ORBITAL FORCING AND CYCLIC SEQUENCES

Orbital Forcing and Cyclic Sequences

EDITED BY P.L. DE BOER
AND D.G. SMITH

SPECIAL PUBLICATION NUMBER 19 OF THE
INTERNATIONAL ASSOCIATION OF SEDIMENTOLOGISTS
PUBLISHED BY BLACKWELL SCIENTIFIC PUBLICATIONS
OXFORD LONDON EDINBURGH BOSTON
MELBOURNE PARIS BERLIN VIENNA

and published for them by
Blackwell Scientific Publications
Editorial Offices:
Osney Mead, Oxford OX2 0EL
25 John Street, London WC1N 2BL
23 Ainslie Place, Edinburgh EH3 6AJ
238 Main Street, Cambridge
Massachusetts 02142, USA
54 University Street, Carlton
Victoria 3053, Australia

Other Editorial Offices:
Librairie Arnette SA
1, rue de Lille
75007 Paris
France

Blackwell Wissenschafts-Verlag GmbH
Düsseldorfer Str. 38
D-10707 Berlin
Germany

Blackwell MZV
Feldgasse 13
A-1238 Wien
Austria

First published 1994

Set by Excel Typesetters, Hong Kong
Printed and bound in Great Britain
at The Alden Press, Oxford

DISTRIBUTORS

Marston Book Services Ltd
PO Box 87
Oxford OX2 0DT
(*Orders*: Tel: 0865 791155
Fax: 0865 791927
Telex: 837515)

USA
Blackwell Scientific Publications, Inc.
238 Main Street
Cambridge, MA 02142
(*Orders*: Tel: 800 759-6102
617 876-7000)

Canada
Oxford University Press
70 Wynford Drive
Don Mills
Ontario M3C 1J9
(*Orders*: Tel: 416 441-2941)

Australia
Blackwell Scientific Publications Pty Ltd
54 University Street
Carlton, Victoria 3053
(*Orders*: Tel: 03 347-5552)

A catalogue record for this title
is available from the British Library

ISBN 0-632-03736-9

Library of Congress
Cataloging in Publication Data

Orbital forcing and cyclic sequences/
edited by P.L. de Boer and D.G. Smith.
p. cm.
(Special publication no. 19 of the
International Association of Sedimentologists)
Includes bibliographical references
and index.
ISBN 0-632-03736-9
1. Sedimentation and deposition – Congresses.
2. Earth – Orbit – Congresses.
I. Boer, Poppe Lubberts de, 1949– .
II. Smith, David G. (David Graham)
III. Series: Special publication . . . of the
International Association of Sedimentologists; no. 19.
QE571.073 1994
551.3′03 – dc20

Contents

Preface

At the International Sedimentological Congress in Nottingham, August 1990, a symposium on Orbital Forcing and Cyclic Sequences was convened by the editors of this volume. A second meeting on the same topic was held in Utrecht in early 1991. This volume brings together some of the papers presented at those two meetings, together with other, invited papers.

There seems to be no shortage of papers on the orbital forcing theme, and there is a shortage neither of interest in the subject, nor of new and original studies deserving of publication. We are pleased to be able to present a significant addition to the literature on orbital forcing and cyclic sedimentary sequences, in the form of 31 papers that between them span the whole range of topics from the astronomical theory, to field studies dealing with a broad variety of sedimentary environments, to modelling and simulation.

Earth's orbital parameters may influence the character of sedimentary sequences. This was first proposed in the last century (Gilbert, 1894), but until fairly recently the subject was considered by many to be a matter of astrology rather than of serious geology.

After Milankovitch published his ideas, orbital forcing was grudgingly considered to be a possible factor in initiating the changes that led to the waxing and waning of polar ice caps. Early papers on orbitally induced sedimentary cycles from non-glacial periods (e.g. Schwarzacher, 1947; Fischer, 1964) were regarded with considerably more scepticism and many preferred interpretations in terms of tectonics, sea level and autocyclicity.

It was only in 1976, with the publication of the benchmark paper of Hays *et al.* (1976), that it became more generally acceptable to ascribe cyclicity in pelagic and lacustrine successions to orbital forcing of climate and oceanography. This development was greatly stimulated by an increase in the understanding of the influence of the orbital movements on the climate on Earth provided by Berger (1978, etc.), and by the explosive development of modern computational techniques. Orbital cycles have, by now, been recognized in many different sedimentary environments, as exemplified in this volume, and the correlation of sedimentary cycles with the predicted orbital influences now even leads to suggestions for improvement of the geological time scale.

While classically orbital cycles have been recognized in pelagic and lacustrine sequences characterized by quiet sedimentation, not disturbed by tectonics, there now is an increasing recognition of the fact that orbital cycles do influence climate and oceanography in general terms. There is also increasing acceptance of the possibility at least that the effect should be felt over large parts of the Earth's surface and that orbital cycles may well leave signs in other sedimentary environments which are commonly considered to be dominated by tectonics and eustasy.

After an introductory review by the editors, the paper by Berger & Loutre sets out the theoretical framework of astronomical forcing through geological time, with particular attention to long-term changes in the absolute and relative frequencies of the key periodicities. As a counterbalance to the predominant theme of orbital forcing of climatic change, Mörner then considers the possibility of internal response of the Earth to orbital forcing.

Since Milankovitch's time, a quantitative approach has been essential to the study of orbital forcing, and the next group of papers looks at statistical techniques and the identification of orbital periodicities. Melnyk *et al.* discuss filtering and frequency mapping as tools in subsurface cyclostratigraphy, and Yang & Baumfalk present examples of Milankovitch cyclicity in the Rotliegend using a similar sliding window approach for their spectral analysis. Van Echelpoel analyses cycles from the Tertiary Boom Clay Formation in Belgium using Walsh spectral analysis, and Longo *et al.* describe high-frequency cycles in carbonate platform deposits in the Southern Apennines. Schwarzacher presents the results of a field-based study in which he attempts a correlation of nearby pelagic sequences in the Mid-Cretaceous of the Gubbio area, using the orbital cycles which gave them their characteristic cyclic appearance. A similar type of succession from the Southern Alps is discussed by Claps & Masetti.

The astronomical theory provides detailed estimates of the orbital cycles during the past few million years, allowing some unusually refined and precisely calibrated chronostratigraphy. Using the curves generated by Berger's algorithm, Hilgen proposes a re-calibration of the geological time scale for the Plio-Pleistocene, and Hooghiemstra & Melice discuss the evolution of orbital periodicities on the basis of a high-resolution pollen record from South America.

From the deep-marine sedimentary record ten Kate *et al.* discuss Late Quaternary monsoonal variations in the western Arabian Sea, and Boyd *et al.* Late Cretaceous cycles from the Exmouth Plateau (Australia). Rhythmic sedimentation related to third-order sea-level variations in the Late Cretaceous Western Interior Basin (USA) is the subject of the paper of Ricken.

Orbitally influenced ichnofacies and biological signals are discussed in the papers of Savrda & Bottjer and of Erba & Premoli Silva. Grötsch's study extends the range of parameters that have been analysed successfully for orbital forcing, by looking at 'guilds' of associated biota and their changes through time, in a Mid-Cretaceous reef setting in Slovenia.

Goldhammer *et al.* show examples of high-frequency, glacio-eustatic cyclicity in Pennsylvanian shelfal carbonates of the Paradox Basin. Cycles in shallow-marine carbonates in the French Jura mountains are discussed by Strasser, and Haas presents new data on the Lofer cycles of the Late Triassic Dachstein platform in Hungary. Reijmer *et al.* analyse periodicities in the composition of Triassic calciturbidites which were shed from the platforms in this region in order to test for extension of sensitivity to orbital forcing from the platform deposits to the deeper water areas of the adjacent basin.

There is increasing awareness that the influence of orbital signals is not confined to marine carbonates and lacustrine deposits formed under very quiet conditions, but that siliciclastic systems can also reflect orbital influences. This recognition is reflected in the papers of van Buchem *et al.* (siliciclastic shelf deposits in the Lias in Yorkshire), Abbott & Carter (mid-Pleistocene shelfal to shoreface sediments in New Zealand formed during a period of orbital control of sea level), van Tassell (Late Devonian Catskill Delta, USA), and Read (Namurian coal-bearing fluvio-deltaic deposits in central Scotland). Cyclic fluvial sequences are described and interpreted by Olsen (lacustrine and fluvial cyclicity in the Devonian of East Greenland). Palaeozoic and Mesozoic examples of orbital control on desert sedimentation is the subject of the contribution of Clemmensen *et al.*

The technical advances of the last decade have made it possible to analyse and calculate very large data sets, leading to the rapid evolution of computer modelling and simulation of the complicated processes which are involved in the transfer of the effects of astronomical processes into the sedimentary record. Examples of this field of research are given by Matthews & Perlmutter who discuss their global cyclostratigraphy model, by Herbert presenting models and examples of deterministic distortion of orbital signals by sedimentation, and by Park & Oglesby applying a general circulation model to the Middle and Late Cretaceous climate. In the final contribution Smith takes a look at the newly emerging theory of chaos, and attempts to judge its potential importance to the understanding and therefore to the modelling of stratigraphic processes.

In the process of reviewing, modifying and (in a few cases) rejecting papers for this volume, the assistance of our external reviewers was invaluable. Apologizing if there is anyone we have forgotten to mention here, we are very grateful to the following for their conscientious help and support:

Kourosh Amiri
Eric Barron
André Berger
David Bottjer
John Bridge
Chris Clayton
Lars Clemmensen
Ed Clifton
Pierre Cotillon
Gerhard Einsele
Al Fischer
Peter Friend
Andrew Gale
Kenneth Glennie
Stephen Hesselbo
Tim Herbert
Linda Hinnov
Michael House
Hugh Jenkyns
George Klein
Kurt Lambeck
J. Laskar
Ole J. Martinsen
David Melnyk
Gerard Middleton
Niels-Axel Mörner
Paul E. Olsen
Chris Paul
George Postma
Arthur Reymer
Werner Ricken
Maurizio Ripepe
Peter Roth
Graham Shimmield
Wolfgang Schlager
Walther Schwarzacher
Bruce Selwood
Nick Shackleton
Herbert Shaw
Antoinette Sprenger
Richard Steele
André Strasser

Colin Summerhayes
Warner ten Kate
Paul Valdes
Erna van Echelpoel
Graham Weedon
Robin Whatley
Toine Wonders
Yang Chang-su
Hans Zijlstra

As with many new developments, the idea of orbital forcing may (and does) lead to overshoots and wishful interpretations of cyclic sequences in terms of orbital forcing. However, we feel sure that the examples presented in this book will withstand future challenge.

POPPE L. DE BOER
Utrecht, The Netherlands

DAVID G. SMITH
London, UK

REFERENCES

BERGER, A. (1978) Long-term variations of caloric insolation resulting from the Earth's orbital elements. *Quat. Res.* **9**, 139–167.

FISCHER, A.G. (1964) The Lofer cyclothems of the Alpine Triassic. *Kansas Geol. Surv. Bull.* **169**, 107–149.

GILBERT, G.K. (1894) Sedimentary measurement of Cretaceous time. *J. Geol.* **3**, 121–127.

HAYS, J.D., IMBRIE, J. & SHACKLETON, N.J. (1976) Variations in the Earth's orbit: pacemaker of the ice ages. *Science* **194**, 1121–1132.

SCHWARZACHER, W. (1947) Uber die sedimentäre Rhythmik des Dachsteinkalkes von Lofer. *Verh. Geol. Bundesanstalt Wien*, **H10–12**, 175–188.

Spec. Publs Int. Ass. Sediment. (1994) **19**, 1–14

Orbital forcing and cyclic sequences

P.L. DE BOER* *and* D.G. SMITH†

* *Comparative Sedimentology Division, Institute of Earth Sciences, P.O. Box 80.021, NL-3508 TA Utrecht, The Netherlands; and*
† *Petroconsultants (UK) Ltd, Europa House, 266 Upper Richmond Road, Putney, London SW15 6TQ, UK*

ABSTRACT

Over periods of tens of thousands to millions of years, periodic changes of climate due to astronomically defined variations in the distribution of solar energy over the Earth are of influence upon the climate and upon oceanic circulation systems. This in turn may affect the production of carbonate and organic matter in marine surface waters and the oxidation and dissolution of these products in the deep ocean. Rhythmic variations in the flux of the resulting pelagic sediments are readily preserved in the deep oceanic record. However, similar reflections of regular, astronomically induced fluctuations of climate have been widely observed in other sedimentary environments, including glacial, lacustrine, fluvial, aeolian, deltaic, shallow marine, and submarine fan.

Periodic climatic and oceanographic changes and the resulting rhythmic sedimentation patterns can be preserved in the sedimentary record, providing that the sediment-supplying mechanisms (weathering and transport on the land, production of biogenic sediment in the ocean) as well as the relative sedimentary facies are sufficiently sensitive. Pelagic and lacustrine, and especially evaporitic, settings have proved to be sensitive to the influence of orbital forcing of climate. It is in such facies that, at the turn of the century, cyclicity in sedimentary sequences was recognized to be the result of orbital forcing (Gilbert, 1894; Bradley, 1929). Other sedimentary environments can be sensitive to such astronomically induced climatic variations in a similar way, but in many cases disturbing factors, such as variations in the rate of sedimentation and intermittent erosion, will obscure an otherwise clear record in the sedimentary column. Recognition of orbital cycles in environments other than the pelagic and lacustrine ones is of more recent date, but even coarse-grained alluvial fan and submarine fan environments can now be proven to bear the marks of orbitally induced climatic changes. Once demonstrated in the record, Milankovitch cycles have applications in assessing rates and durations of geological processes, and in the analysis of the palaeoclimate.

INTRODUCTION

This paper sets the scene of this book by reviewing the subject, and, without attempting to be comprehensive, at least indicating the range of studies that have been undertaken in recent years as part of the general revival of interest in the hypothesis that bears Milankovitch's name.

Astronomical forcing of climate and sedimentary facies depends upon the changing position of the Earth's axis in its varying path around the Sun (Fig. 1; Berger, 1977). Such variations depend on the interaction of gravitational forces in the rotating Sun–Earth–Moon system, and on the influences of the other planets in our solar system. The *precession* (Fig. 1c) is the spinning of the Earth's axis due to the combined effects of the solar and lunar attraction on the equatorial bulge of the Earth. The absolute period of this cycle is of the order of 26 ka*. However, the elliptical orbit of the Earth also rotates, and the average main periods observed from the Earth are about 19 ka and 23 ka, with extremes of

* ka = 1000 years.

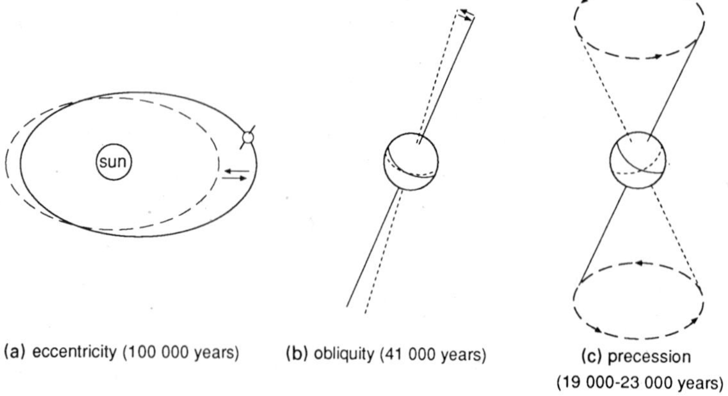

Fig. 1. Schematic representation of the astronomical variables influencing the climate on Earth. From de Boer (1983).

14 ka and 28 ka (Berger, 1988). As a result, the equinoxes, i.e. the position of the Earth in its orbit around the Sun at the moment when the Sun is exactly above the equator and day and night have equal length (21 March and 23 September), move around the elliptical orbit of the Earth. This leads to regular and predictable changes in the distribution of insolation over the Earth, and hence changes in the contrast between summer and winter. The effect of the precession is 180° out of phase between the Northern Hemisphere and the Southern Hemisphere (Figs 1a and 2).

The *obliquity* of the Earth's axis is its angle of tilt (Fig. 1b) with respect to the perpendicular of the ecliptic, the plane in which the Earth rotates around the Sun. Obliquity varies between 22 and 24.5° with a mean period of about 41 ka, and it modulates seasonality, especially at high latitudes.

Finally, the path of the Earth around the Sun is not a perfect circle but an ellipse with varying *eccentricity* (Fig. 1a). The period over which the eccentricity varies from maximum (about 0.06) to minimum (close to zero) and back is, on average, about 100 ka, with major components at 99 ka and 123 ka. Superimposed regular variations of the eccentricity occur over about 400 ka, 1300 ka and 2 Ma, and these longer periods are also recognized in sedimentary successions (e.g. Schwarzacher & Fischer, 1982; de Boer, 1983; Anderson, 1984; Olsen, 1984; Cottle, 1989; Melnyk & Smith, 1989).

The astronomical parameters cause variations in the spatial distribution of solar energy reaching the Earth's surface, and thus they influence the distribution of climatic zones and the variation of the receipt of solar energy with latitude over the year (Berger, 1978a, 1979, 1981). For example, the caloric equator, which is the latitude at which the energy received from the Sun over the year is maximal, moves with varying amplitude and frequency, largely defined by the cycle of precession and its modulation by the varying eccentricity, within limits of about 10° N and 10° S (Fig. 2; Berger, 1978b). At mid-latitudes (20–40°) the astronomical parameters produce different effects, e.g. changes of summer–winter contrast and of monsoon intensity, of the position of the different climate belts and of the boundaries between them (Berger, 1978b; Kutzbach & Otto-Bliesner, 1982).

Development of ideas

Starting early in the nineteenth century, interest in the origin of the growth and melting of ice caps induced astronomers to make calculations of astronomical frequencies, and led to suggestions about the way in which changes of the volume of ice caps could be related to astronomical influences (see Imbrie & Imbrie, 1979). Berger (1988) provides a concise review of the work of pioneers of the astronomical theory during the last two centuries.

At the end of the last century, Gilbert (1894) proposed that the deposition of Late Cretaceous hemi-pelagic carbonate–marl cycles in Colorado had been the result of astronomical forcing. Based on this assumption he calculated the length of part of the Late Cretaceous, and his result is in good agreement with modern geological time scales (Fischer, 1980). Thirty years later, Bradley (1929) recognized precessional cycles in oil shale–dolomite sequences of the Eocene Green River Formation in

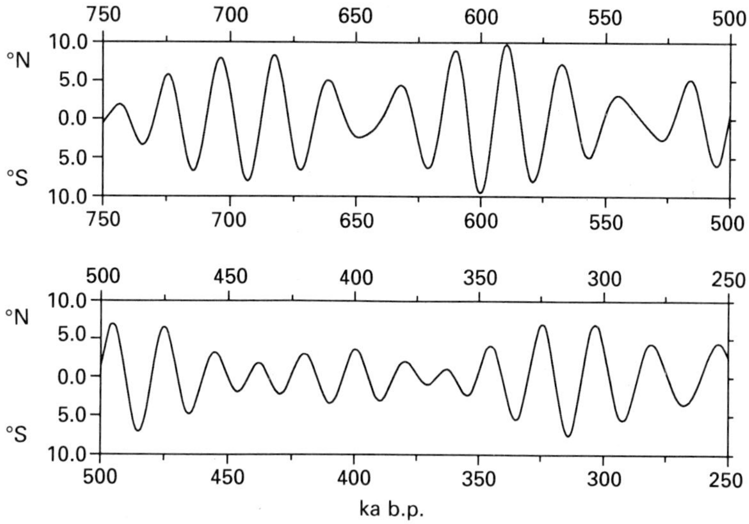

Fig. 2. Changing position of the caloric equator between 750 ka and 250 ka BP. Note the varying frequency. Periods between extreme (N or S) latitudinal positions of the caloric equator vary between 14 ka and 28 ka. From Berger (1978b).

Colorado, Wyoming and Utah, using varves as an unusually precise measure of sedimentation rates.

While the work of Gilbert and Bradley was largely unnoticed, Milankovitch revived the debate about the possible influence of astronomical parameters upon the Earth's climate. In 1941 Milankovitch published his *magnum opus* about the causal relationship between astronomical parameters and the ice ages (see the review by Berger, 1988). With the advance of analytical methods and the possibility of analysing large quantities of data using advanced statistical and computing techniques, and thanks also to the theoretical astronomical basis outlined and made accessible to non-astronomers by Berger, increasing numbers of examples have been documented showing the relation between astronomical influences and rhythmicities in the pre-Pleistocene sedimentary record (Schwarzacher, 1947, 1954; van Houten, 1962, 1964; Fischer, 1964; Arthur, 1979; Anderson, 1982; de Boer & Wonders, 1984; and many others, e.g. *Terra Nova* 1989, vol. 1, issue 5, and *Journal of Sedimentary Petrology* 1991, vol. 61, issue 7). Today the idea of astronomical variables leaving traces in the sedimentary record is becoming widely accepted (see Fischer *et al.*, 1990), although it has to be said that acceptance is not yet universal.

With the development of analytical techniques, the effect of increases and decreases of the ice volume upon the stable oxygen isotope composition of seawater and of biogenic carbonates could be measured, and the marine sedimentary record of $\delta^{18}O$ of carbonate could be related to changes in the volume of global ice caps (Emiliani, 1955). This allowed detailed reconstructions of the growth and melting of ice caps during the Pleistocene, the history of which is related in turn to astronomical influences (cf. Hays *et al.*, 1976; Imbrie *et al.*, 1984).

In much of pre-Pleistocene geological history, polar and other ice caps have been of minor or no importance, and astronomical influences must therefore have had a more direct influence on sedimentation patterns and the environment of deposition, thus allowing their recognition.

Recognition of astronomically induced cyclicities in the sedimentary record

Theoretical astronomy provides us with predictions of the periodicities at which orbitally forced sedimentation rhythms should occur in the stratigraphic record. Stratigraphy provides us with an extremely imperfect, unreliable, noisy and poorly time-calibrated recording of the outcome of a number of variables, of which climate is just one. Analysis of the record to demonstrate control by orbital forcing clearly has to be undertaken with considerable care and rigour. The hypothesis of orbitally forced climatic change has been most clearly validated in the case of the Pleistocene, where (i) the orbitally influenced climatic variation is particularly pronounced, and (ii) the numerical age control is relatively precise. Examples from the pre-Neogene for which sufficiently accurate time control is available are rare and essentially depend on the presence of varves (Bradley, 1929; Anderson, 1984; Olsen, 1984). In the more usual absence of such control on the absolute length of individual rhythms, a common approach is to estimate an average cycle duration by dividing the stratigraphic timespan analysed by the number of sedimentary cycles; numerous published examples provide figures which fit the Milankovitch periodicities, but see Algeo & Wilkinson (1988) for a critique of this approach. The inverse of this procedure is beginning to suggest corrections and improvements to the various geological time scales that are largely based on radiometric methods, with their limitations of precision (e.g. Hilgen & Langereis, 1989; Shackleton, 1989; Hilgen, pp. 109–116, this volume).

Because the theory of orbital forcing predicts periodic phenomena in stratigraphic successions, it is appropriate to apply the methods of spectral analysis to field and borehole data and/or to analytical data obtained in the laboratory. All that is needed is a 'time series' (strictly, a stratigraphic depth series) of measurements of one or more parameters taken at regularly spaced or well-defined intervals along a stratigraphic section. Borehole logs, for example, provide ready-made data sets for this type of analysis. Interpretation of the series in terms of time (rather than stratigraphic intervals) is an important step whose potential for misuse must not be underestimated, though Sander (1936) demonstrated long ago that rhythmicity in stratal thickness is unlikely without rhythmicity in time, while lack of stratal rhythmicity does not preclude rhythmicity in time.

In the case that the most obvious (and readily measured) feature of a sedimentary succession is largely defined by astronomical forcing, and if time control is good, spectral analysis should yield significant peaks at the main astronomical frequencies (see the various examples in this volume). For late Tertiary and Quaternary examples, time control, based on magneto-, bio-, and isotope-stratigraphy may be sufficient to put an adequate number of datum points along sequences in which the variation of certain characteristics (fossil content, petrographical composition, etc.) are analysed. In many cases, however, especially in the pre-Tertiary sedimentary record, time control is limited and depends on interpolations and extrapolations, generally with the assumption that the rate of sedimentation has been constant. It is clear that such assumptions may blur the data set to be analysed if the rate of sedimentation was not constant for example, or if it was influenced or defined by astronomical influences, as in the case of carbonate productivity cycles and dissolution cycles.

If there is a good reason to postulate that particular sedimentary rhythms represent individual precession cycles, then another analytical approach is possible in which precise estimates of time are unnecessary. This is the measurement of the thicknesses of successive rhythms or of the amplitudes of successive deviations from a long-term mean value of easily measured sedimentary characteristics, such as the content of certain microfossils, of carbonate, organic matter, or wind-blown sediment. In such cases, spectral analysis yields the modulation of the precessional cycle (having a variable frequency) by lower frequency astronomical cycles, mainly the eccentricity cycle which, in turn, also has a variable frequency *and* amplitude. In an example in which the parameter analysed was the thickness of successive precession-induced rhythmic beds in a pelagic environment (de Boer, 1983), spectral curves reflected the modulation of the precession effect by longer term cycles that were interpreted as components of the eccentricity cycle and the obliquity cycle. This appeared to be equivalent to sampling and analysing the successive amplitudes or the successive maximal latitudinal positions of the caloric equator. In such cases, where sedimentary rhythms are defined by the position of the caloric equator (which is largely determined by the precessional cycle) or where they are related to climatic variations having similar frequencies and amplitudes as the caloric equator (monsoon intensity, precipitation–evaporation ratios), their spacing in time is irregular, with time intervals varying mainly between 19 ka and 23 ka (e.g. Fig. 2), but with extremes of 14 ka and 28 ka. Successive maxima (carbonate content, dilution by the influx of terrestrial sediments, content of certain fossils, oxygenation state, etc.), unequally spaced in time between 14 ka and 28 ka (present-day values), then reflect the modulation of the effect of the cycle of precession by the varying eccentricity. This explains why many examples of sedimentary cyclicities, which are attributed to the influence of the 21 ka precession cycle, typically exhibit a bundling of cycles in groups of four to five (see de Boer, 1982; Schwarzacher & Fischer, 1982).

Relative influence of different orbital cycles

In many cases sedimentary rhythms seen by the eye, in the field or in wireline logs, clearly show precession-induced rhythms and a subordinate effect/modulation by the eccentricity cycle. Time series analyses, however, often suggest the reverse, i.e. a strong dominance of the eccentricity cycle over the cycle of precession. The obvious explanation is that the variability in time of the frequency of the precession cycle is much greater than that of the eccentricity cycle (see Berger, 1978a, b). This explains *why in many cases mathematical analyses of orbital signals in sedimentary sequences reveal a signal for the eccentricity cycle which is significantly stronger than that of the precession cycle* (see various papers in this volume). However, *such differences in amplitude of peaks, produced by different types of time series analysis, are not the result of the influence of the eccentricity cycle on climate and oceanography being stronger than that of the cycle of precession, but rather result from the fact that the eccentricity cycle causes variations which are much more regular in time than those of the precession cycle, whose frequency varies by a factor of two* (14–28 ka).

Changes of periodicities through geological time

Although the principal periodicities of orbital variation can be calculated very precisely for the present day, it cannot be assumed that the same periodicities have applied for all of geological time. In particular, tidal friction is slowing down the Earth's rotation rate and this is bound to have an effect on both precession and obliquity. However, Berger & Loutre (pp. 15–24, this volume) have shown that the eccentricity periods are not changed by this process, and therefore the ratios between precession : eccentricity and obliquity : eccentricity will have changed through geological time. The effects of tidal friction are not known with sufficient accuracy to make projections back into the remote geological past, but evidence of the number of days in a year from, for example, the growth bands of corals and other fossils provides a limited number of calibration points (Panella, 1972; Brosche & Sündermann, 1978, 1982; Williams, 1991a). Using these as a constraint, Berger and his colleagues estimate that the two main periods of the precession have slowed down from about 16 ka and 18.5 ka in the Early Palaeozoic to their present values of 19 ka and 23 ka. The number of precession cycles per primary eccentricity cycle of about 100 ka should therefore have averaged about six at that time, as compared to less than five at the present day. This should be evident in the bundling of primary sedimentary rhythms. Often there is too much spread to recognize visually such a difference in the field or in wireline logs, but statistical analysis of sufficiently long data sets may reveal such changes of ratio.

Chaos?

Given the complexity of the interaction among the numerous bodies of the solar system, it has been suggested that long-term extrapolation of the present system of motions back into the distant past may not be possible, because of non-linear or 'chaotic' effects. Laskar (1989) has suggested that extrapolation of the present equations of motions beyond about 10 Ma into the past may be unjustified. On the other hand it is a matter of common observation that stratal patterns such as bundling are not obviously different, even for very old successions (Grotzinger, 1986), so it is perhaps reasonable to assume that the orbital parameters have not suffered major qualitative changes through the Earth's history.

MECHANISMS OF ORBITAL FORCING OF CLIMATE AND SEDIMENTARY FACIES

Insolation (the receipt of solar energy at the surface of the Earth per unit area) depends on latitude, on the orientation of the Earth's axis, and on the orientation of the Earth in its eccentric path around the Sun (Fig. 1). Long-term (20–100 ka) variations of these parameters lead to small variations of climate which can, however, have a disproportionately pronounced effect at certain latitudes (Berger, 1978a, b). At low latitudes, close to the equator, the influence of the cycle of precession, modulated by the varying eccentricity of the Earth's orbit, is dominant and causes latitudinal shifts of the caloric equator (Fig. 2). In turn, this causes significant shifts of the boundaries between adjacent climate zones. At mid-latitudes (20–40°) the orbital variations affect the relative length of the seasons and the contrast between summer and winter, and hence of monsoon intensity (see Kutzbach & Otto-Bliesner, 1982). Toward higher latitudes (>40°) the effect of the varying obliquity becomes more prominent (van Woerkom, 1953; Berger, 1978b; Kemper, 1987). Interference of high-latitude obliquity effects and low-latitude precessional dominance, e.g. upon the abundance of different fossil species, may occur when different controlling mechanisms are influenced or dominated by climatic effects generated at different latitudes (see Cottle, 1989; Premoli Silva *et al.*, 1989).

Insolation, and its variation with latitude and through time, can be calculated using the algorithms developed by Berger and his colleagues. However, such calculations yield only the input energy, the energy received at the top of the Earth's atmosphere; they make no predictions about the ways in which the atmosphere–ocean system will respond to such changes. Processes of sedimentation clearly depend on many more variables than insolation alone, and variations in insolation must be propagated through the complexities of the system before they can be encoded in the sediment. The mechanisms for this are not yet clear and they may be many and various. Furthermore, analysis of geological time-series data from very young strata indicates that quite significant time lags may be involved between a particular insolation change and the climatic effect that is predicted to result from it. For instance Hilgen (1991a, b) showed time lags of several thousand years for Pliocene carbonate–marl and anoxic–aerobic cycles in Mediterranean (hemi-)pelagic deposits in Southern Italy, and suggested that the monsoonal system may take that length of time to adjust to changes in the pattern of insolation. Another process which may produce a time lag between orbital parameters and climate response was recently proposed by Adams *et al.* (1992). These authors drew attention to the fact that peat formation is slow, and that the role of peat as a producer of methane may not reach its maximum potential until thousands of years after the related insolation peak has passed.

A comprehensive review of transfer mechanisms would be premature at such a rapidly advancing stage of the science. We therefore attempt to cover only a selection of quite well-established variables interpreted as being an important or critical influence on the development of sedimentary rhythms in the cases cited.

EXAMPLES

The number of examples of stratigraphic phenomena ascribed to orbital forcing is growing rapidly, and only a few examples are given here. We aim to show the diversity of the environments from which examples can now be drawn, and we start with the classic example of pelagic deposits.

Pelagic deposits, chalk

Pelagic sedimentary environments have perhaps the best potential for regular, uninterrupted sedimentation, relatively little disturbed by those tectonic processes that have their strongest expression upon and close to the continents. Therefore they are well able to record changes of climate and oceanography. Examples of orbital forcing have been given from the late Tertiary and Quaternary when polar ice caps exerted their influence (e.g. Imbrie, 1982), as well as from older deposits from periods without (large) polar ice caps. For summaries see, for example, Hay (1988), Berger (1988) and Fischer *et al.* (1990).

Outside the sphere of influence of mass transport from continental shelf areas, the character of the sediment in the deep ocean mainly depends on factors such as the supply of fine-grained sediments derived from the continents, in suspension or through the atmosphere, and on the biogenic production of mainly carbonate, silica and organic matter. Differences in the contribution of each of these constituents depends on geographical factors, such as the distance from the land; on atmospheric circulation systems, the contribution of wind-blown terrigenous sediment and terrestrial organic matter; and on oceanic circulation patterns influencing the productivity in surface waters. Pelagic cyclic sequences are the result of variations in the relative proportions of these features, to which must be added *dissolution* of carbonate below a fluctuating calcite compensation depth (CCD) (e.g. Dean *et al.*, 1981).

Redox cycles result from varying flux rates, or the intermittent absence, of oxygen in deep water (e.g. Arthur, 1979). This may be caused by variations both in organic carbon supply and in supply of oxygen to deep water. The first occurs in areas of very high organic production (e.g. zones of upwelling) and in areas with abundant supply of terrestrial organic matter, the second in areas with poor deep-water circulation (stagnant basins). Moreover, relatively high rates of sedimentation favour the preservation of organic matter (Müller & Suess, 1979; Degens *et al.*, 1986). Direct causal relationships between climate and anoxicity in the Mediterranean during the Quaternary were discussed by, amongst others, Rossignol-Strick (1985) and Rohling & Hilgen (1991).

With respect to carbonate–marl rhythms, regular variations of carbonate productivity induced by changing oceanographic conditions are commonly suggested to have produced variations of sedimentation of carbonate with a more or less constant background sedimentation of non-carbonate sediment (*productivity cycles*; de Boer, 1983; Ditchfield & Marshall, 1989). In some examples, especially from the Western Interior Seaway, it has been suggested that intermittent stratification of the water column has occurred, caused by variations in the supply of fresh water from the continent, producing redox cycles and also involving a variable input of clay (*dilution cycles*; Arthur *et al.*, 1984; Bottjer *et al.*, 1986). Of course, variations of carbonate content can be the result for example of a combination or of a complex interference of features such as production, dissolution and dilution (see Diester-Haass, 1991).

Cool water produced at high-latitudes is an important driving force for currents in the deep ocean, and, obviously, regular variations of high-latitude temperatures, especially in times of polar ice caps, may produce erosion levels (*winnowing cycles*). Changes in the pattern and intensity of these deep currents also may affect carbonate dissolution and the oxygenation state. In periods free of ice caps, currents driven by differences in salinity may have had a similar though less intensive effect.

Changing environmental conditions may influence life at the ocean floor and thereby influence the character of the *ichnofacies*, which thus may be used to distinguish the above processes in the production of rhythmic bedding (e.g. Savrda & Bottjer, 1986 and pp. 195–210, this volume; Erba & Premoli Silva, pp. 211–225, this volume).

In many examples of orbitally induced cyclicities of carbonate-richer and carbonate-poorer beds in the deep marine sedimentary record, $\delta^{18}O$ values of carbonate are reported to show a systematic correlation with the carbonate content. $\delta^{18}O$ of biogenic carbonate depends on the temperature and on the $\delta^{18}O$ of the water in which the carbonate is formed. Due to fractionation during evaporation and precipitation, fresh (rain) water has a low $\delta^{18}O$ value, and runoff of fresh water from the land lowers the $\delta^{18}O$ of seawater. In addition to temperature and freshwater runoff, vital effects may have a subordinate influence upon $\delta^{18}O$ of biogenic carbonate (Goodney *et al.*, 1980). In periods of waxing and waning of polar ice-caps the oceanic $\delta^{18}O$ value is strongly influenced by the amount of fresh water stored in the ice.

Observed $\delta^{18}O$ variations of carbonate in deep marine rhythmic successions from ice-cap-free periods include positive as well as negative correlations between carbonate content and $\delta^{18}O$. For the Mid-Cretaceous in the Apennines and in the Western Interior Seaway, carbonate content and $\delta^{18}O$ of $CaCO_3$ in pelagic deposits show a positive correlation. For the first example this was explained in terms of changes of temperature of surface waters, with warm surface water temperatures in times of slow circulation and low organic productivity (de Boer, 1983; Herbert & Fischer, 1986). For Cretaceous pelagic deposits in Colorado, a similar positive correlation between carbonate content and $\delta^{18}O$ was explained in terms of regular influxes of isotopically light fresh water from the continent, leading to stratification of the water column. This would have led to reduced organic productivity and increases of supply of terrigenous sediment in periods of low surface water salinity (Pratt, 1984; Barron *et al.*, 1985). Eicher & Diner (1989) also discussed the possible causes of carbonate–marl cycles in the Western Interior Seaway. Based on studies of planktonic foraminifera, calcispheres, clay minerals, stable isotopes and quartz content, they concluded that a dilution model is not consistent, but that instead fluctuations in productivity, characteristic for other parts of the Tethyan realm, have been the origin of the rhythms.

Cenomanian chalk–marl cycles in southern England show an opposite trend, the carbonate-rich intervals having the lower $\delta^{18}O$ values. Leary *et al.* (1989) and Ditchfield & Marshall (1989) conclude that in this case temperature rather than nutrients was a limiting factor.

Ice volume and sea-level changes

Milankovitch's original aim was to demonstrate a link between the influence of the orbital parameters and the ice ages (cf. Berger, 1988). Because of the difficulty of dating terrestrial glaciations he was not fully successful in this aim, and it was only later that the link between the predicted insolation changes and cycles of glaciation/deglaciation became established. This was through the Quaternary pelagic record, with its much greater continuity, and through the parameter of oxygen isotope ratios in the shells of pelagic foraminifera. The $^{18}O/^{16}O$ ratio (expressed as $\delta^{18}O$) was used initially as a palaeothermometer, but it became apparent that this ratio is much more strongly affected by the volume of water locked up in the ice caps, as the lighter ^{16}O isotope is more prevalent in evaporation and precipitation. The $^{18}O/^{16}O$ ratio has therefore come to be used as a measure of ice volume, and thus as a slightly less direct proxy for climate. Its sensitivity is nevertheless sufficiently good for it to have become the foundation for the scheme of standard stages for the oceanic Quaternary record. Its variations are sufficiently well dated numerically, mainly by the excellent ties to the palaeomagnetic record, that the match between the glacial cycles and the predicted pattern of insolation is not in doubt. Milankovitch's hypothesis and his life's work can be considered to have been vindicated on the date of publication of the landmark paper of Hays *et al.* (1976).

Ice caps, however, are a rare rather than a common phenomenon in the history of the Earth, and the establishment of a link between ice volume and orbitally forced insolation variations poses as many questions as it solves. The ice caps are widely thought to play an important linking role between orbital forcing and stratigraphically significant effects such as climate and sea-level change. However, during the long parts of the stratigraphic record which were almost certainly non-glacial, rhythmicities in the sedimentary column must have been imposed along other pathways.

Lake deposits

Bradley (1929), van Houten (1962, 1964), Anderson (1982, 1984), Olsen (1984, 1986) and Williams (1991b) have all given examples of lake deposits exhibiting cycles induced by orbital forcing. The calculations of frequencies by these authors are based on the extrapolation of annual cycles or, more roughly, on estimates of annual rates of sedimentation and the ratios between different astronomical frequencies, but all are based on unusually good time control. Anderson's (1982, 1984) example from the Permian Castile Formation in the Delaware Basin (Texas, New Mexico) is based on detailed counts of annual evaporite (gypsum) varves. In the Triassic Lockatong Formation (van Houten, 1962, 1964; Olsen, 1984, 1986), cycles are characterized by alternating fine-grained clastics and intervals with evaporitic minerals or pseudomorphs after them. Bradley (1929) described cyclicities of inferred astronomical origin of alternating organic C-rich and dolomitic intervals from the Green River Formation.

These examples were formed at low palaeolatitudes, where astronomical theory predicts precession-induced changes of climate. Alternations of periods of dominance of precipitation over evaporation and vice versa, related to shifting boundaries of climate zones for example (see Kutzbach & Street-Perrott, 1985) are a logical consequence and may influence the character of the sediment in cases of favourable basin morphology.

Terrigenous clastics: alluvial to submarine fan systems

All models explaining the reflection of orbital cycles in the sedimentary record depend upon the transfer of heat through the Earth's atmosphere. Thus, *if we were not preoccupied by the existing knowledge of orbital rhythms in aqueous domains, and if we were aware only of the possible influence of orbital motions on the Earth's climate, the logical first place to search for signals in the sedimentary record would be among terrestrial deposits.* Sediment transport in alluvial systems depends on the equilibrium between rainfall and evaporation. Indeed both depend on climate and may vary, in the long run, in relation to the effects of the orbital variables. Changes in the ratio between precipitation and evaporation in terrestrial source areas should lead to cyclic fluctuations in sediment yield. Nevertheless, an example of orbital cycles in fluvial deposits, from the Devonian in East Greenland, was described only recently (Olsen, 1990 and pp. 429–430, this volume). Consequently alluvial fan and fan delta systems should also feel the effect of orbitally driven fluctuations in sediment yield (see de Boer *et al.*, 1991). *In the case of examples of inferred Milankovitch cycles in deltaic and shallow marine deposits* (e.g. Goodwin & Anderson, 1985; Heckel, 1986), fluctuations of sea level, as commonly suggested, would be a logical cause, but on the other hand *fluctuations of sediment supply need serious consideration as well.* Explanations in terms of fluctuating sediment yield need special attention in the case of examples of orbitally induced rhythmicity formed at palaeolatitudes of about 30–40°, i.e. at latitudes where orbital changes have been shown to strongly influence the annual precipitation and evaporation (Berger, 1978a, b; Kutzbach & Otto-Bliesner, 1982).

Apart from deltaic and shallow-marine depositional systems, siliciclastic submarine fans also depend on the supply of sediment from the continent. Often such sediments reside in shelf seas until a fall in sea level induces further transport towards submarine fans (e.g. Posamentier & Vail, 1988; Posamentier *et al.*, 1988; van Wagoner *et al.*, 1988). However, if shelf seas are absent and submarine fans are fed directly by sediment supply from the land, submarine fan systems may well directly reflect orbitally induced climatic changes and their effect on sediment yield (Foucault *et al.*, 1987; Weltje & de Boer, 1993).

Nevertheless, terrestrial and marine depositional systems fed by terrigenous sediment supply commonly are highly energetic, and they are often subject to autocyclic processes, to tectonic activity and to intermittent erosion. This implies that orbitally induced climatic changes may well occur and also that they may influence the sediment yield, but that a recognizable signal in the sedimentary record may be absent in many cases.

Aeolian deposits

While considering the fluctuations of climate discussed above, the intensity and direction of aeolian transport may also vary, and in cases of regular subsidence and an overall net sediment supply, orbital cycles may therefore be recorded in aeolian deposits (cf. Clemmensen *et al.*, pp. 439–457, this volume; and Yang & Baumfalk, pp. 47–61, this volume). Factors involved are the character of the wind system, precipitation and humidity, and the level of the groundwater related to the local climate and/or to the height of sea level in the adjacent marine realm.

Palaeosols

The formation of soils is sensitive to climatic conditions, especially to the rate of precipitation and evaporation. Thus, palaeosols represent a sedimentary facies that invites inspection for Milankovitch cycles. However, only a few examples of astronomically induced rhythmicities in palaeosols have as yet been reported or even suggested (see Goodfriend & Magaritz, 1988; Lehman, 1989). This is probably because rates of palaeosol sedimentation are particularly variable and prone to the formation of hiatuses. Also, palaeosols are especially unlikely to form successions which are long enough for time-series analysis of periodicity.

Shallow-marine carbonates

Astronomical cycles in shallow-marine carbonates have been described, e.g. from the Mesozoic in central Europe (Schwarzacher, 1947, 1954; Fischer, 1964; Hardie *et al.*, 1986; Schwarzacher & Haas, 1986; Strasser, 1988). Many of the examples described date from non-glacial periods. In this case conceivable mechanisms for changes of sea level are, amongst others, cyclic evaporation of isolated oceanic basins and thermal expansion of ocean water (see Strasser, 1988) or even orbitally forced vertical movements of the Earth's crust (Mörner, 1981). The subject, however, remains controversial.

Calciturbidites

In many cases carbonate platforms feed calcareous submarine fan systems and in such cases the environmental changes on the platform can be reflected in the composition of calciturbidites (Haak & Schlager, 1989; Reijmer *et al.*, 1991; Reijmer *et al.*, pp. 323–343, this volume).

Storm intensity

The bottom of shallow seas can be subject to storm activity, and it has been suggested that orbitally induced variations in storm intensity produced distinct cycles in shallow-marine mudstones in the Lower Lias in Yorkshire (van Buchem & McCave, 1989). For the hemi-pelagic Chalk deposits in NW Europe, varying rates of reworking of the sediment due to orbitally induced variations in storm intensity and related early diagenetic processes have been proposed as a mechanism which may produce orbitally induced sedimentary cycles such as those seen in the classic Maastrichtian exposures near Maastricht (The Netherlands) and Stevns Klint in Denmark (Zijlstra, in preparation).

Palaeomagnetics

The magnetic field of the Earth is thought to be generated by velocity differences in the outer liquid core and/or at the boundary of core and mantle. The Earth's orbital parameters must lead to changes in the torque exerted by the Sun, the Moon and by other planets and therefore they may well influence perturbations in the core–mantle dynamics. Any variation in coupling will be reflected directly in the properties of the Earth's magnetic field. For the Quaternary, long-term variations in the palaeointensity of the Earth's magnetic field were tentatively correlated with the orbital parameters (Wollin *et al.*, 1977, 1978). Vandenberg *et al.* (1983) and Napoleone & Ripepe (1989) showed a correlation between the palaeomagnetic signal observed in the Mid-Cretaceous sequence at Moria (Italy) and astronomical frequencies.

APPLICATIONS OF THE MILANKOVITCH CONCEPT

Absolute time and cycle calibration

It is generally possible to make only very rough estimates of the periods of presumed orbitally forced sedimentary rhythms, because routine stratigraphic dating methods do not have the necessary degree of precision. An exception is the work of Anderson (1982, 1984). Based on detailed counting of annual evaporite varves in the Permian Castile Formation in Texas and New Mexico, he calculated a mean period of 19.4 ka for the precessional cycle, i.e. significantly shorter than the present-day mean value of about 21 ka. Since Anderson's work was first published, Berger *et al.* (1989a, b) have presented the results of their calculations of the probable changes in the astronomical frequencies over the last 500 Ma due to the decreasing velocity of the Earth's rotation. Their results show that the precession and obliquity cycles have also slowed down through geological time, i.e. their periods were shorter in the past. They did not give estimates of the precession and obliquity periods for the exact date of the Castile Formation, but one can interpolate between their estimates for the Early Permian and the Cretaceous to arrive at an estimate of 19.4 ka for the Late Permian precession cycle, remarkably close to Anderson's estimate from varve counts. This as yet unique result confirms the calculations of Berger *et al.*, and their conclusion that orbital frequencies have indeed slowed down during the Earth's history. It is perhaps unlikely that many more examples will be found where such precision is possible so far back in time. However, the slowing down of the precession cycle should also be evident in the ratio of its period with that of the eccentricity cycle, which is not related to the same astronomical factors as the precession and the obliquity and has therefore not changed its period to the same degree. Frequency analysis of sedimentary rhythms from other periods also often suggests such a change in period ratios through the Earth's history (e.g. various papers in this volume).

Estimating the duration of stratigraphic stages

At the rather longer time scale of a few million years, sedimentary rhythms that can be positively associated with particular orbital frequencies clearly have considerable potential for estimating the lengths of the intervals of the traditional stratigraphic time scale. On the assumption that pelagic carbonate–marl rhythms in Colorado (USA) were the result of the influence of the precession cycle, Gilbert (1894) gave an estimate of the length of part of the Late Cretaceous, which fits well to the results of modern radiometric dating (Fischer, 1980). By counting carbonate–marl rhythms in the Cretaceous of the Apennines, and assuming that the inferred changes in oceanic circulation were driven by the precession cycle, which was assumed to cover 21 ka, de Boer & Wonders (1984) estimated a period of about 7 Ma for the length of the Cenomanian. This value is at the upper limit of the range of estimates based on radiometric and biostratigraphic dating (Obradovich and Cobban, 1975: 3.5 Ma; van Hinte, 1976: about 7 Ma; Odin and Kennedy, 1982: 4 Ma; Hallam *et al.*, 1985: 4.5 Ma; Harland *et al.*, 1990: 6.6 Ma). In this context the value of 5934 ka suggested by Hart (1987), calculated on the basis of supposed obliquity cycles in the English Chalk, also fits well. Gale (1989) gives an estimate of 4.4 Ma based on the assumption that Cenomanian carbonate–marl cycles in southern Britain represent the *c.*21-ka cycle of precession. At present, such estimates based on the counting of orbitally induced sedimentary cycles are clearly tentative; overlooked hiatuses, beds supplied by mass transport, etc. may easily lead to miscalculations, as is well illustrated by fig. 2 of Gale (1989). Although the general principles of the formation of astronomically induced rhythmic bedding are more or less understood, much more work has to be done to understand fully the exact mechanisms, as well as basic sedimentological studies aiming to distinguish pure orbital cycles and features which blur the signal. It will take some time before a cyclostratigraphy for the pre-Quaternary is developed so well that reliable numerical estimates can be given.

For more recent sedimentary successions with good bio- and magneto-stratigraphic control, some convincing examples have been given (Hilgen & Langereis, 1989; Shackleton, 1989; Hilgen, 1991a, b and pp. 109–116, this volume) where field-based astro-cyclostratigraphy together with the extrapolation of present-day astronomical frequencies and amplitudes can be confidently used to improve stratigraphic time scales.

Once having been established, characteristic patterns of astronomical signals in the sedimentary column may serve to correlate sedimentary successions regionally or even over the globe to within

20 000 years (see de Boer, 1983; Cotillon, 1987; Kaufmann, 1988; Gale, 1989; Melnyk & Smith, 1989; Smith, 1989).

Understanding of past climates

Examples of orbital rhythmicities in sediments are concentrated, in respect of both latitude and age (de Boer, 1991). The concentration of examples at palaeolatitudes of 20–40° is probably related to the sensitivity of climate to orbital forcing at that latitude. Moreover, certain (tectonically quiet) periods in Earth history have apparently been more sensitive to the reflection of astronomical forcing of climate in sediment than others. There is no reason to assume that the astronomical influences have fluctuated in intensity during geological time; rather, other features, which may blur or reinforce the astronomical signal, have fluctuated.

CONCLUSIONS

Regular variations in the orientation of the Earth's axis and its distance from the Sun produce small climatic changes. Though small, these changes of climate do influence climate, oceanography, and sedimentary systems if these are sufficiently sensitive and if disturbance by other features is minimal. Statistical proof for the occurrence of astronomically defined rhythms depends on good time control of successions formed under conditions of continuous deposition and/or on the recognition of characteristic patterns in the astronomically induced cyclicity. For many sedimentary environments which are characterized by irregular rates of deposition and/or in which hiatuses are a common feature, astronomically defined variations of lithology, if indeed present, are difficult to prove on a quantitative basis.

With increasing understanding of the mechanism of orbital forcing it may be possible to recognize Milankovitch rhythms in more places and sedimentary facies. Their recognition may help to further refine stratigraphic resolution. The good time control available for the late Tertiary and Quaternary, and detailed studies such as that by Hilgen & Langereis (1989), offer the possibility of arriving at a detailed cyclostratigraphy, to be expanded eventually towards earlier parts of the Earth's history.

REFERENCES

Adams, J.M., Faure, H. & Petit-Maire, N. (1992) Methane and Milankovitch cycles. *Nature* **355**, 214.

Algeo, T.J. & Wilkinson, B.H. (1988) Periodicity of mesoscale Phanerozoic sedimentary cycles and the role of Milankovitch orbital modulation. *J. Geol.* **96**, 313–322.

Anderson, R.Y. (1982) A long geoclimatic record from the Permian. *J. Geophys. Res.* **87**, 7285–7294.

Anderson, R.Y. (1984) Orbital forcing of evaporite sedimentation. In: *Milankovitch and Climate*, Part 1 (Eds Berger, A.L., Imbrie, J., Hays, J., Kukla, G. & Saltzman, B). Reidel Publ. Co., Dordrecht, pp. 147–162.

Arthur, M.A. (1979) Origin of Upper Cretaceous multicoloured claystones of the Western Atlantic. *Init. Rep. DSDP* **47**, 417–420.

Arthur, M.A., Dean, W.E. & Stow, D.A.V. (1984) Models for the deposition of Mesozoic fine-grained organic-carbon-rich sediment in the deep sea. In: *Fine-grained Sediments: Deep-water Processes and Facies* (Eds Stow, D.A.V. & Piper, D.J.W.). Geol. Soc. Lond. Spec. Publ. 15, pp. 527–560.

Barron, E.J., Arthur, M.A. & Kauffman, E.G. (1985) Cretaceous rhythmic bedding sequences: a plausible link between orbital variations and climate. *Earth Planet. Sci. Lett.* **72**, 327–340.

Berger, A. (1977) Long-term variations of the Earth's orbital elements. *Celestial Mechanics* **15**, 53–74.

Berger, A. (1978a) Long-term variations of daily insolation and Quaternary climatic changes. *J. Atmos. Sci.* **35**, 2362–2367.

Berger, A. (1978b) Long-term variations of caloric insolation resulting from the Earth's orbital elements. *Quat. Res.* **9**, 139–167.

Berger, A. (1979) Insolation signatures of Quaternary climatic changes. *Il Nuovo Cimento* **2C**, 63–87.

Berger, A. (1981) Spectrum of climatic variations and possible causes. In: *Climatic Variations and Variability: Facts and Theories* (Ed. Berger, A.). Reidel Publ. Co., Dordrecht, pp. 411–432.

Berger, A. (1988) Milankovitch theory and climate. *Rev. Geophys.* **26**, 624–657.

Berger, A., Loutre, M.F. & Dehant, V. (1989a) Influence of the changing lunar orbit on the astronomical frequencies of pre-Quaternary insolation patterns. *Paleoceanography* **4**, 555–564.

Berger, A., Loutre, M.F. & Dehant, V. (1989b) Astronomical frequencies for pre-Quaternary palaeoclimate studies. *Terra Nova* **1**, 474–479.

Bottjer, D.J., Arthur, M.A., Dean, W.E., Hattin, D.E. & Savrda, C.E. (1986) Rhythmic bedding in Cretaceous pelagic carbonate environments: sensitive recorder of climatic cycles. *Paleoceanography* **1**, 467–481.

Bradley, W.H. (1929) The varves and climate of the Green River Epoch. *US Geol. Surv. Prof. Pap.* **158-E** (Shorter Contributions to General Geology, 1929), 87–110.

Brosche, P. & Sündermann, J. (Eds) (1978, 1982) *Tidal Friction and the Earth's Rotation I and II*. Springer Verlag, Berlin.

Cotillon, P. (1987) Bed-scale cyclicity of pelagic Cretaceous successions as a result of world-wide control. *Mar. Geol.* **78**, 109–123.

Cottle, R.A. (1989) Orbitally mediated cycles from the Turonian of southern England: their potential for high-resolution stratigraphic correlation. *Terra Nova* **1**, 426–431.

Dean, W.E., Gardner, J.V. & Cepek, P. (1981) Tertiary carbonate-dissolution cycles on the Sierra Leone Rise, Eastern Equatorial Atlantic Ocean. *Mar. Geol.* **39**, 81–101.

De Boer, P.L. (1982) Cyclicity and the storage of organic matter in Middle Cretaceous pelagic sediments. In: *Cyclic and Event Stratification* (Eds Einsele, G. & Seilacher, A.). Springer Verlag, Berlin, pp. 456–475.

De Boer, P.L. (1983) Aspects of Middle Cretaceous pelagic sedimentation in S. Europe. *Geol. Ultraiectina* **31**.

De Boer, P.L. (1991) Pelagic black shale–carbonate rhythms: orbital forcing and oceanographic response. In: *Cycles and Events in Stratigraphy* (Eds Einsele, G., Ricken, W. & Seilacher A.). Springer Verlag, Berlin, pp. 63–78.

De Boer, P.L. & Wonders, A.A.H. (1984) Astronomically induced rhythmic bedding in Cretaceous pelagic sediments near Moria (Italy). In: *Milankovitch and Climate*, Part 1 (Eds Berger, A., Imbrie, J., Hays, J., Kukla, G. & Saltzman, B.). Reidel Publ. Co., Dordrecht, pp. 177–190.

De Boer, P.L., Pragt, J.S.J. & Oost, A.P. (1991) Vertically persistent sedimentary facies boundaries along growth anticlines and climatic control in the thrust-sheet-top south Pyrenean Tremp-Graus foreland basin. *Basin Res.* **3**, 63–78.

Degens, E.T., Emeis, K.-C., Mycke, B. & Wiesner, M.G. (1986) Turbidites, the principal mechanism yielding black shales in the early deep Atlantic Ocean. In: *North Atlantic Paleoceanography* (Eds Summerhayes, C.P. & Shackleton, N.J.). Geol. Soc. Lond. Spec. Publ. 21, pp. 361–376.

Diester-Haass, L. (1991) Rhythmic carbonate content variations in Neogene sediments above the oceanic lysocline. In: *Cycles and Events in Stratigraphy* (Eds Einsele, G., Ricken, W. & Seilacher, A.). Springer Verlag, Berlin, pp. 94–109.

Ditchfield, P. & Marshall, J.D. (1989) Isotopic variation in rhythmically bedded chalks: Paleotemperature variation in the Upper Cretaceous. *Geology* **17**, 842–845.

Eicher, D.L. & Diner, R. (1989) Origin of the Cretaceous Bridge Creek cycles in the Western Interior, United States. *Palaeogeogr. Palaeoclimatol. Palaeoecol.* **74**, 127–146.

Emiliani, C. (1955) Pleistocene temperatures. *J. Geol.* **63**, 538–578.

Fischer, A.G. (1964) The Lofer Cyclothems of the Alpine Triassic. *Kansas Geol. Surv. Bull.* **169**, 107–149.

Fischer, A.G. (1980) Gilbert-bedding rhythms and geochronology. In: *The Scientific Ideas of G.K. Gilbert* (Ed. Yochelson, E.L.). Geol. Soc. Am. Spec. Pap. 183, pp. 93–104.

Fischer, A.G., de Boer, P.L. & Premoli Silva, I. (1990) Cyclostratigraphy. In: *Cretaceous Resources, Events and Rhythms* (Eds Ginsburg, R.N. & Beaudoin, B.). Kluwer Academic Publishers, Dordrecht, pp. 139–172.

Foucault, A., Powichrowski, L. & Prud'homme, A. (1987) Le controle astronomique de la sedimentation turbiditique: exemple du Flysch a Helminthoides des Alpes Ligures (Italie). *C.R. Acad. Sci. Paris* **305/II**, 1007–1011.

Gale, A.S. (1989) A Milankovitch scale for the Cenomanian time. *Terra Nova* **1**, 420–425.

Gilbert, G.K. (1894) Sedimentary measurement of Cretaceous time. *J. Geol.* **3**, 121–127.

Goodfriend, G.A. & Magaritz, M. (1988) Paleosols and late Pleistocene rainfall fluctuations in the Negev Desert. *Nature* **332**, 144–146.

Goodney, D.E., Margolis, S.V., Dudley, W.C., Kroopnick, P. & Williams, D.F. (1980) Oxygen and carbon isotopes of recent calcareous nannofossils as paleoceanographic indicators. *Mar. Micropaleontol.* **5**, 31–42.

Goodwin, P.W. & Anderson, E.J. (1985) Punctuated aggradational cycles: a general hypothesis of episodic stratigraphic accumulation. *J. Geol.* **93**, 515–533.

Grotzinger, J.P. (1986) Cyclicity and paleoenvironmental dynamics, Rocknest platform, northwest Canada. *Geol. Soc. Am. Bull.* **97**, 1208–1231.

Haak, A.B. & Schlager, W. (1989) Compositional variations in calciturbidites due to sea-level fluctuations, late Quaternary, Bahamas. *Geol. Rundsch.* **78**, 477–486.

Hallam, A., Hancock, J.M., Labrecque, J.L., Lowrie, W. & Channell, J.E.T. (1985) Jurassic to Paleogene: part 2, Jurassic and Cretaceous geochronology and Jurassic to Paleogene magnetostratigraphy. In: *The Chronology of the Geological Record* (Ed. Snelling, N.J.). Mem. Geol. Soc. London 10, pp. 118–140.

Hardie, L.A., Bosellini, A. & Goldhammer, R.K. (1986) Repeated subaerial exposure of subtidal carbonate platforms, Triassic, Northern Italy: evidence for high frequency sea level oscillations on a 10^4 year time scale. *Paleoceanography* **1**, 447–457.

Harland, W.B., Armstrong, R.L., Cox, A.V., Craig, L.E., Smith, A.G. & Smith, D.G. (1990) *A Geologic Time Scale 1989*. Cambridge University Press, Cambridge.

Hart, M.B. (1987) Orbitally induced cycles in the chalk facies of the United Kingdom. *Cret. Res.* **8**, 335–348.

Hay, W.W. (1988) Paleoceanography: a review for the GSA Centennial. *Geol. Soc. Am. Bull.* **100**, 1934–1956.

Hays, J.D., Imbrie, J. & Shackleton, N.J. (1976) Variations in the Earth's orbit: pacemaker of the Ice Ages. *Science* **194**, 1121–1132.

Heckel, P.H. (1986) Sea-level curve for Pennsylvanian eustatic marine transgressive–regressive depositional cycles along midcontinent outcrop belt, North America. *Geology* **14**, 330–334.

Herbert, T.D. & Fischer, A.G. (1986) Milankovitch climatic origin of mid-Cretaceous black shale rhythms in central Italy. *Nature* **321**, 739–743.

Hilgen, F.J. (1991a) Astronomical calibration of Gauss to Matuyama sapropels in the Mediterranean and implication for the geomagnetic polarity time scale. *Earth Planet. Sci. Lett.* **104**, 226–244.

Hilgen, F.J. (1991b) Extension of the astronomically calibrated (polarity) time scale to the Miocene/Pliocene boundary. *Earth Planet. Sci. Lett.* **107**, 349–368.

Hilgen, F.J. & Langereis, C.G. (1989) Periodicities of

$CaCO_3$ cycles in the Pliocene of Sicily: discrepancies with the quasi-periods of the Earth's orbital cycles? *Terra Nova* **1**, 409–415.

IMBRIE, J. (1982) Astronomical theory of the Pleistocene ice ages: a brief historical review. *Icarus* **50**, 408–422.

IMBRIE, J. & IMBRIE, K.P. (1979) *Ice Ages: Solving the Mystery*. Enslow Publ., Short Hills, NJ.

IMBRIE, J., HAYS, J.D., MARTINSON, D.G. MCINTYRE, A., MIX, A.C., MORLEY, J.J. *et al.* (1984) The orbital theory of Pleistocene climate: support from a revised chronology of the marine $\delta^{18}O$ record. In: *Milankovitch and Climate*, Pt 1 (Eds Berger, A., Imbrie, J., Hays, J., Kukla, G. & Saltzman, B.). Reidel Publ. Co., Dordrecht, pp. 269–305.

KAUFFMAN, E.G. (1988) Concepts and methods of high-resolution event stratigraphy. *Annu. Rev. Earth Planet. Sci.* **16**, 605–654.

KEMPER, E. (1987) Das Klima der Kreide-Zeit. *Geol. Jb.* **A96**, 5–185.

KUTZBACH, J.E. & OTTO-BLIESNER, B.L. (1982) The sensitivity of the African–Asian monsoonal climate to orbital parameter changes for 9000 years BP in a low-resolution general circulation model. *J. Atmos. Sci.* **39**, 1177–1188.

KUTZBACH, J.E. & STREET-PERROTT, F.A. (1985) Milankovitch forcing in the level of tropical lakes from 18 to 0 kyr BP. *Nature* **317**, 130–134.

LASKAR, J. (1989) A numerical experiment on the chaotic behaviour of the solar system. *Nature* **338**, 237–238.

LEARY, P.N., COTTLE, R.A. & DITCHFIELD, P. (1989) Milankovitch control of foraminiferal assemblages from the Cenomanian of southern England. *Terra Nova* **1**, 416–419.

LEHMAN, T.M. (1989) Upper Cretaceous (Maastrichtian) paleosols in Trans-Pecos Texas. *Geol. Soc. Am. Bull.* **101**, 188–203.

MELNYK, D.H. & SMITH, D.G. (1989) Outcrop to subsurface cycle correlation in the Milankovitch frequency band: Middle Cretaceous, central Italy. *Terra Nova* **1**, 432–436.

MÖRNER, N.A. (1981) Revolution in Cretaceous sea-level analysis. *Geology* **9**, 344–346.

MÜLLER, P.J. & SUESS, E. (1979) Productivity, sedimentation rate, and sedimentary organic matter in the oceans. I. Organic carbon preservation. *Deep-sea Res.* **26A**, 1347–1362.

NAPOLEONE, G. & RIPEPE, M. (1989) Cyclic geomagnetic changes in Mid-Cretaceous rhythmites, Italy. *Terra Nova* **1**, 437–442.

OBRADOVICH, J. & COBBAN, W.A. (1975) A time-scale for the Late Cretaceous of the Western Interior of North America. In: *The Cretaceous System in the Western Interior of North America* (Ed. Caldwell, W.G.E.). Geol. Assoc. Can. Spec. Pap. 13, pp. 31–54.

ODIN, G.S. & KENNEDY, W.J. (1982) Géochimie et géochronologie isotopique-mise à jour de l'échelle des temps Mesozoique. *C.R. Acad. Sci. Paris (II)* **294**, 383–386.

OLSEN, H. (1990) Astronomical forcing of meandering river behaviour: Milankovitch cycles in Devonian of East Greenland. *Palaeogeogr. Palaeoclimatol. Palaeoecol.* **79**, 99–115.

OLSEN, P.E. (1984) Periodicity of lake-level cycles in the Late Triassic Lockatong Formation of the Newark Basin (Newark Supergroup, New Jersey and Pennsylvania). In: *Milankovitch and Climate*, Part 1 (Eds Berger, A.L., Imbrie, J., Hays, J., Kukla, G. & Saltzman, B.). Reidel Publ. Co., Dordrecht, pp. 129–146.

OLSEN, P.E. (1986) A 40-million year lake record of Early Mesozoic orbital climatic forcing. *Science* **234**, 842–848.

PANELLA, G. (1972) Paleontological evidence on the Earth's rotational history since early Precambrian. *Astrophys. Space Sci.* **16**, 212–237.

POSAMENTIER, H.W. & VAIL, P.R. (1988) Eustatic controls on clastic deposition II – Sequence and systems tract models. In: *Sea-level Changes – an Integrated Approach* (Eds Wilgus, C.K., Posamentier, H., Ross, C.A. & Kendall, Chr. G. St C.). Soc. Econ. Paleont. Mineral. Spec. Publ. 41, pp. 125–154.

POSAMENTIER, H.W., JERVEY, M.T. & VAIL, P.R. (1988) Eustatic controls on clastic deposition I – conceptual framework. In: *Sea-level Changes – an Integrated Approach* (Eds Wilgus, C.K., Posamentier, H., Ross, C.A. & Kendall, Chr. G. St C.). Soc. Econ. Paleont. Mineral. Spec. Publ. 41, pp. 109–124.

PRATT, L.M. (1984) Influence of paleoenvironmental factors on the preservation of organic matter in Middle Cretaceous Greenhorn Formation, Pueblo, Colorado. *Bull. Am. Assoc. Petrol. Geol.* **68**, 1146–1159.

PREMOLI SILVA, I., RIPEPE, M. & TORNAGHI, M.E. (1989) Planktonic foraminiferal distribution record productivity cycles: evidence from the Aptian–Albian Piobicco core (central Italy). *Terra Nova* **1**, 443–448.

REIJMER, J.J.G., TEN KATE, W.G.H.Z., SPRENGER, A. & SCHLAGER, W. (1991) Calciturbidite composition related to exposure and flooding of a carbonate platform (Triassic, Eastern Alps). *Sedimentology* **38**, 1059–1074.

ROHLING, E.J. & HILGEN, F.J. (1991) The eastern Mediterranean climate at times of sapropel formation: a review. *Geol. Mijnbouw* **70**, 253–264.

ROSSIGNOL-STRICK, M. (1985) Mediterranean Quaternary sapropels, an immediate response of the African monsoon to variations in insolation. *Palaeogeogr. Palaeoclimatol. Palaeoecol.* **49**, 237–263.

SANDER, B. (1936) Beiträge zur Kenntnis der Ablagerungsgefüge. *Mineral. Petrogr. Mitt.* **48**, 27–139.

SAVRDA, C.E. & BOTTJER, D.J. (1986) Trace fossil model for reconstruction of paleo-oxygenation in bottom water. *Geology* **14**, 3–6.

SCHWARZACHER, W. (1947) Uber die sedimentäre Rhythmik des Dachsteinkalkes von Lofer. *Verh. Geol. Bundesanstalt Wien*, **H10–12**, 175–188.

SCHWARZACHER, W. (1954) Die Grossrhythmik des Dachsteinkalkes von Lofer. *Tschermaks Mineral. Petrograph. Mitt.* **4**, 44–54.

SCHWARZACHER, W. & FISCHER, A.G. (1982) Limestone-shale bedding and perturbations of the Earth's orbit. In: *Cyclic and Event Stratification* (Eds Einsele, G. & Seilacher, A.). Springer Verlag, Berlin, pp. 72–95.

SCHWARZACHER, W. & HAAS, J. (1986) Comparative statistical analysis of some Hungarian and Austrian Upper Triassic peritidal carbonate sequences. *Acta Geol. Hung.* **29**, 175–196.

SHACKLETON, N.J. (1989) ODP Site 677: A case for revising the astronomical calibration for the Brunhes–Matuyama and Jaramillo boundaries. *Terra Abstracts* **1**, 185.

SMITH, D.G. (1989) Stratigraphic correlation of presumed Milankovitch cycles in the Blue Lias (Hettangian to earliest Sinemurian), England. *Terra Nova* **1**, 457–460.

STRASSER, A. (1988) Shallowing upward sequences in Purbeckian peritidal carbonates (lowermost Cretaceous, Swiss and French Jura Mountains). *Sedimentology* **35**, 369–383.

VAN BUCHEM, F.S.P. & MCCAVE, I.N. (1989) Cyclic sedimentation patterns in Lower Lias mudstones of Yorkshire (GB). *Terra Nova* **1**, 461–467.

VANDENBERG, J., DE BOER, P.L. & KREULEN, R. (1983) Longterm secular variations of the magnetic field recorded in Late Albian pelagic sediments. *Geol. Utraiectina* **31**, 105–111.

VAN HINTE, J.E. (1976) A Cretaceous time scale. *Bull. Am. Assoc. Petrol. Geol.* **60**, 498–516.

VAN HOUTEN, F.B. (1962) Cyclic sedimentation and the origin of analcime-rich Upper Triassic Lockatong Formation, west-central New Jersey and adjacent Pensylvania. *Am. J. Sci.* **260**, 561–576.

VAN HOUTEN, F.B. (1964) Cyclic lacustrine sedimentation, Upper Triassic Lockatong Formation, Central New Jersey and adjacent Pennsylvania. *Kansas Geol. Surv. Bull.* **169**, 497–531.

VAN WAGONER, J.C., POSAMENTIER, H.W., MITCHUM, R.M., VAIL, P.R., SARG, J.F., LOUTT, T.S. & HARDENBOL, J. (1988) An overview of the fundamentals of sequence stratigraphy and key definitions. In: *Sea-level Changes – an Integrated Approach* (Eds Wilgus, C.K., Posamentier, H., Ross, C.A. & Kendall, Chr. G. St C.). Soc. Econ. Paleont. Mineral. Spec. Publ. 41, pp. 39–45.

VAN WOERKOM, A.J.J. (1953) The astronomical theory of climatic changes. In: *Climatic Change, Evidence, Causes and Effects* (Ed. Shapley, H.). Harvard University Press, Cambridge, MA, pp. 147–157.

WELTJE, G.J. & DE BOER, P.L. (1993) Astronomically induced paleoclimatic oscillations reflected in Pliocene turbidite deposits on Corfu (Greece): implications for the interpretation of higher order cyclicity in fossil turbidite systems. *Geology* **21**, 307–310.

WILLIAMS, G.E. (1991a) Upper Proterozoic tidal rhythmites, South Australia: sedimentary features, deposition, and implications for the Earth's paleorotation. In: *Clastic Tidal Sedimentology* (Eds Smith, D.G. *et al.*). Can. Soc. Petrol. Geol. Mem. 16, pp. 161–178.

WILLIAMS, G.E. (1991b) Milankovitch-band cyclicity in bedded halite deposits contemporaneous with Late Ordovician–Early Silurian glaciation, Canning Basin, Western Australia. *Earth Planet. Sci. Lett.* **103**, 143–155.

WOLLIN, G., RYAN, W.B.F., ERICSON, D.B. & FOSTER, J.H. (1977) Paleoclimate, paleomagnetism and the eccentricity of the Earth's orbit. *Geophys. Res. Lett.* **4**, 267–270.

WOLLIN, G., RYAN, W.B.F. & ERICSON, D.B. (1978) Climatic changes, magnetic intensity variations and fluctuations of the eccentricity of the Earth's orbit during the past 2 000 000 years and a mechanism which may be responsible for the relationship. *Earth Planet. Sci. Lett.* **41**, 395–397.

ZIJLSTRA, J.J.P. (1993) Sedimentology of the Maastrichtian Chalk near Maastricht, The Netherlands. *Geol. Ultraiectina* (in preparation).

Spec. Publs Int. Ass. Sediment. (1994) **19**, 15–24

Astronomical forcing through geological time

A. BERGER *and* M.F. LOUTRE
Institut d'Astronomie et de Géophysique G. Lemaître, 2 Chemin du Cyclotron, 1348 Louvain-la-Neuve, Belgium

ABSTRACT

The sensitivity of the amplitudes and frequencies in the development of the Earth's orbital elements involved in the astronomical theory of palaeoclimates (obliquity and climatic precession) to the Earth–Moon distance and consequently to the length of the day and to the dynamical ellipticity of the Earth is investigated for the last few billions of years. The influence of the stability of the fundamental frequencies of planetary motion is also discussed.

The value of the amplitudes and frequencies for the most important terms of obliquity and climatic precession, as well as the value of the independent term for obliquity, have been computed for the last 2500 Ma with the assumption that the chaotic behaviour of the solar system may be disregarded, which is realistic for the last 200 Ma, and that the general solution of the planetary system has kept its general form over the last billion years. Moreover the time scale presented here assumes only two different lunar recession rates, while it could have varied more continuously.

The shortening of the Earth–Moon distance and of the length of the day, as well as the lengthening of the dynamical ellipticity of the Earth back in time, induce a shortening of the fundamental astronomical periods for precession (the 19-ka and 23-ka quasi-periods becoming respectively 11.3 and 12.7 ka at 2500 Ma BP) and for obliquity (the 41-ka and 54-ka quasi-periods becoming respectively 16.7 and 18.5 ka at 2500 Ma BP). At the same time, one observes for obliquity at 2500 Ma BP a relative enlargement of about 60% of the amplitudes, but a very weak increase (less than 0.1%) of the independent term. On the other hand, the amplitudes of precession change very little. These changes in the frequencies and amplitudes for both obliquity and climatic precession are larger for longer period terms.

The periods in eccentricity development are not influenced by variation of the lunar distance. Only the long-term changes of the fundamental planetary frequencies seem to induce very small variations in these periods, of the order of 1.5% over the last 200 Ma, although the chaotic behaviour of the planetary system complicates the problem particularly before 100 Ma BP.

INTRODUCTION

The aim of the astronomical theory of palaeoclimates, a particular version of which comes from Milankovitch (1920, 1941), is to study the relationship between insolation and climate on the global scale. It comprises four different parts: the theoretical computation of the long-term variations of the Earth's orbital parameters and related geometrical insolations; the design of climatic models to transfer the insolation into climate; the collection of geological data and their interpretation in terms of climate; and the comparison of these proxy data to the simulated climatic variables (Berger, 1988).

The purpose of this paper is to analyse the stability, over the whole history of the Earth, of the orbital elements required for the computation of the insolations used to force the climatic models (i.e. the eccentricity e, the obliquity ε and the climatic precession $e \sin \tilde{\omega}$). This indeed has proved to be of the utmost interest, given that more and more geological data are becoming available to test the validity of the astronomical theory of palaeoclimate over a broader and broader time range. Spectral analysis of Quaternary palaeoclimatic records has provided substantial evidence (Imbrie *et al.*, 1984,

1989; Berger, 1989a) that, at least near the obliquity and precession frequencies, a considerable fraction of the climatic variance is driven in some way by insolation changes forced by changes in the Earth's orbit (Hays *et al.*, 1976; Berger, 1977).

For pre-Quaternary times, it was nearly 100 years ago that Gilbert (1895) suggested that the astronomical frequencies can be found with some degree of confidence in geological records beginning with the Mesozoic (de Boer & Wonders, 1984; Fischer & Schwarzacher, 1984; Fischer, 1986; Olsen, 1986; Berger, 1989b). The sensitivity of climate to astronomical forcing is also becoming clear, in particular for the Cretaceous (Barron *et al.*, 1985), so that there is an imperative need now to understand how and why the orbital frequencies change over the geological time scale.

In the first part of this paper we will describe the astronomical elements used in the astronomical theory of palaeoclimate; then the stability of their trigonometrical expansion (period, amplitude and independent term) will be analysed. The influence of the precessional parameter will be studied first as it is the major source of variation for the astronomical parameters.

ORBITAL ELEMENTS

The solar energy available at any given latitude ϕ on the Earth, on the assumption of a perfectly transparent atmosphere and of a constant solar output, depends upon the semi-major axis of the ecliptic (a), its eccentricity (e), its obliquity (ε) (tilt of the axis of rotation) and the longitude of the perihelion measured from the moving vernal equinox ($\tilde{\omega}$), which combines with e to define the climatic precession parameter $e \sin \tilde{\omega}$ (Berger, 1978a).

The semi-major axis (a) and the eccentricity (e) specify the size and shape of the Earth's orbit around the Sun. In order to measure the angles used to locate the orbit in space and the Earth on it, a reference frame is fixed with a reference plane, usually the ecliptic (orbital plane) at a particular fixed date of reference called the ecliptic of epoch (Ec_0), and a point in that plane, the vernal equinox (γ_0) or First Point of Aries indicating the position of the Sun when it crosses the celestial equator from the austral to the boreal hemisphere. The orientation of the orbital plane in space with regard to the reference plane is specified by two angles (Fig. 1): the longitude (Ω) of the ascending node (N) measured on the ecliptic of epoch between γ_0 and the intersection between the ecliptics of date (Ec) and of epoch (Ec_0), and the inclination (i) which is the angle between these two planes. The position of the perihelion is measured by the longitude of the perihelion (π) which is the sum of two angles lying in different planes: Ω and the angular distance in the orbital plane from the ascending node (N) to the perihelion.

Due to the attraction of the Moon and of the Sun,

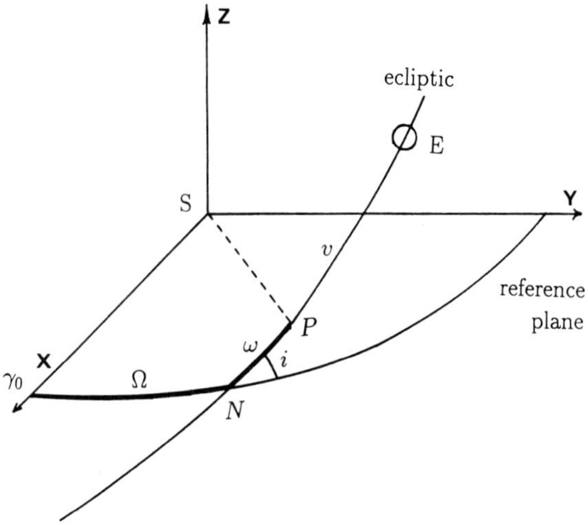

Fig. 1. Position of the Earth (E) around the Sun (S). In astronomy, it is usual to define an orbit and the position of the body describing that orbit by six quantities called the elements. Three of them define the orientation of the orbit with respect to a set of axes, two of them define the size and the shape of the orbit (a and e respectively), and the sixth defines the position of the body within the orbit at a given time. In the case of a planet moving in an elliptic orbit about the Sun, it is convenient to take a set of rectangular axes in and perpendicular to the plane of reference, with the origin at the centre of the Sun. The x-axis may be taken towards the ascending node N, the y-axis being in the plane of reference and 90° from x, while the z-axis is taken to be perpendicular to this reference plane so that the three axes form a rectangular coordinate system. γ_0 is the reference point from where the angles are measured. As the reference plane is usually chosen to be the ecliptic at a particular fixed date of reference (named epoch of reference in celestial mechanics; Woolard & Clemence, 1966), γ_0 is, in such a case, the vernal equinox at that fixed date (the vernal equinox is also referred to as the First Point of Aries indicating the position of the Sun when it crosses the celestial equator from the austral to the boreal hemisphere). P is the perihelion; Ω, the longitude of the ascending node; ω, the argument of the perihelion; $\pi = \Omega + \omega$, the longitude of the perihelion; i, the inclination; v, the true anomaly; $\lambda = \pi + v$, the longitude of the Earth in its orbit.

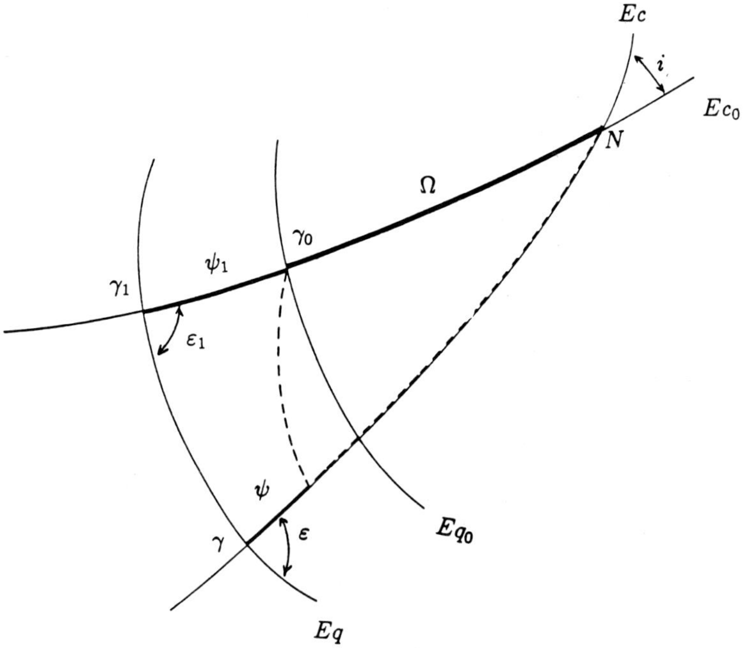

Fig. 2. Precession and obliquity. γ is the vernal equinox of date; γ_0 the vernal equinox of reference; $\gamma_0\gamma_1 = \psi_1$ the luni-solar precession in longitude; ψ the general precession in longitude which provides the longitude of the moving perihelion $\tilde{\omega}$ through $\tilde{\omega} = \pi + \psi$; ε_1 the angle between the equator of date and the ecliptic of reference and ε the obliquity (Berger, 1984); as we are interested in the long-term variations of the astronomical elements, their short-term variations are removed and γ and γ_0 are more adequately referred to as mean vernal equinoxes.

the equatorial plane is not fixed in space, nor is the vernal equinox. Their positions are given respectively by the obliquity, which is the angle between the ecliptic (Ec) and the equator (Eq), and by the general precession (ψ) (Fig. 2). This last angle provides the longitude of the moving perihelion $\tilde{\omega}$ through $\tilde{\omega} = \pi + \psi$.

The expansion of the astro-climatic parameters (e, ε and $e \sin \tilde{\omega}$) can be obtained from the trigonometrical expansion of (e, π) and (i, Ω):

$$e \sin \pi = \sum_{j=1}^{m} M_j \sin(g_j t + \beta_j) \tag{1}$$

$$\sin \frac{i}{2} \sin \Omega = \sum_{i=1}^{n} N_i \sin(s_i t + \delta_i) \tag{2}$$

where M_j, g_j, β_j, and N_i, s_i, δ_i are the fundamental amplitudes, frequencies and phases, which can be found in the development of the astro-climatic parameters:

$$e = e_0 + \Sigma E_i \cos(\lambda_i t + \phi_i) \tag{3}$$

$$\varepsilon = \varepsilon^* + \Sigma A_i \cos(\tilde{f}_i t + \tilde{\delta}_i) \tag{4}$$

$$e \sin \tilde{\omega} = \Sigma P_i \sin(\alpha_i t + \zeta_i) \tag{5}$$

where the frequencies λ_i are linear combinations of g_i only and $\tilde{f}_i$ and α_i are of the following form:

$$s_i + k,\ 2(s_i + k),\ s_i + s_j + 2k,\ s_i + g_j + 2k,\ g_i + k,\ s_i - s_j,\ s_i - g_j \tag{6}$$

The precessional frequency, k, whose present-day value is 50.439 273″/yr, arises through the solution of the Poisson equations describing the Earth–Moon system. Its analytical expression will be given in the next section in equation (7). The amplitudes A_i and P_i in equations (4) and (5) are functions of the N_i, M_j, s_i, g_j and k; the largest of them are respectively of the form $-c_i N_i$ and M_i, c_i being given by equation (11). The expression for the independent term ε^* is given in equations (12) and (13).

SENSITIVITY OF THE PRECESSIONAL FREQUENCY k TO THE 'SLOW' AND 'FAST' VARYING PARAMETERS

The frequency k, introduced through the Poisson equations for precession, can be written as follows:

$$k = \frac{3}{2} \frac{n^2}{\omega} \frac{C - A}{C} \left((1 - e^2)^{-1.5} + \frac{m_{☾}}{m_{\odot}} \frac{a^3}{a_{☾}^3} (1 - e_{☾}^2)^{-1.5} \left(1 - \frac{3}{2} \sin^2 i_{☾}\right) \right) \cos h \tag{7}$$

where n is the mean motion of the Sun in a geocentric reference frame, ω is the rotational angular velocity of the Earth, A and C are the Earth's moments of inertia around the equatorial and polar principal axes of inertia (they combine to define the dynamical ellipticity of the Earth: $H = (C - A)/C$), $a_☾$ and $e_☾$ are the semi-major axis and the eccentricity of the Moon's orbit around the Earth, $i_☾$ is the inclination of the lunar orbit on the ecliptic, $m_☾$ and $m_\odot$ are the masses of the Moon and the Sun respectively, and h is a constant whose value is 23.401 09°. Thus a single change in any one of these parameters influences the value of k.

A set of accurate constants (Table 1) has been used here and by Berger *et al.* (1989a, b) and Loutre & Berger (1989) to reproduce correctly the present-day value of k. Following the sensitivity analysis given by Berger *et al.* (1989b), we can say that the relative impact of the variations of the fast varying parameters $e_☾$ and $i_☾$ on k is at maximum 1% over the last 2000 Ma. Consequently, the value of the eccentricity and inclination of the lunar orbit will be kept constant and equal to their present-day values for all further calculations:

$$e_☾ = 0.055$$

$$i_☾ = 5.15°$$

The sensitivity of k to the variation of the eccentricity of the Earth's orbit has also been tested, showing a maximum relative impact of less than 1% over the same interval of time. Therefore, the value of the eccentricity of the Earth's orbit is also kept constant to its present-day value in this paper:

$$e = 0.01675$$

The impact of the slow-varying parameters, i.e. the semi-major axis of the lunar orbit ($a_☾$), the dynamical ellipticity (H) and the rotational angular velocity of the Earth (ω) has been estimated for the last 440 Ma by Berger *et al.* (1989a, b) and Berger *et al.* (1989c): the values of $a_☾$ have been computed with the expression given by Lambeck (1980)

$$a_☾ = a_0\left[1 - \frac{13}{2}t\frac{\dot{a}}{a_0}\right]^{2/13} \quad (8)$$

where time, t, is positive back from the present, $a_0 =$ 384 000 km is the present value of the semimajor axis of the lunar orbit and $\dot{a}$ is the present average rate of the lunar recession. Prior to 590 Ma, its value has been taken as 0.43×10^{-9} m/s according to Walker & Zahnle (1986); its present-day value (10^{-9} m/s) was chosen from published ones deduced from both indirect proxy data and recent observations. Indeed, evidence for the rate of lunar recession is available on several different time scales: the most important source of long-term information derives from fossil corals, bivalves and stromatolites which have recorded astronomical cycles in their growth rhythms (Lambeck, 1980). Historical records of eclipses and other astronomical phenomena give information over the past few millenia (Stephenson, 1972). More recently, the lunar orbit and the rotation of the Earth have been monitored by lunar laser ranging (Calame & Mulholland, 1978). With these values of $\dot{a}$, the Moon remains well beyond the Roche limit during the time considered. At the Roche limit of about three Earth radii the self attraction of the satellite would be exceeded by the tidal attraction of the planet, and the satellite would become gravitationally unstable and break up (Stacey, 1977, quoted by Lambeck, 1980). The discontinuity in the lunar recession rate at 590 Ma BP is artificial and has been introduced by Walker & Zahnle to link the data of the last 500 Ma and of the Phanerozoic rate. In fact, the rate of lunar recession has been assumed to be constant over each of the two time intervals (before and after 590 Ma BP), but it could have varied within these time intervals, so that the time values tested in Table 2 are only a best guess. The acquisition of new geological data and/or the building up of a precise model of the Earth–Moon system over the whole history of the Earth would allow $\dot{a}$ to be estimated much more precisely. On the other hand, Sonett *et al.* (1988), studying the Elatina Sandstone laminae, estimated the mean rate of retreat of the Moon to be about 2.02 cm/yr during the late Precambrian, which is quite different from the value of 1.36 cm/yr used here.

Table 1. The set of constants used in the computation

Constants used in the computation	
Length of the year	31 471 949.5 s
Mass of the Moon/mass of the Sun	1/26 925 170
Semi-major axis of the orbit of the Earth	149 729 668 km

For the last 440 Ma, the value of the length of the day (l.o.d. $= 2\pi/\omega$) and of $H = (C - A)/C$ have been taken from Stoyko (1970) and Denis (1986). In the model used by Denis to estimate H, the Earth is assumed to have remained in a nearly hydrostatic state throughout geological history, and the inner structure is considered not to have undergone important modifications. The computation assumes the

Table 2. Computed value of k for different geological times taking into account:
1 the variation of the Earth–Moon distance (Lambeck, 1980), of the length of the day (Stoyko, 1970) and of the dynamical ellipticity of the Earth (Denis, 1986) over the last 440 Ma (Berger *et al.*, 1989b, c); and
2 the variation of the Earth–Moon distance (Lambeck, 1980), of the length of the day (Lambeck, 1980) and of the dynamical ellipticity of the Earth (Walker & Zahnle, 1986) until 2500 Ma BP

	Geological period	Date (Ma BP)	$a_{☾}$ (km)	l.o.d. (s)	$\frac{C-A}{C}$ 10^{-6}	k ("/yr)
1	Holocene	0	384 000	86 164	3241	50.439 273
	Late Cretaceous	72	381 690	85 055	3326	51.738 855
	Early Permian	270	374 908	81 962	3581	55.705 000
	Late Carboniferous	298	373 892	81 119	3656	56.623 819
	Middle Devonian	380	370 828	78 888	3866	59.262 787
	Early Silurian	440	368 493	77 342	4022	61.256 931
2		0	384 000	86 164	3240	50.431 631
		72	381 690	84 918	3336	51.800 236
		270	374 908	81 441	3627	56.070 946
		298	373 892	80 942	3672	56.742 065
		380	370 828	79 468	3810	58.818 901
		440	368 493	78 378	3916	60.432 739
		500	366 075	77 275	4029	62.173 118
		1000	356 460	73 151	4496	69.585 861
		1500	348 580	70 050	4903	76.356 342
		2000	339 578	66 776	5395	84.964 091
		2500	329 035	63 263	6011	96.444 174

density distribution of the Preliminary Earth Reference Model (PREM) (Dziewonski & Anderson, 1981) and is based on the classical theory of the Earth's figure expanded to order 3 in the flattening. The equatorial (A) and the polar (C) principal moments of inertia are computed for each geological interval by integration over the whole volume of the Earth, taking into account the time-variable rotation and the flattening of the Earth layers. For periods prior to 440 Ma, for which we do not have the values of H computed through an Earth model, we used an approximate formula deduced from Walker & Zahnle (1986):

$$H = \frac{C - A}{C} = 6.094 \times 10^5 \times \omega^2 \qquad (9)$$

an expression which fits quite correctly the model values for $0 < t < 440$ Ma. Moreover, in this case, we used the tidal rotational angular velocity of the Earth given by Lambeck (1980):

$$\omega = \omega_0\left[-4.87\left(\frac{a_{☾}}{a_0}\right)^{0.5} + 5.87\right] \qquad (10)$$

with

$$\omega_0 = 7.292\,124 \times 10^{-5}\,\text{rad/s}$$

It is expected that much more data will become available for these oldest periods, such as the Precambrian (Sonett *et al.*, 1988; Williams, 1989), so that our computation could reflect as closely as possible what the Milankovitch cyclicities would be during these remote times of the past.

The present-day value of k is estimated to be about 50.44″/yr, which means that γ takes about 25 700 years to circle the ecliptic. Its value increases in the past to reach about 61″/yr at the Early Silurian (corresponding to a 21 250-year period) and more than 95″/yr at 2500 Ma BP (corresponding to a period of 13 438 years). The values of k displayed in Table 2 for the last 440 Ma show good agreement (a departure of less than 2%) between the two sets of parameter values used.

VARIATIONS OF THE PERIODS

The variation of the periods in the trigonometrical expansions (4) and (5) comes from k and from the fundamental frequencies s_i and g_j of the planetary point-mass systems (1) and (2). The stability of these frequencies s_i and g_j has been questioned by Laskar (1989). According to his computations, it is not possible to express his numerical solution for the motion of the planets as a quasi-periodic function which remains valid over the whole 200 Ma of his integration without letting s and g vary with time. When applied to different smaller time intervals, his harmonic analysis shows significant variations of the main frequencies of the inner planets suggesting that the motion of the solar system is chaotic on the geological time scale (Laskar, 1990), which has two consequences. First, it is impossible to compute the exact motion of the solar system over more than about 100 Ma and the solution given over 200 Ma by Laskar (1990) can be considered as only a qualitative possibility before 100 Ma BP. Second, the fundamental frequencies of the planetary system, the s and g frequencies, are definitely not fixed quantities, which would be the case if the motion were not chaotic. However, it seems reasonable to assume that the diffusion of the frequencies was qualitatively the same before 100 Ma BP as over the time interval from 100 Ma BP until now for which the calculation by Laskar (1990) is assumed to be reliable enough. This means that the stability of the main frequencies of e, $e \sin \tilde{\omega}$ and ε over the last 200 Ma can now be tested using this assumption. Over these last 200 Ma, the maximum deviation from the present-day value of the planetary frequencies is 0.20″/yr for g_4, 2×10^{-5}″/yr for g_5, 0.06″/yr for s_3 and 8×10^{-5}″/yr for s_6; these frequencies correspond to the largest amplitude terms in expansions 4 and 5 of ε and $e \sin \tilde{\omega}$ (Berger & Loutre, 1990). Over the same interval of time, the maximal variation of k is of the order of 5″/yr. So, the changes in g frequencies and s frequencies do not seem to be able to counteract the variations of k in the obliquity and climatic precession periods. Therefore, we may consider that the time variation of k (due to the effect of the changing lunar orbit) represents the most important part of the time variation of the frequencies of obliquity and climatic precession over the whole history of the Earth. Assuming that the general solution of the planetary system keeps its general form over the last 1000 Ma, we will therefore consider the sensitivity, to the precessional parameter k only of the frequencies of the main terms (with the largest amplitudes) in the development of ε and $e \sin \tilde{\omega}$.

The frequencies of the most important terms of the obliquity and of the precessional parameter are of the form $s_i + k$ or $g_j + k$ (Sharaf & Boudnikova, 1967; Berger & Loutre, 1990). This shows clearly that any single variation of k has an impact on these frequencies but also that the largest relative variations in the periods will correspond to the longest periods (the lowest frequencies).

For the last 440 Ma, the impact of k on the astronomical periods has been discussed in detail by Berger *et al.* (1989a, b, c) where the relative impact of each of the fast- and slow-varying parameters has been estimated. At 440 Ma BP, according to the value of k, the astronomical periods 19 000, 23 000, 41 000 and 54 000 years become respectively 16 400, 19 300, 30 500 and 37 200 years, i.e. a relative change of between 13 and 30%.

The periods for the obliquity and climatic precession can be computed for times prior to 500 Ma BP assuming that approximation (9) stays valid over that timespan and using the values of $a_{\subset}$ and ω calculated from equations (8) and (10). Table 3 shows that the relative influence of these 'slow'-varying elements together can amount to a decrease of about 65% for the 54 000-year period of the obliquity over the last 2500 Ma. Moreover, at 2500 Ma the four periods considered are very close to each other.

Finally it is interesting to determine whether or not, and eventually to what extent, the eccentricity period is affected by these changes. From equation (1), it can be seen that the eccentricity frequencies depend only on the fundamental frequencies (g),

Table 3. Estimated values of the main periods of the palaeoclimatic parameters taking into account the two points described in Table 2

	Date (Ma BP)	Precession periods (yr)		Obliquity periods (yr)	
		19 000	23 000	41 000	54 000
1	0	19 000	23 000	41 000	54 000
	72	18 645	22 481	39 381	51 226
	270	17 638	21 034	35 145	44 284
	298	17 421	20 725	34 291	42 936
	380	16 824	19 886	32 053	39 484
	440	16 399	19 296	30 546	37 222
2	0	19 000	23 000	41 000	54 000
	72	18 622	22 449	39 280	51 055
	270	17 545	20 902	34 778	43 703
	298	17 387	20 678	34 163	42 736
	380	16 916	20 015	32 390	39 997
	440	16 567	19 529	31 134	38 099
	500	16 207	19 029	29 885	36 245
	1000	14 832	17 162	25 522	30 021
	1500	13 765	15 750	22 520	25 951
	2000	12 612	14 258	19 590	22 136
	2500	11 345	12 659	16 693	18 507

which is also evident from the analytical expansion of e given in Berger (1978b) and Berger & Loutre (1990). This is related to the fact that in the solution of the planetary system, the Earth is represented by the Earth–Moon system at its barycentre, so that the precessional parameter, k, cannot appear. However, in a planetary system where the Earth and the Moon were considered explicitly, the change of $a_{☾}$ would affect the frequencies of the eccentricity. This effect is about 0.02″/yr and varies with $a_{☾}$, but these variations are much smaller than the variations due to the chaotic motion. Therefore, since the largest deviations for the g frequencies related to the chaotic motion of the planets are less than 0.2″/yr over the last 200 Ma, the period of the five largest amplitude terms in the development of eccentricity can change only by roughly 1.5% over this time interval. This seems to be confirmed by recent results obtained by G. Bond (Lamont Doherty Geological Observatory, Columbia University, New York; personal communication) analysing proxy-palaeoclimatic data for the early Palaeozoic, Triassic and Jurassic.

VARIATIONS OF THE AMPLITUDES

The amplitudes of the most important terms in expressions (4) for the obliquity and (5) for the climatic precession are respectively of the form $-c_iN_i$ and M_i (Berger, 1978b; Berger & Loutre, 1990). The values for N_i and M_i are given by solutions (1) and (2) of the Earth's motion in the solar system and c_i is given by Sharaf & Boudnikova (1967), Berger (1976) and Dehant *et al.* (1987) as

$$c_i = \frac{-s_i}{s_i + k} \tag{11}$$

where s_i represents the frequencies in equation (2). The numerical values considered here are those of Berger (1978a), i.e. the numerical values for the development of equations (1) and (2) are those taken from Bretagnon (1974). As we are mostly interested in the relative amplitude of the changes, our conclusions will remain the same if we use the improved solution of Laskar (1988). Consequently, as long as this solution keeps its general form as given by equations (1) and (2) back in time, the amplitude of the most important terms in the expansion of climatic precession will not be influenced by the variation of k, while the amplitude of the first terms in the expansion of the obliquity will change.

A sensitivity analysis of the individual impact of $a_{☾}$, H and ω on the amplitude of the terms in equations (4) and (5) has been made by Loutre & Berger (1989). Because of the form of equation (11), the relative decrease (in absolute value) of the amplitudes in equation (5) will be the same as that of their corresponding period. This decrease amounts to 30% for the amplitude of the 54 000-year period term over the last 440 Ma; it may reach more than 60% at 2500 Ma BP (Table 4).

Table 4. Estimated amplitude of the four most important terms of the expansion of the obliquity taking into account the two points described in Table 2

	Date (Ma BP)	Amplitudes (″)			
		41 000	39 730	53 615	40 521
1	0	−2462.22	−857.32	−629.32	−414.28
	72	−2364.99	−824.48	−597.22	−398.10
	270	−2110.62	−738.17	−516.75	−355.72
	298	−2059.31	−720.69	−501.11	−347.15
	380	−1924.91	−674.80	−461.03	−324.70
	440	−1834.44	−643.82	−434.76	−309.58
2	0	−2462.82	−857.52	−629.52	−414.38
	72	−2360.59	−822.99	−595.78	−397.37
	270	−2089.88	−731.10	−510.41	−352.26
	298	−2052.89	−718.50	−499.17	−346.08
	380	−1946.27	−682.10	−467.32	−328.27
	440	−1870.78	−656.27	−445.25	−315.65
	500	−1795.66	−630.52	−423.67	−303.09
	1000	−1533.41	−540.23	−351.17	−259.15
	1500	−1352.94	−477.75	−303.70	−228.84
	2000	−1176.85	−416.50	−259.16	−199.23
	2500	−1002.78	−355.69	−216.77	−169.90

VARIATIONS OF THE INDEPENDENT TERM OF THE OBLIQUITY

The expansion of the obliquity is given by the deviation around a value ε^* which is dependent upon k and therefore can vary in time; ε^* has the following form:

$$\varepsilon^* = h - \Sigma N_i^2\left[\frac{1}{2}a_i(a_i - s)\tan h + \frac{1}{4}(2a_i - 1)\cot h\right] \tag{12}$$

where

$$a_i = \frac{k}{s_i + k} \tag{13}$$

However, the relative impact on ε^* of the variation of the lunar distance, of the dynamical ellipticity of

Table 5. Estimated amplitudes of the most important terms of the expansion of the obliquity taking into account the two points described in Table 2

	Date (Ma BP)	Independent term (°)
1	0	23.320 556
	72	23.321 688
	270	23.324 620
	298	23.325 206
	380	23.326 733
	440	23.327 754
2	0	23.320 549
	72	23.321 739
	270	23.324 857
	298	23.325 279
	380	23.326 491
	440	23.327 345
	500	23.328 190
	1000	23.331 115
	1500	23.333 104
	2000	23.335 027
	2500	23.336 910

the Earth and of the length of the day amounts only to about 0.03% at the Early Silurian and 0.07% at 2500 Ma BP (Table 5), which is negligible for our geological studies.

CONCLUSIONS

The sensitivity of the amplitudes, periods and independent term for the development of obliquity and of climatic precession to the Earth–Moon distance, the length of the day and the dynamical ellipticity of the Earth has been analysed for the last 2500 Ma. Assuming that the chaotic behaviour of the solar system may be neglected, which is realistic for the last 200 Ma, and that the general solution of the planetary system keeps its general form over the last 1000 Ma, the variation in the past of the fundamental frequencies of the planetary system (s and g) is far less important than the variation of the precessional parameter (k). Therefore we may assume that the frequencies of the obliquity and climatic precession are mainly sensitive to variations of k rather than of s and g. Nevertheless, our results merit further testing taking into account the chaotic behaviour of the planetary system and using improved models of the Earth's interior and of the Earth–Moon system. In particular, we have adopted for the numerical value of the mean retreat rate for the Moon the current Apollo laser value of 3.16 cm/yr for the last 590 Ma and 1.36 cm/yr for earlier times, whereas Sonett *et al.* (1988) found a mean late Precambrian value of 2.02 cm/yr from the interpretation of the Elatina Sandstone laminae of South Australia.

With the values of $a_{☾}$ given by Lambeck (1980), of ω given by Stoyko (1970) and of H given by the model of Denis (1986) for the last 440 Ma, it is concluded that, over that timespan, k has increased back in time by 21%, ε^* has scarcely changed, the amplitudes of the larger terms of the obliquity have decreased in absolute value by about 30% and their frequencies have increased by about 35–45% (which means a decrease of the periods by 25–31%), while those of the precession also increased by 15–20% (a decrease of the periods by 14–16%). It must be stressed that it is only over that time interval that a full model of the Earth's interior provides us with the most accurate values of H as a function of ω and $a_{☾}$.

Using a simplified formula of Walker & Zahnle (1986) where the dynamical ellipticity of the Earth is varying along with the angular rotational speed of the Earth, which itself varies with the Earth–Moon distance, we have extended our computations to 2500 Ma BP: over that timespan, k has increased back in time by more than 45%, ε^* has scarcely changed, the amplitudes of the larger terms of the obliquity have decreased in absolute value by about 60% and their frequencies have increased by a factor of 2.5 to 3, while those of the precession also increased by 70–80%. This means that the 41 000-year period becomes 16 700, the 23 000 becomes 12 700 and the 19 000 roughly 11 300, at 2500 Ma BP.

As far as the eccentricity is concerned, as there are only small variations in time of the fundamental frequencies (g), which are the only ones to be taken into account, the main periods are seen to vary by only 1.5% over 200 Ma.

Finally our computation deals with the past variations of the obliquity and climatic precession. However, if the development of these orbital elements is studied for the distant future, the fact that the Moon will tidally drive the Earth into resonance, as discussed by Ward (1982), will have to be taken into account.

ACKNOWLEDGEMENTS

This research was partly supported by the Agence Nationale Française pour la Gestion des Déchets

Radioactifs (Contract ANDRA 7-0164-A-00-A) and by the Commissariat Français à l'Energie Atomique (Contract CEA/BC-4561).

REFERENCES

BARRON, E.J., ARTHUR, M.A. & KAUFFMAN, E.G. (1985) Cretaceous rhythmic bedding sequences: a plausible link between orbital variations and climate. *Earth Planet. Sci. Lett.* **72**, 327–340.

BERGER, A. (1976) Obliquity and precession for the last 5,000,000 years. *Astron. Astrophys.* **51**, 127–135.

BERGER, A. (1977) Support for the astronomical theory of climatic change. *Nature* **269**, 44–45.

BERGER, A. (1978a) Long-term variations of daily insolation and Quaternary climatic changes. *J. Atmos. Sci.* **35**, 2362–2367.

BERGER, A. (1978b) *A simple algorithm to compute long term variations of daily or monthly insolation.* Contribution No 18, Institut d'Astronomie et de Géophysique G. Lemaître, Université Catholique de Louvain, Louvain-la-Neuve.

BERGER, A. (1984) Accuracy and frequency stability of the Earth's orbital elements during the Quaternary. In: *Milankovitch and Climate* (Eds Berger, A., Imbrie, J., Hays, J., Kukla, G. & Saltzman, B.). Reidel Publ. Co., Dordrecht, pp. 3–39.

BERGER, A. (1988) Milankovitch theory and climate. *Rev. Geophys.* **26**, 624–657.

BERGER, A. (1989a) Pleistocene climatic variability at astronomical frequencies. *Quat. Int.* **2**, 1–14.

BERGER, A. (1989b) The spectral characteristic of pre-Quaternary climatic records, an example of the relationship between the astronomical theory and geo-sciences. In: *Climate and Geo-sciences, a Challenge for Science and Society in the 21st Century* (Eds Berger, A., Schneider, S. & Duplessy, J.C.). Kluwer Academic Publishers, Dordrecht, pp. 47–76.

BERGER, A. & LOUTRE, M.F. (1990) Origine des fréquences des éléments astronomiques intervenant dans le calcul de l'insolation. *Bulletin de la Classe des Sciences. Académie Royale de Belgique, 6ᵉ serie* **1**(1/3), 45–106.

BERGER, A., LOUTRE, M.F. & DEHANT, V. (1989a) Astronomical frequencies for pre-Quaternary palaeoclimate studies. *Terra Nova* **1**, 474–479.

BERGER A., LOUTRE, M.F. & DEHANT, V. (1989b) Influence of the changing lunar orbit on the astronomical frequencies of the pre-Quaternary insolation patterns. *Paleoceanography* **4**, 555–564.

BERGER, A., LOUTRE, M.F. & DEHANT, V. (1989c) Milankovitch frequencies for pre-Quaternary. *Nature* **342**, 133.

BRETAGNON, P. (1974) Termes à longues périodes dans le système solaire. *Astron. Astrophys.* **30**, 141–154.

CALAME, O. & MULHOLLAND, J.D. (1978) Effect of the tidal friction on the lunar orbit and the rotation of the Earth and its determination by laser ranging. In: *Tidal Friction and Earth's Rotation* (Eds Brosche, P. & Sündermann, J.). Springer Verlag, Berlin, pp. 45–53.

DE BOER, P.L. & WONDERS, A.A.H. (1984) Astronomically induced rhythmic bedding in Cretaceous pelagic sediments near Moria (Italy). In: *Milankovitch and Climate* (Eds Berger, A., Imbrie, J., Hays, J., Kukla, G. & Saltzman, B.). Reidel Publ. Co., Dordrecht, pp. 177–190.

DEHANT, V., LOUTRE, M.F. & BERGER, A. (1987) *Les variations à court et à long terme de la rotation de la Terre et de la précession astronomique.* Scientific Report 87/12, Institut d'Astronomie et de Géophysique G. Lemaître, Université Catholique de Louvain, Louvain-la-Neuve.

DENIS, C. (1986) On the change of the kinetical parameter of the Earth during geological times. *Geophys. J. R. Astron. Soc.* **87**, 559–568.

DZIEWONSKI, A.M. & ANDERSON, D.L. (1981) Preliminary Reference Earth Model. *Phys. Earth Planet. Inter.* **25**, 297–356.

FISCHER, A.G. (1986) Climatic rhythms recorded in strata. *Annu. Rev. Earth Planet. Sci.* **14**, 351–376.

FISCHER, A.G. & SCHWARZACHER, W. (1984) Cretaceous bedding rhythms under orbital control? In: *Milankovitch and Climate* (Eds Berger, A., Imbrie, J., Hays, J., Kukla, G. & Saltzman, B.). Reidel Publ. Co., Dordrecht, pp. 163–175.

GILBERT, J.K. (1895) Sedimentary measurement of geologic time. *J. Geol.* **3**, 121–127.

HAYS, J.D., IMBRIE, J. & SHACKLETON, N.J. (1976) Variations in the Earth's orbit: pacemaker of the ice ages. *Science* **194**, 1121–1132.

IMBRIE, J., HAYS, J., MARTINSON, D.G., MCINTYRE, A., MIX, A.C., MORLEY, J.J. *et al.* (1984) The orbital theory of Pleistocene climate: support from a revised chronology of the marine $\delta^{18}O$ record. In: *Milankovitch and Climate* (Eds Berger, A., Imbrie, J., Hays, J., Kukla, G. & Saltzman, B.). Reidel Publ. Co., Dordrecht, pp. 269–305.

IMBRIE, J., MCINTYRE, A. & MIX, A. (1989) Oceanic response to orbital forcing in the late Quaternary: observational and experimental strategies. In: *Climate and Geo-sciences, a Challenge for Science and Society in the 21st Century* (Eds Berger, A., Schneider, S. & Duplessy, J.Cl.). Kluwer Academic Publishers, Dordrecht, pp. 121–164.

LAMBECK, K. (1980) *The Earth's Variable Rotation: its Origin, History and Physical Constitution*, 6th edn. Cambridge University Press, Cambridge.

LASKAR, J. (1988) Secular evolution of the solar system over 10 million years. *Astron. Astrophys.* **198**, 341–362.

LASKAR, J. (1989) A numerical experiment on the chaotic behaviour of the solar system. *Nature* **338**, 237–238.

LASKAR, J. (1990) The chaotic motion of the solar system. A numerical estimate of the chaotic zones. *Icarus* **88**, 266–291.

LOUTRE, M.F. & BERGER, A. (1989) *Pre-Quaternary amplitudes in the expansion of the obliquity and climatic precession.* Scientific report 1989/4, Institut d'Astronomie et de Géophysique G. Lemaître, Université Catholique de Louvain, Louvain-la-Neuve.

MILANKOVITCH, M. (1920) *Théorie Mathématique des Phénomènes Thermiques Produits par la Radiation Solaire.* Académie Yougoslave des Sciences et des Arts de Zagreb, Gauthier-Villars, Paris.

MILANKOVITCH, M. (1941) *Kanon der Erdbestrahlung und seine Anwendung auf das Eiszeitenproblem.* Royal

Serbian Sciences, Spec. Pub. 132, Section of Mathematical and Natural Sciences, Vol. 33, Belgrade. ('Canon of Insolation and the Ice Age Problem.' English Translation by Israël Program for Scientific Translation and published for the US Department of Commerce and the National Science Foundation, Washington, DC, 1969.)

OLSEN, P.E. (1986) A 40-million-year lake record of Early Mesozoic orbital climatic forcing. *Science* **234**, 842–848.

SHARAF, S.G. & BOUDNIKOVA, N.A. (1967) Secular variations of elements of the Earth's orbit which influences the climates of the geological past (in Russian). *Trudy Inst. Theor. Astron. Leningrad* **11**, 233–262.

SONETT, C.P., FINNEY, S.A. & WILLIAMS, C.R. (1988) The lunar orbit in the late Precambrian and the Elatina Sandstone laminae. *Nature* **335**, 806–808.

STACEY, F.D. (1977) *Physics of the Earth.* John Wiley & Sons, New York.

STEPHENSON, F.R. (1972) *Some geophysical, astronomical and chronological deductions from early astronomical records.* Thesis, University of Newcastle.

STOYKO, A. (1970) Les variations séculaires de la rotation de la Terre et les problèmes connexes. *Ann. Guébhard, Neuchâtel* **46**, 293–316.

WALKER, J.C.G. & ZAHNLE, K.J. (1986) Lunar nodal tide and distance to the Moon during the Precambrian. *Nature* **320**, 600–602.

WARD, W.R. (1982) Comments on the long-term stability of the Earth's obliquity. *Icarus* **50**, 444–448.

WILLIAMS, G.E. (1989) Late Precambrian tidal rhythmites in South Australia and the history of the Earth's rotation. *J. Geol. Soc. Lond.* **146**, 97–111.

WOOLARD, E.W. & CLEMENCE, G.M. (1966) *Spherical Astronomy*. Academic Press, New York.

Spec. Publs Int. Ass. Sediment. (1994) **19**, 25–33

Internal response to orbital forcing and external cyclic sedimentary sequences

N.-A. MÖRNER

Department of Paleogeophysics & Geodynamics, University of Stockholm, S-10691 Stockholm, Sweden

ABSTRACT

Orbital Milankovitch forcing is well established for the late Cenozoic period of continental ice caps. Insolation variations affecting Earth's climate, glacial volume and hence ocean volume are usually advocated to explain corresponding cyclic patterns in climate, sea level, different geochemical variables and sedimentary cycles. For pre-glacial/non-glacial times, this causal mechanism cannot be used to explain sea-level changes and cyclic sedimentary and palaeoenvironmental changes. The orbital forcing variables do not only affect insolation, however. They also affect different fundamental internal mechanisms and conditions: the gravity potential (and hence sea level), differential rotation (and hence sea level, ocean circulation, climate, ocean/atmosphere interaction) and palaeomagnetics (and hence the atmospheric shielding of the Earth). These internal factors are likely to be responsible for the majority of the cyclic changes with characteristic orbital forcing frequencies that are recorded in pre-glacial and non-glacial times. This internal response to orbital forcing probably also played a significant, maybe even dominant, role in Quaternary time.

INTRODUCTION

The effect of orbital forcing upon terrestrial systems is generally considered to act via insolation, i.e. changes in the strength and distribution of incoming solar radiation. In the late Cenozoic, the insolation changes led to climatic changes that generated expansions and contractions of the ice, giving rise to corresponding glacial eustatic falls and rises in sea level. This caused simultaneous cyclic changes in various biological, sedimentary and geochemical parameters.

Whilst this explanation may well be true of the late Cenozoic period of continental glaciations, it cannot apply to cyclic sea-level changes in non-glacial periods. Nor does insolation seem sufficient to generate palaeoenvironmental changes significant enough to be the dominant cause of other cyclic changes in non-glacial times (the rhythmicity model in fig. 5 of de Boer, 1990, seems, for example, much more strongly dependent on differential rotation than on actual insolation variations). I have drawn attention many times already to the fact that some fundamental internal parameters will be significantly affected by the orbital forcing variables, or so-called Milankovitch variables (Mörner, 1976, 1978, 1980a, 1984a, b, 1988). Other means of terrestrial response to orbital forcing are, therefore, advocated and analysed in this paper.

CYCLIC CHANGES IN PRE-GLACIAL AND NON-GLACIAL TIMES

There are numerous records of cyclic changes in pre-glacial and non-glacial times that have frequencies that fit well with those of the orbital forcing parameters, and hence are likely to be causally related to them (e.g. Bradley, 1929; Einsele, 1982; Anderson, 1984; Arthur *et al.*, 1984; de Boer & Wonders, 1984; Fischer & Schwarzacher, 1984; Olsen, 1984; Fischer, 1986; Grotzinger, 1986; Schwarzacher, 1987; Goldhammer *et al.*, 1990).

These cycles are usually ascribed to corresponding high-frequency, low-amplitude sea-level changes. It has been customary to try (in the absence of better explanations) to explain these cyclic changes by analogy with the well-established Pleistocene cyclicities (Anderson, 1984; Arthur *et al.*, 1984; de Boer & Wonders, 1984; Fischer & Schwarzacher, 1984; Olsen, 1984; Read *et al.*, 1986; Goldhammer *et al.*, 1990; Walkden & Walkden, 1990). This requires the presence of continental glaciations which are, however, restricted to a few exceptional periods during Earth's Phanerozoic and Proterozoic evolution. In general, there is a total absence of supporting data on corresponding occurrences of glaciations during most of the times under consideration.

It seems inescapable that cyclic changes in palaeoenvironment and sea level also occurred during non-glacial periods. Instead of assuming glaciations contrary to other observational data, efforts should be directed to alternative explanations.

Introducing the theory of sea-level changes due to geoid deformation with time (Mörner, 1976), I also called attention to the multiple effects of the so-called Milankovitch variables: 'the Milankovitch parameters have so far only been discussed in relation to global climatic changes, though they also must have affected the core/mantle coupling, the magnetic field, the ionization layers, and the gravitational forces leading to tidal and geoidal changes'. This was further expanded on in subsequent papers (e.g. Mörner, 1978, 1984a) with graphic illustrations of the multiple effects on palaeogeodesy, palaeomagnetism and palaeoclimate (Fig. 1). I later formulated the theory of differential rotation with interchange of angular momentum between the 'solid' Earth and the hydrosphere and the redistribution of heat (climate) and mass (sea level) via responses in the

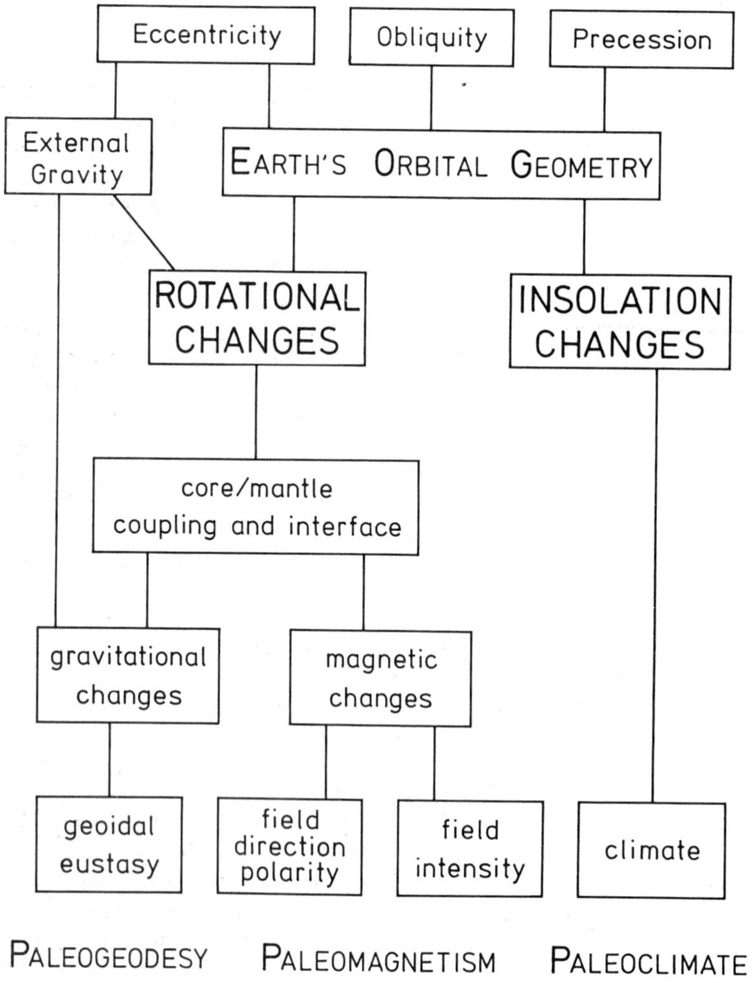

Fig. 1. The multiple effects of the so-called Milankovitch variables according to Mörner (1978, 1984a).

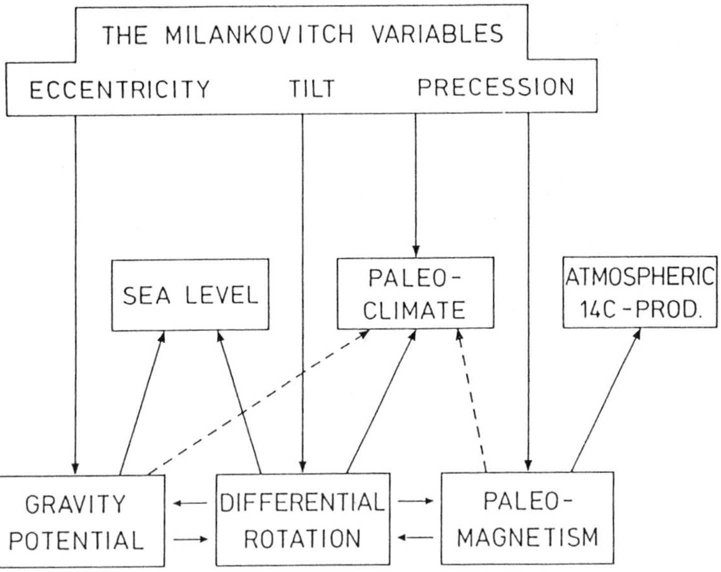

Fig. 2. The multiple effects of orbital forcing. From Mörner (1989a).

ocean current system (Mörner, 1984b, 1987c, 1988).

Today, there are numerous high-quality records (e.g. in this book) of all three Milankovitch variables which refer to non-glacial periods, and therefore have to be explained in terms of a non-Pleistocene response mechanism. As demonstrated by Berger *et al.* (1989), the amplitude and frequency of the obliquity and precession cycles should have been greater with increasing age back through the Phanerozoic as a consequence of Earth's deceleration due to tidal friction.

MULTIPLE EFFECTS OF THE MILANKOVITCH VARIABLES

Figure 2 gives an illustration of the multiple effects of orbital forcing via the Milankovitch variables (Mörner, 1988). In this paper we will discuss the three main internal groups of response to orbital forcing, i.e. gravity potential, differential rotation and geomagnetism (cf. Mörner, 1984b, 1989a). The insolation changes have been fully discussed elsewhere (e.g. Berger, 1988).

Deformation of the gravity potential surface

The Earth's gravity potential surface, the geoid, is controlled partly by internal gravity forces and their irregular distribution over the globe, and partly by the centrifugal forces and the external gravity 'constant' (i.e. the combined external gravity forces from the Universe, the Sun, the planets and the Moon). All these factors are subject to time variations (Mörner, 1976, 1984b).

The internal gravity and its irregular distribution over the globe will change with any changes in the mass distribution and/or density (Mörner, 1980a, 1984b).

The centrifugal forces will be affected by time variations in the Earth's rate of rotation (e.g. changes in the radius due to sea-level changes or mountain building). The Quaternary glacial/interglacial alternations, with sea-level changes in the order of 100–120 m, changed the spin velocity by about 1500–2000 milliseconds (ms). The formation of the Tibetan Plateau implied a considerable deceleration of the Earth (in the order of 6000 ms according to Mörner, 1991).

The external gravity forces will change with changes in the Earth's orbit around the Sun (the Milankovitch variables), the adjustments of the Sun and the planets to the centre of mass (e.g. Landscheidt, 1983), and the Earth–Moon relationship (e.g. Marsden & Cameron, 1966).

We must therefore assume that the gravity potential surface, the geoid, is constantly deforming with time, as indicated by a rapidly increasing amount of observational data (Mörner, 1976, 1980a, 1981, 1983, 1987b; Newman *et al.*, 1980, 1981; Bostrom, 1985; Hopley & Pirazzoli, 1986; Martin *et al.*, 1986; Nunn, 1986, 1987; Wang, 1991).

In the long-term perspective, geoid deformations must have played a significant, probably dominant, role (e.g. Mörner, 1980b, 1983, 1987a). In connection with the glacial eustatic rise of about 100–120 m after the last glaciation maximum at 20 ka, the corresponding deformation of the geoid relief had a variability of up to 50–60 m, i.e. 50% of the glacial eustatic rise (Mörner, 1987b, in prep.). Because the major rise occurred between 18 ka and 6 ka, we can estimate the mean rate of glacial eustatic rise at about 10 mm/yr and the mean rate of maximum geoidal eustatic deformation at about 5 mm/yr. For the Holocene, geoidal deformations in the order of 10 mm/yr and amplitudes of some 2–3 m have been reported (Mörner, 1981, 1983). A main Holocene cyclic geoid deformation in the order of 5–8 m has been reported (Mörner, 1971, 1976, 1980a, 1984b, 1988).

The variability between different 'eustatic' sea-level curves for the last 150 000 years seems to form a cyclic pattern between high and low variability closely related to the precession cycle (Mörner, 1976, fig. 14). The maximum variability ranges between about 30 and 90 m suggesting large-scale geoid deformations. During the last (20 ka) glaciation maximum, the eustatic variability amounted to about 60 m (Table 1). Periods of maximum sea-level differences (i.e. geoid deformations) seem, in general, to correlate with times of reported geomagnetic excursions (Mörner, 1978, 1980a), providing another means of sensing possible internal responses to the precession cycle. This data set needs further investigation, however.

The tides represent daily deformations of the geoid (Mörner, 1981, 1983). Melchior (1978) gave the lunar tidal range at 53.52 cm or 0.1650 mGal, and the solar tidal range at 24.61 cm or 0.0759 mGal (the total combined tidal range being 78.13 cm). He also noted that the gravity potential surface and its derivatives 'exhibit very complicated time variations due to the complexity of the orbital motions of the Earth around the Sun and of the Moon around the Earth'. This multi-body complexity prevents reliable quantifications of the geoid deformations on a long-term basis, i.e. on the time scale of the Milankovitch variables.

Table 1. Sea-level variables and quantities (m)

Present tides	
geoid tides	0.78
ocean tides	up to max. 18
Present geoid relief	
maximum differences	180
Present dynamic sea surface	
low harmonics	up to max. ≈2
major currents	≈5
Earth radius differences	
equator/pole	21 385
Present stored glacial volume (= metres sea level)	
Antarctica	60
Greenland	6
all mountain glaciers	0.5
Glacial eustasy	
Ice Age amplitudes	100–200
Geoid deformations	
on Ma time scale	≈50–250
last 150 ka; 20 ka cycles	≈30–90
last 20 ka	≈60
last 8 ka	≈5–8
Holocene oscillations	≈2
Differential rotation	
Holocene Gulf Stream pulsations	≈1.0–0.1
Holocene Equatorial Current E–W pulses	≈1
El Niño/ENSO events	≈0.3
major current topography	≈5
Rotation rate and sea level	
metres sea level per 15 ms rotation (LOD)	1.0
Ocean thermal expansion	
metres sea level per °C per 1000 m	≈0.2

Earth's differential rotation

The Earth is a multi-layered system consisting of the gaseous atmosphere, the liquid hydrosphere, the solid lithosphere, the viscous asthenosphere, the solid mantle, the liquid outer core and the solid inner core. These layers experience a differential rotation and constantly interchange angular momentum in order to conserve total angular momentum.

Earth's rotation is also subject to changes in the total angular momentum budget. The Earth experiences a continual deceleration of about 2.4 ms per 100 yrs due to tidal friction. This loss of angular momentum is transferred within the Earth–Moon system and compensated by an increase in the Earth–Moon distance. Changes in radius and mass will also affect the total spin velocity, e.g. sea-level falls and rises (like those of the glacial/interglacial alternations) and major mountain building (like the formation of the Tibetan Plateau) as mentioned above (Mörner, 1980a, figs 13 and 14).

Long-term changes in the major oceanic circulation paths, e.g. the successive closing of the circum-equatorial circulation, will affect both the differential rotation and the total rotation (Mörner, 1980a, fig. 15, 1984b).

Accelerations and decelerations of the spin velocity will affect the shape of the Earth's rotational ellipsoid, accelerations driving the water towards the equator, and decelerations driving the water pole-wards. The relation between spin velocity and sea level is about 15 ms in rotation to 1 m in sea level (Table 1).

Changes in the tilt-angle will also deform the rotational ellipsoid. The maximum effects will be in the equatorial and polar regions. The Earth's tilt changes with a *c*.40-ka cycle between 21° 39′ and 24° 36′. Theoretically, the corresponding displacement of the Earth's rotational ellipsoid would lead to cyclic sea-level changes in the order of 700 m in the equatorial and polar regions (see Fairbridge, 1961; Weyer, 1978). Because of the simultaneous interaction between geoidal and ellipsoidal adjustments, the actual effects are only a fraction of this, however, and cannot amount to more than a few metres. (The sea-level fall in many equatorial and low Southern Hemisphere areas during the last 5000–6000 yrs might be an example of this effect.)

The Earth's differential rotation has multiple atmospheric (e.g. Hide & Dickey, 1990), hydrospheric (Mörner, 1984b, fig. 10, 1988), lithospheric (Mörner, 1989a) and geomagnetic (e.g. fig. 3) effects.

The global differentiation of Holocene eustatic sea-level data between 8000 BP and the present (Mörner, 1971) records a cyclic geoid deformation with an amplitude of 5–8 m and a frequency of about 5500 years (Mörner, 1976, 1980a, figs 9 and 10, 1984b, fig. 5). This main Holocene geoid cycle seems to correlate with the archaeomagnetic cycle of Bucha (1970) and the atmospheric ^{14}C productivity cycle (e.g. Suess, 1970) indicating a common origin in differential rotation and core/mantle dislocations affecting the flow in the outer core (and thus the geodynamo), and the mass distribution (and thus the geoid surface). It is interesting that this major geoid deformation occurred in mid-Holocene time, i.e. after the vanishing of the continental ice caps and after the end of the main glacial eustatic rise in sea level. These major Holocene cyclic changes in geoid configuration, geomagnetic field strength and related differential rotation may be a compensational function of the main deceleration in response to the glacial eustatic rise in sea level up till about 6000 BP. Orbital forcing may also be involved (Mörner, 1980a).

If the various 'eustatic' sea-level curves from different parts of the globe were dominated by the glacial eustatic factor up till about 6000 BP, they have been dominated by geoid deformations and differential rotation (besides tectonics) since 6000 BP (Mörner, 1989a, fig. 4). The effects comprise partly a main cyclic trend as discussed above, and partly differential and compensational rises and falls, usually in the order of 1 m or less but sometimes even up to 2–3 m. The corresponding rates are in the order of 10 mm/yr. These short-term differential/compensational oscillations seem generally to represent the redistribution of water masses due to differential rotation and the interchange of angular momentum between the 'solid' Earth and the hydrosphere (Mörner, 1984b, 1988), although in some cases they may represent geoid deformations (Mörner, 1981, 1983).

The Gulf Stream and the Kuroshio Current, bringing warm equatorial water up to high northern latitudes, are specially important for the interchange of angular momentum between the 'solid' Earth and the hydrosphere (Mörner, 1984b, 1987c, 1988). The regional northwest European eustatic curve includes 16 low-amplitude oscillations during the Holocene. These oscillations correlate with continental temperature changes and North Atlantic pulses in Gulf Stream activity (Mörner, 1984b). They provide evidence of the redistribution of heat and mass by the major oceanic current systems. Inevitably, these mass redistributions must be linked to corresponding changes in differential rotation and interchanges of angular momentum. This theory (Mörner, 1984b) is testable by analysis of the last few centuries' instrumental records of the changes in rotation, temperature in northwest Europe and sea level in the North Sea region and adjacent sea areas. This 'test' (e.g. Mörner, 1987c, 1988, fig. 7) shows that there is a very close correlation between recorded changes in the Earth's rotation (in length of the day, LOD) and the temperature and sea-level changes in northwest Europe for the last 300 years. This indicates that the Gulf Stream plays a central role in the interchange of angular momentum on the decadal time scale during this period. Even the El Niño events can be successfully analysed in terms of differential rotation with interchange of angular momentum both with the atmosphere (e.g. Eubanks *et al.*, 1986; Chao, 1989) and with the hydrosphere (Mörner, 1989b).

Differential rotation and the interchange of angular momentum between the 'solid' Earth and the hydrosphere affect the ocean circulation system – the main surface current system as well as the deep-water circulation – giving rise to significant changes in climate (the redistribution of heat) and sea level (the redistribution of water masses) that are of compensational type over the globe, i.e. they represent the redistribution of energy and mass within a given budget (Mörner, 1984b, 1987c, 1988). This mechanism represents a powerful way to create cyclic climatic–eustatic changes as a function of orbital forcing. It seems much more important than direct insolation effects and is here proposed to be the main driving force in many proposed schemes (e.g. de Boer, 1990, fig. 5) to explain observed climatic–eustatic cycles in non-glacial times. For example, the passage south of Africa of the main deep-water circulation is very sensitive to any change in the Earth's rate of rotation. Because this deep-water circulation plays a central role in the whole integrated oceanic palaeoenvironmental system, any changes in its intensity are likely to give rise to significant changes in oceanography, in climate and in sea level on a global scale. Many of the recorded changes in association with the Quaternary glacial/interglacial alternations may, in fact, primarily be caused by ocean circulation changes in response to the spin velocity accelerations and decelerations in the order of 1500–2000 ms. This is precisely why internal responses to orbital forcing may be as important as, or even more important than the effects via insolation changes (Fig. 2; Mörner, 1978, 1984b, 1988).

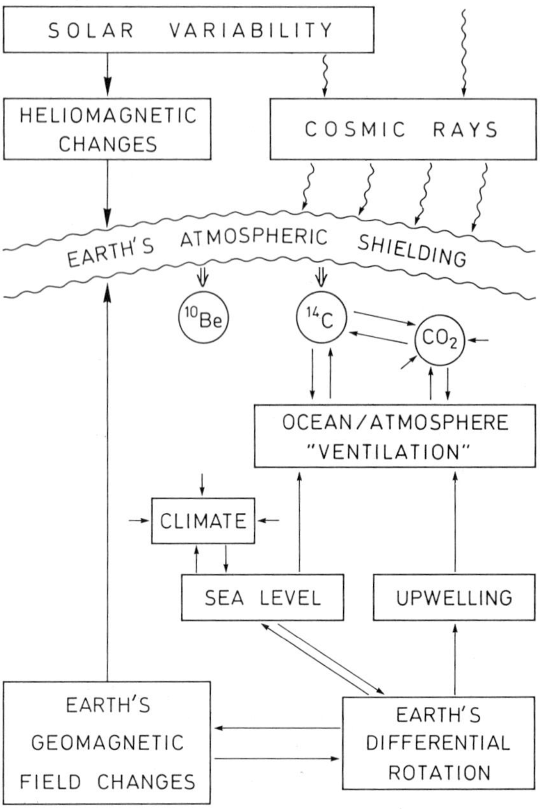

Fig. 3. Illustration of the interaction between Earth's differential rotation and geomagnetic field changes, the geomagnetic/heliomagnetic control of the Earth's atmospheric shielding, its controlling function on incoming ^{10}Be and ^{14}C production, and the multiple origin of climatic changes and atmospheric CO_2 variations.

Cyclic palaeomagnetic changes

The Earth's geomagnetic field is a function of the internal geodynamo. The geodynamo, in turn, is a function of the Earth's rotation and the motions in the liquid outer core, and is hence strongly influenced by any changes in the rate of rotation and interchange of angular momentum between the core and the mantle.

The Earth's geomagnetic field has a fundamental shielding function for incoming radiation. This affects Earth's climate (King, 1974; Wollin *et al.*, 1977; Mörner, 1978, 1984b). It also affects the atmospheric production of ^{14}C and the penetration of ^{10}Be, both of which isotopes may hence have an external (incoming radiation changes) or internal (geomagnetic shielding changes) cause for their variations with time (see Mörner, 1978, 1984a, b).

Figure 3 illustrates the shielding effects of the Earth's geomagnetic field and its interrelationship with Earth's differential rotation and related oceanic changes.

The correlation established (Tric *et al.*, 1990) between variations in the Earth's geomagnetic field and the ^{10}Be changes during the last 100 ka indicates that the primary variable is the Earth's geomagnetic field and not the solar radiation. The geomagnetic field changes are likely to be closely linked to the rotational changes resulting from the glacial eustatic rises and falls in association with the main stadial/interstadial alternations during the last Ice Age period as a whole (Pal & Mörner, 1991).

Palaeomagnetic records of cyclic changes in intensity and/or direction with frequencies in the order of the eccentricity, obliquity and precession cycles have been reported from a number of high-sedimentation

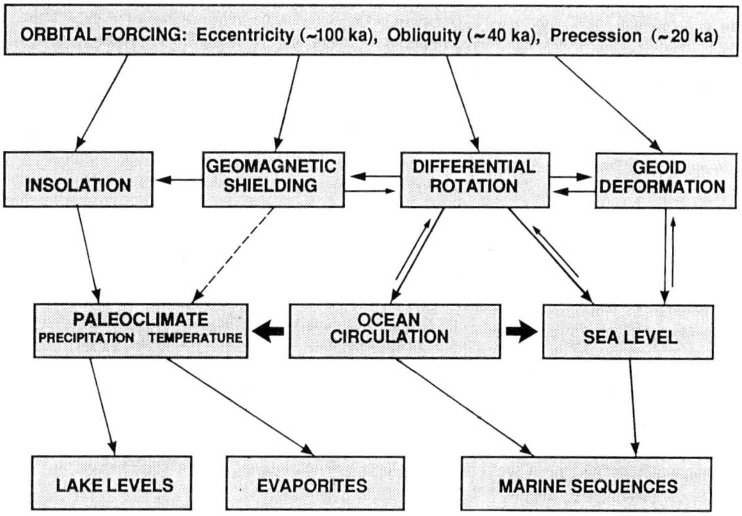

Fig. 4. The multiple effects of orbital forcing leading to cyclic changes in terrestrial lake levels and evaporites, and in various marine environments and sequences. The internal responses play a dominant role, at least, in non-glacial times, and offer a novel means of explaining cyclic records of orbital forcing frequencies in pre-Quaternary times.

rate sequences (e.g. Bucha, 1973; Kent & Opdyke, 1976; Wollin *et al.*, 1977; Bloemendal & de Menocal, 1989; Heller *et al.*, 1990).

Furthermore, it may be significant that reported geomagnetic excursions fall chronologically within periods of maximum geoidal sea-level deformation (see above). The general course of the latter seems to have followed a cyclic pattern during the last 150 000 years that closely correlates with the precession cycle (Mörner, 1976, 1978, 1980a).

ORBITAL FORCING IN NON-GLACIAL TIMES

In Fig. 4, I summarize the multiple effects of orbital forcing. Sea-level changes are generated by geoid deformation, differential rotation and ocean circulation changes. Differential rotation generates ocean circulation changes that affect both sea level and palaeoclimate. Palaeoclimatic changes can be generated from insolation changes as well as from ocean circulation changes, geomagnetic field changes and also deformations of the atmospheric gravity potential surface. This means that orbital forcing can easily produce cyclic changes in terrestrial lake levels (e.g. Olsen, 1984) and evaporites (e.g. Anderson, 1984) as well as in different marine environments and sequences (e.g. Arthur *et al.*, 1984; de Boer & Wonders, 1984; Fischer & Schwarzacher, 1984; Grotzinger, 1986; Goldhammer *et al.*, 1990).

With respect to sea-level changes in non-glacial times (e.g. Mélières, 1979; Grotzinger, 1986; Goldhammer *et al.*, 1990), there are no problems in generating cyclic sea-level changes in the order of 10 m (Grotzinger, 1986), or less (Goldhammer *et al.*, 1990) via geoid deformations and rotational effects. There is no need to introduce unknown (and unlikely) glacial volume changes. Table 1 gives some pertinent sea-level variables.

In conclusion, I propose that the pre-glacial and non-glacial cycles of orbital forcing frequencies are primarily driven by internal variables (with insolation changes playing a subordinate role). These internal forces probably also played a significant (maybe even dominant) role during the Quaternary, which calls for a revision or moderation of the straightforward correlation between insolation variations and major Quaternary changes in palaeoenvironment, climate, glaciation, sea level, etc. Internal responses by the Earth to orbital forcing seem to be a reality, and should be the focus of more intensive analyses and quantifications in the future.

REFERENCES

Anderson, R.Y. (1984) Orbital forcing of evaporite sedimentation. In: *Milankovitch and Climate* (Eds Berger, A., Imbrie, J., Hays, J., Kukla, G. & Saltzman, B.). Reidel Publ. Co., Dordrecht, pp. 147–162.

Arthur, M.A., Dean, W.E., Bottjer, D. & Scholle, P.A. (1984) Rhythmic bedding in Mesozoic–Cenozoic pelagic carbonate sequences: The primary and diagenetic origin of Milankovitch-like cycles. In: *Milankovitch and*

Climate (Eds Berger, A., Imbrie, J., Hays, J., Kukla, G. & Saltzman, B.). Reidel Publ. Co., Dordrecht, pp. 191–222.

Berger, A. (1988) Milankovitch theory and climate. *Rev. Geophys.* **26**, 624–657.

Berger, A., Loutre, M.F. & Dehant, V. (1989) Influence of the changing lunar orbit on the astronomical frequencies of pre-Quaternary insolation patterns. *Paleoceanography* **4**, 555–564.

Bloemendal, J. & de Menocal, P. (1989) Evidence for a change in the periodicity of tropical climate cycles at 2.4 Myr from whole-core magnetic susceptibility measurements. *Nature* **342**, 897–900.

Bostrom, R.C. (1985) Neotectonics of Africa and the Indian Ocean: Development of the geoid low. *Tectonophysics*, **119**, 1–21.

Bradley, W.H. (1929) The varves and climate of the Green River epoch. *US Geol. Surv. Prof. Pap.* **645** (108), 1–45.

Bucha, V. (1970) Influence of the Earth's magnetic field on radiocarbon. In: *Radiocarbon Variations and Absolute Chronology* (Ed. Olsson, I.). Almquist & Wiksell, Stockholm; John Wiley & Sons, Chichester, pp. 501–511.

Bucha, V. (1973) The continuous pattern of variations of the geomagnetic field in the Quaternary and their causes. *Studia Geophys. Geod.* **17**, 218–231.

Chao, B.F. (1989) Length-of-day variations caused by El Niño–Southern Oscillation and Quasi-Biennial Oscillation. *Science* **243**, 923–925.

De Boer, P.L. (1990) Astronomical cycles reflected in sediments. *Zbl. Geol. Paläont. Teil I* **8**, 911–930.

De Boer, P.L. & Wonders, A.A.H. (1984) Astronomically induced rhythmic bedding in Cretaceous pelagic sediments near Moria (Italy). In: *Milankovitch and Climate* (Eds Berger, A., Imbrie, J., Hays, J., Kukla, G. & Saltzman, B.). Reidel Publ. Co., Dordrecht, pp. 177–190.

Einsele, G. (1982) Limestone–marl cycles (periodites): Diagenesis, significance, causes – a review. In: *Cyclic and Event Stratigraphy* (Eds Einsele, G. & Seilacher, A.). Springer Verlag, Berlin, pp. 8–53.

Eubanks, T.M., Steppe, J.A. & Dickey, J.O. (1986) The El Niño, the Southern Oscillation and changes in the length of the day. In: *Earth Rotation: Solved and Unsolved Problems* (Ed. Cazenave, A.). NATO Adr. Inst., Reidel, Dordrecht, pp 163–166.

Fairbridge, R.W. (1961) Eustatic changes in sea level. *Physics Chemist. Earth* **4**, 99–185.

Fischer, A.G. (1986) Climatic rhythms recorded in strata. *Annu. Rev. Earth Planet. Sci.* **14**, 351–376.

Fischer, A.G. & Schwarzacher, W. (1984) Cretaceous bedding rhythms under orbital control? In: *Milankovitch and Climate* (Eds Berger, A., Imbrie, J., Hays, J., Kukla, G. & Saltzman, B.). Reidel Publ. Co., Dordrecht, pp. 163–175.

Goldhammer, R.K., Dunn, P.A. & Hardie, L.A. (1990) Depositional cycles, composite sea-level changes, cycle stacking patterns, and hierarchy of stratigraphic forcing: Examples from Alpine Triassic platform carbonates. *Geol. Soc. Am. Bull.* **102**, 535–562.

Grotzinger, J.P. (1986) Cyclicity and paleoenvironmental dynamics, Rocknest platform, northwest Canada. *Geol. Soc. Am. Bull.* **97**, 1208–1231.

Heller, F., Rüegg, F., Liu, X., Liu, T. & Sun, J. (1990) Magnetic properties of loess sediments in China. In: *27th International Geological Congress, Washington DC 1989*, Vol. 2, pp. 48–49.

Hide, R. & Dickey, J.O. (1990) Earth's variable rotation. *JPL Geod. Geophys. Prepr.* No. 205.

Hopley, D. & Pirazzoli, P. (1986) Late Quaternary and present sea-level changes: magnitude, causes, future applications. In: *Proc. Fifth Coral Reef Congr., Tahiti 1985*, Vol. 3, pp. 77–78.

Kent, D.V. & Opdyke, N.D. (1976) Paleomagnetic field intensity variation recorded in a Brunhes Epoch deep-sea sediment core. *Nature* **166**, 555–573.

King, J.W. (1974) Weather and the Earth's geomagnetic field. *Nature* **247**, 131–134.

Landscheidt, T. (1983) Solar oscillations, sunspot cycles, and climatic changes. In: *Weather and Climate Response to Solar Variations* (Ed. MacCormac, B.M.). Colorado Associated University Press, Colorado, pp. 293–308.

Marsden, B.G. & Cameron, A.G.W. (Eds) (1966) *The Earth–Moon System*. Plenum Press, New York.

Martin, L., Flexor, J.M., Blitzkov, D. & Suguio, K. (1986) Geoid change indications along the Brazilian coast during the last 7000 years. In: *Proc. Fifth Coral Reef Congr., Tahiti 1985*, Vol. 3, pp. 85–90.

Melchior, P. (1978) *The Tides of the Planet Earth*. Pergamon Press, Oxford.

Mélières, F. (1979) Mineralogy and geochemistry of selected Albian sediments from Bay of Biscay, Leg. 48, DSDP, eastern Atlantic Ocean. *Init. Rep. DSDP* **48**, 855–875.

Mörner, N.-A. (1971) The Holocene eustatic sea level problem. *Geol. Mijnbouw* **50**, 699–702.

Mörner, N.-A. (1976) Eustasy and geoid changes. *J. Geol.* **84**, 123–151.

Mörner, N.-A. (1978) Paleoclimatic, paleomagnetic and paleogeoidal changes: interaction and complexity. In: *Proceedings Centre National Etudes Spatiales Colloque Intern., Toulouse 1978*, pp. 221–232.

Mörner, N.-A. (1980a) Eustasy and geoid changes as a function of core/mantle changes. In: *Earth Rheology, Isostasy and Eustasy* (Ed. Mörner, N.-A.). John Wiley & Sons, Chichester, pp. 535–553.

Mörner, N.-A. (1980b) Relative sea-level, tectono-eustasy, geoidal-eustasy and geodynamics during the Cretaceous. *Cret. Res.* **1**, 329–340.

Mörner, N.-A. (1981) Space geodesy, paleogeodesy and paleogeophysics. *Ann. Géophys.* **37**, 69–74.

Mörner, N.-A. (1983) Sea levels. In: *Mega-Geomorphology* (Eds Gardner, R.A.M. & Scoging, H.). Oxford University Press, Oxford, pp. 73–91.

Mörner, N.-A. (1984a) Terrestrial, solar and galactic origin of the Earth's geophysical variables. *Geogr. Ann.* **66A**, 1–9.

Mörner, N.-A. (1984b) Planetary, solar, atmospheric, hydrospheric and endogene processes as origin of climatic changes on the Earth. In: *Climatic Changes on a Yearly to Millennial Basis* (Eds Mörner, N.-A. & Karlén, W.). Reidel Publ. Co., Dordrecht, pp. 483–507.

Mörner, N.-A. (1987a) Pre-Quaternary long-term changes in sea level. In: *Sea Surface Studies. A Global View* (Ed. Devoy, R.J.N.). Croom Helm, Beckenham, pp. 233–241.

Mörner, N.-A. (1987b) Quaternary sea level changes:

Northern Hemisphere data. In: *Sea Surface Studies: A Global View* (Ed. Devoy, R.J.N.). Croom Helm, Beckenham, pp. 242–263.

MÖRNER, N.-A. (1987c) Short-term paleoclimatic changes. Observational data and a noval causation model. In: *Climate, History, Periodicity and Predictability* (Eds Rampino, M.R., Sanders, J.E., Newman, W.S. & Königsson, L.K.), pp. 256–269.

MÖRNER, N.-A. (1988) Terrestrial variations within given energy, mass and momentum budgets: Paleoclimate, sea level, paleomagnetism, differential rotation and geodynamics. In: *Secular Solar and Geomagnetic Variations in the last 10,000 years* (Eds Stephenson, F.R. & Wolfendale, A.W.). Kluwer Academic Publishers, Dordrecht, pp. 455–478.

MÖRNER, N.-A. (1989a) Global changes: The lithosphere: Internal processes and Earth's dynamicity in view of Quaternary observational data. *Quat. Int.* **2**, 55–61.

MÖRNER, N.-A. (1989b) Changes in the Earth's rate of rotation on an El Niño to century basis. In: *Geomagnetism and Palaeomagnetism* (Eds Lowes, F.J. *et al.*). Kluwer Academic Publishers, Dordrecht, pp. 45–53.

MÖRNER, N.-A. (1991) Uplift of the Tibetan Plateau: A short review. *INQUA Spec. Proc. Rew. Rep.* pp. 78–81.

NEWMAN, W.S., MARCUS, L.F., PARDI, R.R., PACCIONE, J.A. & TOMACEK, S.M. (1980) Eustasy and deformation of the geoid: 1000–6000 radiocarbon years BP. In: *Earth Rheology, Isostasy and Eustasy* (Ed. Mörner, N.-A.). John Wiley & Sons, Chichester, pp. 555–567.

NEWMAN, W., MARCUS, L. & PARDI, R. (1981) Paleogeodesy, Late Quaternary geoidal configurations as determined by ancient sea levels. In: *Sea Level, Ice and Climatic Changes*. IAHS Publ. 131, pp. 263–275.

NUNN, P.D. (1986) Implication of migrating geoid anomalies for the interpretation of high-level fossil coral reef. *Geol. Soc. Am. Bull.* **97**, 946–952.

NUNN, P.D. (1987) Late Cenozoic tectonic history of Lau Ridge, southwest Pacific, and associated shoreline displacements: Review and analysis. *N. Z. J. Geol. Geophys.* **30**, 241–260.

PAL, P. & MÖRNER, N.-A. (1991) Rotational and geomagnetic changes during the last glacial period. In: *New Approaches in Geomagnetism and Earth's Rotation* (Ed. Flodmark, S.). World Scientific Publications Co., Singapore, pp. 139–144.

OLSEN, P.E. (1984) Periodicity of lake-level cycles in the Late Triassic Lockatong Formation of the Newark Basin (Newark supergroup, New Jersey and Pennsylvania). In: *Milankovitch and Climate* (Eds Berger, A., Imbrie, J., Hays, J., Kukla, G. & Saltzman, B.). Reidel Publ. Co., Dordrecht, pp. 129–146.

READ, J.F., GROTZINGER, J.P., BOVA, J.A. & KOERSCHNER, W.F. (1986) Models for generation of carbonate cycles. *Geology* **14**, 107–110.

SCHWARZACHER, W. (1987) Astronomical cycles for measuring geological time. *Modern Geology* **11**, 375–381.

SUESS, H.E. (1970) The three causes of the secular C14 fluctuations, their amplitude and time constant. In: *Radiocarbon Variations and Absolute Chronology* (Ed. Olsson, I.). Almquist & Wiksell, Stockholm; John Wiley & Sons, Chichester, pp. 595–605.

TRIC, E., MAZAUD, A., VALET, J.-P., TUCHOLKA, P. & LAJ, C. (1990) The geomagnetic field intensity during the past 120 ky: Spectral analysis and calibration of 14C ages. In: *Reversals, Secular Variations and Dynamo Theory, SEDI Symposium, Santa Fe 1990*, Abstracts, p. 65.

WALKDEN, G.M. & WALKDEN, G.D. (1990) Cyclic sedimentation in carbonate and mixed carbonate–clastic environments: four simulation programs for a desktop computer. In: *Carbonate Platforms*. Spec. Publ. Int. Ass. Sediment. 9, pp. 55–78.

WANG, J. (1991) Late-glacial geoid changes and geoidal eustasy in eastern China. *Chinese Sci. Bull.* **36**, 735–737.

WEYER, E.M. (1978) Pole movements and sea level. *Nature* **273**, 18–21.

WOLLIN, G., RYAN, W.B.F., ERICSON, D.B. & FOSTER, J.H. (1977) Paleoclimate, paleomagnetism and eccentricity of the earth's orbit. *Geophys. Res. Lett.* **4**, 267–270.

Spec. Publs Int. Ass. Sediment. (1994) **19**, 35–46

Filtering and frequency mapping as tools in subsurface cyclostratigraphy, with examples from the Wessex Basin, UK

D.H. MELNYK*, D.G. SMITH *and* K. AMIRI-GARROUSSI

BP Research Centre, Sunbury-on-Thames, TW16 7LN, UK

ABSTRACT

Recent regional studies of the Upper Jurassic (Kimmeridgian through to Portlandian Stages), onshore UK, have demonstrated the advantage of using digitally filtered gamma-ray logs to analyse subsequence-scale depositional packages, or cycles. The low-frequency component of the well-log trace (≫20 m wavelength) tends to be the most correlatable at a regional scale. This low-frequency component can be enhanced by: (i) altering the displayed aspect ratio of the trace by squeezing the vertical scale and stretching the horizontal scale, and changing the horizontal axis to accommodate only the numerical range of log values (i.e. normalization); and (ii) digital filtering of the log values with an appropriate low-pass filter. High-frequency cyclicity is studied by mapping the frequency content of the digital log data as a function of depth, after high-pass filtering.

In the Wessex Basin, gamma-ray logs treated in this manner readily reveal the longer wavelengths (>20 m) associated with major decreases and increases in gamma-ray log activity (i.e. cleaning-up and dirtying-up trends). These cycles can be recognized across changes in lithofacies between wells, and can often be correlated over long distances (*c.*100 km). Missing cycles testify to hiatuses, and biostratigraphic and seismic calibration suggests that cycles are chronostratigraphic.

Mapping the high-frequency component of the well-log trace as a function of depth in the borehole shows that, in general, the high-frequency content of the gamma-ray logs changes abruptly at hiatuses and varies systematically within sequences. Wavelengths are of a similar magnitude to cycles observed at outcrop (i.e. 1–5 m), which have been attributed to the 40-ka Milankovitch obliquity cycle. This permits an estimation of sedimentation rates to be made from the high-frequency maps.

INTRODUCTION

Digital filtering of wireline log data is not new. Geophysicists have been routinely filtering sonic logs for some time as a prelude to the manufacture of synthetic seismograms; principally during the analysis of interval velocities, when low-frequency components from sonic log signatures are added to the high-frequency seismic signal to facilitate comparison with real sonic logs.

However, in stratigraphic studies, in spite of the increasing availability of digital wireline data, mathematical and statistical manipulation of wireline data is, in general, the exception rather than the rule. In this paper, we illustrate the effectiveness of digitally filtering and redisplaying gamma-ray log data. Specifically, we demonstrate that: (i) the low bandpass filtering and redisplay of gamma-ray logs enhances the correlatability of low-frequency cyclic components within the log, cyclicities related to stratigraphic phenomena at the sequence and parasequence scale (≫10 m in thickness and ≫100 ka in duration); and (ii) high bandpass mapping of high-frequency gamma-ray log signal components facilitates the recognition of high-frequency trends and anomalies, cyclicities related to stratigraphic phenomena at the Milankovitch (usually <10 m in thickness and <100 ka in duration) time scale.

In order to demonstrate these techniques, we present data from the Late Jurassic succession of the Wessex Basin (Fig. 1), specifically the Kimmeridgian

* Present address: Scott Pickford PLC, 256 High Street, Croydon, CR0 1NF, UK.

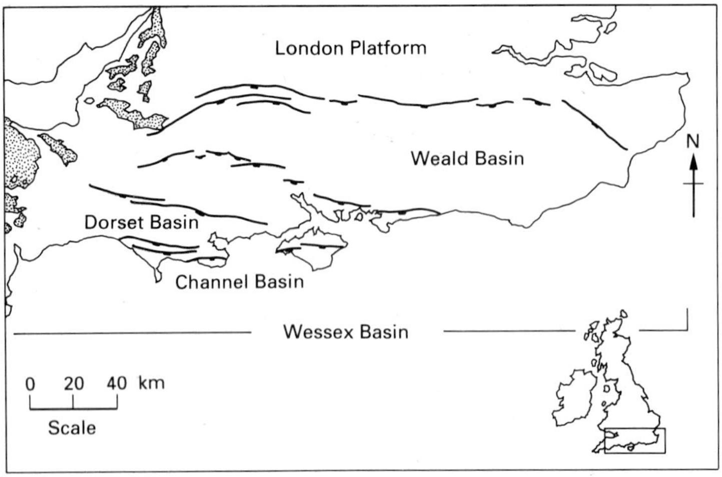

Fig. 1. Map of the areal extent of the Wessex Basin. After Whittaker (1985).

and Portlandian Stages. This mudstone-dominated, mixed siliciclastic and carbonate succession, which at outcrop shows cyclicities at all scales from millimetres to tens of metres, consists of alternating calcareous mudstones, organic-rich mudstones, siltstones, limestones and sandstones (see Fig. 6). Data from 33 wells were analysed during the course of this study using the techniques described below.

METHODS

Gamma-ray logs consist of discrete measurements, taken at equal intervals (usually 15.4 cm, or 0.5 ft) down a well bore. As such, the data are voluminous, and ideally suited to manipulation using the standard techniques of time series analysis. In this paper, we utilize two such methods: low-frequency bandpass filtering and high-frequency bandpass mapping.

Low-frequency bandpass filtering

Fourier theory states that the variance in any time series can be entirely accounted for by the mathematical superposition of sinusoidal waveforms of different amplitudes, frequencies and phases. In the low bandpass filtering of gamma-ray log data, our purpose is to attenuate the high-frequency waveforms (by reducing their amplitudes to zero) without significantly affecting the amplitudes and phases of the low-frequency components. This attenuation can be achieved in either of two ways, namely filtering in the time (or in our case depth) domain, or filtering in the frequency domain. These two methods are mathematically quite distinct, but can be shown to produce equivalent results. We prefer to achieve low bandpass filtering in the time (or depth) domain by the multiple application of moving-average filters; however, high bandpass mapping is best achieved in the frequency domain, as discussed below.

Figure 2 illustrates the philosophy of low bandpass filtering utilizing multiple moving averages. Figure 2a shows a gamma-ray log which has been displayed at a typical (as used in the petroleum industry) aspect ratio. That is, the ratio of the horizontal to vertical scales is fairly small, and the details of the low-frequency signature, although much easier to see than on a typical 1 : 500 scale log display (the display favoured by industry for composite log presentation), are not very clear. Simply lengthening the horizontal axis (Fig. 2b and c) enhances the low-frequency component, although the patterns are now obscured by high-frequency excursions. Finally, low bandpass filtering to remove all variance at wavelengths less than 2 m (Fig. 2d) and 4 m (Fig. 2e) displays low-frequency patterns (i.e. trends and anomalies at parasequence and sequence scale) much more clearly (van Wagoner *et al.*, 1990).

Filtering in the time (or depth) domain by multiple applications of moving averages can be thought of as a process of trend-fitting (Schwarzacher, 1975). The filter is symmetrical, and consists of an odd number of equal weights which sum to unity. In the specific case of the three-point moving average, data points are averaged in sets of threes to produce a new time (or depth) series, which is itself subjected to the same process, and so on. Taken as a

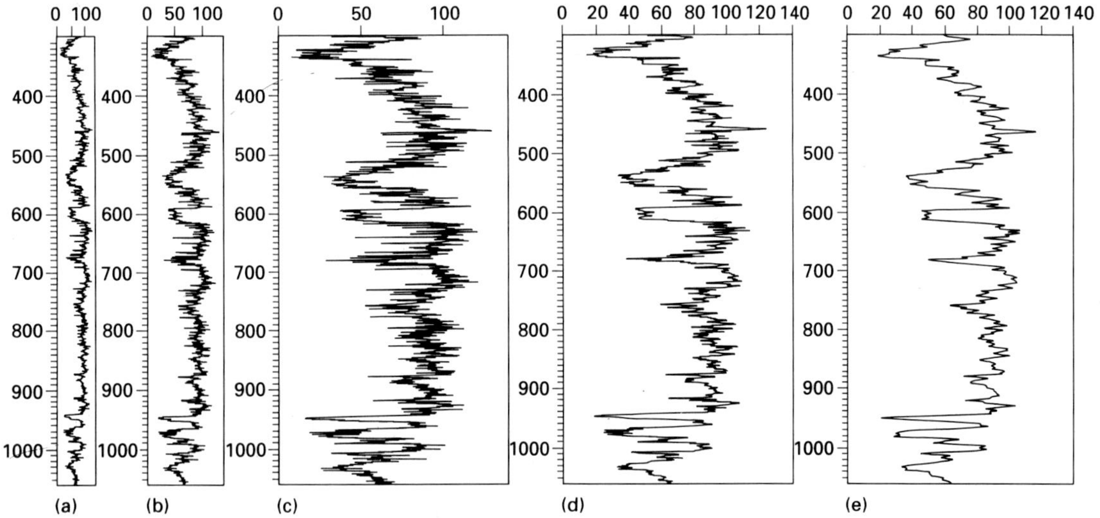

Fig. 2. Stages in gamma-ray log enhancement. (a) A log displayed at a typical aspect ratio, with a low ratio of horizontal to vertical scales. Lengthening the horizontal axis amplifies the low-frequency component of the log signature (b, c), while low bandpass filtering removes variance at wavelengths <2 m (d) and <4 m (e).

whole, the procedure is equivalent to subjecting the original time (or depth) series to a single symmetrical filter composed of normally distributed weights (Holloway, 1958; Schwarzacher, 1975). In many algorithms, the averaging process leads to the loss of ($N/2 - 1$) data points at both ends of the time series (where N is the length of the moving average) for each pass of the moving average. This can be readily remedied by systematically reducing the filter length at the beginning and end of the time series. At the end of the time series the filter length becomes 1, so no data points are lost.

The frequency response of such filters is entirely predictable (Holloway, 1958), as their transfer functions are readily calculated. For example, Fig. 3 compares the frequency responses of single and multiple applications of an 11-point moving average filter by plotting the proportion of variance retained by the filter against frequency (cycles per sample distance). Note how the single application of a moving average is not an ideal filter in that it introduces spurious polarity reversals between frequencies of 0.08 and 0.18 cycles per sample distance, but that the multiple applications of the same moving average produce low-pass filters which are closer to the ideal (i.e. they pass all the low-frequency variance below the specified cut-off frequency, and remove all of the high-frequency variance without introducing systematic noise above the cut-off frequency). As a general rule, it appears that several applications of a short moving average are preferable to the single application of a long moving average. Returning to Fig. 3, to read off the frequency response (i.e. the proportion of the original variance at a particular frequency retained by the filter), project a vertical line from the abscissa to the ap-

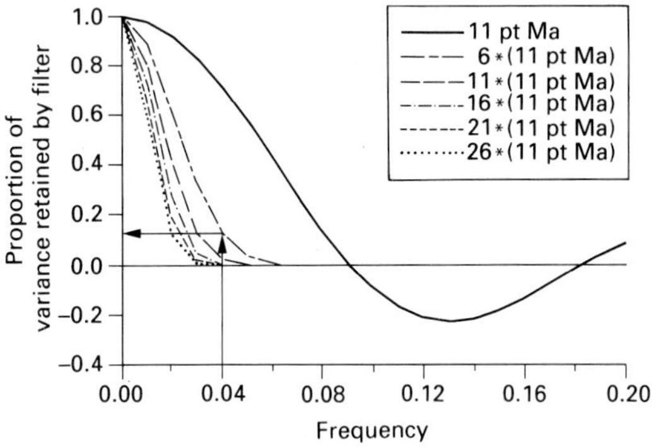

Fig. 3. Transfer functions of 11-point moving average filters. The curves represent the frequency responses (the proportion of the original variance at a particular frequency retained by the filter) of single and multiple applications of 11-point moving average filters after applying multiple smoothing operations.

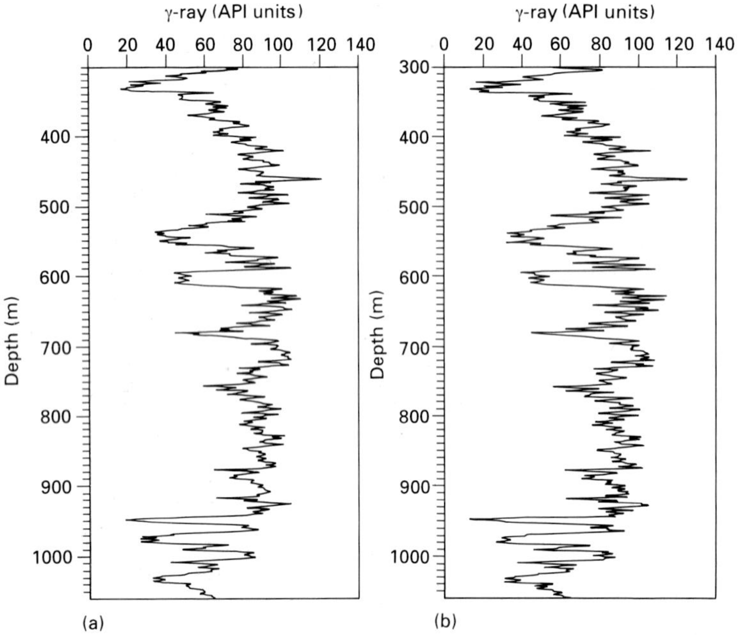

Fig. 4. Gamma-ray log filtered in the time domain (a), using multiple applications of moving averages, and the frequency domain (b), using a discrete Fourier transform followed by an inverse Fourier transform.

propriate curve, and then a horizontal line to the ordinate. For example, six iterations of an 11-point moving average will pass approximately 15% of any variance occurring at a frequency of 0.04 cycles per sample distance in the old time series into the new time series.

Filtering in the frequency domain is mathematically and computationally more complex. Figure 4b depicts a gamma-ray log filtered using a discrete Fourier transform algorithm followed by its inverse. All wavelengths longer than 4 m have been retained in the filtered log. By way of a comparison, Fig. 4a shows the same log filtered by multiple passes of a three-point moving average. In general the results are quite similar. However, the frequency domain transformed log has a perceptibly different quality about it, brought about by the sharp cut-off frequency employed (Gonzalez & Wintz, 1977). On applying the inverse Fourier transform, the sharp cut-off frequency results in the overweighting of the highest frequencies retained by the low-pass filter. If preferred, this can be remedied by tapering the frequency values close to the cut-off frequency, thereby producing a filter with a transfer function similar to those depicted in Fig. 3. The computational cost, however, becomes higher than the multiple moving average strategy and we have therefore opted for the latter.

High-frequency bandpass mapping

The variance removed by low-frequency bandpass filtering is obviously not all noise (by any definition), and will contain stratigraphically important information. In order to study these data, the low-frequency bandpass curve can be subtracted from the original data set, leaving what is usually termed a residual data set.

It is to this residual data set that we apply the technique of high-frequency bandpass mapping (Melnyk & Smith, 1989). The method involves taking successive subsets, or windows, of the residual data set, detrending (linearly), and standardizing (to a mean of 0 and a standard deviation of 1); the subset of data is then transformed to its Fourier power spectrum, the window moved down the residual series by a preset lag (i.e. a preset number of samples), and the process is repeated all the way down the well. The end-product is a set of spectra, each spectrum representing a sample from the residual time series at a given lag. The spectra are digitally stacked side by side to form a grid, smoothed with a two-dimensional Gaussian filter (this is analogous to smoothing a single power spectrum with a Hanning filter, except that with a two-dimensional filter spectra from adjacent lags contribute to the smoothing), and contoured like a

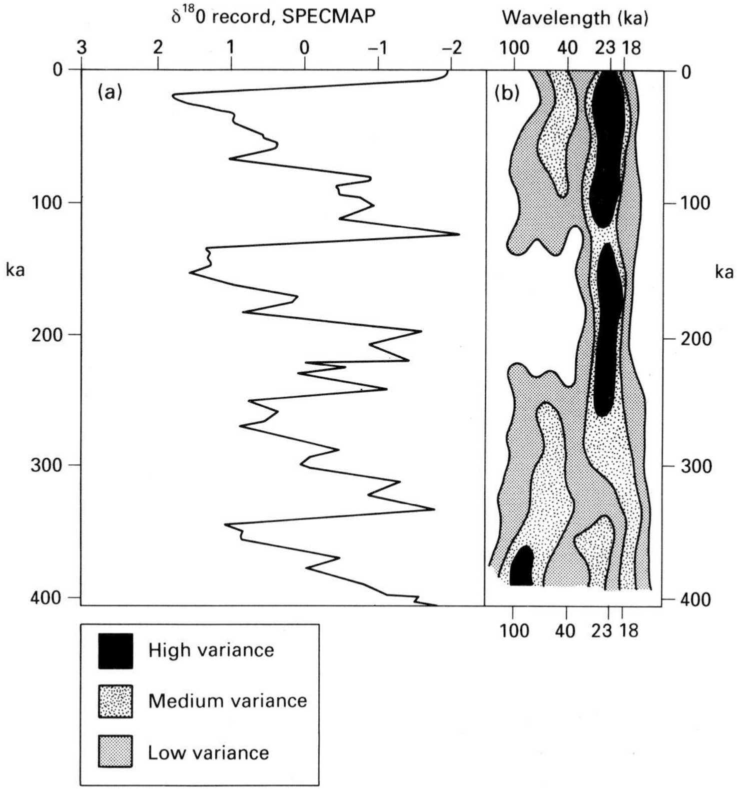

Fig. 5. SPECMAP $\delta^{18}O$ record and accompanying high-frequency bandpass map. The original curve (a) is dominated by the 100-ka Milankovitch frequency, while the high-frequency bandpass map (b) illustrates the temporal variations at the 40-ka, 23-ka and 18-ka Milankovitch frequencies.

map. This gives a visual impression of the changes in the frequency composition of the residual gamma-ray values as a function of depth in the well (Melnyk & Smith, 1989).

Figure 5 is an illustration of this method. We have used the well-known SPECMAP $\delta^{18}O$ record as a test data set (Imbrie *et al.*, 1984). Due to the excellent stratigraphic calibration, the oxygen isotope time series has been tuned to the predicted Milankovitch signal and is plotted as a function of absolute geological time. In the time series in Fig. 5a, note how the large variance attributable to the 100-ka eccentricity cycle, which reflects the major periods of glaciation, obscures the variance associated with the *c*.40-ka obliquity cycle, and the *c*.23-ka and *c*.19-ka precession cycles respectively.

In order to study the distribution of these higher frequencies, low bandpass filtering was used to remove the variance associated with the 100-ka eccentricity signal, and a contour map of the frequency content of overlapping windows of the residual data (Fig. 5b) was plotted adjacent to the original curve (Fig. 5a), as a function of depth. A window length of 80 ka and an overlap of 70 ka were used in the construction of the high-frequency map. The three principal Milankovitch wavelengths (*c*.40 ka *c*.23 ka and *c*.19 ka) are readily apparent on the map, but what is more relevant is their relative importance as a function of time. The *c*.40-ka obliquity cycle, for example, appears to have been relatively strong between 0 and *c*.100 ka BP, and between *c*.275 ka and *c*.350 ka BP; and relatively weak between *c*.100 ka and *c*.275 ka BP. Similarly, much of the precession variance was concentrated at the *c*.23-ka waveband between 0 and *c*.300 ka BP, and at the *c*.19-ka waveband prior to that.

It is evident that high-frequency bandpass mapping permits the recognition and study of subtle trends in frequency composition more readily than simple spectral analysis. Equally important, displays such as Fig. 5 permit the systematic portrayal of the entire variance of the time (or depth) series by displaying patterns in low- and high-frequency content.

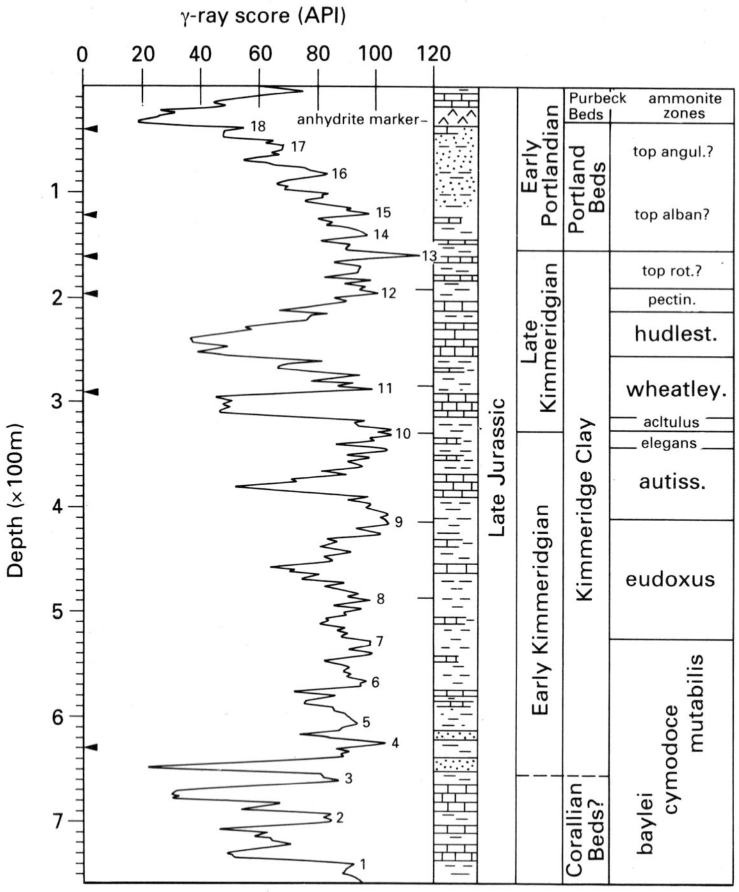

Fig. 6. Type gamma-ray log of the Wessex Basin, with chronostratigraphic, lithostratigraphic and biostratigraphic nomenclature. Correlatable cycle boundaries are numbered 1 through to 18. Possible hiatuses are labelled as black triangles. Kerogen-rich shale intervals are labelled as horizontal dashes. Vertical divisions equal 10 m.

RESULTS

Low-frequency bandpass filtering

The Wessex Basin (Fig. 1) is principally an onshore basin covering an area of over 250 000 km^2 (Chadwick, 1986). The sediment fill, which is mainly Mesozoic and Cenozoic in age, is locally over 3 km thick. Sediment distribution has been explained in terms of polyphase extensional faulting and accelerated subsidence in the Early Triassic, Early Jurassic and Late Jurassic (Chadwick, 1986). During the Kimmeridgian and Portlandian, a dominantly mudstone succession was deposited (Fig. 6). At outcrop, this succession displays cyclicity (reflecting relatively clay-rich and relatively clay-poor intervals) at a number of scales (Cox & Gallois, 1981; House 1986; Oschmann, 1988; Wignall, 1989). The small-scale cyclicity has been utilized as a correlation tool in the subsurface (Penn *et al.*, 1986). In this section we demonstrate the utility of large-scale cyclicity as a correlation tool. We have adopted a flexible definition of the term 'cycle', by which we mean any oscillatory component in a gamma-ray log profile (such as may be identified through frequency analysis).

Large-scale cyclic patterns in gamma-ray log character were enhanced by amplification and digital filtering, thus increasing the ratio of the correlatable low-frequency signals to high-frequency signals. This facilitates the viewing of the gamma-ray log at a much smaller scale (the stratigraphic scale of seismic sequences, parasequences and depositional systems).

We interpret our filtered and amplified gamma-ray logs (Figs 2, 6–8) as facies indicators, with the filtered log reflecting the continuous variation in the trend between relatively radioactive and relatively non-radioactive facies. Large-scale systematic reversals in the facies trend-line reflect basin-wide changes in deposition and can be interpreted as time lines. These time lines can correspond to flooding surfaces, parasequence boundaries and sequence boundaries. Hence, it is not unreasonable to correlate cycles, or packages, as chronostratigraphic units, despite obvious internal variations in lithology. Additionally, Milankovitch theory predicts that systematic variations in a facies tract should occur with definite periodicities, the quasi-periodic Milankovitch forcing mechanisms thereby lending a unique signature to all climatic events. Therefore, if climatic change is proxied by the gamma-ray log (van Buchem *et al.*, 1992) a cyclic pattern should emerge, wherein each cycle has unique character. Moreover, Milankovitch theory predicts that we should be able to recognize individual cycles over a large geographical area and across facies changes, and that these cycles should be datable.

Figures 7 and 8 illustrate our regional correlations. Each well has been oriented relative to a regional marker and correlative gamma-ray highs (or cycle boundaries) have been numbered 1 (oldest) through to 18 (youngest). Only the Lower Kimmeridgian through to the Lower Portlandian succession is illustrated for each well. Figure 7 illustrates a line of correlation from the northern margins of the Channel Basin (left) to the centre of the Weald subbasin (Fig. 1). Similarly, Fig. 8 shows the correlation from the northern margin (left) to the southern margin of the Weald Basin. The striking correlatability of the gamma-ray log cycles is immediately apparent.

Generally speaking, two major trends of decreasing radioactivity (or cleaning-upwards trends) extending over hundreds of metres are apparent: the earlier trend (gamma-ray log markers 4 through to 11/12) comprises much of the Kimmeridge Clay Formation and reflects the gradual upward increase in carbonate content; and the later trend (gamma-ray log markers 12–18) containing the upper part of the Kimmeridge Clay Formation, the Portland Beds and the Purbeck Beds reflects an upward decrease in clay content (Fig. 6). Note that the cycles have been calibrated to the standard ammonite zonation using the Warlingham Borehole (Melnyk *et al.*, 1992).

Contained within these very long wavelength cycles or trends are cycles of shorter wavelengths (usually less than 50 m) (Figs 7 and 8). The lowermost cycles containing the *cymodoce* through to the *baylei* ammonite biozones (gamma-ray log markers 1–4) are developed in limestones and calcareous sandstones. They are overlain by cycles of dark, occasionally bituminous, siltstones and sandstones (gamma-ray log markers 4–7) of the *mutabilis* ammonite biozone; these cycles show condensation over palaeostructures, perhaps implying an onlap surface. Interpretation of unpublished seismic data suggests that a seismic sequence boundary coincides with the vicinity of gamma-ray log markers 3 and 4.

Cyclicity is particularly pronounced between gamma-ray log markers 8 and 13, containing the *eudoxus* through to the *pallasoides* ammonite biozones (Figs 7 and 8). These cycles can be correlated with confidence from Well 6 to Well 7 (Fig. 7), and can be shown to parallel the most reliable biostratigraphic indices available (Melnyk *et al.*, 1992). It is within this interval that the kerogen-rich shales are best developed and the cyclic character of the well logs reflects long-wavelength (50–100 m) transitions from dominantly oxygenated, micritic mudstones to dominantly reducing, black shales. Lithological evidence shows the interval between gamma-ray log markers 12 and 13 to change character, from limestones and mudstones at Well 7, through calcareous siltstones and mudstones at Well 5, to glauconitic mudstones and sandstones at Well 4 (Fig. 7). Cycles maintain their character in spite of facies changes.

Condensation and loss of cycles at the basin margin and on palaeohighs suggests the presence of several hiatuses between gamma-ray log markers 11 and 15 (Figs 7 and 8). A minor hiatus is present at gamma-ray log marker 11, within the *wheatleyensis* ammonite biozone; significant hiatuses also occur at gamma-ray log markers 12 and 13 at the top of the *pectinatus* and *rotunda* ammonite biozones, an observation supported by a basin-wide seismic marker at this level. Cycle loss and condensation also occurs at gamma-ray log marker 15, suggesting an onlap surface close to this level, and significant variation in the thickness of the cycle separating markers 17 and 18 points to a significant hiatus at this level. Generally speaking, markers 13–18 are biostratigraphically poorly constrained, but it seems that markers 16 and 18 lie close to the top of the *albani* and *anguiformis* ammonite biozones respectively.

The preceding discussion, based on Figs 6 and 7, indicates that low-frequency bandpass filtering of gamma-ray data does indeed enhance the correlatability of signals at the parasequence and sequence scale.

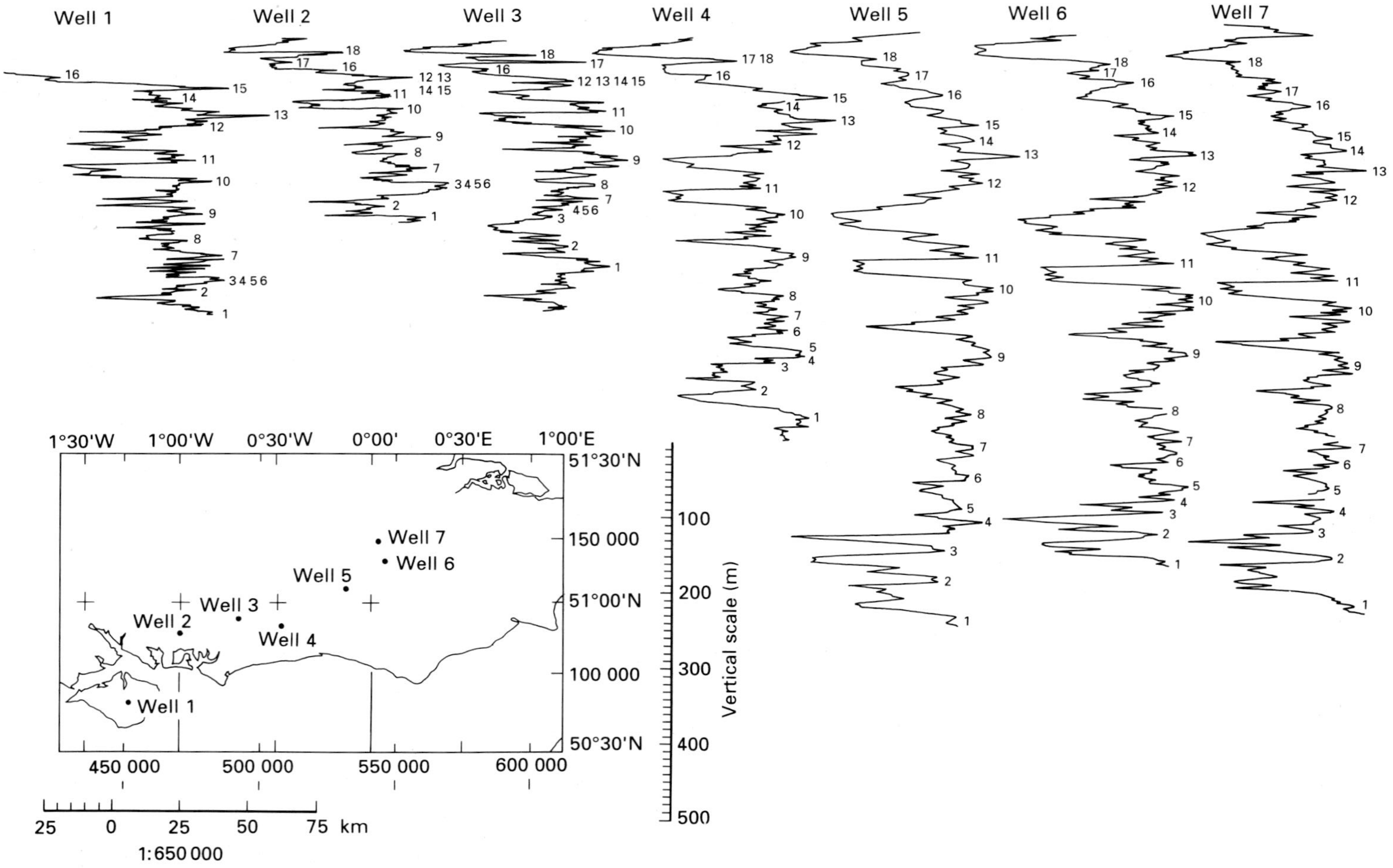

Fig. 7. A wireline log transect, Wessex Basin, emphasizing the west to east regional correlatability of low-frequency cycles. Individual traces have been filtered and hung on a stratigraphic marker horizon (base Purbeck anhydrite). The numbers (1–18) refer to the cycle boundaries illustrated on the type log (see Fig. 6). Where numbers occur at the same horizon a loss of cycles, and therefore a hiatus, is implied.

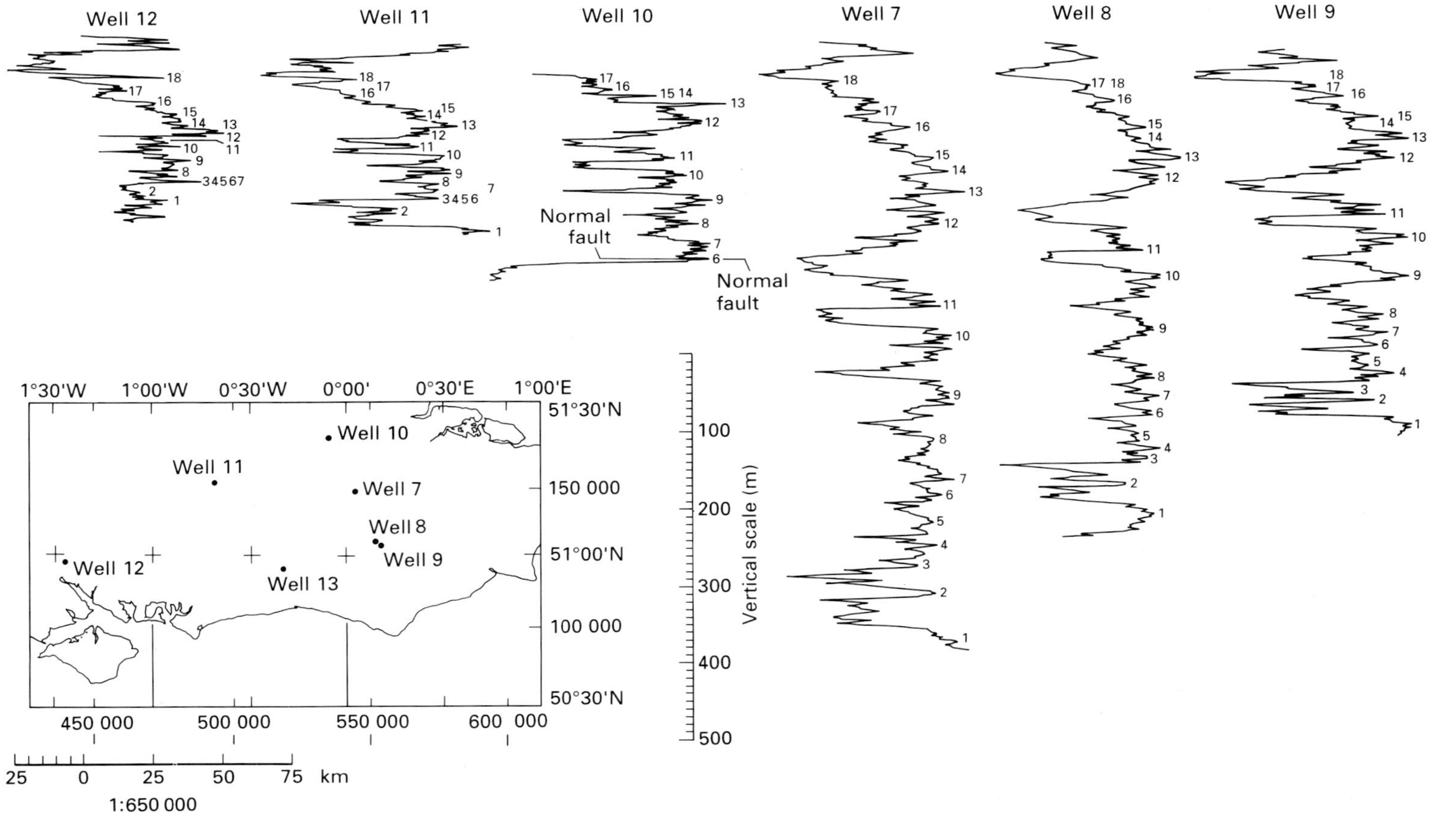

Fig. 8. A wireline log transect, Wessex Basin, emphasizing the west to east regional correlatability of low-frequency cycles. Individual traces have been filtered and hung on a stratigraphic marker horizon. The numbers (1–18) refer to the cycle boundaries illustrated on the type log (see Fig. 6). Where numbers occur at the same horizon a loss of cycles, and therefore a hiatus, is implied.

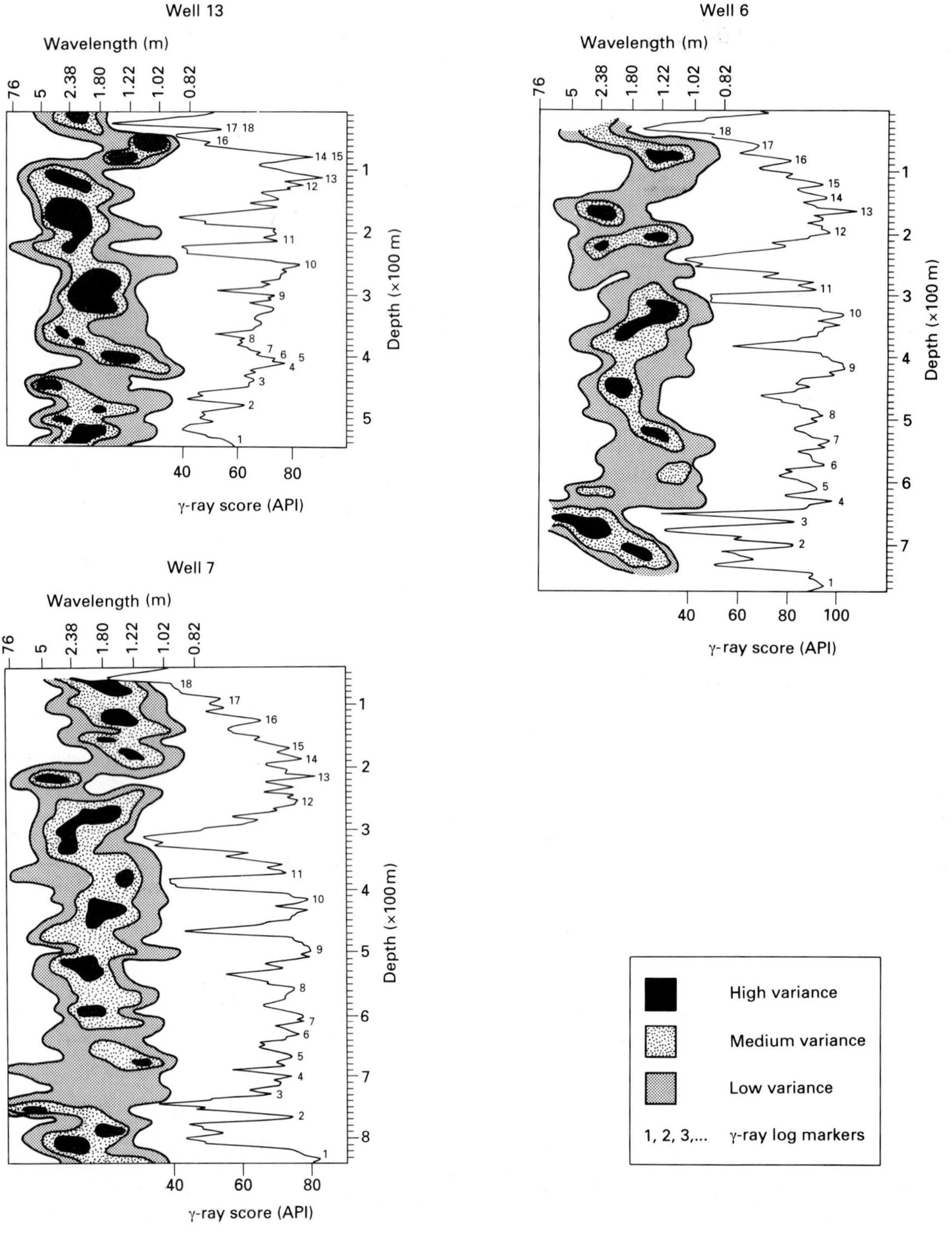

Fig. 9. Low-frequency bandpass curve and high-frequency bandpass maps for Well 6, Well 7 and Well 13. Systematic fluctuations in high-frequency content provide a correlatable signal which reflects the sequence stratigraphy of the Upper Jurassic. The correlatability of the maps is particularly useful where the low-frequency curve does not provide an unambiguous correlation; as, for example, between markers 1 and 8, Well 13.

High-frequency bandpass mapping

Frequency maps have been computed and drawn for three wells: Well 7, Well 6 and Well 13 (Fig. 9). The data on which the maps were drawn are the residuals from an 11-point moving average applied 10 times, and they represent most of the variance in the data set with wavelengths of between 4 m and 0.5 m. No wavelengths below 0.5 m can be expected as this is the effective limit of resolution of most logging tools. A window length of 10 m was used to generate the individual spectra. As the data were standardized (i.e. each data value was divided by the standard deviation of the window, after having the first-order trend of the window subtracted), and for the sake of simplicity and direct comparability, contours were confined to three levels; these levels can be thought of as representing high (dark tone), medium and low (light tone) levels of variance.

The three maps from the three wells are quite similar in terms of trends and anomalies (Fig. 9). Each shows a gradual decrease in frequency (increasing wavelengths) up through the Corallian Beds of gamma-ray log marker 4. At this stratigraphic level, close to the base of the Kimmeridge Clay, there is a marked decrease in wavelength. Wavelengths then increase systematically up-section to the middle of the *eudoxus* biozone (markers 8/9), before decreasing slightly within the *elegans* and *scitulus* biozones (marker 10). The upper *wheatleyensis* through to *hudlestoni* biozones (markers 11/12) appear to be characterized by the longest wavelengths of the entire Kimmeridge Clay. At the base of the Portland Beds (marker 13), wavelengths again decrease dramatically at a major hiatus, before increasing systematically up to the base of the Purbeck Beds (marker 18).

In general, the peaks (dark tone areas) on the contour map correspond to the cleanest parts of the section, as indicated by the lower API scores on the corresponding part of the low-frequency curve (Fig. 9); this is probably a reflection of the greater contrast in radioactivity between the cleaner and shalier beds at these levels. Also noteworthy is the increase in variance associated with short wavelengths (*c*.1.00 m) within the shalier parts of the sections, regardless of the general background trends (i.e. the higher API scores on the low-frequency curve correspond, in general, to the higher frequencies on the map). Finally, the apparent overall similarity in wavelengths between the wells shown in Fig. 9 suggests that that section is lost through the loss of short wavelength cycles, rather than through their thinning and condensation, again implying an onlap surface at this level.

What do these short-wavelength cycles mean geologically? They are certainly of a small enough scale to be recognized at outcrop, and numerous authors have described rhythms at this scale. For example, House (1986) described rhythms of light and dark mudstones from the Kimmeridge Clay Formation of Dorset; he convincingly attributes these rhythms, in the 1–3-m waveband, to the Milankovitch 40-ka obliquity cycle (House, 1985, 1986). If we are observing the equivalent cycles on our frequency maps, then we have at our disposal a potential sedimentation-rate indicator and chronometer. For example, cycle wavelengths in the upper *mutabilis* to upper *scitulus* interval range from just under 1 m to just over 2 m, with an average of around 1.8 m. Assuming a duration of 40 ka/cycle, this would suggest that sedimentation rates varied from 25 to over 50 m/Ma throughout this period of 4.5 Ma. The highest sedimentation rates appear to have occurred during deposition of the *wheatleyensis* through to the *pectinatus* interval.

CONCLUSIONS

Low bandpass filtering of gamma-ray logs of the Upper Jurassic in the Wessex Basin enhances their correlatability, thereby facilitating the rapid and objective subdivision of the succession into chronostratigraphically significant packages or cycles. Correlation of these cycles, which are chronostratigraphic within the resolution of biostratigraphy and seismic stratigraphy, permits the recognition of up to six hiatuses.

High-frequency bandpass mapping emphasizes variations in the short wavelengths present in the residual gamma-ray log curve. Equating these wavelengths to Milankovitch signals facilitates the estimation of sedimentation rates as a function of stratigraphy, and provides a chronometer.

REFERENCES

Chadwick, R.A. (1986) Extension tectonics in the Wessex Basin, southern England. *J. Geol. Soc. Lond.* **143**, 465–488.

Cox, B.M. & Gallois, R.W. (1981) *The stratigraphy of the Kimmeridge Clay of the Dorset type area and its correlation with some other Kimmeridgian sequences.*

Reports of the Institute of Geological Sciences of the United Kingdom, 80.

Gonzalez, R.F. & Wintz, P. (1977) *Digital Image Processing*. Addison-Wesley Publishing Co., Reading, MA.

Holloway, J.L. (1958) Smoothing and filtering of time series and space fields. *Adv. Geophys.* **4**, 351–388.

House, M.R. (1985) A new approach to an absolute timescale from measurements of orbital cycles and sedimentary microrhythms. *Nature* **315**, 712–725.

House, M.R. (1986) Are Jurassic sedimentary microrhythms due to orbital forcing? *Proc. Ussher Soc.* **6**, 299–311.

Imbrie, J., Hays, J.D., Martinson, D.G., McIntyre, A., Mix, A.C., Morley, J.J. *et al.* (1984) The orbital theory of the Pleistocene climate: Support from a revised chronology of the marine delta ^{18}O record. In: *Milankovitch and Climate*, Part 1 (Eds Berger, A., Imbrie, J., Hays, J., Kukla, G. & Saltzman, B.). Reidel Publ. Co., Dordrecht, pp. 269–305.

Melnyk, D.H. & Smith, D.G. (1989) Outcrop to subsurface correlation in the Milankovitch frequency band: Middle Cretaceous, central Italy. *Terra Nova* **1**, 432–436.

Melnyk, D.H., Athersuch, J. & Smith, D.G. (1992) Measuring the dispersion of biostratigraphic events in the subsurface using graphic correlation. *Mar. Petrol. Geol.* **9**, 602–607.

Oschmann, W. (1988) Kimmeridge Clay sedimentation – a new cyclic model. *Palaeogeogr. Palaeoclimatol. Palaeoecol.* **65**, 217–251.

Penn, I.E., Cox, B.M. & Gallois, R.W. (1986) Towards precision in stratigraphy: geophysical log correlation of Upper Jurassic (including Callovian) strata of the Eastern England shelf. *J. Geol. Soc. Lond.* **143**, 381–410.

Schwarzacher, W. (1975) *Sedimentation Models and Quantitative Stratigraphy*. Developments in Sedimentology, 19. Elsevier, Amsterdam.

Van Buchem, F.S.P., Melnyk, D.H. & McCave, I.N. (1992) Chemical cyclicity and correlation of the Lower Lias mudstones using gamma-ray logs, Yorkshire, United Kingdom. *J. Geol. Soc. Lond.* **149**, 991–1002.

Van Wagoner, J.C., Mitchum, R.M., Campion, K.M. & Rahmanian, V.D. (1990) *Siliciclastic Sequence Stratigraphy in Well Logs, Cores, and Outcrops: Concepts for High-resolution Correlation of Time and Facies*. AAPG Methods in Exploration Series, No. 7. Am. Assoc. Petrol. Geol., Tulsa, Oklahoma.

Whittaker, A. (1985) *Atlas of the Onshore Sedimentary Basins in England and Wales: Post-Carboniferous Tectonics and Stratigraphy*. Blackie, Glasgow.

Wignall, P.B. (1989) Sedimentary dynamics of the Kimmeridge Clay: tempests and earthquakes. *J. Geol. Soc. Lond.* **146**, 273–284.

Spec. Publs Int. Ass. Sediment. (1994) **19**, 47–61

Milankovitch cyclicity in the Upper Rotliegend Group of The Netherlands offshore

C.S. YANG *and* Y.A. BAUMFALK

InterGeos B.V., Statenhof, Reaal 5Q, 2353 TK Leiderdorp, The Netherlands

ABSTRACT

The stratigraphic subdivision and correlation of the Upper Rotliegend Group in The Netherlands offshore still provides difficulties, despite the extensive hydrocarbon exploration activities in the last two decades. One of the problems is the lack of biostratigraphic data.

Several kinds of cycles can be discerned in the Upper Rotliegend Group. Third-order sequences are distinct in the wireline logs of the Upper Rotliegend Group. They consist of parasequences or parasequence sets. Based on an analysis of wireline logs of 150 released wells in The Netherlands offshore, 12 third-order sequences, which can be grouped into five supersequences, are discerned in the Upper Rotliegend Group.

In addition to the third-order sequences, a number of higher-order cycles can also be observed in the wireline logs of the Upper Rotliegend Group. Using spectral analysis, a limited number of frequencies can be detected in the wireline log data. The corresponding wavelengths are usually in the range of 1–10 m. The ratios between these wavelengths are very close to the ratios between Milankovitch periods for the Early Permian. Therefore the conclusion is that they are linked to the Milankovitch cycles. Using this approach, the time interval and the net sediment accumulation rates of each third-order sequence can be estimated. Most third-order sequences have a duration of 0.7–1.4 Ma. The net sediment accumulation rates show a long-term trend, which may reflect a long-term climate change and/or the tectonic evolution of the basin. Such a pattern can be used in sequence correlation.

The results of this study show:

1 The upper part of the Upper Rotliegend Group (i.e. Silverpit Formation) contains well-preserved cyclicities. Cyclicities in the lower part are less clear.

2 Based on the sequence and cyclicity analyses, we estimate the total time represented in the Upper Rotliegend Group sediments of The Netherlands to be 10.7 Ma.

Therefore it can be assumed that the Upper Rotliegend of The Netherlands was deposited at approximately 266.8–256.1 Ma, i.e. within the Artinskian and the Kungurian Stages.

INTRODUCTION

The Upper Rotliegend Group of The Netherlands offshore is an Early Permian lithostratigraphic unit (van Wijhe *et al.*, 1980; Glennie, 1990). It consists mainly of clastics, with evaporites in the basin centre. The Rotliegend unconformably overlies the Carboniferous, and is conformably overlain in most cases by the Copper Shale, which marks the beginning of the Zechstein sedimentation, and is understood to represent an approximately isochronous layer just above the Lower/Upper Permian boundary (Haubold & Katzung, 1972).

The Upper Rotliegend Group in The Netherlands offshore has been the target of extensive hydrocarbon exploration activities for the past two decades. Although abundant geological data have been acquired, the stratigraphic subdivision and correlation of the Upper Rotliegend Group is not satisfactory. The classical Upper Rotliegend Group of The Netherlands offshore was subdivided into the Ten Boer Claystone, the Upper Slochteren Sandstone, the Ameland Claystone and the Lower Slochteren Sandstone in the northern part of the sand belt.

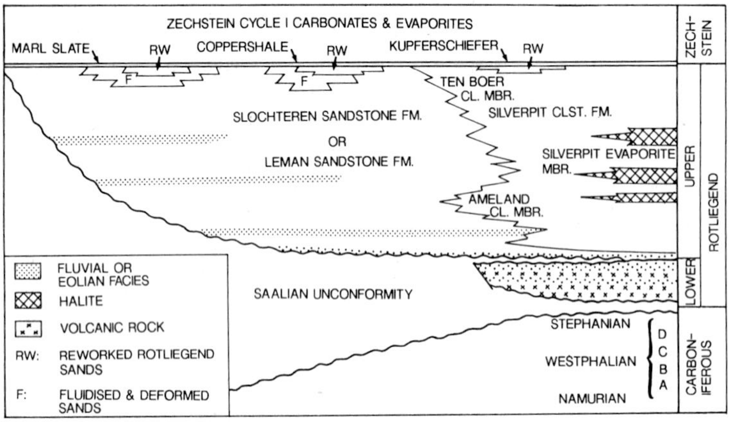

Fig. 1. Classical stratigraphic subdivision of the Upper Rotliegend Group in The Netherlands offshore (modified after Glennie, 1990, and K.W. Glennie, personal communication). This diagram has been modified to show the Carboniferous/Lower Rotliegend unconformity and the Lower Rotliegend/Upper Rotliegend unconformity.

More basinward it was subdivided into the Silverpit Claystone and the Slochteren Sandstone (fig. 1) (Nederlandse Aardolie Maatschappij B.V. & Rijks Geologische Dienst, 1980; Van Wijhe *et al.*, 1980; Glennie, 1990). The boundaries of these classical stratigraphic subdivisions of the Upper Rotliegend Group, however, are defined at the transition from one dominant lithology into another. As a result, the boundaries between members are largely diachronous (Nederlandse Aardolie Maatschappij B.V. & Rijks Geologische Dienst, 1980; Van Wijhe *et al.*, 1980). The absence of fossil control on the age of the Upper Rotliegend Group adds further difficulty to the chronostratigraphic subdivision of the Upper Rotliegend Group.

Based on new developments in sequence stratigraphy, a basin-scale sequence stratigraphy study of the Upper Rotliegend Group in The Netherlands offshore has been conducted using an integrated approach (InterGeos, 1991; Yang & Nio (in press)). This study consisted of two phases.

1 A basin-wide sequence stratigraphic analysis and correlation was carried out over 150 released wells throughout The Netherlands offshore.

2 This was followed by a high-resolution cyclicity analysis to calibrate each sequence. This paper will concentrate on the second aspect of the study, i.e. the recognition of Milankovitch cycles in the Upper Rotliegend Group and its application in high-resolution sequence stratigraphy analysis.

THIRD-ORDER SEQUENCES OF THE UPPER ROTLIEGEND GROUP

Recently, concepts of sequence stratigraphy have been successfully applied to siliciclastic as well as carbonate sediments deposited in marine basins. Analysis of the third-order sequences and systems tracts results in a better understanding of the relationship among eustatic sea-level fluctuations, basin subsidence and sedimentary facies development in a continental shelf and marine basin setting. The Upper Rotliegend Group, however, was deposited in a continental basin, which differs from the continental shelf and marine basin setting in that the base level in the Rotliegend continental basin was not directly controlled by global eustatic sea level, and at times this base level was about 250 m below the ocean level (Smith, 1970, 1979; Ziegler, 1982; Glennie, 1990). Despite this difference, the basic principles of sequence stratigraphy can also be applied to the Upper Rotliegend Group since the climate fluctuations played an important role in controlling both the eustatic sea-level changes and the base-level fluctuations in the Rotliegend Basin. Such base-level fluctuations, together with the basin subsidence, controlled the sequence stratigraphic development and sedimentary facies distribution in the Rotliegend continental basin.

With the application of the principles of sequence stratigraphy, a sequence stratigraphic framework has been established for the Upper Rotliegend Group of The Netherlands offshore (InterGeos, 1991). Based on an analysis of lithology and wireline logs of 150 released wells in The Netherlands offshore, five groups of 12 third-order sequences can be discerned in the Upper Rotliegend Group (Figs 2 and 3). The sequences can be compared to the third-order sequences of Vail *et al.* (1977) and Haq *et al.* (1987). Each sequence consists of parasequences or parasequence sets (the term defined by van Wagoner *et al.*, 1990). The sequence boundaries and

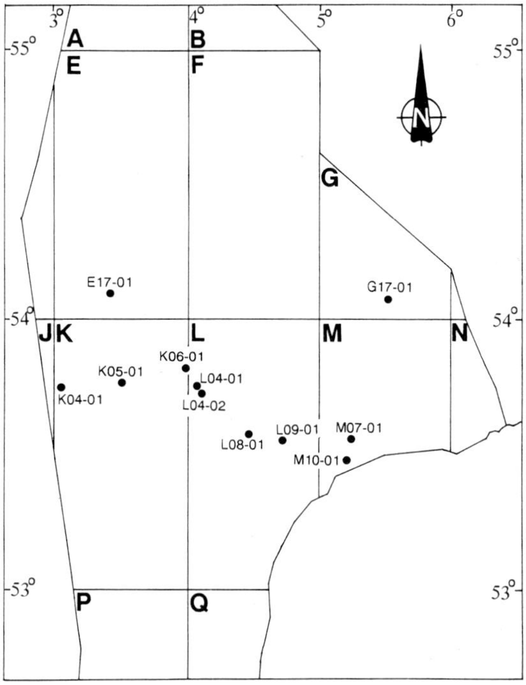

Fig. 2. Location map of the wells in The Netherlands offshore used in this study.

maximum flooding surfaces are regionally persistent and permit reliable sequence stratigraphic correlation throughout The Netherlands offshore. In general, the 12 recognized third-order sequences can be grouped into five supersequences. The distribution of supersequence RO1 is limited in the central part of the basin, while supersequences RO2, RO3, RO4 and RO5 show a successive onlapping pattern on to the basin flank (Fig. 3).

MILANKOVITCH CYCLES

In addition to the third-order sequences, there are also a number of higher-order cycles which can be observed in the wireline logs of the Upper Rotliegend Group. There is convincing evidence that these higher-order cyclic patterns can be identified as Milankovitch cycles.

Milankovitch cycles reflect variations in the orbital parameters of the Earth, which are caused by the gravitational effects of the other bodies in the solar system. As a consequence, the amount of solar radiation reaching a certain point on Earth varies significantly over periods of 10–100 ka. From studies in the Quaternary we know that these astronomical parameters have a major effect on global climate and are the main control on the fluctuation of continental ice caps and the changes in depositional systems.

The astronomical parameters such as the eccentricity of Earth's orbit, the obliquity of Earth's rotation axis, and the precession of the equinoxes have periods of 100 ka, 54 ka and 41 ka, and 23 ka and 19 ka respectively (Berger *et al.*, 1989).

PRESERVATION OF MILANKOVITCH CYCLES IN THE SEDIMENTARY RECORD

The orbital parameters of the Earth result in changes in insolation (solar radiation) at the surface of the Earth, both in absolute terms and relatively over the seasons. These changes must have an important effect on global and local climate respectively. How these climatic changes are reflected in the sedimentary record will depend on the general depositional setting, the latitude, the importance of seasonality in the sediment supply and the depositional processes in the receiving basin.

In the marine pelagic realm the changes in insolation may influence the production (and therefore sedimentation) directly or indirectly (e.g. mediated by changes in availability of nutrients derived from the land). In terrestrial sedimentary settings, prevailing wind direction, weathering and erosion in the hinterland and type of fluvial systems may all change with climate. Whatever the cause, the deposits may show rhythms with the same periods, although the intensity of the contribution of each orbital parameter may vary from place to place.

There is convincing evidence that orbital forcing cycles of climate, causing truly cyclic patterns in sedimentation, existed during the pre-Quaternary. Truly cyclic patterns in pre-Quaternary deposits which are related to precisely measurable orbital events have been recognized, especially from wireline data in subsurface studies (Smith, 1989; Worthington, 1990; Y.A. Baumfalk and A.J. Nederbragt, personal communication).

The recent cyclicities of orbital parameters mentioned above are valid for the last few hundred thousand years. There are strong indications that in the more remote past these values were different,

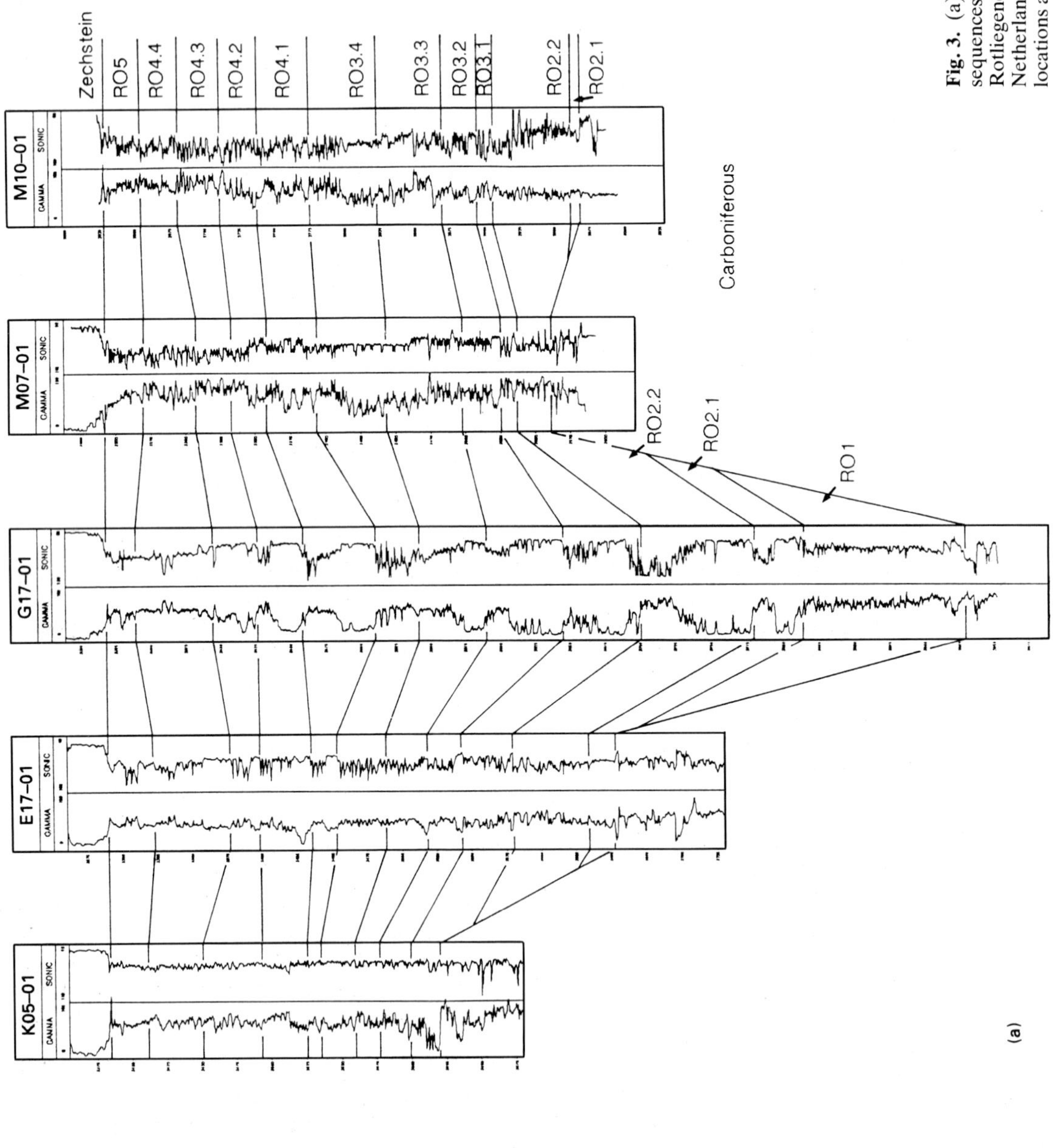

Fig. 3. (a) and (b) Third-order sequences recognized in the Upper Rotliegend Group of The Netherlands offshore. Well locations are indicated in Fig. 2.

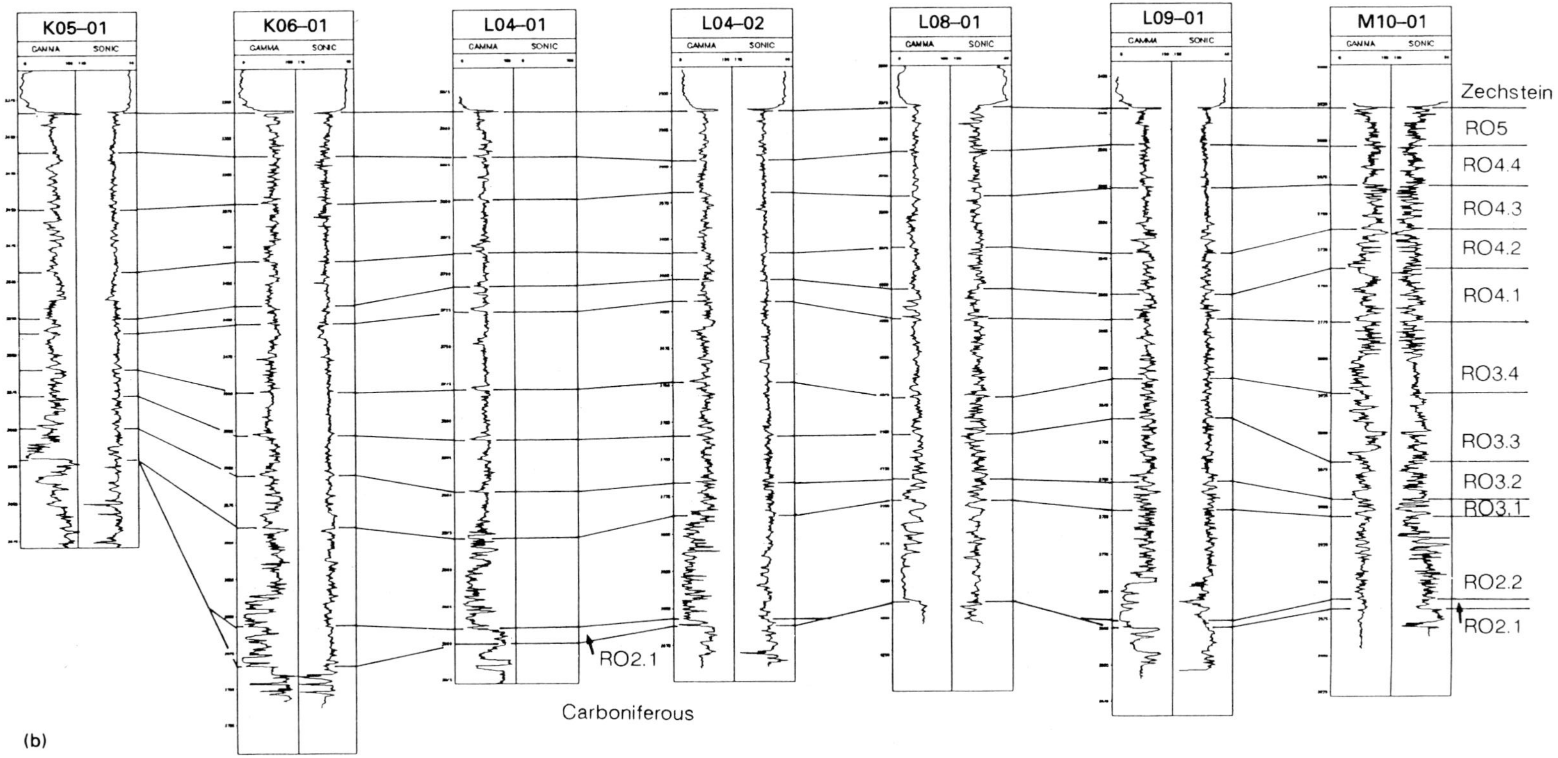

Fig. 3. *Continued.*

Table 1. Estimated values of the periods of the Milankovitch cycles (in ka) for the Early Permian. After Berger *et al.* (1989)

A	100.0	Eccentricity
A′	67.0	
B	44.3	Obliquity
C	35.1	Obliquity
D	30.0	
E	21.0	Precession
F	17.6	Precession

although of the same order of magnitude. Calculations for several periods in geological time have been given by Berger *et al.* (1989). The periods of the Milankovitch cycles in the Early Permian were estimated as 100 ka (eccentricity), 44.3 ka and 35.1 ka (obliquity) and 21 ka and 17.6 ka (precession) (Berger *et al.*, 1989). This study indicates that a few more cycles (or interference frequencies) exist in the Upper Rotliegend sequences (especially in the order of 67 ka and 30 ka). For convenience we designate the Early Permian cycles 100, 67, 44.3, 35.1, 30, 21 and 17.6 ka as A, A′, B, C, D, E and F (Table 1).

HIGHER-ORDER CYCLIC PATTERNS IN THE UPPER ROTLIEGEND GROUP

For a detailed cyclicity analysis of the third-order sequence, gamma-ray logs of the Upper Rotliegend Group from eight wells with well-preserved cyclicities were selected. The orbital forcing cycles of climate can be recorded as cyclic patterns in wireline logs only when: (i) the depositional process is continuous; and (ii) the sediment accumulation rate is constant. Of course these requirements are not always met. As a result, some discrepancies may occur between the ratio of the Milankovitch cycles and that of the cycles discerned in the spectrogram of the Upper Rotliegend sequences.

In order to detect higher-order cyclicities, a sliding-window spectral analysis is applied to the gamma-ray logs of the Upper Rotliegend Group in selected wells (Fig. 4). Sliding-window spectral analysis has been applied to wireline log and outcrop data (e.g. Melnyk & Smith, 1989). In this study, a window of fixed size (256 equally spaced data points in most cases) is moved down the wireline data. In each step a power spectrum is calculated for the window. The left-hand side of the spectrogram is positioned at the middle depth point of the window. Only peaks with their amplitudes above a certain threshold are shown. Spectrograms of successive windows may partly overlap. This way of presentation results in a pattern of parallel bands in intervals where cyclicities are pronounced and a blank pattern in intervals where cyclicities are missing or blurred. When the window includes a discontinuity, the pattern is interrupted for some length.

The size of the window (in number of data points) can be selected at a power of 2 if a fast Fourier transform algorithm is used. A larger window contains more points, and may therefore give a more reliable spectrum, but has a greater likelihood of containing discontinuities. A smaller window does not hold enough points to contain the lower-frequency cycles in this type of deposit. Generally, a window size of 256 points (approximately 40 m at 0.15 m sample spacing) tends to have the best trade-off between interval accuracy and spectrum accuracy.

Figure 4 shows the results of sliding-window spectral analysis of the gamma-ray log in well L09-01. The upper part of the Upper Rotliegend Group (i.e. sequence RO5, 4.4, 4.3, 4.2, 4.1, 3.4, 3.3, 3.2, 3.1 and upper part of RO2.2) displays well-preserved cyclicities. However, cyclicities in the lower part (i.e. lower part of sequence RO2.2 and RO2.1) are less clear. Figure 4 also shows that the boundaries of the third-order sequences are usually characterized by apparent breaks in cyclicity patterns. This feature can be used to draw and check sequence boundaries.

IDENTIFICATION OF MILANKOVITCH CYCLES

For the identification of high-frequency cycles within the third-order sequence, a spectrogram is constructed for each third-order sequence. Figure 4 shows the spectrum diagrams of the third-order sequences in well L09-01. The vertical axis of this graph represents the relative amplitudes of the frequencies. The horizontal axis represents the harmonic numbers of the frequencies: harmonic 1 corresponds to the frequency with a wavelength of half the timespan (or half the thickness of analysed sequence), harmonic 2 to the frequency with a wavelength of one-third the sequence thickness, and in general, harmonic k to the frequency with a wavelength of $1/(k + 1)$ of the sequence thickness.

In Fig. 4 a limited number of frequencies can be discerned. The corresponding wavelengths are

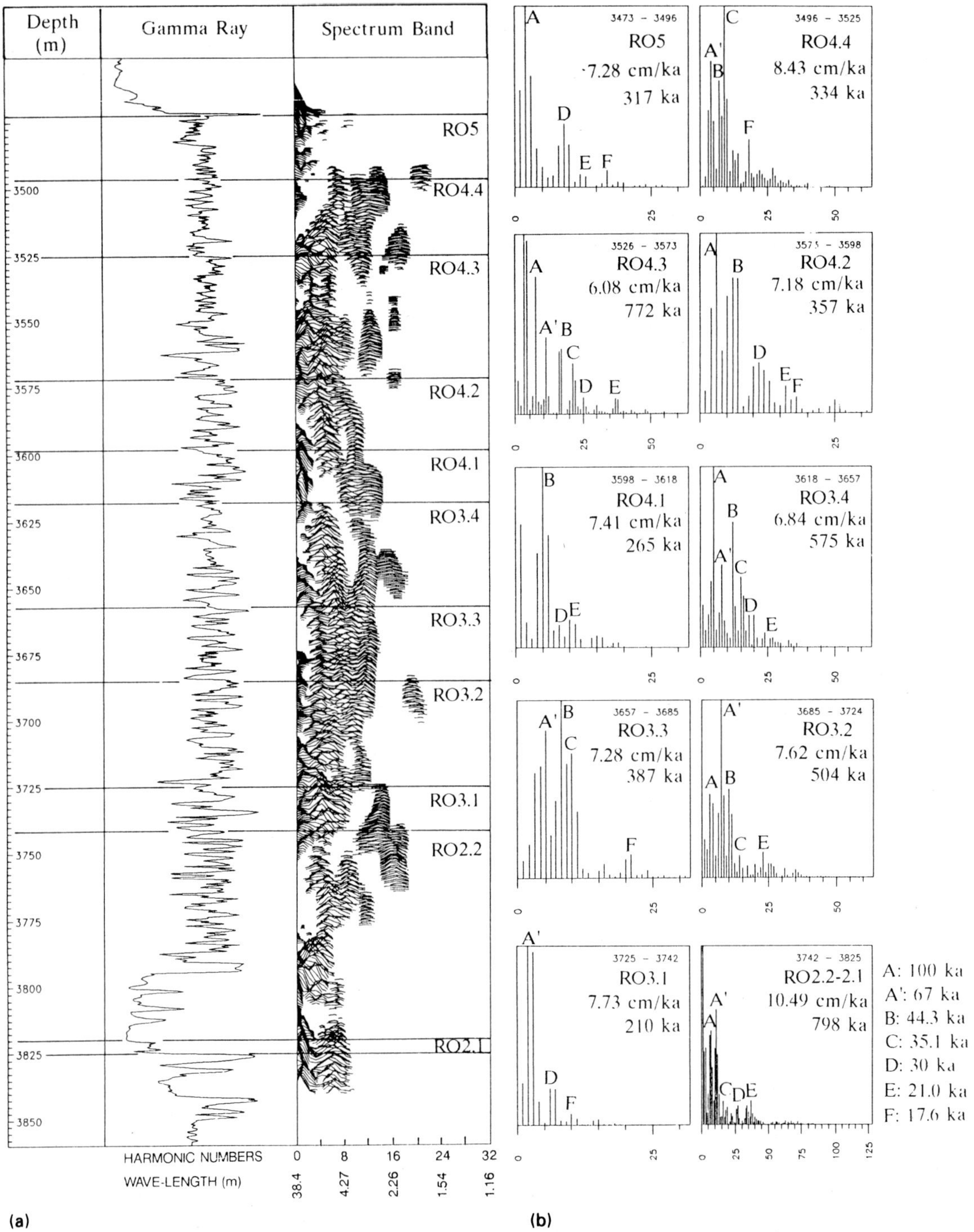

Fig. 4. Higher-order cyclicities in the Upper Rotliegend Group of The Netherlands offshore (well L09-01). (a) Sliding-window spectral analysis of the gamma-ray log; (b) spectrum diagram of the third-order sequences. The vertical axis of the diagram represents the relative amplitudes of the frequencies. The horizontal axis represents the harmonic numbers of the frequencies. Well location is indicated in Fig. 2.

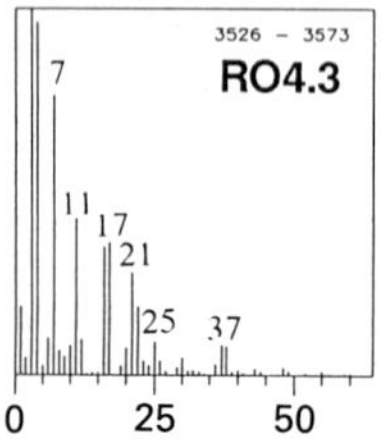

(a)

Harmonic numbers of spectrum peaks: 7 11 17 21 25 37

Ratio of corresponding wavelengths: 1/(7+1):1/(11+1):1/(17+1):1/(21+1):1/ 25+1):1/37+1)=
100 : 66.7 : 44.4 : 36.3 : 30.7 : 21.0

Ratio of Milankovitch periods: 100 : 67 : 44.3 : 35.1 : 30 ; 21

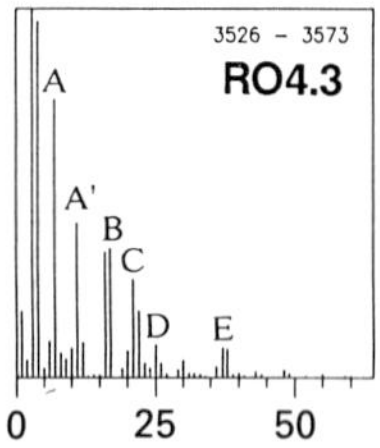

(b)

Peak identification and time duration estimation:
Peak at harmonic no. 7 as 100 ka cycle (A): 100 ka × (7+1) = 800 ka
Peak at harmonic no. 11 as 67 ka cycle (A'): 67 ka × (11+1) = 804 ka
Peak at harmonic no. 17 as 44.3 ka cycle (B): 44.3 ka × (17+1) = 797 ka
Peak at harmonic no. 21 as 35.1 ka cycle (C): 35.1 ka × (21+1) = 772 ka
Peak at harmonic no. 25 as 30 ka cycle (D): 30 ka × (25+1) = 780 ka
Peak at harmonic no. 37 as 21 ka cycle (E): 21 ka × (37+1) = 798 ka

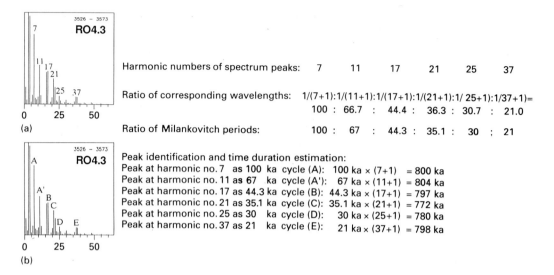

Fig. 5. Identification of Milankovitch cycles in sequence RO4.3 of the Upper Rotliegend Group, well L09-01, The Netherlands offshore. (a) Spectrogram from gamma-ray logs; (b) Milankovitch cycles identified in the same spectrogram.

usually in the range of 1–10 m. The ratios between these wavelengths are very close to the ratios between Milankovitch periods for the Early Permian (Table 1). It is from this similarity that the cyclicities can be linked to the Milankovitch cycles. For example, the spectrogram of sequence RO4.3 shows some major peaks located at harmonics 7, 11, 17, 21, 25 and 37 (Fig. 5). The ratios between their wavelengths are therefore: $1/(7+1):1/(11+1):1/(17+1):1/(21+1):1/(25+1):1/(37+1)$, which are approximately 100:66.7:44.4:36.3:30.7:21.0. These are close to the ratios between the Milankovitch cycles of respectively 100, 67, 44.3, 35.1, 30 and 21 ka (estimates for the Permian, see Berger *et al.*, 1989). These peaks are therefore identified as the 100, 67, 44.3, 35.1, 30 and 21 ka cycles respectively (Fig. 5).

Based on this approach, some of the major peaks discerned in the spectrogram of third-order sequences in the studied well are interpreted as Milankovitch cycles. The results from well L09-01 are summarized in Fig. 6. In some third-order sequences a more or less complete set of Milankovitch cycles (cycles A–F) can be recognized, including a cycle of around 30 ka (cycle D), which is also known from studies in marine Cretaceous deposits from Trinidad (Y.A. Baumfalk and A.J. Nederbragt, personal communication). In some other third-order sequences, however, only a few Milankovitch cycles can be detected. Most peaks show a ratio very close to that of Milankovitch cycles. Some discrepancies, however, do occur. Such discrepancies may reflect the combined effect of several factors.

1 The spectral analysis resolves the serial data into its harmonic components at discrete harmonic numbers. If the frequency of a harmonic component does not correspond to one of the harmonic numbers (positive integer numbers), the round-up error may introduce some discrepancies.

2 The Milankovitch cycles (in terms of time) can be recorded as cyclic patterns in the sedimentary sequences (in terms of depth) only when the sedimentary process is continuous and the sediment accumulation rate is constant. Any changes in sediment accumulation rates, or any discontinuity in the sequence, may cause the observed discrepancies between the cyclic patterns recorded in the sedimentary sequence and the Milankovitch cycles.

3 There is a certain amount of noise during the deposition of the Upper Rotliegend.

4 Not all Milankovitch cycles can be recorded everywhere.

ESTIMATION OF TIME INTERVAL AND NET SEDIMENT ACCUMULATION RATE

With the identification of Milankovitch cycles, it is straightforward to estimate the time duration of each sequence. For example, six peaks have been

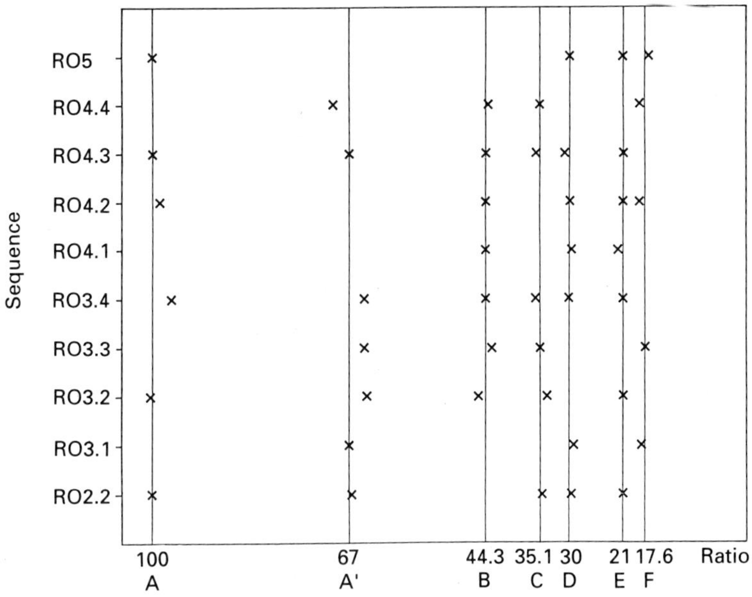

Fig. 6. Ratios between the wavelengths of harmonic peaks discerned in the spectrogram of the third-order sequences in well L09-01. For comparison, the ratios between Milankovitch periods for the Early Permian are also indicated.

identified in the spectrogram of sequence RO4.3 (Fig. 5). The peak at harmonic 21 is identified as the 35.1-ka cycle, which gives an estimated duration for this sequence of (21 + 1) × 35.1 ka = 772 ka. The estimated time durations for the third-order sequences of all selected wells are summarized in Fig. 7.

The net sediment accumulation rate can be estimated from the thickness and time duration of the sequence. In sequences where no pronounced peaks can be identified reliably, the net sediment accumulation rates of similar sequences above or below the interval are extrapolated to get a rough estimate. Figure 8 summarizes the net sediment accumulation rates estimated for third-order sequences of all selected wells. It is notable that the wells studied show a similar trend of changes in net sediment accumulation rate (Fig. 9). Sequence RO1 is characterized by a lower rate (6.9 cm/ka). Supersequence RO2 shows a stage of the highest accumulation rate of the Upper Rotliegend Group (average net sediment accumulation rate 10.4 cm/ka in RO2.1 and 10.8 cm/ka in RO2.2). Supersequence RO3 shows a stage of relatively high accumulation rate (average net rate 8.9–8.1 cm/ka in RO3.1 and RO3.2), followed by a stage of relatively low rate (average net rate 7.4–7.1 cm/ka in RO3.3 and RO3.4). Supersequence RO4 is characterized by a similar pattern with the average net sediment accumulation rate first increasing from 7.7 cm/ka (RO4.1) to 7.9 cm/ka (RO4.2), and then decreasing to 6.1 cm/ka (RO4.3) and 6.7 cm/ka (RO4.4). Finally the average net sediment accumulation rate increases slightly to 7.0 cm/ka in RO5. This pattern may reflect the influence of the long-term climate change on sediment supply and transportation, as well as the influence of tectonics on basin evolution. Such a pattern can be used for the correlation of the third-order sequences.

CHRONOSTRATIGRAPHY OF THE UPPER ROTLIEGEND GROUP

It has been very difficult to estimate the time duration and age of the Upper Rotliegend Group. This is due to the absence of biostratigraphic data in the Upper Rotliegend Group, which was deposited in an arid to semi-arid setting. The identification of higher-order cyclicities in the Upper Rotliegend Group with Milankovitch cycles, however, provides a means for the age determination of the Upper Rotliegend Group. Based on the time duration estimated from cyclicity analysis, an attempt is made to reconstruct the tectonostratigraphic chart of the Upper Rotliegend Group for selected wells (Fig. 10). This chart displays a considerable amount of non-deposition and/or erosion in the stratigraphic record.

It has been estimated that deposition of the

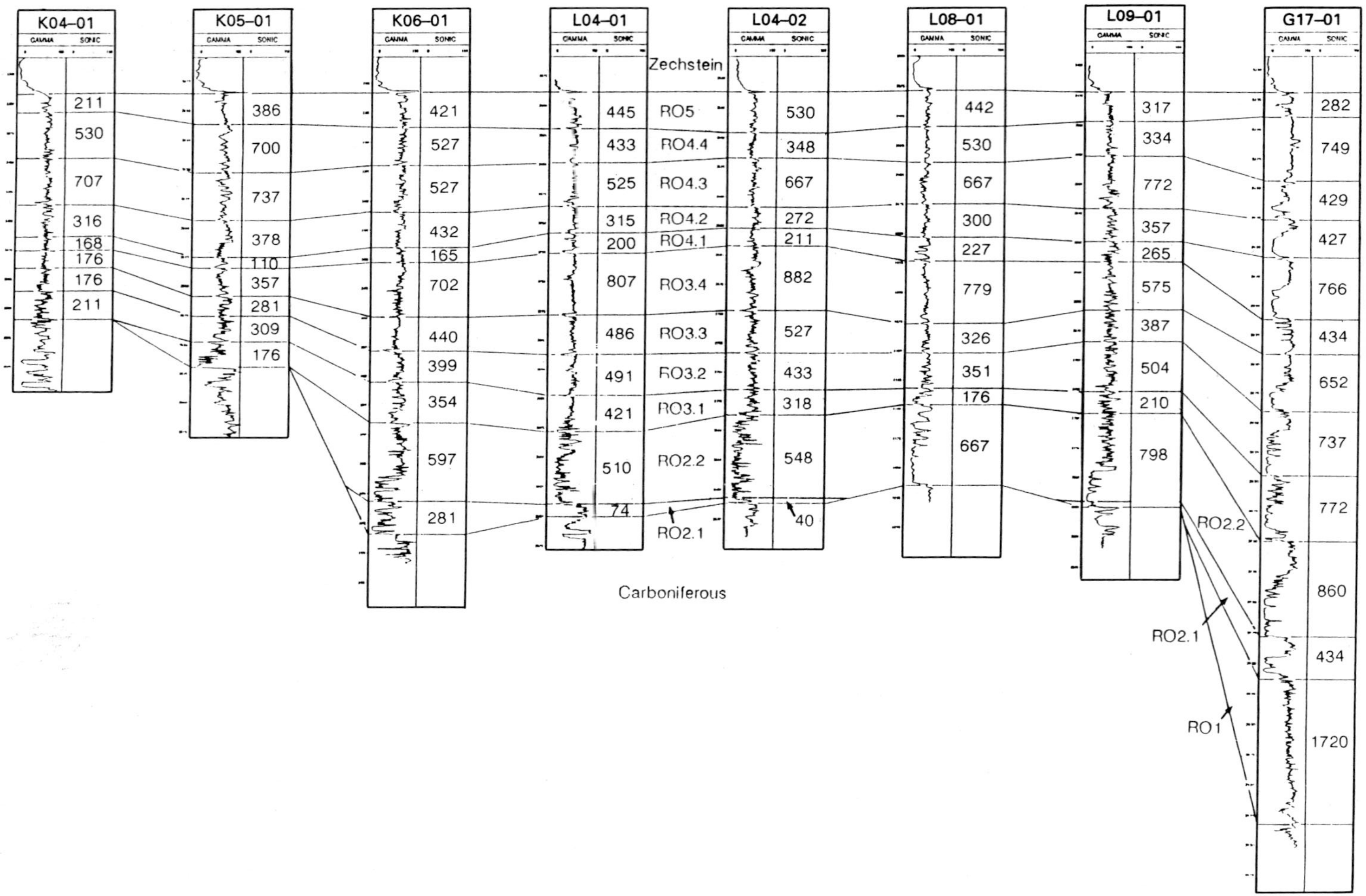

Fig. 7. Estimated time duration (ka) of the third-order sequences in the Upper Rotliegend Group of studied wells, The Netherlands offshore. Well locations are indicated in Fig. 2.

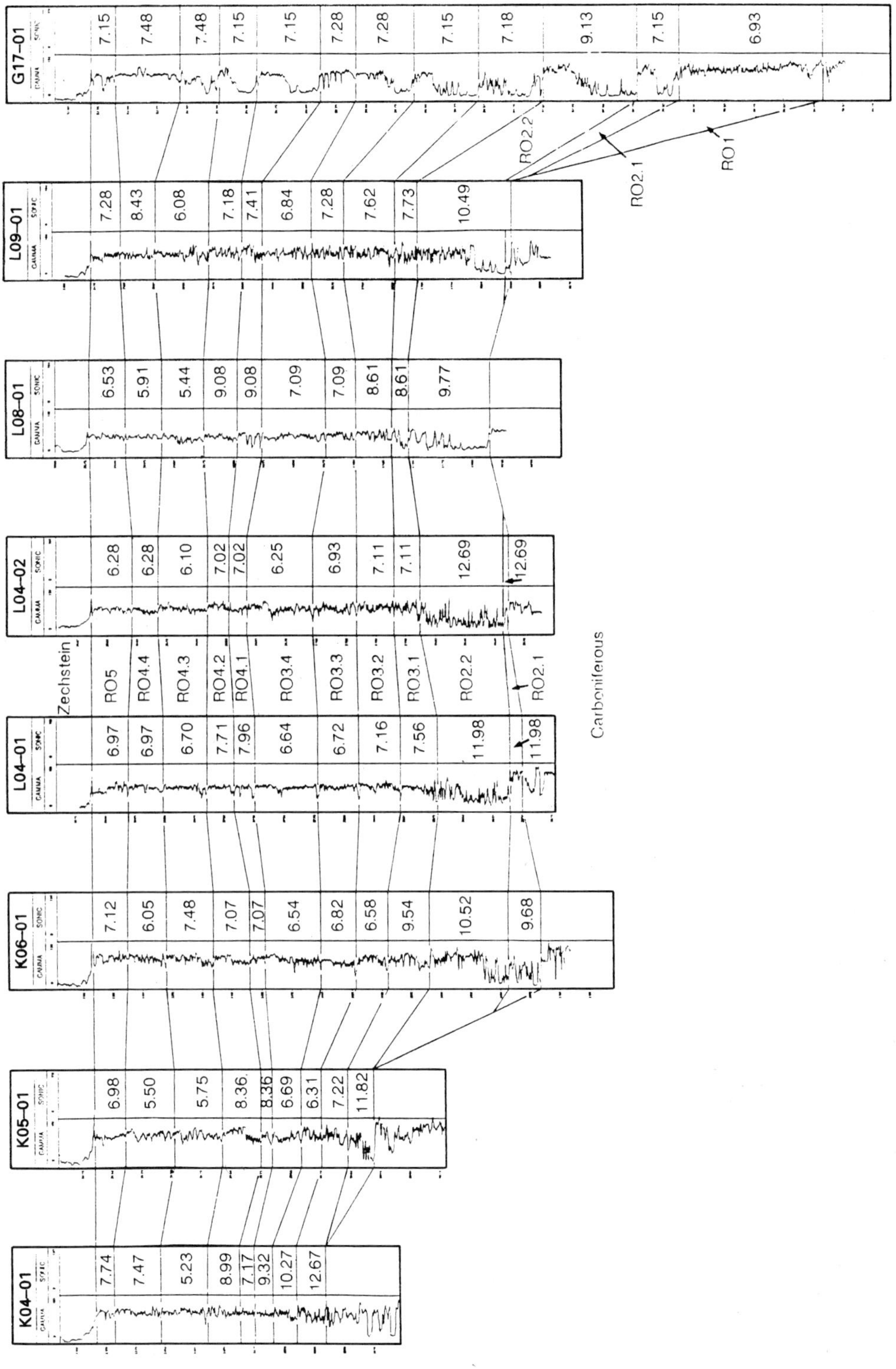

Fig. 8. Estimated net sediment accumulation rate (cm/ka) of the third-order sequences in the Upper Rotliegend Group of studied wells, The Netherlands offshore. Well locations are indicated in Fig. 2.

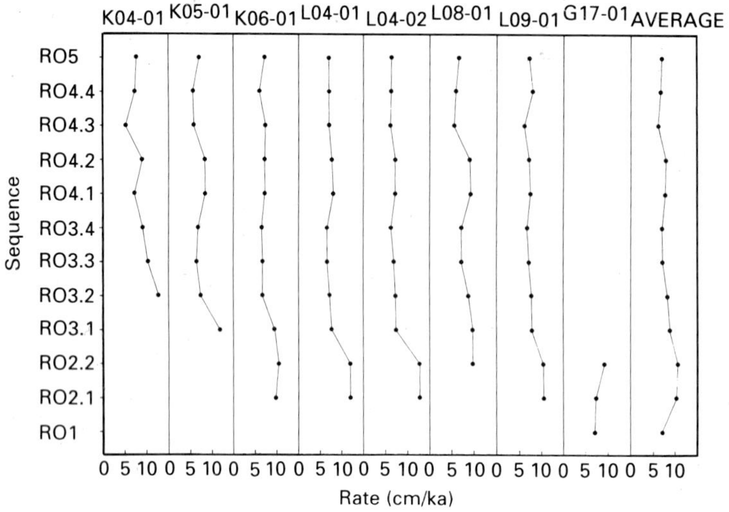

Fig. 9. Trend of changes in net sediment accumulation rates estimated for the third-order sequences in the Upper Rotliegend Group of studied wells, The Netherlands offshore. Note that most wells show a similar trend of changes in net sediment accumulation rate. This pattern may reflect the influence of the long-term climate change on sediment supply and transportation, and the influence of tectonics on basin evolution. Such a pattern can be used for the correlation of the third-order sequences. Well locations are indicated in Fig. 2.

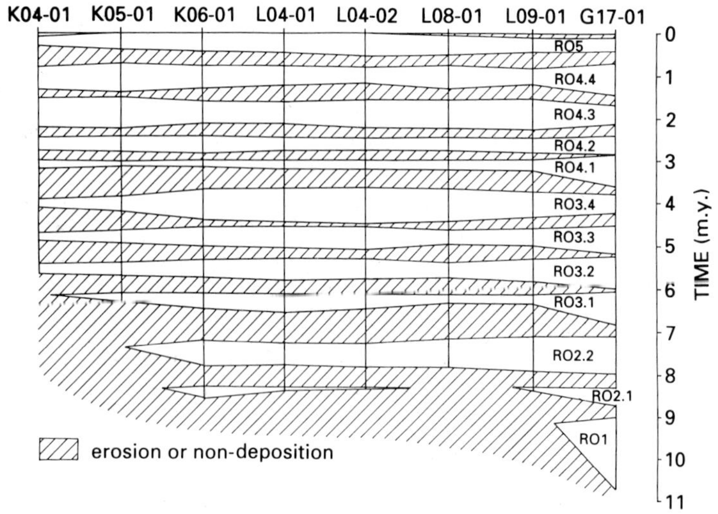

Fig. 10. Tectonostratigraphic chart of the Upper Rotliegend Group based on sequence and cyclicity analyses of selected wells. Well locations are indicated in Fig. 2.

Rotliegend (i.e. the Upper and Lower Rotliegend Groups) extended from approximately 300 to 258 Ma (Gast, 1988; Glennie, 1990). In this study, the maximum estimated duration found for each third-order sequence has been taken as its duration. This estimation may introduce some error if the analysed sequence is not complete. Since the wells studied have relatively complete Upper Rotliegend sequences, and the maximum duration is considered to be the duration of the sequence, we can assume that the error would be small (within about 10%). The estimated duration and age of each third-order sequence are shown in Fig. 11. For comparison the classical lithostratigraphic subdivision of the Upper Rotliegend in The Netherlands offshore (northern part of the sand belt) is also indicated. By adding up the maximum duration of each third-order sequence, we estimate the total time represented in the Upper Rotliegend Group of The Netherlands offshore to be about 10.7 Ma. It has been estimated that the Rotliegend/Zechstein boundary is at 256.1 Ma (Harland *et al.*, 1990). Therefore it can be assumed that the Upper Rotliegend of The Netherlands was deposited between approximately

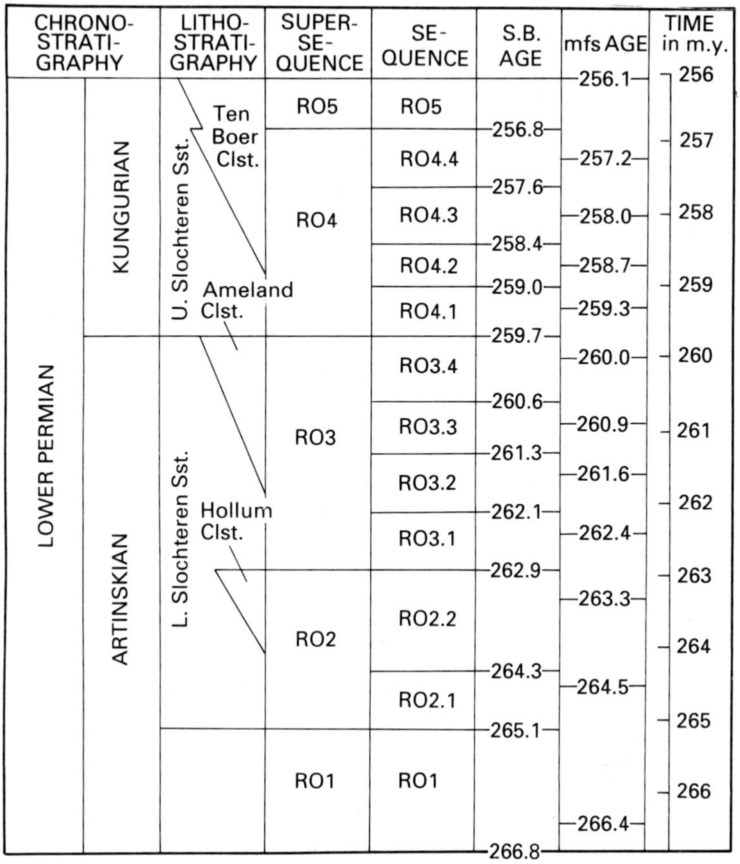

CHRONO-STRATI-GRAPHY
LITHO-STRATI-GRAPHY
SUPER-SE-QUENCE
SE-QUENCE
S.B. AGE
mfs AGE
TIME in m.y.
LOWER PERMIAN
KUNGURIAN
ARTINSKIAN
U. Slochteren Sst.
L. Slochteren Sst.
Ten Boer Clst.
Ameland Clst.
Hollum Clst.
RO5
RO4
RO3
RO2
RO1
RO5
RO4.4
RO4.3
RO4.2
RO4.1
RO3.4
RO3.3
RO3.2
RO3.1
RO2.2
RO2.1
RO1
256.8
257.6
258.4
259.0
259.7
260.6
261.3
262.1
262.9
264.3
265.1
266.8
256.1
257.2
258.0
258.7
259.3
260.0
260.9
261.6
262.4
263.3
264.5
266.4
256
257
258
259
260
261
262
263
264
265
266

Fig. 11. Chronostratigraphy of the Upper Rotliegend Group in The Netherlands offshore based on the sequence stratigraphy and cyclicity analyses of selected wells. The age of the Rotliegend/Zechstein boundary and the ages of the stages follow Harland *et al.* (1990).

266.8 and 256.1 Ma, i.e. within the Artinskian and the Kungurian Stages. This time interval is considerably shorter than the total timespan assigned to the Rotliegend. The difference is attributed to the following factors.

1 Sequence RO1 is incomplete in The Netherlands offshore. It becomes much thicker with additional intervals towards the German sector (Hedemann *et al.*, 1984; Gast, 1988; 1991; Glennie, 1990). These additional intervals may represent a time duration of 5–6 Ma.

2 The Lower Rotliegend Group represents a considerable timespan (Drong *et al.*, 1982; Glennie, 1990).

3 Several important unconformities during the latest Carboniferous and earliest Permian (e.g. those between the Westphalian and Stephanian, between the Stephanian and Lower Rotliegend and between the Lower Rotliegend and Upper Rotliegend) also represent a considerable timespan (Nederlandse Aardolie Maatschappij B.V. & Rijks Geologische Dienst, 1980; K.W. Glennie, personal communication).

CONCLUSIONS

1 The analysis of various cycles observed in the Upper Rotliegend Group may provide a useful tool in the sequence stratigraphic study of the Upper Rotliegend Group.

2 The third-order sequences, which can be compared to the third-order sequences of Vail *et al.* (1977) and Haq *et al.* (1987), are distinct in the wireline logs of the Upper Rotliegend Group. Based on an analysis of wireline logs of 150 released wells in The Netherlands offshore, five groups of 12 third-order sequences can be discerned in the Upper Rotliegend Group.

3 In addition to the third-order sequences, there are

also a number of higher-order cycles in the Upper Rotliegend Group. Sliding-window spectral analysis shows that the upper part of the Upper Rotliegend Group (i.e. Silverpit Formation) contains well-preserved cycles. Cycles in the lower part are less clear.

4 Using spectral analysis of the wireline data, a limited number of frequencies can be detected in each third-order sequence. The corresponding wavelengths are usually in the range of 1–10 m. The ratios between these wavelengths are very close to the ratios between Milankovitch periods for the Early Permian. Based on this similarity, the cycles can be identified as Milankovitch cycles.

5 Using this approach, the time interval and the net sediment accumulation rates of each third-order sequence can be estimated. Most third-order sequences have a duration 0.7–1.4 Ma. The net sediment accumulation rates show a trend of long-term change, which may reflect a long-term climatic change and the tectonic evolution of the basin. Such a pattern can be used in sequence correlation.

6 Based on the sequence and cyclicity analyses, we estimate the total time represented in the Upper Rotliegend Group sediments of The Netherlands offshore to be about 10.7 Ma. Assuming that the Rotliegend/Zechstein boundary is at 256.1 Ma (Harland *et al.*, 1990), it can be inferred that the Upper Rotliegend of The Netherlands was deposited between approximately 266.8 and 256.1 Ma, i.e. within the Artinskian and the Kungurian Stages.

ACKNOWLEDGEMENTS

The Geological Survey of The Netherlands (Rijks Geologische Dienst) provided the wireline logs of released wells for this study. We wish to thank Rijks Geologische Dienst for their help and support. This study is part of the InterGeos study on 'Rotliegend of The Netherlands offshore'. We wish also to thank InterGeos B.V. for permission to publish this paper. Appreciation and thanks are due to our colleagues in InterGeos, especially S.D. Nio, W. Kouwe, H. Jonkman and J.J. van den Hurk for helpful discussions in this study. P.L. de Boer, D.H. Melnyk and K.W. Glennie kindly reviewed the manuscript and made valuable comments. K.W. Glennie also provided interesting figures of the Rotliegend lithostratigraphy.

REFERENCES

Berger, A., Loutre, M.F. & Dehant, V. (1989) Astronomical frequencies for pre-Quaternary palaeoclimate studies. *Terra Nova* **1**, 474–479.

Drong, H.J., Plein, E., Sannemann, D., Schuepbach, M.A. & Zimdars, J. (1982) Der Scheveningen-Sandstein des Rotliegenden – eine äolische sedimentfüllung alter graben strukturen. *Z. Deutsches Geol. Gesampt.* **133**, 699–725.

Gast, R.E. (1988) Rifting im Rotliegenden Niedersachsens. *Die Geowissenschaften* **6**, 115–122.

Gast, R.E. (1991) The perennial Rotliegend saline lake in NW Germany. *Geol. Jb.* **A119**, 25–59.

Glennie, K.W. (1990) Lower Permian-Rotliegend. In: *Introduction to the Petroleum Geology of the North Sea* (Ed. Glennie, K.W.). Blackwell Scientific Publications, Oxford, pp. 120–152.

Haq, B.U., Hardenbol, J. & Vail, P.R. (1987) Chronology of fluctuating sea levels since the Triassic (250 million years ago to present). *Science* **235**, 1156–1167.

Harland, W.B., Armstrong, R.L., Cox, A.V., Craig, L.E., Smith, A.G. & Smith, D.G. (1990) *A Geologic Time Scale 1989*. Cambridge University Press, Cambridge.

Haubold, H. & Katzung, G. (1972) Die Abgrenzung des Saxon. *Geologie* **21**, 883–914.

Hedemann, H.A., Mascheck, W., Paulus, B. & Plein, E. (1984) Mitteilung zur lithostratigraphischen Gleiderung des Oberrotliegenden im nordwestdeutschen Becken. *Nachschr. Deutsches Geol. Gesampt.* **30**, 100–107.

InterGeos (1991) Rotliegend of The Netherlands offshore. InterGeos Report no. EP91046.

Melnyk, D.H. & Smith, D.G. (1989) Outcrop to subsurface cycle correlation in the Milankovitch frequency band: Middle Cretaceous, central Italy. *Terra Nova* **1**, 432–436.

Nederlandse Aardolie Maatschappij B.V. & Rijks Geologische Dienst (1980) *Stratigraphic Nomenclature of The Netherlands*. Het Koninklijk Nederlands Geologisch Mijnbouwkundig Genootschap, Deel 32.

Smith, D.B. (1970) The palaeogeography of the British Zechstein. In: *Third Symposium on Salt* (Eds Rau, J.L. & Dellwig, L.F.). Northern Ohio Geol. Soc., 1, pp. 20–23.

Smith, D.B. (1979) Rapid marine transgressions and regressions of the Upper Permian Zechstein Sea. *J. Geol. Soc. Lond.* **136**, 155–156.

Smith, D.G. (1989) Stratigraphic correlation of presumed Milankovitch cycles in the Blue Lias (Hettangian to earliest Sinemurian), England. *Terra Nova* **1**, 457–460.

Vail, P.R., Mitchum, R.M., Jr & Thompson, S. III (1977) Seismic stratigraphy and global changes of sea level, part four: global cycles of relative changes of sea level. In: *Seismic Stratigraphy – Applications to Hydrocarbon Exploration* (Ed. Payton, C.E.). Am. Assoc. Petrol. Geol. Memoir 26, pp. 83–98.

Van Wagoner, J.C., Mitchum, R.M., Campion, K.M. & Rahmanian, V.D. (1990) *Siliciclastic Sequence Stratigraphy in Well Logs, Cores and Outcrops: Concepts for High-resolution Correlation of Time and Facies*. AAPG

Methods in Exploration Series, 7. Am. Assoc. Petrol. Geol., Tulsa.
VAN WIJHE, D.H., LUTZ, M. & KAASSCHIETER, J.P.H. (1980) The Rotliegend in The Netherlands and its gas accumulations. *Geol. Mijnbouw* **59**, 3–24.
WORTHINGTON, P.F. (1990) Sediment cyclicity from well logs. In: *Geological Applications of Wire-line Logs* (Eds Hurst, A., Lovell, M.A. & Morton, A.C.). Geol. Soc. Lond. Spec. Publ. 48, pp. 123–132.
YANG, C.S. & NIO, S.D. (in press) Applications of high-resolution sequence stratigraphy to the Upper Rotliegend in The Netherlands offshore. In: *Siliclastic sequence stratigraphy: recent developments and applications* (Eds Weimer, P. & Posamentier, H.). Am. Assoc. Petrol. Geol. Memoir (in press).
ZIEGLER, P.A. (1982) *Geological Atlas of Western and Central Europe*. Elsevier, Amsterdam.

Spec. Publs Int. Ass. Sediment. (1994) **19**, 63–76

Identification of regular sedimentary cycles using Walsh spectral analysis with results from the Boom Clay Formation, Belgium

E. VAN ECHELPOEL

Instituut voor Aardwetenschappen, K.U. Leuven, Redingenstraat 16, B-3000 Leuven, Belgium

ABSTRACT

The examination of stratigraphic successions for cyclicity related to orbital variations or Milankovitch cycles relies a great deal upon spectral analysis of selected bio- or lithostratigraphic parameters in measured sections. Recently, Walsh spectral analysis has been introduced for the analysis of digitized sections. Although digitizing is a convenient way to generate long time series, the choice of the coding system used is limited by the nature of the Walsh functions. The effects of coding on the distribution and intensity of powers in Walsh spectra are demonstrated by means of simple models. In some circumstances the bundling of sedimentary cycles, often used to prove Milankovitch-type cycles, cannot be detected by the Walsh approach.

The windowing Walsh analysis gives an impression of the frequency composition of the time series through time. The nature of changes in stratigraphic thickness can be deduced from the set of spectra obtained and important information regarding variations in sedimentation rate can be gained from this scanning technique. The application of the technique is outlined and illustrated for the Boom Clay Formation. This Oligocene siliciclastic shelf deposit is characterized by alternating, laterally persistent beds of silt and clay. The digitized sections, based on grain size and organic matter content, are non-stationary due to the presence of long-term trends. Power spectral analysis of these lithological variations indicates two regular cycles for the Boom Clay Formation. The scale of the cycles and their regularity suggest an indirect link to the 100-ka and 41-ka orbital cycles.

INTRODUCTION

The idea that the geological record contains evidence of cyclic phenomena has been with us for over a century, as has the idea that some cyclic phenomena are truly or nearly periodic (Gilbert, 1895). Hays *et al.* (1976) demonstrated via spectral analysis of stable oxygen isotopes (a climate-sensitive indicator) in Pleistocene deep-sea records that a document of the climatic forcing by orbital variations is contained in the sedimentary record. Since then the Milankovitch concept has been widely applied to cyclic sedimentary successions of various facies and ages. The recent revival of interest in ancient sedimentary cycles is reflected by the large number of recent papers. Comprehensive reviews are provided by, for example, Einsele & Seilacher (1982), Berger *et al.* (1984), Berger (1989), Fischer *et al.* (1990) and papers accompanying Smith (1989).

For Pre-Pleistocene times the identification of stratigraphic rhythmicities with orbital cycles is based largely upon their regularity and their estimated periods. Therefore, spectral analysis is now widely used as an objective tool to demonstrate regular cyclicity in stratigraphic successions. Even when the dating of the studied sections is poor, the detection of regular cycles in thickness is a useful piece of supporting evidence that sedimentary cycles are indeed related to orbital cycles (Fischer & Schwarzacher, 1984; Schwarzacher, 1985; Weedon, 1989). This paper is a contribution to the application of spectral analysis of sedimentary cycles and their relation with orbital cycles.

Different techniques of spectral analysis are available (Pestiaux & Berger, 1984). The choice of the analytical technique strongly depends on the nature of the time series (Beauchamp, 1984; Schwarzacher, 1987). Walsh spectral analysis has been introduced for the analysis of digitized sections (Schwarzacher, 1985; Weedon, 1985, 1989). The first part of this paper attempts to illustrate and to explain the effects of digitization on a power spectrum.

For the majority of Pre-Pleistocene sections a good time control is lacking and the created time series are based on lithostratigraphic thicknesses rather than on a time scale. Due to the presence of long-term trends or oscillations caused by sea-level fluctuations, changes in sediment supply and/or tectonic movements for example, these time series are often non-stationary. In the second part of this paper, windowing Walsh spectral analysis is introduced to examine non-stationarity (Beauchamp, 1984). This technique is especially appropriate for the examination of geological (digitized) time series. An illustration is given in a case study of the Boom Clay Formation.

DIGITIZATION AND SPECTRAL ANALYSIS

All analytical techniques are based on numerical data. Converting lithologies into numerical values by coding is a convenient way to generate long time series. To obtain a meaningful time series certain conditions have to be fulfilled. It is beyond the scope of this paper to discuss all the conditions in detail; in this paper I will concentrate on the more mathematical conditions. For a discussion of the geological limitations the reader is referred to Schwarzacher (1987) and Weedon (1989).

In an analogous manner to Fourier analysis any given set of discrete data (the time series) can be decomposed into a set of Walsh functions. Their relative importance with respect to the total time series is given by the power estimates (Beauchamp, 1984; Negi & Tiwari, 1984). Because Walsh functions are square waves which assume only the two amplitude values +1 and −1, two questions need to be answered. The first concerns the value of the codes: what happens if one uses other values? The second question is even more important: what are the effects if more than two states are distinguished during digitizing? In order to answer these questions some model sections were simulated. The intention was not to be entirely realistic but rather to demonstrate the effects of transforming descriptive geological parameters into numerical values.

Two-component time series

Sections characterized by an alternation of two lithologies provide the simplest cyclic pattern one can imagine. The digitization of such sections is based on two codes, one for each lithology. The amplitude of the created time series is defined by the numerical value of the codes. Changing these values will change the amplitudes of *all* Walsh functions by the same amount and the position of the zero-crossings will be the same. Although the Walsh representation of the time series varies, the normalized power spectrum remains unaffected. So, if we are dealing with a two-component system, the choice of the codes is unlimited, as long as the same value for the positive and the negative code is used. Any stratigraphic section that can be reduced to such a system can be digitized without influencing its Walsh spectrum.

More complex time series

In reality most geological sections are much more complicated. In pelagic and hemi-pelagic environments, for instance, the sedimentary successions may consist of an alternation of dark and light limestones (or marls) and dark and light shales (or clays) (e.g. de Boer, 1983; Weedon, 1985; Herbert & Fischer, 1986). Besides the time series distinguishing only limestones and shales, one can create another time series in which the colour of the sediments is also taken into account (Fig. 1).

The position of the zero-crossings

Time series A and B of Fig. 1 are examples of what we call reduced two-component systems. Time series A separates limestones from shales and in time series B a distinction between light and dark layers is made. Note that the Walsh spectra of A and B (Fig. 2) depend only on the position of the zero-crossings, which depends directly on the coding system applied. The asymmetric shape of time series B produces additional frequencies, called harmonics (Fig. 2) (Schwarzacher, 1987).

Combining sediment type and colour into one time series results in time series C or D, depending on the coding system chosen. The important point is that both time series are based on the same geological information, but their spectra differ appreciably (Fig. 2). In fact spectra C and A are much more similar although time series C contains the additional colour information. The same similarity is

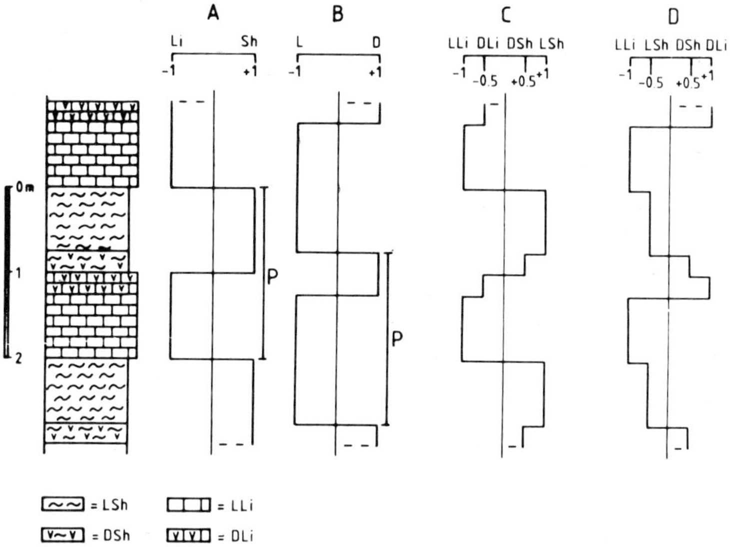

Fig. 1. Diagram illustrating a digitized succession containing two superimposed cycles: a limestone/shale cycle of 2 m (called the 'lithological' cycle in text) and an asymmetric dark/light cycle of 2 m ('colour' cycle). The position of the zero-crossings within the time series is controlled by the coding system. (A) Time series consisting of a limestone/shale classification of the sequence. (B) Time series expressing the colour cyclicity. (C) and (D) contain both cyclic variations but coded differently. Li, limestone/marl; Sh, shale/clay; D, dark; L, light. The corresponding spectra are illustrated in Fig. 2.

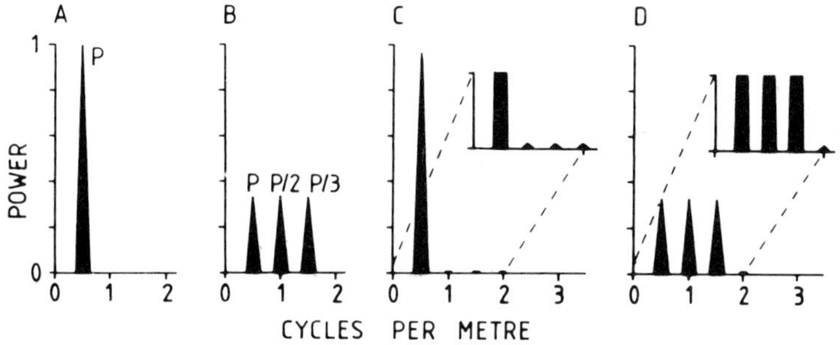

Fig. 2. Walsh power spectra for the time series of Fig. 1.

observed for time series D and B. This is partly explained by the similarity of the positions of the zero-crossings within time series A and C. Simultaneously, there is also the influence of the amplitude of the time series, which is minimized or even neglected (e.g. see Weedon, 1985, 1989). In this experiment the colour variation was assigned to 1 and 0.5, positive codes for one lithology and negative codes for the other. These numerical values +0.5 and −0.5 lead automatically to an underestimate of the frequencies associated with the additional parameter.

The amplitude of the time series: the numerical value of the codes

The influence of the amplitude on a Walsh spectrum is clearly seen in Figs 3 and 4. The succession consists again of an alternation of limestones and shales with a superimposed variation of colour. The position of the zero-crossings within the various time series remains the same. The only difference between time series B, C and D is the value of the codes. This model proves the dominance of certain frequencies as previously suggested. The relative importance of the frequencies detected strongly depends on the numerical values of the codes. It is obvious that the effects generated by the amplitude of the time series can no longer be neglected. Both the amplitude *and* the position of the zero-crossings control the shape of the Walsh power spectrum.

Wavelength ratios have been used as a time-independent criterion to invoke a Milankovitch origin for certain sedimentary cycles (e.g. Olsen, 1984, 1986). However, in some circumstances hiatuses alter the wavelength ratios of the detected cycles (Weedon, 1989). In addition, it appears from Fig. 4

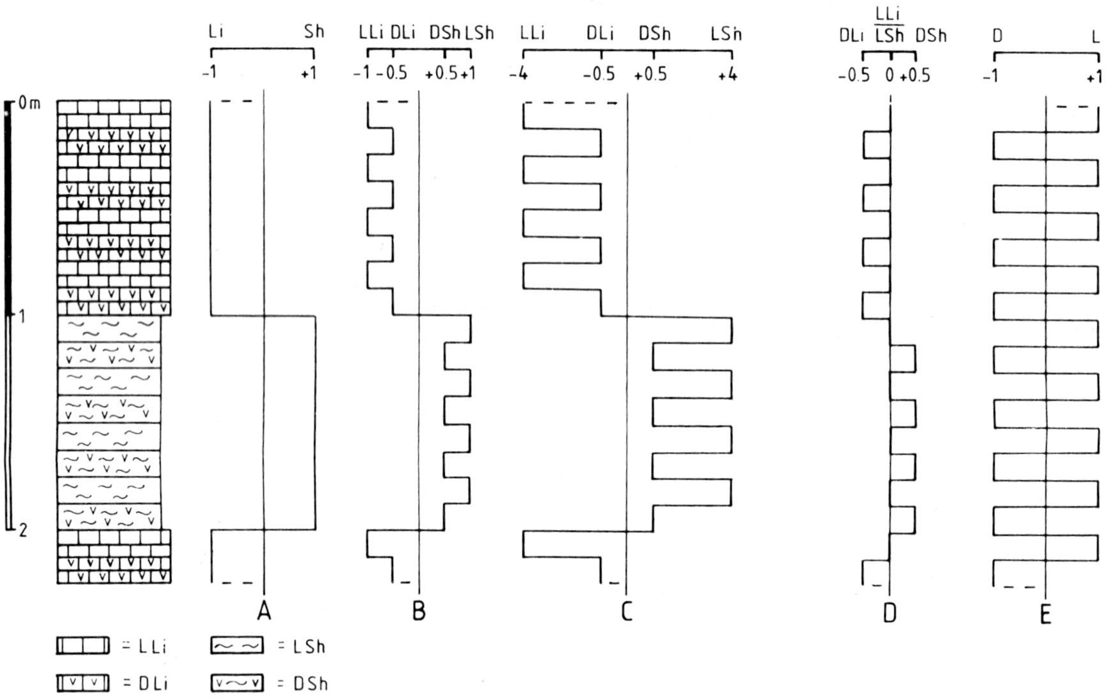

Fig. 3. Diagram illustrating the effects of the amplitude on Walsh power spectra. The model contains two cyclicities: a lithological cycle of 2 m, superimposed on a 25-cm colour cycle. (A) and (E) are reduced two-component systems based on variations in lithology and colour respectively. In time series (B), (C) and (D) only the numerical values of the codes differ. Time series (D) clearly expresses the amplitude modification of the colour cycle by the larger one. The corresponding Walsh spectra are given in Fig. 4.

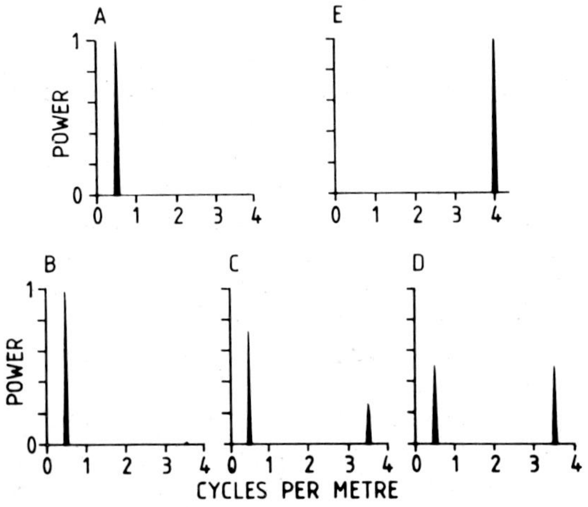

Fig. 4. Walsh power spectra for the time series of Fig. 3 Note the difference in power for the detected cycles and the shift of the colour cycle towards 3.5 cycles/m.

that the method of digitizing is also able to change these ratios. If we examine the colour information separately, as we did in time series E of Fig. 3, we obtain exactly the expected frequency of 4 cycles/m. In all the other spectra (Fig. 4B, C, D), however, the colour cycle is shifted to 3.5 cycles/m. The displacement of this peak is caused by the presence of the lithological cycle in the amplitude of the colour cycle (Fig. 3B, C, D). The lithological cycle acts here as an amplitude modulation of the smaller colour cycle, thereby changing its frequency. Note that any amplitude modulation is described mathematically as a multiplication of the corresponding functions, from which the new function with the 'unexpected' frequency of 3.5 cycles/m originates (Fig. 5) (see also Appendix). Thus, the wavelength ratio of the two cycles, detected by using a complex coding system, changes.

One of the consequences of the coding effects is that the bundling of cycles (Herbert & Fischer,

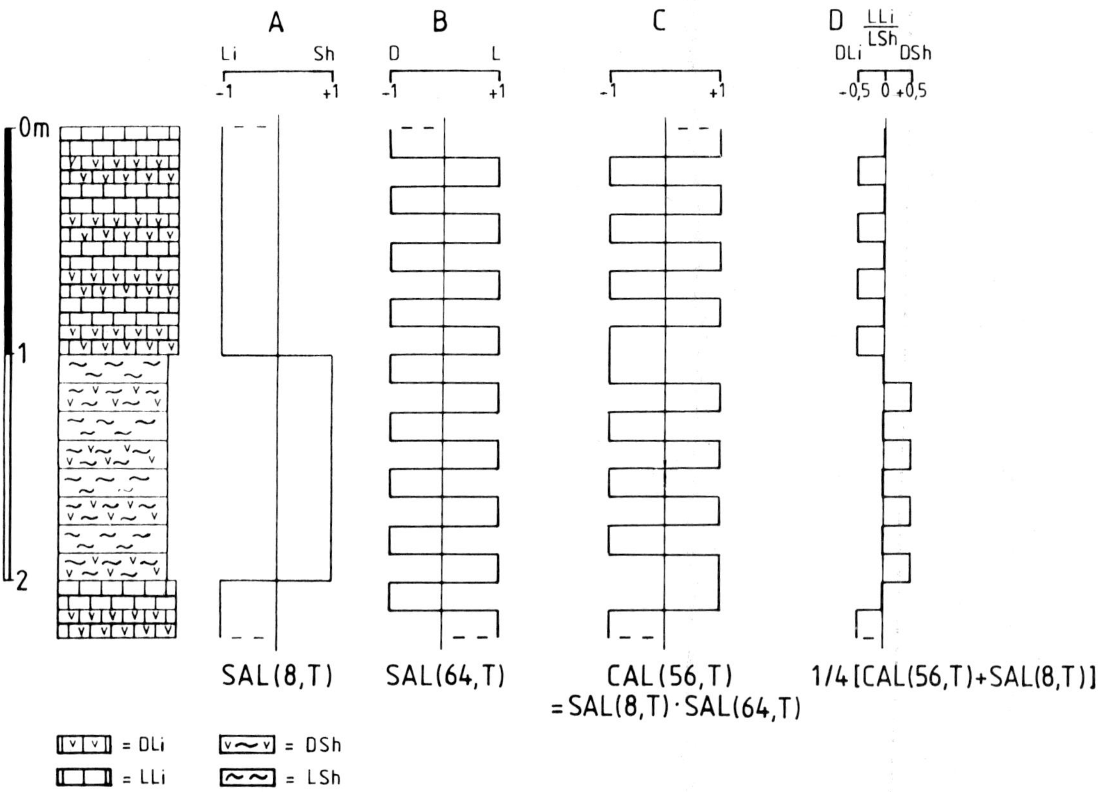

Fig. 5. Graphical explanation of the shift observed in Fig. 4. Time series (A) and (B) correspond to A and E respectively of Fig. 3. The lithology (time series A or the Walsh function SAL(8,T)) controls the amplitude of the colour cyclicity (time series E or the Walsh function SAL(64,T)). (C) Multiplication of these Walsh functions causes the appearance of an unexpected frequency at 3.5 cycles/m, which corresponds in this simulation to CAL(56,T). (D) Time series (D) of Fig. 3 is a combination of this new component and the original lithological cycle. Their relative importance is the same, as indicated in Fig. 4 spectrum D.

1986; Fischer & Schwarzacher, 1984; Fischer *et al.*, 1990) can be detected only in certain circumstances. It is obvious from Fig. 6 that the time series contain two cycles, P1 and P2 or P1 and P3. However, we detect only the smallest cycle (P2 or P3), the bundle P1 not being visible in the power spectrum (Fig. 6B, C). This cycle is observed only if another coding system is used (Fig. 6A, D). The point is that in situations where the packaging of cycles is expressed only by variations in amplitude, the bundle is not present in the spectrum. The grouping of cycles, which is often used as an argument for Milankovitch-type cycles, can be detected by the Walsh approach only if the thickness of the layers shows some bundling.

In conclusion, the number of coded states distinguished during digitization determines the influence of the coding system on the power spectra. As mentioned previously, using two codes is perfect because the numerical value of the codes has no effect on the power spectrum. However, if more than two codes are used, the presence of a dominant factor is automatically induced in the spectrum, because we need to apply values other than ± 1. Secondly, the position of the zero-crossings is now directly linked to the applied coding system. A third reason why a complex coding system should be avoided is the possibility that important frequencies may change. This, in turn, alters the wavelength ratios of the detected regular cycles.

The use of Walsh spectral analysis in this context is thus severely limited by the nature of these functions. In cases of complex geological cyclic patterns we suggest looking for sedimentological criteria that

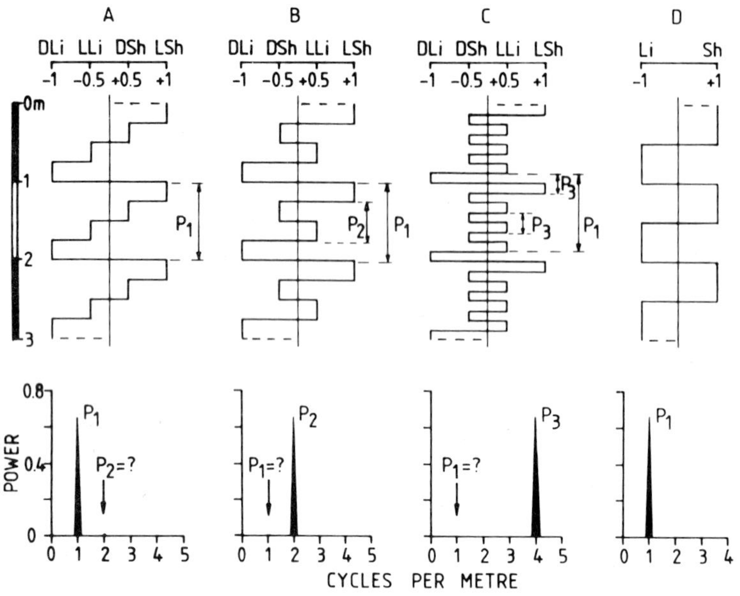

Fig. 6. A complex coding system with more than two states affects the Walsh spectrum. Two small P2 cycles and four P3 cycles are grouped into a bundle (cycle P1). The corresponding spectra underneath the time series only show P2 and P3. Symbols are as in previous models (Figs 1–4). Time series B and C, however, can also be the result of digitizing a $CaCO_3$-record applying different threshold levels.

classify the different rock types into two groups. Such a two-state digitized section can be examined by the Walsh approach.

THE WINDOWING WALSH SPECTRAL ANALYSIS

In the majority of cases where a thickness rather than a time scale is used, the time series created are non-stationary. In fact we are dealing with a distorted time scale, because the stratigraphic thickness is affected by depositional and post-depositional processes. It is very unlikely that any stratigraphic record will be free from long-term trends, caused by sea-level fluctuations, changes in sediment supply or tectonic movements for example. These affect not only the recorded parameter but in most cases also the sedimentation rate and thereby the time scale of the time series (Schwarzacher, 1985, 1989).

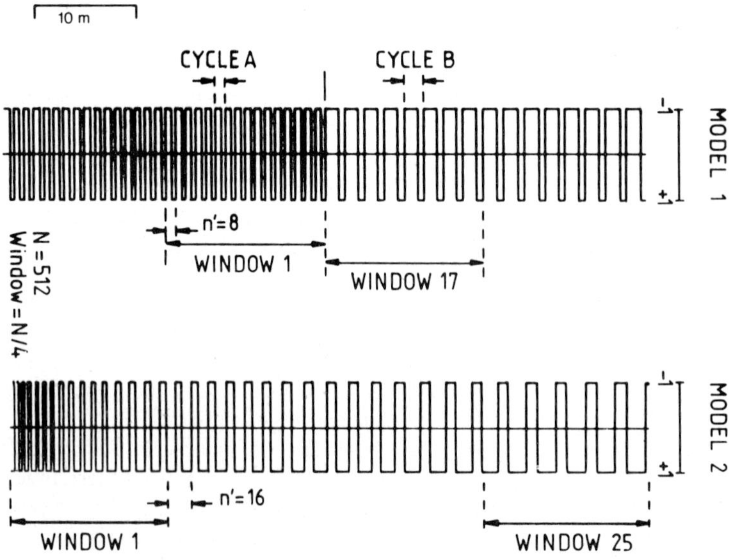

Fig. 7. Simulations of changes in sedimentation rates, expressed as variations in wavelength. Model 1 illustrates an abrupt change. Cycle A has a frequency of 1 cycle/m (8 data points), that of cycle B is 0.5 cycle/m (16 data points). Model 2 simulates a trend-like variation. The time series consists of 512 data points and was sampled with 8 samples/m. The lag n' and the window used for the windowing analysis are indicated. The corresponding spectra are illustrated in Fig. 8.

Because the sedimentation rate is not constant the wavelength of a particular cycle (in time) varies. The power spectrum of such a disturbed time series incorporates all the variations in sedimentation rates. Consequently, the frequency maxima are poorly defined (e.g. see Model 2 of Figs 7 and 8). For non-stationary time series the frequency composition changes in time and thus the power spectrum of the complete data set has to be considered as a composite result. In order to examine the variability of cycles the windowing Walsh spectral analysis was introduced by Beauchamp (1984). Instead of analysing the whole sequence at once, this method employs a set of subsections. This set of (overlapping) segments is obtained by using a window, being moved along the original time series by a preset lag. Each of these short time series is transformed into the frequency domain. The end-product of this procedure is a set of successive spectra, illustrating the variations in frequency composition through time. The accurate location of these changes in time, especially the abrupt changes, requires a short window, whilst the resolution of this analysis is proportional to the number of data points in the window. Note that the resolution of the spectrum controls the accuracy of detected frequency maxima, but the non-stationary character of the data can only be overcome by shortening the window. In practice, therefore, some compromise is necessary in order to secure acceptable results.

Two extreme situations are simulated in Figs 7 and 8. A rather smooth increase in wavelength (Model 2) gives a poorly defined frequency maximum, which moves slowly through the successive windows; whereas an abrupt change in the time series (Model 1) is characterized by a distinct redistribution of power towards other frequencies. Both models, although extreme situations, demonstrate that the nature of the changes in thickness (wavelength) of the cycles can be deduced from the set of spectra. On the assumption that the sedimentary cycles represent an approximately equal time interval, important information regarding the variations in sedimentation rate can be gained from this scanning spectral technique.

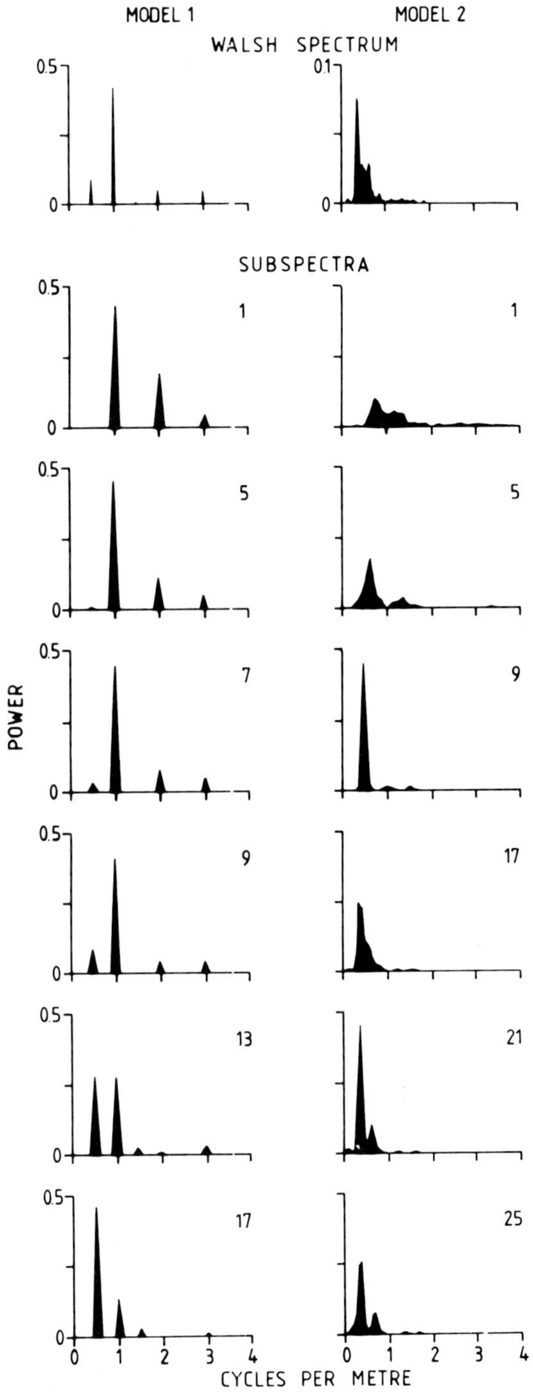

Fig. 8. Averaged Walsh power spectra and the results of the windowing Walsh analysis for time series in Fig. 7. Numbers of the subspectra indicate the position of the window.

THE BOOM CLAY FORMATION

Introduction

The shallow-marine Boom Clay Formation, as exposed in NE Belgium, forms part of the type

Rupelian (Vinken, 1988). It represents the deeper water phase of a major transgression/regression cycle, which started with the deposition of sands in the earliest Oligocene (the Ruisbroek Sands and the Berg Sands) and terminated with the erosion of the top of the Boom Clay in the mid to lȧte Oligocene (Haq *et al.*, 1988). Palaeogeographically, the Boom Clay was deposited in an embayment at the southern end of the North Sea Basin (map 4 in Vinken, 1988).

The decimetre-scale rhythmicity in grain size has been attributed to both tectonic and glacio-eustatic processes under orbital/climatic control (Vandenberghe, 1978; Gullentops & Vandenberghe, 1985). Walsh and Fourier power-spectral analysis has been applied, using data measured in different outcrops, in order to investigate the possibility that the cyclicity was controlled by climate (van Echelpoel & Weedon, 1990). In addition, windowing Walsh spectral analysis is used to examine in detail the non-stationarity of the digitized (composite) section.

Background sedimentology

The sedimentology and the stratigraphy have been described elsewhere (Vandenberghe, 1978; Hooyberghs, 1983; Vandenberghe & van Echelpoel, 1987; Moorkens, in Vinken, 1988). Only the main sedimentological characteristics will be given here.

The Boom Clay Formation consists of a sequence of alternating, laterally persistent beds of detrital silty clay and clayey silt. The chemical and mineralogical properties are fairly constant throughout the whole deposit. The characteristic variables are grain size, carbonate and organic matter content. The grain-size rhythmicity is the most striking feature in the field and bedding is mainly defined by decimetre–metre scale variations in grain size. At the millimetre to centimetre scale, the sediment usually appears homogeneous and bed contacts are gradational over a few centimetres. Pyritized burrows are common.

The range of grain sizes is the same at all levels. The mean grain size, however, is consistently larger in what are termed 'silt' beds and smaller in the so-called 'clay' beds (Fig. 9). The variations in grain size are gradual, and the coarsest and finest fractions are located in the centres of the silt and clay layers respectively. The sediment grains were transported as suspended discrete particles or in floccules (Vandenberghe, 1978). The succession of silty clays and clayey silts is interpreted in terms of alternating periods of more and less important turbulence due to changing wave activity near the sea floor (Vandenberghe, 1978). Bottom-water turbulence sorted a constant initial grain-size population to various degrees through time (Fig. 9).

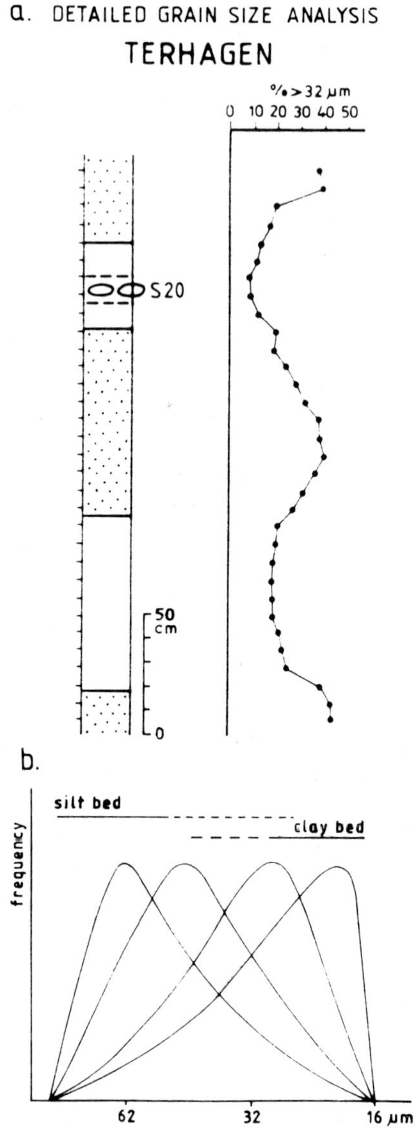

Fig. 9. (a) Detailed grain-size analysis of a section at Terhagen showing two complete cycles. The grain size (expressed as % > 32 μm) changes smoothly. 'Silt' beds are stippled, clay beds are left blank. The position of the samples is indicated on the left of the lithological column. (b) Typical grain-size frequency distributions in the silt and clay beds.

Besides the variations in grain size, a rhythmic alternation of black bituminous and grey (low organic carbon content) beds is observed. All bituminous beds occur just above a silty bed. Most organic matter is allochthonous, comprising Oligocene plant remains and clasts of Carboniferous coals. Organic carbon contents range from 0.3 to 6.7%, maximum concentrations occurring just above the levels of maximum grain size. The origin of the bituminous beds has been attributed to a selective sedimentation due to the difference in density between the organic and inorganic particles (Vandenberghe, 1978).

Most beds have undergone total carbonate dissolution, but at a few horizons fossils survived and carbonate content can still reach up to 25%. The carbonate-rich layers represent periods of an original marl sedimentation. These beds mostly contain widely spaced septarian nodules, a result of diagenetic processes (Vandenberghe, 1978; Vandenberghe & Laga, 1986).

The nodule layers with particular characteristics can be used as marker-levels for correlation. Bed-by-bed correlation has been established over an area of more than 500 km^2 in NE Belgium (Vandenberghe, 1978; Vandenberghe & Van Echelpoel, 1987). Some horizons have been traced from Belgium, across The Netherlands, and into the Lower Rhine area of W Germany (Van Den Bosch & Hager, 1984). In general, beds gradually thicken to the north of Terhagen, away from the palaeoshoreline.

Time series and power spectra

As mentioned above, the Boom Clay Formation possesses three superimposed characteristic variables: grain size, carbonate and organic matter. Obviously, it is important that any parameter used for analysis of cyclicity should represent depositional conditions. The general carbonate dissolution led to spurious carbonate content values and septaria occurring in otherwise decalcified clay (Vandenberghe, 1978). Due to this post-depositional modification the carbonate record no longer reflects depositional conditions. Therefore, carbonate content is excluded here; only variations in grain size and in organic carbon content are analysed. Grain size is here expressed as the percentage of grains larger than 32 μm.

The Fourier spectrum of grain-size data computed for the entire composite section of the Boom Clay reveals two significant spectral peaks denoting regular sedimentary cycles with wavelengths of 100 cm and 46 cm (Fig. 10). In the corresponding Walsh spectrum (van Echelpoel & Weedon, 1990), the peaks are much wider than the peaks in the Fourier spectrum, even though the frequency resolution of the Walsh spectrum is much higher (narrower bandwidth). Windowing Walsh spectral analysis has been used in order to obtain insight into the variability of the cycles. The aim of this additional investigation was to test and amend previous conclusions concerning the non-stationary character of the digitized composite time series (van Echelpoel & Weedon, 1990).

The data used for this analysis are directly based on field observations. At Kruibeke and Terhagen (Fig. 11) each bed was measured to the nearest 4 cm as bed contacts are gradational over a few centimetres. Both sections were digitized twice. Firstly,

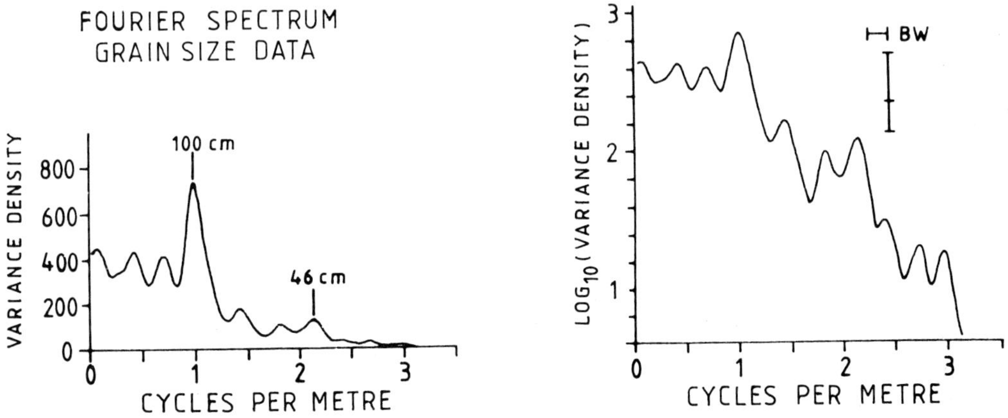

Fig. 10. Fourier spectrum of grain-size data for the entire composite Boom Clay section. The 80% confidence interval (vertical line) and bandwidth (BW) are indicated on the $\log_{10}$ version of the spectrum.

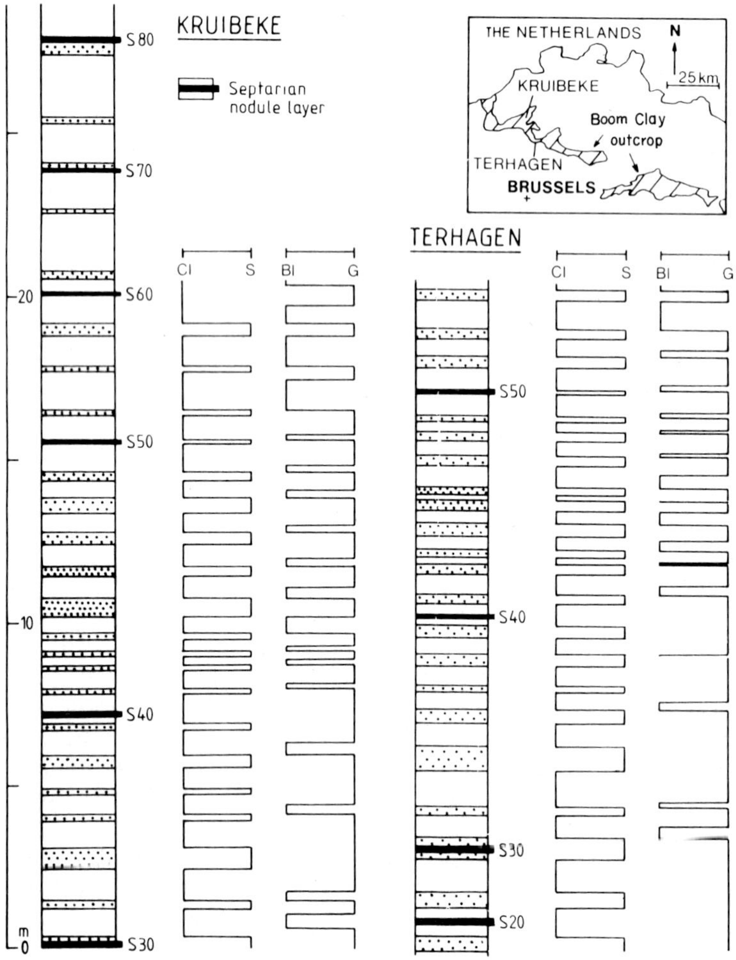

Fig. 11. Time series with a total length of 20.48 m and comprising 512 data points of Kruibeke and Terhagen, digitized as discussed in text. Cl, clay = −1; S, silt = +1; Bl, black or 'high' organic matter content = −1; G, grey or 'low' organic matter content = +1. The location of the sections is indicated on the map.

the beds were classified into silt and clay, representing the grain-size cyclicity. Secondly, a differentiation into black and grey beds was made in order to digitize the organic carbon content variations (Fig. 11). Note that the time series are based on a two-way classification in order to avoid the effects of coding, discussed in the first part of this paper.

The stratigraphic position of the sections, the position of the foraminiferal zones P18–P21 (Hooyberghs, 1983), and also the absolute ages of the boundaries P18/P19 and P20/P21 are indicated in Fig. 12. As these boundaries are located close to the bottom and the top of the Terhagen section, an average sedimentation rate of 0.81 m/100 ka is obtained for this succession. This value has been used to estimate the periods of the cycles detected.

The Walsh spectra given in Fig. 13 are very similar to the Walsh spectrum computed for the entire composite section (van Echelpoel & Weedon, 1990). All the Walsh spectra are characterized by rather poorly defined spectral peaks with wavelengths of 146–98 cm and 47–43 cm. The estimated periods of the cycles are given in Fig. 13.

The time series representing the grain-size variations have been scanned using a window. The time series each contain 512 sampling points and a window of 256 points is used. It is moved upwards at intervals of 32 points or 128 cm. The nine subspectra, from which only the first (lower part), the middle and the last (upper part) spectrum are given in Fig. 14, clearly demonstrate that the poorly defined peaks in the spectra computed for the total time series (Fig. 14) are the result of slow changes in cycle thickness through time. This is supported by

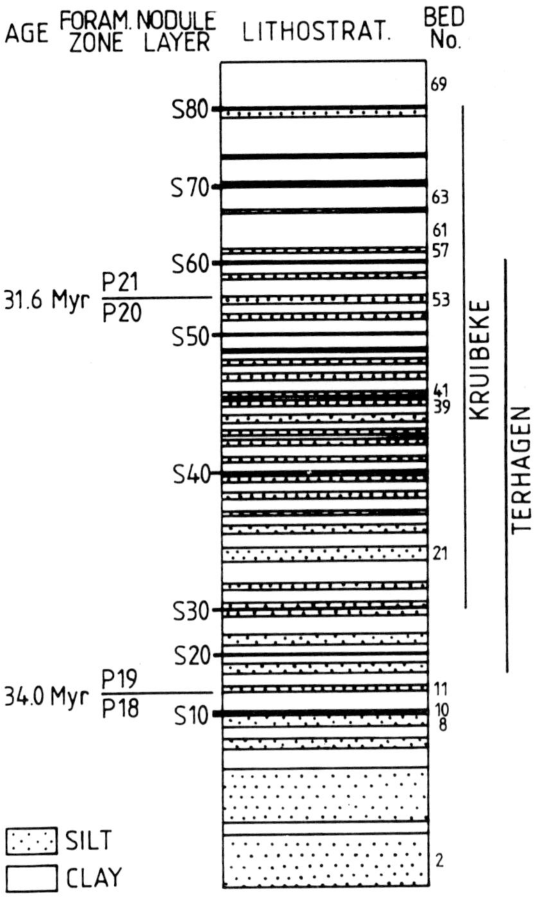

Fig. 12. Composite lithological column of the Boom Clay Formation. The position of the P18/P19 and P20/P21 zonal boundaries are indicated with ages derived from Berggren *et al.* (1986). Using the Haq *et al.* (1987) time scale the respective ages are 31.7 and 34.2 Ma, or from the Harland *et al.* (1990) time scale the values are 30.7 and 33.1 Ma. The lithostratigraphic position of the studied sections are indicated with vertical bars.

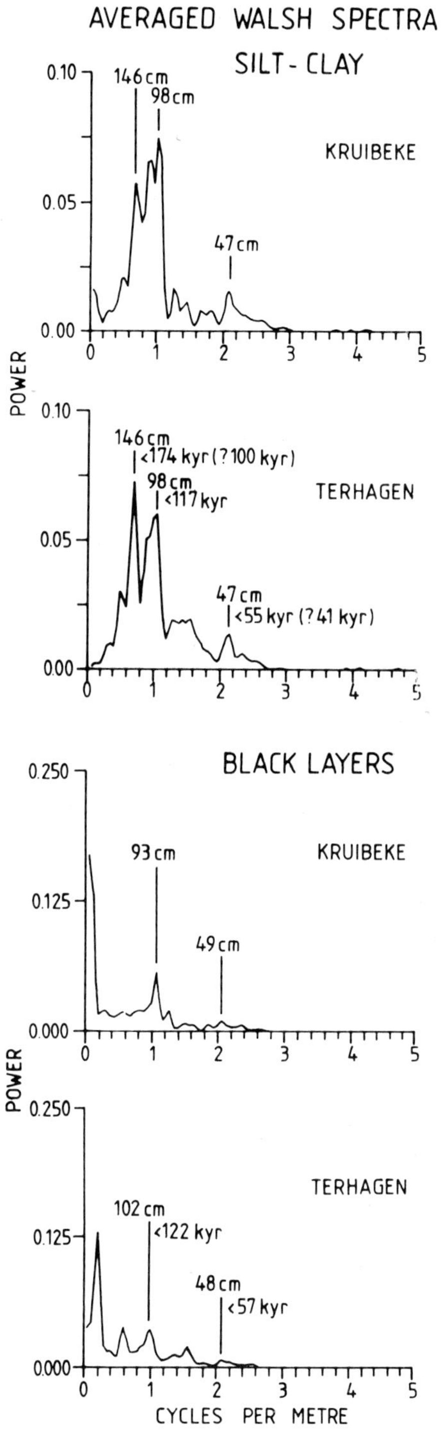

Fig. 13. Walsh power spectra of the total time series given in Fig. 11.

the model, where slow changes in thickness have been simulated (Model 2 in Figs 7 and 8).

Therefore, we conclude that the lateral variations in thickness have had only a minor effect on the spectrum calculated for the composite time series (van Echelpoel & Weedon, 1990). The spread of the Walsh spectral peaks is produced mainly by long-term variations through time. The similarity of the results obtained for the Kruibeke and Terhagen time series suggests that this trend in thickness, which is superimposed on the regular cyclicity, must have been caused by a phenomenon which affected

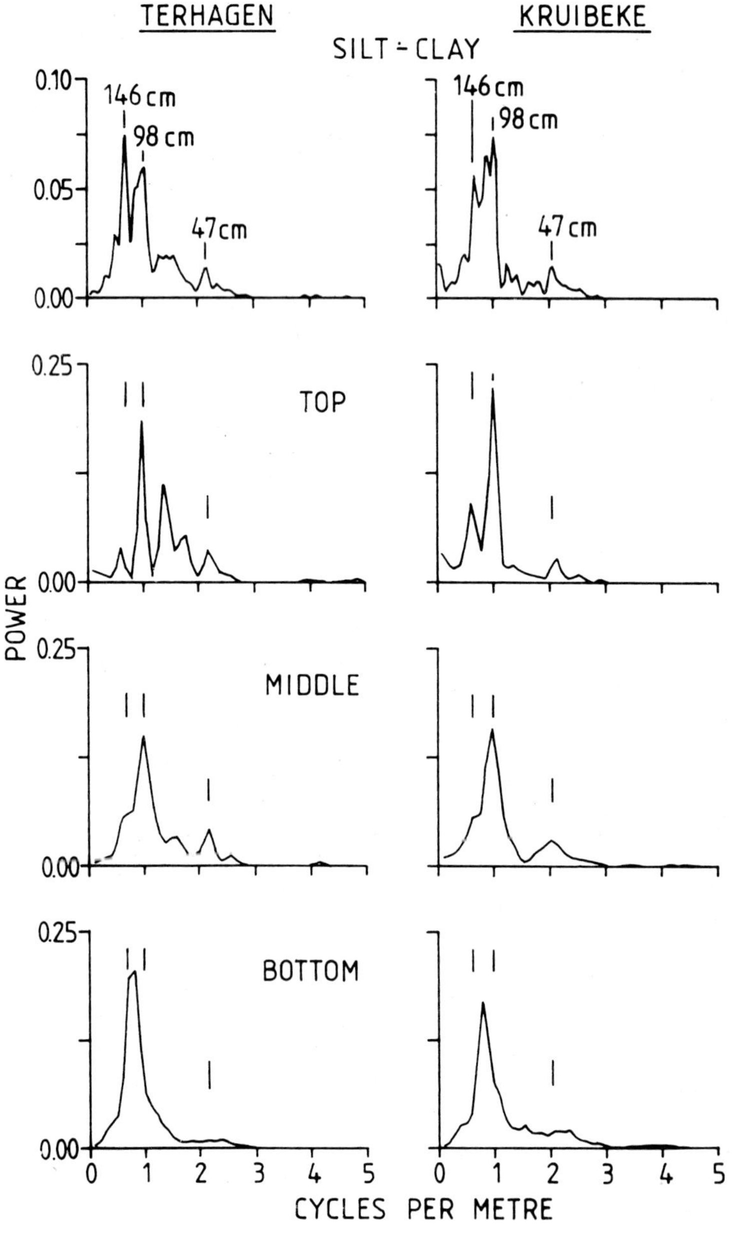

Fig. 14. Walsh spectra of successive overlapping windows, along the silt/clay digitized section of Fig. 10 (window width ± 10 m). Small vertical lines in the top, middle and bottom spectra indicate the position of the frequency maxima in the spectra of the total time series, shown at the top.

this part of the Boom Clay basin over a broad area at the same time. It is assumed that the long-term variation in the thickness of the beds mainly reflects changes in sediment supply (van Echelpoel, 1991). As the Boom Clay Formation is a relatively shallow-marine deposit (Vandenberghe & van Echelpoel, 1987), it is not very surprising that such trend-like variations or large-scale rhythms are present.

Based on the lateral persistence of the beds (regularity in spacing) and their sedimentological characteristics, combined with the regularity in time and the estimated periods of the sedimentary cycles (Fig. 13), we conclude that this siliciclastic sequence does indeed provide a proxy record of palaeoclimatic variations. The 'regular' sedimentary cycles have been attributed to the 100-ka eccentricity cycle and the 41-ka obliquity cycle (van Echelpoel & Weedon, 1990). Due to the superimposed trend or long-term variation in thickness of individual silt and clay beds, the wavelength of these regular cycles varies slowly through time. Consequently, the environment envisaged for the Boom Clay Formation is a shallow marine setting where a delicately balanced deposition system responded to orbitally induced climatic changes, registered here as small-scale variations in grain size and organic matter content.

A Milankovitch origin for the decimetre-scale variations in grain size and organic matter content implies that each silt/clay couplet represents an approximately equal time interval. The thickness of the couplets is thus a measure of the accumulation rates, and changes in wavelength indicate relative changes in sedimentation rates. This, in turn, can be used to refine the sedimentation model of the Boom Clay Formation (van Echelpoel, 1991).

CONCLUSIONS

Walsh spectral analysis is an appropriate technique for examining regular cyclicity in digitized stratigraphic sections provided that the time series generated is a binary type of signal. The appropriate digitization of geological information is dictated by the nature of the Walsh functions. Complex coding systems with more than two codes affect the spectrum in different ways, as demonstrated by the experiments. The amplitude of the time series *and* the position of the zero-crossings within the time series control the shape of the power spectrum. Therefore, in real situations it will often be necessary to reduce the number of states within the sequence, so that the data finally consist of a two-way classification. Thus, digitization does involve a degree of personal judgement.

The windowing Walsh spectral analysis has been introduced. It gives an impression of the variability in the frequency composition of a signal over time. As geological time series are often non-stationary, it is a convenient way to illustrate the variability of the wavelength of cycles. In addition, this method can be applied to sections containing cycles of approximately constant time interval, in order to deduce information on the sedimentation rate.

The spectra from the Boom Clay Formation reveal two regular cycles. The digitized sections, based on a thickness scale, reveal weak non-stationarity. The windowing Walsh spectral analysis indicates that smooth changes in sedimentation rates over time are responsible for the poorly defined Walsh spectral peaks of the Boom Clay Formation. Since windowing analysis can help to detect variations in cycle thickness and hence in sedimentation rates, this technique will probably be useful to unravel objectively the structure of a time series which, in turn, may lead to a refinement of the sedimentation model of the Boom Clay Formation as well as for examples of rhythmic successions elsewhere.

ACKNOWLEDGEMENTS

I thank Dr N. Vandenberghe for supervising my doctoral research, upon which this paper is based. Thanks also to Dr P.L. de Boer, Dr W. Schwarzacher and Dr W. ten Kate for their comments and constructive suggestions for improvements to the manuscript.

APPENDIX

The set of Walsh functions $\mathrm{WAL}(n,T)$ can be classified in two subsets $\mathrm{SAL}(k,T)$ and $\mathrm{CAL}(k,T)$ (Beauchamp, 1984),

$$\mathrm{CAL}(k,T) = \mathrm{WAL}(2k,T)$$
$$\mathrm{SAL}(k,T) = \mathrm{WAL}(2k-1,T) \qquad k = 0, 1, 2, \ldots$$

The product of two Walsh functions is defined as follows

$$\mathrm{WAL}(n,T) \times \mathrm{WAL}(m,T) = \mathrm{WAL}(n \# m,T)$$

(where # indicates modulo-2 addition for the binary representations of n and m) (Beauchamp, 1984, pp. 53–54). Using the SAL- and CAL-notations the equation is replaced by the following set of equations:

$$\begin{aligned}
\mathrm{CAL}(k,T) \times \mathrm{CAL}(p,T) &= \mathrm{CAL}(k \# p,T)\\
\mathrm{SAL}(k,T) \times \mathrm{CAL}(p,T) &= \mathrm{SAL}((k-1) \# p,T)\\
\mathrm{CAL}(k,T) \times \mathrm{SAL}(p,T) &= \mathrm{SAL}((p-1) \# k,T)\\
\mathrm{SAL}(k,T) \times \mathrm{SAL}(p,T) &= \mathrm{CAL}((k-1) \# (p-1),T)
\end{aligned} \qquad (1)$$

In the model shown in Fig. 3 the function $\mathrm{SAL}(8,T)$ represents the lithological variation, and $\mathrm{SAL}(64,T)$ describes the colour variation. Using equation (1), the product of these two functions is a new Walsh function (Fig. 5),

$$\mathrm{CAL}(7 \# 63,T) = \mathrm{CAL}(56,T)$$

having a frequency of 3.5 cycles/m here.

REFERENCES

Beauchamp, K.G. (1984) *Application of Walsh and Related Functions, with an Introduction to Sequence Theory*. Academic Press, London.

Berger, A. (1989) The spectral characteristics of Pre-Quaternary climate records, an example of the relationship between astronomical theory and geosciences. In: *Climate and Geosciences* (Eds Berger, A., Schneider, S. & Duplessy J.C.). Kluwer Academic Publishers, Dordrecht, pp. 47–76.

Berger, A., Imbrie, J., Hays, J., Kukla, G. & Saltzman, B. (1984) (Eds) *Milankovitch and Climate*. Reidel Publ. Co., Dordrecht.

Berggren, W.A., Kent, D.V. & Flynn, J.J. (1986) Palaeogene geochronology and chronostratigraphy. In: *The Chronology of the Geological Record* (Ed. Snelling, N.J.). Mem. Geol. Soc. Lond. 10, pp. 141–198.

De Boer, P.L. (1983) Aspects of middle Cretaceous pelagic sedimentation in Southern Europe. *Geol. Ultraiectina* **31**, 1–112.

Einsele, G. & Seilacher, A. (1982) (Eds) *Cyclic and Event Stratification*. Springer Verlag, Berlin.

Fischer, A.G. & Schwarzacher, W. (1984) Cretaceous bedding rhythms under orbital control? In: *Milankovitch and Climate* (Eds Berger, A., Imbrie, J., Hays, J., Kukla, G. & Saltzman, B.). Reidel Publ. Co., Dordrecht, pp. 164–177.

Fischer, A.G., de Boer, P.L. & Premoli Silva, I. (1990) Cyclostratigraphy. In: *Cretaceous Resources, Events and Rhythms* (Eds Ginsburg, R.N. & Beaudoin, B.). Kluwer Academic Publishers, Dordrecht, pp. 139–172.

Gilbert, G.K. (1895) Sedimentary measurement of Cretaceous time. *J. Geol.* **3**, 121–127.

Gullentops, F. & Vandenberghe, N. (1985) Rhythmicity in the Boom Clay (Rupelian) sedimentation. *Terra Cognita* **5**, 245.

Haq, B.U., Hardenbol, J. & Vail, P.R. (1988) Mesozoic and Cenozoic chronostratigraphy and cycles of sea-level change. In: *Sea-level Changes: An Integrated Approach* (Eds Wilgus, C.K., Hastings, B.S., Posamentier, H., Ross, C.A. & Kendall, Chr. G. St C.). Soc. Econ. Paleont. Mineral. Spec. Publ. 42, pp. 71–108.

Harland, W.B., Armstrong, R.L., Cox, A.V., Craig, L.E., Smith, A.G. & Smith, D.G. (1990) *A Geologic Time Scale 1989*. Cambridge University Press, Cambridge.

Hays, J.D., Imbrie, J. & Shackleton, N.J. (1976) Variations in the Earth's orbit: pacemaker of the ice ages. *Science* **194**, 18–22.

Herbert, T.D. & Fischer, A.G. (1986) Milankovitch climatic origin of mid Cretaceous black shale rhythms in central Italy. *Nature* **321**, 739–743.

Hooybergh, H. (1983) Contribution to the study of planktonic foraminifera in the Belgian Tertiary. *Aardkundige Mededelingen* **2**, 1–131.

Negi, J.G. & Tiwari, R.K. (1984) Periodicities of palaeomagnetic intensity and palaeoclimatic variations: a Walsh spectral approach. *Earth Planet. Sci. Lett.* **70**, 139–147.

Olsen, P.E. (1984) Periodicity of lake-level cycles in the Late Triassic Lockatong Formation of the Newark Basin (Newark Supergroup, New Jersey and Pennsylvania). In: *Milankovitch and Climate* (Eds Berger, A., Imbrie, J., Hays, J., Kukla, G. & Saltzman, B.). Reidel Publ. Co., Dordrecht, pp. 129–147.

Olsen, P.E. (1986) A 40-million-year lake record of Early Mesozoic orbital climatic forcing. *Science* **234**, 842–848.

Pestiaux, P. & Berger, A. (1984) An optimal approach to spectral characteristics of deep-sea climatic records. In: *Milankovitch and Climate* (Eds Berger, A., Imbrie, J., Hays, J., Kukla, G. & Saltzman, B.). Reidel Publ. Co., Dordrecht, pp. 417–445.

Schwarzacher, W. (1985) Principles of quantitative lithostratigraphy – The treatment of single sections. In: *Quantitative Stratigraphy* (Eds Gradstein, F.M. *et al.*). Reidel Publ. Co., Dordrecht, pp. 361–386.

Schwarzacher, W. (1987) The analysis and interpretation of stratification cycles. *Paleoceanography* **2**, 79–95.

Schwarzacher, W. (1989) Milankovitch cycles and the measurement of time. *Terra Nova* **1**, 405–408.

Smith, D.G. (1989) The Milankovitch cyclicity and the stratigraphic record – a review. *Terra Nova* **1**, 402–404.

Vandenberghe, N. (1978) Sedimentology of the Boom Clay (Rupelian) in Belgium. *Verhand. Koninklijke Acad. Wetenshappen België* **40**, 1–137.

Vandenberghe, N. & Laga, P. (1986) The septaria of the Boom Clay (Rupelian) in its type area in Belgium. *Aardkundige Mededelingen* **3**, 229–238.

Vandenberghe, N. & Van Echelpoel, E. (1987) A field guide to the Rupelian stratotype. *Bull. Belg. Vereniging Geol.* **96**, 325–337.

Van Den Bosch, M. & Hager, H. (1984) Lithostratigraphic correlation of Rupelian deposits (Oligocene) in the Boom Area (Belgium), the Winterswijk Area (The Netherlands) and the Lower Rhine District (F.R.G.). *Meded. Werkgr. Tert. Kwart. Geol.* **21**, 123–138.

Van Echelpoel, E. (1991) *Kwantitatieve cyclostratigrafie van de Formatie van Boom (Rupeliaan, België). De methodologie van het onderzoek van sedimentaire cycli via Walsh analyse*. Unpublished PhD thesis, K.U. Leuven, Belgium.

Van Echelpoel, E. & Weedon, G.P. (1990) Milankovitch cyclicity and the Boom Clay Formation: an Oligocene siliciclastic sequence in Belgium. *Geol. Mag.* **127**, 599–604.

Vinken, R. (1988) (Ed.) *The Northwest European Tertiary Basin (Results of IGCP 124). Geol. Jb.* **A100**, 1–508.

Weedon, G.P. (1985) Hemipelagic shelf sedimentation and climatic cycles: the basal Jurassic (Blue Lias) of South Britain. *Earth Planet. Sci. Lett.* **76**, 321–335.

Weedon, G.P. (1989) The detection and illustration of regular sedimentary cycles using Walsh power spectra and filtering, with examples from the Lias of Switzerland. *J. Geol. Soc. Lond.* **146**, 133–144.

Spec. Publs Int. Ass. Sediment. (1994) **19**, 77–85

Fourier evidence for high-frequency astronomical cycles recorded in Early Cretaceous carbonate platform strata, Monte Maggiore, southern Apennines, Italy

G. LONGO*, B. D'ARGENIO†, V. FERRERI† *and* M. IORIO‡

* *Osservatorio Astronomico, Napoli, Italy;*
† *Dipartimento di Scienze della Terra, Università Federico II, Napoli, Italy; and*
‡ *Geomare, Istituto di Geologia Marina del CNR, Napoli, Italy*

ABSTRACT

Carbonate peritidal deposits of Early Cretaceous age, widely outcropping in the carbonate platform sequence of southern Italy, carry distinct signals of cyclicity in the Milankovitch band. We have studied the depositional and diagenetic facies organization of a *c.*100-m-thick sequence of Barremian age, at Monte Raggeto (Monte Maggiore Mountains, near Naples), where, from a total of 60 m analysed at centimetre scale, two sedimentary modules have been recognized.

1 *Depositional cyclothems*: rare, made of one or more subtidal–supratidal couplets and topped by supratidal intervals.

2 *Diagenetic cyclothems*: very common, made of dominantly subtidal intervals which show emersion-generated features (karst, reddened surfaces) at the top.

Cyclothems along the sequence tend to group into fairly regular intervals, each about 10 m thick, formed by sets of seven to nine cyclothems. This trend is confirmed by Fourier analysis of the data, showing periodicities at 105 and 950 cm. Moreover, the mathematical processing of the total data set shows also two shorter periodicities at 40 and 72 cm. The algorithm used for the analysis is a modified version of the Deeming code, first written for astrophysical applications.

The set of periodicities obtained (40, 72, 105 and 950 cm) can be related to the variations of the insolation constant computed for the Cretaceous; moreover the ratios between the two sets of periodicities, expressed in centimetres and in years respectively, show a very high degree of correlation with those predicted for the main orbital periods in the Cretaceous. We propose that the observed cyclicities indicate Barremian sea-level oscillations induced by high-frequency eustatic control under climatic forcing.

Moreover the link between the ratios of time and depositional periodicity sets appears to be a useful method of assigning a duration to a given periodicity in a sequence, quite independently from a precise determination of biostratigraphic age and/or total thickness of a stage.

INTRODUCTION

From the late 1970s studies concerning the rhythms of sedimentation with frequencies in the Milankovitch band brought about new insights into the Cretaceous pelagic sequences of Italy (Fischer & Arthur, 1977; Schwarzacher & Fischer, 1982; de Boer, 1983; Arthur *et al.*, 1984; Premoli Silva *et al.*, 1989). In the Central Apennines A.G. Fischer and his co-workers first recognized sedimentary rhythms expressed by couplets of strata, or groups of couplets, which, on account of their sedimentary and biostratigraphic features, suggest for each couplet or group of couplets intervals of time ranging between 2×10^4 and 4×10^5 years (de Boer, 1983; Herbert & Fischer, 1986; Park & Herbert, 1987).

More recently, high-frequency cyclicity has been shown in carbonate platform sequences of Aptian age outcropping in the Matese Mountains of southern Italy (D'Argenio *et al.*, 1989, 1992). Here metre-scale

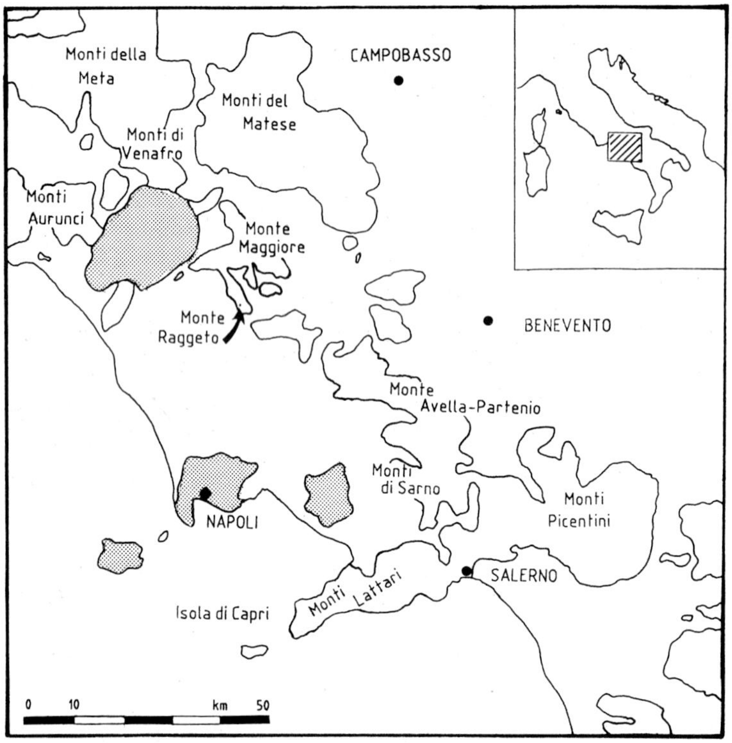

Fig. 1. Location of Monte Raggeto in the area studied. The solid contour encloses mountains predominantly formed by Mesozoic carbonate platform strata; the shading indicates volcanic areas.

stacked cycles exhibit a well-developed supratidal unit passing up into a subtidal level affected by palaeokarstic solution in its topmost part and/or covered by discontinuous varicoloured clayey veneers (asymmetrical peritidal cyclothems). Peritidal cyclicity is also shown by many other Early Cretaceous intervals of southern Italy at present under study (e.g. Matese, Monte Maggiore, Monti di Sarno; D'Argenio & Ferreri, 1991).

The general organization of the above cyclothems, their individual thickness and number per stage suggest that cyclicity was induced by sea-level fluctuations under high-frequency eustatic control.

We present here the main results of a study carried out at Monte Raggeto, Monte Maggiore Mountains, north of Naples (Fig. 1), where from a 100-m-thick interval analysed in detail, 60 m have been investigated at centimetre scale. The mathematical treatment of the data collected will be described in some detail here, while the sedimentary characteristics of this sequence will be only briefly outlined, a more detailed treatment being in preparation.

LITHOFACIES ASSOCIATIONS AND CYCLIC ORGANIZATION

The succession studied is well exposed along quarry walls opened on the northwestern side of Monte Raggeto (Fig. 2) and is composed of dolomitic-calcareous strata whose average thickness is about 110 cm; almost every stratum corresponds to a cyclothem. Fossil associations (green algae and benthic forams, among which are *Salpingoporella muelhbergii*, *S. melitae*, *Debarina* sp., and *Cuneolina laurentii*) suggest a Barremian age.

The strata are sometimes separated by thin (1–10 cm) greyish to greenish clayey horizons, resting on erosional surfaces. Discontinuities in the depositional processes and subaerial exposure are also suggested by karstic features, vadose diagenesis and erosional or bioerosional truncations.

Three main lithofacies associations have been identified (D'Argenio *et al.*, 1989).

1 Gastropod dolomitic limestones (mollusc, benthic foram, green alga and ostracod wackestone and

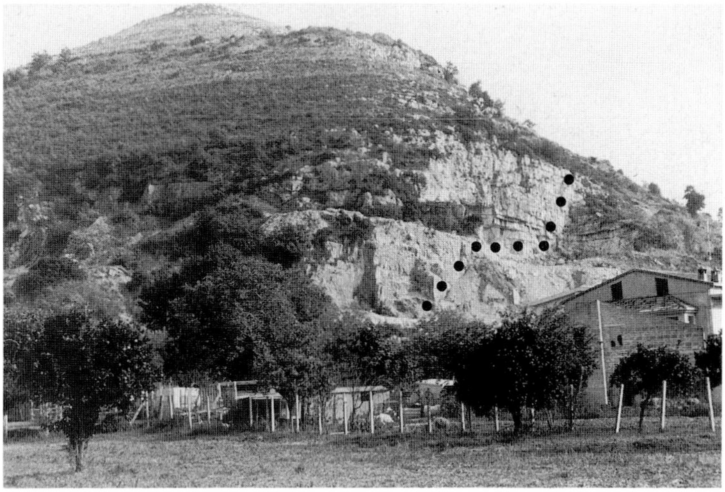

Fig. 2. Quarry wall on the north-western side of Monte Raggeto. Dots indicate part of the measuring and sampling route.

subordinate mudstone and packstone), affected by variable intensity of dolomitization.

2 Stromatolitic and loferitic dolomites (dolomitic bindstone).

3 Intrabioclastic dolomitic calcarenites and calcirudites (dolomitic grainstone and rudstone generally graded, often channelized).

Based on textural characteristics, lime mud content, fossils and diagenetic features, the first lithofacies association is interpreted as the product of subtidal (shallow infralittoral), more or less protected environments, while stromatolitic and loferitic dolomites represent typical tidal flat deposits, as suggested by frequent desiccation and ephemeral emersion features as well as by bioerosion and dolomitization.

While the infralittoral and tidal deposits and related emersion features show a cyclic alternation, the third lithofacies association intercalates episodically in the above two, and is characterized by a clastic texture and by coarse arenitic to ruditic grain size, which suggest higher energy events (*tempestites*, accumulated in tidal to infralittoral settings; Aigner, 1985).

As far as the general organization of these lithofacies associations is concerned, two sedimentary modules have been recognized in the sequence studied.

1 *Depositional cyclothems*: made up of one or more subtidal–supratidal couplets and topped by a loferitic interval.

2 *Diagenetic cyclothems* (as originally defined by D'Argenio, 1976, p. 151, and later by Hardie *et al.*, 1986; see also Tucker & Wright, 1990, p. 163): composed of basal intercalations of stromatolites or calcarenites (tempestites) and followed by subtidal deposits which predominate and show emersion features at the top (vadose features, reddened surfaces).

The depositional cyclothems are rare and reveal a shallowing-upward trend; the diagenetic cyclothems are frequent and also shallow upwards, as indicated by vadose caps and erosional surfaces directly superimposed on subtidal deposits; they also show less intense dolomitization.

Moreover, recurrence every *c.*10 m of a diagenetic cyclothem composed only of a relatively thin subtidal interval, slightly dolomitized to non-dolomitized and capped by palaeokarst, allows recognition of groups of cyclothems, each formed by sets of seven to nine units.

EARLIER INTERPRETATIONS AND CHOICE OF TIME SCALE

A first interpretation of the cyclicity of the Monte Raggeto Barremian, relating the studied cyclothems to one of the Milankovitch orbital parameters, was attempted by D'Argenio *et al.* (1989), who considered that the 100-m sequence basically corresponds to the Barremian or at least to the greatest part of it.

The authors used the time scale of Palmer, 1983, which reports 5-Ma duration for the Barremian, and considered the cyclothems measured at Monte

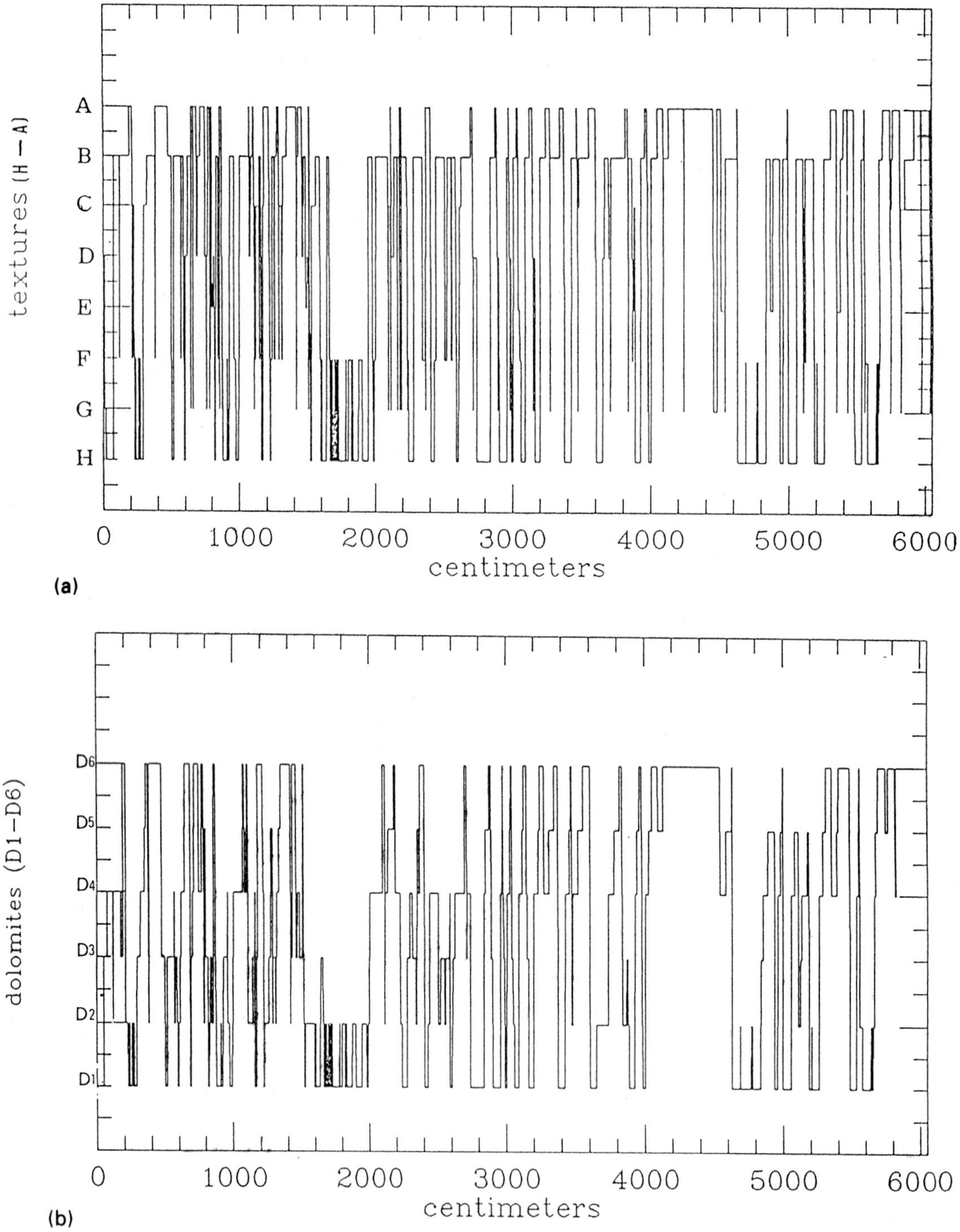

Fig. 3. Time-series plots of the textures (a) and of dolomitization (b). (a) Textures are organized as follows: A, mudstone; B: wackestone; C, packstone; D, grainstone; E, rudstone; F, bindstone; G, clayey levels; H, dolomites (D1). (b) On the vertical axis D1–D6 indicate density variation of dolomitic crystals (number of crystals per square centimetre). Six steps of dolomitization have been empirically distinguished by means of visual estimates: D1 indicating that the dolomitic crystals have completely obliterated the original textures and D6 indicating that there is no dolomitization at all.

Raggeto to be related to the obliquity (50 ka per cyclothem; D'Argenio *et al.*, 1989). However, in the last few years several different time scales have been proposed, acccording to which the Barremian may range from 2.0 (Cowie & Basset, 1989) and 3.5 (Haq *et al.*, 1987) to 7.3 (Harland *et al.*, 1990) Ma. If the 100-m sequence of Monte Raggeto falls entirely within the Barremian and a constant rate of sedimentation is assumed, the 60 m analysed in detail must span at least 1.2–4.4 Ma (60% respectively of 2.0 and 7.3 Ma), implying a sedimentation rate ranging from 50 to 14 m/Ma (i.e. 200–700 years/cm).

To decide which scale fits best with our data we attempt a more sophisticated treatment of them through a Fourier analysis.

PARAMETERS FOR ANALYSIS

The Fourier analysis requires numerical time-series data. The data used here to investigate the systematic recurrence of certain parameters are listed below and briefly commented on. They have been acquired directly in the field at a centimetre scale, using a ×10 hand lens and checked on large oriented samples, with numerous thin sections and acetate peels. For each parameter the time series consists of either the thickness, or simply the presence/absence, of the particular sedimentary feature against its position in the sequence.

Textural parameters

Textural parameters have been considered in terms of thickness (i.e. evaluating sequentially the thickness of each prevailing texture). The terminology used conforms to Embry & Klovan's (1971) nomenclature (Fig. 3a).

A Calcilutites with rare green algae and ostracods (*mudstone*).

B Calcilutites with benthic forams, green algae, gastropods and oncoids (*wackestone*).

C Calcilutites–calcarenites with benthic forams, green algae, micritized grains, gastropods and intraclasts (*packstone*).

D Calcarenites with intraclasts, lithoclasts and rare bioclasts (*grainstone*).

E Calcirudites with intraclasts and lithoclasts (*rudstone*).

F Stromatolites, locally with 'fenestrae', sheet cracks and microteepees, more or less deeply dolomitized (*bindstone*).

G Greyish to greenish clayey horizons, resting on erosional surfaces (*clayey levels*).

H Dolomites whose original textural features have been completely obscured (D1).

Dolomitization intensity has been computed in terms of thickness, evaluating empirically the number of crystals per square centimetre. Six steps of dolomitization have been distinguished by means of visual estimates, D1 indicating that the dolomitic crystals have completely obliterated the original textures (and corresponding to the parameter **H** reported above) and D6 indicating that there is no dolomitization at all (Fig. 3b).

Other sedimentary features

Other sedimentary features have been expressed as linear parameters, i.e. counting only their position in the sequence.

I *Graded bedding*: gradation, from coarser particles at the base to finer particles at the top (normal grading), is generally found in storm deposits.

L *Parallel lamination.*

M *Oblique (hummocky) lamination*: parallel to oblique lamination may be found in the upper part of storm deposits.

N *Erosional surfaces*: more or less continuous, not very 'deep' subaerial unconformities, normally marked by clayey horizons.

O *Bioerosional surfaces*: shallow hardgrounds, marked by boring and microboring, over which new lagoon deposits develop.

P *Cut and fill*: erosional features underlying storm deposits.

Also some diagenetic features have been expressed as linear parameters (even though their expression may reach some centimetres).

Q *Pervasive dissolution and individual small cavities* filled by geopetal crystal silt (microkarst), very frequent in the upper part of subtidal intervals (vadose cap).

R *Paleokarstic cavities*: larger karstic cavities (centimetre to decimetre in size) parallel to the bedding and mechanically filled by polychrome silts may be found in the upper part of subtidal intervals.

DISCUSSION OF DATA

The above data have been mathematically processed to evaluate numerically the observed periodicities.

Usually, the absolute dating of sedimentary sequences is affected by large uncertainties in the radiometric time scales, as well as by time-scale drifts and random errors introduced by long- and medium-term variations in the sedimentation rate. In the present case, the sedimentation rate is assumed to have been quasi-constant for the following combination of conditions:

1 regional regime of thermal subsidence in its mature stage;
2 rate of subsidence $\simeq$ rate of sedimentation (due to the shallowness of the depositional areas and to the high sedimentation potential of peritidal carbonates);
3 no substantial compaction (pervasive early cementation, generally moderate pressure-solution features);
4 no evidence of synsedimentary tectonics (suggesting that emersions which interrupt the sedimentation are evenly distributed over the whole sequence).

The uncertainties mentioned above affect the data treatment in different ways:

1 zero point errors affect the absolute and not the relative scale of ages, and they can therefore be neglected for period estimates;
2 time-scale drifts displace the maxima and may introduce low-frequency fluctuations in the periods. In the case of sequences covering a very limited range in time, they may be considered as a second-order effect whose amplitude may be evaluated by studying the residuals;
3 interruptions in the sequence introduce high-frequency components in the Fourier spectrum which can be kept under control either by filtering the data or by following the more complex procedure described below;
4 random errors are of course unavoidable and affect the overall accuracy.

Within this framework the data d (where d means any of the measured quantities) may be represented as:

$$d_{\text{obs}}(x') = 0 \text{ outside the sampled interval}$$
$$d_{\text{obs}}(x') = d(x') + N(x'),$$

where $x' \in [x_1, x_2]$, i.e. inside the sampled interval, and

$$x' = \alpha(x + x_0)$$

where α is a time drift factor, x_0 is the unknown zero-point offset, $N(x')$ are the random errors, and x_1 and x_2 are the initial and final data of the 60-m sequence.

The code introduced by Deeming (1975) is a Fourier spectrum technique which allows analysis of unequally spaced data which can be represented in the above form. Even though the present data are equally spaced, Deeming's method offers two advantages with respect to more traditional approaches:

1 it minimizes the effects of the aliasing;
2 it discriminates between spurious and true harmonics by estimating the confidence level of each one.

The aliasing is due to the fact that the finite width of the data sampling function introduces spurious harmonics. In order to reduce the aliasing we produced a fictitious set of data defined as:

$$d'(x') = \text{MOD}_{\Delta x}[d(x')]\ \Omega\ (x')$$

where MOD is a simple mathematical operator duplicating the sequence over an infinite interval in order to transform the function $d(x')$ into a periodic function $d'(x')$ having period equal to $\Delta X = x_2 - x_1$. Ω is a very broad Gaussian function having full half-width maximum (FHWM) given by:

$$\text{FHWM} \geqslant 200\Delta X$$

thus obtaining the suppression of all higher order harmonics introduced by the aliasing except the $\mu \simeq$ FHWM one.

Overcoming the problem of chronological interruptions requires the assumption that variations in the sedimentation rates are negligible over a short interruption in the depositional process. The first step consists of using the whole sequence to obtain a first approximation Fourier spectrum. The main harmonics are then used to 'reconstruct' the missing part of the sequence, i.e. the data following an interruption are shifted to minimize the deviations with respect to the lower frequency harmonics (Fig. 4). The reconstructed sequence is then used to investigate higher frequency fluctuations and to evaluate the confidence levels of the harmonics. Figure 5 shows the resulting Fourier spectra obtained for the various parameters used in the analysis.

Expressed in centimetres, the periods found in the Monte Raggeto sequence are at 40, 72, 105 and 950 cm. We note that while 105 and 950 cm are periods which correspond well with the average thickness of cyclothems and groups of cyclothems, as observed in the field, the 40-cm and 72-cm periods are less evident in the field and reflect, among other parameters, the dolomitization intensity (D1) and the presence of dolomitic bindstones.

As a next step we have compared the ratios of the

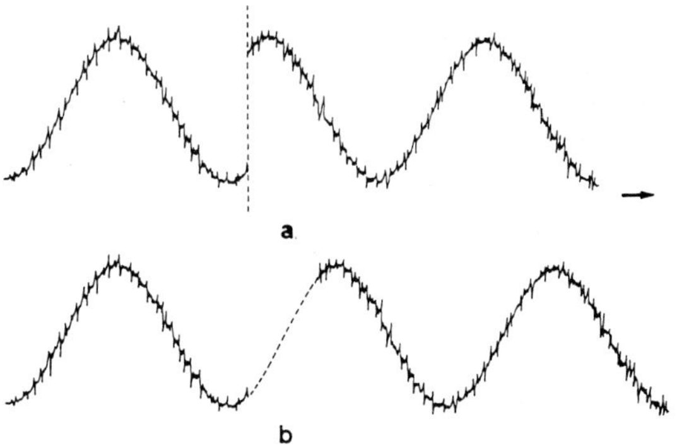

Fig. 4. Schematic illustration of the 'shifting' procedure. (a) An idealized interrupted sequence; (b) the sequence reconstructed on the basis of the extrapolation of the underlying main harmonics, preserving the integrity of the shorter periodicities.

periodicities expressed in years and in centimetres (see Cottle, 1989), as reported in Table 1. In this table, column 1 lists the orbital forcing periodicities predicted for the Cretaceous by Berger *et al.* (1989) on the basis of numerical modelling of the dynamical evolution of the inner solar system. Columns 2 and 3 give the relative ratios of the above periods in units of 19 and 22.5 ka respectively. The small difference between the first two periods in column 1 is very unlikely to be resolved in the present data set, so column 4 gives the orbital forcing periods in units of 21 ka, where 21 ka is assumed to be the equivalent length of the unresolved combination of the 19 and 22.5 ka periods. Column 5 gives the observed periodicities from our data (expressed in centimetres), while column 6 gives the relative ratios of the values in column 5 expressed in units of 40 cm, the shortest period revealed by the Fourier analysis.

Statistical correlation shows that the best agreement is between the sets of values in columns 4 and 6 (which have a linear correlation factor of *c*.0.9), thus arguing in favour of a close coupling between the periodicity of the depositional process and the orbital periods listed in column 1. Therefore we assume that the 40-cm periodicity corresponds to a 21-ka cycle, the 72 cm and the 105 cm to 39.3-ka and 51.2-ka cycles respectively, and the 950 cm to a 413-ka cycle.

From the above discussion it also follows that our *c*.100-m thick sequence, which falls entirely within the Barremian (reliable Aptian was recognized only a few tens of metres above), required a minimum of *c*.5.3 Ma to form. We therefore conclude that a 7.3-Ma duration assigned to the Barremian by Harland *et al.* (1990) is the value which best fits our data.

Lastly the non-detection of the 100-ka period is somewhat puzzling, given that both the 100-ka and the 413-ka periods originate from fluctuations in the same orbital parameter (namely the eccentricity). Either this may be a result of the arbitrary cut in the confidence level introduced in order to separate 'spurious' from 'real' harmonics, or the 100-ka period may be hidden by the stronger *c*.50-ka signal.

We conclude that the periodicities in the depositional record of the Monte Raggeto shallow-water Barremian show strong links with the periodic fluctuations of the insolation constant extensively discussed by Berger *et al.* (1989).

Conversely a precise biostratigraphic calibration may offer a control to the radiometric evaluations of stage duration. This method appears reliable due to the very high linear correlation factor between the observed and predicted values for the period ratios.

ACKNOWLEDGEMENTS

The authors thank Filippo Barattolo from Naples and Raika Radoičič from Belgrade for their help in the age determination. The authors also thank L. Errico from Naples for his help in implementing the

Table 1. Relationships between time and thickness periodicities discussed in the text. See text for details of columns 1–6

Orbital forcing (ka)	Relative ratio (19.0 unit)	Relative ratio (22.5 unit)	Relative ratio (21.0 unit)	Observed periodicities (cm)	Relative ratio (40-cm unit)
19.0	1.00	0.84			
22.5	1.18	1.00	1.00	40	1.00
39.3	2.07	1.78	1.87	72	1.80
51.2	2.70	2.35	2.43	105	2.65
100.0	5.26	4.44	4.76		
413.0	21.73	17.96	19.66	950	23.75

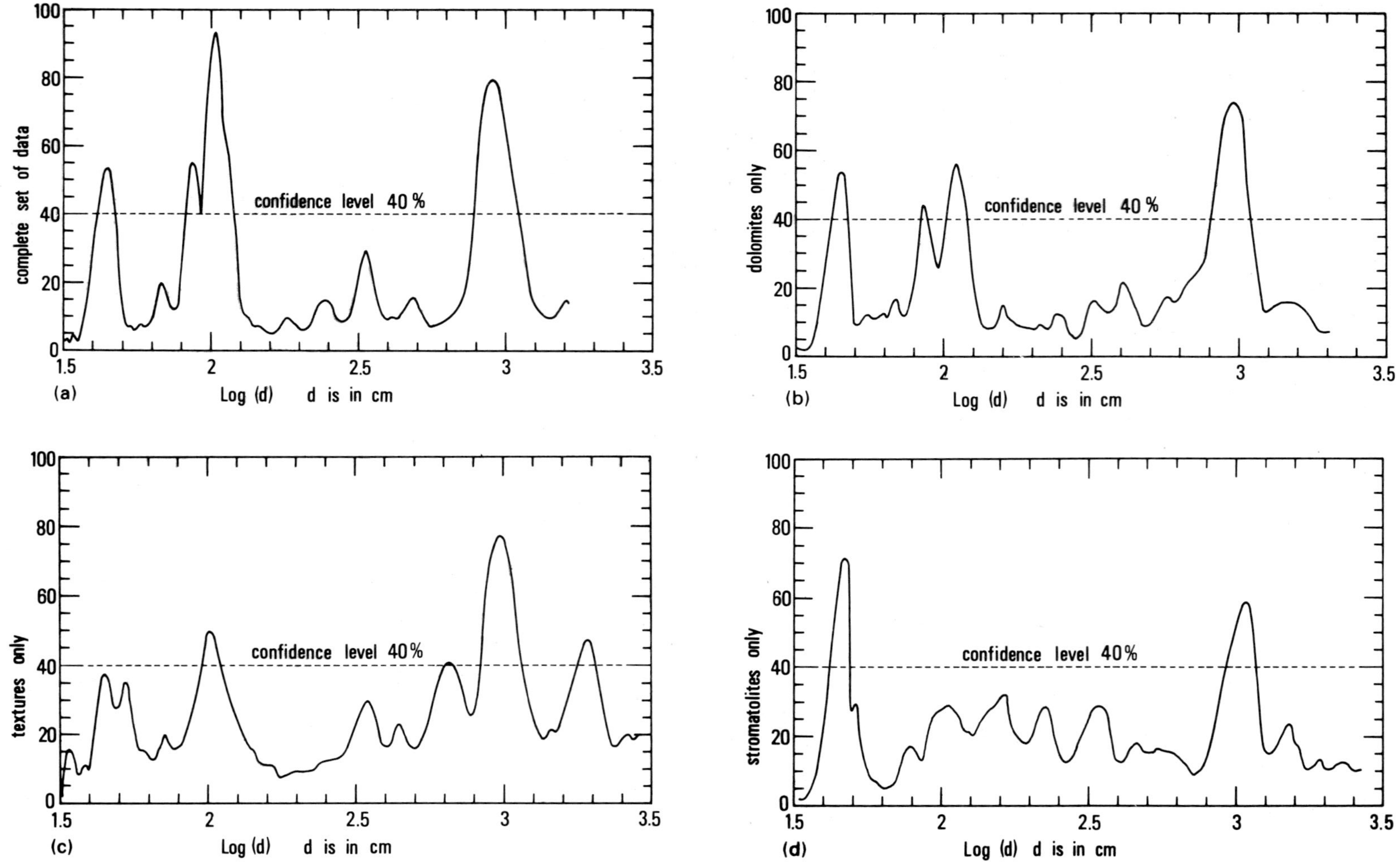

Fig. 5. Fourier spectrum of the analysed data: (a) the Fourier spectrum of the whole data set; (b–d) the Fourier spectrum of (b) dolomitization intensity, (c) textures and (d) stromatolitic levels.

Fourier analysis codes. We especially thank one of the referees (L. Hinnov) for suggesting important references and the editors for improving the manuscript. This research was supported by a 1989 CNR grant for marine geology studies (Geomare program) to B. D'Argenio.

REFERENCES

AIGNER, T.H. (1985) *Storm Depositional Systems*. Lecture Notes in Earth Sciences Vol. 3. Springer Verlag, Berlin.

ARTHUR, M.A., DEAN, W.E., BOTTJER, D. & SCHOLLE, P.A. (1984) Rhythmic bedding in Mesozoic–Cenozoic pelagic carbonate sequences: the primary and diagenetic origin of Milankovitch-like cycles. In: *Milankovitch and Climate* (Eds Berger, A., IMBRIE, J., HAYS, J., KUKLA, G. & SALTZMAN, B.). Reidel Publ. Co., Dordrecht, pp. 191–222.

BERGER, A., LOUTRE, M.F. & DEHANT, V. (1989) Astronomical frequencies for pre-Quaternary paleoclimate studies. *Terra Nova* **1**, 474–479.

COTTLE, R.A. (1989) Orbitally mediated cycles from the Turonian of southern England: their potential for high-resolution stratigraphic correlation. *Terra Nova* **1**, 426–431.

COWIE, J.W. & BASSET, M.G. (1989) Global stratigraphic chart. *Supplement to Episodes* **12**(2).

D'ARGENIO, B. (1976) Le piattaforme carbonatiche periadriatiche. Una rassegna di problemi nel quadro geodinamico mesozoico dell'area mediterranea. *Mem. Soc. Geol. It.* **13**, 137–160.

D'ARGENIO, B. & FERRERI, V. (1991) High frequency cyclicity in carbonate platform sequences, Lower Cretaceous of Southern Italy. E.U.G. VI Meeting, Strasbourg, March 24–28, 1991. *Terra Abstracts* **30**, 278.

D'ARGENIO, B., FERRERI, V., IORIO, M. & RUBERTI, D. (1989) Ritmi sedimentari con frequenze di Milankovitch nel Cretacico inferiore dell'Appennino Meridionale. Dati preliminari. Conf. Scient. 1989, Dipartimenti Geofisica e Vulcanologia Paleontologia e Scienze della Terra, Università di Napoli, 93–95, Napoli.

D'ARGENIO, B., FERRERI, V. & RUBERTI, D. (1992) Cicli ciclotemi e tempestiti nei depositi carbonatici aptiani del Matese (Appennino Campano). *Mem. Soc. Geol. It.* **41**, 761–773.

DE BOER, P.L. (1983) Aspects of Middle Cretaceous pelagic sedimentation in Southern Europe. *Geol. Ultraiectina* **31**.

DEEMING, T.J. (1975) Fourier analysis with unequally-spaced data. *Astrophys. Space Sci.* **36**, 137–158.

EMBRY, A.F. & KLOVAN, J.E. (1971) A Late Devonian reef tract on northeastern Banks Island, N.W.T. *Bull. Can. Petrol. Geol.* **19**, 730–781.

FISCHER, A.G. & ARTHUR, M.A. (1977) Secular variation in the pelagic realm. In: *Deep-water Carbonate Environments* (Eds Cook, H.E. & Enos, P.). Soc. Econ. Paleont. Mineral., Spec. Publ. 25, pp. 19–50.

HAQ, B.U., HARDENBOL, J. & VAIL, P.R. (1987) Chronology of fluctuating sea level since Triassic (250 million years ago to present). *Science* **235**, 1156–1167.

HARDIE, L.A., BOSELLINI, A. & GOLDHAMMER, R.K. (1986) Repeated subaerial exposure of subtidal carbonate platforms, Triassic, Northern Italy: evidence for high frequency sea level oscillations on a 10^4 year scale. *Paleoceanography* **14**, 447–457.

HARLAND, B.W., ARMSTRONG, R.L., COX, A.V., CRAIG, L.E., SMITH, A.G. & SMITH, D.G. (1990) *A Geologic Time Scale 1989*, Cambridge University Press, Cambridge.

HERBERT, T.D. & FISCHER, A.G. (1986) Milankovitch climatic origin of Mid-Cretaceous black shale rhythms in Central Italy. *Nature* **321**, 739–743.

PALMER, A.R. (1983) The decade of North American Geology, 1983 Geologic Time Scale. *Geology* **11**, 503–504.

PARK, J. & HERBERT, T.D. (1987) Hunting for paleoclimatic periodicities in a geologic time series with an uncertain time scale. *J. Geophys. Res.* **92**, 14027–14040.

PREMOLI SILVA, I., RIO, M. & TORNAGHI, M.E. (1989) Planktonic foraminiferal distribution record productivity cycles: evidence from the Aptian–Albian Piobbico core (Central Italy). *Terra Nova* **1**, 443–448.

SCHWARZACHER, W. & FISCHER, A.G. (1982) Limestone shale bedding and perturbation of the Earth's orbit. In: *Cyclic and Event Stratification* (Eds Einsele, G. & Seilacher, A.). Springer Verlag, Berlin, pp. 72–95.

TUCKER, M.E. & WRIGHT, V.P. (1990) *Carbonate Sedimentology*. Blackwell Scientific Publications, Oxford.

Spec. Publs Int. Ass. Sediment. (1994) **19**, 87–97

Cyclostratigraphy of the Cenomanian in the Gubbio district, Italy: a field study

W. SCHWARZACHER

School of Geosciences, Queen's University Belfast, Belfast, UK

ABSTRACT

The Cenomanian of the Central Apennines in Italy shows well-developed stratification cycles. Groups of four to seven beds form a 'bundle'; adjacent bundles are often separated by a marl layer. Two sections in the Scaglia Bianca (top Albian to Cenomanian) approximately 20 km apart have been correlated and in both sections 65–70 cycles have been recognized. Absolute time estimates make it likely that the cycles are controlled by the 100-ka eccentricity cycle. The length of the Cenomanian excluding the Bonarelli horizon is estimated to be 5.8–6.2 Ma. Spectral analysis and filtering indicates evidence also of a 400-ka cycle. The quality of data and the lengths of the sections were insufficient to prove the existence of longer cycles.

INTRODUCTION

The Cenomanian in the Umbrian–Marchian sequence of the Central Apennines in Italy is represented by the Scaglia Bianca. These are well-bedded pelagic limestones with only occasional thicker marl layers. The Cenomanian is underlain by the Aptian–Albian Scisti a Fucoidi and terminated by the black shales of the Bonarelli horizon. The latter represents an anoxic event which can be traced worldwide. The sequence, which in its higher parts contains the Cretaceous–Tertiary (K–T) boundary, has been extremely well studied by palaeontologists and stratigraphers (Tornaghi *et al.*, 1989 for references).

Cyclic sedimentation has been recognized in the sequence from the Neocomian up to the Eocene (Arthur, 1976; de Boer, 1982, 1983; Schwarzacher & Fischer, 1982). The best data are from a borehole through the Aptian–Albian Scisti a Fucoidi at Piobbico (province of Pessaro-Urbino) and these were studied in great detail by a group of Italian and American scientists (Fischer & Herbert, 1988; Premoli Silva *et al.*, 1989; Tornaghi *et al.*, 1989).

The time-series analysis of the Piobbico data (Premoli Silva *et al.*, 1989) confirms that the cycles are Milankovitch related and, when assuming an average sedimentation rate of 5 mm/ka, the major periods of the Milankovitch spectrum (20, 44 and 111 ka) were found. The Scaglia Bianca which follow the Scisti show an essentially identical but less well developed cyclicity (de Boer, 1982, 1983).

This paper deals with the interval of the Scaglia Bianca Formation below the Bonarelli horizon; it is a study that had to rely entirely on field observations without any detailed petrographic or faunal examinations. It was undertaken to demonstrate the extent to which a cyclostratigraphic approach can be used during the 'mapping stage' of a stratigraphic investigation. Thus it is important not only to prove that a cycle of a certain period exists by using statistical methods such as spectral analysis, but also to identify each cycle in the stratigraphic sequence. The cycle boundaries should be identified using objective methods, if possible.

THE SECTIONS

The following sections have been measured to provide the basic data.

1 A section (Contessa) along the Contessa highway leading from Gubbio to Cantiano.

2 A section (Petrano) near the top of Monte Petrano along the small road leading from the summit to the village of Palecano.

3 The classical section in the Bottacione gorge near Gubbio.

4 A section at the base of Monte Petrano along the road leading from Moria to the summit.

Sections 3 and 4 are incomplete. The Bottacione section cannot be reached in its lower part because the wire netting which covers it is filled with loose material. The top 40 m are partly overgrown but have been measured. The section near Moria provides very good exposures in the Scisti a Fucoidi but only 25 m of the Cenomanian are at present sufficiently well exposed to be measured.

All the sections used in the present study have been measured before by a number of geologists. In the Contessa and Bottacione sections, A.G. Fischer and his colleagues painted large black spots on a white background (bull's-eyes) at intervals of 250 cm. Unfortunately, some of these are already missing and particularly in the Contessa section the numbering is no longer legible. The positions of bull's-eyes which are still visible have been recorded with the measurements. The measurements were carried out bed by bed and their reproducibility can be roughly judged by comparing the new measurements with the existing marks (bull's-eyes). Using all the available data, an average discrepancy of 1.9% was found. It is of course impossible to tell whether the present measurements are better than the previous measurements, but the deviations are surprisingly small and are not systematic. (A copy of the original measurements can be obtained by sending a $5\frac{1}{4}$-inch floppy disk to the author.)

The first few metres of the Monte Petrano and Moria sections were measured by Fischer and Ripepe (Premoli Silva & Erba, 1988). They were able to correlate these sections, which contain the top part of the Albian, with the sequence in the Piobbico borehole. This correlation is important as it provides the connection between the cyclostratigraphy of the Aptian–Albian with that of the Cenomanian. The base of the newly measured sections in this study at Moria, Monte Petrano and Contessa was taken at the top of the first limestone of the Piobbico borehole at the depth of 70 cm in Unit 1 (Tornaghi *et al.*, 1989). At this level, the lithology changes from the Fucoid marl into the typical limestone sequence of the Scaglia. All subsequent data are expressed in centimetres above this level. Using this zero reference, the Bonarelli horizon was found at 6509 cm in the Contessa section and at 6537 cm in the Petrano section.

De Boer (1982) measured a section near Moria which is no longer fully exposed. He recorded the first appearance of *Planomalina buxtorfi* at 10 m above the base of his section. The *P. buxtorfi* datum was recorded at 19 cm in the Piobbico borehole which would mean that de Boer's zero is at −949 cm on the scale used in this paper. According to this tentative correlation the Bonarelli horizon in de Boer's section would occur at 6592 cm using the present scale and the Albian–Cenomanian boundary at approximately 750 cm (Fig. 1).

Apart from the distinct changes in lithology at the beginning and end of the sections, several events are recorded which can be used to correlate the sections in more detail. At 1186 cm in the Contessa section and at 1204 cm in the Petrano section, a 10–13-cm thick shale band, which is partly developed as black shale, occurs and indicates a short return to the anoxic conditions which are more common in the Albian. At 2424 and 2508 cm in the Contessa section

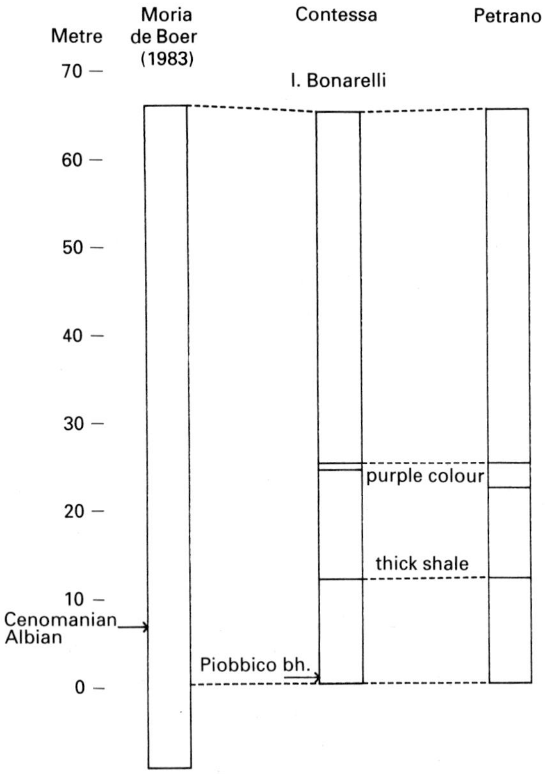

Fig. 1. Correlation between de Boer's Moria section and the Contessa and Monte Petrano sections with the Piobbico core. The zero of the Piobbico borehole is indicated by an arrow; de Boer's Albian–Cenomanian boundary is also shown. The base datum for the Contessa and Petrano sections (zero on the metre scale on the left) is as described in the text.

and at 2255 and 2527 cm in the Petrano section, a distinct colour change is seen for a short interval. The normally white limestone becomes red to purple; the change is seen in all the measured sections and can be used for local correlation. Finally a definite change in the colour of cherts occurs. Cherts in the lower part of the sections are grey brown or red in colour and become distinctly black in the higher parts of the sequence. The first black chert band was recorded at 4527 cm in the Petrano section and at 4869 cm in the Contessa section. The two levels correlate only approximately, but it is possible that with a petrographic study more detailed correlation could be obtained from the distribution of chert bands.

THE STRATIFICATION CYCLES

The term stratification cycle has been defined as a group of beds which are regularly repeated (Schwarzacher, 1987). The number of beds composing such cycles can be constant or regularity may be due to repeated sequences of bed-thicknesses or composition.

The term bundle (German: *Bündle*) was first used to describe Triassic platform limestones which are grouped into stratification cycles of five beds (Schwarzacher, 1952). The term is a useful and expressive way of describing such groups but it is not, as some geologists believe, a technical term. The number of beds making a bundle can change either by the fusing of beds or by incomplete preservation, which can reduce the bundle to a bundle of one!

Considerably more difficult is the definition and classification of bedding planes. Such planes must be parallel to the original sedimentation surface and therefore they are always related to changes in primary sedimentation. An approximate classification of bedding planes can be obtained by considering their lateral persistency (Schwarzacher, 1958). In the present study, two types of bedding planes (and therefore of beds) were recorded. Beds which can be traced throughout the extent of an outcrop (usually tens of metres) are called normal beds and less persistent beds are called subsidiary beds. Subsidiary beds are always thinner than normal beds, their bedding planes often occur near normal planes and they are particularly numerous near stratification cycle boundaries. Cycle boundaries can be correlated throughout the field area and they are therefore master bedding planes. Stylolitic seams are often associated with bedding planes and they may occur close to and subparallel to bedding planes. Stylolites, however, have not been recorded, although they too are ultimately an indication of changes in primary sedimentation if they are parallel to sedimentary bedding.

In the following, a distinction is made between cycles and beds. The cycles are stratification cycles, thought to have been caused by the orbital eccentricity cycle, and their duration is estimated to be 100 ka. It is very possible that bed formation also involves cyclic processes and indeed it is likely that the 21-ka precession cycle was responsible for normal bedding, but the present evidence provides no proof for this.

At the base of the sections towards the top of the Scisti a Fucoidi, stratification cycles are 50–70 cm thick. The cycles consist of four to seven limestone beds which increase in thickness towards the top of each cycle. Shale is concentrated at the base of the cycle and the marl layers between the beds are at most only 1–2 cm thick. In the lowest part of the section, one can still define limestone–marl couplets, even if they are often subdivided by subsidiary bedding so that the 'couplet' may contain several limestone beds to one marl band. From about 9 m above the base of the section, shale layers between limestones are below the accuracy of measurement and such layers become simply bedding planes. At this stage, the stratification cycle may still contain a measurable thickness of marl at its base and the grouping of beds is still accentuated by this lithological change. From about 21 m above the base, shale layers separating the stratification cycles become very thin or disappear. Despite this, there is still the impression that beds occur in packets of around 1 m in thickness, but it is difficult to fix precise stratification cycle boundaries. From approximately 3500 to 4400 cm, the cyclic classification is uncertain. Stratification cycles are again very obvious with the appearance of black chert. Such chert bands are more likely to occur near the base of cycles and are often associated with marl bands, which may in fact have been partly replaced by chert.

When measuring the section in the field and without any analysis, 73 stratification cycles were recorded in the Contessa section and 75 in the Petrano section. These numbers are likely to be too high. Cyclicity close to the Bonarelli horizon is very badly developed and bedding in the anoxic interval of the Bonarelli is much thinner and very irregular. No evidence was found to suggest that the Bonarelli contains cycles that are comparable to the cycles in the lower part of the Scaglia. However, de Boer (1982) was able in his sections to find between 23 and 33 couplets of different lithology or colour and was able to estimate the duration of the Bonarelli interval as 0.5 Ma.

THE ANALYSIS OF THE STRATIFICATION PATTERN

One of the major problems in the study of suspected Milankovitch cycles is finding a suitable method to describe sediment variation.

The first 9 m of the sections could be analysed by using marl percentages based on field estimates of the composition of fixed intervals. However, in the compositionally quite monotonous higher part of the succession, the only immediately measurable quantity is the bed thickness. If bedding represents at least approximately equal time intervals, then thickness would be proportional to sedimentation rates and this would be a very valuable parameter. Unfortunately, the bedding in the Scaglia may contain several orders of thickness and it also appears that the intensity of the process which eventually led to the formation of bedding planes was quite variable. The heterogeneity of bedding is seen in the frequency distribution of bed thicknesses, which is bimodal with a mode at 8–10 cm and a less clearly developed mode at 15–20 cm. The variances of bed thickness are therefore relatively high (see Table 1).

The classification of bedding planes into normal and subsidiary ones is clearly subjective and also depends on the state of weathering in the exposed sections. It is not therefore possible to construct a time scale or indeed any scale using bed numbers as units. The alternative to bed-by-bed plotting of thicknesses is to consider bed-thickness as a function of stratigraphic thickness. This can be done either by averaging thicknesses over some defined constant interval or, as has been done in this study, by taking the measuring points so close that not a single bed is missed. In its simplest form, one can consider each layer, regardless of its lithology, as a 'bed'. Different sedimentation models can be introduced by giving different weights to different lithologies. For example, the deposition of shale can be accentuated by giving the marl bed negative thickness values or by multiplying limestone sedimentation by a constant factor. The effect of such manipulations can alter the shape of the resulting curves considerably.

In the present analysis, thickness values have been taken at 1-cm intervals, and since this is the accuracy of the original measurements no information has been lost. Thickness indices have been constructed by treating both normal and subsidiary bedding planes as equal. Thicknesses of shale or marl layers greater than 1 cm have been given negative values of half their thickness and the resulting thickness index can vary between +20 and −20. This method was adopted after considerable experimentation and its advantage is that the resulting curve is a step function which fairly accurately mimics the actual exposure (Fig. 2). The sedimentological meaning of this index, however, is not very well defined. If the beds did actually represent equal time intervals, then the curve would show sedimentation rates as a function of stratigraphical position and not, as in the hypothetical bed-by-bed plot, as a function of time. Since the time values are unknown, the thickness indices are used in the hope that they somehow represent the conditions of sediment formation. Since the recognition of

Table 1. Variation of bed and cycle thicknesses

	Mean thickness (cm)	Coefficient of variation (%)	Number
Combined limestones and underlying marl beds			
Contessa highway	12.0	67.0	543
Monte Petrano	13.1	60.4	499
Cycle thicknesses			
Contessa highway	99.8	29.7	64
Monte Petrano	100.6	28.5	64

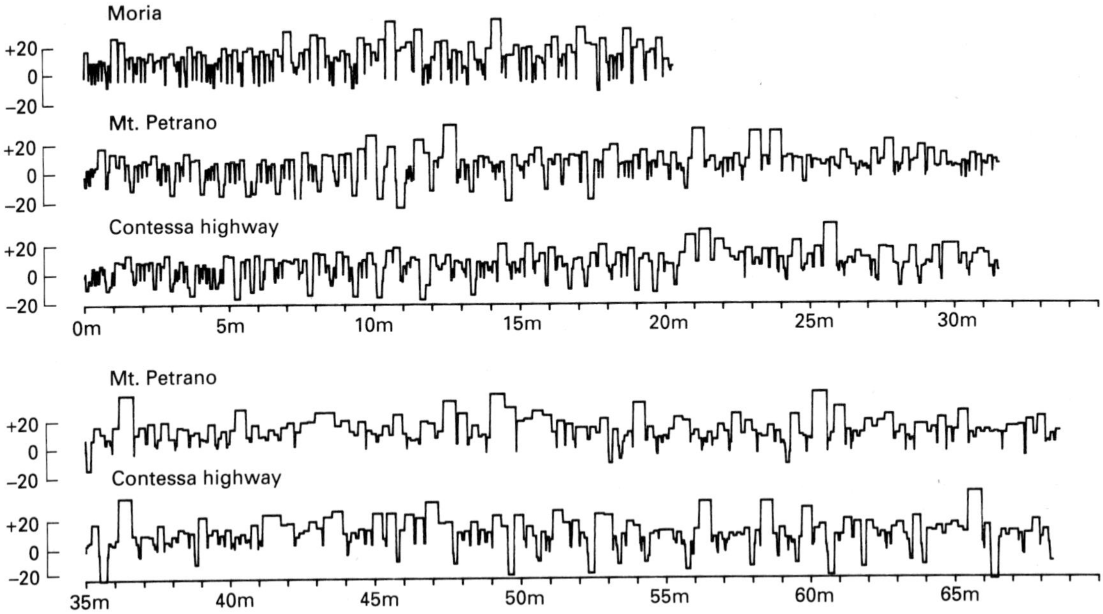

Fig. 2. Thickness index curves of the measured sections. Sampling points are at 1-cm intervals.

bedding depends partly on weathering and other uncontrolled fluctuations, the stratification cycle cannot be clearly defined by the shape of the thickness index curve. In parts of the section with upward-thickening beds, cycles are indicated by an asymmetric curve with a maximum near the top of the cycle boundaries. In other parts, the central part of a cycle is more massive and the index curve can be nearly symmetrical. In other cases, two peaks may occur in one cycle. This variability of the thickness index curve must be taken into account in any further analysis.

POWER SPECTRAL ANALYSIS

A short inspection of the sections in Fig. 2 will show that the sequences are far from stationary. Not only does the marl content decrease rapidly towards the younger parts of the section, but the pattern of bedding changes several times and the cycle length increases. The effect of this is that the power spectral analysis of the complete section gives average results which are difficult to interpret and which have only limited geological meaning.

Since the data in this study are like many stratigraphic time series in consisting of step functions, Walsh transforms were used in the analysis. Power spectra were averaged to minimize phase bias. Walsh spectral analysis is based on sequency rather than frequency. The term sequency (Beauchamp, 1984) is used to describe a periodic repetition rate which is independent of the waveform. In the spectral analysis, it is the sequence of zero-crossings of a bed thickness curve after subtracting the mean. The positions of zero-crossings within a cycle depend on the shape of the curve and if such shapes are variable, a number of peaks can be generated in the power spectrum.

A reasonably representative spectrum of the sequence can be obtained by averaging the periodogram of four non-overlapping subsections, each with a length of 16.25 m (Fig. 3). To obtain a better scaling, the lowest two frequencies have been omitted. This leaves the highest frequency peak at 18 cycles per subsection which corresponds to 72 cycles for the complete Scaglia Bianca below the Bonarelli horizon. This peak clearly represents the basic stratification cycle, which has been interpreted as the 100-ka eccentricity cycle. Two further maxima are found at 5 and 42 cycles per subsection. When the basic cycle is set equal to 1.0 then the ratios of the two

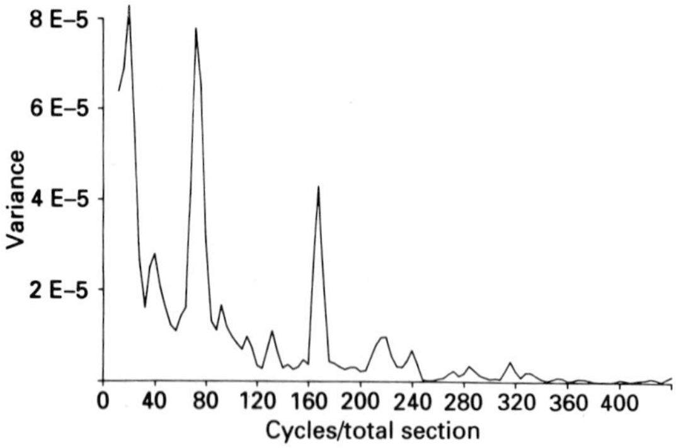

Fig. 3. Walsh power spectrum of the thickness indices of the Contessa highway section. Average of four 16.25-m long subsections.

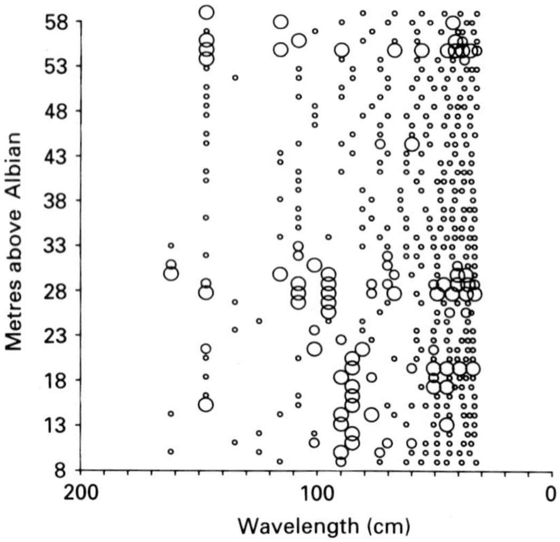

Fig. 4. Maxima of power spectra based on a 16-m wide moving window. The vertical scale gives the position of the window's centre and the horizontal scale is calibrated in wavelengths. Large circles represent maxima with 50% of the red noise, intermediate circles represent maxima with 25–50% variance of the red noise and small circles represent maxima with 10–25%.

additional maxima are at 3.6 and 0.42. These come close to the periods for the 400-ka eccentricity cycle and the 41-ka obliquity cycle. However, the latter, which is approximately one-half the basic cycle, is very uncertain. The variability of bed development makes the thickness index used in this analysis more and more unreliable with higher frequencies. The earlier mentioned tendency of thickness maxima to shift from the centre to the top of cycles could be partly responsible for the higher frequency maximum.

It was found that spectra based on Fourier transforms gave essentially the same maximum for the stratification cycle, but here the higher frequencies are dominated by the harmonics of this cycle and no evidence for the problematic obliquity cycle was found using this method.

Examining the high-frequency spectra on an expanded scale, maxima corresponding to wavelengths of 13–25 cm and maxima in the range of 5–8 cm are found. These represent clearly the thicknesses of the beds, and the two groups of maxima reflect the bimodality of the thickness distribution which was found earlier. The two groups roughly represent the two types of bed caused by normal and subsidiary bedding planes as classified during field observations.

To investigate the changes of cyclicity in the section, a window 16-m wide (1024 measuring points) was used to calculate spectra at 1-m intervals. The maxima of these staggered spectra are given in Fig. 4. In this diagram, spectral power is plotted against wavelength and not frequency to illustrate the changes in sedimentation rates which probably caused the frequency shifts. The basic stratification cycle near the base of the section is found at a thickness of 85 cm. The thickness of this cycle is between 80 and 90 cm up to about 20 m above base and at this point changes fairly abruptly to a thickness of 100–120 cm. From 33 to 53 m above the base, cyclicity is not well developed and the maxima are much lower compared with the red noise component. In the higher parts of the section, cycle thicknesses increase to 120 cm, representing an increase in sedimentation rate which is also found by direct measurement of bed thicknesses (see later).

THE FILTERING OF SECTIONS

Decomposing the time series into Walsh functions has the great advantage that filtering methods are made very simple and effective. The method used in this study is known as sequence-based vector filtering (Beauchamp, 1984). Walsh transforms are used to obtain the full number of coefficients which represent the series and filtering is simply achieved by setting coefficients corresponding to unwanted frequencies to zero. Renewed transformation will yield the filtered signal. In contrast to frequency

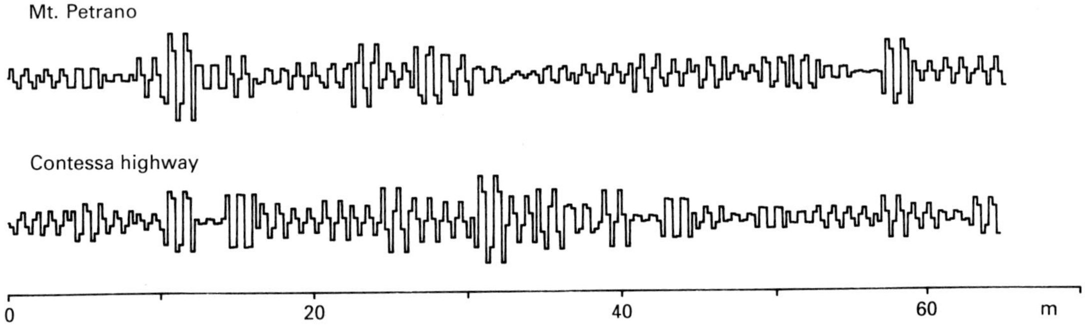

Fig. 5. Contessa highway and Monte Petrano measurements filtered with a bandwidth of 60–70 cycles.

filtering with trigonometric functions, there is always a finite number of Walsh functions which fully represents the series.

It is an essential feature of such filtering that it involves a restriction in the cycle range which can be recognized. Because the waveform in Walsh functions does not mark unit spaces, very narrow filters can produce results where square waves are arranged in regular groups resembling higher order cycles or bundles. Such effects can be recognized simply by generating Walsh functions of similar orders and comparing them with the filter results.

Figure 5 shows the two sections of Petrano and Contessa after filtering with a bandpass filter of 60–80 cycles. Depending on the interpretation, between 65 and 67 cycles can be counted for the complete interval. If one accepts the tentative correlation with de Boer's section, then the Albian–Cenomanian boundary is at cycle 7. Large amplitudes in the filtered data indicate that the series is well represented by the chosen band. Irregular cycles or cycles which are outside the filtered band have low amplitudes. Such variations could be part of the original signal, in which case both sections should behave in a similar way. The variation can also be caused by the degree of preservation in the outcrops and by errors of measurement. The two sections are similar to a limited degree. The agreement is quite good between 0 and 20 m and again reasonable from 50 m upwards. The middle part of the section, which also showed very low spectral maxima (see Fig. 4), appears to have some quite divergent developments. In this part of the section, recognition and classification into cycles is difficult.

Using a much wider bandpass filter allows a wider variation of cycle lengths and it is possible to show the sedimentary variation without imposing a definite cycle period. Choosing a filter of 20–150 cycles per total section gives in this case a very good representation of the sequence (Fig. 6), and the cycles which can be deduced from this record coincide well in most cases with cycles which were recognized during measuring in the field. The latter are indicated by vertical lines in the diagram. The total number of cycles in the filtered sections is 64, but it is possible that some cycle boundaries have been missed or misinterpreted.

When examining the low-frequency component of the thickness index, it is again important not to impose the number of 'cycles' by choosing a filter of too narrow a frequency range. In the present data, a good 'trend' was obtained by using a low-pass filter with a cut-off frequency of 15 cycles and by giving the resulting step function a smooth appearance using running averages (Fig. 7). The two resulting curves are certainly similar overall but once again not similar in very great detail. The similarity is largely due to two maxima which are approximately in the same position but which are caused by different sedimentological events. The lower maximum is largely caused by thicker marl beds in this part of the section, whereas the higher maximum is due to thicker limestone beds and is due particularly to one exceptionally thick bed which may be turbiditic in origin. The similarity of the thickness index trends with smoothed SP logs of wells in the Adriatic (Melnyk & Smith, 1989) is only very slight (Fig. 7) and is largely caused by the decreasing frequency of marl beds in the higher parts of the sequence. It would be difficult to associate any of the observed trend variations with some low-frequency orbital

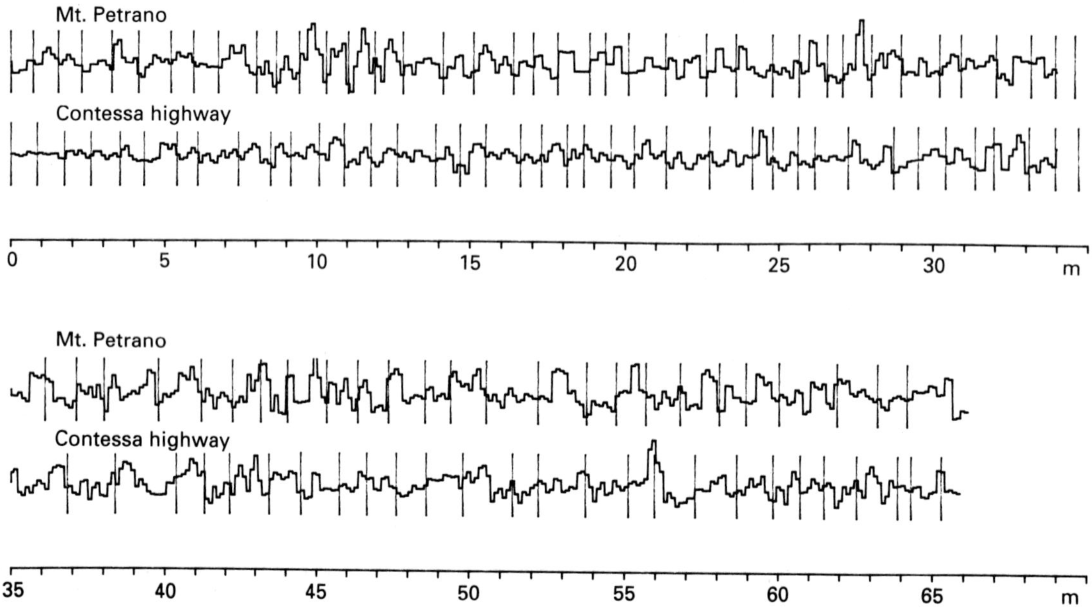

Fig. 6. Contessa highway and Monte Petrano measurements filtered with a bandwidth of 20–150 cycles. The vertical lines represent the positions of cycle boundaries as determined from field notes and from the comparison of sections.

cyclicity. For example, the trend of the Contessa section contains three maxima which appear at roughly equal intervals. However, if the basic eccentricity cycle is used as a time measure, the length between the maxima is found to be 2.8, 2.0 and 1.4 Ma. This hardly fits into the concept of an astronomically controlled cycle.

CYCLOSTRATIGRAPHIC CORRELATION

Since the analysis of both the Contessa and Petrano sections led to a division into almost exactly the same number of cycles and since the beginnings and the ends of the sections are well correlated, a cycle-by-cycle correlation should be possible. This can indeed be done with some confidence for the first 35 m of the section. Between 35 and 41 m, the identification of cycles is more uncertain, and although the interval yielded the same number of cycles in both sections, the interpretation may be wrong. This particular interval is also disturbed by a small fault in the Petrano section, which again may have led to some slight errors in measuring the section. In the higher part of the section, from 41 to 65 m, the correlation is again good.

The lower part of the section contains a very well-defined shale band at 1204 cm (Petrano) and 1186 cm (Contessa) which permits a very accurate correlation. Several prominent chert bands and a shale band at 4759 cm (Petrano) and 4712 cm (Contessa) provide additional checks in the higher part of the section.

De Boer (1983) showed that sections which are as much as 80 km apart can be accurately correlated on a bed-by-bed basis. This was not achieved for the two sections in the present study. Although the number of beds in correlated cycles is often similar, there are sometimes considerable discrepancies. It is often found that the bed thickness index has a maximum in one cycle corresponding to a minimum in the correlated cycle. This may be partly due to different weathering in the outcrop but could also be a genuine difference in the sequences.

The cycle thicknesses of both sections again show several similarities (Fig. 8) and the plots could be used for an approximate correlation. In both sections, there is an increase in cycle length towards the younger parts of the sequence. Fitting a straight line to the data, a cycle thickness of 80 cm (Petrano) and 86 cm (Contessa) is found at the base and 120 cm (Petrano) and 113 cm (Contessa) at the top of the sections. This corresponds to an increase of 3–4 mm/ka throughout the Cenomanian.

The coefficient of variation of cycle length is 29%. The variability of the sedimentation rates can be roughly estimated by comparing the variance of the 65 stratification cycles with the variance of 65 eccentricity cycles which were calculated using Berger's algorithm (1978). The time variance of the astronomical cycles is 374 ka, and the variance of the combined sections is 848 cm. Assuming a constant average sedimentation rate of 1 cm/ka it is found that a variance of 473 cm has to be explained by the irregularities of the sedimentation rates. The fluctuations of the sedimentation rates therefore account for 21% of the observed thickness variation. Such fluctuations of sedimentation rates prohibit the resolution of cyles like the 100-ka eccentricity cycle into its 126-ka and 98-ka components.

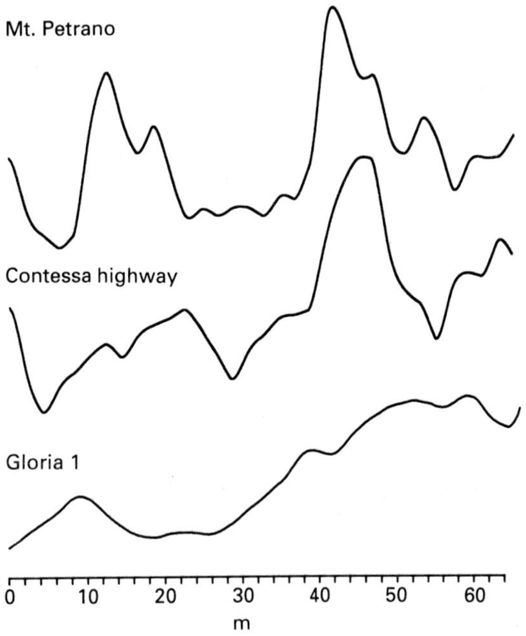

Fig. 7. The 'trend' of the measured sections compared with the smoothed SP curve of the offshore Adriatic well Gloria 1 (see text). The vertical axes have been arbitrarily scaled, for better comparison.

CONCLUSIONS

A study of cyclic limestone sedimentation without palaeontological and petrographic analysis is far from ideal but certain conclusions can be drawn from the field evidence.

The interval which has been studied possibly includes the very top of the Albian and contains the complete Cenomanian. Absolute dates suggest a duration of the Cenomanian of 6.6 Ma (Harland *et al.*, 1990). Assuming a constant time of 100 ka for the basic cycle gives a duration of 6.5–7.2 Ma for the interval investigated. Using the tentative correlation with the section given by de Boer (1983), the Albian–Cenomanian boundary is at 720 cm. According to this, the top Albian is represented by seven cycles and the timespan for the Cenomanian would therefore be 5.8–6.2 Ma. The uncertainty of the cyclostratigraphic estimate is due to the difficulty in recognizing and correctly identifying eccentricity cycles. This applies in particular to the middle part of the section. More detailed investigations may help

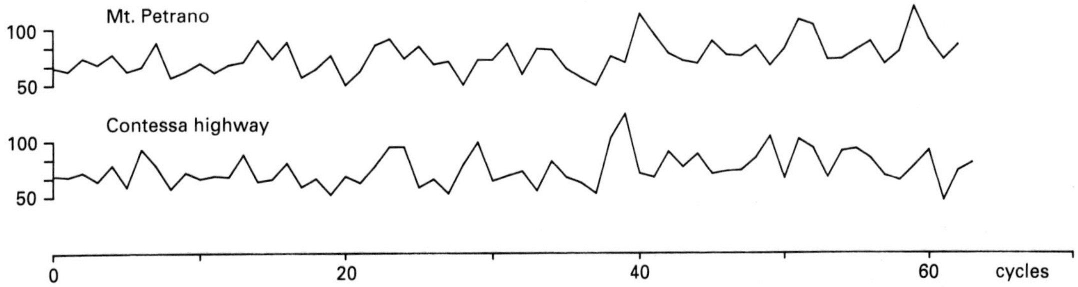

Fig. 8. Thicknesses of the successive cycles in the two measured sections. The vertical scale is thickness of cycles in centimetres, and the horizontal scale is the cycle number and can be regarded as an 'average' time scale. The displacements of some maxima could indicate errors in correlation.

to improve this situation. It is, however, possible that the middle Cenomanian was a time in which the eccentricity cycle was less effective. The poor evidence for an effective obliquity cycle may be largely due to the inferior observational material of this study.

A possible hypothesis which could explain the available evidence is that the climatic signal controlled the influx of non-carbonate material and that the formation of bedding planes is more likely when this influx is high. If the supply of non-carbonate material pulsates with the 21-ka precessional cycle and in addition fluctuates with the 100-ka eccentricity cycle, then a pattern similar to the one found would be generated. Again, if the formation of subsidiary bedding planes is favoured by high clay content, more bedding planes would be found at the base of the cycle, which generally starts with a marl layer. It is envisaged that there is no sharp and well-defined threshold which leads to the formation of bedding but that there are levels which fluctuate at random in relation to the non-carbonate fraction. Whilst the non-carbonate supply is regarded as being the most important factor, variable carbonate production and even dissolution could have contributed. However, fluctuating carbonate production cannot explain the distribution of marl throughout the section. If the cycles with no marls near the top part of the section resulted from dilution of the non-carbonate fraction, they would have to have enormous thicknesses. Although it has been suggested that tectonic movement could have removed the clay in the top parts of both sections, this seems unlikely.

It is impossible, without petrographic examination, to decide how much diagenesis has contributed towards the formation of bedding planes. However, it is certain that the much-discussed migration of carbonate from the marls into the centre of the limestone beds was not effective (if it happened at all), since there is no positive correlation between the thicknesses of marls and the thicknesses of succeeding limestone beds. Furthermore, limestones in the practically marl-free higher part of the sequence are just as well cemented as in the lower parts, without having any marl reservoirs from which they could derive the carbonate.

As far as methods of stratigraphic time-series analysis are concerned, power spectral analysis was found to be of little use when it came to the examination of long sequences. In particular, the recognition of cycles in the one to several million year range is difficult because sedimentation rates change over such time intervals. More important than strict spectral analysis is filtering. Indeed, a kind of spectral analysis can be obtained by filtering long sequences with narrow bandpass filters in which the pass frequency is systematically increased. In such filter procedures, Walsh transforms were found to be vastly superior to Fourier transforms.

A very important part of this study has been the parallel examination of two sections which are geographically separated by approximately 20 km. In comparing such sections, random variations can be eliminated and what is possibly even more important, any wrong conclusions which might be based on them can also be discounted.

Comparison of individual cycles has shown that the thickness index curves will reflect the cyclic behaviour of the sediment in space, but that the actual shape of the curve can be highly variable. For example, the maximum thickness value may be at the base of a cycle in one section but could be in the centre or at the top of the cycle in the second section. The symmetry or asymmetry of such curves therefore does not tell us anything about the mechanism of cycle formation. The idea that the symmetry of a lithological cycle reflects the symmetry of astronomical cycles is too simplistic to be considered. The trend variation of the two sections could be correlated only very approximately and no evidence of an astronomical cycle longer than the 400-ka eccentricity cycle has been found.

REFERENCES

ARTHUR, M.A. (1976) Sedimentology of Gubbio sequence and its bearing on paleomagnetism. *Mem. Soc. Geol. It.* **15**, 9–20.

BEAUCHAMP, K.G. (1984) *Application of Walsh and Related Functions with an Introduction to Sequence Theory*. Academic Press, London.

BERGER, A.L. (1978) *A simple algorithm to compute long term variations of daily or monthly insolation*. Contrib. 18, Inst. Astronomie Géophysique G. Lemaître, Univ. Catholique de Louvain.

DE BOER, P.L. (1982) Cyclicity and the storage of organic matter in Middle Cretaceous sediments. In *Cyclic and Event Stratification* (Eds Einsele, G. & Seilacher, A.). Springer Verlag, Berlin, pp. 456–475.

DE BOER, P.L. (1983) Aspects of Middle Cretaceous pelagic sedimentation in southern Europe. *Geol. Ultraiectina* **31**.

FISCHER, A.G. & HERBERT, T.D. (1988) Stratification rhythms: Italo-American studies in the Umbrian facies. *Mem. Soc. Geol. It.* **31**, 45–51.

Harland, W.B., Armstrong, R.L., Cox, A.V., Craig, L.E., Smith, A.G. & Smith, D.G. (1990) *A Geologic Time Scale 1989*. University Press, Cambridge.

Melnyk, D.H. & Smith, D.G. (1989) Outcrop to subsurface cycle correlation in the Milankovitch frequency band: middle Cretaceous, central Italy. *Terra Nova* **1**, 432–436.

Premoli Silva, I. & Erba, E. (1988) Field guide of Umbria-Marche area September 16, 1988. Working group 3 – Cyclostratigraphy, Perugia, Italy.

Premoli Silva, I., Ripepe, M. & Tornaghi, M.E. (1989) Planktonic foraminiferal distribution record productivity cycles: evidence from the Aptian–Albian Piobbico core (central Italy). *Terra Nova* **1**, 443–445.

Schwarzacher, W. (1952) Zum kartieren mit sedimentären Rhythmen. *Verh. Geol. Bundesanstalt (Austria)* 2.

Schwarzacher, W. (1958) The stratification of the Great Scar Limestone in the Settle district of Yorkshire. *Liverpool Manchester Geol. J.* **72**, 124–142.

Schwarzacher, W. (1987) The analysis and interpretation of stratification cycles. *Paleoceanography* **2**, 79–95.

Schwarzacher, W. & Fischer, A.G. (1982) Limestone–shale bedding and perturbations in the earth's orbit. In *Cyclic and Event Stratification* (Eds Einsele, G. & Seilacher, A.). Springer Verlag, Berlin, pp. 72–95.

Tornaghi, M.I., Premoli Silva, I. & Ripepe, M. (1989) Lithostratigraphy and planktonic foraminiferal biostratigraphy of the Aptian–Albian 'Scisti a Fucoidi' in the Piobbico core, Marche, Italy: background for cyclostratigraphy. *Riv. It. Paleont. Strat.* **95**, 223–264.

Spec. Publs Int. Ass. Sediment. (1994) **19**, 99–107

Milankovitch periodicities recorded in Cretaceous deep-sea sequences from the Southern Alps (Northern Italy)

M. CLAPS *and* D. MASETTI

Dipartimento di Scienze Geologiche e Paleontologiche dell'Università di Ferrara, Corso Ercole I d'Este 32, 44100 Ferrara, Italy

ABSTRACT

Power spectral analysis has been applied for the first time to Cretaceous pelagic and hemi-pelagic sections in the Southern Alps (Venetian Prealps and Dolomites), to look for orbital–climatic (Milankovitch) forcing cycles. At the Castagnè section (middle Cenomanian), bed thickness, bedding plane position and codified lithological-class time series have been analysed to obtain spectra showing periodicities in the range of the orbital eccentricity, obliquity and precession cycles. At the Ra Stua section (Barremian), a bed thickness time series has been investigated which shows variable main peak values relative to all the three Milankovitch cycles; the ichnogenera presence/absence time series shows a dominant frequency corresponding to the short eccentricity cycle. In the Cismon Valley section (late Cenomanian), analysis of bed thickness and calcimetric time series show main spectral peaks with wavelength ratios similar to those existing between the short eccentricity, obliquity and precession cycles (100, 41 and 21 ka). This allows us to propose a sedimentation rate (about twice the value estimated by the existing time scale), calculated by matching the three most significant ranges of peaks to these three cycle time values.

INTRODUCTION

The spectral analysis technique has been applied, for the first time in the Cretaceous of the Southern Alps, to three pelagic and hemi-pelagic sections, located in the Venetian Prealps and in the Dolomites (Fig. 1). These sections were deposited during the Barremian–Cenomanian and show, as a general feature, an alternation of carbonate-rich and carbonate-poor layers.

This rhythmicity is common in a variety of pelagic and hemi-pelagic sequences and is usually evidenced by a cyclic thickness pattern with a hierarchy of couplets and bundles (Schwarzacher & Fischer, 1982; Fischer & Schwarzacher, 1984), a regular oscillation in the $CaCO_3$ content, and/or a variation in the colour of the successive beds (Herbert & Fischer, 1986). The possible forcing mechanisms that may influence this kind of cyclicity are productivity, dilution and redox cycles. These can be reflected in regular variation of $CaCO_3$ and terrigenous supply, and in variable availability of oxygen at the sea floor (Einsele, 1982; Arthur *et al.*, 1984; Premoli Silva *et al.*, 1989). It has been demonstrated (Fischer, 1986; Fischer *et al.*, 1990) that the time values proposed for these various hierarchical ranks often fall in the range from 10 to 400 ka. This suggests a link between the stratigraphic cycles and the periods of the orbital perturbations predicted in the Milankovitch theory (Berger, 1984). The cycles of precession (19–23 ka), obliquity of the Earth's rotation axis (41–54 ka), and eccentricity of the orbit around the Sun (short term: 98–123 ka; long term: 413 ka) (Berger, 1984) result in cycles in caloric insolation. These become climatic cycles by the interaction of various parameters in the Earth's climatic system and through amplification by feedback and threshold effects (Fischer *et al.*, 1990). The stratigraphic response of orbital–climatic cycles can be detected in various depositional environments, provided that these are sufficiently sensitive to record the changing environmental conditions.

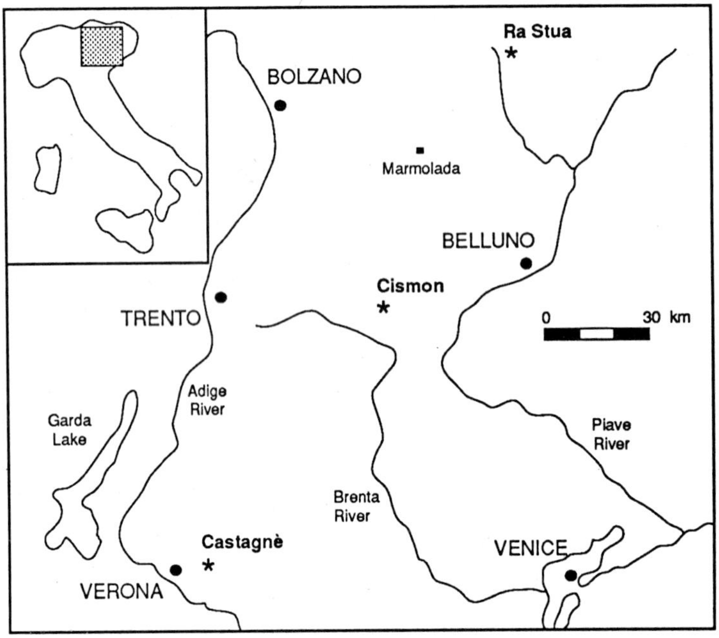

Fig. 1. Map showing the location of the three studied sections described in the text: Castagnè, Ra Stua and Cismon Valley.

Spectral analysis is used to detect lithological periodicities in the measured sections, in order to search for orbital–climatic Milankovitch-type cycles. Two different processing routines have been applied: the fast Fourier transform (FFT) algorithm (Bendat & Piersol, 1971; Pestiaux & Berger, 1984; Ripepe, 1988) has been used to process bed thickness, bedding plane position and calcimetric time series; the fast Walsh transform (FWT) algorithm (Weedon, 1985) has been used to analyse codified time series, such as coded lithotype or the presence/absence of some ichnogenera. Figure 2 shows the bed thickness time series of the three sections studied. The results of the data processing are amplitude and power spectra, showing frequency peaks with different levels of energy. Only those passing the 95% confidence level are indicated in the diagrams and later examined in order to identify the periodicities. They are computed first in terms of space measurements and are later converted into time values, using the average sedimentation rate inferred for each section by the time duration of stratigraphic units (stages or biozones) and their thickness in the section studied. To prevent distortion related to the form of the input signal a triangular smoothing filter was used as a way of obtaining a more sinusoidal-like wave (Bendat & Piersol, 1971; Schwarzacher, 1989). The stability of the most relevant peaks can be tested by splitting the original time series into two or more subsections (Weedon, 1985, 1989). They are then processed to generate spectra with different resolution levels. Since we currently lack specific data for interpretation in terms of productivity or dilution cycles, we will refer to other authors on this point (de Boer, 1982; Einsele, 1982; Arthur *et al.*, 1984; Fischer, 1986; Herbert & Fischer, 1986; Premoli Silva *et al.*, 1989).

MILANKOVITCH CYCLES IN THE CRETACEOUS OF THE SOUTHERN ALPS

Figure 3A (from Channell *et al.*, 1979) represents the simplified stratigraphy of the Cismon section, and can be roughly applied to the Castagnè section, apart from thickness. As is clearly shown by the stratigraphic column (applicable to the main part of the Venetian Prealps), the whole Cretaceous succession is characterized by pelagic micritic deposits containing a variable amount of clay. On the basis of colour and clay content, this succession has been divided into the Biancone, Scaglia Variegata and Scaglia Rossa Formations. The first section was measured in the Venetian Prealps (Castagnè,

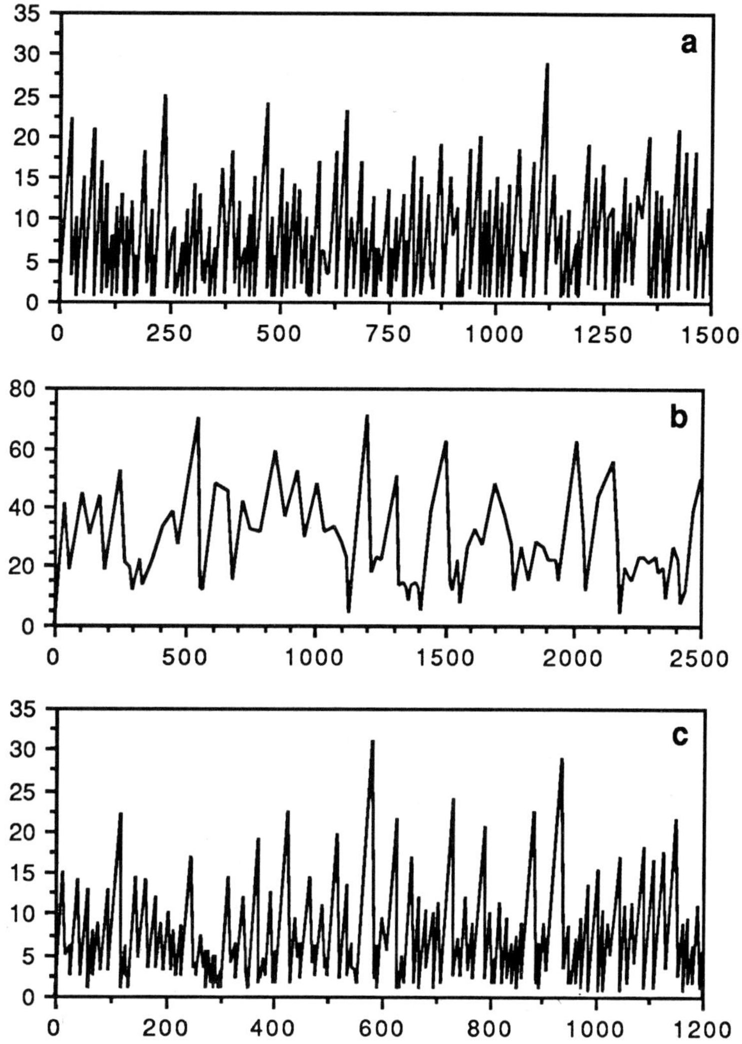

Fig. 2. Bed thickness time series for the three studied sections: (a) Castagnè; (b) Ra Stua; (c) Cismon Valley. The vertical axes indicate the bed thickness (cm) and the horizontal axes refer to the cumulative thickness (cm).

Lessini Mountains) and belongs to the Middle Cenomanian (Scaglia Variegata Formation). An average sedimentation rate of 1.7 cm/ka was computed, on the basis of the presence of the whole *Rotalipora reicheli* Zone and its estimated time duration (about 0.5 Ma; Haq *et al.*, 1987). It is characterized by an alternation of micritic white and dark green marl beds (Fig. 4). Bed thickness and bedding plane positions have been processed by FFT and lithological class time series by FWT (Fig. 5a–c). The spectra obtained exhibit periodicities in the range of eccentricity (111–83 ka), obliquity (60–41 ka) and precession of the equinoxes (24–18 ka). Figure 5c reveals the presence of another frequency group with an average value of 12 ka. This sub-Milankovitch frequency range is also common to the next sections and was observed by Fischer (1986), Napoleone & Ripepe (1989) and Premoli Silva *et al.* (1989). It does not derive from data processing artefacts, but may represent a real frequency higher than those commonly expected in the Milankovitch band. We suggest, following de Boer (1982) and Herbert & Fischer (1986), that these lithological rhythms are the sedimentological result of the fluctuating productivity of calcareous nannofossils. For example, high-amplitude precession cycles may influence the development of oceanic upwelling zones. In such a phase the nannofossil fauna will flourish,

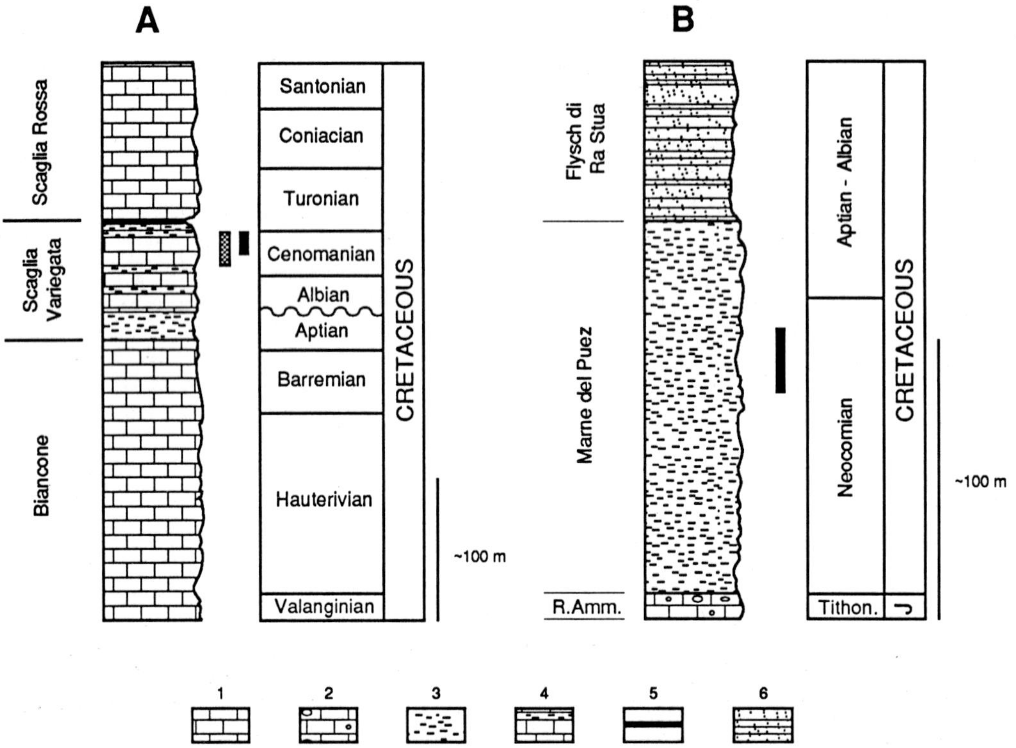

Fig. 3. Column A represents the simplified stratigraphy of the Cismon Valley (modified from Channell *et al.*, 1979), which is also applicable to the Castagnè section. The black bar indicates the stratigraphic range of the Cismon section. The grey bar refers to the Castagnè section. Column B is the generalized stratigraphy of the Ra Stua area (modified from Scudeler Baccelle & Semenza, 1974). The black bar shows the stratigraphic range of the measured section. Symbols: 1, limestone; 2, nodular limestone; 3, marls; 4, limestone with thin marl intercalations; 5, black shales; 6, sandstone.

with the deposition of the carbonate-rich semi-couplet as a consequence.

In the Dolomites area, both the Biancone and Scaglia Variegata Formations are replaced by the Marne del Puez Formation. Figure 3B, modified from Scudeler Baccelle & Semenza (1974), shows the generalized stratigraphy of the Ra Stua section (Barremian). Here the rhythms investigated are characterized by lithological couplets of limestone and marly limestone, changing in colour from white to light grey (Fig. 6). The estimated average sedimentation rate of 2.5 cm/ka is based on the biostratigraphical evidence (R. Busnardo, personal communication, 1986) that the whole section spans one ammonite zone (*Barremense* Zone). Its duration is estimated at about 1 Ma (Van Hinte, 1976; Harland *et al.*, 1982, 1990). Spectra were computed from bed thickness and *Zoophycos* presence/absence time series (Fig. 7a, b, respectively). Both time series show spectral peaks at 113–101, 56–36 and 21–17 ka (Fig. 7). Figure 7a shows a well-identified periodicity peak with a time-converted value of about 14 ka, rather less than the precession range components (19–23 ka). The spectrum computed on the presence/absence of *Zoophycos* indicates one dominant frequency in the eccentricity range, probably produced by periodic variation in the substrate firmness (Ekdale *et al.*, 1984). Based on the Haq *et al.* (1987) time scale, the duration of the Barremian is about one-half of the value estimated by other authors. The values shown in Fig. 7 would become half of the previous estimates, making them shorter than the Milankovitch periodicities. As suggested for the Cismon section, the wavelength ratios are in tune with those of the Milankovitch frequencies, supporting the longer estimates for the duration of Barremian time. The effects of all three Milankovitch orbital–climatic

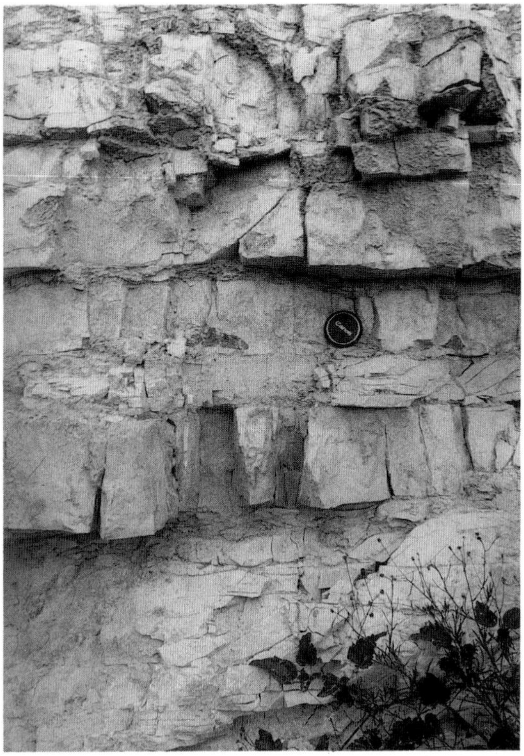

Fig. 4. Limestone–marl couplets of the Castagnè section.

cycles are evident in the bed thickness distribution spectrum. We suggest that they have been caused by the combined actions of $CaCO_3$ productivity and dilution cycles. A high precession index, which determines a stronger nutrient flux to the surface, is also responsible for global ice-mass reduction (high-stand phase), and a consequent decrease of terrigenous input. Lack of direct evidence of glaciation in Cretaceous time does not necessarily exclude development of restricted ice caps in the polar region (Fischer, 1986). These effects may thus account for the deposition of the carbonate part of each couplet.

The third section sampled is located in the Cismon Valley (Venetian Prealps) and belongs also to the Scaglia Variegata Formation (late Cenomanian, Fig. 3A). An average sedimentation rate of 0.6 cm/ka has been calculated on the basis of the chronological scale of Haq *et al.* (1987) and the presence of the whole *Rotalipora cushmani* Zone (estimated time duration: 1.8–1.9 Ma). It consists of a cyclic repetition of white bioturbated limestone and grey to black, fissile marly limestone (Fig. 8), corresponding to cyclic biogenic carbonate production and variable aerobic to anaerobic conditions on the sea floor (de Boer, 1982; Herbert & Fischer, 1986). The periodicities of the couplets and bundles are detected by FFT analysis of the bed thickness and calcimetric time series (Fig. 9); both spectra show stable and regular frequency peaks. Bearing that in mind although we are not able to find any obvious hiatus in the measured section, some intraformational truncation surfaces are locally present in under- and overlying portions of the same section. So we can assume that the mean sedimentation rate may not reflect good time conversion of the space periodicities, because some parts of the succession may have been lost at hiatuses. Nevertheless the computed spectra show wavelength ratios of spectral peaks similar to those existing between the known Milankovitch cycles in the Pleistocene records (1 : 2 : 5). We propose a different sedimentation rate (1.4 cm/ka) for the complete section, evaluated by matching the three most relevant groups of peaks to the time values of the short eccentricity, obliquity and precession cycles (Weedon, 1985). The values obtained in this way are about 100, 43–41 and 23–17 ka. The proposed time conversion shows in addition another frequency group in the sub-Milankovitch band (from 14 to 11 ka), as outlined in the previous sections and observed by other authors (Fischer, 1986; Napoleone & Ripepe, 1989; Premoli Silva *et al.*, 1989). In this way it may be possible to bypass the application of the existing geological time scales, which are subject to a variety of uncertainties. We assume that these Milankovitch cycles are recorded in the Cismon section by the effect of productivity cycles of calcareous nannofossils, in the same way as in the other sections studied.

SUMMARY AND CONCLUSION

In summary the spectral analysis of these Cretaceous sedimentological records, investigated in the Southern Alps, has provided evidence that the time values of the rhythmicities probably contain three main frequencies. These are of the same order of magnitude as the Milankovitch cycles (21, 41 and 100 ka), recognized widely as a consequence of climatic changes and variable solar insolation, forced by the orbital parameters (Berger, 1984). The oscillations in the climatic behaviour of the atmosphere–hydrosphere–biosphere system, greatly amplified

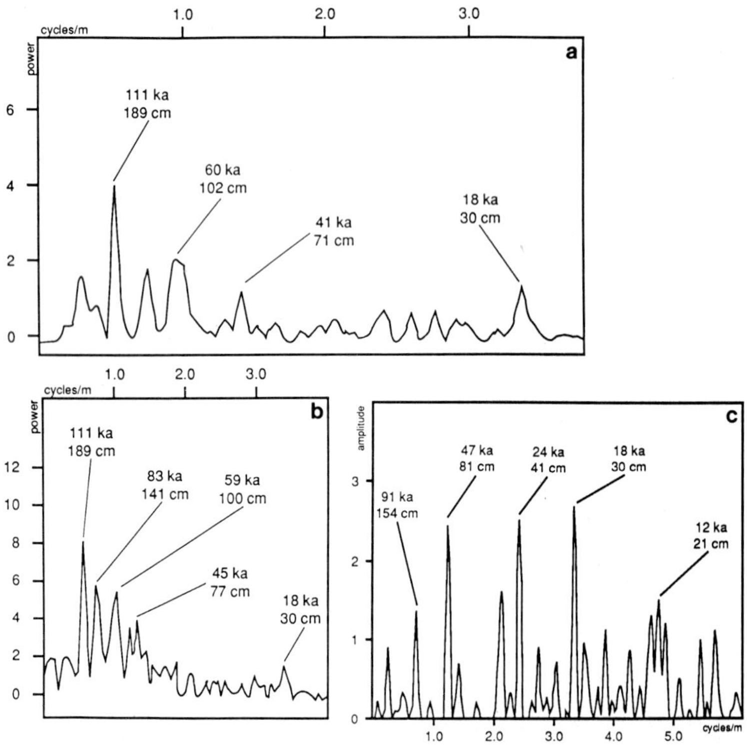

Fig. 5. Spectra of the Castagnè section computed from time series of (a) bedding plane positions, (b) bed thickness and (c) coded lithology; (a) and (b) are computed by FFT, (c) by FWT. Estimated sedimentation rate is 1.7 cm/ka. The periodicities shown are in the range of the orbital eccentricity (111–83 ka), obliquity (60–41 ka) and precession (24–18 ka) cycles.

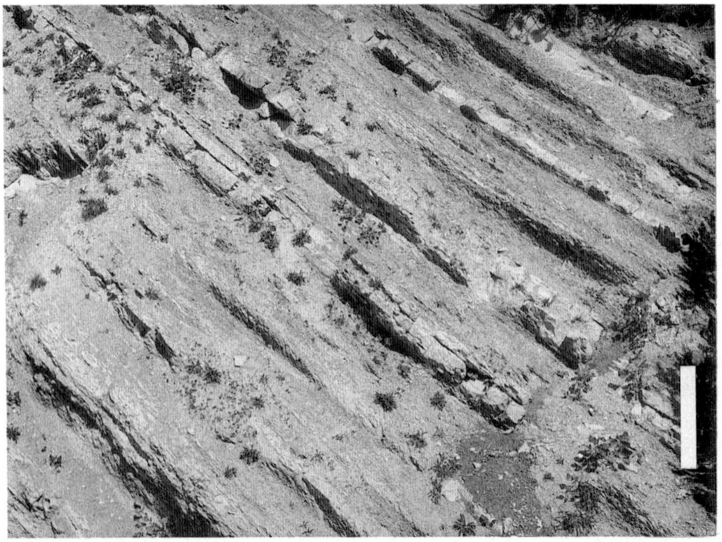

Fig. 6. Lithological rhythms of carbonate-rich and carbonate-poor beds in the Ra Stua section. The white scale bar on the right is about 1 m.

by feedback and non-linear effects (Fischer *et al.*, 1990), can be reflected in the sedimentological rhythms by productivity, dilution and redox cycles (Arthur *et al.*, 1984; Premoli Silva *et al.*, 1989). As outlined in previous works (Premoli Silva *et al.*, 1989; Ripepe, 1990), the eccentricity cycle shows in most cases the highest degree of relevance in the processed spectra. Another frequency, in the sub-Milankovitch range, occurs in the higher frequency part of some of the spectra. Similar periodicities are found in Pleistocene records and probably depend on non-linear responses of the system, forced by the known Milankovitch frequencies (Pestiaux *et al.*, 1988). In one of the cases studied we suggest a different way of interpreting the spectrum, apart from the sedimentation rate computed through the thickness–time ratio. In this case the frequency ratios between the most stable spectral peaks were found to be similar to those of the standard 100-, 41- and 21-ka Milankovitch cycles. Matching the three Milankovitch frequencies to the appropriate spectral peaks yields an alternative estimate of the sedimentation rate for the section. In this way new estimated values for the rate of sedimentation may be obtained, thus avoiding possible errors in the existing time scales.

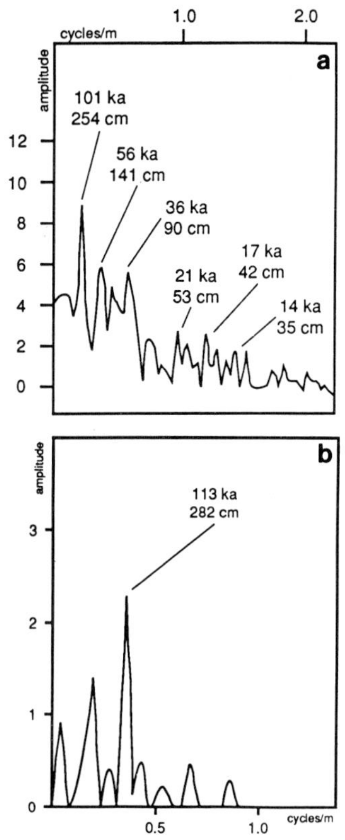

Fig. 7. Spectra of the Ra Stua section: time series based on (a) bed thickness and (b) presence/absence of the ichnogenus *Zoophycos*, processed using FFT and FWT respectively. The estimated sedimentation rate is 2.5 cm/ka. The results indicate periodicity peaks in a similar range to those of the Castagnè section: 113–101 ka (eccentricity), 56–36 ka (obliquity) and 21–17 ka (precession).

ACKNOWLEDGEMENTS

Work on this project has been supported by grants from the MURST (60%) and the CNR (40%). We are grateful to Francesco Pedrielli for assistance with setting up the spectral analysis and to Linda

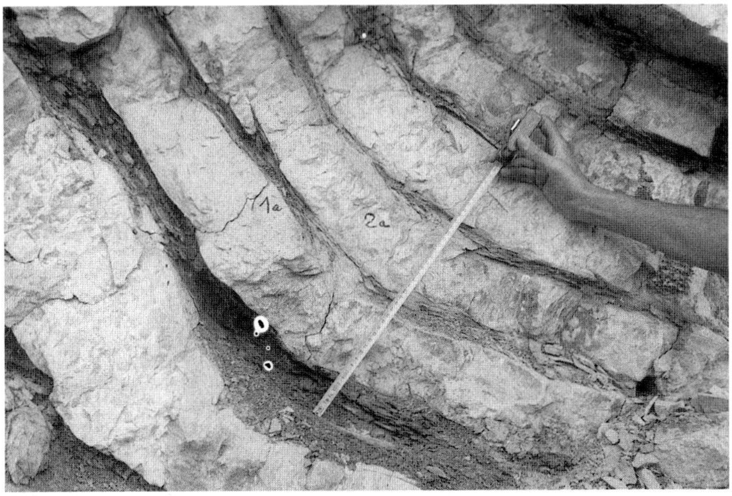

Fig. 8. Limestone–marl couplets, corresponding to precession cycles, in Cismon Valley section.

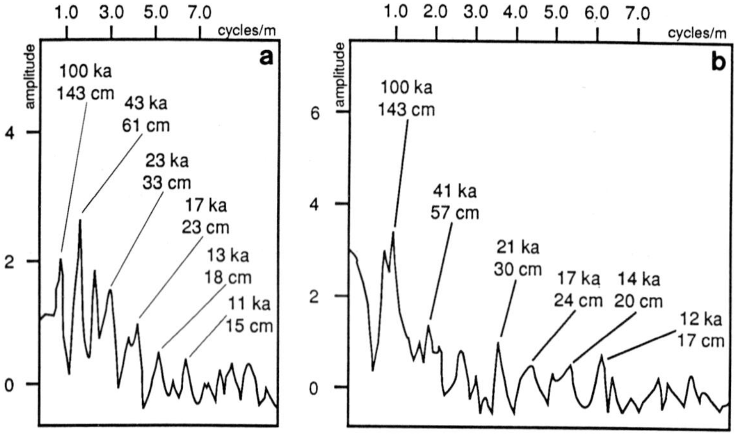

Fig. 9. Spectra of the Cismon Valley section: (a) bed thickness and (b) $CaCO_3$ time series analysed by FFT. The main groups of spectral peaks are in tune with the three principal Milankovitch periodicities (wavelength ratios of 1 : 2 : 5). As explained in the text a sedimentation rate of about 1.4 cm/ka has been proposed.

Hinnov for technical review. The English text was reviewed by William Zempolich. We also thank Poppe de Boer and Graham Weedon for useful reviewing and suggestions.

REFERENCES

Arthur, M.A., Dean, W.E., Bottjer, D. & Scholle, P.A. (1984) Rhythmic bedding in Mesozoic–Cenozoic pelagic carbonate sequences – the primary and diagenetic origin of Milankovitch-like cycles. In: *Milankovitch and Climate*, Part 1 (Eds Berger, A.L., Imbrie, J., Hays, J., Kukla, G. & Saltzman, B.). Reidel Publ. Co., Dordrecht, pp. 191–222.

Bendat, J.S. & Piersol, A.G. (1971) *Random Data: Analysis and Procedure*. John Wiley & Sons Inc., New York.

Berger, A. (1984) Accuracy and frequency stability of the Earth's orbital elements during the Quaternary. In: *Milankovitch and Climate*, Part 1 (Eds Berger, A.L., Imbrie, J., Hays, J., Kukla, G. & Saltzman, B.). Reidel Publ. Co., Dordrecht, pp. 1–39.

Channell, J.E.T., Lowrie, W. & Medizza, P. (1978) Middle and early Cretaceous magnetic stratigraphy from the Cismon section, Northern Italy. *Earth Planet. Sci. Lett.* **42**, 153–166.

De Boer, P.L. (1982) Cyclicity and storage of organic matter in Middle Cretaceous pelagic sediment. In: *Cyclic and Event Stratification* (Eds Einsele, G. & Seilacher, A.). Springer Verlag, New York, pp. 456–475.

Einsele, G. (1982) Limestone–marl cycles (periodites): diagnosis, significance, causes, a review. In: *Cyclic and Event Stratification* (Eds Einsele, G. & Seilacher, A.). Springer Verlag, New York, pp. 8–53.

Ekdale, A.A., Bromley, R.G. & Pemberton, S.G. (1984) *Ichnology – the use of trace fossils in sedimentology and stratigraphy*. Short Course 15. Soc. Econ. Paleont. Mineral., Tulsa, pp. 214–231.

Fischer, A.G. (1986) Climatic rhythms recorded in strata. *Annu. Rev. Earth Planet. Sci.* **14**, 351–376.

Fischer, A.G. & Schwarzacher, W. (1984) Cretaceous bedding rhythms under orbital control? In: *Milankovitch and Climate*, Part 1 (Eds Berger, A.L., Imbrie, J., Hays, J., Kukla, G. & Saltzman, B.). Reidel Publ. Co., Dordrecht, pp. 163–175.

Fischer, A.G., De Boer, P.L. & Premoli Silva, I. (1990) Cyclostratigraphy. In: *Cretaceous Resources, Events and Rhythms – Background and Plans for Research* (Eds Ginsburg, R.N. & Beaudoin, B.). Kluwer Academic Publishers, Dordrecht, pp. 139–172.

Haq, B.W., Hardenbol, J. & Vail, P.R. (1987) Chronology of fluctuating sea levels since the Triassic. *Science* **235**, 1156–1167.

Harland, W.B., Cox, A.V., Llewellyn, P.G., Pickton, C.A.G., Smith, A.G. & Walters, R. (1982) *A Geologic Time Scale*. Cambridge University Press, Cambridge.

Harland, W.B., Armstrong, R.L., Cox, A.V., Craig, L.E., Smith, A.G. & Smith, D.G. (1990) *A Geologic Time Scale 1989*. Cambridge University Press, Cambridge, 263 pp.

Herbert, T.D. & Fischer, A.G. (1986) Milankovitch climatic origin of mid-Cretaceous black shale rhythms in central Italy. *Nature* **321**, 739–743.

Napoleone, G. & Ripepe, M. (1989) Cyclic geomagnetic changes in Mid-Cretaceous rhythmites, Italy. *Terra Nova* **5**, 437–442.

Pestiaux, P. & Berger, A. (1984) An optimal approach to the spectral characteristics of deep-sea climatic records. In: *Milankovitch and Climate*, Part 1 (Eds Berger, A.L., Imbrie, J., Hays, J., Kukla, G. & Saltzman, B.). Reidel Publ. Co., Dordrecht, pp. 417–445.

Pestiaux, P., van der Mersch, I. & Berger, A.L. (1988) Paleoclimatic variability at frequencies ranging from 1 cycle per 10 000 years to 1 cycle per 1000 years: evidence for nonlinear behaviour of the climate system. *Clim. Change* **12**, 9–37.

Premoli Silva, I., Ripepe, M. & Tornaghi, M.E. (1989) Planktonic foraminiferal distribution record productivity cycles: evidence from the Aptian–Albian Piobbico core (Central Italy). *Terra Nova* **1**, 443–448.

RIPEPE, M. (1988) Stratabase: a stratigraphic database and processing program for microcomputers. *Computers & Geosciences* **14**, 369–375.

RIPEPE, M. (1990) Risposta del sistema geologico alle variazioni orbitali. In: *75° Congresso Nazionale S.G.I.: La Geologia italiana degli anni '90, riassunti relazioni a invito*, pp. 122–128.

SCHWARZACHER, W. (1989) Milankovitch cycles and the measurement of time. *Terra Nova* **1**, 405–408.

SCHWARZACHER, W. & FISCHER, A. (1982) Limestone–shale bedding and perturbations of the Earth's orbit. In: *Cyclic and Event Stratification* (Eds Einsele, G. & Seilacher, A.). Springer Verlag, New York, pp. 73–95.

SCUDELER BACCELLE L. & SEMENZA, E. (1974) Flysch terrigeno con strutture contornitiche nel Cretaceo delle Dolomiti ampezzane. Caratteristiche sedimentologiche e significato geodinamico. *Ann. Univ. di Ferrara* **5**, 165–199.

VAN HINTE, J.E. (1976) A Cretaceous time scale. *Bull. Am. Assoc. Petrol. Geol.* **60**, 498–516.

WEEDON, G.P. (1985) Hemipelagic shelf sedimentation and climatic cycles: the basal Jurassic (Blue Lias) of South Britain. *Earth Planet. Sci. Lett.* **76**, 321–335.

WEEDON, G.P. (1989) The detection and illustration of regular sedimentary cycles using Walsh power spectra and filtering with examples from the Lias of Switzerland. *J. Geol. Soc. Lond.* **146**, 133–144.

Spec. Publs Int. Ass. Sediment. (1994) **19**, 109–116

An astronomically calibrated (polarity) time scale for the Pliocene–Pleistocene: a brief review

F.J. HILGEN

Department of Geology, Institute of Earth Sciences, State University of Utrecht, Budapestlaan 4, 3584 CD Utrecht, The Netherlands

ABSTRACT

This paper gives a brief review of the development of an astronomically calibrated (polarity) time scale (A(P)TS) with emphasis on the recent extension of this time scale from the late Pleistocene down to the Miocene/Pliocene boundary. This time scale is based on the calibration of sedimentary cycles, and other cyclic variations in climatic proxy records, to the astronomical time series of the quasi-periodic variations in the Earth's orbit and is independent of radiometric dating. The A(P)TS gives new ages for the Brunhes/Matuyama boundary (0.78 Ma vs. a conventional age of 0.73 Ma after Berggren *et al.*, 1985), top Jaramillo (0.99 vs. 0.92), bottom Jaramillo (1.07 vs. 0.98), Cobb Mountain (1.19 vs. 1.10), top Olduvai (1.77 or 1.79 vs. 1.66), bottom Olduvai (1.95 vs. 1.88), Gauss/Matuyama boundary (2.60 vs. 2.47), top Kaena (3.04 vs. 2.92), bottom Kaena (3.11 vs. 2.99), top Mammoth (3.22 vs. 3.08), bottom Mammoth (3.33 vs. 3.18), Gilbert/Gauss boundary (3.58 vs. 3.40), top Cochiti (4.18 vs. 3.88), bottom Cochiti (4.29 vs. 3.97), top Nunivak (4.48 vs. 4.10), bottom Nunivak (4.62 vs. 4.24), top Sidufjall (4.80 vs. 4.40), bottom Sidufjall (4.89 vs. 4.47), top Thvera (4.98 vs. 4.57) and bottom Thvera (5.23 vs. 4.77). New ages for the Pliocene/Pleistocene and Miocene/Pliocene boundaries arrive at 1.81 (vs. 1.68) and 5.32 (vs. 4.86) Ma, respectively. Discrepancies with other astronomical time scales and existing conventional time scales will be briefly reviewed and discussed in the light of new radiometric datings based on the single-crystal laser-fusion technique. Finally, the importance and applications of the new time scale will be dealt with.

INTRODUCTION

It has long been realized that sedimentary cycles may reflect climatic oscillations that are ultimately controlled by the Earth's orbital cycles and, hence, that they can be used to estimate the amount of time represented (Gilbert, 1895; see also Fischer, 1980; Berger, 1989). Preceding the invention and application of radiometric dating at the turn of this century, these estimates provided an argument in favour of a much older age of our planet than the then widely, although certainly not generally, accepted age of 100 (or even 20) Ma, based on calculations of a thermodynamic cooling model of the Earth (see Hallam, 1983).

However, an important shortcoming of this use of astronomically induced sedimentary cycles is that only the amount of time can be determined; no absolute ages can be assigned to cyclically bedded sequences and their associated geological events. This shortcoming can be avoided for the youngest part of the geological record by correlating sedimentary cycles – or any other astronomically controlled cyclic variation in climatic proxy records – directly to the calculated orbital time series using the Recent as a fixed calibration point. This procedure was initially used to calibrate the classical succession of (Alpine) glaciations to the summer insolation curve of 65° N Lat supposed to be critical to the expansion and retraction of (Northern Hemisphere) ice sheets (Köppen & Wegener, 1924; Milankovitch, 1941; see also Imbrie & Palmer Imbrie, 1979). More systematically, it was applied for the first time by Hays *et al.* (1976) using high-resolution records from continuous deep-sea sequences recovered by piston coring, $\delta^{18}O$ as a proxy of ice volume and refined statistical methods for time-series analysis. This work has subsequently been elaborated by Morley & Hays (1981), Imbrie *et al.* (1984) and Martinson *et al.* (1987) to further evaluate the astronomical theory of the ice ages and to provide an astronomically calibrated time scale for the last 780 ka, which, in view of the accuracy of the astronomical solution (Berger, 1984), is more accurate and has a higher resolution than existing conventional time scales. Such a time scale is not only instrumental to a better understanding of the relationships between orbital forcing and ocean and climate response (Imbrie *et al.*, 1989; McIntyre *et al.*, 1989), but it is also fundamental to the study of the youngest part of the Earth's history as a whole. Since the mid 1970s, this is indeed feasible for the last million years, the Milankovitch (1941) time scale having been largely improved by Berger (1976). More recently, the accuracy of that astronomical time scale was further extended back in time to roughly 3–5 Ma (Berger & Loutre, 1991), owing to a refinement of the astronomical solution of the planetary system and of the Earth's rotation. It is pertinent therefore to apply the same approach to successively older parts of the geological record – in the first instance to the early Pleistocene and Pliocene – to extend the astronomically calibrated (geological) time scale. In combination with detailed magnetostratigraphical records, such an extension would in addition provide a totally independent check on the accuracy of existing and widely used conventional (polarity) time scales, which are essentially based on radiometric dating (Berggren *et al.*, 1985; Harland *et al.*, 1989).

EXTENSION OF THE ASTRONOMICALLY CALIBRATED (POLARITY) TIME SCALE

This extension was initially strongly hampered by problems encountered in recovering older, undisturbed and continuous deep-sea sequences which, except in areas with extremely low sedimentation rates, were beyond the reach of ordinary piston cores. Sedimentary sequences from such low sediment accumulation areas are less suitable for detailed studies, however, because the resolution of deep-sea records is generally inversely proportional to the sedimentation rate. This major problem concerning the recovery of uninterrupted, high-resolution deep-sea records was ultimately solved by the application of advanced piston coring of multiple holes. Such parallel sites are necessary to avoid stratigraphic problems encountered at core breaks (Ruddiman *et al.*, 1986). Similarly good sequences of Pliocene and early Pleistocene age, however, can also be found on land in areas that underwent a strong uplift resulting in the subaerial exposure of relatively young deep-sea sediments. One of these areas, undoubtedly, is the Mediterranean, which has the additional advantage that its latitudinal position in combination with its semi-enclosed, land-locked configuration make its sedimentary systems particularly sensitive to recording astronomically forced oscillations in climate.

Initial attempts to extend the astronomical time scale yielded ages of geomagnetic reversal boundaries that were essentially consistent with the K/Ar radiometric datings (Pisias & Moore, 1981; Raymo *et al.*, 1989; Ruddiman *et al.*, 1989). Serious doubt was nevertheless expressed as to the correctness of the astronomical calibration of the Brunhes/Matuyama (B/M) boundary interval (Johnson, 1982; Ruddiman *et al.*, 1989) and Johnson (1982) even arrived at an age of 0.79 Ma for the B/M boundary. Based on the high-resolution oxygen isotope record from ODP Site 677, Shackleton *et al.* (1990) provided an alternative calibration of this particular interval, which resulted in an age of 0.78 Ma for the B/M boundary. The astronomical calibration of the isotope record from ODP Site 677, in combination with a close re-examination of the standard isotope record from DSDP Site 607, in addition led to the conclusion that the interval between the B/M boundary and the top of the Olduvai must have lasted significantly longer than previously assumed (Shackleton *et al.*, 1990; see also Hilgen, 1991a). The resulting age of 1.77 Ma for the top of the Olduvai was essentially confirmed by the independent work of Hilgen (1991a; 1.79 Ma), who established an astronomical time scale by correlating late Pliocene to early Pleistocene sapropel (brownish coloured, often well-laminated interbeds) patterns in marine sequences exposed in Italian land-based sections to the orbital time series of precession and eccentricity (solution Ber90 in Berger & Loutre, 1991). For this purpose, phase relationships were first established between these astronomical cycles and the youngest, late Pleistocene sapropels recovered in numerous piston cores from the eastern Mediterranean. These relationships, where individual sapropels correspond to minima in the precession index and sapropel clusters to maxima in eccentricity, were then employed to astronomically calibrate the older sapropels. This (alternative) astronomical calibration of the isotope records from Sites 607 and 677 (Shackleton *et al.*, 1990) and the Mediterranean sapropel sequence (Hilgen, 1991a) resulted in new ages for the bottom of the Olduvai (1.95 Ma), the Reunion (2.14–2.15), the Gauss/Matuyama (2.60 or 2.59/2.62) and the top of the Kaena (3.04). The age of the Pliocene/Pleistocene boundary, as defined at the base of the homogeneous claystone which conformably overlies the sapropeletic layer coded e in the Vrica section (Aquirre & Pasini, 1985), is thus 1.81 Ma (Hilgen, 1991a; see also Table 1).

It was further anticipated that this new time scale could even be extended to the Miocene/Pliocene boundary by taking the $CaCO_3$ cycles in the early Pliocene Trubi Formation on Sicily into account. Hilgen & Langereis (1989) initially established an alternative (polarity) time scale for the major part of the Gilbert and Gauss Chrons using: (i) an average periodicity of 21.7 ka of the precession cycle as the periodicity of small-scale $CaCO_3$ cycles in the Trubi Formation; and (ii) the Gilbert/Gauss boundary dated radiometrically at 3.40 Ma as a fixed calibration point. This procedure was adopted in an attempt to correct for the observed shift of significant peaks in the $CaCO_3$ spectra to slightly, but consistently higher frequencies than those of the Earth's orbital cycles; this shift appeared in the spectrum if conventional ages of reversal boundaries were used to generate the $CaCO_3$ time series. To establish a time scale totally independent from radiometric dating, Hilgen (1991a) provisionally extended his astronomically calibrated time scale for the late Gauss to early Matuyama by adjusting the time scale of Hilgen & Langereis (1989) to the age difference of 180 ka found for the top of the Kaena. In a subsequent paper, Hilgen (1991b) correlated the $CaCO_3$ cycles of the Trubi directly to the astronomical record using: (i) phase relationships between these $CaCO_3$ cycles and precession and eccentricity (grey beds in small-scale $CaCO_3$ cycles correspond to minima in the precession index and $CaCO_3$ minima in larger-scale $CaCO_3$ cycles to eccentricity maxima; these relationships were established indirectly, namely by establishing first the relationships between $CaCO_3$ and sapropel cycles); and (ii) fixed astronomical calibration points provided by the sapropels which occur in the topmost part of the $CaCO_3$ record. This correlation, in combination with the high-resolution magnetostratigraphy, resulted in astronomically calibrated ages for the reversal boundaries of the Kaena (3.04–3.11 Ma) and Mammoth (3.22–3.33), the Gilbert/Gauss Chron boundary (3.58) and the reversal boundaries of the Cochiti (4.18–4.29), Nunivak (4.48–4.62), Sidufjall (4.80–4.89) and Thvera (4.98–5.23). The age of the Miocene/Pliocene boundary, as defined by the base of the Trubi marls (Cita, 1975), is thus 5.32 Ma (see Fig. 1; Table 1). This older part of the new time scale is still preliminary and awaits further confirmation. In particular, the underlying astronomical calibration needs to be duplicated in another, preferably extra-Mediterranean sequence.

COMPARISON WITH OTHER (POLARITY) TIME SCALES

The resulting astronomically calibrated time scale of Shackleton *et al.* (1990) and Hilgen (1991a, b) deviates markedly from previously established astro nomically calibrated time scales, which are essentially consistent with conventional radiometric datings of reversal boundaries (Fig. 1; Table 1). The latter time scales were constructed using a different approach, namely by employing the rather uniform variations of the obliquity cycle, whereas we used the more variable pattern of precession instead. This distinctive pattern results from the modulation of precession by the variations in eccentricity. Close re-examination of the isotope record from DSDP Site 607 provided strong evidence that two rather weakly developed obliquity-related cycles have not been recognized by Ruddiman *et al.* (1989) in the B/M to top Olduvai interval (Hilgen, 1991a), whereas the B/M boundary itself proved to be 50 ka older than previously assumed (Shackleton *et al.*, 1990).

Most surprisingly, however, the new time scale departs significantly from existing conventional time scales which for this interval are either based on K/Ar radiometric dating of lavas of known polarity (Mankinen & Dalrymple, 1979) or on linear interpolation between radiometrically dated calibration points in marine anomaly sequences (Berggren *et al.*, 1985; Harland *et al.*, 1989). Age discrepancies with the most widely used conventional time scales range from 3 to 10% for individual reversal boundaries. These discrepancies cannot be explained by errors in the astronomical solution because the most recent solution is considered to be accurate in the time domain over the last 5.0 Ma (Berger & Loutre, 1991). A consistent error in the decay constants used in K/Ar dating can also be excluded as a

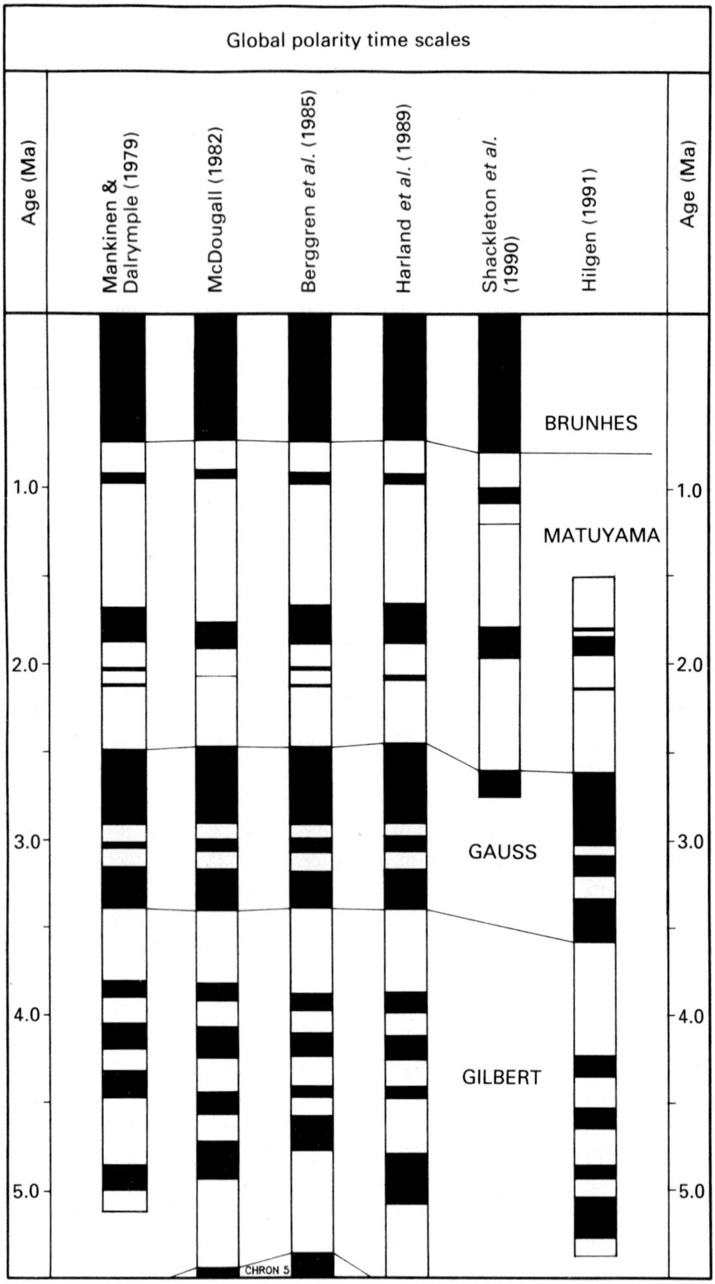

Fig. 1. Comparison of the astronomically calibrated polarity time scale (A(P)TS) for the Pliocene–Pleistocene with other polarity time scales.

possible cause because their inaccuracy is assumed to be in the order of 0.5% (Gale, 1982).

At present, the most likely explanation for these discrepancies is the loss of radiogenic argon from lavas. This would result in aberrant young radiometric ages (Hilgen, 1991b). This satisfactorily explains why in several cases the oldest K/Ar age for a certain polarity event or reversal boundary is in good agreement with its astronomical age (see Hilgen, 1991b). New techniques in radiometric dating, such as (single crystal) laser fusion $^{40}Ar/^{39}Ar$ dating, may provide a definite answer to the question

Table 1. Ages for reversal boundaries according to the different time scales. M&D 79, Mankinen & Dalrymple (1979); McD 79, McDougall (1979); Ber 85, Berggren *et al.* (1985); Har 89, Harland *et al.* (1989); Sha 90, Shackleton *et al.* (1990); Hil 91, Hilgen (1991a, b). Ages for epoch boundaries marked by an asterisk were calculated by linear interpolation or extrapolation of magnetostratigraphic datum planes in the Vrica and Eraclea Minoa sections (Hilgen & Langereis, 1988; Zijderveld *et al.*, 1991) using ages for reversal boundaries after the various time scales

	K/Ar dated		Sea-floor anomaly		Astronomical	
	M&D 79	McD 79	Ber 85	Har 89	Sha 90	Hil 91
Reversal boundary						
Brunhes/Matuyama	0.73	0.72	0.73	0.72	0.78	
Jaramillo top	0.90	0.89	0.92	0.91	0.99	
Jaramillo bottom	0.97	0.94	0.98	0.97	1.07	
Cobb Mountain	—	—	1.10	—	1.19	
Olduvai top	1.67	1.76	1.66	1.65	1.77	1.79
Olduvai bottom	1.87	1.91	1.88	1.88	1.95	1.95
Reunion 1 top	2.01	2.07	—	2.06	—	2.14
Reunion 1 bottom	2.04	2.07	—	2.09	—	2.15
Reunion 2 top	2.12	—	—	—	—	—
Reunion 2 bottom	2.14	—	—	—	—	—
Gauss/Matuyama	2.48	2.47	2.47	2.45	2.60	2.59/2.62
Kaena top	2.92	2.91	2.92	2.91		3.04
Kaena bottom	3.01	3.00	2.99	2.98		3.11
Mammoth top	3.05	3.07	3.08	3.07		3.22
Mammoth bottom	3.15	3.17	3.18	3.17		3.33
Gilbert/Gauss	3.40	3.41	3.40	3.40		3.58
Cochiti top	3.80	3.82	3.88	3.87		4.18
Cochiti bottom	3.90	3.92	3.97	3.99		4.29
Nunivak top	4.05	4.07	4.10	4.12		4.48
Nunivak bottom	4.20	4.25	4.24	4.26		4.62
Sidufjall top	4.32	4.44	4.40	4.41		4.80
Sidufjall bottom	4.47	4.57	4.47	4.48		4.89
Thvera top	4.85	4.72	4.57	4.79		4.98
Thvera bottom	5.00	4.94	4.77	5.08		5.23
Epoch boundary						
Pliocene/Pleistocene	1.69*	1.78*	1.68*	1.67*		1.81
Miocene/Pliocene	5.07*	5.04*	4.86*	5.21*		5.32

of whether conventional K/Ar ages are too young. Preliminary results for the B/M boundary and the Reunion event suggest that this is indeed the case (Baksi *et al.*, 1991a, b). Also the age of the top of the Jaramillo Subchron may be somewhat older than in the currently accepted time scale (Tauxe *et al.*, 1991). New datings from the type locality of the Olduvai further indicate that the base of the Olduvai is also considerably older (between 1.98 and 2.01 Ma) than presently assumed, although the inferred maximum age of 1.74 Ma for the top of the Olduvai (Walter *et al.*, 1991) is certainly not in agreement with the astronomical ages. The radiometric ages of the lavas and tuffs in the Olduvai Gorge represent mean ages based on a number of age determinations of single grains. The oldest age of a single grain in the youngest tuffs still seems to be consistent with the astronomical ages for the Olduvai (fig. 2b in Walter *et al.*, 1991). This suggests that differential diffusion of ^{40}Ar with respect to ^{39}Ar may be responsible for the shift of mean ages to relatively young values as compared with the astronomical estimates.

Literature in addition provides important arguments for the need to adjust also the conventional ages of early Pliocene polarity reversals. An age of 10.17 Ma is obtained if the observed offset between the conventional (of Berggren *et al.*, 1985) and astronomically calibrated (Hilgen, 1991b) ages for successively older polarity reversals in the Gilbert Chron is extrapolated back to the younger end of Chron 11. After the Gilbert/Gauss boundary, this reversal boundary represents the next older age calibration point in marine anomaly sequences initially used by Berggren *et al.* (1985) to construct their polarity time scale. The resulting 1.3-Ma discrepancy is in fair, although certainly not complete, agreement with the work of McDougall *et al.* (1984) and Baksi (1988), who provided independent evidence that the age of Chron 11 should be increased by approximately 0.9 Ma. Evidence for the existence of similar errors in the conventional polarity time scale between 15 and 17 Ma has finally been provided by results of $^{40}Ar/^{39}Ar$ dating of basalts from the northwest Pacific (Baksi & Farrar, 1990). Based on the references cited above, it is proposed that the time scale based on astronomical calibration is more accurate than previous ones, which are essentially based on (K/Ar) radiometric dating.

APPLICATIONS AND FUTURE RESEARCH

The high resolution and degree of accuracy of the A(P)TS is crucial to the study of complex relationships between astronomical forcing and response in the climate–ocean system, as has been shown convincingly for both the late Pleistocene (Imbrie *et al.*, 1989; McIntyre *et al.*, 1989) and for the late Pliocene (Lourens *et al.*, 1992). In addition, however, it provides a perfect framework:

1 To analyse in detail the synchronism or diachronism of biostratigraphic datum planes in order to determine their usefulness for chronostratigraphic correlations to sequences lacking a distinct cyclicity;

2 To determine the exact number and duration of short polarity events and, moreover, to reveal whether polarity reversals are recorded synchronously in different sedimentary environments or diachronously as a consequence of delayed acquisition of remanent magnetization (see van Hoof & Langereis, 1991);

3 To establish temporal relationships between such widely varying geological phenomena as (glacio-)eustatic sea-level changes, hiatuses, tectonic activity and rotations, changes in sedimentary conditions, faunal and floral extinctions, etc.;

4 To calculate (changes in) sea-floor spreading rates much more precisely, etc.

Future research should specifically aim at extending the A(P)TS to older parts of the geological record, in the first instance to the middle to late Miocene. This is especially relevant because phase relationships between orbital cyclicity and changes in the climate–ocean system during the Miocene may well differ from (late) Pliocene to Pleistocene relationships. For instance, this is suggested by the fact that essentially monopolar ice build-up (i.e. Antarctic) occurred during the Miocene and bipolar ice build-up during the (late) Pliocene and Pleistocene.

Further extension of the A(P)TS based on a direct calibration of cyclic variations in climatic proxy records to the orbital time series is at present not feasible because the accuracy of the most recent astronomical solution is assumed to be restricted to the last 5 Ma in the time domain and to the last 10 Ma in the frequency domain (Berger & Loutre, 1991). Theoretical arguments further oppose an unlimited extension of the accuracy of future solutions (in the time domain) over much longer periods due to the inferred chaotic behaviour of the solar system (Laskar, 1989). Moreover, quasi-periods of precession and obliquity will be reduced with increasing age as a consequence of a shortening of the Earth–Moon distance (Berger *et al.*, 1989). The A(P)TS can nevertheless be extended by using as a first stage only eccentricity-related variations in proxy records since the eccentricity periods are not affected by changes in the Earth–Moon distance (Berger & Loutre, 1992). For this purpose, the 400-ka eccentricity cycle is most suitable because of its longer periodicity than the 100-ka cycle and the fact that it is not composed of different frequency components. In this way, the A(P)TS can in principle be extended very accurately into the Mesozoic and even the Palaeozoic. By such an approach also the character of precession and obliquity in the more remote geological past may be better understood in the future.

ACKNOWLEDGEMENTS

P.L. de Boer critically read a first draft of the manuscript. The reviews of A. Berger and N. Shackleton are gratefully acknowledged.

REFERENCES

AQUIRRE, E. & PASINI, G. (1985) The Pliocene–Pleistocene boundary. *Episodes*, **8**, 116–120.

BAKSI, A.K. (1988) Estimation of lava extrusion and magma production rates for two flood basalt provinces. *J. Geophys. Res.* **93**, 11809–11815.

BAKSI, A.K. & FARRAR, E. (1990) Evidence for errors in the geomagnetic polarity time-scale at 17–15 Ma; results of $^{40}Ar/^{39}Ar$ dating of basalts from the Pacific Northwest, USA. *Geophys. Res. Lett.* **17**, 1117–1120.

BAKSI, A.K., HOUGHTON, B., MCWILLIAMS, M., TANAKA, H. & TURNER, G. (1991a) What is the age of the

Brunhes–Matuyama polarity transition. *Eos* **72**, 44, 135 (Abstract).

Baksi, A.K., Hoffman, K.A. & McWilliams, M.O. (1991b) Preliminary results of $^{40}Ar/^{39}Ar$ dating studies directed to testing the accuracy of the geomagnetic polarity time-scale (GPTS) at 2–5 Ma. *Eos* **72**, 44, 135 (Abstract).

Berger, A. (1976) Obliquity and precession for the last 5 000 000 years. *Astron. Astrophys.* **51**, 127–135.

Berger, A.L. (1984) Accuracy and frequency stability of the Earth's orbital elements during the Quarternary. In: *Milankovitch and Climate* (Eds Berger, A., Imbrie, J., Hays, J., Kubla, G. & Salzman, B.). Reidel Publ. Co., Dordrecht, pp. 3–39.

Berger, A. (1989) The spectral characteristics of pre-Quaternary climatic records, an example of the relationship between the astronomical theory and geosciences. In: *Climate and Geosciences* (Eds Berger, A., Schneider, S. & Duplessy, J.C.). Kluwer Academic Publishers, Dordrecht, pp. 47–76.

Berger, A. & Loutre, M.F. (1991) Insolation values for the climate of the last 10 million years. *Quat. Sci. Rev.* **10**, 297–317.

Berger, A.L., Loutre, M.F. & Dehant, V. (1989) Influence of the changing lunar orbit on the astronomical frequencies of pre-Quaternary insolation patterns. *Paleoceanography* **4**, 555–564.

Berger, A., Loutre, M.F. & Laskar, J. (1992) Stability of the astronomical frequences over the Earth's history for paleoclimatic studies. *Science* **255**, 560–566.

Berggren, W.A., Kent, D.V., Flynn, J.J. & van Couvering, J.A. (1985) Cenozoic geochronology. *Geol. Soc. Am. Bull.* **96**, 1407–1418.

Cita, M.B. (1975) The Miocene/Pliocene boundary. History and definition. In: *Late Neogene Epoch Boundaries* (Eds Saito, T. & Burckle, L.H.). Micropaleontol. Press, Spec. Publ. 1, pp. 1–30.

Fischer, A.G. (1980) Gilbert-bedding rhythms and geochronology. In: *The Scientific Ideas of G.K. Gilbert* (Ed. Yochelson, E.L.). Geol. Soc. Am. Spec. Pap. 183, pp. 93–104.

Gale, N.H. (1982) The physical decay constants. In: *Numerical Dating in Stratigraphy* (Ed. Odin, G.S.). John Wiley & Sons–Interscience, Chichester, pp. 107–122.

Gilbert, G.K. (1895) Sedimentary measurement of geologic time. *J. Geol.* **3**, 121–127.

Hallam, A. (1983) *Great Geological Controversies*. Oxford University Press, Oxford.

Harland, W.B., Armstrong, R.L., Cox, A.V., Craig, L.E., Smith, A.G. & Smith, D.G. (1989) *A Geologic Time Scale 1989*. Cambridge University Press, Cambridge.

Hays, J.D., Imbrie, J. & Shackleton, N.J. (1976) Variations in the earth's orbit: pacemaker of the ice ages. *Science* **194**, 1121–1132.

Hilgen, F.J. (1991a) Astronomical calibration of Gauss to Matuyama sapropels in the Mediterranean and implication for the Geomagnetic Polarity Time Scale. *Earth Planet. Sci. Lett.* **104**, 226–244.

Hilgen, F.J. (1991b) Extension of the astronomically calibrated (polarity) time scale to the Miocene/Pliocene boundary. *Earth Planet. Sci. Lett.* **107**, 349–368.

Hilgen, F.J. & Langereis, C.G. (1988) The age of the Miocene–Pliocene boundary in the Capo Rossello area (Sicily). *Earth Planet. Sci. Lett.* **91**, 214–222.

Hilgen, F.J. & Langereis, C.G. (1989) Periodicities of $CaCO_3$ cycles in the Pliocene of Sicily: Discrepancies with the quasi-periods of the Earth's orbital cycles? *Terra Nova* **1**, 409–415.

Imbrie, J. & Palmer Imbrie, K. (1979) *Ice Ages: Solving the Mystery*. Harvard University Press, Cambridge, MA.

Imbrie, J., Hays, J.D., Martinson, D.G., McIntyre, A., Mix. A.C., Morley, J.J. *et al.* (1984) The orbital theory of Pleistocene climate: Support from a revised chronology of the marine $\delta^{18}O$ record. In: *Milankovitch and Climate* (Eds Berger, A., Imbrie, J., Hays, J., Kukla, G. & Saltzman, B.). Reidel Publ. Co., Dordrecht, pp. 269–305.

Imbrie, J., McIntyre, A. & Mix, A. (1989) Oceanic response to orbital forcing in the late Quaternary: observational and experimental strategies. In: *Climate and Geosciences* (Eds Berger, A., Schneider, S. & Duplessy, J.C.). Kluwer Academic Publishers, Dordrecht, pp. 121–164.

Johnson, R.G. (1982) Brunhes–Matuyama dated at 790 000 yr B.P. by marine astronomical correlations. *Quat. Res.* **17**, 135–147.

Köppen, W. & Wegener, A. (1924) *Die Klimate der geologischen Vorzeit*. Gebruder Borntraeger, Berlin.

Laskar, J. (1989) A numerical experiment on the chaotic behaviour of the solar system. *Nature* **338**, 237–238.

Lourens, L.J., Hilgen, F.J., Zachariasse, W.J. & Gudjonsson, L. (1992) Late Pliocene–early Pleistocene astronomically forced surface water temperatures and productivity variations in the Mediterranean. *Mar. Micropaleontol.* **19**, 49–78.

Mankinen, E.A. & Dalrymple, G.B. (1979) Revised geomagnetic polarity time scale for the interval 0–5 m.y. B.P. *J. Geophys. Res.* **84**, 615–626.

Martinson, D.G., Pisias, N.G., Hays, J.D., Imbrie, J., Moore, T.C. & Shackleton, N.J. (1987) Age dating and the orbital theory of the ice ages: Development of a high-resolution 0 to 300 000-year chronostratigraphy. *Quat. Res.* **27**, 1–29.

McDougall, I. (1979) The present status of the Geomagnetic Polarity Time Scale. In: *The Earth, its origin, structure and evolution* (Ed. McElhinney, N.W.). Academic Press, London, pp. 543–565.

McDougall, I., Kristjansson, L. & Saemundsson, K. (1984) Magnetostratigraphy and geochronology of Northwest Iceland. *J. Geophys. Res.* **89**, 7029–7060.

McIntyre, A., Ruddiman, W.F., Karlin, K. & Mix, A.C. (1989) Surface water response of the equatorial Atlantic to orbital forcing. *Paleoceanography* **4**, 19–55.

Milankovitch, M. (1941) *Kanon der Erdbestrahlung und seine Anwendung auf das Eiszeitenproblem*. Royal Serb. Acad. Sci., Spec. Publ. 133, Belgrade.

Morley, J.J. & Hays, J.D. (1981) Towards a high-resolution, global, deep-sea chronology for the last 750 000 years. *Earth Planet. Sci. Lett.* **53**, 279–295.

Pisias, N.G. & Moore, T.C. (1981) The evolution of Pleistocene climate: A time series approach. *Earth Planet. Sci. Lett.* **52**, 450–458.

Raymo, M.E., Ruddiman, W.F., Backman, J., Clement, B.M. & Martinson, D.G. (1989) Late Pliocene variation in Northern Hemishere ice sheets and North Atlantic deep water circulation. *Paleoceanography* **4**, 413–446.

RUDDIMAN, W.F., CAMERON, D. & CLEMENT, B.M. (1986) Sediment disturbance and correlation of offset holes drilled with hydraulic piston corer: Leg 94. *Init. Rep. DSDP* **94**, 615–634.

RUDDIMAN, W.F., RAYMO, M.E., MARTINSON, D.G., CLEMENT, B.M. & BACKMAN, J. (1989) Pleistocene evolution: Northern Hemisphere ice sheets and north Atlantic Ocean. *Paleoceanography* **4**, 353–412.

SHACKLETON, N.J., BERGER, A. & PELTIER, W.R. (1990) An alternative astronomical calibration of the lower Pleistocene time scale based on ODP Site 677. *Trans. R. Soc. Edinburgh: Earth Sci.* **81**, 251–261.

TAUXE, L., DEINO, A., POTTS, R. & BEHRENSMEYER, A.K. (1991) Magnetostratigraphy and single crystal Ar–Ar dating: implications for the age of the Jaramillo (abstract). *International Association of Geomagnetism and Aeronomy Program and Abstracts, XX-International Union of Geodesy and Geophysics Congres, Vienna.*

VAN HOOF, A.A.M. & LANGEREIS, C.G. (1991) Reversal records in marine marls and delayed acquisition of remanent magnetization. *Nature* **351**, 223–225.

WALTER, R.C., MANEGA, P.C., HAY, R.L., DRAKE, R.E. & CURTIS, G.H. (1991) Laser-fusion $^{40}Ar/^{39}Ar$ dating of Bed I, Olduvai Gorge, Tanzania. *Nature* **354**, 145–149.

ZIJDERVELD, J.D.A., HILGEN, F.J., LANGEREIS, C.G., VERHALLEN, P.J.J.M. & ZACHARIASSE, W.J. (1991) Integrated magnetostratigraphy and biostratigraphy of the upper Pliocene–lower Pleistocene from the Monte Singa and Crotone areas in Calabria, Italy. *Earth Planet. Sci. Lett.* **107**, 697–714.

Spec. Publs Int. Ass. Sediment. (1994) **19**, 117–126

Pleistocene evolution of orbital periodicities in the high-resolution pollen record Funza I, Eastern Cordillera, Colombia

H. HOOGHIEMSTRA* *and* J.L. MELICE†

**Hugo de Vries-Laboratory, Department of Palynology and Paleo-/Actuo-ecology, University of Amsterdam, Kruislaan 318, 1098 SM Amsterdam, The Netherlands; and*
†ORSTOM, 74, route d'Aulnay, F-93143 Bondy, France and
Institut d'Astronomie et de Géophysique G. Lemaître, Université Catholique de Louvain, 2, Chemin du Cyclotron, 1348 Louvain-la-Neuve, Belgium

ABSTRACT

The evolution of orbital periodicities during the Pleistocene has been studied in the 30–1450-ka Funza I pollen record from the sedimentary basin of Bogotá, Eastern Cordillera, Colombia. Based on the improved time control of the Bogotá sediments (fission-track datings of Funza II tephra beds) the 1178 pollen samples analysed from between 2.90 and 340.00 m core depths give an average time resolution of *c.*1205 years. Regular sampling was achieved with cubic spline interpolation and subsequently several time series, related to different variables of the palaeoclimate, were analysed by maximum entropy spectral analysis and Thomson multi-taper spectral analysis using a moving 400-ka window with steps of 20 ka.

The presence of eccentricity periods is restricted to the last *c.*800 ka of the Pleistocene, whereas periodicities of the precession band are present throughout the record. Three frequency bands of high palaeoclimatic variability centred at 10.1–10.6, 11.4–11.9 and 14–15 ka are also detected. These periods are very close to those predicted by the climatic non-linear model of Ghil and Le Treut & Ghil. Temperature- and precipitation-related changes of the altitudinal zonation of the vegetation belts on the tropical mountains of Colombia thus respond to orbital forcing, testifying to the global scale of climatic change.

INTRODUCTION

Pollen records from lake sediments, peat bogs and ocean cores are well known for their potential in documenting the local or regional (subcontinental) vegetation history. Based on the relationship between modern vegetation and modern climatic conditions, the climatic history can be inferred from the vegetation history.

Intermontane basins have a favourable position in accumulating sediments. Long sediment sequences from sedimentary basins often preserve a continuous document of the vegetation history based on the intercepted and conserved pollen grains. Tropical mountains especially seem to be in a favourable position because a change in climatic conditions results mainly in a vertical shift of vegetation belts over the mountain slopes. Thus, the different vegetation belts stay in the vicinity and are registered continuously by their intercepted pollen (Fig. 1). It is the relative contribution of each vegetation belt to the total intercepted pollen rain that forms the basis for the reconstruction of climatic change.

The basin of Bogotá at 2550 m altitude represents a key site among long continental records of palaeoclimate (Fig. 1). The pollen rain that fell into the former lake of Bogotá was provided principally by the vegetation on the surrounding mountains, which rise up to *c.*3000–3200 m in the near vicinity and to 3600–3800 m in the more distant mountain chains (33–40 km away). Accordingly, high percentages of arboreal pollen indicate that the surrounding slopes were entirely or almost entirely covered with forest. Increasing percentages of elements of open equa-

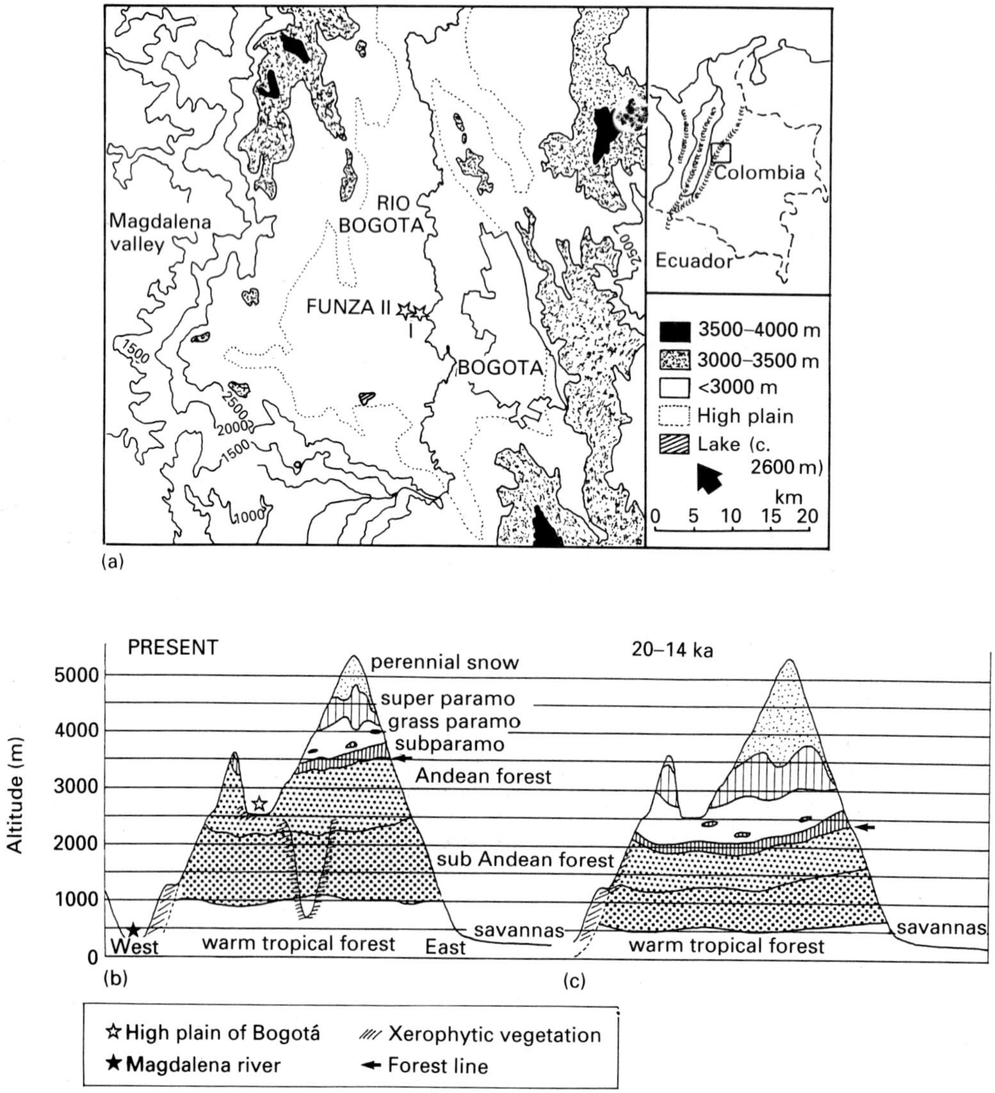

Fig. 1. Location map of boreholes Funza I and Funza II in the high plain of Bogotá (Eastern Cordillera, Colombia) (a) and cross-section through the Eastern Cordillera, indicating the modern (b) and Last Glacial (c) altitudinal position of the vegetation belts. After van der Hammen (1974).

torial alpine vegetation (paramo vegetation) correspond with a progressively downward shift of the upper forest line. This repeating process is the basis of the palaeoclimatic reconstruction.

Long continental records of climatic change are scarce but are of great importance in facilitating the comparison of land-based and ocean-based climatic histories. A high resolution in time is necessary to be able to detect climatic oscillations of at least Milankovitch frequencies. Some pollen records have provided evidence of Milankovitch cycles, such as the records of Greece (Wijmstra & Groenhart, 1983), East Atlantic off Mauritania (Dupont & Hooghiemstra, 1989), France (Molfino *et al.*, 1984) and Japan (Kanari *et al.*, 1984).

The impact of the different frequency bands of orbital forcing depends on the geographical position (Crowley & North, 1991). The Colombian pollen records from the Eastern Cordillera represent up to the present the first almost equatorial records that have been analysed within this scope.

DATA AND ANALYTICAL METHODS

The evolution of orbital periodicities during the Pleistocene has been studied in the Funza I pollen record of the high plain of Bogotá, Eastern Cordillera, Colombia. Difficulties in dating the intercalated tephra beds of the Funza I borehole (Hooghiemstra, 1984, 1989; see also Andriessen *et al.*, in press) resulted in poor time control for the pollen record obtained in 1976 and published in 1984 and 1989. In 1988, a new deep borehole probably penetrated the complete basin infill and undisturbed sediments were recovered to a depth of 586 m below the surface of the high plain (2550 m, 4° N, 74° W), which represents the bed of a drained lake. Pollen record Funza II (first results published in Hooghiemstra & Sarmiento, 1991; Hooghiemstra & Cleef, submitted; Hooghiemstra & Ran, submitted) correlates closely with pollen record Funza I. This correlation was based on the number of recorded major climatic cycles, the first appearance dates of important Andean forest trees (*Alnus, Quercus*), characteristic changes in the pollen assemblage throughout the core, characteristic down-core changes in the frequency of climatic oscillations, characteristic down-core changes in the lithology of the sediments, and absolute dating results of intercalated tephra beds. The lithology of the upper 320 m of both Funza cores consists of lacustrine clays, only occasionally interrupted by (light) sandy clays, lignites and volcanic ash horizons. Sandy intervals are more frequent down-core (Hooghiemstra, 1984). Time control of the Funza II sediments was based on fission-track datings of selected zircon crystals (Andriessen *et al.*, submitted) from eight inter-

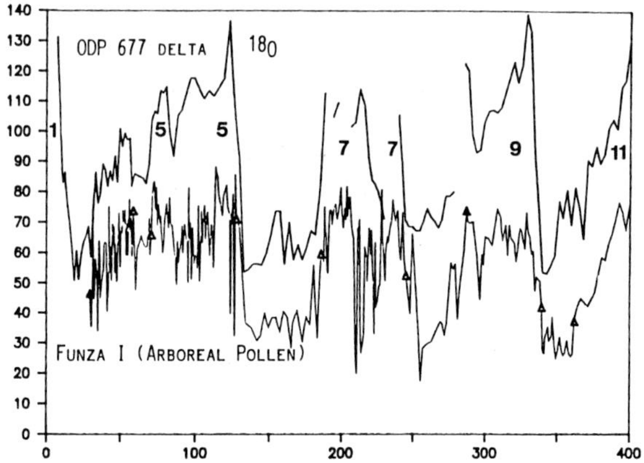

Fig. 2. Experimental visual correlation of the pollen record of Funza I (Hooghiemstra, 1984, 1989) and the oxygen-isotope record of ODP Site 677 (Shackleton *et al.*, 1990). By using the time control of Funza II and ODP Site 677, the intervals representing stage 22 in both records and the core tops were correlated. Subsequently, the pollen record was stretched or compressed over 35 short intervals between 36 control points. The parallel records show the tentative correlation for oxygen-isotope stages 3 through to 25 (some uncertainty exists at stages 5, 7 and 16). The correlation of oxygen-isotope stages 26 through to 35 has to be regarded as experimental. From N.J. Shackleton and H. Hooghiemstra (unpublished data).

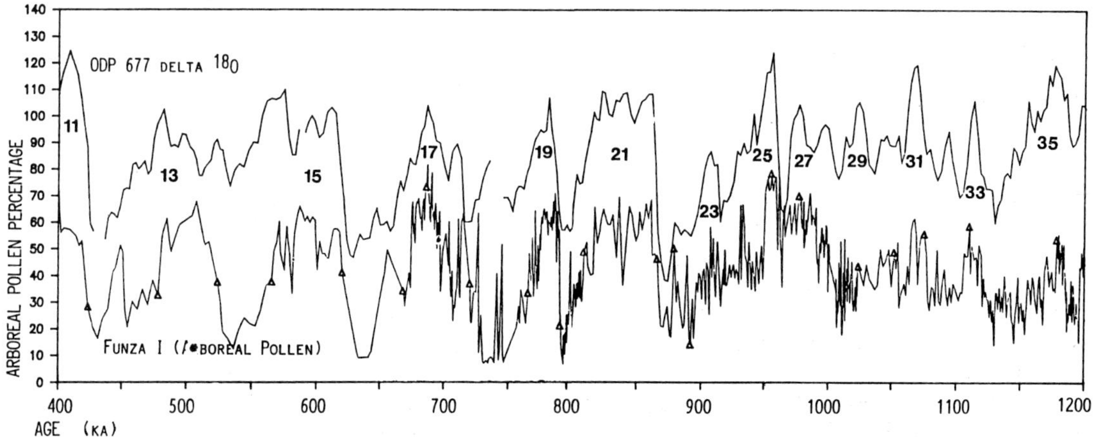

calated volcanic ash horizons. Based on this improved absolute time control of the sediments of the Bogotá basin, the 1178 Funza I pollen samples analysed from core depths of 2.90 to 340.00 m give an average time resolution of *c.*1205 years. A simple three-point time-control model based on absolute datings is provided by the top of the record (3.00 m: 30 ka BP, estimated on the basis of a radiocarbon date), a cluster of three dated tephra beds between 298 and 322 m core depth (selected point 302.00 m: 1020 ka BP) and the tephra bed at 506.20 m core depth (2740 ka BP). Correlation of the pollen record Funza I with the ODP 677 $\delta^{18}O$ record (Fig. 2) (Hooghiemstra & Sarmiento, 1991; Shackleton & Hooghiemstra, unpublished data) has generated a 36-point time-control model. For this correlation the pollen record was stretched and squeezed between 36 control points (indicated in Fig. 3) in such a way that the oscillations of the mainly temperature-related 'arboreal pollen' record ran as parallel as possible to the oxygen isotope curve.

With the objectives (i) of identifying significant periodicities in this high-resolution Funza I pollen record and (ii) of detecting changes in the periodicities and their significance throughout the Pleistocene period, we used a moving 400-ka window with steps of 20 ka. A combination of different spectral techniques, namely maximum entropy spectral analysis (MESA; Burg, 1972) and Thomson multitaper spectral analysis (TMT; Thomson, 1982), has been applied to five different data sets of the pollen record. We show here selected results of the 'arboreal pollen' record, the '*Alnus*' record and the 'subparamo elements' record. Use of the 36-point time-control model was selected because it apparently accounts more for changes in the accumulation rate of the lake sediments. On the other hand, this model introduces some bias as the oxygen isotope record has been orbitally tuned. In the discussion we will consider our procedure.

RESULTS

The 'arboreal pollen' record (Fig. 3) reflects the altitudinal shifts of the upper forest line over the slopes of the Eastern Cordillera, which are mainly related to changes in temperature. The altitudinal position of the forest line (at present at *c.*3200 m) shifted between a maximum of *c.*3500 m (during inter-glacial periods) and a minimum of *c.*1800 m (during glacial periods), rendering the sediments in

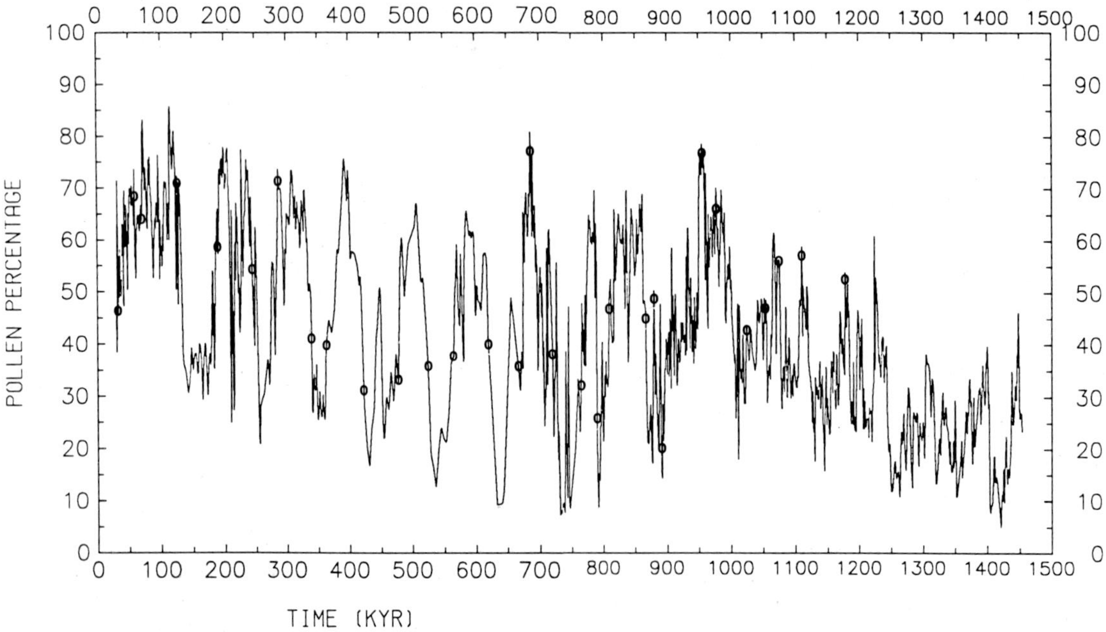

Fig. 3. Graph of Funza I pollen record 'arboreal pollen' percentages (*Alnus* included in the pollen sum) versus time (36-point time-control model). The position of the 36 control points is indicated by dots.

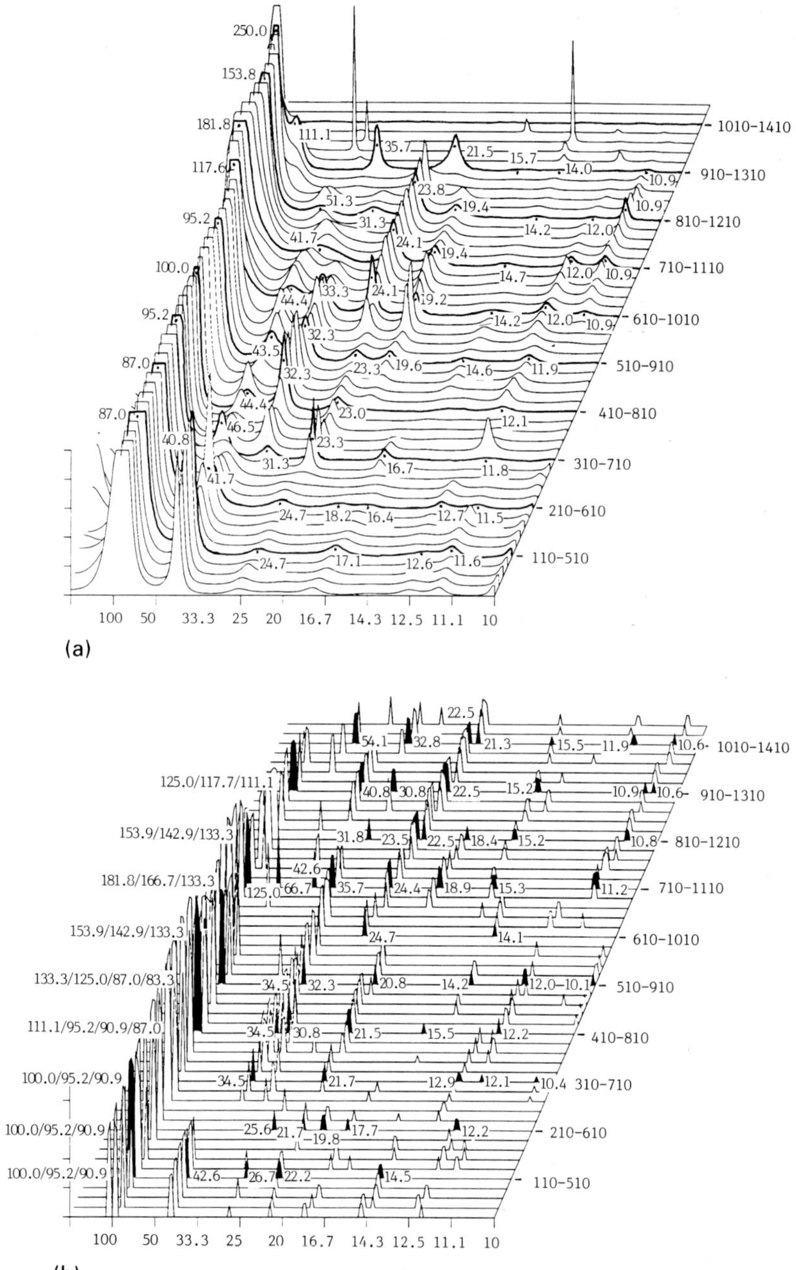

Fig. 4. MESA and TMT with significance level above 80% of 'arboreal pollen' percentages (*Alnus* included in the pollen sum) of Funza I pollen record, using a 36-point time-control model. A moving 400-ka window with steps of 20 ka provides 52 spectra. The first spectrum concerns the interval 30–430 ka BP. The maxima of the periodicities are expressed in ka and are indicated at every fifth spectrum.

the basin of Bogotá at 2550 m a sensitive recorder of climatic change. An arboreal pollen percentage of c.40% (*Alnus* included in the pollen sum) reflects an altitudinal position of the upper forest line at the level of the high plain of Bogotá. The results of the MESA and TMT analyses (Fig. 4) show that during the last 0.8 Ma, in the eccentricity band, periods of 87–100 ka are significant, shifting to periods of 117–180 ka during the interval of 0.8–1.2 Ma BP. The last interval also shows significant periods of 21–24 and 19 ka (precession band) and c.11 ka.

The tree *Alnus* is an important element included

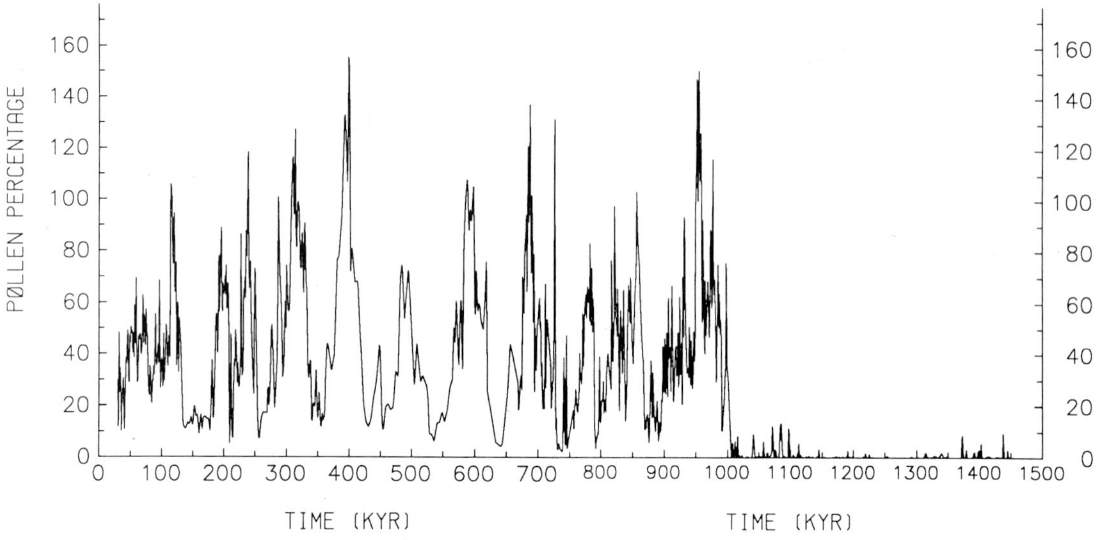

Fig. 5. Graph of Funza I pollen record *Alnus* percentages (*Alnus* not included in the pollen sum) versus time (36-point time-control model).

in the 'arboreal pollen' record. *Alnus* forms extensive stands of azonal swamp forest around the former lake of Bogotá. As *Alnus* does not occur above the forest line, this tree is only properly registered when the forest line is situated above 2550 m (during interglacial intervals). Therefore this record (Fig. 5) is very sensitive to transitions from glacial to interglacial conditions (and vice versa) when the forest line (corresponding to the annual *c*.9.5°C isotherm) passes the altitudinal level of the high plain of Bogotá. To a minor extent lake-level fluctuations are also reflected by the *Alnus* record (Hooghiemstra, 1984, 1989). During the last 0.8 Ma, in the eccentricity band, periods of 87–100 ka are significant (Fig. 6). A 41–44-ka period is present with changing significance (TMT) and the interval of 0.5–0.8 Ma BP also shows significant periods of 31–33 and 23 ka (MESA). Before 0.7 Ma BP a 19-ka (precession band) peak is also significant.

The group of subparamo elements represents the belt of open (alpine) scrub vegetation above the upper forest line. Many pollen diagrams from sites between altitudes of 2500 and 4000 m in the Eastern and Central Cordillera of Colombia suggest distinct changes in the vertical expansion of the subparamo vegetation belt (at the present day *c*.300 m wide), namely from *c*.3200 to *c*.3500 m. During the last 30 ka BP a very wide subparamo belt has been registered at *c*.27 ka BP (e.g. in pollen diagram Laguna de Fuquene: van Geel & van der Hammen, 1973) and *c*.6 ka BP (e.g. in pollen diagram TPN 21B: Salomons, 1986; and pollen diagram La Primavera: Melief, 1985). Thus, these intervals characterized by a wide subparamo belt are separated by about 21 ka in time, corresponding to one precession cycle. A relation to oscillations in climatic humidity is suggested by these pollen records. The record of the subparamo elements (Fig. 7) has been analysed by MESA and TMT techniques (Fig. 8). Results show that the presence of a significant precession band is most characteristic: periods of 21–24 and 29–31 ka during the most recent 0.8 Ma, preceded by periods of 19 and 22–23 ka during the interval of *c*.0.8–1.4 Ma BP. Other significant periods in the subparamo data set are 14–15 ka (0–0.9 Ma BP), 12.5–13 ka (0.8–1.4 Ma BP) and 10.1–10.6 ka throughout the record.

DISCUSSION

The ODP 677 $\delta^{18}O$ record (Shackleton *et al.*, 1990) has been astronomically tuned, so that it could be argued that the significant oscillations detected in the pollen records are artificially forced to match the Milankovitch frequencies. On the other hand, the same spectral analyses performed after using a linearly interpolated age model based on the three-

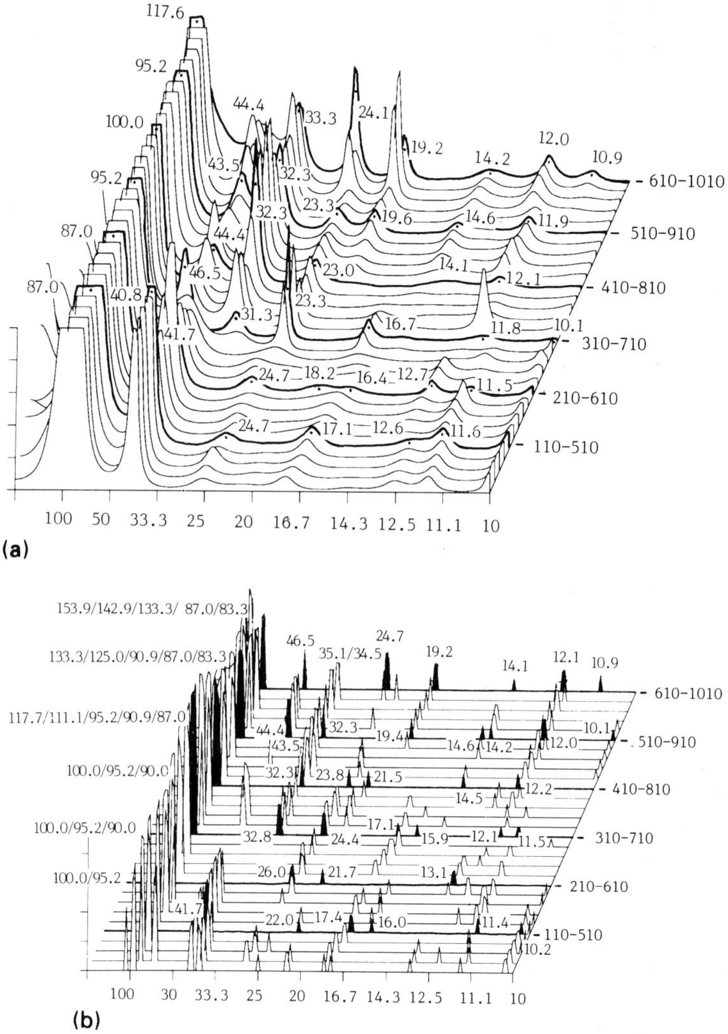

Fig. 6. MESA and TMT with significance level above 80% of *Alnus* percentages of Funza I pollen record, using a 36-point time-control model. A moving 400-ka window with steps of 20 ka provides 30 spectra. The first spectrum concerns the interval 30–430 ka BP. The maxima of the periodicities are expressed in ka and are indicated at every fifth spectrum. Note: The genus *Alnus* is originally a Northern Hemisphere element which migrated to South America after the closure of the Panamanian Isthmus (*c*.4 Ma BP). The record of *Alnus* starts in cores Funza I and Funza II at 257 m core depth. For this reason the *Alnus* record is shorter than the record of 'arboreal pollen'.

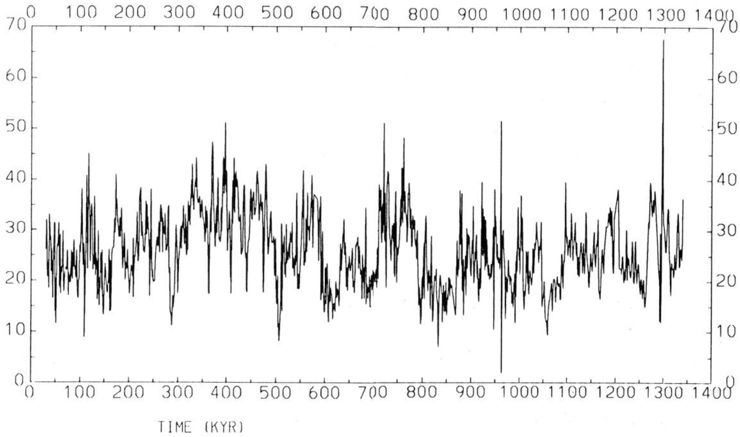

Fig. 7. Graph of subparamo element percentages of Funza I pollen record versus time (36-point time-control model).

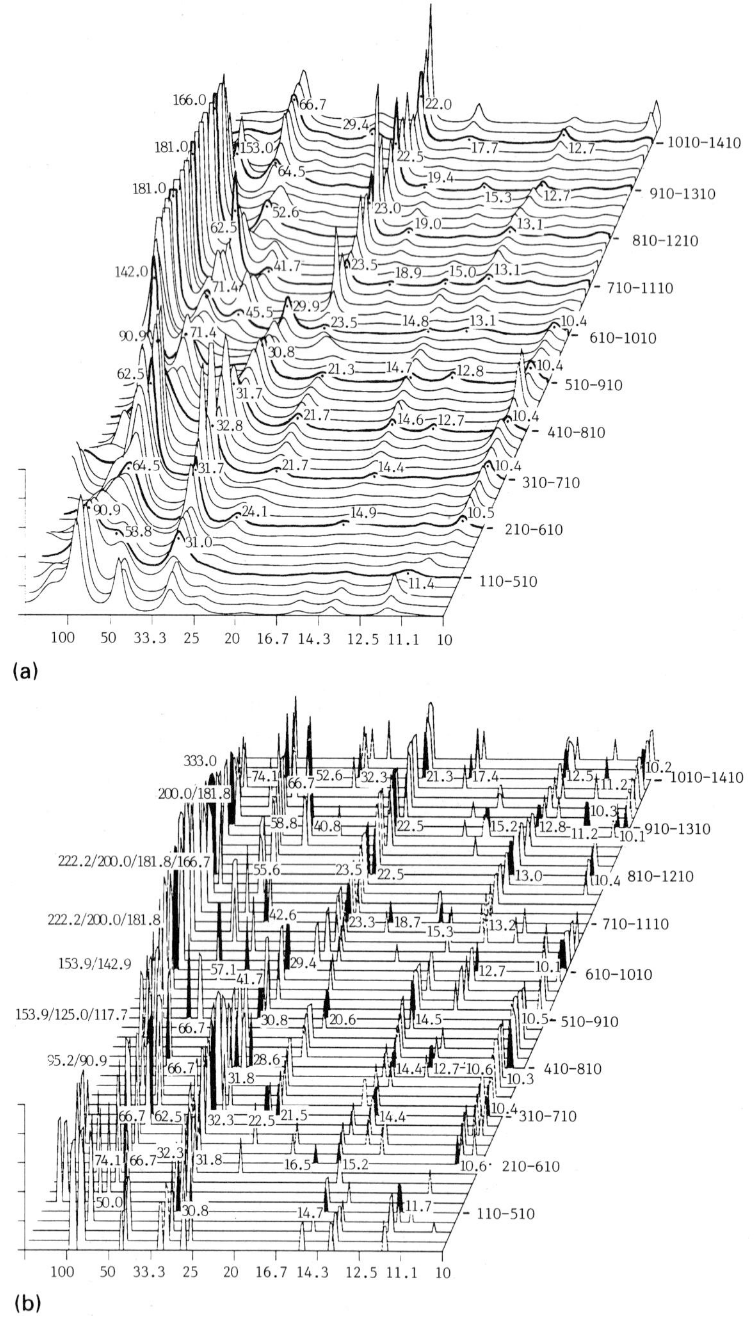

Fig. 8. MESA and TMT with significance level above 80% of subparamo element percentages of Funza I pollen record, using a 36-point time-control model. A moving 400-ka window with steps of 20 ka provides 52 spectra. The first spectrum concerns the interval 30–430 ka BP. The maxima of the periodicities are expressed in ka and are indicated at every fifth spectrum.

point time-control model (thus independent from orbital tuning) were able to extract oscillations in good agreement with the results of the biased 36-point time-control model. Especially when stretching the second control point from 1020 to 1206 ka, we obtained an alternative linear age model which is quite close to the 36-point model. The frequencies extracted when using such a three-point simplified chronology match almost perfectly those produced with the more sophisticated 36-point model. It has to be added that the age of 1206 ka used for the second control point lies within the uncertainty bar of the absolute dating.

The significant peaks found at *c*.15, *c*.13 and *c*.10 ka could be explained in terms of the climatic system's non-linear response to variations in the insolation available at the 'top of the atmosphere'. The climatic system can indeed be modelled as a non-linear oscillator forced by the periodic components of the orbital forcing periods (equal to 19, 23 and 41 ka respectively) (Ghil, 1989). More precisely, when the climatic non-linear model of Le Treut & Ghil (1983) is forced by these three periodic components it generates oscillations not only at the same periods, but also at 109, 14.7, 13.0, 11.5 and 10.4 ka. These predicted periodicities are very close to the ones obtained from our data.

CONCLUSIONS

The analysed data sets of this equatorial Andean pollen record reflect different variables of the palaeoclimate. Orbital forcing has been evidenced by the presence of Milankovitch periods in all analysed data sets. Significant periods of the eccentricity band have been registered particularly in the late and middle Pleistocene, and periods of the precession band are present throughout the record. Also, periodicities of higher palaeoclimatic variability centred at 14–15, 11.4–11.9 and 10.1–10.6 ka are detected throughout the record. The most important change in climatic variability occurred around 800 ka BP.

ACKNOWLEDGEMENTS

Professor Dr A. Berger is acknowledged for stimulating this research. The first author thanks the Netherlands Organization for Scientific Research (NWO, The Hague, grant H 75-284) for financial support. This paper was prepared on the occasion of the symposium 'Orbital forcing and cyclic sedimentary sequences', Utrecht, The Netherlands.

REFERENCES

Andriessen, P.A.M., Helmens, K.F., Hooghiemstra, H., Riezebos, P.A. & van der Hammen, T. (in press) Pliocene–Quaternary chronology of the sediments of the high plain of Bogotá, Eastern Cordillera, Colombia. Quat. Sci. Rev. (in press).

Burg, J.R. (1972) The relationship between maximum entropy spectra and maximum likelihood spectra. *Geophysics* **37**, 375–376.

Crowley, T.J. & North, G.R. (1991) *Paleoclimatology*. Oxford Monographs on Geology and Geophysics 16. Oxford University Press, Oxford.

Dupont, L.M. & Hooghiemstra, H. (1989) The Saharan–Sahelian boundary during the Brunhes chron. *Acta Bot. Neerl.* **38**, 405–415.

Ghil, M. (1989) Deceptively simple models of climatic change. In: *Climate and Geosciences* (Eds Berger, A., Schneider, S. & Duplessy, J.C.). Kluwer Academic Publishers, Dordrecht, pp. 211–240.

Hooghiemstra, H. (1984) *Vegetational and Climatic History of the High Plain of Bogotá, Colombia: a Continuous Record of the Last 3.5 Million Years*. Dissertaciones Botanicae 79. J. Cramer, Vaduz. (Also in: *The Quaternary of Colombia*, Vol. 10.)

Hooghiemstra, H. (1989) Quaternary and Upper-Pliocene glaciations and forest development in the tropical Andes: evidence from a long high-resolution pollen record from the sedimentary basin of Bogotá, Colombia. *Palaeogeogr. Palaeoclimatol. Palaeoecol.* **72**, 11–26.

Hooghiemstra, H. & Cleef, A.M. (in press) Lower Pleistocene and Upper-Pliocene climatic change and forest development in Colombia: pollen record Funza II (205–540 m core interval) *Palaeogeogr. Palaeoclimatol. Palaeoecol.* (in press).

Hooghiemstra, H. & Ran, E.T.H. (in press) Upper and Middle Pleistocene climatic change and forest development in Colombia: pollen record Funza II (2–158 m core interval) *Palaeogeogr. Palaeoclimatol. Palaeoecol.* (in press).

Hooghiemstra, H. & Sarmiento, G. (1991) Long continental pollen record from a tropical intermontane basin: Late Pliocene and Pleistocene history from a 540-meter core. *Episodes* **14**, 107–115.

Kanari, S., Fuji, N. & Horie, S. (1984) The paleoclimatological constituents of paleotemperature in Lake Biwa. In: *Milankovitch and Climate* (Eds Berger, A.L., Imbrie, J., Hays, J., Kukla, G. & Saltzman, B.). Reidel Publ. Co., Dordrecht, pp. 405–414.

Le Treut, H. & Ghil, M. (1983) Orbital forcing, climatic interactions and glaciation cycles. *J. Geophys. Res.* **88**, 5167–5190.

Melief, A.B.M. (1985) *Late Quaternary paleoecology of the Parque Nacional Natural los Nevados (Cordillera Central), and Sumapaz (Cordillera Oriental) areas, Colombia*. Thesis, University of Amsterdam. (Also in: *The Quaternary of Colombia*, Vol. 12.)

Molfino, B., Heusser, L.H. & Woillard, G.M. (1984) Frequency components of a Grande Pile pollen record: Evidence of precessional orbital forcing. In: *Milankovitch and Climate* (Eds Berger, A.L., Imbrie, J., Hays, J., Kukla, G. & Saltzman, B.). Reidel Publ. Co., Dordrecht, pp. 391–404.

Salomons, J.B. (1986) *Paleoecology of Volcanic Soils in the Colombian Central Cordillera (Parque Nacional Natural de los Nevados)*. Dissertaciones Botanicae 95. J. Cramer, Gebr. Borntraeger, Berlin-Stuttgart. (Also in: *The Quaternary of Colombia*, Vol. 13.)

Shackleton, N.J., Berger, A. & Peltier, W.R. (1990) An alternative astronomical calibration of the lower

Pleistocene timescale based on ODP Site 677. *Trans. R. Soc. Edinburgh, Earth Sci.* **81**, 251–261.

THOMSON, D.J. (1982) Spectrum estimation and harmonic analysis. *IEEE Proc.* **70**, 1055–1096.

VAN DER HAMMEN, T. (1974) The Pleistocene changes of vegetation and climate in tropical South America. *J. Biogeogr.* **1**, 3–26.

VAN GEEL, B. & VAN DER HAMMEN, T. (1973) Upper Quaternary vegetational and climatic sequence of the Fuquene area (Eastern Cordillera, Colombia). *Rev. Paleobot. Palynol.* **14**, 9–92.

WIJMSTRA, T.A. & GROENHART, M.C. (1983) Record of 700 000 years vegetational history in Eastern Macedonia (Greece). *Rev. Acad. Colomb. Cienc. Exactas, Fis. Nat.* **15**(58), 87–98.

Spec. Publs Int. Ass. Sediment. (1994) **19**, 127–143

Late Quaternary monsoonal variations in the western Arabian Sea based on cross-spectral analyses of geochemical and micropalaeontological data (ODP Leg 117, core 728A)

W.G.H.Z. TEN KATE, A. SPRENGER, T.N.F. STEENS *and* C.J. BEETS

Geomarine Centre, Institute of Earth Sciences, Free University, de Boelelaan 1085, 1081 HV Amsterdam, The Netherlands

ABSTRACT

ODP Hole 728A (Leg 117), located at a semi-enclosed slope basin on the continental margin of Oman, underlies a seasonally monsoon-enhanced upwelling zone. The sediment composition reflects past environmental fluctuations which are linked to variations in the monsoon system. These include seasonal upwelling, palaeoproductivity, variation in the extent of the OMZ and aeolian input. The top-most Pleistocene (21.5 m; 524 ka) was sampled at 25-cm intervals for micropalaeontological, geochemical and stable isotope analyses.

Twelve of the micropalaeontological and geochemical variables determined have been used as proxy indicators. These are assumed to be directly related to wind strength (Ti, Cr), biological productivity ($CaCO_3$, TOC, U, Ba, Br and I), intensity of the OMZ (planktonic foraminiferal tests whole–fragments ratio) and upwelling-induced productivity ($\delta^{13}C$ *Globigerina bulloides* – $\delta^{13}C$ *Neogloboquadrina dutertrei*). Examination for coherence and phase relationship with the SPECMAP stacked $\delta^{18}O$ record, using cross-spectral analyses, has revealed that productivity proxies vary in concert with OMZ proxies, and are coherent with the orbital cycles of the Milankovitch model. However, upwelling and aeolian intensity proxies may not be related directly to the orbital parameters, but show instead a more complicated behaviour.

INTRODUCTION

ODP Hole 728A, Leg 117 was drilled in an upper slope basin on the continental margin of Oman at 17° 40.790′ N latitude and 57° 49.553′ E longitude at a water depth of 1428 m (Fig. 1; Prell *et al.*, 1989a). One of the major objectives of this Leg was the palaeoenvironmental reconstruction of the Indian Ocean monsoon system, as recorded in the sediments of the western Arabian Sea.

The monsoon in this part of the world is dynamic, driven by atmospheric interactions between the Asian continent and the Indian Ocean. Recent hydrography of the southeast continental margin of Oman is typified by three interrelated phenomena: (i) seasonal upwelling caused by the SW summer monsoon; (ii) stable intermediate waters of high salinity derived from the marginal seas; and (iii) a strong oxygen minimum zone (OMZ) between 300 and 1500 m (Prell *et al.*, 1989a).

Heating of the Asian continent in summer and cooling in winter creates a summer low-pressure cell and a winter high-pressure cell above the Himalayas and the Tibetan Plateau (Fig. 2; Prell, 1984). In winter this results in near-surface winds of moderate strength that blow from Asia to the Arabian Peninsula. In summer the wind direction is reversed, and much stronger southwesterlies blow parallel to the coast of Oman. Not only do these winds carry an appreciable amount of dust, they also cause an offshore flow of surface waters by Ekman transport. The surface waters are replenished by nutrient-rich, oxygen-poor cold waters from deeper levels. Variations in intensity of this summer upwelling off Oman and Somalia have a profound effect on the nutrient content, temperature and biogenic productivity in the surface waters, and consequently on the intensity of the OMZ (e.g. Currie *et al.*, 1973; Bruce, 1974; Prell *et al.*, 1990).

Knowledge of the palaeoceanography of the

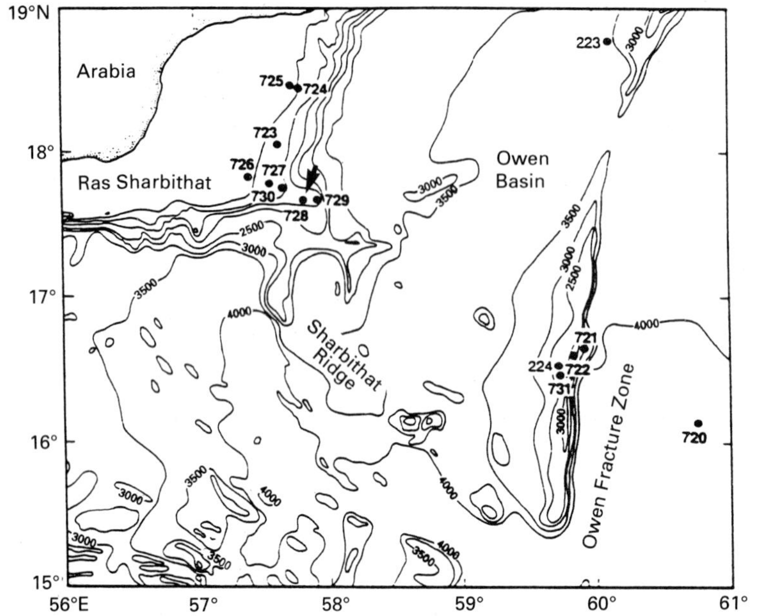

Fig. 1. Physiography of the Oman margin and the location of ODP Site 728 (water depth 1427.8 m). From Prell *et al.* (1989a).

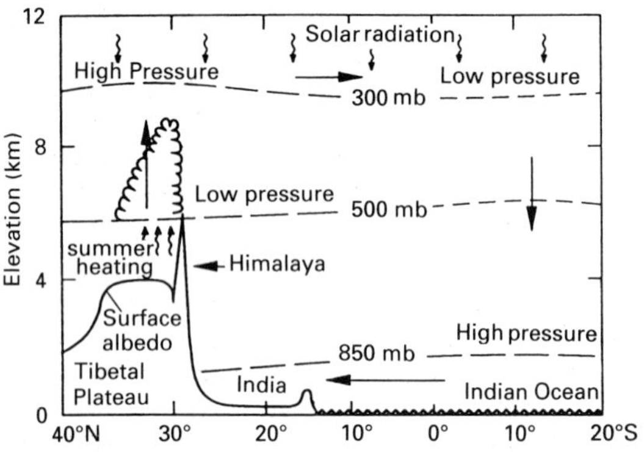

Fig. 2. Schematic boundary conditions of monsoonal atmospheric circulation in the northwest Indian Ocean area during the summer monsoon. Heating of air masses over the highlands of Asia creates a low-pressure cell which induces intense southwesterly winds in the lower atmosphere over the northwest Indian Ocean. mb, atmospheric pressure surfaces (exaggerated) in millibars; arrows denote wind direction. Redrawn after Prell (1984).

Arabian Sea and the monsoon system stems from the sedimentary record and from simulations of the monsoonal circulation with a general atmospheric circulation model (Prell & Kutzbach, 1987). During the past 150 ka, variations in seasonal distribution of solar radiation induced by changes in the Earth's orbital parameters appear to have forced the monsoon system. This caused the strength of the SW monsoon to increase during interglacial periods coinciding with precession-induced maxima in solar radiation (e.g. Prell & Kutzbach, 1987; Anderson & Prell, 1991; Clemens & Prell, 1991; Steens *et al.*, 1991).

Hole 728A, which is situated directly below the upwelling zone, provides a long record, back into the Miocene, of past environmental fluctuations coupled to variations in the monsoon system. The main emphasis of this paper lies in the spectral analysis of geochemical data and stable isotope ratios, determined for 87 samples at 25-cm intervals from late Pleistocene sediments (0–21.5 m). Altogether about 30 geochemical and microfaunal parameters were determined and inferences were made about their proxy value for palaeoenvironmental reconstruction. From these we selected the following: the $\delta^{18}O$ variations in *Neogloboquadrina dutertrei* (d'Orbigny) as an indicator for global palaeoice volume; calcium carbonate ($CaCO_3$) and total organic carbon (TOC) (not corrected for loss by dissolution or for dilution effects) as rough proxies for biogenic productivity; the trace elements barium (Ba) (Schmitz, 1987; Bishop, 1988; Shimmield & Mowbray, 1991), uranium (U) (ten Haven *et al.*, 1988), bromine (Br) (Fenical, 1975;

ten Haven *et al.*, 1988) and iodine (I) as indicators of biogenic productivity; the ratio of whole foraminiferal tests and the number of whole plus fragmented tests as proxy for the intensity of the OMZ; and titanium (Ti) (Boyle, 1983) and chromium (Cr) (Shimmield & Mowbray, 1991), which are thought to represent palaeowind strength and possibly sediment winnowing by (contour) currents. Upwelling intensity is thought to be documented by the $\delta^{13}C$ gradient between *Globigerina bulloides* (d'Orbigny) and *N. dutertrei*. Classical spectral analyses help to determine whether the proxy variables are wholly or partly controlled by orbital variations. Cross-spectral, coherence, and phase analyses in relation to SPECMAP stack $\delta^{18}O$ show deviations from global palaeocirculation and ice volume.

Shimmield & Mowbray (1991) and Weedon & Shimmield (1991) studied the shallow Oman Margin Site 724, and the pelagic Owen Ridge Site 722 (Fig. 1). Oman Margin Site 728 is deeper than Site 724, and therefore probably less influenced by fluvial material. They used basically the same methods, but our series is about 200 ka longer. A foraminiferal fragmentation index and the magnetic susceptibility record are also investigated. Our Ba/Al record comes from the Margin Site, whereas their data were from the Ridge Site.

METHODS

Analytical methods

To isolate foraminiferal tests, samples were washed over a 60-µm sieve and dried at 50°C. We counted whole planktonic foraminiferal tests and fragments in the 125–250-µm and >250-µm fractions. The ratio of whole tests (W) versus fragments (F) is expressed as: $(W/(W + F)) \times 100\%$ (data in Steens *et al.*, 1991).

Stable oxygen and carbon isotope ratios were measured on 50 specimens per sample of *N. dutertrei* from the >250-µm fraction, and 120 specimens of *G. bulloides* from the 125–250-µm fraction. The tests were cleaned ultrasonically in analytical grade methanol for 2 minutes, and dissolved in 100% phosphoric acid under vacuum at 50°C. Isotopic analyses were performed off-line on a MAT 251 mass-spectrometer. Analytical precision of a continuously run standard ('Merck' 100% calcite) was 0.05‰ for $\delta^{18}O$ and 0.04‰ for $\delta^{13}C$ during the measuring period. The isotope values are referred to PDB using the standard notation (Craig, 1957) and are calibrated to the NBS 19, 20 standard. The $\delta^{18}O$ values are listed in full in Steens *et al.* (1991), the $\delta^{13}C$ values in Steens *et al.* (1992).

Volume magnetic susceptibility was measured on board ship at intervals of 6–8 cm (Prell *et al.*, 1989b).

The coulometrically determined $CaCO_3$ content of the samples was calibrated with a standard solution of sodium carbonate. Reproducibility is within 2.5% of the measured values (data in Steens *et al.*, 1991).

Major and trace element contents of the bulk sediments (desalted by dialysis) were determined using standard instrumental neutron activation analyses with a reproducibility of 5–10% for U, Th, Ba, Br and I and 10–15% for Al (de Bruin, 1983).

Numerical methods

Figure 3 shows $\delta^{18}O$ values determined on *N. dutertrei* at 25-cm intervals down-core, and the volume magnetic susceptibility at intervals of 6–8 cm (Prell *et al.*, 1989b). *Neogloboquadrina dutertrei* is a thermocline dweller, therefore the oxygen isotope ratios mainly reflect the global ice volume signal. The isotope data were visually matched with SPECMAP stack-11 (Imbrie *et al.*, 1984) down to isotopic stage 13 (Steens *et al.*, 1991). This results in 28 match points over the timespan of 12–524 ka (Table 1). These are used to transform the data linearly from depth to time. The results of this transformation are shown in Fig. 4.

The number of datum planes is high, resulting in a straightforward transformation from depth-in-core into time. We decided to use the time-series data as input for spectral analyses. Figure 4 shows that accumulation rates were higher during glacial stages than during interglacials, hence the data points are unequally spaced in time. A new set of equally spaced data was generated by fitting cubic splines through the existing data points. In order to avoid an excessive increase of interpolated values and to counteract a change in resolution of the spectra, the interpolation interval was fixed at 6 ka and 86 interpolation points were calculated.

The input for spectral analysis needs to be stationary; at least the mean and variance of the data should not change with time. All variables were tested for the presence of a linear trend; only minor trends were found and these were removed from the series. Stationarity in variance was inferred from the time-series curves and has been assumed (Figs 6, 8, 10 and 12).

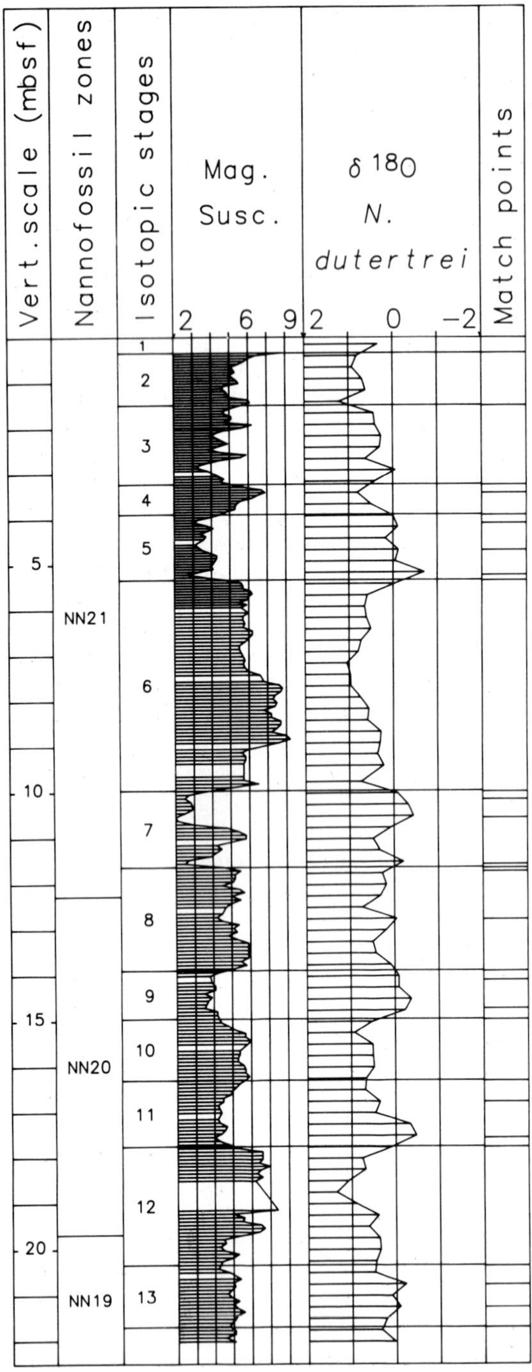

Table 1. List of data used to construct the time–depth relationships. Depth (metres below sea floor), age and events (SPECMAP) of the 28 points used to visually match the $\delta^{18}O$ profile of *N. dutertrei* with the SPECMAP curve (Imbrie *et al.*, 1984)

Depth (mbsf)	Age (ka)	Event
0.35	12	2.0
1.49	24	3.0
3.23	59	4.0
3.41	64	4.2
3.89	71	5.0
4.07	80	5.1
4.66	100	5.3
5.21	122	5.5
5.33	128	6.0
9.96	186	7.0
10.13	195	7.1
10.53	216	7.3
11.55	238	7.5
11.63	245	8.0
11.71	250	8.2
12.76	286	8.5
13.89	303	9.0
14.09	310	9.1
14.71	330	9.3
14.95	339	10.0
16.30	362	11.0
16.76	368	11.1
17.55	405	11.3
17.75	423	12.0
20.35	478	13.0
20.76	482	13.11
21.26	500	13.13
21.70	524	14.0

Univariate power spectra were calculated by the classical Blackman & Tukey (BT) method, in which the autocorrelation function is estimated from a sampled time series and then, by taking the Fourier transform, the power spectral density is obtained. The BT-spectrum is continuous and each power is given by the value of the integral of the power function over a small interval about the corresponding frequency. Because the sampled time series are finite, the Fourier transform creates a series of side lobes around every spectral peak. These are

Fig. 3 (*left*). Overview of the $\delta^{18}O$ values (‰ PDB), determined on *N. dutertrei*, at 25-cm intervals down-core, measured in the top 21.5 m of ODP Leg 117 Hole 728A, and the magnetic susceptibility data (μSI), determined at intervals of 6–8 cm. Horizontal lines mark measuring points. Odd numbers refer to interglacial periods, even numbers to glacials. Data from Steens *et al.* (1991).

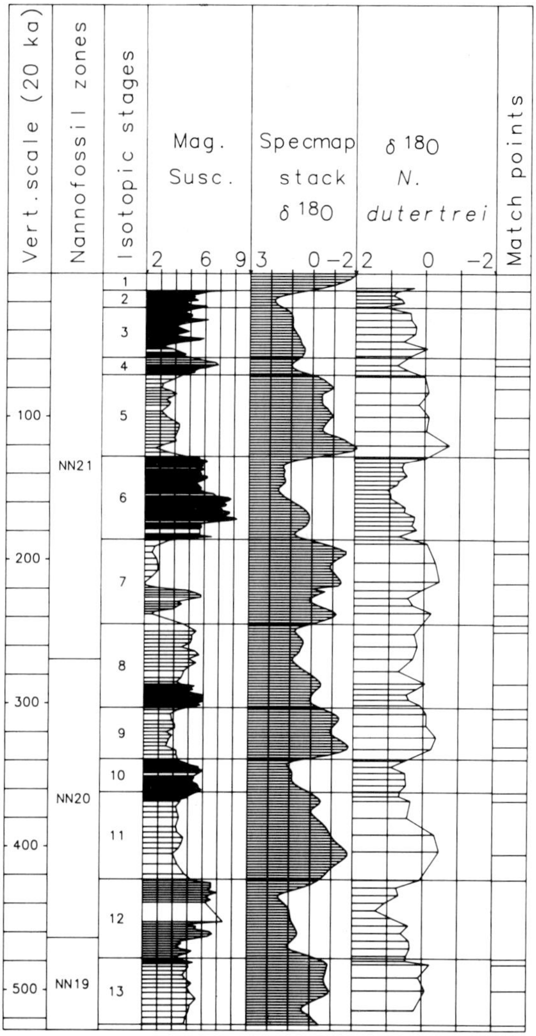

Fig. 4. Overview of data used to construct the time–depth relationships. The curve in the middle is the time series of the SPECMAP stack-11 (Imbrie *et al.*, 1984), sampled from 12 to 524 ka at intervals of 2 ka. The stack is reversed; high negative values correspond to interglacial conditions. The $\delta^{18}O$ profile of *N. dutertrei* (Fig. 3) is visually matched with the SPECMAP-curve, down to isotopic stage 13. This results in 28 match points over a timespan of 12–524 ka (Table 1). These were used to transform the data linearly from depth to time.

eliminated to a large extent by weighting the auto-correlation function with a proper lag window, which smooths the power over adjacent frequencies. The degree of smoothing is controlled by the width of this filter and is expressed in a spectrum by the bandwidth. Estimated power values at two frequencies separated by more than the bandwidth may be taken to be independent. Increasing the lag window reduces the bandwidth and gives a more detailed spectrum, but on the other hand the standard error of the estimated spectral power is increased and fewer power values will differ significantly from zero at a particular confidence level. For the theory the reader is referred to Blackman & Tukey (1958) and Jenkins & Watts (1968). We used the Tukey–Hanning lag window with 43 parameters. All time series were standardized, which does not affect the shape of the spectra, but allows a direct comparison of power estimates.

The SPECMAP stack was sampled at the same intervals of 6 ka and cross-spectra were made with all the proxies. Squared coherency at a particular frequency measures the correlation of the power between two time series at that frequency. Its value varies between 0, i.e. no correlation at all, and 1, i.e. a perfect correlation, which means either a perfect direct or a perfect inverse relationship. The extent to which one of the series leads or lags the other at a certain frequency is measured by the phase angle of that particular period in degrees. Phase angles are given in the range −180° to +180° and a whole number of multiples of 360° may be added or subtracted. A phase estimate is reliable only at a frequency at which the coherence estimate is significant.

DISCUSSION

At the top of Fig. 5 the power spectra of the SPECMAP (solid curve) and of the *N. dutertrei* oxygen isotope data (dashed curve) are shown. The numbers plotted above the peaks of the SPECMAP spectrum are periods in ka. The orbital frequencies are shown with a range of ±1 ka. The eccentricity cycle around 100 ka dominates the spectrum. The 41-ka obliquity cycle and 23-ka precession cycle are represented by clear peaks; the 19-ka precession peak is subdued.

In the centre of the figure the spectrum of squared coherencies is shown with the 80% confidence level. All coherencies below this level do not differ significantly from zero. As expected, both series are coherent at the orbital frequencies 100, 41 and 23 ka.

The spectrum at the bottom shows the phase angles between both series together with a band of 6-ka lead or lag of the SPECMAP curve with respect to the isotope curve. Phase angles are nearly zero at the orbital frequencies 100, 41 and 23 ka. The 19-ka

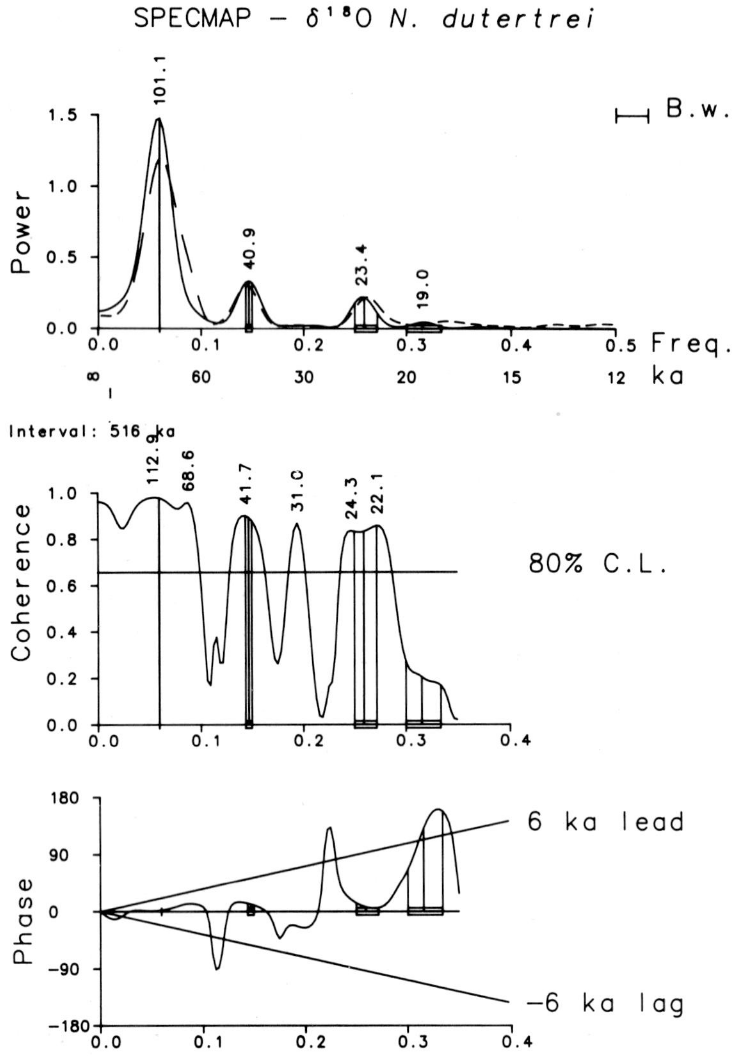

Fig. 5. Top: univariate Blackman–Tukey spectra of SPECMAP stack $\delta^{18}O$ (solid curve) and of $\delta^{18}O$ *N. dutertrei* (dashed curve) are shown. The spectra are based on time series of 86 data points, which are sampled at 6-ka intervals and detrended for small linear trends. Both spectra are smoothed with a Tukey–Hanning filter with the bandwidth as indicated. The numbers plotted above the peaks of the SPECMAP spectrum are periods in ka. The orbital frequencies at 100, 41, 23 and 19 ka are indicated with a band of ±1 ka. Middle: squared coherence spectrum of the bivariate cross-spectrum together with the 80% confidence level. Bottom: phase spectrum together with a band of 6-ka lead or lag.

cycle is not coherent and not in phase and its power is subordinate. This might be ascribed to the combined effect of insufficient resolution of the depth–time relationship at that scale, sampling distance, and subordinate power. Although 28 match points over the range 12–524 ka is high for geological data, the average distance of match points is 19 ka. The SPECMAP curve is sampled at constant time intervals, whereas our isotope record of *N. dutertrei* has much wider spaced time intervals in interglacial than in glacial periods. Comparing isotopic stage 6 to stage 7 in Fig. 4 shows the undersampling of the latter. The SPECMAP spectrum has a minor peak at 19 ka, whereas the power in the isotope spectrum of *N. dutertrei* is a subdued broad peak from 16 to 20 ka, with a maximum near 17–18 ka (Fig. 5).

The relatively low number of sampling points (87) limits the resolution of the spectra. Nevertheless, we may conclude that the isotope record of *N. dutertrei* is forced by changes in global ice volume and thus by orbitally induced climatic variation in time.

Using geochemical proxies to quantify palaeoproductivity

The Arabian Sea is characterized by strong seasonal upwelling caused by the SW summer monsoon. The enhanced nutrient supply stimulates biological productivity. This results in high $CaCO_3$% and TOC% in the sediments. The organic matter produced in the euphotic zone decays while it is sinking towards the sea floor and helps to create an OMZ between 300 and 1500 m water depth. The reason for the existence of this OMZ lies in the stable stratification of the saline intermediate water and the low lateral and vertical exchange of oxygen. The sources for the intermediate water are the marginal basins, the Arabian Gulf, the Red Sea, the open Arabian Sea and the Gulf of Aden (Wyrtki, 1973).

The most variable parameter governing the short-term duration and strength of the OMZ is the seasonal biological productivity. Over a longer period global ice volume, and thus sea level, plays an important role (considering the various shallow sill depths of the marginal seas).

With the high primary productivity Ba becomes enriched in the sediments (Dymond, 1981; Bishop, 1988) provided that no redox distribution occurs. High U, Br and I concentrations are often associated with high organic carbon content in the sediments (e.g. ten Haven *et al.*, 1988). Using these trace elements as proxies for biological productivity, the input from other (mainly terrigenous) sources has to be excluded. Shimmield *et al.* (1990) suggested normalizing against Al, where Al stands for the terrigenous aluminosilicates.

Ba/Al and U/Th in our 524-ka sedimentary record show peaks during the warmer, interglacial periods and lows during the colder, glacial periods (Fig. 6). These glacial–interglacial variations indicate a strong dependence on global changes in palaeocirculation and ice volume (Shimmield & Mowbray, 1991).

The TOC and bulk $CaCO_3$ contents are selected as rough estimates of biogenic palaeoproductivity. The time series of TOC fluctuates strongly, but on average higher values dominate in the interglacials and low values in the glacials (Fig. 6). The cross-spectrum reveals high coherences at orbital frequencies (Fig. 7a). The phase spectrum shows slight lags with respect to the SPECMAP spectrum. Bulk $CaCO_3$ content is diluted by terrigenous material of aeolian or fluviatile origin and by dissolution in the OMZ. Visually the time series of SPECMAP and $CaCO_3$ content show a remarkable correspondence (Fig. 6). The cross-spectrum varies coherently at the orbital frequencies and SPECMAP leads the $CaCO_3$ signal by a few thousand years (Fig. 7b).

The main variation in the series of normalized Ba and U are peaks during warm periods and lows

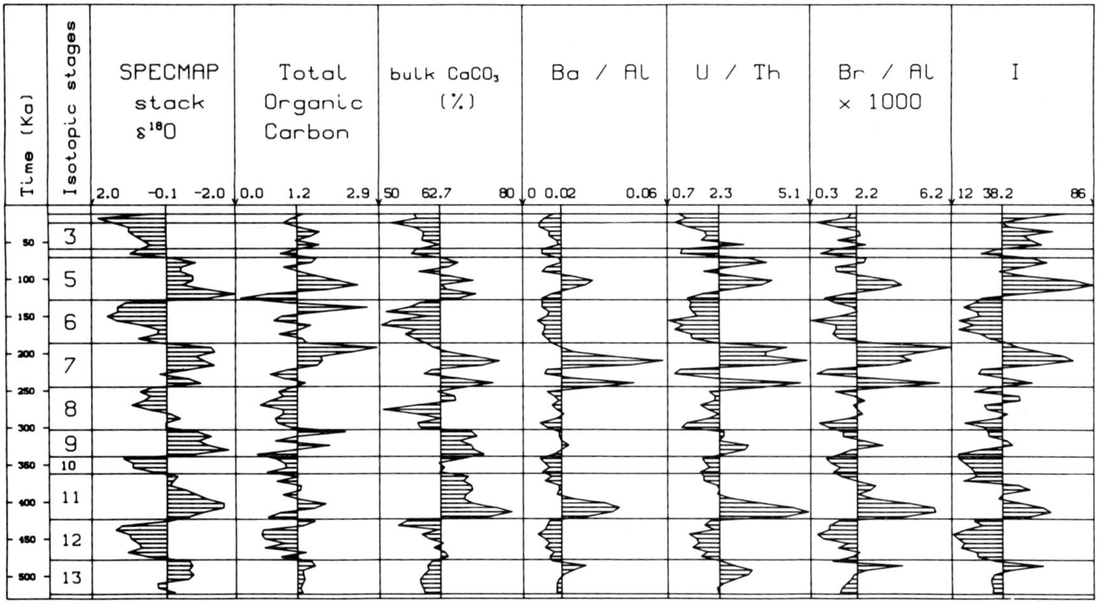

Fig. 6. Time series of the SPECMAP stack $\delta^{18}O$, sampled from 12 to 522 ka at intervals of 6 ka in relation to the cubic spline interpolated (lag = 6 ka) time series of geochemical variables which quantify paleoproductivity, TOC, bulk $CaCO_3$ content, Ba/Al, U/Th, (Br/Al) × 1000, and I. Each data series consists of 86 data points.

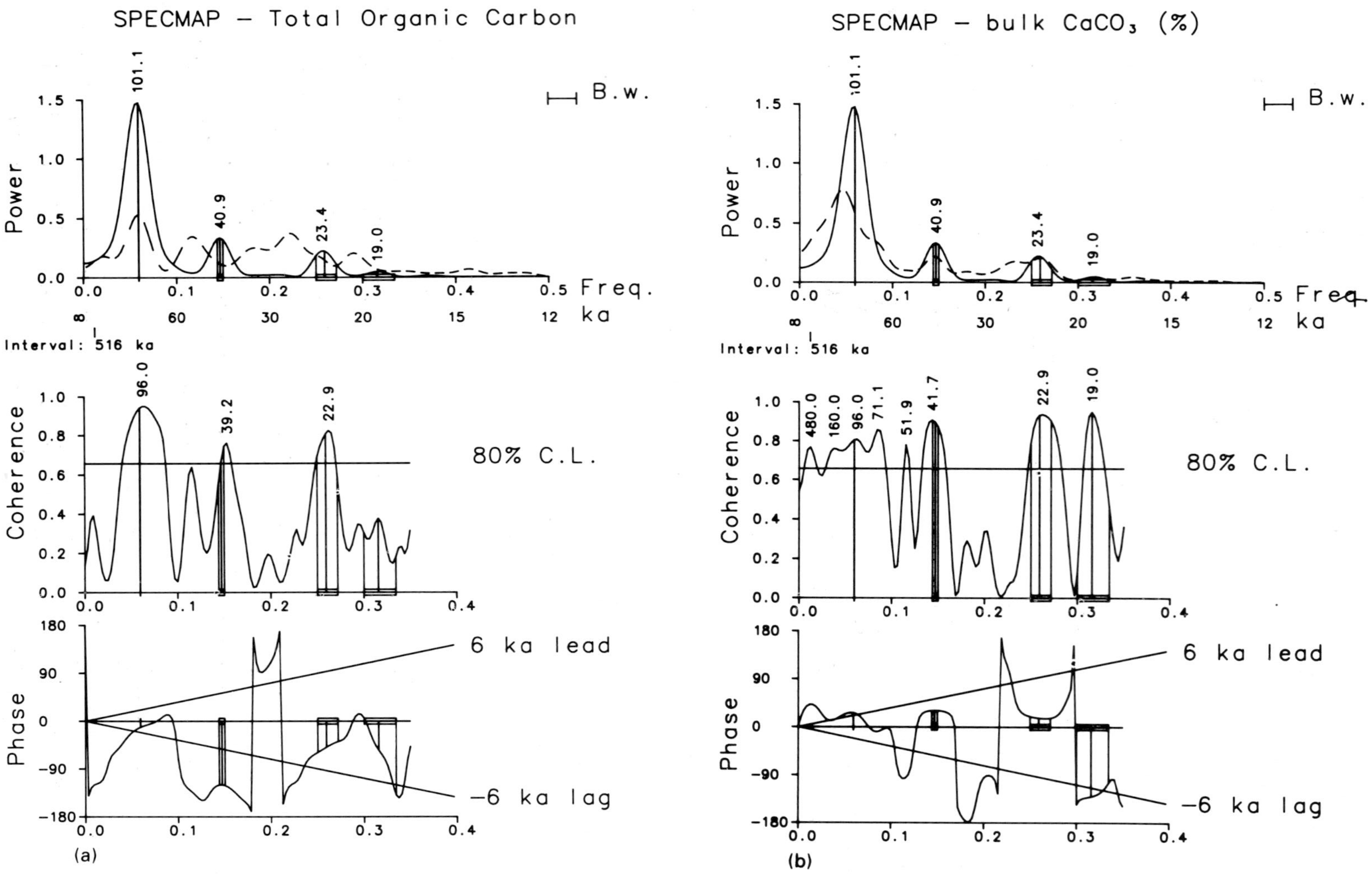
SPECMAP – Total Organic Carbon
B.w.
Power
Freq.
ka
Interval: 516 ka
Coherence
80% C.L.
Phase
6 ka lead
–6 ka lag
(a)
SPECMAP – bulk CaCO₃ (%)
B.w.
Power
Freq
ka
Interval: 516 ka
Coherence
80% C.L.
Phase
6 ka lead
–6 ka lag
(b)

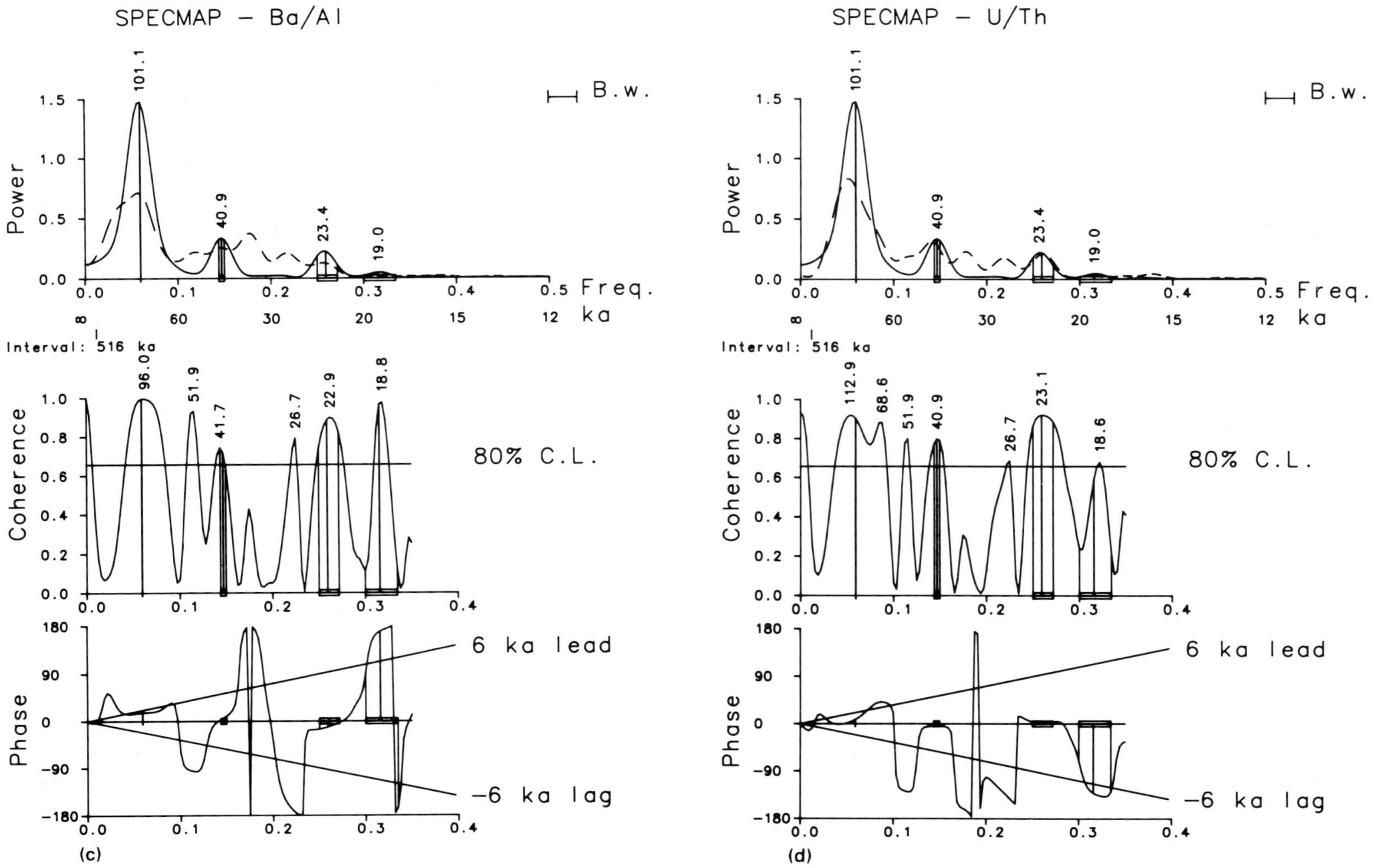

Fig. 7. BT, coherence and phase spectra of the SPECMAP data with geochemical variables approximating palaeoproductivity: (a) TOC, (b) bulk $CaCO_3$ content, (c) Ba/Al, (d) U/Th, (e) (Br/Al) ×1000. For further explanation see Fig. 5.

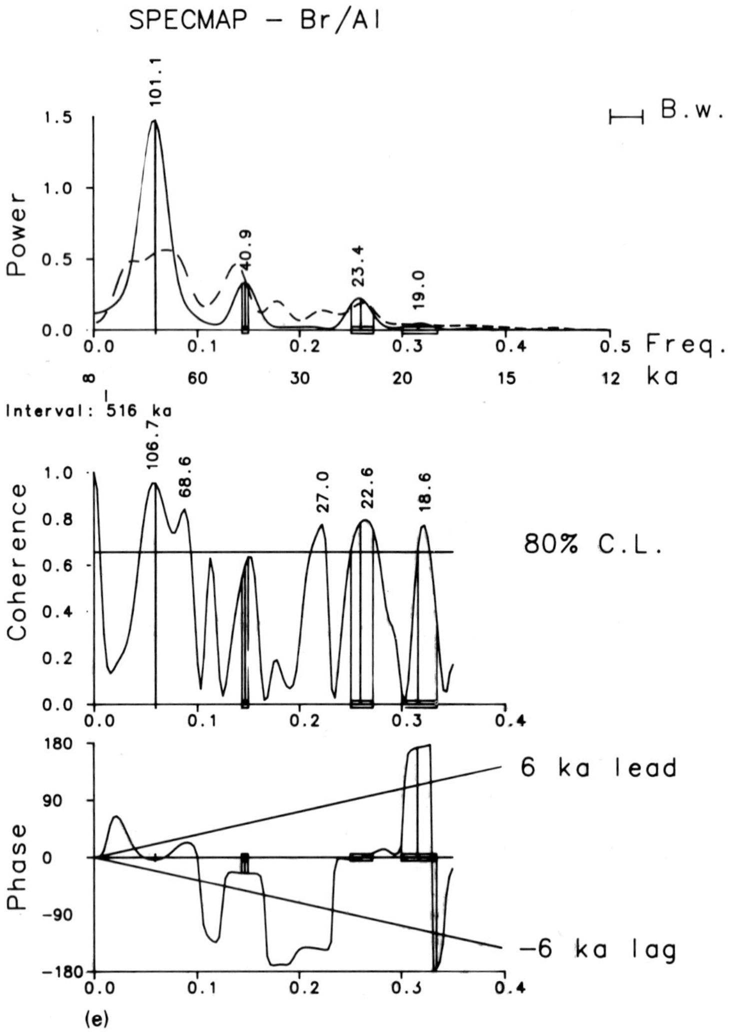

Fig. 7. *Continued.*

during glacials (Fig. 6). This is confirmed by a high coherence and a small phase angle at the eccentricity frequency (Fig. 7c, d). At higher frequencies the records show more variation, but both are coherent and in phase at the precession and obliquity frequencies.

The fact that certain algae accumulate Br and I (Fenical, 1975; ten Haven *et al.*, 1988; Pedersen & Shimmield, 1990) can be used as an indicator for the standing crop of organisms. High Br and I concentrations should coincide with periods of high primary productivity. The time series (Fig. 6) and the spectra of Br (Fig. 7e) and I are very similar to those of Ba and U. Moreover, this similarity indicates that no migration of I occurred due to diagenetic processes, which in turn points to a stable OMZ.

Intensity of the OMZ

High productivity results in a decreased oxygen concentration in the OMZ and enhanced acidity, which in turn enhances the dissolution of $CaCO_3$. Steens *et al.* (1991) proposed using the ratio of whole planktonic foraminiferal tests/(whole tests + fragments) (W/(W + F)) × 100% as a proxy for the intensity of dissolution and therefore of the strength of the OMZ. The perfect correlation between high productivity (high Ba/Al and U/Th) and a low (W/(W + F)) × 100% ratio confirms the direct link between strong upwelling, causing high biological productivity, inducing low oxygen and increased acidity and thus fragmentation of foraminiferal tests.

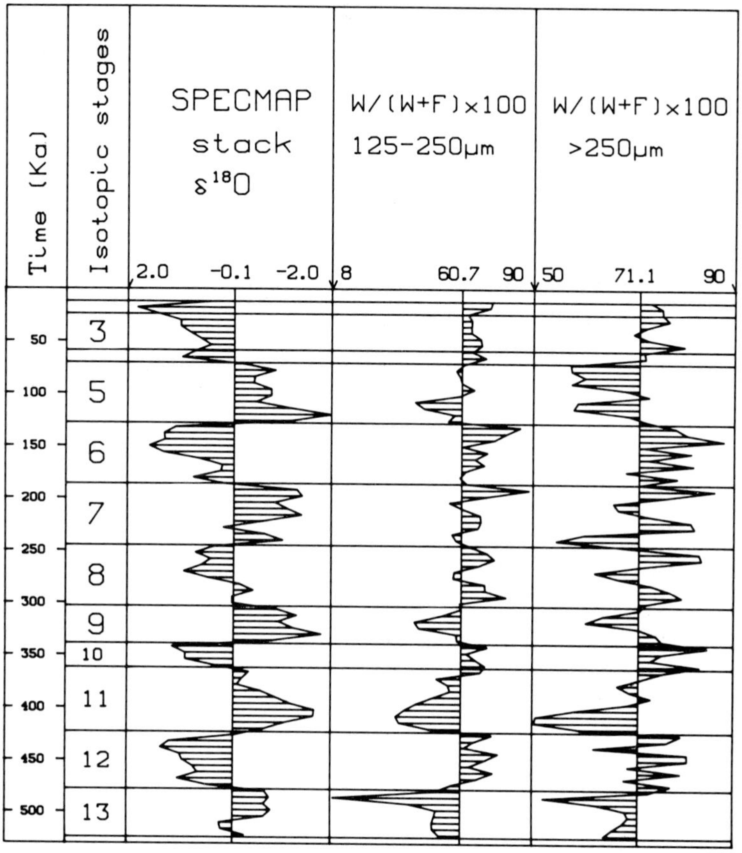

Fig. 8. Overview of the time series of SPECMAP and proxies for the intensity of the OMZ: the ratio whole planktonic foraminiferal tests/whole tests + fragments expressed as $(W/(W + F)) \times 100\%$ (*W*, whole tests; *F*, fragments), in the fractions 125–250 μm and >250 μm (data in Steens *et al.*, 1991).

The fragmentation is mainly due to dissolution and not to mechanical breaking (Steens *et al.*, 1991)

The time series of this proxy (Fig. 8) is practically a mirror image of the productivity factors, especially those of U/Th, Ba/Al and $CaCO_3$. The cross-spectrum of the fragmentation index varies coherently with SPECMAP at the eccentricity and obliquity frequencies, but not at the precession frequency (Fig. 9). Due to the antisymmetric relationship the phase angle at these frequencies is 180°.

Upwelling-related factors

According to Kroon & Ganssen (1989) the carbon isotope composition in various modern species of planktonic Foraminifera reflects different phases of upwelling. During the initial stages of upwelling, *N. dutertrei* becomes enriched in ^{12}C, while in the final phase of upwelling *G. bulloides* shows a depletion of ^{12}C. They suggest that under upwelling conditions the difference in carbon isotope values between the two becomes smaller and may serve as an indicator of upwelling.

However, both the time series (Fig. 10) and coherence spectrum (Fig. 11) show that this proxy has no simple relationship with the SPECMAP curve. Furthermore, the obliquity peak (41 ka) is not coherent because of a shift in phase angle. The lack of correlation can be explained by a change from an inverse to a direct relationship in the upwelling indicator with $\delta^{18}O$ in isotopic stage 11 (Steens *et al.*, 1992).

Terrigenous input factors

Average accumulation rates were considerably higher during glacial stages than during the interglacials (Fig. 4). This could be the effect of lowstands of sea level during which a higher input of terrigenous material occurred. The exposed shelf area is enlarged, enhancing transport in the nepheloid layer (Sirocko & Sarnthein, 1989) and input of

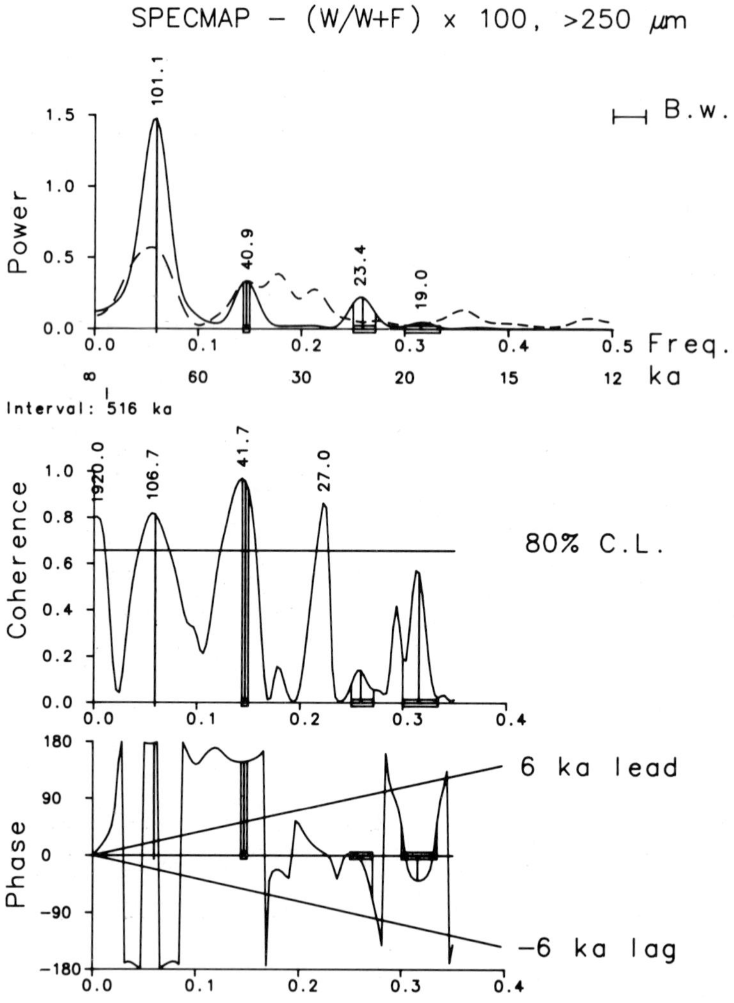

Fig. 9. BT, coherence and phase spectra of the SPECMAP data with the ratio of whole planktonic foraminiferal tests/whole test + fragments in the >250-μm fraction. For further explanation see Fig. 5.

aeolian dust, which also increased due to greater overall aridity on the continent (Clemens & Prell, 1991).

Magnetic susceptibility, Ti/Al and Cr/Al are regarded as proxies for terrigenous input (e.g. Shimmield & Mowbray, 1991). The particles that cause the magnetic susceptibility are mainly restricted to the clay-sized fraction of the sediments, and are brought in by dust-storms and rivers (de Menocal *et al.*, 1991). The time series of the magnetic susceptibility mirrors the SPECMAP $\delta^{18}O$ curve (Fig. 12). It is systematically above average during glacials and below average during interglacials. The possibility that this is due to dilution with $CaCO_3$ during interglacials is ruled out because accumulation rates during glacial stages are much higher than in interglacial stages.

Coherence around the precession and eccentricity frequencies is high, but at the obliquity frequency not significant (Fig. 13a). Phase differences at orbital frequencies are nearly 180° due to the negative correlation between $\delta^{18}O$ and magnetic susceptibility.

From Core 722B on the Owen Ridge, Shimmield & Mowbray (1991) and Weedon & Shimmield (1991) reported evidence that Ti and Cr are especially concentrated in the heavy mineral fraction of the aeolian input. They proposed to use the Al-normalized values of Ti and Cr as proxies for aeolian input. The time series for Hole 728A are shown in Fig. 12. The cross-spectra of both proxies are very similar, and the coherence between SPECMAP $\delta^{18}O$ and the Ti/Al ratio is effectively zero (Fig. 13b). This is probably due to input of heavy minerals from fluvial and aeolian sources, coupled with the strong im-

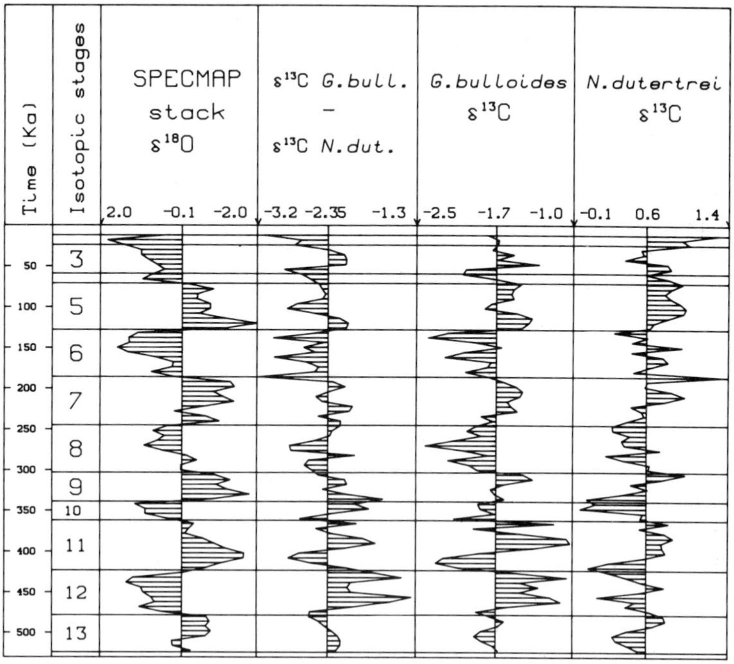

Fig. 10. The leftmost curve is the time series of the SPECMAP stack $\delta^{18}O$, sampled from 12 to 522 ka at intervals of 6 ka. The curve to the right of this column is the cubic spline interpolated (lag = 6 ka) time series of $\delta^{13}C$ *G. bulloides* − $\delta^{13}C$ *N. dutertrei*. Each data series consists of 86 data points.

pact of sea-level change on Oman Margin sediments (Weedon & Shimmield, 1991).

Our findings at Margin Site 728 differ from those described for Site 722 on the Owen Ridge (e.g. Shimmield *et al.*, 1990; Clemens & Prell, 1991). Both Sites 728 and 722 are greatly influenced by the monsoon. The non-carbonate mass flux at Site 722 on the Owen Ridge reflects largely aeolian dust, with precession as the ruling factor. At Site 728 the highest peak is found at the eccentricity frequency. Most likely this is due to the position of 728, on the margin of Oman. Sea-level fluctuations related to ice volume changes are the most important factor, resulting in exposure of the shelves and cut-off from marginal seas, leaving their imprint on the sediments.

CONCLUSIONS

The various proxies used and their coherence with SPECMAP stack $\delta^{18}O$ at the orbital quasi-periodicities 100, 41, 23 and 19 ka ± 1 ka, and significant at the 80% level, are summarized in Table 2.

The proxies used to quantify the global ice volume ($\delta^{18}O$ of *N. dutertrei*) and biological palaeo-productivity (TOC content, bulk $CaCO_3$ content, Ba/Al, U/Th, Br/Al and I) vary coherently with the orbital frequencies of eccentricity (100 ka), obliquity (41 ka) and precession (23 and 19 ka). The 19-ka

Table 2. Overview of the various proxies used and their coherence with SPECMAP stack $\delta^{18}O$ at the orbital quasi-periodicities 100, 41, 23 and 19 ka ± 1 ka and significant at the 80% level

Global ice volume				
$\delta^{18}O$ *N. dutertrei*	100	41	23	—
Productivity				
TOC (%)	100	41	23	—
bulk $CaCO_3$ (%)	100	41	23	19
Ba/Al	100	41	23	19
U/Th	100	41	23	19
Br/Al	100	—	23	19
Upwelling intensity				
$\delta^{13}C$ *G. bulloides* − $\delta^{13}C$ *N. dutertrei*	—	—	—	19
Intensity OMZ				
whole/(whole + fragments), >250 μm	100	41	—	—
Terrigenous input				
magnetic susceptibility	100	—	23	—
Ti/Al	—	41	—	—
Cr/Al	—	—	—	—

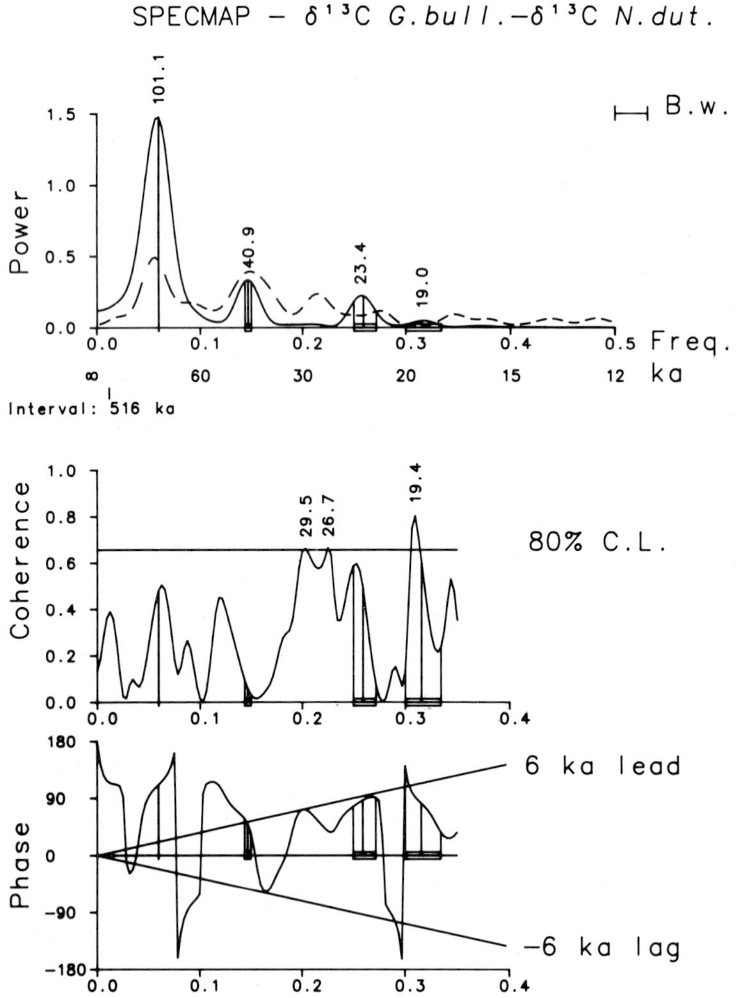

Fig. 11. BT, coherence and phase spectra of the SPECMAP data of $\delta^{13}C$ *G. bulloides* – $\delta^{13}C$ *N. dutertrei*. For further explanation see Fig. 5.

cycle is subdued in power in all spectra due to undersampling of the interglacial periods.

No distinct relationship between the upwelling proxy ($\delta^{13}C$ *G. bulloides* – $\delta^{13}C$ *N. dutertrei*) and orbital frequencies of SPECMAP could be established. Up to stage 10 the variation in both curves is synchronous. Beyond stage 10 this relationship is inverted (Fig. 10).

On the other hand the proxy for intensity of the OMZ (ratio of whole planktonic foraminiferal tests and whole tests plus fragments) shows a direct relationship at the eccentricity (100 ka) and obliquity (41 ka) frequencies with the SPECMAP curve.

Of all proxies reflecting terrigenous input, magnetic susceptibility varies coherently with eccentricity (100 ka) and precession (23 ka), but not with obliquity. Ti/Al and Cr/Al ratios in our data are not coherent with the Milankovitch cycles at all.

The fact that the Owen Ridge sites are dominated by precession and the Oman Margin sites by eccentricity can be explained by the location of these sites. The non-carbonate sediments of the Owen Ridge originate mainly through aeolian deposition, while at the Oman Margin site they are influenced by the combination of aeolian flux, fluvial deposition, sediment reworking and, very important, changes in sea level, which combine to modify the sediment accumulation and leave a much more complicated record.

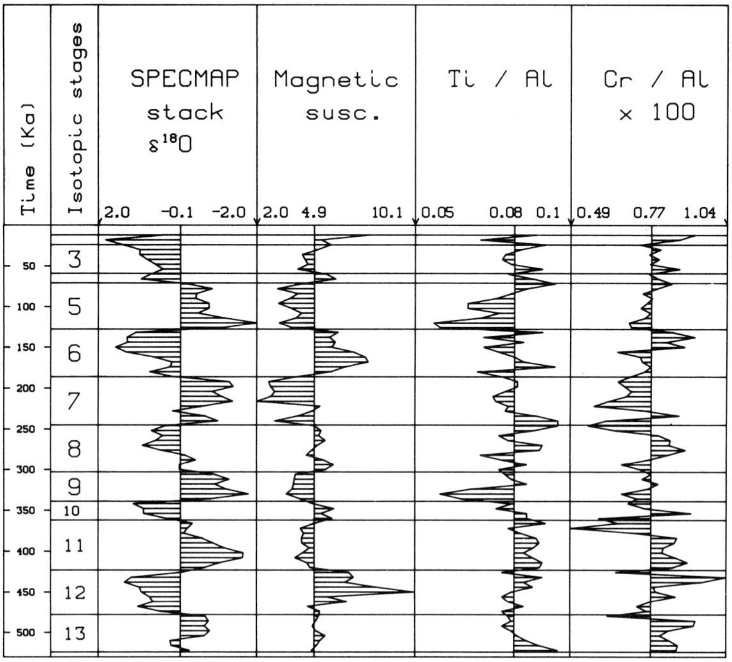

Fig. 12. The leftmost curve is the time series of the SPECMAP stack $\delta^{18}O$, sampled from 12 to 522 ka at intervals of 6 ka. The curves to the right of this column are the cubic spline interpolated (lag = 6 ka) time series of geochemical variables of aeolian or fluvial input, magnetic susceptibility, Ti/Al and (Cr/Al) ×100. Each data series consists of 86 data points.

ACKNOWLEDGEMENTS

We thank Gerald Ganssen for his advice on stable isotopes, Dick Kroon for making his data available and for his many useful suggestions, Thea van Meerten for performing neutron activation analyses, Gerard Klaver for helpful suggestions, and Peter Willekes, Roel van Elsas and Martin Koonert who determined the calcium carbonate content and prepared many washed residues. We thank Jan E. van Hinte for initiating this project and together with Poppe L. de Boer for critically reviewing the first draft of the manuscript. We thank G. Shimmield and G.P. Weedon for their thorough reviews of the manuscript. This study has been carried out on material of ODP Leg 117 and was supported by the Nederlandse Organisatie voor Wetenschappelijk Onderzoek (NWO) (T.N.F. Steens and C.J. Beets, grant 751-356-018) and the Stichting Onderzoek der Zee (SOZ). This is a publication of the Geomarine Centre Amsterdam.

REFERENCES

Anderson, D.M. & Prell, W.L. (1991) The coastal upwelling off Oman gradient during the Late Pleistocene. In: *Proc. Ocean Drilling Program Scientific Results, 117* (Eds Prell, W.L., Niitsuma, N. *et al.*). College Station, TX (Ocean Drilling Program), pp. 265–276.

Bishop, J.K.B. (1988) The barite–opal–organic carbon association in oceanic particulate matter. *Nature* **332**, 341–343.

Blackman, R.B. & Tukey, J.W. (1958) *The Measurement of Power Spectra.* Dover Publications Inc., New York.

Boyle, E.A. (1983) Chemical accumulation variations under the Peru current during the past 130 000 yrs. *J. Geophys. Res.* **88**, 7667–7680.

Bruce, J.G. (1974) Some details of upwelling off the Somali and Arabian coast. *J. Mar. Res.* **8**, 419–423.

Clemens, S.C. & Prell, W.L. (1991) Late Pleistocene variability of Arabian Sea summer monsoon winds and continental aridity: Eolian records from the lithogenic component of deep-sea sediments. *Paleoceanography* **5**, 109–145.

Craig, H. (1957) Isotopic standards for carbon and oxygen and correction factors for mass spectrometric analysis of CO_2. *Geochim. Cosmochim. Acta* **12**, 133–149.

Currie, R.I., Fisher, A.E. & Hardgraves, P.M. (1973) Arabian Sea upwelling. In: *The Biology of the Indian Ocean* (Eds Zeitzschel, B. & Gerlach, S.A.). Springer-Verlag, New York, pp. 37–52.

De Bruin, M. (1983) *Instrumental Neutron Activation Analysis – a Routine Method.* Delftse Universitaire Pers.

De Menocal, P., Bloemendal, J. & King, J. (1991) A rock-magnetic record of monsoonal dust deposition to the Arabian Sea: evidence for a shift in the mode of deposition at 2.4 Ma. In: *Proc. Ocean Drilling Program Scientific Results, 117* (Eds Prell, W.L., Niitsuma, N.

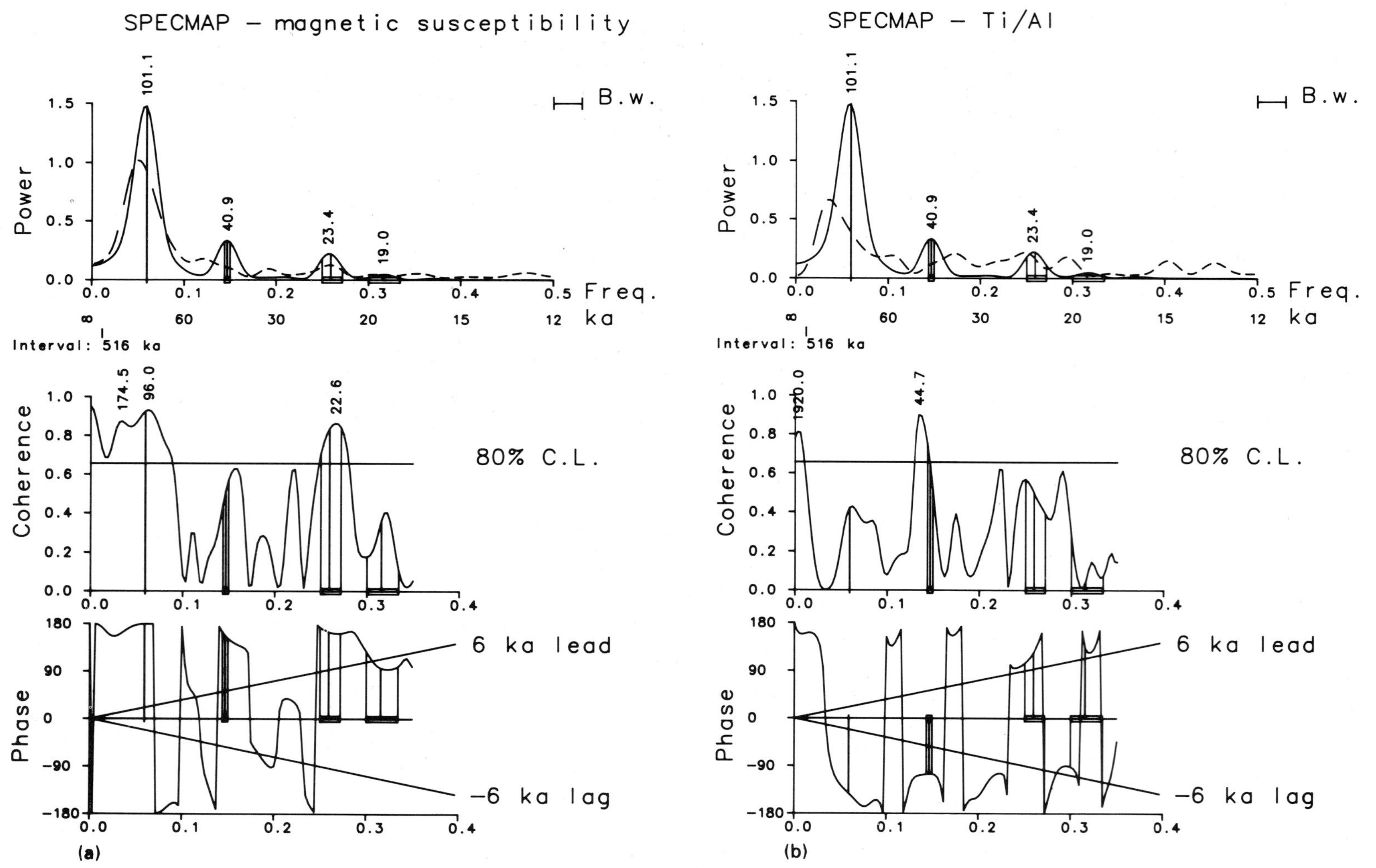

Fig. 13. BT, coherence and phase spectra of the SPECMAP data with geochemical variables of terrigenous input: (a) magnetic susceptibility; (b) Ti/Al. For further explanation see Fig. 5.

et al.). College Station, TX (Ocean Drilling Program), pp. 389–407.

DYMOND, J. (1981) Geochemistry of Nazca plate surface sediments: an evaluation of hydrothermal, biogenic, detrital, and hydrogenous sources. *Mem. Geol. Soc. Am.* **154**, 133–173.

FENICAL, W. (1975) Halogenetion in the Rhodophyta: a review. *J. Phycol.* **11**, 245–259.

IMBRIE, J., HAYS, J.D., MARTINSON, D.G., MCINTYRE, A., MIX, A., MORLEY, J.J. *et al.* (1984) The orbital theory of the Pleistocene climate: support from a revised chronology of the marine $\delta^{18}O$ record. In: *Milankovitch and Climate* (Eds Berger, A., Imbrie, J., Hays, J., Kukla, G. & Saltzman, B.). Reidel Publ. Co., Dordrecht, pp. 269–305.

JENKINS, G.M. & WATTS, D.G. (1968) *Spectral Analysis and its Applications*. Holden-Day, San Francisco.

KROON, D. & GANSSEN, G. (1989) Northern Indian Ocean upwelling cells and the stable isotope composition of living planktonic foraminifers. *Deep-Sea Res.* **36**, 219–236.

PEDERSEN, T.F. & SHIMMIELD, G.B. (1990) The geochemistry of reactive trace metals and halogens in hemipelagic continental margin sediments. *Rev. Aquatic Sci.* **3**, 255–279.

PRELL, W.L. (1984) Variation of monsoonal upwelling: a response to changing solar radiation. In: *Climatic Processes and Climate Sensitivity* (Eds Hansen, J.E. & Takahashi, T.). Am. Geophys. Union Geophys. Monogr. 29, Maurice Ewing Series 5, pp. 48–57.

PRELL, W.L. & KUTZBACH, J.E. (1987) Monsoon variability over the past 150 000 years. *J. Geophys. Res.* **92**, 8411–8425.

PRELL, W.L., NIITSUMA, N. & Shipboard Scientific Party (1989a) Explanatory Notes. In: *Proc. Ocean Drilling Program Init. Repts, 117* (Eds Prell, W.L., Niitsuma, N. *et al.*). College Station, TX (Ocean Drilling Program), pp. 5–10.

PRELL, W.L., NIITSUMA, N. & Shipboard Scientific Party (1989b) Site 728. *Proc. Ocean Drilling Program Init. Repts, 117* (Eds Prell, W.L., Niitsuma, N. *et al.*). College Station, TX (Ocean Drilling Program), pp. 495–545.

PRELL, W.L. & Shipboard Party of ODP Leg 117 (1990) Neogene tectonics and sedimentation of the SE Oman continental margin: results from ODP Leg 117. In: *The Geology and Tectonics of the Oman Region* (Eds Robertson, A.M.F., Searle, M.P. & Ries, A.C.). Geol. Soc. Lond. Spec. Publ. 49, pp. 745–758.

SCHMITZ, B. (1987) Barium, equatorial high productivity, and the northward wandering of the Indian continent. *Paleoceanography* **2**, 63–77.

SHIMMIELD, G.B. & MOWBRAY, S.R. (1991) The inorganic geochemical record of the northwest Arabian Sea: a history of productivity variation over the last 400 k.y. from sites 722 and 724. In: *Proc. Ocean Drilling Program Init. Repts, 117* (Eds Prell, W.L., Niitsuma, N. *et al.*). College Station, TX (Ocean Drilling Program), pp. 409–431.

SHIMMIELD, G.B., PRICE, N.B. & PEDERSEN, T.F. (1990) The influence of hydrography, bathymetry and productivity on sediment type and composition of the Oman margin and in the Northwest Arabian Sea. In: *The Geology and Tectonics of the Oman Region* (Eds Robertson, A.M.F., Searle, M.P. & Ries, A.C.). Geol. Soc. Lond. Spec. Publ. 49, pp. 759–769.

SIROCKO, F. & SARNTHEIN, M. (1989) Wind-borne deposits in the northwest Indian Ocean: record of Holocene sediments versus modern satellite data. In: *Paleoclimatology and Paleometeorology: Modern and Past Patterns of Global Atmospheric Transport* (Eds Leinen, M. & Sarnthein, M.). Kluwer Academic Publishers, Dordrecht, NATO ASI Ser. C, **282**, 401–433.

STEENS, T.N.F., KROON, D., TEN KATE, W.G. & SPRENGER, A. (1991) Late Pleistocene periodicities of oxygen isotope ratios, calcium carbonate and magnetic susceptibilities of western Arabian Sea margin Hole 728A (ODP Leg 117). In: *Proc. Ocean Drilling Program Scientific Results, 117* (Eds Prell, W.L., Niitsuma, N. *et al.*). College Station, TX (Ocean Drilling Program), pp. 309–320.

STEENS, T.N.F., GANSSEN, G. & KROON, D. (1992) Oxygen and carbon isotopes in planktonic foraminifera as indicators of upwelling intensity and upwelling induced high productivity in sediments from the northwestern Arabian Sea. In: *Upwelling Systems, Evolution since the Early Miocene.* (Eds Summerhayes, C.P., Prell, W.L. & Emeis, K.C.). Geol. Soc. Lond. Spec. Publ. 64, pp. 107–119.

TEN HAVEN, H.L., DE LEEUW, J.W., SCHENCK, P.A. & KLAVER, G.T. (1988). Geochemistry of Mediterranean sediments. Bromine/organic carbon and uranium/organic carbon ratios as indicators for different sources of input and post-depositional oxidation, respectively. *Org. Geochem.* **13**, 255–261.

WEEDON, G.P. & SHIMMIELD, G.B. (1991) Late Pleistocene upwelling and productivity variations in the Northwest Indian Ocean deduced from spectral analyses of geochemical data from ODP Sites 722 and 724. In: *Proc. Ocean Drilling Program Scientific Results, 117* (Eds Prell, W.L., Niitsuma, N. *et al.*). College Station, TX (Ocean Drilling Program), pp. 431–445.

WYRTKI, K. (1973) Physical oceanography of the Indian Ocean. In: *The Biology of the Indian Ocean* (Eds Zeitschel, B. & Gerlach, S.A.). Springer Verlag, New York, pp. 18–36.

Spec. Publs Int. Ass. Sediment. (1994) **19**, 145–166

Milankovitch cyclicity in Late Cretaceous sediments from Exmouth Plateau off northwest Australia

R. BOYD*, Z. HUANG† *and* S. O'CONNELL‡

** Centre for Marine Geology, Dalhousie University, Halifax, Nova Scotia, Canada B3H 3J5;*
† Atlantic Geoscience Centre, Geological Survey of Canada, Bedford Institute of Oceanography, Box 1006, Dartmouth, Nova Scotia, Canada B2Y 4A2; and
‡ Department of Earth and Environmental Sciences, Wesleyan University, Middletown, CT 06457, USA

ABSTRACT

Well-developed early Campanian to early Maastrichtian pelagic cyclic sediments were recovered from Hole 762C on the Exmouth Plateau, off northwest Australia during Ocean Drilling Program (ODP) Leg 122. The cycles consist of nannofossil chalk (light beds) and clayey nannofossil chalk (dark beds), which are strongly to moderately bioturbated, alternating on a decimetre scale, and exhibit gradual boundaries. Trace fossils, which introduced material from a bed of one colour into an underlying bed of another colour, and differences in composition between the light and dark beds indicate that the cycles in these sediments are a depositional feature rather than a diagenetic outcome.

Walsh spectral analysis was applied to the upper Campanian–lower Maastrichtian cyclic sediments to examine the regularity of the cycles. With an average sedimentation rate of 1.82 cm/ka in this interval, the most predominant wavelengths of the colour cycles yield periods of around 21 ka and 41 ka, respectively, comparable to the precession and obliquity cycles, strongly suggesting an orbital origin for the cycles.

On the basis of sedimentological evidence and plate tectonic reconstructions, we propose the following mechanism for the formation of the cyclic sediments during the Late Cretaceous in this region. The cyclic variations in insolation in response to periodic orbital changes controlled the alternation of two prevailing climates in the area. During the wetter, equable and warmer climatic phases under high insolation, more clay minerals and other terrestrial materials were produced on land and supplied by higher runoff to an ocean with low bioproductivity, and the dark clayey beds were deposited. During the drier and colder climatic phases under low insolation, fewer clay minerals were produced and supplied to the ocean, bioproductivity was increased, and the light beds were deposited.

INTRODUCTION

Since Hays *et al.* (1976) successfully related the climatic cyclicity in pelagic Quaternary sediments to the orbital cycles proposed by Milankovitch, many researchers have attempted to relate pre-Pleistocene cyclic sediments, both from land outcrop and from deep-sea drilling, to orbital cycles (e.g. Schwarzacher & Fischer, 1982; Arthur *et al.*, 1984; Cotillon & Rio, 1984; Barron *et al.*, 1985; Bottjer *et al.*, 1986; Herbert & Fischer, 1986; Research on Cretaceous Cycles (ROCC) Group, 1986; Weedon, 1986; Hart, 1987; Ogg *et al.*, 1987; Schwarzacher, 1987; Jarrard & Arthur, 1989) with various data sets and approaches. However, controversy often followed. The question of a diagenetic versus climatic origin for the basal Jurassic cyclic sediments (Blue Lias) of south Britain is one example (Hallam, 1986; Weedon, 1986). This case is the result of the relatively great geological age of the strata, insufficient stratigraphic control, errors in the time scale, and evident diagenetic overprint. Another example is the short-term sea-level fluctuation origin versus a climatic fluctuation origin for the Late Cretaceous marl–limestone sequences in Alabama (Bottjer *et al.*, 1986; ROCC Group, 1986; King, 1990). A sea-level change origin here seems plausible because the sequences are the product of shelf sedimentation. However, the sequence contains insufficient beds for a valid statistical test. Pelagic cyclic sediments younger than Early Cretaceous are better suited for the purpose of testing the theory of orbital origin of sediment cyclicity in pre-Pleistocene time in a quantitative fashion. In this paper we report our study of a Late Cretaceous cyclic sequence from the Indian Ocean (Exmouth Plateau), recovered during Ocean Drilling Program (ODP) Leg 122. Early Cretaceous cyclic sequences from the Atlantic–Tethyan Region have been previously studied (Cotillon & Rio, 1984; Barron *et al.*, 1985; Herbert & Fischer, 1986; Ogg *et al.*, 1987; ten Kate & Sprenger, 1989), and also Late Cretaceous cyclic sequences from the South Atlantic (Herbert & D'Hondt, 1990) and North America (ROCC Group, 1986). However, Late Cretaceous cyclic sequences from the Indian Ocean were not reported before ODP Leg 122.

The Late Cretaceous cyclic sediments of this study were recovered from Hole 762C, which is located on the western part of the central Exmouth Plateau (19° 53.24′ S and 112° 15.24′ E) in water depth of 1360 m, and about 260 km offshore (Fig. 1). The estimation of the palaeolatitude for the Australian continent during the Maastrichtian is between 45° S and 65° S (Barron, 1987). A more specific palaeolatitude estimation for the Exmouth region by J.G. Ogg (personal communication, 1990) is 53° S. The burial history analysis by F. Gradstein & U. von Rad (personal communication) estimates that the palaeowater depth in Late Cretaceous time was at least 550 m. The Cretaceous succession from Hole 762C starts with black to very dark grey silty claystone and clayey siltstone deposited in a shelf-margin prodelta environment during Berriasian to Valanginian time. The latest Valanginian to Barremian is represented by a major hiatus (Haq *et al.*, 1990). Early Aptian sediments are black to dark grey calcareous claystones deposited in a hemipelagic epicontinental shelf environment. The rest of the Cretaceous sediments are mostly pelagic nannofossil chalk and clayey nannofossil chalk with minor claystone and they record a mature ocean phase of passive margin development (Boyd *et al.*, 1992). Cyclicity in the form of colour banding, mainly on a decimetre scale, begins in Late Cretaceous sediments (younger than Cenomanian), but is best developed in Cores 122-762C-47X through to 122-762C-56X, 592.5–687.5 m below sea floor (mbsf) (late Campanian to early Maastrichtian). The presence of nannofossil chalk in this interval indicates a pelagic environment. Similar cyclic sediments and environments were also recovered in Holes 761 and 763 in this region during ODP Leg 122 (Haq *et al.*, 1990). With high core recovery (78% for Cores

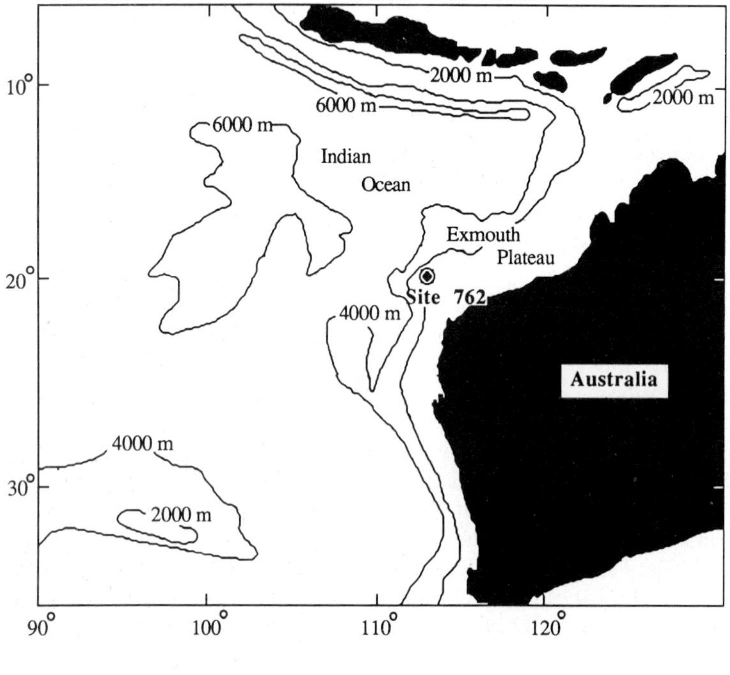

Fig. 1. Location of ODP Hole 122-762C.

43X–76X), this interval of cyclic sediments provides us with a good opportunity for a quantitative test of the orbital origin of sediment cyclicity in pre-Pleistocene time.

Recently, Berger & Loutre (1989) calculated the long-term variations of the main orbital parameters during the past 500 Ma, and demonstrated that the periods of these parameters were shorter in the geological past. According to their results, the difference in the periods between the present and the Cretaceous is not very significant. For example, the presently 41-ka obliquity cycle was about 39.5 ka in the Late Cretaceous (around 74 Ma). The difference for obliquity periodicity between the Cretaceous and the present is 2000 years at most. The difference for precession periodicity is smaller. The precession periodicities were 22.5 and 18.6 ka in the Late Cretaceous (around 74 Ma). These small differences do not pose a difficulty for testing the Milankovitch theory in Late Cretaceous strata.

In this study we examined the sedimentological features of the cyclic sediments. Then time series were built on the basis of core inspection. The time series were analysed with the Walsh spectral analysis method (Beauchamp, 1975) to reveal possible periodicity in the sediments and thus to provide insight into the process of cyclic sedimentation and to search for the ultimate origin of the cyclicity.

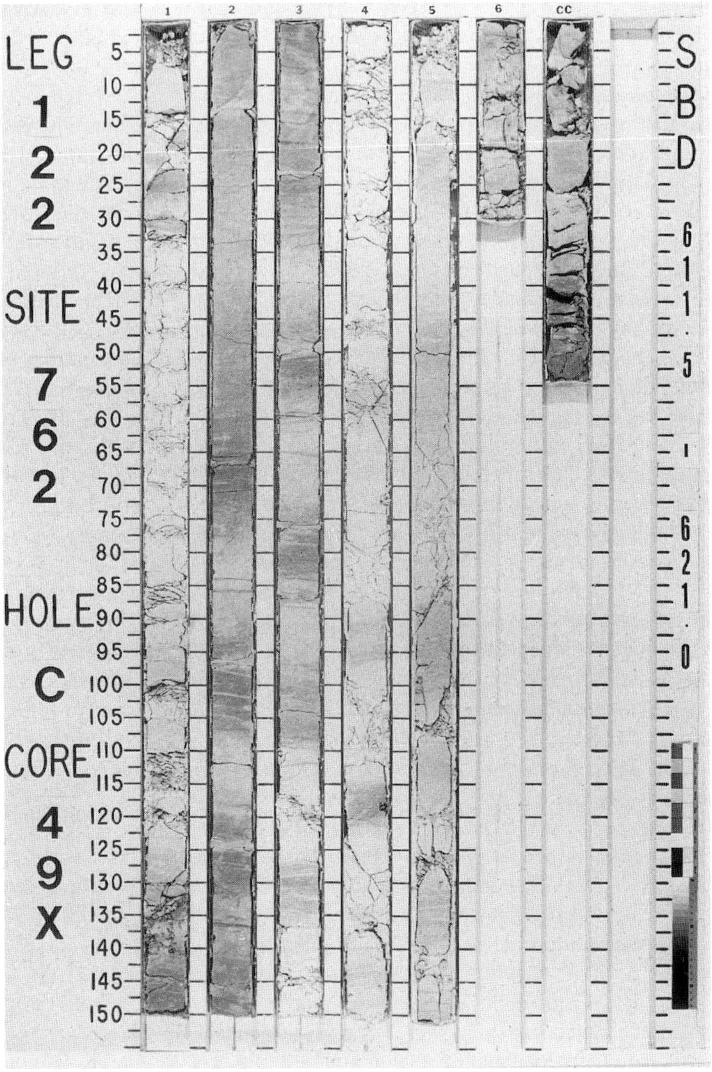

Fig. 2. Core 122-762C-49X (lower Maastrichtian), displaying alternation of light nannofossil chalk and dark, clayey nannofossil chalk. The boundaries between the light and dark beds are gradual.

THE FEATURES OF THE CYCLIC SEDIMENTS

Cyclicity in the form of colour variation is observed in the Late Cretaceous pelagic sediments of Cores 43X–76X, 554.5–819.5 mbsf. The cyclicity is best developed in Cores 47X–56X (592.5–687.5 mbsf, late Campanian to early Maastrichtian). Figure 2, a photograph of Core 49X, is an example of late Campanian to early Maastrichtian cyclic sediments. The sediments have been classified on the basis of smear slide examination and $CaCO_3$ abundance using the ODP sediment classification scheme (Haq *et al.*, 1990). Samples from Cores 47X–53X were used for coulometric determination of $CaCO_3$ content, X-ray diffraction (XRD) analysis and geochemical analyses. Preprocessing was carried out on subsamples for the latter two analyses. The samples (nannofossil chalk and clayey nannofossil chalk) were dried at a temperature lower than 80°C, ground to powder, and then rinsed with distilled water three times to remove salt after being sieved with a 100-mesh sieve. The concentrations of Ca, Fe, Mn, Mg, Sr, Si, Al and Ti were determined with an atomic emission spectrometer and those of Na and K with an atomic absorption spectrometer at the University of Hawaii laboratory. There was at least one replication of each sample (analysis on a separate dilution of the same digested sample solution), to check the reproducibility of the method. For the element Ca, a number of samples of different dilutions were re-run because the results of the first run were over the high end of the calibration curve. Average and standard deviation were calculated for all measurements for each element and, separately, for the measurements on samples from the light beds and those from dark beds. For each sample the difference between the replicate was taken as the analytical error. In the cases where there were more than two measurements on a sample, the largest difference was taken as the analytical error.

The Late Cretaceous age in this hole was determined with nannofossils which are more abundant and better preserved than other minor fossil groups. The nannofossil events observed by Bralower & Siesser (1992), together with lithological records, indicate a stratigraphically complete Late Cretaceous succession in Hole 762C. The biostratigraphy of the interval of interest is shown in Fig. 3. The Santonian–Campanian boundary is placed at the first occurrence of *Broinsonia parca* (744.74 mbsf). The early Campanian–late Campanian boundary is placed at the first occurrence of *Quadrum gothicum* (687.80 mbsf). The Campanian–Maastrichtian boundary is placed at the last occurrence of *Eiffellithus eximius* (611.50 mbsf). The early Maastrichtian–late Maastrichtian boundary is placed at the first occurrence of *Lithraphidites quadratus* (578.30 mbsf). There is a hiatus at the Cretaceous–Tertiary boundary. The top of Late Cretaceous strata is placed at the top of Core 43X (554.80 mbsf), where the last Cretaceous species occurs.

The numerical ages of the stage boundaries in millions of years shown in Fig. 3 are from the time scale by Kent & Gradstein (1985). For the interval of well-developed cyclicity, we used the early Maastrichtian–late Maastrichtian boundary (71 Ma) and the early Campanian–late Campanian boundary (77 Ma) as the age control to estimate the average sedimentation rate. The sediment thickness between these two boundaries is 109.50 m and the duration is 6 Ma. The estimated average sedimentation rate for this interval is therefore 1.82 cm/ka. This average sedimentation rate may contain some errors from both the placement of the age boundaries and from the time scale; the actual sedimentation rate may have differed in different parts of this interval.

The completeness in both stratigraphy and recovery for the interval of well-developed colour alternations enables a detailed study on the nature of the cyclic sediments. Because of high gas pressure in the drilling area, the cores were moderately disturbed and slightly expanded, resulting in recovery higher than 100%, but this disturbance was not serious enough to prevent recognition of colour variation couplets in most of the cores. Our statistical analysis of the cycles, however, was affected by whole-round sampling for organic geochemistry (OG) analysis and interstitial water (IW) analysis. We were unable to determine if a colour variation boundary occurs in a whole-round sample (25 cm long for OG sample and 10 cm long for IW sample) and, if so, where the boundary is.

General character of the cycles

The 95-m-thick interval (Cores 47X–56X) of well-developed cyclic sediments displays alternations of light-coloured and dark-coloured beds (Fig. 2). The colour tends to become darker down-hole in this interval. In this interval there are two subunits, with the boundary at 603.5 mbsf (top of Section 2 of Core 48X). Above 603.5 mbsf, the light beds are white (2.5Y 8/0) to very light greenish grey (10Y 8/1)

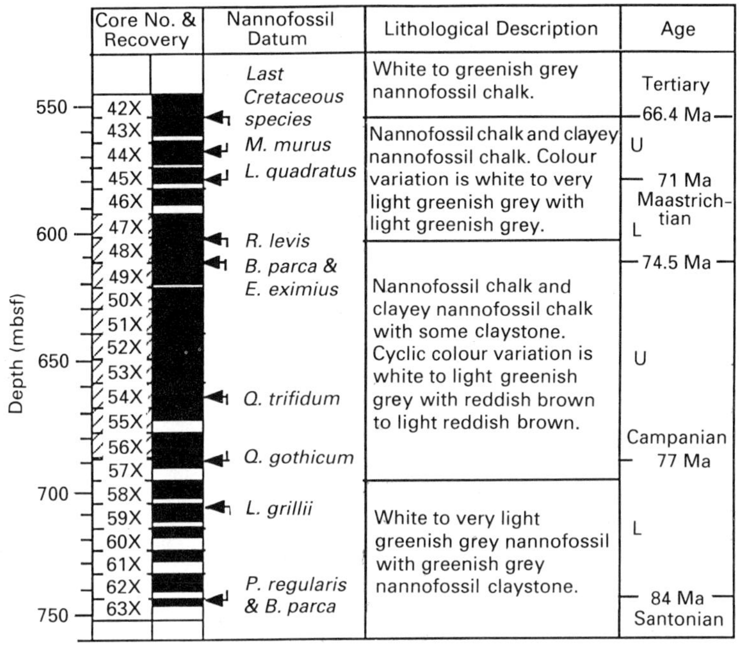

Fig. 3. Lithological and stratigraphic summary of the Campanian to Maastrichtian cyclic sequence in Hole 122-762C, showing core number, core recovery (in black), nannofossil data (after Bralower & Siesser, 1992), lithology and the age. Shaded area indicates the interval of well-developed cyclic colour variations, which is composed of two subunits with the boundary at 603.5 mbsf. The numerical ages at the stage boundaries are from the time scale by Kent & Gradstein (1985).

nannofossil chalk, whereas the dark beds are light greenish grey (5GY 7/1) clayey nannofossil chalk. Below 603.5 mbsf the light beds are white (10YR 8/1) to light greenish grey (10YR 7/1) nannofossil chalk and the dark beds are light reddish brown (5YR 6/3) to reddish brown (5YR 5/3) clayey nannofossil chalk with some nannofossil chalk.

The thickness of colour alternations ranges from 10 to 120 cm, but most of the alternations are on a decimetre scale. Boundaries between the light and dark beds are generally gradual with transitional zones varying from 2 to 12 cm. It is feasible to judge the regularity of the cycles visually.

The thickness of the light and dark beds was measured in Cores 47X–56X, but beds which are not complete (e.g. those from which whole-round samples were removed) were not included. Light beds (n = 195) vary in thickness from 4.0 to 80 cm, with an average of 18 cm and a standard deviation of 14 cm. Dark beds (n = 194) vary from 3.5 to 89 cm in thickness, with an average of 22 cm and a standard deviation of 18 cm, indicating that the dark beds are thicker and more variable than the light beds. In addition, only part of the dark beds show lamination, and where present it is very faint. The lamination is not caused by current activity as indicated by the absence of ripple cross-bedding and cut-and-fill features.

Bioturbation

Both the light and the dark beds which comprise the cycles are moderately to strongly bioturbated with many recognizable trace fossils such as *Planolites*, *Zoophycos*, *Chondrites*, *Teichichnus* and *Helminthoida*. *Inoceramus* fragments are more common in light beds. *Teichichnus* and *Helminthoida* are relatively rare and the latter is only found above 603.5 mbsf. The burrows are predominantly horizontal. However, downward burrowing has locally introduced lighter sediments into the darker underlying beds. Section 56X-2 contains a typical example (Fig. 4). These burrows indicate that the colour variation is primarily a depositional product rather than the result of diagenesis.

$CaCO_3$ and total organic carbon content of the dark and light beds

$CaCO_3$ measurements were carried out on 186 samples from Cores 47X–56X (detailed results can be obtained from the authors). The 92 $CaCO_3$ measurements from light beds vary from 74.71 to 93.72%, with an average of 86.99% and a standard deviation of 3.17%. The 94 measurements from dark beds vary from 57.22 to 91.5%, with an average of 74.46% and a standard deviation of 7.03%.

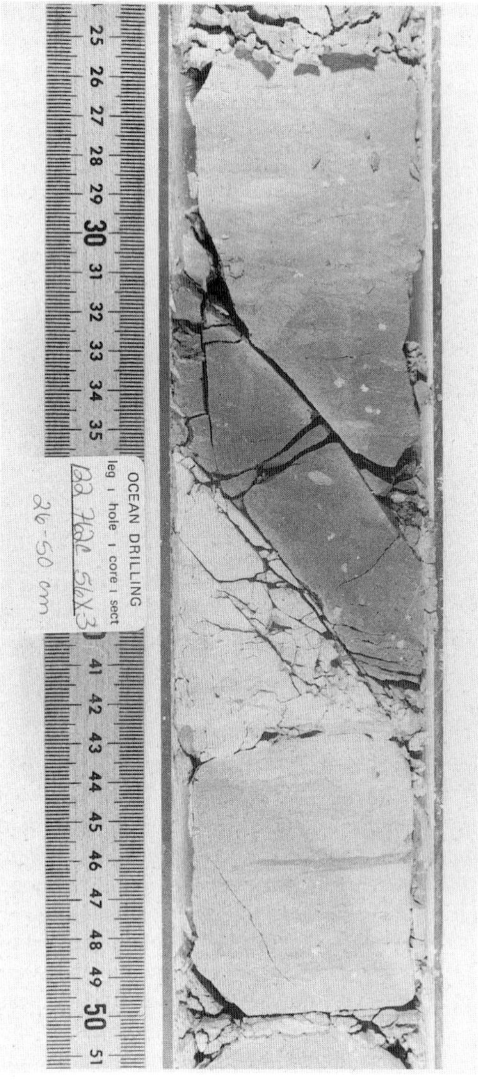

Fig. 4. Typical bioturbation which introduces the upper light sediments into the lower dark bed. The lower sharp inclined boundary represents a microfault plane (sample 122-762C-56X-3, 26–50 cm).

Therefore, on average, the $CaCO_3$ content is about 13% lower and more variable in the dark beds than in the light beds. Detailed measurements in Cores 48X, 49X, 51X and 52X show that the $CaCO_3$ content of light beds is always higher than that of the nearby dark beds (Fig. 5A), and the $CaCO_3$ content is more variable in the dark beds than in the light beds. Overall, both light and dark beds above 603.5 mbsf are more calcareous than those below 603.5 mbsf.

Total organic carbon (TOC) content is low in both light and dark beds, according to shipboard analysis. There is no significant difference in TOC between dark and light beds. The TOC values vary from 0.0 to 0.05% in eight dark beds and from 0.0 to 0.02% in four light beds. In contrast, in cyclic sequences deposited at greater depths, e.g. the Lower Cretaceous sequence from the northwest Atlantic Ocean, the dark beds have much higher TOC values than the light beds (Ogg *et al.*, 1987).

Other differences between the dark and light beds

X-ray diffraction (XRD) analysis was carried out on Cores 47X–53X. Only quartz and calcite consistently show reliable peaks on the X-ray spectra. Peaks of clay minerals and other accessory minerals, such as feldspar, zeolite and pyrite, were identified when possible. Content of total clay minerals was calculated, and with rare exceptions more clay minerals were present in dark beds than in the adjacent light beds (Fig. 5B). The same tendency was found in the variation of quartz content (Fig. 5C). We also observed that more kaolinite and illite peaks than smectite peaks appear in dark beds, more zeolite and feldspar peaks are detected in dark beds, and, conversely, more smectite peaks than kaolinite and illite peaks appear in light beds.

Composition differences between the light and dark beds were also detected in the shipboard smear slide examinations of Cores 47X–56X (Haq *et al.*, 1990). The differences presented by the smear slide record can be outlined as follows: (i) quartz, clay and opaques are more common in dark beds than in light beds; (ii) zeolites and dolomite are more common in dark beds than in light beds; (iii) sponge spicules and ostracods are more abundant in dark beds than in light beds; (iv) biogenic carbon grains and bioclasts are more common in dark beds than in light beds; and (v) there are more abundant nannofossils in light beds than in dark beds.

The smear slide observation that there are more nannofossils in the light beds is in agreement with what has been observed in similar cyclic sequences (i.e. Atlantic Early Cretaceous cyclic sequences, Ogg *et al.*, 1987; and Early Cretaceous cyclic sequences from the Vocontian Basin, Darmedru *et al.*, 1982). This observation can be explained by higher bioproductivity in the surface water during the phase in which the light beds were deposited.

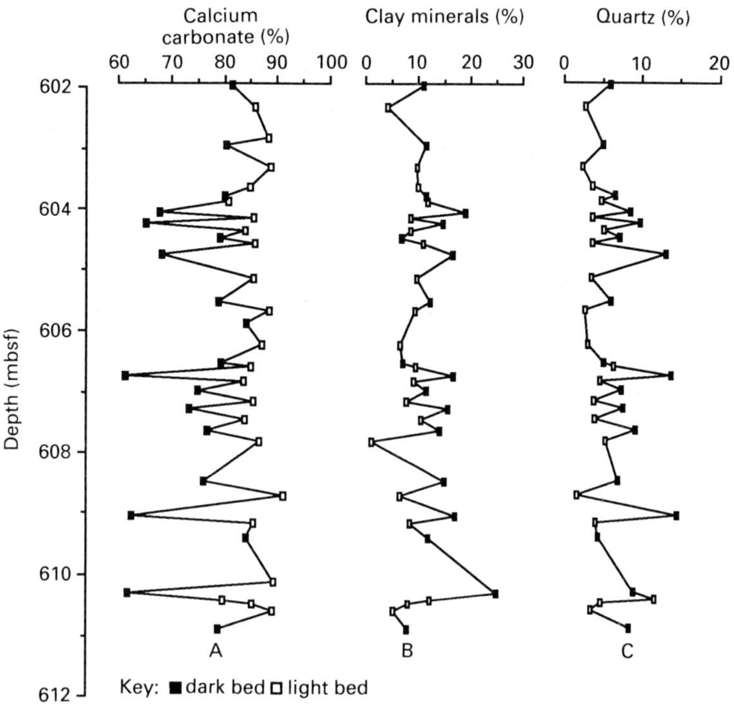

Fig. 5. Composition differences in a succession of cyclic colour alternations, Core 122-762C-48X. A, Content of $CaCO_3$ (%) measured with coulometric method. The content of $CaCO_3$ in a light bed is always higher than that of the adjacent dark beds. B, Content of total clay minerals (%) from XRD analysis. There are more clay minerals in dark beds than in light beds, with rare exceptions. C, Content of quartz (%) from XRD analysis. Quartz is more abundant in dark beds than in light beds, also with rare exceptions.

Geochemical differences between the dark and light beds

On average, there is more Fe, Si, Al, Ti, Na and K in the dark beds than in the light beds, whereas there is more Mn in the light beds than in the dark beds. As there are more samples analysed in Core 48X than in Core 47X, the variations of these elements are plotted for Core 48X (Figs 6, 7, 8 and 9). In producing these plots, the replicates with higher concentration were used, as the lower concentration measured is because of an incomplete solubilization of the aluminosilicate fraction (E. de Carlo, personal communication).

A systematic difference in the amounts of Si, Al, Ti, Fe, Mn, Na and K between the dark and light beds can be seen (Figs 7, 8 and 9). Figure 7 shows that generally there is more Si, Al and Ti in the dark beds than in the light beds in Core 48X. Figure 8 shows that generally there is more Mn in the light beds than in the dark beds, but there is more Fe in the dark beds than in the light beds. Figure 9 shows that generally there is more Na and K in the dark beds than in the light beds.

The variations of Si, Al, Ti, Na and K are essentially in phase with the difference in clay content (the aluminosilicate fraction) between the dark and the light beds (Fig. 5). The higher clay and quartz content in the dark beds may explain why there is more Si, Al, Ti, Na and K in the dark beds. Similarly, in the Early Cretaceous cyclic sequences from the North Atlantic Ocean, there is more TiO_2 and Al_2O_3 in the laminated limestones (Dean & Arthur, 1987). It was also reported that there is more TiO_2 in chalky marl than in chalk in Cenomanian rhythmic sequences of southeastern England, a phenomenon reflecting a greater fine-grained detrital component (Bottjer *et al.*, 1986).

For Ca, Mg and Sr, no systematic variations between the dark and light beds appear on the plots (Fig. 6). The erratic variation of Ca in Fig. 6 may be explained by the fact that the average analytical error of the inductively coupled plasma method for this element is greater than the average Ca difference between the dark and light beds. The apparent non-matching behaviour between the Ca concentration and $CaCO_3$ content can also be explained by the fact that Ca (also Mg and Sr to a lesser extent) is present in minerals from both the carbonate fraction and the detrital fraction. Most of the Ca comes from the $CaCO_3$, but a significant amount of these elements can be associated with clays and feldspar

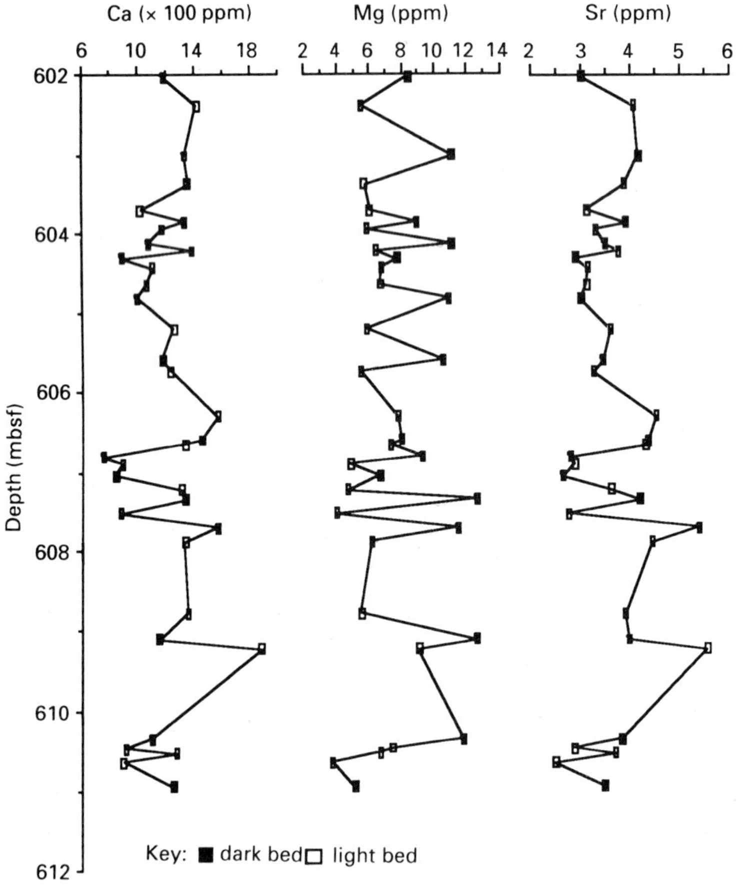

Fig. 6. Variations in Ca, Mg and Sr in Core 122-762C-48X. The variations of the three elements appear to be erratic.

minerals if the latter are sufficiently abundant. Variations in Mn concentrations may be related to slight variations of oxygen content in the bottom water (Balzer, 1982).

Samples from Cores 47X and 48X for geochemical analysis cover the brownish hue/greenish hue boundary at 605.5 mbsf. The concentration of Si, Al, Fe and Ti decreases upwards (Fig. 10). This trend is more obvious in measurements of the dark beds than of the light beds. Since these elements are mainly derived from terrigenous sediments such as clay and quartz (radiolaria are not abundant in this interval), the upward decrease in the concentration of these elements suggests a gradual decrease of terrestrial input toward the end of the Maastrichtian.

Diagenetic overprint

The sediments recovered in the interval studied are relatively soft. Diagenetic alteration appears as thin seams cutting trace fossils. These seams occur irregularly and more often in dark beds than in light beds. They cut trace fossils and result from early diagenetic compaction, local $CaCO_3$ dissolution, and reprecipitation during burial (Fig. 11). Examination by scanning electron microscopy (SEM) on seven samples from 3.5 cycles in Sections 49X-2 through to 49X-3 shows etched nannofossils and some secondarily deposited $CaCO_3$ in samples from dark beds, and less dissolved nannofossils also with some secondarily deposited $CaCO_3$ in samples from light beds (Fig. 12). The effects of the diagenetic process are not profound enough to alter the overall cyclic colour alternation produced by depositional processes, as redeposition of $CaCO_3$ is not restricted to the light beds.

Physical property measurements (Haq *et al.*, 1990) show that light beds have higher water content, lower bulk density, slightly lower grain density and higher porosity than dark beds. SEM examination

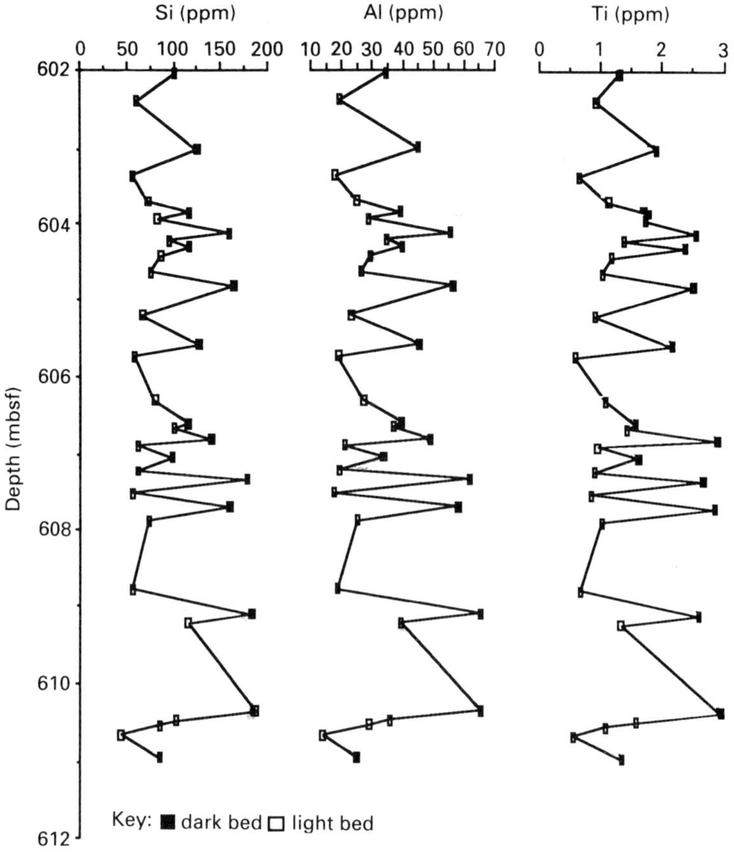

Fig. 7. Variations in Si, Al and Ti in Core 122-762C-48X. There is more Si, Al and Ti in the dark beds than in the light beds.

also indicates that the samples from light beds are more porous. Porosity in both light and dark beds tends to decrease with depth. As the cyclic sequences on the Exmouth Plateau are not fully indurated, the influence from cementation is small and the dissimilarity in physical properties (e.g. porosity) may be more directly related to small differences in compaction between the light and dark beds. The mean values of difference in water content and porosity are less than 4.5%. The small differences in compaction between the dark and light beds as reflected by the physical properties are unlikely to interfere with our statistical analysis.

Geological implications of cyclic properties

A summary of the differences between the light and dark beds is given in Table 1. The major geological implications from sedimentological examination of the cyclic interval is that the cyclic colour alternation in the sediments is a depositional feature and not a diagenetic product. Strong to moderate bioturbation throughout the cyclic interval and the generally low TOC value in both light and dark beds indicate that the bottom water had enough oxygen for benthic organisms to live. The occasional faint lamination in dark beds may imply a bottom water with slightly less oxygen, as is confirmed by the small variations in Mn concentrations, and hence a decrease in benthic organisms. The changes in oxygen supply to the bottom water were subtle and are not crucial to the formation of the cycles. Differences in both mineral composition and such chemical elements as Si, Al, Ti, Na and K between the light beds and dark beds imply that the change lies mainly in sediment supply. When the light beds were deposited, biogenic material was the major component, with less terrestrial input. When the dark beds were deposited, terrestrial input was increased, with some relative decrease in biogenic supply. The smaller

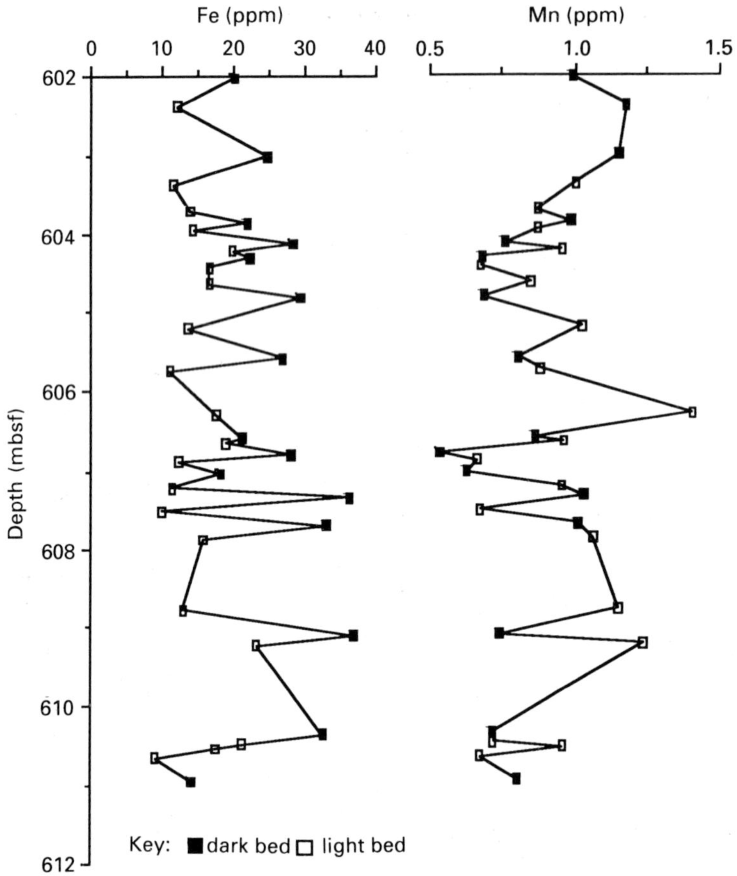

Fig. 8. Variations in Fe and Mn in Core 122-762C-48X. There is more Fe in the dark beds than in the light beds, but there is more Mn in the light beds than in the dark beds.

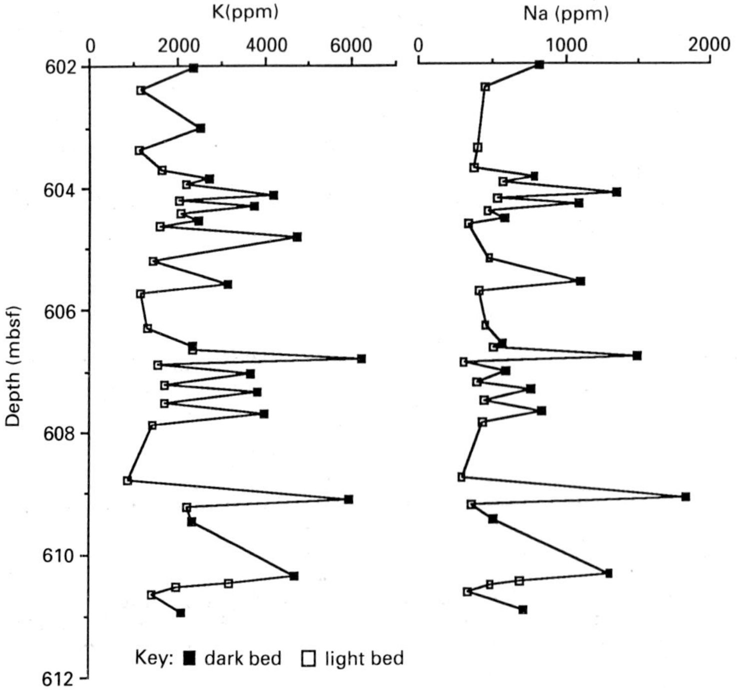

Fig. 9. Variations in K and Na in Core 122-762C-48X. There is more K and Na in the dark beds than in the light beds.

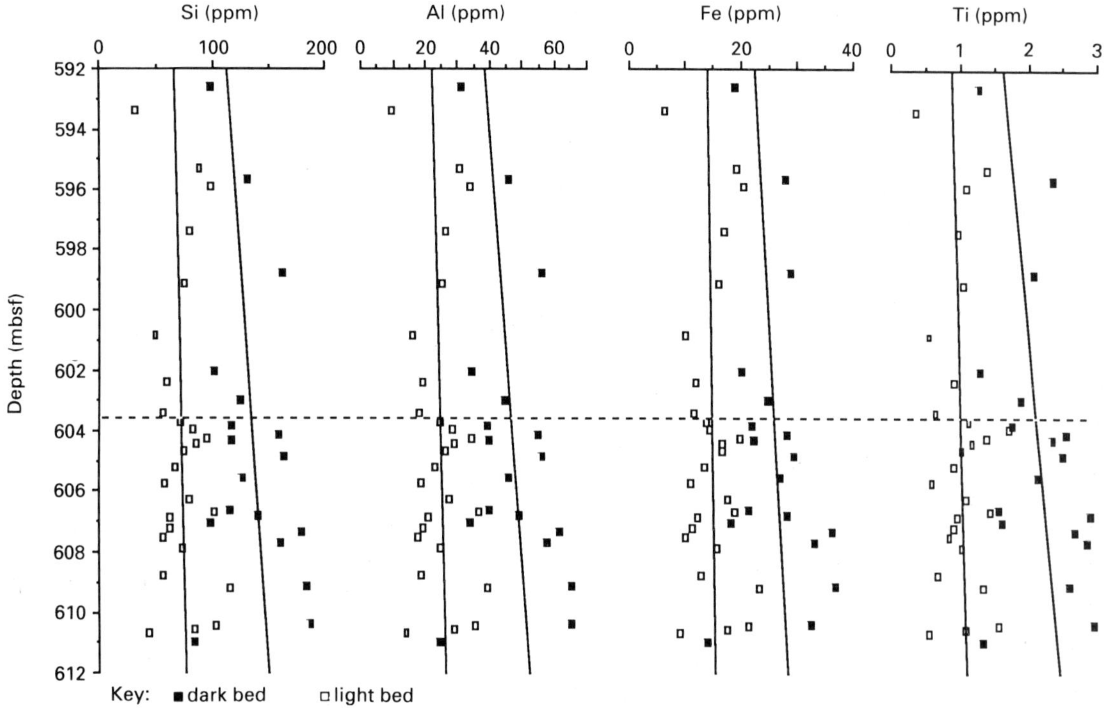

Fig. 10. Depth–concentration plots of elements Si, Al, Fe and Ti, in Cores 122-762C-47X and 122-762-48X. A dashed line is drawn at 603.5 m, where the brownish hue changes upward to a greenish hue. The lines through the data points are the linear fitting for measurements in the dark and light beds, respectively.

standard deviation of bed thickness and less variable $CaCO_3$ content of the light beds may suggest that the supply of biogenic material is less variable than the supply of terrestrial material. More variability in the terrestrial supply indicates that land-based changes may have played a more important role in cyclic sedimentation. Diagenetic processes may have enhanced the colour cyclicity, producing small seams of $CaCO_3$ dissolution/reprecipitation which locally cut trace fossils.

A background change in the cyclic sequence is indicated by the overall colour change from a brownish hue interval below 603.5 mbsf to a greenish hue interval with pyrite crystals above that depth. This background change is also reflected by exclusive occurrence of the trace fossil *Helminthoida*

Table 1. A summary of some important differences between the light and dark beds

	Light beds	Dark beds
Average thickness and variance (cm)	18 ± 14	22 ± 18
$CaCO_3$ content and variance (%)	86.99 ± 3.17	74.46 ± 7.03
Clay content	Low	High
Clay minerals	More smectite than kaolinite and illite	More kaolinite and illite than smectite
Quartz	Less	More
Nannofossils	More	Less
Sponge spicules and ostracods	Less	More
Physical properties	Higher water content, higher porosity	Lower water content, lower porosity
Geochemical characteristics	Less Si, Al, Ti, Fe, K and Na, more Mn	More Si, Al, Ti, Fe, K and Na, less Mn

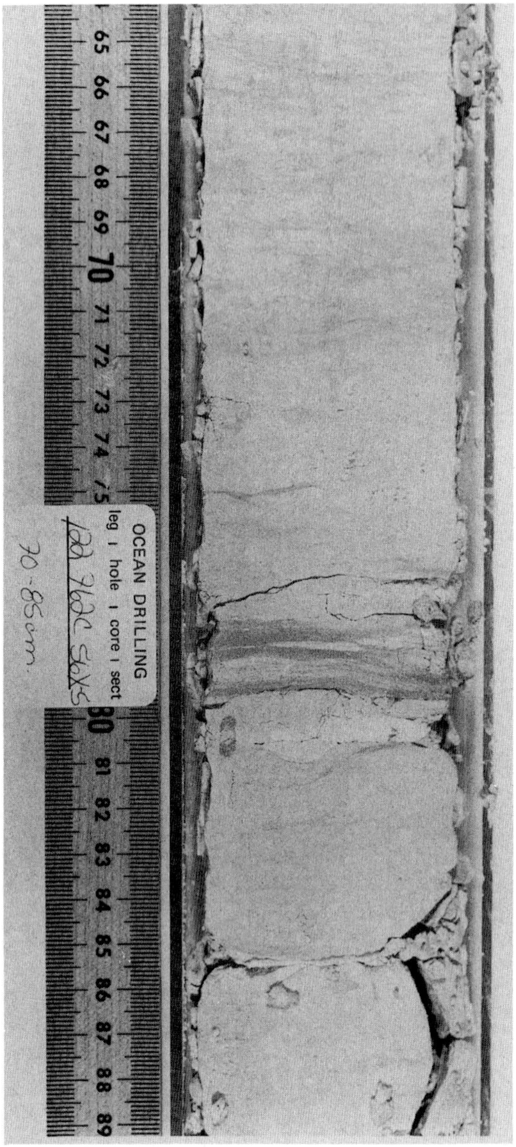

Fig. 11. Diagenetic alteration of the sediments: small and thin seams cutting trace fossils (Sample 122-762C-56X-5, 70–85 cm).

above this depth. The background change has been correlated with a relative sea-level change from high-stand to low-stand (Haq *et al.*, 1990). Whatever the cause is, the background change does not seem to have affected the formation of the cycles, and it is not a controlling factor.

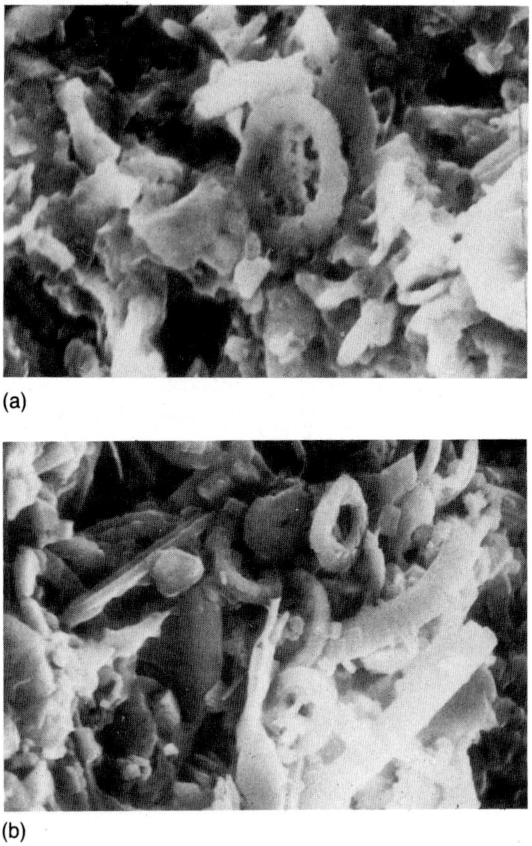

Fig. 12. SEM photos of dark clayey nannofossil chalk and light nannofossil chalk (×5000). (a), In the dark beds, the nannofossils are etched, with some secondary-deposited $CaCO_3$ (Sample 122-762C-49X-3, 20–21 cm). (b), In the light beds, the nannofossils are less dissolved, also with some secondary-deposited $CaCO_3$. The sediments appear more porous (Sample 122-762C-49X-3, 65–67 cm).

QUANTITATIVE SEARCH FOR THE PERIODICITY OF CYCLIC SEDIMENTS

The method and time-series generation

As it is difficult to estimate the regularity of the colour cycles visually, a statistical method should be utilized for this purpose. Spectral analysis is frequently employed in the study of Quaternary cyclic sediments. However, the conventional Fourier spectral analysis method is most suitable for smooth time series such as geophysical logging data or curves built by a large number of closely spaced laboratory-derived data points, for example oxygen/carbon isotopes, $CaCO_3$ content and other geochemical measurements. A typical geophysical logging tool of high vertical resolution used by ODP is the natural gamma-ray intensity log (0.3 m vertical resolution; Schlumberger, 1987). It would be unable to resolve colour alternations thinner than 30 cm. Although we have one sample from every bed for $CaCO_3$ measurements in some of the cores, the number of data points is still too sparse to build a smooth time series. In this situation it seems better to build step-like (discrete) time series, on the basis of core inspection and measurement of bed thickness, and to apply the Walsh spectral analysis method to the discrete time series (Schwarzacher, 1987). As with Fourier analysis, any time series can be resolved into Walsh functions with different sequences (zero-crossings of the time base), equivalent to frequency in Fourier spectral analysis (Weedon, 1986). Walsh functions are discontinuous waves which take only two values, +1 or −1. Therefore, Walsh analysis is well suited to analysing step-like time series (for more details on the method, refer to Beauchamp, 1975). Walsh spectral analysis was applied to geoscience for studying the cyclicity of magnetic reversals (Negi & Tiwari, 1983). Later Weedon (1986) and Schwarzacher (1987) introduced this technique to study cyclic sediments. One advantage of this method is the speed in transform computation as only addition and subtraction is required.

The application of spectral analysis to geological problems involves four aspects: time-series generation, transform operation, spectrum generation and interpretation. In our time-series generation for ODP Site 762 sediments, we coded the light beds with a value of +1 and the dark beds with a value of −1. Figure 13 is an example of the step-like time series generated. Time series generated in this way are no more than an approximation of the cycles, since in reality the boundary between light and dark beds is gradual, with a transitional zone varying from 2 to 12 cm, a gradation which may also be reflected in $CaCO_3$ content. For the purposes of this study we assumed that a single bed was produced under relatively constant depositional conditions. To generate the step-like time series, we placed the colour boundary in the middle of the transition zone, the most suitable choice for an interval without enough laboratory measurements but with a good lithological record.

High recovery in all of the cores in the interval, except Core 55X (5 m long), aided time-series generation. Other factors, however, prevented time-series analysis of every core. Microfault planes were observed in Sections 47X-2 and 56X-3 (Fig. 4). The microfault planes, which offset the cyclic succession, can distort the time series and this distortion is difficult to correct. Whole-round samples were taken in Sections 47X-5 (115–140 cm, 140–150 cm) and 53X-4 (115–140 cm, 140–150 cm). Core 54X is not only more expanded (>100% recovery) but also more disturbed than the other cores. For these reasons, Sections 47X-1 through to 47X-5, Cores 53X–56X were rejected for time-series generation. The three time series we have built in the cyclic interval are:

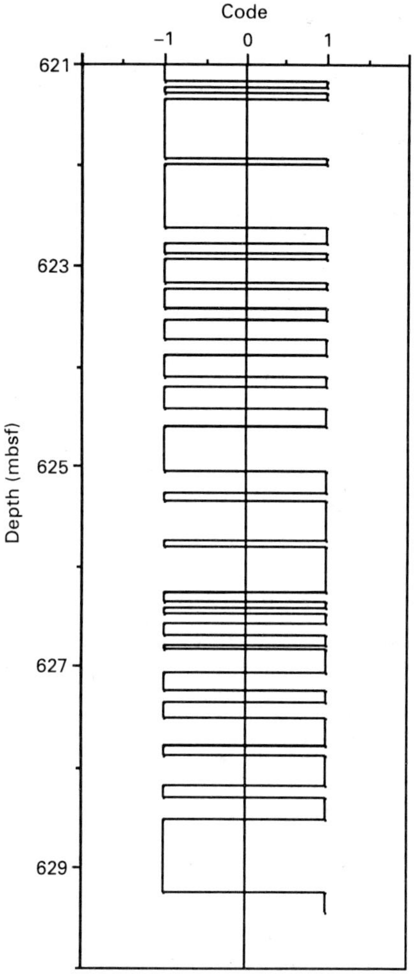

Fig. 13. An example of the time series generated from core inspection and measurement record (time series B). The value +1 stands for the light bed and −1 for the dark bed. Sections 122-762C-50X-1 to 122-762C-50X-6.

1 Sections 47X-6 through to 49X, core catcher (600.0–621.0 mbsf);
2 Sections 50X-1 through to 50X-6 (621.0–629.74 mbsf);
3 Sections 51X-1 through to 52X, core catcher (630.5–649.5 mbsf).

In Section 50X-2, there is a 10-cm-long IW whole-round sample, but we interpolated the colour of the whole-round sample based on the surrounding colour. Two cores were merged in time series 1 and 3 to increase the length of time base for analysis. An 18-m time series is more than 10 times longer than the longest cycle length observed in the cores. Longer time series would improve the accuracy of the position of frequency peaks. With longer time series it is also possible to examine the stability of the frequency peaks by dividing the time series into two subsections and applying spectral analysis to each of them. However, the cost of using longer time series is the introduction of an additional complication caused by unknown changes in sedimentation rate.

To examine further and interpret the power spectra we also used wavelength ratios of the frequency peaks. To test the hypothesis that the cyclic sediments resulted from processes with Milankovitch frequencies, the wavelength ratios were examined and compared with the ratios of present-day periods of the orbital elements (Table 2). According to Berger & Loutre (1989), the periods of orbital elements in the Late Cretaceous were only slightly shorter than those of the present day (2000 years shorter at most for the obliquity and precession). The use of ratios for interpretation is independent of absolute time and can avoid some errors brought in by sedimentation rate estimation. However, the ratio comparison can be affected by several factors (Molinie & Ogg, 1990). Therefore, the time duration in thousands of years of the frequency peaks converted from estimated average sedimentation rate, which may also have errors, should also be examined. Mutual constraint of ratio comparison and time duration examination is expected to achieve more certain insight into the periodicities of the cyclic sequence.

The power spectra

In spectral analysis computation, the step-like time series is first sampled with a user-defined uniform interval. The number of data points for a fast Walsh transform must be equal to a power of 2. In this study the sampling interval varies slightly with the length of every time series, but is less than 1 cm, to guarantee sampling the thinnest bed observed in the cores. The program subsequently performed the Walsh transform and power value calculations. The power values were normalized for an application of Fisher's test for white noise (Nowroozi, 1967), and the Walsh spectrum was smoothed using a three-point Hanning window. The Walsh power spectrum illustrates the relative importance of the frequency peaks. The 90% confidence level was used to inspect the validity of the frequency peaks. The use of a 90% confidence level implies that in case no periodicity is present in the time series, 1 out of every 10 peaks will exceed this level.

Walsh spectral analysis was carried out on the three time series. The stability of the frequency peaks in a spectrum was tested by dividing a time series into two or more subsections and calculating subsection spectra from them.

The power spectra calculated from time series 1, 2 and 3 are shown in Figs 14, 15 and 16 respectively. On each spectrum the lengths of the time series are indicated by a time base, and the bandwidth by a short horizontal bar. A narrower bandwidth indicates higher resolution. The 90% confidence level is drawn as a horizontal line on whole-section spectra. To help interpretation, the ratios of the

Table 2. Ratios of the modern periods (in ka) of the precession (P1, P2), obliquity (O) and eccentricity (E1, E2) cycles to each other. Orbital periods are according to Berger (1977)

	P1 (19.0)	P2 (23.7)	O (41.0)	E1 (94.9)	E2 (123.3)
Ratio of precession period (P1) to other orbital periods	1.00	0.80	0.46	0.20	0.15
Ratio of precession period (P2) to other orbital periods	1.25	1.00	0.58	0.25	0.19
Ratio of obliquity period (O) to other orbital periods	2.15	1.73	1.00	0.43	0.33
Ratio of eccentricity period (E1) to other orbital periods	4.99	4.07	2.31	1.00	0.77
Ratio of eccentricity period (E2) to other orbital periods	6.49	5.20	3.01	1.30	1.00

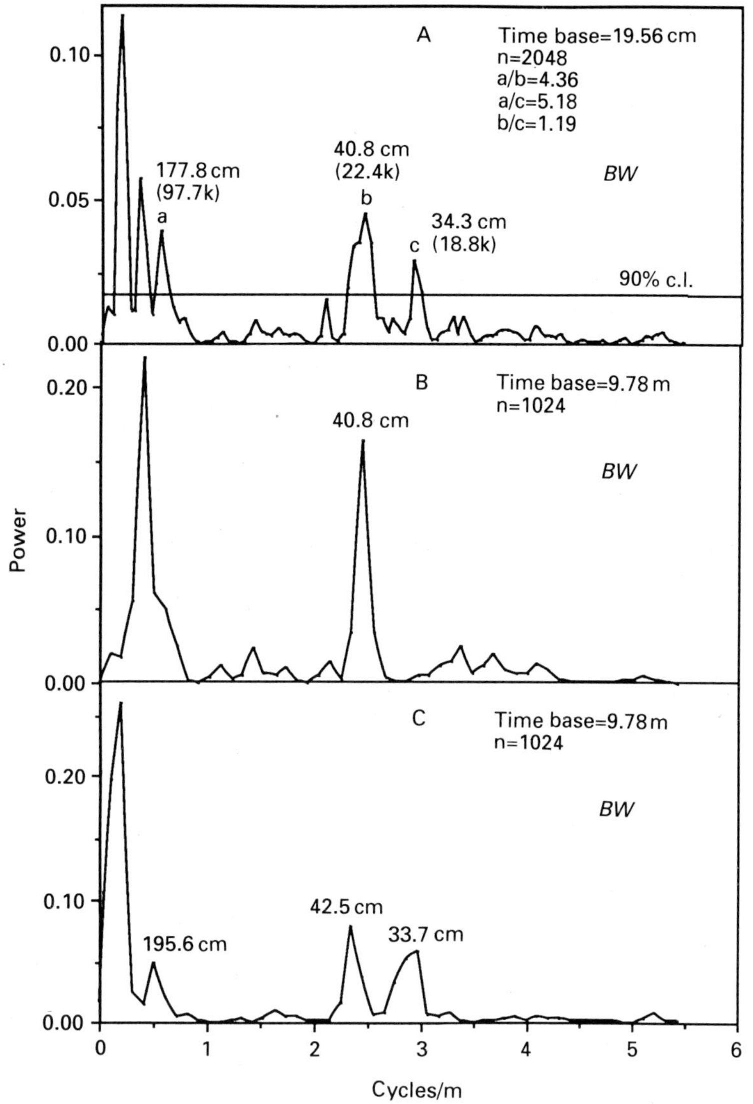

Fig. 14. Power spectra for time series 1. The number of sampling points is indicated by n. The bandwidth is expressed by a short horizontal bar. A, Whole-section spectrum. The time base is the length of the entire time series. The horizontal line is the 90% confidence level (c.l.) from Nowroozi (1967). The wavelengths of the stable-frequency peaks are marked in centimetres and corresponding durations from the average sedimentation rate of 1.82 cm/ka are shown in parentheses. B, Subsection spectrum for the upper half of time series 1. C, Subsection spectrum for the lower half of time series 1.

wavelengths of significant frequency peaks and the time duration of the frequency peaks are also indicated on the whole-section spectra. As the subsection is only one-half of the length of the whole section, the peaks in very low-frequency regions of the whole-section spectra are difficult to compare with those in subsection spectra in Figs 14 and 16. Therefore, peaks in the low-frequency region (less than 0.5 cycles/m) have not been interpreted and discussed in this study. We regard a frequency peak in a whole-section spectrum to be stable if a strong counterpart is found in one or both of the subsection spectra. In Figs 14 and 16, the location of the frequency peaks in the spectra from the subsections deviate slightly from their corresponding stable peaks. The subsection spectra of time series 3 display more variation. The variations may be attributed to changes in sedimentation rate and influence from differential compaction. The location of the frequency peaks in Fig. 15 is comparable to the location of those appearing in Figs 14A and 16A. Thus the frequency peaks with wavelength 84.0, 38.2 and 35.0 cm are interpreted to be stable. The sampling intervals in analysing these three time series are less

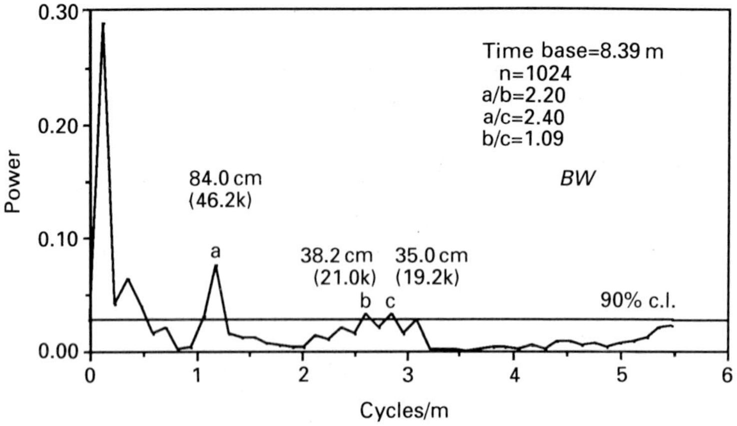

Fig. 15. Power spectra for time series 2. The number of sampling points is indicated by n. The bandwidth is expressed by a short horizontal bar. The time base is the length of the time series. The horizontal line is the 90% confidence level (c.l.) from Nowroozi (1967). The wavelengths of the stable-frequency peaks are marked in centimetres and corresponding durations from the average sedimentation rate of 1.82 cm/ka are shown in parentheses.

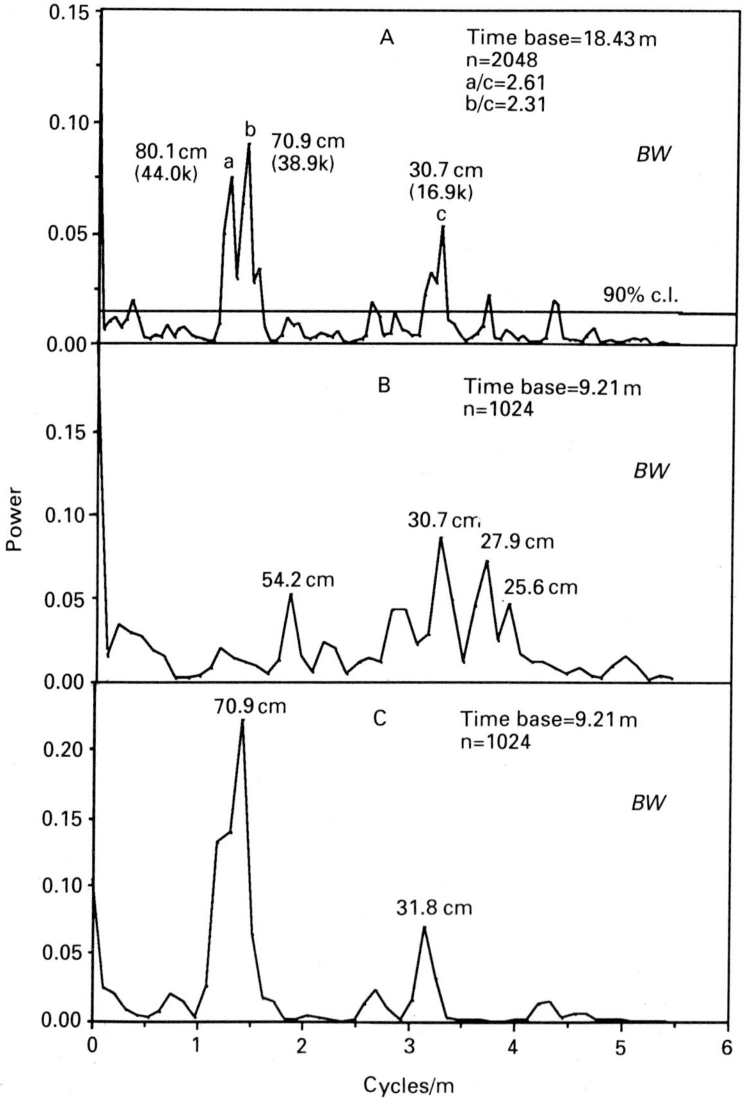

Fig. 16. Power spectra for time series 3. The number of sampling points is indicated by n. The bandwidth is expressed by a short horizontal bar. A, Whole-section spectrum. The time base is the length of the entire time series. The horizontal line is the 90% confidence level (c.l.) from Nowroozi (1967). The wavelengths of the stable-frequency peaks are marked in centimetres and corresponding durations from the average sedimentation rate of 1.82 cm/ka are shown in parentheses. B, Subsection spectrum for the upper half of time series 3. C, Subsection spectrum for the lower half of time series 3.

than 1 cm, much smaller than the thickness of the thinnest bed. The Nyquist frequency is at about a 10-ka period, judging from the estimated average sedimentation rate (1.82 cm/ka). Therefore, the frequency peaks in the spectra are not subjected to an aliasing problem (see Pisias & Mix, 1988). The structure of the spectra does not suggest there is a harmonic problem. We therefore conclude that the stable-frequency peaks reveal the predominant frequencies of the colour alternations in the cyclic sediments.

In Fig. 14A, the wavelength ratio of peak a to peak b is 4.36, near the E1/P2 ratio (4.07, see Table 2), that of peak a to peak c is 5.18, near the E1/P1 ratio (4.99, see Table 2) and that of peak b to peak c is 1.19, near the P2/P1 ratio (1.25, see Table 2). The time durations of peaks a, b and c, from the estimated average sedimentation rate (1.82 cm/ka), are 97.7, 22.4 and 18.8 ka. These time durations are very near the modern periodicity of one of the eccentricity cycles (94.9 ka) and the two main precession cycles (23.7 and 19.0 ka). Therefore, peak a in Fig. 14A (time series 1) is comparable to the short eccentricity cycle and peaks b and c to the precession cycles. In this case the wavelength ratios fit very well with the time duration from estimated sedimentation rate in interpreting the spectrum.

In Fig. 15, the wavelength ratio of peak a to peak b is 2.20, displaced from the O/P2 ratio (1.73, see Table 2), and that of peak a to peak c is 2.40, near the O/P1 ratio (2.15, see Table 2). The wavelength ratio of peak b to peak c is 1.09, near the P2/P1 ratio (1.25, see Table 2). The time durations, from the estimated average sedimentation rate, for peaks a, b and c are 46.2, 21.0 and 19.2 ka. Combining both the wavelength ratio and the estimated time duration, peak a is interpreted to represent the obliquity cycle and peaks b and c possibly the precession cycles.

In Fig. 16A, the wavelength ratio of peak a to peak c (2.61) is near the O/P1 ratio (2.15, see Table 2) and the ratio of peak b to peak c (2.31) is also near to the O/P1 ratio (2.15, see Table 2). The time durations of peaks a, b and c are 44.0, 38.9 and 16.9 ka. Therefore, peaks a and b may relate to the obliquity cycle and peak c possibly to the precession cycle.

It was found that wavelength ratios fit less well with the time durations at increasing depth (i.e. from Figs 14–16). This may be attributed to the increasing effect of differential compaction on the sediment column with depth, and also on the wavelength ratio method. Overall, wavelength ratios were within 13% of the orbital ratios predicted from the spectral peaks in Figs 14–16 except for the O/P2 ratio in time series 2 (21% error) and the O/P1 ratio in time series 3 (18% error).

The examination and interpretation of the power spectra with both wavelength ratios and time duration converted from the estimated average sedimentation rate demonstrates that cyclic sequences from Exmouth Plateau can be best explained in terms of cyclic variations in orbital elements, rather than other processes. The obliquity, precession and eccentricity signals are all found in this cyclic sequence, though the last is detected in only one of the time series.

MODEL FOR THE ORIGIN OF CYCLICITY

With the establishment of Milankovitch-like cycles in Hole 762C and with the available sedimentological evidence, we now attempt to describe the origin of the cyclicity. The mechanism considered to produce cyclic sediments at Site 762 is summarized in Fig. 17. We suggest that orbital variation may have controlled climate and oceanography and thus induced the observed cyclic sedimentation on the Exmouth Plateau during the Late Cretaceous. Previous studies have documented that the Late Cretaceous was globally warmer than present and free of large-scale continental glaciation (Frakes, 1979; Barron, 1983). According to the Cretaceous plate tectonic reconstructions by Barron (1987), during Maastrichtian time the Australian continent, stretching west to east, was between 45° S and 65° S. J.G. Ogg (personal communication, 1990) gives an estimate of 53° S for the palaeolatitude of the Exmouth region. In a latitudinal basin such as the Exmouth Plateau and with cyclic change in insolation caused by the obliquity and precession variations, we suggest a cyclic alternation of two different climatic regimes and the operation of the following model. We suggest that this model is one possible solution for the generation of cyclic sediments by orbital processes and recognize that alternative models may also be suggested.

Given that the proportion of ocean and land areas is constant in a region, in higher insolation phases both the ocean and land receive more energy from the Sun. Because of the difference in the thermal inertia of water and land, the temperature increase in the ocean is much smaller than on the land. On

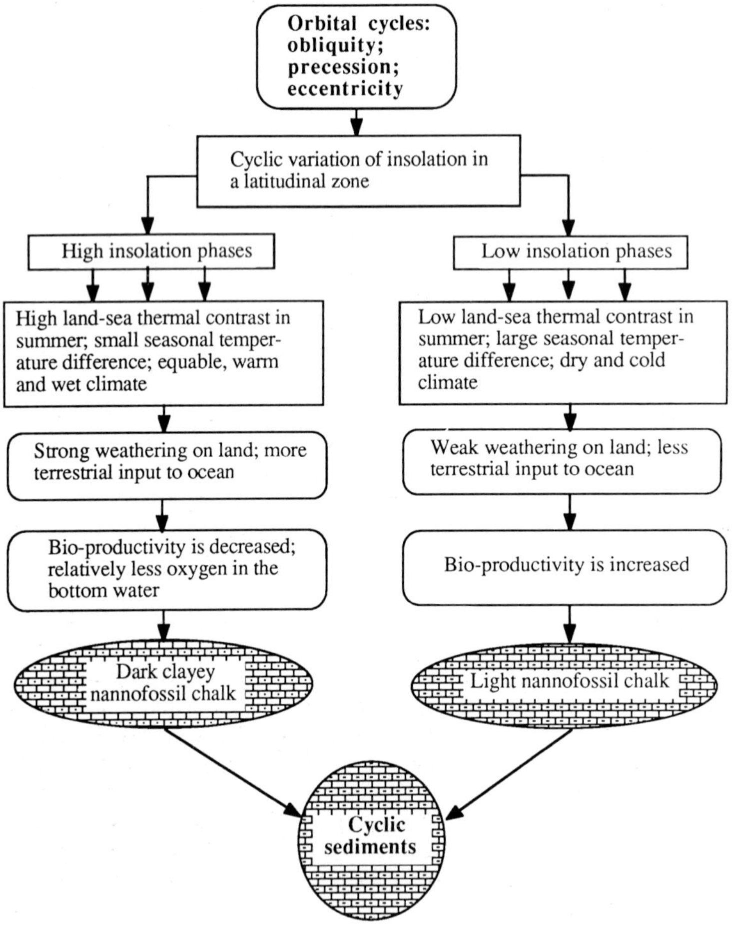

Fig. 17. A proposed mechanism for the cyclic sedimentation under the control of orbitally induced cyclic climatic changes on the Late Cretaceous Exmouth Plateau (see text for detailed explanation).

the other hand, the ocean, which has great thermal inertia, may store some solar energy and serve as a thermal reservoir. In this period, land–sea thermal contrast is great in summer time, resulting in higher precipitation when more hot air rises over the continent to be replaced by moist cooler marine air, but the summer–winter temperature difference is relatively small because the ocean serves as an efficient thermal reservoir and diminishes the temperature contrast. In this phase, the climate has a smaller seasonal contrast in temperature and is rainy. An equable, warm, rainy climate favours the weathering of rocks, especially chemical weathering, and more clay minerals are formed on land. The high runoff, because of higher precipitation, increases summer terrestrial input of clay minerals (kaolinite and illite) and siliciclastic silt to the ocean. The rainy climate and high runoff may also lead to more precipitation than evaporation in the sea area and produce a surface water of lower salinity which may show seasonal variation. Warm surface water with varying and decreased salinity may not be suitable for most planktonic species to 'bloom'. Excessive suspension of sediments in the upper water column may also affect the development of a planktonic community. The reduced equator–pole temperature contrasts during increased solar insolation may cause a decrease in wind strength, upwelling and deep circulation, as the Earth's circulation system as a whole is slowed down. Total bioproductivity may be reduced in this phase of reduced nutrient supply, as 90% of the nutrients in marine organic production are recycled from deep water through ocean circulation (de Boer, 1986). It was under such a Cretaceous climate at Site 762 that we consider the dark beds with more clay to have been deposited.

In the low insolation phase, the total energy from the Sun is decreased. There is a smaller land–sea thermal contrast in the summer, making it less rainy in comparison with the summer in high-insolation phases. The seasonal contrasts (winter, summer) in temperature become pronounced as the function of the ocean as a thermal reservoir is reduced. The winter also becomes drier. Consequently, climate in the low-insolation phase is colder and drier. A colder and drier climate hinders the weathering of rock, especially chemical weathering, and fewer clay minerals are formed on land. The dilution of terrestrial input is reduced because of low precipitation and resulting low runoff. Planktonic species increase their abundance because the surface waters have less stratification caused by freshwater influx and experience greater nutrient replenishment from deeper levels in a period of increased wind strength and probably more upwelling. It was under such a climate that we interpret the light carbonate-rich beds to have been deposited at Site 762.

This mechanism can be supported by the quantitative climatic modelling results for Cretaceous time by Oglesby & Park (1989). Although in their study an uncoupled climate model is employed, and only insolation variations caused by the precession cycle are considered, the results support our explanation for the climatic-induced cyclic sedimentation on the Exmouth Plateau. Their results for South proto-Asia show that the January–July difference of land surface temperature is greater in a low-insolation phase than in a high-insolation phase, and in both January and July the specific humidity and precipitation are higher in a high-insolation phase than in a low-insolation phase.

In our model we have integrated the factors causing the variation in terrestrial material input and the change in bioproductivity. However, during the Late Cretaceous, under the control of orbital variations, the changes on land may have played a leading role in the cyclic sedimentation on Exmouth Plateau. The changes in the sea area of the Exmouth Plateau are more or less dependent on the changes that happened on the land. The change in planktonic microfauna is a response not only to water temperature changes (which may not be very substantial owing to the great thermal inertia of water), but also to input of terrestrial material and the surface salinity. Nutrient supply is another important factor for bioproductivity. As both dark and light beds are bioturbated, in both climatic phases the bottom water was well oxygenated. The oxygen levels in the bottom water may have been variable, as indicated by the variations of Mn, but apparently never diminished enough to eliminate bottom life. Therefore, the pattern and rate of circulation could not have changed substantially from one climatic phase to the other in this region. The fertility of the surface water may have been more affected by the influx of fresh water than upwelling or deep circulation, according to the palaeowater depth (550 m) estimated by L. Gradstein & U. von Rad (personal communication) and the fact that the Exmouth Plateau is not far from the terrestrial source (260 km offshore).

DISCUSSION

Both split-power spectra and wavelength ratios suggest the presence of eccentricity (around 100 ka), obliquity (41 ka) and precession (23.7 and 19 ka) cycles in the data from Site 762 on the Exmouth Plateau. In the interval between 630.5 and 649.5 mbsf (time series 3) there are strong peaks most likely related to the obliquity and one less-strong peak possibly associated with precession (Fig. 16A). Between 621.0 and 630.0 mbsf (time series 2), the peak related to the obliquity is still strong (Fig. 15). In the upper interval from 600.0 to 621 mbsf (time series 1), there are peaks possibly related to the eccentricity and precession cycles (Fig. 14A). However, neither wavelength ratios nor time duration (from the average sedimentation rate) indicate a peak in the whole-section spectrum (Fig. 14A) that could be possibly related to the obliquity cycle. This implies a possible transition from obliquity-controlled climatic variations to eccentricity–precession-controlled climatic variations. This also suggests that the response of the pelagic sedimentation system to the controlling effects of the long-term orbital variations at Site 762 may have changed through the Late Cretaceous. This change could have been due to plate motion into lower latitudes, where, theoretically, the precession is dominant over the obliquity (Berger, 1984). However, the motion is of small scale; using a spreading rate of 2.0 cm/year we calculated that the Australian plate could only have moved less than 100 km during the time of deposition investigated here. Another possibility is that falling sea level during the end of the late Campanian (Haq *et al.*, 1987) caused some changes in the response of climate to orbital cycles and some changes in cyclic sedimentation. It appears that the response of pelagic sedimentation to the controlling effects of long-term orbital variations is complex and does not adhere to any simple relationship in the geological past. The development of cycles in Pleistocene sediments involved changes in ice volume controlled by orbital variation (Pisias & Moore, 1981). The study by Hagelberg & Pisias (1990) indicates that the climatic response to orbital forcing during the Pliocene is totally different from that during the Pleistocene. Similarly, Mayer (in press) notes clear shifts from eccentricity to obliquity frequencies in high-resolution Pleistocene cycles. Investigators of Upper Cretaceous cyclic sequences (ROCC Group, 1986) suggested that the sedimentary response to climatic forcing is variable.

In Figs 14–16, the time durations converted from the average sedimentation rate for the peaks discussed are near to the periodicities (modern values, compare with Table 2) of the orbital elements to which they are interpreted to respond. This may imply that a total 6-Ma duration for the late Campanian to early Maastrichtian used for estimating the sedimentation rate is near the actual duration.

Long-term variations in sedimentation rate could explain some variability in the data. For example, in the spectrum from time series 2 (Fig. 15) the peak associated with obliquity is longer (46.2 ka) than the obliquity periodicity (41.0 ka). Therefore, the actual sedimentation rate for this section (time series 2) could be higher than 1.82 cm/ka. In the spectrum from time series 3 (Fig. 16A) there are two peaks whose time durations (44.0 and 38.9 ka) are both near the obliquity periodicity (41 ka). They are the product of one dominant obliquity cycle (41 ka). Two sedimentation rates (1.95 and 1.73 cm/ka) are derived when the wavelengths of these peaks are divided by the obliquity periodicity. These two values may represent the fluctuating range of sedimentation rate in this interval (time series 3).

The brown to green change in background hue at 603.5 mbsf is considered by us to have resulted from the decreasing intensity of weathering on land. The background change has been correlated with a relative sea-level change from high-stand to low-stand (Haq *et al.*, 1990). Our explanation is not in contradiction with this correlation. When the interval of brownish hue was deposited, the global climate was warmer and more humid, corresponding to a higher relative sea level, and the weathering products (clayey minerals, etc.) were brownish because of complete oxidation. When the interval of greenish hue was deposited, the background climate became colder or at least less humid, corresponding to a lower relative sea level, and the weathering products (clayey minerals, etc.) were not fully oxidized and appeared greenish. It is also possible that the terrestrial input of iron occurred in a reduced state.

CONCLUSIONS

1 The cyclic colour variations in late Campanian to early Maastrichtian pelagic sediments in the Indian Ocean are interpreted to have resulted from depositional processes due to astronomically induced variations in the quantity of terrigenous sediment input to the ocean. Diagenetic processes at most enhanced the appearance of the cyclicity.

2 Spectral analysis reveals that the predominant wavelengths of the colour cycles are 34–41 and 71–84 cm. With an average sedimentation rate of 1.82 cm/ka, the time durations of the colour cycles are around 41 and 21 ka, which correlate to the obliquity and precession periods and strongly suggest an orbital origin for the cycles. A weak low-frequency signal that correlates to the eccentricity cycle is found in the upper part of the sequence, which agrees with the dominant precession signal there (time series 1).

3 During the Late Cretaceous, when there was no continental glaciation, cyclic sedimentation on the Exmouth Plateau of the eastern Indian Ocean was primarily caused by cyclic changes on land (soil formation and runoff) in response to orbitally controlled insolation variations. Bioproductivity varied 180° out of phase with changes in terrestrial input into the ocean, but no stagnant and oxygen-depleted bottom water was involved in the process. Warm climatic phases resulted in the production of dark beds, whereas cooler climatic phases produced light beds.

4 Spectral analysis indicates that the orbital variations controlling cyclic sedimentation may vary with time. In the older part of the sequence (time series 2 and 3), there are strong obliquity signals together with precession signals, but in the upper parts of the sequence (time series 1) there are only precession and eccentricity signals. A transition from obliquity-controlled climatic variations to eccentricity–precession-controlled climatic variations is implied. The change from brownish hue to greenish hue in the sequence may indicate a major shift of the background climatic pattern, from very humid to less humid.

ACKNOWLEDGEMENTS

We thank the Shipboard Scientific Party of Leg 122 for the cooperative scientific effort that made this study possible, G. Hampt for calcium carbonate measurements, J. Tribble for XRD analysis and E. DeCarlo and W. Shibata for the geochemical analyses. SEM examination was assisted by F. Thomas. This work benefited from discussions with Dr R. Wilkens. Our manuscript was improved by comments and suggestions from Drs F.M. Gradstein and J.G. Ogg. We are grateful to Drs P.L. de Boer and D. Smith for their critical review. Financial support for R.B. and Z.H. was provided by a Canadian NSERC Collaborative Special Project Grant, a University Research Grant from Imperial Oil (Canada), and for S.O. by USSAC. Sabbatical facilities and support for R.B. were provided by the Alexander von Humboldt Foundation and Geomar Institute, Kiel.

REFERENCES

ARTHUR, M.A., DEAN, W.E., BOTTJER, D.J. & SCHOLLE, P.A. (1984) Rhythmic bedding in Mesozoic–Cenozoic pelagic carbonate sequences: the primary and diagenetic origin of Milankovitch-like cycles. In: *Milankovitch and Climate*, Part 1 (Eds Berger, A., Imbrie, J., Hays, J., Kukla, G. & Saltzman, B.). Reidel Publ. Co., Dordrecht, pp. 191–222.

BALZER, W. (1982) On the distribution of iron and manganese at the sediment/water interface: thermodynamic versus kinetic control. *Geochim. Cosmochim. Acta* **46**, 1153–1161.

BARRON, E.J. (1983) A warm, equable Cretaceous: the nature of the problem. *Earth Sci. Rev.* **19**, 305–338.

BARRON, E.J. (1987) Cretaceous plate tectonic reconstructions. *Palaeogeogr. Palaeoclimatol. Palaeoecol.* **59**, 3–29.

BARRON, E.J., ARTHUR, M.A. & KAUFFMAN, E.G. (1985) Cretaceous rhythmic bedding sequences: a plausible link between orbital variations and climate. *Earth Planet. Sci. Lett.* **72**, 327–340.

BEAUCHAMP, K.G. (1975) *Walsh Functions and Their Applications*. Academic Press, Orlando, FA.

BERGER, A.L. (1977) Support for the astronomical theory of climatic change. *Nature* **269**, 44–45.

BERGER, A.L. (1984) Accuracy and frequency stability of the Earth's orbital elements during the Quaternary. In: *Milankovitch and Climate*, Part 1 (Eds Berger, A., Imbrie, J., Hays, J., Kukla, G. & Saltzman, B.). Reidel Publ. Co., Dordrecht, pp. 3–39.

BERGER, A.L. & LOUTRE, M.F. (1989) Pre-Quaternary Milankovitch frequencies. *Nature* **342**, 133.

BOTTJER, D.J., ARTHUR, M.A., DEAN, W.E., HATTIN, D.E. & SAVRDA, C.E. (1986) Rhythmic bedding produced in Cretaceous pelagic carbonate environments. *Paleoceanography* **1**, 467–481.

BOYD, R., WILLIAMSON, P. & HAQ, B.U. (1992) Seismic stratigraphy and passive margin evolution of the southern Exmouth Plateau, Exmouth Plateau. In: *Proc. Ocean Drilling Program, Sci. Repts, 122*. College Station, TX (Ocean Drilling Program), pp. 39–59.

BRALOWER, T.J. & SIESSER, W.G. (1992) Cretaceous calcareous nannofossil biostratigraphy of ODP Leg 122 Sites 751, 762 and 763, Exmouth and Wombat Plateau, NW Australia. In: *Proc. Ocean Drilling Program, Sci. Repts, 122*. College Station, TX (Ocean Drilling Program), pp. 529–556.

COTILLON, P. & RIO, M. (1984) Cyclic sedimentation in the Cretaceous of Deep Sea Drilling Project Sites 535 and 540 (Gulf of Mexico), 534 (Central Atlantic) and in the Vocontian Basin (France). *Init. Rep. DSDP* **77**, 339–376.

DARMEDRU, C., COTILLON, P. & RIO, M. (1982) Climatic and biological rhythms in marine pelagic realm: their relationships in Cretaceous alternating beds from the Vocontian Basin (SE France) (in French). *Bull. Soc. Géol. Fr.* Ser. **7**, **24**, 672–640.

DEAN, W.E. & ARTHUR, M.A. (1987) Inorganic and organic geochemistry of Eocene to Cretaceous strata recovered from the lower continental rise, North American Basin, Site 603, Deep Sea Drilling Project Leg 93. *Init. Rep. DSDP* **93**, 1093–1138.

DE BOER, P.L. (1986) Changes in the organic carbon burial during the Early Cretaceous. In: *North Atlantic Palaeoceanography* (Eds Summerhayes, C.P. & Shackleton, N.J.). Geol. Soc. Lond. Spec. Publ. 21, pp. 321–331.

FRAKES, L. (1979) *Climate Through Geological Time*. Elsevier, New York.

HAGELBERG, T.K. & PISIAS, N. (1990) Nonlinear response of Pliocene climate to orbital forcing: evidence from the eastern equatorial Pacific. *Paleoceanography* **5**, 595–617.

HALLAM, A. (1986) Origin of minor limestone–shale cycles: climatically induced or diagenetic? *Geology* **14**, 609–612.

HAQ, B.U., HARDENBOL, J. & VAIL, P.R. (1987) Chronology of fluctuating sea levels since the Triassic. *Science* **235**, 1156–1167.

HAQ, B.U., VON RAD, U. & O'CONNELL, S. *et al.* (Eds) (1990) *Proc. Ocean Drilling Program. Init. Results, 122*. College Station, TX (Ocean Drilling Program).

HART, M.B. (1987) Orbitally induced cycles in the chalk facies of the United Kingdom. *Cret. Res.* **8**, 335–348.

HAYS, J.D., IMBRIE, J. & SHACKLETON, N.J. (1976) Variations in the Earth's orbit: pacemaker of the ice ages. *Science* **194**, 1121–1132.

HERBERT, T.D. & D'HONDT, S.L. (1990) Precession climate cyclicity in Late Cretaceous–Early Tertiary marine sediments: a high resolution chronometer of Cretaceous–Tertiary boundary events. *Earth Planet. Sci. Lett.* **99**, 263–275.

HERBERT, T.D. & FISCHER, A.G. (1986) Milankovitch climatic origin of Mid-Cretaceous black shale rhythms in Central Italy. *Nature* **321**, 739–742.

JARRARD, R. & ARTHUR, M.A. (1989) Milankovitch paleoceanographic cycles in geophysical logs from ODP Leg 105, Labrador Sea and Baffin Bay. In: *Proc. Ocean*

Drilling Program, Sci. Results, 105 (Eds Srivastava, S.P., Arthur, M.A., Clememt, B. *et al.*). College Station, TX (Ocean Drilling Program), pp. 757–772.

KENT, D.V. & GRADSTEIN, F.M. (1985) A Cretaceous and Jurassic geochronology. *Geol. Soc. Am. Bull.* **96**, 1419–1427.

KING, D.T. Jr (1990) Upper Cretaceous marl–limestone sequences of Alabama: possible products of sea-level change, not climate forcing. *Geology* **18**, 19–22.

MAYER, L.A. (1991) Extraction of high resolution carbonate records for paleoclimate reconstruction. *Nature* **352**, 148–150.

MOLINIE, A.J. & OGG, J.G. (1990) Sedimentation-rate curves and discontinuities from sliding-window spectral analysis of logs. *The Log Analyst* **31**, **No. 6**, 370–374.

NEGI, J.G. & TIWARI, R.K. (1983) Matching long term periodicities of geomagnetic reversals and galactic motions of the solar system. *Geophys. Res. Lett.* **10**, 713–716.

NOWROOZI, A.A. (1967) Table for Fischer's test of significance in harmonic analysis. *Geophys. J. R. Astron. Soc.* **12**, 512–520.

OGG, J.G., HAGGERTY, J. & SARTI, M. (1987) Lower Cretaceous pelagic sediments of Deep Sea Drilling Project Site 603, Western North Atlantic: a synthesis. *Init. Rep. DSDP* **93**, 849–880.

OGLESBY, R. & PARK, J. (1989) The effect of precessional insolation changes on Cretaceous climate and cyclic sedimentation. *J. Geophys. Res.* **94**, 14973–14816.

PISIAS, N.G. & MIX, A.C. (1988) Aliasing of the geologic record and the search for long-period Milankovitch cycles. *Paleoceanography* **3**, 613–619.

PISIAS, N.G. & MOORE T.C. Jr (1981) The evolution of Pleistocene climate: a time series approach. *Earth Planet. Sci. Lett.* **52**, 450–458.

RESEARCH ON CRETACEOUS CYCLES (ROCC) GROUP (ARTHUR, M.A., BOTTJER, D.J., DEAN, W.E., FISCHER, A.G., HATTIN, D.E., KAUFFMAN, E.G., PRATT, L.M. & SCHOLLE, P.A.) (1986) Rhythmic bedding in Upper Cretaceous pelagic carbonate sequences: varying sedimentary response to climatic forcing. *Geology* **14**, 153–156.

SCHLUMBERGER (1987) *Log Interpretation Principles/Applications*. Schlumberger Education Services, Houston, Texas.

SCHWARZACHER, W. (1987) The analysis and interpretation of stratification cycles. *Paleoceanography* **2**, 79–95.

SCHWARZACHER, W. & FISCHER, A.G. (1982) Limestone–shale bedding and perturbation of the Earth's orbit. In: *Cyclic and Event Stratification* (Eds Einsele, G. & Seilacher, A.). Springer Verlag, New York, pp. 72–95.

TEN KATE, W.G. & SPRENGER, A. (1989) On the periodicity in a calcilutite–marl succession (SE Spain). *Cret. Res.* **10**, 1–31.

WEEDON, G.P. (1986) Hemipelagic shelf sedimentation and climatic cycles: the basal Jurassic (Blue Lias) of South Britain. *Earth Planet. Sci. Lett.* **76**, 321–335.

Spec. Publs Int. Ass. Sediment. (1994) **19**, 167–193

Complex rhythmic sedimentation related to third-order sea-level variations: Upper Cretaceous, Western Interior Basin, USA

W. RICKEN

University of Cologne, Geological Institute, Zülpicher Strasse 49a, 50674 Köln, Germany

ABSTRACT

Rhythmic sedimentation in Upper Cretaceous carbonaceous marls and chalks deposited in the Western Interior Basin is quantified in order to assess the original flux variations governing these rhythms. Investigations are carried out using carbonate mass balance calculations, isotope studies and numerical flux models for three sediment components. Sections are investigated from mid-basin, reflecting various positions on two major third-order transgressive–regressive (TR) cycles (i.e. Greenhorn TR cycle and Niobrara TR cycle). Estimation of the dominant flux pattern and quantification of relative depositional fluxes is performed by evaluating different C_{org}–$CaCO_3$ dilution and concentration styles.

The shallow seaway of the Western Interior Basin is sensitive to small-scale climatic cycles, due to combined variations in carbonate productivity, clastic supply from continental sources and redox conditions. Rhythmic deposition is expressed differently for different portions of each of the two third-order TR cycles. Section intervals reflecting maximum transgression show combined variations in carbonate productivity and redox rhythms, while the more regressive units show variations in calcareous productivity, clastic dilution and dysaerobic bottom waters. For most of the cycles investigated, persistent dysaerobic bottom waters are largely uninfluenced by variations in pelagic carbonate productivity, except for section intervals representing maximum transgression. This suggests the existence of a circulation pattern in which a mixed surface water zone was underlain by more stratified, oxygen-depleted water masses with slow circulation.

Only the carbonate-rich units, which are formed during sea-level high-stands, have substantial diagenetic overprint; this overprint is medium to low in the more marly units. Differential carbonate dissolution and cementation processes lower the carbonate content in the marl beds and augment the carbonate content in the limestone beds. The organic carbon content is passively enriched in the marl beds and reduced in the limestone beds. These processes, along with differential compaction, augment the original bedding phenomena.

INTRODUCTION

Marl–limestone cycles are a prominent bedding feature in both deep-sea and epicontinental settings. In most of these settings, Milankovitch orbital variations have been proposed as the mechanism governing these cycles, based on time-series analyses, and isotopic and palaeontological arguments (see reviews in Fischer, 1986, 1991; Fischer *et al.*, 1990). The investigated chalks, black shales and marls of the US Western Interior Basin (Fig. 1) provide a classic example of rhythmically bedded sequences formed in the middle of a foreland basin which is epicontinental in character (Pratt *et al.*, 1985). Bedding rhythms are best developed in pelagic and chalky sections formed during sea-level high-stands (e.g. Kauffman, 1984). Rhythms represent repeated changes in carbonate productivity, clastic supply and redox conditions, indicating that the Western Interior Basin is a sensitive system transforming climatic signals into bedding rhythms (Hattin, 1971; Pratt, 1984; Barron *et al.*, 1985; Eicher & Diner, 1989, 1991; Pratt *et al.*, 1991). The aim of this paper is to quantify the complex depositional styles which led to different expressions of bedding rhythms over the course of two third-

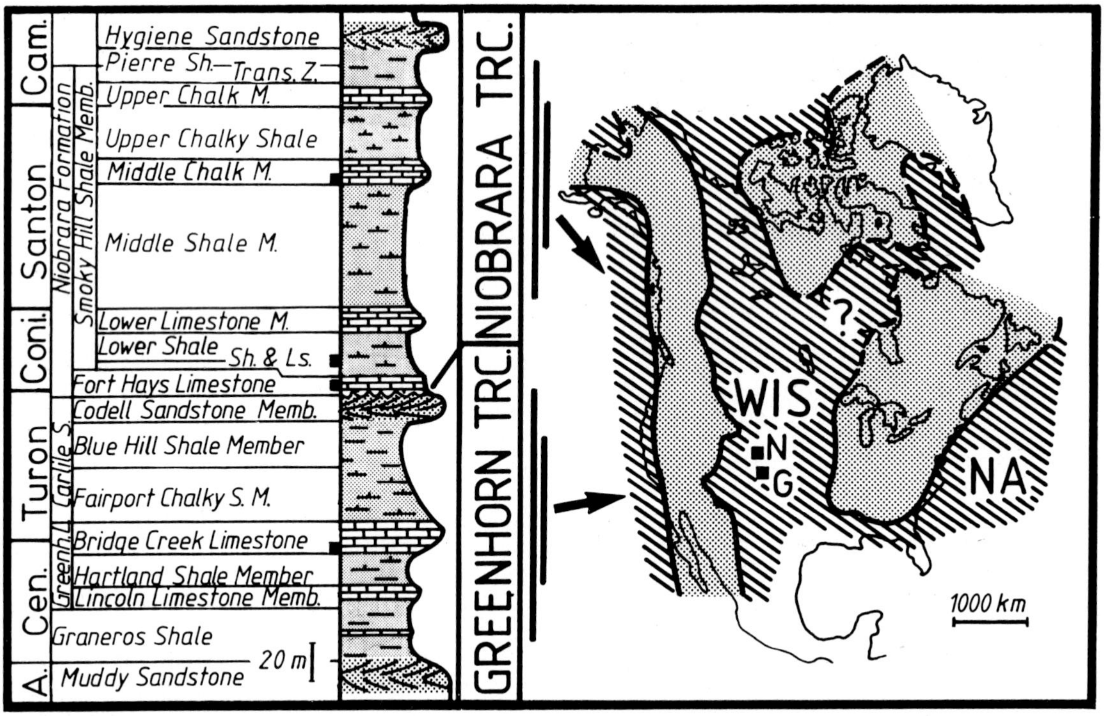

Fig. 1. Western Interior Basin during maximum transgression at the Cenomanian/Turonian boundary. Locations and stratigraphic levels of studied sections are depicted by black symbols. G, Greenhorn TR cycle; N, Niobrara TR cycle; WIS, Western Interior Basin; NA, North Atlantic. Palaeogeography after Kauffman (1984).

order sea-level variations. A detailed quantitative assessment of the depositional processes cannot be performed unless diagenetic overprint is first determined (Hallam, 1986; Ricken, 1986; Bathurst, 1987). Diagenetic overprint is assessed through compaction studies and carbonate mass balance calculations, while the original depositional processes are investigated through numerical flux determination, based on evaluation of characteristic C_{org}–$CaCO_3$ relationships.

RHYTHMIC BEDDING IN THE WESTERN INTERIOR BASIN

The Upper Cretaceous sediment fill of the Western Interior foreland basin shows an overall pattern of several third-order TR cycles (Weimer, 1960; Kauffman, 1984; Vail *et al.*, 1991). Only section intervals interpreted as representing sea-level highstands are characterized by well-developed rhythmic stratification (e.g. Pratt *et al.*, 1985). The first major descriptions of rhythmic bedding were made by Gilbert (1895), who invoked climatic cycles to explain these variations. Other significant contributions were made by Hattin (1971, 1985) and Elder (1987), who showed that individual limestone layers of the Bridge Creek Limestone are isochronous and can be traced for several thousands of kilometres across the basin, aided by the presence of closely intercalated bentonite beds. Pratt (1984) made the first detailed analyses of the Bridge Creek Limestone by documenting changing styles of bioturbation between beds, which correlated with organic matter types and changes in composition in the clastic fraction. Oxygen isotope data for the Bridge Creek Limestone documented by Pratt (1984), Pratt & Threlkeld (1984), Arthur *et al.* (1985) and Pratt *et al.* (1991) denote lighter oxygen isotopes in the carbonate-rich beds, which Pratt interpreted as an effect of reduced salinity in the surface water. Diagenetic influences on isotope composition were addressed by Arthur *et al.* (1984, 1985) and Arthur & Dean (1991). Climate and water mass models were introduced by various authors to explain the rhythmicity, including Barron *et al.* (1985), Eicher &

Diner (1985, 1989, 1991), Laferriere *et al.* (1987) and Watkins (1989). These models show that variations in water circulation and productivity are essential factors controlling cyclic sedimentation. The response of infauna to changing water stratification and oxygen content, which thereby creates changing bioturbation styles, was investigated by Savrda & Bottjer (1986, 1987, 1989), Sageman (1989), Kauffman & Sageman (1990) and Sageman *et al.* (1991); Elder (1987) meanwhile documented macrofossil changes for the Bridge Creek Limestone. Important summaries of concepts and data are presented in the volumes edited by Pratt *et al.* (1985) and Eicher & Diner (1989).

This study is an attempt to quantify diagenetic and primary processes in rhythmically bedded strata of the Western Interior Basin, in order to demonstrate different styles of deposition, which are interpreted as reflecting different environmental processes related to third-order sea-level stands. Three basic aspects are addressed here.

1 Evaluation of the magnitude of diagenetic overprint in order to assess the 'original' (i.e. reconstructed) sediment composition; this is performed by determining differential compaction and by performing carbonate mass balance calculations.

2 Determination of the changing sediment flux between succeeding beds of the reconstructed 'original' sediment succession by using a flux model for three components (i.e. carbonate, clastics and organic matter). Different styles of deposition are recognized and quantified by different correlation patterns between carbonate and organic carbon contents.

3 Connection of the different depositional processes to the well-established transgressive–regressive history of the Western Interior Basin.

Investigated sections

Four short sections were investigated (Figs 2–5) which displayed the most prominent rhythmic bedding horizons in the Greenhorn and the Niobrara TR cycles. The Cenomanian/Turonian Bridge Creek marl–limestone alternation (Greenhorn TR cycle) represents the maximum transgression associated with deposition of carbonate ooze and rhythmically changing bottom-water stratification. Three sections of the Coniacian to Santonian Niobrara TR cycle were studied: the extremely well bedded Fort Hays Limestone, deposited during peak transgression; the Shale and Limestone Member; and the Middle Chalk Member (Barlow & Kauffman, 1985). Outcrop localities represent mid-basin positions (see Fig. 1). The Bridge Creek Limestone was sampled in the type section near Pueblo, Colorado, while the Niobrara TR cycle was sampled in well-exposed quarries near Lyons, north of Boulder, Colorado (see Pratt *et al.*, 1985 for detailed outcrop maps).

In the Niobrara cycle, the investigated sections were continuously sampled, cut into slabs, and polished in order to map sedimentological details and bioturbation patterns. This was not performed for the Bridge Creek Limestone, as detailed petrological studies had already been carried out by Pratt (1984). All sections were analysed for organic carbon and carbonate contents, compaction and porosity, which are the basic data needed to perform carbonate mass balance calculations and to carry out relative flux determinations. These methods are described in detail in the following sections of this paper.

DIAGENETIC OVERPRINT

Diagenetic overprint in rhythmically bedded marly and calcareous sequences is associated with cementation processes controlled by early organic matter decomposition (Raiswell, 1987, 1988), or with differential cementation and dissolution processes (Arthur *et al.*, 1984; Hallam, 1986; Ricken, 1986; Bathurst, 1987, 1991; Ricken & Eder, 1991). In fine-grained limestone–marl alternations, carbonate is dissolved in the marl beds and reprecipitated in the pore spaces of the limestone layers. As a consequence, limestone layers are considerably cemented, dense and micritic, with burrows and fossils slightly reduced in original thickness, while marl beds show the products of chemical compaction, namely seams, fitted fabrics and stylolites (Wanless, 1979; Buxton & Sibley, 1981; Hattin, 1981; Precht & Pollastro, 1985; Bathurst, 1987, 1991). In the marl beds, considerable flattening of burrows and fossils can be observed (Gaillard & Jautee, 1985; Ricken, 1986, 1987).

Cemented limestone layers have a higher carbonate content compared to the original composition, but a lower organic carbon content, because precipitation of diagenetic carbonate 'dilutes' the initial organic matter contained in the limestone bed precursors (Ricken & Eder, 1991). On the other hand, dissolution-affected marl beds show lower carbonate contents than the original beds, while organic matter becomes passively concentrated. Other chemical changes associated with these differential

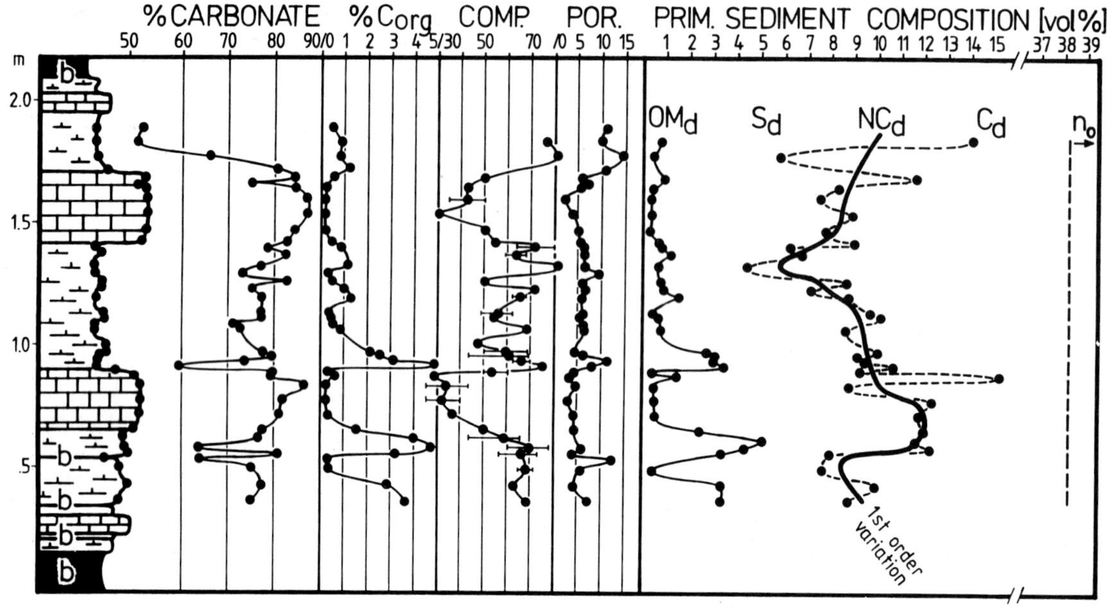

Fig. 2.

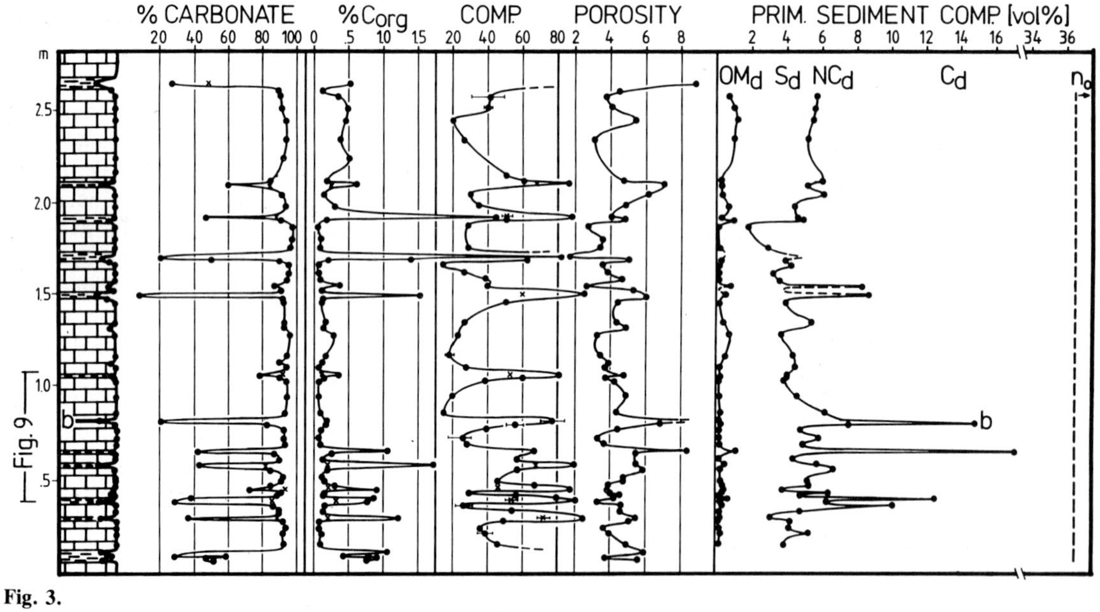

Fig. 3.

Figs 2–5. The present and calculated 'original' compositions of four typical sections with rythmic bedding. One section is from the Greenhorn TR cycle at Pueblo, Colorado (Bridge Creek Limestone, Fig. 2), the other sections are from the Niobrara TR cycle at Lyons, Colorado (Fort Hays Limestone, Fig. 3; Shale and Limestone Member, Fig. 4; Middle Chalk Member, Fig. 5). Curves on the right show the carbonate content (wt%), organic carbon content (C_{org}, wt%), compaction (vol% of original thickness), and porosity (vol%). Error bars display the total range of several compaction measurements. The reconstructed, 'original' sediment depicts bulk contents of organic matter (OM_d, vol% of the total original sediment), clastic sediment (S_d, vol%), and non-carbonate (NC_d, vol%). Note that detailed variation of the primary bulk carbonate content (C_d, vol%) cannot be assessed, because an average value of decompaction porosity (n_o, vol%) is determined in carbonate mass balance calculations. Black beds are bentonites (b). For stratigraphic position see Fig. 1.

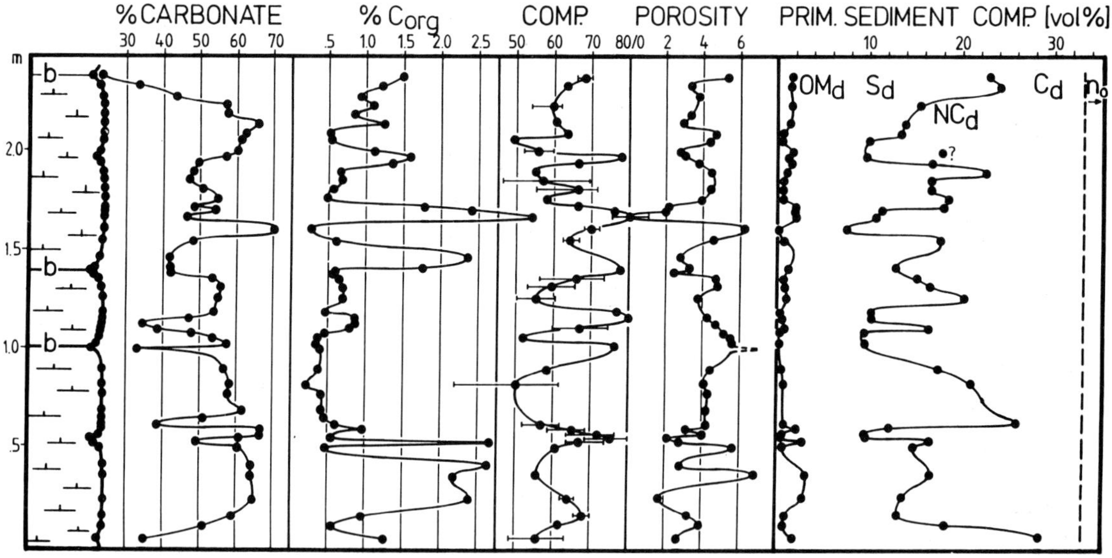

Fig. 4.

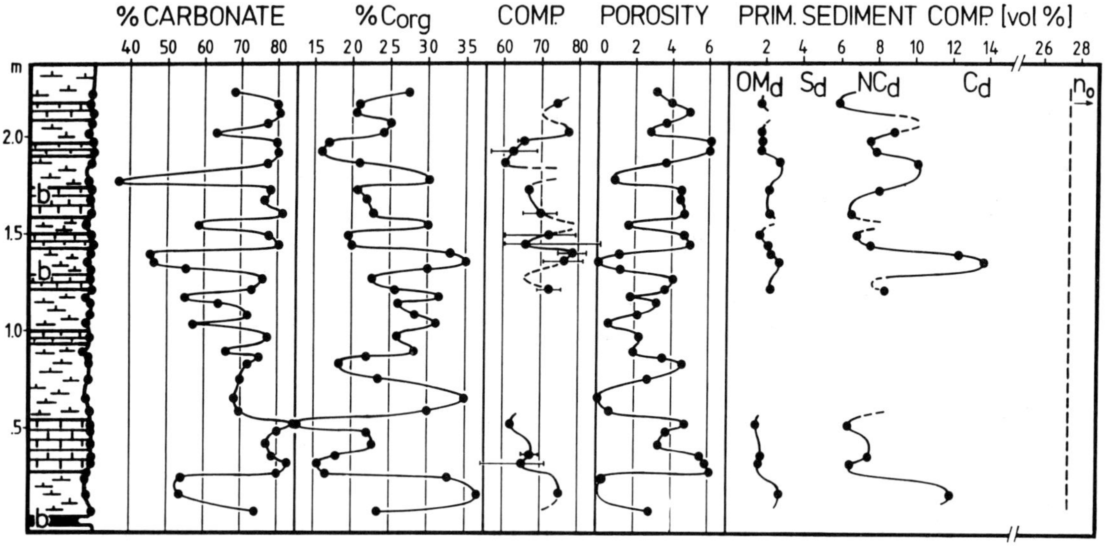

Fig. 5.

dissolution–cementation processes are changes in trace element associations, and in carbon and oxygen isotopes (Arthur *et al.*, 1984; Ricken, 1986, 1992; Arthur & Dean, 1991). One of the most conspicuous features associated with differential diagenesis is the accentuation of the original bedding rhythm. Limestone layers are only slightly reduced from their original thickness, while the highly compacted marl beds are significantly diminished. After severe diagenetic overprint, a rhythmic sequence may look like a brick wall, with equally thick limestone layers interbedded with highly compacted, thin marl beds (see the Fort Hays section, Fig. 3; Simpson, 1985; Ricken, 1986). The number of original beds is influenced by differential diagenesis, selective cementation and weathering; in addition, the intervals between beds and the bundling of a sequence is affected (Ricken, 1986; Bathurst, 1987; Ricken & Eder, 1991).

Methods for determining diagenetic overprint: decompaction and carbonate mass balance calculations

Two principal methods are commonly applied in quantifying diagenetic overprint. They can be grouped into chemical evaluation and volume assessment strategies. The chemical evaluation strategy is the more classical approach described by Brand and Veizer (1980), Veizer (1983), Arthur *et al.* (1985) and Arthur & Dean (1991). The degree of overprint can be qualitatively assessed by plotting element ratios which are sensitive to diagenetic changes. It is difficult, however, to estimate the transferred (dissolved and reprecipitated) amounts of carbonate. Such questions are addressed in the volume assessment strategy. Based on Bathurst (1976) and Shinn *et al.* (1977), the volume assessment strategy involves the indirect determination of cemented and dissolved amounts of carbonate by using the degree of compaction measured in the flattening of trace fossils. This is performed by numerical decompaction and mass balance calculations.

A detailed presentation of the carbonate mass balance method has already been given in various publications by Ricken (1986, 1987, 1992). Therefore, only a brief overview of the procedures is presented here. The first four steps calculate the bulk organic carbon and siliciclastic contents for the carbonate-rich and carbonate-poor beds of the reconstructed and decompacted 'original' sediment. The balance calculation is then carried out to determine the original porosity, original mean composition and amounts of dissolved and cemented carbonate (Steps 5 and 6).

Step 1

Compaction is measured using the degree of deformation of originally circular or nearly circular burrows of *Thalassinoides*, *Chondrites* and suitable types of *Planolites*. Compaction is determined by measuring the deformed and undeformed axes in horizontally emplaced burrows perpendicular to the long axis of the burrow (Ricken, 1987). Whenever possible, several measurements are made to determine an average compaction value. In the depicted diagrams (Figs 2–5) the lowest and highest compaction values from several measurements are denoted by horizontal error bars, which commonly span less than ±10% of the mean value.

Step 2

For each rock sample with measured compaction, the carbonate and organic carbon contents are determined after Pratt (1984) using an LECO induction furnace. Additional samples are analysed in order to get a complete, detailed carbonate–organic carbon record of the studied sections. For a few cemented burrow tubes, where rock compaction is significantly higher than burrow compaction, an indirect technique can be applied. Rock compaction is determined by using the difference in carbonate content between the cemented burrow and the compacted, surrounding rock, as described in Ricken (1986, 1987).

Step 3

The porosity of the existing rocks is derived by sealing dried samples with shellac, and by determining their weight in water and in air. To calculate the porosity, a grain density of 2.7 g/cm^3 is used for the non-organic fraction and 1.01 g/cm^3 for the organic fraction. A factor of 1.3 is taken to transform the weight percent organic carbon into weight percent organic matter (Tissot & Welte, 1984).

Step 4

From the data determined so far, the bulk amount of the non-carbonate fraction in the original sediment can be calculated. This non-carbonate fraction is expressed as a percentage of the primary bulk (or decompacted) sediment volume (NC_d); it is composed of the 'primary' bulk organic matter (OM_d) and the bulk clastic fraction (S_d). Calculation follows the 'carbonate compaction law', a well-established and theoretically grounded relationship describing the compaction of the original sediment volume and associated changes in carbonate content in argillaceous, calcareous sediments and rocks (Ricken, 1987):

$$NC_d(\text{vol\%}) =$$

$$OM_d + S_d = \frac{(100 - K)(100 - n)(100 - C)}{10\,000} \quad (1)$$

where C is the volume percentage of carbonate, K is the percentage of compaction and n is the porosity (vol%). As the grain densities of the non-carbonate and carbonate fractions are similar (approximately 2.7 g/cm^3), volume percentage carbonate is essentially equivalent to weight percentage carbonate for organic carbon contents below 1.5% (Ricken, 1993). The compaction law depends on the assumption that the non-carbonate fraction is for the most part not involved in carbonate dissolution and reprecipitation processes. Determination of the primary bulk volumes of the organic and clastic fractions allows important conclusions to be drawn on the composition of the primary sediment.

Step 5

Numerical decompaction is performed throughout the entire section to reconstruct the original sediment volume, and to determine the degree of closure of the diagenetic carbonate system. First, compaction is calculated for small intervals with a thickness of 1 cm by solving equation (1) for K. Stepwise decompaction with $n_o = \Sigma[(1 - 0.01\,K)n + K]$ then results in a restored original sediment, in which the carbonate fraction is still affected by differential diagenesis. The mean porosity of such a decompacted section is the decompaction porosity, n_o. Its value represents a critical measure for the degree of closure of the diagenetic carbonate system. The diagenetic system was open when the decompaction porosity is substantially above or below the mean porosity for freshly deposited calcareous ooze or mud. Under such conditions, diagenetic carbonate was transported by diffusion or advection either out of, or into, the system of the considered section, enhancing or reducing the decompacted porosity, respectively. The original porosity of fine-grained calcareous sediment is reported to range mostly from 60 to 75% (Hamilton, 1976; Enos & Sawatzky, 1981; Osmond, 1981; Moore, 1989). These values are identical to the mean decompaction porosities found in this investigation (62–73%). Consequently, this indicates that the diagenetic carbonate systems of the sections investigated must have been largely closed. Therefore, the dissolved and reprecipitated carbonate masses are largely redistributed between beds and thus can be quantified in carbonate mass balance calculations.

Step 6

The analysed data show that the decompaction porosity of marl beds is relatively high because of carbonate loss during diagenetic dissolution. In contrast, limestone layers have a relatively low decompaction porosity because their pore space is partly occluded by carbonate cement. The cemented and dissolved carbonate masses are balanced by assuming that dissolution-affected and cemented parts originally had an equivalent pore space (n_o). The effects of smaller porosity differences between original beds are neglected. From the balance calculation, the percentage of cement and the mean primary composition for the precursors of the marl and limestone layers are obtained. It must be emphasized that the resulting 'original' organic carbon content (C_{org0}) in the mass balance calculation represents the residual C_{org} content under burial conditions. This C_{org} content is lower than the C_{org} content in the primary sediment (see pp. 181–182).

Results of carbonate mass balance calculations

Carbonate mass balance calculations show that diagenetic overprint in the interbedded calcareous shales, marls and limestone sequences is generally low to medium in degree (Fig. 6, Table 1). Average cement contents (Z_c) in the limestone layers range from 11 to 21% of the total carbonate fraction. Only the Fort Hays Limestone underwent substantial carbonate redistribution, with a Z_c in the limestone layers of 45%. The degree of diagenetic redistribution is expressed through enhancement factors determining the former and present differences in composition between the carbonate-rich and carbonate-poor beds. In the shale and marl sequences, this factor ranges from 1.8 to 2.8 for carbonate and organic carbon contents, whereas in the Fort Hays Limestone the enhancement factor is approximately 5 (Table 1). Thus, in the interbedded Upper Cretaceous shale and marl sequences studied, the difference in $CaCO_3$ and C_{org} content between beds is clearly enhanced compared to the original sediment. Without diagenetic overprint, rhythmic sequences would exhibit weaker bedding features.

In the cemented limestone layers, the organic carbon content is diagenetically lowered as a result of 'dilution' through cement precipitation, whereas organics are enriched in the marl beds relative to the amount of exported carbonate. Thus, an originally reversed C_{org}–$CaCO_3$ relationship between marl and limestone beds, associated with carbonate deposition, is further accentuated (see pp. 179–181). Such accentuation of originally reversed C_{org}–$CaCO_3$ contents is expressed in the perfectly

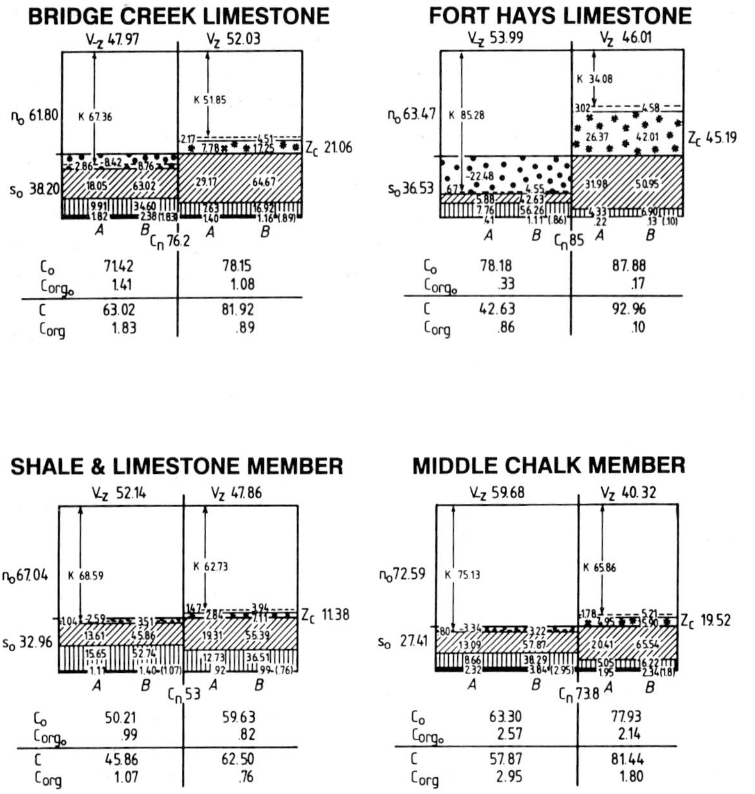

Fig. 6. Box models to assess differential carbonate diagenesis as well as the 'original' sediment composition. Models show the reconstructed, 'original' sediment for dissolution-affected (left side) and cemented (right side) beds with mean decompacted porosity (n_o) and original amounts of solids (s_o). Original solids are composed of carbonate (diagonally hatched), clastic (vertically hatched) and organic matter (solid). Carbonate redistribution is shown by dissolved (dotted) and cemented (starred) volumes. K is the average amount of compaction. Z_c is the cement content expressed as a percentage of the total carbonate fraction. Small numbers within diagrams: A, sediment fractions as a percentage of dissolution-affected and cemented sediment volumes (V_{-Z} and V_Z); B, percentages of solid sediment fractions. C_n is the statistically 'neutral' carbonate content between cementation and dissolution. Lower numbers show average contents in 'original' and present carbonate and organic carbon contents (C_o and C_{org0} and C and C_{org}, respectively); the C_{org0} value is significantly reduced by early diagenetic processes.

Table 1. Parameters describing diagenetic enhancement of carbonate and organic carbon contents in the sections investigated. Average values for composition, cement content and porosity (per cent)

	Bridge Creek Limestone	Fort Hays Limestone	Shale & Limestone Member	Middle Chalk Member
'Primary' composition marl beds				
$CaCO_3$	71.4	78.2	50.2	63.3
C_{org}	1.4	0.3	1.0	2.6
'Primary' composition limestones				
$CaCO_3$	78.2	87.9	59.6	77.9
C_{org}	1.1	0.2	0.8	2.1
Present composition marl beds				
$CaCO_3$	63.0	42.6	45.9	57.9
C_{org}	1.8	0.9	1.1	3.0
Present composition limestones				
$CaCO_3$	81.9	93.0	62.5	81.4
C_{org}	0.9	0.1	0.8	1.8
Factor (F) of diagenetic enhancement				
F-$CaCO_3$	2.8	5.2	1.8	1.6
F-C_{org}	2.9	4.8	1.8	2.7
Cement content limestone	21.1	45.2	11.4	19.5
Mean decompaction porosity	61.8	63.5	76.0	72.6

opposite covariation of carbonate and organic carbon contents in some of the alternations investigated (see Figs 3 and 5).

The diagenetic histories of the rocks investigated are characterized by a phase of mechanical compaction succeeded by cementation and chemical compaction (i.e. pressure dissolution; Scholle, 1977). Sediments with the highest carbonate contents (e.g. the precursor of the Fort Hays Limestone) have the earliest onset of differential cementation and dissolution, compared to shaly sediments with a much longer phase of mechanical compaction. In the Fort Hays Limestone much of the original pore space in the limestone layer was available for cement precipitation, which was provided by equally intense carbonate dissolution in the adjacent marl beds. In the shale sequences investigated, mechanical compaction and pore-space reduction before the onset of diagenetic redistribution were much larger, resulting in a low degree of diagenetic overprint.

The amount of mechanical compaction is an important criterion for the amount of overburden and timespan (since deposition) required to initiate cementation. According to Ricken (1986), mechanical compaction is equivalent to the compaction in the middle of the cemented, carbonate-rich bed. Cementation starts in the middle of the limestone beds and continues with increasing compaction towards the limestone layer margins. The amount of overburden relative to the onset of cementation can be derived from standard porosity reduction curves (e.g. Hamilton, 1976; Baldwin & Butler, 1985; Allen & Allen, 1990) and the mean decompaction porosities obtained from the carbonate mass balance calculations. In these curves, porosity (n) can be transformed into compaction (K) with the equation:

$$n = \frac{n_o - K}{1 - 0.01K} \qquad (2)$$

where n_o is the porosity of the original sediment (i.e. the decompaction porosity). In the limestone sequences studied, the equivalent overburden to the onset of cementation is 450 m for the Bridge Creek Limestone and 70 m for the Fort Hays Limestone (according to the Baldwin–Butler equation). Using the porosity–overburden relationship for deep-sea carbonates given by Hamilton (1976), the appropriate overburden required to initiate carbonate redistribution in the Fort Hays Limestone is 120 m (Table 1). Similar values are also observed for European limestone–marl alternations (Ricken, 1986) as well as for pelagic and hemi-pelagic carbonates, where the ooze–chalk transition occurs between 100 and 400 m sub-bottom depth (e.g. Schlanger & Douglas, 1974; Garrison, 1981; Scholle *et al.*, 1983). In contrast, mechanical compaction in the laminated and C_{org}-rich sediments, at 55–64%, is substantially higher than in the European marl–limestone alternations in bioturbated C_{org}-poor sediments with the same carbonate contents (Ricken, 1986). Presumably, this is related to the parallel alignment of elongated sediment particles and clay minerals in laminated, non-bioturbated sediments, which may promote mechanical compaction. When the Baldwin–Butler equation is applied, the overburden necessary to initiate differential carbonate diagenesis would be approximately 1000 m. It is not clear whether the Baldwin–Butler equation describing pore-space reduction in average shales gives too high an overburden value for the carbonaceous sediments investigated.

Compactional enrichment of organic matter

One of the most important processes associated with differential diagenesis is the compactional enrichment of organic matter (Shinn *et al.*, 1984). In the marl beds, organic carbon concentrations increase with increasing compaction, while in the cemented limestone layers the organic carbon concentrations are reduced. Compactional enrichment of organic matter is a widely occurring process which partly controls the formation of calcareous source rocks. A primary sediment with low organic carbon content and weak bedding features can be transformed into a well-bedded sequence with elevated organic carbon levels in the interbeds. Solution-affected, compacted interbeds may expel substantial quantities of hydrocarbons when passing through the oil window (Shinn *et al.*, 1984).

Compactional enrichment can be observed for insoluble or less-soluble sediment constituents, such as bioclastic grains, dolomite crystals and organic matter, which become concentrated at the same time as the non-carbonate fraction is augmented due to dissolution processes (Wanless, 1979; Eder, 1982; Pollastro & Martinez, 1985; Ricken, 1986; Bathurst, 1987, 1991). For example, concentration of organic matter in dissolution seams and stylolites is widely found in argillaceous calcareous rocks (Shinn *et al.*, 1984; Füchtbauer, 1988).

In the rhythmically bedded sequences investigated here, compactional enrichment can be illustrated by plotting the organic carbon content

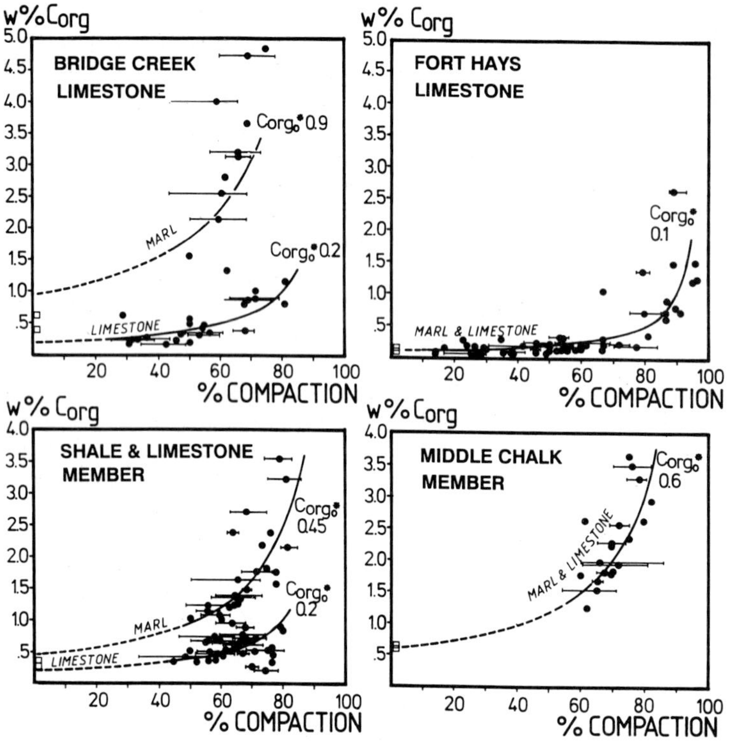

Fig. 7. Compactional enrichment of organic carbon. Weight percent organic carbon (C_{org}) is plotted against measured compaction. Error bars indicate the whole range of several compaction measurements. Note fit between measured data and theoretical enrichment curves (equation 3). Marl and limestone layers display either one or two major enrichment curves, depending on the primary difference in organic carbon. 'Primary' organic carbon contents resulting from equation (3) (C_{org0}*) are approximately equivalent to carbonate contents determined for a theoretical carbonate rock with zero compaction but complete cementation (see squares, derived from average C_{org0} values of carbonate mass balance calculations performed, in Fig. 6).

against the measured degree of burrow compaction (Fig. 7). Despite scattering, the weight percentage of organic carbon can be seen to increase statistically with increasing compaction. The maximum organic carbon content is 2–10 times larger than the amount of organic carbon in cemented beds with the least compaction. Two individual enrichment curves for marl and limestone beds are presented for the Bridge Creek Limestone and the Shale and Limestone Member. In the Fort Hays Limestone and the Middle Chalk Member, the primary differences in organic carbon content between the two types of beds are so low that essentially one major enrichment curve can be formed (Fig. 7).

The process of compactional enrichment is illustrated by concentrating sediment constituents in a sediment volume which becomes increasingly smaller due to compaction (Fig. 8). Water is expelled, while solids become enriched. This mechanical compaction does not influence the composition of the solid fraction, unless differential cementation and dissolution processes are involved. In calcareous rocks, mechanical compaction is followed by cementation within the decreasing sediment volume, incorporating ever higher amounts of organic matter. In addition, compaction may be so extreme that carbonate dissolution (i.e. chemical compaction) causes further enrichment of organic matter. In any case, whether carbonate rock underwent cementation or pressure dissolution, compaction influences the amount of organic carbon contained in a lithified rock volume (Fig. 8). This is expressed in the basic equation:

$$N = \frac{N_o}{1 - 0.01K} \qquad (3)$$

where N is the amount of present constituents, N_o the initial amount of constituents in the non-compacted sediment volume, and K is the percentage of compaction. Using this equation, the swarm of points was simulated; the overall fit between the theoretical curves according to equation (3) and the actual data is excellent. The initial concentration of organic matter (equation (3)) is approximately equivalent to the 'original' organic carbon content (derived from mass balance calculations) when complete cementation of the original pore space is assumed (see Fig. 7). The greatest degree of enrichment occurs in the Fort Hays Limestone, which also shows the greatest diagenetic overprint. Other

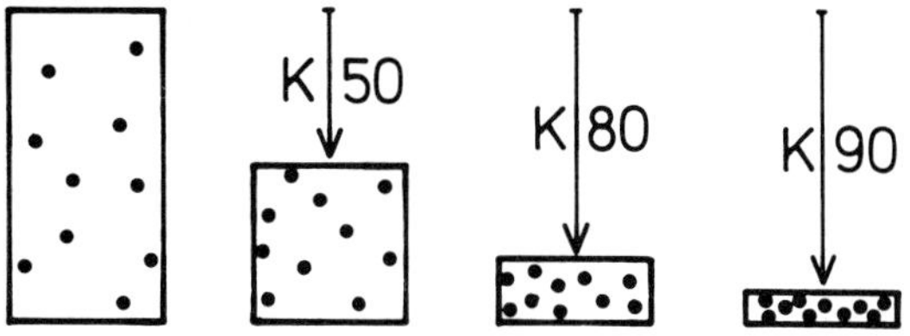

$$N = \frac{N_o}{1 - 0.01K}$$

Fig. 8. Schematic depiction of compactional enrichment. Original number of particles (N_o, dots) is not affected by increasing compaction (K, % of original thickness), but note increasing particle number (N) contained in a constant sample unit. In rhythmically bedded rocks with differential diagenesis, various levels of compactional enrichment are related to either cementation in the pore space (low degrees of compaction) or pressure dissolution (high degrees of compaction).

investigated sequences primarily rich in clay and organic matter underwent a longer phase of mechanical compaction. They have less diagenetic overprint with moderate organic carbon enrichment.

Oxygen and carbon isotopes

Oxygen and carbon isotopes provide valuable clues to salinity, temperature and water mass changes, as well as to diagenetic processes involved in the formation of interbedded shale–limestone sequences (e.g. Arthur *et al.*, 1984; Pratt *et al.*, 1991). The isotopic record of the most pelagic unit of the Greenhorn TR cycle (i.e. the Bridge Creek Limestone) has been extensively studied by Arthur *et al.* (1984, 1985), Pratt (1984), Pratt & Threlkeld (1984), Barron *et al.* (1985), Eicher & Diner (1985, 1989), and Pratt *et al.* (1991). Oxygen isotopes show an overall strongly negative (diagenetic) signature, but smaller $\delta^{18}O$ differences between marl and limestone beds are thought to represent primary salinity or temperature changes (e.g. Pratt, 1984); $\delta^{13}C$ differences are taken as indicators of productivity and water-mass variations (e.g. Eicher & Diner, 1989). This study focuses on the most pelagic unit of the Niobrara TR cycle, the Fort Hays Limestone, which underwent the highest degree of diagenetic overprint. Detailed $\delta^{13}C$ and $\delta^{18}O$ measurements were performed for five limestones and their enclosing marl beds, using the carbonate fraction of the whole rock (Fig. 9). Analyses were carried out in the Oil and Gas Branch of the United States Geological Survey in Denver, with the collaboration of Lisa Pratt.

The isotope curves (Fig. 9) have a conspicuous 'blocky' character, indicating different levels of carbon and oxygen isotopes per bed. As already observed by Arthur *et al.* (1985), both carbon and oxygen isotopes are heavier in the marl beds and lighter in the limestones; this pattern is even more pronounced in $\delta^{13}C$ values. An additional factor comes into play with the alteration processes of small bentonite beds, which create strongly positive excursions (Pratt, 1984).

Carbon isotopes show a pattern comparable to that observed in most of the studied rhythmically bedded sequences of the Western Interior Basin. In the Bridge Creek Limestone, the more positive marl beds were interpreted as reflecting organic matter which was produced in ^{13}C-enriched, stratified water masses (Eicher & Diner 1985, 1989, 1991). Alternatively, cements could incorporate light CO_3 carbon from decomposed organic matter, which would decrease the $\delta^{13}C$ content in the limestones (Hudson, 1977; Pratt, 1984; Arthur *et al.*, 1985; Raiswell,

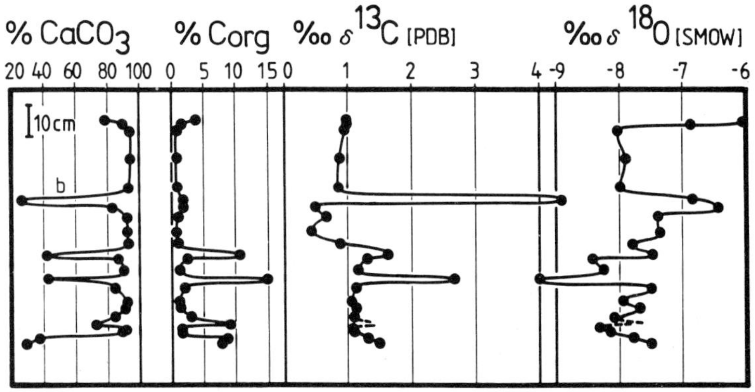

Fig. 9. Carbon and oxygen isotope record for five limestone and enclosing marl beds. Bentonitic marl bed is indicated by 'b'. Fort Hays Limestone. See Fig. 3 for location.

1987). Unlike carbon isotopes, oxygen isotopes in marl and limestone beds are reversed compared to the Bridge Creek Limestone. In the Bridge Creek Limestone, $\delta^{18}O$ values are heavier in the limestones, reflecting higher salinities in the limestone beds (Pratt, 1984), or incursions of mixed oceanic water masses into the Western Interior Basin (Eicher & Diner 1989, 1991). In the Fort Hays Limestone, however, $\delta^{18}O$ values are lower in the carbonate-rich beds. This might be associated with warmer climates during deposition (see pp. 185–186) or may be an effect of cementation under burial elevated temperatures and pore waters (Arthur *et al.*, 1984; Arthur & Dean, 1991).

Here, an attempt is made to relate carbon isotopes to the volume of cemented or dissolved carbonate (Z_d) using the equation:

$$Z_d(\text{vol\%}) = n_o - n + K(0.01n - 1) \qquad (4)$$

(Ricken, 1986), where n_o and n are the primary (i.e. decompacted) and present porosities (vol%), respectively, and K the percentage of compaction. Statistically, the samples with the highest cement contents have lighter isotopes, whereas dissolution-affected marls contain heavy isotopes (Figs 9 and 10). Several diagenetic effects may overlap, such as precipitation of light carbon cements in the limestone layers (related to bacterial sulphate reduction and fermentation), and enrichment of an isotopically heavy carbonate fraction in the marl beds, when the carbonate is composed of two fractions with both different solubility and different isotopic composition. Commonly found *Inoceramus* prisms and shell frag ments may form such solution-resistant calcitic carbonate embedded in the micritic matrix. *Inoceramus* shell fragments are up to 4‰ heavier in $\delta^{13}C$ than the enclosing fine-grained carbonate (Pratt & Barlow, 1985).

Carbon isotopes seem to be influenced by cementation, as suggested by decreasing $\delta^{13}C$ values with increasing cement contents; see the relatively clean correlation curve in Fig. 10. This is also illustrated by estimating the $\delta^{13}C$ content of the cement, which is based on carbon isotopes representing the bulk composition, as well as on the 'residual' composition, which is thought to indicate the $\delta^{13}C$ content before the onset of cementation. The 'residual' $\delta^{13}C$ value can be derived from the carbon isotope plot when the cement content is zero. According to this model, carbon isotopes in the cement are isotopically lighter by 0.5‰ compared to the $\delta^{13}C$ value in the 'residual' carbonate fraction. Early cements (in samples with the highest cement volumes) contain relatively light $\delta^{13}C$ values, which become slightly heavier with increasing diagenesis and overburden (Fig. 10). This light organic carbon cement results when about 2%

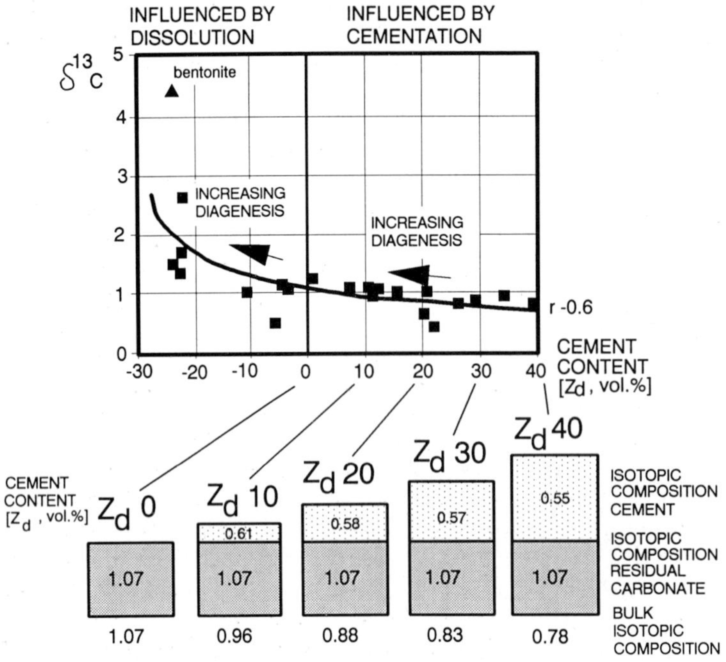

Fig. 10. Estimate of the $\delta^{13}C$ content in the cement (Z_d, vol%), Fort Hays Limestone. The cement content is derived using equations (1) and (4), and the mean decompaction porosity (n_o) from carbonate mass balance calculations. Positive values of Z_d indicate cementation, while negative values indicate carbonate dissolution. The $\delta^{13}C$ content of the carbonate fraction not involved in dissolution and cementation processes (i.e. 'residual' carbonate) is taken from the upper diagram ($Z_d = 0$) with $\delta^{13}C = 1.07$‰. The cement $\delta^{13}C$ is estimated using the $\delta^{13}C$ content of the 'residual' carbonate (assumed to be constant) and the bulk isotopic composition read on the regression line of the upper diagram.

of the carbon is derived from the decomposition of light organic matter with −25‰. If these assumptions are correct, then carbon isotopes are only slightly influenced by the early decomposition of organic matter, associated with sulphate reduction and methane production processes occurring in the upper tens of metres of the sediment (e.g. Claypool & Kaplan, 1974; Gautier & Claypool, 1984; Raiswell, 1987). In addition, oxygen isotopes indicate burial cementation with the incorporation of light cements. These results agree with the compaction studies performed here, which point to a largely closed diagenetic carbonate system and an overburden load between 70 and 120 m thick at the onset of lithification processes in the Fort Hays Limestone (see p. 175).

ASSESSMENT OF DEPOSITIONAL PROCESSES

The carbonate mass balance calculations presented here determine the average primary composition of marl and limestone beds. But primary differences in sediment composition are not easily translated into the associated changes in depositional flux. In the literature, attempts have been made to estimate the clastic or calcareous flux variation of rhythmic bedding by expressing bed thicknesses versus carbonate contents in various ways (e.g. de Boer, 1983, 1991; Arthur *et al.*, 1984; Cotillon, 1985, 1991; Eicher & Diner, 1989, 1991; ten Kate & Sprenger, 1989; Arthur & Dean, 1991; Einsele & Ricken, 1991; Ricken, 1991a, b). Herbert *et al.* (1986) performed flux calculations by assuming Milankovitch periodicities for beds in Mid-Cretaceous chalks in central Italy. Arthur & Dean (1991) have introduced methods to determine depositional changes between carbonate-rich and carbonate-poor beds by plotting the contents of the three major sediment fractions in ternary diagrams, thereby generating 'mixing curves' among the three major sediment inputs. Slightly different from their procedure, the concept followed here is as described in Ricken (1991a, 1993). It is based on evaluating carbonate–organic carbon relationships plotted in *xy* diagrams. For a more comprehensive treatment of the methodology, which cannot be adequately presented here, the reader is referred to Ricken (1993).

Three component system: carbonate, clastic and organic matter deposition

Deposition is viewed as a flux system with three components: carbonate, clastics and organic matter. It is assumed that the style of deposition can be described by idealized types of deposition, where the flux of one component varies, while the other two fluxes are more constantly delivered, forming the background sediment. This idealized flux pattern is described as carbonate, clastic and organic matter deposition. The principles of the three component system are as follows.

1 The three major types of deposition (carbonate, clastic and organic matter deposition) are associated with, and recognized by, characteristic C_{org}–$CaCO_3$ relationships, which result from depositional dilution and concentration processes.

2 Relative sedimentation rates can be derived by comparing the difference between the C_{org} and $CaCO_3$ values of two succeeding sediment or rock samples, as expressed in dilution equations.

3 Relative sedimentation rates can be transformed into associated changes in relative sediment flux for carbonate, clastic and organic matter deposition.

As schematically depicted in Fig. 11, the basic types of deposition are idealized and expressed by different C_{org}–$CaCO_3$ relationships.

1 *Carbonate deposition*: a significant increase in the carbonate supply (by keeping the deposition of the two other fractions constant) would ideally result in diluting the organic carbon content to lower weight percentages, causing an increase in the carbonate content and simultaneous decrease in the organic carbon content in the sediment. When plotting the weight percentages of organic carbon and carbonate in an *xy* diagram, carbonate deposition would result in a negative correlation between these two concentrations if the deposition of organic matter and siliciclastic sediment is largely constant (Fig. 11a). The sedimentation rate increases with rising carbonate and declining organic matter concentrations.

2 *Clastic deposition*: when ideally increasing silt and clay deposition by keeping organic carbon and carbonate deposition unchanged, the carbonate concentration of the sediment decreases, while the organic carbon content becomes simultaneously diluted to lower weight percentages. This leads to a positive C_{org}–$CaCO_3$ correlation pattern (Fig. 11b). The sedimentation rate increases with decreasing amounts of organic carbon and carbonate in the sediment.

3 *Organic matter deposition*: changes in organic

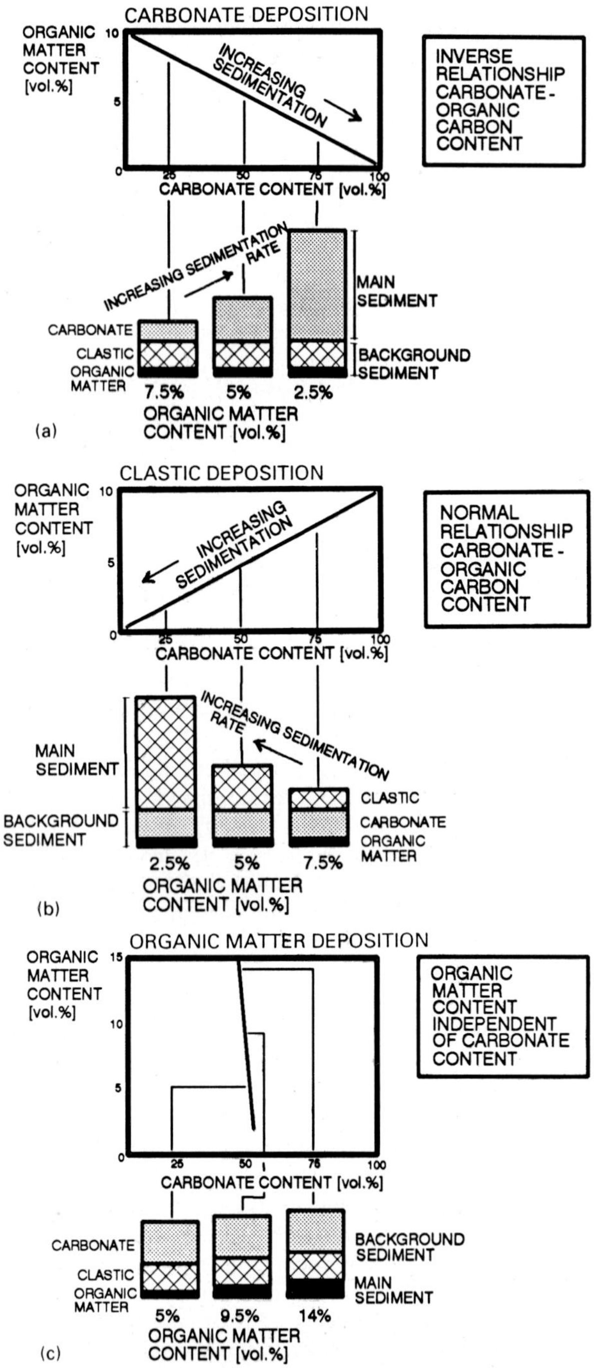

Fig. 11. Organic carbon–carbonate relationships for (a) carbonate, (b) clastic and (c) organic matter deposition, according to the three component system. Characteristic C_{org}–$CaCO_3$ relationships are generated when only the flux of the main component varies, while the fluxes of the two other components are constant, forming the background sediment. Different C_{org}–$CaCO_3$ relations reflect different dilution styles. Carbonate and clastic deposition are associated with inverse and normal C_{org}–$CaCO_3$ correlations, respectively, while organic matter deposition is associated with C_{org} variations largely independent of the carbonate content.

matter deposition correlate with a C_{org}–$CaCO_3$ pattern which is largely independent of the carbonate concentration, because depositional changes in the relatively small organic fraction have a low potential for changing the much larger carbonate and clastic fractions (Fig. 11c). Therefore, organic matter deposition has only a weak influence on the sedimentation rate.

Evaluation of a large data set from the DSDP/ODP and other sources shows that the three basic C_{org}–$CaCO_3$ relationships are indeed observed in many rhythmically bedded sediments (Ricken, 1991a). It must be stressed that composite depositional variations can also be recognized by the superposition of two or more C_{org}–$CaCO_3$ regression curves.

When one of the basic depositional styles is documented by a characteristic C_{org}–$CaCO_3$ curve, one can numerically transform a carbonate-poor bed into a thicker, carbonate-richer bed representing a higher rate of deposition and a higher $CaCO_3$ input. Such transformations, which result in relative sedimentation rate changes, are described for the different depositional styles in organic carbon dilution equations. The relative sedimentation rate between two samples (s_r) is expressed by the ratio of changing organic matter or carbonate volumes of these samples, $s_r = OM_{1vol}/OM_{2vol}$ for carbonate and siliciclastic deposition, and $s_r = C_{1vol}/C_{2vol}$ for organic matter deposition. When these volume ratios are transformed into weight percentages, one obtains for carbonate or clastic deposition the following dilution equation (based on grain densities of 1.01 and 2.7 g/cm^3 for the organic matter and the carbonate and clastic fractions, respectively, and a conversion factor of $OM = 1.3\,C_{org}$):

$$s_{r[C,\ S\ dep.]} = \frac{C_{org1}}{C_{org2}} \times \frac{1 + 0.0218\,C_{org2}}{1 + 0.0218\,C_{org1}} \times \frac{100 - n_1}{100 - n_2} \quad (5)$$

while the relative sedimentation rate for organic matter deposition is expressed through

$$s_{r[OM\ dep.]} = \frac{C_1}{C_2} \times \frac{1 + 0.0218\,C_{org2}}{1 + 0.0218\,C_{org1}} \times \frac{100 - n_1}{100 - n_2} \quad (6)$$

C_{org1} and C_{org2} are the organic carbon contents in wt%, C_1 and C_2 the carbonate contents in wt%, and n_1 and n_2 the porosities (in vol%) of the first and second samples, respectively. Equations (5) and (6) become simplified in lithified rock, where the last factor can be neglected because porosity differences are only small. A detailed explanation of these equations is given in Ricken (1993).

A succession of several (relative) sedimentation rates ($s_{r1}, s_{r2}, s_{r3}, \ldots, s_{rn}$) determined between two samples can be combined by multiplication:

$$s_r = s_{r1} * s_{r2} * s_{r3} \ldots * s_{rn} \quad (7)$$

For example, rhythmic sedimentation of the Middle Chalk Member shows a complex C_{org}–$CaCO_3$ pattern suggesting simultaneous changes in carbonate and clastic deposition (see Figs 12 and 16(b)). Such combined deposition is simulated by starting with the average, diagenetically corrected composition of the marl beds, which is defined as having a relative sedimentation rate of $s_{r1} = 1$. In order to determine the limestone layer deposition, this composition must shift along a carbonate dilution line towards a second point located between the marl and limestone beds, in which the relative sedimentation rate increases from $s_{r1} = 1$ to $s_{r2} = 1.12$ (equation (5)). Here, C_{org}–$CaCO_3$ values indicate decreasing clastic deposition; sedimentation rates decline along a clastic dilution line from $s_{r2} = 1$ to $s_{r3} = 0.92$. The resulting composition represents a third sample that is again located on a carbonate dilution line; increasing carbonate deposition leads to rising deposition from $s_{r3} = 1$ to $s_{r4} = 1.14$. The entire change in sedimentation rate from point 1 to point 4 is $s_r = 1 \times 1.12 \times 0.92 \times 1.14 = 1.17$. Thus, the diagenetically corrected, average sedimentation rate in the limestones is approximately 1.2 times as high as in the marl beds.

When the relative sedimentation rates in the marl and limestone bed precursors are determined, one can derive relative sediment inputs for the carbonate, clastic and organic fractions; these are expressed here as the relative sedimentation rates for these fractions (s_{rC}, s_{rS}, s_{rOM}):

$$s_{rC} = \frac{s_r \times C(1 - 0.01\,n)}{100 + 2.18\,C_{org}} \quad (8)$$

$$s_{rS} = \frac{s_r \times (100 - C - 1.3\,C_{org})(1 - 0.01\,n)}{100 + 2.18\,C_{org}} \quad (9)$$

$$s_{rOM} = \frac{s_r \times 3.475\,C_{org}(1 - 0.01n)}{100 + 2.18\,C_{org}} \quad (10)$$

where s_r is the relative sedimentation rate, C and C_{org} are the carbonate and organic carbon contents, respectively, and n is the porosity (in vol%). The flux patterns of marl and limestone beds (see Figs 13, 14 and 16) are derived using these equations. Note that the reconstructed organic matter flux (s_{rOM}) is substantially reduced by diagenetic processes compared to the original sediment (i.e. 'apparent' organic matter flux; see following section).

Diagenetic influences on the three component system

Flux calculations performed using the three component system are influenced by diagenetic processes affecting the organic and carbonate fractions. As shown on pp. 169–179, the effects of differential carbonate diagenesis can be quantified by carbonate mass balance methods. But the diagenetic influences on the organic fraction are difficult to quantify.

In aerobic bottom waters, most of the easily degradable organic matter is already oxidized in the surface sediments. In the equatorial Pacific Ocean, Emerson *et al.* (1985) found a C_{org} decrease of more than 50% in the uppermost, 1-cm-thick layer of the sediment. Stein (1991) calculated that the degradation of organic matter in the surface sediment is between 60 and 70% for coastal upwelling environments, and between 70 and 80% for non-upwelling coasts, whereas in the open ocean 90–98% of the organic carbon flux which reaches the sediment surface is decomposed. Thus, compared to the primary production of organic matter in the surface waters, only a minute percentage is commonly embedded in the sediment (between 0.001 and 10%; for most aerobic environments below 1%; Bralower & Thierstein, 1984, 1987). Below the sulphate reduction and fermentation zones (below 10 to several hundreds of metres of overburden), degradation of organic matter continues. Using S/C ratios, Raiswell & Berner (1987) have estimated that the organic matter loss between the end of the sulphate-reduction zone and the onset of the oil window is approximately 30%.

All these diagenetic losses of organic carbon, due to various early and late degradation processes, are thought to have affected the different rock types in one section similarly, thus systematically flattening the slope of all types of C_{org}-$CaCO_3$ dilution curves, and lowering the apparent organic carbon input. This lowering, however, has little influence on sedimentation rates as calculated with equations (5) and (6), because these sedimentation rates are determined by relative means. When the absolute amount of organic carbon contained in a time-equivalent sediment interval is lowered equally for lithologies rich and poor in carbonate content, then the calculation of relative sedimentation rates is little affected for both carbonate and clastic deposition.

An additional factor is the 'sealing effect' related to greater preservation of organic matter with increasing sedimentation rates (e.g. Heath *et al.*, 1977; Toth & Lerman, 1977; Müller & Suess, 1979; Ibach, 1982; Betzer *et al.*, 1984; Berger *et al.*, 1989). Stein (1986, 1991) showed that the sealing effect is much lower than suggested by Müller & Suess (1979) because their data set was also influenced by productivity and water depth changes. A varying supply of non-organic sediment mainly causes dilution of organic carbon, as established in numerical models performed by Emerson (1985), Emerson *et al.* (1985) and Emerson & Hedges (1988). Additionally, the sealing effect is strongly reduced in anoxic to dysaerobic bottom waters (Stein, 1986, 1991). Such oxygen-depleted bottom waters are assumed for most of the sections investigated, based on the occurrence of lamination or restricted bioturbation (see below). Therefore, no correction for a possible sealing effect is performed here.

The sections investigated generally show an inverse carbonate–organic carbon correlation pattern (Fig. 12), denoting carbonate deposition. This pattern is further amplified through diagenetic dissolution and reprecipitation of $CaCO_3$. Organic carbon is diagenetically concentrated in the carbonate-poor beds, and diluted in the carbonate-rich layers (see pp. 173–177). Compared to the original sediment, diagenetically affected correlation curves span a wider C_{org}–$CaCO_3$ range (i.e. they become longer), while the slopes of carbonate dilution curves remain unchanged. A correction is performed by reducing the lengths of the diagenetically affected correlation curves, according to the diagenetic enhancement factor derived through carbonate mass balance calculations (see Table 1). Note that carbonate mass balance calculations exhibit the mean 'original' composition per bed whether it be rich or poor in carbonate; they do not give the maximum difference in composition (i.e. the amplitude of original $CaCO_3$ variations). Despite these restrictions with the method, it is still possible to derive a reasonable pattern of relative sediment fluxes.

Complex rhythmic sedimentation

Rhythmic bedding is associated with either simple or composite flux types (Arthur *et al.*, 1986; Ricken 1991a). Simple flux types reflect varying deposition of only one major fraction, whereas composite rhythms show simultaneous variations in several components. Most of the sections investigated have composite rhythms which show that the depositional system is sensitive to climatic cycles by the interaction of different sediment sources.

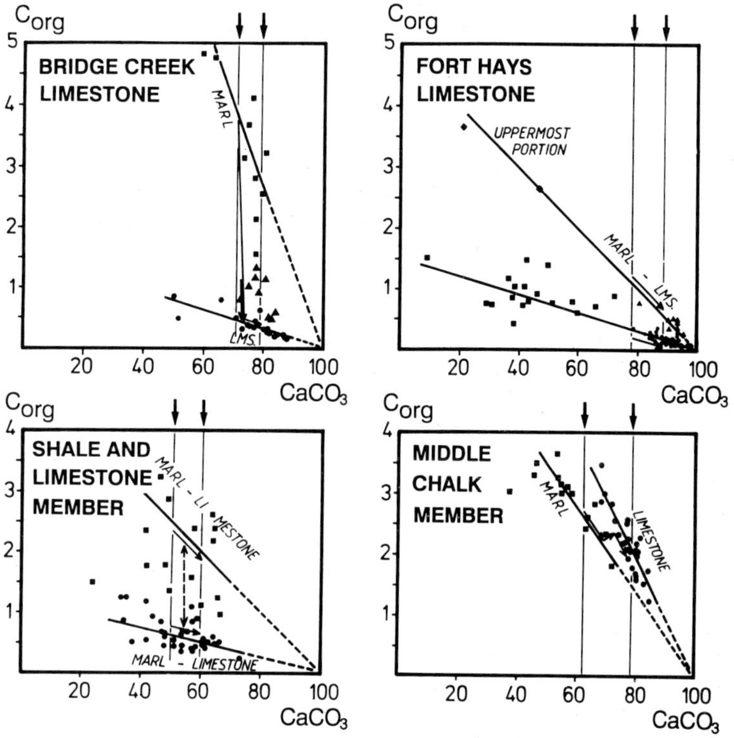

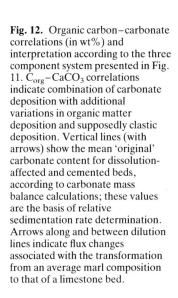

Fig. 12. Organic carbon–carbonate correlations (in wt%) and interpretation according to the three component system presented in Fig. 11. C_{org}–$CaCO_3$ correlations indicate combination of carbonate deposition with additional variations in organic matter deposition and supposedly clastic deposition. Vertical lines (with arrows) show the mean 'original' carbonate content for dissolution-affected and cemented beds, according to carbonate mass balance calculations; these values are the basis of relative sedimentation rate determination. Arrows along and between dilution lines indicate flux changes associated with the transformation from an average marl composition to that of a limestone bed.

Bridge Creek Limestone

The original sediment of the Cenomanian/Turonian Bridge Creek Limestone was composed of a rhythmically bedded, nannoplankton–foraminiferal ooze (Pratt, 1984; Elder, 1987; Eicher & Diner 1989). The original difference in $CaCO_3$ ($CaCO_3$ amplitude) in the section investigated was approximately 14% between the middles of carbonate-rich and carbonate-poor beds, with average decompacted porosities of 62% (see Fig. 6). Carbonate-rich beds were entirely bioturbated, while the interbedded C_{org}-rich layers were slightly affected by (micro)bioturbation. The Bridge Creek Limestone exhibits distinct intervals with higher organic matter contents (see Fig. 2). Obviously, water-mass stratification and oxygen content fluctuated during the rhythmic deposition of the Bridge Creek Limestone (Pratt, 1984; Barron *et al.*, 1985; Elder, 1987; Eicher & Diner 1989; Kauffman & Sageman, 1990; Arthur & Dean, 1991; Pratt *et al.*, 1991).

Organic carbon and carbonate contents generally show an inverse relationship, resulting in negatively sloping dilution lines (Fig. 12). As discussed in the previous section, this indicates that oscillating carbonate input was the dominant type of deposition. Two major dilution lines occur, one flatly sloping and representing the limestones, and the other steeply sloping with higher C_{org} contents denoting the marl beds. Between these curves, transitional values exist in a pattern parallel to the C_{org} axis, which points to increasing organic carbon deposition in the marl layers. Thus, the Bridge Creek marl–limestone alternation indicates simultaneous variation in surface water carbonate productivity (Eicher & Diner, 1989, 1991) and redox conditions at the sediment/water interface (Pratt, 1984; Barron *et al.*, 1985; Arthur & Dean, 1991). Periods with high carbonate production are dominantly associated with higher oxygenation at the sediment surface, expressed in the decreasing supply of organic matter (Fig. 13).

Statistically, carbonate-rich beds represent a relative $CaCO_3$ flux which is 1.2 times higher than in the carbonate-poor beds. When the entire carbonate sediment fraction is related to surface water productivity, as assumed by Eicher & Diner (1989), distinct productivity changes must be assumed in order to explain the reconstructed original carbonate differences between beds (Fig. 13). The carbonate–organic carbon data investigated do not suggest that

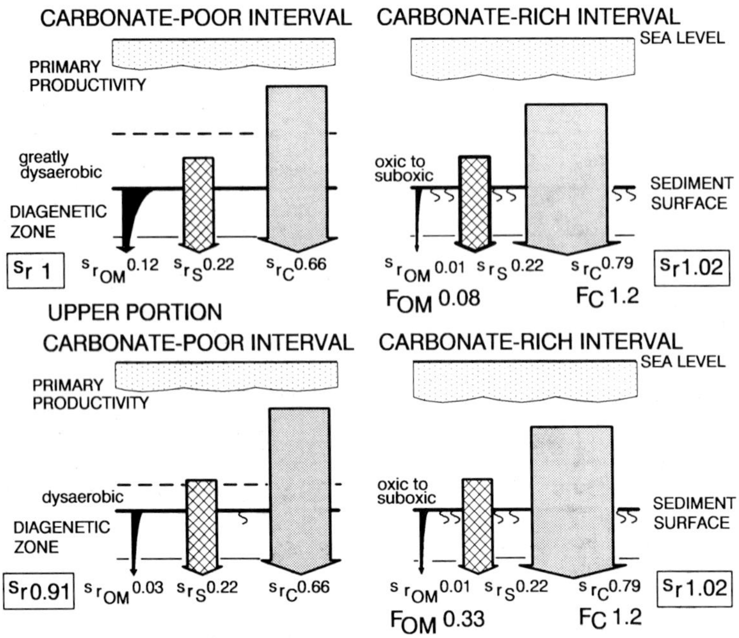

Fig. 13. Average relative flux variations for carbonate-rich and carbonate-poor beds of the Bridge Creek Limestone. Upper two diagrams show fluxes for the lowermost part of the sequences depicted in Fig. 2, while lower diagrams show estimated fluxes for the upper portion of this sequence. Bedding rhythm is caused by changing carbonate and organic matter input. Fluxes pass the sediment surface and a diagenetic zone in which the organic matter flux is greatly reduced. Dashed line above sediment surface indicates onset of dysaerobic bottom water masses, while stippled zone below water surface denotes pelagic carbonate productivity. The various relative fluxes are expressed as the relative sedimentation rates of the individual organic matter (s_{rOM}), clastic (s_{rS}) and carbonate (s_{rC}) sediment fractions, forming the bulk sedimentation rate (s_r). Input changes, relative to the marl beds, are expressed by factors F_{OM} and F_C. Calculation is based on evaluation of C_{org}–$CaCO_3$ data represented in Fig. 12. Carbonate flux is corrected for diagenetic changes according to the carbonate mass balance calculations, which determine the 'original' $CaCO_3$ contents averaged for intervals rich or poor in carbonate content.

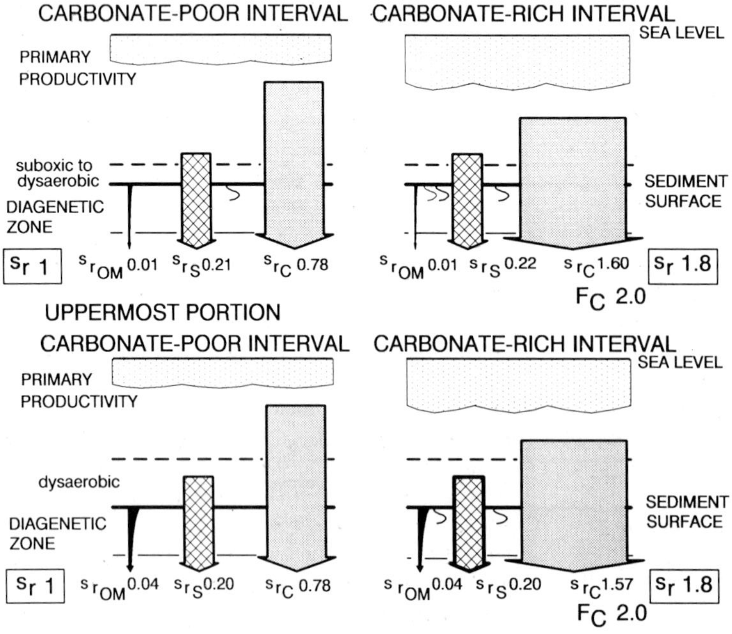

Fig. 14. Schematic depiction of relative sediment flux of the Fort Hays Limestone. Marl and limestone precursors are related to variations in carbonate supply, while the supplies of clastic and organic matter are approximately constant. The uppermost portion of the investigated section denotes a generally higher organic matter supply (lower diagrams). For symbols and numbers see Fig. 13.

a changing supply of clastic sediment plays a dominant role in forming these cycles, as also documented by approximately equal Na/K ratios for marl and limestone beds (Arthur & Dean, 1991).

Carbonate-poor intervals are associated with a larger organic matter supply and less bioturbation, interpreted as reflecting stratified, dysaerobic bottom-water masses. However, the degree of oxygen depletion in the carbonate-poor beds varies in the section investigated (see Fig. 2). In the lowermost metre the organic matter flux is large in the marl beds over several cycles, while in the following metre the organic matter supply is reduced by a factor of four. The transition between these changes is rather sharp. Thus, the Bridge Creek Limestone represents a combination of productivity and redox cycles with larger-scale shifts in the level of oxygen depletion in the carbonate-poor intervals.

Fort Hays Limestone

The Coniacian Fort Hays Limestone shows well-developed rhythmicity associated with the maximum flooding of the Niobrara cycle. This conspicuous alternation provides an example of varying carbonate deposition which underwent the highest degree of diagenetic carbonate redistribution in the sequences investigated. Originally, this alternation was composed of nannoplankton ooze (Hattin, 1981; Barlow & Kauffman, 1985) interbedded with laminated C_{org}-rich beds slightly thinner than the calcareous beds. The sediment displayed an original carbonate difference (i.e. $CaCO_3$ amplitude between carbonate-rich and carbonate-poor beds) of approximately 13% between the middles of succeeding beds, with an average porosity of about 63%. Bioturbation was largely restricted to the carbonate-rich intervals (Savrda & Bottjer, 1989).

In the reconstructed original sediment one can see a covariation between the bulk organic matter and bulk clastic contents (see Fig. 3). This covariation is partly interrupted by deposition of thin bentonite clays. The reconstructed sediment differs substantially from the present sequence when C_{org}–$CaCO_3$ values are compared. The currently high organic carbon content in the marl beds has been greatly amplified due to compactional enrichment (see pp. 175–177).

Carbonate–organic carbon correlation curves scatter around two distinct negatively sloping regression lines, indicating carbonate deposition (see Fig. 12). The lower regression line shows C_{org}–$CaCO_3$ data from the lower two-thirds of the investigated section depicted in Fig. 3, while the upper curve represents the upper third of the section with generally higher organic carbon values. It must be stressed that each regression line represents values from both marl and limestone beds. Here, rhythmic deposition occurs when the marl bed sediment is diluted by greater carbonate deposition, thereby creating a limestone bed, and vice versa; thus the C_{org}–$CaCO_3$ values shift along one regression line. If this interpretation is correct, carbonate-poor and carbonate-rich beds had largely equivalent organic carbon inputs, while the average carbonate supply changed by about a factor of two (Fig. 14). In the uppermost portion of the section investigated, the degree of oxygen depletion decreased substantially for both marl and limestone beds.

These findings, of a largely equal organic matter supply for both the marl and limestone precursors, apparently conflict with the bioturbation changes observed by Savrda & Bottjer (1986, 1987, 1989). Savrda & Bottjer found that limestone layers show trace fossil assemblages of *Thalassinoides* (for some beds), *Planolites*, *Teichichnus*, *Zoophycos* and *Chondrites*, whereas little to no bioturbation was found in the marl beds. This general change in bioturbation was interpreted by Savrda & Bottjer as representing changing degrees of oxygenation, but flux determination indicates continued oxygen deficiency in the water mass overlying the sediment surface. An interpretation which combines the trace fossil pattern with the observations here is to assume moderately dysaerobic bottom waters and concurrent, but intensely varying carbonate deposition, diluting or concentrating the organic matter in the sediment. Changing C_{org} concentrations affect the sedimentary oxygen content that is thought to control the intensity of bioturbation.

During increased calcareous deposition, the organic carbon content in the Fort Hays Limestone was diluted to concentrations low enough to allow restricted infaunal activity in a dysaerobic environment. On the other hand, when carbonate deposition was low, organic matter became concentrated to such high values that oxygen consumption in the surface sediment caused a further reduction of dissolved oxygen in the pore water (e.g. Demaison & Moore, 1980). It is thought that by this process burrowers were prevented from penetrating the sediment (Fig. 15). The oxygenation curves presented by Savrda & Bottjer therefore seem to indicate the oxygen content in the upper layer of the sediment,

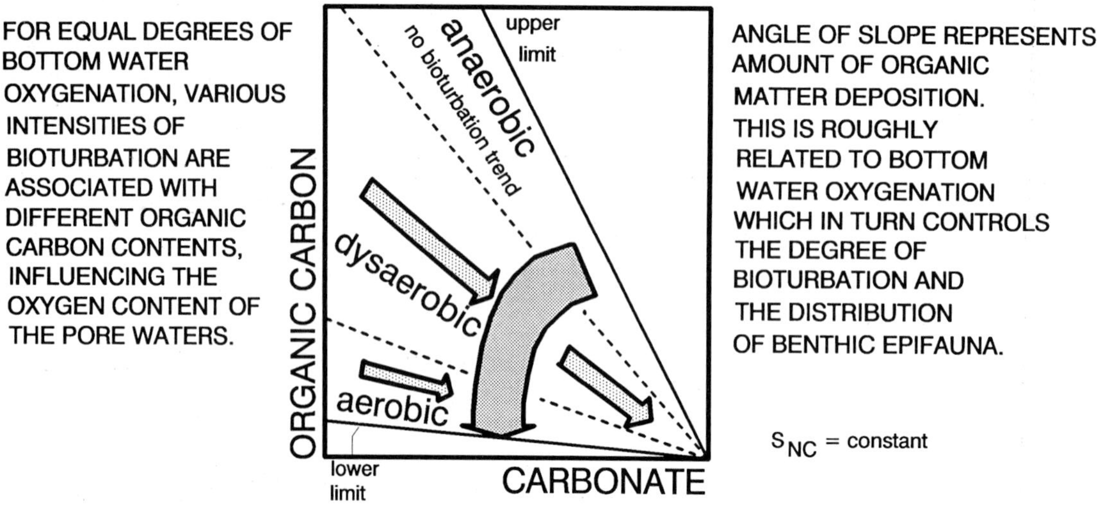

FOR EQUAL DEGREES OF BOTTOM WATER OXYGENATION, VARIOUS INTENSITIES OF BIOTURBATION ARE ASSOCIATED WITH DIFFERENT ORGANIC CARBON CONTENTS, INFLUENCING THE OXYGEN CONTENT OF THE PORE WATERS.
ORGANIC CARBON
anaerobic
no bioturbation trend
upper limit
dysaerobic
aerobic
lower limit
CARBONATE
ANGLE OF SLOPE REPRESENTS AMOUNT OF ORGANIC MATTER DEPOSITION. THIS IS ROUGHLY RELATED TO BOTTOM WATER OXYGENATION WHICH IN TURN CONTROLS THE DEGREE OF BIOTURBATION AND THE DISTRIBUTION OF BENTHIC EPIFAUNA.
S_{NC} = constant

Fig. 15. Bioturbation trends associated with inversely correlated C_{org}–$CaCO_3$ relationships denoting carbonate deposition (see Fig. 11a). Steeply sloping correlation curves represent high organic matter input, indicating oxygen deficiency in the bottom waters, while shallowly sloping curves represent more aerobic conditions. Note two trends of increasing bioturbation, one towards smaller slopes of C_{org}–$CaCO_3$ correlation lines (curved arrow), and one towards decreasing C_{org} and increasing $CaCO_3$ in the sediment (small arrows, parallel to carbonate dilution line). In the Fort Hays Limestone, the latter trend is thought to reflect bioturbation changes with approximately constantly dysaerobic bottom waters. Further explanation in text.

instead of the oxygen content in the bottom waters. Note that the maximum changes in carbonate supply are larger than those depicted in Fig. 14 because the average 'original' carbonate contents (for beds rich or poor in $CaCO_3$) were utilized instead of the maximum and minimum carbonate contents; this is a consequence of the carbonate mass balance calculations (see pp. 172–173). Thus, the primary sediment of the Fort Hays Limestone is thought to represent a surface water productivity cycle with substantial changes in the pelagic carbonate supply, underlain by largely stagnant, dysaerobic water masses.

Shale and Limestone Member

The interbedded Coniacian Shale and Limestone Member (Niobrara TR cycle) is related to simultaneous variations in carbonate productivity and redox conditions in the bottom waters. The original sediment is composed of laminated to slightly bioturbated marl rich in organic carbon (Barlow & Kauffman, 1985). Differences in $CaCO_3$ between the middles of originally carbonate-rich or carbonate-poor beds were around 21%, with average decompacted porosities of 67%.

The bulk organic matter and clastic contents vary randomly to a certain degree in the reconstructed sediment. This is also documented in the relationship between the organic carbon and carbonate contents (see Fig. 12). Data show two major regression lines with inverse C_{org}–$CaCO_3$ relationships, indicating carbonate deposition under conditions of alternating low and high organic carbon input. Both lines generally span the same carbonate content. Thus, redox conditions in the bottom waters varied, uninfluenced by periods of low and high carbonate production (Fig. 16). Unlike the Bridge Creek Limestone, no association between bottom-water stratification and calcareous surface-water productivity is found. The irregular stratification, with sudden changes in C_{org} content and trace fossil assemblages from one bed to another, supports this interpretation. It seems doubtful whether this type of stratification actually represents true rhythmic bedding resulting from Milankovitch cycles. The lowermost bed in the section investigated shows covariation between the carbonate and organic carbon contents which could even indicate clastic deposition for this bed (see Fig. 4). Sedimentation rates for the diagenetically corrected sediment are on average 1.1–1.2 times higher in the carbonate-rich beds. These values reflect

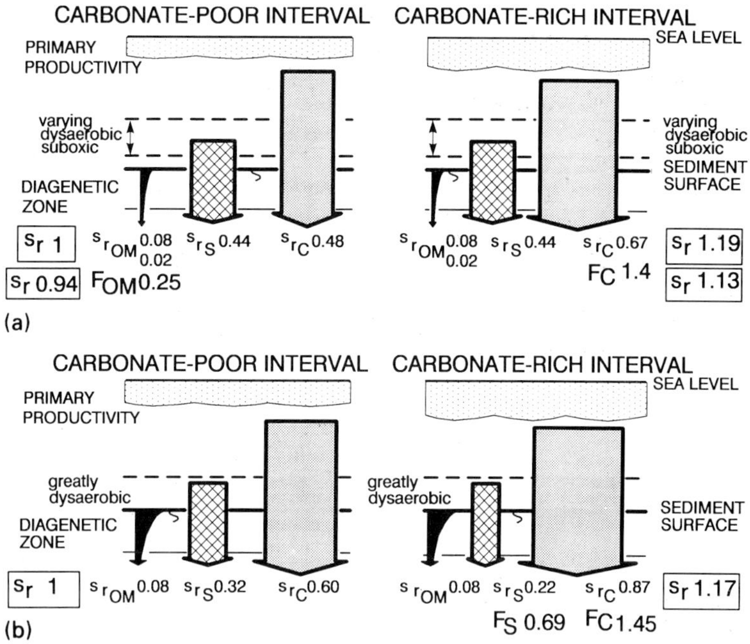

Fig. 16. Schematic depiction of average sediment flux for original beds rich or poor in carbonate. Shale and Limestone Member is characterized by carbonate deposition with a randomly varying organic matter supply, which is thought to indicate oscillations in the degree of oxygen depletion (a). Middle Chalk Member is characterized by a combination of carbonate and clastic deposition, with constantly dysaerobic bottom waters (b). For symbols and numbers see Fig. 13.

pelagic carbonate productivity elevated by a factor of 1.4 compared to the carbonate-poor beds.

Consequently, the Shale and Limestone Member documents bedding processes related to moderately changing carbonate productivity in the surface water, which is underlain by a water mass with fluctuating oxygen content (dysaerobic to subaerobic). Oxygen fluctuations were largely uninfluenced by the variations in carbonate supply.

Middle Chalk Member

The Santonian Middle Chalk Member shows well-developed rhythmic bedding, with an inverse relationship between the carbonate and organic carbon contents (see Fig. 5). Sedimentological data suggest composite bedding cycles related to simultaneous deposition of the carbonate and the clay fractions. Of the bedding cycles investigated, the Middle Chalk Member underwent the smallest degree of diagenetic overprint.

The primary sediment in the Middle Chalk Member was composed of a laminated, nannofossil, marly chalk containing a relatively high organic carbon content of 1–3.5% (Hattin, 1981). The estimated $CaCO_3$ difference between the maximum and minimum $CaCO_3$ values in the beds was approximately 24%, with a relatively high average decompacted porosity of 73%. The present alternation shows an inverse covariation between carbonate and organic matter, which was less developed in the (reconstructed) 'original' sediment. Organic matter is diagenetically enriched in the marl beds of the present alternation.

The inverse C_{org}–$CaCO_3$ correlation pattern indicates carbonate variation as the dominant process in deposition (see Fig. 12). Thus, rhythmically increasing and decreasing carbonate deposition dilutes and concentrates the organic carbon content in the sediment, resulting in the perfectly opposite covariation between carbonate and organic carbon (see Fig. 5). This general process is similar to that observed in the Fort Hays Limestone, although the Middle Chalk Member is more organic carbon-rich. Compared to the limestones, marl beds plot on a slightly displaced C_{org}–$CaCO_3$ regression line representing carbonate deposition. It is difficult to judge which depositional processes shift the sediment composition from one carbonate deposition curve to the other. Data points in Fig. 12 are too scattered to provide clear clues. The two regression lines can be theoretically connected through a C_{org}–$CaCO_3$ curve representing organic carbon or clastic deposition (see Fig. 11). The former would indicate higher organic matter input during deposition of the carbonate-rich beds, which is difficult to interpret but which may be an effect of elevated productivity. The latter would indicate decreased clay input during deposition of the carbonate-rich bed, which may be a result of a more arid climate and lower river runoff (Pratt, 1984). The more symmetrical shape of the bedding cycle may point to such simultaneous deposition of calcareous and clastic sediment (see Fig. 5). Ideally, carbonate deposition is reflected by thicker carbonate-rich beds, whereas clastic deposition is associated with thicker carbonate-poor beds (Arthur *et al.*, 1984; Einsele & Ricken, 1991; Ricken, 1991a, 1993).

The carbonate supply is 1.5 times higher in the carbonate-rich bed, while clay deposition is thought to decrease by a factor of 0.7. The consistent repetition of the C_{org}–$CaCO_3$ pattern suggests actual rhythmic bedding, possibly orbital–climatic in origin. Hence, rhythmic bedding in the Middle Chalk Member represents a combination of carbonate productivity and clastic dilution, with stable, highly dysaerobic bottom-water masses. Periods of greater carbonate supply are thought to be associated with lower input of clastic sediment.

CONCLUSIONS

Sensitivity of a shallow basin in transforming climatic cycles into rhythmic bedding

The Western Interior Basin represents a sensitive system which is capable of transforming weak climatic cycles into rhythmic stratification. In the sections investigated, representing the more transgressive units in mid-basin, the ratio of clastic to carbonate inputs is 1:1 to 1:4 (and 1:7 in the carbonate-rich beds of the Fort Hays Limestone). Consequently, original sediments with moderate carbonate contents are mostly generated. For such sediments, varying inputs are expressed in the relatively large differences in carbonate content between alternating beds (Ricken, 1991a). The Western Interior Basin represents a foreland basin with clastic shorelines on the western side (towards the Sevier orogenic front), but is generally epicontinental in character (e.g. Weimer, 1960; Kauffman, 1984). Climatic cycles are represented by productivity changes in the surface waters and changes in the redox conditions of the bottom waters. The land masses of the Sevier orogenic belt were influenced by changing runoff (probably related to changing precipitation, evaporation and vegetation cover), creating a varying clastic supply into the basin. Consequently, climatic cycles are generally well expressed in the marine strata of the Western Interior Basin, related to characteristic carbonate–clastic supply ratios and to a complex interaction of different sediment sources.

In some of the regressive units, however, orbital–climatic signals are little documented if at all. Rhythmic sedimentation is either not developed due to disturbing 'noise' (e.g. non-cyclic events and bioturbation) or cannot be recognized in carbonate-poor shales, as substantial changes in clastic input are necessary to have an effect on the carbonate content of beds (e.g. Graneros Shale, Blue Hill Shale; calcarenitic event beds in the Hartland Shale, etc.). Thus, the achievement of a critical water depth well below storm wave base, the realization of fully marine conditions to allow pelagic carbonate production, and the existence of a diluting clastic flux approximately balancing carbonate deposition are the most important factors controlling the occurrence of climatic cycles in the Western Interior Basin (Fig. 17).

Rhythmic deposition controlled by third-order sea-level stands

Rhythmic bedding has developed differently for different portions of the low-frequency third order sea-level variations. Section intervals with the highest carbonate contents, which are thought to represent maximum transgressions (Bridge Creek Limestone, Fort Hays Limestone), document calcareous productivity rhythms or a combination of these and redox rhythms. Rhythms are little influenced by varying clastic deposition and are thought to reflect periods of generally more mixed water masses. In the Bridge Creek Limestone, periods of water mixing are related to higher pelagic carbonate productivity. However, this coupling has not always developed. In the Fort Hays Limestone, productivity changes in the surface waters are probably not accompanied by fluctuating oxygen contents in the bottom waters. Instead, variously mixed surface waters are thought to be underlain by stagnant, dysaerobic bottom waters.

Section intervals with medium carbonate contents, which represent early regression (e.g. Middle Chalk Member), are thought to indicate rhythms combining calcareous productivity and clastic dilu-

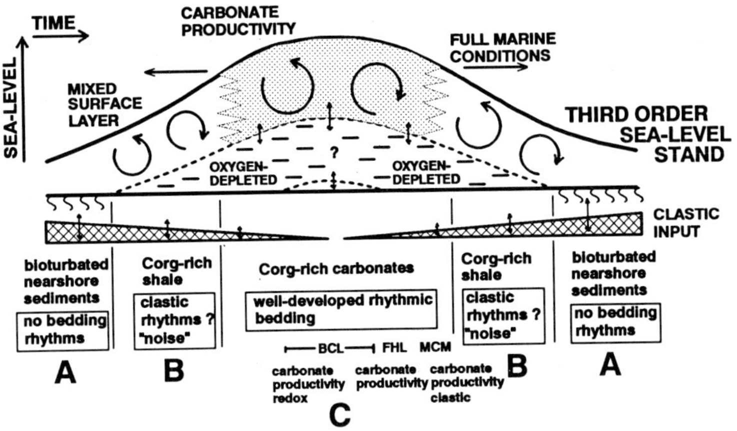

Fig. 17. Bedding rhythms related to different third-order sea-level stands; simplified scheme. Upper curve shows changing sea level with time. A mixed surface-water zone (curved arrows) is underlain by oxygen-depleted bottom waters (parallel bars). Carbonate productivity in the upper water zone is related to maximum transgression with fully marine conditions (shaded). Decreasing and then increasing clastic input with rising and falling sea level, respectively, is schematically indicated by two sideways-orientated arrows. Small vertical arrows denote oscillations of various sediment supplies. Three general facies zones are distinguished, a disturbed and bioturbated zone free of rhythms (A), a zone with oxygen-depleted bottom waters and possibly clastic rhythms (B), and a zone of well-developed bedding rhythms (C). Within the latter zone, the relative positions of the bedding rhythms investigated here are indicated; productivity–redox rhythms occur during maximum sea-level stands, and productivity–clastic rhythms occur during moderately high sea-level stands (Bridge Creek Limestone, BCL; Fort Hays Limestone, FHL; Middle Chalk Member, MCM).

tion. Periods with low calcareous productivity and possibly less surface water mixing are associated with higher runoff from the Sevier orogenic belt. Water-mass stratification might be intensified by this runoff producing slightly lowered salinities in the surface waters (e.g. Pratt, 1984).

Permanence of dysaerobic water masses

Dysaerobic bottom waters occurred during deposition of most of the sections investigated, as documented by high organic matter deposition and by generally low degrees of bioturbation. Oxygen depletion of these water masses was little influenced by changing carbonate productivity of the surface waters. Only the Bridge Creek Limestone shows such a coupling between declining carbonate and increasing organic matter flux, which might be explained by an alternation between periods of more stagnant and more mixed water masses during a prominent sea-level high-stand. In all the other rhythms investigated, the organic matter flux is largely equal for intervals high or low in carbonate deposition, or the organic matter flux varies randomly. Consequently, two major water masses must be assumed for most of the marine sequences in the Western Interior Basin: a mixed surface zone, underlain by slowly circulating oxygen-depleted bottom waters (Fig. 17). Only during maximum transgressions was this pattern partly replaced by complete mixing.

Diagenetic overprint

Diagenetic overprint is most developed in the carbonate-rich, transgressive sections, but is low to moderate in the marly sections investigated. Differential dissolution and cementation processes between beds augment primary compositional differences. The original carbonate content is partly dissolved in the carbonate-poor beds and reprecipitated as carbonate cement in the pore spaces of the limestone layers. As an effect of these redistribution processes, organic carbon becomes passively concentrated in the marl beds, whereas it is diluted in the limestones. Consequently, differential compaction distorts the primarily sinusoidal rhythms by greatly reducing the thicknesses of the marl beds, while only slightly compacting the cemented limestone layers.

ACKNOWLEDGEMENTS

I would like to acknowledge friends and colleagues who helped me in innumerable ways. Robin Bathurst made important comments regarding the role of diagenesis in rhythmic carbonate rocks; Gerhard Einsele and Philip A. Sandberg helped to interpret the three component model; and John Hudson made important written contributions to the isotope data. Erle G. Kauffman and Bradley B. Sageman discussed the role of oxygen-depleted bottom waters and supported me in the field and in many other ways. Lisa M. Pratt and Christopher Shaw provided me with analytical work and were always available for discussion. Alfred G. Fischer, Don L. Eicher and Richard Diner made helpful comments on the oceanographic processes of cyclic sedimentation. Charles Savrda and David Bottjer provided tools for understanding the style of bioturbation in rhythmic bedding. Poppe de Boer, Linda Hobert, Chris Paul and David Smith critically reviewed this text. To all of them I offer my grateful thanks.

REFERENCES

Allen, P.A. & Allen, J.R. (1990) *Basin Analysis, Principles and Applications*. Blackwell Scientific Publications, Oxford.

Arthur, M.A. & Dean, W.E. (1991) A holistic geochemical approach to cyclomania: examples from Cretaceous pelagic sequences. In: *Cycles and Events in Stratigraphy* (Eds Einsele, G., Ricken, W. & Seilacher, A.). Springer Verlag, Berlin, pp. 126–166.

Arthur, M.A., Dean, W.E., Bottjer, D. & Scholle, P.A. (1984) Rhythmic bedding in Mesozoic–Cenozoic pelagic carbonate sequences: The primary and diagenetic origin of Milankovitch-like cycles. In: *Milankovitch and Climate* (Eds Berger, A.L., Imbrie, J., Hays, J., Kukla, G. & Saltzman, B.). Reidel Publ. Co., Dordrecht, pp. 191–222.

Arthur, M.A., Dean, W.E., Pollastro, R.M. & Scholle, P.A. (1985) Comparative geochemical and mineralogical studies of two cyclic transgressive pelagic limestone units, Cretaceous Western Interior Basin, U.S. In: *Fine-grained Deposits and Biofacies of the Cretaceous Western Interior Seaway* (Eds Pratt, L.M., Kauffman, E.G. & Zelt, F.). Soc. Econ. Paleont. Mineral. Fieldtrip Guidebook 4, pp. 16–27.

Arthur, M.A., Bottjer, D.J., Dean, W.E., Fischer, A.G., Hattin, D.E., Kauffman, E.G. *et al.* (1986) Rhythmic bedding in Upper Cretaceous pelagic carbonate sequences: varying sedimentary response to climate forcing. *Geology* **14**, 153–156.

Baldwin, B. & Butler, C.O. (1985) Compaction curves. *Bull. Am. Assoc. Petrol. Geol.* **69**, 622–626.

Barlow, K.L. & Kauffman, E.G. (1985) Depositional cycles in the Niobrara Formation, Colorado, Front Range. In: *Fine-grained Deposits and Biofacies of the Cretaceous Western Interior Seaway* (Eds Pratt, L.M., Kauffman, E.G. & Zelt, F.). Soc. Econ. Paleont. Mineral. Fieldtrip Guidebook 4, pp. 199–208.

Barron, E.J., Arthur, M.A. & Kauffman, E.G. (1985) Cretaceous rhythmic bedding sequences: a plausible link between orbital variations and climate. *Earth Planet. Sci. Lett.* **72**, 327–340.

Bathurst, R.G.C. (1976) *Carbonate Sediments and Their Diagenesis*. Developments in Sedimentology. Elsevier, Amsterdam.

Bathurst, R.G.C. (1987) Diagenetically enhanced bedding in argillaceous platform limestones: stratified cementation and selective compaction. *Sedimentology* **34**, 749–778.

Bathurst, R.G.C. (1991) Pressure-dissolution and limestone bedding: the influence of stratified cementation. In: *Cycles and Events in Stratigraphy* (Eds Einsele, G., Ricken, W. & Seilacher, A.). Springer Verlag, Berlin, pp. 450–463.

Berger, W.H., Smetacek, V.S. & Wefer, G. (1989) Ocean productivity and paleoproductivity – an overview. In: *Productivity of the Ocean: Present and Past* (Eds Berger, W.H., Smetacek, V.S. & Wefer, G.), Dahlem Konferenzen, 1–34. John Wiley & Sons.

Betzer, P.R., Showers, W.J., Laws, E.A., Winn, C.D., di Tullio, G.R. & Kroopnick, P. (1984) Primary productivity and particle fluxes on a transect of the equator at 153° W in the Pacific Ocean. *Deep-Sea Res.* **31**, 1–11.

Bralower, T.J. & Thierstein, H.R. (1984) Low productivity and slow deep-water circulation in mid-Cretaceous oceans. *Geology* **12**, 614–618.

Bralower, T.J. & Thierstein, H.R. (1987) Organic carbon and metal accumulation rates in Holocene and mid-Cretaceous sediments: palaeoceanographic significance. In: *Marine Petroleum Source Rocks* (Eds Brooks, J. & Fleet, A.J.). Geol. Soc. Lond. Spec. Publ. 26, pp. 345–369.

Brand, U. & Veizer, J. (1980) Chemical diagenesis of a multicomponent carbonate system – I. Trace elements. *J. Sedim. Petrol.* **50**, 1219–1236.

Buxton, T.M. & Sibley, D.F. (1981) Pressure solution features in shallow buried limestone. *J. Sedim. Petrol.* **51**, 19–26.

Claypool, G.E. & Kaplan, I.R. (1974) The origin and distribution of methane in marine sediments. In: *Natural Gases in Marine Sediments* (Ed. Kaplan, I.R.). Plenum Press, New York, pp. 99–139.

Cotillon, P. (1985) Les variations à différentes échelles du taux d'accumulation sédimentaire dans les séries pélagiques alternantes du Crétacé inférieur. *Bull. Soc. Géol. France* **8**, 59–68.

Cotillon, P. (1991) Varves, beds, and bundles in pelagic sequences and their correlation (Mesozoic of SE France and Atlantic). In: *Cycles and Events in Stratigraphy* (Eds Einsele, G., Ricken, W. & Seilacher, A.). Springer Verlag, Berlin, pp. 820–839.

De Boer, P.L. (1983) Aspects of Middle Cretaceous pelagic sedimentation in southern Europe. *Geol. Ultraiectina* **31**.

De Boer, P.L. (1991) Pelagic black shale–carbonate rhythms: orbital forcing and oceanographic response.

In: *Cycles and Events in Stratigraphy* (Eds Einsele, G., Ricken, W. & Seilacher, A.). Springer Verlag, Berlin, pp. 63–78.

Demaison, G.J. & Moore, G.T. (1980) Anoxic environments and oil source bed genesis. *Bull. Am. Assoc. Petrol. Geol.* **64**, 1179–1209.

Eder, F.W. (1982) Diagenetic redistribution of carbonate, a process in forming limestone–marl alternations. In: *Cyclic and Event Stratification* (Eds Einsele, G. & Seilacher, A.). Springer Verlag, Berlin, pp. 98–112.

Eicher, D.L. & Diner, R. (1985) Foraminifera as indicators of warm water mass in the Cretaceous Greenhorn Sea, Western Interior. In: *Fine-grained Deposits and Biofacies of the Cretaceous Western Interior Seaway* (Eds Pratt, L.M., Kauffman, E.G. & Zelt, F.). Soc. Econ. Paleont. Mineral. Fieldtrip Guidebook 4, pp. 60–71.

Eicher, D.L. & Diner, R. (1989) Origin of the Cretaceous Bridge Creek cycles in the Western Interior, United States. *Palaeogeogr. Palaeoclimatol. Palaeoecol.* **74**, 127–146.

Eicher, D.L. & Diner, R. (1991) Environmental factors controlling Cretaceous limestone–marlstone rhythms. In: *Cycles and Events in Stratigraphy* (Eds Einsele, G., Ricken, W. & Seilacher, A.). Springer Verlag, Berlin, pp. 79–93.

Einsele, G. & Ricken, W. (1991) Limestone–marl alternation – an overview. In: *Cycles and Events in Stratigraphy* (Eds Einsele, G., Ricken, W. & Seilacher, A.). Springer Verlag, Berlin, pp. 23–47.

Elder, W.P. (1987) The paleoecology of the Cenomanian–Turonian stage boundary extinctions at Black Mesa, Arizona. *Palaios* **2**, 24–40.

Emerson, S. (1985) Organic carbon preservation in marine sediments. In: *The Carbon Cycle and Atmospheric CO_2* (Eds Sundquist, E.T. & Broecker, W.S.). Am. Geophys. Union Geophys. Monogr. 32, pp. 78–87.

Emerson, S. & Hedges, J.I. (1988) Processes controlling the organic carbon content of open ocean sediments. *Paleoceanography* **3**, 621–634.

Emerson, S., Fischer, K., Reimers, C. & Heggie, D. (1985) Organic carbon dynamics and preservation in deep sea sediments. *Deep-Sea Res.* **32**, 1–21.

Enos, P. & Sawatzky, L.H. (1981) Pore networks in Holocene carbonate sediments. *J. Sedim. Petrol.* **51**, 961–985.

Fischer, A.G. (1986) Climatic rhythms recorded in strata. *Annu. Rev. Earth Planet. Sci.* **14**, 351–376.

Fischer, A.G. (1991) Orbital cyclicity in Mesozoic strata. In: *Cycles and Events in Stratigraphy* (Eds Einsele, G., Ricken, W. & Seilacher, A.). Springer Verlag, Berlin, pp. 48–62.

Fischer, A.G., de Boer, P.L. & Premoli Silva, I. (1990) Cyclostratigraphy. In: *Cretaceous Resources, Events and Rhythms* (Eds Ginsburg, R.N. & Beaudoin, B.). Kluwer Academic Publishers, Dordrecht, pp. 139–172.

Füchtbauer, H. (1988) *Sedimente und Sedimentgesteine*. Schweizerbart, Stuttgart.

Gaillard, C. & Jautee, E. (1985) Compaction et déformation des structures de bioturbation. *Ass. Sed. Fran., Paris* (Abstract).

Garrison, R.E. (1981) Diagenesis of oceanic carbonate sediments: a review of the DSDP perspective. In: *The deep sea drilling project: a decade of progress* (Eds Warme, J.E., Douglas, R.G. & Winterer, E.L.). Soc. Econ. Paleontol. Mineral. Spec. Publ. 32, pp. 181–207.

Gautier, D.L. & Claypool, G.E. (1984) Interpretation of methanic diagenesis in ancient sediments by analogy with processes in modern diagenetic environments. In: *Clastic Diagenesis* (Eds McDonald, D.A. & Surdam, R.C.). Mem. Am. Assoc. Petrol. Geol. 37, pp. 111–123.

Gilbert, G.K. (1895) Sedimentary measurement of geologic time. *J. Geol.* **3**, 121–127.

Hallam, A. (1986) Origin of minor limestone–shale cycles: climatically induced or diagenetic? *Geology* **14**, 609–612.

Hamilton, E.L. (1976) Variations of density and porosity with depth in deep-sea sediments. *J. Sedim. Petrol.* **46**, 280–300.

Hattin, D.E. (1971) Widespread, synchronously deposited, burrow-mottled limestone beds in the Greenhorn Limestone (Upper Cretaceous) of Kansas and central Colorado. *Bull. Am. Assoc. Petrol. Geol.* **55**, 412–431.

Hattin, D.E. (1981) Petrology of Smokey Hill Member, Niobrara Chalk (Upper Cretaceous) in type area, western Kansas. *Bull. Am. Assoc. Petrol. Geol.* **65**, 831–849.

Hattin, D.E. (1985) Distribution and significance of widespread, time-parallel pelagic limestone beds in the Greenhorn Limestone (Upper Cretaceous) of the central Great Plains and southern Rocky Mountains. In: *Fine-grained Deposits and Biofacies of the Cretaceous Western Interior Seaway* (Eds Pratt, L.M., Kauffman, E.G. & Zelt, F.). Soc. Econ. Paleont. Mineral. Fieldtrip Guidebook 4, pp. 28–37.

Heath, G.R., Moore, T.C. & Dauphin, J.P. (1977) Organic carbon in deep sea sediments. In: *The Fate of Fossil Fuel CO_2 in the Oceans* (Eds Anderson, N.R. & Malahoff, A.). Plenum, New York, pp. 605–625.

Herbert, T.D., Stallard, R.F. & Fischer, A.G. (1986) Anoxic events, productivity rhythms and the orbital signature in a mid-Cretaceous deep-sea sequence from central Italy. *Paleoceanography* **1**, 495–506.

Hudson, J.D. (1977) Stable isotopes and limestone lithification. *J. Geol. Soc. Lond.* **133**, 637–660.

Ibach, L.E. (1982) Relationship between sedimentation rate and total organic carbon in ancient marine sediments. *Bull. Am. Assoc. Petrol. Geol.* **66**, 170–188.

Kauffman, E.G. (1984) Paleobiogeography and evolutionary response dynamic in the Cretaceous Western Interior Seaway of North America. In: *Jurassic–Cretaceous Biochronology and Paleogeography of North America* (Ed. Westermann, G.E.G.). Geol. Assoc. Can. Spec. Pap. 27, pp. 273–306.

Kauffman, E.G. & Sageman, B.B. (1990) Biological sensing of benthic environments in dark shales and related oxygen-restricted facies. In: *Cretaceous Resources, Events and Rhythms* (Eds Ginsburg, R.N. & Beaudoin, B.). Kluwer Academic Publishers, Dordrecht, pp. 121–138.

Laferriere, A.P., Hattin, D.E. & Archer, A.W. (1987) Effects of climate, tectonics and sea-level changes on rhythmic bedding patterns in the Niobrara Formation (Upper Cretaceous), U.S. Western Interior. *Geology* **15**, 233–236.

Moore, C.H. (1989) *Carbonate Diagenesis and Porosity*. Developments in Sedimentology 46. Elsevier,

Amsterdam.
MÜLLER, P.J. & SUESS, E. (1979) Productivity, sedimentation rate, and sedimentary organic carbon content in the oceans. 1. Organic carbon preservation. *Deep-Sea Res.* **26**, 1347–1362.
OSMOND, J.K. (1981) Quaternary deep-sea sediments: Accumulation rates and geochronology. In: *The Oceanic Lithosphere* (Ed. Emiliani, C.), pp. 1329–1371.
POLLASTRO, R.M. & MARTINEZ, C.J. (1985) Whole-rock, insoluble residue, and clay mineralogies of marl, chalk and bentonite, Smokey Hill Shale Member, Niobrara – depositional and diagenetic implications. In: *Fine-grained Deposits and Biofacies of the Cretaceous Western Interior Seaway* (Eds Pratt, L.M., Kauffman, E.G. & Zelt, F.). Soc. Econ. Paleont. Mineral. Fieldtrip Guidebook 4, pp. 215–222.
PRATT, L.M. (1984) Influence of paleoenvironmental factors on preservation of organic matter in Middle Cretaceous Greenhorn Formation, Pueblo, Colorado. *Bull. Am. Assoc. Petrol. Geol.* **68**, 1146–1159.
PRATT, L.M. & BARLOW, L.K. (1985) Isotopic and sedimentological study of the lower Niobrara Formation, Lyons, Colorado. In: *Fine-grained Deposits and Biofacies of the Cretaceous Western Interior Seaway* (Eds Pratt, L.M., Kauffman, E.G. & Zelt, F.). Soc. Econ. Paleont. Mineral. Fieldtrip Guidebook 4, pp. 199–208.
PRATT, L.M. & THRELKELD, C.N. (1984) Stratigraphic significance of $^{13}C/^{12}C$ ratios in Mid-Cretaceous rocks of the Western Interior, U.S.A. Mem. Can. Soc. Petrol. Geol. 9, pp. 305–312.
PRATT, L.M., KAUFFMAN, E. & ZELT, F. (Eds) (1985) *Fine-grained Deposits and Biofacies of the Cretaceous Western Interior Seaway*. Soc. Econ. Paleont. Mineral. Fieldtrip Guidebook 4.
PRATT, L.M., ARTHUR, M.A., DEAN, W.E. & SCHOLLE, P.A. (1991) Paleoceanographic cycles and events during the Late Cretaceous in the Western Interior Seaway of North America. In: *Evolution of the Western Interior Basin* (Eds Caldwell, W.G.E. & Kauffman, E.G.). Geol. Assoc. Can. (in press).
PRECHT, W.F. & POLLASTRO, R.M. (1985) Organic and inorganic constituents of the Niobrara Formation in Weld County, Colorado. In: *Fine-grained Deposits and Biofacies of the Cretaceous Western Interior Seaway* (Eds Pratt, L.M., Kauffman, E.G. & Zelt, F.). Soc. Econ. Paleont. Mineral. Fieldtrip Guidebook 4, pp. 223–233.
RAISWELL, R. (1987) Non-steady state microbiological diagenesis and the origin of concretions and nodular limestones. In: *Diagenesis in Sedimentary Sequences* (Ed. Marshall, J.D.). Geol. Soc. Lond. Spec. Publ. 36, pp. 41–54.
RAISWELL, R. (1988) Chemical model for minor limestone–shale cycles by anaerobic methane oxidation. *Geology* **16**, 641–644.
RAISWELL, R. & BERNER, R.A. (1987) Organic carbon losses during burial and thermal maturation of normal marine shales. *Geology* **15**, 853–856.
RICKEN, W. (1986) *Diagenetic Bedding: a Model for Marl–Limestone Alternations*. Lecture Notes Earth Science 6. Springer Verlag, Berlin.
RICKEN, W. (1987) The carbonate compaction law: a new tool. *Sedimentology* **34**, 571–584.
RICKEN, W. (1991a) Variation of sedimentation rates in rhythmically bedded sediments – distinction between depositional types. In: *Cycles and Events in Stratigraphy* (Eds Einsele, G., Ricken, W. & Seilacher, A.). Springer Verlag, Berlin, pp. 167–187.
RICKEN, W. (1991b) Time span assessment. In: *Cycles and Events in Stratigraphy* (Eds Einsele, G., Ricken, W. & Seilacher, A.). Springer Verlag, Berlin, pp. 723–794.
RICKEN, W. (1992) A volume and mass approach to carbonate diagenesis: the role of compaction and cementation. In: *Diagenesis, III* (Eds Wolf, K.H. & Chilingarian, G.V.). Developments in Sedimentology 47. Elsevier, Amsterdam, pp. 291–315.
RICKEN, W. (1993) *Sedimentation as a Three-component System: Organic Carbon, Carbonate, Noncarbonate*. Lecture Notes in Earth Science (in press).
RICKEN, W. & EDER, F.W. (1991) Diagenetic modification of calcareous beds. In: *Cycles and Events in Stratigraphy* (Eds Einsele, G., Ricken, W. & Seilacher, A.). Springer Verlag, Berlin, pp. 430–449.
SAGEMAN, B.B. (1989) The benthic boundary biofacies model: Hartland Shale Member, Greenhorn Formation (Cenomanian), Western Interior, North America. *Palaeogeogr. Palaeoclimatol. Palaeoecol.* **74**, 87–110.
SAGEMAN, B.B., WIGNALL, P.B. & KAUFFMAN, E.G. (1991) Biofacies models for oxygen-deficient facies in epicontinental seas: tool for paleoenvironmental analysis. In: *Cycles and Events in Stratigraphy* (Eds Einsele, G., Ricken, W. & Seilacher, A.). Springer Verlag, Berlin, pp. 542–564.
SAVRDA, C.E. & BOTTJER, D.J. (1986) Trace-fossil model for reconstructions of paleo-oxygenation in bottom waters. *Geology* **14**, 3–6.
SAVRDA, C.E. & BOTTJER, D.J. (1987) The exaerobic zone, a new oxygen-deficient marine biofacies. *Nature* **327**, 54–56.
SAVRDA, C.E. & BOTTJER, D.J. (1989) Development of a trace fossil model for the reconstruction of paleo-bottom water redox conditions: evaluation and application to Upper Cretaceous Niobrara Formation, Colorado. *Palaeogeogr. Palaeoclimatol. Palaeoecol.* **74**, 49–74.
SCHLANGER, S.O. & DOUGLAS, R.G. (1974) The pelagic ooze–chalk limestone transition and its implications for marine stratigraphy. In: *Pelagic sediments: on land and under the sea* (Eds Hsü, K.J. & Jenkyns, H.C.). Spec. Publ. Int. Ass. Sediment. 1, pp. 117–148.
SCHOLLE, P.A. (1977) Chalk diagenesis and its relation to petroleum exploration – oil from chalks, a modern miracle? *Bull. Am. Assoc. Petrol. Geol.* **61**, 982–1009.
SCHOLLE, P.A., ARTHUR, M.A. & EKDALE, A.A. (1983) Pelagic sediments. In: *Carbonate Depositional Environments* (Eds Scholle, P.A., Bebout, D.G. & Moore, C.H.). Mem. Am. Assoc. Petrol. Geol. 33, pp. 620–691.
SHINN, E.A., HALLEY, R.B., HUDSON, J.H. & LINDZ, B.H. (1977) Limestone compaction – an enigma. *Geology* **5**, 21–24.
SHINN, E.A., ROBBIN, D.M. & CLAYPOOL, G.E. (1984) Compaction of modern carbonate sediments: implications for generation and expulsion of hydrocarbons. In: *Petroleum Geochemistry and Source Rock Potential of Carbonate Rocks* (Ed. Palacas, J.G.). Am. Assoc. Petrol. Geol. Stud. Geol. 18, pp. 197–203.
SIMPSON, J. (1985) Stylolite-controlled layering in an homogeneous limestone: pseudo-bedding produced by

burial diagenesis. *Sedimentology* **32**, 495–505.

STEIN, R. (1986) Organic carbon and sedimentation rate – further evidence for anoxic deep-water conditions in the Cenomanian/Turonian Atlantic Ocean. *Mar. Geol.* **72**, 199–209.

STEIN, R. (1991) *Accumulation of Organic Carbon in Marine Sediments*. Lecture Notes in Earth Science 34. Springer Verlag, Berlin.

TEN KATE, W.G. & SPRENGER, A. (1989) On the periodicity in a calcilutite–marl succession (SE Spain). *Cret. Res.* **10**, 1–31.

TISSOT, B.P. & WELTE, D.H. (1984) *Petroleum Formation and Occurrence*. Springer Verlag, Berlin.

TOTH, D.J. & LERMAN, A. (1977) Organic matter reactivity and sedimentation rates in the ocean. *Am. J. Sci.* **277**, 465–485.

VAIL, P.R., AUDEMARD, F., BOWMAN, S.A., EISNER, P.N. & PEREZ-CRUZ, C. (1991) The stratigraphic signatures of tectonics, eustasy and sedimentology. In: *Cycles and Events in Stratigraphy* (Eds Einsele, G., Ricken, W. & Seilacher, A.). Springer Verlag, Berlin, pp. 617–659.

VEIZER, J. (1983) Trace elements and isotopes in sedimentary carbonates. In: *Carbonates: Mineralogy and Chemistry* (Ed. Reeder, R.J.). Min. Soc. Am. Rev. Min. 11, pp. 265–299.

WANLESS, H.R. (1979) Limestone response to stress: pressure solution and dolomitization. *J. Sed. Petrol.* **49**, 437–462.

WATKINS, D.K. (1989) Nannoplankton productivity fluctuations and rhythmically-bedded carbonates of the Greenhorn Limestone (Upper Cretaceous). *Palaeogeogr. Palaeoclimatol. Palaeoecol.* **74**, 75–85.

WEIMER, R.J. (1960) Upper Cretaceous stratigraphy, Rocky Mountain area. *Bull. Am. Ass. Petrol. Geol.* **44**, 1–20.

Spec. Publs Int. Ass. Sediment. (1994) **19**, 195–210

Ichnofossils and ichnofabrics in rhythmically bedded pelagic/hemi-pelagic carbonates: recognition and evaluation of benthic redox and scour cycles

C.E. SAVRDA* *and* D.J. BOTTJER†

* *Department of Geology, Auburn University, Auburn, AL 36849-5305, USA; and*
† *Department of Geological Sciences, University of Southern California, Los Angeles, CA 90089-0740, USA*

ABSTRACT

Provided that their production and preservation are understood, discrete trace fossils and general ichnofabrics within pelagic and hemi-pelagic strata may aid in the recognition of depositional cyclicity and the interpretation of the palaeoenvironmental mechanisms responsible. Trace fossils are particularly sensitive to periodic benthic redox variations and scour events.

Variations in benthic oxygenation levels are indicated by systematic changes in the diversity, diameter and penetration depth of burrows. These parameters can be employed to recognize oxygen-related ichnocoenoses (ORI), the vertical stacking patterns of which can be translated to palaeo-oxygenation curves. As demonstrated by applications to the lower Bridge Creek Limestone Member (Cenomanian–Turonian) of the Greenhorn Limestone (Colorado), such curves can help to refine reconstructions of cyclicity. Curves for the Bridge Creek, constructed on the basis of vertical disposition of laminated strata and four recurring ORI, exhibit short-term redox cycles arranged in bundles of four to six cycles. Bundling patterns and estimated cycle periodicities suggest forcing by axial precession modulated by orbital eccentricity. Although palaeoceanographic implications can be drawn from trace-fossil-derived redox histories, the complex feedback relations among bottom-water oxygenation, bioturbation, physical and chemical properties of substrates, and other parameters necessitate caution in interpretation.

Scour cycles, reflecting periodic fluctuations in bottom current strength, are indicated by the ichnological record of the response of organisms to changes in substrate consistency. Intervals characterized by normal softground ichnofabrics alternate with intervals upon which firmground prelithification omission suites dominated by *Thalassinoides* and, possibly, hardground postlithification omission suites of borings are superimposed. Rhythmic occurrences of hardgrounds, nodular limestones and/or omission surfaces that are suggestive of orbital control are common in European Late Cretaceous chalks but are typically lacking in North American examples.

INTRODUCTION

Cyclic sedimentation and rhythmic bedding in pelagic and hemi-pelagic sequences have been the subject of considerable study in recent years. Most of this work has focused on Cretaceous chalk and marl sequences, which commonly exhibit rhythmic variations in calcium carbonate content on a decimetre to metre scale (e.g. Einsele, 1982; Arthur *et al.*, 1984, 1986; Bottjer *et al.*, 1986). Although carbonate cyclicity in some sequences has been diagenetically enhanced to varying degrees (Ricken, 1986, pp. 167–193, this volume; Ricken & Eder, 1991), most rhythmites undoubtedly reflect a primary depositional signal (Fischer *et al.*, 1985; Fischer, 1986). Bedding rhythms in pelagic sediments have been attributed to one or a combination of several palaeoenvironmental mechanisms, including fluctuations in: (i) the level of benthic oxygenation and, hence, the extent of preservation of organic matter (redox cycles); (ii) the degree of saturation of the water column with respect to calcium carbonate (dissolution cycles); (iii) the rate of input of terrigenous sediments (dilution cycles); (iv) the production of calcareous biogenic sediments by plankton (productivity cycles); and (v) the intensity of bottom currents (scour cycles). On the basis of estimated cycle periods, these fluctuations are often attributed to climate rhythms induced by cyclic changes in the Earth's orbital parameters (i.e. orbital eccentricity, axial obliquity and axial precession), the so-called Milankovitch cycles (see de Boer, 1982; Schwarzacher & Fischer, 1982; de Boer & Wonders, 1984; Fischer & Schwarzacher, 1984; Fischer *et al.*, 1985; Fischer, 1986, 1991; de Boer, 1991; Einsele & Ricken, 1991).

Differentiation of the aforementioned cycle types in ancient strata is not always easy and typically requires a multidisciplinary approach that includes aspects of sedimentology, palaeontology and organic and inorganic geochemistry (see de Visser *et al.*, 1989; Arthur & Dean, 1991; Ricken, 1991). The purpose of this paper is to review the potential role of trace fossils in cycle recognition and differentiation. Because infaunal organisms and the ichnofossils they produce are most responsive to redox and scour cycles, emphasis will be placed on these. The responses of infauna to dilution, productivity, and dissolution cycles are very poorly known. We will comment briefly on these cycle types in the context of redox cyclicity.

TRACE-FOSSIL PRESERVATION IN RHYTHMICALLY BEDDED STRATA

In order to employ trace fossils effectively in the assessment of the palaeoenvironmental mechanisms responsible for cyclicity, it is crucial to have a sound understanding of the production of ichnofabrics and the preservation of discrete ichnofossils in pelagic sediments. Here we review the burrow stratigraphy of modern pelagic sediments and use this stratigraphy to explain the typical ichnofabrics and disposition of trace fossils observed in rhythmically bedded units in the rock record.

Burrow stratigraphy in modern pelagic sediments

On the basis of the extent and character of bioturbation, modern fine-grained pelagic substrates can be divided into three general levels (Fig. 1): a surface mixed layer, a transition layer and a historical layer (Ekdale *et al.*, 1984). The surface mixed layer represents an interval of rapid and complete biogenic homogenization. Under relatively static

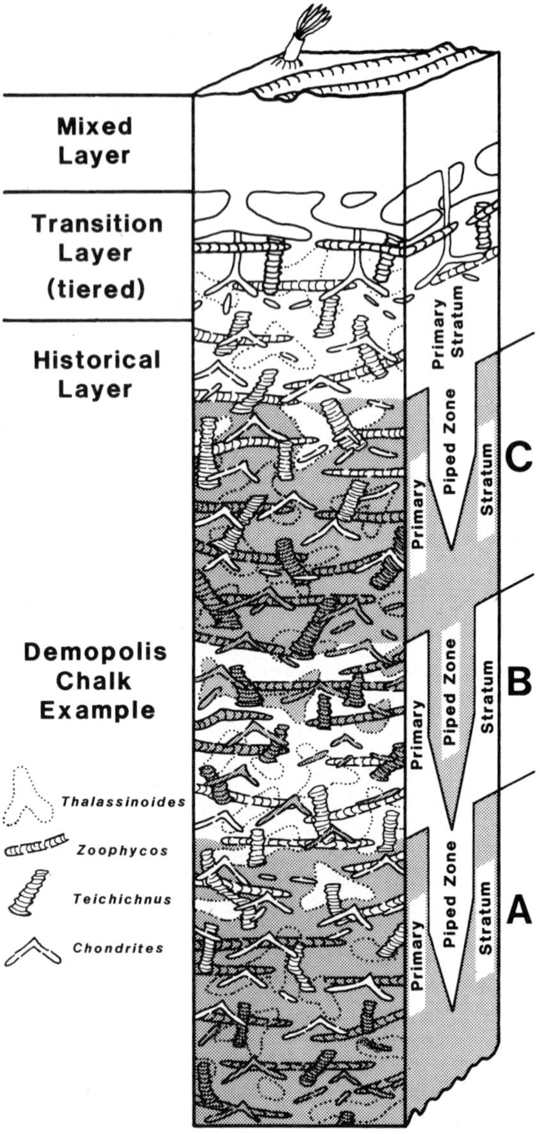

Fig. 1. General burrow stratigraphy of fine-grained pelagic sediments. Schematic ichnofabrics shown in historical layer, modelled after those in the Demopolis Chalk (Campanian) in western Alabama, illustrate typical preservation of trace fossils and development of primary strata and piped zones in rhythmically bedded sequences. Note that the tiering structure of transition-layer burrows is reflected by cross-cutting relations among historical-layer burrows in general and by the order of appearance of ichnogenera upward through piped zones. Recurring Demopolis Chalk ichnogenera depicted are, in order of increasing tier depth, *Thalassinoides suevicus* (dotted branching burrows), *Teichichnus* (vertical spreite), *Zoophycos* (horizontal spreite) and *Chondrites* (small branched burrows). A, B and C indicate beds discussed in text.

depositional conditions, the burrows and surface tracks and trails produced in this zone are normally not preserved due to low sediment shear strengths, continued rapid biogenic stirring and subsequent overprinting by transition-layer burrows with ongoing sediment accretion (Berger *et al.*, 1979).

The transition layer, a zone of heterogeneous mixing, is characterized by burrows produced by organisms that live or feed at greater depths in the substrate (i.e. below the mixed layer), where sediments are soft but considerably less fluid than those of the mixed layer. Because infaunal organisms typically exhibit tiering (i.e. different organisms preferentially occupy different depth zones below the sediment–water interface), burrows produced within the transition layer are also typically tiered (Bottjer & Ausich, 1986) (Fig. 1).

With continued sediment accretion and associated upward migration of mixed and transition layers, sediments pass out of the actively bioturbated zone into the historical layer, where no new disruptive burrowing takes place. Historical-layer ichnofabrics, which are essentially those that will be preserved in the stratigraphic record, consist of two temporally disjunct components: (i) homogeneous to vaguely burrow-mottled backgrounds produced during passage through the surface mixed layer; and (ii) superimposed discrete burrows emplaced during subsequent passage through the transition layer (Fig. 1). The tiering structure of transition-layer communities can be inferred from cross-cutting relationships between burrows in historical-layer sediments, i.e. the shallowest emplaced transition-layer burrows are progressively overprinted by the burrows of progressively deeper-dwelling organisms (Fig. 1) (Wetzel, 1983; Ekdale, 1985; Bromley & Ekdale, 1986; Savrda & Bottjer, 1987, 1989a, b). The deeper emplaced burrows also typically have the better defined boundaries and are less compacted, owing to progressive dewatering with depth in the sediment column (Bromley, 1990).

Character and genesis of ichnofabrics in rhythmically bedded sequences

For the purpose of simplicity, we initially restrict our discussion of ichnofabrics in rhythmically bedded sequences to stratigraphic intervals wherein cycles bear little evidence of the influence of scour or redox cycles, i.e. substrate consistency is considered to have been constant (soup to softgrounds), and sedimentation and bioturbation were continuous. The Campanian Demopolis Chalk of western Alabama serves as a good example.

The Demopolis Chalk is characterized by a well-developed decimetre-scale rhythmic alternation of chalks and marly chalks that have undergone little burial diagenesis. Bedding couplets, and the bundles into which they are arranged, are interpreted as recording the effects of axial precession and orbital eccentricity, respectively (Savrda *et al.*, 1991). The specific mechanisms responsible for this carbonate rhythmicity are the subject of an ongoing investigation. Current efforts are focused on evaluating the relative roles of dilution and productivity cycles; scour and redox cycles have been eliminated as important mechanisms, at least for the sections examined thus far. Minor scour features are limited stratigraphically to a few widely spaced intervals and occur in association with both chalks and marly chalks. Influence by bottom-water redox variations is dismissed on the basis of continuous bioturbation and the lack of significant changes in trace-fossil assemblages across cycle boundaries.

As is typical for most rhythmically bedded pelagic sequences, carbonate rhythmicity in the Demopolis Chalk is clearly expressed as alternating light and dark beds (Figs 1 and 2). Colour differences between adjacent beds, which vary depending on the degree of contrast in carbonate and organic carbon contents (72–91% and 0.2–1.0%, respectively), are important for understanding the genesis of ichnofabrics therein. On the basis of general ichnofabrics and the disposition of discrete burrows, two components are recognized for each light and dark interval: a primary stratum and a piped zone (Fig. 1). In order to facilitate description of the character and genesis of these components, we will initially consider only ichnofabrics of dark beds.

Ichnofabrics of dark-coloured beds reflect the superposition of three distinct components that were produced in the following order:

1 homogeneous to vaguely burrow-mottled dark backgrounds;
2 discrete, dark-coloured burrows; and
3 well-defined, light-coloured burrows (see beds A and C, Fig. 1).

Dark-coloured components (1 and 2) define the primary stratum of each dark interval whereas component 3 defines the piped zones of the superjacent light intervals (Fig. 1).

Components 2 and 3 are characterized by the same ichnofossil assemblage. Our observations on assemblages and tiering structure support those pre-

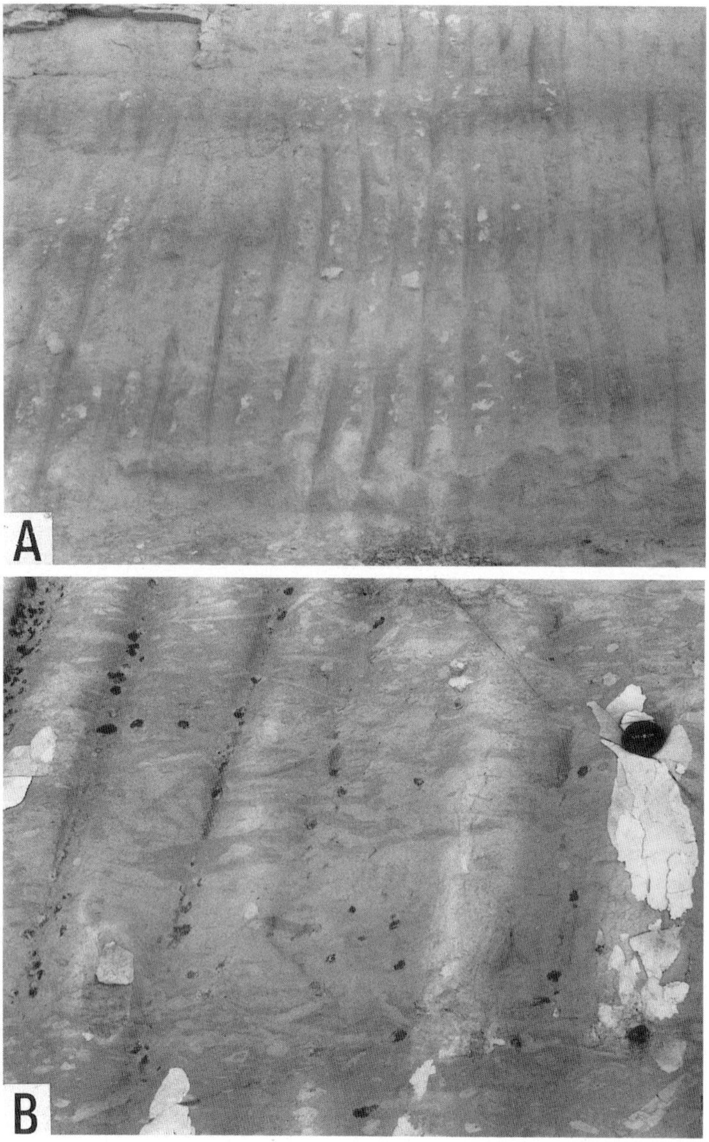

Fig. 2. (A) Rhythmic alternation of bioturbated chalks (light) and marly chalks (dark) in the Demopolis Chalk exposed at Port of Epes, western Alabama (dark beds in this bundle are *c*.30 cm thick). Note diffuse bed boundaries. (B) Close-up of Demopolis Chalk cycles showing primary strata and overlapping piped zones of alternating light chalk and darker marly chalk. Note density of piped zone burrows increases upward towards bases of associated primary strata. Lens cover is 5 cm wide.

viously reported by Frey and Bromley (1985). Dominant ichnogenera are, in order of increasing tier depth as indicated by cross-cutting relationships, *Thalassinoides*, *Zoophycos*, *Teichichnus* and *Chondrites*. Discrete dark burrows (component 2) are typically best defined in the lower portions of dark primary strata. Discrete light burrows (component 3) are typically restricted to the upper parts of dark beds, particularly where beds are thick, and increase in density upwards (Fig. 1). Vertical zonation among these piped-zone burrows typically mimics the tiering structure inferred from cross-cutting relationships. The order of initial occurrence of ichnogenera from the base of the piped zones upward is as follows: *Chondrites*, *Teichichnus*, *Zoophycos* and *Thalassinoides* (Fig. 1).

Each of the ichnofabric components can be explained in terms of normal historical-layer preservation. The homogeneous to burrow-mottled backgrounds (component 1) were generated by mixed-layer bioturbation during the deposition of darker sediments. The superimposed dark burrows (component 2) represent subsequent progressive upward migration of the tiered transition layer during continued accumulation of dark-coloured sediments. Light-coloured piped-zone burrows (component 3) also represent the emplacement of transition-layer burrows into dark sediment but only after the initiation of light-sediment deposition. Production of piped-zone burrows continued until the superjacent light sediment reached a thickness necessary for migration of the transition layer upward into light sediments. This upward migration is responsible for the progressive increase in density of pipedzone burrows towards dark-bed tops. Because the emplacement of the deepest piped-zone burrows roughly corresponds to the onset of light-sediment deposition, the thickness of piped zones provides a minimum estimate of maximum burrow-penetration depth (post-compaction).

The character and genesis of ichnofabrics in light beds are essentially the same. However, colours are obviously inverted; light primary strata are overprinted by dark piped-zone burrows (e.g. bed B, Fig. 1). Boundaries between light and dark primary strata are typically diffuse (Fig. 2). This may be related to gradual changes in sediment composition during each cycle. However, mixing of light and dark sediments by bioturbation, particularly in the mixed layer, during transitions in sedimentation must also contribute to this diffusion.

In summary, the ichnofabric of each bed in the Demopolis Chalk and other rhythmically bedded units can be divided into both primary and secondary components. Their distinction is critical: the secondary components, which define the piped zone of the superjacent interval, reflect conditions during deposition of the overlying bed and, in some cases as will be described below, are not at all representative of conditions during deposition of the sediments that they cross-cut.

TRACE FOSSILS AND REDOX CYCLICITY

Trace-fossil response to redox variations

Studies of modern marine environments (e.g. Rhoads, 1975; Pearson & Rosenberg, 1978; Savrda *et al.*, 1984) have shown that the character of infaunal communities and their associated bioturbation vary systematically with changes in bottom-water oxygenation. A brief synopsis of observed trends, which are outlined in detail elsewhere (Savrda & Bottjer, 1987, 1989a), is provided below.

Although the general burrow stratigraphy remains unchanged as oxygenation varies, faunal composition changes predictably as bottom-water oxygen concentrations decline. Infaunal assemblages become less diverse due to the progressive loss of larger and more active organisms, which typically have higher oxygen requirements. The aeration of sediments is consequently reduced, and the associated rise of the redox potential discontinuity (RPD) within the substrate results in progressive reduction in depths of residence and feeding of most infauna. The responses of bioturbating infauna are preserved and reflected by ichnological features. As bottom-water oxygenation decreases, either along lateral sea-floor redox gradients or through time, the thickness of the surface mixed layer decreases and the diversity, diameter and depth of penetration of transition-layer burrows all generally decline (Fig. 3). Eventually, when oxygen concentrations drop

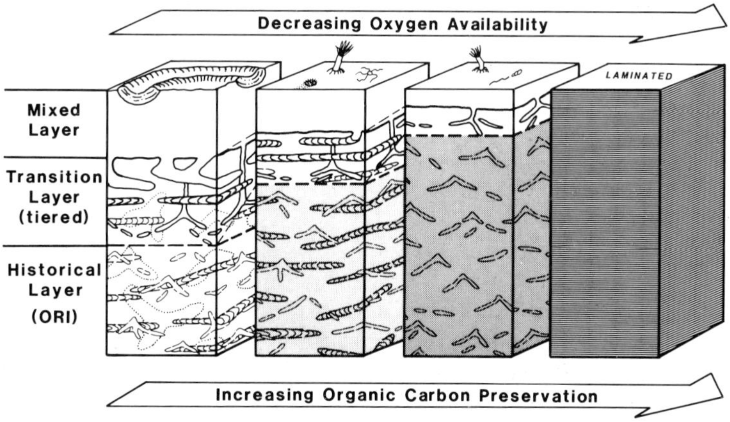

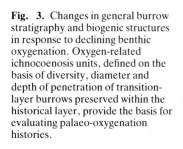

Fig. 3. Changes in general burrow stratigraphy and biogenic structures in response to declining benthic oxygenation. Oxygen-related ichnocoenosis units, defined on the basis of diversity, diameter and depth of penetration of transition-layer burrows preserved within the historical layer, provide the basis for evaluating palaeo-oxygenation histories.

below a critical threshold level (*c*.0.1 ml/l dissolved oxygen), bioturbation virtually ceases altogether and laminated, organic-rich sediments accumulate.

Oxygen-related ichnocoenoses

The ichnological trends summarized above form the basis for reconstructing palaeo-oxygenation histories during the deposition of ancient fine-grained marine strata (Savrda & Bottjer, 1986, 1987, 1989a, b, 1991). Relative degree of oxygenation can be evaluated on the basis of the diversity, diameter and vertical extent of well-defined burrows within historical-layer sediments. Taken together, these criteria permit the recognition and delineation of oxygen-related ichnocoenosis (ORI) units (Fig. 3). ORI units representing the low end of the benthic oxygenation spectrum are characterized by low-diversity assemblages of small-diameter, relatively shallowly penetrating burrows. ORI units generated under progressively higher oxygen concentrations are recognized by progressively higher diversity ichnofossil assemblages that include progressively larger and more vertically extensive burrows.

Redox cyclicity and palaeo-oxygenation curves

In order to evaluate temporal changes in palaeo-oxygenation during the deposition of a stratigraphic package, it is necessary to decipher the vertical stacking pattern of ORI units therein (Fig. 4). Each interval in a rhythmically bedded sequence can be assigned to an ichnocoenosis on the basis of ichnogeneric diversity, maximum diameter and depth of penetration of primary burrows within the primary stratum (see e.g. Savrda & Bottjer, 1989a). However, better-defined components within the associated piped zone may also be useful, particularly if visibility of primary burrows in primary strata is low. This is especially true for burrow-penetration depth, which can be assessed readily on the basis of piped-zone thickness (see above).

Cyclic changes in oxygenation typically result in partial overprinting of different ichnocoenoses (Fig. 4). Except where overlain directly by a thick laminated interval, each ichnocoenosis unit is overprinted by piped-zone burrows of the superjacent ichnocoenosis, which reflect different conditions from those recorded by the primary ichnofabric that they cross-cut. An extreme example is provided by laminated sediments (deposited under highly oxygen-deficient conditions) that have been overprinted by piped-zone burrows as a result of an improvement in oxygenation. Recognition of this superposition of ichnocoenoses is crucial for accurate interpretation of palaeoredox conditions.

Overprinting by mixed-layer bioturbation may also be important but is not universal across all ichnocoenosis transitions (Fig. 4). This type of overprinting is not considered significant when oxygen levels are declining, owing to the associated decrease in the depth of bioturbation. However, during increases in oxygenation, the associated increase in bioturbation depth can result in significant overprinting by mixed-layer bioturbation and may destroy altogether a portion of the previous record of oxygenation. The potential for loss of record generally increases with increased bioturbation depth and decreased sedimentation rate (Savrda & Bottjer, 1987, 1991).

Vertical ORI stacking patterns can be used to construct oxygenation curves that reflect not only relative degree of oxygenation for each stratigraphic interval, but also the rates and magnitudes of longer-term temporal changes in oxygenation. The construction of oxygenation curves, illustrated in Fig. 4 (see Savrda & Bottjer, 1987, 1989a for details), is therefore important for recognizing and/or clarifying redox cyclicity. An example of the use of trace fossils in this capacity is provided below.

Example application: Bridge Creek Limestone

The utility of ichnofabric analyses in the characterization and interpretation of redox cyclicity in pelagic carbonates has been demonstrated in several previous studies. These include studies of the Late Cretaceous Niobrara Formation, US Western Interior (e.g. Savrda & Bottjer, 1989a), and the Late Albian Scisti a Fucoidi in Italy (Erba & Premoli Silva, pp. 211–225, this volume). Here we focus on new ichnological data generated from studies of the Bridge Creek Limestone Member (latest Cenomanian–earliest Turonian) of the Greenhorn Formation, Colorado.

General background

The Bridge Creek Limestone, which represents the peak transgressive phase of the Greenhorn marine cycle (Elder & Kirkland, 1985), is characterized by a decimetre-scale rhythmic alternation of highly bioturbated, organic-poor micritic limestones and laminated to weakly bioturbated, organic-rich marlstones

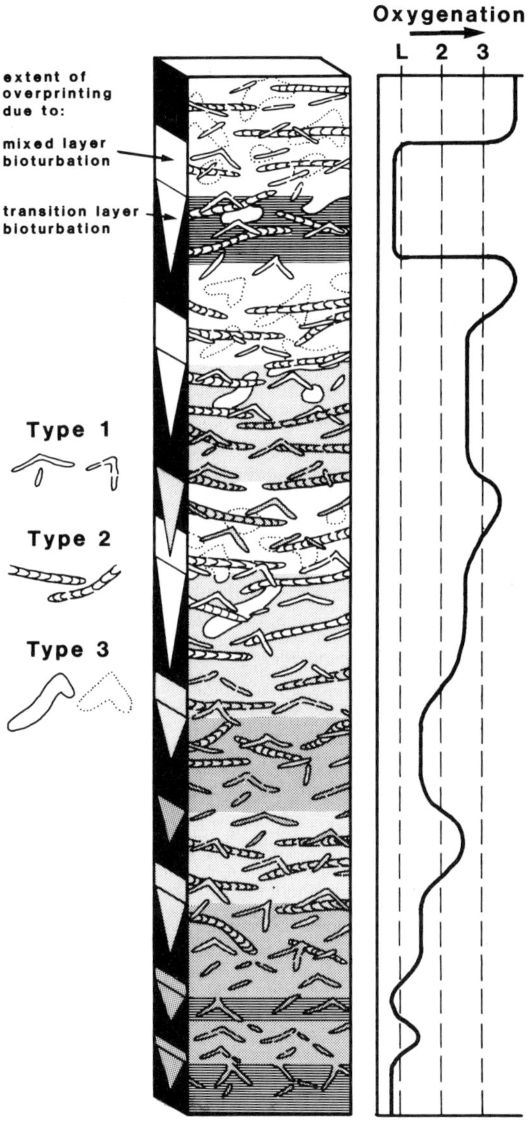

Fig. 4. Stratigraphic column showing hypothetical vertical stacking pattern of ORIs resulting from temporal redox variations, predicted extent of ichnocoenosis overprinting related to mixed-layer and transition-layer bioturbation, and construction of palaeo-oxygenation curve. Note that extent of overprinting increases as oxygen availability and, consequently, depth of bioturbation increase. Palaeo-oxygenation curve is constructed using three reference lines. The L line represents oxygen levels below which lamination is preserved and above which the producers of small type-1 burrows can survive. Lines 2 and 3 represent threshold oxygen levels required for inhabitation of substrates by producers of the larger type-2 and type-3 burrows, respectively. Modified after Savrda & Bottjer (1987, 1989a).

and calcareous shales. Minor lithological components include thin (<2 cm) calcarenites and bentonites (altered ashes) of variable thickness (Pratt, 1984; Elder, 1985; and references therein). Many of the limestone beds and bentonitic seams can be correlated over a large area of the Western Interior (Hattin, 1985).

Rhythmicity in the Bridge Creek has been interpreted to be the product of Milankovitch-driven climatic cycles. However, there is some disagreement among previous authors regarding the specific mechanisms responsible for the observed covariation of bioturbation and organic carbon and carbonate contents. For example, Pratt (1984) attributes these bedding rhythms to combined redox–dilution cycles related to wet–dry climatic oscillations, whereas Eicher & Diner (1989, 1991) argue for combined redox–productivity cycles. Owing to poorly constrained ages for the Bridge Creek interval, estimates of cycle periods have been variable. Hence, there are also differences of opinion regarding the specific orbital parameter(s) responsible. Kauffman (1977) calculated a period of 40–80 ka for cycles defined by prominent limestones. Fischer (1980) estimated a carbonate oscillation period of 27 ka, which is suggestive of control by axial precession. However, he later modified his estimate to *c*.40 ka, and hence implicated obliquity cycles, on the basis of relatively even spacing of limestones and the apparent lack of the well-expressed bundling that typically characterizes precession cycles modulated by eccentricity (Fischer *et al.*, 1985; Fischer, 1986). Employing an updated time scale and an estimated average sedimentation rate for the member (0.5–1.0 cm/ka), Elder (1985) estimated a period of *c*.100 ka for cycles defined by the most prominent and laterally continuous limestones, suggesting that eccentricity played a role. Elder's (1985) suggestion implies that shorter cycles expressed by intervening less prominent limestones or marlstones may reflect a precession signal. Below, we will present ichnological evidence that supports this contention.

The relationships between bioturbation and benthic palaeoredox conditions have been recognized for the Bridge Creek Limestone by various authors (Hattin, 1971, 1975; Pratt, 1984; Elder, 1985; Elder & Kirkland, 1985; Glenister & Kauffman, 1985). However, ichnological observations have been generalized and qualitative. Of these previous studies, Pratt's (1984) analysis of a core drilled near Pueblo, Colorado is the most detailed. With the aid of acetate-peel replicates, Pratt identified four general ichnofabrics (laminated, microburrowed, moderately macroburrowed and highly macroburrowed) and integrated these with sedimentological, mineralogical and organic geochemical data to assess the palaeoceanographic and palaeoclimatic mechanisms responsible for depositional cyclicity.

The study described herein involves a more detailed analysis of ichnofabrics in the Bridge Creek using the acetate peels prepared by Pratt (1984). We restrict discussion to the lower Bridge Creek Limestone Member. This interval of the Pueblo core is approximately 5.5 m thick and, with the exception of a few minor bentonite-rich intervals, is represented by a virtually continuous and complete set of acetate peels.

Results

The lower Bridge Creek Limestone contains both laminated and bioturbated intervals. Initial examination demonstrated that the latter contain one or more of four recurring trace-fossil assemblages. These assemblages, identified on the basis of the presence or absence of recurring ichnogenera and named after the ichnogenera with the largest burrow diameters therein, are referred to herein as the: (i) *Chondrites*; (ii) *Planolites*; (iii) *Zoophycos/Teichichnus*; and (iv) *Thalassinoides* ichnocoenoses. Although the *Chondrites* ichnocoenosis may correspond in part to Pratt's (1984) 'microburrowed' fabrics, most of the *Chondrites*, *Planolites* and *Zoophycos/Teichichnus* ichnocoenoses probably correspond to Pratt's 'moderately macroburrowed' fabrics. The *Thalassinoides* ichnocoenosis, and perhaps some occurrences of the *Zoophycos/Teichichnus* ichnocoenosis, correspond to Pratt's 'highly macroburrowed' fabrics.

After the vertical distribution of ichnocoenoses was established, maximum burrow diameters and, where possible, depth of burrow penetration were logged for each occurrence. Burrow diameters were typically recorded from the primary strata of bioturbated beds. Burrow penetration depths were more difficult to assess in the peels and, in most cases, were estimated from the thickness of the associated piped zones. Diameter and depth data are plotted along with general ichnogeneric diversity for each ichnocoenosis in Fig. 5. The systematic increases in ichnogeneric diversity, maximum burrow diameter and depth of burrow penetration from the *Chondrites* ichnocoenosis through to the *Thalassinoides* ichnocoenosis suggest that these are ORIs. This interpretation is supported by strong relationships among burrow diameter, ichnocoenosis type and available organic carbon data (Fig. 6).

Using the ORI unit concept, an interpreted palaeo-oxygenation curve was constructed for the lower Bridge Creek Limestone (Fig. 7). The curve shows 27 low-order cycles that are grouped into bundles of four to six cycles each. As delineated in

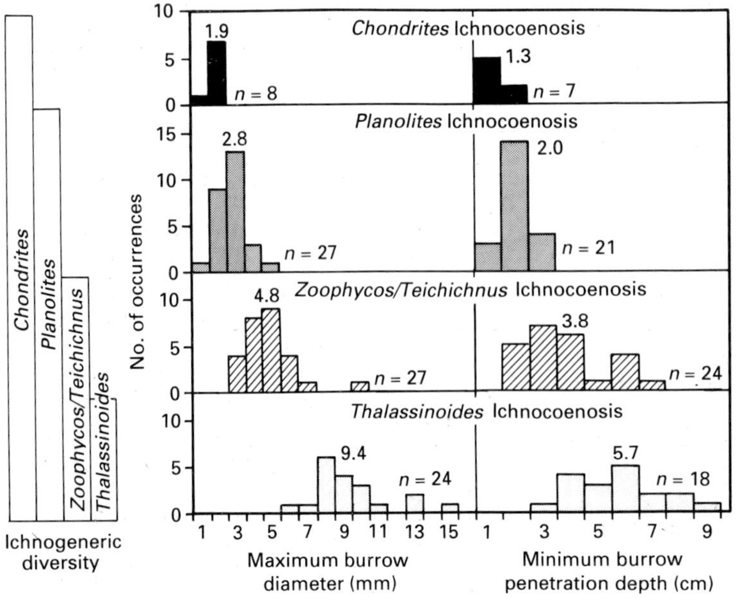

Fig. 5. Ichnogeneric diversity, maximum burrow diameters and burrow-penetration depths recorded for the *Chondrites*, *Planolites*, *Zoophycos/Teichichnus* and *Thalassinoides* ichnocoenoses observed in the Lower Bridge Creek Member of the Greenhorn Formation, Pueblo, Colorado. Numbers above each histogram reflect average values.

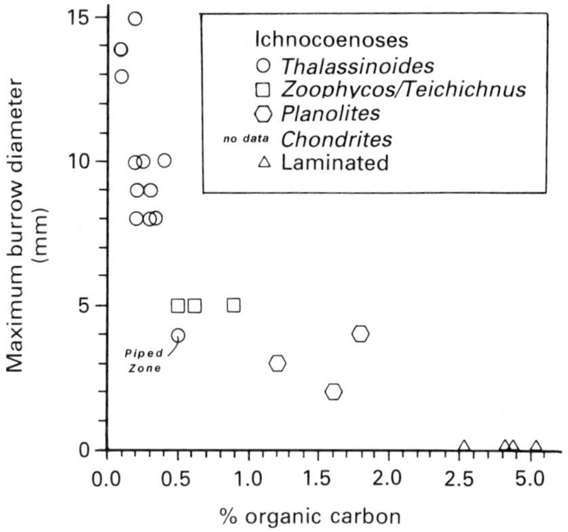

Fig. 6. Relationships among organic carbon content (data from Pratt, 1984), burrow diameter and ORIs in the Lower Bridge Creek Limestone Member. No organic carbon data were available for intervals characterized by the *Chondrites* ichnocoenosis.

Fig. 7, bundles are defined by occurrences of the *Thalassinoides* ichnocoenosis, which generally corresponds to prominent limestone beds. This bundling, previously unrecognized or reported as poorly expressed, is in accord with combined precession/eccentricity cycles. Previous failures to recognize bundling in this interval may be related to the overemphasis placed on prominent limestones and lack of attention paid to more subtle lithological and fabric variations.

Accurate estimates of periodicities are difficult to derive without knowledge of short-term sedimentation rates. Using the time scale of Kauffman (1977), Elder (1985) has calculated an average sedimentation rate for the Bridge Creek Limestone Member as a whole of 0.5–1.0 cm/ka. If the upper limit of 1.0 cm/ka is applied to the oxygenation curve, the short- and long-term cycles have average periodicities of 20.5 and 105 ka, respectively. This further supports the interpretation that carbonate rhythmicity was forced by axial precession and orbital eccentricity.

The example summarized above demonstrates the utility of ichnofabric studies in recognizing and evaluating the role of redox cyclicity in the development of bedding rhythms. However, trace fossils by themselves cannot discriminate between the dilution and productivity mechanisms previously proposed

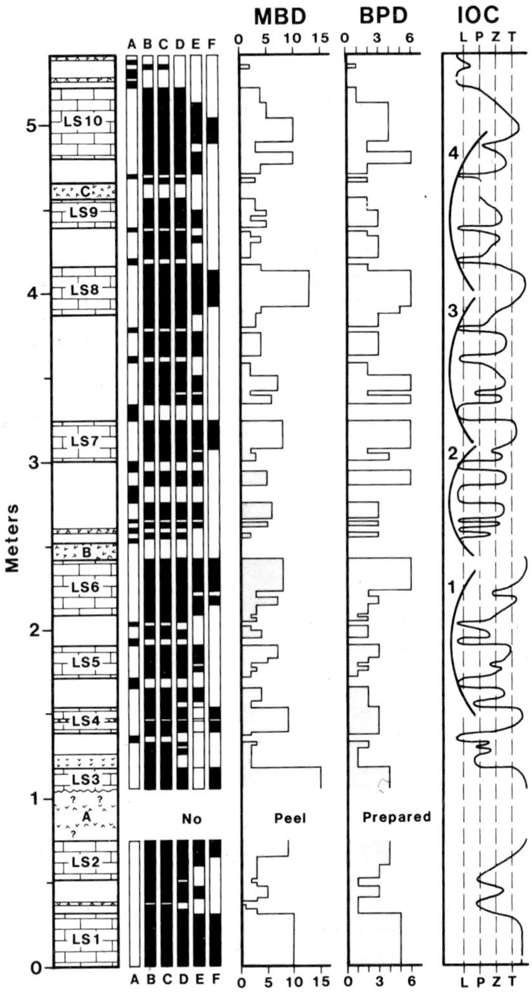

Fig. 7. Distribution of general sediment fabrics (A, laminated; B, bioturbated) and recurring ichnogenera (C, *Chondrites*; D, *Planolites*; E, *Zoophycos/Teichichnus*; F, *Thalassinoides*; piped zones not shown), maximum burrow diameter (MBD, in mm), burrow penetration depth (BPD, in cm), and interpreted oxygenation curve (IOC) for lower Bridge Creek Limestone. IOC is constructed using four reference lines. L line represents oxygen levels below which laminations are preserved and above which *Chondrites* producers can survive. P, Z and T lines represent threshold oxygen levels required for occupation by the producers of *Planolites*, *Zoophycos/Teichichnus* and *Thalassinoides*, respectively. Redox cycles are arranged in at least four bundles (labelled 1–4), which are defined by prominent limestone beds. Bed designations for limestones (LS1–LS10) and thicker bentonites (A, B and C) in the stratigraphic column are after Elder (1985). Intervening strata are characterized by smaller scale, relatively weakly defined alternations of marly limestones, marlstones and calcareous shales.

to explain the carbonate oscillations (see Eicher & Diner, 1991).

Interpretation of redox trends: bottom-water or pore-water oxygenation?

It is commonly assumed that the dominant control of pore-water oxygenation and bioturbation in fine-grained sediments is bottom-water oxygenation (see e.g. Pratt, 1984; Arthur *et al.*, 1986; Savrda & Bottjer, 1986, 1987, 1989a). Hence, bioturbation trends such as those described above are typically interpreted to reflect redox changes associated with regional changes in circulation or water-mass stratification (see e.g. Pratt, 1984; Arthur *et al.*, 1986; Savrda & Bottjer, 1989a). However, the cause–effect relationships and feedback mechanisms among bottom-water oxygenation, infauna and the physical and chemical properties of substrates may be more complex than assumed by these and other previous workers (Fig. 8; see Tyson, 1987; Savrda & Bottjer, 1991).

Pore-water oxygenation is probably affected by other factors, including the amount of organic matter and the metabolic rates of bacteria. The amount of sedimentary organic matter is controlled, in part, by oxygenation, but also by independent factors such as water depth, sedimentation rate, sediment texture and productivity (Tyson, 1987). Similarly, rates of bacterial metabolism are probably governed more by water temperature than by oxygenation (Jannasch *et al.*, 1971). Considering these complexities, it has been suggested that, in some cases, bioturbation trends in mudrocks may reflect changes in pore-water oxygenation that were independent of bottom-water conditions (Savrda & Bottjer, 1991). In ancient shallow epeiric seas, where a greater proportion of organic matter produced in the water column could reach the sea floor and elevated water temperatures could support rapid bacterial metabolisms, pore-water conditions may have actually regulated bottom-water oxygenation (Tyson, 1987; Savrda & Bottjer, 1991).

Physical sedimentological parameters must also be considered. Although benthic oxygenation is a principal control on bioturbation trends in fine-grained marine sediments, additional parameters such as substrate consistency are also important. The fluidity and shear strength of sediments is controlled, in part, by bioturbation (see e.g. Rhoads & Young, 1970; Rhoads & Boyer, 1982). However, other factors that are independent of bioturbation,

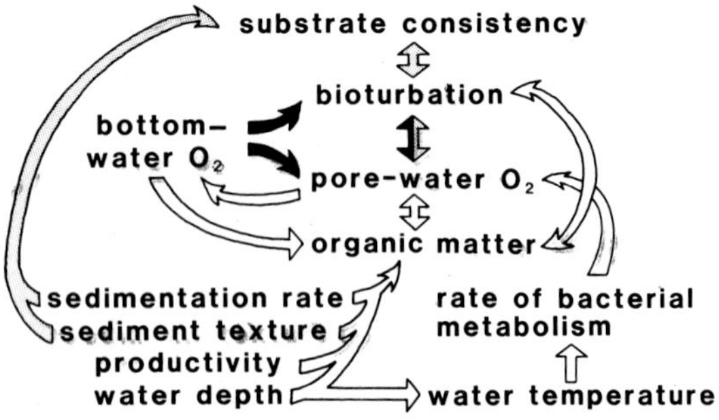

Fig. 8. Complex feedback relationships among bottom-water oxygenation, bioturbation, physical and chemical properties of substrates, and other parameters. Simple relationships depicted by black pathways are often assumed when interpreting bioturbation trends. However, other pathways, some of which involve little change in bottomwater oxygenation (e.g. stippled), may have been important in the generation of redox cyclicity as well.

such as changes in sedimentation rate and sediment texture and composition, may also result in changes in substrate consistency that, in turn, could affect burrowing infauna (e.g. Harrison & Wass, 1965) and pore-water oxygenation. Although not yet adequately documented in modern settings, it seems likely that the ability of organisms to irrigate and aerate substrates would decline as sediments become progressively more fluid, owing to difficulties in maintaining open burrows. Hence, pore-water oxygenation could decline and, through feedback, further affect infaunal organisms and their bioturbation without being regulated by a decline in water-column oxygenation (Savrda & Bottjer, 1991).

The above discussion underscores the need for caution in drawing palaeoenvironmental and/or palaeoceanographic inferences solely on the basis of tracefossil derived redox reconstructions. Rhythmic bedding in the Bridge Creek Limestone and the Niobrara Chalk illustrates potential problems in interpretation. In both of these units, ORIs are strongly correlated not only with organic carbon content, but also with carbonate content (Fig. 9). The redox variations and carbonate oscillations are typically treated as two distinct responses to the same climateforcing mechanism. The wet–dry climate model proposed by Pratt (1984) and others (Barlow & Kauffman, 1985; Arthur *et al.*, 1986; Savrda & Bottjer, 1989a) serves as an example. In this model, periods of increased regional surface runoff resulted in both the dilution of carbonate by clastic sediments and the formation of a brackish surface-water mass that reduced vertical advection

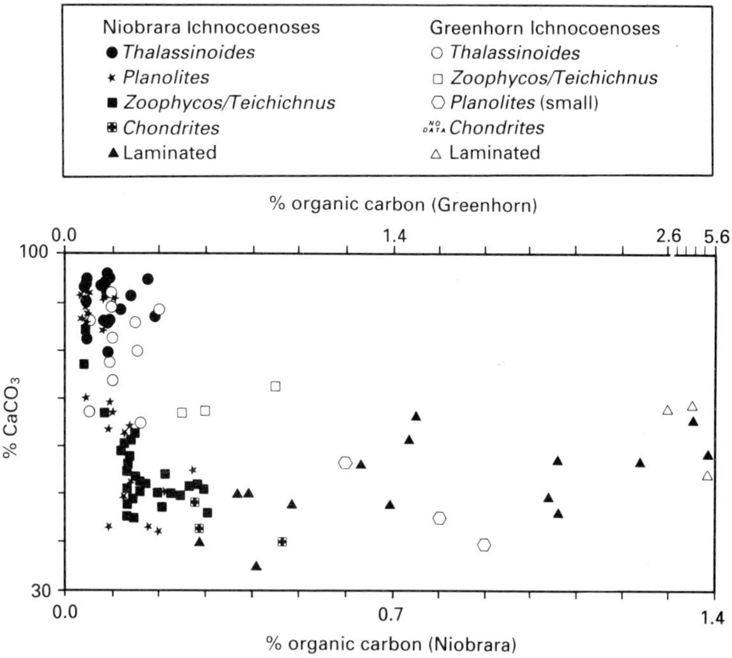

Fig. 9. Relationships among $CaCO_3$, organic carbon and ORIs in Greenhorn (lower Bridge Creek Member) and Niobrara (Fort Hays and lowermost Smoky Hill Member) rhythmites. Increases in oxygenation reflected by ichnocoenoses are generally associated with decreasing organic carbon preservation and increasing carbonate content. Greenhorn $CaCO_3$ and organic data is from Pratt (1984). No data were available for beds characterized by the *Chondrites* ichnocoenosis in the Greenhorn. Niobrara ichnocoenoses are described in detail in Savrda & Bottjer (1989a).

and induced oxygen deficiency in deeper parts of the water column. During drier periods, clastic input was reduced and the deterioration of the brackish water cap led to the restoration of vertical circulation. Other explanations, where redox variations are not considered solely as an independent response, are feasible if additional feedback pathways (Fig. 8) were important. For example, Hattin (1982, 1986) and Sageman (1989) have suggested that inferred redox changes may reflect a feedback response to physical substrate changes induced by climate-mediated dilution and/or productivity. The palaeoceanographic implications of the latter model, where no significant changes in water-column stratification or vertical advection are necessary, differ markedly from those of the water-column stagnation model.

TRACE FOSSILS AND SCOUR CYCLES

Changes in ichnofabrics associated with redox cycles primarily reflect the *biochemical* response of infauna to variations in benthic (bottom- and/or pore-water) oxygenation. The ichnological response to scour cycles is equally pronounced. However, changes in ichnofabric associated with scour cycles primarily reflect the reaction of infaunal organisms to variations in the *physical* character of the sediments, specifically substrate consistency.

Substrate changes associated with scour

Intensification of current activity at the sea floor results in non-deposition, winnowing of previously deposited sediments and the production of omission surfaces. Substrate consistency is thereby modified in several ways. Erosion of the sea floor results in the exhumation of sediments that, owing to previous dewatering associated with overburden pressure, are firm or overconsolidated. Depending on the duration of omission (and hence sea-floor exposure) and chemical conditions, substrate consistency may be altered further by growth of early diagenetic carbonate cements. This cementation typically commences at isolated nuclei in the shallow subsurface to form dispersed lithified nodules. If cementation proceeds, nodules may coalesce, resulting in the complete lithification of the uppermost sediment column. The latter stages of this continuum are typically associated with the precipitation of phosphatic and/or glauconitic crusts.

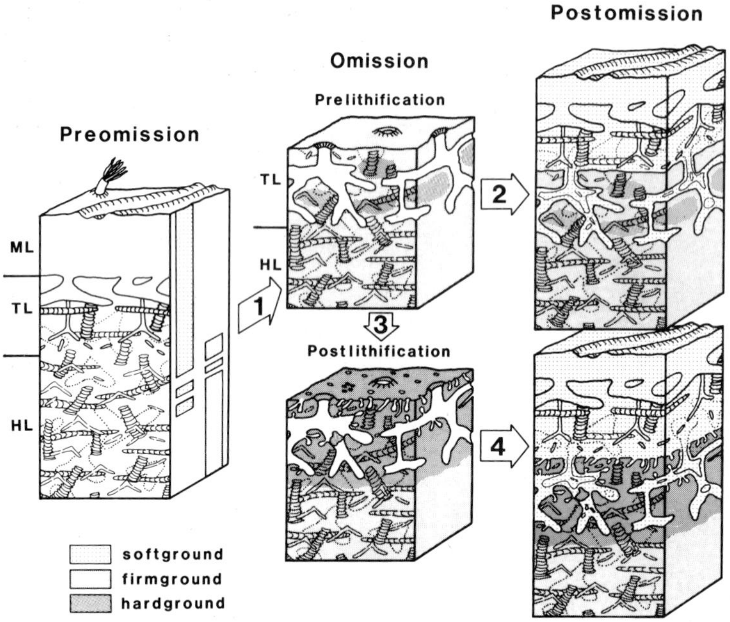

Fig. 10. Progressive development of composite ichnofabrics associated with scour events (based on Bromley, 1967, 1975). Erosive exhumation (1) results in overprinting of softground pre-omission suite by a firmground prelithification omission suite of burrows dominated by *Thalassinoides*, which may exhibit stenomorphism if early diagenetic cementation is initiated. If normal sedimentation resumes prior to sea-floor lithification (2), the production of softground ichnofabrics will eventually resume. If hiatus is extended and sea-floor cementation progresses to completion (3), a hardground postlithification omission suite of borings may be emplaced before resumption of sedimentation (4). Production of softground ichnofabrics will resume only after normal sedimentation is initiated. Post-omission burrows emplaced within soft fills of firmground burrows or hardground borings may exhibit distorted, xenomorphic configurations.

Trace-fossil response to changes in substrate consistency

The response of trace-producing organisms to omission and the associated progressive firming and/or lithification of substrates was first documented in detail by Bromley (1967, 1975) through studies of omission surfaces in European chalks. Bromley demonstrated the complexity of omission-related ichnofabrics, which are the product of the superposition of as many as four distinct suites of trace fossils: the pre-omission, prelithification omission, postlithification omission and post-omission suites. The emplacement of each of these suites in the progression of changes in substrate induration and the resulting production of composite ichnofabrics are summarized below and in Fig. 10. Bromley's (1975) summary should be consulted for details.

Initial scour significantly disrupts the normal burrow stratigraphy via the removal of the pre-existing surface mixed layer, the transition layer and, potentially, softer parts of the historical layer. Exhumed historical-layer ichnofabrics are characterized by a pre-omission suite of trace fossils that are morphologically undistorted, or idiomorphic (e.g. *Thalassinoides suevicus*; Bromley, 1967), reflecting production in originally soft substrates. Exhumation of these now overconsolidated sediments (firmgrounds) generally precludes extensive occupation by organisms that require soupy to soft substrates to burrow. As a result, the development of a new surface mixed layer is virtually precluded altogether and, hence, the upper part of the substrate becomes a zone of heterogeneous mixing that resembles more a normal transition layer. However, burrowers in this modified transition layer are restricted to those organisms that are capable of excavating in firm substrates. For this reason, the burrows of decapod crustaceans, particularly *Thalassinoides*, tend to dominate the prelithification omission suite of trace fossils that overprint the pre-omission suite (Bromley, 1967; Kennedy & Garrison, 1975). *Thalassinoides* therein typically exhibit well-defined burrow-wall ornamentation (e.g. scratch marks). Moreover, in cases where early diagenetic cementation is initiated, dispersed lithified nodules restrict the movement of burrowers, resulting in distorted or stenomorphic burrow configurations (e.g. *Thalassinoides paradoxicus*; Bromley, 1967, 1975). Cramping and degree of distortion of burrows increases as cementation progresses.

If cementation proceeds further, a hardground develops. At this stage, endolithic organisms become important and a postlithification omission suite of borings (e.g. *Trypanites*, *Gastrochaenolites*, etc.) is overprinted on the already complex composite ichnofabric. Borings are emplaced not only from the sea-floor surface but also into the hardened, mineralized walls of prelithification suite burrows.

With the resumption of normal sedimentation, postlithification borings and/or prelithification burrows are filled with unconsolidated sediments. Burrows emplaced within these fills, however, are distorted (xenomorphic; Bromley, 1967) as a result of the irregular geometry of soft-sediment burrowing space. A normal burrow stratigraphy will be re-established only after sediment accumulates to a level where the zone of active bioturbation is above the omission surface.

Stratigraphic expression of scour cycles

The ichnological expression of scour cycles in the stratigraphic record will vary depending on a variety of factors including the magnitude and frequency of scour events and the duration of omission. If after each scour event sedimentation resumes before early diagenetic cementation begins, the record will be one of a normal historical-layer softground ichnofabric punctuated at intervals by a firmground prelithification omission suite dominated by *Thalassinoides*. If early diagenetic processes come into play, as would be expected if hiatuses are prolonged, the sequence will be punctuated by nodular limestones with associated stenomorphic *Thalassinoides* burrow systems. Where cementation proceeds to completion, the sequence will be punctuated by more complex composite ichnofabrics that include later-stage postlithification borings.

Obviously stratigraphic expression may be complicated if the degree of early lithification varies from one scour phase to the next. Moreover, more than one scour phase may be represented in a single composite hardground if successive erosion events repeatedly exhume the same indurated horizon. Composite hardgrounds may be recognized on the basis of ichnological criteria, such as periodic filling and re-excavation of prelithification burrows, boring of fills of earlier borings, and superposition of borings (for detailed summary see Bromley, 1975).

Although hardgrounds, nodular chalks and omission surfaces have been observed in various lithologies of a broad age range, decimetre- to metre-scale alternation of these features is most common in Late Cretaceous shallow-water chalk facies. These scour cycles are characteristic of many European chalk occurrences (e.g. Kennedy & Garrison, 1975), but are typically lacking in North American examples (Bottjer, 1986). The various forms of scour cycles in European chalk, all of which Kennedy & Garrison (1975) attribute to the same mechanism, were interpreted by Arthur *et al.* (1986) to be the result of changes in oceanographic circulation mediated by Milankovitch cycles. A preliminary step towards analysing the duration of these cycles was taken by Hart (1987). His analyses of rhythmically bedded intervals of the Lower Chalk (Cenomanian; England) that are dominated by omission surfaces suggest forcing by axial obliquity. Obviously, attempts at cyclostratigraphy in sequences dominated by nodular limestones and hardgrounds will be inhibited by the effects of extended phases of non-deposition and/or erosion, which could have resulted in non-preservation of an unknown number of cycles.

CONCLUSIONS

Ichnofabric analysis has proved to be a uniquely useful source of information for understanding the origin of decimetre- to metre-scale rhythmic bedding in pelagic carbonates. This is so for a variety of reasons. For example, differences in these beds, such as compositional variations that lead to alternating light and dark beds, or differences in cementation, caused by firmground or hardground formation, enhance the visibility of trace fossils in these strata. Alternating beds in these rhythmic sequences were commonly formed under different environmental conditions, so that different trace fossils and hence different ichnofabrics occur from one bed to the next, commonly repetitively, allowing detailed palaeoenvironmental reconstruction for each individual bed.

For the very reason that these beds are bioturbated, analyses of ichnofabric and trace-fossil content are potentially superior to approaches that rely on analysis of small samples, particularly those done with no accompanying analysis of bioturbation structures. This is because, in these pelagic carbonate environments, bioturbation commonly causes variable amounts of vertical mixing on the decimetre scale, so that microfossils or sediment analysed for compositional or other geochemical parameters may not have originated from the sampled bed, potentially leading to a considerable amount of noise in the signal that each of these approaches provides.

Because they are primary, *in situ* sedimentary structures, trace fossils may also provide a more reliable source of information on rhythmic bedding in pelagic carbonates that have undergone extensive diagenesis. As an example, trace fossils typically survive episodes of rhythmic unmixing of carbonate from carbonate-poor to carbonate-rich beds.

Fortunately, in many sequences, the depth of extensive bioturbation did not exceed 10–12 cm, so that beds with a decimetre-scale thickness are not destroyed by biogenic activity. In these cases, trace fossils and ichnofabric act as valuable tools in unravelling the primary environmental signal. However, this bioturbation also has the potential of completely mixing and, hence, of destroying beds with a thickness of only a few centimetres. It is likely that this destructive process may not be recognized, so that knowledge of the previous existence of these relatively thin beds is typically lost. Because of the extensive mixing, other methods may also be unable to detect the prior existence of these beds.

Thus, analyses of ichnofabric and trace fossils have assumed a particularly important role in determining the primary environmental signal in rhythmic beds of pelagic and hemi-pelagic carbonates. This evidence from bioturbation structures provides a separate line of evidence that can be used to enhance and test data retrieved through use of other approaches, and that can be used to further understand the role of orbital forcing in the generation of these sedimentary sequences.

ACKNOWLEDGEMENTS

Greenhorn peels from the Pueblo core were made by Lisa M. Pratt, and access for our use of these peels was facilitated by A.G. Fischer, W.E. Dean, and Lisa M. Pratt. Lisa M. Pratt also graciously allowed us to use Greenhorn organic carbon and carbonate data that she generated. Studies of the Demopolis Chalk were initiated under the support of an Auburn University Grant-in-Aid of Research to C.E.S. We thank Poppe de Boer, Chris Paul, David Smith and Hans Zijlstra for critiques of an earlier version of this paper.

REFERENCES

Arthur, M.A. & Dean, W.E. (1991) An holistic geochemical approach to cyclomania: examples from Cretaceous pelagic limestone sequences. In: *Cycles and Events in Stratigraphy* (Eds Einsele, G., Ricken, W. & Seilacher, A.). Springer Verlag, Berlin, pp. 126–166.

Arthur, M.A., Dean, W.E., Bottjer, D.J. & Scholle, P.A. (1984) Rhythmic bedding in Mesozoic–Cenozoic pelagic carbonate sequences: the primary and diagenetic origin of Milankovitch-like cycles. In: *Milankovitch and Climate*, Part 1 (Eds Berger, A.L., Imbrie, J., Hays, J., Kukla, G. & Saltzman, B.). Reidel Publ. Co., Dordrecht, pp. 191–222.

Arthur, M.A., Bottjer, D.J., Dean, W.E., Fischer, A.G., Hattin, D.E., Kauffman, E.G. *et al.* (1986) Rhythmic bedding in Upper Cretaceous pelagic carbonate sequences: varying sedimentary response to climatic forcing. *Geology* **14**, 153–156.

Barlow, L.K. & Kauffman, E.G. (1985) Depositional cycles in the Niobrara Formation, Colorado Front Range. In: *Fine-grained Deposits and Biofacies of the Cretaceous Western Interior Seaway: Evidence for Cyclic Sedimentary Processes* (Eds Pratt, L.M., Kauffman, E.G. & Zelt, F.B.). Soc. Econ. Paleont. Mineral. Fieldtrip Guidebook 4, Midyear Meeting, Golden, Colorado, pp. 209–214.

Berger, W.H., Ekdale, A.A. & Bryant, P.F. (1979) Selective preservation of burrows in deep-sea carbonates. *Mar. Geol.* **32**, 205–230.

Bottjer, D.J. (1986) Campanian–Maastrichtian chalks of southwestern Arkansas: Petrology, paleoenvironments and comparison with other North American and European chalks. *Cret. Res.* **7**, 161–196.

Bottjer, D.J. & Ausich, W.I. (1986) Phanerozoic development of tiering in soft substrata suspension-feeding communities. *Paleobiology* **12**, 400–420.

Bottjer, D.J., Arthur, M.A., Dean, W.E., Hattin, D.E. & Savrda, C.E. (1986) Rhythmic bedding produced in Cretaceous pelagic carbonate environments: sensitive recorders of climatic cycles. *Paleoceanography* **1**, 467–481.

Bromley, R.G. (1967) Some observations on burrows of thalassinidean Crustacea in chalk hardgrounds. *Q. J. Geol. Soc. Lond.* **123**, 157–182.

Bromley, R.G. (1975) Trace fossils at omission surfaces. In: *The Study of Trace Fossils* (Ed. Frey, R.W.). Springer Verlag, New York, pp. 399–428.

Bromley, R.G. (1990) *Trace Fossils: Biology and Taphonomy*. Unwin-Hyman, London.

Bromley, R.G. & Ekdale, A.A. (1986) Composite ichnofabrics and tiering in burrows. *Geol. Mag.* **123**, 59–65.

De Boer, P.L. (1982) Cyclicity and the storage of organic matter in Cretaceous pelagic sediments. In: *Cyclic and Event Stratification* (Eds Einsele, G. & Seilacher, A.). Springer Verlag, Berlin, pp. 456–475.

De Boer, P.L. (1991) Pelagic black shale–carbonate rhythms: Orbital forcing and oceanographic response. In: *Cycles and Events in Stratigraphy* (Eds Einsele, G., Ricken, W. & Seilacher, A.). Springer Verlag, Berlin, pp. 63–78.

De Boer, P.L. & Wonders, A.A.H. (1984) Astronomically induced rhythmic bedding in Cretaceous pelagic sediments near Moria (Italy). In: *Milankovitch and Climate*, Part 1 (Eds Berger, A.L., Imbrie, J., Hays, J., Kukla, G. & Saltzman, B.). Reidel Publ. Co., Dordrecht, pp. 177–190.

De Visser, J.P., Ebbing, J.H.J., Gudjonsson, L.,

Hilgen, F.J., Jorissen, F.J., Verhallen, P.J.J.M. & Zevenboom, D. (1989) The origin of rhythmic bedding in the Pliocene Trubi Formation of Sicily, southern Italy. *Palaeogeogr. Palaeoclimatol. Palaeoecol.* **69**, 45–66.

Eicher, D.L. & Diner, R. (1989) Origin of the Cretaceous Bridge Creek cycles in the western interior, United States. *Palaeogeogr. Palaeoclimatol. Palaeoecol.* **74**, 127–146.

Eicher, D.L. & Diner, R. (1991) Environmental factors controlling Cretaceous limestone–marlstone rhythms. In: *Cycles and Events in Stratigraphy* (Eds Einsele, G., Ricken, W. & Seilacher, A.). Springer Verlag, Berlin, pp. 79–93.

Einsele, G. (1982) Limestone–marl cycles (periodites)–diagnosis, significance, causes: a review. In: *Cyclic and Event Stratification* (Eds Einsele, G. & Seilacher, A.). Springer Verlag, Berlin, pp. 8–53.

Einsele, G. & Ricken, W. (1991) Limestone–marl alternation – An overview. In: *Cycles and Events in Stratigraphy* (Eds Einsele, G., Ricken, W. & Seilacher, A.). Springer Verlag, Berlin, pp. 23–47.

Ekdale, A.A. (1985) Paleoecology of the marine endobenthos. *Palaeogeogr. Palaeoclimatol. Palaeoecol.* **50**, 63–81.

Ekdale, A.A., Bromley, R.G. & Pemberton, S.G. (1984) *Ichnology*. Soc. Econ. Paleont. Mineral. Short Course Notes **15**.

Elder, W.P. (1985) Biotic patterns across the Cenomanian–Turonian extinction boundary near Pueblo, Colorado. In: *Fine-grained Deposits and Biofacies of the Cretaceous Western Interior Seaway: Evidence for Cyclic Sedimentary Processes* (Eds Pratt, L.M., Kauffman, E.G. & Zelt, F.B.). Soc. Econ. Paleont. Mineral. Fieldtrip Guidebook 4, Midyear Meeting, Golden, Colorado, pp. 157–169.

Elder, W.P. & Kirkland, J.I. (1985) Stratigraphy and depositional environments of the Bridge Creek Limestone Member of the Greenhorn Limestone at Rock Canyon Anticline near Pueblo, Colorado. In: *Fine-grained Deposits and Biofacies of the Cretaceous Western Interior Seaway: Evidence for Cyclic Sedimentary Processes* (Eds Pratt, L.M., Kauffman, E.G. & Zelt, F.B.). Soc. Econ. Paleont. Mineral. Fieldtrip Guidebook 4, Midyear Meeting, Golden, Colorado, pp. 122–134.

Fischer, A.G. (1980) *Gilbert-bedding Rhythms and Geochronology*. Geol. Soc. Am. Spec. Pap. **183**, pp. 93–104.

Fischer, A.G. (1986) Climatic rhythms recorded in strata. *Annu. Rev. Earth Planet. Sci.* **14**, 351–376.

Fischer, A.G. (1991) Orbital cyclicity in Mesozoic strata. In: *Cycles and Events in Stratigraphy* (Eds Einsele, G., Ricken, W. & Seilacher, A.). Springer Verlag, Berlin, pp. 48–62.

Fischer, A.G. & Schwarzacher, W. (1984) Cretaceous bedding rhythms under orbital control? In: *Milankovitch and Climate*, Part 1 (Eds Berger, A.L., Imbrie, J., Hays, J., Kukla, G. & Saltzman, B.). Reidel Publ. Co., Dordrecht, pp. 163–175.

Fischer, A.G., Herbert, T.D. & Premoli Silva, I. (1985) Carbonate bedding cycles in Cretaceous pelagic and hemipelagic sequences. In: *Fine-grained Deposits and Biofacies of the Cretaceous Western Interior Seaway: Evidence for Cyclic Sedimentary Processes* (Eds Pratt, L.M., Kauffman, E.G. & Zelt, F.B.). Soc. Econ. Paleont. Mineral. Fieldtrip Guidebook 4, Midyear Meeting, Golden, Colorado, pp. 1–10.

Frey, R.W. & Bromley, R.G. (1985) Ichnology of American chalks: the Selma Group (Upper Cretaceous), western Alabama. *Can. J. Earth Sci.* **22**, 801–828.

Glenister, L.M. & Kauffman, E.G. (1985) High-resolution stratigraphy and depositional history of the Greenhorn regressive hemicyclothem, Rock Canyon Anticline, Pueblo, Colorado. In: *Fine-grained Deposits and Biofacies of the Cretaceous Western Interior Seaway: Evidence for Cyclic Sedimentary Processes* (Eds Pratt, L.M., Kauffman, E.G. & Zelt, F.B.). Soc. Econ. Paleont. Mineral. Fieldtrip Guidebook 4, Midyear Meeting, Golden, Colorado, pp. 170–183.

Harrison, W. & Wass, M.L. (1965) Frequencies of infaunal invertebrates related to water content of Chesapeake Bay sediments. *Southeastern Geol.* **6**, 177–187.

Hart, M.B. (1987) Orbitally induced cycles in the chalk facies of the United Kingdom. *Cret. Res.* **8**, 335–348.

Hattin, D.E. (1971) Widespread, synchronously deposited, burrow-mottled limestone beds in Greenhorn Limestone (Upper Cretaceous) of Kansas and central Colorado. *Bull. Am. Assoc. Petrol. Geol.* **55**, 412–431.

Hattin, D.E. (1975) Stratigraphy and depositional environment of Greenhorn Limestone (Upper Cretaceous) of Kansas. *Kansas Geol. Surv. Bull.* **209**.

Hattin, D.E. (1982) Stratigraphy and depositional environments of the Smoky Hill Chalk Member, Niobrara Chalk (Upper Cretaceous), in the type area, western Kansas. *Kansas Geol. Surv. Bull.* **255**.

Hattin, D.E. (1985) Distribution and significance of widespread, time-parallel pelagic limestone beds in Greenhorn Limestone (Upper Cretaceous) of the central Great Plains and southern Rocky Mountains. In: *Fine-grained Deposits and Biofacies of the Cretaceous Western Interior Seaway: Evidence for Cyclic Sedimentary Processes* (Eds Pratt, L.M., Kauffman, E.G. & Zelt, F.B.). Soc. Econ. Paleont. Mineral. Fieldtrip Guidebook 4, Midyear Meeting, Golden, Colorado, pp. 28–37.

Hattin, D.E. (1986) Carbonate substrates of the late Cretaceous sea, central Great Plains and southern Rocky Mountains. *Palaios* **1**, 347–367.

Jannasch, H.W., Elmhjellen, K. & Wirsen, K.O. (1971) Microbial degradation of organic matter in the deep sea. *Science* **171**, 672–675.

Kauffman, E.G. (1977) Geological and biological overview: Western Interior Cretaceous basin. *Mountain Geologist* **14**, 75–99.

Kennedy, W.J. & Garrison, R.E. (1975) Morphology and genesis of nodular chalks and hardgrounds in the Upper Cretaceous of southern England. *Sedimentology* **22**, 311–386.

Pearson, T.H. & Rosenberg, R. (1978) Macrobenthic succession in relation to organic enrichment and pollution of the marine environment. *Oceanography Mar. Biol. Annu. Rev.* **16**, 229–311.

Pratt, L.M. (1984) Influence of paleoenvironmental factors on preservation of organic matter in middle Cretaceous Greenhorn Formation, Pueblo, Colorado. *Bull. Am. Assoc. Petrol. Geol.* **68**, 1146–1159.

Rhoads, D.C. (1975) The paleoecologic and paleoenvironmental significance of trace fossils. In: *The Study of Trace Fossils* (Ed. Frey, R.W.). Springer Verlag, New York, pp. 147–160.

RHOADS, D.C. & BOYER, L.F. (1982) The effects of marine benthos on the physical properties of sediments: a successional perspective. In: *Animal–Sediment Relations* (Eds McCall, P.L. & Tevesz, J.C.). Plenum, New York, pp. 3–52.

RHOADS, D.C. & YOUNG, D.K. (1970) The influence of deposit-feeding organisms on sediment stability and community trophic structure. *J. Mar. Res.* **28**, 150–178.

RICKEN, W. (1986) *Diagenetic Bedding – A Model for Marl–Limestone Alternations*. Lecture Notes in Earth Science 6. Springer Verlag, New York.

RICKEN, W. (1991) Variation of sedimentation rates in rhythmically bedded sediments: Distinction between depositional types. In: *Cycles and Events in Stratigraphy* (Eds Einsele, G., Ricken, W. & Seilacher, A.). Springer Verlag, Berlin, pp. 167–187.

RICKEN, W. & EDER, W. (1991) Diagenetic modification of calcareous beds – An overview. In: *Cycles and Events in Stratigraphy* (Eds Einsele, G., Ricken, W. & Seilacher, A.). Springer Verlag, Berlin, pp. 430–449.

SAGEMAN, B.B. (1989) The benthic boundary biofacies model: Hartland Shale Member, Greenhorn Formation (Cenomanian), Western Interior, North America. *Palaeogeogr. Palaeoclimatol. Palaeoecol.* **74**, 87–110.

SAVRDA, C.E. & BOTTJER, D.J. (1986) Trace fossil model for reconstruction of paleo-oxygenation in bottom water. *Geology* **14**, 3–6.

SAVRDA, C.E. & BOTTJER, D.J. (1987) Trace fossils as indicators of bottom-water redox conditions in ancient marine environments. In: *New Concepts in the Use of Biogenic Sedimentary Structures for Paleoenvironmental Interpretation* (Ed. Bottjer, D.J.). Soc. Econ. Paleont. Mineral., Pacific Sect., Volume and Guidebook 52, pp. 3–26.

SAVRDA, C.E. & BOTTJER, D.J. (1989a) Trace fossil model for reconstructing oxygenation histories of ancient marine bottom waters: application to Upper Cretaceous Niobrara Formation, Colorado. *Palaeogeogr. Palaeoclimatol. Palaeoecol.* **74**, 49–74.

SAVRDA, C.E. & BOTTJER, D.J. (1989b) Anatomy and implications of bioturbated beds in 'black shale' sequences: Examples from the Jurassic Posidonienschiefer (southern Germany). *Palaios* **4**, 330–342.

SAVRDA, C.E. & BOTTJER, D.J. (1991) Oxygen-related biofacies in marine strata: an overview and update. In: *Modern and Ancient Continental Shelf Anoxia* (Eds Tyson, R.V. & Pearson, T.H.). Geol. Soc. Lond. Spec. Publ. 58, pp. 201–219.

SAVRDA, C.E., BOTTJER, D.J. & GORSLINE, D.S. (1984) Development of a comprehensive oxygen-deficient marine biofacies model: evidence from Santa Monica, San Pedro, and Santa Barbara basins, California continental borderland. *Bull. Am. Assoc. Petrol. Geol.* **68**, 1179–1192.

SAVRDA, C.E., DEMKO, T.M. & SMITH, M. (1991) Cyclicity in the Demopolis Chalk (Upper Cretaceous), western Alabama (Abstract). *J. Alabama Acad. Sci.* **62**, 95.

SCHWARZACHER, W. & FISCHER, A.G. (1982) Limestone–shale bedding and perturbations of the earth's orbit. In: *Cyclic and Event Stratification* (Eds Einsele, G. & Seilacher, A.). Springer Verlag, Berlin, pp. 72–95.

TYSON, R.V. (1987) The genesis and palynofacies characteristics of marine petroleum source rocks. In: *Marine Petroleum Source Rocks* (Eds Brooks, J. & Fleet, A.J.). Geol. Soc. Lond. Spec. Publ. 26, pp. 47–68.

WETZEL, A. (1983) Biogenic sedimentary structures in a modern upwelling regime: northwest Africa. In: *Coastal Upwelling and its Sediment Record, Part B, Sedimentary Records of Ancient Coastal Upwelling* (Eds Thiede, J. & Suess, E.). Plenum, New York, pp. 123–144.

Spec. Publs Int. Ass. Sediment. (1994) **19**, 211–225

Orbitally driven cycles in trace-fossil distribution from the Piobbico core (late Albian, central Italy)

E. ERBA *and* I. PREMOLI SILVA

Dipartimento di Scienze della Terra, University of Milan, via Mangiagalli 34, Milan 20133, Italy

ABSTRACT

The Aptian–Albian Scisti a Fucoidi Formation from central Italy consists of a pelagic rhythmic varicoloured sequence of marls, marly clays, marly limestones and subordinate limestones. Black shale layers are abundant and cyclically modulated in several intervals.

A multidisciplinary study performed on the late Albian 'Amadeus Segment' of the Piobbico core (Marche) shows that the rhythmically alternating beds of grey to black marls and whitish limestones (marl–limestone bedding couplets) record the precession cycle, and grouping of these couplets into bundles records the short and long cycles in eccentricity.

In order to reconstruct the bottom-water palaeo-oxygenation history, a semiquantitative study of the trace-fossil distribution was performed from the 'Amadeus Segment'. Based on absence/occurrence of laminations, absence/occurrence of bioturbation, type of trace fossils (ichnogenera), density of bioturbation, and maximum burrow diameter and penetration into sediments, three assemblages were recognized: (i) *Chondrites*-1 assemblage represented by intervals with only 'small' *Chondrites*; (ii) *Chondrites*-1/*Chondrites*-2 assemblage represented by intervals with 'small' and 'large' *Chondrites*; and (iii) *Teichichnus*/*Zoophycos* assemblage represented by intervals with 'small' and 'large' *Chondrites* along with *Teichichnus* and/or *Zoophycos*. The constructed palaeo-oxygenation curve shows that sedimentation occurred mainly under dysaerobic conditions. However, a trend was detected from a more oxygenated environment in the lower portion of the segment to overall less oxygenated conditions, which also alternate with real anoxia, followed by a reverse trend towards more oxygenated conditions again in the uppermost portion of the interval investigated.

Fast Fourier spectral analysis was performed on palaeo-oxygenation and bioturbation density using the Stratabase program. Regular and cumulative spectra of both parameters show two main peaks with periodicities around 100 and 41 ka, which are correlatable with the orbital eccentricity and obliquity frequencies. Spectral evidence for precession (19–23 ka) is weaker, though present in both palaeo-oxygenation curve and bioturbation density spectra, which show a double peak at 28 and 19 ka and at 26 and 20 ka, respectively. Weakness of precessional peak was previously recorded in calcium carbonate, light transmission and planktonic foraminiferal spectra from the same interval, and suggests the presence of distortions in all curves.

Cross-correlation of palaeo-oxygenation and carbonate-content curves shows that these two parameters oscillate with the same frequencies and are perfectly in phase. This means that the increased seasonality, leading to a better stirring of the oceans, is responsible for both the increase in carbonate primary productivity and the enhancement of bottom-water ventilation. The resulting sediments are intensively bioturbated whitish limestone. In contrast, during times of weak seasonality, primary productivity is lower and bottom waters are less oxygenated, as represented by the darker marls showing very slight bioturbation or being barren of trace fossils. The anoxic episodes were cyclic and were controlled by precession, modulated by eccentricity (both short and long) cycles, with a discrete influence of obliquity, at least at the palaeolatitude of the Piobbico core.

INTRODUCTION

Mid-Cretaceous pelagic sequences, in a variety of palaeoenvironments and palaeodepths, record unusual organic carbon-rich facies (Schlanger & Jenkyns, 1976; Schlanger & Cita, 1982; Arthur *et al.*, 1990). These pelagic sequences are commonly characterized by a well-developed cyclicity expressed by limestone–marlstone couplets and redox rhythms. Several authors have tried to identify the mechanisms responsible for the accumulation and preservation of such a large amount of organic matter, and to decipher the origin of the rhythmicity. Interpretations include productivity cycles, changes in bottom-water oxygenation, dissolution cycles, dilution cycles due to terrigenous input, rhythmic input of terrestrial organic matter, and winnowing cycles (e.g. ROCC Group, 1986; Arthur *et al.*, 1990 and references therein; Fischer *et al.*, 1990 and references therein).

One of the most representative formations of pelagic rhythmic sedimentation is the Scisti a Fucoidi (Fucoid Marls) of Aptian–Albian age, cropping out widely in the Umbrian–Marchean Basin of central Italy, an analogue of the deep-sea sequences recovered in the Atlantic, Pacific and Indian Oceans (Arthur & Premoli Silva, 1982; Arthur *et al.*, 1984). The Fucoid Marls have attracted the attention of several scientists with their potential clues to the accumulation of organic-rich sediments in an open marine environment, and with the evident rhythmicity of their varicoloured strata apparently related to orbital forcing (e.g. de Boer, 1982; Schwarzacher & Fischer, 1982; de Boer & Wonders, 1984; Fischer *et al.*, 1985; Fischer & Herbert, 1988).

To gain a better understanding of these problems, a core was drilled for that purpose through the Aptian–Albian Scisti a Fucoidi Formation at Piobbico (Marche, central Italy). A detailed multidisciplinary study was carried out on the Piobbico core, including sedimentology, organic and inorganic geochemistry, calcareous nannofossil and planktonic foraminiferal biostratigraphy and assemblage composition, and palaeomagnetism (Fischer *et al.*, 1985; Erba, 1986, 1988, 1992; Herbert & Fischer, 1986; Herbert *et al.*, 1986; Pratt & King, 1986; Herbert, 1987; Erba *et al.*, 1989; Napoleone & Ripepe, 1989; Premoli Silva *et al.*, 1989a, b; Tornaghi *et al.*, 1989; Ripepe, 1992).

The aim of the present paper is to examine the type, abundance and penetration of ichnogenera in the late Albian Scisti a Fucoidi of the Piobbico core in order (i) to reconstruct the fluctuations in oxygen content at the sea floor during the investigated interval, and (ii) to demonstrate that changes in palaeo-oxygenation are partly driven by orbital forcing, as already inferred for the light transmission, colour changes, calcium carbonate content and planktonic foraminiferal distribution (Herbert & Fischer, 1986; Herbert *et al.*, 1986; Herbert, 1987; Premoli Silva *et al.*, 1989a, b; Tornaghi *et al.*, 1989).

LITHOSTRATIGRAPHIC AND BIOSTRATIGRAPHIC BACKGROUND

An 84-m section was cored at Le Brecce, located 3 km west of the town of Piobbico, Marche, Central Italy (Fig. 1). The core penetrated the entire Scisti a Fucoidi Formation, including the upper and lower transitions and the uppermost portion of the Maiolica Formation. The recovery was 98%, and the total thickness, corrected for the average dip of 23°, is 77.70 m, of which 75.83 m correspond to the Scisti a Fucoidi, and 1.87 m to the underlying Maiolica Formation (Fig. 2).

The Scisti a Fucoidi Formation consists of a cyclically bedded pelagic sequence of marlstone, marly limestone, marly claystone and black shale (black sediment, usually with a low carbonate content and an organic carbon content higher than 1%) with subordinate limestone. On the basis of detailed analyses of colours, carbonate content and occurrence or absence of black shale, 18 lithological units were distinguished within the Scisti a Fucoidi; the oldest unit of the core corresponds to the Maiolica (Unit 19) (Fig. 2). For a complete description of the Piobbico core lithostratigraphy see Erba (1988) and Tornaghi *et al.* (1989). It is worth mentioning that the entire sequence is fully pelagic, lacks recognizable resedimented or turbiditic layers and is devoid of apparent hiatuses.

A total of 154 black shale layers are recorded in the cored sequence: 143 layers occur within the Scisti a Fucoidi Formation, and 11 within the Maiolica Formation. Black shales are not evenly distributed through the core. They are frequent, usually cyclically modulated, in units dominated by greenish-grey, olive-grey and maroon lithotypes. Black-shale layers are absent in units characterized by red to reddish-brown colours. In units consisting of varicoloured reddish and greenish lithotypes, black shale is extremely rare. Unit 18 (Fig. 2), whose lower portion is correlatable with the regional

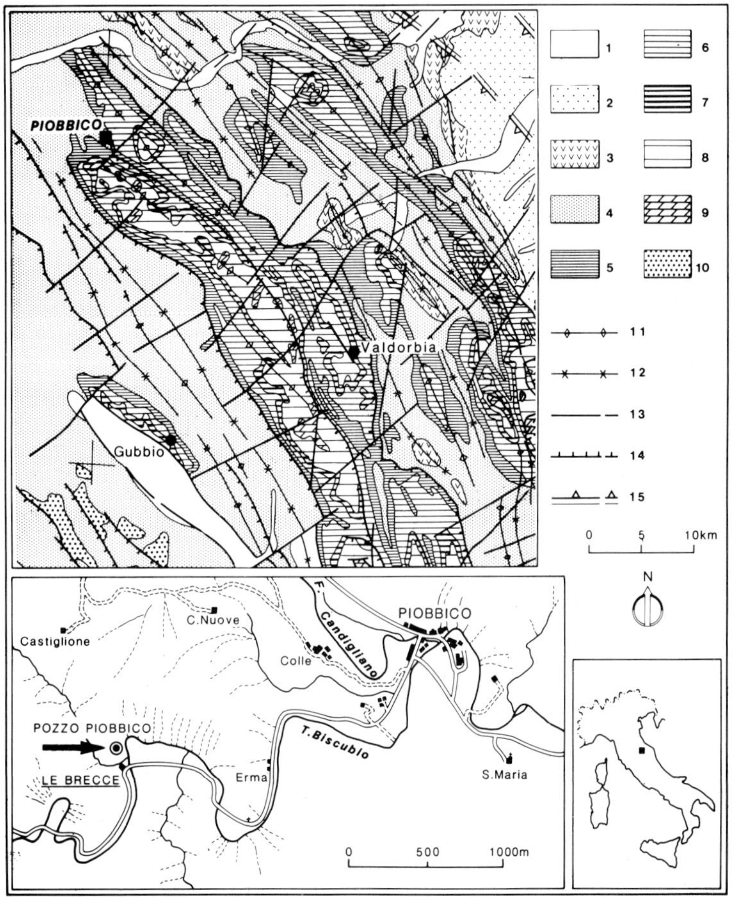

Fig. 1. Geological map of the area investigated and location of the Piobbico core (after Erba, 1988). 1, Quaternary–Pliocene continental facies; 2, Quaternary–Pliocene marine terrigenous facies; 3, Messinian evaporitic facies; 4, Miocene marine terrigenous facies; 5, Oligocene–Palaeocene pelagic facies; 6, Upper Cretaceous pelagic facies; 7, Mid-Cretaceous pelagic facies (Scisti a Fucoidi Formation); 8, Lower Cretaceous–Lower Jurassic pp. pelagic facies; 9, Lower Jurassic pp.–Upper Triassic carbonate platform facies; 10, allochthonous units; 11, anticline; 12, syncline; 13, fault; 14, thrust; 15, buried inverse fault.

marker Livello Selli (Wezel, 1985; Coccioni *et al.*, 1987, 1989) contains several black-shale layers occasionally characterized by common fish remains and pyrite nodules. Frequent black shales are recorded in the Maiolica. Radiolarians occur in numerous discrete layers, which occasionally appear to be affected by winnowing. However, their occurrence does not show any evident cyclic pattern (Tornaghi *et al.*, 1989).

Biostratigraphically, the core spans the interval from the *Globigerinelloides blowi* Zone to the *Planomalina buxtorfi* Zone (*Rotalipora appenninica* Zone of Caron, 1985) (planktonic Foraminifera) and from the *Chiastozygus litterarius* Zone to the *Eiffellithus turriseiffelii* Zone (calcareous nannofossils) of early Aptian to latest Albian age (Erba, 1988; Tornaghi *et al.*, 1989). Biostratigraphy is based on 10 nannofossil biohorizons and seven interval zones, and 11 planktonic foraminiferal biozones (Fig. 2). In the Piobbico core the Aptian–Albian boundary was estimated to be located at about −43 m, within the lower portion of the peculiar maroon claystone of Unit 11, and within the lower portion of the *Predis-cosphaera columnata* nannofossil Zone (Bréhéret *et al.*, 1986; Delamette *et al.*, 1986; Erba, 1988). The Barremian–Aptian boundary was not recovered and occurs below the base of the core (Coccioni *et al.*, 1992).

The Scisti a Fucoidi at Piobbico represents most of the Aptian–Albian interval. Fischer *et al.* (1985) estimated a duration of 18 Ma for the cored sequence and they calculated therefore a mean accumulation rate of about 5 m/Ma.

MODELLING OF LATE ALBIAN CYCLIC ANOXIA

Previous, detailed work on an 8-m-long segment from the Piobbico core, belonging to the late Albian

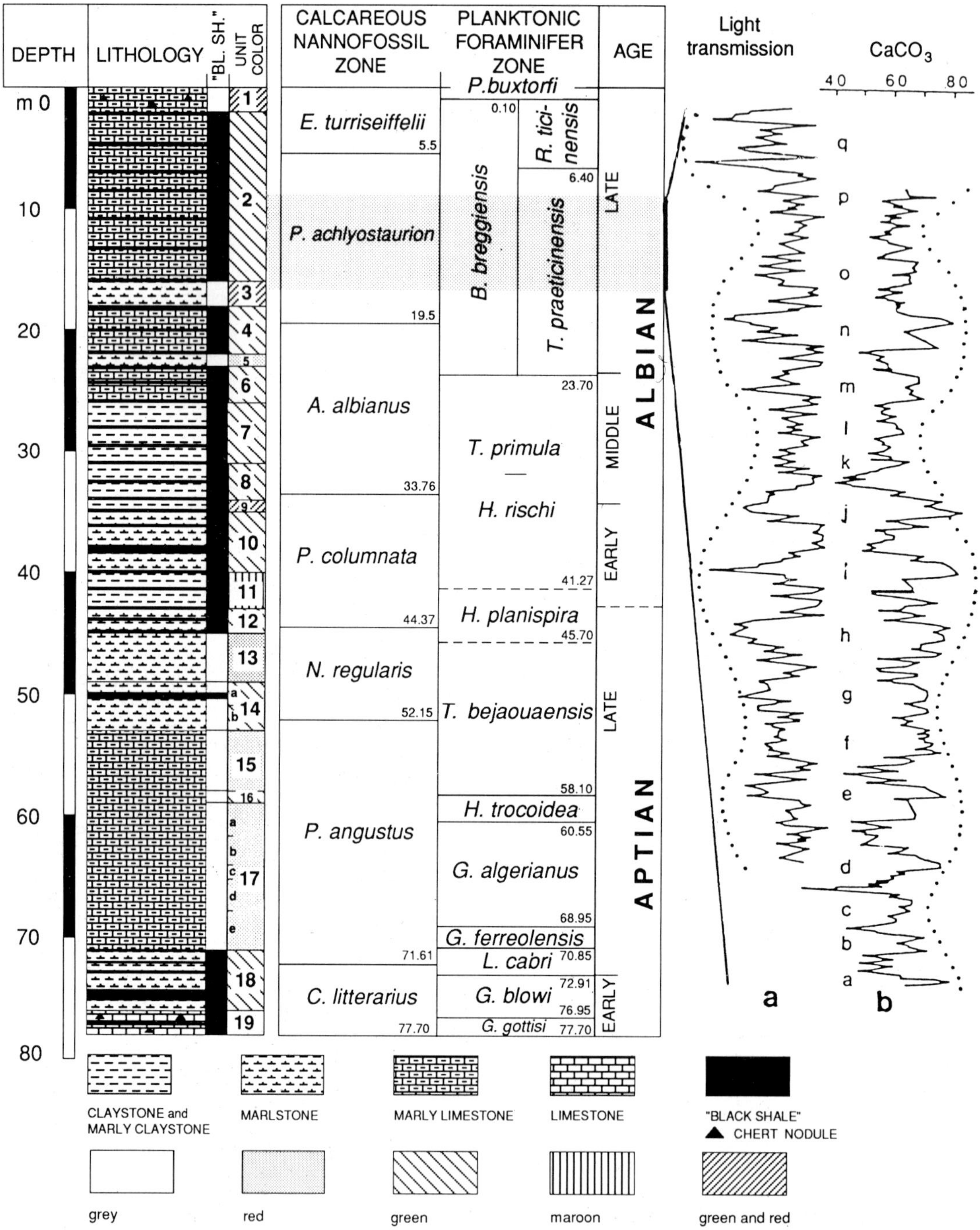

Fig. 2. Lithology and biostratigraphy of the Piobbico core plotted against metres below ground surface and cycles from the 'Amadeus Segment' (9–17 m). Identified cycles and bundles in light transmission (a) and carbonate content (b) are indicated by letters; dotted line marks the 400-ka envelopes. From Herbert & Fischer (1986).

Ticinella praeticinensis foraminiferal Subzone and *Parhabdolithus achlyostaurion* nannofossil Zone, suggested that limestone–marlstone couplets are the sedimentary expression of orbitally induced (precessional) cycles modulated by eccentricity cycles, as a result of changes in carbonate flux (Herbert & Fischer, 1986; Herbert *et al.*, 1986). As proposed earlier by de Boer (1982, 1983) and based on the fact that mainly marlstone is rich in organic matter, these authors concluded that during periods with a low precession index carbonate flux was low, and ventilation at the sea floor was poor, because of the lack of a vigorous current system. In contrast, during periods with a high precession index carbonate flux was higher and water circulation was more vigorous, thus preventing the accumulation of organic matter at the bottom. Calcareous nannofossil fertility indices and planktonic foraminiferal distribution and composition corroborate this hypothesis (Erba, 1986, 1992; Premoli Silva *et al.*, 1989a; Tornaghi *et al.*, 1989).

Herbert & Fischer's (1986) interpretation implies short-term (10 ka) temporal changes in bottom-water oxygenation. This hypothesis can be tested by quantifying the trace-fossil distribution, following the model proposed by Savrda & Bottjer (1986, 1989). Savrda & Bottjer's model is based on trace-fossil diversity, assemblage composition, burrow-size and penetration depth, which are influenced by the oxygen availability in bottom waters. In particular, these parameters decrease when oxygenation diminishes, and therefore the qualitative and quantitative study of palaeoichnocoenoses may be used to evaluate the changes in redox conditions through a given stratigraphic interval. Although the resulting palaeo-oxygenation curves are only qualitative, they can reflect the relative degree of oxygenation and rates and magnitude of such a change.

Savrda & Bottjer (1986, 1989) applied this model to Upper Cretaceous rhythmically bedded sequences (Niobrara Formation) and recognized oxygen-related ichnocoenoses. The palaeo-oxygenation curves based on these trace-fossil assemblages provide evidence for Milankovitch-like climatic cycles triggering palaeoceanographic perturbations.

We will apply the trace-fossil model to the same segment of the Piobbico core previously studied in detail (see above). Recently Coccioni & Galeotti (1991) named this portion of the Scisti a Fucoidi the 'Amadeus Segment' to celebrate the bicentennial anniversary of the death of Wolfgang Amadeus Mozart. The lithology of this segment, from 9 to 17 m, consists of alternating whitish limestone and black shale between 9 and 14 m and is visibly cyclic, whereas between 14 and 17 m black shale is rare and marlstone is the dominant lithology.

TRACE-FOSSIL ASSEMBLAGES

The study of trace-fossil distribution was performed on polished core surfaces and acetate peels through the 8-m-thick segment (9–17 m) (Fig. 3). Identification of ichnogenera is based on the description of Ekdale *et al.* (1984), Savrda & Bottjer (1987) and Bromley (1990). We first checked the absence/occurrence of lamination and bioturbation, and then recorded the assemblage composition, maximum size of burrows, penetration depth and the density

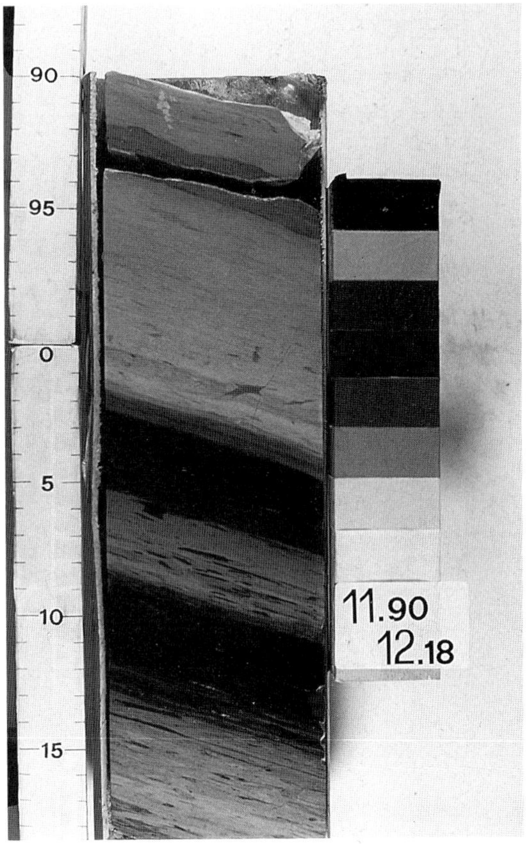

Fig. 3. Bioturbations (*Chondrites*-1, *Chondrites*-2 and *Zoophycos*) in the Piobbico core, segment from −10.95 to −11.20 m (11.90–12.18 m uncorrected depth), late Albian.

of bioturbation using the reference table of Baccelle & Bosellini (1965). Figure 4 illustrates the trace-fossil distribution in the interval studied. Bioturbations occur throughout the segment with the exception of black-shale layers, which are laminated in most cases. Slight bioturbation was observed in only three black-shale layers.

Three ichnogenera were identified: *Chondrites*, *Teichichnus* and *Zoophycos*. *Chondrites* is the dominant form and may be the only type of bioturbation present in several stratigraphic intervals, especially within or close to the black-shale layers. This ichnogenus is common in bioturbated intervals sedimented under dysaerobic conditions (Bromley &

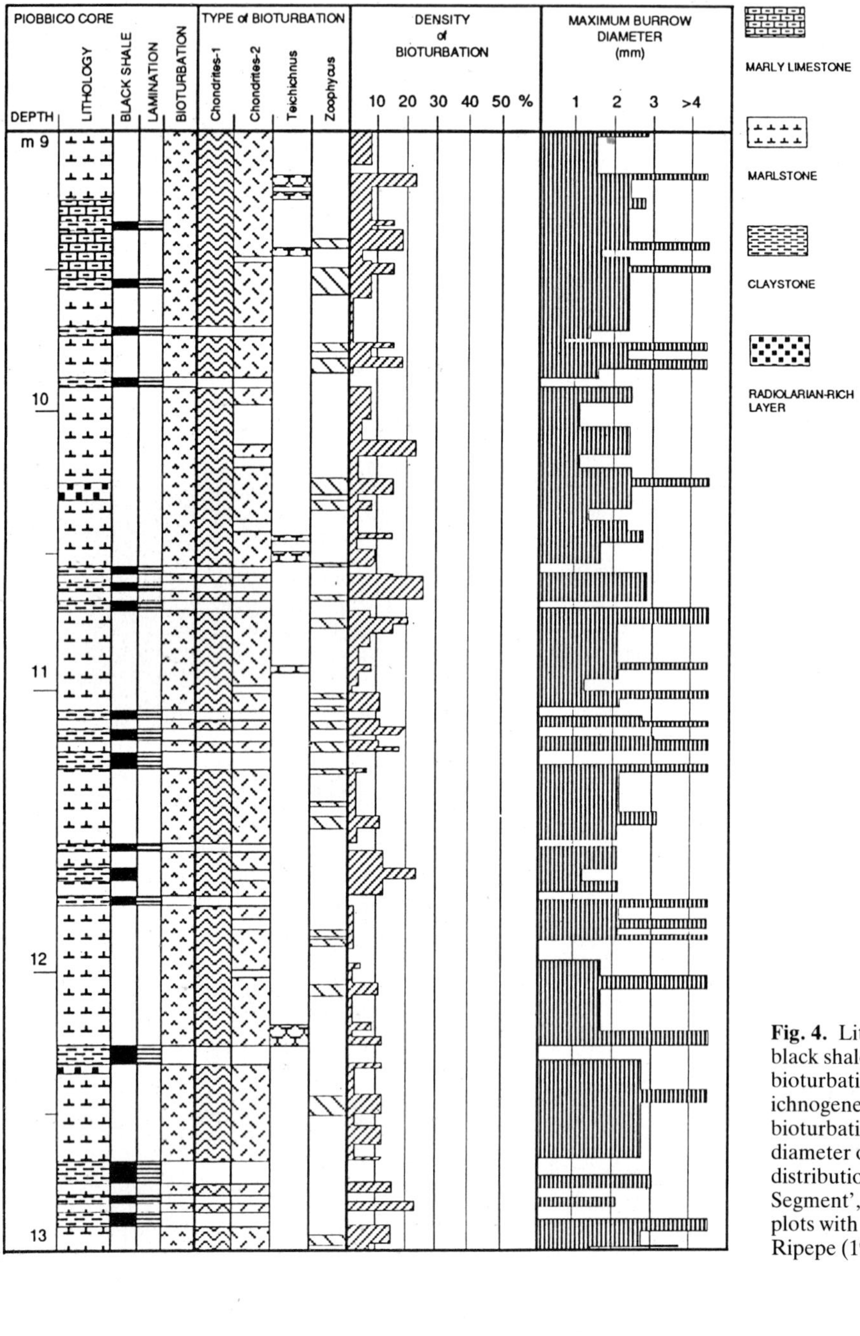

Fig. 4. Lithology, occurrence of black shales, lamination and bioturbation plotted against ichnogenera assemblages, density of bioturbation and maximum diameter of bioturbation distribution from the 'Amadeus Segment', Piobbico core. Graphic plots with Stratabase program by Ripepe (1988).

Ekdale, 1984; Sandberg & Gutschick, 1984; Savrda & Bottjer, 1986, 1987; Bromley, 1990), and is usually the only type of bioturbation present just before the onset of anoxia.

We subdivided the *Chondrites* trace-fossils into two groups based on maximum burrow diameter. The *Chondrites*-1 group comprises burrows with a diameter between 0.5 and 1.5 mm, whereas the *Chondrites*-2 group consists of burrows with a diameter between 1.5 and 2.5 mm (after compaction). The penetration depth of all *Chondrites* is usually 1–2 cm below their apparent origin point. The tiny trace fossils of the *Chondrites*-1 group are occasionally the only type of bioturbation present, indicating

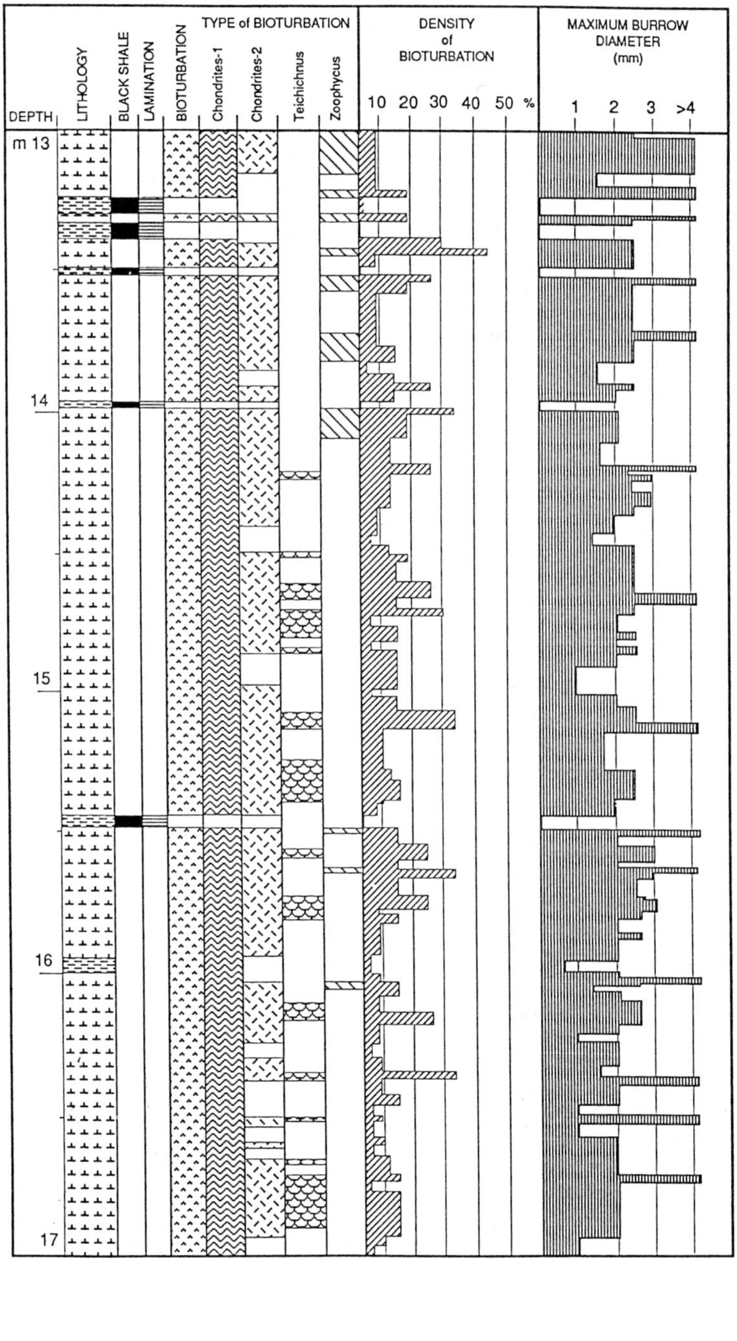

Fig. 4. *Continued.*

a very low oxygen content in bottom waters or the lowest tier in the vertical profile of the endobenthic community (Bromley, 1990). However, since there is no evidence of interrupted or extremely low accumulation rates, the first interpretation is more likely.

Teichichnus and *Zoophycos* are very similar and might have been produced by the same organism, as suggested by Savrda & Bottjer (1989). In the interval studied, these two ichnogenera are largely mutually exclusive and co-occur rarely. *Teichichnus* is more common in the lower portion of the segment studied up to approximately 14 m, whereas *Zoophycos* is more abundant in the upper portion. Both ichnogenera are always observed associated with both *Chondrites*-1 and *Chondrites*-2; they never occur alone or with *Chondrites*-2 only. The maximum burrow diameter measured from both *Teichichnus* and *Zoophycos* varies between 2 and 10 mm, but most values are restricted to between 3 and 5 mm. Due to their morphology, it is difficult to estimate the depth of penetration. Discernible penetration depth fluctuates between 3 and 6 cm. The plot of maximum burrow diameter (Fig. 4) follows the distribution of the ichnogenera.

With regard to the cross-cutting relationship, we noted that *Chondrites*-1 and *Chondrites*-2 are usually cut by *Teichichnus* and *Zoophycos*, but not vice versa, suggesting a fairly stable assemblage composition/burrow-size/tier-depth relationship. These data are in agreement with the results obtained by previously by Bromley & Ekdale (1984) and Savrda & Bottjer (1986, 1987, 1989) from Mesozoic pelagic sequences.

Another parameter that we took into account is the density of bioturbation reflecting the intensity of organism activity through the sediments. This parameter represents the amount of discrete, identifiable burrows and is partly correlatable to the ichnofabric indices proposed by Droser & Bottjer (1986, 1991). The density of bioturbation observed corresponds to Droser & Bottjer's ichnofabric 1 (no bioturbation), 2 (up to 10% bioturbation) and 3 (10–40% bioturbation). Within this range, however, we further distinguished bioturbation density using classes with a 2.5% increase.

Ichnofabrics 4–6 (strongly bioturbated to totally homogenized sediments) of Droser & Bottjer (1986, 1991) were not observed in the interval studied, suggesting that biogenic reworking never succeeded in blurring all primary sedimentary structures. It is possible that the oxygen availability acted as a major limiting factor for biological activity at the bottom. The maximum values of bioturbation density, close to 40% of surface cut, were usually observed close to the black shales, possibly suggesting higher food concentration stores in the C_{org}-rich layers, or relative peaks of oxygen content sharply interrupted by anoxic conditions. However, peaks were recorded also in intervals without black-shale layers, especially between 14 and 17 m. The relatively low values of bioturbation density suggest that dysaerobic bottom waters are likely; this is especially true for the *Chondrites* trace fossils, which are always represented by densities lower than 25%.

The association and areal relationship of the trace-fossil genera permit the recognition of three assemblages named: (i) *Chondrites*-1 assemblage represented by intervals with only 'small' *Chondrites*; (ii) *Chondrites*-1/*Chondrites*-2 assemblage represented by intervals with 'small' and 'large' *Chondrites*; and (iii) *Teichichnus*/*Zoophycos* assemblage represented by intervals with 'small' and 'large' *Chondrites* along with *Teichichnus* and/or *Zoophycos*. These assemblages are comparable with those recognized by Savrda & Bottjer (1989), but we further subdivide their *Chondrites* assemblage into two assemblages, as discussed above.

PALAEO-OXYGENATION CURVES

The detailed analysis of trace-fossil assemblage composition and diversity, cross-cutting relationships, density of bioturbation and maximum burrow diameter constitutes the basis for the reconstruction of oxygen-related ichnocoenoses, following the procedure of Savrda & Bottjer (1986, 1989) and similar to that used by Bromley & Ekdale (1984). The relative degree of oxygenation is based on the estimated tolerance of organisms to low-oxygen conditions. Bromley (1990) summarized and discussed the opportunistic trace fossils, and pointed out that opportunistic ichnotaxa are the pioneers and forms flourishing in extreme environments. *Chondrites* has been regarded as an opportunistic ichnogenus characteristic of oxygen-depleted environments (Bromley & Ekdale, 1984; Ekdale, 1985; Savrda & Bottjer, 1986). Similarly, *Zoophycos* has been considered as an opportunistic form when low-oxygen conditions prevail (Bromley, 1990). As suggested by Savrda & Bottjer (1987, 1989), *Chondrites*, *Zoophycos* and *Teichichnus* may represent variable conditions within the dysaerobic belt since they seem to

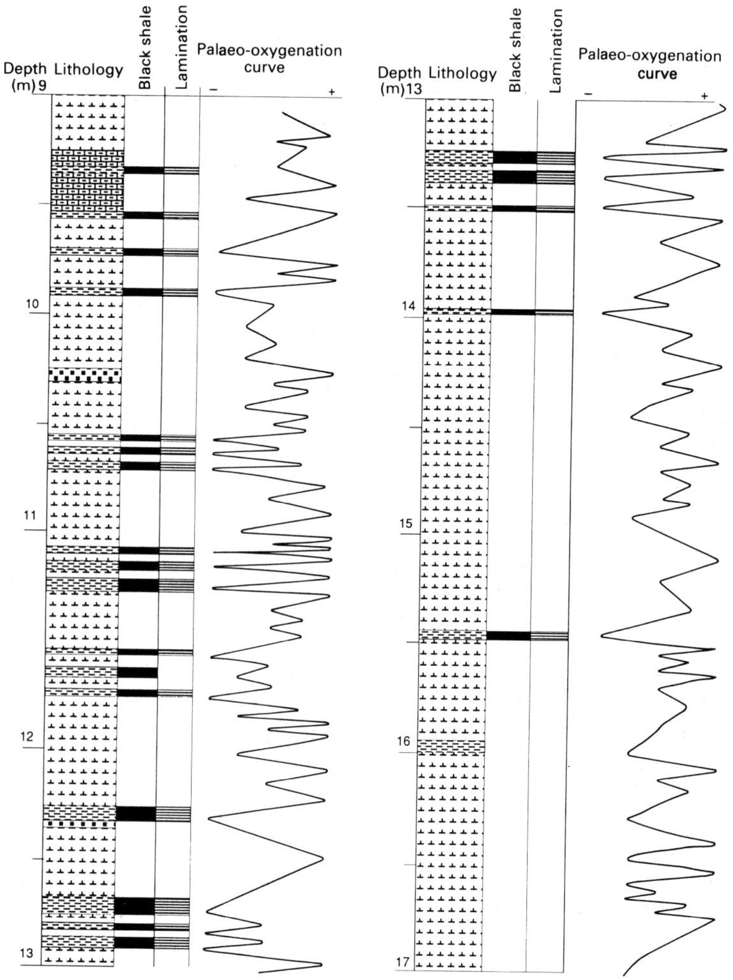

Fig. 5. Palaeo-oxygenation curve based on trace-fossil assemblages, density of bioturbation and maximum burrow diameter plotted against lithology, 'Amadeus Segment'. After Savrda & Bottjer's (1987, 1989) model.

colonize increasingly more oxygenated niches. Conversely, progressively impoverished palaeoichnocoenoses would suggest progressively lower palaeo-oxygen content in the bottom waters.

In the interval studied, *Teichichnus* and *Zoophycos* are considered less tolerant than *Chondrites*, and probably the *Chondrites*-2 group is less tolerant with respect to low-oxygen content than the *Chondrites*-1 group. Based on this supposition, we constructed qualitative palaeo-oxygenation curves (Fig. 5). Discriminations between step-like or gradual changes are not possible and therefore we only interpolated from maxima to minima and vice versa to compile a simplified picture of the changes in palaeo-oxygenation of the bottom waters. We were conscious that fluctuations in the rate and magnitude of oxygenation and deoxygenation, and changes in the sedimentation rate may strongly affect the shape of the curves.

The palaeo-oxygenation curve shows that sedimentation mainly occurred under dysaerobic bottom waters, but with fluctuations from anoxia to relatively more oxygenated conditions. Low-amplitude fluctuations are observed in the lower portion up to 14 m, with minimum (laminations) occurring every 1.5 m (Fig. 5). Above that level, up to 10 m, fluctuations increase remarkably in amplitude and number, with several anoxic laminated levels, which are rare again in the uppermost part of the curve. Excursions of the palaeo-oxygenation curve seem to be symmetrical, suggesting a gradual alternation of oxygenation and deoxygenation phases. However, some bioturbation peaks immediately precede the black-shale layers and consequently the palaeo-oxygenation curve is asymmetrical with respect to the black shale. This may record a more rapid phase of deoxygenation than of oxygenation. Then, the distribution of palaeoichnogenera might simply reflect tiering profiles frozen after very rapid deoxygenation, as testified by laminated black shales.

MILANKOVITCH CYCLES

Fast Fourier spectral analyses were performed to investigate the trace-fossil parameter and carbonate fluctuations, and to see if their rhythmicities are correlatable to the Earth's orbital cycles.

The data were stored at a 1-cm scale in the Stratabase program (Ripepe, 1988) and standard time-series analyses were applied. The functions were smoothed, applying two iterations by a five-point triangular filter and spectral analyses were carried out on the entire segment. The resulting spectra for both palaeo-oxygenation and density of bioturbation show peaks correlatable to eccentricity, obliquity and precession frequencies using an estimated average sedimentation rate of 5 mm/ka (Fig. 6).

Spectral analysis was also performed using a 2-m moving window in 50-cm steps through the curve. The same procedure was used for lithological and palaeontological spectra (Premoli Silva *et al.*, 1989b). The derived spectra database comprises 14 normalized spectral functions, which were subsequently stacked into a single cumulative spectrum. The application of these techniques results in: (i) an increase in the degrees of freedom and, consequently, in the stationarity of spectra; and (ii) an attenuation of distortions introduced by very short-term fluctuations (noise) and by variations in sedimentation rate.

The results of the latter spectral analyses of the palaeo-oxygenation curve and density of bioturbation from the 8-m 'Amadeus Segment' are shown

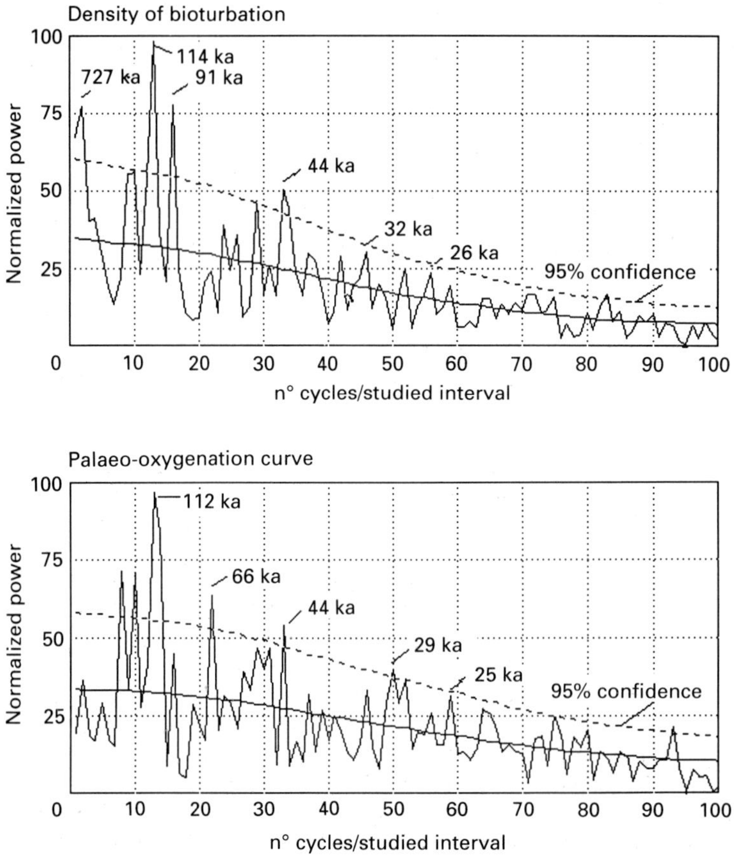

Fig. 6. Spectral functions of density of bioturbation and palaeo-oxygenation curve from the 'Amadeus Segment', Piobbico core, with frequencies in thousands of years using a sedimentation rate of 5 mm/ka. Stratabase program by Ripepe in Premoli Silva *et al.* (1989b).

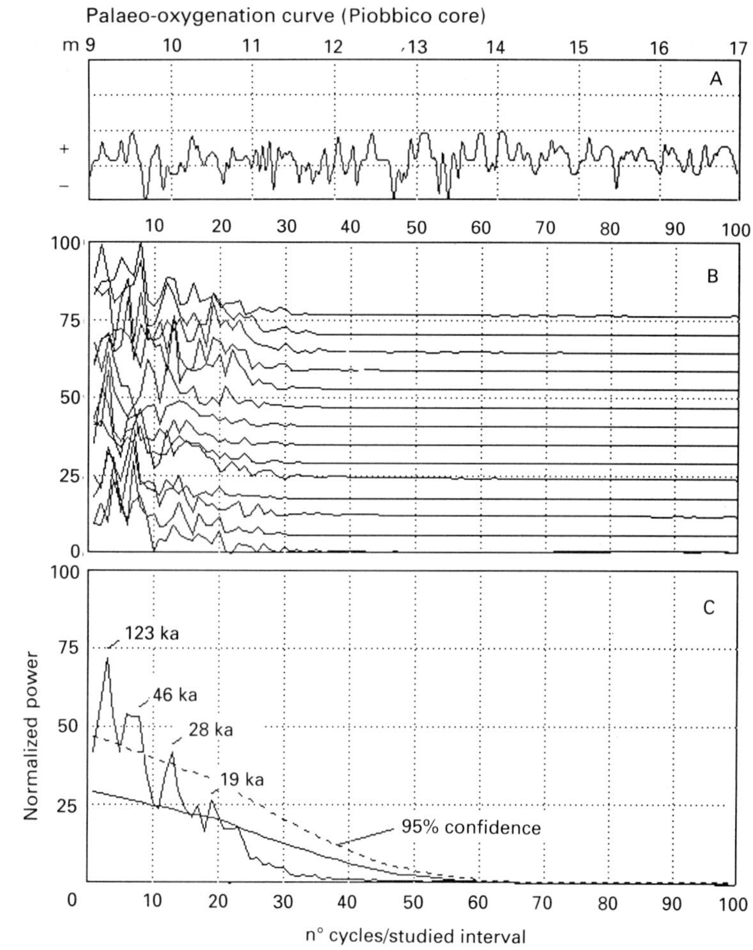

Fig. 7. (A) Smoothed palaeo-oxygenation curve of the 'Amadeus Segment' of the Piobbico core (−9 to −17 m) sampled at 1-cm scale. (B) Spectra database assembling the 14 spectra calculated via a 200-cm wide window moved in 50-cm steps. (C) Cumulative spectrum obtained by stacking the 14 spectra in (B) with frequencies in thousand years using a sedimentation rate of 5 mm/ka. Stratabase program by Ripepe in Premoli Silva *et al.* (1989b).

in Figs 7 and 8. The cycle periodicities (ka) were based on the same estimated average sedimentation rate of 5 mm/ka as in the previous analysis. Both cumulative spectra show a dominant cyclicity around 123 and 92 ka respectively, which is interpreted as the short (98–126 ka) eccentricity frequency. Both spectra show a second cyclicity, still above the 95% confidence level, around 41 ka in the spectrum of bioturbation density and 46 ka in the palaeo-oxygenation curve spectrum; these values are close to the obliquity (40–53 ka) frequency. Spectral evidence for precessional (23–19 ka) frequencies is weaker, though present in both palaeo-oxygenation and density spectra, which show a double peak at 28 and 19 ka and at 26 and 20 ka, respectively. Both bioturbation spectra show a remarkable coherence with the other parameters analysed, such as lithology, colour index, planktonic Foraminifera and calcium carbonate (Herbert & Fischer, 1986; Premoli Silva *et al.*, 1989b). In fact, these new spectra show a strong eccentricity peak and a weak precessional signal. In addition, both spectra show a powerful obliquity frequency peak, even stronger than that observed in the microforam spectrum (Premoli Silva *et al.*, 1989b, Fig. 4). We conclude that bioturbation is not only cyclically driven by orbital forcing, but seems less affected by distortion due to diagenesis than other parameters. It would thus better reflect the primary signal of climatic–oceanographic changes caused by orbital forcing. Recently, Thierstein & Roth (1991) suggested that Cretaceous carbonate cycles in deep-sea sediments seem to be related to bioturbation intensity indicative of varying deep-water and oxygen renewal rates, although they also stated that the lithological cycles studied are dominated by diagenetic effects.

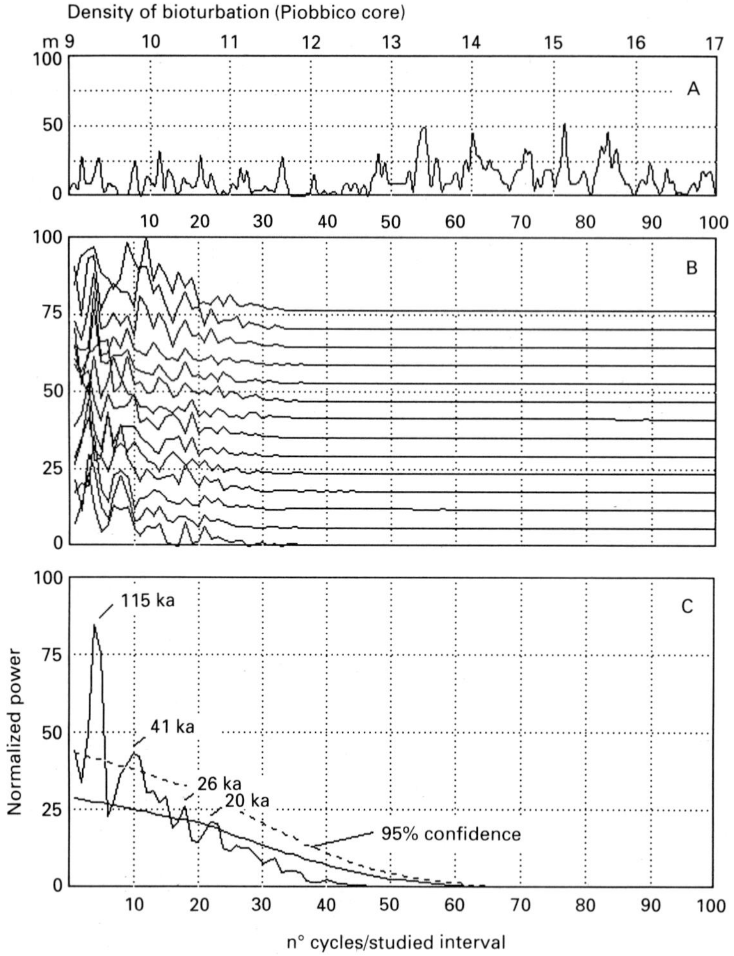

Fig. 8. (A) Smoothed bioturbation density curve of the 'Amadeus Segment' of the Piobbico core (−9 to −17 m) sampled at 1-cm scale; (B) and (C) obtained as in Fig. 6.

Cross-correlation of the carbonate and palaeo-oxygenation curves simultaneously compares the spectral behaviour at all the levels of orbital cycles, and indicates that these two curves oscillate with the same frequencies and are perfectly in phase (Fig. 9). This is evidenced by the maximum value of the cross-correlation function that occurs at time-lag zero, showing that the two functions contain all the frequencies with the same phase. Conversely, if two functions contain the same frequency range but the phases are opposite, the maximum peak occurs on the negative axis. Cross-correlation is, therefore, a powerful tool when two functions have to be compared in terms of frequency and phase content simultaneously.

The palaeo-oxygenation and density of bioturbation curves, which fluctuate in phase, are compared with the calcium carbonate curve (Herbert & Fischer, 1986; Premoli Silva *et al.*, 1989b) in Fig. 10. Higher palaeo-oxygenation and especially bioturbation density correspond to calcium carbonate peaks except at 13.40 m. At this level, a peak in the density curve occurs in a carbonate low, which is possibly related to a higher food concentration stored in the black-shale layers. Figure 10 also shows that higher density occurs in correspondence to the greatest eccentricity shown in the 400-ka envelope, whereas the minimum density corresponds to lows in the 400-ka eccentricity cycle as reconstructed by Herbert & Fischer (1986). In addition, the 100-ka cycles are visually recognizable in the density curve, especially in the lower portion of the segment, where maximum bioturbation density coincides with the thickest, carbonate-rich layers. The general trend

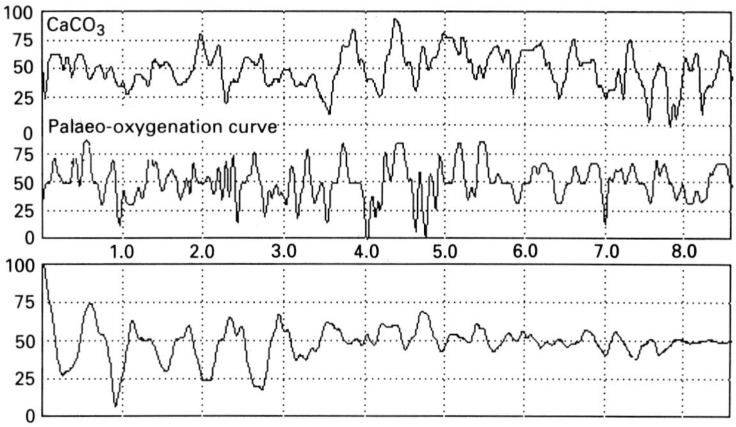

Fig. 9. Cross-correlation function of palaeo-oxygenation and calcium carbonate from the 'Amadeus Segment' of the Piobbico core. Note that the maximum value occurs at time-lag zero indicating that the two functions contain all the frequencies with the same phase. From Ripepe in Premoli Silva *et al.* (1989b).

ranges from a relatively well-oxygenated environment to medium to poorly oxygenated conditions intermittently replaced by anoxic conditions.

CONCLUSIONS

Based on a quantitative analysis of trace-fossil distribution, a palaeo-oxygenation curve was constructed for the late Albian 'Amadeus Segment' of the Piobbico core. In this interval, sedimentation mainly occurred in dysaerobic bottom waters. A trend in palaeo-oxygenation was, however, detected from a more oxygenated environment in the lower portion to overall less oxygenated conditions, which also alternate with real anoxia, followed by a reverse trend towards more oxygenated conditions again in the uppermost portion of the interval investigated.

Spectral analysis performed on changes in palaeo-oxygenation and density of bioturbation shows a dominant cyclicity around 100 ka and a less marked, but still important one around 41 ka correlatable with Milankovitch eccentricity (short) and obliquity cycles, respectively. Spectral evidence for precessional frequencies is weak as in previous calcium carbonate, light transmission and planktonic foraminiferal spectra, where such a weak precessional signal was obscured by errors and failed to rise above the noise level in Fourier spectra (Fischer *et al.*, 1991).

Cross-correlation of the carbonate and palaeo-oxygenation curves indicates that these two parameters are fluctuating not only with the same frequencies but also in phase. The highest density of bioturbation occurs not only within the richest car-

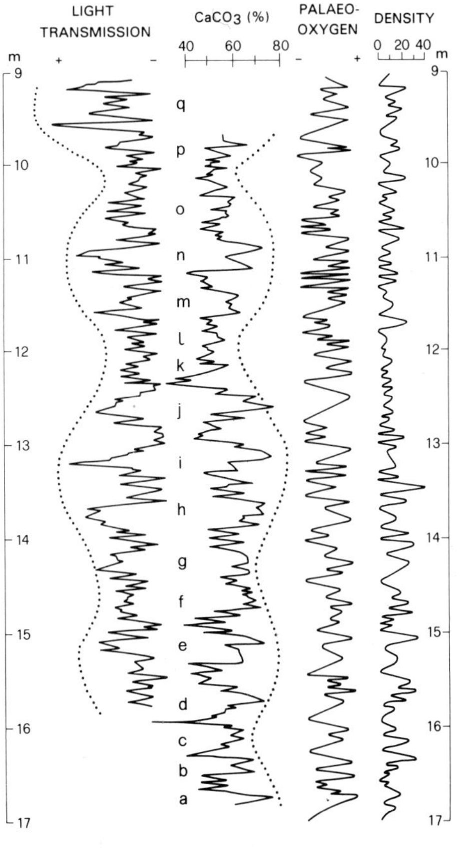

Fig. 10. Plots of the light transmission, carbonate content, palaeo-oxygenation and bioturbation density curves of the 'Amadeus Segment' of the Piobbico core. Identified cycles and bundles in light transmission and carbonate content are reported with letters; dotted lines mark the 400-ka envelopes. From Herbert & Fischer (1986).

bonate layers, but also in the highest amplitude part of the 400-ka eccentricity envelope, whereas the lowest density is recorded in the lows of the envelope. This suggests that bioturbation reflects real climatic changes driven by orbital forcing, which resulted in fluctuations in current intensity and oxygenation at the sea floor. That is, the increased seasonality, leading to a better stirring of the oceans, is responsible for both the increase in carbonate productivity and the enhancement of bottom-water ventilation. The resulting sediments are intensively bioturbated whitish limestones. In contrast, during times of weak seasonality carbonate productivity was lower and bottom waters were less oxygenated, as represented by the greyish marls showing very slight or no bioturbation. Accordingly, anoxic episodes might be cyclic and controlled by precession modulated by eccentricity (both short and long) cycles with a discrete influence of obliquity at least at the latitude of the Piobbico core.

Previous interpretations of black shales as the expression of stagnation, and limestones of increased circulation, seem therefore to be substantially correct.

ACKNOWLEDGEMENTS

The authors would like to thank Poppe de Boer for inviting them to contribute to this volume. Their warm thanks go to their former students, Elia Conti and Nadia Schaweker, who collected part of the data, and to Maurizio Ripepe for continuing support in applying his Stratabase program. They also would like to thank David Bottjer, Graham Weedon and Poppe de Boer for reviewing the present paper.

This is a contribution to the CRER Program, WG 3 – Cyclostratigraphy. The work was supported by MURST 40% to I.P.S.

REFERENCES

Arthur, M.A. & Premoli Silva, I. (1982) Development of widespread organic carbon-rich strata in the Mediterranean Tethys. In: *Nature and Origin of Cretaceous Carbon-rich Facies* (Eds Schlanger, S.O. & Cita, M.B.). Academic Press, London, pp. 7–54.

Arthur, M.A., Dean, W.E., Bottjer, D. & Scholle, P.A. (1984) Rhythmic bedding in Mesozoic–Cenozoic pelagic carbonate sequences: the primary and diagenetic origin of Milankovitch-like cycles. In: *Milankovitch and Climate*, Part 1 (Eds Berger, A., Imbrie, J., Hays, J., Kukla, G. & Saltzman, B.). Reidel Publ. Co., Dordrecht, pp. 191–222.

Arthur, M.A., Jenkyns, H.C., Brumsack, H.J. & Schlanger, S.O. (1990) Stratigraphy, geochemistry, and paleoceanography of organic carbon-rich Cretaceous sequences. In: *Cretaceous Resources, Events and Rhythms* (Eds Ginsburg, R.N. & Beaudoin, B.). Kluwer Academic Publishers, Dordrecht, pp. 75–119.

Baccelle, L. & Bosellini, A. (1965) Diagrammi di stima visiva della composizione percentuale nelle rocce sedimentarie. *Annali Univ. Ferrara* IX, **1/3**, 59–62.

Bréhéret, J.G., Caron, M. & Delamette, M. (1986) Niveaux riches en matérie organique dans l'Albien vocontien; quelques caractères du paléoenvironment; essai d'interprétation génétique. *Bur. Rech. Géol. Min.* **110**, 141–191.

Bromley, R.G. (1990) *Trace Fossils. Biology and Taphonomy*. Unwin-Hyman, London.

Bromley, R.G. & Ekdale, A.A. (1984) *Chondrites*: a trace fossil indicator of anoxia in sediments. *Science* **224**, 872–874.

Caron, M. (1985) Cretaceous planktonic foraminifera. In: *Plankton Stratigraphy* (Eds Bolli, H.M., Saunders, J.B. & Perch-Nielsen, K.). Cambridge University Press, Cambridge, pp. 17–86.

Coccioni, R. & Galeotti, S. (1991) Orbitally induced cycles in benthonic foraminiferal assemblage distribution from the Aptian–Albian organic-rich Scisti a Fucoidi (Central Italy). In: *Abstracts, Meeting IGCP 262 – Pelagic Facies, Grenoble, May 24–26, 1991*, pp. 1–2.

Coccioni, R., Nesci, O., Tramontana, M., Wezel, C.F. & Moretti, E. (1987) Descrizione di un livello guida 'Radiolaritico-Bituminoso-Ittiolitico' alla base delle Marne a Fucoidi nell'Appennino Umbro-Marchigiano. *Bull. Soc. Geol. Ital.* **106**, 183–192.

Coccioni, R., Franchi, R., Nesci, O., Wezel, C.F., Battistini, F. & Pallecchi, P. (1989) Stratigraphy and mineralogy of the Selli Level (early Aptian) at the base of the Marne a Fucoidi in the Umbrian–Marchean Apennines (Italy). In: *Cretaceous of the Western Tethys* (Ed. Wiedmann, J.). Proc. 3rd Int. Symp. Tubingen 1987. E. Schweizerb. Verlagsbuch., Stuttgart, pp. 563–584.

Coccioni, R., Erba, E. & Premoli Silva, I. (1992) Barremian–Aptian calcareous plankton biostratigraphy from the Gorgo Cerbara section (Marche, central Italy) and implications for plankton evolution. *Cret. Res.* **13**, 517–537.

De Boer, P.L. (1982) Cyclicity and storage of organic matter in middle Cretaceous pelagic sediments. In: *Cyclic and Event Stratification* (Eds Einsele, G. & Seilacher, A.). Springer Verlag, New York, pp. 456–474.

De Boer, P.L. (1983) Aspects of Middle Cretaceous pelagic sedimentation in S. Europe. *Geol. Ultraiectina* **31**.

De Boer, P.L. & Wonders, A.A.H. (1984) Astronomically induced rhythmic bedding in Cretaceous pelagic sediments near Moria (Italy). In: *Milankovitch and Climate*, Part 1 (Eds Berger, A., Imbrie, J., Hays, J., Kukla, G. & Saltzman, B.). Reidel Publ. Co., Dordrecht, pp. 177–190.

Delamette, M., Caron, M. & Bréhéret, J.G. (1986) Essai d'interprétation génétique des faciès euxiniques de l'Eo-Albien du Bassin Vocontien (SE France) sur la base de données macro- et microfauniques. *C.R. Acad.*

Sci. Paris **302**, 1085–1090.
DROSER, M.L. & BOTTJER, D.J. (1986) A semiquantitative field classification of ichnofabric. *J. Sedim. Petrol.* **56**, 558–559.
DROSER, M.L. & BOTTJER, D.J. (1991) Trace fossils and ichnofabric in Leg 119 cores. In *Proc. Ocean Drilling Program, Sci. Results, 119* (Eds Barron, J., Larsen, B. *et al.*). College Station, TX (Ocean Drilling Program), pp. 635–641.
EKDALE, A.A. (1985) Paleoecology of marine endobenthos. *Palaeogeogr. Palaeoclimatol. Palaeoecol.* **50**, 63–81.
EKDALE, A.A., BROMLEY, R.G. & PEMBERTON, S.G. (1984) *Ichnology. The Use of Trace Fossils in Sedimentology and Stratigraphy*. Soc. Econ. Paleont. Mineral. Short Course 15.
ERBA, E. (1986) *I nannofossili calcarei nell'Aptiano-Albiano (Cretacico inferiore): biostratigrafia, paleoceanografia e diagenesi degli Scisti a Fucoidi del Pozzo Piobbico (Marche)*. PhD dissertation, Univ. Milano.
ERBA, E. (1988) Aptian–Albian calcareous nannofossil biostratigraphy of the Scisti a Fucoidi cored at Piobbico (central Italy). *Riv. Ital. Paleont. Strat.* **94**, 249–284.
ERBA, E. (1992) Calcareous nannofossil distribution in pelagic rhythmic sediments (Aptian–Albian Piobbico core, central Italy). *Riv. Ital. Paleont. Strat.* **97**, 455–484.
ERBA, E., PREMOLI SILVA, I., PRATT, L.M. & TORNAGHI, M.E. (1989) Aptian–Albian Black Shales from 'Scisti a Fucoidi' of Piobbico Core (Marche, Italy). In: *Abstract 28th Int. Geol. Congr.*, Vol. 1, p. 455.
FISCHER, A.G. & HERBERT, T.D. (1988) Stratification rhythms: Italo-American studies in the Umbrian facies. *Mem. Soc. Geol. Ital.* **31**, 45–51.
FISCHER, A.G., HERBERT, T.D. & PREMOLI SILVA, I. (1985) Carbonate bedding cycles in Cretaceous pelagic and hemipelagic sequences. In: *Fine-grained Deposits and Biofacies of the Cretaceous Western Interior Seaway: Evidence for Cyclic Sedimentary Processes* (Eds Pratt, L.M., Kauffman, E.G. & Zelt, F.B.). Soc. Econ. Paleont. Mineral. Fieldtrip Guidebook 4, pp. 1–10.
FISCHER, A.G., DE BOER, P.L. & PREMOLI SILVA, I. (1990) Cyclostratigraphy. In: *Cretaceous Resources, Events and Rhythms* (Eds Ginsburg, R.N. & Beaudoin, B.). Kluwer Academic Publishers, Dordrecht, pp. 139–172.
FISCHER, A.G., HERBERT, T.D., NAPOLEONE, G., PREMOLI SILVA, I. & RIPEPE, M. (1991) Albian pelagic rhythms (Piobbico core). *J. Sedim. Petrol.* **61**, 1164–1172.
HERBERT, T.D. (1987) *Eccentricity and precessional orbital periodicities in a mid-Cretaceous deep-sea sequence: identification and application to quantitative paleoclimatology*. PhD dissertation, University of Princeton.
HERBERT, T.D. & FISCHER, A.G. (1986) Milankovitch climatic origin of mid-Cretaceous black shale rhythms in central Italy. *Nature* **321**, 739–743.
HERBERT, T.D., STALLARD, R.F. & FISCHER, A.G. (1986) Anoxic events, productivity rhythms and the orbital signature in a mid-Cretaceous deep sea sequence from central Italy. *Paleoceanography* **1**, 495–506.
NAPOLEONE, G. & RIPEPE, M. (1989) Cyclic geomagnetic changes in Mid-Cretaceous rhythmites, Italy. *Terra Nova* **1**, 437–442.
PRATT, L.M. & KING, J.D. (1986) Variable marine productivity and high eolian input recorded by rhythmic black shales in Mid-Cretaceous pelagic deposits from central Italy. *Paleoceanography* **1**, 507–522.
PREMOLI SILVA, I., ERBA, E. & TORNAGHI, M.E. (1989a) Paleoenvironmental signals and changes in surface fertility in Mid-Cretaceous C_{org}-rich pelagic facies of the Fucoid Marls (central Italy). *Geobios, Mém. sp.* **11**, 225–236.
PREMOLI SILVA, I., RIPEPE, M. & TORNAGHI, M.E. (1989b) Planktonic foraminiferal distribution record productivity cycles: evidence from the Aptian–Albian Piobbico core (central Italy). *Terra Nova* **1**, 443–448.
RIPEPE, M. (1988) Stratabase: a stratigraphical database and processing program for microcomputers. *Computers Geosciences* **14**, 369–375.
RIPEPE, M. (1992) Risposta del sistema geologico alle variazioni orbitali. *Mem. Soc. Geol. Ital.* **45**, 687–695.
ROCC (Research on Cretaceous cycles) GROUP (1986) Rhythmic bedding in Upper Cretaceous pelagic carbonate sequences: varying response to orbital forcing. *Geology* **14**, 153–156.
SANDBERG, C.A. & GUTSCHICK, R.C. (1984) Distribution, microfauna and source-rock potential of Mississippian Della Phosphatic member of Woodman Formation and equivalents, Utah and adjacent states. In: *Hydrocarbon Source Rocks of the Greater Rocky Mountain Region* (Eds Woodward, J., Meissner, F.F. & Clayton, J.L.). Rocky Mountain Assoc. Geologists, Denver, pp. 135–178.
SAVRDA, C.E. & BOTTJER, D.J. (1986) Trace-fossil model for reconstruction of paleo-oxygenation in bottom waters. *Geology* **14**, 3–6.
SAVRDA, C.E. & BOTTJER, D.J. (1987) Trace fossils as indicators of bottom-water redox conditions in ancient marine environments. In: *New Concepts in the Use of Biogenic Sedimentary Structures for Paleoenvironmental Interpretation* (Ed. Bottjer, D.J.). Soc. Econ. Paleont. Mineral., Pacific Section Guidebook 52, pp. 3–26.
SAVRDA, C.E. & BOTTJER, D.J. (1989) Trace-fossil model for reconstructing oxygenation histories of ancient marine bottom waters: application to Upper Cretaceous Niobrara Formation, Colorado. *Palaeogeogr. Palaeoclimatol. Palaeoecol.* **74**, 49–74.
SCHLANGER, S.O. & CITA, M.B. (Eds) (1982) *Nature and Origin of Cretaceous Carbon-rich Facies*. Academic Press, London.
SCHLANGER, S.O. & JENKYNS, H.C. (1976) Cretaceous oceanic anoxic sediments: causes and consequences. *Geol. Mijnbouw* **55**, 179–184.
SCHWARZACHER, W. & FISCHER, A.G. (1982) Limestone–shale bedding and perturbations in the Earth's orbit. In: *Cyclic and Event Stratification* (Eds Einsele, G. & Seilacher, A.). Springer Verlag, New York, pp. 72–95.
THIERSTEIN, H.R. & ROTH, P.H. (1991) Stable isotopic and carbonate cyclicity in Lower Cretaceous deep-sea sediments: Dominance of diagenetic effects. *Mar. Geol.* **97**, 1–34.
TORNAGHI, M.E., PREMOLI SILVA, I. & RIPEPE, M. (1989) Lithostratigraphy and planktonic foraminiferal biostratigraphy of the Aptian–Albian 'Scisti a Fucoidi' in the Piobbico core, Marche, Italy: background for cyclostratigraphy. *Riv. Ital Paleont. Strat.* **95**, 223–264.
WEZEL, C.F. (1985) Facies anossiche ed episodi geotettonici globali. *Giorn. Geologia, S. 3* **47**, 281–286.

Spec. Publs Int. Ass. Sediment. (1994) **19**, 227–242

Guilds, cycles and episodic vertical aggradation of a reef (late Barremian to early Aptian, Dinaric carbonate platform, Slovenia)

J. GRÖTSCH*

Institut für Paläontologie, Universität Erlangen-Nürnberg, Loewenichstr. 28, D-8520 Erlangen, Germany

ABSTRACT

A new approach to high-resolution stratigraphy of massive reef limestones is presented. On the northern edge of the Dinaric carbonate platform next to the Slovenian trough (Kanalski Vrh, Slovenia), an 'Urgonian' type reef of late Barremian to early Aptian age has been investigated. The section comprises a reef flat to back reef environment. Guilds of reef organisms (constructor, binder, baffler, dweller, destroyer) in 267 samples spaced at equal distances have been quantified together with sediment parameters (primary content of biogenic aragonite, facies). The first component of a principal component analysis of the data set, which is controlled by the process of construction in the reef, reveals cyclic changes in the variables. Cycles are composed of an alternation of baffler and detrital framework facies. The shallow subtidal baffler facies is characterized by fine-grained packstones or wackestones with *Tubiphytes* as the main baffling organism. Other components are miliolid foraminifera, debris of molluscs, bryozoans, sponges, echinoderms and dasyclads. In contrast, the intertidal detrital, rarely inplace framework facies contains grainstones, rudstones and framestones with reef-building organisms (scleractinian corals, hydrozoans, calcisponges), molluscs, agglutinating and miliolid foraminifera, algae and, less important, bryozoans and echinoderms.

Even though rhythmic sedimentation cannot be traced directly in the field, cycles do show up in the spectral analysis. The periodogram of the time series shows that the reef has been built up by the stacking of episodic, vertical, aggradational cycles with an average thickness of about 3.2 m, representing the 20-ka period of the Milankovitch band. Cyclostratigraphic results suggest that the 133.5-m section was built up during roughly 800 ka. The high-frequency, low-amplitude Milankovitch cycles are superimposed on a third-order rise of sea level which took place during the latest Barremian to early Aptian.

INTRODUCTION

The evolution of reefs in the Earth's history has been a periodic process (Talent, 1988), interrupted by pronounced breaks which are in many instances linked to global drowning events of carbonate platforms (Schlager, 1981). Reefs are very sensitive indicators of environmental (Hallock & Schlager, 1986), climatic and sea-level changes (MacIntyre, 1988). Thus, they can be used as palaeoclimatic recorders. In order to achieve this, differentiation should be made between times of long-term, vertical aggradation of reefs and periods of erosion and/or drowning which are intercalated in the growth record or mark the final stage of platform evolution.

In this paper, a period of worldwide vertical aggradation of platforms from the latest Barremian to early Aptian is investigated (see also Bebout & Loucks, 1977; Groupe Français du Crétacé, 1979; Masse & Philip, 1981; Moldovanyi & Lohmann, 1984; Scott, 1984; Velic *et al.*, 1987). This time interval is equivalent to the so-called 'Urgonian Facies'. In order to reduce environmental influences (e.g. detrital input) to a minimum, an example of an open ocean carbonate platform reef is used as a 'dip stick' to record relative changes of sea level. Up to now, one of the most important problems in reef

* Present address: Shell Research B.V., Volmerlaan 6, 2288 GD Rijswijk, The Netherlands.

analysis is the evaluation of time. As biostratigraphy in such an ecologically highly diversified environment gives only very limited time resolution, a new approach to high-resolution stratigraphy is tested. From inner platform settings, examples of cyclic changes in sedimentation have been known since the early works of Sander (1936), Schwarzacher (1948, 1954) and Fischer (1964). For reef ecosystems no such analysis is available yet.

Reef ecosystems are extremely sensitive to environmental changes, such as those caused by short-term fluctuations of relative sea level. Following such changes, there should also be a change in the community structure of a reef. In order to trace these variations, the guild concept of Fagerstrom (1987) is applied. A guild is defined as a group of species that exploits the same class of environmental resources in a similar way and overlap significantly in their niche requirements, without regard to taxonomic position (Fagerstrom, 1987, p. 193). Therefore, the proportions of the different guilds in a reef (constructor, binder, baffler, dweller, destroyer) should vary according to changes of external parameters. An additional indicator, which can only be used qualitatively, is the diagenetic history of the sediment, especially where it preserves early meteoric cements suggesting short-term exposure.

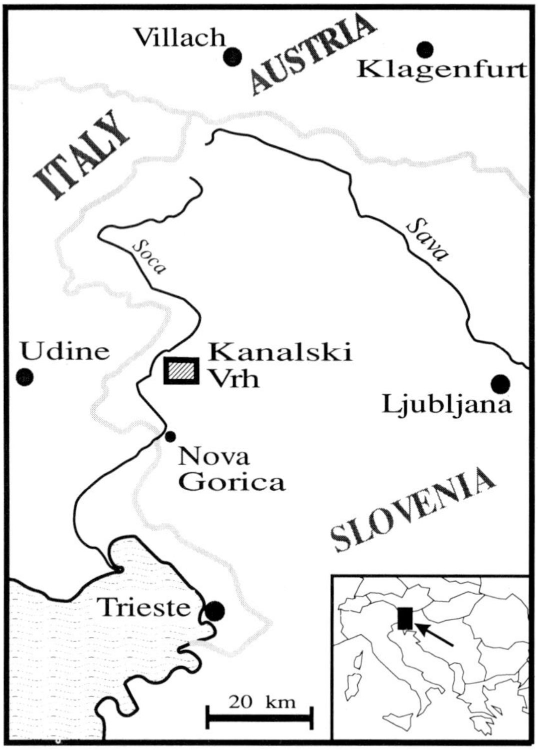

Fig. 1. Location map of Slovenia with the investigated area (hatched) just E of the small village of Kanalski Vrh in NW Slovenia.

GEOLOGICAL SETTING

The area investigated is located on the NW edge of the Dinaric carbonate platform in Slovenia, 1 km E of the small village of Kanalski Vrh (Fig. 1). Detailed mapping of the region has been carried out by the Geoloski Zavod of Slovenia during the campaign for the Geological Map of Yugoslavia 1:100 000 by Stanko Buser and his colleagues. In the tectonic subdivision of the Dinarides proposed by Herak (1986), the platform is part of the Dinaricum. It is bounded to the N by the Slovenian Trough and to the S by the autochthonous Adriatic Platform (Buser, 1987). The units are part of the Trnovo Nappe, which has been overthrust towards the SW (Placer, 1981).

Dinaric carbonate platform

The platform is part of a wide area of Mesozoic, open ocean carbonate platforms striking in a NW–SE direction from NE Italy along the Adriatic Sea through Apulia, the former Yugoslavia and Albania down to the western part of Greece. It is subdivided by large-scale intra-platform basins (Belluno Trough, Slovenian Trough, Cukali Trough, Ionian Trough, Pindos Trough). In the now compressed stage, the platforms and troughs cover an area of about 1400 × 350 km. The carbonate development of the Dinaric platform began in the Late Permian and ended in the Eocene. It was interrupted by several events, e.g. in the earliest Jurassic (partial drowning), in the Late Jurassic (karstification), in the late Albian (karstification and partial drowning), at the end of the Cenomanian (partial drowning) and at the Cretaceous/Tertiary boundary (bauxite formation). Beginning in the Campanian the platform broke up and was progressively covered by southward migrating flysch sedimentation (Buser, 1987).

Stratigraphy

The abundant occurrence of *Palorbitolina lenticularis* and *Palaeodictyoconus arabicus* allows new

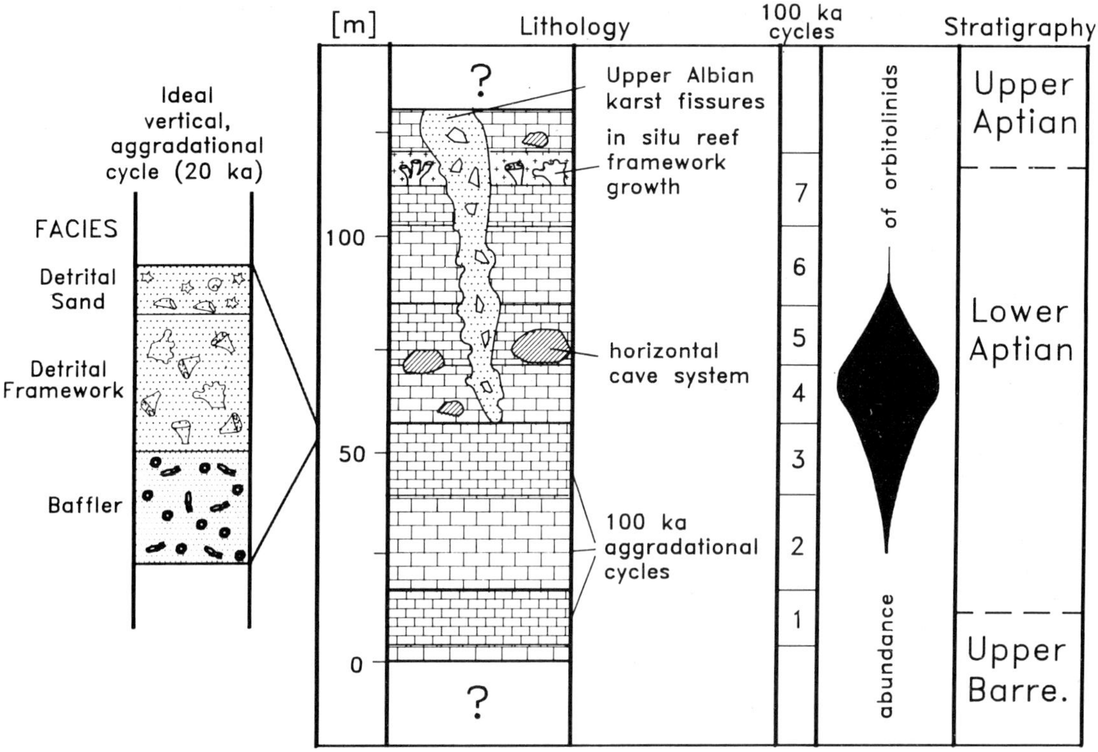

Fig. 2. Log of the Kanalski Vrh II section with thickness (m), lithology, cyclostratigraphic subdivision, abundance of orbitolinid foraminifera (comprising *Palorbitolina lenticularis* and *Paleodictyoconus arabicus*) and chronostratigraphic correlation. Note that the platform was karstified in the late Albian. Vertical fissures and horizontal, biconvex cave systems were filled during the *Rotalipora appenninica* Zone of the late Albian by shallow-water reef carbonate and pelagic wackestones with abundant planktonic Foraminifera. On the left, a detail of an ideal, vertical aggradational cycle of the reef is illustrated, showing the associated changes in facies.

biostratigraphic dating of the Kanalski Vrh II section (Fig. 2). It comprises a time interval from the latest Barremian to the early Aptian (R. Schroeder, personal communication, 1989). Thus, the reef growth stage is time equivalent to the 'Urgonian Facies' known from many locations in the Tethys and the Gulf of Mexico. On the Dinaric and Adriatic Platform, this time interval is equivalent to the lithostratigraphic unit called the 'Lower *Orbitolina* Limestone'.

Lithology

The Kanalski Vrh II section comprises a massive, reefal limestone, of which 133.5 m have been investigated in detail (Fig. 2). In the coarse-grained, highly fossiliferous sediment, mostly packstones, grainstones and rudstones occur. Signs of bedding are rare. Only occasionally can flat encrusting reef-building organisms such as corals or hydrozoans be seen *in situ*. Between 110 and 120 m from the base of the section they become noticeably more abundant (Fig. 2). The transition from coarse-grained reefal sediments into lagoonal wackestones can be studied at an outcrop 750 m towards the SW of the studied section (see map in Grötsch, 1991, Fig. 3). It shows a cyclic alternation of the bioclastic and mud-dominated facies. Further to the SW only lagoonal, intraplatform carbonates crop out.

From the surrounding outcrops, irregular vertical karst fissures are known which are filled with eroded underlying reefal sediment, late Albian shallow-water carbonate and pelagic wackestones from the *Rotalipora appenninica* Zone (Grötsch, 1991, pp. 45ff.). In addition, cross-bedded, pelagic wackestones are found in biconvex, horizontal cave systems (Fig. 2).

CHARACTERISTICS OF THE REEF

Quantitative application of the guild concept (Fagerstrom, 1987) to the 'Urgonian' reef on the northern edge of the Dinaric Platform reveals a rhythmic change in the occurrence of baffling organisms on the one hand and constructing and binding organisms on the other. Although the average percentage of the constructor guild throughout the section is high (16%), it is mostly detrital components that form the reef. Knowing this, it is important to analyse not only initial carbonate production, but also biological and physical degradation on the reef surface (e.g. by bioerosion, wave action, episodic storms) as well as the redistribution of the resulting sediment inside and outside the reef (Hubbard *et al.*, 1990, p. 336). Thus, in fossil reefs one is more likely to encounter evidence of secondary processes of redistribution than the original inplace frame builders. Hence, it is necessary to recognize not only the primary producers but also the sediment forming processes. In agreement with Hubbard *et al.* (1990), the associated facies dominated by reworked constructors is therefore called 'detrital framework facies' in the remainder of this paper.

Guild structure of the reef

The organisms in the reef are classified not only according to common taxonomic concepts but also by applying the guild structures proposed by Fagerstrom (1987, pp. 193ff., 421–432) for the Early Cretaceous. Although there are some differences, especially concerning baffling and constructing organisms (see Fagerstrom, 1987, Fig. 14.9), the basic idea remains the same. Assignment to the individual guilds in the case of Kanalski Vrh II is made by using taxonomic criteria, growth form and distribution in the different facies (Table 1). Therefore, it is possible that one species or genus is a member of more than one guild. Membership and abundance of an organism in a guild varies through time. It is inferred that the qualitative and quantitative composition of guilds is influenced by external parameters such as changes in relative sea level, ecological factors, and preservation of the original biotope in the actual sediment (the 'diagenetic sieve' of Dullo, 1983).

According to Fagerstrom (1987, p. 201) organisms of the *constructor* guild are characterized by their colonial habit, rigid skeleton, rapid vertical growth and high volumetric share in the internal sediment of the reef (Table 1). As in many other fossil examples, the proportion of *in situ* preserved reef builders is low. Nevertheless, together with the binders, they are the most abundant guild in the Kanalski Vrh II section (16%, averaged over all samples), followed by the dwelling organisms (11%).

In many reefs a considerable overlap of function between constructor and *binder* guilds can be observed. This overlap is the reason that the two guilds are grouped together as one for later statistical analysis. Binders (Table 1) are the group most often preserved in life position as they are extremely wave resistant (Fagerstrom, 1987, pp. 202, 210).

Table 1. Guild structure of the 'Urgonian' reef at Kanalski Vrh II. The organisms or groups of organisms in each column are listed in decreasing order of their relative abundance. Assignment to individual guilds is made by using taxonomic criteria, growth form and distribution in the different facies. Note that certain groups or genera can be part of more than one guild. Organisms listed in parentheses occur only occasionally

Constructor	Binder	Baffler	Destroyer	Dweller
Scleractinia	Problematica	Problematicum	Echinoidea	Bivalvia
Hydrozoa	*Lithocodium*	*Tubiphytes*	'Borers'	Hippuritacea
Actinarea	*Baccinella*	Dasycladaceae	Algae	Requeiniidae
Calcispongea	Rhodophyta	Chlorophyta	Fungi	Monopleuridae
Peronidella	*Ethelia*	Rhodophyta	Porifera	Radiolitidae
Actinostromaria	*Lithothamnium*	Bryozoa	Bivalvia	Caprinidae
Barroisia	*Solenopora*		Gastropoda	Gastropoda
Tabulozoa	Foraminiferida		Cirripedia	Nerinea
Sclerospongia	*Placopsilina*		Fish?	Bivalvia
(Bryozoa)	(Chlorophyta)		*Lithocodium*?	Foraminiferida
(Serpulida)	(Hydrozoa)			Miliolina
	(Tabulozoa)			Textulariina
	(Scleractinia)			(Brachiopoda)

According to Fagerstrom (1987, pp. 199ff.) there are two groups of *baffling* organisms:
1 bafflers that reduce wave velocities along the platform edge, thus protecting reefs from physical destruction by wave action;
2 bafflers that reduce current velocities, inducing deposition of reef-derived sediment.
Here only the latter group will be regarded as the actual baffling one, whereas the first group is added to the constructor guild. This is necessary not only because of the limited area used for quantification of the guilds ($16\,cm^2$ per 0.5 m of section), but also because their environments of maximum abundance are considerably different in respect of water energy and therefore position in the reef. None of the members of the baffler guild has a strong, heavy skeleton; instead they are rather flexible, only partly mineralized, fragile and sometimes have an organic skeleton with low preservation potential. They are mostly colonial. In the example investigated, *Tubiphytes* (Problematicum) is the most important preserved member of that guild.

The *destroyer* guild comprises, taxonomically and morphologically, a wide variety of organisms that includes bivalves, gastropods, echinoderms, sponges, fish, polychaetes, balanides, algae, bacteria and fungi. In fossil examples a quantitative analysis of this guild is extremely difficult, as most of its members are preserved only as traces or else they cannot be differentiated from members of other guilds (e.g. bivalves, gastropods). Therefore only echinoderms are quantified as representatives of the destroyer guild, as their preservation potential is very good. Additionally, in the environment investigated, the echinoderms belong almost completely to this guild and are easy to quantify. Reducing the destroyer guild to one member is a simplification but the correlation coefficients and component loadings of the principal component analysis (PCA) show good results, justifying this procedure (Table 2).

The *dweller* guild is formed by organisms not belonging to any of the other guilds (Table 1). In contrast to Scott (1981), rudists are not considered to belong to the constructors but rather to

Table 2. Results of the principal component analysis (PCA): correlation matrix of the quantified variables, eigenvalues, component loadings, and the variance explained by them. Component 1 represents the process of construction (framework building), whereas component 2 is linked to erosion of the reef. Components 3 and 4 represent the processes of dwelling and baffling, respectively. The latter two combined explain only 20% of the total variance, making them less suitable for quantitative analysis and usage as an index parameter

Correlation matrix						
	FAC	CONBIN	DWE	DES	BAF	ARA
FAC	1.000					
CONBIN	0.718	1.000				
DWE	0.183	−0.139	1.000			
DES	−0.090	−0.395	0.200	1.000		
BAF	−0.599	−0.485	−0.290	−0.017	1.000	
ARA	0.703	0.758	0.358	−0.294	−0.542	1.000
Latent roots (eigenvalues)						
	1	2	3	4	5	6
	3.035	1.404	0.778	0.444	0.244	0.094
Component loadings						
	1	2	3	4	5	6
	Construc	Erosion				
FAC	0.872	0.079	−0.228	0.199	0.374	−0.039
CONBIN	0.864	−0.375	−0.160	0.130	−0.159	0.213
DWE	0.254	0.792	0.539	0.047	0.044	0.117
DES	−0.306	0.724	−0.557	0.225	−0.141	0.004
BAF	−0.738	−0.322	0.229	0.546	0.038	0.016
ARA	0.908	0.041	0.218	0.191	−0.237	−0.183
Variance explained by components						
	1	2	3	4	5	6
	3.035	1.404	0.778	0.444	0.244	0.094
Percentage of total variance explained						
	1	2	3	4	5	6
	50.579	23.406	12.971	7.399	4.074	1.570

the dwellers (see Skelton & Gili, 1990). They are, next to the gastropods, the most important group in this guild, but they are not able to form any wave-resistant, colonial or framework structures, as for example scleractinian corals do. They have only a limited vertical growth potential, even though their contribution to carbonate production is high. Their distribution is always biostromal, and only occasionally can they be found in growth position.

Parameterization of reef characteristics

In order to evaluate production of carbonate on a platform, it is necessary to choose an environment and a time interval with a high preservation potential. The latter is strongly dependent on relative sea level and therefore on accommodation space.

In the following paragraphs the method of quantitative facies analysis will be explained. The aim is to obtain a relative index for in-place framework production (RIFP), which can be a useful device in the classification of fossil reefs and a very sensitive parameter in high-resolution stratigraphy.

The continuously exposed section of 133.5 m was sampled with an equal-distance spacing of 0.5 m. From each sample a peel with an area of 16 cm^2 was prepared. Additionally, in some instances thin sections for diagenetic or special palaeontological purposes were made. In order to set up a semi-quantitative index that will trace carbonate production in the fossil record per time unit (Δt) on a platform, several different variables characteristic of the environment have to be considered. The variables used are as follows.

1 The relative abundance of reef guilds (constructor, CON; binder, BIN; baffler, BAF; dweller, DWE; destroyer, DES).

2 The primary content of biogenic aragonite (ARA), comprising completely or partly aragonitic and now recrystallized organisms. Periplatform sediments in the Tongue of the Ocean (Bahama Platform) have revealed a different aragonite content for glacials and interglacials. Turbidite frequency and aragonite content of sediment produced during interglacials is higher, which can be explained by diagenetic alteration, dissolution cycles or changing primary production of aragonite (Droxler *et al.*, 1983; Droxler & Schlager, 1985; Reymer *et al.*, 1988; Haak & Schlager, 1989).

3 As an additional parameter, facies classification, according to Dunham (1962). In order to use this qualitative variable for later statistics, it was transformed into a facies number (FAC), which increases with increasing sparitic cement, grain size and therefore inferred wave energy. The following arbitrary values were assigned to the facies classes: mudstone = 0, wackestone = 2, packstone = 6, grainstone = 10, rudstone = 14 and (detrital) framestone = 15, allowing all possible transitions from 0 to 15. Even though this assumption of an essentially linear progression is not valid for certain boundstones, this is not very important in the present case, as most samples with facies number 15 belong to a detrital and not to an in-place framework facies (see Hubbard *et al.*, 1990 and explanation above).

After determining the guild members and the facies (FAC) for each of the 267 samples under the microscope, the other six variables (CON, BIN, BAF, DWE, DES, ARA) were estimated from framed acetate peels projected on a screen using estimation charts from Flügel (1982, fig. 40, see also for discussion of errors). As in many instances a clear distinction between constructor and binder was not possible, especially because of the limited area quantified, both variables were added together to form one (CONBIN; for data see Grötsch, 1991, Table 1).

Statistics

The variables do not always fluctuate in parallel; thus it is advantageous to reduce the multivariate data set by applying principal component analysis (Jöreskog *et al.*, 1976; Guillaume, 1977). As will be shown in the following sections, in the case of Kanalski Vrh II the component loadings of the principal components allow the reconstruction of the mechanisms controlling sedimentation (see Table 2), which are baffling, constructing and destroying. Each principal component is a weighted, linear composite of the observed variables. Usage of a correlation matrix gives equal weight to each of the parameters.

The correlation coefficients of the variables in Table 2 show a positive correlation of CONBIN with ARA and FAC, whereas there is a negative relation to BAF and to a lesser extent to DES and DWE. The loadings indicate the strength of the influence of the variables on each component. The first component, which accounts for 50.6% of the total variance, shows similar results, as seen in the correlation coefficients. In Fig. 3 the scores of this component are plotted against the variables CONBIN, ARA,

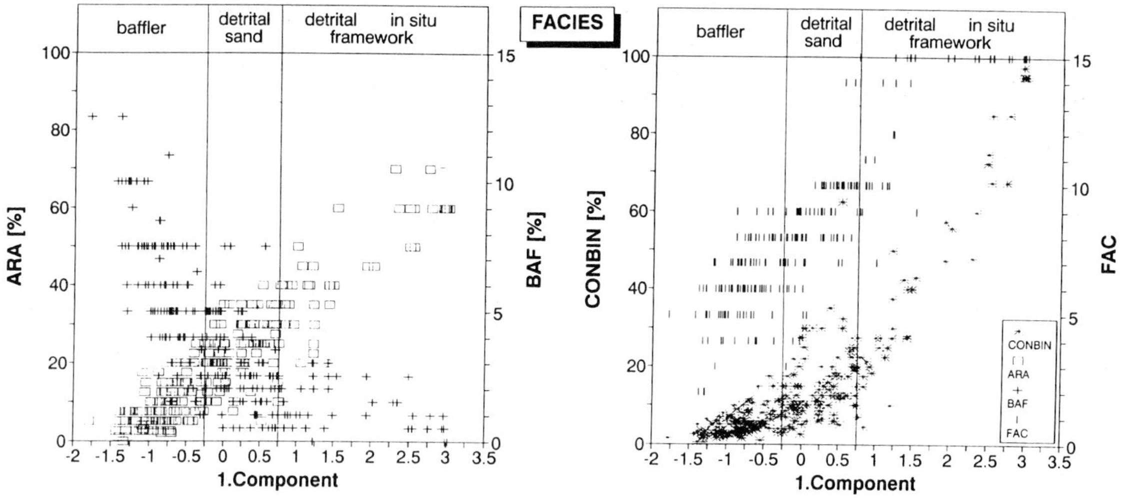

Fig. 3. Plot of first principal component (PC1) against (a) ARA (in %) and BAF (in %) and (b) CONBIN (in %) and FAC (in facies units from 1 to 15), showing respectively the exponential (CONBIN, BAF) and linear (FAC, ARA) relationships of the variables to the different reef-building processes (see legend on the lower right). The approximate facies transitions are indicated by vertical lines in the graphs.

FAC and BAF, showing the linear or slight exponential relation of the parameters in respect to component 1.

Facies and guilds

From thin-section analysis three different types of the facies can be distinguished which can also be seen in the triangle of reef sedimentation in Fig. 4. There, component 1 is plotted against component 2 using the facies number (FAC) to label the data points. The corners of the triangle are attributed to the main reef-forming processes. The three processes are well discriminated by FAC, by the distribution of the guilds and their reflection in the first two components of the PCA, and also by common facies analysis.

Baffler facies

Baffler facies, whose components 1 and 2 are always $\leqslant 0$ and whose FAC is $\leqslant 7$, is mostly composed of wackestones to packstones, rarely of floatstones. Thus, it is characterized by a higher content of muddy matrix. The important components are fine-grained detritus, *Tubiphytes*, miliolid foraminifera and fragments of molluscs, bryozoans, sponges, echinoderms and dasyclads. Because of its fossil content and the implied quiet sedimentary environment, this facies is interpreted to be of subtidal, back-reef origin (see Flügel, 1982).

It is deposited during the beginning of an episodic, vertical aggradational cycle (see Anderson *et al.*, 1984) of the platform rim, during which the reef core is keeping (or catching) up with a short-term relative sea-level rise ($\Delta RSL/\Delta t > 0$), thus protecting the slowly accreting back reef from wave energy and leaving it for a short time in a subtidal regime (Fig. 5). Therefore the baffler facies can also be called a lag facies. This condition prevails until relative sea-level rise approaches 0 ($\Delta RSL/\Delta t \approx 0$) and framework production is migrating laterally over the back reef.

Detrital or in-place framework facies

This facies has a FAC between 12 and 15, which indicates higher *in situ* carbonate production. Component 1 is always $\geqslant 0.5$, whereas component 2 is mostly negative. This facies is characterized by grainstones, rudstones and framestones and by a dominance of the constructor guild. Most of the framestones match a detrital framework, described by Hubbard *et al.* (1990) from Holocene reefs in the Caribbean, rather than boundstones after Dunham (1962). The sedimentation in the 'Urgonian' reef of Kanalski Vrh II can therefore be compared with

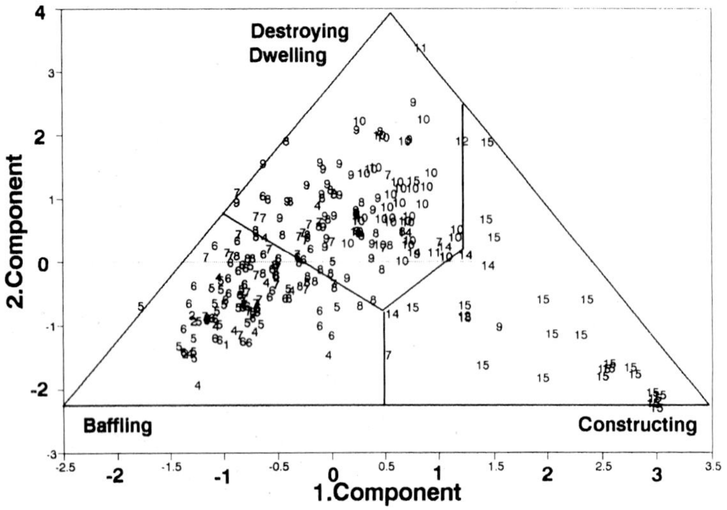

Fig. 4. Graph showing principal components 1 (constructing) and 2 (destroying); data points are labelled by facies number (FAC). The corners of this 'triangle of reef sedimentation' mark the processes of baffling, destroying/dwelling and constructing. The three sectors in the triangle represent the three major facies types of the reef flat to back reef environment (see Fig. 5).

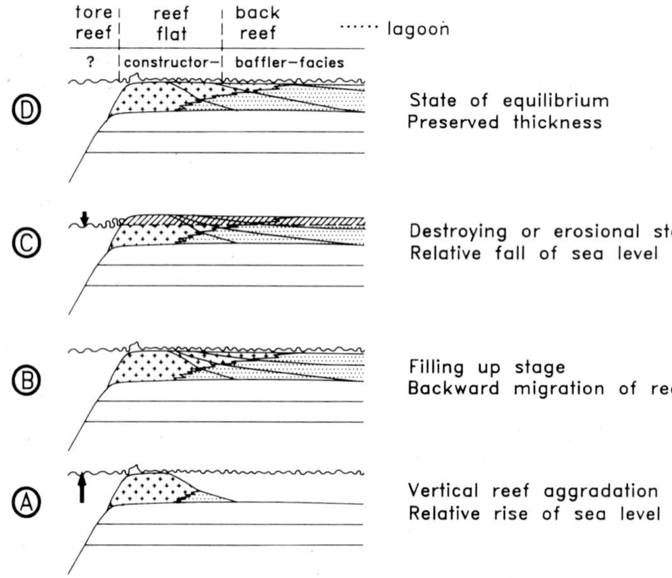

Fig. 5. Schematic two-dimensional facies model of an episodic, vertical, aggradational cycle with facies boundaries (ragged line) and time lines (curved). (A) Due to a pulse-like rise of relative sea level ($\Delta RSL/\Delta t > 0$) vertical growth of platform-rimming reefs (crosses) is initiated, whereas the baffler facies (dots) covers only a limited area in the reef. (B) After the end of sea-level rise ($\Delta RSL/\Delta t \approx 0$) the detrital or in-place framework facies migrates over former back reef sediments, and thus the reef flat area expands (filling-up stage). (C) Depending on amplitude of subsequent fall of sea level ($\Delta RSL/\Delta t < 0$), biogenic destruction and/or physical erosion by waves begins. The duration and depth of meteoric influence in the reef are major factors controlling preservation of cycles in the sediment column. (D) After the end of sea-level fall ($\Delta RSL/\Delta t \approx 0$) a short-term state of equilibrium is reached before the beginning of another aggradational cycle.

sediment found in Recent reefs, which is dominated by reef-derived detritus and which resembles more a 'garbage pile' than an 'in-place framework' (Hubbard *et al.*, 1990, p. 351).

The environment of the section investigated is not located along the platform margin, but rather represents cyclic changes from reef flat to back reef. Hence, the in-place carbonate production of this facies is not due to the actual, short-term rise of sea level, but is rather an intermediate stage of an individual cycle (Fig. 5). After the end of the short-term rise of sea level ($\Delta RSL/\Delta t \approx 0$) and resulting vertical accretion of the reef core, the detrital framework facies migrates towards the back reef due to lack of accommodation space, resulting in a lateral expansion of the reef flat. The constructors growing there have a much better preservation potential than the frame builders along the platform margin (Hubbard *et al.*, 1990).

Detrital sand facies

This facies is characterized by an FAC between 8 and 11 and a second component which is mostly positive. Component 1 can vary from -1 to $+1$. So, DWE and DES dominate, whereas CON is less important. The detrital framework facies can be replaced or followed by the detrital sand facies. The latter is composed of grainstones with strongly reworked detritus of reef dwellers (DWE), boring and destroying organisms (DES) and to a lesser extent of constructors (CON) in a sparitic cement. As component 2 of the PCA suggests, this facies is related to a different process of sediment formation. Transition to the other facies is gradual (Fig. 5). Hence, rudists and gastropods can replace frame builders due to lack of vertical accommodation space. The detrital sand facies, with a peak occurrence of DES, can also follow a detrital framework facies, terminating the cycle with biological destruction (Fig. 5).

If change in relative sea level is constant or negative over some time ($\Delta RSL/\Delta t \leqslant 0$), biological and physical destruction and/or chemical dissolution starts, as vertical accommodation space diminishes. Preservation of the destroying or erosional stage of the cycles in the sediment column is not as good as in the other two cases, depending on $\Delta RSL/\Delta t$ (Fig. 5C). Nevertheless, this stage could also be formed during a transgressive lag, when sedimentation rate is low and DES can attack subtidally exposed sediment. In that case one would expect good preservation of a bored surface at the beginning of a cycle, which is not the case.

In some instances early vadose overprinting and emersion is indicated by cements (Grötsch, 1991, plate 6/3–5). Unfortunately preservation of this diagenetic signal is scarce, probably due to physical erosion and fast destruction of early cement generations in a sediment with high porosity and permeability. In addition, it is not possible to see the depth of intrusion of vadose overprinting as is possible in the case of lagoonal, mud-dominated intraplatform cycles of the same age (Grötsch, 1991, pp. 124ff.). The shallowing-upward cycles in the Kanalski Vrh II section could be analogous to what Walker & Alberstadt (1975) described schematically as short-term reef successions, especially the stage from colonization (analogous to baffler facies) to diversification (analogous to detrital framework facies), although reef organisms are not preserved in growth position but rather as reefal detritus. Besides that, there is no indication of a 'stabilization' or 'domination' phase.

Carbonate production and sediment deposition

After differentiating facies types reflected in the rock record as well as in the PCA, we will see which processes are important for sediment production and deposition. As indicated in Fig. 3 by vertical lines, the first component is related to facies and reflects the process of construction. Figure 6 shows the simultaneous fluctuation of CONBIN and the first component, implying a high in-place framework production for high PC1 scores. The high aragonite content is very well correlated to pulse-like occurrences of the detrital framework facies (Fig. 3). Hence, rhythmic sedimentation is interpreted to be due to repeated, short-term rises of relative sea level ($\Delta RSL/\Delta t > 0$; Δt is the time interval corresponding to 0.5 m thickness, assuming a constant sedimentation rate throughout the section) and therefore to change in accommodation potential, which results in episodic, vertical aggradation of the reef. Primary production on the platform rim might therefore also be the reason for higher aragonite contents in interglacial turbidites shed from the Bahama Platform (see Droxler & Schlager, 1985).

The second important process, reflected in component 2 and explaining 23.4% of the total variance, might be due to intercalated short periods of bioerosion (by DES) or low carbonate production (by DWE). Component 2 correlates very well with

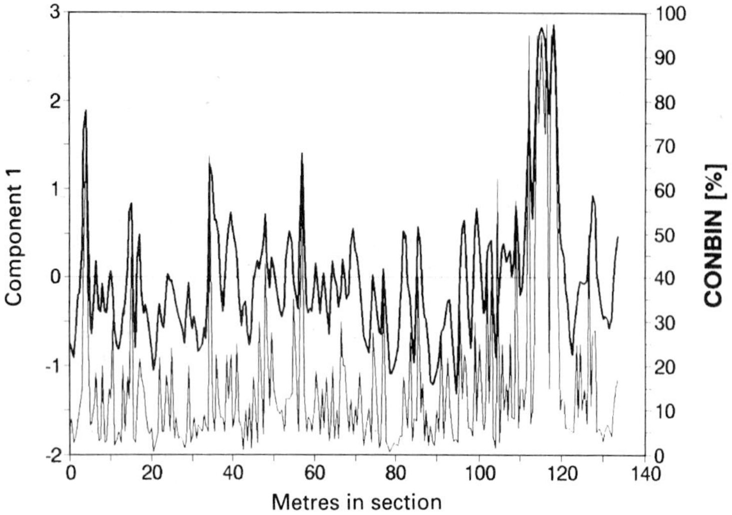

Fig. 6. Plot of principal component 1 (thick line) and the sum of CONstructor and BINder guilds (thin line) against position in section (in metres above base). PC1 has been smoothed with a (weight = 1,2,1) moving average filter. There is a good visual correlation between the two curves, which confirms the control of PC1 by detrital or in-place framework production in the reef.

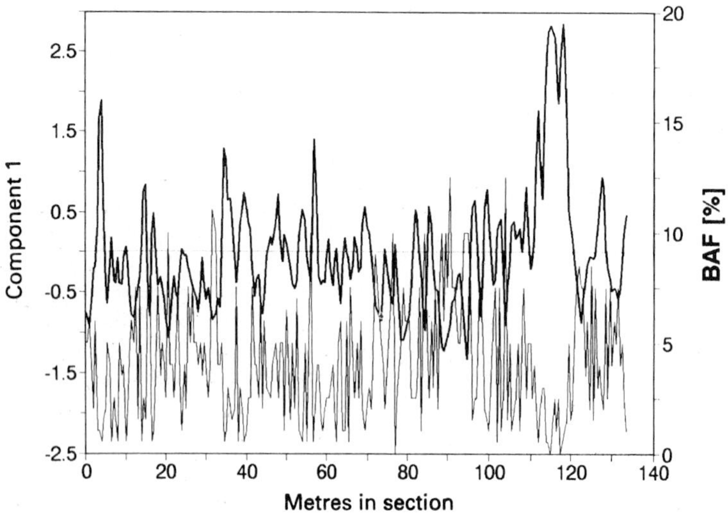

Fig. 7. Principal component 1 (thick line) plotted with the percentage of the baffler guild (thin line) against position in section. PC1 has been smoothed with a (weight = 1,2,1) moving average filter. Note that maxima of bafflers coincide with minima of PC1. Thus, periods of high detrital or in-place framework production by reef-building organisms are interrupted by a facies type with maximum appearance of bafflers. Therefore, the vertical aggradation in the reef near Kanalski Vrh II is not continuous but rather cyclic.

the sum of DES and DWE. Thus, these intervals of sedimentation might indicate lowering ($\Delta RSL/\Delta t < 0$) or still-stands of relative sea level ($\Delta RSL/\Delta t \approx 0$), suggested also for the present state of Holocene reefs (Hopley, 1986, pp. 204ff.; Pirazzoli *et al.*, 1988; Woodroffe *et al.*, 1990). If one compares the relative change of component 1 from one sample to the next ($\Delta PC1/\Delta t$) and the scores of the second component, a lag of about two samples (1 m) of the latter in respect to $\Delta PC1/\Delta t$ can be observed in many instances. Hence, the maximum occurrence of DES or DWE in many cases follows the maxima of constructing and binding organisms (CONBIN). In contrast, the minima of destroyers and dwellers are linked with the maximum increase in frame-building organisms (Grötsch, 1991, fig. 12). If this is a general feature, the maxima of component 2 could be assigned to the end of a vertical aggradational cycle on the platform (Fig. 5C).

The third process that is important for carbonate production and deposition is baffling, i.e. the slowing-down of current velocities by organisms and the trapping thereby of transported sediment grains. Figure 7 shows the negative correlation between

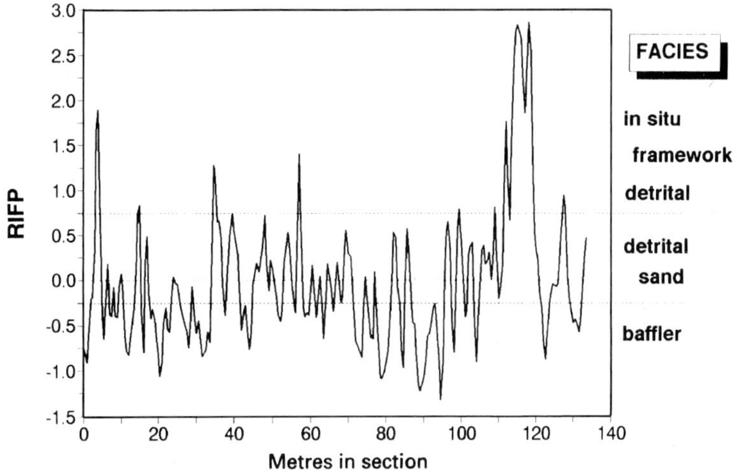

Fig. 8. Plot of PC1, here referred to as RIFP (see text) smoothed with a (weight = 1,2,1) moving average filter against position in section. The approximate distribution of the three different facies types is indicated by dashed lines. Note the high proportion of baffler facies from 0–35 and 70–105 m.

component 1 and baffling organisms. So, in addition to the previously discussed factors of constructing and destroying/dwelling, a third factor, here called baffling, is important. It actually marks the beginning of an aggradational cycle on the platform, during which the reef flat environment lags behind vertical growth of the reef core itself (Fig. 5A). In summary, whereas the reef core is keeping up with relative sea level, the reef flat is filled up later on by reefal debris and backward migration of frame-building organisms (Fig. 5B).

VERTICAL AGGRADATION OF THE REEF

Relative index of framework production

The results from the PCA indicate that the first component is controlled by high positive loadings for CONBIN, ARA, FAC and negative ones for BAF. Therefore, it can be assigned to the process of construction in the reef. This suggests the possibility of using the first component as a relative index of framework production (RIFP). Figure 8 shows the RIFP (PC1 scores smoothed with a moving average filter) with the facies transitions indicated by the dashed horizontal lines. The strongly positive values of the RIFP show a dominance of abundant, mostly destroyed skeletons of frame builders. Despite this, it is not possible to use the RIFP as a direct indicator of vertical platform growth or productivity. Rather it is an index for in-place production of carbonate by frame builders, which migrate over the back reef towards the end of relative sea-level rises. The actual short-term rises of sea level are characterized by the baffler or lag facies at the beginning of a cycle.

Cyclostratigraphic analysis

Periodic changes in climate and the resulting rhythmic sedimentation are a well-known phenomenon (Hays *et al.*, 1976; Einsele & Seilacher, 1982; Berger, 1988). Cyclic sedimentation has been described, for example, from deep-sea carbonates of Pleistocene (Arrhenius, 1952; Grötsch *et al.*, 1991) and Cretaceous age (Herbert & Fischer, 1986), as well as from internal platform environments in the earliest Cretaceous (Strasser, 1988), the Mid-Cretaceous (Grötsch, 1991, NW Greece), the Triassic (Sander, 1936; Fischer, 1964; Goldhammer *et al.*, 1987) and the Carboniferous (Schwarzacher, 1975). For massive reefal limestones, such as Kanalski Vrh, no cyclostratigraphic analyses have been available up to now.

The Kanalski Vrh II section was sampled with an equal-distance spacing of 0.5 m to facilitate later spectral analysis, as described in Davis (1986, pp. 248ff.). Figure 9 shows a periodogram (plot of magnitude^2 against frequency) for a fast Fourier transform (FFT) of the RIFP (thick line, emphasizing lower frequencies) and its first-order differences (thin line, emphasizing higher frequencies). Significant peaks can be seen at cycle thicknesses of 64, 16, 5.2, 3.6 and 3.2–3.0 m.

The ratios of cycle thickness in the periodogram (20:5:2:1) are very similar to those between the

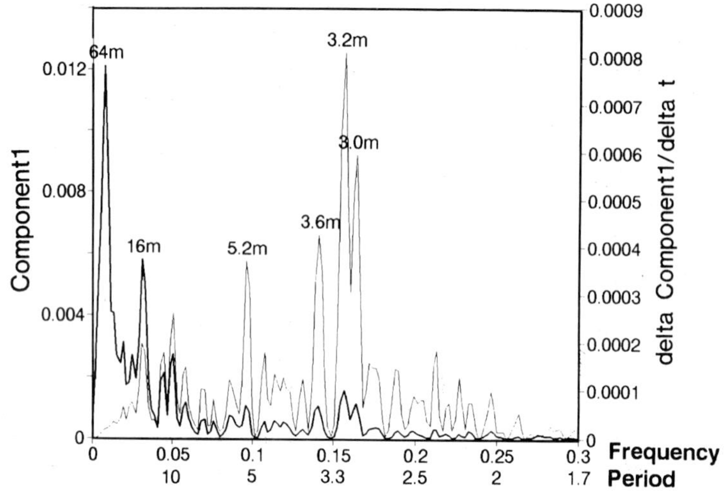

Fig. 9. Periodogram (spectrum) of component 1 (thick line, left y-axis) and its relative change (ΔComponent1/Δt with Δt equal to the sample spacing of 0.5 m; thin line, right y-axis). The x-axis shows frequency and period (m/cycle), whereas the y-axis gives the square of magnitude. The component 1 data were first smoothed with a (weight = 1,2,1) moving average filter, then detrended and tapered, before performing a fast Fourier transform. In the first case (component 1) low frequencies (64 m and 16 m) are emphasized; in the second case (Δcomponent1/Δt) high frequencies (5.2 m, 3.6 m, 3.2–3.0 m) are emphasized. The frequency ratios correlate very well with those of the orbital cycles predicted by the Milankovitch theory, i.e. 400, 100, 41, 23 and 19 ka. The very prominent appearance of the cycle peaks in the periodogram implies a constant vertical aggradation of the platform during the timespan represented in the section (800 ka).

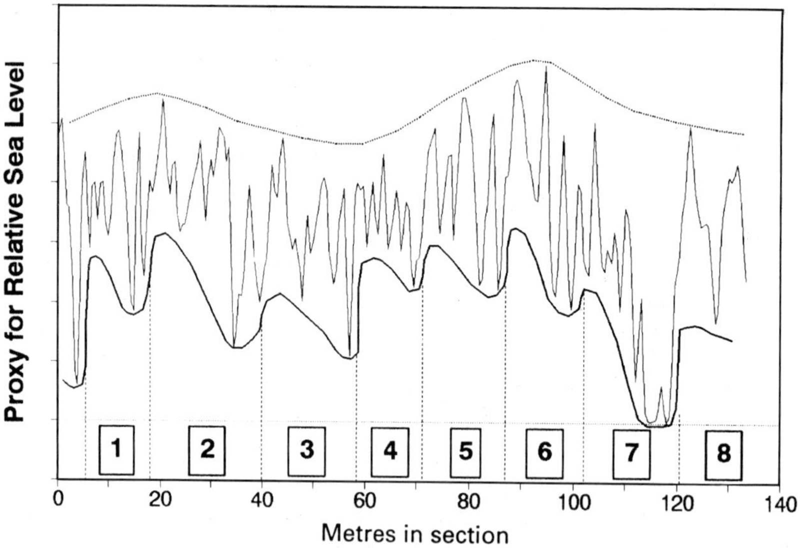

Fig. 10. Plot of proxy for relative sea level (PRSL) against position in section. The PRSL index is generated using results from facies and spectral analysis and by applying a transform function on the scores of component 1 of the principal component analysis. Lower, thicker envelope and large numbers in the lower part of the graph indicate the inferred 100-ka eccentricity (E1) cycle (see Fig. 2) and show a distinct bundling of four or five smaller cycles representing the 23- or 19-ka precession cycle. Upper envelope might be a reflection of the 400-ka E2 eccentricity cycle, which is also reflected in the absolute abundance of foraminifera species *Palorbitolina lenticularis* and *Paleodictyoconus arabicus* (see Fig. 2). The scaling of the y-axis is arbitrary as there is probably no strict linear relation between RIFP and relative sea level. Additionally, with the method proposed here, changes resulting in emersion of the reef cannot be quantified but can only be seen qualitatively in early diagenetic features. The index gives high-frequency fluctuations of sea level in the Milankovitch band, but not long-term, third-order changes.

main orbital frequencies (400, 100, 40 and 20 ka; see Berger, 1988). These ratios together with the facies analysis and the biostratigraphic framework are interpreted as indicating cyclicity at Milankovitch frequencies. According to this interpretation, sedimentation of the massive reef limestone of Kanalski Vrh in the early Aptian, despite showing no distinct bedding and only minor variation in facies, was controlled by orbital forcing.

As Algeo & Wilkinson (1988) point out, independent time control in cyclic sequences is a major problem because the time resolution of biostratigraphy is low and gaps in sedimentation can add up to long time intervals. However, in this study thicknesses of cycles and highly variable average durations of cycles were not used as indicators for Milankovitch band cyclicity, but rather frequency analysis and the resulting ratios of cycle thickness. These ratios from spectral analysis (20:5:2:1) may be used to calculate the sedimentation rate over the whole section (133 m $\approx$ 800 ka). Individual 3.2 m-cycles are thus inferred to represent 20 ka. Nevertheless these individual cycles might still have highly variable sedimentation rates, as suggested by facies analysis (see Fig. 5).

If this interpretation is right, the 64-m cycle reflects the rarely documented 400-ka E2 cycle (Fischer, 1986). In Figure 10 this cycle is indicated by the upper, enveloping line. It can also be seen in the absolute abundance of *Palorbitolina lenticularis* and *Palaeodictyoconus arabicus* (Fig. 2; Grötsch, 1991, fig. 18). The second important peak at 16-m thickness would represent the 100-ka eccentricity period E1. In Fig. 10 it is marked by the lower envelope. As in other areas (e.g. Goldhammer *et al.*, 1987), the 100-ka cycle comprises bundles of four or five smaller cycles. These are the cycles in the periodogram with thicknesses of about 3.6 or 3.2–3.0 m. They are interpreted as representing the precessional cycles with periods of 23 (22) and 19 (18.5) ka. The numbers in parentheses give the estimated periods for the Mid-Cretaceous (100 Ma) after calculations made by Berger *et al.* (1989); the eccentricity periods are unchanged from their present lengths. Surprisingly, the spectrum even splits up the precessional cycle into the two periods as it does for Pleistocene deep-sea $\delta^{18}O$ records (Hays *et al.*, 1976; Imbrie *et al.*, 1984). However, one should keep in mind the possible effects of noise in the data set. Finally, the 41 (38.8)-ka cycle can be identified with the 5.2-m peak in the periodogram.

Although rhythmic sedimentation cannot be observed directly in the field, after densely spaced sampling and quantification of the sedimentary parameters, cycles show up clearly in the spectrum (Fig. 9). This indicates an overall high potential of cycle preservation with only minor effects of erosion during periods of falling sea level ($\Delta RSL/\Delta t \leq 0$). Hence, the high-frequency cycles in the Milankovitch band have to be combined with high rates of subsidence and/or overlain by a third-order rise of sea level to allow such preservation. Following the cyclostratigraphic sub-division based on facies analysis, the RIFP and the periodogram of the data set, the 133.5 m thickness of the Kanalski Vrh II section comprises a time interval of only about 800 ka, which is equivalent to a rate of platform aggradation of 167 Bubnoff ($\approx$ m/Ma). This 800 ka of aggradation is located in the timespan from latest Barremian to early Aptian. Thickness and stratigraphic data from the Early Cretaceous Trnovo section about 15 km SE of Kanalski Vrh (Koch *et al.*, 1989) suggest a subsidence rate of about 20–60 Bubnoff for the NW part of the Dinaric Platform. This is in agreement with Schlager (1981, fig. 5), according to whom subsidence rates vary in the range of 10–100 Bubnoff (rarely up to 300 Bubnoff). Therefore, the reef aggradation is probably linked to a third-order eustatic sea-level rise as there are no indications for increased subsidence.

If the sequence of the Kanalski Vrh II section was deposited as part of a transgressive or high-stand systems tract, the question arises whether it is correlatable to the sequence chart of Haq *et al.* (1988). In order to do so, it would be preferable to analyse a longer section with good biostratigraphic control. According to the available data, the investigated sequence might be compared to LBZ-4.1 (Haq *et al.*, 1988), as has also been proposed for the Urgonian of the Basque–Cantabrian Basin in northern Spain (Garcia-Mondéjar, 1990). However, the eustatic sea-level curve of Haq *et al.* (1988) does not agree with the data presented here.

Inferred high-frequency changes of sea level

Using the precessional cycles (22 and 18.5 ka) with an average thickness of about 3.2 m (−0.8 m subsidence per 20 ka) as an example, the average eustatic rise of sea level above the platform edge is about 2.4 m for one cycle, not taking compaction into account. This amplitude might be responsible for the change from subtidal baffler facies to intertidal detrital (or in-place) framework facies.

As can also be seen in the facies model (Fig. 5), the RIFP is negatively correlated to sea level, pulse-like increases being recorded as low values and high-frequency high-stands being reflected as high values. This is because the section is located in a reef flat environment. Thus, the RIFP has a negative relation to sea level that is not linear. So, in order to get an approximate sea-level curve for the time interval recorded by the platform sediments, a transfer function has to be applied to the RIFP, which is:

$$\text{PRSL} = \text{RIFP} \times (-1) + C$$

where PRSL represents a proxy for relative sea level (first approximation to change of relative sea level), the constant (-1) reverses the curve to allow for the negative correlation between RIFP and sea level, and C is a constant to make RIFP values positive (here 3). Figure 10 shows the resulting approximation of a sea-level curve for the NW part of the Dinaric Platform. This curve only takes Milankovitch band cycles of fourth (400 ka, 100 ka) and fifth order (41 ka, 23 ka, 19 ka) into account. For the Aptian the three latter cycles have periods of about 38.8, 22 and 18.5 ka, whereas the eccentricity cycle (100 ka) remains constant (Berger *et al.*, 1989). The curve does not trace third-order changes. Because of the non-linear transferability of facies and palaeontological data to sealevel changes, units on the *y*-axis are omitted but might be in the order of metres.

CONCLUSIONS

The evolution of late Barremian to early Aptian reefs

A wide variety of papers on the evolution of reefs through time has been published (for further references see Fagerstrom, 1987; Talent, 1988). Therein the episodic occurrence of reefs and the extinction of reef-building organisms in Earth's history is discussed. One of the decisive problems in interpreting the development of fossil reefs is the missing high-resolution stratigraphy now available for Pleistocene and Holocene examples (see MacIntyre, 1988). The paper presented here is an attempt to achieve a roughly similar resolution for a Mid-Cretaceous (latest Barremian to early Aptian) example. The results show that in this time interval it is misleading to investigate the general composition of the reef, but instead it is necessary to analyse the organisms in 'time slices' in order to get an idea of the mechanism of vertical aggradation of the platform rim.

In the case of Kanalski Vrh the results of this study support cyclic, vertical aggradation of the reef, controlled by regional subsidence rate, third-order change of eustatic sea level and superimposed high-frequency fluctuations in the Milankovitch band. The latter control facies and facies change, while the long-term variations (subsidence and third-order sea-level change) influence the preservation potential of the episodic reef aggradational cycles. Figure 5 gives a facies model of such a cycle. After an initial stage of equilibrium and the following low-amplitude rise of sea level during a high-frequency fluctuation, vertical growth of the reef core (A) and subsequent filling up of the back reef and lagoon takes place (B). An intermediate state of equilibrium is succeeded by lowering of sea level and accompanying erosion (C). The latter controls the degree of cycle preservation (D).

Thus, the construction of the detrital (and in-place) framework in the reef core (Hubbard *et al.*, 1990), as well as the change from baffler to constructor facies in the reef flat, is controlled by rate of subsidence, third-order sea-level change and Milankovitch band fluctuations. These three factors are the major influence on the accommodation potential of the reef and, consequently, of the whole platform, if other factors such as nutrient input do not prohibit reef growth (Hallock & Schlager, 1986). Thus, it might not in many cases be necessary or possible for the reef to form a rigid, elevated framework (e.g. if change in relative sea level is constant or slightly positive). This could be a reason why only a few, rigid, in-place frameworks are preserved in the fossil record, and why the volume of reef detritus (produced by bioerosion, wave and storm action) is in many cases much higher than the volume of the *in situ* framework.

Reefs and sea level

Reefs might therefore be classified by the parameter carbonate production per unit time (Δt), whereby a high accommodation potential results in high carbonate production, if reefs do not drown. The time unit (Δt) is of major importance. Comparing Pleistocene reef growth and latest Barremian to early Aptian reef growth per 20 ka, the first is considerably higher, as amplitudes of Pleistocene sea-level fluctuations in the Milankovitch band are higher (about 130 m vs. 2–5 m). If we use a Δt of 1 Ma, the difference is small, as in both cases net change in sea level is about 100 m ($\Delta SL/\Delta t \approx 100$ m) and the net difference in accommodation space is therefore zero. Nevertheless, even if the proposed method of high-resolution stratigraphy in reefs is applicable to other examples, there is still the question of obtaining a satisfactory index for carbonate productivity in the fossil record. Reefs might not be different regarding carbonate production for long units of time, but rather in quality, quantity and preservation of frame-building organisms (constructor) and the associated guilds (binder, baffler, destroyer, dweller). Hence, there should be fundamental differences in mode of construction and in composition of the organisms in a reef between a greenhouse period (see Fischer, 1981) with high-frequency, low-amplitude sea-level fluctuations (e.g. latest Barremian to early Aptian) and an icehouse period with high-frequency, high-amplitude fluctuations of sea level (e.g. Pleistocene).

ACKNOWLEDGEMENTS

I thank Erik Flügel and Roman Koch for extremely stimulating discussions and assistance in the field. Stanko Buser from the Geoloski Zavod in Ljubljana/Slovenia who not only helped in the field, but also provided his know-how from many years of mapping in the investigated area and its surroundings. Rolf Schroeder generously determined the orbitolinid foraminifera, providing the biostratigraphic framework. I am also grateful to Wolfgang Schlager and John Reymer for useful criticism as well as to André Strasser, Robert Goldhammer and Poppe de Boer for reviewing this paper. Work was financed by a scholarship of the Studienstiftung des deutschen Volkes.

REFERENCES

Algeo, T.J. & Wilkinson, B.H. (1988) Periodicity of mesoscale Phanerozoic cycles and the role of Milankovitch orbital modulations. *J. Geol.* **96**, 313–322.

Anderson, R.J., Goodwin, P. & Sobieski, T.H. (1984) Episodic accumulation and the origin of formation boundaries in the Helderberg Group of New York. *Geology* **12**, 120–123.

Arrhenius, G. (1952) Sediment cores from the East Pacific: Swedish Deep-Sea Expedition (1947–1948). *Repts* **5**(1–4), 1–228.

Bebout, D.G. & Loucks, R.B. (Eds) (1977) *Cretaceous Carbonates of Texas and Mexico: Applications to Subsurface Exploration*. University of Texas at Austin Bur. Econ. Geol. Rep. Inv. 89.

Berger, A. (1988) Milankovitch theory and climate. *Rev. Geophys.* **26**, 624–657.

Berger, A., Loutre, M.F. & Dehant, V. (1989) Astronomical frequencies for pre-Quaternary paleoclimate studies. *Terra Nova* **1**, 474–479.

Buser, S. (1987) Development of the Dinaric and the Julian carbonate platforms and the intermediate Slovenian basin (NW-Yugoslavia). *Mem. Soc. Geol. Ital.* **40**, 313–320.

Davis, J.C. (1986) *Statistics and Data Analysis in Geology*, 2nd edn. John Wiley & Sons, New York.

Droxler, A.W. & Schlager, W. (1985) Glacial versus interglacial sedimentation rates and turbidite frequency in the Bahamas. *Geology* **13**, 799–802.

Droxler, A.W., Schlager, W. & Whallon, C.C. (1983) Quaternary aragonite cycles and oxygen-isotope record in Bahamian carbonate ooze. *Geology* **11**, 235–239.

Droxler, A.W., Bruce, C.H., Sager, W. & Watkins, D.H. (1988) Pliocene–Pleistocene variations in aragonite content and planktonic oxygen-isotope record in Bahamian periplatform ooze, Hole 633A. In: *Proc. Ocean Drilling Program, Scientific Results, 101* (Eds Austin, J.A., Schlager, W. *et al.*). College Station, TX (Ocean Drilling Program), pp. 221–236.

Dullo, W.C. (1983) Fossildiagenese im miozänen Leitha-Kalk der Paratethys von Österreich: ein Beispiel für Faunenverschiebungen durch Diageneseunterschiede. *Facies* **8**, 1–112.

Dunham, R.J. (1962) Classification of carbonate rocks according to depositional texture. *Am. Assoc. Petrol. Geol. Mem.* **1**, 108–121.

Einsele, G. & Seilacher, A. (Eds) (1982) *Cyclic and Event Stratification*. Springer Verlag, Berlin.

Fagerstrom, J.A. (1987) *The Evolution of Reef Communities*. John Wiley & Sons, New York.

Fischer, A.G. (1964) The Lofer cyclothems of the Alpine Triassic. *Kansas Geol. Surv. Bull.* **169**, 107–149.

Fischer, A.G. (1981) Climate in Earth history. In: *Climate in Earth History* (Ed. Berger, W.). National Academy of Sciences, Washington DC, pp. 97–104.

Fischer, A.G. (1986) Climate rhythms recorded in strata. *Annu. Rev. Earth Planet. Sci.* **14**, 351–376.

Flügel, E. (1982) *Microfacies Analysis of Limestones*. Springer Verlag, Berlin.

Garcia-Mondejar, J. (1990) The Aptian–Albian episode of the Basque–Cantabrian Basin (northern Spain): general characteristics, controls and evolution. In: *Carbonate Platforms: Facies, Sequences and Evolution* (Eds Tucker, M.E., Wilson, J.L., Crerollo, P.D., Sang, J.R. & Read, J.F.). Spec. Publ. Int. Assoc. Sediment. 9, pp. 257–290.

Goldhammer, R.K., Dunn, P.A. & Hardie, L.A. (1987) High frequency glacio-eustatic sea level oscillations with Milankovitch characteristics recorded in middle Triassic platform carbonates in N-Italy. *Am. J. Sci.* **287**, 853–892.

Grötsch, J. (1991) *Die Evolution von Karbonatplattformen des offenen Ozeans in der mittleren Kreide (NW-Jugoslawien, NW-Griechenland, NW-Pazifik): Möglichkeiten zur Rekonstruktion von Meeresspiegel-*

veränderungen verschiedener Größenordnungen. PhD thesis, Universität Erlangen-Nürnberg.

GRÖTSCH, J., WU. G. & BERGER, W.H. (1991) Carbonate cycles in the Pacific: Reconstruction of saturation fluctuation. In: *Cycles and Events in Stratigraphy* (Eds Ricken, W., Einsele, G. & Seilacher, A.). Springer Verlag, Berlin, pp. 110–125.

GROUPE FRANÇAIS DU CRETACE (1979) *L'Urgonien des Pays Méditerranéens*. Geobios Mem. Spec. 3.

GUILLAUME, A. (1977) *Introduction à la Géologie Quantitative*. Masson, Paris.

HAAK, A. & SCHLAGER, W. (1989) Compositional variations in calciturbitides due to sea-level fluctuations, late Quaternary, Bahamas. *Geol. Rundsch.* **78**, 477–486.

HALLOCK, P. & SCHLAGER, W. (1986) Nutrient excess and the demise of coral reefs and carbonate platforms. *Palaios* **1**, 389–398.

HAQ, B.U., HARDENBOL, J. & VAIL, P. (1988) Mesozoic and Cenozoic chronostratigraphy and cycles of sea-level change. In: *Sea-Level Changes: An Integrated Approach* (Eds Wilgus, C.K., Hastings, B.S., Kendall, C.S., Posamentier, H.W., Ross, C.A. & van Wagoner, J.C.). Soc. Econ. Paleont. Mineral. Spec. Publ. 42, pp. 71–108.

HAYS, J.D., IMBRIE, J. & SHACKLETON, N.J. (1976) Variations in the Earth's orbit: pacemaker of the ice ages. *Science* **194**, 1121–1132.

HERAK, M. (1986) A new concept of geotectonics of the Dinarids. *Acta Geol.*, **16**, 1–42.

HERBERT, T.D. & FISCHER, A.G. (1986) Milankovitch climatic origin of mid-Cretaceous black shale rhythms in central Italy. *Nature* **321**, 739–743.

HOPLEY, D. (1986) Corals and reefs as indicators of paleo-sea level with special reference to the Great Barrier Reef. In: *Sea-Level Research* (Ed. Van de Plassche, O.). Geo Books, Norwich, pp. 195–228.

HUBBARD, D.K., MILLER, A.I. & SCATURO, D. (1990) Production and cycling of calcium carbonate in a shelf-edge reef system (St. Croix, U.S. Virgin Islands): applications to the nature of reef systems in the fossil record. *J. Sedim. Petrol.* **60**, 335–360.

IMBRIE, J., HAYS, J.D., MARTINSON, D.G., MCINTYRE, A., MIX, A.C., MORLEY, J.J. *et al.* (1984) The orbital theory of Pleistocene climate: support from a revised chronology of the marine $\delta^{18}O$ record. In: *Milankovitch and Climate*, Part I (Eds Berger, A., Imbrie, J., Hays, J., Kukla, G. & Saltzman, B.). Reidel Publ. Co., Dordrecht, pp. 269–305.

JÖRESKOG, K.G., KLOVAN, J.E. & REYMENT, R.A. (1976) *Methods in Geomathematics, 1. Geological Factor Analysis*. Elsevier, Amsterdam.

KOCH, R., OGORELEC, B. & OREHEK, S. (1989) Microfacies and diagenesis of Lower and Middle Cretaceous carbonate rocks of NW-Yugoslavia (Slovenia, Trnovo Area). *Facies* **21**, 135–170.

MACINTYRE, I.G. (1988) Modern coral reefs of Western Atlantic: New geological perspective. *Bull. Am. Assoc. Petrol. Geol.* **72**, 1360–1369.

MASSE, J.-P. & PHILIP, J. (1981) Cretaceous coral-rudist buildups of France. In: *European Fossil Reef Models* (Ed. Toomey, D.F.). Soc. Econ. Paleont. Mineral. Spec. Publ. 30, pp. 399–426.

MOLDOVANYI, E.P. & LOHMANN, K. (1984) Isotopic and petrographic record of phreatic diagenesis: Lower Cretaceous Sligo and Cupido formation. *J. Sedim. Petrol.* **54**, 972–985.

PIRAZZOLI, P.A., MONTAGGIONI, L.F., SALVAT, B. & FAURE, G. (1988) Late Holocene sea level indicators from twelve atolls in the central and eastern Tuamotus (Pacific Ocean). *Coral Reefs* **7**, 51–68.

PLACER, L. (1981) Geologic structure of southwestern Slovenia (in Serbo-Croatian). *Geologija* **24**, 27–60.

REYMER, J.J., SCHLAGER, W. & DROXLER, A.W. (1988) Site 632: Pliocene–Pleistocene sedimentation cycles in a Bahamian basin. In: *Proc. Ocean Drilling Program, Scientific Results, 101* (Eds Austin, J.A., Schlager, W. *et al.*). College Station, TX (Ocean Drilling Program), pp. 213–220.

SANDER, B. (1936) Beiträge zur Kenntnis der Anlagerungsgefüge (rhythmische Kalke und Dolomite aus Tirol). *Tschermaks Mineral. Petrogr. Mitt.* **46**, 27–209.

SCHLAGER, W. (1981) The paradox of drowned reefs and carbonate platforms. *Geol. Soc. Am. Bull.* **92**, 197–211.

SCHROEDER, R. (1975) General evolutionary trends in Orbitolinas (1). *Rev. Espan. Micropaleont. Num. Esp.* 117–128.

SCHWARZACHER, W. (1975) *Sedimentation Models and Quantitative Stratigraphy*. Developments in Sedimentology 19. Elsevier, Amsterdam.

SCOTT, R.W. (1981) Biotic relations in Early Cretaceous coral–algal–rudistid reefs, Arizona. *J. Paleontol.* **55**, 463–478.

SCOTT, R.W. (1984) Evolution of Early Cretaceous reefs in the Gulf of Mexico, *Paleontograph. Am.* **54**, 406–412.

SKELTON, P.W. & GILI, E. (1990) Did rudists build reefs? In: *13th Int. Sedimentol. Congr. Abstr.*, p. 501.

STRASSER, A. (1988) Shallowing upward sequences in Purbeckian peritidal carbonates (lowermost Cretaceous, Swiss and French Jura Mountains). *Sedimentology* **35**, 369–383.

TALENT, J.A. (1988) Organic reef-building: episodes of extinction and symbiosis? *Senkenbergiana lethaea* **69**, 315–368.

VELIC, I., TISLJAR, J. & SOKAC, B. (1987) The variability of thickness of the Barremian, Aptian and Albian carbonates as a consequence of changing depositional environments and emersion in Western Istria (Croatia, Yugoslavia). *Mem. Soc. Geol. Ital.* **40**, 209–218.

WALKER, K.R. & ALBERSTADT, L.P. (1975) Ecological successions as an aspect of structure in fossil communities. *Paleobiology* **1**, 238–257.

WOODROFFE, C., MCLEAN, R., POLACH, H. & WALLENSKY, E. (1990) Sea level and coral atolls: Late Holocene emergence in the Indian Ocean. *Geology* **18**, 62–66.

Spec. Publs Int. Ass. Sediment. (1994) **19**, 243–283

High-frequency, glacio-eustatic cyclicity in the Middle Pennsylvanian of the Paradox Basin: an evaluation of Milankovitch forcing

R.K. GOLDHAMMER, E.J. OSWALD
and P.A. DUNN

Exxon Production Research Co., P.O. Box 2189, Houston, TX 77252-2189, USA

ABSTRACT

Middle Pennsylvanian (Desmoinesian) shelf carbonates in the SW Paradox Basin display three superimposed orders of stratigraphic cyclicity with a systematic, vertical succession of facies, cycle and sequence stacking patterns. Fifth-order cycles (34 cycles in 196-m section; average 6 m thick; mean period of 29 ka) are grouped into fourth-order sequences (average 35 m thick; mean period 257 ka), which in turn vertically stack to define a third-order sequence (196+ m thick; 2–3 Ma duration). Fifth-order cycles are composed of shallowing-upward packages of dominantly subtidal shelf carbonates with sharp cycle boundaries (either exposure or flooding surfaces). Fifth-order cycles are packaged into fourth-order sequences bounded by regionally correlative subaerial exposure surfaces. These type 1 sequences contain a down-dip, restricted low-stand wedge of evaporites and quartz clastics in topographic lows on the Paradox shelf (intra-shelf depressions). The low-stand systems tract is overlain by a regionally correlative transgressive shaley mudstone (condensed section), and a high-stand systems tract composed of thinning-upward, aggradational, fifth-order cycles. Systematic variation in the thickness of fourth-order sequences (thinning-upward followed by thickening-upward), as well as systematic variations in the number of fifth-order cycles/fourth-order sequences (decreasing followed by increasing number), define a third-order accommodation trend which is also regionally correlative.

High-frequency cycles and sequences are interpreted as dominantly aggradational allocycles generated in response to composite fourth- and fifth-order glacio-eustatic sea-level fluctuations. Two different orbital forcing (Milankovitch) scenarios are evaluated to explain the composite stratigraphic cyclicity of the Paradox sequences, each of which is plausible given Desmoinesian age date estimates. The cycle, sequence and facies stacking patterns have been replicated via computer by superimposing composite, high-frequency glacio-eustasy atop regional subsidence using depth-dependent sedimentation.

INTRODUCTION: PROBLEMS IN INTERPRETING PLATFORM CARBONATE CYCLICITY

The fundamental building blocks of ancient shallow-marine, platform carbonates are high-frequency (metre-scale), shallowing-upward depositional cycles. Cycles are composed of a relatively conformable succession of genetically related subtidal facies bounded by peritidal facies, subaerial exposure surfaces, and/or marine flooding surfaces. Typical cycle durations range from 10 to 500 ka, and a hierarchy of high-frequency cycle periods may be designated (Table 1), with fifth-order cycles (0.01–0.1 Ma) and fourth-order cycles (0.1–1.0 Ma) grouped into third-order sequences (1.0–10 Ma; Goldhammer *et al.*, 1990).

The origin of high-frequency (i.e. fifth- and fourth-order) carbonate cyclicity has received considerable attention (Grotzinger, 1986; Hardie & Shinn, 1986; Goldhammer *et al.*, 1987) and essentially three mechanisms have been proposed to create the successive stacks of fifth/fourth-order cycles that characterize carbonate deposits of all ages: (i) high-frequency glacio-eustatic oscillations forced by Milankovitch climatic rhythms (Fischer, 1964); (ii) autocycles

Table 1. Orders of stratigraphic and eustatic cyclicity. Data summarized from Sloss (1963), Rona (1973), Pitman (1978), Donovan & Jones (1979), Schlager (1981), Kendall & Schlager (1981), Hine & Steinmetz (1984), Miall (1984), Haq *et al.* (1987), Goldhammer *et al.* (1987, 1990), Ross & Ross (1987), Crevello *et al.* (1989)

Sequence strat. terminology	Eustatic cycles (orders)	Duration (Ma)	Amplitude (m)	Rise/fall rates (cm/ka)
	First	>100		<1
Sepersequence	Second	10–100	50–100	1–3
Sequence	Third	1–10	50–100	1–10
Sequence, cycle	Fourth	0.1–1	1–150	40–500
Parasequence, cycle	Fifth	0.01–0.1	1–150	60–700

driven by seaward progradation of tidal flat complexes (Ginsburg, 1971); and (iii) tectonic pulsing of carbonate platforms related to successive downdropping of platform tops or reversal tectonics (Cisne, 1986; Hardie *et al.*, 1990).

Milankovitch cycles

The general acceptance of the Milankovitch theory for Pleistocene cycles has been provided by the correlation of the Pleistocene deep-sea and coral reef record with Milankovitch climatic rhythms (Hays *et al.*, 1976; Berger, 1984; Imbrie, 1985). An outgrowth of the Milankovitch revival is the question of whether orbital variations have produced a legible stratigraphic signal in pre-Pleistocene cyclic carbonate platforms. This question was initially addressed by Fischer (1964), who, following Sander (1936), called upon small-scale, glacio-eustatic sea-level fluctuations induced by orbital variations to account for the Alpine Triassic Lofer cycles. In particular, Fischer (1964) favoured the 41-ka obliquity cycle as the 'eustatic dictator' responsible for the fundamental fifth-order Lofer cyclothem. Pursuit of the feasibility of pre-Pleistocene Milankovitch forcing in carbonate systems is exemplified by studies published over the last several years.

For example, Grotzinger (1986) called upon a Milankovitch allocyclic mechanism to explain the cyclicity of the Precambrian Rocknest Formation of Canada. The pervasive metre-scale peritidal cycles of the Cambrian passive margin of North America have been attributed to Milankovitch forcing by Koerschner & Read (1989) and Bond *et al.* (1991). Goodwin & Anderson (1985) have invoked glacio-eustasy under orbital control to interpret the origin of punctuated aggradational cycles (PACs) of the Devonian of the Appalachian Basin. In the Middle Triassic of the Dolomites (Italy), Goldhammer *et al.* (1987) reported a hierarchy of fifth-order platform carbonate cycles grouped into fourth-order thinning-upward megacycles. Time-series analyses of fifth-order cycle stacking patterns complemented with forward stratigraphic modelling led them to infer a Milankovitch glacio-eustatic mechanism.

To establish the existence of Milankovitch forcing, some workers have simply compared calculated cycle periods or a range of periods with Milankovitch values (e.g. Heckel, 1986; Koerschner & Read, 1989 – see discussion of this paper by Kozar *et al.*, 1990). Others have noted the bundling of small-scale, fifth-order cycles into larger scale fourth-order cycles and compared this stratigraphic hierarchy to the Milankovitch hierarchy of superimposed orders of climatic cyclicity (Goodwin & Anderson, 1985; Goldhammer *et al.*, 1987; Strasser, 1988).

In the light of errors involved in age determinations and uncertainties in the pre-Pleistocene time scale, the calculation of cycle periods and a comparison with predicted Milankovitch frequencies is fraught with hazards (Hardie *et al.*, 1986; Algeo & Wilkinson, 1988). Analysis of cycle stacking patterns and comparison of cycle bundles with Milankovitch ratios alone provides ambiguous results. Time-series analysis provides the most reliable means of demonstrating Milankovitch frequencies within platform carbonates. For example, Goldhammer *et al.* (1987) assigned equal durations to fifth-order cycles, each of which consists of a thin subtidal unit with a vadose diagenetic cap resulting from high-frequency allocyclic sea-level oscillations. Using successive cycle thicknesses, the resulting time-series analyses yielded a spectrum of superimposed orders of relative sea-level change with the same dimensionless frequencies as Milankovitch rhythms (Hinnov & Goldhammer, 1988). The system studied by Goldhammer *et al.* (1987) and Hinnov & Goldhammer (1988) is unique in that progradational subfacies were absent from the cyclic record, thereby eliminating autocyclicity as a viable mechanism. The aggradational stacks of high-frequency cycles could be attributed directly to relative sea-level changes.

In systems complicated by progradation, peritidal subfacies with mudcracked cryptalgal laminites typically cap cycles, rendering the interpretation of direct time-series analyses more difficult. An approach to this problem has been outlined in Bond *et al.* (1991) for the Cambrian of Utah.

However, a more serious problem regarding time-series analyses of ancient platform cyclic sequences pertains to the completeness of the cyclic succession. Inherent in traditional time-series analysis is the assumption that the cyclic record is complete and that each recorded stratigraphic rhythm equates directly to one rhythmic pulse of the cycle-producing mechanism (e.g. glacio-eustasy; Duff *et al.*, 1967, p. 13). The approach with the rock record is to scan stratigraphic data sets for cyclic signatures and compare the extracted frequencies with known Milankovitch frequency ratios. In examining the relationship between cyclic events in time (i.e. rhythms) and the resulting stratigraphic record, Schwarzacher (1975) pointed out that carbonate platforms may depict at best a rather 'faulty recording mechanism'. Despite a regular, rhythmic time pattern of cycle-producing events (e.g. Milankovitch-driven glacio-eustasy), the corresponding stratigraphic record may not preserve each and every potential cycle-producing pulse due to complexities of sedimentation and erosion.

This is exemplified by the latest Pleistocene record of the Florida and Bahama shallow-water carbonate platform record, where although sea level was oscillating with a 20-ka rhythm induced by the Earth's precessional cycle, sea level stayed well below the top of these platforms for about 120 ka (Hardie & Shinn, 1986; Goldhammer *et al.*, 1990). This resulted in loss of platform *depositional cycles* despite the existence of 20-ka *eustatic cycles* related to Milankovitch glacio-eustasy. These 'missed beats' of deposition leading to time gaps in the stratigraphic succession are recognized in the Florida and Bahama platforms only because of the independent external

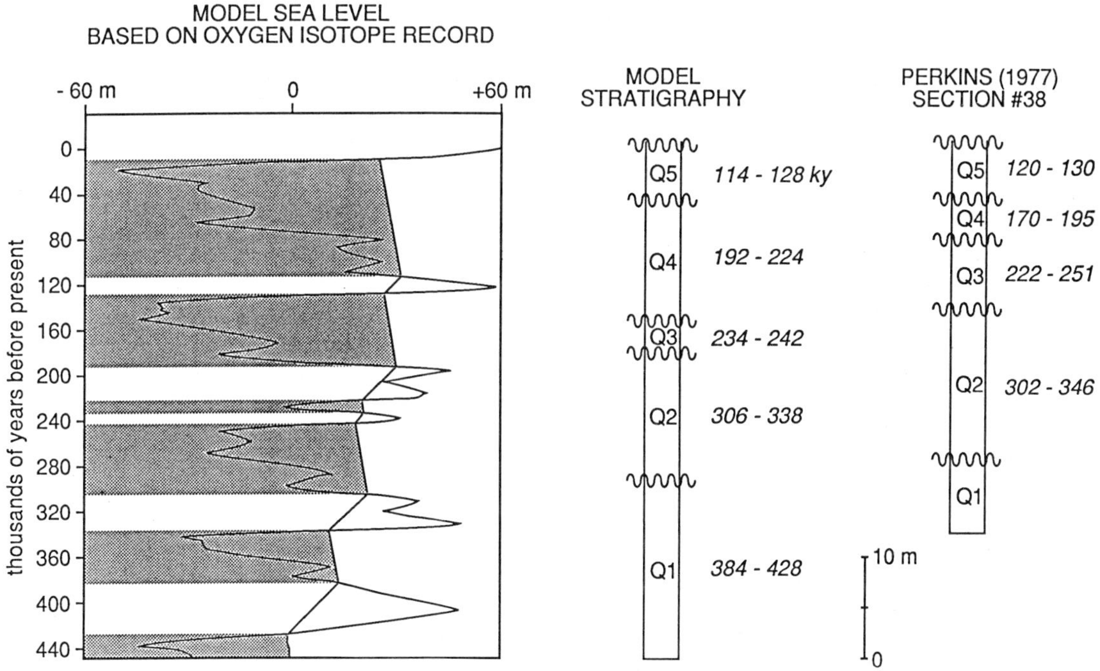

Fig. 1. Computer simulation of the south Florida Pleistocene platform carbonate stratigraphy described by Perkins (1977, section 38). The Pleistocene record consists of five marine subtidal units (Q1–Q5) capped by subaerial exposure surfaces. Ages of Pleistocene units (in thousands of years BP) are given to the right of stratigraphic columns. Combined subsidence–sedimentation–sea-level history shown to the left. Input parameters utilized are as follows: run length = 440 ka; subsidence rate = 0.06 m/ka; sedimentation rate = 0.40 m/ka; rate of subaerial exposure development = 0.01 m/ka. Temporal variations in oxygen isotope data in Quaternary deep-sea sediments used as a proxy for sea level scaled to 120-m amplitude based on Pleistocene coral reef data. From Goldhammer *et al.* (1990).

Pleistocene sea-level history provided by deep-sea oxygen isotope curves and radiometrically dated uplifted Pleistocene coral reef terraces (e.g. Barbados and New Guinea; data summarized in Goldhammer *et al.*, 1987, 1990). The Pleistocene succession of south Florida suggests that this is not only true for the last 120 ka, but apparently was the case back to about 400 ka. Insight regarding the significance of 'missed beats' in interpreting the Pleistocene platform stratigraphy of south Florida was reviewed by Goldhammer *et al.* (1990) and is summarized below.

Perkins (1977) has identified five marine subtidal units (Q1 to Q5) separated by subaerial unconformities recording vadose diagenesis in the Pleistocene carbonates of south Florida (Fig. 1). Pleistocene unit Q5 is dated at 120–130 ka (Broecker & Thurber, 1965; Osmond *et al.*, 1965). Mitterer (1975) has assigned ages of 170–195, 222–251 and 302–346 ka to units equivalent to Q4, Q3 and Q2 respectively (Fig. 1). Perkins (1977) has interpreted these Pleistocene unconformity-bounded units as the result of alternating shallow-marine deposition and subaerial exposure controlled by Pleistocene glacio-eustatic sea-level oscillations.

In an effort to further understand the cyclicity of the south Florida Pleistocene carbonates Goldhammer *et al.* (1990) simulated the Q1 to Q5 vertical succession through forward computer modelling. As a proxy for glacio-eustatic sea-level changes over the last 400 ka, they used the temporal variations in oxygen isotope concentrations in Quaternary deep-sea sediments (the smoothed data of Imbrie, 1985, his fig. 17). Sea-level amplitudes were scale based on Pleistocene coral reef data (Steinen *et al.*, 1973; Aharon, 1984). Their simulated cyclic stratigraphy is compared with the Pleistocene succession in Fig. 1. What the simulation reveals is a basic theme of 'missed beats' and underscores the dangers of literal interpretation of time series of platform carbonates. Despite the rhythmic oscillation of sea level in tune to composite Milankovitch astronomical cycles (sinusoidal *c.*20-ka precessional beats superimposed on asymmetrical *c.*100-ka eccentricity beats, Fig. 1) only a fraction of the basic *c.*20-ka oscillations are recorded. The large amplitude of the steeply asymmetrical *c.*100-ka component carried Pleistocene sea level high enough above the top of the south Florida platform to submerge it for one or more of the initial *c.*20-ka precessional beats of each *c.*100-ka eccentricity cycle (Fig. 1). In turn, during the *c.*100-ka low-stands sea level dropped far enough below the platform top to keep it exposed for two or more of the last *c.*20-ka beats of each *c.*100-ka cycle (Fig. 1). Therefore, the stratigraphic record is one of incomplete fourth-order megacycles with single or amalgamated subtidal units capped by extensive soils. While it is clear from the computer simulation that the basic *c.*20-ka precessional rhythm of sea-level rises and falls has been missed, the stratigraphic succession of single and amalgamated subtidal units with their karstic soil caps has essentially caught the higher order *c.*100-ka eccentricity rhythm (Fig. 1). In essence, the Q1, Q2, Q3–Q4 (combined), Q5, and the Holocene unconformity-bounded units are the stratigraphic records of the high-stands of the *c.*100-ka component of the composite Pleistocene sea-level changes. However, the actual durations of the periods of subtidal deposition and the intervening periods of subaerial exposure are quite irregular (Fig. 1) and without the aid of accurate age dating of each cycle it would not be possible from the cycle thickness patterns to demonstrate unequivocally that the cyclicity was controlled by externally driven rhythmic processes. Average cycle duration of this south Florida Pleistocene cyclic succession is *c.*67 ka (six cycles in 400 ka), a value that does not reflect any of the actual Milankovitchian components that have shaped the Pleistocene sea-level history.

As demonstrated by the south Florida example, 'missed beats' of stratigraphic cyclicity are a potential problem in evaluating platform carbonate stratigraphy. In the Pleistocene example, missed beats stemmed from the relative amplitude ratios of the superimposed orders of fourth- and fifth-order eustasy. The corresponding deep-sea oxygen isotope record provided the key ingredient in unravelling the Pleistocene platform carbonate stratigraphy. In short, the deep-sea organism wears the 'eustatic glasses'. Optimum situations in interpreting carbonate 'platform' stratigraphy exist when the shallow-water stratigraphy is paired with coeval 'basinal' stratigraphy.

In a similar way Goldhammer *et al.* (1990) also demonstrated that lower-frequency, third-order relative sea-level changes can lead to 'missed beats' of higher-frequency fourth- and fifth-order eustatic cycles. During third-order relative rise higher-frequency eustatic cycles may not 'touch down' and expose the top of the platform, leading to subtidal amalgamation. During third-order relative fall, higher-frequency cycles may oscillate beneath the platform top, leading to subaerial condensation.

In summary, despite sophisticated time-series techniques, completeness of the section due to 'missed beats' and/or erosion needs to be considered. The problem is best alleviated where two-dimensional platform-to-basin control exists such that one can 'see' the 'missed platform beats' in the down-dip position.

Autocycles

Based on Holocene insight, Ginsburg (1971) formulated a progradation-driven autocyclic mechanism that readily accounts for repetitious stacks of metre-scale fifth- and fourth-order platform carbonate cycles. This model, summarized by Hardie & Shinn (1986), couples continuous carbonate sedimentation and progradation of tidal flat complexes across gently sloping carbonate platforms under conditions of constant subsidence and stationary sea level. The subtidal carbonate factory provides the sediment which is transported shoreward, accreting against exposed land masses or islands. The tidal flat complex progrades out and over the subtidal deposits, yielding a shallowing-upward cycle capped by progradational tidal flat facies. As pointed out by Hardie & Shinn (1986), the key aspect of the model is that carbonate production rates will be unable to keep pace with subsidence as tidal flat progradation diminishes the size of the 'subtidal factory' and eventually the progradational cycle will halt. With passive subsidence, a minimum 'lag depth' (the depth required to resume carbonate production) of a metre or two is achieved, initiating carbonate sedimentation, and hence a new cycle.

As shown by Hardie & Shinn (1986) and Hardie *et al.* (1991), Ginsburg's autocyclic model can readily produce metre-scale cycles with periodicities that fall within the Milankovitch band. Dunn (1991) has developed a two-dimensional forward computer model, incorporating tidal flat progradation, which elegantly simulates successive stacks of Ginsburg autocycles (Dunn *et al.*, 1991). Time-series analyses of simulated autocyclic successions do not reveal Milankovitch frequency ratios (Dunn, 1991), even when high-frequency Milankovitch eustasy is built into the simulations (Hardie *et al.*, 1991), testifying to the complicating effects of autocyclic progradation.

In any cyclic carbonate platform succession composed of metre-scale cycles capped by progradational tidal flat facies, Ginsburg's (1971) autocyclic model must be considered. In geological scenarios involving autocyclic progradation and superimposed Milankovitch glacio-eustasy, care must be taken to separate the effects of each, and stratigraphers need to search the rock record for appropriate signs of each. For example, Milankovitch-style stacking patterns of tidal flat cycles, or extensive vadose diagenesis superimposed atop tidal flat capping facies indicating relative drops in sea level, would suggest a glacio-eustatic component.

Tectonic cycles

Another viable cycle-producing mechanism is that of tectonically dictated cycles. Repeated tectonic pulses of down-dropping and uplift (reversal or 'yo-yo' tectonics) could induce metre-scale cyclicity by alternately submerging and exposing (subaerially) carbonate platforms (Hardie & Shinn, 1986). Episodic, metre-scale down-dropping driven by faulting could also create accommodation space, leading to shallowing-upward cycles (Hardie *et al.*, 1991). An important difference between the two types of tectonic models is the internal facies architecture which will be recorded in the cycle. For example, reversal tectonics will create subtidal facies capped by subaerial exposure surfaces (caliches, etc.). Depending on competing rates of sedimentation and uplift, the cycles may not contain a classic shallowing-upward internal facies arrangement. Episodic, non-uniform subsidence would probably create a typical shallowing-upward succession that may or may not fill to sea level, again a function of sedimentation rates versus periodicity and rate of down-faulting. Likewise, cycles created by episodic subsidence would not be capped by subaerial exposure surfaces containing evidence for vadose diagenesis. With both models, the accommodation space required for each cycle would be controlled by metre-scale pulses that might be somewhat rhythmic if a threshold stress must be achieved to induce crustal rupturing (Hardie & Shinn, 1986).

Modern data concerning reversal tectonics is scarce and information regarding recurrence frequency, amount of vertical displacement and the size of areas involved is lacking. Cisne (1986) has outlined a model of stick-slip faulting at the edges of carbonate platforms that demands a fault-bounded shelf-edge. In this model, the platform edge and the up-dip area located within 100 km of the shelf edge would experience high-frequency crustal oscillations due to reversal tectonics (Cisne, 1986). While this model is intriguing, there is no way of estimating recurrence frequencies and it necessitates block faulting coincident with a platform shelf edge.

Presently, there are sufficient data regarding tectonic forcing through step-wise subsidence to render episodic subsidence a plausible mechanism for the generation of metre-scale sedimentary cycles and larger sequences (Hardie *et al.*, 1991). For example, modern manifestations of coseismic subsidence over large areas (equal to or greater than the size of many carbonate platforms) associated with very large earthquakes are reported along convergent margins and in intra-plate regions (Plafker, 1965; Atwater, 1987). Although large areas can be down-dropped a few to several metres, data regarding recurrence frequencies are sparse, suggesting only that recurrence frequencies span the 10–100-ka band (Hardie *et al.*, 1991).

Complicating the interpretation of platform carbonate sequences is the fact that all three mechanisms (autocyclic progradation, glacio-eustasy and high-frequency tectonic forcing) may operate in concert.

Pennsylvanian cyclicity: an example of multiple working hypotheses

The Pennsylvanian period is renowned throughout the world for the numerous examples of cyclic sedimentation, involving carbonates, clastics and evaporites. With regard to studies of cyclic sediments, Middle and Upper Pennsylvanian cycles of the mid-continent USA and Europe (or 'cyclothems', Weller, 1930) have historically probably received more attention than any other cyclic deposits (e.g. Weller, 1964; Duff *et al.*, 1967; Klein & Willard, 1989; Read & Forsyth, 1989). Interpretations of Pennsylvanian cycles have varied but they essentially fall into three categories, similar to the models reviewed above.

1 *Glacio-eustatic control.* Mixed clastic–carbonate cyclothems of the mid-continent USA and Europe have been inferred to result from the waxing and waning of large Gondwanan ice caps (Wanless & Shepard, 1936; Wheeler & Murray, 1957; Wanless & Cannon, 1966; Crowell, 1978; Ramsbottom, 1979). More recently, Heckel (1986, 1989) and Ross & Ross (1987) have advocated a direct Milankovitch control on the cycles characteristic of the interior and southwestern regions of the USA. As their main line of evidence was calculated cycle periodicities, Klein (1990) recently challenged a Milankovitch control based on estimates of cycle periods utilizing different time scales. Commenting on Klein's (1990) paper, de Boer (1991) drew attention to the fact that mid-continent Pennsylvanian cyclothems are grouped into bundles of five, supporting a precession-driven origin, regardless of the particular time scale used. Original glacial proponents (Wanless & Shepard, 1936) drew upon the widespread distribution of cyclothems, extreme lateral continuity of lithofacies and the occurrence of age-equivalent cyclothems on different continents to support their viewpoint.

2 *Autocyclic controls.* Autocyclic influences were suggested by van der Heide (1950), who speculated that compaction of peat layers in clastic portions of cyclothems in The Netherlands could have induced sudden marine transgressions, contributing to cyclic deposition. More recently, autocyclic models of delta-lobe shifting have been applied to Upper Pennsylvanian mixed clastic–carbonate cyclothems (Galloway & Brown, 1973; Elliot, 1976). Recognizing that most of the deltaic deposits occur within the regressive portions of cyclothems, Heckel (1989) viewed autocyclic delta-lobe switching as a subordinate process operating within an overall eustatic framework.

3 *Tectonic controls.* Stout (1931) interpreted mid-continent cyclothems in Ohio as records of periodic down-dropping of the basin, inducing transgression, followed by steady sediment infill (i.e. a form of episodic subsidence). Weller's (1930, 1956, 1964) 'diastrophic control theory' centred upon alternate pulsating uplift and depression (reversal tectonics) of large continental areas, with the magnitude being greatest in the uplands, declining basinward. Periodic episodes of uplift resulted in termination of marine deposition (inferred to occur at the top of a cyclothem), erosion and influx of clastic (non-marine) material.

Middle Pennsylvanian cyclicity of the Paradox Basin

The Middle Pennsylvanian cyclic stratigraphy of the Paradox Basin in southeast Utah provides an ideal laboratory for evaluating models of cyclic carbonate deposition. In the light of the controversy regarding the origin of the mid-continent-type 'cyclothems', the Paradox system offers an independent assessment of the nature and origin of Pennsylvanian cyclicity. The Paradox system is characterized by cyclic shelfal carbonates and basinal evaporites that reflect cyclic and reciprocal accumulation patterns between the Paradox shelf and a deeper restricted basin (Fig. 2).

A great deal of attention has focused on the facies

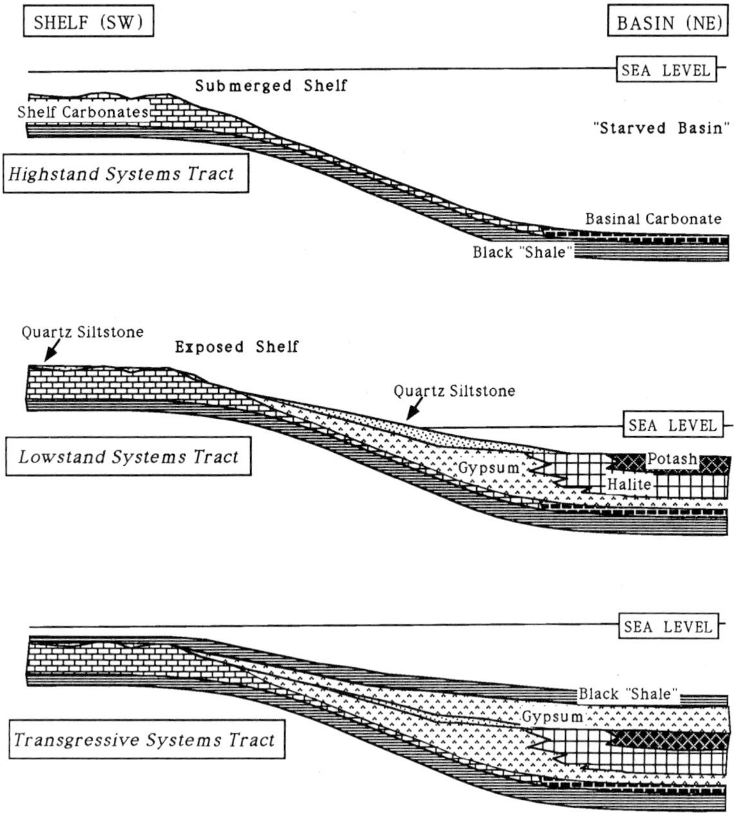

Fig. 2. Simplified depositional model for the development of fourth-order depositional sequences illustrating the cyclic and reciprocal style of shelf vs. basin accumulation. Modified from Hite (1970).

relationships within the cyclic Paradox strata as well as the correlation of shelfal carbonate cycles to basinal evaporite cycles (see Goldhammer *et al.*, 1991 for a review). The idealized Paradox shelf 'cycle' (Fig. 2) averages 35 m in thickness (ranging from 10 to 40 m) and contains a thin, basal transgressive 'black shale' or dark carbonate mudstone overlain by a shallowing-upward succession of shelf carbonate interpreted as the regressive phase of the cycle (e.g. Choquette & Traut, 1963; Elias, 1963; Peterson & Ohlen, 1963; Hite, 1970). The 'black shale' unit is actually a dark, argillaceous sapropelic dolomite which has historically been termed a 'shale' because of its characteristic 'hot' gamma-ray response in well logs. The regressive carbonate portion typically contains a thick phylloid algal bioherm facies overlain by a capping ooid grainstone facies. Unconformities with evidence for subaerial exposure typically cap the shelf cycles (Peterson & Hite, 1969; Choquette, 1983). In some of the shelf cycles a low-stand deposit of siliciclastic sandstone occurs between the subaerial unconformity and the overlying black 'shale' (Goldhammer *et al.*, 1991). The basal 'shale' member of the shelf cycles can be traced basinward into the equivalent basinal cycles (Peterson & Ohlen, 1963; Hite, 1970).

An idealized Paradox basinal evaporite 'cycle' averages near 45–60 m in thickness and consists of the following from bottom to top (Peterson & Hite, 1969; Hite, 1970): (i) disconformity in the form of a subaqueous dissolution surface atop halite and/or potash salts; (ii) nodular and laminated anhydrite; (iii) silty dolomite; (iv) black sapropelic, silty, calcareous shale, equivalent to the basal 'shale' of the shelf cycles; (v) dolomite as before; (vi) anhydrite as before; (vii) halite, often grading vertically into potash salts; and (viii) disconformity surface. Hite (1970) interprets the evaporite–shale cycles as responses to decreasing and increasing salinites resulting from alternating periods of relative rise and fall of sea level in a restricted, barred evaporite basin.

Interpretations regarding the origin of the Paradox 'cycles' generally favour a glacio-eustatic mechanism. Proponents of a glacio-eustatic mechanism (Elias, 1963; Peterson & Ohlen, 1963; Peterson & Hite, 1969; Hite & Buckner, 1981) suggested eustatic sea-level oscillations triggered by Southern Hemisphere glaciation of Gondwana. Extensive lateral correlation of cycles, the asymmetrical nature of the internal facies architecture (i.e. basal thin transgressive shale), and subaerial exposure with inferred meteoric leaching of carbonate have all been cited in support of glacio-eustasy (e.g. Peterson & Hite, 1969). Whereas tectonic influences have been suggested as significant in controlling localization of phylloid algal bioherm development (Baars & Stevenson, 1982), tectonic mechanisms, as well as autocyclic processes, have not been viewed as significant in generating the Paradox cyclicity.

Although previous workers have suggested a glacio-eustatic control, no attempt has been made to evaluate the role of Milankovitch forcing as the driving agent. Based on sedimentation rates, Hite & Buckner (1981) calculated a period of 110 ka for one of the Paradox 'cycles' and suggested that this value reflected predicted periods of continental ice surge. Other workers have noted the presence of smaller cycles or 'subcycles' within the larger Paradox 'cycles' (e.g. Choquette & Traut, 1963; Peterson & Hite, 1969; Hite & Buckner, 1981), but no published work has addressed this possible hierarchy of cyclicity in any detail. Indeed, essentially all published subsurface and outcrop descriptions and correlations only address the larger scale shale-based 'cycles'. If glacio-eustasy was operating, this hierarchy of cyclicity takes on added significance, as it may reflect superimposed orders of high-frequency glacio-eustasy oscillating in tune to a Milankovitch beat (see for example, de Boer, 1991).

In an attempt to unravel the effects of Milankovitch glacio-eustasy upon the origin of the Paradox shelf cycles, we have integrated outcrop and subsurface data sets within the Paradox shelf and applied cycle stacking pattern analysis to establish a hierarchy of depositional cyclicity. This analysis includes incorporation of the vertical facies successions that make up the highest frequency cycles. Cycle stacking pattern analysis has been complemented by time-series analysis to search for Milankovitch rhythms. The scale of the outcrop and the correlation to the subsurface also provides a basis for applying sequence stratigraphic concepts to the problem, and integrating sequence stratigraphy with models of cyclic carbonate deposition. Additionally, the variation of facies and (stratigraphic) cycle hierarchy, combined with the possibility of glacio-eustatic forcing, render the Paradox shelf a stratigraphic package well-suited to forward modelling. The forward stratigraphic simulation techniques allow us to demonstrate the mechanics of cycle hierarchy development and to test a Milankovitch model.

GENERAL SETTING

The Paradox Basin is a northwest–southeast trending, asymmetric, evaporite basin of Pennsylvanian age located in southeastern Utah and southwestern Colorado (Figs 3 and 4). The regional geology and tectonic evolution of this area have been reviewed by numerous authors (Peterson & Hite, 1969; Baars, 1988; Stevenson & Baars, 1988). Stevenson & Baars (1988) suggest the Paradox Basin is a pull-apart basin formed primarily by right-lateral extension along two northwest–southeast oriented master faults.

Subsidence analysis of the Palaeozoic section from the Paradox shelf illustrates the area's tectonic development (Fig. 5). This analysis employs a simple one-dimensional Airy model in which the lithosphere responds to sediment loads by local isostatic adjustments. The backstripping procedure used here was outlined by Steckler & Watts (1978) and Bond & Kominz (1984). Following a pre-Pennsylvanian phase of general stability, the Desmoinesian Paradox shelf experienced increased rates of subsidence in response to the development of the Marathon–Ouachita convergent orogenic front located to the south (Pindell & Dewey, 1982; Ross; 1986). This trend continued through the Wolfcamp stage. Despite this pulse of increased subsidence, Paradox shelfal carbonates easily kept pace with subsidence.

The pre-Pennsylvanian stratigraphy across the Colorado Plateau records widespread stable shelf conditions (Ohlen & McIntyre, 1965). Late in Mississippian time regional uplift led to exposure and karstification of Mississippian carbonates and to the development of extensive red palaeosols. The overlying Pennsylvanian section is 900–1200 m thick on the Paradox shelf (Ohlen & McIntyre, 1965). The initial marine transgression is recorded in reworked soil material of the Molas Formation (Atokan) and mixed clastics and carbonates of the Pinkerton Trail Formation (Figs 6 and 7).

The Hermosa Group comprises the Pinkerton Trail Formation (Atokan), the Paradox Formation (Desmoinesian) and the Honaker Trail Formation (Desmoinesian–Virgilian; Fig. 6). The Paradox

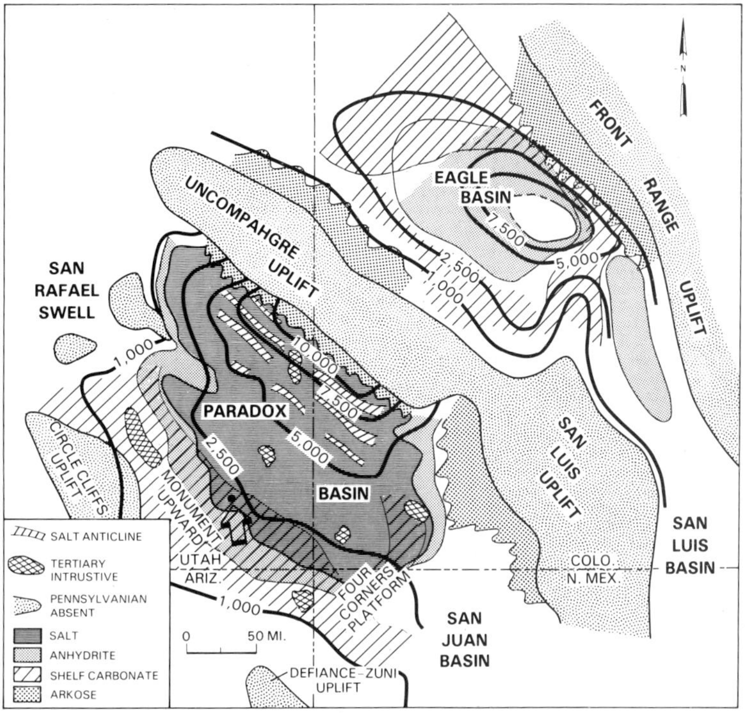

Fig. 3. Isopach and facies map of the Pennsylvanian System of the Paradox and Eagle basins. Isopachs in feet (0.304 m). Modified from Peterson & Hite (1969). Large arrow in southwest quadrant of figure marks location of the Honaker Trail.

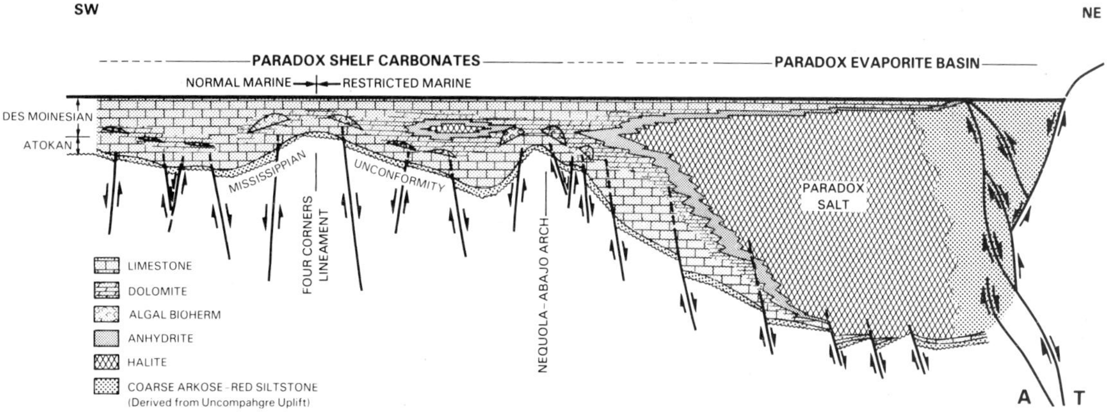

Fig. 4. Generalized SW–NE cross-section across the Paradox Basin illustrating the gross facies relations between Middle Pennsylvanian shelf carbonates, basinally restricted evaporites, and coarse clastics proximal to the Uncompahgre Uplift. Modified from Baars (1988).

Formation includes both the Desmoinesian basinal evaporites and their shelfal equivalent, which consists of carbonate, and secondary interbedded siliciclastics. The basinal evaporite section contains 29 salt–shale cycles (Peterson & Hite, 1969), which decrease in number laterally as the lower cycles onlap up-dip into the shelfal sections, where equivalent carbonate–shale cycles are recognized (Fig. 7; Peterson & Hite, 1969).

In Desmoinesian time, up to 1500 m of bedded salt accumulated in the rapidly subsiding basin centre (Figs 3 and 7). The halite facies changes to anhydrite

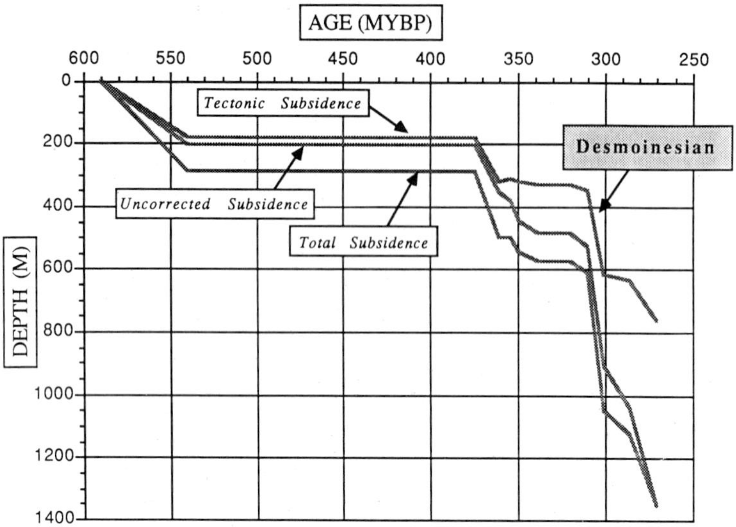

Fig. 5. Subsidence analysis of Paradox shelf based on age–depth pairs from the Texaco Johns Canyon no. 1 well located in San Juan County, Utah (R. 18 E, T. 41 S.; sec. 6). Palaeozoic time scale is that of Harland *et al.* (1982). There are three curves on the plot: the top curve is the tectonic subsidence curve, which depicts non-isostatic basement subsidence; the middle curve is a plot of present-day rock thickness vs. time; the lower curve is the total subsidence curve, which includes the contributions to subsidence by sediment loading, compaction, palaeobathymetry and basement subsidence. Originating in the Desmoinesian, the Paradox shelf experienced increased tectonic subsidence in response to the Marathon–Ouachita orogenic event to the south.

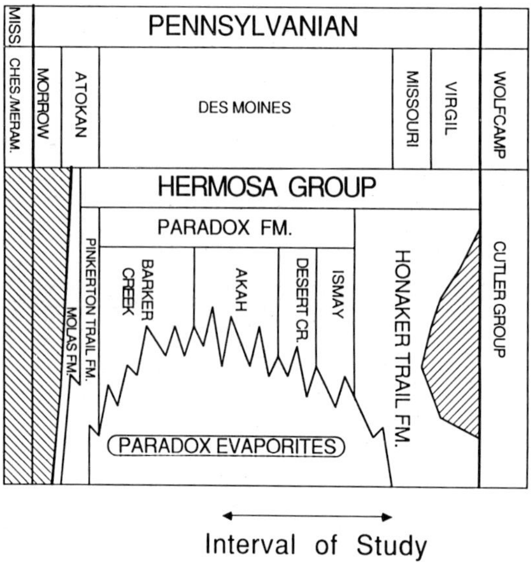

Fig. 6. Pennsylvanian chronostratigraphy of the Paradox Basin. Modified after Baars & Stevenson (1982).

facies at the up-dip limit of the restricted basin. These evaporites onlap the flanks of an extensive carbonate shelf that developed along the southern and western margins of the basin. Depositional strike within the southwest Paradox shelf is northwest–southeast, with very gentle dips to the northeast. The restricted nature of the Paradox Basin resulted from the existence of uplifts that bounded the basin (Fig. 3). The only accessible, unrestricted pathways to the open sea were to the southeast and the southwest (Peterson & Hite, 1969). The cyclic nature of the Paradox shelf carbonates and the basinal evaporite–shale deposits resulted from oscillations in relative sea level (Peterson & Hite, 1969). During relative sea-level rise the narrow seaways of the Paradox shelf were flooded, resulting in shelf carbonate deposition and basin starvation. With relative sea-level fall, the carbonate shelf was exposed, causing restriction of the basin and evaporite deposition in the basin centre. Thus during times of relative lowered sea level the Paradox carbonate shelf acted as a topographic barrier restricting influx of normal sea water into the evaporite basin (Peterson & Hite, 1969; Hite, 1970).

METHODS

A continuous Middle Pennsylvanian section is exposed along the Honaker Trail in the San Juan Canyon of southeast Utah (Fig. 8). Our section begins in the upper part of the Akah interval of the Paradox Formation (middle Desmoinesian) and ends within the Honaker Trail Formation at the top of the Desmoinesian (Figs 9 and 11). The outcrop section was correlated to the subsurface with well-logs, which provided a regional stratigraphic framework (Figs 8, 12 and 13). The outcrop strata are well-exposed (Figs 17, 18 and 19), facilitating the tie to the subsurface. From the McElmo Creek oil field,

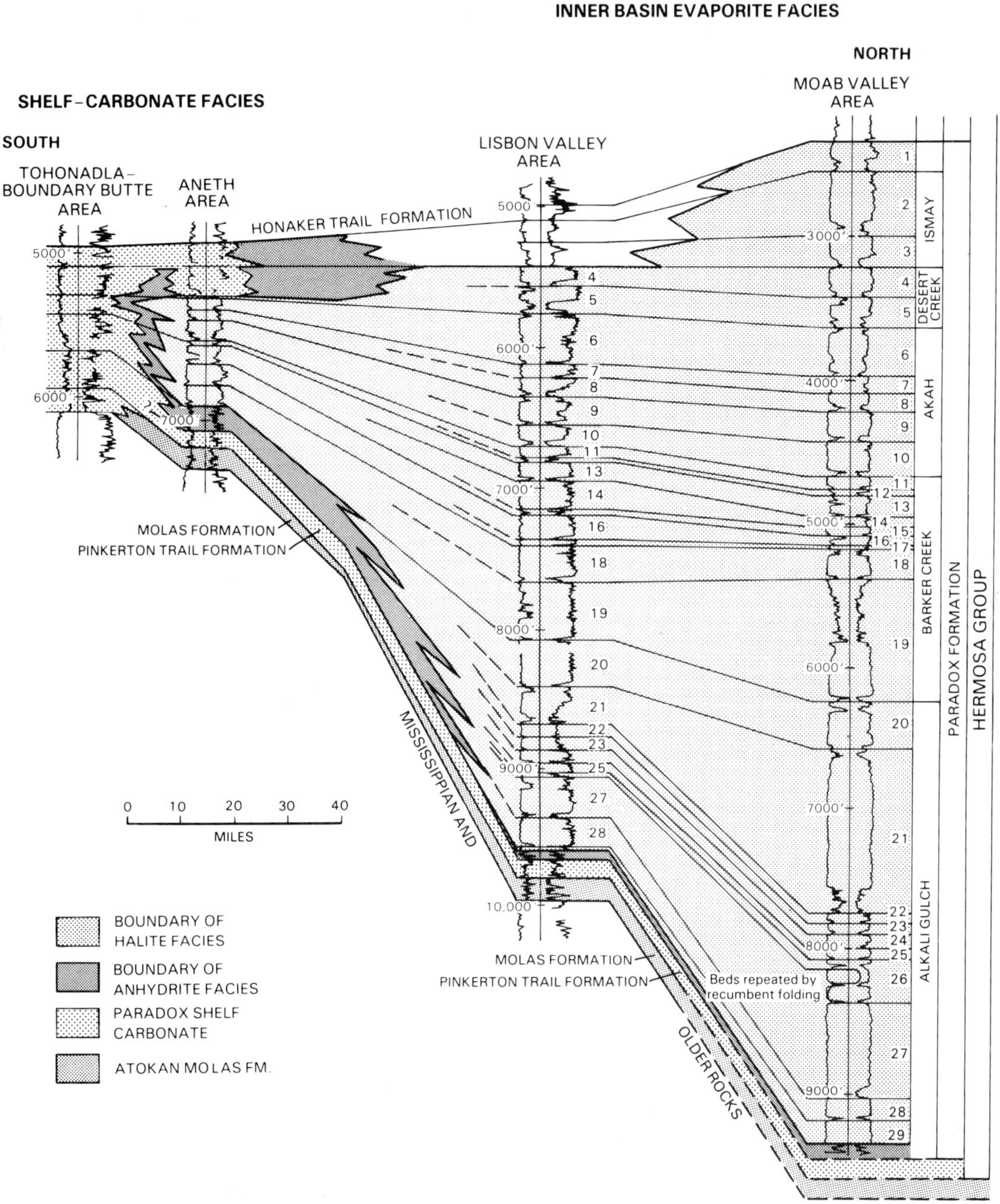

Fig. 7. Correlation of well logs across the Paradox Basin showing relations between shelfal carbonate facies and basinal anhydrite and halite facies. Note the 29 evaporite–shale, basinal cycles of the Paradox Formation. These regionally correlative cycles equate to the fourth-order depositional sequences recognized at the Honaker Trail section. Modified from Peterson & Hite (1969).

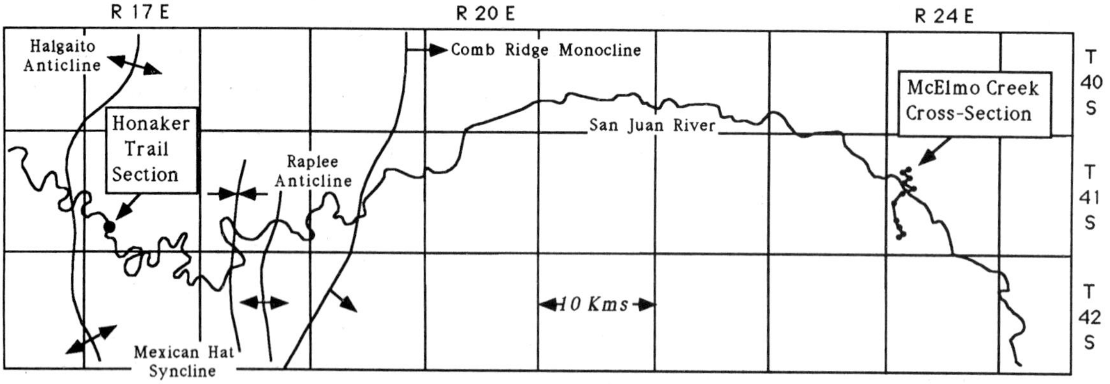

Fig. 8. Location map of Honaker Trail section and well-logs used in construction of McElmo Creek cross-section, displayed in Fig. 12. Refer to Goldhammer *et al.* (1991) for well-log cross-sections linking outcrop locality to subsurface stratigraphy at McElmo Creek.

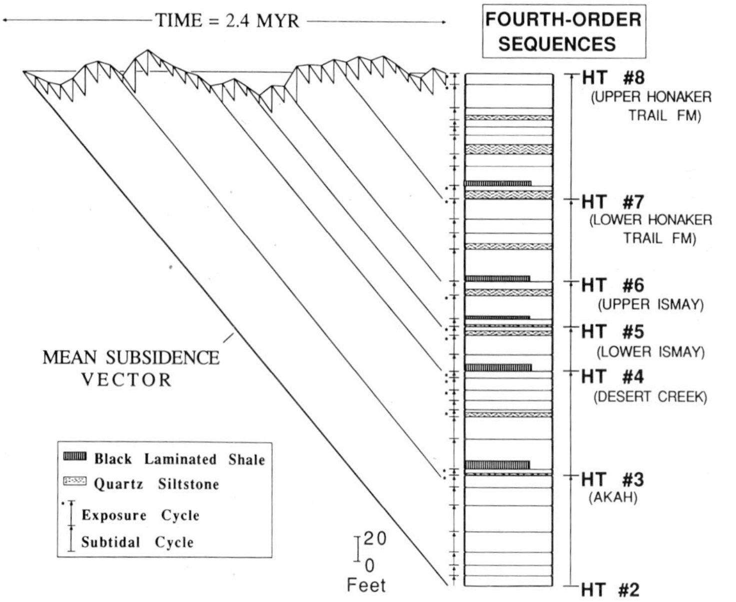

Fig. 9. Simplified stratigraphy of the Honaker Trail section and 'Fischer plot', emphasizing fifth-order cycle stacking patterns within fourth-order sequences. Fifth-order cycles marked with an asterisk are exposure cycles (see text). Black laminated mudstone facies depicted by even parallel line pattern, quartz sandstone facies by dashed pattern. On the 'Fischer plot' the horizontal axis represents relative time and individual fifth-order cycles are plotted as triangles evenly spaced along the axis, on the assumption that each cycle is of equivalent duration. The vertical axis is the Honaker trail measured section, and the straight line that connects the base of the stratigraphic section to time zero is the mean subsidence vector, depicting net subsidence. The changes in slope over the entire string of cycles reflect deviations in long-term accommodation.

75 km east of the Honaker Trail section (Fig. 8), a series of cores were described in detail from one of the Paradox 'cycles' (Desert Creek interval of the Paradox Formation). The stratigraphic cross-section constructed from the core descriptions enables us to delineate the transition from shelf carbonates to low-stand evaporites and siliciclastic deposits that accumulated in topographic lows (or intra-shelf depressions) on the Paradox shelf (Fig. 13). This provided some two-dimensional control and allowed us to visualize 'missed beats' of platform cyclicity preserved in a down-dip position, essential for constructing a proper cycle hierarchy. Such 'missed beats' are obviously not recorded on top of the shelf (i.e. at the Honaker Trail locality).

DEPOSITIONAL FACIES

The Paradox shelfal carbonate facies have been studied in much detail by numerous workers because of their significance as petroleum reservoirs (e.g.

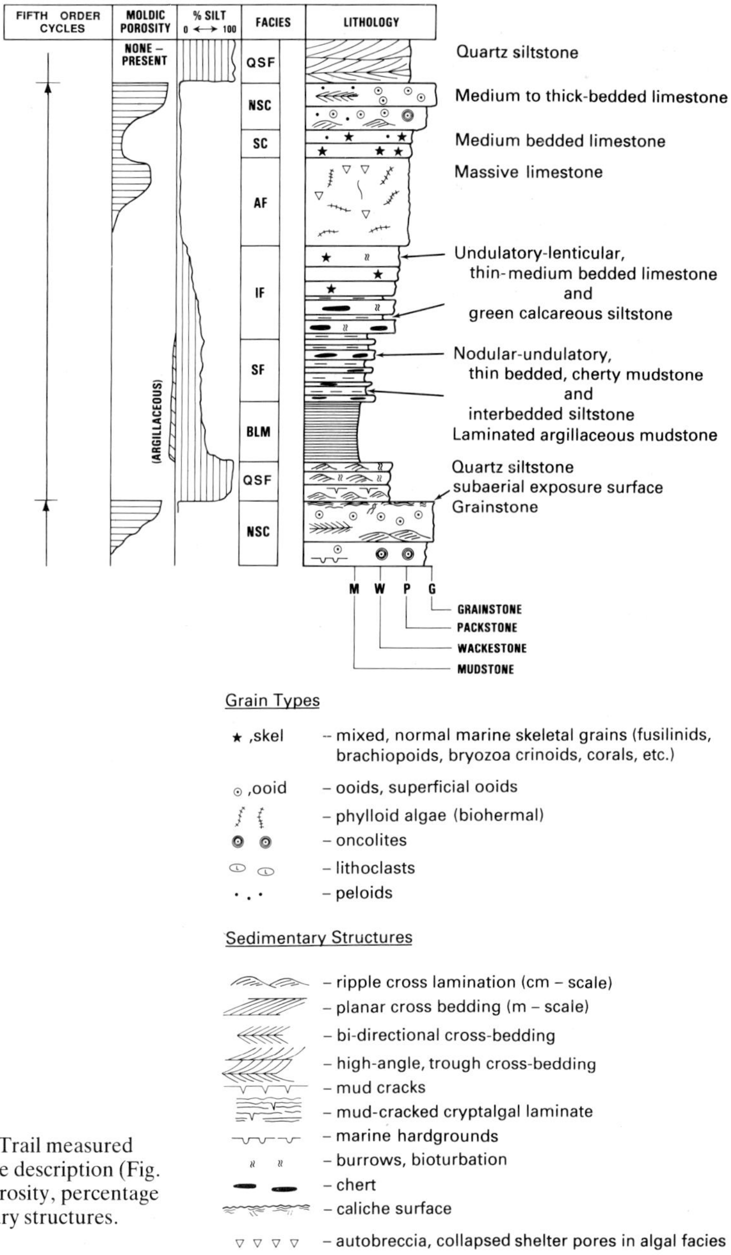

Fig. 10. Facies and legend for Honaker Trail measured section (Fig. 11) and McElmo Creek core description (Fig. 13). Attributes shown include moldic porosity, percentage silt, lithology, grain types and sedimentary structures. Refer to text for facies code.

Choquette & Traut, 1963; Elias, 1963; Peterson & Ohlen, 1963; Pray & Wray, 1963; Peterson & Hite, 1969; Hite, 1970; Hite & Buckner, 1981; Choquette, 1983). The depositional facies, their inferred palaeo-water depths, and early diagenetic attributes of the Honaker Trail section and the McElmo Creek cores discussed in this study (Fig. 10) have been described and interpreted in detail by Goldhammer *et al.* (1991; Table 2) and only a brief summary is provided here. We recognize the following depositional facies.

1 *Quartz sandstone facies* (QSF): trough cross-bedded, calcareous sandstone and laminated, mudcracked siltstone; shallow marine to tidal flat,

Table 2. Summary of Paradox shelf carbonate facies

Facies designation	Quartz sandstone facies (QSF)	Black laminated mudstone (BLM)	Sponge facies (SF)	Intermediate facies (IF)
Bedding and sedimentary structures	Laterally discontinuous lenses to continuous sand sheets (<1–5 ft thick); medium-scale, wedge-shaped, low-angle trough cross-bedding; ripple cross-lamination; burrows; mudcracks	Thin (<1–2 inch thick), varve-like, even, planar lamination	Undulatory-nodular thin bedding; fine (<1–2 inches thick), wispy-planar lamination; slump features	Undulatory, thin–medium bedding; burrow mottling
Rock type and depositional textures	Calcareous siltstones to sandstones; coarse silt to very fine sand-sized quartz; very well sorted, angular–subrounded; no clay fraction	Dolomitic shales to silty carbonate mudstones	Argillaceous, cherty carbonate mudstone to calcareous siltstone	Silty mudstone–wackestone, with local coarse packstone lenses
Grain types	Quartz sand; 10–30% carbonate peloids, ooids, skeletal material (worn, abraded, well-sorted); angular, carbonate lithoclasts; traces of glauconite	High organic content (sapropelic); quartz silt (35%), clays (15–25%), carbonate peloids (20%); rare conodonts, bone fragments, fish teeth, thin-shelled phosphatic brachiopods	Quartz silt (<5–40%); clays (<5–10%); carbonate peloids; siliceous sponge spicules; rare echinoids, phosphatic brachiopods, conodonts	Normal marine skeletal grains (crinoids, brachiopods, bryozoa, fusilinids, corals, forams) and peloids; up to 30% quartz silt
Conspicuous early diagenetic phenomena		Mechanical compaction	Mechanical compaction	
Depositional environment	Transported during low-stands of sea level by aeolian processes, modified with marine overprint with ensuing transgression; shallow-marine, high-energy, shoreface deposition	Deeper subtidal (>35-m water depths) deposition under marine reducing conditions; minimum water turbulence; quiet, toxic bottom conditions	Deep subtidal (25–35-m water depths), anaerobic to dysaerobic deposition under restricted, hypersaline conditions	Moderately shallow (5–25-m water depths), well-oxygenated, normal marine deposition

probably reworked aeolian siliciclastics delivered during sea-level low-stands.

2 *Black laminated mudstone* (BLM): non-fossiliferous, black, shaley mudstones with centimetre-scale parallel laminae; low-energy, anaerobic conditions, >35-m water depth (arguments for palaeowater depths discussed in full by Goldhammer *et al.* 1991).

3 *Sponge facies* (SF): silty, argillaceous, cherty carbonate mudstone with abundant sponge spicules and rare phosphatic debris; low-energy, dysaerobic conditions, 25–35-m water depth).

4 *Intermediate facies* (IF): thin-bedded, silty, mixed skeletal wackestone–packstone full of normal marine fossils (crinoids, brachiopods, Bryozoa, etc.) and burrows; moderate energy, normal marine subtidal, 5–25-m water depth.

5 *Algal facies* (AF): phylloid algal mound and mound-flank wackestone–packstone; normal marine, subtidal phylloid algal bioherms, 5–25-m water depth).

6 *Skeletal cap facies* (SC): well-sorted, abraded, mixed skeletal packstone–grainstone; shoaling subtidal, 0–5-m water depth.

7 *Non-skeletal cap facies* (NSC): trough cross-bedded and ripple cross-laminated oolitic–peloidal grainstone; high-energy shoaling subtidal; 0–5-m water depth.

8 *Caliches/subaerial exposure surfaces*: surfaces containing evidence for vadose/meteoric exposure

Table 2. *Continued*

Facies designation	Algal facies (AF)	Skeletal cap facies (SC)	Non-skeletal cap facies (NSC)	Caliches/subaerial exposure surfaces
Bedding and sedimentary structures	Flat-based, convex-upward mounds (20–40 ft thick; 30–90 ft long); internally massive core; mound flanks display low-angle accretionary bedding: laterally adjacent to IF	Medium-thick bedding: small-scale, low-angle trough cross-stratification; ripple cross-lamination; burrows	Medium-thick bedding; medium-scale trough cross-stratification; planar cross-stratification; wave ripple and current ripple cross-lamination; hardground surfaces	Laterally discontinuous, irregular laminated thin (<2–6 inches thick) crusts; subvertical cracks and subcircular vugs and pipes lined with laminated, brecciated carbonate
Rock type and depositional textures	Grain-supported algal bafflestone and algal-rich wackestones; mound flank to foreslope lithoclastic wackestone to packstone	Wackestone to dominantly packstone; medium to coarse sand-sized allochems	Packstone to dominantly grainstone; well-sorted, medium to coarse sand-sized allochems	Wackestone to packstone
Grain types	Phylloid algae (*Ivanovia*) form framework; admixed peloids and normal marine skeletal grains, intraclasts	Abundant worn, abraded normal marine skeletal grains (crinoids, brachiopods, fusilinids, bryozoa); encrusting forams (opthalmidids); micritized grains	Abundant non-skeletal allochems (ooids, peloids, oncolites, intraclasts, coated grains); admixed abraded skeletal debris; 5% rounded quartz sand	Micritic peloids, angular, black lithoclasts and micritic lumps; admixed quartz silt; altered primary carbonate (micritized)
Conspicuous early diagenetic phenomena	Syndepositional 'autobrecciation' of algal wackestone; fabric-selective dissolution of phylloid algal plates and solution-enlarged porosity; meteoric, phreatic cementation	Minor fabric-selective, solution porosity	Solution-enlarged, moldic porosity (leached ooids, peloids, molluscs); freshwater phreatic cementation and neomorphic recrystallization	Moldic, solution-enlarged porosity; neomorphic recrystallizaton; extensive micritization; geopetal, internal sediment; primary grain alteration
Depositional environment	Moderately shallow (5–20-m water depths), well-circulated, oxygenated, normal marine deposition	Shallow subtidal to near shoal-water deposition (0–5-m water depths)	High-energy, very shallow subtidal to lower intertidal shoal-water deposition (0–5-m water depths)	Subaerial exposure horizons with alteration of primary carbonate; caliche formation

characterized by moldic porosity, vugs, neomorphic recrystallization, and thin laminated caliche crusts.

In the Desert Creek subsurface interval at Aneth Field, two facies occur restricted solely to the most basinward well (Carter Navajo 115-1; Figs 12 and 13). This well occupies a depositional low, occurring within an intra-shelf depression on the greater Paradox shelf. These facies are an *evaporite facies*, consisting of nodular mosaic anhydrite and a *foreslope facies* consisting of poorly sorted angular litho-clastic-skeletal wackestone–packstone. The evaporite is interpreted as subaqueous because it shallows into mudcracked cryptalgal laminite. The foreslope debris beds are largely composed of allochthonous, phylloid algal material and muddy clasts, interpreted as mound-flank deposits.

SEQUENCE STRATIGRAPHY AND STRATIGRAPHIC CYCLICITY

Low-frequency, fourth-order depositional sequences

Based on our interpretation of the cyclic stratigraphy and the age model presented below, we divided the Desmoinesian outcrop and subsurface stratigraphy into regionally correlative, fourth-order depositional sequences (Figs 9, 11 and 12). At the Honaker Trail section, eight sequences are recognized from the base of the outcrop (base Akah) to the top of the

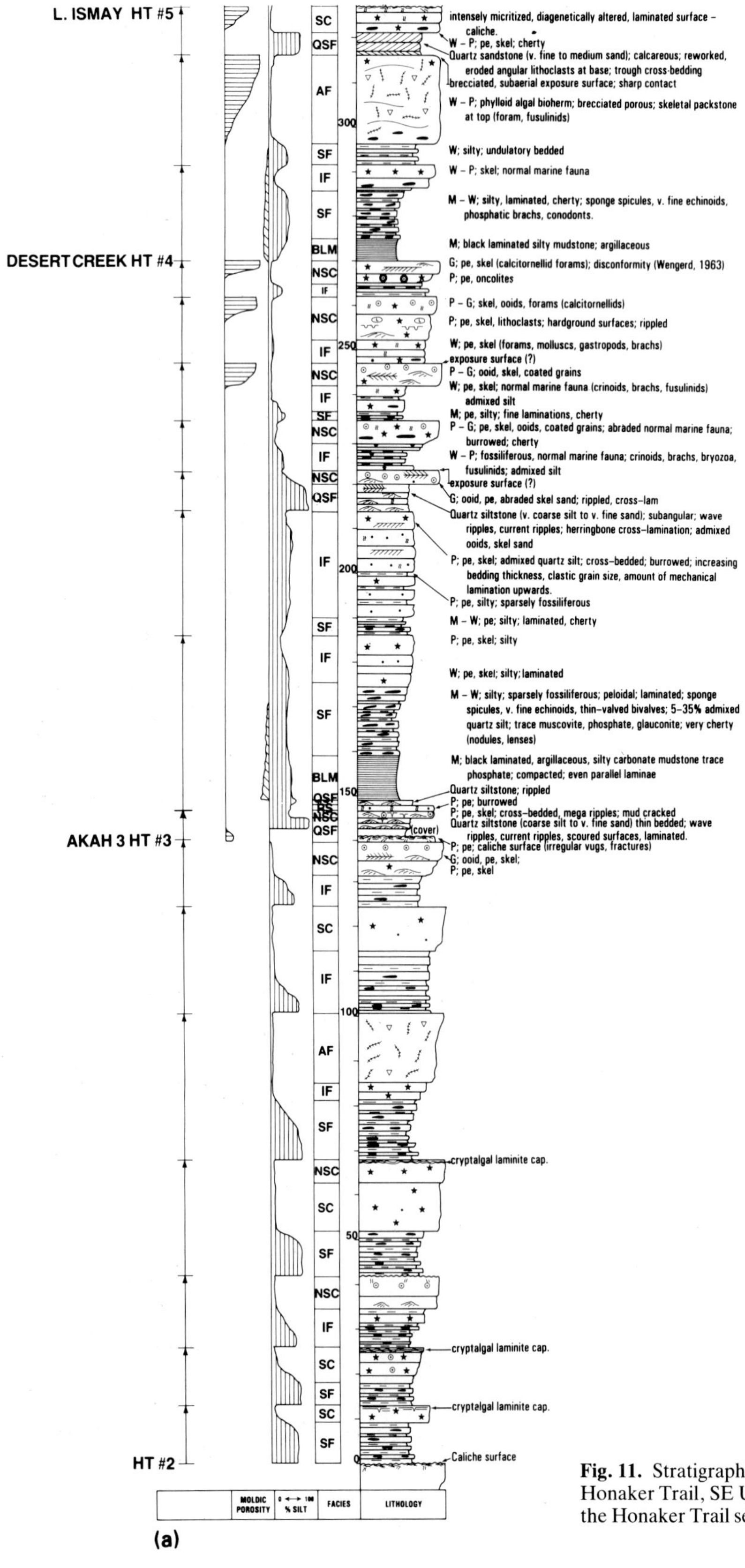

Fig. 11. Stratigraphic section measured along the Honaker Trail, SE Utah. Refer to Fig. 14 for a view of the Honaker Trail section.

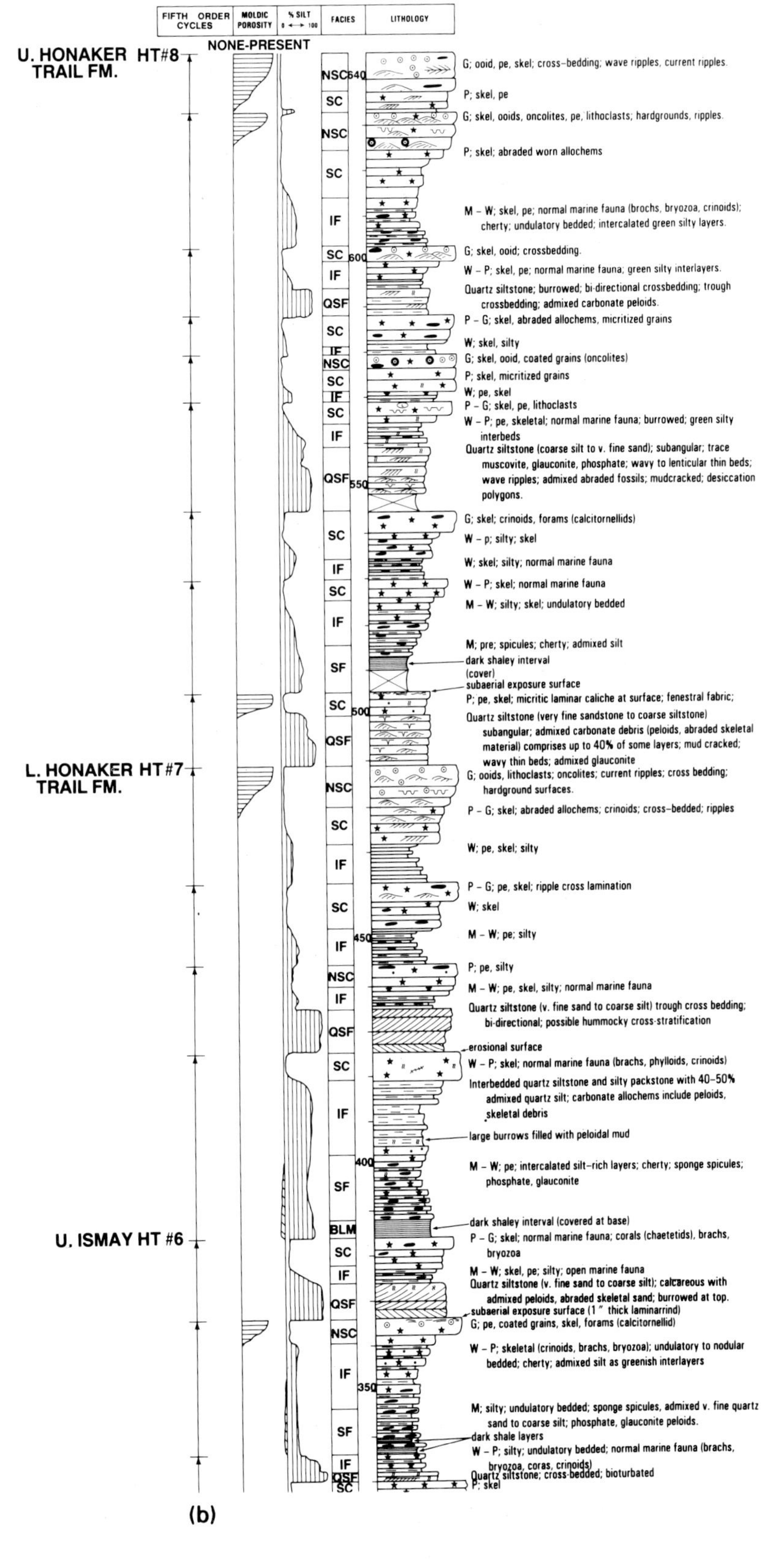

Fig. 11. *Continued.*

Desmoinesian (Figs 9, 11, 14–16). They average 35 m in thickness, ranging from 17 to 48 m thick. The sequences are labelled Honaker Trail sequence #1 through to #8 and they are correlated to the subsurface productive intervals. At McElmo Creek, the cored interval occupies one complete sequence, sequence HT#4 (Desert Creek interval), which ranges in thickness from 45 to 60 m (Fig. 13).

The following criteria were used in recognizing the sequences.

1 All sequences to be bounded by regionally correlative surfaces (sequence boundaries) that contain evidence for subaerial exposure. These surfaces are traceable from outcrop to the subsurface, where they can be correlated over 100–120 km across the Paradox shelf (Goldhammer *et al.*, 1991). At the Honaker Trail, the best-developed caliches occur at these fourth-order boundaries (e.g. top HT#2, HT#3, HT#5). These boundaries to be also marked by significant moldic porosity, recrystallization and inferred freshwater cementation (Goldhammer *et al.*, 1991).

2 Sequence boundaries to be onlapped or draped by evaporite deposits which are interpreted as low-stand deposits (Goldhammer *et al.*, 1991). The anhydrites were deposited during periods of restriction within intra-shelf depressions associated with sea-level falls to beneath the Paradox shelf (Peterson & Ohlen, 1963; Peterson & Hite, 1969; Hite, 1970). Low-stand evaporites restricted to such intra-shelf topographic lows onlap the top of sequence HT#3 (Goldhammer *et al.*, 1991), the top of sequence HT#4 (Fig. 12) and the tops of sequences HT#5 and #6 (Fig. 12).

3 Sequences to be marked by the occurrence of the BLM facies near or at their bases. The BLM facies is interpreted as the superposition of the fourth-order condensed interval atop thin, shelfal low-stand–transgressive deposits (Figs 9 and 12). The BLM facies, which is thin over the shelf, thickens down-dip into more basinal or topographically lower settings.

4 High-frequency cycle stacking patterns within the sequences are also diagnostic. Two types of cycles are recognized in this study, based on physical bounding surfaces: cycles bounded by marine flooding surfaces, termed *subtidal cycles*, and those bounded by subaerial exposure surfaces, *exposure cycles*. In general, fifth-order cycles tend to thin upward within a sequence, as is well illustrated in the HT#3 and HT#4 sequences (Figs 9 and 15). The use of 'Fischer plots' (Fischer, 1964) clarifies this pattern, as groups of cycles plot as relative rising and falling 'waves' within sequences (Fig. 9). The most 'complete' sequences (HT#2, HT#3, HT#8) exhibit these stacking patterns fairly well. Cycles thin upward within a sequence due to the increase and subsequent decrease in long-term accommodation associated with each sequence (e.g. Read and Goldhammer, 1988). Additionally, subtidal cycles tend to cluster at the base of sequences, whereas exposure cycles dominate the upper portions (Fig. 9).

We use the term 'complete' sequence to describe the thickest sequences that contain the highest number of higher-frequency, small-scale cycles. On 'Fischer Plots' these sequences plot as well-defined rise–fall waves. In contrast, 'incomplete sequences' are thinner, contain fewer higher-frequency cycles, and do not plot as well-defined rise–fall waves on 'Fischer plots'. The master sequence boundaries which correlate regionally define the sequences.

Systems tract development of fourth-order sequences

The fourth-order sequences can be subdivided into component systems tracts in both the outcrop (Fig. 16) and subsurface (Fig. 12). All of the Paradox sequences examined in this study are type 1 sequences bounded by type 1 sequence boundaries (Vail *et al.*, 1984; Vail, 1987) interpreted to have formed when fourth-order sea level fell beneath the Paradox shelf.

The low-stand systems tract (LST) consists of wedges of evaporite and quartz siliciclastics restricted to topographically low intra-shelf depressions (Figs 12 and 13). Low-stand wedges range in thickness from 6 to 18 m and contain a lower evaporite-rich portion composed of mosaic anhydrite and an upper siliciclastic-rich portion. The wedges occupy depositional troughs and laterally thin by onlap against carbonate depositional highs on the shelf. In the McElmo Creek core cross-section (Fig. 13), the low-stand wedge (Carter Navajo 115-1 well) contains two onlapping high-frequency cycles that are restricted within a topographic low. These cycles indicate the persistence of fifth-order relative sea-level oscillation during times of fourth-order low-stand. In detail (Fig. 13) the lower cycle shoals from subaqueous nodular, mosaic anhydrite to mud-cracked cryptalgal laminite. The depositional relief at the top of this cycle indicates a minimum relative fourth-order sea fall of about 25–35 m.

The upper cycle shoals from restricted muddy carbonate (lagoon) to laminated calcareous siltstone with desiccation cracks, scours and starved ripples (QSF facies). This siltstone can be traced up-dip to the shelf edge (Carter Navajo 114-7 well), where it thins from about 6 m to less than 1 m. Here the siltstone sharply overlies shelfal carbonate and contains within it reworked cobbles of lithified oolitic grainstone full of moldic porosity. This indicates the following sequence of events: (i) deposition of oolitic grainstone on the shelf as the capping phase of a high-frequency cycle (cycle #7); (ii) exposure of the shelf and subaerial leaching developing moldic porosity, probably contemporaneous with deposition of the lower onlapping cycle in the low-stand wedge; (iii) siliciclastic transport across the shelf feeding the upper onlapping cycle, with erosion of leached grainstone and formation of oolitic lithoclasts.

At the shelfal, Honaker Trail section, fourth-order low-stands are marked mainly by low-stand subaerial diagenesis (caliches, moldic porosity, etc.). At a few of the boundaries (e.g. HT#4) siliciclastic sands fill minor incisions in the underlying carbonate, and are interpreted as fourth-order low-stand drapes on the shelf. However, as alluded to earlier, not all sequence boundaries are overlain by quartz siltstones.

The transgressive systems tract (TST) deposited during fourth-order rise in sea level is composed primarily of shales and shaley carbonate mudstones (BLM facies) that represent the condensed section of the sequence (Figs 11 and 12). These shaley intervals thicken into the basin (up to 10–13 m thick) and thin up-dip onto the shelf (less than 1–3 m thick), essentially mantling the underlying sequence boundaries. Typically, however, there is a thin (less than 2–3 m thick) transgressive oolitic grainstone between the underlying fourth-order sequence boundary and the condensed section. The transgressive systems tract may contain one or two recognizable thin high-frequency cycles (e.g. base of sequence HT#4). The top of the transgressive systems tract, termed the maximum flooding surface (MFS; Vail, 1987) is always marked by the vertical facies change from BLM to SF, and in outcrop it lies within the recessive weathering part of the sequence (Fig. 16). No downlap of strata against this surface is observed.

The high-stand systems tract (HST), which records a decline in the rate of fourth-order rise and the early portion of the fourth-order fall, is characterized by an aggradational stack of shelfal carbonate (up to 65 m thick) and a thinner carbonate section within intra-shelf depressions (Figs 12 and 13). The depositional topography and facies make-up of the McElmo Creek cross-section (Fig. 13) can be used to divide the high-stand systems tract into two phases, an early high-stand and a late high-stand. The early high-stand phase consists of stacked, aggradational, high-frequency subtidal cycles composed primarily of sponge facies shoaling into algal facies, with a thin skeletal cap. Shelf-to-basin depositional relief progressively increases with each successive cycle. The late high-stand phase consists of stacked, aggradational, high-frequency exposure cycles dominated by intermediate facies shoaling to skeletal or non-skeletal capping facies. Shelf-to-basin depositional relief is maintained. High-frequency exposure cycles are also characteristic of the late high-stand in the sequence at the Honaker Trail section.

High-frequency, fifth-order depositional cycles

Fourth-order sequences are constructed of higher-frequency cycles that are fifth order based on the age model developed below. At the Honaker Trail section, 34 fifth-order cycles are recognized (average 6 m/cycle; Figs 9, 11 and 15). In the subsurface interval of sequence HT#4, seven shelfal cycles occur (average 7 m/cycle; Fig. 13). The high-frequency cycles are apparently regionally correlative, at least from the Honaker Trail section to McElmo Creek within sequence HT# 4. Overall, the cycles form dominantly aggradational, laterally continuous stacks which internally contain little evidence for lateral progradation, such as offlapping lateral facies changes (e.g. Fig. 13).

Cycles are recognized on the basis of vertical facies associations that shallow upward (Figs 11 and 13). Internally, boundaries between facies components are transitional. Evidence for shoaling within individual cycles includes: (i) tendency for grain-supported textures and grain size of carbonate allochems to increase upwards; (ii) upward increase in mechanically generated sedimentary structures, such as ripples and cross-bedding within the carbonate units of a cycle; (iii) upward increase in non-skeletal high-energy components, such as ooids, superficial ooids, oncolites and intraclasts; (iv) upward increase in normal marine fauna, which may be abraded and well sorted towards the top of the cycle; (v) occurrence of caliches and other indicators of subaerial exposure at the tops of some cycles; (vi) vertical decrease in quartz silt content upward; and (vii) overall vertical change in colour, Fe-sulphide and organic matter content, from dark and pyritic (e.g. BLM, SF) to light and non-pyritic (e.g. SC, NSC, etc.).

Over half of the cycles at the Honaker Trail and in the McElmo Creek example are *exposure cycles* that have upper boundaries marked by subaerial exposure features (caliches, moldic porosity, recrystallization, and minor mudcracked peritidal laminites). Exposure cycles may have erosional tops (e.g. cycle #17 at the Honaker Trail, cycle #7 at McElmo Creek) and are always overlain by deeper-water carbonate facies. The remainder of the cycles are *subtidal cycles* that have upper boundaries marked by an abrupt deepening.

Age model for Desmoinesian cyclicity

Prior to any interpretation as to the origin of the cyclicity, it is essential to attempt to establish the durations of the various orders of cyclicity. As pointed out previously, the thick evaporite depocentre of the Paradox basin contains 29 regionally correlative 'shale'–evaporite 'cycles' of Desmoinesian age (Peterson & Hite, 1969; Hite, 1970; Fig. 7). These 'cycles', which equate to the shelfal 'sequences', decrease in number towards the Paradox shelf due to onlap of the lower cycles up-dip. The 'shale' members of these cycles correlate to the shaley, BLM interval that occupies the lower parts of the equivalent fourth-order, shelfal sequences (Peterson & Hite, 1969; Hite, 1970; Hite & Buckner, 1981). Due to differences in relative subsidence between the Paradox shelf and the evaporite basin, the shelfal section recorded only a few 'shale'-based sequences (Fig. 4).

In order to ascertain the correct duration of the 'shale'-based sequences, the total number of *basinal* sequences (29) is simply divided by the total amount of time represented by the Desmoinesian (Table 3). If one takes the total number of *shelfal* sequences (8) and divides by the total time, an incorrect value for the duration of the sequences would be obtained, as the shelf did not record the deposition of all the sequences. From the number of low-frequency sequences (29) and the radiometric age determinations for the Desmoinesian, the average periodicity of the sequences is between 138 and 345 ka per sequence (Table 3). Where errors are included in the assessment of age determinations (e.g. Harland *et al.*, 1990), minimum and maximum values can be calculated. Minimum values fall in the fifth-order range (Table 3), but on average the low-frequency sequences are *fourth-order* (0.1–1.0 Ma duration; Table 1).

The low-frequency, fourth-order sequences consist of stacks of higher-frequency cycles which form the fundamental units of the sequences (Figs 9, 11 and 15). At the Honaker Trail, the number of high-frequency cycles per sequence ranges from three to nine. Because of the likelihood of missed fifth-order beats up on top of the shelf (similar to the Pleistocene example illustrated in Fig. 1), it is feasible that not every fourth-order sequence will record the maximum number of higher-frequency fifth-order sea-level oscillations, hence the variation in number of higher-frequency cycles per sequence. This is

Table 3. Summary of Desmoinesian age estimates (duration in Ma) and calculated periodicities of low-frequency (fourth-order) sequences and high-frequency (fifth-order) cycles. Minimum and maximum values are given for accompanying age determinations that include errors. Calculation of high-frequency cycle periods assumes a maximum ratio of nine cycles per sequence (see text)

Source	Duration (Ma)	Periodicity of low-frequency sequences			Periodicity of high-frequency cycles		
		Minimum	Average	Maximum	Minimum	Average	Maximum
van Eysinga (1975)	10	—	345 000	—	—	38 333	—
Harland *et al.* (1982)	6.25 ± 6	8600	215 527	422 414	955	23 946	46 934
Harland *et al.* (1989)	7 ± 10.3	0	241 379	596 551	0	26 819	66 283
Palmer (1983)	9 ± 10	0	310 345	655 172	0	34 482	72 796
Ross & Ross (1987)	8.5	—	293 100	—	—	32 566	—
Odin & Gale (1982) Lippolt *et al.* (1984) Klein (1990)	4	—	137 931	—	—	15 325	—

demonstrated in our subsurface McElmo Creek example, in which sequence HT#4 contains nine higher-frequency cycles, seven in the shelf and two within the low-stand wedge occupying an intrashelf depression (Fig. 13). In this regard sequence HT#4 is 'incomplete' when viewed only from the perspective of the shelf record. Recall that we use the term 'complete' sequence to describe the thickest sequences that contain the highest number of higher-frequency, small-scale cycles. On the other hand, 'incomplete sequences' are thinner, and contain fewer higher-frequency cycles. The master sequence boundaries which correlate regionally define the sequences, not the number of higher-frequency cycles.

Based on all available data we believe that the maximum ratio of higher-frequency cycles per fourth-order sequence is 9:1 (as indicated by the most 'complete' sequences). Assuming a ratio of 9:1, the periodicity of a high-frequency cycle is derived by dividing the calculated periodicities of the fourth-order sequences by the maximum number of cycles (9) (Table 3). From this calculation, the average periodicity of the high-frequency cycles ranges from 15 325 to 38 333 years per cycle. Taking into account age date errors, even maximum values fall within the realm of fifth-order sequences. Thus, in all likelihood, the higher-frequency cycles are *fifth-order* cycles (Table 3).

The assumption that all fourth-order sequences are of equal duration is a critical one, an assumption that we cannot prove unequivocally at this point. This assumption is somewhat precarious when one considers that there are large climatic variations with periods of tens to a few hundreds of thousands of years which are *irregular* and cannot be accounted for by the Milankovitch theory alone (G. Weedon, personal communication; see for example Ruddiman *et al.*, 1989, and Weedon & Jenkyns, 1990). Additionally, whereas we believe that the fifth-order cycles are of equal duration (i.e. rhythmic), this assumption can also be challenged. For example, Berger (1989) has shown that the precessional cycle is not of constant periodicity, such that over the last 5 Ma he estimates that the number of precessional cycles per short eccentricity cycle varies from three

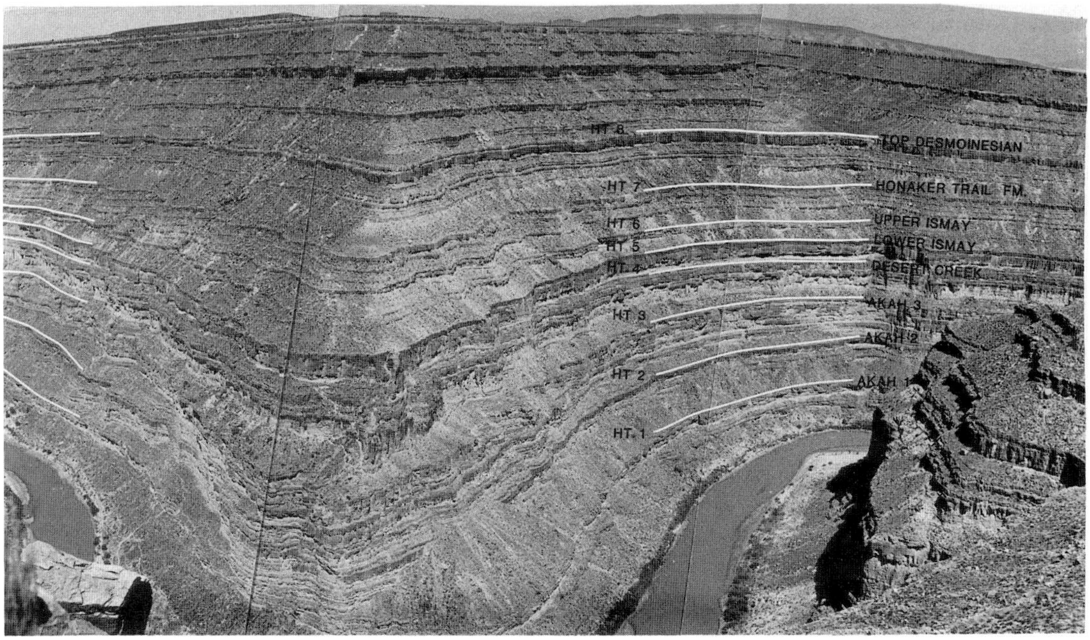

Fig. 14. Panoramic view of the Middle Pennsylvanian section exposed at the Honaker Trail, on the east side of the San Juan River. Fourth-order sequence boundaries are marked by white lines. Note the third-order accommodation cycle marked by progressive thinning upward of fourth-order sequences (from HT#1 to HT#5) followed by progressive thickening upward of fourth-order sequences (HT#6 to HT#8). Total vertical section exposed equal to approximately 400 m.

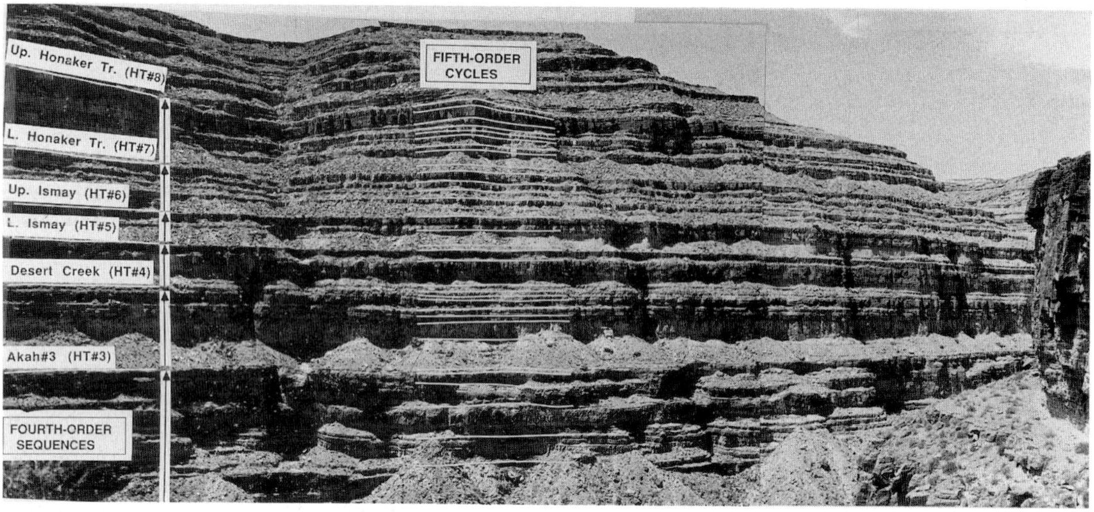

Fig. 15. Panoramic view of the canyon wall on the west side of the San Juan River, directly opposite the Honaker Trail measured section. Fourth-order sequence boundaries and fifth-order cycles are marked and have been correlated to the Honaker Trail measured section.

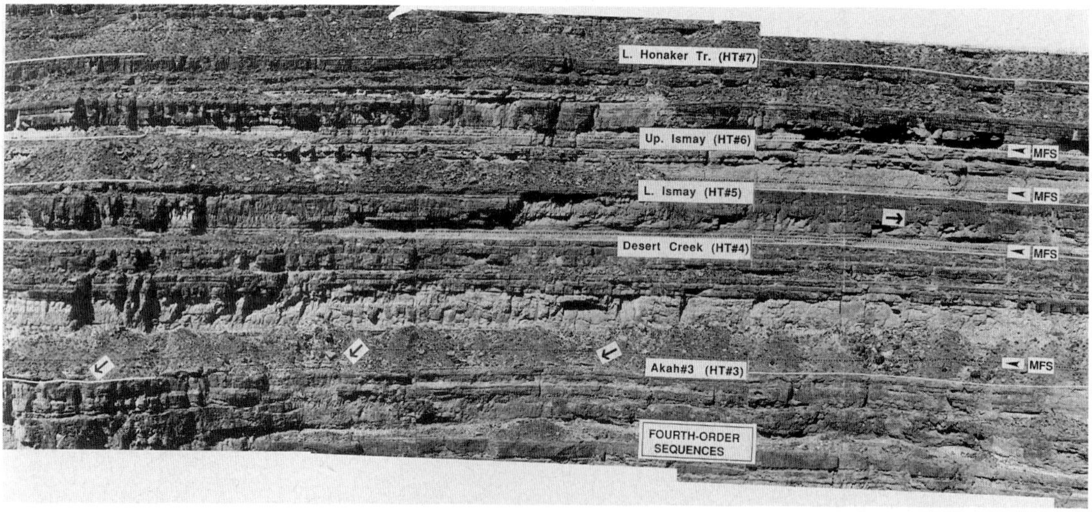

Fig. 16. Panoramic view to the southwest of that shown in Fig. 15. Fourth-order sequence boundaries and maximum flooding surfaces (MFS) are labelled. Note laterally discontinuous drape of quartz sandstone, filling irregular, incised topography atop the HT#3 sequence boundary (arrows, lower left). Note also truncated tops to algal bioherms overlain by quartz sandstone (QSF facies) in sequence HT#5 (arrow, centre right).

to seven (G. Weedon, personal communication). Such sobering messages from the better understood Pleistocene record need to be kept in mind in the interpretation of ancient systems, such as the Pennsylvanian.

The shelfal fourth-order sequences stack to yield a lower-frequency accommodation cycle (Figs 14–16) defined by systematic thinning and thickening of fourth-order sequences. Based on the age model for the fourth-order sequences, this accommodation cycle is third-order in duration.

Interpretation of fourth-order sequence development

Evaluation of autocyclicity

In evaluating criteria for distinguishing allocyclic versus autocyclic mechanisms for generating platform carbonate cycles and sequences, Hardie & Shinn (1986) and Goldhammer *et al.* (1987) pointed out that cycles or sequences composed of subtidal deposits capped by subaerial exposure surfaces must result from *relative sea-level oscillations*, either high-frequency eustasy or tectonic 'yo-yoing'. As reviewed above, autocyclic mechanisms involving progradation must produce shallowing-upwards cycles or sequences capped by a progradational peritidal facies. All Paradox fourth-order sequences are bounded by regionally correlative sequence boundaries that contain evidence for subaerial exposure (laminar caliches plus significant moldic porosity, recrystallization, etc.). The surfaces are superimposed directly on subtidal deposits and thus indicate relative drops in fourth-order sea level, ruling out autocyclicity. In the McElmo Creek oilfield example recall the evidence for erosion at the top of the Desert Creek sequence. Also recall the existence of the subaerially exposed fifth-order cycle (cycle #8) within the low-stand wedge, located about 30 m down-dip from the fourth-order boundary on the shelf top (Fig. 13). This constrains the amount of relative sea-level fall.

Additionally, all fourth-order sequence boundaries are traceable from outcrop to the subsurface, where they can be correlated over 100–120 km across the Paradox shelf (Goldhammer *et al.*, 1991). Not one of the fourth-order sequences is capped by an intertidal–supratidal deposit (e.g. mudcracked laminite) suggestive of progradational autocyclicity. This, coupled with the aggradational stacking of facies, diminishes the feasibility that the fourth-order sequences were generated through some progradational autocyclic mechanism. Thus we are left with a choice between the two allocyclic mechanisms: tectonic forcing or fourth-order eustasy.

Evaluation of tectonic forcing

Because the sequences are capped by subaerial diagenetic caps, then, if tectonics is to be invoked, reversal tectonics or 'yo-yoing' of the Paradox shelf is demanded, as opposed to simple non-uniform subsidence driven by faulting (episodic down-dropping). The palaeostructural interpretation of the Paradox Basin is that of an extensional rhombochasm bounded by right-lateral master faults that originated in Atokan time (Baars & Stevenson, 1982). This model is supported by the presence of thick alluvial/fluvial red bed sequences deposited along the southwest flank of the Uncompahgre uplift, testifying to this active movement (Fig. 3; Stevenson & Baars, 1988). Clearly, tectonic forcing of the Paradox shelf deserves serious consideration.

As stated earlier, modern data concerning 'yo-yo' tectonics is scarce and information regarding recurrence frequency, amount of vertical displacement and the size of areas involved is lacking. With regard to Cisne's (1986) tectonic model, estimating recurrence frequencies of reversal tectonics is imprecise. Additionally, application of this model to the Paradox cyclic shelf strata would necessitate Desmoinesian-age block faulting coincident with the Desmoinesian Paradox shelf edge.

Baars & Stevenson (1982) suggest that 'small-scale oscillating vertical movements recurred along fault blocks throughout early Palaeozoic time' (i.e. the Early Mississippian) across the 'Four Corners Platform of northwestern New Mexico' (i.e. the San Juan Basin, Fig. 3). They suggest that in the Pennsylvanian the same 'small-scale basement structures . . . actively controlled the distribution of Middle Pennsylvanian carbonate lithofacies' in the Paradox Basin by providing positive fault blocks favourable to shallow-water carbonate production (Baars & Stevenson, 1982, p. 154). They do not, however, suggest tectonic oscillations of the shelf, nor do they present data, such as isopachs of the Desmoinesian shelf or regional subsurface cross-sections, to indicate faults that cut through the Desmoinesian section on the southwestern Paradox shelf (Baars & Stevenson, 1981, 1982; Baars, 1988; Stevenson & Baars, 1988). In fact our own regional subsurface cross-sections (Fig. 12; see also Goldhammer *et al.*, 1991) and our inspection of unpublished seismic data across the shelf do not indicate any syn-Desmoinesian faulting of the southwestern Paradox shelf. It is more likely that localized tectonics, in the form of gentle uplifts, acted to influence the distribution of major algal mound development (Choquette & Traut, 1963).

If in fact reversal tectonics was operative it is unlikely that the entire Paradox shelf would have experienced synchronous and uniform amounts of uplift and down-drop, a consideration demanded by the lack of abrupt lateral thickness changes of the regionally correlative sequences. Additionally, the regional correlation of fifth-order cycles and the repetitive stacking of facies within successive fifth-order cycles (e.g. compare the repetitious facies stacking in cycles 5–7 at McElmo Creek, Fig. 13) are features unlikely to have developed from tectonic oscillation of the Paradox shelf. Furthermore, a comparison of the relatively small deviation in fifth-order cycle thickness (a range of 2.5–11 m/cycle) with the average cycle thickness (average about 6 m/cycle) argues against repeated tectonic oscillations, as this would require uniform amounts of repetitive displacement.

Evaluation of fourth-order eustasy

The fourth-order sequences are bounded by regionally correlative subaerial exposure surfaces that indicate alternating submergence and subaerial exposure of the Paradox shelf with significant relative drops in sea level to ensure effective leaching and caliche formation. In this respect, the fourth-order Paradox sequences are similar to late Quaternary fourth-order 'cycles' of various carbonate platforms (e.g. the Plio-Pleistocene of the Great Bahama Bank, Beach & Ginsburg, 1978; the Pleistocene of Caicos Platform, Wanless & Dravis, 1989; the Pleistocene of south Florida Platform, Perkins, 1977) in that both are essentially subtidal deposits capped by subaerial exposure surfaces ('diagenetic caps' in the form of laminar caliches, karstic dissolution pipes, soil breccias, etc.). Associated with both are freshwater, near-surface, early diagenetic features (e.g. leached ooids, neomorphic recrystallization, etc.). These and other late Quaternary 'cycles' of alternating shallow-marine deposition and subaerial exposure are interpreted to have been controlled by late Quaternary glacio-eustatic sea-level oscillations (see Hardie & Shinn, 1986; Goldhammer *et al.*, 1987, 1990 for a review).

The calculated periods of late Quaternary 'cycles' are similar to those derived for the Paradox fourth-order sequences. For example, the Upper Pliocene to Holocene platform carbonates of the Great Bahama Bank (Beach & Ginsburg, 1978) contain 15 subtidal cycles with diagenetic soil caps representing the last 3.2 Ma, yielding an average cycle duration of 213 ka/cycle (Hardie & Shinn, 1986).

Thus by analogy with late Quaternary cycles, which are widely accepted as eustatic in origin, we favour eustatic sea-level oscillations as the mechanism responsible for the development of the fourth-order Paradox sequences. Recall from Table 3 that the periodicities of the Paradox sequences range from 138 to 345 ka per sequence. Ignoring for the time being the 4-Ma estimate for the Desmoinesian (Odin & Gale, 1982; Lippolt *et al.*, 1984), the range of fourth-order periodicities varies from 216 to 345 ka per sequence, that is, *medial* fourth-order.

This is significant in that numerous other examples of Pennsylvanian cyclicity show that the major cycles are medial fourth-order in duration (Table 4). For example, Heckel (1983, 1986, 1989) in a synthesis of mixed clastic–carbonate cyclothems from the midcontinent of the USA, calculated periodicities of between 235 and 400 ka for regionally extensive, major fourth-order cycles ('cyclothems'). For analogous transgressive–regressive coal-bearing cycles from the Appalachian Basin, Busch & Rollins (1984) derived periodicities of between 400 and 450 ka/cycle for the most obvious fundamental cycles (ranging in thickness from 5 to 30 m) that correlate for over 300 km (Busch & Rollins, 1984). In their recent compilation of depositional sequences in the southwestern USA, Ross & Ross (1987) reported 23 regionally significant, fourth-order sequences for the Desmoinesian, with an average duration of 390 ka. Ramsbottom (1979) has recognized a cyclic hierarchy within the British Carboniferous containing a fundamental cycle (his 'cyclothem'), with a period in the 200–500 ka range. Goldstein (1988) reported mixed clastic–carbonate cycles capped by palaeosols from the Virgilian Holder Formation of New Mexico that have an average periodicity of 300 ka. Driese & Dott (1984) calculated a range in cycle period for the 17 mixed sandstone–carbonate cycles of the Middle Pennsylvanian Morgan Formation of between 470 and 880 ka. Finally, Algeo & Wilkinson (1988) in a review of Phanerozoic cycle periods found that for Late Mississippian through to Late Pennsylvanian cycles a statistically significant number of mesoscale cycles (1–100 m thick) clustered about a 400-ka period.

Without exception, all of the above examples of Pennsylvanian cyclicity are interpreted by the various authors as eustatic in origin, caused by the waxing and waning of glaciers as the Gondwana supercontinent drifted near and across the South Pole (Crowell, 1978). Crowell (1978) pointed out that the timespan of Gondwanan glaciation (about 90 Ma) in the Southern Hemisphere coincides with the Carboniferous interval of stratigraphic cyclicity in the Northern Hemisphere.

Many of the above examples of Pennsylvanian cyclicity, including the Paradox sequences, have

Table 4. Summary of Pennsylvanian, fourth-order stratigraphic cyclicity

Source	Stratigraphic unit	Cycle periodicity (ka)
This study	Desmoinesian Paradox Fm; SE Utah, SW Colorado; 29 regionally correlative, basinal shale–evaporite cycles	230–385
Driese & Dott (1984)	Middle Pennsylvanian Morgan Fm; northern Utah and Colorado; 17 mixed clastic–carbonate cycles	470–800
Heckel (1986, 1989)	Middle Pennsylvanian 'cyclothems' from the mid-continent of the USA; 25 shale-based mixed clastic–carbonate cycles	235–400
Busch & Rollins (1984)	Pennsylvanian 'cyclothemic PAC sequences' from the northern Appalachian Basin; 12 mixed clastic–carbonate regionally extensive cycles with coal horizons	400–450
Ross & Ross (1987)	Desmoinesian depositional sequences of the southwestern USA; 23 regionally significant sequences	390
Ramsbottom (1979)	Carboniferous 'cyclothems' of Britain	200–500
Goldstein (1988)	Virgilian Holder Fm of New Mexico; mixed clastic–carbonate cycles	300
Algeo & Wilkinson (1988)	Compilation of Carboniferous mesoscale cycles	400

other characteristics consistent with a glacio-eustatic mechanism. First, several of the examples are lithologically similar to the late Quaternary fourth-order cycles from the Bahamas and south Florida reviewed above. That is, they are diagenetic 'cycles' that consist of shallow-marine carbonates (phylloid algal bioherms, ooid grainstones, etc.) that are capped by subaerial caliches (see also Goldhammer & Elmore, 1984; Goldstein, 1988). Second, the sequences tend to be 'regressive-skewed', that is, there is a tendency toward thin, argillaceous, condensed transgressive phases and thick regressive limestones. This asymmetric trend in sequence components reflects the mechanics of the actual build-up and decay of ice caps. The Pleistocene sea-level curve has a 'sawtooth' shape inferred from dated coral reef terraces (Steinen *et al.*, 1973; Aharon, 1984) or oxygen isotope data (Imbrie, 1985) with relatively rapid rises and slow falls, reflecting the slow build-up and rapid decay of ice sheets.

Amplitudes of fourth-order Pennsylvanian eustasy, as published in the literature, are believed to be fairly high. Ross & Ross (1987) indicate 50–150-m amplitudes and Wilson (1967) speculated minimum eustatic drops in sea level for the cyclic Virgilian Holder Formation of 30–50 m. Heckel's (1989) interpretation of the black shale, core facies of the midcontinent cyclothems requires water depth changes on the order of 100 m. The high-amplitude fluctuations and the relative shape of the inferred eustatic curve are both consistent with a glacio-eustatic interpretation. Our stratigraphic simulations (see below) suggest that these estimates are too high.

Interpretation of composite stratigraphic cyclicity and cycle stacking patterns

If, as we believe, the fourth-order sequences are eustatic in origin, then we still must explain the origin of the higher-frequency, fifth-order cycles that make up the sequences. One of the central themes that emerges from the stratigraphy of the Paradox shelf is that of composite stratigraphic cyclicity, in which small depositional cycles build into larger sequences in an ordered hierarchy. Recognition of vertical stacking patterns (thickness, facies character, early diagenetic attributes) of the high-frequency, fifth-order cycles is the key to unravelling composite stratigraphies.

In the Paradox example the number of fifth-order cycles per sequence ranges from three to nine. In the thickest fourth-order sequences, which have the highest number of fifth-order cycles (e.g. sequences HT#3, HT#4, HT#8), this value ranges from seven to nine (Fig. 11). However, most of the shelfal sequences may actually be 'incomplete' with respect to the true number of fifth-order sea-level cycles per sequence. This is due to the fact that the fourth-order low-stand wedges contain a few fifth-order cycles that onlap below the shelf top in topographic depressions. Thus the up-dip, shelfal sections may not contain every fifth-order stratigraphic cycle, and thus do not record the maximum number of fifth-order cycles.

Below we address the possible mechanisms that might have generated the composite stratigraphic cyclicity of the Paradox shelf.

Composite reversal tectonics

An explanation for composite stratigraphic cyclicity calling upon solely tectonics would require superimposed, multiple orders of reversal tectonics. If we view the fourth-order sequences as eustatic and call upon fifth-order reversal tectonics to generate the component cycles, then we must place strict limitations on the timing and amount of high-frequency 'yo-yoing' to generate the repetitive stacks of fifth-order cycles. Based on uncertain recurrence frequencies of major earthquakes (see Hardie *et al.*, 1991), and lacking a tectonic model that would predict superimposed, high-frequency, relative crustal oscillations, we have no way of evaluating the model at this time.

Composite eustasy

The concept of high-frequency composite eustasy is an established one (Goldhammer *et al.*, 1990), exemplified by the Plio-Pleistocene isotopic records of deep-sea cores of the Caribbean (Broecker & van Donk, 1970) and the Indian Ocean (Hays *et al.*, 1976), as well as by published Pleistocene sea-level curves derived from uplifted successions of dated coral reef terraces (Steinen *et al.*, 1973; Aharon, 1984). In these and other examples, composite eustasy has resulted from glacio-eustatic sea-level oscillations generated in response to Milankovitch orbital forcing.

Based on: (i) the eustatic interpretation of the fourth-order sequences; (ii) the high probability of glacio-eustasy in the Middle Pennsylvanian (Wanless & Shepard, 1936; Crowell, 1978; Fischer, 1986); and (iii) the composite stratigraphic cyclicity of the Paradox sequences, glacio-eustasy triggered by astronomical (Milankovitch) phenomena deserves consideration. The issue now becomes which orders of stratigraphic cyclicity equate to the various Milankovitch orbital cycles. In effect, how should we link our stratigraphic cycles with the orbital periods? Based on the age determinations and calculation of fourth-order sequence and fifth-order cycle periods (Table 3), there appear to be two viable scenarios: (i) a long eccentricity–obliquity model; and (ii) a short eccentricity–precession model. Below we evaluate both of these cases.

Long eccentricity–obliquity model. Relying on the age determinations of van Eysinga (1975), Harland *et al.* (1982, 1990), Palmer (1983) and Ross & Ross (1987), the duration of fourth-order sequences ranges from 216 to 345 ka per sequence (Table 3). These values fall in the vicinity of the Earth's long eccentricity cycle with a period of 413 ka (Berger, 1977, 1984). Considering age date errors, the Paradox fourth-order sequences could reflect the waxing and waning of glaciers in response to net solar insolation changes driven by the long eccentricity cycle. Indeed, Heckel (1986) and Fischer (1986) have argued that Late Carboniferous large, rhythmic marine transgressions and regressions of the mid-continent USA were driven by the Milankovitch long eccentricity cycle, a view championed by Ross & Ross (1987). A review of the fourth-order cycle periods in Table 4 supports the notion of the dominance of the long eccentricity cycle.

Recalling the 9 : 1 ratio of fifth-order cycles per fourth-order Paradox sequence, the fifth-order periods range from 24 to 38 ka/cycle, with a mean value of 31 230 years (Table 3). Recently, Berger *et al.* (1989) have recalculated the predicted obliquity and precessional periods back through geological time, accounting for the effect of the shortening of the Earth–Moon distance and the concomitant shortening of the length of day back through time (Berger *et al.*, 1989, table 2, p. 558; pp. 15–24, this volume). Because of these phenomena the fundamental astronomical periods for obliquity and precession must have been less than their present values. For obliquity, the value is reduced from 41 ka to about 34 ka, and for precession the values are shortened from 19 and 23 ka to approximately 17 and 21 ka respectively. There is no change, however, in the Earth's short eccentricity cycle, with a quasi-period of about 100 ka, nor in the Earth's long eccentricity cycle, with a fixed period of 413 ka. Within the limitation of available age determinations, it is conceivable to view the Paradox fifth-order cycles as a record of the Earth's obliquity cycle, with a period close to 34 ka/cycle.

Supporting the contention of Milankovitch-forced glacio-eustasy, Heckel (1986, 1989) classified mid-continent Pennsylvanian cycles into minor cycles (with periods from 44 to 118 ka) and major cycles (with periods from 235 to 393 ka). Heckel (1986,

1989) suggested that these periods and the hierarchy of minor and major cycles can best be explained by glacio-eustasy driven by the Earth's orbital cycles in accordance with the Milankovitch insolation theory, similar to the Pleistocene climatic forcing (Hays *et al.*, 1976; Imbrie & Imbrie, 1980; Imbrie, 1985). Minor cycles reflect the obliquity cycle (dominant period near 41 ka) and the short eccentricity cycle (averaging about 100 ka, but ranging from 95 to 125 ka; Berger, 1977), whereas the major cycles record the long eccentricity cycle. Fischer (1986) pointed out that in the Carboniferous, the shorter Milankovitch rhythms, particularly the precessional cycle (two dominant periods presently averaging 19 and 23 ka), appear to have been suppressed in favour of the long eccentricity cycle. Fischer (1986) concluded that this 'red shift' seems more pronounced in the Carboniferous than in the Pleistocene, indicating that Carboniferous climates 'reacted with greater inertia' (Fischer, 1986, p. 371).

A Quaternary and Miocene analogue supporting the existence of the Earth's long eccentricity cycle, as recorded in deep-sea carbonate sediments, is provided by two different cores from the Pacific (Moore *et al.*, 1982). In their variance spectra of the Quaternary core (Moore *et al.*, 1982, fig. 3), peaks corresponding to the 400-ka and 100-ka eccentricity cycle as well as the 41-ka obliquity cycle are illustrated which they relate to Milankovitch phenomena. In their Miocene example, the 400-ka eccentricity cycle is dominant which they again relate to the Earth's long eccentricity cycle.

Thus the composite stratigraphic cyclicity of the Paradox shelf could very well be modelled as two superimposed glacio-eustatic sea-level oscillations, with the fourth-order driver depicting the Earth's long eccentricity cycle (413 ka) and the fifth-order fluctuation responding to the obliquity cycle (approximately 34 ka).

This is somewhat problematic in that if the 400-ka cycle is present within the stratigraphic data set, one would expect the 100-ka cycle to register as well, considering that the 100-ka signal is of comparable size to the 413-ka signal (P.L. de Boer, personal communication). If the 100-ka cycle is present in the Paradox cycle data set, it is not obvious based on cycle stacking patterns. However, it is interesting to note that in 'complete' shelf, fourth-order sequences, quartz clastics are observed every third fifth-order cycle, suggesting still another level of cyclicity, approximately 100 ka in duration, intriguingly close to the Earth's short eccentricity cycle.

Equally problematic is the fact that it is unlikely to have the effects of eccentricity without the linked effects of precession (P.L. de Boer, personal communication). The effects of the precession and eccentricity cycles are intrinsically linked. If the fundamental fifth-order cycles of the Paradox shelf are 34 ka in duration, then the effects of precession are perhaps amalgamated within the fundamental cycle, and we cannot clearly decipher a precessional record.

Short eccentricity–precession model. Utilizing the age determinations of Odin & Gale (1982), Lippolt *et al.* (1984) and Klein (1990), the duration of fourth-order sequences averages 138 ka per sequence (Table 3). This value is close to the Earth's short eccentricity cycle, with a quasi-period approximating 100 ka (the modulation envelope ranges from about 95 to 123 ka; Berger, 1977). Thus, the Paradox fourth-order sequences could conceivably represent the short eccentricity cycle. If this were the case, then the component fifth-order Paradox cycles would average 15.5 ka/cycle and tie more closely to the precessional cycle, which in the Pennsylvanian modulated between 17.4 and 20.7 ka (Berger *et al.*, 1989; pp. 15–24, this volume).

To aid in evaluating this second Milankovitch model we performed relative time-series analysis on the fifth-order cyclic succession measured at the Honaker Trail (Figs 9, 11 and 15). A complete discussion of the application of time series to stratigraphic problems is beyond the scope of this paper and the reader is referred to Schwarzacher (1975), Schwarzacher & Fischer (1982) and Weedon (1989) for more information.

Underlying the spectral analysis of the Paradox record presented below is the assumption that each fifth-order depositional cycle is of equal duration. The limitations of this assumption are discussed above. The Honaker Trail section was analysed using the maximum entropy spectral analysis (MESA) technique of Burg (1967) in an algorithm from Press *et al.* (1986). The MESA technique is discussed in full by Dunn (1991) and here we present only the results of the analysis. Firstly, we used the MESA technique to analyse the raw cycle thickness data, which was converted to oscillate about a zero point by subtracting the average cycle thickness from the thickness value of each individual cycle (the (− average) curve in Fig. 17A). Secondly, we utilized the MESA technique to analyse the log transformed thickness data (the Log 10 transform curve in Fig. 17A).

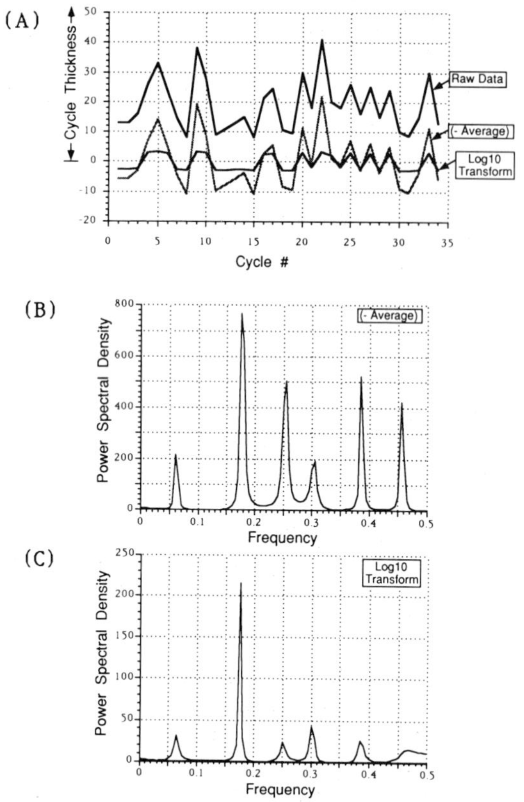

Fig. 17. Relative time-series analysis of fifth-order cycle succession at the Honaker Trail. (A) shows cycle thickness versus cycle # ('Raw Data') and two transformations of the raw data. The '(−average)' transformation is simply cycle thickness minus the average thickness; the 'Log 10 transform' is the log10 transformation of the '(-average)' data. (B) is the frequency spectrum for the '(-average)' data set derived from MESA. (C) is the frequency spectrum for the 'Log 10 transform' derived from MESA. For both data sets the dominant frequency occurs at 0.175 (refer to text).

Application of the MESA technique to these data yields frequency peaks expressed as power spectral density (Fig. 17B, C). These spectral peaks indicate underlying, lower-frequency stratigraphic rhythms, where 1/frequency yields the ratio of fundamental fifth-order cycles to lower-frequency cycles. For both the normal and the log10 transformed data sets, the strongest peak occurred at a frequency of 0.175 indicating a 5.71 ratio of two superimposed orders of stratigraphic cyclicity. Assuming that the fundamental fifth-order cycles represent precession with a period modulating between 17.4 and 20.7 ka, then the 5.71 ratio yields a fourth-order cycle (with values ranging from 99 354 to 118 197 years) suggestive of the short eccentricity cycle.

Intriguing as these results are, this analysis assumes that all fourth-order sequences are 'complete'. That is, all fifth-order eustatic oscillations of sea level were recorded as fifth-order stratigraphic cycles on the Paradox shelf – the shelf does not contain any fifth-order 'missed beats'. This assumption is probably not true based on the existence of fifth-order cycles trapped down-dip in the low-stand wedge of sequence HT#5 on top of sequence HT#4 (Fig. 13). Goldhammer *et al.* (1991) noted similar low-stand fifth-order cycles within sequence HT#4 restricted to intra-shelf depressions overlying sequence HT#3. Thus the application of time series adds little support for a short eccentricity–precession model.

Summary of composite eustasy. Based on our assessment of all possible models, we strongly favour high-frequency composite eustasy as the mechanism responsible for the composite stratigraphic cyclicity of the Paradox shelf. The multiple orders of solar insolation variation predicted by the Milankovitch theory make it the only testable model for composite eustasy at the fourth- and fifth-order time scales. However the bridge from recognition of a composite stratigraphy to proof of Milankovitch forcing is not easily crossed. Uncertainty in age dating, incompleteness of the record, variable subsidence and sedimentation rates as well as errors inherent in projecting astronomical parameters and resulting climatic change back in time can individually, or in combination, render a stratigraphy illegible even if it was laid down at a time when sea level resonated to a Milankovitchian tune.

Utilizing Milankovitch astronomical forcing of glacio-eustasy as a mechanism we have evaluated two scenarios, both of which are viable based on published radiometric age determinations. Both scenarios call upon a composite mixture of fifth-order eustasy and fourth-order eustasy acting in concert. In the light of the observed 9:1 ratio of fifth-order cycles to fourth-order sequences we favour the long eccentricity–obliquity model.

Third-order accommodation trend

In addition to the obvious fifth-order and fourth-order cyclicity, a longer-term lower-frequency trend or accommodation cycle is observed at the Honaker

Trail and in the subsurface. This third-order trend is delineated by the vertical stacking patterns of the fourth-order sequences (Figs 14–16) which progressively thin from sequence HT#2 (approximately 50 m thick) to sequence HT#6 (17.5 m thick), and subsequently thicken upward to sequence HT#8 (47 m thick). In addition, the number of fifth-order cycles per fourth-order sequence progressively decreases and increases coincident with the thinning–thickening trend (i.e. sequence HT#3, seven cycles; HT#4, eight cycles; HT#5, three cycles; HT#6, three cycles; HT#7, four cycles; HT#8, nine cycles). The Honaker Trail 'Fischer plot' (Fig. 9) also illustrates this third-order relative decrease and increase in accommodation.

This long-term trend is probably related to long-term subsidence patterns. However, it is interesting to note that the same overall pattern of third-order change is reported by Ross & Ross (1987) for the upper of several fourth-order sequences of the south-western USA (their Altamont through to Hertha sequences; Ross & Ross, 1987, their fig. 3), as well as by Heckel (1986) for the latest Desmoinesian (his Verdigris through to Lost Branch cyclothems, Heckel, 1986, his fig. 2), suggesting the presence of a third-order eustatic trend. In addition, Berger's (1977) expansion series of the eccentricity actually predicts a lower-frequency forcing of the 400-ka eccentricity cycle. The 400-ka cycles tend to cluster into sets of five and six, suggestive of a 2–2.4 Ma underriding eccentricity component.

STRATIGRAPHIC SIMULATIONS

In order to gain insight pertaining to the mechanics whereby composite eustasy could generate composite stratigraphic cyclicity, the outcrop and subsurface stratigraphy was simulated using the 'Mr. Sediment' computer program (Dunn *et al.*, 1986; details of the model outlined in Dunn, 1991 and Goldhammer *et al.*, 1987). The simulations demonstrate the hierarchy of stratigraphic forcing generated by three superimposed orders of relative sea-level oscillation, and visually portray the relationship between vertical facies successions, sequence and cycle stacking patterns, and the formation of subaerial vs. subtidal cycle boundaries. We performed both one-dimensional and two-dimensional simulations.

Additionally, the simulations allow us to test the hypothesis that the Paradox stratigraphic cyclicity resulted from Milankovitch-driven glacio-eustasy. We have chosen to model the Paradox composite cyclicity by superimposing fifth-, fourth- and third-order eustasy for the sake of graphical presentation. The modelling constrains the range of scenarios likely to be involved. In performing the simulations we have opted to model the high-frequency composite eustasy in accordance with the long eccentricity–obliquity model outlined above to honour the 9 : 1 ratio of fifth- to fourth-order cyclicity observed. The exact values for the input parameters together provide one solution only in that all parameters are interrelated.

One-dimensional simulations

The Paradox sequences and cycles are interpreted as largely aggradational; thus they can be modelled as one-dimensional stacks. One of the prime objectives of the modelling was to attempt to constrain the input variables to be geologically reasonable. The input parameters for the 'Mr. Sediment' program include the following.

Subsidence. The input subsidence value is the rate at which the sediment column subsides (in m/ka). It is the total subsidence that the sediment column experienced, and thus includes the effects of compaction, isostatic loading and tectonic subsidence. For the Paradox shelf a long-term subsidence rate of 0.15 m/ka was calculated by backstripping the Palaeozoic section (Fig. 5). In the simulations this value was kept constant, with the assumption that long-term subsidence rates will change at a significantly slower rate than rates of eustatic change, especially in a glacio-eustatic period.

Sedimentation. A depth-dependent sedimentation function was used which has a maximum sedimentation rate at 0.40 m/ka at water depths between 0 and 5 m. Holocene accumulation rates on shallow-water carbonate platforms vary from about 0.1 to 1.0 m/ka (Stockman *et al.*, 1967; Taft *et al.*, 1968; Neumann & Land, 1975; Harris, 1979; Schlager, 1981). It is well established that shallow-water, platform carbonate production is depth dependent (Wilson, 1975; Schlager, 1981). While the general shape of this depth-dependent function is known (Wilson, 1975; Schlager, 1981), the actual depth-dependent sedimentation values vary as a function of the particular data set.

As outlined in detail by Goldhammer *et al.* (1991), water-depth ranges were assigned to the facies. To simplify the simulations, the algal and intermediate facies were lumped together in the same depth range, as was the skeletal and non-skeletal capping facies. In addition a depth-related sedimentation trigger, termed *lag depth*, was incorporated. This function sets the minimum depth of water required for carbonate sedimentation to resume following subaerial exposure. A lag depth of 1 m was used, following arguments outlined in Goldhammer *et al.* (1987). A rate of caliche formation (0.01 m/ka; Goldhammer *et al.*, 1987) was used when the top of the sediment column was exposed.

Eustasy. Several different orders of eustasy were utilized in accordance with a Milankovitch-forced glacio-eustatic model. For *fourth-order eustasy* we used an asymmetric, sawtooth-shaped curve with a period of 400 ka. The periodicity approximates that of the long eccentricity cycle, and the sawtooth shape reflects the hypothesis that ice caps melt faster than they grow (Hays *et al.*, 1976; Heckel, 1986). This sawtooth variation of the fourth-order driver is also characteristic of the Pleistocene, judging by the shape of Pleistocene sea-level curves (Steinen *et al.*, 1973; Aharon, 1984; Imbrie, 1985). Fourth-order amplitude was varied between 25 and 28 m. This is constrained by the shelf-to-basin cross-section at McElmo Creek, where the top of the lowest onlapping fifth-order cycle within the spatially restricted low-stand wedge is approximately 33 m beneath the underlying sequence boundary up on the shelf (Fig. 13). It must be stressed that the top of this fifth-order cycle is mudcracked and that it sits well beneath the antecedent shelf edge. Decompacting the basinal section and restoring the shelf-to-basin geometry suggests a value of about 25 m for the fourth-order amplitude.

For *fifth-order eustasy* a sinusoidal wave with a period of 40 ka was used, approximating the obliquity cycle. The amplitude of this oscillation (9 m) was constrained by the stacking patterns of the fifth-order cycles, which tend to thin upward toward the top of a sequence, indicating that the relative amplitude had to be less than that of the fourth-order driver. Supporting this is the systematic gradual shift in facies types that comprise fifth-order cycles at the base of a sequence compared with those that make up cycles at the top of a sequence. If the fifth-order oscillation dominated the fourth-order wave, the succession of facies within each fifth-order cycle would not vary greatly depending on cycle position within a sequence. The actual amplitude of 9 m was derived by trial and error, honouring the water-depth limitations for the facies, the fourth-order amplitude, rates of subsidence, etc. The important point is that of the relative amplitude ratio between the two different orders. This relative amplitude ratio, with fourth-order sea level forcing fifth-order eustasy, is characteristic of other Pennsylvanian cyclic strata (see, for example, Heckel, 1986).

For *third-order eustasy* a sinuousform-shaped wave was added to the higher-frequency waves because of the existence of the underlying long-term accommodation cycle discussed above. The form of the wave is not strictly a sine curve, but approaches it in shape. Our intent was to force the fourth-order sequences to progressively thin upward and then thicken upward. The period was set equal to 2.4 Ma, based on the feasibility of a longer-term eccentricity cycle (Berger, 1977). The 'Fischer plot' for the entire outcrop succession indicates a longterm accommodation amplitude of about 15–20 m (Fig. 9). We want to remind the reader that we could have just as readily modelled the third-order component as a long-term subsidence variation, but the effects of the third-order forcing are best visually portrayed by building it in as a eustatic function.

In accordance with a Milankovitchian glacio-eustatic model, we added a minor short eccentricity component (100-ka sinusoidal wave; 1-m amplitude) as well as a minor precessional component (20-ka sinusoidal wave; 2-m amplitude). As our basic fifth-order cycle is believed to be about 40 ka in duration, it would be very difficult to 'read' a 20-ka component in the rock record. There may, however, be some evidence for a subtle 100-ka cycle, as reflected by the occurrence of QSF facies every third cycle within sequence HT#8. The important message is that the interaction of the precession and the short eccentricity (i.e. the climatic precession, Fischer, 1986) may have had a minimal effect in the Middle Pennsylvanian. Climatic precession is clearly dominant, however, in many other cases (Goldhammer *et al.*, 1987; Strasser, 1988).

Results of one-dimensional simulations

Initially, the shelfal section from the Desert Creek subsurface interval (McElmo Creek) was simulated (Figs 18 and 19). The simulation was run over about 880 ka at a time step of 2 ka. Figure 18 illustrates the sea-level–subsidence–sedimentation history. In this diagram, time moves forward from base to top

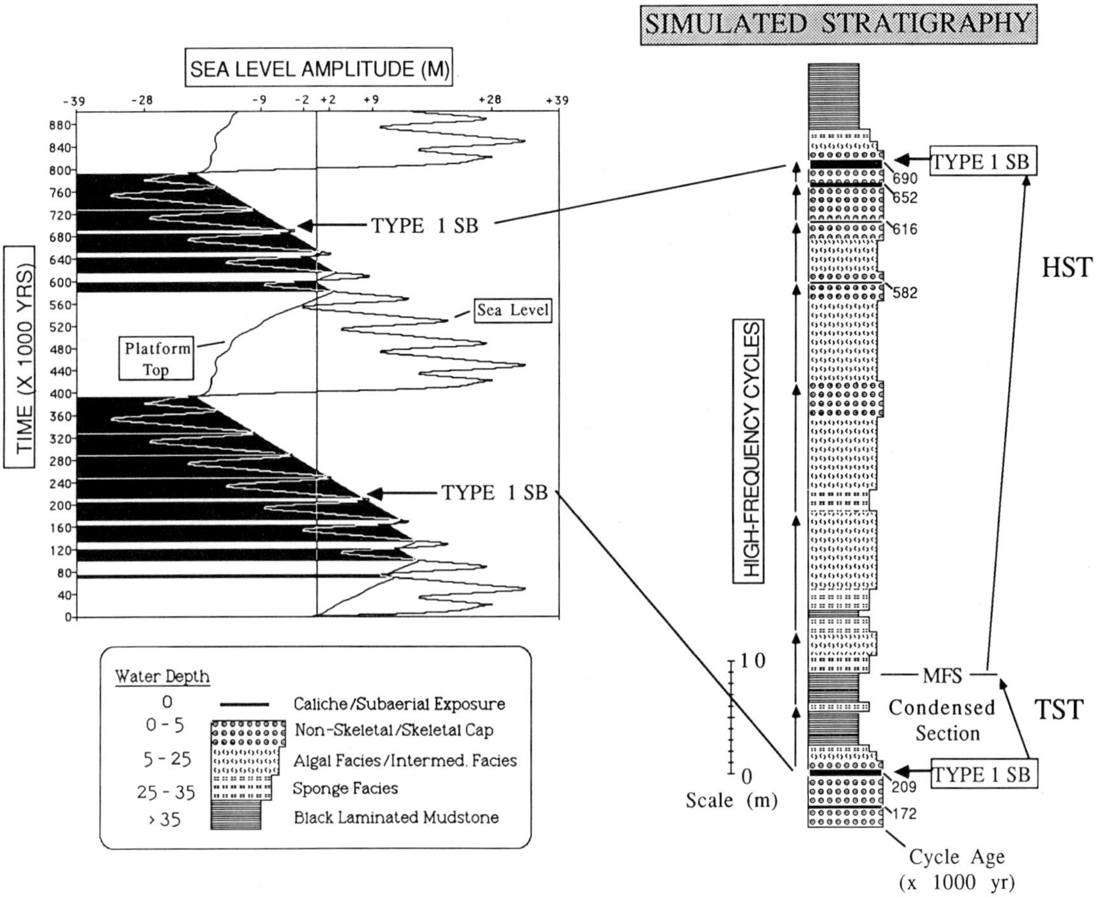

Fig. 18. Stratigraphic simulation of one of the fourth-order Paradox sequences (sequence HT#4) at McElmo Creek, with the sea-level–subsidence–sedimentation history and corresponding simulated stratigraphy. Input parameters, water-depth dependent facies and depth-dependent sedimentation curve are discussed in the text. Numbers to the right of simulated cycles refer to age of cycle formed during simulation, which ran from 0 to 880 ka. These cycles, also marked by thick black lines are exposure cycles. SB refers to sequence boundary, MFS to maximum flooding surface.

(0–880 ka), sea level rises to the right and falls to the left, white depicts marine submergence and sediment aggradation, and black depicts subaerial exposure. The top of the sediment column is represented by the thin black line that intersects the sea-level curve upon exposure. Upon exposure the sediment column subsides at the fixed rate of background subsidence (top of black area sloping to the left). The simulated stratigraphy consists of fifth-order cycles (marked by vertical arrows) composed of various facies. Subtidal cycles are marked by flooding surfaces (as indicated by deepening facies) and exposure cycles are marked by thin caliche caps (dated to the right in thousands of years). The fourth-order sequence stratigraphy interpretation is shown to the right in Fig. 18.

The simulation illustrates several important facets of the Paradox cyclicity (Fig. 18). Firstly, one can observe how the fourth-order oscillation forces the fifth-order waves. The fourth-order fluctuation, together with subsidence, creates the space for sedimentation and controls the overall water-depth profile, thus effectively controlling the vertical facies distribution. The fifth-order fluctuations act to 'fine-tune' the stratigraphy, creating more subtle internal facies changes, and more subtle stratigraphic cycle boundaries.

Secondly, one can observe the delicate interplay

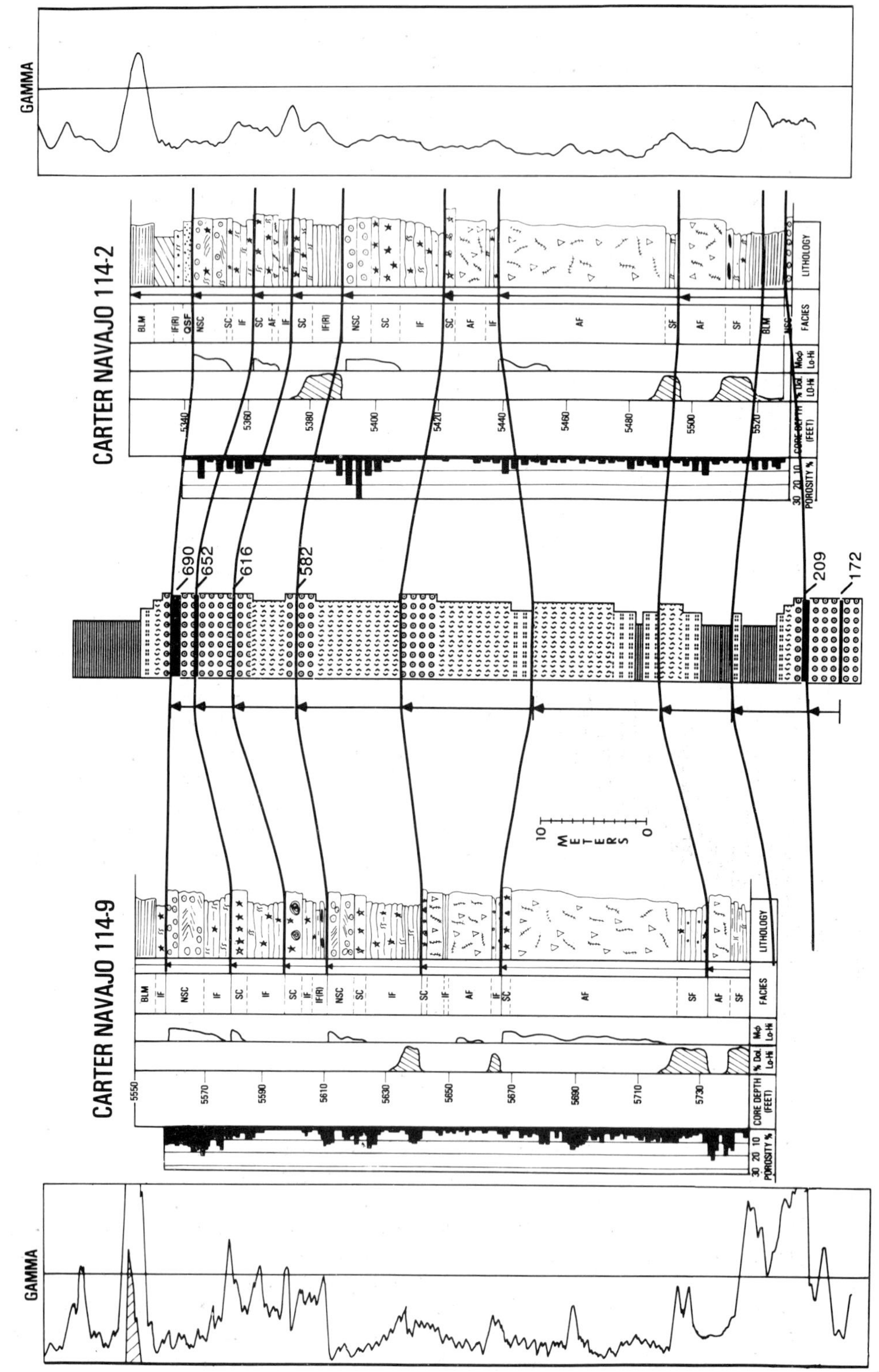

Fig. 19. Comparison of simulated stratigraphy with actual stratigraphy from the shelfal section at McElmo Creek, within sequence HT#4. The fourth-order sequence consists of several fifth-order cycles.

between subsidence, sedimentation and sea level. Note that with the rapid fourth-order rise in sea level, sedimentation rates are reduced substantially due to decline in sediment production rates during deposition of the BLM and sponge facies (water depths >25–35 m). This results in the generation of a fourth-order condensed section due to deeper-water conditions and reduced sedimentation rates, composed of essentially two thin fifth-order cycles that are dominated by deeper-water facies. With fourth-order fall in sea level, water depths are reduced, which causes depth-dependent sedimentation to increase, resulting in sediment aggrading to within the 'optimum window' of sediment production (<25 m). Thus cycles 3 and 4 build up to shallower water depths with successively shallower facies, but still do not become exposed (they are subtidal cycles). These cycles expand in thickness compared to underlying cycles 1 and 2 because they developed within the optimum window for sedimentation and because there was still plenty of previously formed space available to accommodate their aggradation. During cycle 5, sedimentation is operating at maximum efficiency as the fourth-order fall starts to dominate the system, and shallow-water facies (algal and capping facies) easily aggrade to sea level, at which point a subaerial exposure surface is formed.

The remaining fifth-order cycles (cycles 6–8) are increasingly 'regressive-skewed', that is they thin upward, are all capped by exposure surfaces and consist of shallowest-water facies. As these cycles are all subaerially exposed, near-surface meteoric diagenesis (leaching, recrystallization) would occur in the capping facies of these cycles. Due to its position at the very back-end of the fourth-order fall the last fifth-order eustatic oscillation does not flood the platform in the up-dip shelfal position, and thus the last stratigraphic cycle (cycle 8 dated at 690 ka) experiences close to 100 ka of subaerial exposure. This exposure surface capping cycle 8 marks the fourth-order sequence boundary, and is the thickest caliche cap generated. Many of the underlying fifth-order cycles previously exposed on a fifth-order time scale would be subject to yet another prolonged period of meteoric diagenesis. With renewed fourth-order rise, a thin veneer of transgressive capping facies (ooid grainstone) is deposited that subsequently deepens upward into black laminated mudstone.

In terms of sequence stratigraphy, the thick caliches which bound the fourth-order sequences are type 1 sequence boundaries (Vail, 1987), which as the simulation demonstrates, formed because the rate of fourth-order eustatic fall exceeded the rate of subsidence. The first two thin fifth-order cycles, which are dominantly BLM facies, constitute the condensed section at the top of the transgressive systems tract. The one-dimensional model does not create a low-stand deposit, other than the subaerial exposure surface, but it is during the lengthy period of subaerial exposure associated with the sequence boundary that low-stand conditions exist. This surface would correlate to the down-dip wedge of evaporites and siliciclastics. Due to the asymmetry of the fourth-order sea-level curve, the transgressive systems tract is very thin, and the fourth-order transgressive surface lies immediately above a thin deposit of capping facies (ooid grainstone). The high-stand systems tract consists of an early high-stand phase composed of fifth-order, subtidal cycles, and a late high-stand phase characterized by fifth-order, exposure cycles.

The simulated stratigraphy can be directly compared with the core descriptions from the shelfal section at McElmo Creek (Fig. 19). The actual stratigraphy contains seven recognizable fifth-order cycles, but we suggest that in reality the first cycle actually contains two fifth-order beats of eustasy amalgamated in the BLM facies as the simulation demonstrates. In this case the fourth-order sequence at McElmo Creek would contain eight fifth-order cycles. The correlation between the simulated section and the actual section is excellent with respect to fifth-order cycle stacking patterns, vertical facies distribution and the location of exposure surfaces and moldic porosity (subaerial leaching).

The entire measured section at the Honaker Trail was simulated next, with input values identical to those described above (Figs 20 and 21). The simulation ran over about 3.2 Ma at a step size of 2 ka. As above, fourth-order sequences with component fifth-order cycles were produced. Figure 21 compares directly the measured section with the simulated stratigraphy. The effect of adding the third-order 2.4-Ma eustatic component resulted in the progressive thinning upward followed by thickening upward of the fourth-order sequences. Additionally, the number of fifth-order cycles per sequence progressively declines and then increases.

Both of these results are a direct consequence of the hierarchy of stratigraphic forcing. The third-order accommodation cycle forced the fourth-order eustatic oscillations to produce both: (i) *'complete' sequences* with a maximum number of fifth-order cycles/sequence, a well-developed transgressive BLM, a thin subaerial sequence boundary (e.g. HT#2, 3, 8); and (ii) *'incomplete' sequences* with a minimum number of fifth-order cycles/sequence, lacking a well-developed transgressive shaley mudstone, but characterized by a very thick subaerial exposure surface which records several beats of fifth-order eustasy below the top of the shelf (e.g. HT#4, 5, 6, 7). The modelling predicts prolonged subaerial exposure at fourth-order sequence boundaries during the third-order fall. Note that the simulation suggests 200–300 ka of subaerial exposure at these fourth-order boundaries (e.g. top of HT#4, 5, 6, 7).

Two-dimensional simulation

In order to investigate questions regarding the lateral distribution of facies, lateral continuity of individual high-frequency cycles and overall stratal geometries within fourth-order systems tracts, a two-dimensional simulation was performed. Dunn (1991) has written a two-dimensional computer program ('Mr. Sediment') designed to handle the lateral transport of sediment and progradation of the shoreline across shallow carbonate platforms. The details of the program are to be found in Dunn (1991), but in essence the model calculates two-dimensional cyclostratigraphy of either vertically aggrading subtidal platform deposits or prograding tidal flat deposits across the depositional strike of a carbonate platform. The platform may experience simple rotational subsidence about a fixed hinge-line or undergo differential flexural subsidence. An important facet of 'Mr. Sediment' is that despite the fact that two-dimensional simulations are presented graphically as a series of one-dimensional 'boreholes', the program is a 'seamless' one that keeps track in space and time of all points across the entire cross-sectional area being modelled.

We generated a synthetic cross-section (Figs 22–24) of the Paradox shelf assuming an across strike width of 190 km with a very gentle basinward dip and kept track of 20 depositional sites graphically.

(A)

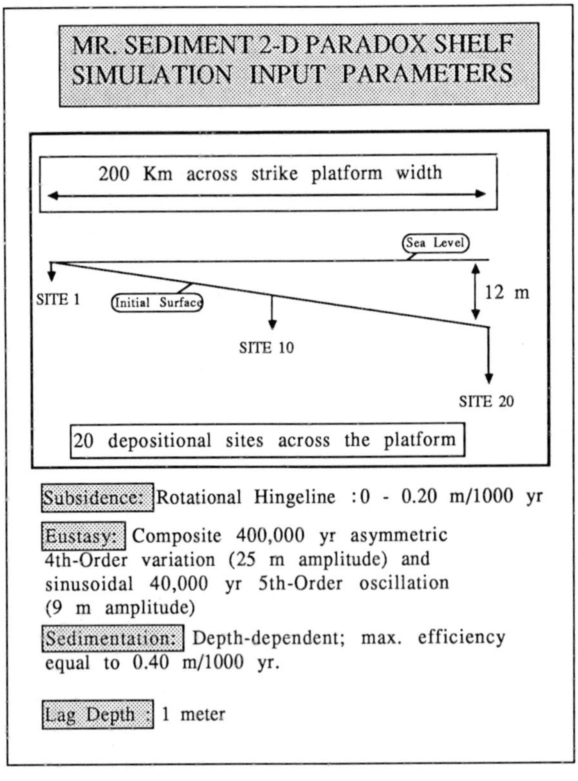

(B)

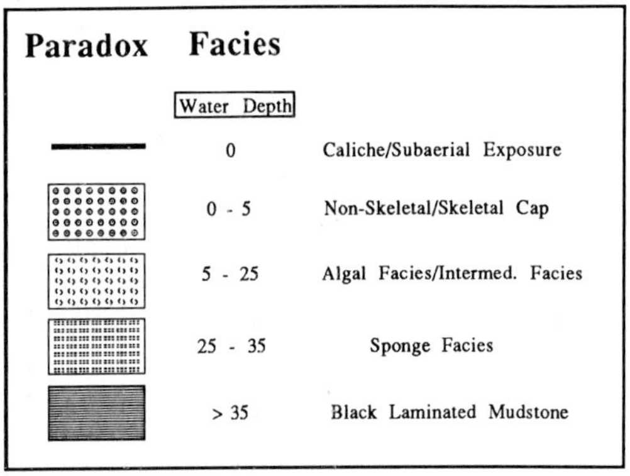

Fig. 22. Two-dimensional forward model of the Paradox shelf cyclostratigraphy generated by the computer program 'Mr. Sediment' (Dunn, 1991). (A) Initial starting conditions and input parameters shown are discussed in the text. (B) Paradox shelfal facies key for two-dimensional simulation shown in Fig. 24.

Initial starting conditions fixed point 1 at zero water depth and zero rate of subsidence, and fixed point 20 at 12-m water depth and a rate of subsidence equal to 0.20 m/ka (Fig. 22A). Initial water depths and subsidence values for intervening sites were linearly interpolated (Fig. 22A). All other variables (depth-dependent sedimentation, depth-dependent facies, eustasy, lag depth) were identical to those utilized in the one-dimensional simulations (Fig. 18). No third-order eustatic component was added. The interaction of sea level, subsidence and sedimentation is recorded graphically at six depositional sites across the profile (Fig. 23).

Results of two-dimensional simulation

The simulated two-dimensional cross-section illustrated in Fig. 24 contains three fourth-order depositional sequences bounded by sequence bounding subaerial unconformities. Each sequence contains a thin transgressive carbonate grainstone (non-skeletal cap facies) that onlaps the underlying fourth-order sequence boundary. This in turn is overlain by a series of offlapping fifth-order cycles (up to a maximum of 10 per sequence) each of which has a fifth-order transgressive component followed by a fifth-order high-stand component. Note the lateral

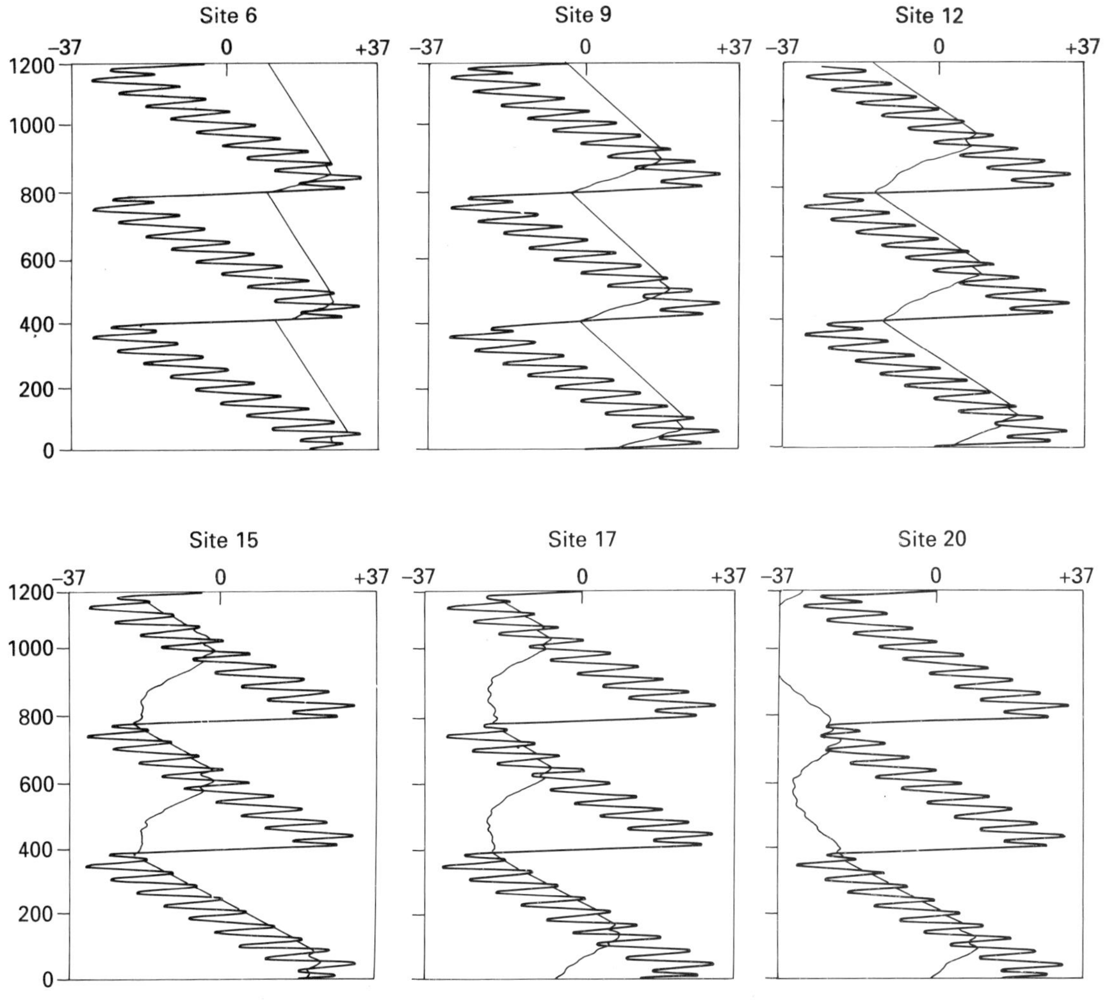

Fig. 23. Combined sea level–subsidence–sedimentation histories of six of the sites corresponding to the two-dimensional simulated cross-section shown in Fig. 24. These six sites illustrate the interaction of sedimentation and accommodation changes.

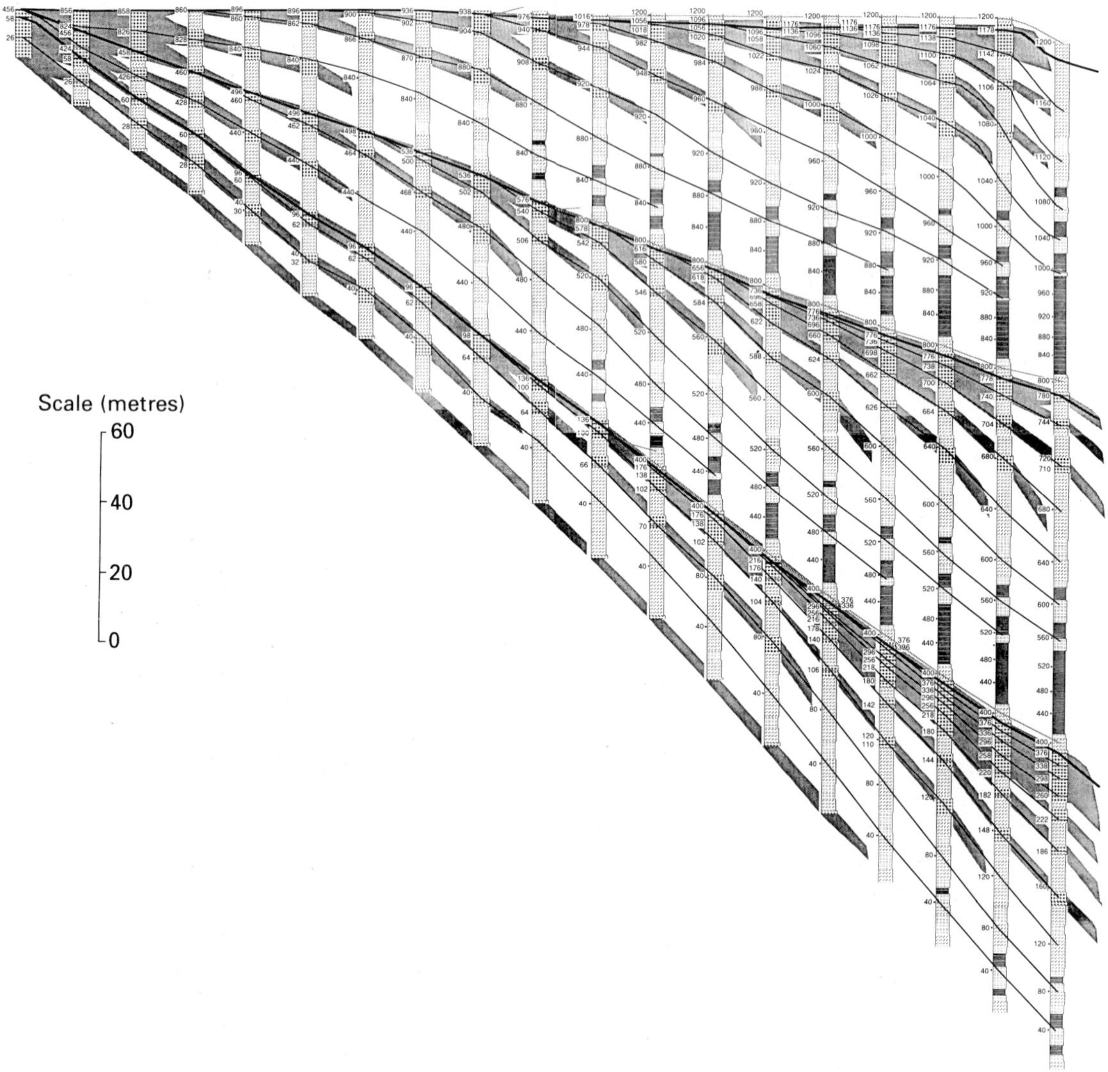

Fig. 24. Two-dimensional, synthetic forward model of the Paradox shelf cyclostratigraphy. Individual 'boreholes' record the facies and fifth/fourth-order cyclicity at each site. Cycle ages (in thousands of years) are shown to the right of 'boreholes'. Fourth-order sequence boundaries denoted by thick black lines. Fifth-order cycle boundaries marked by thin black lines. See text for discussion and refer to Fig. 22B for facies key.

and vertical facies changes that occur both within individual cycles, and within sequences as a whole. Note that the BLM facies is not a homogeneous lithologic entity, but rather in detail is interbedded with both the sponge facies and algal facies, reflecting the fifth-order eustasy. Since the computer dates the individual fifth-order cycle tops at each locality, the cycles are easily correlated laterally to reveal the geometry of stratal surfaces.

In terms of fourth-order systems tracts, the landward thinning wedge of the BLM facies which dominates the basal portion of fourth-order sequences constitutes the TST. The series of offlapping fifth-order cycles above this make up the HST. The down-dip terminations of each individual fifth-order cycle (where it changes facies laterally into BLM) collectively approximate the fourth-order downlap surface or maximum flooding surface of Vail (1987).

CONCLUSIONS

1 The Middle Pennsylvanian (Desmoinesian) shelf carbonates of the SW Paradox basin are an excellent example of *composite stratigraphic cyclicity* composed of fifth-order cycles grouped into fourth-order sequences, which in turn vertically stack to define a third-order accommodation cycle. At the core of this hierarchical scheme are the metre-scale, fifth-order depositional cycles, each of which shoals upward and is bounded by a subaerial or submarine diastem. Within the next level of the hierarchy, the fourth-order sequences are described by the vertical *stacking patterns* (systematic variation in the thickness, facies make-up and early diagenetic features) of the fundamental fifth-order cycles. At the outermost level, at vertical scales of hundreds of metres and at timespans of millions of years, a third-order sequence is defined by thickness trends of the fourth-order packages.

2 After an evaluation of autocyclic, tectonic and eustatic mechanisms for the generation of the Paradox cyclicity, we have called upon high-frequency *composite eustasy* as a means of interpreting the composite stratigraphic cyclicity of the Paradox shelf. This interpretation follows from: (i) the facies architecture and fifth-order cycle stacking patterns within a fourth-order sequence; (ii) regionally correlative subaerial exposure surfaces which cap the fourth-order sequences; (iii) the similarity between the Paradox fourth-order sequences and Plio-Pleistocene analogues known to have a eustatic origin; and (iv) the high probability of glacio-eustasy in the Middle Pennsylvanian. Tectonic control cannot be completely ruled out, but as yet there are no subsidence models which include composite cyclicity at fourth- and fifth-order time scales.

3 Using published radiometric age determinations, we have evaluated two variants of Milankovitch glacio-eustasy as possible mechanisms for producing the observed stratigraphic cyclicity. Both the *long eccentricity–obliquity* and the *short eccentricity–precession* models call upon a Pennsylvanian sea-level history containing a combination of fourth- and fifth-order oscillations. In the light of all available evidence and the uncertainties inherent in Palaeozoic age determinations, we cannot rule one out over the other but can only conclude that Milankovitchian glacio-eustasy was certainly possible.

4 The plausibility of composite eustatic control of fifth-order cycle, fourth-order sequence and third-order sequence formation was demonstrated through computer simulations. An important message which arose from the simulations was that of the '*hierarchy of stratigraphic forcing*' (Goldhammer *et al.*, 1990). Order within the hierarchy is maintained by systematic control of high-frequency cycle characteristics (thickness, facies make-up, diagenetic features, etc.) by longer-term changes in accommodation. Starting with the elemental shoaling-upward cycles and working at progressively larger scales, composite stratigraphies can be constructed even in seemingly monotonous stacks of platformal cycles.

ACKNOWLEDGEMENTS

We would like to thank S. Schutter of Exxon Production Research Co. for useful discussions of the Pennsylvanian stratigraphy and cyclicity of the Paradox Basin, and for carefully reviewing the manuscript. We also acknowledge the thorough and thoughtful reviews of P.L. de Boer, G. Weedon and G. deV. Klein. We are especially grateful to L.A. Hardie and Linda Hinnov for timely and critical reviews of Milankovitch theory and discussions pertaining to the interpretation of the Paradox cyclicity. We are indebted to M.G. Fitzgerald and R.G. Todd of EPR for allowing us to complete this study and publish the results.

REFERENCES

Aharon, P. (1984) Implications of the coral-reef record from New Guinea concerning the astronomical theory of ice ages. In: *Milankovitch and Climate* (Eds Berger, A., Imbrie, J., Hays, J., Kukla, G. & Saltzman, B.). Reidel Publ. Co., Dordrecht, pp. 379–389.

Algeo, T.J. & Wilkinson, B.H. (1988) Periodicity of mesoscale Phanerozoic sedimentary cycles and the role of Milankovitch orbital modulation. *J. Geol.* **96**, 313–322.

Atwater, B.F. (1987) Evidence for great Holocene earthquakes along the outer coast of Washington state. *Science* **236**, 942–944.

Baars, D.L. (1988) Basins of the Rocky Mountain region. In: *Sedimentary Cover – North America Craton U.S.* (Ed. Sloss, L.L.). The Geology of North America, D-2, Geol. Soc. Am., Boulder, Colorado, pp. 109–129.

Baars, D.L. & Stevenson, G.M. (1981) Tectonic evolution of the Paradox Basin, Utah and Colorado. In: *Geology of the Paradox Basin: Rocky Mountain Association of Geologists 1981 Field Conference* (Ed. Wiegand, D.L.). Rocky Mountain Association of Geologists, Denver, Colorado, pp. 23–31.

Baars, D.L. & Stevenson, G.M. (1982) Subtle stratigraphic traps in Paleozoic rocks of Paradox Basin. In: *The Deliberate Search for the Subtle Trap* (Ed.

Halbouty, M.T.). Mem. Am. Assoc. Petrol. Geol. 32, pp. 131–158.

Beach, D.K. & Ginsburg, R.N. (1978) Facies succession of Pliocene-Pleistocene carbonates, northwestern Great Bahama Bank. *Bull. Am. Assoc. Petrol. Geol.* **64**, 1634–1642.

Berger, A.L. (1977) Support for the astronomical theory of climate change. *Nature* **269**, 44–45.

Berger, A.L. (1984) Accuracy and frequency stability of the Earth's orbital elements during the Quaternary. In: *Milankovitch and Climate* (Eds Berger, A., Imbrie, J., Hays, J., Kukla, G. & Saltzman, B.). Reidel Publ. Co., Dordrecht, pp. 3–39.

Berger, A.L. (1989) The spectral characteristics of pre-Quaternary climate records, an example of the relationship between astronomical theory and geosciences. In: *Climate and Geosciences* (Eds Berger, A.L., Schneider, S. & Dupessy, J.C.). Reidel Publ. Co., Dordrecht, pp. 47–76.

Berger, A., Loutre, M.F., & Dehant, V. (1989) Influence of the changing lunar orbit on the astronomical frequencies of pre-Quaternary insolation patterns. *Paleoceanography* **4**, 555–564.

Bond, G.C., & Kominz, M.A. (1984) Construction of tectonic subsidence curves for the early Paleozoic miogeocline, southern Canadian Rocky Mountains: Implications for subsidence mechanisms, age of breakup, and crustal thinning. *Geol. Soc. Am., Bull.* **95**, 155–173.

Bond, G.C., Kominz, M.A. & Beavan, J. (1991) Evidence for orbital forcing of Middle Cambrian peritidal cycles: Wah Wah range, south-central Utah. In: *Sedimentary Modeling: Computer Simulations and Methods for Improved Parameter Definition* (Eds Franseen, E.K., Watney, W.L., Kendall, G.C. St C. & Ross, W.). *Kansas Geol. Surv. Bull.* **233**, 293–317.

Broecker, W.S. & Thurber, D.L. (1965) Uranium-series dating of corals and oolites from Bahaman and Florida Keys limestones. *Science* **149**, 58–60.

Broecker, W.S. & van Donk, J. (1970) Insolation changes, ice volumes, and the oxygen 18 record of deep-sea cores. *Rev. Geophys. Space Phys.* **8**, 169–197.

Burg, J.P. (1967) Maximum Entropy Spectral Analysis (Abstract). In: *37th Annual International Society of Exploration Geophysics Meeting, Oklahoma City, OK, USA*, p. 54.

Busch, R.M. & Rollins, H.B. (1984) Correlation of Carboniferous strata using a hierarchy of transgressive–regressive units. *Geology* **12**, 471–474.

Choquette, P.W. (1983) Platy algal reef mounds, Paradox Basin. In: *Carbonate Depositional Environments* (Eds Scholle, P.A., Bebout, D.G. & Moore, C.H.). Mem. Am. Assoc. Petrol. Geol. 33, pp. 454–462.

Choquette, P.W. & Traut, J.D. (1963) Pennsylvanian carbonate reservoirs, Ismay field, Utah and Colorado. In: *Shelf Carbonates of the Paradox Basin* (Ed. Bass, R.O.). Four Corners Geological Society 4th Field Conference Guidebook, pp. 157–184.

Cisne, J.L. (1986) Earthquakes recorded stratigraphically on carbonate platforms. *Nature* **323**, 320–322.

Crevello, P., Sarg, J.F., Read, J.F. & Wilson, J.L. (Eds) (1989) *Controls on Carbonate Platform to Basin Development*. Soc. Econ. Paleont. Mineral. Spec. Publ. 44.

Crowell, J.C. (1978) Gondwana glaciation, cyclothems, continental positioning, and climate change. *Am. J. Sci.* **278**, 1345–1372.

De Boer, P.L. (1983) Aspects of Middle Cretaceous pelagic sedimentation. *Geol. Ultraiectina* **31**, 112 pp.

De Boer, P.L. (1991) Pennsylvanian time scales and cycle periods – comment. *Geology* **19**, 408–409.

Donovan, D.T. & Jones, E. (1979) Causes of world-wide changes in sea level. *J. Geol. Soc. Lond.* **136**, 187–192.

Driese, S.G. & Dott, R.H. (1984) Model for sandstone–carbonate 'cyclothems' based on upper member of Morgan Formation (Middle Pennsylvanian) of northern Utah and Colorado. *Bull. Am. Assoc. Petrol. Geol.* **68**, 574–597.

Duff, P. McL. D., Hallam, A. & Walton, E.K. (1967) *Cyclic Sedimentation*. Developments in Sedimentology 10. Elsevier, New York.

Dunn, P.A. (1991) *Diagenesis and cyclostratigraphy: an example from the Middle Triassic Latemar platform, Dolomites Mountains, northern Italy*. Unpublished Ph.D. dissertation, The Johns Hopkins University, Baltimore.

Dunn, P.A., Goldhammer, R.K. & Hardie, L.A. (1986) Mr. Sediment – A computer model for carbonate cyclicity (Abstract). *Geol. Soc. Am. Abstracts with Programs* **18**, 590.

Dunn, P.A., Goldhammer, R.K., Hardie, L.A. & Nguyen, C.T. (1991) Two-dimensional forward modelling of Lower Ordovician platform carbonate sequences (Beekmantown Gp, central Appalachians): the search for high-frequency autocyles (Abstract). *Bull. Am. Assoc. Petrol. Geol.* **75**, 565.

Elias, G.K. (1963) Habitat of Pennsylvanian algal bioherms, Four Corners area. In: *Shelf Carbonates of the Paradox Basin* (Ed. Bass, R.O.). Four Corners Geological Society 4th Field Conference Guidebook, pp. 185–203.

Elliot, T. (1976) Upper Carboniferous sedimentary cycles produced by river-dominated, elongate deltas. *J. Geol. Soc. Lond.* **132**, 199–208.

Fischer, A.G. (1964) The Lofer cyclothems of the Alpine Triassic. *Kansas Geol. Surv. Bull.* **169**, 107–149.

Fischer, A.G. (1986) Climatic rhythms recorded in strata. *Ann. Rev. Earth Planet. Sci.* **14**, 351–376.

Galloway, W.E. & Brown, L.F., Jr (1973) Depositional systems and shelf-slope relations on cratonic basin margin, uppermost Pennsylvanian of north-central Texas. *Bull. Am. Assoc. Petrol. Geol.* **57**, 1185–1218.

Ginsburg, R.N. (1971) Landward movement of carbonate mud: new model for regressive cycles in carbonates (Abstract). *Bull. Am. Assoc. Petrol. Geol.* **55**, 340.

Goldhammer, R.K. & Elmore, R.D. (1984) Paleosoils capping regressive carbonate cycles in the Pennsylvanian Black Prince Limestone, Arizona. *J. Sedim. Petrol.* **54**, 1124–1137.

Goldhammer, R.K., Dunn, P.A. & Hardie, L.A. (1987) High frequency glacio-eustatic sea level oscillations with Milankovitch characteristics recorded in Middle Triassic platform carbonates in northern Italy. *Am. J. Sci.* **287**, 853–892.

Goldhammer, R.K., Dunn, P.A. & Hardie, L.A. (1990) Depositional cycles, composite sea level changes, cycle stacking patterns, and the hierarchy of stratigraphic forcing: examples from platform carbonates of the Alpine Triassic. *Geol. Soc. Am. Bull.* **102**, 535–562.

Goldhammer, R.K., Oswald, E.J. & Dunn, P.A. (1991) Hierarchy of stratigraphic forcing: example from Middle Pennsylvanian shelf carbonates of the Paradox Basin. In: *Sedimentary modeling: Computer Simulations and Methods for Improved Parameter Definition* (Eds Franseen, E.K., Watney, W.L., Kendall, G.C. St C. & Ross, W.). *Kansas Geol. Surv. Bull.* **233**, 361–413.

Goldstein, R.H. (1988) Paleosols of Late Pennsylvanian cyclic strata, New Mexico. *Sedimentology* **35**, 777–804.

Goodwin, P.W. & Anderson, E.J. (1985) Punctuated aggradational cycles: a general hypothesis of episodic stratigraphic accumulation. *J. Geol.* **93**, 515–533.

Grotzinger, J.P. (1986) Cyclicity and paleoenvironmental dynamics, Rocknest platform, northwest Canada. *Geol. Soc. Am. Bull.* **97**, 1208–1231.

Haq, B.U., Hardenbol, J. & Vail, P.R. (1987) Chronology of fluctuating sea levels since the Triassic. *Science* **235**, 1156–1166.

Hardie, L.A. & Shinn, E.A. (1986) *Carbonate depositional environments, modern and ancient, 3, Tidal flats*. Colorado School of Mines Quarterly, 81, no. 1.

Hardie, L.A., Bosellini, A. & Goldhammer, R.K. (1986) Repeated subaerial exposure of subtidal carbonate platforms, Triassic, northern Italy: evidence for high frequency sea level oscillations on a 10 000 year scale. *Paleoceanography* **1**, 447–457.

Hardie, L.A., Dunn, P.A. & Goldhammer, R.K. (1991) Field and modelling studies of Cambrian carbonate cycles, Virginia Appalachians: a discussion. *J. Sedim. Petrol.* **61**, 636–646.

Harland, W.B., Cox, A.V., Llewellyn, P.G., Picton, C.A., Smith, A.G. & Walters, R. (1982) *A Geological Time Scale*. Cambridge University Press, New York.

Harland, W.B., Armstrong, R.L., Cox, A.V., Craig, L.E., Smith, A.G. & Smith, D.G. (1990) *A Geologic Time Scale 1989*. Cambridge University Press, New York.

Harris, P.M. (1979) *Facies anatomy and diagenesis of a Bahamian ooid shoal*. Sedimenta VII, University of Miami Comparative Sedimentology Laboratory.

Hays, J.D., Imbrie, J. & Shackleton, N.J. (1976) Variation in the Earth's orbit: pacemaker of the ice ages. *Science* **194**, 1121–1132.

Heckel, P.H. (1983) Diagenetic model for carbonate rocks in Midcontinent Pennsylvanian eustatic cylothems: *J. Sedim. Petrol.* **53**, 733–760.

Heckel, P.H. (1986) Sea-level curve for Pennsylvanian eustatic marine transgressive–regressive depositional cycles along midcontinent outcrop belt, North America. *Geology* **14**, 330–334.

Heckel, P.H. (1989) Current view of Midcontinent Pennsylvanian cyclothems. In: *Middle and Late Pennsylvanian Chronostratigraphic Boundaries in North-central Texas: Glacial–Eustatic Events, Biostratigraphy, and Paleoecology* (Ed. Boardman, D.R.). Texas Tech Univ. Studies in Geology 2, pp. 17–34.

Hine, A.C. & Steinmetz, J.C. (1984) Cay Sal Bank, Bahamas – a partially drowned carbonate platform. *Mar. Geol.* **59**, 135–164.

Hinnov, L.A. & Goldhammer, R.K. (1988) The identification of Milankovitch signals in M. Triassic platform carbonate cycles using a super-resolution spectral technique (Abstract). *Bull. Am. Assoc. Petrol. Geol.* **72**, 197.

Hite, R.J. (1970) Shelf carbonate sedimentation controlled by salinity in the Paradox Basin, southeast Utah. In: *Third Symposium on Salt* (Eds Ron, J.L. & Dellwig, L.F.). Northern Ohio Geologic Society 1, pp. 48–66.

Hite, R.J. & Buckner, D.H. (1981) Stratigraphic correlations, facies concepts, and cyclicity in Pennsylvanian rocks of the Paradox Basin. In: *Geology of the Paradox Basin: Rocky Mountain Association of Geologists 1981 Field Conference* (Ed. Wiegand, D.L.). Rocky Mountain Association of Geologists, Denver, Colorado, pp. 147–159.

Imbrie, J. (1985) A theoretical framework for the Pleistocene ice ages. *J. Geol. Soc. Lond.* **142**, 417–432.

Imbrie, J. & Imbrie, J.Z. (1980) Modelling the climatic response to orbital variations. *Science* **207**, 943–953.

Kendall, G. St C. & Schlager, W. (1981) Carbonates and relative changes in sea level. *Mar. Geol.* **44**, 181–212.

Klein, G. deV. & Willard, D.A. (1989) Origin of the Pennsylvanian coalbearing cyclothems of North America. *Geology* **17**, 152–155.

Klein, G. deV. (1990) Pennsylvanian time scales and cycle periods. *Geology* **18**, 455–457.

Koerschner III, W.F. & Read, J.F. (1989) Field and modelling studies of Cambrian carbonate cycles, Virginia Appalachians. *J. Sedim. Petrol.* **59**, 654–687.

Kozar, M.G., Weber, L.J. & Walker, K.R. (1990) Field and modelling studies of Cambrian carbonate cycles, Virginia Appalachians, a discussion. *J. Sedim. Petrol.* **60**, 790–794.

Lippolt, H.J., Hess, J.C. & Burger, K. (1984) Isotopische Alter von pyroklastischen Sanidinen aus kaolin-Kohlentonstein als Korrelationsmarken fur das mitteleuroaische Oberkarbon. *Forschrift Geologisches, Rheinland und Westfalen* **32**, 119–150.

Miall, A.D. (1984) *Principles of Sedimentary Basin Analysis*. Springer Verlag, New York.

Mitterer, R.M. (1975) Ages and diagenetic temperatures of Pleistocene deposits of Florida based on isoleucine epimerization in *Mercenaria*. *Earth Planet. Sci. Lett.* **28**, 275–282.

Moore, T.C., Jr, Pisias, N.G. & Dunn, D.A. (1982) Carbonate time series of the Quaternary and Late Miocene sediments in the Pacific Ocean: a spectral comparison. *Mar. Geol.* **46**, 217–233.

Neumann, A.C. & Land, L.S. (1975) Lime mud deposition and calcareous algae in the Bight of Abaco, Bahamas: a budget. *J. Sedim. Petrol.* **45**, 763–786.

Odin, G.S. & Gale, N.H. (1982) Mise à jour d'échelles des temps caledoniens et hercyniens. *C.R. Acad. Sci. Paris* **294**, 453–456.

Ohlen, H.R. & McIntyre, L.B. (1965) Stratigraphy and tectonic features of Paradox Basin, Four Corners Area. *Bull. Am. Assoc. Petrol. Geol.* **49**, 2020–2040.

Osmond, J.K., Carpenter, J.R. & Windom, H.L., (1965) $^{230}Th/^{234}U$ age of the Pleistocene corals and oolites of Florida. *J. Geophys. Res.* **70**, 1843–1847.

Palmer, A.R. (1983) The Decade of North American Geology 1983 geologic time scale. *Geology* **11**, 503–504.

Perkins, R.D. (1977) Depositional framework of Pleistocene rocks in south Florida. *Mem. Geol. Soc. Am.* **147**, 131–198.

PETERSON, J.A. & HITE, R.J. (1969) Pennsylvanian evaporite–carbonate cycles and their relation to petroleum occurrence, southern Rocky Mountains. *Bull. Am. Assoc. Petrol. Geol.* **53**, 884–908.

PETERSON, J.A. & OHLEN, H.R. (1963) Pennsylvanian shelf carbonates, Paradox Basin. In: *Shelf Carbonates of the Paradox Basin* (Ed. Bass, R.O.). Four Corners Geological Society 4th Field Conference Guidebook, pp. 65–79.

PINDELL, J.L. & DEWEY, J.F. (1982) Permo-Triassic reconstruction of western Pangaea and the evolution of the Gulf of Mexico/Caribbean region. *Tectonics* **1**, 179–211.

PITMAN, W.C. (1978) Relationship between eustasy and stratigraphic sequences of passive margins. *Geol. Soc. Am. Bull.* **89**, 1389–1403.

PLAFKER, G. (1965) Tectonic deformation associated with the 1964 Alaskan earthquake. *Science* **148**, 1675–1687.

PRAY, L.C. & WRAY, J.L. (1963) Porous algal facies (Pennsylvanian) Honaker Trail, San Juan Canyon, Utah. In: *Shelf Carbonates of the Paradox Basin* (Ed. Bass, R.O.). Four Corners Geological Society 4th Field Conference Guidebook, pp. 204–234.

PRESS, W.H., FLANNERY, B.P., TENKOLSKY, S.A. & VETTERLING, W.T. (1986) *Numerical Recipes: the Art of Scientific Computing*. Cambridge University Press, New York.

RAMSBOTTOM, W.H.C. (1979) Rates of transgression and regression in the Carboniferous of NW Europe. *J. Geol. Soc. Lond.* **136**, 147–153.

READ, J.F. & GOLDHAMMER, R.K. (1988) Use of Fischer plots to define third-order sea-level curves in Ordovician peritidal cyclic carbonates, Appalachians. *Geology* **16**, 895–899.

READ, W.A. & FORSYTH, I.H. (1989) Allocycles and autocycles in the upper part of the Limestone Coal Group (Pendleian E1) in the Glasgow–Stirling region of the Midland Valley of Scotland. *Geol. J.* **24**, 121–137.

RONA, P.A. (1973) Relations between rates of sediment accumulation on continental shelves, sea-floor spreading, and eustasy inferred from the central North Atlantic. *Geol. Soc. Am. Bull.* **84**, 2851–2872.

ROSS, C.A. (1986) Paleozoic evolution of southern margin of Permian basin. *Geol. Soc. Am. Bull.* **97**, 536–554.

ROSS, C.A. & ROSS, J.R.P. (1987) Late Paleozoic sea levels and depositional sequences. In: *Timing and Depositional History of Eustatic Sequences: Constraints on Seismic Stratigraphy* (Eds Ross, C.A. & Haman, D.). Cushman Foundation for Foraminiferal Research, Spec. Publ. 24, pp. 137–149.

RUDDIMAN, W.F., RAYMO, M.E., MARTINSON, D.G., CLEMENT, B.M. & BACKMAN, J. (1989) Pleistocene evolution: northern hemisphere ice sheets and North Atlantic Ocean. *Paleoceanography* **4**, 353–412.

SANDER, B. (1936) Beitrage zur Kenntnis der Anlagerungsgefuge (Rhythmische Kalke und Dolomite aus der Trias). *Mineral. Petrogr. Mitt.* **48**, 27–139.

SCHLAGER, W. (1981) The paradox of drowned reefs and carbonate platforms. *Geol. Soc. Am. Bull.* **92**, 197–211.

SCHWARZACHER, W. (1975) *Sedimentation Models and Quantitative Stratigraphy*. Developments in Sedimentology 19. Elsevier, New York.

SCHWARZACHER, W. & FISCHER, A.G. (1982) Limestone–shale bedding and perturbations of the Earth's orbit. In: *Cyclic and Event Stratification* (Eds Einsele, G. & Seilacher, A.). Springer Verlag, New York, pp. 72–95.

SLOSS, L.L. (1963) Sequences in the cratonic interior of North America. *Geol. Soc. Am. Bull.* **74**, 93–113.

STECKLER, M.S. & WATTS, A.B. (1978) Subsidence of the Atlantic-type continental margin off New York. *Earth Planet. Sci. Lett.* **41**, 1–13.

STEINEN, R.P., HARRISON, R.S. & MATTHEWS, R.K. (1973) Eustatic low stand of sea level between 125 000 and 105 000 BP: evidence from the subsurface of Barbados, West Indies. *Geol. Soc. Am. Bull.* **84**, 63–70.

STEVENSON, G.M. & BAARS, D.L. (1988) Overview: Carbonate reservoirs of the Paradox Basin. In: *Occurrence and Petrophysical Properties of Carbonate Reservoirs in the Rocky Mountain Region* (Eds Goolsby, S.M. & Longman, M.W.). Rocky Mountain Association of Geologists 1988 Carbonate Symposium, 149–162.

STOCKMAN, K.W., GINSBURG, R.N. & SHINN, E.A. (1967) The production of lime mud by algae in south Florida. *J. Sedim. Petrol.* **37**, 633–648.

STOUT, W.E. (1931) Pennsylvanian cycles in Ohio. *Illinois State Geological Survey Bulletin* **60**, 195–216.

STRASSER, A. (1988) Shallowing-upward sequences in Purbeckian peritidal carbonates (lowermost Cretaceous, Swiss and French Jura Mountains). *Sedimentology* **35**, 369–383.

TAFT, W.M., ARRINGTON, F., HAIMOVITZ, A., MACDONALD, C. & WOOLHEATER, C. (1968) Lithification of modern carbonate sediments at Yellow Bank, Bahamas. *Bull. Mar. Sci., Gulf Carib.* **18**, 762–828.

VAIL, P.R. (1987) Seismic stratigraphy interpretation procedure. In: *Atlas of Seismic Stratigraphy, Volume 1* (Ed. Bally, A.W.). Am. Assoc. Petrol. Geol. Stud. Geol. 27, Tulsa, Oklahoma, pp. 1–11.

VAIL, P.R., HARDENBOL, J. & TODD, R.G. (1984) Jurassic unconformities, chronostratigraphy, and sea-level changes from seismic stratigraphy and biostratigraphy. In: *Interregional Unconformities and Hydrocarbon Accumulation* (Ed. Schlee, J.). Mem. Am. Assoc. Petrol. Geol. 36, pp. 129–144.

VAN DER HEIDE, S. (1950) Compaction as a possible factor in upper Carboniferous rhythmic sedimentation. In: *Report of the 18th International Geologic Congress, part 4*, pp. 38–45.

VAN EYSINGA, F.W.B. (1975) *Geologic Time Table*. Elsevier, Amsterdam.

WANLESS, H.R. & CANNON, J.R. (1966) Late Paleozoic glaciation. *Earth Sci. Rev.* **1**, 247–286.

WANLESS, H.R. & DRAVIS, J.J. (1989) *Carbonate environments and sequences of Caicos Platform*. Fieldtrip. Guidebook T 374, 28th Int. Geol. Congr., Am. Geophys. Union.

WANLESS, H.R. & SHEPARD, F.P. (1936) Sea level and climate changes related to late Paleozoic cyclothems. *Geol. Soc. Am. Bull.* **47**, 1177–1206.

WEEDON, G.P. (1989) The detection and illustration of regular sedimentary cycles using Walsh power spectra and filtering, with examples from the Lias of Switzerland. *J. Geol. Soc. Lond.* **146**, 133–144.

WEEDON, G.P. & JENKYNS, H.C. (1990) Regular and irregular climatic cycles and the Belemnite Marls (Pliensbachian, Lower Jurassic, Wessex Basin). *J. Geol. Soc. Lond.* **147**, 915–918.

Weller, J.M. (1930) Cyclical sedimentation of the Pennsylvanian period and its significance. *J. Geol.* **38**, 97–135.

Weller, J.M. (1956) Argument for diastrophic control of late Paleozoic cyclothems. *Bull. Am. Assoc. Petrol. Geol.* **40**, 17–50.

Weller, M.W. (1964) Development of the concept and interpretation of cyclic sedimentation. *Kansas Geol. Soc. Bull.* **169**, 607–619.

Wheeler, H.E. & Murray, H.H. (1957) Base level control patterns in cyclothemic sedimentation. *Bull. Am. Assoc. Petrol. Geol.* **41**, 1985–2011.

Wilson, J.L. (1967) Cyclic and reciprocal sedimentation in Virgilian strata of southern New Mexico. *Geol. Soc. Am. Bull.* **78**, 805–818.

Wilson, J.L. (1975) *Carbonate Facies in Geologic History*. Springer Verlag, New York.

Spec. Publs Int. Ass. Sediment. (1994) **19**, 285–301

Milankovitch cyclicity and high-resolution sequence stratigraphy in lagoonal–peritidal carbonates (Upper Tithonian–Lower Berriasian, French Jura Mountains)

A. STRASSER

Institut de Géologie et Paléontologie, Pérolles, 1700 Fribourg, Switzerland

ABSTRACT

Three sections of the Tidalites-de-Vouglans and Goldberg Formations have been studied in the French Jura. The sedimentary record consists of well-stratified carbonates which represent shallow-lagoonal, intertidal and supratidal depositional environments where salinities ranged from normal marine to hypersaline or fresh water.

The beds display a hierarchical stacking which is probably related to climatically induced sea-level fluctuations in the Milankovitch frequency band. Elementary sequences (commonly corresponding to an individual bed) would represent the 20-ka precession cycle, larger composite sequences the 100- and 400-ka eccentricity cycles.

Elementary and larger sequences can, partly and on a small scale, be analysed in terms of sequence stratigraphy. Sequence boundaries mark the top of the beds and in many cases are erosive. Low-stand deposits comprise calcrete, conglomerates and marls with freshwater fossils, or are missing altogether. Thin transgressive deposits follow a generally well-defined transgressive surface and contain reworked pebbles and mixed marine and freshwater fossils. High-stand deposits make up the bulk of the sequences and generally exhibit a shallowing-upward facies evolution. A large part of the sea-level cycle, however, was dominated by non-deposition, reworking and erosion. The time framework given by the inferred Milankovitch cyclicity permits estimation of rates of sediment accumulation and of diagenetic processes.

Detailed analysis of depositional sequences interpreted to have been induced by Milankovitch cycles suggests a duration of about 3.6 Ma for the two formations studied. Larger sequences are difficult to identify, but partial time control by ammonites and charophyte–ostracod assemblages allows for a tentative correlation with the global sea-level chart of Haq *et al.* (1987). Difficulties with and the validity of such a comparison are discussed.

INTRODUCTION

Most ancient shallow-water carbonates which formed on platforms or ramps are well stratified. The stratification results from stacking of small depositional sequences, which in many cases display a facies evolution with a dominant shallowing-upward trend (James, 1984). Elementary sequences usually correspond to a bed in the stratigraphic record and form part of larger sequences with regressive or transgressive trends of facies evolution (Goodwin & Anderson, 1985; Anderson & Goodwin, 1990). The more or less regular recurrences of depositional sequences imply that cyclic processes controlled carbonate production and/or deposition.

It has been demonstrated by several authors (e.g. Fischer, 1964; Grotzinger, 1986a, b; Hardie *et al.*, 1986; Heckel, 1986; Schwarzacher & Haas, 1986; Goldhammer *et al.*, 1987, 1990; Goldstein, 1988; Read & Goldhammer, 1988; Goldhammer & Harris, 1989; Koerschner & Read, 1989; Read, 1989) that the cyclicity found in ancient shallow-water to peritidal carbonates has time periods in the order of a few tens of thousands to a few hundreds of thousands of years and shows Milankovitch characteristics. It is therefore suggested that cyclic perturbations of the Earth's orbit (Milankovitch, 1941; Berger, 1980; Berger *et al.*, 1989) induced climatic

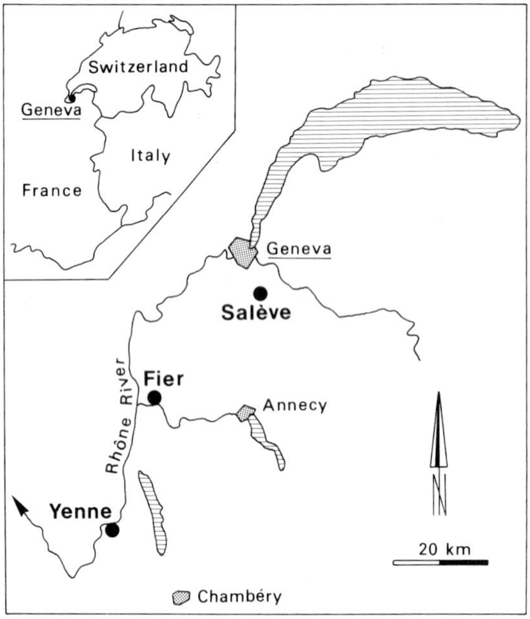

Fig. 1. Location map of the sections studied.

cycles which in turn controlled, directly or indirectly, carbonate productivity and eustatic sea level.

The peritidal carbonate sediments of late Tithonian and early Berriasian age in the Swiss and French Jura Mountains were deposited on the northern passive margin of the Tethys Ocean. They show well-developed cyclic stacking of elementary sequences (Strasser, 1988a). Unfortunately, outcrop conditions are such that only three continuous sections could be sampled, which cover the Upper Tithonian (Portlandian) Tidalites-de-Vouglans Formation (Bernier, 1984) and the Lower Berriasian (Purbeckian) Goldberg Formation (Häfeli, 1966). The section of Salève has been studied along the footpath of Grotte de la Mule on the northwestern face of Mount Salève, close to Geneva (Strasser, 1988b). The section of Fier is situated in the quarry next to regional road D 14, which winds through the canyon of Val du Fier. The third section has been sampled just west of Yenne, in a quarry and along national road N 504, where the Rhône River cuts through the Cluse de la Balme (Fig. 1).

The aim of this paper is to present an analysis of the sections studied in terms of facies evolution and high-resolution sequence stratigraphy, and to discuss the validity and potential applications of a time framework based on Milankovitch cyclicity. Independent time control is given locally by fossils and allows for a tentative correlation with the global eustatic sea-level curve of Haq *et al.* (1987).

DEPOSITIONAL SEQUENCES

The stratigraphic record of the Tidalites-de-Vouglans and Goldberg Formations commonly shows well-defined beds. Bed thicknesses vary between a few centimetres and 4.50 m. Facies indicate shallow-subtidal, intertidal and supratidal depositional environments, comprising marine and restricted lagoons, beaches and sand bars, tidal flats, algal marshes, coastal sabkhas, freshwater lakes, and land (Chevallier & Strasser, 1985; Strasser, 1988a; see also Wright, 1984, for a general description of peritidal facies models). Many sequences display a shallowing-upward trend which is expressed by a gradual change from deeper to shallower facies, or by intertidal to supratidal overprinting of subtidal facies.

The stacking of elementary sequences leads to sequences of a higher order, which equally have a shallowing-upward tendency (Fig. 2). These larger sequences, a few tens of centimetres up to 10 m thick, commonly terminate with marls, conglomerates or pedogenic caps and are easy to pick out in the field. They generally consist of four, five or six elementary sequences.

Continuous shallowing-upward from deeper to shallower facies can be explained by progradation or lateral migration of the depositional system (Ginsburg, 1971; Pratt & James, 1986). Local erosion may be due to tidal or storm-induced currents. Widespread erosional features and pedogenic overprinting of subtidal facies, on the other hand, suggest a drop of relative sea level (Strasser, 1991). It is therefore probable that sea-level fluctuations controlled, at least partly, the deposition of the carbonate sequences.

Sequence stratigraphy defines depositional systems and surfaces related to changes of eustatic sea level. Sequences resulting from high-amplitude sea-level fluctuations are delimited by sequence boundaries and can be decomposed into systems tracts (e.g. Vail *et al.*, 1984; Vail, 1987; Sarg, 1988). Beds and bedsets formed by smaller sea-level changes are defined as parasequences which are bounded by marine flooding surfaces (van Wagoner *et al.*, 1990). On a small scale also, the studied

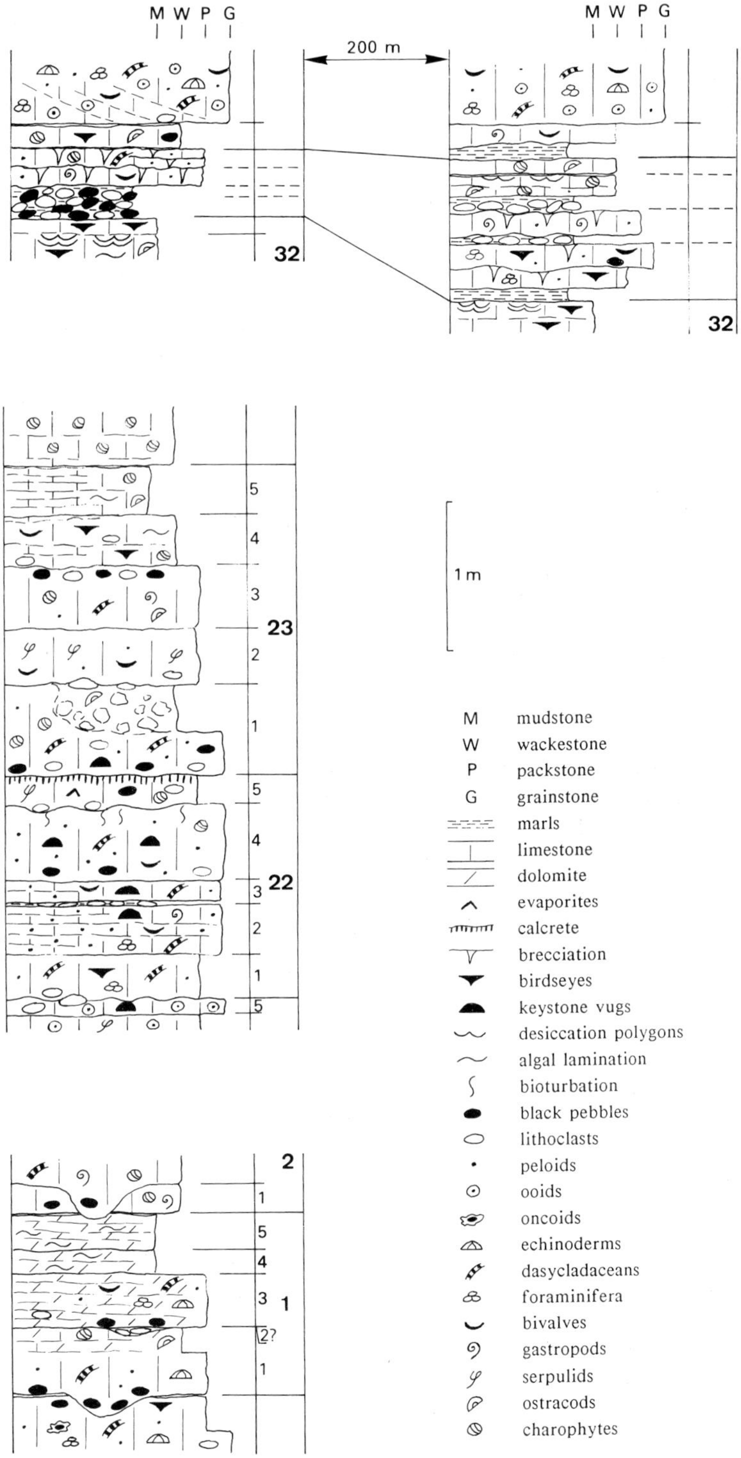

Fig. 2. Detailed sequences of the Salève section. Larger sequences are composed, on average, of five elementary sequences. Sequences of both orders express a shallowing-upward trend of facies evolution. Lateral changes (upper part of figure) demonstrate how some sequences may be reduced by non-deposition, erosion, or reworking. Large numbers in right column correspond to the numbers of supposed 100-ka sequences in Fig. 8. Compare also with Figs 7 and 9.

elementary sequences and the larger composite sequences contain many elements which can be described in terms of sequence stratigraphy. In the very shallow depositional environments of the formations studied the typical geometry of the systems tracts could, of course, not develop. Sediments accumulated basically through vertical aggradation. However, emersion surfaces, transgressive surfaces and facies changes indicating maximum flooding and shallowing-up can be identified in many cases. It is therefore preferable to call the studied beds and bedsets 'small-scale' and 'composite, larger sequences' instead of 'parasequences', and to delimit them with sequence boundaries. This integration of sequence stratigraphy in the description of small-scale sequences provides a more dynamic picture of their depositional history.

Sequence boundaries in the formations studied commonly correspond to the upper bedding surfaces and are characterized by erosion, pedogenetic brecciation, root traces, or desiccation polygons. These features formed during a drop of relative sea level. The overlying low-stand deposits are represented by calcrete crusts, conglomerates and marls. Lithoclasts have been reworked from the underlying sequence, and abundant black pebbles are impregnated by terrestrial organic matter, which in some cases had been burnt by forest fires (Strasser & Davaud, 1983). Green marls contain authigenic illite, which formed during repeated wetting by sea water and drying (Deconinck & Strasser, 1987). Charophytes, ostracods and gastropods in the marls point to the presence of freshwater lakes (Mojon & Strasser, 1987). Low-stand deposits are commonly thin but well developed on the top of the larger sequences, but may be reduced to a film of marls or missing completely between the elementary sequences.

The transgressive surfaces are mostly well defined. Transgressive deposits contain reworked lithoclasts and black pebbles, mixed lagoonal and freshwater fossils, and locally birdseyes or keystone vugs, which indicate intertidal to supratidal conditions. In some cases, the marls below the transgressive surface already contain a mixed fauna. This may indicate reworking by storms and spring tides, and suggests that the observed well-developed transgressive surface does not necessarily correspond to the initial transgressive movement, but to the stage when the location of the section studied was actually flooded. The transgressive deposits of the small-scale sequences are usually thin and in some cases missing,

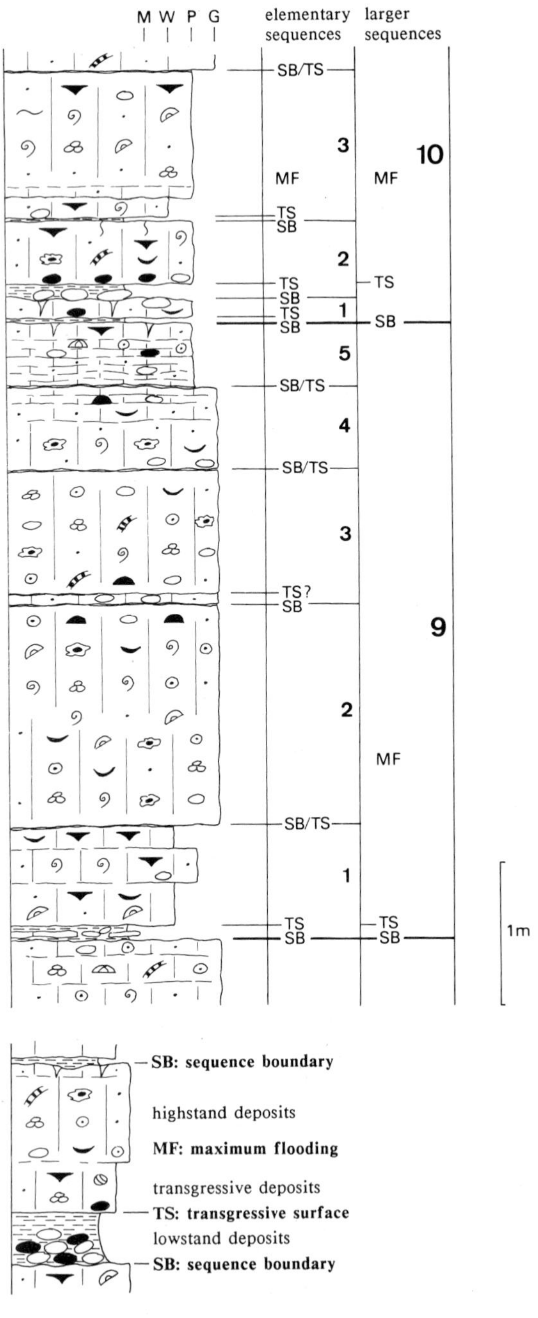

Fig. 3. Larger sequences 9 and 10 (supposed 100-ka sequences) interpreted in terms of sequence stratigraphy. Compare also with Fig. 6. The hypothetical section at the bottom shows positions of sequence boundaries, transgressive surface and approximate maximum flooding, as well as of deposits formed during one transgressive–regressive cycle. Symbols as in Fig. 2.

which is probably due to reduced carbonate production after prolonged intertidal to supratidal exposure.

Once sea level had risen enough to create new lagoons, carbonate production could set in again. During sea-level high-stand, sediments accumulated up to the water surface, which led to a shallowing-upward facies succession. Transgressive deposits commonly pass into regressive high-stand deposits. Actual surfaces of downlap or maximum flooding (Posamentier *et al.*, 1988) can only rarely be recognized in the small-scale sequences studied (in Fig. 3, the approximate level of deepest facies has been labelled MF: maximum flooding). High-stand deposits make up the bulk of the sequences in the formations studied. Sediments of the late high-stand may exhibit birdseyes or keystone vugs indicating intertidal to supratidal exposure, they may be dolomitized or contain evaporites, or be rich in freshwater fossils. A sequence boundary then terminates the sequence.

In many cases, the accommodation potential (defined by the combined rates of subsidence and eustatic sea-level change) was insufficient to permit deposition of complete sequences. Non-deposition, erosion and reworking resulted in incomplete or missing sequences. Elementary sequences may be represented only by a conglomerate or by a brecciated layer. Larger sequences may contain only one or two beds, or may themselves be reduced to a conglomerate (e.g. Fig. 2, top).

SEDIMENTARY RECORD AND TIME CONTROL

The sections studied show a trend from thicker beds in the Tidalites-de-Vouglans Formation to thinner beds in the Goldberg Formation (Fig. 4), which indicates a general decrease of accommodation potential on the platform. Facies evolve from rather marine-lagoonal to rather restricted lagoonal and fresh water. In detail, however, the sedimentary record is a complex stacking of depositional sequences (Figs 5 and 6).

At the base of the Salève section, thick-bedded marine-lagoonal limestones are truncated by an erosion surface (below sequence 1 in Figs 2 and 7). The following two larger depositional sequences (1 and 2) exhibit dolomitized algal mats and charophytes (Fig. 2, bottom). This rapid facies change is interpreted to coincide with the boundary

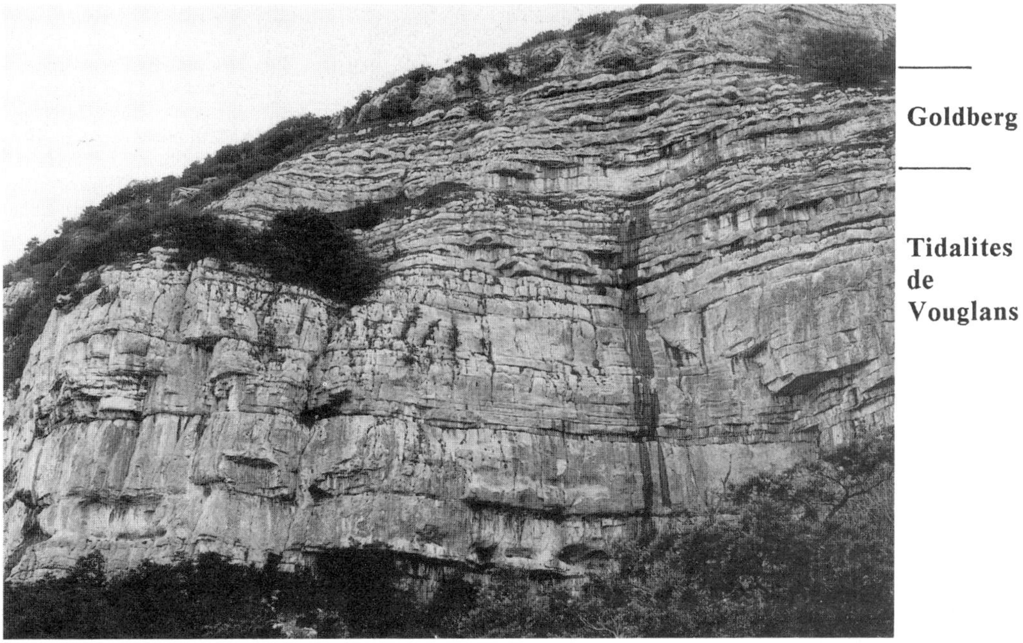

Fig. 4. Section of the northwest face of Salève Mountain, displaying a general trend of thinning-upward in the Tidalites-de-Vouglans and Goldberg Formations. The cliff is about 100 m high.

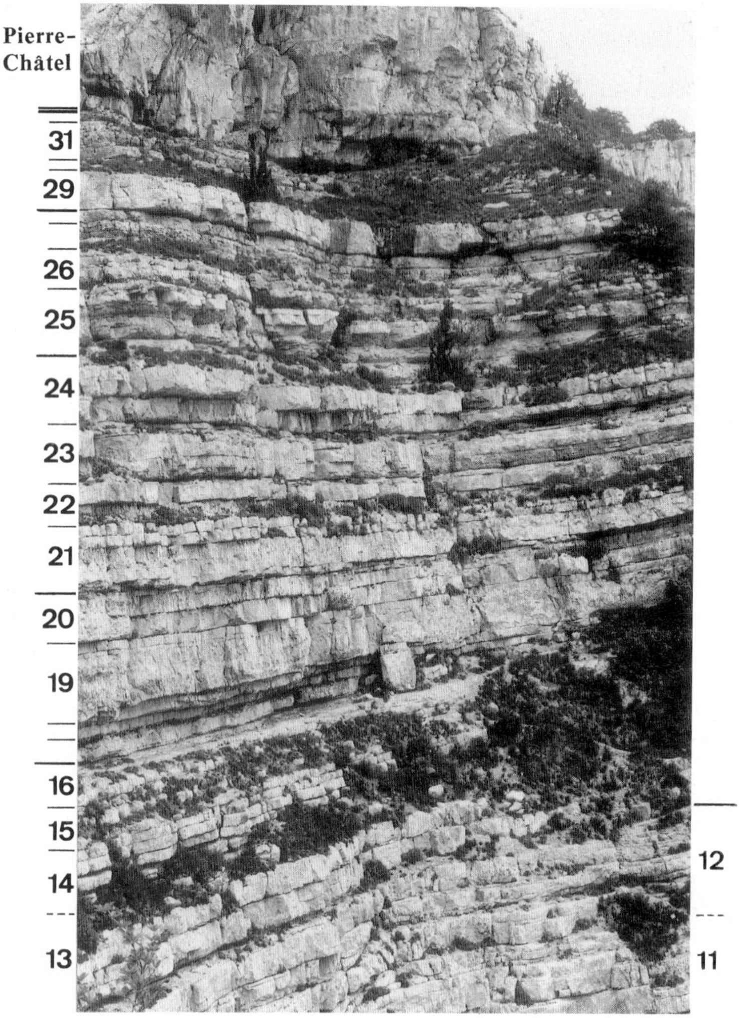

Fig. 5. Goldberg Formation at Salève exhibiting hierarchical stacking of sequences. Inferred 100-ka sequences (numbered as in Fig. 8) show bundling which implies modulation by a 400-ka orbital cycle. Sequence 19 is 2.10 m thick.

between the underlying Couches-du-Chailley Formation and the Tidalites de Vouglans (Bernier, 1984). In the sections of Fier and Yenne, facies changes are less pronounced. Concentration of black pebbles and dolomitization, however, may be the lateral equivalent of the restricted facies encountered at Salève (Fig. 8).

Facies of the overlying sequences are mostly marine (example in Fig. 3), and many of the sequences are partly dolomitized (Fig. 8). Black pebbles commonly occur in low-stand and transgressive deposits. The limit between the Vouglans and Goldberg formations has been set where, in the Salève section, the first well-developed black-pebble conglomerate appears (top of sequence 12, Fig. 8). The sequences of the Goldberg Formation generally show more restricted and lacustrine facies and display abundant calcrete (at Salève, example of Fig. 2, centre) and pedogenic brecciation (at Yenne).

The sequences at the very top of the Goldberg Formation are commonly much reduced (Figs 2, top, and 9). Green marls and black-pebble conglomerates are common. Following a thin transgressive bed, coarse grainstones and packstones of the Pierre-Châtel Formation (Steinhauser & Lombard, 1969) then mark the change to fully marine conditions.

The depositional sequences as defined above show a hierarchical stacking. On average, five elementary sequences compose one larger sequence

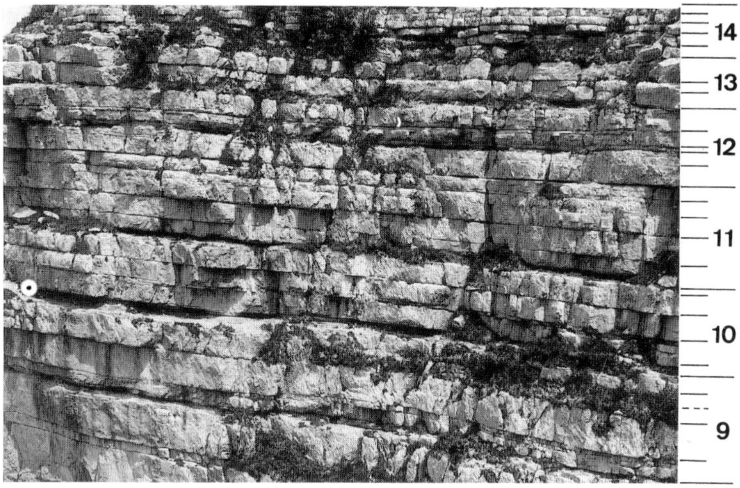

Fig. 6. Detail showing superposition of beds (inferred 20-ka sequences) which form larger sequences corresponding to the 100-ka orbital cycle. Numbers as in Fig. 8. Sequence 10 is 2.70 m thick.

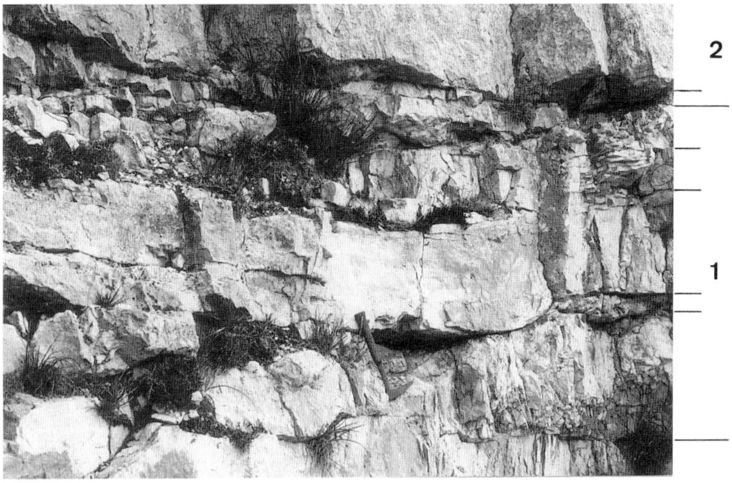

Fig. 7. Base of Tidalites-de-Vouglans Formation at Salève. Note irregular eroded bed surfaces. Sequence numbers correspond to those in Fig. 8. Hammer for scale.

(Figs 2, 6 and 8). Furthermore, especially in the Salève section, four larger sequences are in many cases grouped together to form sequences of a still higher order which display well-marked erosional tops or thinning-upward of beds, or which are followed by well-developed low-stand deposits. The definition of sequences and the correlation between the sections of Salève, Fier and Yenne presented in Fig. 8 is, of course, not unequivocal. It is rather a best-fit solution which, for the upper half of the sections, has been completed by the study of 24 other sections in the Jura Mountains (Strasser, 1988a). In Fig. 8, the larger sequences have been numbered.

Independent timing of the sequences is difficult. Ammonites have been found only in the uppermost part of the Goldberg Formation and in the lower part of Pierre-Châtel (Clavel *et al.*, 1986; Waehry, 1988). The base of Pierre-Châtel is situated in the *privasensis* subzone (Middle Berriasian). An ammonite of the *grandis* subzone (Lower Berriasian) has been cited at Cluse de Chaille by Clavel *et al.* (1986), in a bed that probably corresponds to sequence 33 (Fig. 10). Mojon (in Detraz & Mojon, 1989) established a zonation based on charophyte and ostracod assemblages which is calibrated with ammonites and calpionellids. Some sequences of the sections studied could thus be dated (Fig. 8); some are dated by correlation with other sections.

The erosion surface at the very base of the Salève

section and the important facies changes in all three sections studied imply a sea-level fall and may, in terms of sequence stratigraphy, be interpreted as a sequence boundary of a larger-scale sea-level cycle (short-term cycle of Haq *et al.*, 1987). The dolomitized and freshwater-dominated sequences 1 and 2 (and partly 3) would correspond to low-stand and/or earliest transgressive deposits (Figs 2, bottom, and 7). A well-marked thinning-upward trend at Salève points to diminishing accommodation potential as high-stand deposits fill in the platform, and thus to a possible larger sequence boundary on top of sequence 16 (Fig. 5). Other sections in the Jura, however, show a rapid change from dolomitic to limestone facies at the top of sequence 12. The thick bed of sequence 19 indicates rapid deepening of water and could correspond to a time of maximum flooding of the platform. Other possible large-scale sequence boundaries may occur on top of sequences 24 or 28 (followed by thicker beds and/or

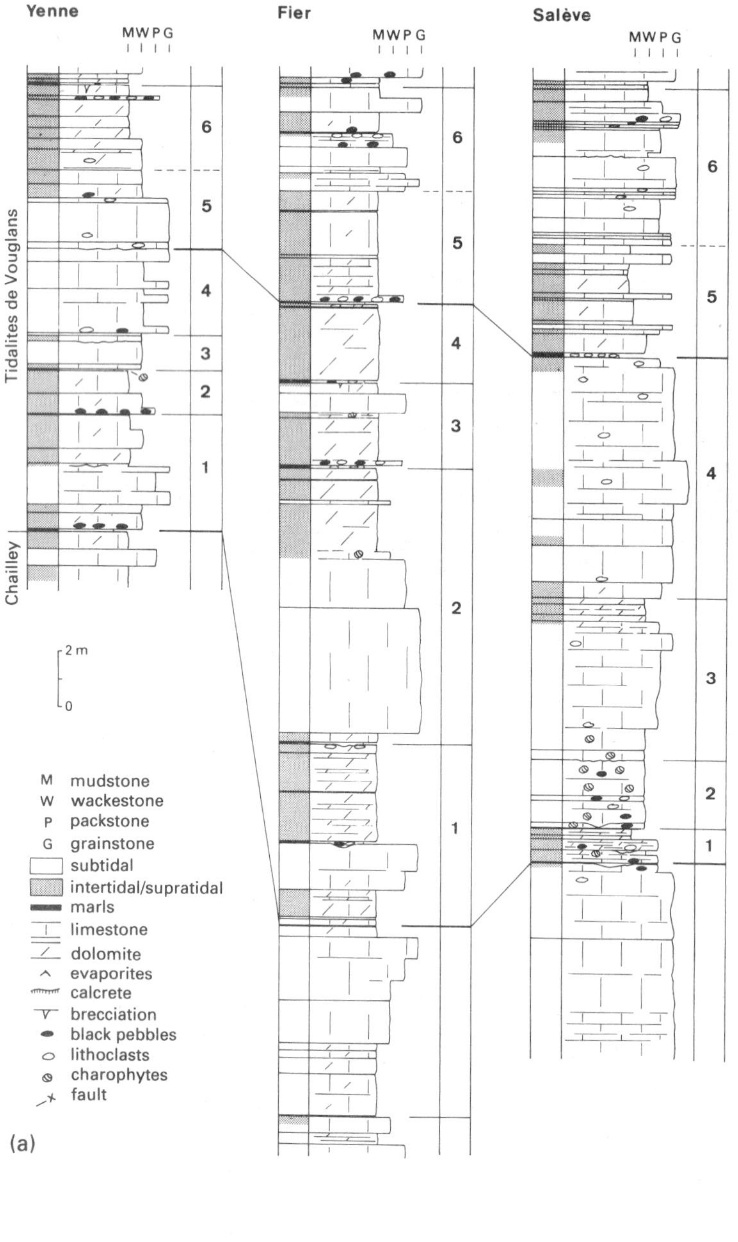

Fig. 8. Studied sections of Salève, Fier and Yenne with simplified facies interpretation. Inferred 100-ka sequences are numbered, limits of 400-ka groups are marked with heavy lines. *M1b* to *M3* are charophyte–ostracod assemblages of P.O. Mojon (personal communication). For discussion refer to text.

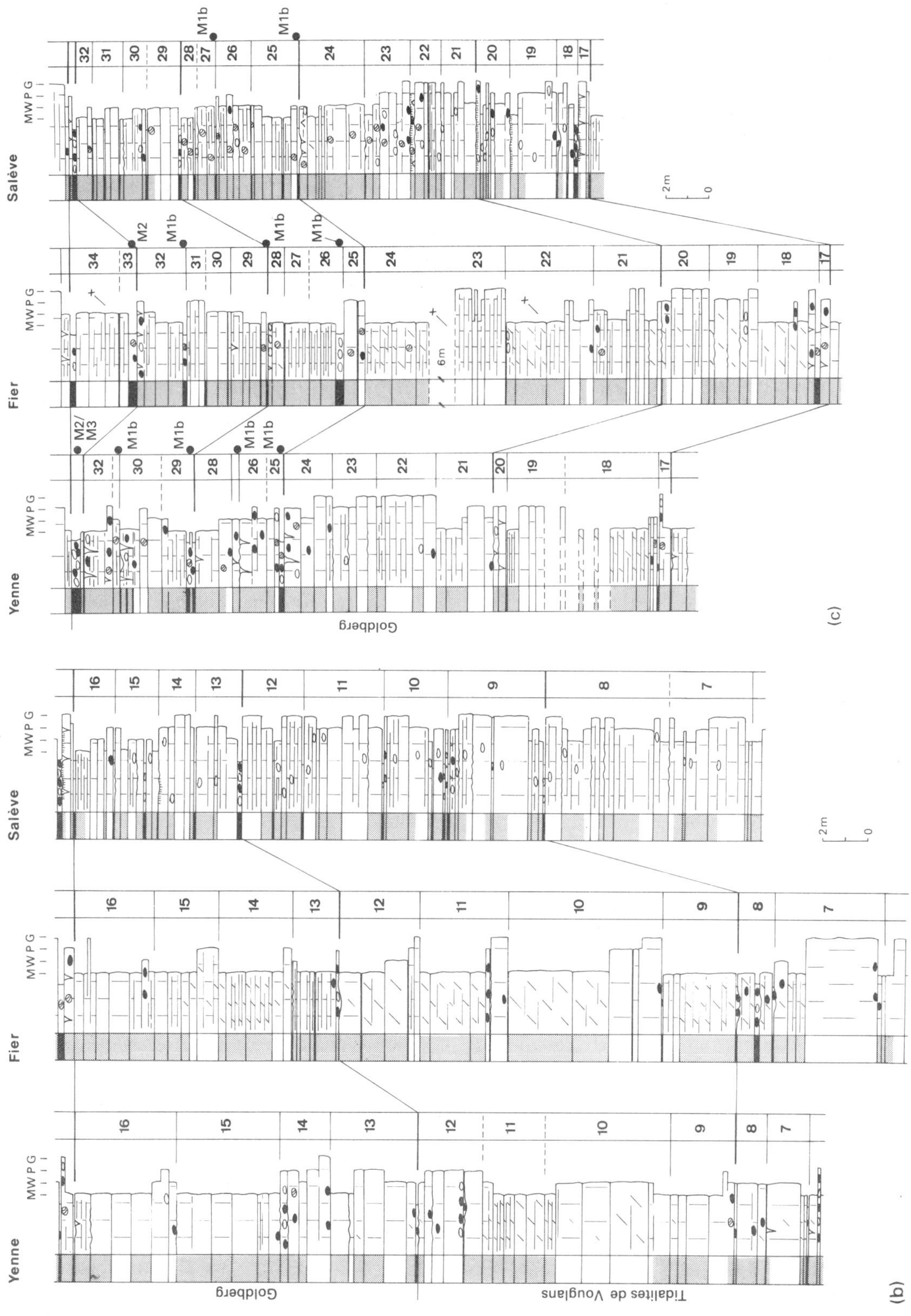

Fig. 8. *Continued.*

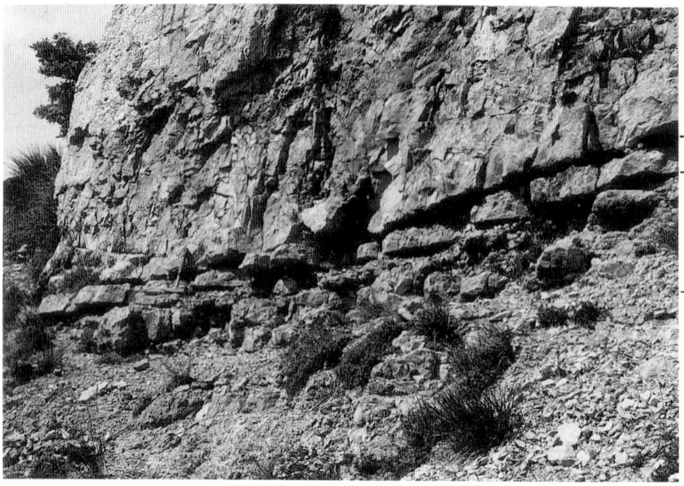

Fig. 9. Top of Goldberg Formation at Salève. Sequences are strongly reduced, reworked, or missing. Hammer for scale.

deeper facies in sequences 29 and 30), and at the top of sequence 32, after which the beds are much reduced and reworked. The rapid change to fully marine conditions at the base of Pierre-Châtel then indicates an important transgression (Figs 9 and 10).

DISCUSSION

Milankovitch cyclicity

Stacked shallowing-upward sequences in shallow-water carbonates can form by autocyclic or allocyclic processes (Strasser, 1991). Autocyclic processes include progradation of shorelines and lateral migration or shifts of tidal channels, tidal inlets and bars, or islands (Ginsburg, 1971; Matti & McKee, 1976; Wong & Oldershaw, 1980; Pratt & James, 1986; Selg, 1988). Such processes are inherent to shallow platforms, and the resulting sequences are limited to specific depositional environments. Allocyclic processes, on the other hand, are controlled by factors such as basin-wide or global changes of sea level or sediment production rates.

The fact that many sequences of the Tidalites-de-Vouglans and Goldberg Formations display pedogenic caps and widespread erosional features implies that drops of relative sea level were involved, which could have been due to periodic tectonic uplifts or to eustatic sea-level falls. It has been suggested by Cisne (1986) that high-frequency tectonic movements at the platform edge can form successive shallowing-upward sequences. Differential subsidence did influence the Jura platform (Wildi *et al.*, 1989), but the observed hierarchical stacking of small sequences can be better explained by climatically controlled eustatic sea-level changes.

Glacial eustasy was very effective during the Quaternary and has been proven to be linked to insolation changes induced by cyclic perturbations of the Earth's orbit (Milankovitch, 1941; Hays *et al.*, 1976). In Late Jurassic and Early Cretaceous times, climate was more equable (Barron, 1983). Nevertheless, Frakes & Francis (1988) suggested the presence of polar ice whose waxing and waning could have influenced sea level. Other climatically tuned factors modifying eustatic sea level are presence or absence of Alpine glaciers (Fairbridge, 1976), changing evaporation rates (Donovan & Jones, 1979), and thermal expansion of the upper layers of the ocean (Gornitz *et al.*, 1982). The thicknesses of the studied small-scale sequences suggest that the high-frequency sea-level fluctuations were in the order of only one to a few metres. Any of the factors mentioned above could have produced such low-amplitude sea-level fluctuations. Climatically controlled changes in the rate of organic and inorganic carbonate production could equally have influenced the sedimentary conditions. Climatically induced presence or absence of terrestrial vegetation may have held back or released clay minerals which could slow down carbonate production. Furthermore, climatically controlled changes in water circulation could lead to nutrient excess and plankton blooms, which reduced water transparency and thus the vitality of carbonate-producing organisms (Hallock & Schlager, 1986).

The hierarchy of the stacked sequences suggests superimposed cycles of eustatic sea-level changes which reflect a Milankovitch cyclicity. As a working hypothesis it is assumed that the elementary sequences correspond to the precession cycle of the equinoxes which, in the Early Cretaceous, had a duration of about 20 000 years (Berger *et al.*, 1989). The larger sequences (composed of an average of five elementary sequences) would represent the 100 000-year eccentricity cycle, and the groups of four larger sequences the second cycle of eccentricity of 400 000 years (Milankovitch, 1941; Berger, 1980; Berger *et al.*, 1989). Obliquity cycles could not be identified. Considering the fact that many beds in the outcrops studied have eroded tops, that some beds may be missing, that compaction rates vary from one facies to another, and that autocyclic processes cannot be excluded, a statistical analysis of bed thicknesses or simple facies alternations (e.g. Schwarzacher & Haas, 1986) has not been undertaken. The recognition of such a hierarchy is therefore based mainly on facies evolution and comparison with other sections.

It has been shown by Weedon & Jenkyns (1990) that bundling of elementary sequences can also be caused by climatic variations not related to orbital eccentricity cycles. Furthermore, Laskar (1989) argued that the chaotic behaviour of the solar system excludes predictability of the orbital parameters of the Earth for times older than a few tens of millions of years. Even though it cannot be proven to what extent the observed stratigraphic record was controlled by orbital cycles, it is practical to assume such a control as a first approximation. This allows a tentative time framework to be established, which can then be compared with the framework given by the larger sea-level variations and by biostratigraphy.

Superposition of Milankovitch cycles on larger sea-level cycles

High-frequency sea-level cycles of the Milankovitch type are superimposed on larger cycles of eustatic sea level with durations of one to several million years and amplitudes of several tens of metres (Goldhammer *et al.*, 1990; Osleger & Read, 1991). The resulting fluctuations of sea level, together with probable cyclic carbonate production and differential subsidence, control water depth and thus facies evolution. Shallow-lagoonal and peritidal depositional environments respond to the slightest modifications of these three parameters, which leads to a very complex sedimentary record.

The most detailed sea-level curve available for the Mesozoic is the one published by Haq *et al.* (1987), although it is controversial in places (Schlager, 1991). In Fig. 10, the supposed Milankovitch cycles are compared with the short-term sea-level cycles of Haq *et al.* (1987). The total number of supposed 100-ka sequences is inferred from the detailed analysis and correlation of the studied sections (Fig. 8). The brecciated beds and conglomerates especially at the top of the Goldberg Formation, and the condensation of the charophyte–ostracod assemblages *M2/M3* (Fig. 8), imply that some sequences are much reduced or missing (Fig. 2, top). The suggested number of 36 is therefore rather a minimum estimate (Fig. 10).

The sequence boundary at the base of the Tidalites-de-Vouglans Formation could not be dated by palaeontological means. However, it represents a major break in the sedimentary record and is tentatively correlated with sequence boundary 134 of Haq *et al.* (1987), which occurs at the base of the *jacobi* subzone. Three other possible sequence boundaries can be proposed (Fig. 10), but it is not known to what extent they are due to a larger cycle or to a strongly expressed superimposed Milankovitch signal. One is placed in the *jacobi* subzone (dated indirectly by correlation with charophyte–ostracod assemblage *M1a*; Mojon, in Detraz & Mojon, 1989). The two other possible sequence boundaries appear in the *grandis* subzone, one of them probably at its very top. In a recent study of deeper-water carbonates in south-eastern France, Jan du Chêne *et al.* (1993) propose two sequence boundaries in the *jacobi grandis* zone, one at the base of the *subalpina* subzone, and one at its top which corresponds to sequence boundary 131.5 in Haq *et al.* (1987). This may suggest that more time needs to be accounted for at the top of the Goldberg Formation, and that sequence boundary 134 in Haq *et al.* could also be represented by the boundary placed at the top of sequences 12 or 16 (Fig. 10). More biostratigraphic data are needed, and detailed platform-to-basin correlations will have to be carried out before the positions of these sequence boundaries can be established with confidence (in Fig. 10 the first possibility has been taken into account). The transgressive surface corresponding to the base of the Pierre-Châtel Formation has been well dated

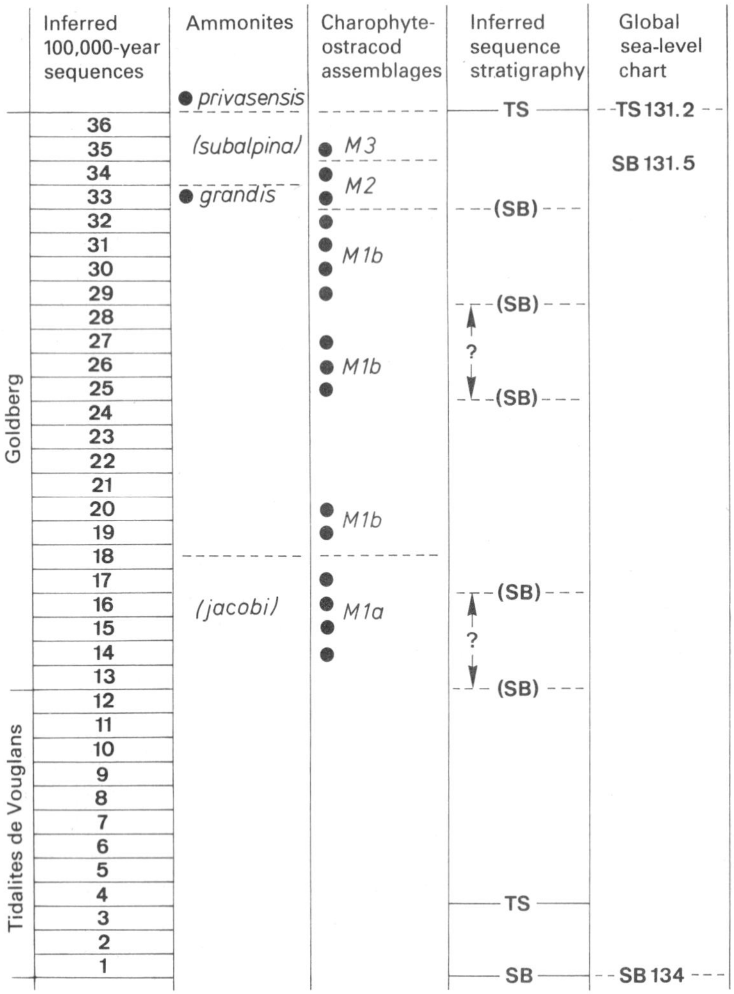

Fig. 10. Tentative comparison between Milankovitch cyclicity and large-scale sequence stratigraphy. Charophyte–ostracod assemblages after Mojon (in Detraz & Mojon, 1989), global sea-level chart after Haq *et al.* (1987). For discussion refer to text.

by ammonites and can be correlated with the large-scale transgression at about 131.2 Ma (Haq *et al.* dated the maximum flooding at 131 Ma).

Thirty-six sequences of possibly 100 ka imply a timespan of 3.6 Ma for the formation of the Tidalites de Vouglans and Goldberg successions. However, according to the sea-level chart of Haq *et al.* (1987) and the proposed correlations, only 2.8 Ma are available. Another approach is counting the beds in the studied sections, and multiplying by 20 ka (assuming each bed to be an elementary sequence). This gives the following values: 3.3 Ma for Salève, 2.8 Ma for Fier, and 2.4 Ma for Yenne. Some beds may, of course, be due to autocyclic processes, many have been eroded, or time may have passed without deposition.

It has to be kept in mind that the top of the Goldberg Formation commonly shows condensation due to non-deposition, reworking or erosion, so that dates furnished by fossils often cannot be attributed to a particular sequence. Interpretation and correlation of supposed 100-ka sequences is in many cases a best-fit solution, and dating by sequence correlation may well be out by 100–200 ka. Furthermore, the ages of sequence boundaries of Haq *et al.* (1987) are indicated with steps of 0.5 Ma for the time period here concerned, and the absolute time scales are not very well established. According to Haq *et al.* (1987) the Berriasian has a duration of 6 Ma, according to Harland *et al.* (1990, p. 58) only 5 Ma. Finally, on a shallow platform, a strong Milankovitch signal can modulate a sea-level drop leading to an important sequence boundary and displace its strongest physical expression by a few hundred thousand years. These uncertainties may explain some of the discrepancies in Fig. 10.

Rates of sedimention and diagenesis

If it is assumed that an elementary sequence comprises about 20 ka, a time framework for estimating rates of sediment production and accumulation, of facies evolution, and of diagenetic processes is given.

In order to evaluate the initial sediment thickness, compaction has to be considered. Dewatering and mechanical compaction may account for a porosity loss of about 10% in grainstones and of 30% or more in shallow-water carbonate muds (Shinn & Robbin, 1983; Moore, 1989, pp. 244–247). For deep burial producing pressure solution and stylolitization, values between 20 and 30% of chemical compaction seem to be reasonable (Moore, 1989, pp. 247–251). Depending on the facies, total compaction can accordingly be estimated to have varied between 30 and 60% or more.

Allowing an average compaction of 50%, no significant erosion and a timespan of about 3.5 Ma for the deposition of the two formations studied, the average subsidence rate can be estimated at about 0.05 m/ka (Grotzinger, 1986b, indicates 0.05–0.1 m/ka for mature passive margins). The varying thicknesses of time-correlated sequences (Fig. 8) can be explained by differential subsidence on the faulted Jura platform (Wildi *et al.*, 1989).

Figure 11 illustrates schematically that, in a peritidal setting, a great part of the 20 ka of an elementary cycle is spent in non-deposition or erosion. Decompacted bed-thicknesses suggest that accommodation space and thus relative sea-level fluctuations were in the order of a few metres. The estimated subsidence of about 0.05 m/ka was greatly outpaced by sedimentation which was first slow during a certain lag time, but which later reached 1 m/ka (Schlager, 1981, indicates this value as an average for ancient carbonate platforms; Hardie & Ginsburg, 1977, suggest 0.3–3 m/ka for Recent tidal-flat accumulation). Sea-level rise is estimated at about 0.2 m/ka on average (as calculated from an average accumulation up to the intertidal zone of 2 m/10 ka of non-compacted sediment), but reached 0.4 m/ka at its fastest point. Using these theoretical values, it is clear from Fig. 11 that the space available for sediment accumulation is soon filled up, even before the sea-level curve reaches its culmination. The base of the resulting sedimentary sequence will consist of a lag deposit and transgressive facies, until deepest water is reached at maximum flooding. Water depth then diminishes and a shallowing-up facies succession is deposited, which soon attains emersion. Falling sea level starts to erode the previously accumulated sediment. If a freshwater lens is present, cementation will occur and stabilize the carbonate, which will then be exposed to pedogenesis, vadose diagenesis and karstification.

This model suggests that, in a peritidal carbonate system, sediment production and accumulation occurs mostly during a rise of eustatic sea level, when accommodation space is created. Transgressive de-

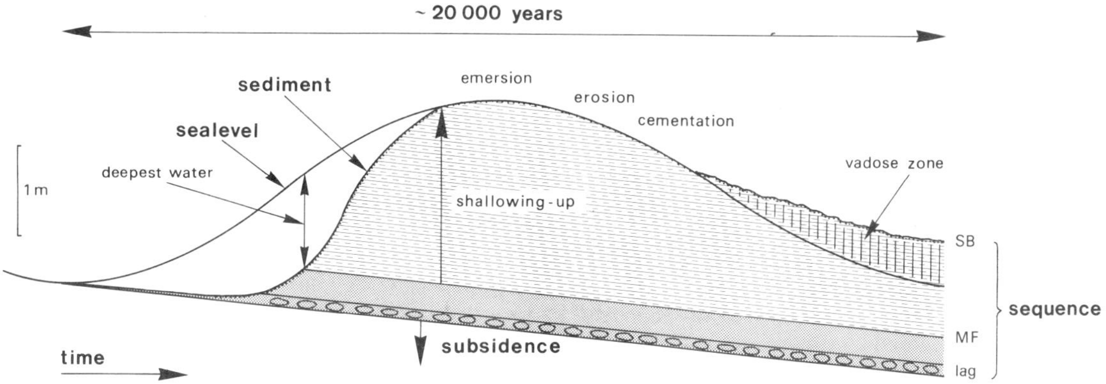

Fig. 11. Model of deposition of one small-scale peritidal carbonate sequence during a sea-level cycle of about 20 ka. Sediment accumulation takes place only during eustatic sea-level rise. Some sediment is exported as sea level starts to fall, but cementation soon prevents further erosion. Sea level drops below the sediment surface, thus creating a vadose diagenetic zone. MF, maximum flooding; SB, sequence boundary. For discussion refer to text.

posits with a deepening-upwards trend develop only at the beginning of eustatic sea-level rise. As soon as sediment accumulation outpaces sea-level rise, a shallowing-upward high-stand deposit is created. It is evident that only a few thousand years are needed to deposit such a sequence, and that much time is available for erosion, diagenesis and reworking.

Taking for example the lagoonal grainstones of elementary sequence 2 (part of sequence 9, Fig. 3), decompacting them by 30% and dividing by 4 ka accumulation time (as suggested in Fig. 11) gives a sedimentation rate of about 0.5 m/ka. The packstones of elementary sequence 5 (Fig. 3), decompacted by 50%, imply a sedimentation rate of 0.17 m/ka. These values are rather small when compared to Recent production rates. They might suggest that even less time is needed to accumulate an elementary sequence. Quinn & Matthews (1990) estimated that in post-Miocene atoll carbonates less than 10% of lapsed time was recorded by sediment accumulation. Some of the sediment may have been exported by currents and storms, as for example has been documented by Wilber *et al.* (1990) for the Holocene of the Great Bahama Bank, where during sea-level high-stand large quantities of bank-derived sediment accumulated on the slope. It is also possible that Tithonian–Berriasian production rates were lower than modern ones. More data are needed to clarify this point.

Cementation in a freshwater lens can be very fast (Halley & Harris, 1979, have shown that oolite was cemented within 1 ka). The reworked clasts at the base of small-scale sequences testify to such rapid consolidation of the underlying sediment. Cathodoluminescence shows complex patterns of early cementation which are difficult to correlate in a vertical section. This is probably due to repeated exposure at the top of small sequences, allowing for the installation of freshwater lenses and early cementation of each sequence, before the following one was deposited. Horbury & Adams (1989) and Sun (1990) have illustrated in detail such cyclic cementation in shallow-water carbonates.

Studies of oxygen isotopes performed on Purbeckian micrites have shown an evolution towards heavier $\delta^{18}O$ values at the top of 100-ka sequences, indicating increased evaporation (Joachimski, 1990). Carbon isotopes generally become lighter towards the top of 100-ka sequences, which is probably due to the enrichment of the pore waters by light carbon from soil gas (Joachimski, 1990).

CONCLUSIONS

The shallow-lagoonal and peritidal carbonates studied display a hierarchical stacking of beds. One bed represents in most cases an elementary depositional sequence. An average of five elementary sequences composes larger sequences, which again form groups of four. The sequences of all three orders generally display a shallowing-upward trend of facies evolution. Recurring pedogenic caps and widespread erosion surfaces imply repeated drops of sea level, and thus sea-level fluctuations.

As the formation of the sequences is, at least partly, controlled by sea level, the concept of sequence stratigraphy may be applied in their description and interpretation. This underlines the dynamic evolution of depositional sequences in response to changing accommodation potential. Bed surfaces are in many cases erosive and are interpreted as sequence boundaries, the overlying calcretes, conglomerates and marls as low-stand deposits. Transgressive surfaces are mostly well defined. Transgressive deposits contain reworked pebbles and mixed marine and freshwater fossils. High-stand deposits commonly exhibit subtidal, marine to restricted facies in their lower part, then shallow up into intertidal and supratidal, hypersaline or freshwater facies.

The hierarchical stacking of sequences is probably due to sea-level fluctuations controlled by orbitally induced climatic cycles in the Milankovitch frequency band. The elementary sequences may correspond to the 20-ka cycle of the precession of the equinoxes, the two larger orders of sequences to the eccentricity cycles with periods of 100 and 400 ka.

Dating by fossils allows correlation of some sequences of the studied sections with the global sea-level curve of Haq *et al.* (1987). Large-scale features of sequence stratigraphy are not always easy to recognize: in the peritidal depositional environments, high-frequency sea-level fluctuations overprinted many of the signals of a general sea-level change.

More biostratigraphical and palaeomagnetic work, very detailed analyses of facies evolution and sequential patterns, and comparisons with other, widely spaced sections are needed to resolve this incoherency and to test the tentative correlation presented in Fig. 10. If it can be demonstrated that the recognized sequences indeed represent Milankovitch cycles with periods of 20, 100 and 400 ka, and if time correlations of the sequences are well established, a very precise framework of absolute time will be available.

Within this framework it becomes possible to estimate rates of sediment production and accumulation, as well as of diagenetic processes. There is the potential that, through very precise studies of the ecological, sedimentological and diagenetic record, the evolution of an ancient sedimentary system can be monitored on a time scale which is comparable to that of the Holocene.

ACKNOWLEDGEMENTS

Material for this work was collected during my stay at the Department of Geology and Palaeontology in Geneva. I would like to thank Eric Davaud who introduced me to the Purbeckian and Alex Waehry who accompanied me on the long and winding path through high-resolution sequence stratigraphy. Pierre-Olivier Mojon provided the charophyte–ostracod dates. Georges Gorin kindly read and improved a first version of the manuscript. I also thank Poppe de Boer, Wolfgang Schlager, David Smith and Hugh Jenkyns for their critical and helpful reviews. This study has been supported by the Swiss National Science Foundation as part of projects No. 2.897.083 and 21–28.988.90.

REFERENCES

Anderson, E.J. & Goodwin, P.W. (1990) The significance of metre-scale allocycles in the quest for a fundamental stratigraphic unit. *J. Geol. Soc. Lond.* **147**, 507–518.

Barron, E.J. (1983) A warm, equable Cretaceous: the nature of the problem. *Earth Sci. Rev.* **19**, 305–338.

Berger, A.L. (1980) The Milankovitch astronomical theory of paleoclimates: A modern review. *Vistas in Astronomy* **24**, 103–122.

Berger, A., Loutre, M.F. & Dehant, V. (1989) Astronomical frequencies for pre-Quaternary palaeoclimate studies. *Terra Nova* **1**, 474–479.

Bernier, P. (1984) Les formations carbonatées du Kimméridgien et du Portlandien dans le Jura méridional. Stratigraphie, micropaléontologie et sédimentologie. *Doc. Lab. Géol. Lyon* **92**.

Chevallier, T. & Strasser, A. (1985) Dépôt de plaine d'estran dans le Portlandien de la montagne de Parves (Jura méridional, Ain, France). *Eclogae Geol. Helv.* **78**, 287–298.

Cisne, J.L. (1986) Earthquakes recorded stratigraphically on carbonate platforms. *Nature* **323**, 320–322.

Clavel, B., Charollais, J., Busnardo, R. & Le Hegarat, G. (1986) Précisions stratigraphiques sur le Crétacé inférieur basal du Jura méridional. *Eclogae Geol. Helv.* **79**, 319–341.

Deconinck, J.F. & Strasser, A. (1987) Sedimentology, clay mineralogy and depositional environment of Purbeckian green marls (Swiss and French Jura). *Eclogae Geol. Helv.* **80**, 753–772.

Detraz, H. & Mojon, P.O. (1989) Evolution paléogéographique de la marge jurassienne de la Téthys du Tithonique–Portlandien au Valanginien: corrélations biostratigraphiques et séquentielles des faciès marins à continentaux. *Eclogae Geol. Helv.* **82**, 37–112.

Donovan, D.T. & Jones, E.J.W. (1979) Causes of worldwide changes in sea level. *J. Geol. Soc. Lond.* **136**, 187–192.

Fairbridge, R.W. (1976) Convergence of evidence on climatic change and ice ages. *Ann. N.Y. Acad. Sci.* **91**, 542–579.

Fischer, A.G. (1964) The Lofer cyclothems of the Alpine Triassic. *Kansas Geol. Surv. Bull.* **169**, 107–149.

Frakes, L.A. & Francis, J.E. (1988) A guide to Phanerozoic cold polar climates from high-latitude ice-rafting in the Cretaceous. *Nature* **333**, 547–549.

Ginsburg, R.N. (1971) Landward movement of carbonate mud: new model for regressive cycles in carbonates (abstract). *Bull. Am. Assoc. Petrol. Geol.* **55**, 340.

Goldhammer, R.K. & Harris, M.T. (1989) Eustatic controls on the stratigraphy and geometry of the Latemar buildup (Middle Triassic), the Dolomites of northern Italy. In: Controls on Carbonate Platform and Basin Development (Eds Crevello, P.D., Sarg, J.F., Read, J.F. & Wilson, J.L.). Soc. Econ. Paleont. Mineral. Spec. Publ. 44, pp. 323–338.

Goldhammer, R.K., Dunn, P.A. & Hardie, L.A. (1987) High frequency glacio-eustatic sealevel oscillations with Milankovitch characteristics recorded in Middle Triassic platform carbonates in northern Italy. *Am. J. Sci.* **287**, 853–892.

Goldhammer, R.K., Dunn, P.A. & Hardie, L.A. (1990) Depositional cycles, composite sea-level changes, cycle stacking patterns, and the hierarchy of stratigraphic forcing: Examples from Alpine Triassic platform carbonates. *Geol. Soc. Am. Bull.* **102**, 535–562.

Goldstein, R.H. (1988) Paleosols of Late Pennsylvanian cyclic strata, New Mexico. *Sedimentology* **35**, 777–804.

Goodwin, P.W. & Anderson, E.J. (1985) Punctuated aggradational cycles: a general hypothesis of episodic stratigraphic accumulation. *J. Geol.* **93**, 515–533.

Gornitz, V., Lebedeff, S. & Hansen, J. (1982) Global sea level trend in the past century. *Science* **215**, 1611–1614.

Grotzinger, J.P. (1986a) Cyclicity and paleoenvironmental dynamics, Rocknest platform, northwest Canada. *Geol. Soc. Am. Bull.* **97**, 1208–1231.

Grotzinger, J.P. (1986b) Upward shallowing platform cycles: a response to 2.2 billion years of low-amplitude, high-frequency (Milankovitch band) sea level oscillations. *Paleoceanography* **1**, 403–416.

Häfeli, C. (1966) Die Jura/Kreide-Grenzschichten im Bielerseegebiet (Kt. Bern). *Eclogae Geol. Helv.* **59**, 565–696.

Halley, R.B. & Harris, P.M. (1979) Fresh-water cementation of a 1000-year-old oolite. *J. Sedim. Petrol.* **49**,

969–988.

Hallock, P. & Schlager, W. (1986) Nutrient excess and the demise of coral reefs and carbonate platforms. *Palaios* **1**, 389–398.

Haq, B.U., Hardenbol, J. & Vail, P.R. (1987) Chronology of fluctuating sea levels since the Triassic. *Science* **235**, 1156–1167.

Hardie, L.A. & Ginsburg, R.N. (1977) Layering: the origin and environmental significance of lamination and thin bedding. In: *Sedimentation on the Modern Carbonate Tidal Flats of Northwest Andros Island, Bahamas* (Ed. Hardie, L.A.). Johns Hopkins University Press, Baltimore, pp. 50–123.

Hardie, L.A., Bosellini, A. & Goldhammer, R.K. (1986) Repeated subaerial exposure of subtidal carbonate platforms, Triassic, northern Italy: evidence for high frequency sea level oscillations on a 10^4 year scale. *Paleoceanography* **1**, 447–457.

Harland, B.W., Armstrong, R.L., Cox, A.V., Craig, L.E., Smith, A.G. & Smith, D.G. (1990) *A Geologic Time Scale 1989*. Cambridge University Press, Cambridge.

Hays, J.D., Imbrie, J. & Shackleton, N.J. (1976) Variations in the Earth's orbit: pacemaker of the ice ages. *Science* **194**, 1121–1132.

Heckel, P.H. (1986) Sea-level curve for Pennsylvanian eustatic marine transgressive–regressive depositional cycles along midcontinent outcrop belt, North America. *Geology* **14**, 330–334.

Horbury, A.D. & Adams, A.E. (1989) Meteoric phreatic diagenesis in cyclic late Dinantian carbonates, northwest England. *Sedim. Geol.* **65**, 319–344.

James, N.P. (1984) Shallowing-upward sequences in carbonates. In: *Facies Models* (Ed. Walker, R.G.). Geoscience Canada Reprint Ser. 1 (2nd edn), pp. 213–228.

Jan du Chêne, R., Busnardo, R., Charollais, J., Clavel, B., Deconinck, J.-F., Emmanuel, L., Gardin, S., Gorin, G., Manivit, H., Monteil, E., Raynaud, J.-F., Renard, M., Steffen, D., Steinhauser, N., Strasser, A., Strohmenger, C. & Vail, P. (1993) Sequence stratigraphic interpretation of Upper Tithonian – Berriasian reference sections in South East France: a multidisciplinary approach. *Bull. Centres Rech. Explor.-Prod. Elf Aquitaine* (in press).

Joachimski, M. (1990) Depositional environments and shallowing-upward cycles in Purbeckian carbonates: evidence from stable isotopes. In: *Abstr. 13th Int. Sedim. Congr. Nottingham*, pp. 253–254.

Koerschner, W.F. III & Read, J.F. (1989) Field and modelling studies of Cambrian carbonate cycles, Virginia Appalachians. *J. Sedim. Petrol.* **59**, 654–687.

Laskar, J. (1989) A numerical experiment on the chaotic behaviour of the Solar System. *Nature* **338**, 237–238.

Matti, J.C. & McKee, E.H. (1976) Stable eustasy, regional subsidence, and a carbonate factory: a self-generating model for onlap–offlap cycles in shallow-water carbonate sequences. *Geol. Soc. Am. Abstr.* **8**, 1000–1001.

Milankovitch, M. (1941) *Kanon der Erdbestrahlung und seine Anwendung auf das Eiszeitenproblem*. Acad. Roy. Serbe spec. ed. 133.

Mojon, P.O. & Strasser, A. (1987) Microfaciès, sédimentologie et micropaléontologie du Purbeckien de Bienne (Jura suisse occidental). *Eclogae Geol. Helv.* **80**, 37–58.

Moore, C.H. (1989) *Carbonate Diagenesis and Porosity*. Developments in Sedimentology 46. Elsevier, Amsterdam.

Osleger, D. & Read, J.F. (1991) Relation of eustasy to stacking patterns of meter-scale carbonate cycles, Late Cambrian. *U.S.A.J. Sedim. Petrol.* **61**, 1225–1252.

Posamentier, H.W., Jervey, M.T. & Vail, P.R. (1988) Eustatic controls on clastic deposition I – conceptual framework. In: Sea-level Changes: an Integrated Approach (Eds Wilgus, C.K., Hastings, B.S., Posamentier, H., Van Wagoner, J., Ross, C.A. & Kendall, C.G. St C.). Soc. Econ. Paleont. Mineral. Spec. Publ. 42, pp. 109–124.

Pratt, B.R. & James, N.P. (1986) The St George Group (Lower Ordovician) of western Newfoundland: tidal flat island model for carbonate sedimentation in shallow epeiric seas. *Sedimentology* **33**, 313–343.

Quinn, T.M. & Matthews, R.K. (1990) Post-Miocene diagenetic and eustatic history of Enewetak Atoll: model and data comparison. *Geology* **18**, 942–945.

Read, J.F. (1989) Controls on evolution of Cambrian–Ordovician passive margin, U.S. Appalachians. In: Controls on Carbonate Platform and Basin Development (Eds Crevello, P., Sarg, J.F., Read, J.F. & Wilson, J.L.). Soc. Econ. Paleont. Mineral. Spec. Publ. 44, pp. 147–165.

Read. J.F. & Goldhammer, R.K. (1988) Use of Fischer plots to define third-order sea-level curves in Ordovician peritidal cyclic carbonates, Appalachians. *Geology* **16**, 895–899.

Sarg, J.F. (1988) Carbonate sequence stratigraphy. In: Sea-level Changes: an Integrated Approach (Eds Wilgas, C.K., Hastings, B.S., Posamentier, H., Van Wagoner, J., Ross, C.A. & Kendall, C.G. St C.). Soc. Econ. Paleont. Mineral. Spec. Publ. 42, pp. 155–181.

Schlager, W. (1981) The paradox of drowned reefs and carbonate platforms. *Geol. Soc. Am. Bull.* **92**, 197–211.

Schlager, W. (1991) Depositional bias and environmental change – important factors in sequence stratigraphy. *Sedim. Geol.* **70**, 109–130.

Schwarzacher, W. & Haas, J. (1986) Comparative statistical analysis of some Hungarian and Austrian Upper Triassic peritidal carbonate sequences. *Acta Geol. Hung.* **29**, 175–196.

Selg, M. (1988) Origin of peritidal carbonate cycles: Early Cambrian, Sardinia. *Sedim. Geol.* **59**, 115–124.

Shinn, E.A. & Robbin, D.M. (1983) Mechanical and chemical compaction in fine-grained shallow-water limestones. *J. Sedim. Petrol.* **53**, 595–618.

Steinhauser, N. & Lombard, A. (1969) Définition de nouvelles unités lithostratigraphiques dans le Crétacé inférieur du Jura méridional (France). *C.R. Soc. Phys. Hist. nat. Genève* **4**, 100–113.

Strasser, A. (1988a) Shallowing-upward sequences in Purbeckian peritidal carbonates (lowermost Cretaceous, Swiss and French Jura Mountains). *Sedimentology* **35**, 369–383.

Strasser, A. (1988b) Enregistrement sédimentaire de cycles astronomiques dans le Portlandien et Purbeckien du Salève (Haute-Savoie, France). *Arch. Sci. Genève* **41**, 85–97.

STRASSER, A. (1991) Lagoonal–peritidal sequences in carbonate environments: autocyclic and allocyclic processes. In: *Cyclic and Event Stratification* (Eds Einsele, G., Ricken, W. & Seilacher, A.). Springer Verlag, Berlin, pp. 709–721.

STRASSER, A. & DAVAUD, E. (1983) Black pebbles of the Purbeckian (Swiss and French Jura): lithology, geochemistry and origin. *Eclogae Geol. Helv.* **76**, 551–580.

SUN, S.Q. (1990) Facies-related diagenesis in a cyclic shallow marine sequence: the Corallian Group (Upper Jurassic) of the Dorset coast, southern England. *J. Sedim. Petrol.* **60**, 42–52.

VAIL, P.R. (1987) Seismic stratigraphy interpretation procedure. In: *Atlas of Seismic Stratigraphy I* (Ed. Bally, A.W.). Am. Assoc. Petrol. Geol. Stud. Geol. 27, pp. 1–10.

VAIL, P.R., HARDENBOL, J. & TODD, R.G. (1984) Jurassic unconformities, chronostratigraphy, and sea-level changes from seismic stratigraphy and biostratigraphy. In: Interregional Unconformities and Hydrocarbon Accumulation (Ed. Schlee, J.S.). *Am. Assoc. Petrol. Geol. Mem.* **36**, 129–144.

VAN WAGONER, J.C., MITCHUM, R.M., CAMPION, K.M. & RAHMANIAN, V.D. (1990) *Siliciclastic Sequence Stratigraphy in Well Logs, Cores, and Outcrops*. Am. Assoc. Petrol. Geol. Methods in Exploration Ser. 7.

WAEHRY, A. (1988) *Faciès et séquences de dépôt dans la Formation de Pierre-Châtel (Berriasien moyen, Jura méridional/France)*. Dipl. Univ. Genève (unpubl.).

WEEDON, G.P. & JENKYNS, H.C. (1990) Regular and irregular climatic cycles and the Belemnite Marls (Pliensbachian, Lower Jurassic, Wessex Basin). *J. Geol. Soc. Lond.* **147**, 915–918.

WILBER, R.J., MILLIMAN, J.D. & HALLEY, R.B. (1990) Accumulation of bank-top sediment on the western slope of Great Bahama Bank: rapid progradation of a carbonate megabank. *Geology* **18**, 970–974.

WILDI, W., FUNK, H., LOUP, B., AMATO, E. & HUGGENBERGER, P. (1989) Mesozoic subsidence history of the European marginal shelves of the Alpine Tethys (Helvetic realm, Swiss Plateau and Jura). *Eclogae Geol. Helv.* **82**, 817–840.

WONG, P.K. & OLDERSHAW, A.E. (1980) Causes of cyclicity in reef interior sediments. Kaybob Reef, Alberta. *Bull. Can. Petrol. Geol.* **28**, 411–424.

WRIGHT, V.P. (1984) Peritidal carbonate facies models: a review. *Geol. J.* **19**, 309–325.

Spec. Publs Int. Ass. Sediment. (1994) **19**, 303–322

Lofer cycles of the Upper Triassic Dachstein platform in the Transdanubian Mid-Mountains, Hungary

J. HAAS

Geological Research Group of Hungarian Academy of Sciences, H-1088 Budapest Múzeum krt. 4/a, Hungary

ABSTRACT

Lofer cycles are lagoonal–peritidal cycles that are characteristic of extremely thick and broad carbonate platforms along the margin of the Upper Triassic Tethys. In the Transdanubian Mid-Mountains borehole sections expose continuous sequences of the cyclic platform carbonates several hundred metres thick. Sedimentological investigations and statistical analyses have revealed that the cycles consist of symmetric and asymmetric sequences 2–5 m thick. The ideal cycle is fairly symmetric but many cycles are condensed or incomplete and truncated. The cycles are related to relatively small-scale sea-level variations, which has resulted in considerable lateral facies migrations on a wide, marginal carbonate platform. Periodicities are estimated to range between 20 and 40 ka, and are suggestive of orbital control.

INTRODUCTION

Early in the Alpine tectonic cycle (Late Permian–Triassic) the Transdanubian Mid-Mountain (TDMM) area was part of the Tethys shelf between the Southern Alps and the Upper Austroalpine nappes, and in the Late Triassic it formed part of the broad marginal carbonate platform (Dachstein platform) (Figs 1 and 2) (Balla, 1982, 1988; Kovács, 1982; Kázmér & Kovács, 1985, 1989; Haas, 1987b; Haas *et al.*, 1990).

On this platform a carbonate sequence 2.5–3 km thick has accumulated. In the inner platform facies area metre-scale cyclicity is the most spectacular feature of the succession. Fischer's (1964, 1975) studies on cycles of similar age and palaeogeographic setting in the Northern Limestone Alps led to the description of the classic Lofer cycles. Fischer interpreted the Lofer cycles as transgressive, upward-deepening cycles reflecting high-frequency sea-level oscillations (Milankovitch cycles). Later Bosselini & Hardie (1985) considered Lofer cycles of the Dolomia Principale in the Southern Alps to be regressive and upward shallowing. The observations of Goldhammer *et al.* (1990) led to a similar conclusion for the Lofer cycles in both the Dolomia Principale and the Dachstein Limestone sequences. They suggested that Lofer facies were largely controlled by short-term variations in subsidence rate. In the Northern Limestone Alps, Schwarzacher (1948, 1954) described a pattern of cycles of lower frequency. There, groups of five or sometimes more beds represent higher order cycles, i.e. megacycles which are separated by comparatively thick 'interbed facies' (intertidal and/or supratidal beds). In surface exposures the bedding planes are laterally persistent (master bedding planes). Fischer (1964) observed that the thickness of individual cycles in a bundle decrease from the base upward (thinning-upward megacycle pattern). In the TDMM area megacycles separated by master bedding planes were described by Haas (1987a) from the Rhaetian segment of the Dachstein Limestone in the Gerecse Mountains.

The aim of the present paper is to summarize our observations on the Lofer cycles in the TDMM and to suggest a revised genetic model.

Upper Triassic carbonates of the TDMM are particularly suited to a detailed study of the Lofer cyclicity, because:

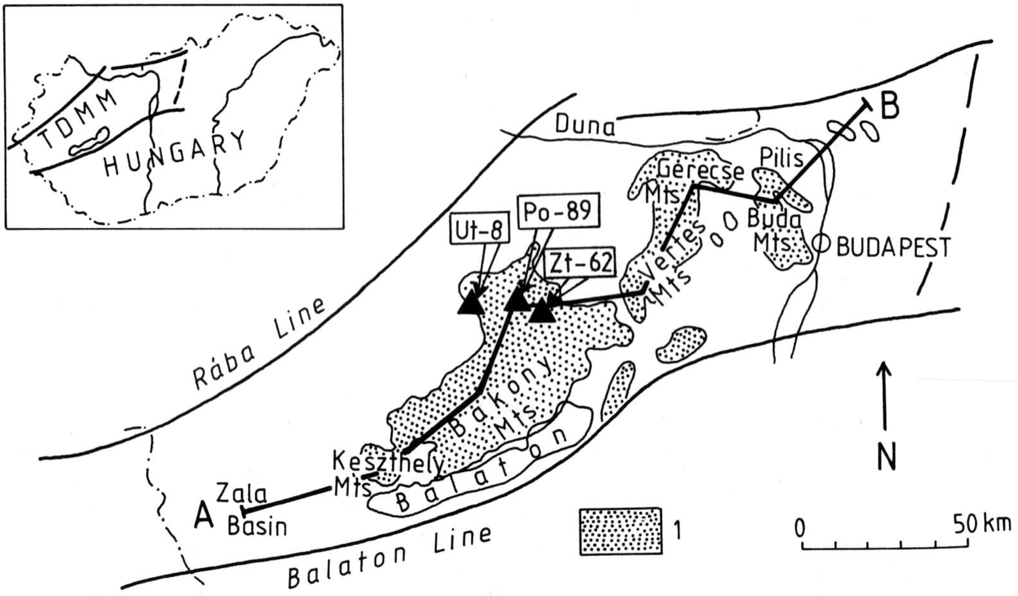

Fig. 1. Sketch map of the Transdanubian Mid-Mountains showing the study area position of key boreholes (Ut-8, Po-89, Zt-62) and general cross-section (A–B) in Fig. 3. 1, Triassic formations on the surface.

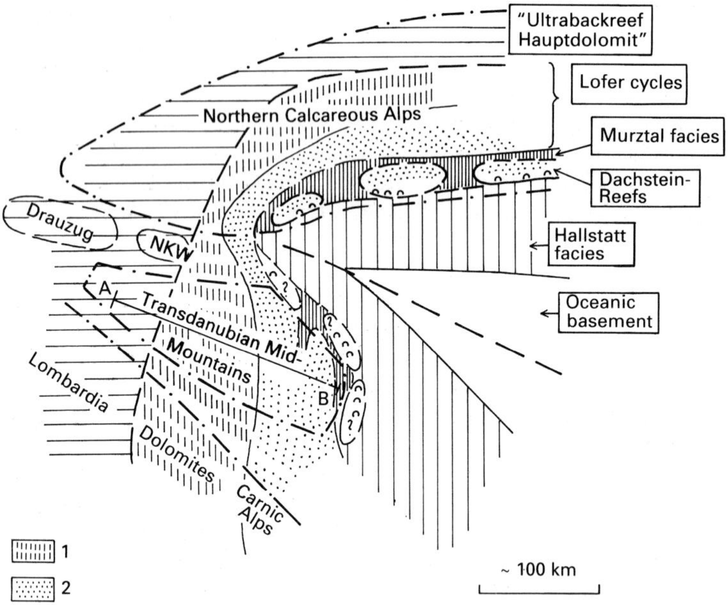

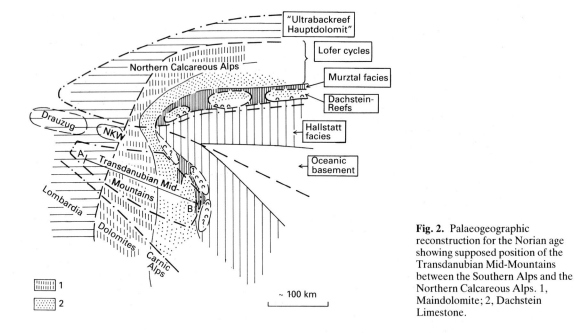

Fig. 2. Palaeogeographic reconstruction for the Norian age showing supposed position of the Transdanubian Mid-Mountains between the Southern Alps and the Northern Calcareous Alps. 1, Maindolomite; 2, Dachstein Limestone.

1 In this area a complete profile of the original carbonate platform has been preserved in a terrane less tectonized than in the Alpine nappes.

2 Lofer cycles developed continuously over a period of 16 Ma from the late Carnian to the Rhaetian.

3 Undolomitized, partially dolomitized and fully dolomitized Lofer cyclic sequences are equally known.

4 Lofer cyclic sections are exposed in huge quarries and penetrated by boreholes. Cut cores made it possible to observe thick continuous sequences over several hundred metres in detail.

METHODS

Numerous continuous drill cores provided the data base for the cycle analysis. To improve accuracy of field observations cores were cut vertically. Boreholes were supplemented by large exposures in quarries, allowing the study of lateral facies relations. For field descriptions a graphic log system was used providing an accuracy of 10 cm in data recording.

Analysis of the cycles was carried out by two different methods: (i) the frequency of the cycle-types in the sequences of the carbonate platform lithological units was determined and (ii) by applying Markov analysis the regularities in the superposition of the different subfacies (cycle members) were investigated. As for the cycle-type analysis, first the logically possible stacking patterns were determined followed by their frequency in the studied successions.

Markov analysis was applied in order to understand the characteristics of the various superimposed facies (Schwarzacher, 1975; Vistelius, 1980). The Markov chain is a series of discrete states in which a given state influences (or determines) the subsequent state. In a chain, the number of transitions from a given state to the same or another state is called transition frequency, while the corresponding probability is expressed in a transition probability matrix. Statistical tests and modelling primarily deal with these frequencies and probabilities (Anderson & Goodmann, 1957; Billingsley, 1961). Lofer cycle members were coded prior to the analysis (A, B, C, B′ and d for the disconformities, Fig. 5). The codes correspond to layers with fixed properties (and different genesis). The method applied represents a special type of Markov chain called embedded Markov-chains, in which the transition from a given state to the same state is impossible by definition (Davis, 1986). We obtained such sequences from all boreholes studied, and analysed their characteristics using the statistical tests of Goodman (1968) and Davis (1986). First, the Markov property was tested in each of the sequences. After that, the most frequent cycle types were determined (Tables 2–4). For each cycle type a simple hypothetical (geological and corresponding Markov chain) model was formed and the expected and observed transition frequences (i.e. cycle types) were compared statistically.

GEOLOGICAL SETTING

The time and space interrelations of the Upper Triassic megasedimentological-(lithostratigraphical) units of the TDMM are schematically demonstrated in Fig. 3.

The carbonate platform evolution started in the late Carnian when the Maindolomite Formations began to form. Its total thickness is 1–1.5 km.

During the Lower Norian the Maindolomite in the northeastern part of the TDMM gradually evolved into the Dachstein Limestone through a thick (400–500 m) partially dolomitized transitional unit. In the Bakony Mountains (in the southwestern part) the deposition of sediments in the Maindolomite facies continued. Consequently, during that time and in different regions of the TDMM, the Dachstein Limestone, the transitional unit and the Maindolomite were formed as coeval facies (Fig. 3).

During the Upper Norian and Lower Rhaetian, sediment accumulation in the Dachstein Limestone facies area continued in the northeastern part of the TDMM while contemporaneously the organic-rich Rezi Dolomite of restricted basin facies and subsequently the anoxic, argillaceous Kössen Formation were formed in the southeast (Fig. 3).

In the southeastern part of the TDMM (Southern Bakony Mountains), in the Upper Rhaetian, above the Kössen Formation, the Dachstein Limestone appears again. Simultaneously, in the Northern Bakony Mountains, and in the Vértes and Gerecse Mountains the formation of the Dachstein Limestone was continuous.

In the Bakony Mountains the formation of the Dachstein Limestone continued up to the latest Triassic (total thickness 600–800 m), and platform carbonate sedimentation prevailed also in the early Liassic (Kardosrét Limestone). In the northeastern part of the TDMM (Vértes and Gerecse Mountains) the uppermost Rhaetian is missing and the Dachstein Limestone is overlain by non-platform Liassic formations.

KEY SECTIONS

In the Bakony Mountains three key boreholes (Fig. 1) penetrated an almost complete sequence of the

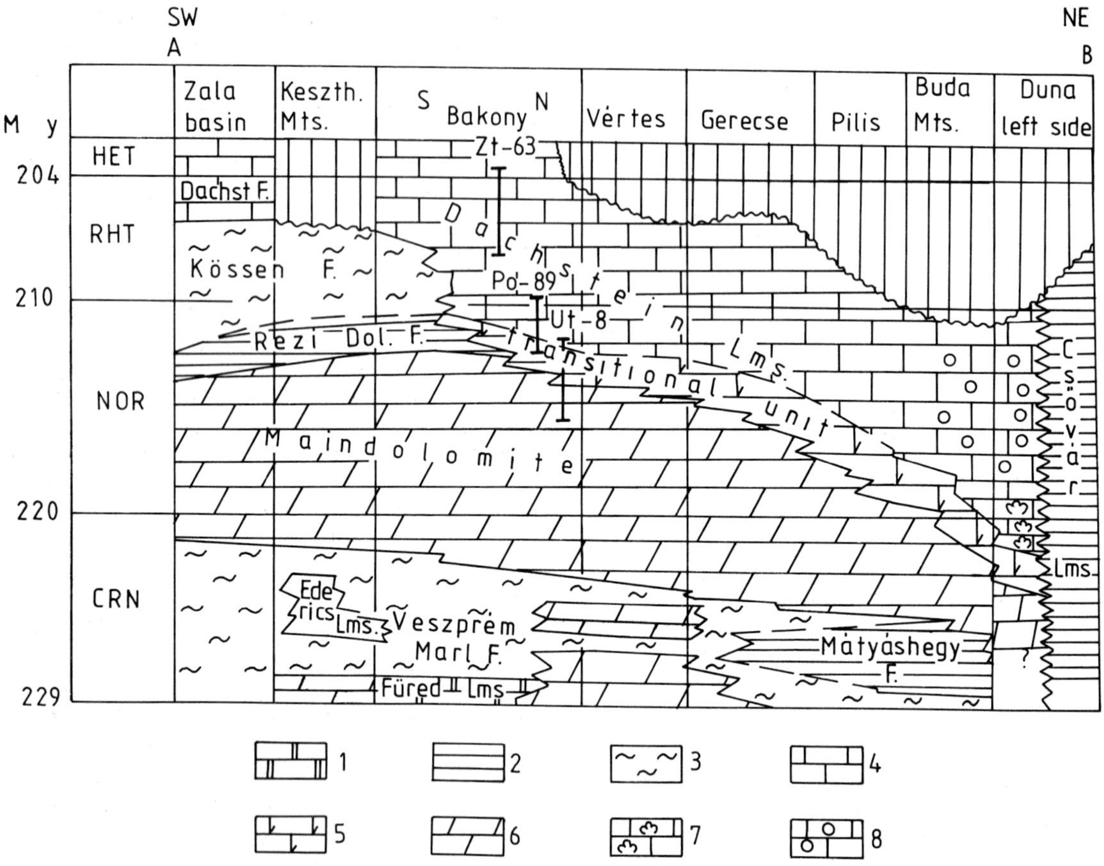

Fig. 3. Upper Triassic lithostratigraphical units of the Transdanubian Mid-Mountains. 1, Pelagic limestone, restricted basin facies; 2, carbonates; 3, marls; carbonate platform facies; 4, limestones; 5, dolomitic limestones, calcareous dolomites; 6, dolomites; 7, reef limestones; 8, oncolite facies.

Norian–Rhaetian Dachstein Limestone Formation. Borehole Ugod Ut-8 (Fig. 4) exposed the topmost part of the Maindolomite (120 m) and the transitional unit between the Maindolomite and the Dachstein Limestone (180 m). Borehole Porva Po-89 (Fig. 4) shows a 400-m-long segment of the lower part of the Dachstein Limestone from the transitional unit upwards. Borehole Zirc Zt-62 (Fig. 4) drilled the topmost part of the Dachstein Formation (220 m) under the Lower Jurassic beds. These borehole sections provided the best data set for analysis of the Lofer cycles in the TDMM. Observations in the Gerecse Mountains were also considered.

LITHOLOGICAL FEATURES AND STACKING PATTERNS OF THE LOFER CYCLES

Our study confirmed Fischer's (1964) observation and interpretations concerning the basic facies pattern of the Lofer cycles, i.e. from base to top:

1 Member A: red to green argillaceous intraclastic mudstone (supratidal).

2 Member B: laminated carbonates containing fenestral pores (intertidal).

3 Member C: massive carbonates with marine biota (subtidal).

Within each member some subfacies could be distinguished (Fig. 5). Their palaeoenvironmental interpretation is presented in Fig. 6. Naturally in any one individual cycle only some of these subfacies are

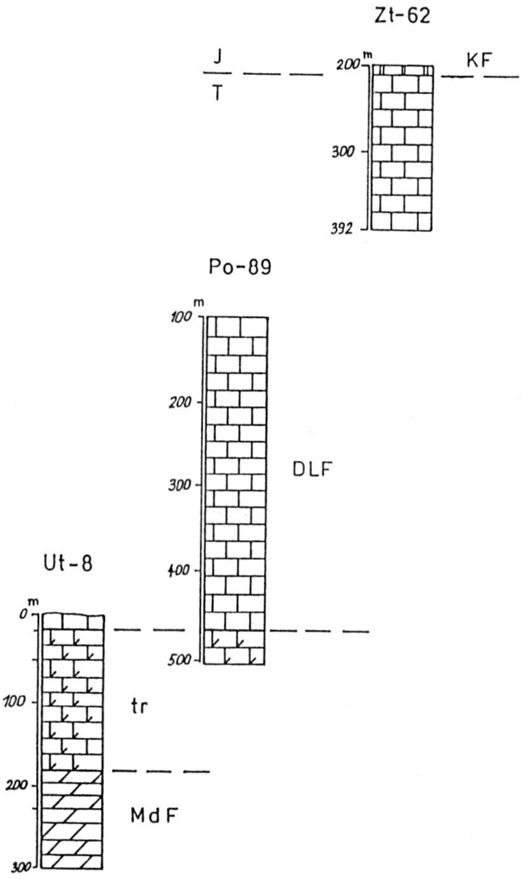

Fig. 4. Correlation of the key sections investigated in the Bakony Mountains. MdF, Maindolomite Formation; tr, transitional unit; DLF, Dachstein Limestone Formation; KF, Kardosrét Limestone Formation (Early Jurassic).

represented. Facies showing intermediate features between the three basic facies (A, B, C) are also common, e.g. black pebbles and reworked palaeosol occur in the Megalodont-bearing subtidal facies (A + C), and large algal mat slabs appear in the C bed (B + C), etc.

Figures 7–10 show the thickness and facies succession of the Lofer cycles in the studied key sections. Systematic study and statistical analysis of these sequences revealed that in many cases cycles contain an additional Member B above the C facies but below the disconformity at the top of the sequence (Member B′; Haas, 1982). In a few cases *in situ* preserved palaeosol layers could also be observed on top of Member B′. Since it belongs to the regressive phase too it may be symbolized by A′.

It must be emphasized here that Member A is generally a mixture of marine fossiliferous carbonate mud, reworked palaeosol and semi-consolidated tidal sediments. That is why it is usually considered as the basal member of the transgressive tract.

Observed stacking patterns of Lofer cycles in the borehole Po-89 are shown in Fig. 11. Some of these cycles are fairly symmetric (close to the ideal A–B–C–B′–(A′) pattern), while others are either: (i) truncated, i.e. the top part is missing (e.g. A–B–C; A–B); (ii) incomplete (anomalous), i.e. basal or internal members are missing (e.g. B–C–B′; A–C–B′); (iii) condensed, i.e. consisting of extremely thin members; or (iv) showing combinations of these features. In some cases B and C members alternate without a disconformity between the cycles and consequently one cannot distinguish between B and B′ (B/B′).

RESULTS OF STATISTICAL ANALYSIS

Frequency analysis

The logically possible combinations in the succession of the cycles (cycle types) are shown in Fig. 12. Some of these combinations were indistinguishable in practice (these are in brackets). Facies of intermediate character were disregarded in this case and the A and A′ facies were not separated.

The frequency of the cycle types in the key sections studied are shown in Fig. 13. The following features were observed.

1 The upper part of the Maindolomite (lower segment of borehole Ugod Ut-8) is characterized by the subordinance of member A, and a dominance of B–C couplets.

2 Within the transitional unit between the Maindolomite and the Dachstein Limestone (upper part of borehole Ut-8) the predominantly dolomitic middle part is also characterized by B–C successions. In the section consisting of alternations of limestone and dolomite, as well as carbonates of transitional composition (calcareous dolomite, dolomitic limestone), member A and the regressive B′ member also appear. Consequently, besides the prevailing B–C couplets, cycles A–C and B–C–B′ are subdominant.

3 In the lower part of the Dachstein Limestone (borehole Porva Po-89) member A is usually present, whereas member B is commonly missing. The A–C cycle type is predominant, and A–B–C and

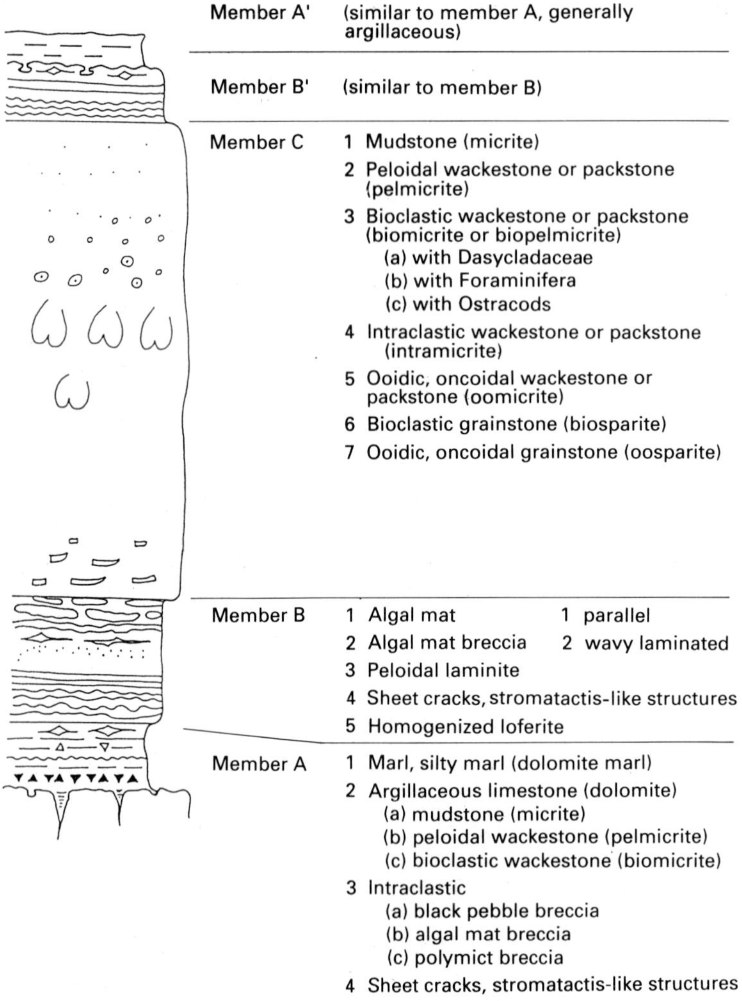

Fig. 5. The ideal Lofer cycle and the most common subfacies of the members.

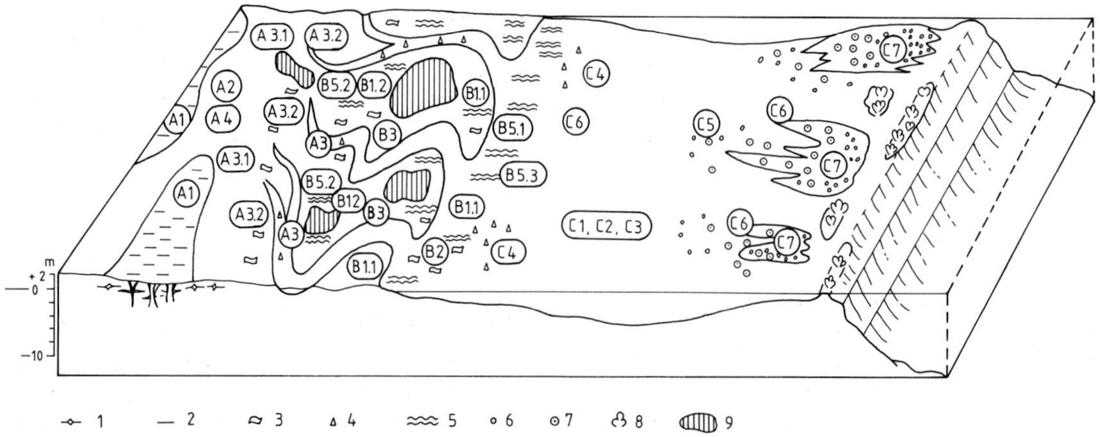

Fig. 6. Schematic palaeogeographic model of the Dachstein-type platform with indication of the component sedimentary environments. For codes of facies and subfacies see Table 1. 1, shrinkage cracks, sheet cracks; 2, palaeosol; 3, flat pebbles (of algal mat origin); 4, intraclasts; 5, algal mat facies; 6, ooidic facies; 7, oncoidal facies; 8, reefs; 9, tidal flat pools.

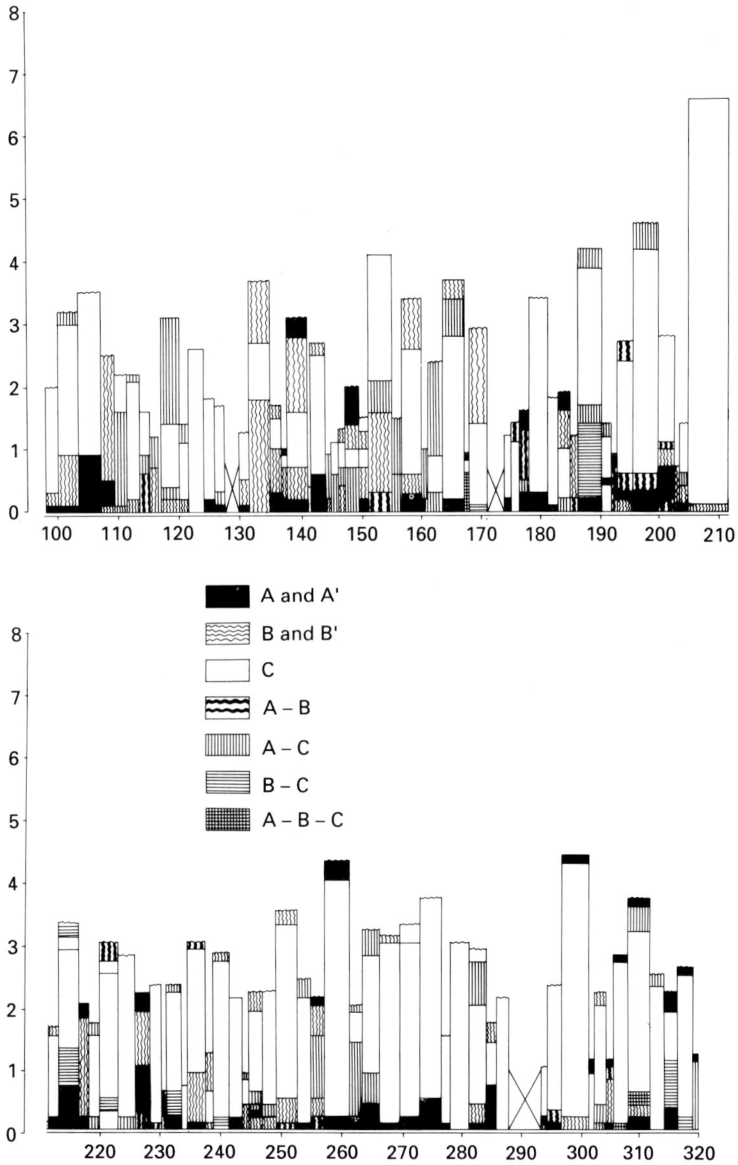

Fig. 7. Thicknesses and facies successions of the individual Lofer cycles in the upper segment of borehole Porva Po-89. Diagram shows the thicknesses of the individual cycles and within them the thicknesses of the members (i.e. facies types) horizontally. Symbols mark the distinguished facies types (e.g. A, B, A + B transitional facies, etc.). Superposition of the cycles in the log is displayed vertically.

A–B–C–B′ successions are subdominant, but almost every combination may occur (A′ is also comparatively common).

4 In the upper part of the Dachstein Limestone (borehole Zirc Zt-62, and in many sections in the Gerecse Mountains) cycles of A–C composition prevail, whereas combinations B–C and A–B–C are less common.

Markov analysis

Results of the Markov analysis are shown in Tables 1–3. Using this analysis, the characteristic successions were summarized in a diagram, where the thickness of the arrows between the codes reflects the frequency of the superposition of the members. In all sequences studied the Markov property (i.e.

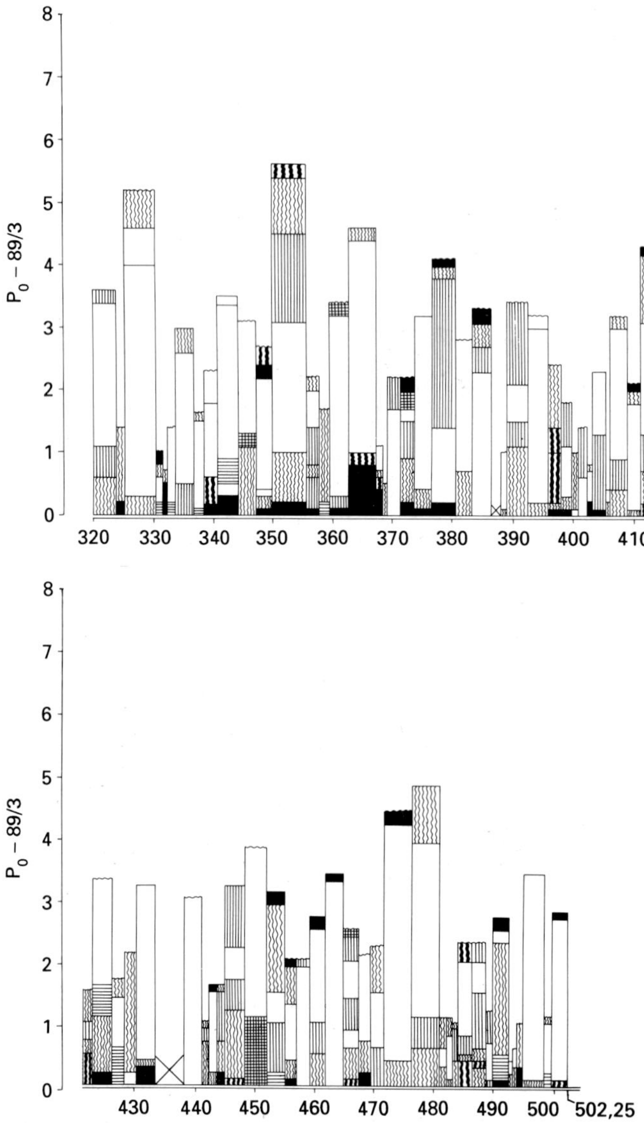

Fig. 8. Thicknesses and facies successions of the individual Lofer cycles in the lower segment of the borehole Porva Po-89. Legend is the same as in Fig. 7.

regular superposition of beds of determined features) was tested statistically. The conclusions can be summarized as follows.

1 In borehole Ugod Ut-8 (Table 1) through the topmost part of the Maindolomite and the transitional unit, the d–B–C–B′/B cycle succession is definitely predominant, although in many cases disconformities were likely not observed due to the drilling technology applied.

2 In borehole Porva Po-89 (Table 2), representing the lower part of the Dachstein Limestone, the build-up of the cycles is particularly varied. The d–A–C–d and d–A–B–C–d cycles are predominant, but d–A–B–C–B′–d and d–A–B–d successions are also common.

3 In borehole Zirc Zt-62 (Table 3), representing the upper part of the Dachstein Limestone, the d–A–C–d stacking pattern is predominant, but many other combinations occur.

The statistical study reveals the following characteristics:

1 In the Maindolomite, B–C couplets prevail.

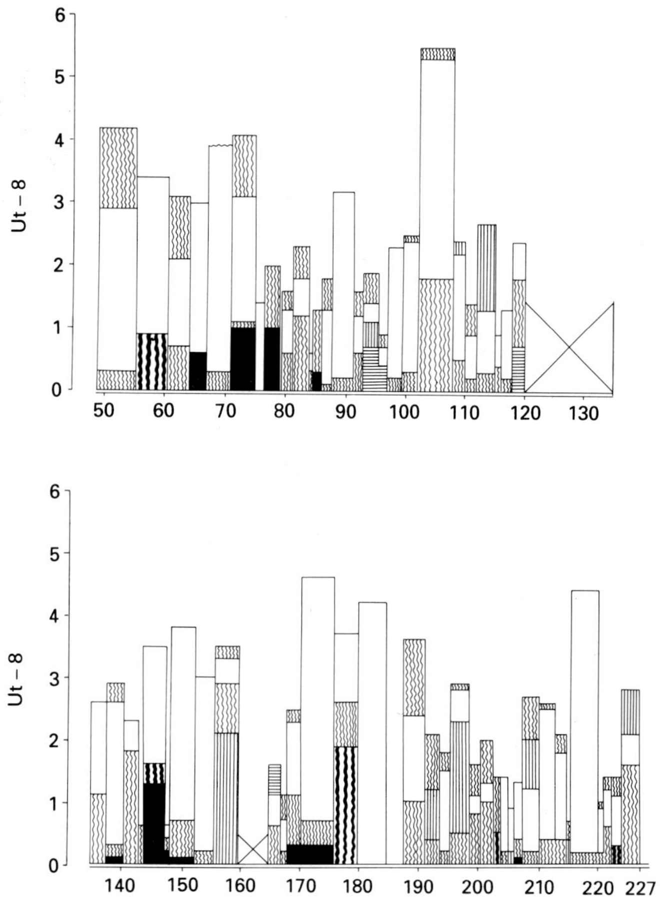

Fig. 9. Thicknesses and facies successions of the individual Lofer cycles in the borehole Ugod Ut-8. Legend is the same as in Fig. 7.

2 In the Dachstein Limestone, numerous cycle types occur from the deepening upward A–C or A–B–C to the symmetric A–B–C–B′ cycles.

3 In the transitional unit the stacking patterns of the cycles are also intermediate.

MEGACYCLES

Based on power spectral analysis, wavelengths of 30–25 m, 12 m and 6–7 m were demonstrated by Schwarzacher & Haas (1986) from the Norian Dachstein Limestone (boreholes Ut-8 and Po-89). Visual inspection of thicknesses (Figs 7–9) of the individual cycles does not show a definite thinning-upward megacyclic pattern, and no characteristic regularity in the internal upbuilding of the cycles seems recognizable.

Thus although there are traces of higher rank cyclicity even in the Norian part of the Dachstein Limestone, random distribution of cycle types and cycle thicknesses appears to dominate the successions. A preliminary Fischer plot of borehole Po-89 (J.F. Read and A. Balog, personal communication) shows bundling of five cycles in a few cases only, but it does indicate a strong third-order relative sea-level signal.

DISCUSSION

In his classic paper on the Lofer cycles Fischer (1964) suggested that individual metre-scale cycles represent a transgressive record related to short-term eustatic sea-level oscillations in the Milankovitch band. Assuming 15 Ma for the Norian–Rhaetian he

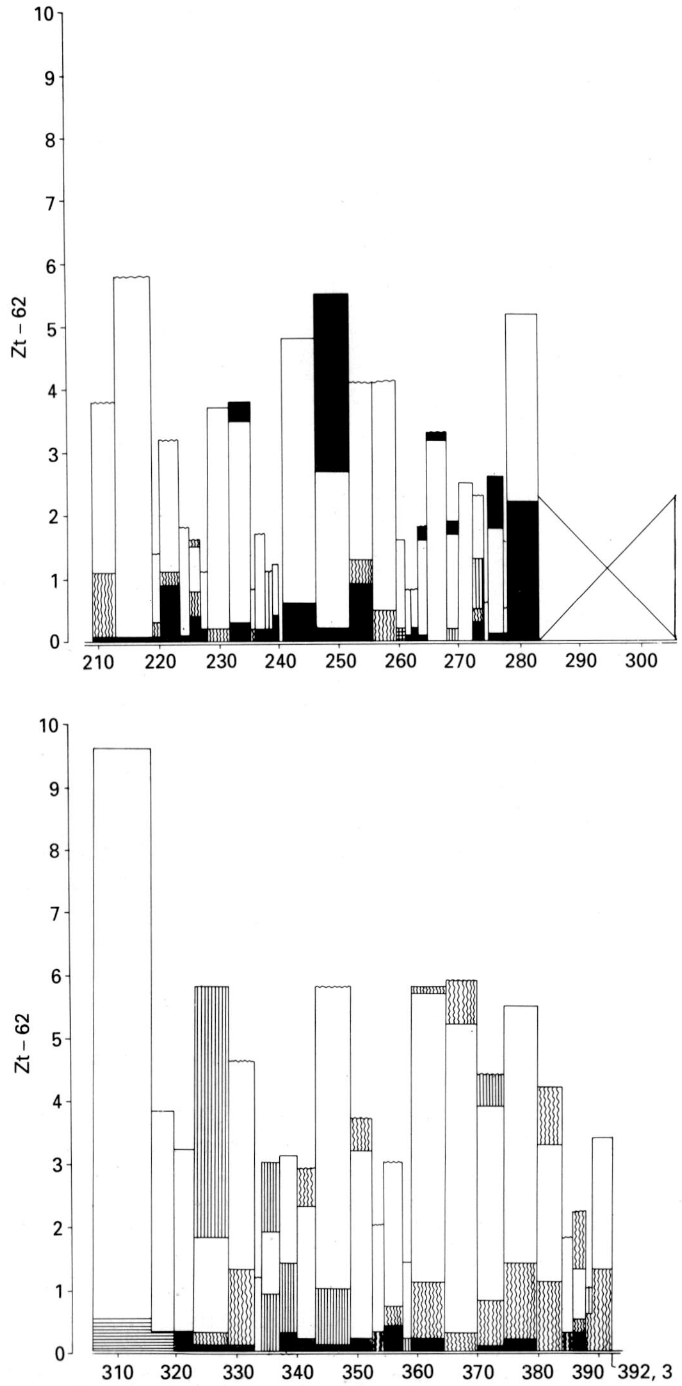

Fig. 10. Thicknesses and facies successions of the individual Lofer cycles in borehole Zirc Zt-62. Legend is the same as in Fig. 7.

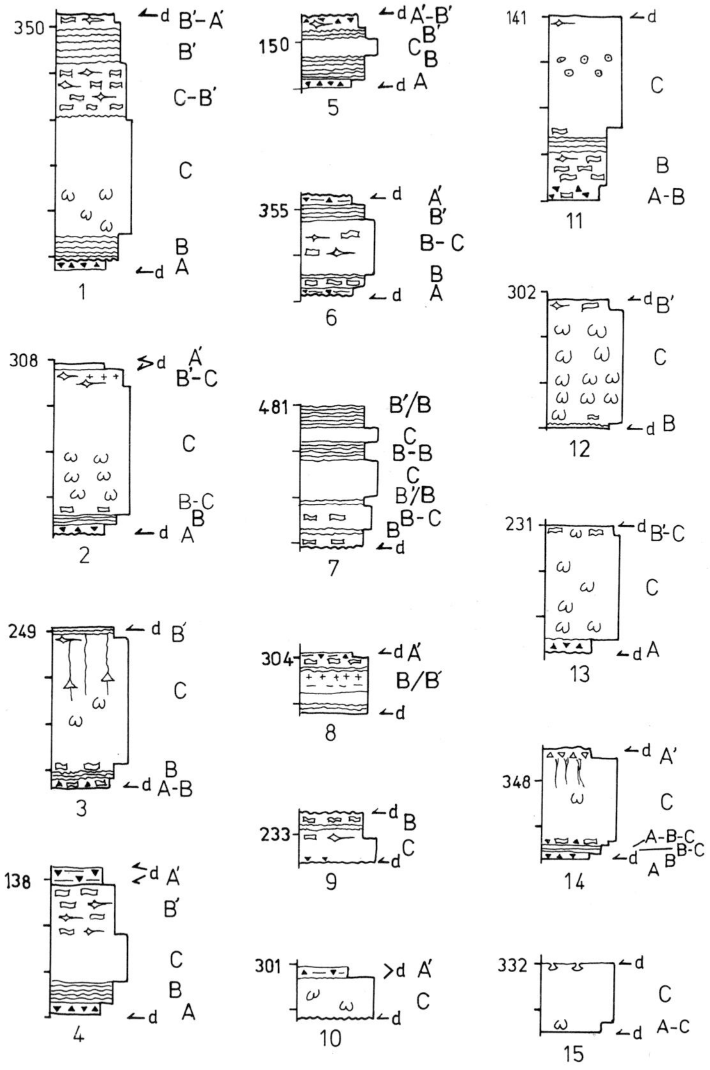

Fig. 11. Typical stacking patterns from the sequence of borehole Po-89.

calculated a period of 50 ka for the accumulation of one individual cycle. In 1975 he described the ideal Lofer cycle as follows: 'A theoretical complete sequence of this kind of transgressive–regressive cycle might be expected to read d (disconformity)–A (soil)–B (intertidal, transgressive)–C (subtidal)–B (intertidal regressive)–d. The Lofer sequence generally lacks the regressive phase of B, probably because of the succeeding erosion' (Fischer, 1975). He proposed three genetic models: a steady-state tidal marsh model (autocyclic), a eustatic oscillation model and a diastrophic oscillation model. The first model was rejected and the eustatic model was preferred.

Studies on Lofer cycles of the TDMM in Hungary revealed that in many cycles a transgressive–regressive record has been preserved and that, as a consequence, symmetric cycles are fairly common (Haas, 1982; Haas & Dobosi, 1982; Schwarzacher & Haas, 1986).

Based on data sets from the Northern Limestone Alps and from the TDMM, Schwarzacher & Haas (1986) calculated a duration of 23 ka for the average basic cycle of about 3 m thickness, showing good agreement with the shortest 21-ka Milankovitch cycle. This would imply that part of the cycle has not been deposited or preserved.

In the Southern Alps various types of metre-scale

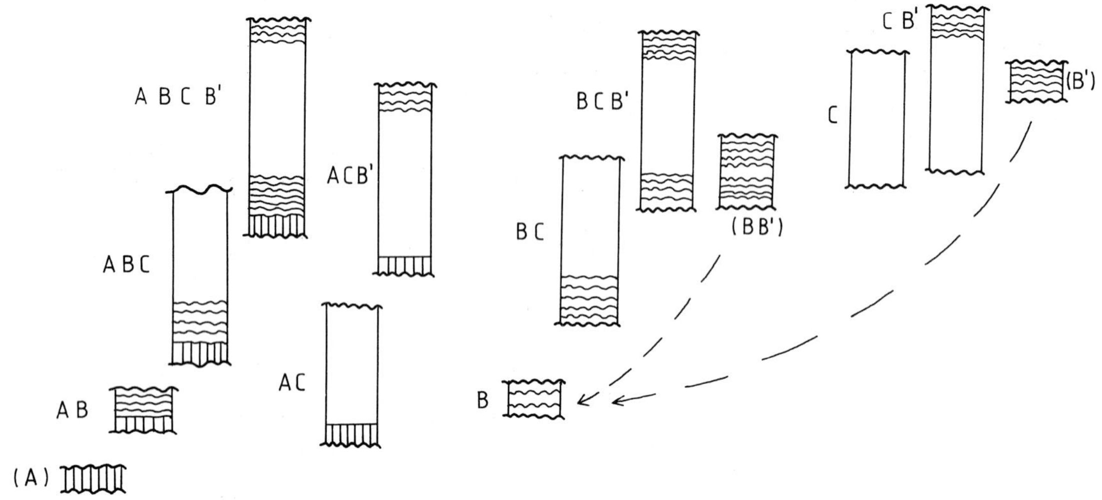

Fig. 12. The logically possible stacking patterns.

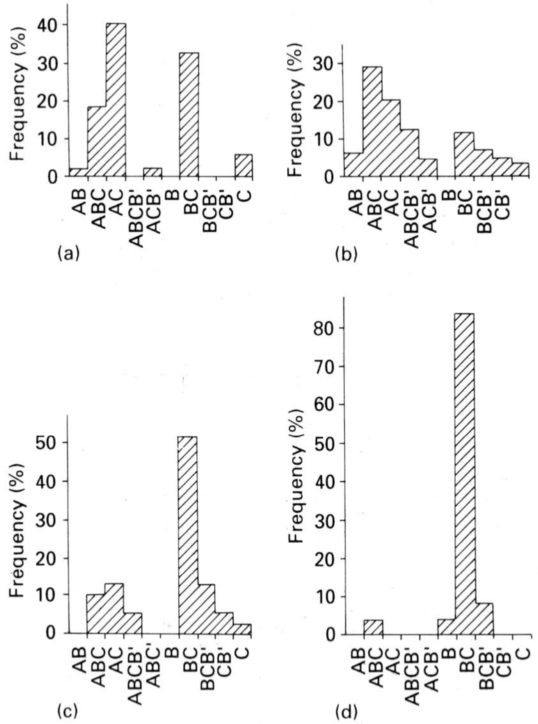

Fig. 13. Frequency of the cycle types in the various lithogenetic units. (a) The upper part of the Dachstein Limestone; (b) the lower part of the Dachstein Limestone; (c) the transitional unit between the Maindolomite and the Dachstein Limestone; (d) the uppermost part of the Maindolomite.

shallowing-upward peritidal cycles were observed in the Norian Dolomia Principale (Bosellini, 1965, 1967; Bosellini & Hardie, 1985). Recently Goldhammer *et al.* (1990) re-examined Lofer cycles in the Northern Limestone Alps and revealed that most were in fact regressive shallowing-upward cycles. Taking into account the apparently random succession of the cycle types and cycle thicknesses they suggested that short-term changes in the subsidence rate significantly modified the eustatic pattern.

Based on observations in the TDMM the main genetic factors of the Lofer cyclic carbonate accumulation are summarized below.

Palaeogeographic conditions, platform evolution

Lofer cycles of the TDMM were formed in the topographically balanced inner part of a huge Dachstein-type carbonate platform. Factors that controlled the platform evolution are (Fig. 14):

1 Subsidence: fast, slightly accelerating from the late Carnian to the Rhaetian (roughly from 0.1 to 0.2 m/ka).

2 Eustatic sea-level changes: long-term sea-level fall from the early Norian to the Hettangian (Haq *et al.*, 1987), short-term oscillation.

3 Carbonate accumulation: could keep pace with subsidence but was discontinuous, and was controlled by fluctuations in accommodation space due to small-scale sea-level changes.

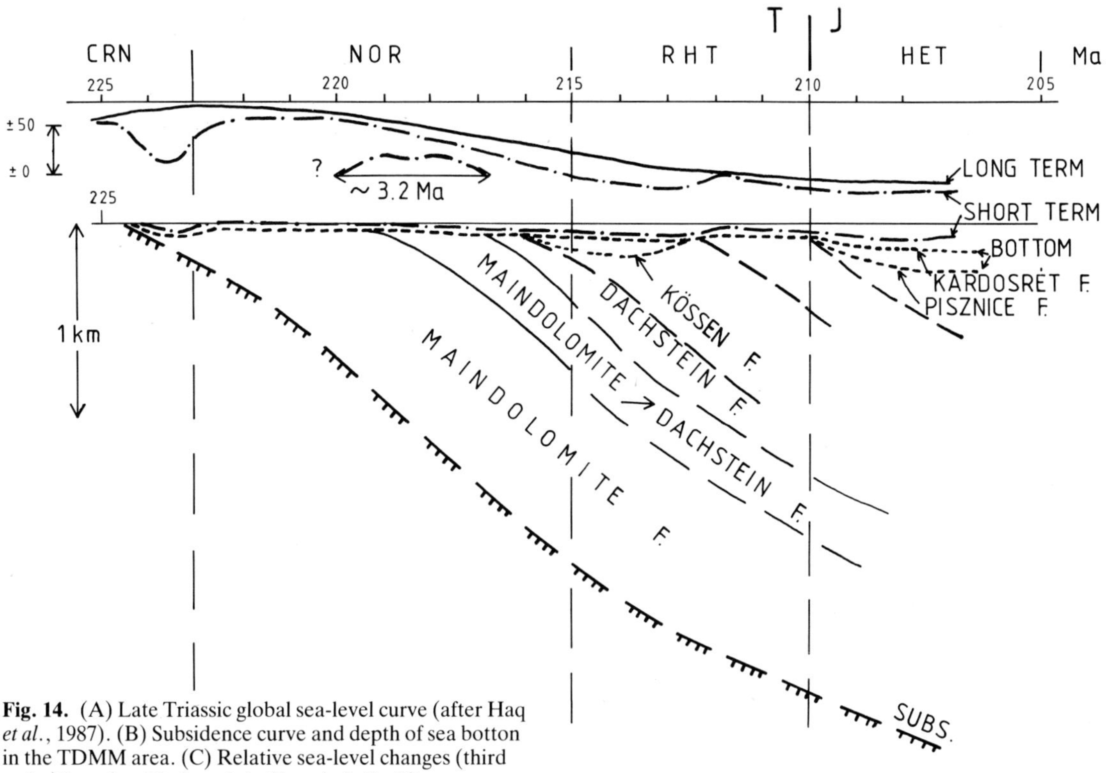

Fig. 14. (A) Late Triassic global sea-level curve (after Haq *et al.*, 1987). (B) Subsidence curve and depth of sea botton in the TDMM area. (C) Relative sea-level changes (third order) based on Fischer plot of borehole Po-89.

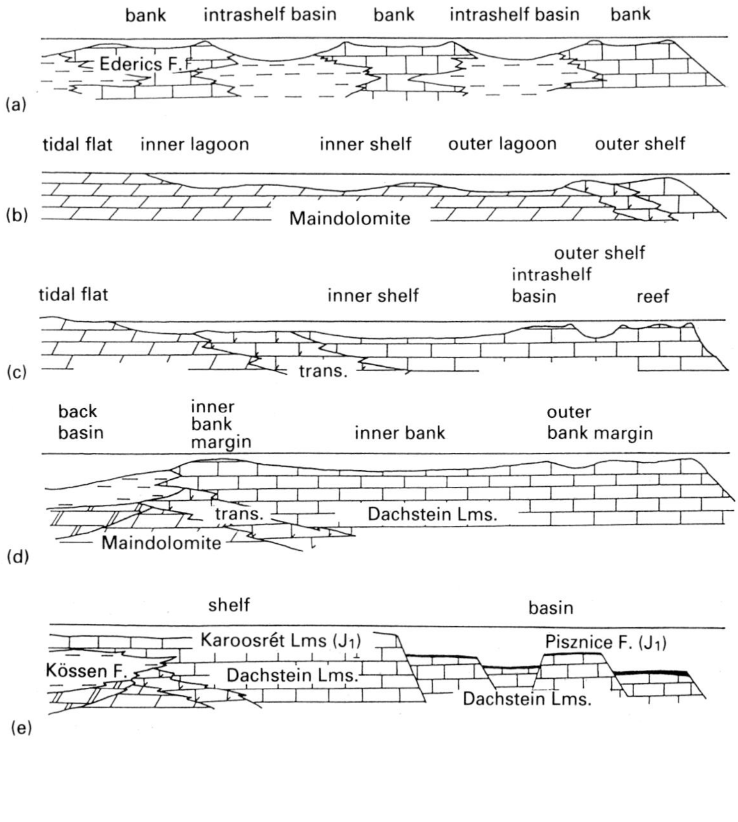

Fig. 15. Carbonate platform evolution phases: (a) basin upfilling–equilibration (late Carnian); (b) homoclinal ramp–shallow lagoon (latest Carnian–early Norian); (c) rimmed shelf (middle Norian); (d) isolated platform (late Norian–Rhaetian); (e) platform disintegration–partial drowning (Early Jurassic).

Table 1. Borehole Ugod Ut-8

1 Test of Markov property (iter. numb. = 7)

		A	B	B′	C	d
Fixed probability		0.05	0.29	0.24	0.31	0.12
Expected frequency matrix	A	0.0	3.2	2.3	3.5	1.0
	B	3.2	0.0	19.4	29.4	8.1
	B′	2.3	19.4	0.0	21.6	5.9
	C	3.5	29.4	21.6	0.0	9.0
	d	1.0	8.1	5.9	9.0	0.0
Observed frequency matrix	A	0.00	0.80	0.00	0.20	0.00
	B	0.00	0.00	0.00	1.00	0.00
	B′	0.00	0.80	0.00	0.00	0.20
	C	0.00	0.00	0.77	0.00	0.23
	d	0.42	0.54	0.55	0.04	0.00

$\chi^2 = 293$, degrees of freedom = 11

2 Cycle model

d→B→C→B′→B
↘ ↗↑ ↘
A→B d

dominant:
d–B–C–B′–B

3 Matrix model

	A	B	B′	C	d
A	0.0	1.0	0.0	0.0	0.0
B	0.0	0.0	0.0	1.0	0.0
B′	0.0	1.0	0.0	0.0	0.0
C	0.0	0.0	0.75	0.0	0.25
d	0.5	0.5	0.0	0.0	0.0

4 Test of observed frequency matrix

	A	B	B′	C	d
A	0.0	8.0	0.0	2.0	0.0
B	0.0	0.0	0.0	60.0	0.0
B′	0.0	40.0	1.0	0.0	9.0
C	0.0	0.0	49.0	0.0	15.0
d	10.0	13.0	0.0	1.0	0.0

$\chi^2 = 2.9$, degrees of freedom = 2

4 Terrigenous input: was generally subordinate but increased probably due to a change in the climate in the late Norian (Kössen basin).

The main phases of carbonate platform evolution are shown in Fig. 15 (for a detailed description of the phases see Haas, 1988). Final drowning of the Dachstein platform (from the Rhaetian to the late Hettangian) was a consequence of block faulting and accelerated sinking in the initial phase of the Jurassic rifting of the Tethys.

Genetic model

The controlling factors of the Lofer cyclic sedimentation on the platform interior were: (i) subsidence rate; (ii) sea-level oscillations; and (iii) carbonate accumulation. Cycles similar to the ideal pattern (Fig. 16) were generated when sea-level change was fairly regular (sine-wave-shaped sea-level curve) and when subsidence rate was moderate (about 0.15 m/ka). Deviations from the ideal pattern may have been caused by irregular sea-level variations (deformed, asymmetric sea-level curve; Fig. 17A, B), or a subsidence rate lower or higher than the mean value (Fig. 17C, D).

Change of subsidence rate should have been slow and gradual (long-term acceleration from the late Carnian to the Rhaetian). There is no evidence for significant tectonic displacements during this period in the TDMM area. Therefore I assume that the observed features of the individual basic Lofer cycles reflect mainly the character of the sea-level curve. If this is correct a sea-level curve can be reconstructed assuming a constant subsidence rate during a comparatively short period. Figure 17 shows an example of a reconstruction based on these principles. If durations of the basic cycles are more or less equal and the subsidence rate is constant, thickness of the cycles indicates the amplitude of the sea-level changes. Succession, thickness and other features of

Table 2. Borehole Porva Po-89

1 Test of Markov property (iter. numb. = 7)

		A	B	B′	C	d
Fixed probability		0.18	0.16	0.12	0.27	0.26
Expected frequency matrix	A	0.0	8.8	6.1	17.3	16.9
	B	8.8	0.0	5.3	15.1	14.8
	B′	6.1	5.3	0.0	10.4	10.2
	C	17.3	15.1	10.4	0.0	29.0
	d	16.9	14.8	10.2	29.0	0.0
Observed frequency matrix	A	0.00	0.47	0.00	0.53	0.00
	B	0.00	0.00	0.02	0.84	0.14
	B′	0.06	0.16	0.00	0.00	0.78
	C	0.00	0.01	0.43	0.00	0.56
	d	0.66	0.21	0.00	0.13	0.00

$\chi^2 = 286.7$, degrees of freedom = 11

2 Cycle model

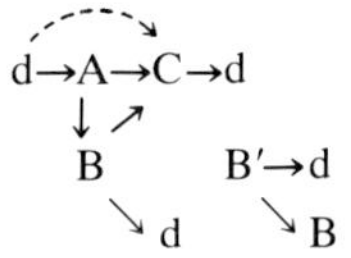

dominant:
d–A–C–d
d–A–B–C–d

3 Matrix model

		A	B	B′	C	d
	A	0.0	0.5	0.0	0.5	0.0
	B	0.0	0.0	0.0	1.0	0.0
	B′	0.0	0.0	0.0	0.0	1.0
	C	0.0	0.0	0.5	0.0	0.5
	d	1.0	0.0	0.0	0.0	0.0

4 Test of observed frequency matrix

		A	B	B′	C	d
	A	0.0	23.0	0.0	26.0	0.0
	B	0.0	0.0	1.0	37.0	6.0
	B′	2.0	5.0	0.0	0.0	25.0
	C	0.0	1.0	31.0	0.0	40.0
	d	47.0	15.0	0.0	9.0	0.0

$\chi^2 = 12$, degrees of freedom = 2

Table 3. Borehole Zirc Zt-62

1 Test of Markov property (iter. numb. = 9)

		A	B	B′	C	d
Fixed probability		0.19	0.15	0.11	0.32	0.23
Expected frequency matrix	A	0.0	5.9	4.1	17.5	10.6
	B	5.9	0.0	2.9	12.6	7.6
	B′	4.1	2.9	0.0	8.7	5.3
	C	17.5	12.6	8.7	0.0	22.7
	d	10.6	7.6	5.3	22.7	0.0
Observed frequency matrix	A	0.00	0.34	0.00	0.58	0.08
	B	0.00	0.00	0.03	0.90	0.07
	B′	0.19	0.52	0.00	0.00	0.29
	C	0.10	0.00	0.34	0.00	0.56
	d	0.61	0.11	0.00	0.28	0.00

$\chi^2 = 159$, degrees of freedom = 11

2 Cycle model

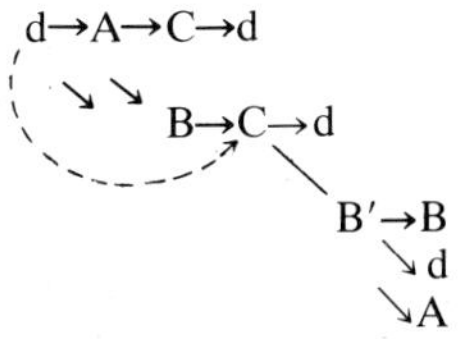

dominant:
d–A–C–d

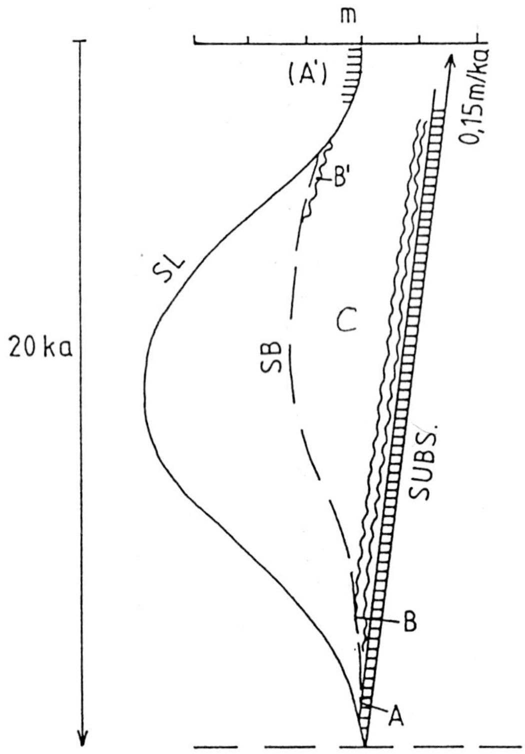

Fig. 16. Genetic model of the ideal Lofer cycle. SUBS, subsidence rate; SB, sea bottom; SL, sea level.

the cycle members provide information on the shape of the curves.

Differences in the stacking patterns of various lithogenetic units (Maindolomite, transitional unit, lower and upper part of the Dachstein Limestone; Fig. 12) should be the consequence mainly of long-term changes in subsidence rate. This factor is particularly important in the case of the Maindolomite. A reduced subsidence rate (about 0.1 m/ka) resulted in the predominance of truncated cycles and pervasive dolomitization during the comparatively long subaerial exposure. On the other hand, in the fast subsiding outer platform zone, supratidal facies were subordinate, and massive oncoidal limestones were formed throughout the Norian.

Periodicity

Based on data sets from the Northern Limestone Alps and TDMM, Schwarzacher and Haas (1986) calculated 23 ka for the duration of a basic cycle. Average cycle thickness in the TDMM is 3.1 m (standard deviation 2.1 m). Since the total sequence represents 16 ± 6 Ma (based on Visscher, 1985) and 2000 ± 500 m a mean duration of 24.7 ka/cycle is calculated. Taking into consideration the uncertainty of the total thickness and common erosion between the cycles, as well as amalgamation or extreme condensation of the cycles, less than 24.7 ka seems an appropriate estimate, representing the frequency band of Milankovitch cycles. The 21-ka precession period may well be responsible for the basic cyclicity (see Berger, 1977; Berger & Loutre, 1987).

CONCLUSIONS

Significant features and conclusions about the origin of the Lofer cycles in the TDMM are summarized below.

1 Lofer cycles were formed in the wide inner belt of a topographically balanced large carbonate platform representing a discrete stage of the Tethys passive margin evolution from the late Carnian to the end of the Rhaetian.

2 The ideal Lofer cycle is essentially symmetric and bounded by disconformities. It is composed of supratidal to intertidal and subtidal members: A–B–C–B′–(A′). In fact there are a lot of variations in the stacking pattern which can be deduced from the ideal cycles (condensed, incomplete, truncated cycle types). Statistical analysis revealed that the individual lithogenetic units of the Dachstein platform (Maindolomite, transitional unit, lower and upper part of the Dachstein Limestone) are characterized by the dominance of certain stacking patterns. A clear megacyclic pattern reflecting orbital influence was observed only in the upper part of the Dachstein Limestone (in the NE part of the TDMM).

3 Main controlling factors in the generation of Lofer cycles were: (i) subsidence rate (0.1–0.2 m/ka); (ii) short-term sea-level oscillations; and (iii) carbonate production (in the subtidal zone), deposition and redeposition.

4 The ideal cycle is a record of a regular sea-level fluctuation (sine-wave-shaped sea-level curve) of 5–10-m amplitude at a subsidence rate of about 0.15 m/ka. In the cases of deviations from the ideal pattern, either the sea-level curves were deformed or the subsidence rate was higher or lower than the mean value. (Carbonate production was controlled

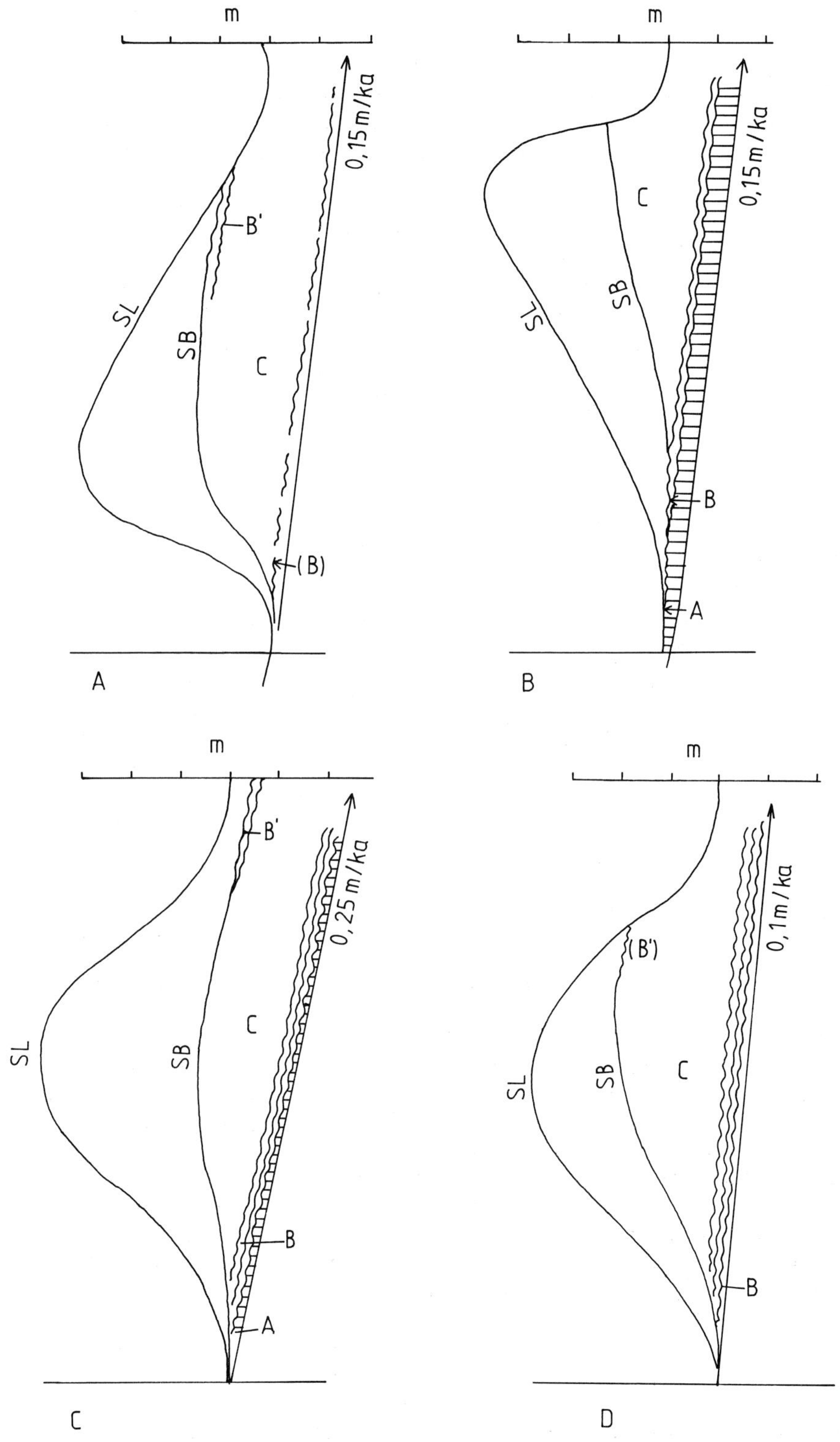

Fig. 17. Genetic models of the anomalous cycle types (end members). (a) Fast transgression–slow regression; (b) slow transgression–fast regression; (c) increased subsidence rate; (d) reduced subsidence rate.

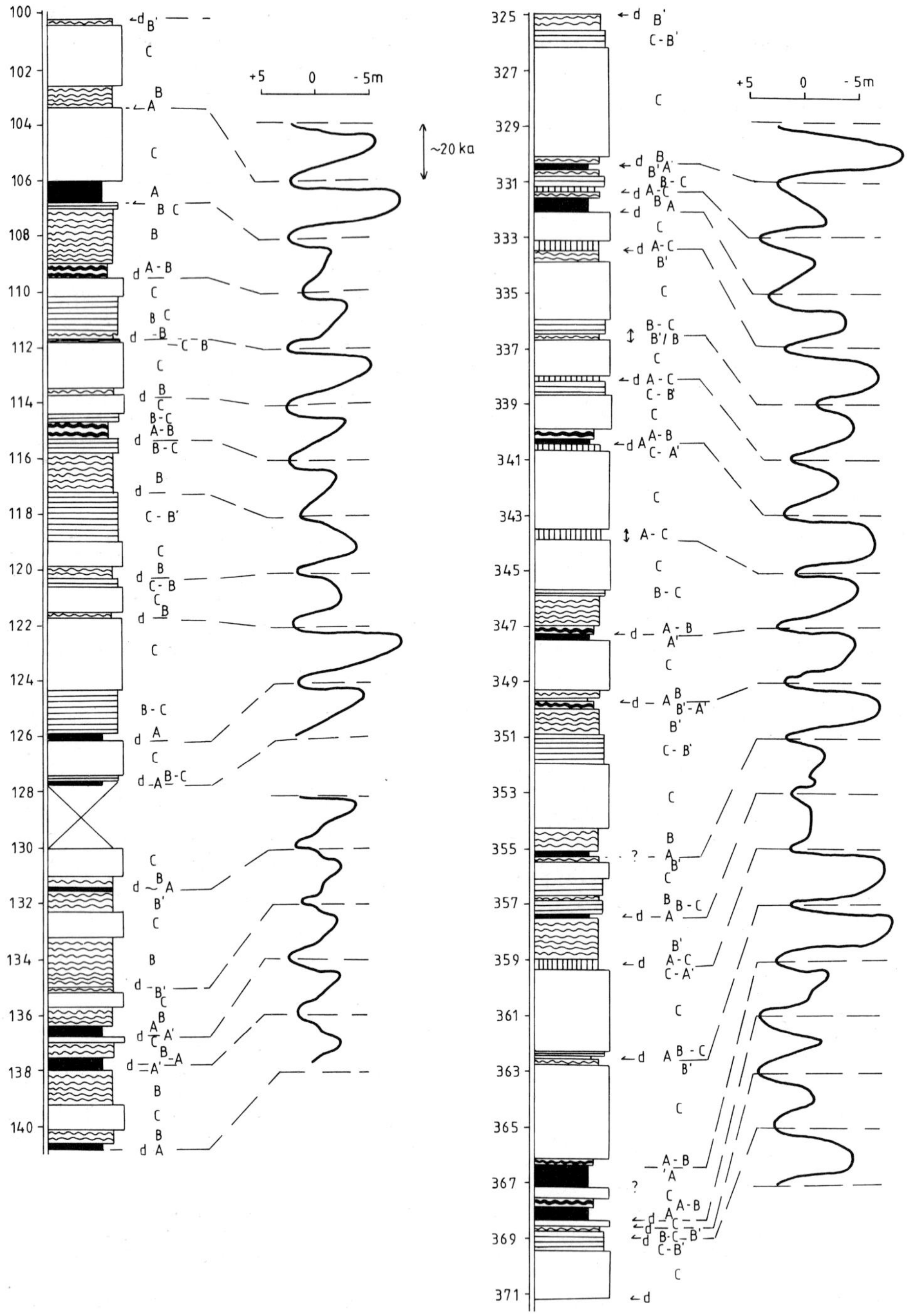

Fig. 18. Reconstructed sea-level curve for two selected segments of borehole section Po-89. Legend is the same as in Fig. 7.

mainly by these two factors, which determined the actual water depth.) Since early diagenetic processes were also controlled by the sea-level changes, relations between the stacking pattern and the diagenetic features are evident. This was a crucial factor in the early dolomitization of the Maindolomite Formation.

5 Basic Lofer cycles can be considered as records of high-frequency Milankovitch cycles (20 ka). Causes of definite or poor manifestation of the megacycles (100 ka) are not known, but local tectonics as well as short-term changes in the subsidence rate do not seem to be critical factors. However, long-term change in the subsidence rate (acceleration from the late Carnian to the Norian) and third-order relative sea-level changes might be important controlling factors.

ACKNOWLEDGEMENTS

This work results from the National Key-Section Program of the Hungarian Geological Institute funded by the Hungarian Central Office of Geology. The computer work for the Markov analysis was carried out by L.Ó. Kovács and G. Turczi (Hungarian Geological Institute) and their assistance was also significant in the interpretation of the results. The valuable help of A. Balog and advice received from J.F. Read (Virginia Polytechnic Institute and State University) is greatly appreciated. The manuscript benefited from reviews by W. Schlager, J.G. Reijmer and P.L. de Boer.

REFERENCES

Anderson, T.W. & Goodman, L.A. (1957) Statistical inference about Markov chains. *Ann. Math. Stat.* **28**, 89–110.

Balla, Z. (1982) Development of the Pannonian basin basement through the Cretaceous–Cenozoic collision: A new synthesis. *Tectonophysics* **88**, 61–102.

Balla, Z. (1988) Clockwise paleomagmatic rotations in the Alps in the light of the structural pattern of the Transdanubian Range (Hungary). *Tectonophysics* **145**, 277–292.

Berger, A. (1977) Support of the astronomical theory of climatic change. *Nature* **268**, 44–45.

Berger, A. & Loutre, M.F. (1987) *Origine des fréquences des éléments astronomiques intervenant dans le calcul de l'insolation*. Scientific Report 1987/13, Inst. d'Astronomie et de Géophys., Louvain-la-Neuve.

Billingsley, P. (1961) Statistical methods in Markov chains. *Ann. Math. Stat.* **32**, 12–40.

Bosellini, A. (1965) Analisi petrografica della 'Dolomia Principale' nel Gruppo di Sella (regione Dolomitica). *Univ. Ferrara, Mem. Geopaleo.* **1**, 49–109.

Bosellini, A. (1967) La tematica deposizionale della Dolomia Principale (Dolomiti e Prealpi Venete). *Soc. Geol. Ital. Boll.* **86**, 133–169.

Bosellini, A. & Hardie, L.A. (1985) Facies e Cicli Della Dolomia Principale Delle Alpi Venete. *Mem. Soc. Geol. Its.* **30**, 245–266.

Davis, J.C. (1986) *Statistics and Data Analysis in Geology*. John Wiley and Sons, New York.

Fischer, A.G. (1964) The Lofer cyclothems of the Alpine Triassic. *Kansas. Geol. Surv. Bull.* **169**, 102–149.

Fischer, A.G. (1975) Tidal deposits, Dachstein Limestone of the North-Alpine Triassic. In: *Tidal Deposits* (Ed. Ginsburg, R.N.). Springer-Verlag, New York, pp. 235–242.

Goldhammer, R.K., Dunn, P.A. & Hardie, L.A. (1990) Depositional cycles, composite sea-level changes, cycle stacking patterns, and the hierarchy of stratigraphic forcing: Examples from Alpine Triassic platform carbonates. *Geol. Soc. Am. Bull.* **102**, 535–562.

Goodman, L.A. (1968) The analysis of cross-classified data: independence, quasi-independence, and interactions in contingency tables with or without missing entries. *J. Am. Statis. Assoc.* **63**, 1091–1131.

Haas, J. (1982) Facies analysis of the cyclic Dachstein Limestone Formation (Upper Triassic) in the Bakony Mts., Hungary. *Facies* **6**, 75–84.

Haas, J. (1987a) Correlation of Upper Triassic profiles on the basis of Lofer cycles (Gerecse Mts.). *Földtani Közlöny, Bull. Hung. Geol. Soc.* **117**, 375–383.

Haas, J. (1987b) Position of the Transdanubian Central Range structural unit in the Alpine evolution phase. *Acta Geol. Hung.* **30**, 243–256.

Haas, J. (1988) Upper Triassic carbonate platform evolution in the Transdanubian Mid-Mountains. *Acta Geol. Hung.* **31**, 299–318.

Haas, J. & Dobosi, K. (1982) Felsö-triász ciklusos karbonátos közetek vizsgálata bakonyi alapszelvényeken (Study of Upper Triassic cyclic carbonate rocks in basic profiles, Bakony Mountains). *Földt. Int. Évi Jel. 1980-ról*, 135–168.

Haas, J., Császár, G., Kovács, S. & Vörös, A. (1990) Evolution of the western part of the Tethys as reflected by the geological formations of Hungary. *Acta Geol. Geoph. Mont. Hung.* **25**, 325–344.

Haq, B.U., Hardenbol, J. & Vail, P.R. (1987) Chronology of fluctuating sea levels since the Triassic. *Science* **235**, 1156–1167.

Kázmér, M. & Kovács, S. (1985) Permian–Paleogene paleogeography along the eastern part of the Insubria-Periadriatic lineament system: evidence for continental escape of the Bakony-Drauzug unit. *Acta Geol. Hung.* **28**, 71–84.

Kámér, M. & Kovács, S. (1989) Triassic and Jurassic oceanic/paraoceanic belts in the Carpathian–Pannonian region and its surroundings. In: *Tectonic Evolution of the Tethyan Region* (Ed. Sengör, A.M.C.). Kluwer Academic Publishers, Dordrecht, pp. 77–92.

Kovács, S. (1982) Problems of the 'Pannonian Median Massif' and the plate tectonic concept. Contribution based on the distribution of Late Paleozoic–Early

Mesozoic isotopic zones. *Geol. Rundsch.* **71**, 617–640.
SCHWARZACHER, W. (1948) Über dies sedimentäre Rhythmik das Dachsteinkalkes von Lofer. *Geol. Bundesanstalt Verhandl.* (p. 11), 175–188.
SCHWARZACHER, W. (1954) Die Grossrhythmik des Dachsteinkalkes von Lofer. *Mineral. Petrogr. Mitt.* **4**, 44–54.
SCHWARZACHER, W. (1975) *Sedimentation Models and Quantitative Stratigraphy*. Elsevier, Amsterdam.
SCHWARZACHER, W. & HAAS, J. (1986) Comparative statistical analysis of some Hungarian and Austrian Upper Triassic peritidal carbonate sequences. *Acta Geol. Hung.* **29**, 175–196.
VISSCHER, H. (1985) Subcommission on Triassic stratigraphy Report 1984. *Albertiana* **3**, 1–2.
VISTELIUS, A.B. (1980) *Osznovü Matematicseszkoj Geologii*. Nauka, Leningrad.

Spec. Publs Int. Ass. Sediment. (1994) **19**, 323–343

Periodicities in the composition of Late Triassic calciturbidites (Eastern Alps, Austria)

J.J.G. REIJMER*, A. SPRENGER†, W.G.H.Z. TEN KATE†, W. SCHLAGER† *and* L. KRYSTYN‡

* *GEOMAR Forschungszentrum für marine Geowissenschaften, Wischhofstr. 1–3, D-24148 Kiel, Germany;*
† *Faculteit der Aardwetenschappen, Vrije Universiteit, De Boelelaan 1085, 1081 HV Amsterdam, The Netherlands; and*
‡ *Institut für Paläontologie der Universität Wien, Universitätsstrasse 7, A-1010 Wien I, Austria*

ABSTRACT

Carbonate platform sediments of the Triassic Dachstein Formation display the so-called Lofer cycles, attributed to sea-level fluctuations. The Lofer cycles have been attributed by previous authors to the Milankovitch variations of the parameters of the Earth's orbit.

The Late Triassic Pedata/Pötschen Schichten represent the basinal equivalent of the Dachstein Formation. The petrographical composition of 810 calciturbidites of the Late Triassic Lacke section (Eastern Alps, Austria) was analysed to see whether the flooding/ exposure cycles of the platform could be traced into the adjacent basin. In each thin section, 200 points were counted. Seven categories of platform-bank or basin-derived grains were distinguished. The platform-bank derived input (e.g. Dasycladaceae, platform Foraminifera, pellets, Bryozoa, corals, calcisponges, microproblematica) displays a clear variation throughout the section. Spectral analysis revealed various cyclicities in the basinal sediments. The effects on the power spectra of individual debris flows and variations in sedimentation rate are evaluated.

The fluctuation in calciturbidite composition along the section is conceived as a response to sea-level controlled variation in platform sediment production. After transformation into time, correspondence of several spectral peaks with Milankovitch quasi-periodicities can be suggested.

INTRODUCTION

The rhythmic bedding in the platform (Lofer) facies of the Late Triassic Dachstein Formation in the Northern and Southern Alps is a characteristic feature of these sediments. Sander (1936) proposed that this rhythmicity is the result of cyclic events in time, arguing that repetitive stratigraphic sequences were unlikely to result from disorder. Schwarzacher (1948, 1954) described the cycles in more detail and noted the bundling of cycles into megacycles. The link between sea-level changes and the facies present in the platform cycles was demonstrated by Schwarzacher (1948, 1954) and Fischer (1964). The classic transgressive Lofer cyclothem (A member: karst/erosion; B member: intertidal; and C member: subtidal) is often found in the sediments of the Dachstein Formation (Haas, 1982; Goldhammer *et al.*, 1990), but is usually capped with a regressive intertidal member B′ (Haas, 1982). The average thickness of a basic Lofer cycle on the platform is 3–4 m and these basic cycles are bundled into cycles of higher order. The duration of deposition for a basic cycle is estimated to be 20–50 ka (Fischer, 1964). Other studies of the Middle and Upper Triassic platform carbonates of the Southern Alps also revealed a close relationship between the quasi-periodicities of the Milankovitch model and the cyclicity present in the Dachstein Formation (Hardie *et al.*, 1986; Schwarzacher & Haas, 1986; Goldhammer & Harris, 1989). Goldhammer *et al.* (1990), however, argued against the presence of cyclicity related to the Milankovitch frequencies in the platform sediments of the Dachstein Formation.

Our analysis of the composition of calciturbidites of the Pedata/Pötschen Schichten, the basinal equivalent (Hallstatt Facies) of the platform sediments of the Dachstein Formation, reveals a close relationship between biota present on the Dachstein carbonate platform and the composition of coeval calciturbidites. The compositional variations could be attributed to fluctuations in sea level that alternately flooded and exposed the platform, creating and destroying shallow-water habitats on the platform-top (Reijmer & Everaars, 1991; Reijmer *et al.*, 1991).

In this paper the following problems will be discussed: (i) Do the basinal Pedata/Pötschen Schichten show cyclicity? (ii) If so, can it be correlated with the cyclicities demonstrated in the platform sediments of the Dachstein Formation? This paper thus aims at studying the effects of platform cycles in the basinal environment and is the first analysis of a pre-Pleistocene calciturbidite sequence in the light of sedimentary cyclicity attributed to Milankovitch cycles (Fischer *et al.*, 1990; de Boer, 1991).

SETTING

The Lacke section is situated in the Eastern Alps (Northern Calcareous Alps) about 60 km southeast of Salzburg in Austria in the Gosau Valley (Fig. 1). The sediments are lithostratigraphically attributed to the Pötschen Kalke (Schlager, 1966) or Pedata Schichten (Tollmann, 1976) and represent the basinal equivalent of the carbonate platform limestones of the Dachstein Formation (Ohlen, 1959; Zapfe, 1960; Fabricius, 1966; Zankl, 1967). The transition from the basinal Pedata/Pötschen Schichten, to the facies of the platform, the Dachstein facies, has been clearly demonstrated in the Gosaukamm area (Ganss *et al.*, 1954; Schlager, 1967) and in the Totes Gebirge area (Schöllnberger, 1973). The 200/210-m-thick succession consists of an alternation of blue–grey mudstones, packstones and grainstones (classification of Dunham, 1962) and green–yellow calcisiltites to marls. In the section analysed (95 m) the thickness of the mudstone to grainstone beds varies between 0.5 and 132 cm, and that of the calcisiltites between 0.1 and 7 cm (Figs 2, 3, 4 and 5A). The mudstone to grainstone beds show grading

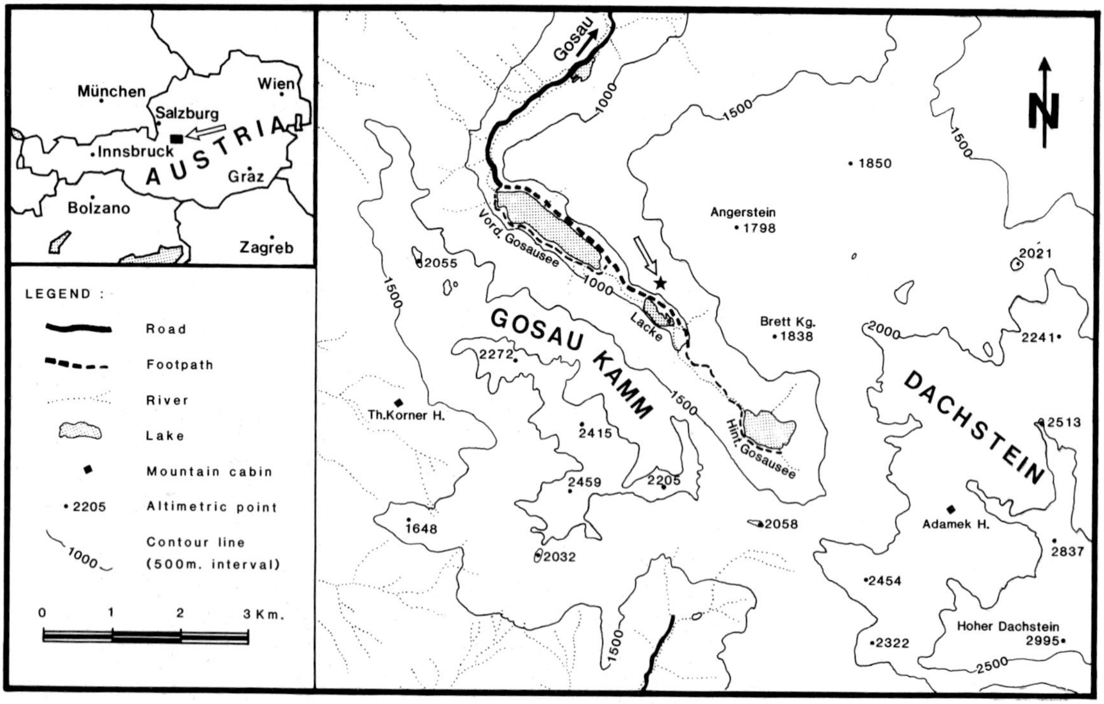

Fig. 1. Location map. The section studied is indicated with an asterisk.

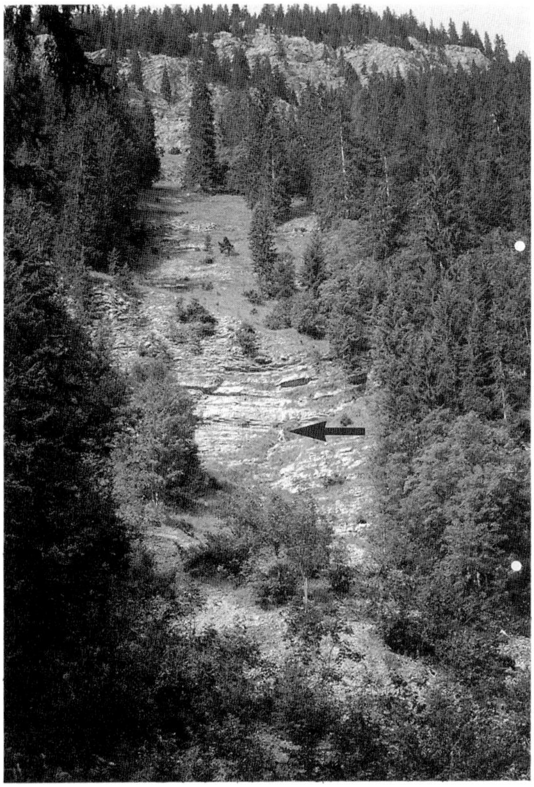

Fig. 2. Photograph of the outcrop located on the eastern side of the Lacke in which the section is situated. For scale, the height of the person, indicated with an arrow, in the centre of the photograph is 1.82 m. The section used in this analysis is situated between the two dots on the picture, 95 m in total.

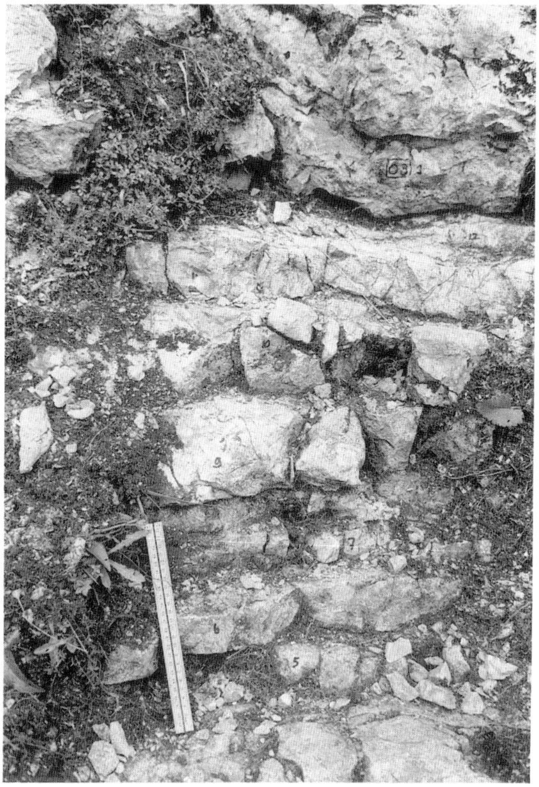

A

Fig. 3A, B. Two details of the lithology in the outcrop, showing calciturbidites alternating with very thin bedded calcisiltites or marls. Measuring-rod and hammer for scale.

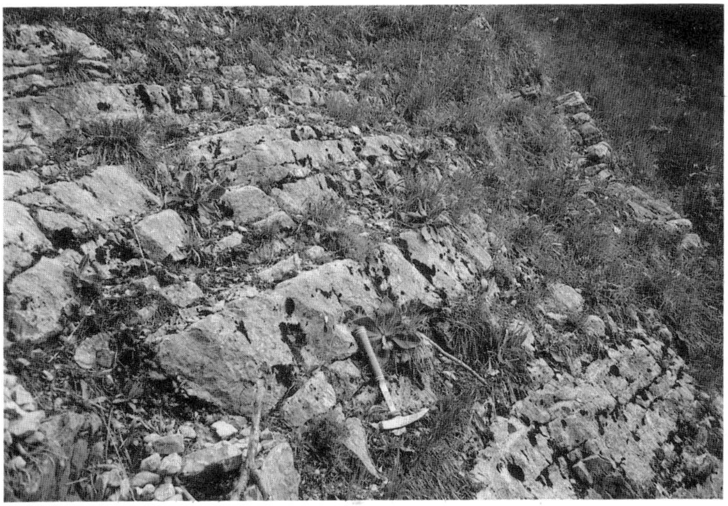

Fig. 3B. B

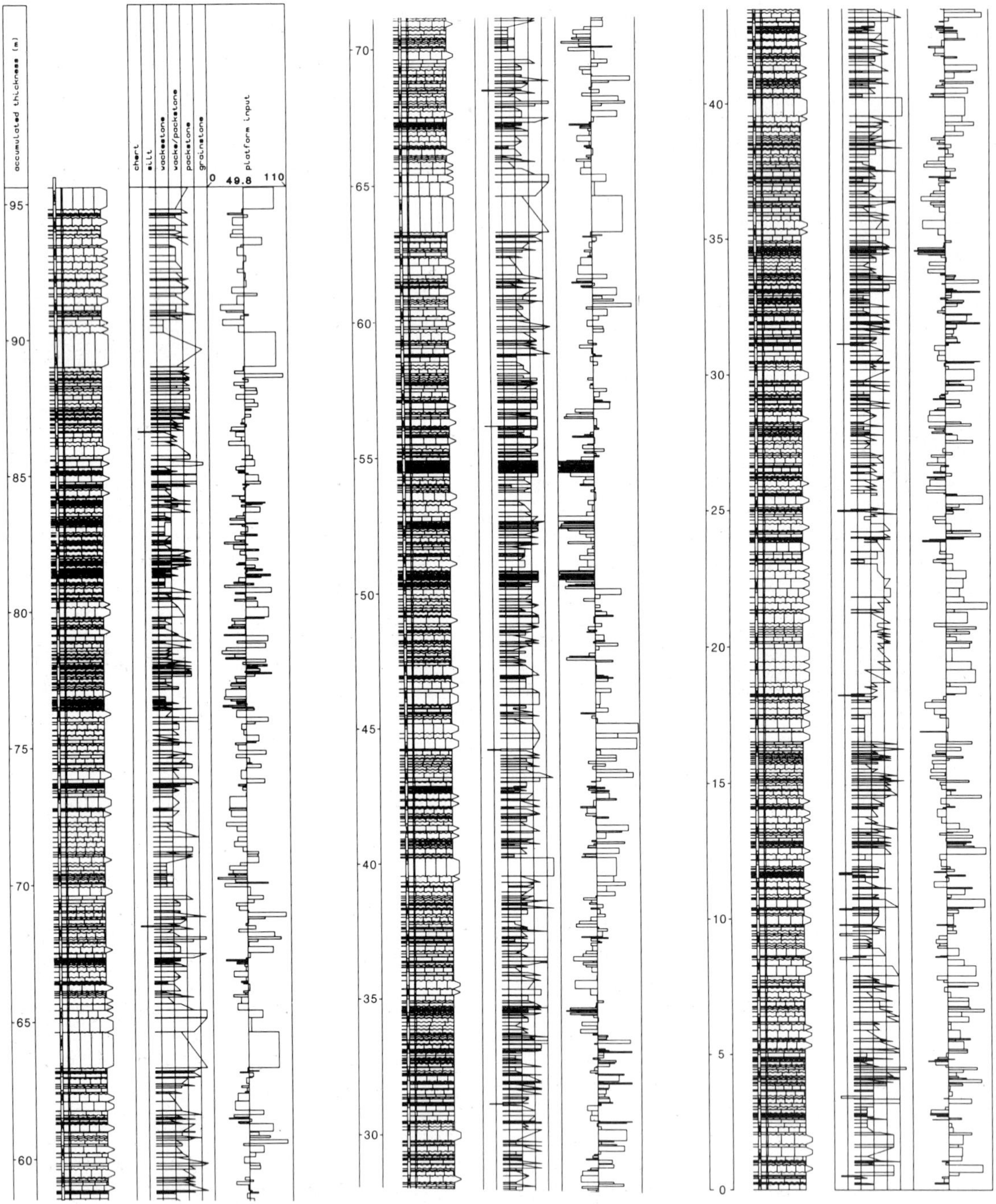

Fig. 4. The vertical distribution of the platform-derived input along the Lacke section. Vertical scale in metres. All calciturbidites are attributed a chalk notation in the second column. In the next column the lithology of the calciturbidites is shown, using the classification scheme of Dunham (1962). The last column displays the counts of the platform input variable within each turbidite. The counts are centred around the mean of the whole sample population, indicated with a vertical line and its value (49.8) at the top of the figure.

and sharp contacts and are interpreted as calciturbidites. The thickness ratio between calcisiltites and calciturbidites over the analysed interval is approximately 1:20. Chert nodules and layers occur throughout the section (Figs 2–4).

Conodont biozones point to a late Norian age for the sediments. The boundary between the Alaunian and Sevatian substages is present 5 m above the base of the analysed succession and at the top of the outcrop (80–90 m above the top of the analysed succession) the Sevatian 1 and 2 transition was found (Fig. 2; Krystyn, 1980).

METHODS

Sampling procedure

The main goal of our investigation was to detect periodicities. Due to the irregular character of the bedding thickness in the section, equally spaced sampling, the ideal basis for this type of analysis, was impossible. Instead we used the following sampling procedure. Bedding thicknesses were measured and lithologies described using the classification scheme of Dunham (1962). The grain size and bedding characteristics were determined and the presence and size of chert layers or nodules noted. The field procedure ended with sampling the top and bottom of each calciturbidite bed. In calciturbidite beds thicker than 30 cm an additional sample was taken from the middle part.

Thin-section analysis

In the 95-m-section analysed, 810 individual calciturbidite beds were present (Reijmer & Everaars, 1991). Thin sections for point-count analysis were made of the coarsest part of the turbidite layers, usually from the base, but occasionally at a higher level. In total 747 thin sections out of 810 were point-counted. Sixty-three beds (7.8%) showing strong chertification or dolomitization were eliminated from the analysis. In each thin section 200 points were counted volumetrically, i.e. large grains were counted several times if succeeding grid points fell upon them. Comparison with a point-count of 400 points of the lower third of the sequence using the same samples showed no detectable differences, thus justifying the point-count method.

Point-count group definition

The various constituents encountered in the thin sections were assigned to point-count groups that characterize particular palaeoenvironments, i.e. lagoon, reef complex and basin. Definition of these groups is inspired by studies on the palaeoecology of reef biota in the Upper Triassic Dachstein Reefs (e.g. Fabricius, 1966; Ott, 1972; Lobitzer, 1975; Flügel, 1981; Buser *et al.*, 1982; Flügel, 1982a, b). For an extensive description of the biota the reader is referred to Reijmer & Everaars (1991) and Reijmer *et al.* (1991).

The groups used in the point-count analysis are as follows.

1 *Biota, non-specific.* Biota not characteristic for a certain palaeoenvironment are incorporated in this point-count group. It includes unspecified skeletal material, echinoderms, crinoids, as well as non-facies diagnostic foraminifera and microproblematica.

2 *Clasts.* The grains incorporated in this group consist of so-called intra-reef clasts (Reid, 1987) and lithoclasts with a variety of biologically produced fragments.

3 *Platform interior biota.* This group comprises platform foraminifera such as *Aulotortus* sp. and *Triasina* sp. and microproblematica *Thaumatoporella*. Dasyclads and the rare ooids have been allotted to this group.

4 *Shallow reef biota.* This group contains frame-building biota, the microproblematica *Tubiphytes*, *Lithocodium*, *Bacinella*, the platform foraminifera *Alpinophragmium* sp., *Glomospirella* sp., *Glomospira* sp. (Senowbari-Daryan, 1980), *Kaeveria* sp., *Sigmoilina* sp., *Galeanella* sp. and other often sessile foraminifera. Pellets *sensu lato* have been counted in this group as well, based on the corresponding environment.

5 *Deep reef and/or forereef biota.* The constituents of this group are the microproblematica *Radiomura*, *Microtubus*, *Baccanella*, *Cheilosporites*, *Muranella*, and *Lamellitubus*.

6 *Open ocean input.* This group includes filaments, Radiolaria, *Globochaete* as well as thin shelled Nodosariids, thin shelled agglutinants and *Lenticulina* sp. and contains quartz grains.

7 *Embedding sediment.* This volumetrically important point-count group includes cement and embedding sediment (matrix).

Definition of the variable for spectral analysis

In the present-day Bahamas a sharp contrast exists between the rim and interior facies on the platform. The non-skeletal facies dominates the interior, while skeletal material prevails in the rim facies (Purdy, 1963). The composition of banktop-derived turbidites varies. Turbidites deposited in interglacial periods, when the platforms are flooded, show a dominance of non-skeletal material (pellets, ooids and grapestones) while in glacial periods, when the banktops are exposed, skeletal material dominates the neritic input (Cartwright, 1985; Droxler & Schlager, 1985; Haak & Schlager, 1989). The composition of the calciturbidites is thus a good indicator of banktop exposure or flooding.

In the Triassic the compositional variation of the calciturbidites is restricted to the contrast between platform and basinal input. A cyclic variation of composition exists such that the system oscillates between abundant platform input from both rim and interior, and little platform input, thus dominance of open-ocean material. The nature of this compositional variation strongly suggests platform conditions that swing back and forth between a state of flooding and prolific production, and a state of exposure with hardly any neritic production (Reijmer & Everaars, 1991; Reijmer *et al.*, 1991). This variation is almost certainly linked to the Lofer cycles of the platform, because these cycles fall within the same frequency range and indicate a history of marine deposition alternating with exposure (Fischer, 1964).

In this analysis we calculated frequency spectra of the periodicities in the basin to test the hypothesis of orbital control. A new variable was created, the input from the platform, combining point-count groups 3 and 4. We assume that the response of platform biota to sea-level fluctuations is represented by this variable. It will be used as the continuous variable in the spectral analysis.

Time-series analysis

The Milankovitch theory establishes the effect of quasi-periodic variations of the orbital parameters of the Earth upon the amount of solar energy entering the atmosphere (Milankovitch, 1941). Tests for possible orbital forcing of bedding rhythms in the sedimentary record are faced with a variety of problems: the number and precision of datum planes (palaeontological markers, radiometric and palaeomagnetic dates); variation in accumulation rates, levels of non-deposition or erosion; selection of an adequate numerical method that analyses the selected stratigraphical and micropalaeontological attributes; and an adequate sampling plan that guarantees an optimal spectral resolution, among others.

We applied various spectral analytical techniques in search of the possible relationship between orbital forcing and rhythmicity in our basinal section. We used the platform-derived input as input variable. The spectral analytic methods used were Blackman–Tukey, harmonic, maximum entropy, Walsh and average Walsh analyses. This multi-method approach is used to distinguish regular cyclicities from spurious peaks that arise by the method used, or by the pulsating character of the platform input variable.

Each method has its own specific (dis)advantages which will be discussed in separate paragraphs. For a summary and mathematical comparison of spectral analytical methods the reader is referred to Box & Jenkins (1976) and Pestiaux & Berger (1984).

The input series for the different spectral techniques were tested for stationarity in the mean by fitting a linear regression line. In the data sets used only minor linear trends were present and these were removed prior to spectral analysis.

The Blackman–Tukey, harmonic and Walsh spectra were smoothed by a Tukey–Hanning filter in order to suppress spurious powers that arise due to the finiteness of the sampled section (Blackman–Tukey and harmonic), to abrupt changes in amplitude (Blackman–Tukey and harmonic) and to irregular and incomplete cycles. In the harmonic and Walsh periodograms a simple first-order autoregressive model is indicated together with its confidence limits at the 80% confidence level.

Blackman–Tukey (BT)

The classic Blackman and Tukey method estimates the autocovariance function from a sampled series and then, by taking the real Fourier (cosine) transform, the power spectral density is obtained (Blackman & Tukey, 1958; Jenkins & Watts, 1968). Advantages of this method are the accuracy of the spectral amplitude estimation and the possibility of testing statistical properties of these estimates. A disadvantage is poor resolution in the low-frequency domain. The BT spectrum is continuous and each power is given by the value of the integral of the power function over a small interval about the corresponding frequency.

All spectral methods used require equally spaced sample points instead of unequal spacing related to the turbidite thickness. To meet this requirement for BT, harmonic and maximum entropy analyses, we fitted cubic spline functions through the data points and subsequently sampled this curve at equally spaced distances. The interpolation interval was fixed at 9.25 cm, which falls in the range between the mean (11.1 cm) and mode (6.9 cm) of the original sample distances (see Fig. 5A). The interpolated data were detrended.

The BT method was also used when calculating spectra of segments of the section. These 'subspectra' were calculated to investigate whether nonstationarity existed in the data, or whether the length of the cycles changed along the section. A sliding window of 47.4 m was used, equal to half of the section. After calculating a spectrum of this segment the window was moved upwards for 10 m and the next spectrum was calculated. This procedure thus creates an overlap of 80% between two succeeding segments. The entire section was covered by a total of five subspectra (see Figs 7 and 8).

Harmonic (HA)

A discrete analogue of the BT model, called harmonic analysis or periodogram spectral analysis was used also. In this analogue the power spectrum is directly evaluated by least squares fit of a set of harmonically related sinusoids, the Fourier harmonics. The resulting periodogram is a Fourier line spectrum that plots the average spectral power of each harmonic versus its frequency. The periodogram is discrete and each power is directly proportional to the height of its peak. We used the same cubic spline interpolated data set as in the BT analysis.

Maximum entropy (ME)

This method computes a set of autocorrelations from the data and calculates the power spectral density of the most random time series with corresponding autocorrelations. A disadvantage of the maximum entropy spectral estimate is the impossibility of gaining general analytical expressions for its statistical properties. Despite this shortcoming, the method is very useful because it provides a more precise identification of the low-frequency peaks, resolves close spectral peaks, has a continuous spectrum and yields an indication of the regularity of the detected quasi-periodicities. Several peaks on a common broad base detected by maximum entropy indicate that the quasi-periodicity present in some part of the record has slightly changed along the section (Pestiaux & Berger, 1984). The same cubic spline interpolated data set as BT analysis was used.

Walsh (WA)

The fourth method applied is Walsh spectral analysis. The model consists of a set of orthonormal square wave functions which take only the values +1 and −1 and are called Walsh functions. Just as a Fourier harmonic consists of a sine and a cosine component, each Walsh function consists of a CAL and an SAL component, but unlike the Fourier harmonics, the Walsh functions are fixed in phase. The data are transformed by a fast Walsh transform followed by periodogram estimation (Beauchamp, 1984). Disadvantages are that the spectrum is discrete and has low resolution in the low-frequency range like the Fourier spectrum, but the main disadvantage is the fixed phase of the Walsh functions, which will affect the Walsh periodogram by creation of spurious peaks when a periodicity in the data and its corresponding Walsh function are not in phase. The advantage of this method is that it is less sensitive to abrupt changes in amplitude and thus compensates for the turbidite-generated, pulsating character of our variable.

The data set for this analysis was created using rectangular interpolation between the sample points. The value of the proxy at the base of a calciturbidite was maintained until the next value was encountered at the base of the next turbidite (see Fig. 4). The fast Walsh transform algorithm used requires the number of data points to be a power of two. The chosen interpolation interval of 9.25 cm covered the succession with a total of 1024 points. Because subtracting a linear trend would destroy the rectangularity of the interpolated values, the original data were detrended before the interpolation.

'Average Walsh' (AV-WA)

This method computes the average of a number of Walsh periodograms. It starts off with the calculation of a Walsh periodogram on the original data set, then the bottom sample is placed on top of the top sample of the succession and a new periodogram is calculated. This sample redistribution procedure followed by the calculation of a periodogram is continued until the original starting position is reached once again. All periodograms are stacked and the overall mean of these periodograms is calculated. In practice only one-quarter of the number of data needs to be circulated in order to arrive at the final average periodogram. The advantages of this procedure are that the spectrum does not change along the section, and spurious peaks, which arise when periodicities present in the data and the corresponding Walsh functions are not in phase, are smoothed to a large extent. Consequently, however, peaks of regular periodicities are smoothed also. The same rectangular interpolated data set was used as in single Walsh spectral analysis.

RESULTS

In the following pages we present the results of the different time-series analyses of the cyclicity in calciturbidite composition. The basinal Pedata/Pötschen sequence offered us an excellent opportunity for a comparison between the cyclicity in the peritidal carbonate platform sequences (Dachstein Limestone) and the cyclicity in the coeval basinal sequence.

In total 747 thin sections were researched out of a total of 810 calciturbidite beds present in the 95-m section analysed. The mean turbidite thickness is 11.1 cm, with a median of 8.9 cm and a mode of 6.9 cm (Fig. 5A). The distribution of thickness through the section is shown in Fig. 5B.

The variable used for spectral analysis (the platform-derived input) is shown in Fig. 4 (far right column). It is defined as the sum of two point-count groups (groups 3 and 4 out of the seven used in the point-count analysis). It combines the point-count groups of platform interior biota and platform rim biota. This variable fluctuates between 0 and 108 counts out of a total of 200 points counted in each thin section. The measures of central tendency over the entire data set are 49.8 counts as mean value, with a modal value of 42.2, a median of 48.9, a kurtosis of −0.071 and a skewness of 0.104 indicating a non-symmetrical frequency distribution. The measures of dispersion occur with the following values: a standard deviation of 21.5, a variance of 463.5 and a coefficient of variation of 43.2.

Blackman–Tukey

The first spectrum (Fig. 6a) resulted from a BT analysis on the cubic spline interpolated (lag = 9.25 cm) data set. The frequencies beyond 0.1 cycles/

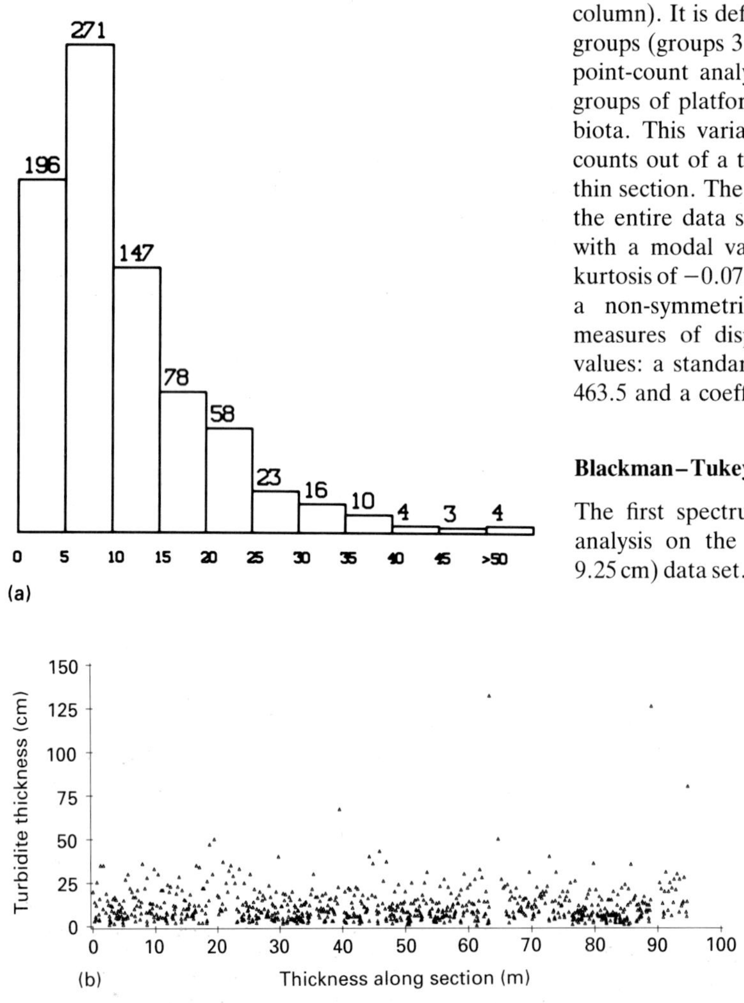

Fig. 5. (a) Histogram of the thicknesses (cm) of the calciturbidite beds of the section analysed (95 m). Number of beds within each class denoted at the top. (b) The distribution of turbidite thickness (cm, vertical axis) along the section (m, horizontal axis). Note the two debris flow beds at 64 and 90 m (132 and 126 cm thick respectively).

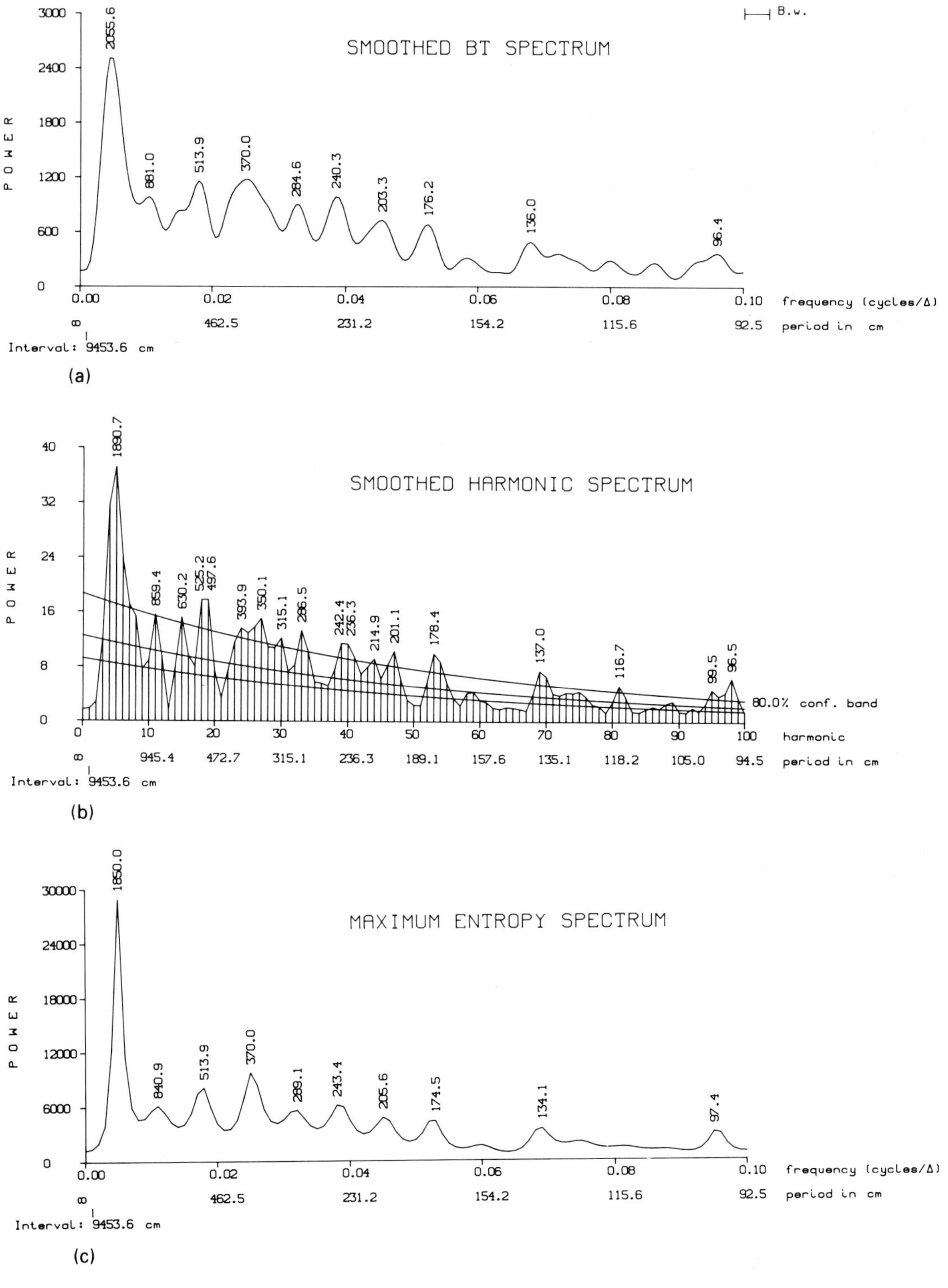

Fig. 6. Power spectra of the section: (a) Blackman–Tukey spectral analysis, smoothed by a Tukey–Hanning lag window; (b) harmonic; (c) maximum entropy; (d) Walsh; (e) average Walsh. The horizontal axis is a frequency axis and for easy reference the corresponding periods (cm) are plotted below; the vertical axis shows power. When possible the fit of the series to a first-order autoregressive process model together with an 80% confidence band are shown. Periods (cm) of significant peaks are indicated with numbers. Table 3 shows the ratios of the orbital frequencies calculated for the Triassic by Berger *et al.* (1989) and the ratios of the peaks.

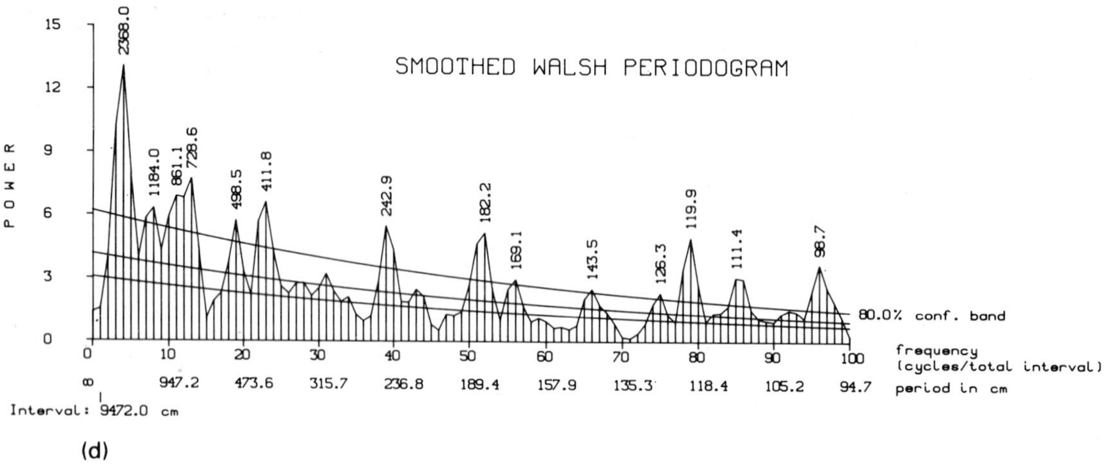

(d)

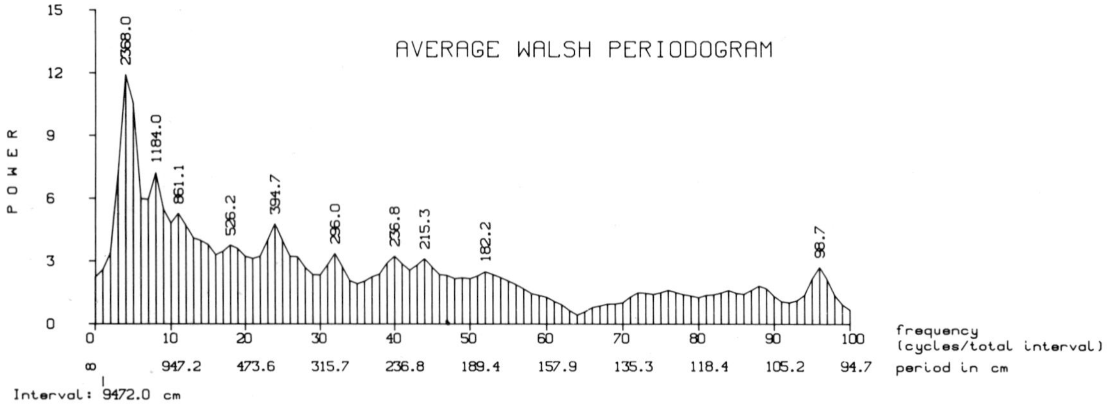

(e)

Fig. 6. *Continued.*

sampling distance (periods less than 93 cm) are not reproduced, because they are reduced in power. Thus, only at the extreme side of the spectrum are peaks present that may be attributed to the mean and modal bedding thicknesses of the calciturbidites. Prominent peaks in the spectrum are: 2055.6, 881.0, 513.9, 370.0, 284.6, 240.3, 203.3, 176.2, 136.0 and 96.4 cm (Fig. 6a and Table 1).

Harmonic

The second periodogram (Fig. 6b) used the same data set. The resulting periodogram differs from the BT periodogram in low frequencies due to the discrete resolution. Both show a good match in the higher frequencies on the right side of the spectrum. The overall distribution of powers shows a gradual decrease from left to right. The significant peaks at the 80% confidence level present in the spectrum are: 1890.7, 859.4, 630.2, 525.2, 497.6, 393.9, 350.1, 315.1, 286.5, 242.4, 236.3, 214.9, 201.1, 178.4, 137.0, 116.7, 99.5 and 96.5 cm (Fig. 6b and Table 1).

Maximum entropy

The third spectrum is a ME spectrum (Fig. 6c). The data used are equal to the data set for the BT analysis. The peaks that are very prominent in this spectrum are: 1850.0, 840.9, 513.9, 370.0, 289.1, 243.4, 205.6, 174.5, 134.1 and 97.4 cm. For comparison the dominant peaks of each spectrum are listed in Table 1.

Table 1. Overview of the wavelengths (cm) of dominant peaks encountered in the power spectra (Fig. 6) of the different spectral analytical techniques used in this analysis

Blackman–Tukey	Harmonic	Maximum entropy	Walsh	Average Walsh
2055.6	1890.7	1850.0	2368.0	2368.0
—	—	—	1184.0	1184.0
881.0	859.4	840.9	861.1	861.1
—	—	—	728.6	—
—	630.2	—	—	—
513.9	525.2	513.9	—	526.2
—	497.6	—	498.5	—
370.0	393.9	370.0	411.8	394.7
—	350.1	—	—	—
—	315.1	—	—	—
284.6	286.5	289.1	—	296.0
240.3	242.4	243.4	242.9	—
—	236.3	—	—	236.8
—	214.9	—	—	215.3
203.3	201.1	205.6	—	—
176.2	178.4	174.5	182.2	182.2
—	—	—	169.1	—
136.0	137.0	134.1	143.5	—
—	—	—	126.3	—
—	116.7	—	119.9	—
—	—	—	111.4	—
—	99.5	97.4	98.7	98.7
96.4	96.5	—	—	—

Walsh

The fourth periodogram is a WA spectrum (Fig. 6d). It is based on the rectangular interpolated data set as described in the methods section. The outcome of this analysis is in good agreement with the results of the BT analysis. The cycle thicknesses that exceed the 80% confidence level are: 2368.0, 1184.0, 861.1, 728.6, 498.5, 411.8, 242.9, 182.2, 169.1, 143.5, 126.3, 119.1, 111.4 and 98.7 cm (see also Table 1).

Average Walsh

The final periodogram is an AV-WA spectrum (Fig. 6e), which is invariant to depth along the section and in which spurious peaks are reduced. The most prominent peaks are: 2368.0, 1184.0, 861.1, 526.2, 394.7, 296.0, 236.8, 215.3, 182.2 and 98.7 cm (Table 1).

Comparison of spectra

When the minor linear trend in both data sets is not removed before the spectra are calculated, all powers are subdued except for one dominant peak with a length of about 95 m, equal to the total section interval (these spectra are not reproduced). We attribute this trend to an overall gradual transgression along the succession upwards. When this trend is removed and the results of the different power spectra (Fig. 6a–e and Table 1) are compared it turns out that several peaks occur with minor variations in all spectra; we attribute them to regular sedimentary cycles with wavelengths of: (i) 23.7–18.9 m, (ii) 11.9 m, (iii) 8.8–8.4 m, (iv) 5.3–5.1 m, (v) 3.9–3.7 m, (vi) 2.9–2.8 m, (vii) 2.4 m, (viii) 2.1–2.0 m, (ix) 1.8–1.7 m, (x) 1.4–1.3 m and (xi) 1.0 m.

As can be observed in Fig. 6 and Table 1, the cyclicities detected in the power spectra show some variation when a comparison is made between the different methods used. The high frequencies show a good match, but the low frequencies differ considerably. These differences can be attributed to a number of factors.

1 Creating a new data set by re-sampling either a rectangular or a cubic spline interpolated curve.
2 Differences in the mathematical approach used in each spectral analysis to quantify the cyclicity present (see Pestiaux & Berger, 1984).
3 The difference that exists in the resolution of a continuous spectrum (BT and ME) versus a discrete spectrum (HA, WA). In a continuous spectrum the positioning of the cycles can be precise, while in a discrete periodogram a period can be resolved only at discrete fixed frequencies.
4 The influence of variations in sedimentation rate on the different spectra which are discussed in the next section (see also Pestiaux & Berger, 1984; Weedon, 1989).

Subspectra

Point-count analysis of the entire 95-m section revealed a change in the overall character of the sedimentation regime (Reijmer & Everaars, 1991). At 50–55 m above the base of the section a succession of three coquina layers is present. Calciturbidites preceding this interval contain a lesser amount of carbonate mud (point-count group 7) and clasts (point-count group 2), and a higher amount of skeletal grains compared to calciturbidites deposited after this interval. To check whether these variations in composition are related to changes in sedimentation rates, we calculated spectra of segments of the section. BT spectral analysis was applied. A sliding

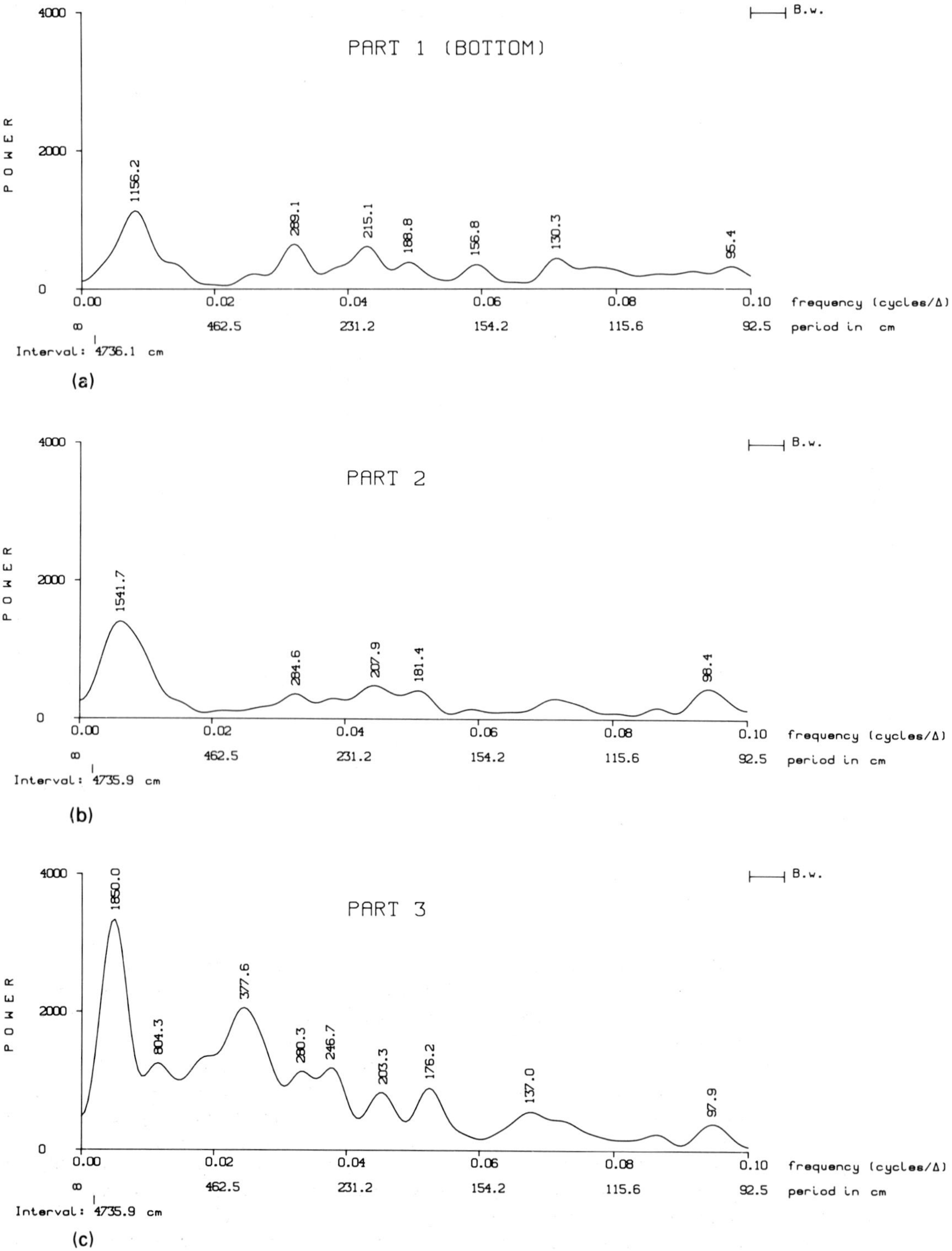

Fig. 7. Blackman–Tukey spectra of successive 47.4-m segments of the section. Succeeding segments overlap each other by 80%. The horizontal axis shows frequency and periods (cm); the vertical axis, power. Periods of significant peaks are indicated with numbers. The orbital frequencies calculated for the Triassic by Berger *et al.* (1989) and the ratios of the peaks are shown in Table 4.

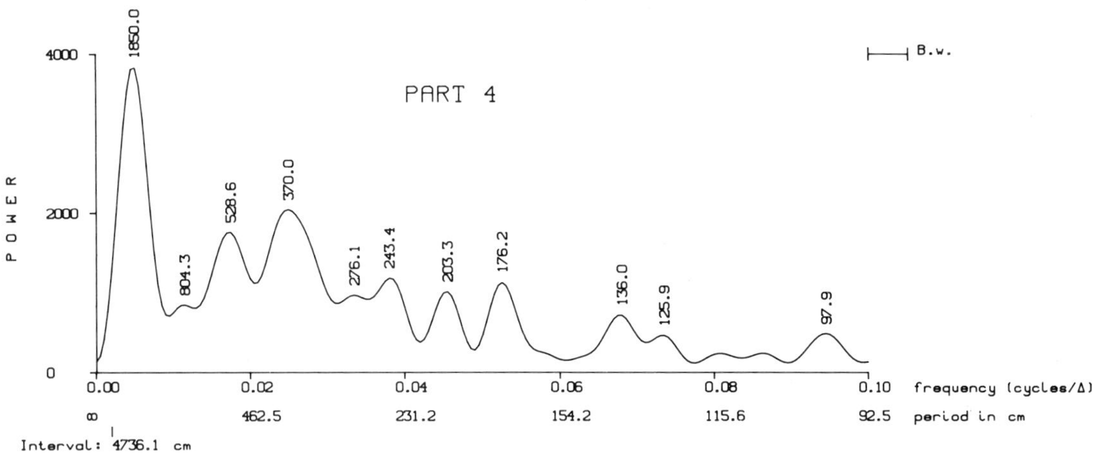

(d)

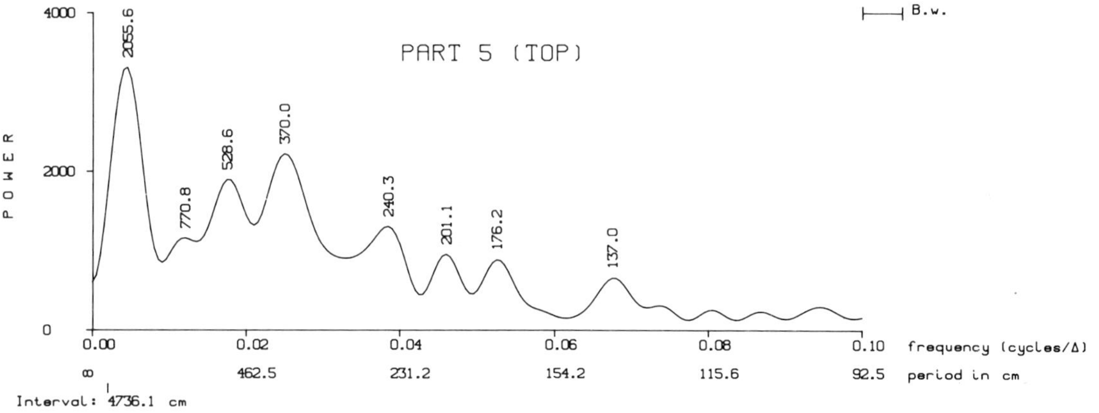

(e)

Fig. 7. *Continued.*

window of 47.4 m was used, equal to the bottom half of the section just beneath the coquina level. After calculating a spectrum of the lowermost segment, the window was moved upwards by 10 m and the next spectrum was calculated. This procedure creates an overlap of 80% between two succeeding segments. The entire section was covered by a total of five 'subspectra' (Fig. 7). The dominant low-frequency peak present in the subspectra shows a gradual increase in length from the bottom to the top, from 11.6 m up to 20.6 m. This explains the variation in the length of the dominant peak(s) found by the different methods used (BT, 20.6 m; HA, 18.9 m; ME, 18.5 m; WA, 23.7 and 11.8 m; AV-WA, 23.7 and 11.8 m). We interpret this as an overall increase in sedimentation rate. Other periods show the same trend, from 2.2 to 3.8 m for example. The 5.3-m peak is present only in subspectra 3, 4 and 5. The peak pattern from the 0.05 frequency upward (periods less than 2 m) is fairly consistent. Because the overall sedimentation rate increases, but the thickness distribution of individual calciturbidites does not change along the section (Fig. 5b), we interpret the peaks 1.9 to 0.97 m and the 5.3-m peak to originate from the thickest beds (seven beds are thicker than 45 cm, see Fig. 5a).

As a proof, another set of subspectra (Fig. 8) was calculated after the removal from the data set of two coarse-grained, clast-rich grainstones at 64 and 90 m (Fig. 4). These two beds, much thicker than all the others (Fig. 5b), deviate sedimentologically and

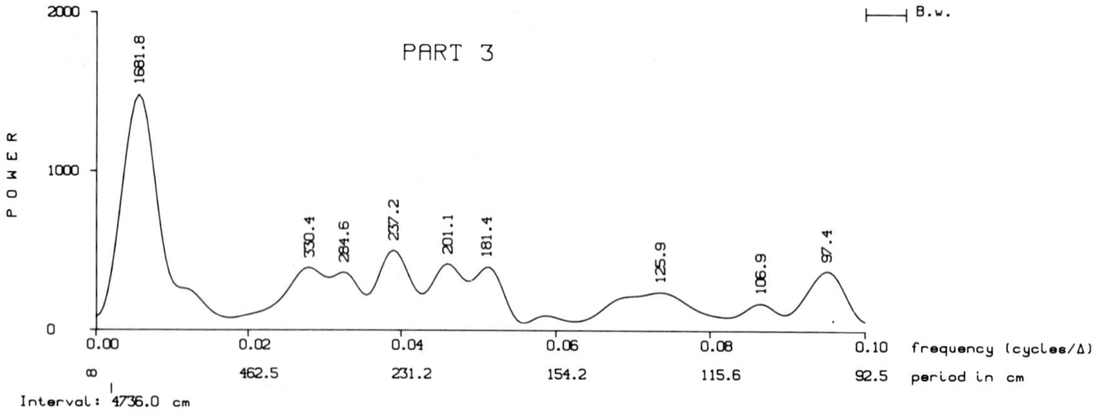

(a)

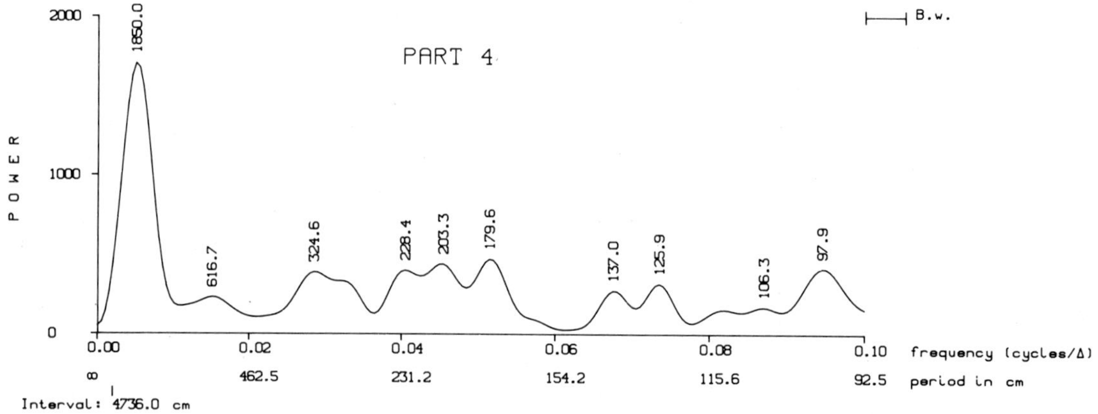

(b)

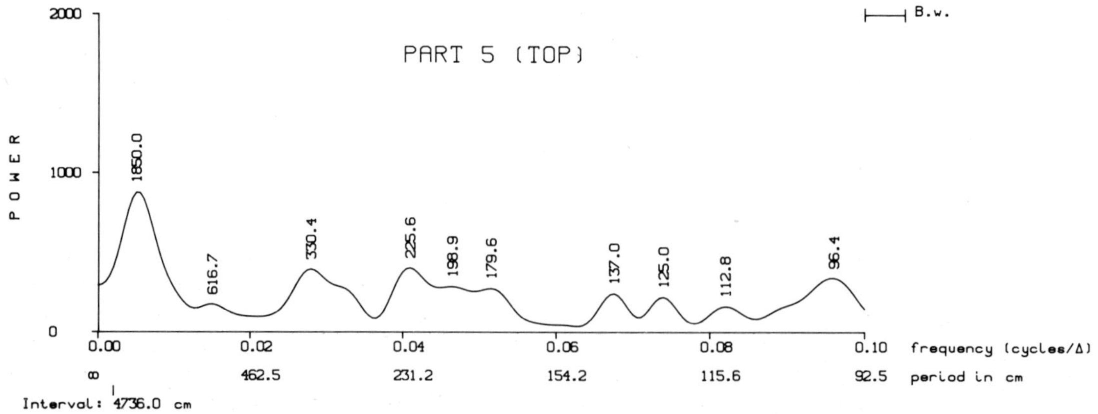

(c)

Fig. 8. As Fig. 7, but after removal of two thick debris flow beds (present in the windows of parts 3–5 only) from the data set.

lithologically from the overall calciturbidite pattern, and are interpreted as debris flow deposits. They are present only in the windows of parts 3, 4 and 5. The same preconditions as previously described for the calculation of subspectra were used. The subspectra 1 and 2 did not alter, but the 5.3-m peak present in subspectra 3, 4 and 5 of the original data set disappears (compare Fig. 7 with Fig. 8).

DISCUSSION

Thickness versus time

An important step in the study of periodicities in stratigraphy is the transition from thickness to time. This step involves various assumptions that will be evaluated in the following paragraphs. In our analysis we assumed that thickness was proportional to time. Obviously this assumption does not hold true for time intervals relative to the spacing between individual turbidite events. However, it has been shown that sedimentation in turbidite sequences is fairly steady when averaged over long timespans and many individual turbidite events (e.g. Lancelot *et al.*, 1980). In our example the grand mean of the turbidite thickness is 11.1 cm for the estimated minimum sedimentation rate of 100 m/Ma (see below). This implies that the shortest wavelength interpreted in the Milankovitch spectrum, the precession cycle, still consists of at least 16–18 individual turbidites. We believe that at this level sedimentation rates are no longer influenced by the pulsating nature of turbidite deposition and that the principle of thickness being proportional to time is a reasonable approximation of reality.

As stated in the literature, differences in sedimentation rates occur in carbonate basinal environments in response to sea-level changes. High-stand shedding characterizes these sediments, high sedimentation rates during high-stands and low sedimentation rates during low-stands (Droxler & Schlager, 1985; Reijmer *et al.*, 1988; Wilber *et al.*, 1990; Glaser & Droxler, 1991). Variations in sedimentation rates also occur in the Lacke sequence as can be seen in the ME spectrum (Fig. 6c) showing several peaks on a common broad base in the 2.0-m range instead of one distinct peak. The direct response in the basin to the production of material on the platform can also be seen when a comparison is made between the BT spectrum of the entire section (Fig. 6a) and the BT subspectra (Fig. 7). The spectrum of the entire section is a composite of variations in accumulation rate due to the pulsating nature of calciturbidites and an overall increase in sediment accumulation by an increased number of calciturbidites per time interval. The result of these variations in sedimentation rates is a number of closely spaced peaks within each spectrum (Fig. 6a–e). These variations will disturb the time–thickness relation and can partly be made visible in the subspectra.

Another critical assumption involves the timespan present in the calcisiltites between the calciturbidites. The calciturbidites make up almost 93% of the thickness of the entire 95-m section, but do not represent an equal percentage of time. They are episodic events that individually occur within a very short timespan. Erosion resulting from the deposition of the calciturbidites may have removed parts of the actual sequence of basinal calcisiltites. This is a further distortion of the ideal 1 : 1 time–thickness relation.

Two other related factors influencing the time–thickness relationship are the diagenetic and compaction history of the sediments. The argillaceous calcisiltites are clearly more affected by diagenesis than the calciturbidites. X-ray diffraction analysis of the calcisiltites reveals the occurrence of dolomite and diagenetically altered micas. In fact, the argillaceous calcisiltites can be considered as bundles of microstylolites. In the calciturbidites diagenetic features like chertification, dolomitization and rare stylolites also occur. The thickness loss caused by pressure solution remains uncertain, but these effects should only modify composition over a scale of centimetres. Compaction must have affected the calcisiltites more than the calciturbidites as is to be expected, and can be concluded from the porosity/depth diagrams of Bond & Kominz (1984). The loss in thickness by these processes cannot be calculated and compensated for without making assumptions.

Datum levels

The accurate estimation of the periods of cycles depends on reliable datum levels. Two time markers based on conodonts were established. At the base of the section, the Alaunian–Sevatian transition is recorded. The Sevatian 1–Sevatian 2 boundary was found 80–90 m above the top of the sampled section. Consequently the section represents exactly one-half of a conodont zone, the Upper *Epigondolella bidentata* Zone. The conodont subzone correlates to the *Rhabdoceras suessi* ammonoid Zone (Krystyn, 1980).

Table 2. Triassic age dates for the Norian and Rhaetian using the different time scales indicated at the top of the table. Duration (Ma) of each stage is shown in parentheses

		Odin & Letolle (1982)	Forster & Warrington (1985)	Harland *et al.* (1989)
Jurassic				
		204 ± 4	205 ± 5	208.0 ± 7.5
Triassic	Rhaetian	(4)	(5)	(1.5)
		208 ± 5	210 ± 5	209.5 ± 8
	Norian	(12)	(10)	(13.9)
		220 ± 8	220 ± 5	223.4 ± 10
	Carnian			

The available radiometric time scales show considerable differences in the duration of the Norian, between 10 and 13.9 Ma. (Table 2).

Using the two time markers and assuming equal duration of Triassic ammonoid zones, we estimate the time (minimum spread) incorporated in the section as 1.4–1.95 Ma depending on the time scale used. Using the stratigraphic thickness, approximately 200 m, the sedimentation rate can be estimated as 100–150 m/Ma (Bubnoffs).

Astronomical cycles

In addition to the preceding restrictions another complication is added, namely the gradual change of duration in the pre-Quaternary of the orbital precession and obliquity cycles, at the present day about 19 and 23 ka, and 41 and 54 ka (Berger *et al.*, 1989). Studies on astronomical frequencies for the pre-Quaternary showed the sensitivity of the Earth's orbital frequencies to the Earth–Moon distance and consequently to the length of the day and to the dynamical ellipticity of the Earth (Berger, 1989). The shortening of the Earth–Moon distance and the length of the day back in time induces a shortening of the obliquity and the precession periods. For the Norian (Late Triassic) this shortening results in a duration for the precession cycles of 18 and 21.5 ka and for the obliquity of 36.5 and 46.5 ka.

Orbital control

Platform

The record observed represents the ultimate response of a complex chain of processes affecting the platform and basinal sediments. Schwarzacher & Haas (1986) showed the connection between the cyclicity observed on the platform and the quasi-periodicities of the Milankovitch model. They proposed two scenarios to explain the cyclic pattern present in the platform sediments of the Leoganger–Steinberge Mountain Range. The first assigns the basic 3.7-m Lofer cycle to the 20-ka period of the precessional cycle (Table 3). The 7.2-m cycle could then be coupled to the mean 43-ka obliquity cycle, leaving the 15.2-m cycle as a candidate for the E1 (100-ka) eccentricity cycle. The 27.7-m cycle remains unexplained in this scenario. The second option presented by Schwarzacher & Haas (1986) proposed the 27.7-m cycle as the 100-ka eccentricity cycle and the 15.2-m cycle as the 43-ka obliquity cycle. This variant does not correlate the 3.7-m cycle with one of the known Milankovitch periods.

Goldhammer *et al.* (1990) suggested a quite different origin for the Lofer facies of the Leoganger–Steinberge Mountain Range. Their analysis on cycle thicknesses did not reveal Milankovitch-related cyclic patterns, but rather a chaotic stratigraphic distribution of cycle thicknesses and diagenetic features probably controlled by short-term variations in subsidence rates.

As shown by Sadler (1981) all sedimentary stratigraphic sections must contain hiatuses at some scale. Presumably the same holds for the succession of the Dachstein Formation. The occurrence of hiatuses will distort spectral shapes and affect wavelength ratios (Weedon, 1989). This may partly explain why wavelength ratios observed on spectra of the Dachstein Formation are not exactly those of the orbital cycles even when orbital forcing has been inferred (Schwarzacher & Haas, 1986). Hiatuses may even obscure Milankovitchian rhythms (Goldhammer *et al.*, 1990).

Basin

When we apply the 100 m/Ma sedimentation rate to the resulting peaks present in the different power

Table 3. Lacke basinal cycle wavelengths (cm) and their ratios as analysed by five spectral methods (see Fig. 6), compared with the quasi-periodicities (ka) and their ratios predicted by Berger's model (1989), and the Lofer platform cycles analysed by Schwarzacher & Haas (1986). Two possible correlations of the Lofer cycles with the predicted quasi-periodicities are given, as (a) and (b). The five spectral methods are Blackman–Tukey (BT), harmonic (HA), maximum entropy (ME), Walsh (WA) and average Walsh (AV-WA)

Berger model	100 ka	46.5 ka	36.5 ka			21.5 ka	18 ka				
	1	2.1:1	2.7:1			4.6:1	5.6:1				
Lofer platform cycles											
(a)	1525 cm	725 cm				367 cm					
	1	2.1:1				4.2:1					
(b)	2770 cm	1525 cm		725 cm						367 cm	
	1	1.8:1		3.8:1						7.5:1	
Lacke basinal cycles											
BT	2056 cm	881 cm		514 cm			370 cm			285 cm	240 cm
	1	2.3:1		4.0:1			5.6:1			7.2:1	8.5:1
HA	1891 cm	859 cm	630 cm	525 cm	498 cm	394 cm	350 cm	315 cm	286 cm	242 cm	
	1	2.2:1	3.0:1	3.6:1	3.8:1	4.8:1	5.4:1	6.0:1	6.6:1	7.8:1	
ME	1850 cm	841 cm		514 cm			370 cm		289 cm	243 cm	
	1	2.2:1		3.6:1			5.0:1		6.4:1	7.6:1	
WA	2368 cm	1184 cm	861 cm	729 cm		499 cm	412 cm				243 cm
	1	2.0:1	2.7:1	3.2:1		4.7:1	5.7:1				9.7:1
AV-WA	2368 cm	1184 cm	861 cm			526 cm		395 cm			296 cm
	1	2.0:1	2.7:1			4.5:1		6.0:1			8.0:1

spectra of our basinal proxy some match with the quasi-periodicities can be reached, but very significant peaks present in all spectra are left unexplained. The peak around 11.9 m present in the Walsh periodograms (see Fig. 6d, e) can be interpreted as the short eccentricity cycle. The subdued peak around 4 m is associated with the obliquity cycle while the closely spaced double peaks with periods of 2.1 and 1.8 m correspond to the precession cycles (Table 1). This interpretation does not explain the peak at 18.5–23.7 m or at 5.2 m. When we use the 150 m/Ma sedimentation rate the results are indefinite and show no cyclicity that can be matched with Milankovitch.

When the ratios of the peaks are calculated and we interpret the dominant peak in the low-frequency domain (wavelength between 18 and 24 m) as the short cycle of eccentricity, the correlation with the quasi-periodicities of Milankovitch is reasonable (Table 3). Considering the other estimated periods, the 8.4–8.8-m cycle (BT, ME and WA) would fall within the range of the obliquity and the 3.5–3.9-m cycle could be attributed to the precession (Fig. 6a–e; Tables 1 and 3). For example, in the BT spectrum (Fig. 6a), the peak around 20.6 m can be interpreted as the short eccentricity cycle. The subdued peak around 8.8 m can be associated with the obliquity cycle and the broad peak around 3.7 m with the precession cycles. The prominent peak at 5.1 m, however, is not explained by orbital forcing, but by the presence of two debris flows (see discussion on the subspectra).

As mentioned before, an overall change in the sedimentation regime occurred in the 95-m section (Reijmer & Everaars, 1991). Spectra of segments of the section using BT analysis showed the influence of the compositional variations which were accompanied by changes in sedimentation rates (Fig. 7). The dominant peak present in the low-frequency domain of all subspectra shows a gradual increase in length from 11.6 m up to 20.6 m, from bottom to top. Other periods display the same trend or remain constant. The 5.3-m peak is visible only in subspectra 3, 4 and 5. The more or less consistent image of the peak pattern from 0.05 frequency upward may be a product of variations in bed thicknesses and the pulsating character of the calciturbidites. Short-term palaeoclimatic fluctuations might also be present within this range (for an overview see Elrick *et al.*, 1991). The match with the quasi-periodicities of Milankovitch is better in these subspectra. When the dominant peak is interpreted as the short eccen-

Table 4. Results of the sliding window Blackman–Tukey spectral analysis of the Lacke section (see Fig. 7). The Berger model quasi-periodicities (ka) and their ratios are shown at the top as in Table 3. A sliding window of half the total section length (47.4 m) was used with 80% overlap between successive windows, giving five windows in all. The analysis was repeated for the upper three windows (3–5) after excluding the debris flows from the data (see Fig. 8). The results are given as cycle wavelengths (cm) and their ratios

Berger model		100 ka	46.5 ka	36.5 ka			21.5 ka	18 ka		
		1	2.1 : 1	2.7 : 1			4.6 : 1	5.6 : 1		
47.4-m window subspectra										
Debris flows included	1	1156 cm			356 cm	289 cm		215 cm	189 cm	157 cm
	(bottom)	1			3.2 : 1	4.0 : 1		5.4 : 1	6.1 : 1	7.4 : 1
	2	1542 cm						285 cm	243 cm	208 cm
		1						5.4 : 1	6.3 : 1	7.4 : 1
	3	1850 cm	804 cm		500 cm		378 cm		280 cm	247 cm
		1	2.3 : 1		3.7 : 1		4.9 : 1		6.6 : 1	7.5 : 1
	4	1850 cm	804 cm		529 cm		370 cm		276 cm	243 cm
		1	2.3 : 1		3.5 : 1		5.0 : 1		6.7 : 1	7.6 : 1
	5	2056 cm		771 cm	529 cm			370 cm		240 cm
	(top)	1		2.7 : 1	3.9 : 1			5.6 : 1		8.6 : 1
Debris flows excluded	3	1682 cm					330 cm	285 cm	237 cm	
		1					5.1 : 1	5.9 : 1	7.1 : 1	
	4	1850 cm		617 cm				325 cm		228 cm
		1		3.0 : 1				5.7 : 1		8.1 : 1
	5	1850 cm		617 cm				330 cm		225 cm
	(top)	1		3.0 : 1				5.6 : 1		8.2 : 1

tricity cycle, other peaks can be attributed to obliquity and precession cycles (Table 4). The match is not perfect, but the discrepancies are within reasonable limits.

Two coarse grained, clast-rich grainstones at 64 and 90 m in the section were interpreted as debris flows and thus as deviations from the normal sedimentation pattern (Fig. 5b). After removal of these beds, another set of subspectra was made. Subspectra 1 and 2 remained the same, but the 5.3-m peak present in subspectra 3, 4 and 5 of the first set of subspectra disappears. The overall character of the subspectra is smoothed and the correlation with the quasi-periodicities of Milankovitch improved (Table 4). This illustrates the distortion of spectra by single events.

The subspectra display an increase in wavelength of the most dominant peaks, from 11.6 to 20.6 m (Fig. 7 and Table 4). This lengthening of the cycles possibly reflects the positioning of the section nearer to the source of sediment, perhaps related to the first stage of progradation of the carbonate platform system. The actual progradation of the platform over the basinal sediments is visible at the top of the outcrop (Reijmer & Everaars, 1991).

Ratios

In order to compare the periodicities from sediments of basin and platform, the ratios between peaks present in the spectra were calculated. The results of all spectra were used in this comparison (Table 3).

To represent the platform the Walsh periodicities reported by Schwarzacher & Haas (1986) which are possibly associated with the orbital periods were used. Two scenarios are possible: either the 1525-cm or the 2770-cm period matches the short cycle of eccentricity of 100 ka (Schwarzacher & Haas, 1986). Both ratio scenarios show a good match with the ratios found in the basin analysis. The first scenario, with the 1525-cm cycle as a basic cycle, demonstrates the better coherency with the basinal ratios. The 2770-cm scenario also shows a tie from the platform to the 2-m cycles in the basin environment. In both cases the cyclicity present in the basin successions indicates a possible relationship to the cyclicity present on the platform, even for those frequencies that cannot be explained by orbital perturbations. We conclude that there is a strong similarity between cycles in the Late Triassic platforms and the basin section of the Lacke. Irrespective of their ultimate cause, this similarity argues for a causal link between platform and basin cycles.

After compensation for variations in sedimentation rate, and disturbances caused by single events, the match between the ratios of peaks observed in the basinal environment and those of the orbital model was improved (Table 4). The sedimentation pattern in the basinal environment thus seems to be a response to periodic changes of climate, related to variations in the distribution of solar energy over the Earth. These variations influence the production of material on the platform and/or the subsequent off-bank transport. Upper Triassic basinal deposits also showed the effect of climate and eustasy on the sedimentation of platform-originated carbonate mud and basinal terrigenous mud (Masetti *et al.*, 1989; Burchell *et al.*, 1990).

CONCLUSIONS

The pronounced cyclicity present in the platform sediments of the Triassic Dachstein Formation resulted from sea-level fluctuations influencing the sedimentation patterns. These sea-level changes influenced the distribution and production potential of certain biota and the off-bank transport of excess material. In the periplatform deposits of the Pedata/Pötschen Schichten, the compositional variation of platform input in calciturbidites displayed a cyclicity analogous to that of the platform sediments. When spectral analyses are applied to the platform-derived input in the basin, several high powers of the spectra can be matched with the quasi-periodicities of the Milankovitch model. The peak at 18–23 m, the subdued peak at 8–9 m and the peak at 3–4 m match the short eccentricity, the obliquity, and the precession cycles, respectively. The peaks at 5.2, 2.0 and 1 m could not be explained with the Milankovitch model. Two debris flows were shown to be responsible for the presence of the 5.2-m peak. In addition, the BT spectral analysis of the sequence using a sliding window revealed an increase in sedimentation rate towards the top of the section through the lengthening of the period of the dominant peaks.

The fluctuation in calciturbidite composition along the section is conceived as a response to variation in platform sediment production as controlled by sea level. Analogous to the Lofer cycles, the compositional variations in the basinal sediments occur within the Milankovitch frequency band.

ACKNOWLEDGEMENTS

The permission of Dr W. Janoschek of the Austrian Geological Survey to investigate this aspect of Austrian geology is greatly appreciated. Dr W. Piller (Wien) and Dr B. Senowbari-Daryan (Erlangen) are thanked for carbonate petrographical advice. We also thank Hemmo Bosscher, Poppe de Boer and Juul Everaars who reviewed earlier versions of the manuscript. Tim Herbert and David G. Smith are thanked for their reviews and suggestions. Partial funding was provided by Amoco Prod. Co., British Petroleum Ltd and Royal Dutch/Shell (KSEPL).

Readers interested in the spectral analytical methods used and the computer programs are referred to A. Sprenger & W. ten Kate, who have an article in preparation on that subject. All programs run on an ES/9000 mainframe with VM/CMS operating system. The overview of the entire section (Fig. 4) was produced by program STRATCOLUMN (Sprenger & ten Kate, 1990).

REFERENCES

Beauchamp, K.G. (1984) *Applications of Walsh and Related Functions, with an Introduction to Sequence Theory*. Academic Press, London.

Berger, A. (1989) The spectral characteristics of Pre-Quaternary climatic records, an example of the relationship between the astronomical theory and geosciences. In: *Climate and Geo-Sciences, a Challenge for Sciences and Society in the 21st Century* (Eds Berger, A., Schneider, S. & Duplessy, J.Cl.). Kluwer Academic Publishers, Dordrecht, pp. 47–76.

Berger, A., Loutre, M.F. & Dehant, V. (1989) Astronomical frequencies for pre-Quaternary paleoclimate studies. *Terra Nova* **1**, 474–479.

Blackman, R.B. & Tukey, J.W. (1958) *The Measurements of Power Spectra*. Dover Publ. Inc., New York.

Bond, G.C. & Kominz, M.A. (1984) Construction of tectonic subsidence curves for the early Paleozoic miogeocline, southern Canadian Rocky Mountains: Implications for subsidence mechanisms, age of breakup, and crustal thinning. *Geol. Soc. Am. Bull.* **95**, 155–173.

Box, G.E.P. & Jenkins, G.M. (1976) *Time Series Analysis: Forecasting and Control*. Holden-Day, San Francisco.

Burchell, M.T., Stefani, M. & Masetti, D. (1990) Cyclic sedimentation in the Southern Alpine Rhaetic: the importance of climate and eustasy in controlling platform-basin interactions. *Sedimentology* **37**, 795–816.

Buser, S., Ramovs, A. & Turnsek, D. (1982) Triassic Reefs in Slovenia. *Facies* **6**, 15–24.

Cartwright, R.A. (1985) *Petrography of turbidites from deep-sea fans in the Bahamas*. PhD dissertation, Univ. of Miami.

De Boer, P.L. (1991) Astronomical cycles reflected in sediments. *Zbl. Geol. Paläont.* **1**, 911–930.

Droxler, A.W. & Schlager, W. (1985) Glacial versus interglacial sedimentation rates and turbidity frequency in the Bahamas. *Geology* **13**, 799–802.

Dunham, R.J. (1962) Classification of carbonate rocks according to depositional texture. In: *Classification of Carbonate Rocks – a Symposium* (Ed. Ham, W.E.). Am. Assoc. Petrol. Geol. Mem. 1, pp. 108–121.

Elrick, M.A., Read, J.F. & Coruh, C. (1991) Short-term paleoclimatic fluctuations expressed in lower Mississippian ramp-slope deposits, southwestern Montana. *Geology* **19**, 799–802.

Fabricius, F.H. (1966) *Beckensedimentation und Riffbildung an der wende Trias/Jura in den Bayerisch-Tiroler Kalkalpen*. International sedimentary petrographical series (Eds Cuvillier, J. & Schürmann, H.M.E.), Vol. IX. E.J. Brill, Leiden.

Fischer, A.G. (1964) The Lofer cyclothems of the Alpine Triassic. *Kansas Geol. Surv. Bull.* **169**, 107–149.

Fischer, A.G., de Boer, P.L. & Premoli Silva, I. (1990) Cyclostratigraphy. In: *Cretaceous Resources, Events and Rhythms* (Eds Ginsburg, R.N. & Beaudoin, B.). Kluwer Academic Publishers, Dordrecht, pp. 139–172.

Flügel, E. (1981) Paleoecology and facies of Upper Triassic reefs in the Northern Calcareous Alps. In: *European Fossil Reef Models* (Ed. Toomey, D.F.). Soc. Econ. Paleont. Mineral. Spec. Publs 30, pp. 291–359.

Flügel, E. (1982a) *Microfacies Analysis of Limestones*. Springer-Verlag, Berlin.

Flügel, E. (1982b) Evolution of Triassic reefs: Current concepts and problems. *Facies* **6**, 297–328.

Forster, S.C. & Warrington, G. (1985) Geochronology of the Carboniferous, Permian and Triassic. In: *The Chronology of the Geological Record* (Ed. Snelling, N.J.). Mem. Geol. Soc. London 10, pp. 99–113.

Ganss, O., Kümel, F. & Spengler, E. (1954) Erläuterungen zur geologischen Karte der Dachsteingruppe. *Wiss. Alpenver.* **15**.

Glaser, K.S. & Droxler, A.W. (1991) High production and highstand shedding from deeply submerged carbonate banks, northern Nicaragua Rise. *J. Sedim. Petrol.* **61**, 128–142.

Goldhammer, R.K. & Harris, M.T. (1989) Eustatic controls on the stratigraphy and geometry of the Latemar buildup (Middle Triassic), the Dolomites of northern Italy. In: *Controls on Carbonate Platforms and Basin Development* (Eds Crevello, P.D., Wilson, J.L., Sarg, J.F. & Read, J.F.). Soc. Econ. Paleont. Mineral. Spec. Publs 44, pp. 323–338.

Goldhammer, R.K., Dunn, P.A. & Hardie, L.A. (1990) Depositional cycles, composite sea-level changes, cycle stacking patterns, and the hierarchy of stratigraphic forcing: Examples from Alpine Triassic platform carbonates. *Geol. Soc. Am. Bull.* **102**, 535–562.

Haak, A.B. & Schlager, W. (1989) Compositional variations in calciturbidites due to sea-level fluctuations, late Quaternary, Bahamas. *Geol. Rundsch.* **78**, 477–486.

Haas, J. (1982) Facies analysis of the cyclic Dachstein Limestone Formation (Upper Triassic) in the Bakony Mountains, Hungary. *Facies* **6**, 75–84.

Hardie, L.A., Bosellini, A. & Goldhammer, R.K. (1986) Repeated subaerial exposure of subtidal carbonate platforms, Triassic, northern Italy: Evidence for high frequency sea level oscillations on a 100 000 year scale. *Paleoceanography* **1**, 447–457.

Harland, W.B., Armstrong, R.L., Cox, A.V., Craig, L.E., Smith, A.G. & Smith, D.G. (1989) *A Geologic Time Scale 1989*. Cambridge University Press, Cambridge.

Jenkins, G.M. & Watts, D.G. (1968) *Spectral Analysis and its Applications*. Holden-Day, San Francisco.

Krystyn, L. (1980) Triassic conodont localities of the Salzkammergut region (Northern Calcareous Alps). *Abh. Geol. B.-A.* **35**, 61–98.

Lancelot, Y., Winterer, J.E. and the Shipboard Scientific Party (1980) Site 416, in the Moroccan Basin, Deep Sea Drilling Project Leg 50. *Init. Rep. DSDP* **50**, 115–302.

Lobitzer, H. (1975) Fazielle Untersuchungen an norischen Karbonatplattform-Beckengesteinen (Dachsteinkalk-Aflenzer Kalk im südöstlichen Hochschwabgebiet, Nördliche Kalkalpen, Steiermark). *Mitt. Geol. Ges. Wien* **66–67**, 75–99.

Masetti, D., Stefani, M. & Burchell, M. (1989) Asymmetric cycles in the Rhaetic facies of the Southern Alps: Platform-basin interactions governed by eustatic and climatic oscillations. *Riv. It. Paleont. Strat.* **94**, 401–424.

Milankovitch, M. (1941) Kanon der Erdbestrahlung und seine Anwendung auf das Eiszeitenproblem. *Ed. Spec. Acad. Royale Serbe Belgrade* **132**.

Odin, G.S. & Letolle, R. (1982) The Triassic time scale in 1981. In: *Numerical Dating in Stratigraphy* (Ed. Odir, G.S.). John Wiley & Sons, Chichester, pp. 523–533.

Ohlen, H.R. (1959) *The Steinplatte Reef Complex of the Alpine Triassic (Rhaetian) of Austria*. PhD thesis, Princeton University.

Ott, E. (1972) Mitteltriadische Riffe der Nördliche Kalkalpen und altersgleiche Bildungen auf Karaburun und Chios (Ägäis). *Mitt. Ges. Geol. Bergbaustud.* **21**, 251–276.

Pestiaux, P. & Berger, A. (1984) An optimal approach to the spectral characteristics of deep-sea climatic records. In: *Milankovitch and Climate* (Eds Berger, A., Imbrie, J., Hays, J., Kukla, G. & Saltzman, B.). Reidel Publ. Co., Dordrecht, pp. 417–445.

Purdy, E.G. (1963) Recent calcium carbonate facies of the Great Bahama Bank. 2. Sedimentary facies. *J. Geol.* **71**, 472–497.

Reid, P.R. (1987) Nonskeletal peloidal precipitates in Upper Triassic reefs, Yukon Territory (Canada). *J. Sedim. Petrol.* **57**, 893–900.

Reijmer, J.J.G. & Everaars, J.S.L. (1991) Carbonate platform facies reflected in carbonate basin facies (Triassic, Northern Calcareous Alps, Austria). *Facies* **25**, 253–278.

Reijmer, J.J.G., Schlager, W. & Droxler, A.W. (1988) Site 632: Pliocene–Pleistocene sedimentation cycles in a Bahamian basin. In: *Proceedings of the Ocean Drilling Program, Scientific Results 101* (Eds Austin, J.A. Jr, Schlager, W. *et al.*). College Station, Texas (Ocean Drilling Program), pp. 213–220.

Reijmer, J.J.G., ten Kate, W.G.H.Z., Sprenger, A. & Schlager, W. (1991) Calciturbidite composition related to exposure and flooding of carbonate platform (Triassic, Eastern Alps). *Sedimentology* **38**, 1059–1074.

Sadler, P.M. (1981) Sediment accumulation rates and the completeness of stratigraphic sections. *J. Geol.* **89**, 569–584.

Sander, B. (1936) Beiträge zur Kentniss der Anlagerungsgefüge (Rhythmische Kalke und Dolomite aus der

Trias). *Mineral. Petrogr. Mitt.* **48**, 27–139, 141–209.

SCHLAGER, W. (1966) Fazies und Tektonik am Westrand der Dachsteinmasse (Österreich). *II. Mitt. Ges. Geol. Bergbaustud.* **17**, 205–282.

SCHLAGER, W. (1967) Hallstätter und Dachsteinkalk-Fazies am Gosaukamm und die Vorstellung ortsgebundener Hallstätter Zonen in den Ostalpen. *Verh. Geol. Bundesanstalt* **1/2**, 50–70.

SCHÖLLNBERGER, W. (1973) Zur Verzahnung von Dachsteinkalk-Fazies und Hallstätter Fazies am Südrand des Toten Gebirges (Nördliche Kalkalpen, Österreich). *Mitt. Geol. Bergbaustud.* **22**, 95–153.

SCHWARZACHER, W. (1948) Sedimentpetrografische Untersuchungen kalkalpiner Gesteine. Hallstätterkalk von Hallstatt und Ischl. *Jb. Geol. B.-A.* **91**, 1–48.

SCHWARZACHER, W. (1954) Die Grossrythmik des Dachsteinkalkes von Lofer. *Tschermaks Mineral Petrogr. Mitt.* **4**, 44–54.

SCHWARZACHER, W. & HAAS, J. (1986) Comparative statistical analysis of some Hungarian and Austrian upper Triassic peritidal carbonate sequences. *Acta Geol. Hung.* **29**, 175–196.

SENOWBARI-DARYAN, B. (1980) Fazielle und paläontologische Untersuchungen in oberrhätischen Riffen (Feichtenstein- und Gruberriff bei Hintersee, Salzburg, Nördlichen Kalkalpen). *Facies* **3**, 1–237.

SPRENGER, A. & TEN KATE, W.G. (1990) A graphical software system to present stratigraphic information of surveyed sections. *Computer & Geosciences* **16**, 517–537.

TOLLMANN, A. (1976) *Analyse des klassischen Nordalpinen Mesozoikums. Stratigraphie, Fauna und Fazies der Nördlichen Kalkalpen.* Deuticke, Vienna.

WEEDON, G.P. (1989) The detection and illustration of regular sedimentary cycles using Walsh power spectra and filtering, with examples from the Lias of Switzerland. *J. Geol. Soc. Lond.* **146**, 133–144.

WILBER, R.J., MILLIMAN, J.D. & HALLEY, R.B. (1990) Accumulation of bank-top sediment on the western slope of Great Bahama Bank: Rapid progradation of a carbonate megabank. *Geology* **18**, 970–974.

ZANKL, H. (1976) Die Karbonatsedimente der Obertrias in den nördlichen Kalkalpen. *Geol. Rundsch.* **56**, 128–139.

ZAPFE, H. (1960) Untersuchungen im obertriadischen Riff des Gosaukammes (Dachsteingebiet, Oberösterreich). I. Beobachtungen über das Verhältnis der Zlambachschichten zu den Rifflalken im Bereich des Grossen Donnerkogels. *Verh. Geol. B.-A.* 236–240.

Spec. Publs Int. Ass. Sediment. (1994) **19**, 345–366

Orbitally induced small-scale cyclicity in a siliciclastic epicontinental setting (Lower Lias, Yorkshire, UK)

F.S.P. VAN BUCHEM*, I.N. McCAVE *and* G.P. WEEDON†

Department of Earth Sciences, University of Cambridge, Cambridge, UK

ABSTRACT

The Yorkshire Lower Lias consists essentially of fine-grained sediments, in the clay to fine silt size range. The occurrence of several types of shellbed, siliciclastic layers and concretionary horizons makes it possible to distinguish five stratigraphically distinct facies, which can be interpreted in terms of specific depositional environments.

One of these facies is the Banded Shales unit, which covers most of the *jamesoni* ammonite zone, and consists of 64 couplets of regularly alternating darker and lighter grey layers (layer thickness 10–70 cm), in general with plane gradational contacts. The layers consist mainly of quartz silt, clay, black and white micas, calcite and organic matter. Regular variation in the relative concentration of these components accounts for the rhythmically bedded nature of the unit. The clay mineral and total organic carbon (TOC) distributions also show evidence for a longer term variation covering four to five couplets.

To investigate the presence of cyclicity, spectral analysis has been applied to three geochemical time series (isothermal remanent magnetization, TOC, Si/Al ratio) all covering the bottom 10 m (=20 couplets) of the Banded Shales unit, and to a gamma-ray log of the adjacent Felixkirk Borehole. Within the dating limitations of the ammonite zonation scheme it seems possible to distinguish the precession, obliquity and eccentricity cycles of the Earth's orbit. Based on this evidence it is suggested that orbitally forced climatic changes influenced the depositional system at two levels: (i) short-term variations expressed on a couplet scale, affecting the storm frequency and perhaps magnitude; and (ii) longer term variations affecting weathering and clay production in the source area.

The temporal control established by the recognition of precession frequencies allows us to suggest that Jurassic ammonite subzones are markedly unequal in duration.

INTRODUCTION

Orbitally induced small-scale cyclicity in Mesozoic rocks has been widely documented in the literature for pelagic and hemi-pelagic settings, marine carbonate platforms, evaporite basins and lake sediments (see overview in Fischer *et al.*, 1990). Few examples have, however, been described from epicontinental, siliciclastic settings, which can be expected to act as accurate recorders of small changes in the sedimentary environment. Their relatively shallow depth and the mixture of mudstones and siltstones/sandstones are ideal conditions to register the particular nature of the hydrodynamic processes involved in sedimentation. The near-land position ensures a direct expression in the sediment of changes in the amount and composition of the terrigenous sediment supply. Additionally, changes in coastal vegetation, marine benthos and micro-fauna and flora might be expected to find direct expression in the sedimentary record.

This paper provides a case study from the Lower Lias in the Cleveland Basin (Yorkshire, UK; Figs 1 and 2) which consists of a series of mudstones and siltstones deposited in a shallow epicontinental sea in a subtropical climate (Wall, 1965; Sellwood, 1970, 1971, 1972; van Buchem & McCave, 1989). The small-scale cyclicity occurs in the *jamesoni*

* Present address: Institut Français du Pétrole, BP 311, 92506 Rueil-Malmaison Cedex, France.

† Present address: School of Geological Sciences, University of Luton, Park Square, Luton, UK.

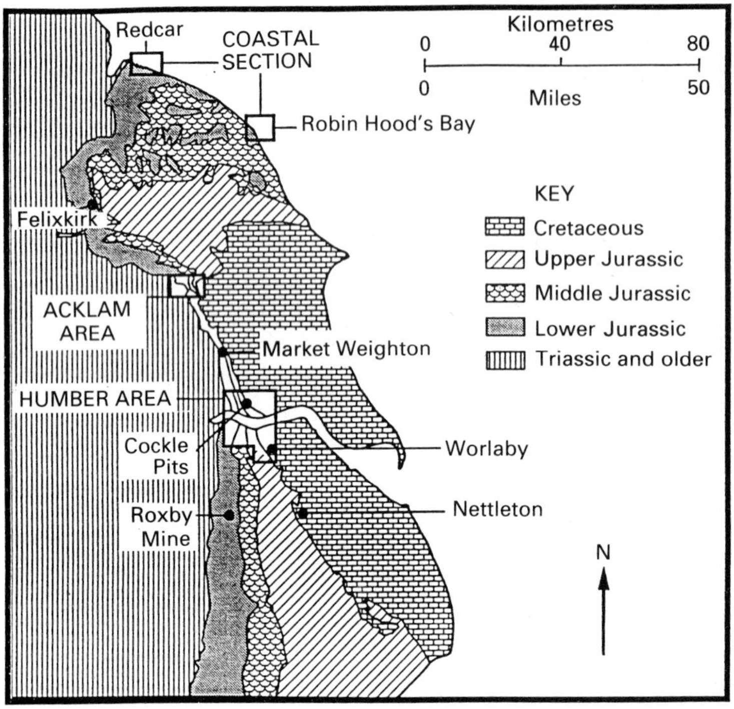

Fig. 1. Geological map of NE Yorkshire. The Cleveland Basin covers the area between Redcar, Felixkirk, the Acklam area and Robin Hood's Bay. The position of the coastal sections and sites of the boreholes have been indicated.

ammonite zone of the Pliensbachian stage, and has also been recorded in other parts of the UK at this stratigraphic level: Dorset (Weedon & Jenkyns, 1990), Oxfordshire (Horton & Poole, 1983) and Lincolnshire (Sellwood, 1972; Gaunt *et al.*, 1980). Because of its characteristic lithological nature, van Buchem & McCave (1989) introduced the lithostratigraphic term Banded Shales for the unit displaying the small-scale cyclicity in Yorkshire.

The first sedimentological work on the Yorkshire Lias cycles was carried out by Sellwood (1970, 1971, 1972). Here a further investigation of the mineralogical, chemical and organic matter composition of the Banded Shales interval is presented in order to establish a better understanding of environmental conditions involved in their formation. The Banded Shales, which cover the *polymorphus*, *brevispina* and *jamesoni* subzones of the *jamesoni* ammonite zone, crop out along the Yorkshire coast in Robin Hood's Bay and at Huntcliff (Saltburn-on-the-sea, just south of Redcar; Fig. 1). A detailed study was carried out on a stratigraphic interval of 10 m that is well exposed in the cliff face of Robin Hood's Bay, just north of Bay Town in Dungeon Hole (NZ 9535 0555) (Fig. 3).

METHODS

Bulk mineralogy and clay mineralogy (for 9 and 40 samples respectively) were determined by X-ray diffractometry (XRD) on smear slides using a Philips PW 1730 X-ray generator. Clay mineral percentages were calculated following Griffin's (1971) semi-quantitative method.

The major (Si, Al, K, Na, P, Mg, Ca, Mn, Fe, Ti) and trace (Zr, Cr, Rb, V, Zn, Ni, Cu, Pb, Y, Sr) element composition of 51 samples was determined by X-ray fluorescence (XRF) analysis on lithium metaborate fusion beads and pressed powder pellets respectively, using a Philips PW-1480.

Isothermal remanent magnetization (IRM), anhysteritic remanent magnetization (ARM) and magnetic susceptibility (MS) were determined for all 214 samples. IRM was induced by a direct current field of 0.3 T (3000 Oe).

The analysis of the organic matter (OM) was approached using both organic geochemical analysis (total organic carbon (TOC); hydrogen index (HI); organic carbon stable isotopes), and optical examination of the isolated OM-fraction. The TOC and HI were determined for 114 samples using a Rock-Eval

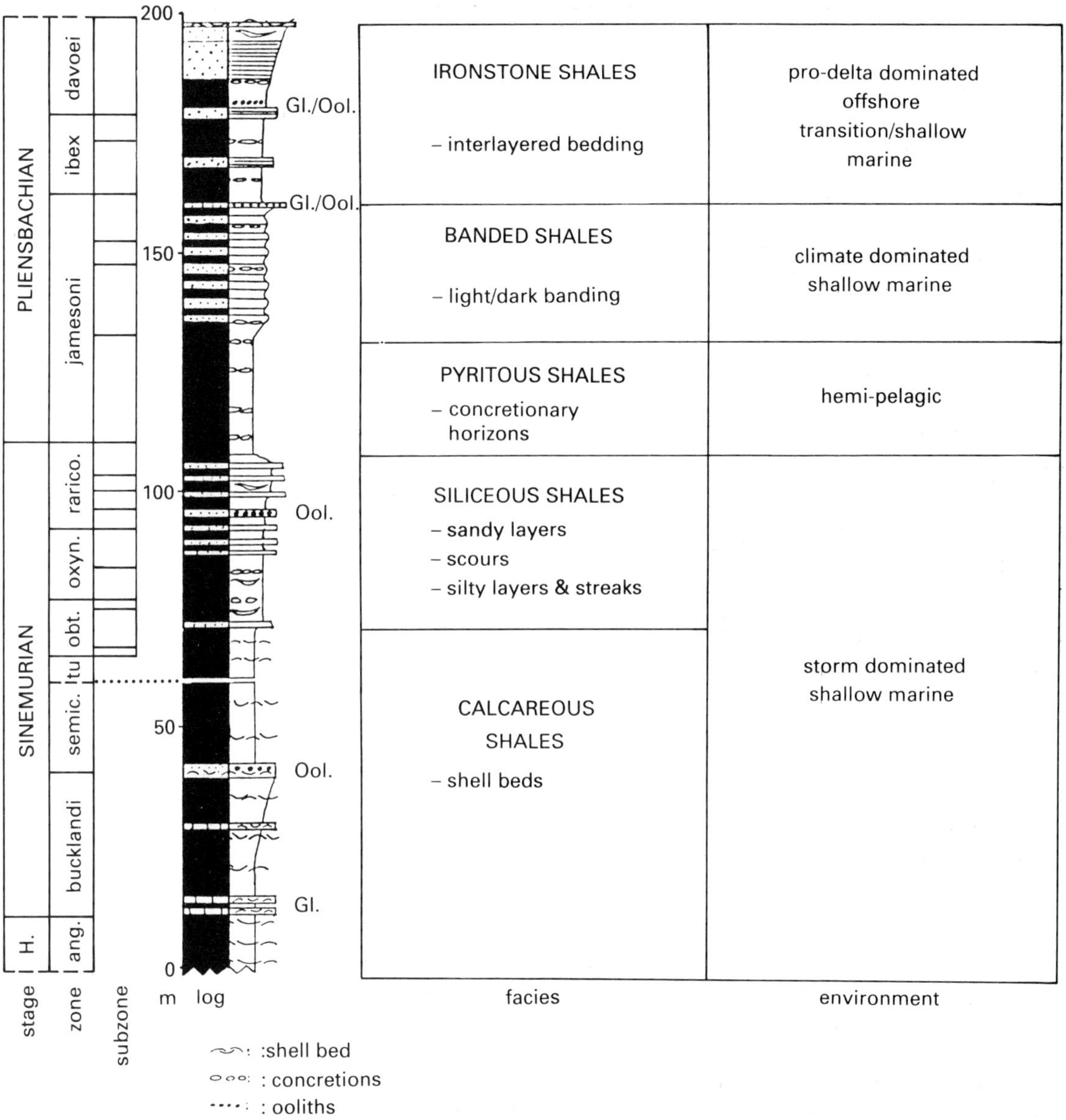

Fig. 2. Summarized lithological log of the Lower Lias Redcar Mudstone Formation in Yorkshire, and interpretation of the facies in terms of depositional environments. The log is composed of the sections exposed in Redcar (0–60 m) and Robin Hood's Bay (60–200 m). Biostratigraphy from Tate & Blake (1876), Bairstow (1969), Sellwood (1970, 1972) and Cope *et al.* (1980). Lithostratigraphy from Buckman (1915), Powell (1984) and van Buchem & McCave (1989). Subzonation follows Cope *et al.* (1980). ang., *angulata*; semic., *semicostatum*; tu., *turneri*; obt., *obtusum*; oxyn., *oxynotum*; rarico., *raricostatum*; Gl., glauconite; Ool., iron-ooliths.

and LECO carbon analyser. The stable organic carbon isotope ratio $\delta^{13}C$ of 20 selected samples has been determined on the extracted kerogen fraction.

Optical examination of the isolated organic matter fraction was carried out on 10 samples following standard palynological methods (HF, HCl, 5–10 min oxidation, 10-μm sieve). Counting was performed on slides studied under the transmitted light microscope; representative areas were selected and at least 200 particles per slide were categorized. Further examination of the organic particles was carried out with a scanning electron microscope (SEM).

PHYSICAL SEDIMENTOLOGY

The banded interval consists of 64 couplets of alternating darker and lighter grey layers (couplet thickness 15–135 cm), usually with planar contacts that are gradational over a few centimetres (Figs 3 and 4). The rock consists mainly of quartz silt, clay, black and white micas and calcite, with minor bioclasts.

The pale colour is caused by a higher concentration of quartz silt (grain size 30–60 μm), and calcite in the form of shell material and cement (backscatter SEM (BSEM) observations). Shell beds typically occur in the light layers, and can be continuous for several metres. They are characterized by a rich suspension feeding macro-fauna (Tate & Blake, 1876; Sellwood, 1970). *Pinna*, *Gryphaea*, thick-shelled pectinids and belemnites can be extremely abundant. *Pinna* normally occurs in a flat-lying position, still articulated, and is found either with other shell material, or at the base of thinly laminated silt/very fine sand quartz layers. In many places, however, it has also been found in life-position. Thin beds of quartzose silt to fine sand, varying in lateral continuity from centimetres to tens of metres, are a common feature in the lighter layers. They show parallel lamination and small-scale cross-lamination, and sometimes have slightly erosional bases. The sediment is generally homogenized by bioturbation. Examination of hand specimens shows the presence of small *Chondrites* traces. X-radiographs of thin slabs demonstrate the overall presence of fine pyritized burrow systems.

The dark layers contain more clay, relatively less silt/very fine sand sized quartz and little shell material. Both light and dark beds have the same range of quartz silt grain size (30–60 μm). The benthic fauna is strongly reduced in numbers and variety in the dark layers, and is characterized by small, thin-shelled, deposit-feeding protobranch and lucinoid bivalves (Tate & Blake, 1876; Sellwood, 1970). Belemnites are much less abundant than in the light layers. Laminated, silty streaks are rare in the dark layers. The sediment is homogenized by a very small type of burrow (mean diameter <0.5 mm).

An important early diagenetic feature of the Banded Shales is the frequent occurrence of siderite concretionary horizons. They are present either as continuous horizons several decimetres in thickness, or as scattered nodules, and occur preferentially at the contact between dark and light layers.

Figure 4 shows a lithological log of the complete Banded Shales unit. Physical sedimentological observations show the gradual transition from a black shale with a sparse fauna (the Pyritous Shales, Fig. 2) into a silty facies (the Banded Shales) displaying an alternation of (i) layers containing a rich fauna, partly *in situ* macrofauna and many wave-induced sedimentary structures (silty streaks), suggestive of favourable living conditions and relatively high turbulent stress at the sea floor, with (ii) layers containing a very sparse fauna, and few wave-induced sedimentary structures suggesting unfavour-

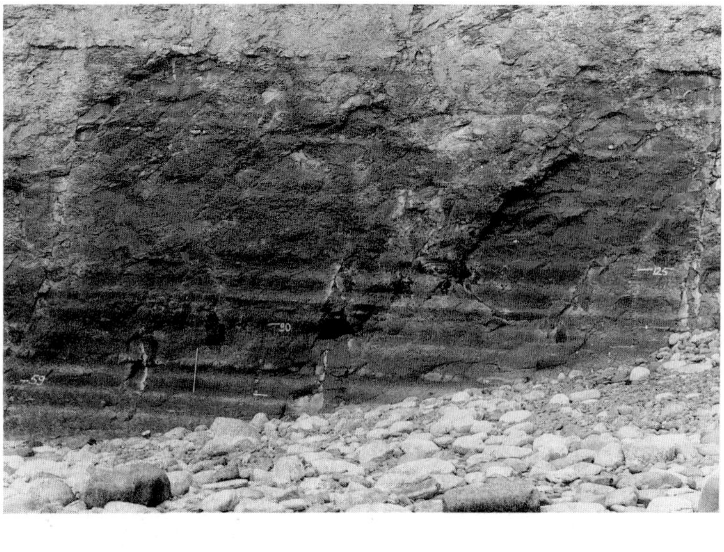

Fig. 3. Lower part of the Banded Shales in the *polymorphus* ammonite subzone (Robin Hood's Bay) showing 20 couplets of darker and lighter grey layers. Measuring rod equals 1 m.

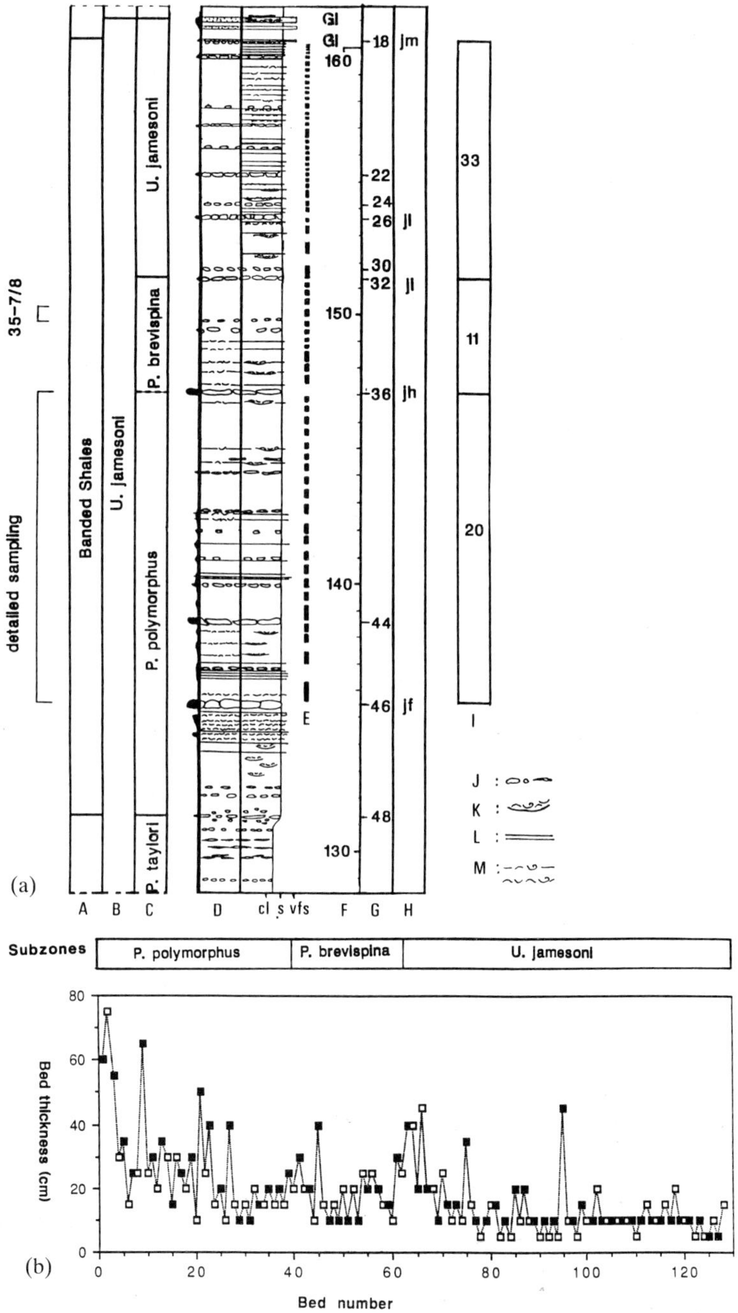

Fig. 4. (a) Detailed lithological log of the Banded Shales. The lower 10 m was sampled at 5-cm intervals. Position of palynological samples shown in Fig. 8 has been indicated (35-7/8). A, lithological unit; B, ammonite zone; C, ammonite subzone; D, lithological log; E, light/dark banding; F, thickness in metres; G, bed numbers of Tate & Blake (1876); H, bed numbers of Sellwood (1970); I, couplets per subzone; J, concretions; K, nests of shell material; L, silty layers and streaks; M, silty shell beds; Gl, glauconite; cl, clay; s, silt; vfs, very fine sand. (b) Bed thickness diagram showing variations in thickness of the light and dark layers up section. A black square represents a dark layer, an open square represents a light layer.

able living conditions and relatively low stress at the sea floor. Superimposed on the alternation of these two 'modes' of deposition, a larger scale mechanism may have been responsible for both the bed thickness variations in the lower part of the sequence and the overall thinning trend (Fig. 4b). The top of the banded interval is marked by two sandy, glauconitic horizons which have been recognized all over the Cleveland Basin (Huntcliff, Felixkirk Borehole, Brown Moor and Howsham Boreholes). This stratigraphic level, marking the top of the *jamesoni* and the base of the *ibex* ammonite zones, has also been found to represent a condensed section in the order of one or two ammonite subzones on a large regional scale (Dorset: Weedon & Jenkyns, 1990; Oxfordshire: Horton & Poole, 1983; Lincolnshire: Sellwood, 1972; Market Weighton: Gaunt *et al.*, 1980; and Felixkirk: Ivimey-Cook & Powell, 1991).

CHEMICAL SEDIMENTOLOGY

In order to quantify the physical sedimentary observations in terms of chemical and mineralogical composition of the sediment, 10.70 m of the cliff section was sampled at 5-cm intervals (214 samples; Fig. 4) and analysed for clay minerals, major and trace elements. The complete data set can be found in van Buchem (1990). Here only those parameters related to grain-size distribution and relative terrestrial sediment influx will be discussed.

IRM, the Si/Al ratio and grain-size distribution

The observations on hand specimen suggesting that light layers contain more silt/sand sized quartz and less clay were confirmed by semi-quantitative XRD analysis (unweighted peak areas) of bulk sediment of a selected number of layers (Fig. 5), and BSEM observations on a number of polished cube samples.

A chemical indicator of grain-size variation is the Si/Al ratio, which reflects the silt/clay content of the sediment (Pye & Krinsley, 1986). Figure 6 shows the Si/Al ratio curve standardized for typical shale concentrations (Krauskopff, 1985). The match with the banding pattern is very good, showing a relative enrichment of Si in the light quartz-rich layers, and a relative depletion of Si in the dark clay-rich layers. Influence of the clay mineral assemblage on the Si/Al ratio can be excluded, as explained in the next section.

Attempts to disaggregate the mudstones and siltstones using ultrasonic and acid treatment gave very unsatisfactory results. Instead a rock magnetic parameter sensitive to magnetic minerals (isothermal remanent magnetization (IRM); Oldfield *et al.*, 1985; Hall *et al.*, 1989) has been used to obtain a more precise insight into the quartz grain-size distribution. This parameter has been measured at 5-cm intervals. It is not sensitive to diagenetic overprint by secondary paramagnetic minerals such as the iron in siderite concretions. The IRM curve shows a very good match with the rhythmic bedding pattern, high values correlating with the light quartz-rich layers (Fig. 6). The fact that the fine-grained magnetic mineral fraction, monitored by the ARM curve, does not follow the bedding pattern (van Buchem, 1990) confirmed that the IRM curve is mainly controlled by variations in the concentration of the coarser, silt sized ferrimagnetic minerals. This coarser grained fraction is not influenced by diagenetic overprint, and is therefore purely detrital and an excellent indicator of changes in the primary depositional environment. The coarse-grained ferrimagnetic mineral fraction is believed to have been brought in as part of the silty quartz-rich fraction, and it thus reflects variations in the influx of its transportation host. Since the relatively coarse-grained fraction of these sediments is mineralogically very mature (Sellwood, 1970), and, as indicated by the high concentration of Zr (average 300 ppm; van Buchem, 1990), contains a relatively large heavy mineral fraction, it is suggested that the heavy mineral assemblage is the most likely carrier of the

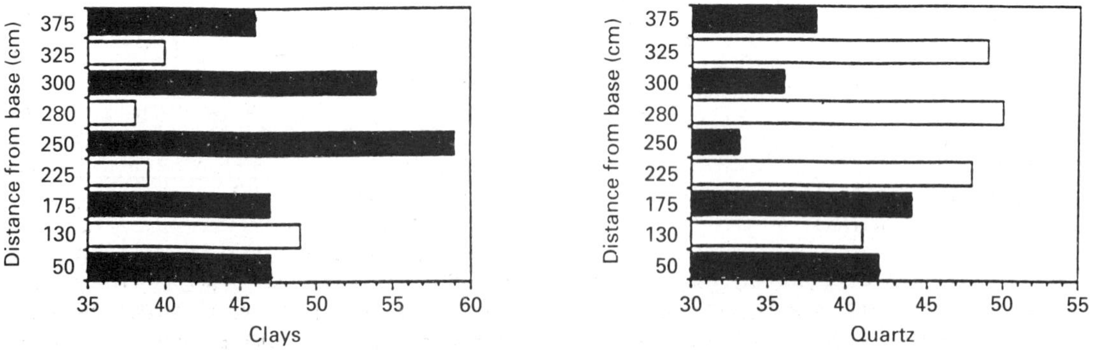

Fig. 5. Mineralogical composition of the first nine dark and light layers of the Banded Shales, given as a percentage of the bulk sample. Black represents a dark layer, white represents a light layer. Based on semi-quantitative XRD analysis, unweighted peak heights.

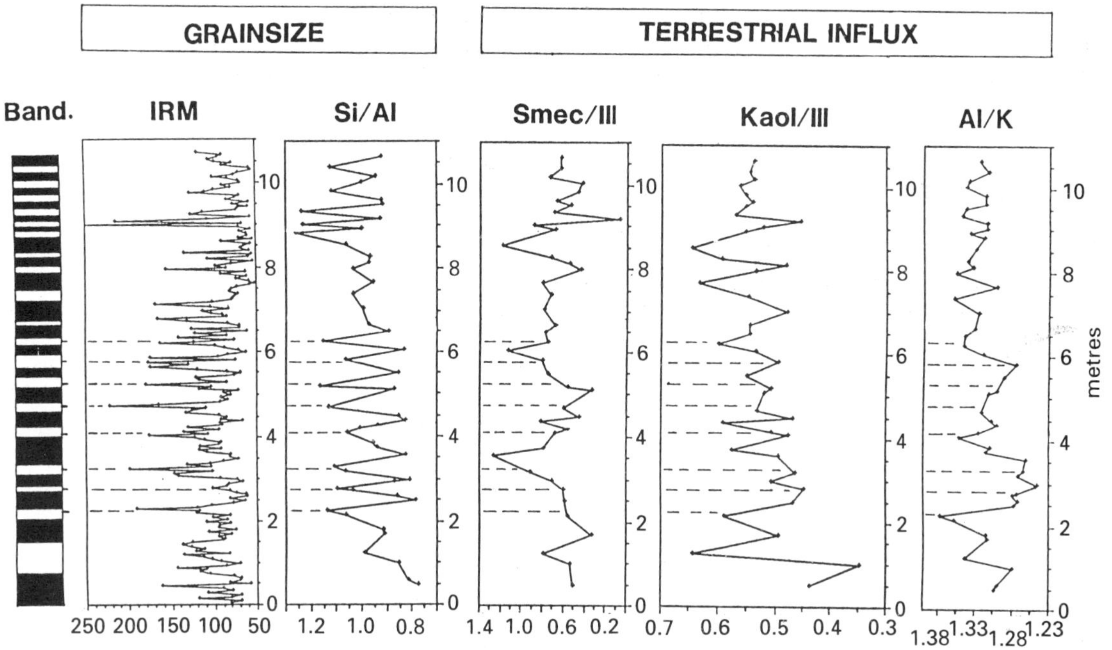

Fig. 6. Variations in grain size and terrestrial influx parameters in the lower 10 m of the Banded Shales unit. Horizontal broken lines represent stratigraphic position of light layers. IRM, isothermal remanent magnetization; Si/Al and Al/K ratios are standardized for average shales (Krauskopf, 1985); for smectite/illite and kaolinite/illite ratios real values have been given. Banding: alternation of dark and light layers as observed in outcrop.

coarse-grained magnetic minerals (i.e. in the form of magnetite or titanomagnetite inclusions). The spiky nature of the IRM signature is a result of the fact that the coarse-grained quartz fraction, containing the magnetic minerals, is often preserved as thin sandy/silty streaks or layers.

Observations on hand specimen, and mineralogical and chemical analyses all show a distinct systematic variation in the relative amount of silt/sand sized quartz contained in the dark and light layers.

Clay mineralogy, the Al/K ratio and the influx of terrestrial material

As an indication of variations in the type of terrestrial material supplied to the basin, the composition of the clay mineral assemblage has been studied. The kaolinite and chlorite fractions show relatively little variation (16–24% and 8.5–16% respectively), whereas the illite and smectite percentages vary considerably (15–42% and 32–56% respectively). From the smectite/illite and kaolinite/illite ratio curves (Fig. 6) it is obvious that the variations in the clay mineral distribution do not show a correlation with the dark/light banding pattern on a bed scale. However, a larger scale variation seems to be present in the smectite/illite ratio showing high values at 3.5, 6 and 8 m (Fig. 6). The mineralogical results are supported by the data obtained from the geochemistry. The Al/K ratio, which is an indication of the relative concentrations of clay minerals, and in particular of K-rich illite and Al-rich kaolinite (Pye & Krinsley, 1986), shows no correlation with the bedding (Fig. 6). It is clear that there is no direct relationship between the clay mineral distribution and the small-scale cyclicity on a layer scale. For the interpretation of the long-term variation present in the smectite/illite ratio, three main factors determining distribution and concentration of clay minerals have to be considered: (i) composition of the sediment load; (ii) segregation by differential flocculation or transportation; and (iii) diagenetic neoformation or alteration.

BSEM observations showed some growth of kaolinite in micas, but no other evidence for important neoformation of clay minerals, ruling out the first factor. Variations in the composition of the clay mineral assemblage may potentially reflect changing climatic conditions. Clay minerals basically express the intensity of weathering, and especially of hydrolysis, in the landmass adjacent to the basin. In the ideal situation, warm and humid climates lead to high hydrolysis and the subsequent pedogenic formation of kaolinite and smectite, and the degradation of illite. Cold and drier climates have a low hydrolysis, and are characterized by the presence of (pre-existing) well crystallized illite, chlorite and smectite (Chamley, 1989). Alternations of kaolinite/illite-rich and smectite-rich layers have been documented in several Cretaceous sequences (e.g. Deconinck *et al.*, 1983; Tribovillard & Cotillon, 1989). The clay mineralogy of these Cretaceous cycles generally displays a systematic increase of smectite abundance in carbonate-rich beds, and an increase of kaolinite and often illite in clay-enriched interbeds. This type of variation in mineralogy is generally attributed to more humid conditions during the deposition of the marly (kaolinite/illite rich) beds. Increased continental runoff produces more hydrolysis, preferential development of pedogenic kaolinite and the increased erosion of both soil- and rock-derived minerals (i.e. upstream supplied kaolinite and illite). By contrast, smectite-enriched calcareous beds are usually considered to reflect drier climate on land, causing a lesser terrigenous influx, and the preferential removal of downstream soil products (Chamley, 1989, p. 454). Following from this, the smectite/illite curve in the Banded Shales may reflect a larger scale variation of the climatic conditions.

It has been observed that a general preference exists for kaolinite to accumulate in continental sediments, illite in transitional to marine, and smectite in deeper marine deposits as a result of differential settling related to grain size and flocculation (review in Chamley, 1989). However, a strong argument against transportation processes being the main factor controlling the clay mineral distribution is the fact that the small-scale cyclicity, which is expressed as grain-size variations, is not reflected in the clay mineral distribution. This leaves climatic factors, determining amount and composition of the sediment supplied from the land, as the main mechanism to explain the long-term variation in the clay mineral distribution.

ORGANIC MATTER

The amount and composition of OM preserved in sediments is determined by three main factors: (i) production of the biomass; (ii) transport process; and (iii) degradation (Huc, 1988). In this section the importance of each of these three factors is investigated.

Organic geochemistry

The TOC content of the interval studied is low (average 0.8%). However, the variation of TOC is consistent with the colour density (values of 0.5–0.8% for the light bands, and 0.7–1.4% for the dark bands). Figure 7 shows a profile of the TOC content next to a lithological log and coloration intensity profile of the layers as observed in the field. In the first 7 m there is a good correlation between percentage TOC and the coloration pattern. In the top 3 m the concentration and variation of OM decreases and correlation is increasingly difficult. Determination of TOC by Rock-Eval pyrolysis in rock samples containing low percentages of OM may be influenced by mineral matrix effects (Espitalié *et al.*, 1977). For this reason a separate analysis of organic carbon was carried out on a LECO apparatus after digestion of the carbonate fraction in HCl. Figure 8A shows the excellent correlation between the values obtained using the two analytical methods, giving extra credibility to the consistency of the observed variations in TOC.

The hydrogen index provides information about the properties of the bulk OM fraction independent of the concentration (it is defined as the quantity of hydrocarbons (HC) released by Rock-Eval pyrolysis and standardized to organic carbon content). The average value of the HI is 65 mgHC/gTOC. This is normally found for type III OM (Tissot & Welte, 1984). The trend in the HI profile follows that of the TOC, showing a good correlation with the banding pattern in the lower part and reduced amplitude variations in the upper part (Fig. 7). The plot of TOC versus HI values shows a weak positive correlation (Fig. 8B). The difference of the quality of the OM in the dark and light layers is not large enough to produce a separation of the two sample groups in a van Krevelen-type diagram (Fig. 9). *T*max, as determined by Rock-Eval pyrolysis, varies between 430 and 440°C, indicating that organic maturity is relatively low (Espitalié *et al.*, 1985/86).

Variation of the quantity and quality of the OM is associated with a small but consistent variation of the isotopic signature of the isolated kerogen fraction (Fig. 7). The stable organic carbon isotope values ($\delta^{13}C$) vary from −25.91 to −26.68‰, PDB (maxi-

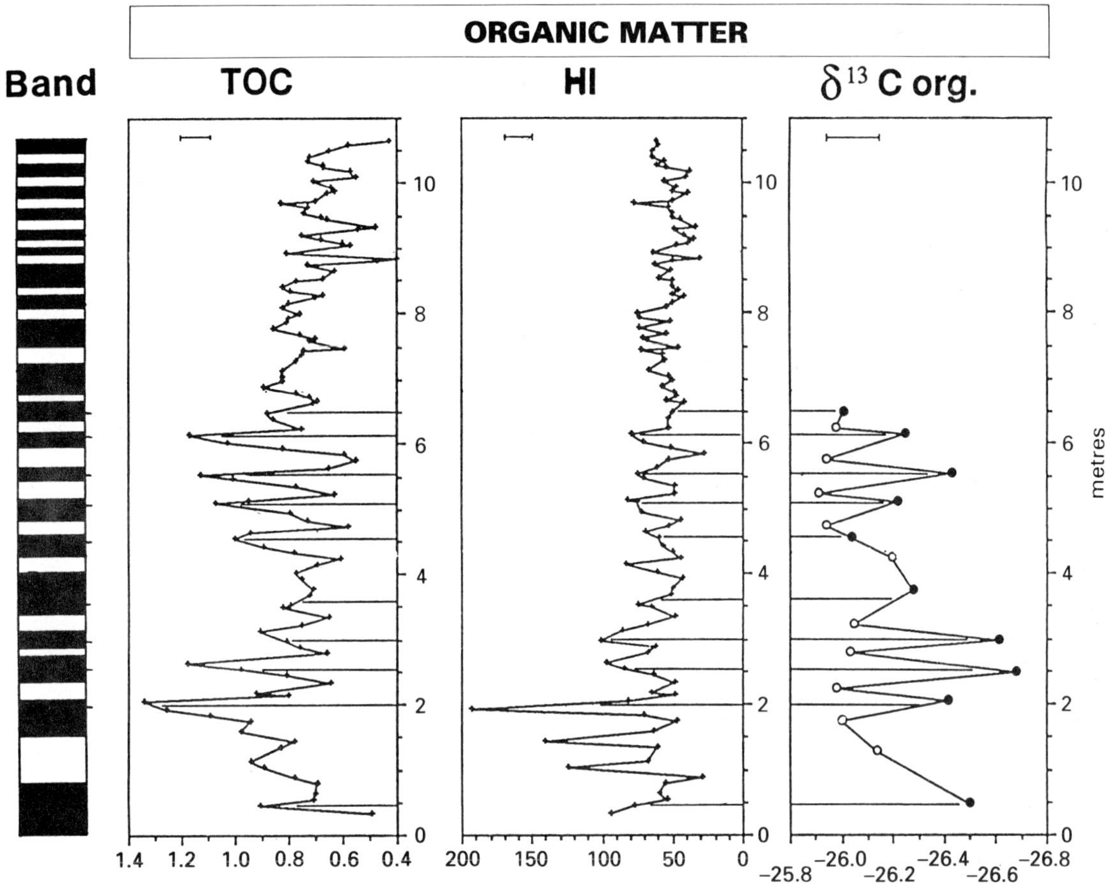

Fig. 7. Variations in quantity and quality of the OM fraction for the lower 10 m of the Banded Shales unit. Horizontal lines represent stratigraphic position of dark layers. Standard error bars have been indicated. TOC, total organic carbon in %; HI, hydrogen index in mg hydrocarbons/gTOC; $\delta^{13}C$ org., organic carbon stable isotopes in ‰; Band., alternation of dark and light layers as observed in outcrop.

mum variation 0.77; standard error of carbon isotope values is ±0.1‰). The more negative values tend to be associated with higher TOC and HI values occurring in the dark layers (Figs 7, 8C, D).

Optical examination

The aim of the optical examination was to give a general characterization of the OM composition with regard to the allochthonous (terrigenous) versus autochthonous (marine) origin, and to investigate whether this composition changes over the sequence. A relatively quick qualitative, semi-quantitative approach has been followed. Four main categories of OM particles have been distinguished. Data tables can be found in van Buchem (1990). The two most important fractions are the opaque particles (60–80%) and the structured transparent particles (6–25%). Transparent woody debris and amorphous OM vary between 1–7% and 1–9% respectively. Intact palynomorphs contribute relatively little to the total OM fraction.

The four main categories are as follows.

1 *Opaque particles.* These fragments are black in transmitted light and vary in size from 5 to 1000 μm; their shape can be rounded, blocky or splinter-like. SEM examination reveals that the majority of these particles show anatomical details of the kind found in carbonized wood fragments of terrestrial plants, like open cellular structures and tracheids (Cope, 1981). Comparison with the charred woody material described from a near-shore Early Jurassic flora in

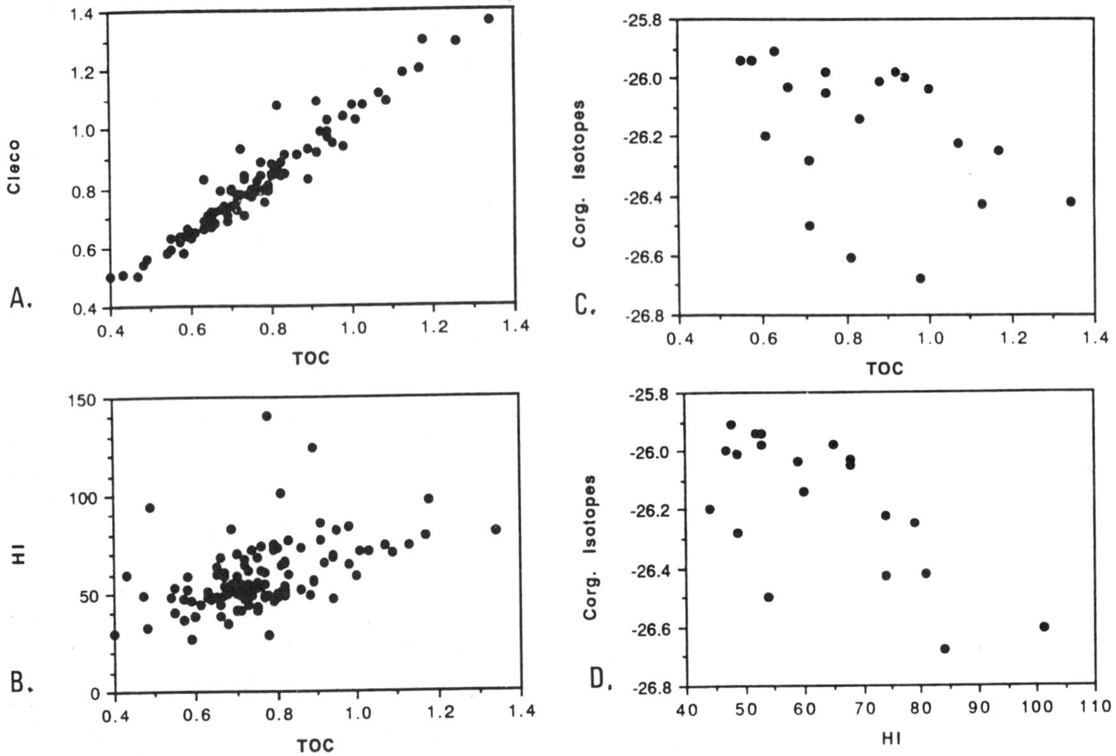

Fig. 8. Organic matter cross plots: (A) TOC-Cleco; (B) TOC-HI; (C) TOC-organic carbon isotopes; (D) HI-organic carbon isotopes. TOC, total organic carbon values determined with a Rock-Eval; Cleco, total organic carbon values determined with a LECO analyser; HI, hydrogen index.

Wales (Harris, 1957) suggests that a (substantial) part of the opaque particles may originate from conifer wood. Almost all the material Harris found was charred, and 79% of it consisted of the carbonized wood and bark of the conifer *Cheirolepis muensteri*.

2 *Structured, transparent particles*. All transparent particles which show structure are assigned to this group. Their size varies from 10 to 450 μm. Dark brown particles of woody origin form a distinct component. The group also includes clearly recognizable particles like leaf cuticles and degraded stem bundles, and a substantial amount of smaller particles whose affinity was not determined (e.g. fragments of palynomorphs).

3 *Palynomorphs*. This group includes both marine and terrestrial palynomorphs. Spores and bisaccate pollen are relatively common. Acritarchs do occur, but are rare. Dinoflagellates and algae have not been observed. Intact marine and terrestrial palynomorphs contribute only a few percent to the total OM.

4 *Amorphous OM*. All translucent masses of OM which do not show cellular detail are included in this group. Size varies and can be up to 100 μm. The affinity is difficult to determine. It is not clear to what extent this fraction is of primary origin or an alteration product. The most common precursor of amorphous material is thought to be marine algal material, although bacterial and thermal degradation can produce amorphous organic matter from many of the other types of palynodebris (Rogers, 1979; Boulter & Riddick, 1986). The contribution of amorphous OM varies from 2 to 9% of the OM fraction. It should, however, be realized that these numbers may be biased by the preparation method, which includes (mild) oxidation and sieving (10 μm).

The general state of preservation of the OM is rather poor. The majority of the particles are in a fully oxidized, carbonized state (opaque particles,

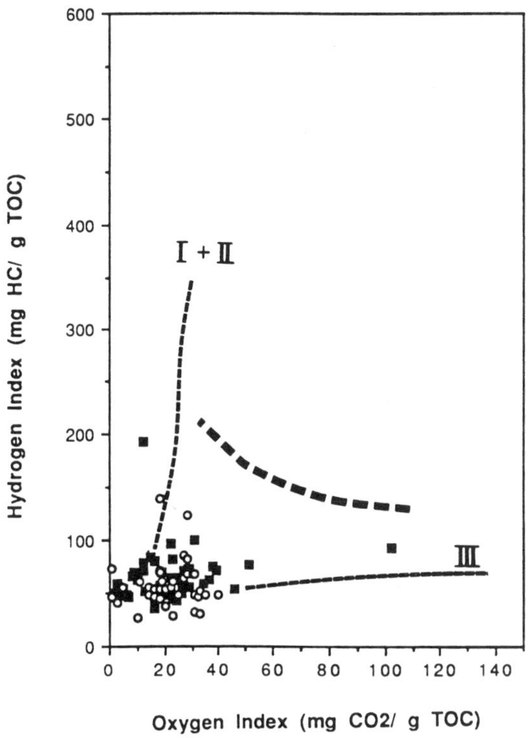

Fig. 9. Van Krevelen diagram showing OM types in the dark layers (black squares) and in the light layers (open circles). I, II, III are standard OM types (Tissot & Welte, 1984).

most probably as a result of woodfires). The relatively low percentage of amorphous OM could be the result of decomposition under generally oxic conditions in the depositional environment. Many of the palynomorphs were affected by pyritization as shown by the impressions of small cubic crystals, formed during early diagenesis in reducing conditions.

The studies of Harris (1957) and Cope (1981) give ample evidence for a charcoal origin of the opaque particles. Together with the dark brown woody particles they serve as an indicator of relative input of terrestrial OM. The palynomorphs, structured/transparent and amorphous OM fractions are of mixed marine and terrestrial origin, which are much more sensitive to oxic conditions and more difficult to transport (opaque particles are considered to be more buoyant; van der Zwan, 1990). The ratio of the woody, carbonized, terrestrial fraction to the mixed, transparent/translucent debris fraction is used here as an indicator of relative composition of the OM. Out of ten light and dark layers, seven layers show a consistent pattern. The dark layers contain more opaque/woody particles and less mixed transparent/translucent debris (terrestrial/mixed ratio of around 5.5). Conversely the light layers contain relatively fewer opaque/woody particles and more mixed transparent/translucent debris (terrestrial mixed ratio of around 2.6). Figure 10 shows characteristic palynofacies assemblages of light and dark layers.

Discussion

The variation in quantity (TOC) and quality (HI, stable organic carbon isotopes and palynofacies) of the OM fraction for the lower 7 m of the sequence studied is in phase with the alternation of light and dark layers. The differences are small, but consistent over the sequence.

Variation in the amount of preserved OM can result from changes in the type supplied (marine and/or terrestrial), nature of preservation, dilution by clastics and carbonate, or a combination of these. The generally poor preservation of OM and low TOC values are a normal feature of well-oxygenated sediments (Demaison & Moore, 1980; Pratt, 1984), and agrees with the interpretation that the Banded Shales facies as a whole represents an aerobic to dysaerobic environment. However, the relatively high HI values in the dark layers, and their impoverished fauna and scarcity of sedimentary structures, suggest deposition under dysaerobic conditions with little turbulence at the sea floor. The light layers contain a rich macrofauna and thin, laminated streaks of silt and very fine sand with low HI values, and are interpreted as being deposited in a fully oxygenated environment (aerobic) as a result of strong mixing of the water column.

A possible explanation for the variation in TOC and HI could thus be that the combination of more oxygen and high biological activity (burrowing, grazing, etc.) caused a stronger decomposition of OM, leaving less for burial in the sediment. A dilutional control of TOC values seems less likely. The HI variations indicate variable oxidation of the OM, being greatest under conditions favouring coarser sediment size. As a third possibility a change in the flux of OM has to be considered, i.e. less material supplied in light layers, more in dark layers. It is necessary to distinguish between the allochthonous fraction, consisting of plant debris brought in from the land, and the autochthonous fraction, consisting

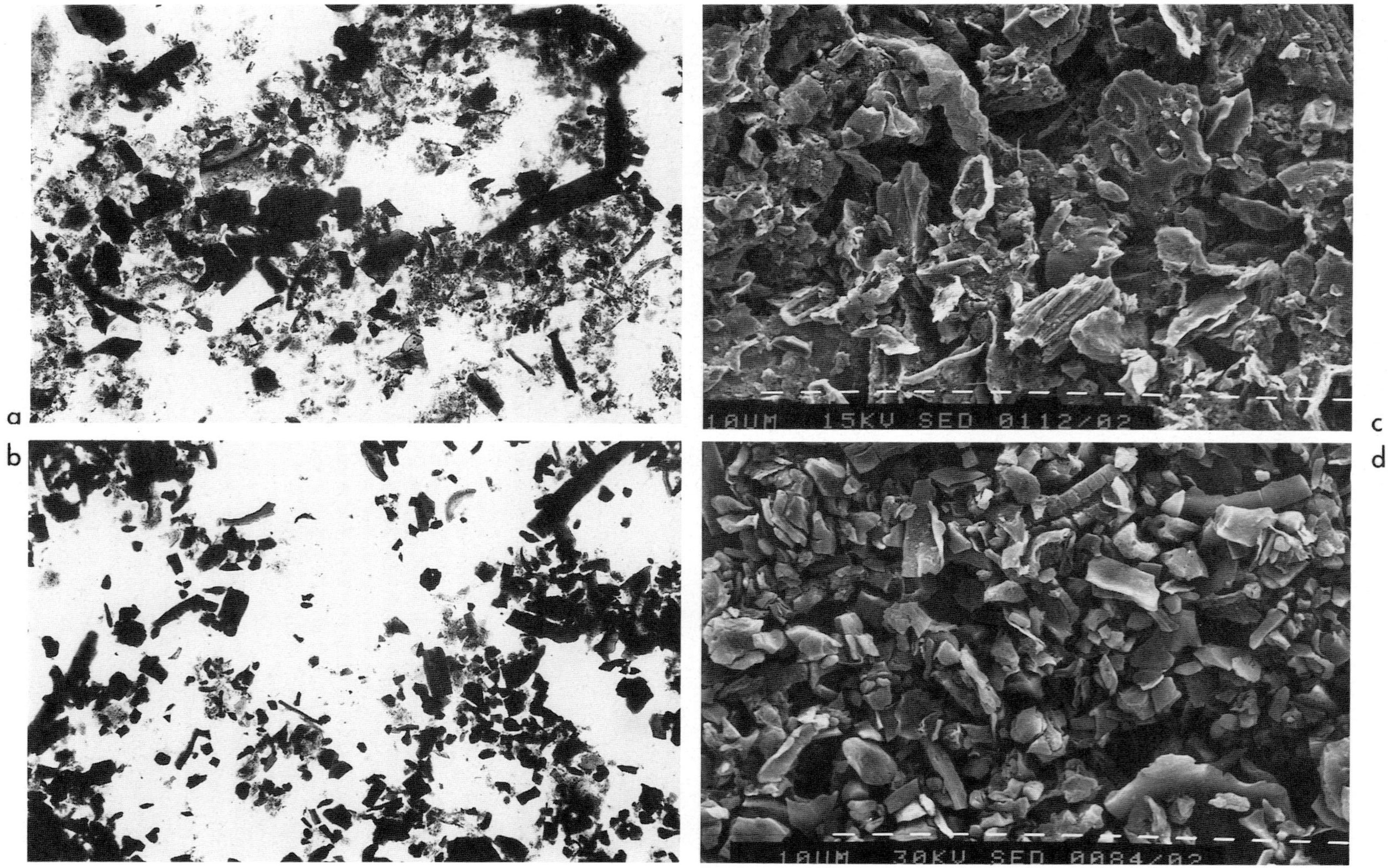

Fig. 10. Isolated OM fraction (after HF treatment) of dark and light layers. The palynofacies of the light layers is characterized by opaque, structured transparent and amorphous particles in the light microscopic picture (a), and by a large variety in size and shape of the particles in the SEM picture (c) (sample 35-7). The palynofacies of the dark layers is characterized by a predominance of opaque, blocky particles in the light microscopic picture (b), and by uniformity in size and shape of the particles in the SEM picture (d) (sample 35-8). White bar in SEM pictures is 10 μm; same bar represents approximately 40 μm in light microscopic picture.

of primary marine organic matter. Organic geochemical characterization (Fig. 9) classified the OM into the type III kerogen group, which is normally considered to be composed of terrestrial plant debris and strongly oxidized type II kerogen (Tissot & Welte, 1984). Optical examination confirmed the dominant presence of woody material from terrestrial plants (75–85%). The OM fractions, which are structured and unstructured/transparent/translucent, are of mixed terrestrial and marine origin. The maximum contribution of marine organic material to this fraction is estimated to be 10–15%, which is based on the assumption that most amorphous OM and part of the unidentifiable tissue is of marine origin.

Extra information about the composition of the OM is provided by the stable organic carbon isotope data. Values vary between −25.91 and −26.68‰. As 75–85% of the OM consists of terrestrially derived woody material, one may assume that the isotopic composition of the kerogen is mainly determined by vascular land plants employing the C3 cycle. They have typical organic carbon values around −26/−27‰, PDB (Deines, 1980; Popp *et al.*, 1989). However, in a series of recent studies it has been shown that marine OM can occur in an isotopically light (−35 to −26‰) and an isotopically heavy form (−24 to −20‰) (Küspert, 1982; Dean *et al.*, 1986; Lewan, 1986; Wenger *et al.*, 1988; Popp *et al.*, 1989). Light forms are mainly recorded from Palaeozoic and Mesozoic sediments, whereas heavy forms have been found in Cenozoic and modern sediments. Layers with enhanced OM preservation and relative hydrogen enrichment are isotopically lighter, contrary to the trend seen in modern sediments. The earlier studies of Mesozoic $\delta^{13}C$ therefore suggest that the dark layers in the Banded Shales contain a slightly higher amount of marine OM than the light layers. If this is true, two alternatives could explain the variation: better preservation, or higher primary productivity. Since the dark layers were formed under dysaerobic conditions, with reduced mixing of the water column, selective preservation seems the most likely cause. The light layers show indications of a very well oxygenated setting, which will certainly have led to enhanced OM decomposition, particularly of the more labile fraction (Pratt, 1984; Savrda *et al.*, 1984; Tyson, 1987).

The semi-quantitative palynofacies analysis showed a relatively high percentage of tissue material in the light layers. In view of the organic geochemical analysis this would imply that the 'tissue' material in the dark layers (including amorphous OM) mainly represents marine OM, while in the light layers a large amount of the 'tissue' fraction would originate from land plants (like cuticles). This would also fit the hypothesis that the light layers were formed during periods with a higher frequency and perhaps greater maximum magnitude of storms, during which larger, low buoyancy plant tissue would have been spread over the basin. Additionally, since variation in storm frequency may have been related to climatic changes, a response of the coastal vegetation to the changing climatic conditions is a possibility.

Control of the TOC by continental runoff has to be considered, since most of the OM is of land plant origin. A widely used indication of terrestrial runoff is the kaolinite content (Chamley, 1989). From the clay mineral distribution (Fig. 6) it is clear that there is no correlation with the variations in TOC on a layer scale, which excludes runoff as a cause for TOC variation. However, a longer term variation is apparent in both curves, as considered below.

In summary, the OM fraction is of dominantly terrestrial vascular C3 plant origin, possibly largely of the conifer *Cheirolepsis* which dominated the homogeneous coastal floras. The contribution of marine micro-organisms to the OM composition is estimated to be smaller than 10–15% of the total OM.

Variations in TOC, HI and stable organic carbon isotopes are most likely caused by selective preservation during alternating fully aerobic and dysaerobic conditions at the sea floor, with aerobic conditions being characterized by lower TOC, lower HI and slightly less marine OM (heavier $\delta^{13}C$).

Variations in the palynofacies seem to reflect changes in the distribution of the higher plant debris over the basin as a result of variations in the storm frequency. It is suggested that also the composition and/or abundance of the coastal floras was affected by the same climatic changes.

STRATIGRAPHY

Biostratigraphy

The biostratigraphy of the Yorkshire Lias is well known (Tate & Blake, 1876; Cope *et al.*, 1980). The highest stratigraphical resolution is realized by the ammonite zonation scheme, of which the latest

Table 1. Ammonite subzones and distribution of dark/light couplets

Subzone	Time (ka)	Couplets	Thickness (m)	Average time (ka)	Average thickness (cm)
jamesoni	433	33	8.85	13.1	27.0
brevispina	433	11	4.50	39.4	41.0
polymorphus (72%)	312	20	10.65	15.6	53.0
Total	1178	64	24.00	18.4	37.5

version is given by Cope *et al.* (1980). The ammonite subzone has been chosen as the basic time unit in this study. Cope *et al.* (1980) list 144 subzones for the Jurassic, and the time scale of Harland *et al.* (1990) gives a duration of 62.4 Ma for the whole of the Jurassic. Thus the *average* duration for a (British) Jurassic ammonite subzone is 433 ka.

The Banded Shales cover almost three subzones of the *jamesoni* ammonite zone. Table 1 shows the distribution of the light/dark couplets over the subzones and gives average values for their duration and thickness. Only 72% of the *polymorphus* subzone shows the banded pattern (Fig. 4), hence the shorter time estimate. The average durations of the couplets that are listed are based on the assumptions that the boundaries of the subzones are at the correct position, and that subzones represent equal periods of time.

The continuity of the sequence can be questioned if there is evidence in the lithology for hiatuses or condensation. Indications of episodically reduced accumulation rates are provided by the glauconitic horizons at the top of the Banded Shales above which the fine banding pattern disappears. Within the banded sequence there are many sideritic concretionary horizons of varying thickness, the formation of which may involve a short period of reduced sedimentation rate (Sellwood, 1971; Bayer *et al.*, 1985; Raiswell, 1987). Although the variation in bed thickness of the dark and light layers has been interpreted in terms of changing degree of winnowing, there is no evidence for major erosional gaps. The Banded Shales terminate with a condensation event, but within the sequence there is only evidence for minor changes in accumulation rate. We believe we are justified in working on the presumption that the sequence is complete at the 10-ka level, and substantially so at 1 ka (Sadler, 1981). This implies that the difference in numbers of couplets found in the subzones cannot be explained in terms of physical sedimentological processes (erosion, condensation). We are then left with two hypotheses, assuming that the location of the ammonite subzones is correct: (i) subzones are of equal duration, which would imply that the observed cycles are basically of two types – one with an average duration of 13.1 to 15.6 ka and a second with a duration of 39.4 ka; (ii) the cycles are all of the same duration, implying that the time represented by the different ammonite subzones varies considerably. The latter possibility will be further investigated in the next section.

Milankovitch cyclicity

Over the last two decades the relationship has been established between regular sedimentary cycles observed in the rock record and the Earth's orbital variations as hypothesized by Milankovitch (1941; Berger, 1978, 1988). Of key importance was the recognition of the sensitivity of climate and ocean behaviour to external forcing (Emiliani, 1955; Hays *et al.*, 1976; Fischer, 1981). Comprehensive reviews of Milankovitch cyclicity found in the pre-Quaternary record are given by Einsele (1982), Fischer (1986), Berger (1989), Fischer *et al.* (1990), and elsewhere in this volume. In a series of recent studies of the British Mesozoic, strong evidence has been found for the presence of orbitally induced cycles (House, 1985; Weedon, 1985; Gale, 1989; Leary *et al.*, 1989; Smith, 1989; Weedon & Jenkyns, 1990). For instance it was argued that cyclicity in the Blue Lias (Hettangian–Sinemurian) was orbitally induced, based on spectral analysis of time series of digitized measured section data (Weedon, 1985). Two frequencies were detected, possibly related to the precession and obliquity cycles. In the Belemnite Marls analysis of similar time series showed the presence of a short wavelength cycle, probably related to the precession or the obliquity cycle, and a longer wavelength oscillation which has been suggested to represent longer-term, irregular climatic changes rather than the eccentricity cycle (Weedon & Jenkyns, 1990). Since the Yorkshire Banded Shales span the same ammonite zone as the Belemnite Marls, the possibility of a similar cause for the observed cyclicity needs to be investigated.

Milankovitch cycles are based on three orbital parameters, which when superimposed produce various patterns that can be expressed and recognized in the rock record (Berger, 1978; Fischer *et al.*, 1985). The exact climatic and sedimentological mechanisms through which the orbital parameters find expression in the rock record remains a matter of study, but generally changes in distribution of insolation over the Earth's surface leading to climatic changes in the order of tens of thousands of years is believed to be a key part of the process. A characteristic feature of these rock sequences is that time series extracted from them are dominated by a few regular oscillations/cycles (Fischer *et al.*, 1990). To test whether the cyclicity observed in the Banded Shales is the indirect product of several, superimposed, orbital–climatic cycles, spectral analysis has been carried out on three time series (see, for example, Weedon, 1991). Since the cyclicity is mainly expressed in the clay/silt ratio, a grain-size dependent parameter such as IRM (see above), which is easy and quick to determine, is ideal to produce a high-resolution time series. The Si/Al data are more widely spaced, and have been added to support the credibility of IRM as a grain-size dependent parameter. TOC is another parameter analysed, believed to be largely independent of grain size. Before spectral analysis can be carried out, the TOC and Si/Al intermediate values need to be interpolated to produce a time series with data points at constant height intervals.

Because the 5-cm sample spacing was unable to record the very fine-scale grain-size variations in the top 3.70 m of the sampled interval, only data from 3.70 to 10.65 m has been used for spectral analysis in order to avoid aliasing (see Pisias & Mix, 1988). Figure 11 shows the three time-series curves with sample positions and the power spectra for this interval. IRM produces two spectral peaks, which indicate the presence of two regular cycles with wavelengths of 50 and 83 cm. The TOC data have peaks at the same frequencies, but 50-cm variations are dominant here, and there is a third spectral peak denoting a regular oscillation with a wavelength of 300 cm. Coherency between IRM and TOC is high at the frequency of the 50-cm cycle, and is just significant at the 80% level at the frequency of the 83-cm cycle. Thus there is a strong relationship between amplitude variations in IRM (grain size) and amplitude variations in TOC. The phase spectrum indicates a phase difference of about 140°, but this is statistically indistinguishable from 180°. Thus variations in IRM and TOC are interpreted to be inversely related (high IRM–low TOC; Fig. 11). The dominant peak of the Si/Al ratio at 50 cm shows good coherence with IRM, and is in phase. The 83-cm peak is small, but does show coherency. Thus using two independent parameters the presence of two wavelengths defining the regular cyclicity has been detected. Support for the consistency of the cyclicity in the Banded Shales within the Cleveland Basin comes from power spectra obtained from a selected interval of a gamma-ray log of the Felixkirk borehole, located 40 km west of Robin Hood's Bay, in which cyclicity was shown at 58, 96 and 270 cm (van Buchem *et al.*, 1992).

To illustrate the regular cycles detected by spectral analysis, Fig. 12 shows the result of three smoothing operations on the IRM curve. The smoothing technique employed is a weighted three point moving average (weights = 0.25, 0.5, 0.25), which removes most of the noise of the measurements. The first curve is the result of three smoothing operations, and the second one has been smoothed 110 times which reveals the background long-wavelength variations. The third curve indicates departure from the mean values of the curve (realized by subtracting the second curve from the first one) and reflects the basic variation in silt/clay content, and as such represents the dark/light layering. The match with the field observations is very good for the lower 7 m, as is clear from the match with the banding pattern, and it is this interval that has been used for the spectral analysis. In the top 3 m variation in the thrice-smoothed IRM curve is much smaller (except for the peak at −1.60 m), and the thin layering observed in the field is not clearly visible. This may be the result of a more intense homogenization by bioturbation. Since IRM is a grain-size sensitive parameter, it can be assumed that the spiky nature of the measurements in the lower part of the section (especially apparent in the light layers) is a result of the fact that there the silt fraction is often preserved as thin streaks.

The translation of the bed thicknesses into the time involved in their formation depends on the accuracy of the time scale (biozonation in this case), and the completeness of the section. Since the ammonite subzonation for this interval is uncertain

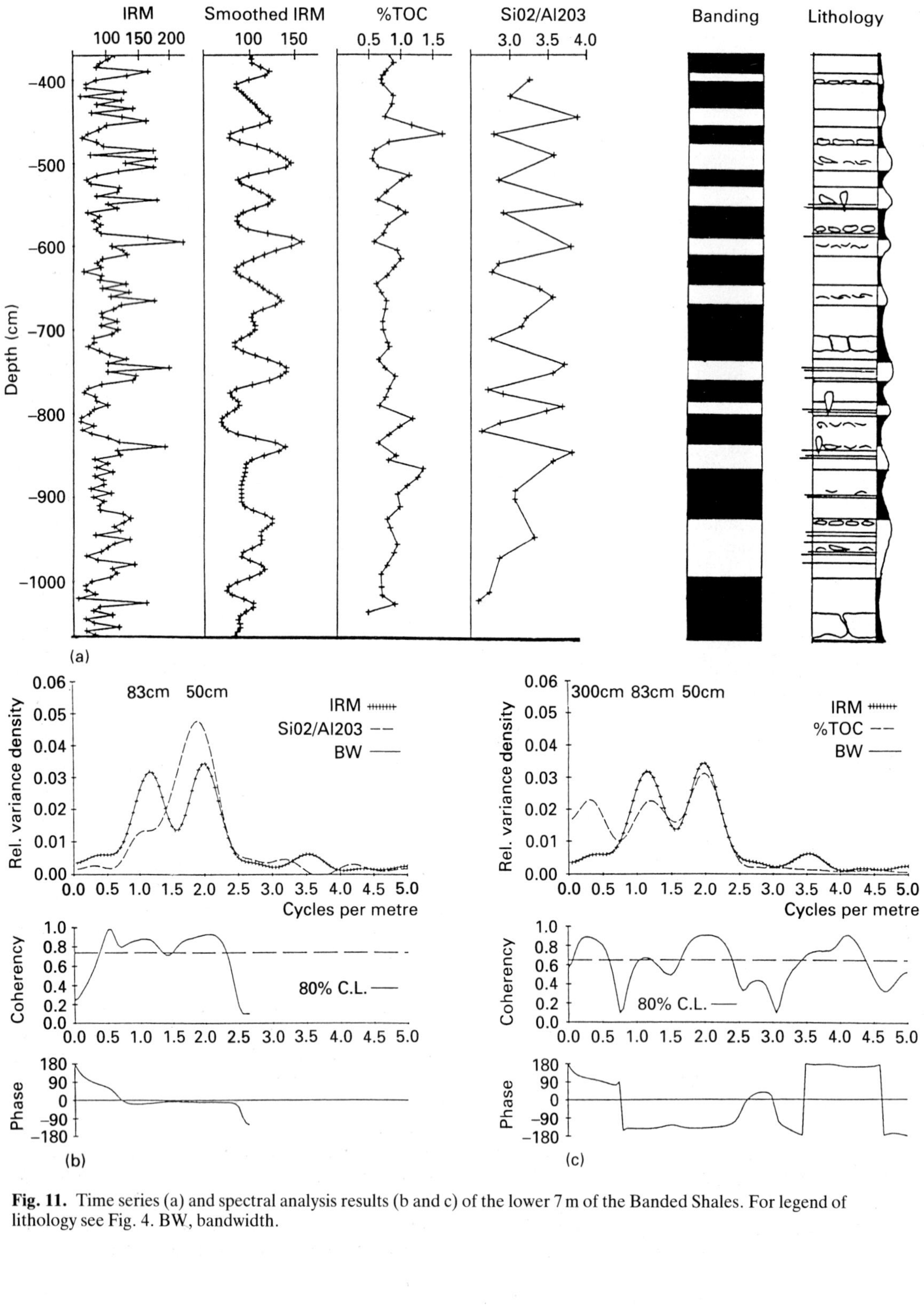

Fig. 11. Time series (a) and spectral analysis results (b and c) of the lower 7 m of the Banded Shales. For legend of lithology see Fig. 4. BW, bandwidth.

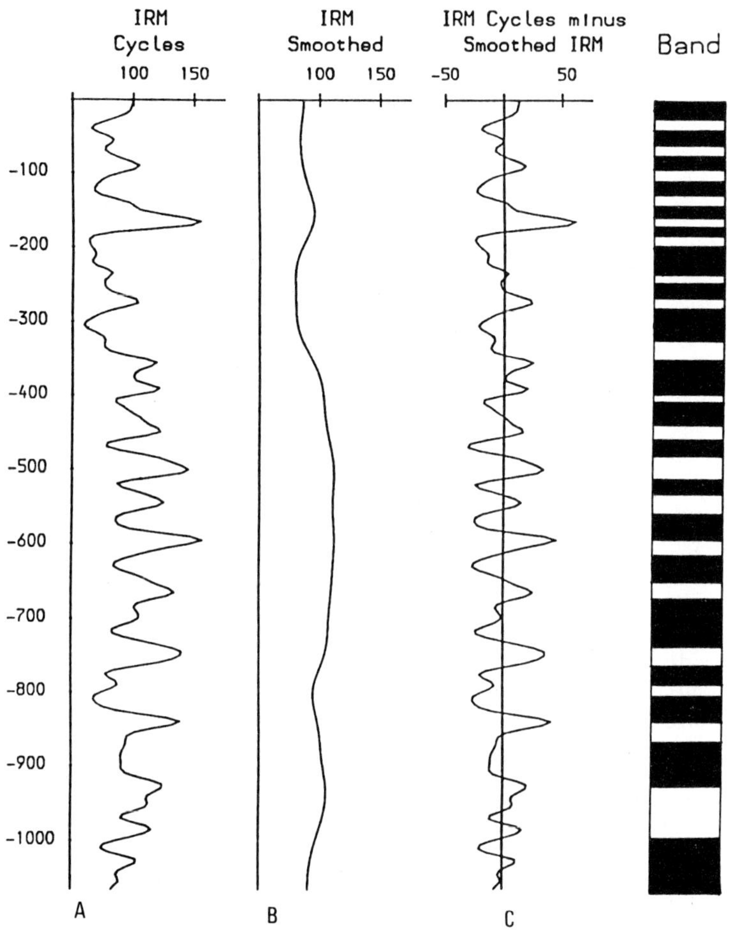

Fig. 12. Smoothed IRM data: (A) after three smoothings; (B) after 110 smoothings; (C) curve obtained by subtracting curve B from curve A, giving the basic variations without the long-term background variation.

(Sellwood, 1978; Wilson & Manning, 1978; Gaunt *et al.*, 1980) wavelength ratios have been used as a method independent of dating, to decide which orbital parameters are present. As has been pointed out by Weedon (1989) there are some risks involved in this method since any incompleteness of the section decreases the ratio values. However, if we compare the wavelength ratios obtained from the two sites in the Yorkshire Basin with the values predicted by Berger *et al.* (1989) for the Early Jurassic (precession 18 and 21.8 ka, and obliquity 36.8 and 47 ka) there is a good correlation (Table 2).

Of related significance are the variance spectra of $CaCO_3$ and TOC from the Belemnite Marls in Dorset (Weedon & Jenkyns, 1990), which showed peaks corresponding to cycles of 37.5 and 300 cm, the latter with high wavelength variability. These results were based on a 10.50-m-long section covering 43% by thickness of the *U. jamesoni* zone (including subzones *P. polymorphus*, *P. brevispina* and *U. jamesoni*). This is exactly the same time interval covered by the Banded Shales in Yorkshire, where, however, a total thickness of 24.0 m is reached. Weedon & Jenkyns (1990) suggest that the 37.5-cm cycle represents the 20-ka precession period. The irregular longer period probably records climatic variations, and the obliquity frequency is not recorded in Dorset though it appears to be represented in Yorkshire in the 83-cm wavelength. This difference may lie in the fact that the Belemnite Marls represent a hemi-pelagic setting, where sedimentation may have been more influenced by ocean currents, water temperature and primary productivity, whereas the shallow marine, siliciclastic Yorkshire setting was more directly influenced by epicontinental features such as sen-

Table 2. Wavelength and wavelength ratios in Liassic small-scale cyclicity intervals. 1, Berger *et al.* (1989); 2, this paper; 3 and 4, van Buchem *et al.* (1992). E, eccentricity; O, obliquity; P, precession

	Wavelength (cm)			Wavelength ratios			
	E	O	P	E/E	E/O	E/P	O/P
1 Calculated numbers				1	2.2–2.7	4.6–5.5	1.68–2.6
Yorkshire Lias, jamesoni *zone, Lower Pliensbachian*							
2 Geochem. time series, cliff	—	83	50	—	—	—	1.66
3 Gamma-ray, cliff	190	85	45	1	2.2	4.2	1.89
4 Gamma-ray, borehole	250	96	58	1	2.6	4.3	1.65

sitivity to storm magnitude and frequency, and by changes on the land affecting vegetation cover, weathering intensity and runoff.

CONCLUSIONS

Physical sedimentological, chemical, mineralogical and organic matter studies of a selected interval of the Banded Shales have shown that the dark and light layering is mainly the result of a difference in primary composition (Table 3).

A depositional model for the Banded Shales should define conditions which led to the formation of the two different facies, and take into account that these facies apparently alternated 64 times on a regular basis. The general scene within which conditions have to be defined can be envisaged as a shallow epicontinental sea in a subtropical climate, which received mainly siliciclastic sediments from a relatively close, densely vegetated, landmass to the north and west (Wall, 1965; Hallam, 1975; Ziegler, 1982).

Small-scale cyclicity on the precession and obliquity scales (50 and 83 cm thickness) is shown by the quartz/clay ratio (also Si/Al), grain size (IRM), sedimentary structures, fauna and OM parameters (TOC, HI, $\delta^{13}C$ and palynodebris). Clay mineralogy, TOC and whatever controls gamma-ray level (clay mineral and organic matter abundance fixing U) show longer period change at the eccentricity scale. Our data do not allow a good separation between parameters controlling the calibre of material supplied to the shallow marine area (weathering and runoff) and those controlling its transport and winnowing (current and storm activity). However the simplest hypothesis is that wind-stress was the primary control, expressed as more frequent (and perhaps correspondingly bigger) storms yielding more efficient offshore dispersal and winnowing of coarse

Table 3. Summary of different parameters which vary consistently in the dark/light layers of the Banded Shales

	Light layer	Dark layer
Physical features	Silty streaks/beds (cm/m) abundant/common	Silty streaks (cm) scarce
Fauna	Suspension feeding macrofauna, abundant	Deposit feeding macrofauna, scarce
Organic matter		
TOC	0.5–0.8%	0.8–1.4%
HI	25–70	70–200
$\delta^{13}C_{org}$ (ppt PDB)	−25.91 to −26.20	−26.00 to −26.68
Palynofacies	Woody 75%	Woody 85%
Lithology		
Quartz	40–50%	32–42%
Clays (total)	38–48%	47–59%
Si/Al	1.05–1.3*	0.7–1.05*
Zr	310–390 ppm	240–310 ppm

* Standardized for average shale values.

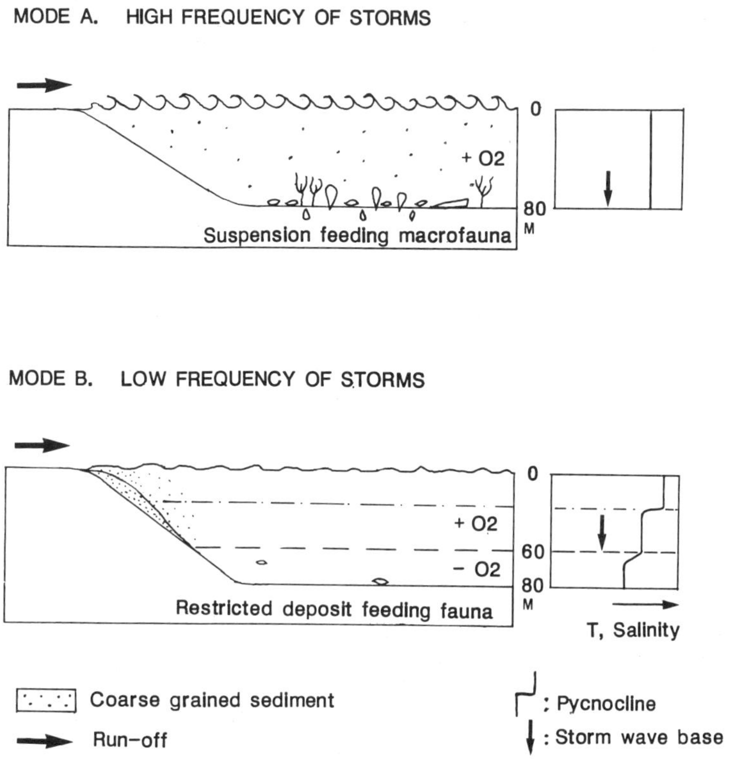

Fig. 13. Depositional model for the Banded Shales. Mode A represents the situation during deposition of the light layers: high frequency of storms causes a thorough mixing of the water column leading to favourable living conditions at the sea floor and the distribution of the coarse silt fraction over the basin. The preservation potential of organic matter is low. Mode B represents the situation during deposition of the dark layers: low frequency of storms enhances the stratification of the water column (pycnocline) leading to dysaerobic conditions at the sea floor and better preservation of the organic matter. The coarse silt fraction is trapped near shore.

silty sediment (Fig. 13). This same agency would also be responsible for water-column mixing and thus oxygenation in the face of oxygen consumption by sinking organic debris. The variations of TOC, HI, $\delta^{13}C$ and palynodebris are explicable by preservation. However, the longer period change (100 ka) in mineralogy requires more than wind, probably humidity and rainfall controlling weathering and clay production.

Given the good evidence for a climatic cause of the cyclicity, the next question to address is what mechanism could have controlled the consistent and regular alternations over a long period of time. Spectral analysis carried out on geochemical time series and gamma-ray logs of the Banded Shales detected the presence of three main wavelengths, which correspond to the three orbital parameters. In the absence of the strong amplifying effects of large polar ice caps (Hallam, 1975; Frakes, 1979), astronomically induced fluctuations of climate are likely to be of a different kind, and possibly smaller. This is also suggested by the fact that Milankovitch cycles found in Mesozoic rocks tend to have been formed between 20 and 40° latitude, close to the boundaries of the (sub-)tropical and temperate climate belts (de Boer, 1983, 1991). In these climatically sensitive 'zones' even small changes in climatic conditions would be likely to change the position of the climate belts and as such the depositional environment. An example of the high sensitivity of the climatic system has been given by Kutzbach & Otto-Bliesner (1982) and Kutzbach & Guetter (1984), who demonstrated the importance of low-latitude solar insolation changes as an important modulator of the monsoon circulation during the last 9000 years.

The Cleveland Basin was located about 41° N (Smith *et al.*, 1981) in Lower Lias times, which, according to the available Jurassic climatic maps (Pomerol, 1975, p. 140; Hallam, 1975, 1985, p. 158; Epshteyn, 1978; Parrish & Curtis, 1982; Parrish *et al.*, 1982), is a few hundred kilometres from the subtropical/temperate climate boundary. In other words, relatively small shifts of the climate belts could easily affect the depositional conditions in the Yorkshire area. The fact that the *jamesoni* zone in the whole of Britain is characterized by regular, small-scale cyclicity suggests that the whole region was influenced by the regular climatic changes. However, the different sediment sources and depositional settings caused a variety of expressions: limestone/marlstone cycles in the hemi-pelagic Dorset setting (Weedon & Jenkyns, 1990); grey, calcareous mudstone/olive grey mudstone cycles in Oxfordshire (Horton & Poole, 1983), and siltstone/mudstone cycles in the shallow-marine setting of Yorkshire.

The possibility of regular sea-level changes is not considered of major importance here because of the lack of evidence for the existence of substantial ice caps in the Early Jurassic (Frakes, 1979). But it should be realized that a change in sea level may produce the same changes as those related to a change in storm *magnitude* (see van Echelpoel, pp. 63–76, this volume), but not frequency. The argument for orbital forcing would, however, still explain: (i) the small-scale cyclicity as a result of the climatic variations causing the growing and waning of the ice cap and subsequent changes in sea level; and (ii) the longer term variation caused by climatic changes affecting conditions on the land and subsequently runoff.

Finally we return to the problem of the duration of ammonite subzones (Table 1). We believe that the couplets in our spectrally analysed record from the *polymorphus* subzone represent a 20-ka precession cycle period. If this is the case for all the couplets then the *jamesoni* subzone represents 660 ka, *brevispina* 220 ka and the banded part of *polymorphus* 400 ka (the whole zone at the same sedimentation rate would be 555 ka), a total duration of 1435 ka, which is close to the value of 1300 ka obtained from assumption of uniform duration for subzones. That assumption should now be seen as unlikely for individual subzones.

ACKNOWLEDGEMENTS

The authors wish to thank the following institutions for the use of their equipment: BP Research, Sunbury-on-Thames (carbon isotopes), Institut Français du Pétrole, Rueil-Malmaison (Rock-Eval), University of East Anglia (rock magnetic measurements), University of Reading (XRF), University of Cambridge (XRD). This text benefited from discussions with A. Huc, C. Jeans, K. Pye and S. Robinson, and critical reading by P.L. de Boer, G. Einsele and B.W. Sellwood. F.S.P.v.B. was sponsored by the British Council (Amsterdam), G.P.W. was a research fellow of Downing College, Cambridge during preparation of this paper. Cambridge Earth Sciences contribution 2307.

REFERENCES

Bairstow, L. (1969) Lower Lias. In: *William Smith bicentenary field meeting in north-east Yorkshire* (Ed. Hemingway, J.E.).

Bayer, U., Altheimer, E. & Deutschle, W. (1985) Environmental evolution in shallow epicontinental seas: sedimentary cycles and bed formation. In: *Sedimentary and Evolutionary Cycles* (Eds Bayer, U. & Seilacher, A.). Springer Verlag, Berlin, Lecture Notes in Earth Science 1, pp. 347–381.

Berger, A.L. (1978) Long term variations of daily insolation and Quaternary climatic changes. *J. Atmos. Sci.* **35**, 2362–2367.

Berger, A.L. (1988) Milankovitch theory and climate. *Rev. Geophys.* **26**, 624–657.

Berger, A.L. (1989) The spectral characteristics of Pre-Quaternary climate records, an example of the relationship between astronomical theory and geosciences. In: *Climate and Geosciences* (Eds Berger, A.L., Schneider, S. & Duplessy, J.C.). Reidel Publ. Co., Dordrecht, pp. 47–76.

Berger, A., Loutre, M.F. & Dehant, V. (1989) Astronomical frequencies for pre-Quaternary palaeoclimate studies. *Terra Nova* **1**, 474–479.

Boulter, M.C. & Riddick, A. (1986) Classification and analysis of palynodebris from the Paleocene sediments of the Forties field. *Sedimentology* **33**, 871–886.

Buckman, S.S. (1915) A palaeontological classification of the Jurassic rocks of the Whitby district; with a zonal table of Lias ammonites. In: *The Geology of the Country between Whitby and Scarborough* (Eds Fox-Strangways, C. & Barrow, G.). Mem. Geol. Surv. UK. **44**, 144.

Chamley, H. (1989) *Clay Sedimentology*. Springer Verlag, Berlin.

Cope, J.C.W., Getty, T.A., Howarth, M.K., Morton, N. & Torrens, H.S. (1980) *A Correlation of Jurassic Rocks in the British Isles. Part one: Introduction and Lower Jurassic*. Geol. Soc. Lond. Spec. Report 14.

Cope, M.J. (1981) Products of natural burning as a component of the dispersed organic matter of sedimentary rocks. In: *Organic Maturation Studies and Fossil Fuel Exploration* (Ed. Brooks, J.). Academic Press, London, pp. 89–109.

Dean, W.E., Arthur, M.A. & Claypool, G.E. (1986) Depletion of ^{13}C in Cretaceous marine organic matter: source diagenetic or environmental signal? *Mar. Geol.* **70**, 119–157.

De Boer, P.L. (1983) Aspects of Middle Cretaceous pelagic sedimentation in southern Europe. *Geol. Ultraiectina* **31**.

De Boer, P.L. (1991) Pelagic black shale–carbonate rhythms: orbital forcing and oceanographic response. In: *Cycles and Events in Stratigraphy* (Eds Einsele, G., Ricken, W. & Seilacher, A.). Springer Verlag, Berlin, pp. 63–78.

Deconinck, J.F., Chamley, H., Debrabant, P. & Colbeaux, J.P. (1983) Le Boulonnais au Jurassique supérieur: données de la minéralogie des argiles et de la géochimie. *Ann. Soc. Geol. Nord* **102**, 145–152.

DEINES, P. (1980) The isotopic composition of reduced organic carbon. In: *Handbook of Environmental Isotope Geochemistry, Vol. 1, The Terrestrial Environment A* (Eds Fritz, P. & Fantes, J.Ch.). Elsevier, Amsterdam, pp. 329–406.

DEMAISON, G.J. & MOORE, G.T. (1980) Anoxic environments and oil source bed genesis. *Bull. Am. Assoc. Petrol. Geol.* **64**, 1179–1209.

EINSELE, G. (1982) Limestone–marl cycles (periodites): diagenesis, significance, cause: a review. In: *Cyclic and Event Stratification* (Eds Einsele, G. & Seilacher, A.). Springer Verlag, Berlin, pp. 8–53.

EMILIANI, C. (1955) Pleistocene temperatures. *J. Geol.* **63**, 538–578.

ESPSHTEYN, O.G. (1978) Mesozoic–Cenozoic climates of northern Asia and glacial marine deposits. *Int. Geol. Rev.* **20**, 49–58.

ESPITALIÉ, J., LAPORTE, J.L. MADEC, M., MARQUIS, F., LEPLAT, P., PAULET, J. & BOUTEFEU, A. (1977) Méthode rapide de caractérisation des roches mères de leur potentiel pétrolier et de leur degré d'évolution. *Rev. Inst. Fr. Petrole.* **32**, 23–42.

ESPITALIÉ, J., DEROO, G. & MARQUIS, F. (1985/86) La Pyrolyse Rock-Eval et ses applications. *Rev. Inst. Fr. Pétrole* 1985, 5/6; 1986, 1.

FISCHER, A.G. (1981) Climatic oscillations in the biosphere. In: *Biotic Crises in Ecological and Evolutionary Time* (Ed. Nitecki, M.H.). Elsevier, New York, pp. 103–131.

FISCHER, A.G. (1986) Climatic rhythms recorded in strata. *Annu Rev. Earth Planet. Sci.* **14**, 351–376.

FISCHER, A.G., HERBERT, T. & PREMOLI SILVA, I. (1985) Carbonate bedding cycles in Cretaceous pelagic and hemi-pelagic sequences. In: *Fine Grained Deposits and Biofacies of the Cretaceous Western Interior Seaway: Evidence of Cyclic Sedimentary Processes* (Eds Pratt, L.M., Kaufmann, E.G. & Zelt, F.B.). Soc. Econ. Paleont. Mineral. Field Trip no 9, pp. 1–10.

FISCHER, A.G., DE BOER, P.L. & PREMOLI SILVA, I. (1990) Cyclostratigraphy. In: *Cretaceous Resources, Events and Rhythms* (Eds Ginsburg, R.N. & Beaudoin, B.). Reidel Publ. Co., Dordrecht, pp. 139–172.

FRAKES, L.A. (1979) *Climates Throughout Geologic Time.* Elsevier, Amsterdam.

GALE, A.S. (1989) A Milankovitch scale for Cenomanian time.*Terra Nova* **1**, 420–425.

GAUNT, G.D., IVIMEY-COOK, H.C., PENN, I.E. & COX, B.M. (1980) *Mesozoic rocks proved by IGS boreholes in the Humber and Acklam areas.* Rep. Inst. Geol. Sci. 79/13.

GRIFFIN, G.M. (1971) Interpretation of X-ray diffraction data. In: *Procedures in Sedimentary Petrology* (Ed. Carver, R.E.). John Wiley & Sons Interscience, New York, pp. 541–569.

HALL, F.R., BUSCH, W.H. & KING, J.W. (1989) The relationship between variations in rock-magnetic properties and grain size of sediments from ODP Hole 645C. In: *Proc. Ocean Drilling Program, Scientific Results 105* (Eds Srivasta, S.P., Arthur, M., Clement, B. *et al.*). College Station, TX (Ocean Drilling Program), pp. 837–841.

HALLAM, A. (1975) *Jurassic Environments.* Cambridge University Press, Cambridge.

HALLAM, A. (1985) A review of mesozoic climates. *J. Geol. Soc. Lond.* **142**, 433–455.

HARLAND, W.B., ARMSTRONG, R.L., COX, A.V., CRAIG, L.E., SMITH, A.G. & SMITH, D.G. (1990) *A Geologic Time Scale 1989.* Cambridge University Press, Cambridge.

HARRIS, T.M. (1957) A Liasso-Rhaetic flora in S. Wales. *Proc. R. Soc. Lond.* **147B**, 289–308.

HAYS, J.D. IMBRIE, J. & SHACKLETON, N.J. (1976) Variations of the earth's orbit: pacemaker of the ice ages. *Science* **194**, 1121–1132.

HORTON, A. & POOLE, E.G. (1983) The lithostratigraphy of three geophysical marker horizons in the Lower Lias of Oxfordshire. *Bull. Geol. Surv. GB* **62**, 13–24.

HOUSE, M.R. (1985) A new approach to an absolute timescale from measurements of orbital cycles and sedimentary microrhythms. *Nature* **315**, 721–725.

HUC, A.Y. (1988) Sedimentology of organic matter. In: *Humic Substances and their Role in the Environment* (Eds Frimmel, F.H. & Christeman, R.F.). John Wiley & Sons Ltd, Dahlem Konferenzen, pp. 215–243.

IVIMEY-COOK, H. & POWELL, J. (1991) Late Triassic and early Jurassic biostratigraphy of the Felixkirk Borehole North Yorkshire. *Proc. Yorks. Geol. Soc.* **48**, 367–374.

KRAUSKOPF, K.B. (1985) *Introduction to Geochemistry,* 2nd edn. McGraw-Hill, London.

KÜSPERT, W. (1982) Environmental changes during oil shale deposition as deduced from isotope ratios. In: *Cyclic and Event Stratification* (Eds Einsele, G. & Seilacher, A.). Springer Verlag, Berlin, pp. 482–501.

KUTZBACH, J.E. & GUETTER, P.J. (1984) The sensitivity of monsoon climates to orbital parameter changes for 9000 years BP: experiments with the NCAR general circulation model. In: *Milankovitch and Climate* (Eds Berger, A., Imbrie, J., Hays, J., Kukla, G. & Saltzman, B.). Reidel Publ. Co., Dordrecht, pp. 801–820.

KUTZBACH, J.E. & OTTO-BLIESNER, B.L. (1982) The sensitivity of the African–Asian monsoonal climate to orbital parameter changes for 9000 yr BP in a low resolution general circulation model. *J. Atmos. Sci.* **39**, 1177–1188.

LEARY, P.N., COTTLE, R.A. & DITCHFIELD, P. (1989) Milankovitch control of foraminiferal assemblages from the Cenomanian of southern England. *Terra Nova* **1**, 416–419.

LEWAN, M.D. (1986) Stable carbon isotopes of amorphous kerogens from Phanerozoic sedimentary rocks. *Geochim. Cosmochim. Acta* **50**, 1583–1591.

MILANKOVITCH, M. (1941) *Kanon der Erdbestrahlung und seine Anwendung auf das Eiszeiten problem.* Akad. Royale Serbe 133.

OLDFIELD, F., MAHER, B.A., DONAGHUE, J. & PIERCE, J. (1985) Particle-size related mineral magnetic source sediment linkages in the Rhode River catchment, Maryland, U.S.A. *J. Geol. Soc. Lond.* **142**, 1035–1046.

PARRISH, J.T. & CURTIS, R.L. (1982) Atmospheric circulation, upwelling and organic-rich rocks in the Mesozoic and Cenozoic Eras. *Palaeogeogr. Palaeoclimatol. Palaeoecol.* **40**, 31–66.

PARRISH, J.T., ZIEGLER, A.M. & SCOTESE, C.R. (1982) Rainfall patterns and the distribution of coals and evaporites in the Mesozoic and Cenozoic. *Palaeogeogr.*

Palaeoclimatol. Palaeoecol. **40**, 67–101.

Pisias, N.G. & Mix, A.C. (1988) Aliasing of the geological record and the search for long-period Milankovitch cycles. *Paleoceanography* **3**, 613–619.

Pomerol, Ch. (1975) *Ere Mésozoique, stratigraphie et paléogéographie*. Doin, Editeurs, Paris.

Popp, B.N., Takiguki, R., Hayes, J.M., Louda, J.W. & Baker, E.W. (1989) The post-Paleozoic chronology and mechanism of ^{13}C depletion in primary marine organic matter. *Am. J. Sci.* **289**, 436–454.

Powell, J. (1984) Lithostratigraphical nomenclature of the Lias Group in the Yorkshire Basin. *Proc. Yorks. Geol. Soc.* **45**, 51–57.

Pratt, L. (1984) Influence of paleoenvironmental factors on preservation of organic matter in Middle Cretaceous Greenhorn Formation, Pueblo, Colorado. *Bull. Am. Assoc. Petrol. Geol.* **68**, 1146–1159.

Pye, K. & Krinsley, D.H. (1986) Microfabric, mineralogy and early diagenetic history of the Whitby Mudstone Formation (Toarcian), Cleveland Basin, U.K. *Geol. Mag.* **123**, 191–203.

Raiswell, R. (1987) Non-steady state microbiological diagenesis and the origin of concretions and nodular limestones. In: *Diagenesis of Sedimentary Sequences* (Ed. Marshall, J.D.). Geol. Soc. Lond. Spec. Pub. 36, pp. 41–54.

Rogers, M.A. (1979) Application of organic facies concepts to hydrocarbon source rock evaluation. In: *Proc. of 10th World Petr. Congress, Bucharest*, pp. 23–30.

Sadler, P.M. (1981) Sediment accumulation rates and the completeness of stratigraphic sections. *J. Geol.* **89**, 569–584.

Savrda, C.E., Bottjer, D.J. & Gorsline, D.S. (1984) Development of a comprehensive oxygen-deficient marine biofacies model: evidence from Santa Monica, San Pedro, and Santa Barbara Basins, California Continental Borderland. *Bull. Am. Assoc. Petrol. Geol.* **68**, 1179–1192.

Sellwood, B.W. (1970). The relation of trace fossils to small scale sedimentary cycles in the British Lias. In: *Trace Fossils* (Eds Crimes, T.P. & Harper, J.C.). Geol. J. Spec. Issue, 3, pp. 489–504.

Sellwood, B.W. (1971) The genesis of some sideritic beds in Yorkshire Lias (England). *J. Sedim. Petrol.* **41**, 854–858.

Sellwood, B.W. (1972) Regional environmental changes across a lower Jurassic Stage-boundary in Britain. *Palaeontology* **15**, 125–157.

Sellwood, B.W. (1978) Jurassic. In: *The Ecology of Fossils, an Illustrated Guide* (Ed. McKerrow, W.S.). Duckworth, London, pp. 204–279.

Smith, A.G., Hurley, A.M. & Briden, J.C. (1981) *Phanerozoic Paleocontinental World Maps*. Cambridge University Press, Cambridge.

Smith, D.G. (1989) Stratigraphic correlation of presumed Milankovitch cycles in the Blue Lias (Hettangian to earliest Sinemurian), England. *Terra Nova* **1**, 457–460.

Tate, R. & Blake, J. (1876) *The Yorkshire Lias*. J van Voorst, London.

Tissot, B.P. & Welte, D.H. (1984) *Petroleum Formation and Occurrence*, 2nd edn. Springer Verlag, Berlin.

Tribovillard, N.P. & Cotillon, P. (1989) Relationships between climatically influenced sedimentation and salt diapirism in the French Western Alps based on evidence from organic and inorganic geochemistry. *Palaeogeogr. Palaeoclimatol. Palaeoecol.* **71**, 271–280.

Tyson, R.V. (1987) The genesis and palynofacies characteristics of marine petroleum source rocks. In: *Marine Petroleum Source Rocks* (Eds Brooks, J. & Fleet, A.). Geol. Soc. Lond. Spec. Publ. 26, pp. 47–68.

Van Buchem, F.S.P. (1990) *Sedimentology and diagenesis of Lower Lias mudstones in the Cleveland Basin, Yorkshire, UK*. Unpubl. PhD thesis, University of Cambridge.

Van Buchem, F.S.P. & McCave, I.N. (1989) Cyclic sedimentation patterns in Lower Lias Mudstones of Yorkshire (GB). *Terra Nova* **1**, 461–467.

Van Buchem, F.S.P., Melnyk, D.H. & McCave, I.N. (1992) Chemical cyclicity and correlation of Lower Lias mudstones using gamma ray logs, Yorkshire, UK. *J. Geol. Soc. Lond.* **149**, 991–1002.

Wall, D. (1965) Microplankton, pollen and spores from the Lower Jurassic of Britain. *Micropaleontology* **11**, 151–190.

Weedon, G.P. (1985) Hemipelagic shelf sedimentation and climatic cycles: the basal Jurassic (Blue Lias) of South Britain. *Earth Planet. Sci. Lett.* **76**, 321–335.

Weedon, G.P. (1989) The detection and illustration of regular sedimentary cycles using Walsh power spectra and filtering, with examples from the Lias of Switzerland. *J. Geol. Soc. Lond.* **146**, 133–144.

Weedon, G.P. (1991) The spectral analysis of stratigraphic time series. In: *Cycles and Events in Stratigraphy* (Eds Einsele, G., Ricken, W. & Seilacher, A.). Springer Verlag, Berlin, pp. 840–854.

Weedon, G.P. & Jenkyns, H.C. (1990) Regular and irregular climatic cycles and the Belemnite Marls (Pliensbachian, Lower Jurassic, Wessex Basin). *J. Geol. Soc. Lond.* **147**, 915–918.

Wenger, L.M., Baker, D.R., Chung, H.M. & McCulloh, T.H. (1988) Environmental control of carbon isotope variations in Pennsylvanian black-shale sequences, Midcontinent, U.S.A. *Org. Geochem.* **13**, 765–771.

Wilson, H.E. & Manning, P.I. (1978) *Geology of the Causeway Coast*. Mem. Geol. Surv. Northern Ireland, Belfast.

Ziegler, P.A. (1982) *Geological Atlas of Western and Central Europe*. Elsevier, Amsterdam.

Zwan, C.J. van der (1990) Palynostratigraphy and palynofacies reconstruction of the Upper Jurassic to lowermost Cretaceous of the Draugen Field, Offshore Mid Norway. *Rev. Palaeobot. Palynol.* **62**, 157–186.

Spec. Publs Int. Ass. Sediment. (1994) **19**, 367–394

The sequence architecture of mid-Pleistocene (*c*.1.1–0.4 Ma) cyclothems from New Zealand: facies development during a period of orbital control on sea-level cyclicity

S.T. ABBOTT *and* R.M. CARTER

Department of Geology, James Cook University of North Queensland, Townsville, Q4811, Australia

ABSTRACT

South Wanganui Basin, located in western North Island, New Zealand, and about 200 × 200 km in area, contains a fill up to 4 km thick of mainly shelf facies sediment of Plio-Pleistocene age. The eastern edge of the basin has undergone gentle upwarping along the adjacent Pacific/Indo-Australian plate boundary, resulting in excellent coastal exposures through late Pliocene to early Pleistocene strata. The markedly cyclic mid-Pleistocene (*c*.1.1–0.4 Ma) section exposed at Wanganui comprises 10 superposed cyclothems with a total thickness of *c*.160 m corresponding to fifth (100 ka)- and sixth (40 ka)-order sequences equivalent to odd-numbered oxygen isotope stages 11–31. Each cyclothem comprises three parts: (i) a basal suite of shoreface and innermost shelf sediments containing intertidal and shallow subtidal molluscan faunas and, more commonly, transported shell gravels (type A shellbeds); (ii) a mid-cycle shellbed which includes *in situ* offshore molluscs in a matrix of muddy fine sandstone or siltstone (type B shellbeds); and (iii) an upper unit of terrigenous siltstone, either bedded and barren of macrofossils, or massive, bioturbated and containing a sparsely scattered *in situ* fauna, commonly similar to that of the subjacent shellbed. The threefold subdivision of each cyclothem occurs within both fifth- and sixth-order sequences, and corresponds to the transgressive systems tract, mid-cycle condensed shellbed and high-stand systems tract of the Exxon sequence stratigraphic model, respectively. Compared with sixth-order cyclothems, fifth-order cyclothems may contain thicker transgressive systems tracts and mid-cycle condensed shellbeds, deeper water faunas in their mid-cycle shellbeds, and deeper water sedimentary facies in their high-stand systems tracts. Because the Wanganui coastal cyclothems were deposited many tens of kilometres east (inshore) from the contemporary shelf-edge, the stratigraphic record at Wanganui contains no marine low-stand systems tract sediment. The section therefore comprises dominantly periods of interglacial deposition, representing the late transgressive and high-stand portions of each sea-level cycle. Glacial periods are represented only at surfaces of marine planation and bioerosion at the base of each cyclothem, which mark the sequence boundaries. Marine ravinement and *in situ* boring pholad bivalves (type C shellbeds; equivalent to *Glossifungites* ichnofacies) have removed any former traces of non-marine conditions at the sequence boundaries in the coastal section, but subaerial surfaces (including soils) are preserved inland towards the eastern edge of the basin. The facies architecture of the Wanganui Basin is of compelling interest because (i) the sediments were deposited during a period of known sea-level fluctuation driven by Milankovitch orbital controls, and (ii) the invertebrate faunas used to assist the sequence interpretation comprise living species, i.e. are of known habitat significance. For these reasons, the sedimentary interpretation at Wanganui possesses a degree of certainty which is lacking from sequence stratigraphic studies of ancient, pre-Pliocene sections, for which dating, palaeoecologic reconstruction and the role of sea-level controls are of necessity highly inferential.

INTRODUCTION

The study of orbital controls on cyclic sedimentary sequences has generally been approached using theoretical calculations of orbital periodicities (Milankovitch cycles; Vernekar, 1968), and their retrodiction to the geological period of interest (Berger, 1989; Berger *et al.*, 1989). Power spectrum analysis is then used to identify any comparable periodicities present within a suite of stratigraphic data, typically oxygen isotope ratios, percentage calcium carbonate or inferred sea surface palaeotemperature (Hays *et al.*, 1976; Imbrie *et al.*, 1984). This approach has been particularly successful as applied to Plio-Pleistocene deep-sea sediments, for which the climatic forcing functions are now well established, and which possess a limited variety of facies and quasi-constant sedimentation rates (e.g. Raymo *et al.*, 1989; Ruddiman *et al.*, 1989). Much of the sedimentary record preserved on the continents, however, is of shallow marine or non-marine origin, its deposition resulting from a complex interplay of local palaeogeography, varying sediment supply, tectonics, climate and sea level. Despite some outstanding attempts (e.g. Kamp, 1978; Weedon, 1985), it is a far from simple exercise to identify correctly the orbital signals in such sections, even when the sediments are dominantly fine grained and therefore potentially of reasonably constant sedimentation rate (Hesselbo *et al.*, 1990a).

Analysis of sedimentary rhythms in continental margin (predominantly coastal plain-shelf) sedimentary basins during the 1980s focused on the identification of seismic sequences, believed by the Exxon school of seismic stratigraphers to be fundamentally sea-level controlled (e.g. Vail *et al.*, 1977; Vail, 1987; van Wagoner *et al.*, 1988). This sequence stratigraphic approach has now been applied to sedimentary sections of a wide range of ages, from Precambrian (Christie-Blick *et al.*, 1988; Eriksson & Simpson, 1990), through Palaeozoic (Cotter, 1988; Boardman & Heckel, 1989) and Mesozoic (Hubbard, 1988), to Tertiary (Kidwell, 1984; Plint, 1988) and Quaternary (Clifton *et al.*, 1988). Fulthorpe & Carter (1989) and Weedon & Jenkyns (1990) have discussed similar but high-resolution sequences respectively of Miocene and Liassic age, whilst expressing caution on the matter of a primary orbital or sea-level control on deposition of all cycles present.

It is clearly important to identify accurately any climatic or sea-level periodicity that is contained within the stratigraphic record (e.g. Kominz & Bond, 1990). Fulthorpe (1991) has shown how a stratigraphic column can be viewed as a variable frequency response recorder whose contained record has to be inverted to obtain the pattern and frequency of any input cyclicity. In thus analysing ancient sediments, it is vital to locate correctly the sequence boundaries and maximum flooding surfaces for each cycle. Yet, even for well-studied sections, it is often difficult to designate a particular diastem unequivocally in sequence stratigraphic terms.

Ambiguities intrinsic to sequence stratigraphy

The well-known Lower Lias of Dorset (e.g. House, 1989) contains conspicuous cyclic bedding, particularly in the Pliensbachian Belemnite Marls (House, 1985; Weedon & Jenkyns, 1990). The section is mud dominated, contains frequent banded concretionary layers, and possesses a refined ammonite biostratigraphy. Particular concretionary horizons can be recognized as potential markers of sequence boundaries using evidence such as firmground or hardground features, condensed shellbed character and the absence of particular faunal zone(s) (Hesselbo *et al.*, 1990a). One such horizon, the Coinstone, is marked by a pyritized top with bored, encrusted and mineralized exhumed concretions, and bored belemnites, some of which are preserved within vertical cracks; two Sinemurian ammonite zones are missing across the bed.

The Dorset Lias section was an important input into the Jurassic part of the Exxon global sea-level curve (Vail *et al.*, 1977; Haq *et al.*, 1988), in which the concretionary Coinstone horizon, identified as a diastem by Lang *et al.* (1928), was considered to mark a 196.5-Ma mid-cycle maximum flooding surface. Hallam (1969, 1988), however, mindful of evidence for exposure of the Coinstone, argued instead that the bed marks a sea-level fall and therefore a sequence boundary.

Similar ambiguity in interpreting demonstrable stratigraphic breaks as either sequence boundaries or maximum flooding surfaces has been described for estimated 103, 106 and 107.5 Ma levels in the Aptian–Albian of Folkestone (Hesselbo *et al.*, 1990b), and from a *c.*29-Ma level in the mid-Oligocene of New Zealand (Carter, 1985; Loutit *et al.*, 1988). In sequence terms ambiguity cannot be more profound than these cases exemplify, since the alternative interpretations are 180° out of phase, and no simple observational tests are available to resolve the matter.

A major difficulty in applying sequence stratigraphic analysis, particularly in circumstances such as those just described for the Coinstone, is the present lack of sedimentological description of an unequivocal set of transgressive, high-stand and low-stand systems tracts. With the exception of recent work on the Gulf Coast shelf and slope (which is necessarily based on an analysis of seismic and core materials, e.g. Lowrie & McDaniel-Lowrie, 1985; Loutit *et al.*, 1988; Boyd *et al.*, 1989; Feeley *et al.*, 1990), and that of Collier (1990) on Pleistocene high-stand systems tracts exposed in the Corinth Canal, there is little published data regarding *known* as opposed to inferred systems tracts. We therefore present here an analysis of a well-exposed set of mid-Pleistocene strata from the Wanganui Basin, New Zealand. Coastal platforms and cliffs afford unique three-dimensional exposure through a set of 10 cyclothems, and display with great clarity the succession of sedimentary facies characteristic of the transgressive, mid-cycle flooding and high-stand phases of the sea-level cycle (Carter *et al.*, 1991; Abbott, in preparation).

REGIONAL BACKGROUND

There are three particular attractions to studying the sequence stratigraphy of the Wanganui coastal section. First, the sediments were deposited during a time when it is known that glacio-eustatic fluctuations of sea level of amplitude greater than 100 m were occurring regularly (e.g. Shackleton & Opdyke, 1973). Second, the sediments are richly fossiliferous throughout, containing molluscan faunas which are closely similar to modern faunas and thereby greatly aid accurate sedimentary environmental analysis (e.g. Fleming, 1953). And third, the section contains a wide variety of evidence suitable for age determination, thus allowing accurate correlation to the global isotope curve (e.g. macro- and micro-fossils, Beu & Edwards, 1984; fission track analysis, Seward, 1976; oxygen isotopes, Stevens & Vella, 1981; amino acid racemization, B. Pillans, personal communication; and palaeomagnetic analysis, Turner & Kamp, 1990; Pillans *et al.*, in preparation).

Stratigraphic and tectonic setting

The South Wanganui Basin is a 200 × 200 km, ovoid sedimentary basin situated in a back-arc position with respect to the subducting plate boundary between the Pacific and Indo/Australian plates in North Island, New Zealand (Fig. 1A). Sediments deposited in the eastern part of the basin outcrop in hills on the western side of North Island, and have been upwarped at rates which increase eastwards from *c.*0.3–0.5 m/ka at the coast to 1–3 m/ka along the forearc axial mountain range (Pillans, 1986). The sediments of the western side of the basin, described herein, are particularly well exposed in coastal cliffs to the northwest of Wanganui.

The modern continental shelf west of North Island is underlain from east to west by sediments of the Wanganui Basin, Taranaki Basin and Western Platform (Fig. 1B). Plio-Pleistocene subsidence, and sedimentation, has been concentrated in the vicinity of the Wanganui Basin, and is attributed by Stern & Davey (1989) to the presence of a subjacent, locked, subducting plate interface, and by Stern & Davey (1990) to foreland basin thrust loading. Nonetheless, the characteristics of the western North Island shelf have changed little since the late Pliocene, the age of a conspicuous seismic reflection mapped by Anderton (1981) within the Wanganui Basin and westwards to the shelf edge.

This particular set of circumstances presents a unique opportunity to study the outcrop stratigraphy of cyclothems deposited during the known mid-Pleistocene fluctuations of sea level (Fig. 2). Whilst the cyclothems deposited at the Wanganui coast comprise exclusively latest glacial and interglacial transgressive and high-stand systems tracts, the intervening glacial low-stand systems tracts are also well preserved offshore in the Giant Foresets Formation (Pilaar & Wakefield, 1978; Beggs, 1990) which underlies the Western Platform and the shelf edge (see Fig. 1B).

Correlation to the mid-Pleistocene isotope scale

In describing the stratigraphy of western Pacific core V28-238, Shackleton & Opdyke (1973) established the use of the oceanic Pleistocene oxygen isotope curve as a surrogate for global ice volume, and hence for eustatic sea level. A wealth of detailed stratigraphic and sedimentary data now supports this conclusion, and the corollary that repeated fluctuations of sea level of amplitude greater than 100 m have characterized much of the last 2.5 Ma, driven by Milankovitch variations in the Earth's orbital parameters.

The most detailed versions of the Pleistocene oxygen isotope curve currently available are those of Williams *et al.* (1988), Ruddiman *et al.* (1989), Joyce *et al.* (1990) and Shackleton *et al.* (1990), based upon hydraulic piston coring by the Ocean Drilling Program (ODP). High-quality records of climatic cycles from locations near the Wanganui Basin include those of ODP holes 593 (Dudley & Nelson, 1989) and 594 (Nelson *et al.*, 1985). Figure 2 sum-

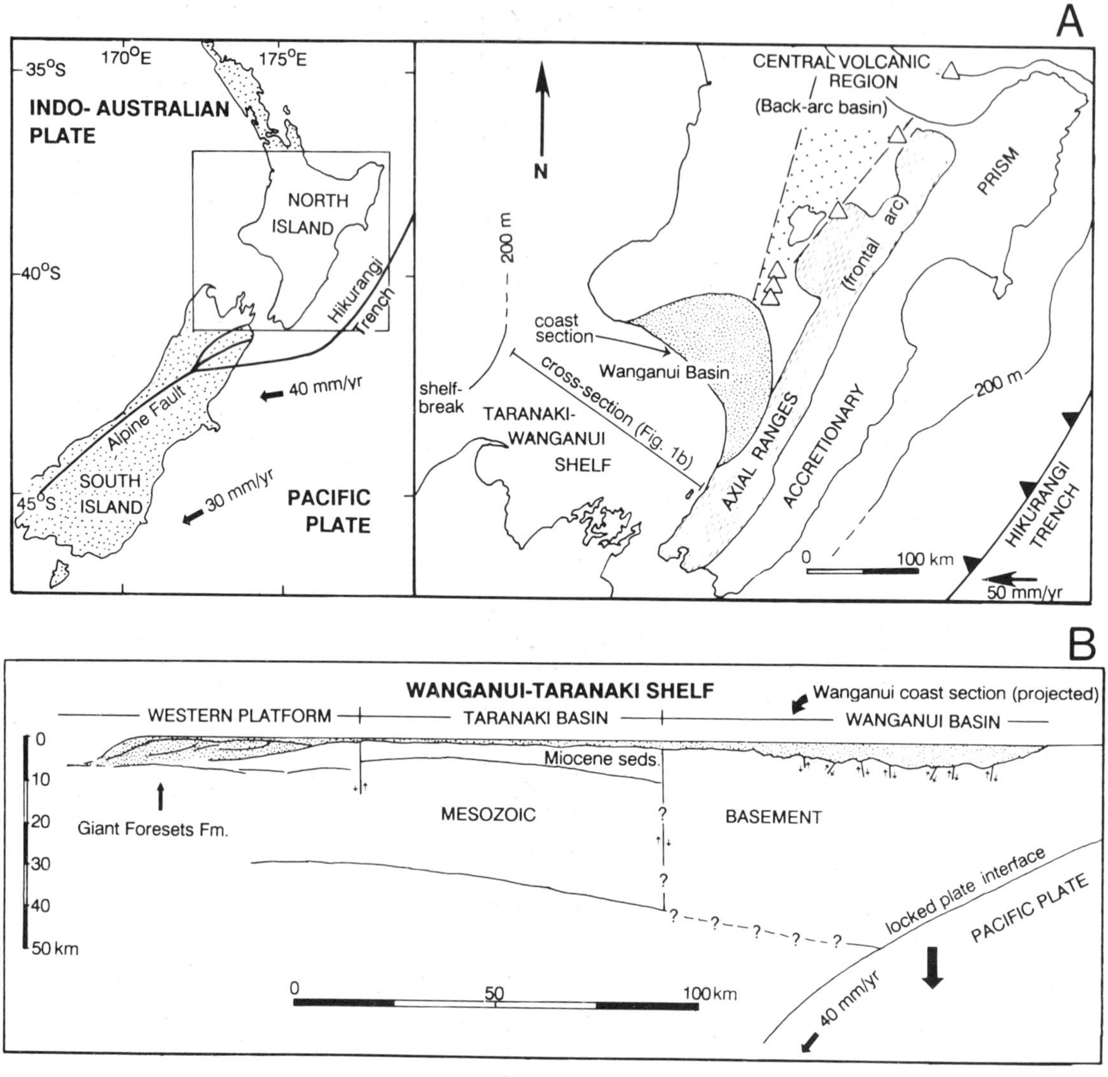

Fig. 1. (A) Location and regional geological setting of the Wanganui Basin. (B) Schematic section across the Wanganui–Taranaki shelf showing the Wanganui Basin–Taranaki Basin–Western Platform, with Pleistocene sediments stippled. The locked plate interface beneath the Wanganui Basin is a possible driving mechanism for Quaternary subsidence. After Stern & Davey (1989).

marizes the correlation between the cyclothems of the Wanganui coastal section and the isotope record, using the Milankovitch-cycle tuned time scale of Shackleton *et al*. (1990) which indicates stage boundaries about 5% older than previously assumed. The precise correlation also differs from that of Beu & Edwards (1984) and Turner & Kamp (1990), in that it is based on detailed sedimentological analysis as well as traditional correlation tools. Nonetheless, the palaeomagnetic tie-points provided by Turner & Kamp (1990), and particularly the equivalence between the Kaikokopu Shellbed (cylothem 6) and the Brunhes/Matuyama boundary (0.78 Ma), and the Lower Okehu Siltstone (cyclothem 2) and the top-Jaramillo reversal (0.99 Ma), are the lynchpins of the interpretation, and are given more weight than

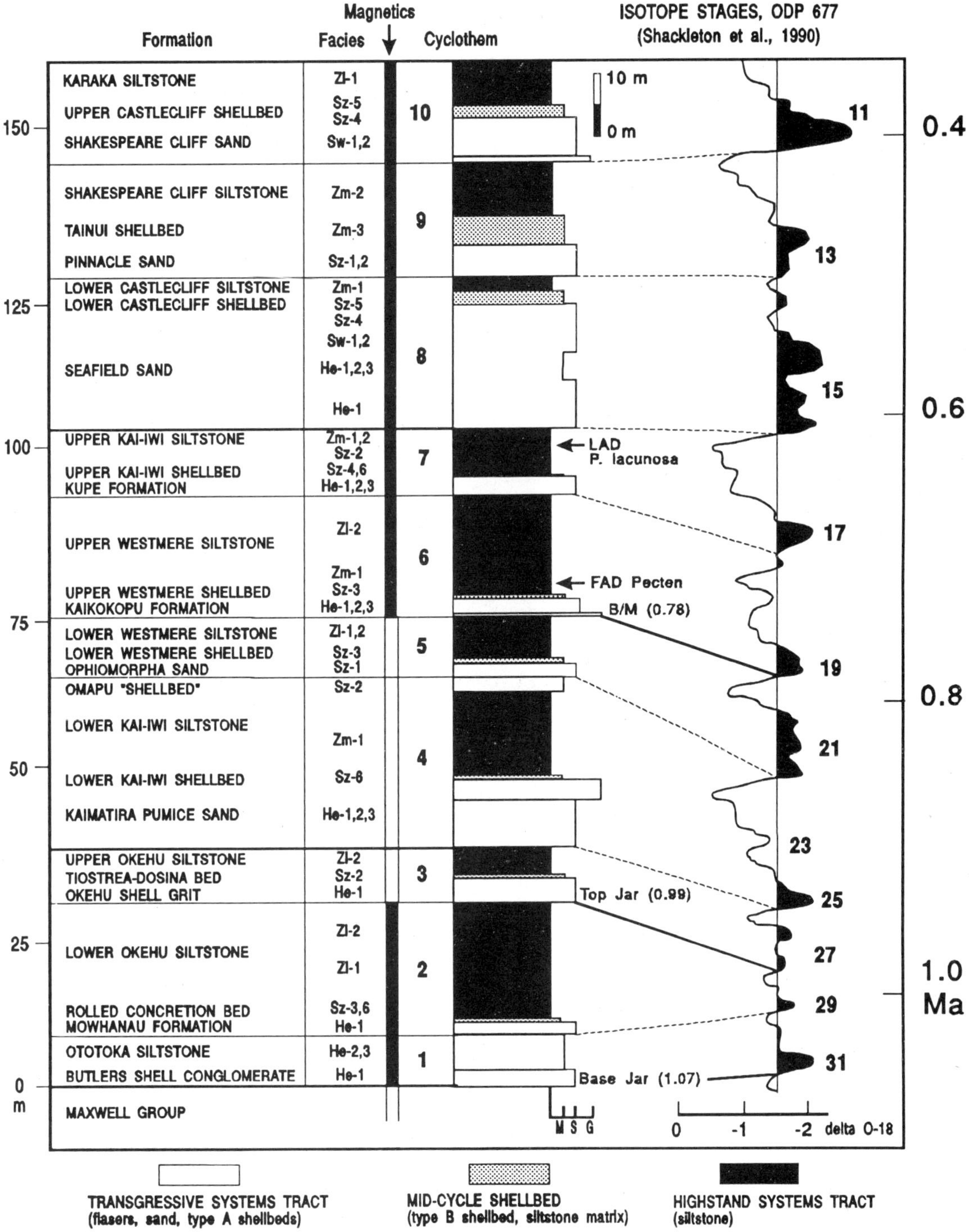

Fig. 2. Summary of the stratigraphic column exposed on the Wanganui coast between Otokoka Stream and Castlecliff. Ten cyclothems are distinguished and correlated to the mid-Pleistocene global isotope curve and Milankovitch-tuned age model of Shackleton *et al.* (1990). Biostratigraphic data after Beu & Edwards (1973). Palaeomagnetic data after Turner & Kamp (1990). Jar, Jaramillo event; B/M, Brunhes/Matuyama boundary. For explanation of column of abbreviations for sediment facies, see Table 2.

Table 1. Alternative correlation schemes between the Wanganui cyclothems and the oxygen isotope scale, based respectively on palaeomagnetic and faunal criteria. The most satisfactory match is based upon columns A and B, which is therefore adopted in Figs 2 and 3

Sequence/ cyclothem	Correlation fixed by isotope stages			
	A	B	C	D
10	11	11	9	7
9	13	13	11	9
8	15	15	13	11
7	17	17	15	13
6	19	19	17	15
5	21	21	19	17
4	25	25	21	19
3	27	27	25	21
2	29	29	27	25
1	31	31	29	27

A, Top Jaramillo at stage 28/27 boundary.
B, Brunhes/Matuyama at stage 20/19 boundary.
C, Brunhes/Matuyama at stage 19/18 boundary.
D, LAD of *Pseudemiliana huxleyi* at stage 13/12 boundary.

the probably less accurate fission track ages and microfaunal data published earlier.

A number of sources of minor uncertainty exist in the precise correlation of the Wanganui cyclothems with the standard isotope scale (see Table 1):

1 Recently, deMenocal *et al.* (1990) have argued that the Brunhes/Matuyama (B/M) reversal falls about 6 ka after the peak of isotope (interglacial) stage 19, rather than at or just before the peak as adopted hitherto. Inner shelf cyclothemic shelf sections such as that at Wanganui are deposited mostly during the final stages of post-glacial transgression and at the subsequent interglacial sea-level high. Therefore, and if the B/M boundary in fact post-dates the peak transgression associated with interglacial stage 19, the first normally magnetized cyclothem (= cyclothem 6 at Wanganui) should correlate with isotope stage 17 rather than stage 19 as would have previously been assumed (see Table 1). We do not adopt this interpretation because such a position for the B/M boundary (column C, Table 1) conflicts with a correlation based on the position of the top-Jaramillo boundary (column A, Table 1).

2 The LAD of *Pseudemiliana lacunosa* occurs within the Upper Kai-iwi Siltstone of cyclothem 7 (Beu & Edwards, 1984), i.e. near the isotope stage 16/15 boundary, despite the belief that the global extinction of this form occurred at the stage 13/12 boundary (e.g. Joyce *et al.*, 1990). Recent resampling of the Lower Castlecliff Siltstone (cyclothem 8, correlated with stage 15) confirms that it contains a diverse nannoflora consistent with deposition from normal marine waters, but without *P. lacunosa* (A.R. Edwards, personal communication). Apparently *P. lacunosa* became extinct in parts of the southwest Pacific up to two isotope stages earlier than elsewhere.

3 Virtually all recent published climatic records show interglacial isotope stage 23 to correspond to a weakly developed interglacial, and therefore to a lower than usual sea-level high. It is therefore likely that the Wanganui coast section contains no record, or only a cryptic record, of deposition during stage 23. It may be that a fossiliferous siltstone at the top of cyclothem 4 (the hitherto enigmatic Omapu 'shellbed') represents the relatively minor sea-level fluctuation associated with stage 23. In any case, our overall correlation indicates that, unless it is missing altogether at the superjacent sequence boundary, stage 23 must correlate with the upper part of cyclothem 4.

We conclude that the Wanganui section can be correlated with the standard isotope scale with an uncertainty of no more than one full glacial/interglacial cycle. However, any chosen correlation must depend upon judgements about the relative validity of particular correlation tools, and on the interplay of the absolute magnitude of the eustatic sea-level signature and the rate of subsidence or uplift at any point in the section. The main alternative correlations are summarized in Table 1.

It is noteworthy also that the Wanganui section overlaps in time with a fundamental shift of behaviour of the isotope curve. Isotope stage couplets 31–22 (1.07–0.85 Ma) are characterized by lesser amplitude cycles (<100 m eustatic) dominated by the 40-ka spectral signature of axial tilt, whereas stage couplets 21/20 and younger are dominated by the high amplitude (*c.*135 m eustatic) 100-ka cycle characteristic of orbital eccentricity (Ruddiman *et al.*, 1989). Thus, given our preferred correlation to the isotope scale (Figs 2 and 3), Wanganui cyclothems 1–3 and 5 (Butler's Shell Conglomerate to Lower Westmere Siltstone) were deposited under the influence of a sixth-order 40-ka tilt cycle, whereas cyclothems 4 and 6–10 (Kaikokopu Formation to Karaka Siltstone) correspond to the fifth-order 100-ka eccentricity cycle.

Table 2. Major sedimentary facies associations distinguished within the 10 cyclothems of the Wanganui coastal section. A specified interval within a particular formation is nominated as the type for each facies. The distribution of each facies within the Wanganui systems tracts (transgressive systems tract, condensed shellbed and high-stand systems tract) and, where relevant, the equivalent shellbed type, is summarized in the three columns towards the right (see Table 3 for more detail on shellbed types)

Facies		Description	Type	TST	MCS	HST	Environment
Siltstone association							
(massive)	Zm-3	Bands, clumps and scattered shells (including common paired *Tiostrea*, *Chlamys*) in a gritty siltstone matrix	Tainui shellbed (all 4.5 m)		B		
	Zm-2	Massive siltstone; barren to sparsely fossiliferous; bioturbated	Shakespeare Cliff Siltstone (all 9 m)			*	
	Zm-1	Massive fine sandy siltstone, scattered sparse fossils; bioturbated	Lower Kai-Iwi Siltstone (0.2–5.5 m grades to Sz upwards)			*	Inner-mid shelf
(laminated)	Zl-2	Cm-dm interbedded, streaky-laminated and bioturbated siltstone; rare fossils	Lower Westmere Siltstone (4.5–8.5 m)			*	
	Zl-1	Massive silty fine sandstone to fine sandy siltstone with thin streaky laminated zones; scattered fossils	Lower Westmere Siltstone (1.2–4.5 m)	*		*	
Silty sandstone association							
(shell-rich)	Sz-6	Parallel-bedded shell conglomerate, muddy sand matrix	Kupe Fmn. (*Mactra tristis* layer) (all 40 cm)		*		†
	Sz-5	Clumps and bands of shells (*Tiostrea*) in gritty, silty sandstone	Lower Castlecliff Shellbed (0.5–1.5 m)		B		
	Sz-4	Close-packed shells (bored, encrusted) in bioturbated, very gritty, silty sandstone matrix	Lower Castlecliff Shellbed (0–0.5 m)		B		
(matrix-rich)	Sz-3	Disarticulated, bedding-parallel shells in bioturbated gritty, muddy fine sand matrix	Lower Westmere Shellbed (all 10 cm)		B		Inner shelf
	Sz-2	Massive, intensely bioturbated, fine sandy siltstone; scattered fossils	Omapu 'Shellbed' (all 3 m)			*	
	Sz-1	Massive, fossiliferous, intensely bioturbated muddy fine sandstone; abundant scattered shells, shell lags	Pinnacle Sand (0–3 m)	*			
Well-sorted sand association	Sw-3	Intensely burrowed/bioturbated gritty shelly slightly muddy fine sand; occasional mud drapes	Shakespeare Cliff Sandstone (6.5–8 m)	*			
	Sw-2	Parallel-laminated & trough cross-bedded, gritty well-sorted fine sand with shelly lags	Shakespeare Cliff Sandstone (1.5–6.5 m)	A			Intertidal–subtidal shoreface
	Sw-1	As He-1, but with sandy matrix and no associated mud drapes, pebbles or coarse sand	Shakespeare Cliff Sandstone (0–1.5 m)	A			
Heterolithic association	He-3	Lenticular and streaky bedded fine sands and silts; mud dominated	Kaimatira Pumice Sand (0.25–3.5 m)	*			
	He-2	Wavey-flaser-lenticular bedded fine sands and silts; sand or coarse silt dominated	Kaimatira Pumice Sand (3.5–8.5 m)	*		*	Muddy nearshore shelf
	He-1	Metre-scale cross-bedded shelly conglomerate with mud drapes (dune facies), coarse sand matrix	Butlers Shell Conglomerate (all 10 m)	A			

A/B, shellbed type.
* Occurs at Wanganui.
† Palimpsest shell lag on flooding surface.

SEDIMENTARY FACIES ASSOCIATIONS

The Wanganui cyclothems encompass a wide variety of sediment facies representative of coastal and inner-middle shelf environments. Individual sedimentary facies can be grouped into five major facies associations, which are based primarily on lithological characteristics. Each association corresponds also to a broad environment of deposition. We describe briefly below the main characteristics of each of these facies associations, which are summarized in Figs 3 and 12 and Tables 2 and 3.

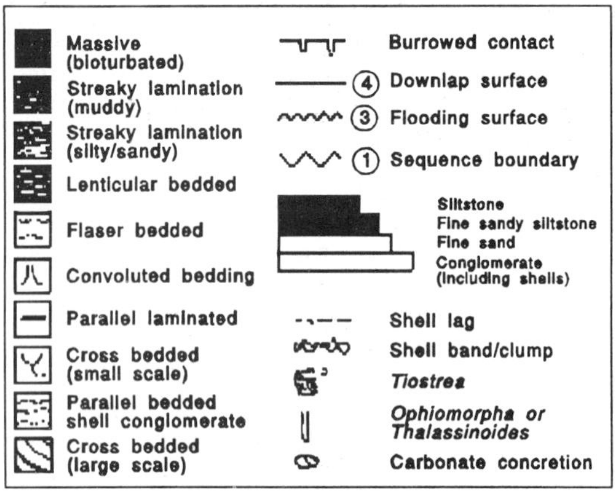

Fig. 4. Key to Figs 5–8.

Heterolithic association (nearshore shelf)
(Figs 4 and 5)

Typical occurrence: Kaimatira Pumice Sand, cyclothem 4

The heterolithic association is characterized by 1–3-m-thick, trough cross-bedded pebbly shell conglomerate (facies He-1), often with foreset mud-drapes. The shell conglomerates contain a mixed fauna of variously abraded and rounded shells and shell fragments from estuarine, open shoreface and numerous offshore environments in a matrix of coarse terrigenous sand and granule-sized shell fragments. Facies He-1 occurs intimately interbedded with finer grained lithologies, which comprise a spectrum of subfacies between lenticular and streaky-bedded siltstone and flaser-bedded and parallel-laminated fine sand or silt (facies He-2, He-3), which often occur interbedded on a centimetre–decimetre scale. Macrofossils are rare in facies He-2 and He-3, but burrows or more pervasive bioturbation are common at some horizons.

Table 3. Major shellbed types distinguished within the 10 cyclothems of the Wanganui coastal section. Type A shellbeds are those deposited in zones of relatively high terrigenous sediment input and within the influence of tidal and storm currents. Type B shellbeds are deposited in sediment starved locations, mainly below storm wave base. Type C shellbeds are associated with the marine ravinement surface at sequence boundaries

Shellbed	Characteristic species	Texture	Matrix	Environment	Typical	Systems tract
Type A1	Largely reworked shells, especially *Mactra tristis*, *Paphies subtriangulata*; *in situ* (?) *Pseudechinus*	Sedimentological (palimpsest)	Coarse sand and shell gravel	Nearshore shelf	Kaimatira Pumice Sand	Transgressive
Type A2	Intertidal and shallow subtidal species; *Dosinia*, *Paphies*, *Mactra*, *Fellaster*, *Divaricella*, *Antisolarium*	*In situ* and near *situm*	Sand and silty sand	Ocean beach, subtidal	Shakespeare Cliff Sand Pinnacle Sand	Transgressive
Type B	Soft-bottom and shell-bottom species; *Dosina*, *Kereia*, *Tawera*, *Pecten*, *Chlamys*, *Austrofusus*, *Tiostrea*, Bryozoa	*In situ* and near *situm*	Sandy silt and silt	Inner shelf Inner-mid shelf	Lower Westmere Shbd. Lower Castlecliff Shbd. Tainui Shellbed	Mid-cycle (between local flooding and downlap surfaces)
Type C	*Barnea similis*, *Pholadidea* sp., *Leptomya retaria*	*In situ* (borings)	Siltstone	Intertidal rock platform	Kupe/Upper Westmere Formation contact	Sequence boundary (ravinement)

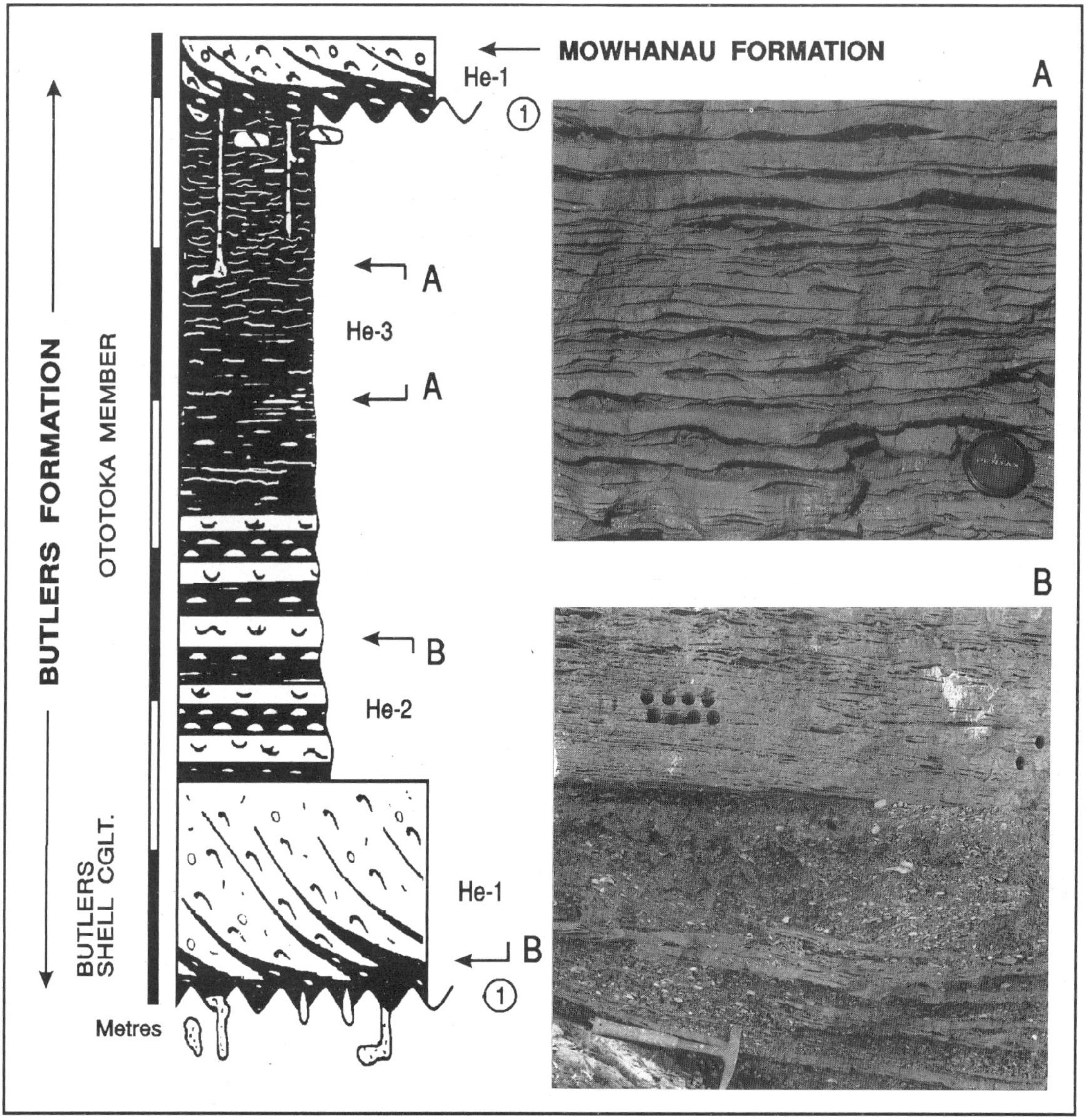

Fig. 5. Heterolithic (He) facies developed within the transgressive systems tract of cyclothem 1 (Butlers Formation) and base of cyclothem 2 (Mowhanau Formation). Inset photos: (A) lenticular and flaser-bedded mud and coarse silt showing some symmetrical (wave oscillation) and unidirectional (?tidal) ripples of coarse silt (facies He-2) (4.5-cm diameter lens cap for scale); (B) cross-bedded foresets of shell gravel (facies He-1, shellbed type A1) with mud drapes, overlain by lenticular and streaky-laminated siltstone (swale facies He-3) (hammer for scale). For detailed key, see Fig. 4.

Interpretation

Small-scale rhythmically bedded structures such as flasers and cross-bedded conglomerates with larger scale mud-drapes are common in modern subtidal to intertidal estuarine environments (e.g. Reineck & Singh, 1986), and on muddy inner shelves subjected to storm activity (e.g. Swift & Rice, 1984). We conclude that the heterolithic association represents a shallow shelf mosaic of migrating pebbly shell–gravel dunes and shoals, composed of transported shell material derived from contemporary paralic

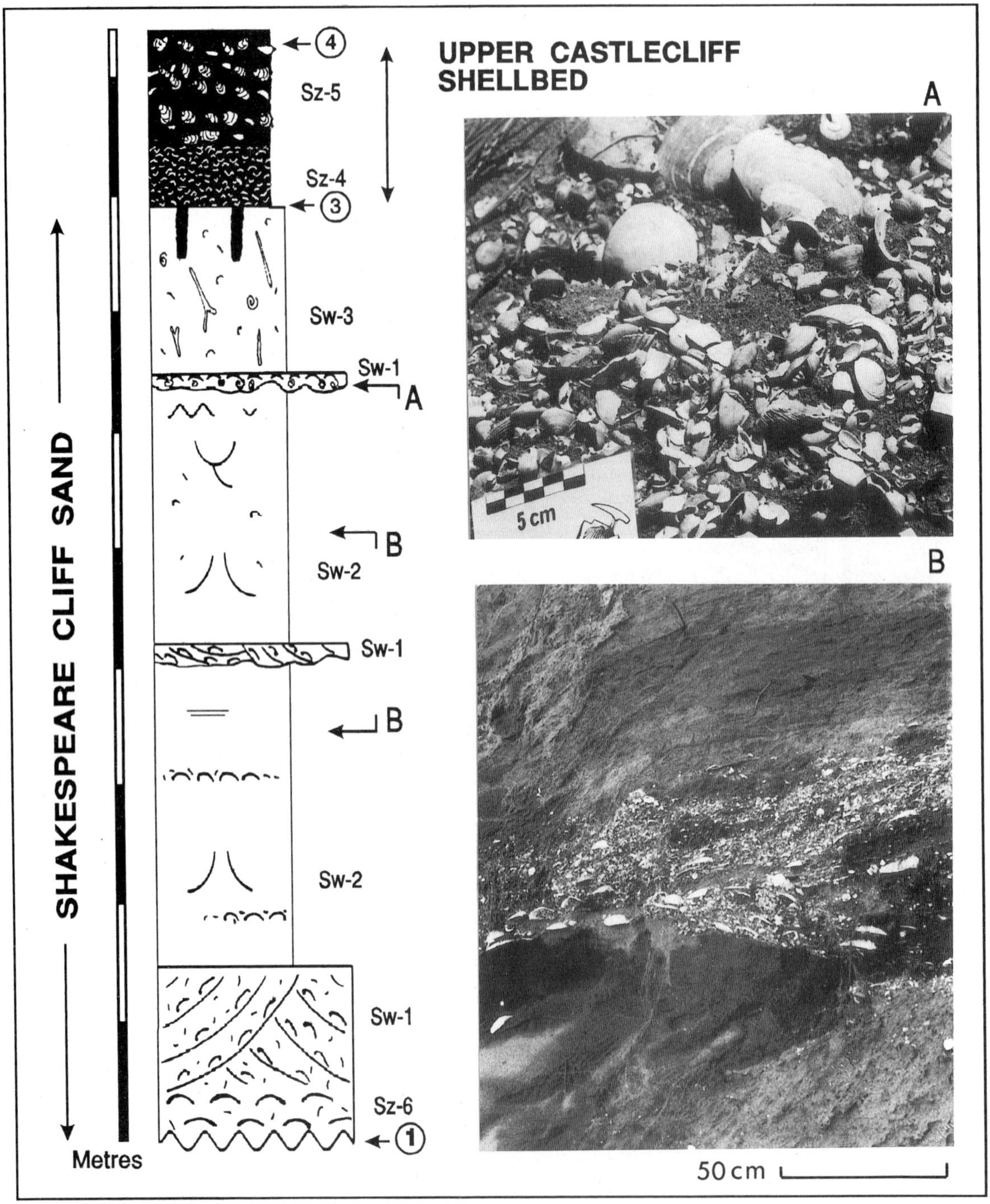

Fig. 6. Well-sorted shoreface sand (Sw) facies developed within the transgressive systems tract of cyclothem 10 (Shakespeare Cliff Sand). Inset photographs: (A) detailed view of shell lens (facies Sw-1), showing the characteristic intertidal–shallow subtidal forms *Paphies* and *Zethalia* (scale in cm); (B) well-sorted, laminated beach sands (facies Sw-2) enclosing a current-emplaced shellbed (facies Sw-2) containing molluscs of open shoreface and surf zone habitat (notably *Paphies*). For detailed key, see Fig. 4.

and inner shelf environments, or eroded from slightly older strata in coastal cliffs, intermixed with river-derived greywacke sand and gravel. These mobile gravel dunes (facies He-1) interfingered with, and passed offshore into, the sand and mud substrate represented by heterolithic facies He-2 and He-3.

Well-sorted sand association (shoreface)
(Figs 4 and 6)

Typical occurrence: Shakespeare Cliff Sand, cyclothem 10

The well-sorted sand facies association contains well-sorted, parallel-laminated to trough cross-bedded, bioturbated, gritty fine sand (facies Sw-2). The association contains a diagnostic sandy beach macrofossil assemblage, but includes also common transported species from nearby estuarine or offshore habitats. Shell material is well preserved to variously rounded and fragmented, and occurs scattered or concentrated into thin lags in mainly convex-up, bedding-parallel orientation; paired bivalves are unusual (facies Sw-2). Sandy, cross-bedded shell conglomerate is also present (facies Sw-1).

Interpretation

The texture, sedimentary structures and fauna of this facies association are consistent with deposition in the shoreface–intertidal zone of a relatively high-energy, exposed coastline, similar to that of the present Wanganui coast. The presence of rare *in situ* bivalves, and highly bioturbated and burrowed facies, indicates an associated lower shoreface, subtidal environment (Howard & Frey, 1973). The presence of rare mud-drapes, and transported estuarine species, indicates also that paralic environments occurred nearby.

Silty sandstone association (inner shelf)
(Figs 4 and 7)

Typical occurrence: Pinnacle Sand, cyclothem 9 (matrix-rich subassociation); Lower Castlecliff Shellbed, cyclothem 8 (shell-rich subassociation)

The silty sandstone association is the most diverse facies grouping recognized, and is subdivided into two major subassociations based upon the respective dominance of terrigenous matrix or shells (Table 2). The Pinnacle Sand (facies Sz-1) comprises massive, intensely bioturbated muddy fine sandstone with abundant scattered shells and shell lags. The fauna includes abundant *Antisolarium*, *Zenatia*, *Serratina*, *Gari* and other species characteristic of inner shelf stations. The Omapu 'Shellbed' (a misnomer, since the shells are widely scattered in matrix support) (facies Sz-2) has a similar but finer grained matrix, and sparser fossils which are of slightly more offshore aspect (*Austrofusus*, *Poirieria*, *Scalpomactra*). In cyclothems 2, 5 and 6, the heterolithic facies is overlain by a mid-cycle, 20–50 cm thick, intensely bioturbated, silty fine sandstone with abundant, scattered fossils of inner shelf habitat (facies Sz-3, Fig. 7B). Shells are well preserved, and most bivalves occur disarticulated, scattered in bed-parallel or bed-oblique orientations.

Silty sand also forms the matrix in the shell-supported mid-cycle shellbed facies of cyclothems 7, 8 and 10. Two different facies are distinguished. The lower part of the Lower Castlecliff Shellbed (0–0.5 m, facies Sz-4, Fig. 7A) comprises a diverse fauna of close-packed, unoriented shells, including common double-valved bivalves (*Purpurocardia*, *Tawera*). The upper part of the shellbed (0.5–1.5 m, facies Sz-5) has a less diverse faunal composition, and is dominated by *in situ* clumps and bands of the oyster *Tiostrea*.

A third shell-rich facies is recognized in cyclothems 4 and 7, within which a thin shell-lag occurs on the inferred mid-cycle flooding surface (facies Sz-6).

Interpretation

The matrix-rich silty sandstone of the Pinnacle Sand occurs within the upper parts of the transgressive systems tract, whereas the majority of facies within this association – shell-rich silty sandstones – occur within the mid-cycle shellbed, i.e. above the transgressive systems tract (see Fig. 9).

Overall, the silty sandstone facies association

Fig. 7. (*Overleaf*) Mid-cycle condensed shellbeds (type B shellbeds) of the silty sandstone (Sz) and siltstone (Zm) facies. Inset photographs: (A) detail of offshore shelf molluscs concentrated in matrix of poorly sorted muddy fine sand (facies Sz-4) (cyclothem 7, Upper Kai-iwi Shellbed) (2-cm coin for scale); (B) inner shelf molluscs, including *in situ* bivalves, within poorly sorted muddy fine sandstone (facies Sz-3) (cyclothem 6, Upper Westmere Shellbed) (photograph of entire shellbed, pencil for scale); (C) scattered *in situ* epifaunal clumps and individuals, mainly *Tiostrea* and Bryozoa, in siltstone matrix (facies Zm-3) (cyclothem 9; Tainui Shellbed, photograph of entire unit, hammer for scale). For detailed key, see Fig. 4.

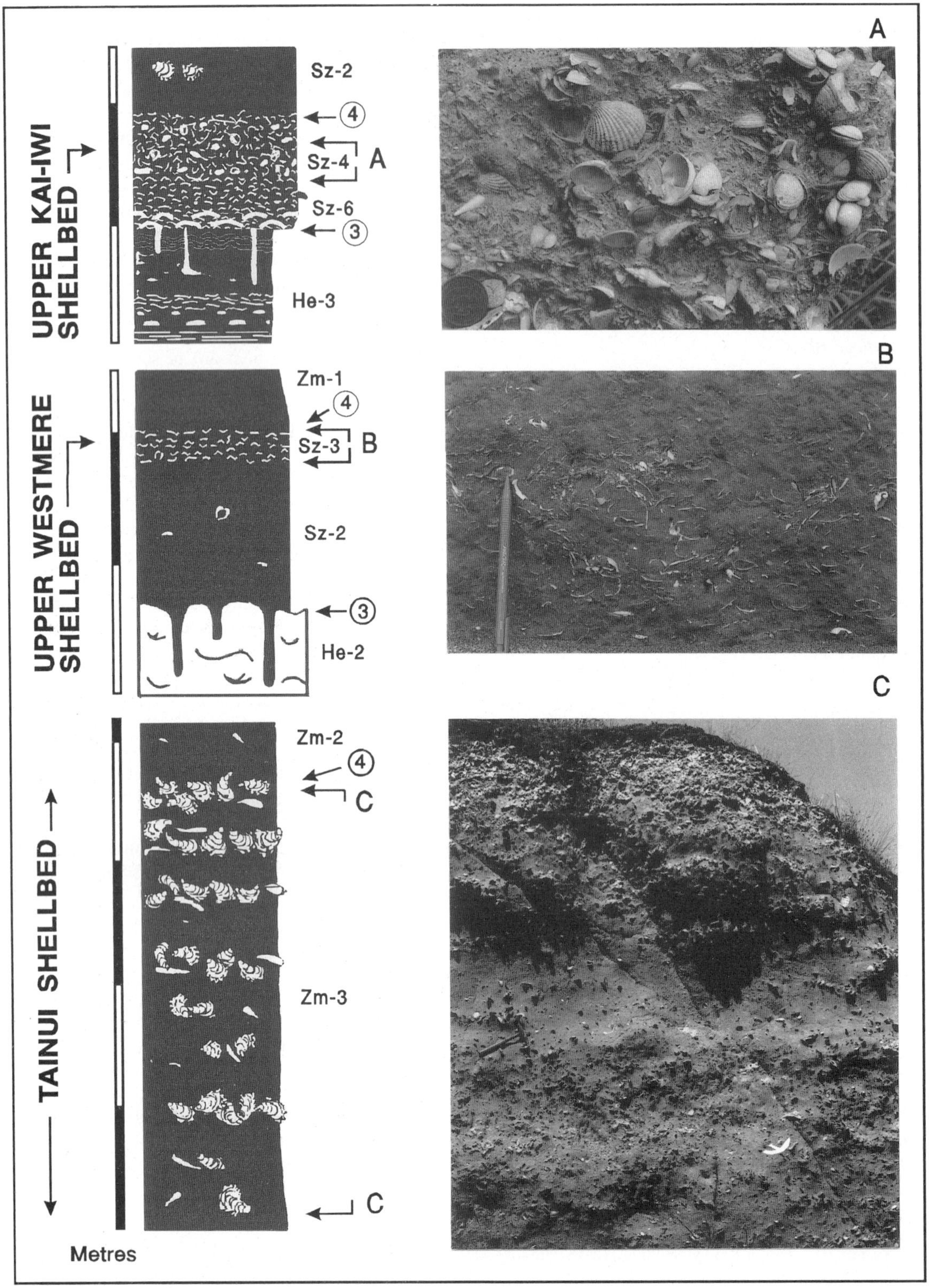
A
B
C
UPPER KAI-IWI SHELLBED
Sz-2
Sz-4
Sz-6
He-3
4
3
UPPER WESTMERE SHELLBED
Zm-1
Sz-3
Sz-2
He-2
TAINUI SHELLBED
Zm-2
Zm-3
Metres

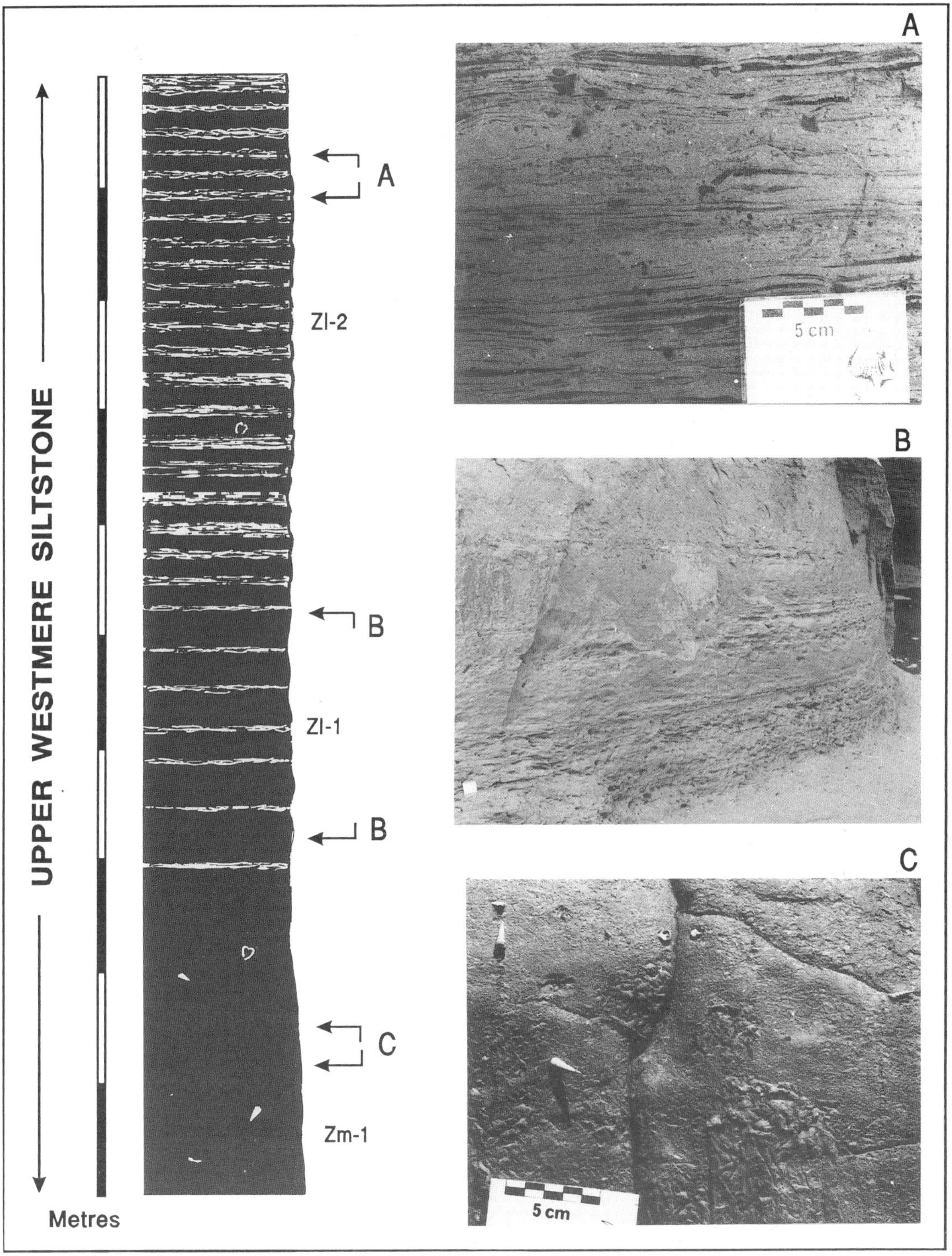

Fig. 8. Shelf siltstone (Zm, Zl) facies developed within the high-stand systems tract of cyclothem 6 (Upper Westmere Siltstone); note that the alternation shown on the column between zones of massive and streaky lamination is schematized. Inset photographs: (A) alternating streaky-laminated and bioturbated siltstone (facies Zl-2); (B) faintly bedded sandy siltstone with mud partings (facies Zl-1) (8-cm-long card for scale at bottom left); (C) offshore shelf molluscan shells (*Chlamys*, *Nemocardium*) concentrated in pods within bioturbated siltstone (facies Zm-1). For detailed key, see Fig. 4.

corresponds to the transition zone of Reineck & Singh (1986, p. 373), i.e. an area of high biological productivity and intermediate sediment texture located between a sand-rich coast and offshore muds, an interpretation consistent with the presence of thin shelly lags and rare shellbeds which represent occasional substrate reworking by storm-waves (facies Sz-1, Sz-2). The mid-cycle condensed shellbed facies grouped with the silty sandstone association contain a characteristic 'shelf' fauna, but their close-packed and in some cases slightly current-modified textures (facies Sz-4) suggest that they accumulated within the reach of occasional storm wave-base.

Siltstone association (inner to middle shelf)
(Figs 4 and 8)

Typical occurrence: Lower Kai-iwi Siltstone, cyclothem 4 (fossiliferous massive siltstone, facies Zm-1); Lower Westmere Siltstone, cyclothem 5 (laminated barren siltstone, facies Zl-1, Zl-2); Tainui Shellbed, cyclothem 9 (shellbed, facies Zm-3)

The shelf siltstone association is composed of blue–grey, terrigenous siltstones (locally termed 'papa') and dominates the coastal succession. Siltstone units are up to 19.75 m thick (cyclothem 2), and consist of at least one of three main facies:

1 Unfossiliferous, centimetre to decimetre alternations of streaky and bioturbated (massive) siltstone, including black reduced layers (facies Zl-1, Zl-2);

2 Intensely bioturbated, fossiliferous siltstones, with scattered soft-bottom fossil assemblages sometimes similar to those in the subjacent shelf shellbed or transition facies (facies Zm-1, Zm-2);

3 Shellbeds comprising a close-packed accumulation of broken and whole shells, including *in situ* occurrences of both soft-bottom infauna and firm-bottom epifaunal forms (*Tiostrea*, *Chlamys gemmulata*, brachiopods, Bryozoa). Bivalve shells are frequently paired, and concentrations of *Tiostrea–Chlamys–*Bryozoa are arranged into crude bands and clumps within an intensively bioturbated siltstone matrix (facies Zm-3).

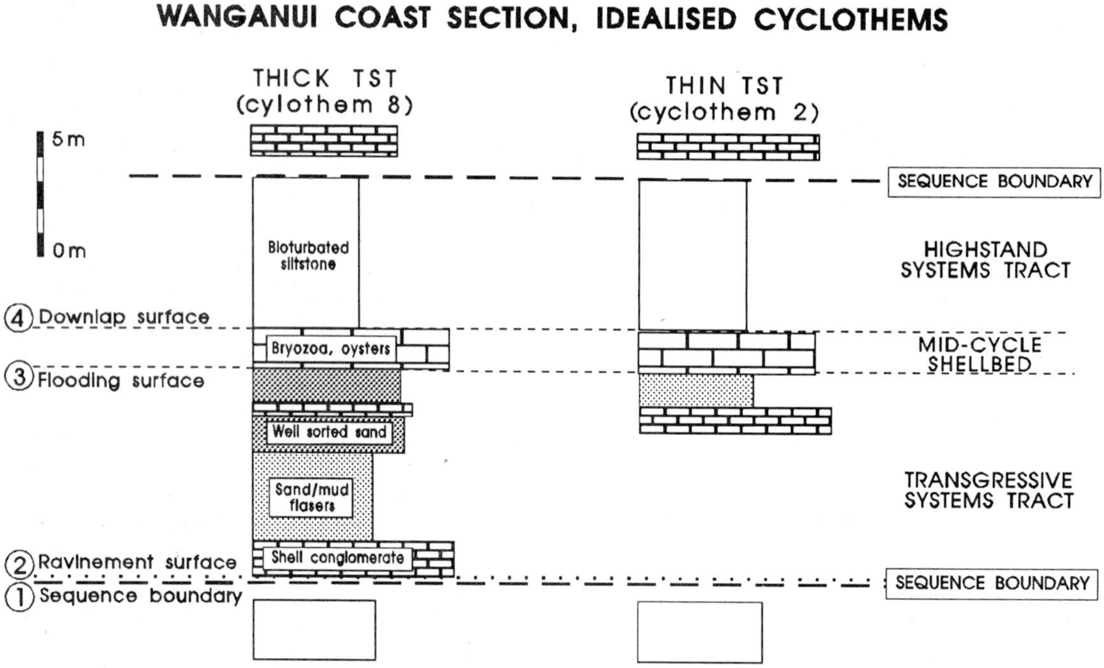

Fig. 9. Summary characteristics of the two main types of cyclothem developed in the Wanganui coast section. Left: cyclothem with a thick transgressive systems tract, including well-sorted sands and shellbeds of the well-sorted sand (shoreface) facies association, indicative of proximity to riverine source or of a pause in the rate of sea-level rise, or both. Right: cyclothem with a thin transgressive systems tract, comprised of heterolithic (tide and storm-influenced) facies, possibly derived exclusively from coastal erosion and reworking during transgression. Thin TST-type cyclothems dominate the Wanganui section, comprising seven of the ten cyclothems present.

Interpretation

The bioturbated siltstone facies includes common *Echinocardium* spines and burrows together with other macrofossils indicative of offshore shelf habitats in water depths of *c.*15 m to greater than 100 m (Fleming, 1953; McKnight, 1969). The foraminiferal faunas are slightly more depth specific, and are consistent with deposition in the shallower parts of that range (N. de B. Hornibrook, personal communication). In several cyclothems, the fossiliferous and massive siltstones occur in a gradational stratigraphic order consistent with a shoaling-upward sequence. Intensely bioturbated fossiliferous siltstone at the base passes up-section into fossiliferous silty fine sandstone (facies Zm-1 to Zm-2; cyclothem 4), or to barren, streaky-laminated siltstone (facies Zm-2 to Zl-2; cyclothems 5 and 6). The first case represents transition into higher energy, relatively shallow water. The second also represents transition upwards into relatively nearshore environments, but where a high rate of mud sedimentation inhibited biological activity, resulting in the preservation of primary bedding features.

The mid-cyclothem Tainui Shellbed (facies Zm-3) of cyclothem 9 is interpreted as the most offshore (i.e. deepest) facies in the Wanganui section, and as accumulating in an offshore low energy environment. A low input of terrigenous sediment allowed the growth of a flourishing epibenthic fauna and the concentration and preservation of shells in a relatively condensed shellbed. The abundant shell detritus, and occurrence of paired bivalves in other than life position, probably mostly reflects biological activities such as bioerosion, burrowing, scavenging and predation. Together with some shellbeds of facies Sz-4 and Sz-5, the Tainui Shellbed displays no sign of physical hydraulic sorting or cross-bedding, indicating that it accumulated below the contemporary storm wave-base.

CYCLE ARCHITECTURE AND SEQUENCE ANALYSIS

The Castlecliff section has been divided into 10 cyclothems, 8.3–28.2 m in thickness, on the basis of the repetitive facies sequence which occurs between cycle-bounding unconformities (Figs 3 and 4). Typically, a cyclothem consists of the following elements in ascending stratigraphic order (Fig. 9):

1 basal erosional unconformity (sequence boundary = ravinement surface);
2 heterolithic and/or well-sorted sand facies association, most commonly including a cross-bedded shell conglomerate (type A shellbed) lying on the basal unconformity;
3 mid-cycle local flooding surface;
4 mid-cycle, condensed shell-rich or shellbed facies (type B shellbed) belonging to either the silty sandstone or siltstone facies association;
5 downlap surface;
6 shelf siltstone facies association, sometimes coarsening upwards.

Relationship between facies associations and cyclothem stratigraphy

In terms of sequence stratigraphy, each cyclothem at Wanganui represents a sequence, and the major intra-cyclothem lithological subdivisions correspond to a transgressive systems tract, mid-cycle condensed shellbed and high-stand systems tract respectively (Fig. 9); low-stand systems tracts are absent. The four major facies associations described above are the heterolithic, well-sorted sand, silty sandstone and siltstone and siltstone associations, and broadly represent successively more offshore depositional environments which could have coexisted across a single shore-shelf transect. Waltherian reasoning allows, therefore, that all facies could be superposed (in the listed order) in a transgressive systems tract, and then superposed again (in reverse order) in a high-stand systems tract. In reality, however, the particular nature of the mid-Pleistocene sea-level cycles (short, fast transgressions and more extended high-stand still-stands or gradual falls), together with the average palaeobasinal position of the sediments preserved in the Wanganui section, combine to produce an asymmetric facies signature. Sediments of transgressive systems tracts overlie an erosional unconformity and mainly encompass the heterolithic and well-sorted sand associations, whereas high-stand systems tracts are dominantly made up of the shelf siltstone association. The mid-cycle condensed shellbeds are slightly more complex, with particular shellbeds belonging to the silty sandstone (either matrix or shell rich) or siltstone associations depending upon the particular position of each shellbed on the palaeoshelf (see Figs 3 and 12 and Tables 2 and 3).

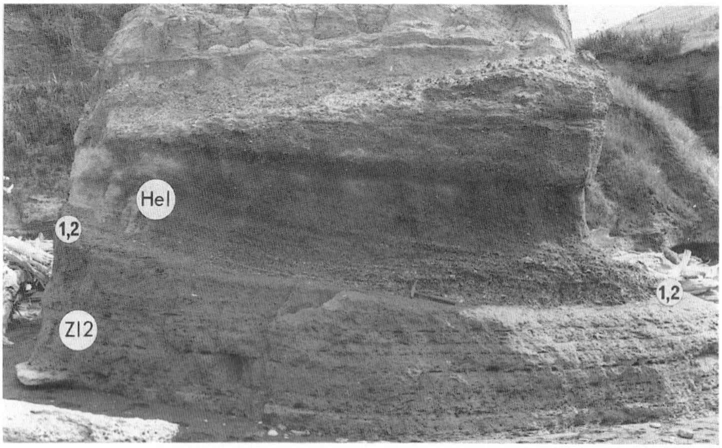

(a)

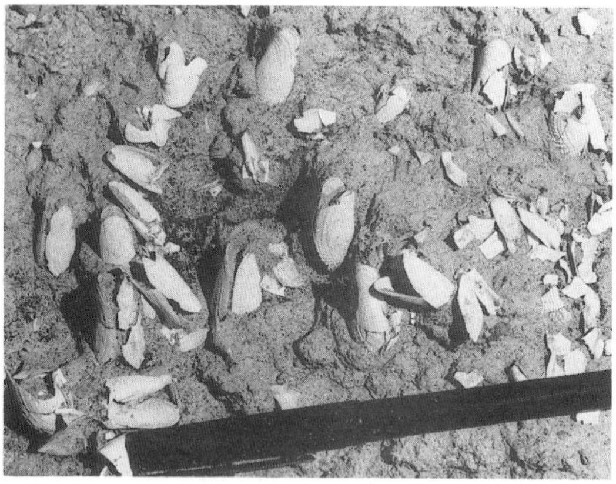

(b)

Fig. 10. (a) Outcrop view of sediments in the vicinity of the sequence boundary between cyclothems 2 and 3 (hammer for scale). Parallel-bedded streaky silts (facies Zl-2) of the Lower Okehu Siltstone (high-stand systems tract of cyclothem 2), overlain sharply at the sequence boundary by large-scale cross-bedded shell conglomerate (facies He-1) of the Okehu Shell Grit (transgressive systems tract of cyclothem 3). (Outcrop capped by surficial, late Quaternary, terrace tread which is not part of the cyclothemic sequence.) (b) Bedding plane view of the sequence boundary between cyclothems 7 and 8, showing *in situ* double valves of the rock-boring pholad *Barnea similis* penetrating into the Upper Kai-iwi Siltstone (14-cm-long pen for scale).

Sequence boundaries

Bases of sequences are marked by erosional unconformities, with traction-emplaced sandy or gravelly sediment resting on a sharp, unweathered, eroded, planar surface cut into the siltstone association of the underlying cycle. The sequence boundary at the base of cyclothem 1 exhibits broadly undulating relief of up to 10 m on a wavelength greater than 100 m, reflecting the presence of erosive gutters or channels which cut across the contemporary shelf bathymetry. For all other cyclothems, the sequence boundaries are essentially horizontal, with barely perceptible relief up to a few decimetres. Erosion at the unconformity is demonstrated by the common presence of small irregularities in the top of the underlying unit, and by the occurrence on and just above the sequence boundary of mud-clasts or concretions derived from the underlying siltstone (Fig. 10).

Sequence boundaries are penetrated by grit-filled thalassanidian burrows up to 2 m deep, and by preservation within their burrows of double valves of the rock-boring marine pholad *Barnea similis* (type C shellbeds; Table 3). Should the pholad shells not have been fossilized, the trace-fossil assemblage that would be present is identical with the *Glossifungites* ichnofacies of Pemberton & Frey (1985).

The evidence summarized above is consistent with each cycle-bounding unconformity representing a surface of marine planation, cut by the landward passage of the shoreface of a transgressing sea (Curray, 1964; Swift, 1968). Besides marking sequence boundaries between successive cyclothems, each unconformity therefore also coincides with a ravinement surface (Demarest & Kraft, 1987;

Nummedal & Swift, 1987). In addition, the sequence boundary at the top of cyclothem 1 (Butler's Formation) has superposed on it the mid-cycle local flooding surface (equivalent to the submarine unconformity of Embry, 1990), because it is onlapped laterally by the base of a rolled concretion layer which further south forms the mid-cycle condensed shellbed of cyclothem 2 (see Fig. 12).

The extent of erosion at the Wanganui sequence boundary unconformities is such that soils or nonmarine sediments marking the subaerial exposure during the preceding glacial low-stand have been removed, together with an unknown thickness of the topmost high-stand systems tract of the underlying cycle. However, soils are known to occur on sequence boundaries on the eastern side of the basin in the Rangitikei Valley (e.g. Abbott, 1992).

Transgressive systems tracts

The transgressive systems tract for eight out of the ten Wanganui cyclothems comprises mainly the heterolithic facies association, including at the base cross-bedded shell conglomerate (type A shellbeds; Table 2, Figs 2 and 5). The two exceptions are cyclothems 9 (Pinnacle Sand) and 10 (Shakespeare Cliff Sand).

Transgressive sequences are mostly thin, generally between 0.3 and 8 m, suggesting a high rate of sea-level rise relative to sediment supply. Possibly the sediment was exclusively derived from coastal reworking and erosion. Despite the abundance of sedimentary features consistent with deposition from rapidly alternating currents, there exists little evidence for true intertidal deposition (for instance, edgewise mud-chip conglomerates are absent). We conclude, therefore, that facies He-1 to He-3 represent deposition on a fully marine, nearshore, shallow, tide and/or storm-dominated, current-swept shelf.

The atypically thick Seafield Sand of cyclothem 8 comprises 15.5 m of heterolithic association followed gradationally by 1.5 m of well-sorted sand association. In contrast to the transgressive models of Demarest & Kraft (1987) and Nummedal & Swift (1987), no sharp ravinement surface of marine erosion occurs within the transgressive systems tract of this cyclothem. We conclude that this facies succession represents a small seaward-prograding paracycle deposited during a slow-down or pause in the overall transgression.

Mid-cycle condensed shellbeds

Mid-cycle shellbeds include one or more of the shell-rich/shellbed facies Sz-3 to Sz-5, Zm-3. Note that a mid-cycle shellbed may include shell-rich (Sz-3) as well as shellbed facies, and that it comprises only one part of a complete mid-cycle 'condensed section' (Fig. 12). The mid-cycle shellbed is bounded above and below by the downlap and local flooding surfaces respectively. However, relatively condensed sedimentation may be present also in the uppermost part of the underlying transgressive systems tract and the lower parts of the overlying high-stand systems tract (Baum & Vail, 1988).

A condensed shellbed results from the drowning of the shelf by a rapid sea-level rise, such that portions of the offshore shelf come to be located seawards of the main shore-connected sediment bodies. The development of a typical condensed shellbed therefore requires both the absence of a major supply of sediment and the presence of an *in situ* fauna. The omission surface at the base of the shellbed marks the local marine flooding surface (unconformity 3 of Fig. 12), and is coincident with the surface of maximum starvation (Baum & Vail, 1988) and with the submarine unconformity of Embry (1990). The condensed shellbed generally grades rapidly (over a few centimetres) up into terrigenous silts of the high-stand systems tract, this contact marking the downlap surface of Vail *et al.* (1984) and Embry (1990) (unconformity 4 of Fig. 12). Van Wagoner *et al.* (1988, p. 44) equate the downlap surface with the maximum flooding surface. However, the maximum flooding surface is seldom able to be located in outcrop, and observed local flooding surfaces rarely equate with the downlap surface (see under heading Maximum Flooding Surfaces below).

Baum & Vail (1988) recognize two parts to a mid-cycle condensed shellbed, the lower coinciding in ancient examples with a lithified, glauconitic, and/or burrowed omission surface or hardground, and the upper being the shellbed itself. Except for the glauconitized rim which occurs on boulders in the rolled concretion layer of cycle 2, no such alteration is observed on the Wanganui mid-cycle flooding surfaces, probably because they were not exposed at the sea floor for long periods of time. At Wanganui, the mid-cycle condensed shellbeds therefore correspond only to the 'upper condensed section' of Baum & Vail (1988). The shellbeds comprise a rich *in situ* and near *situm* fauna (autochthonous and

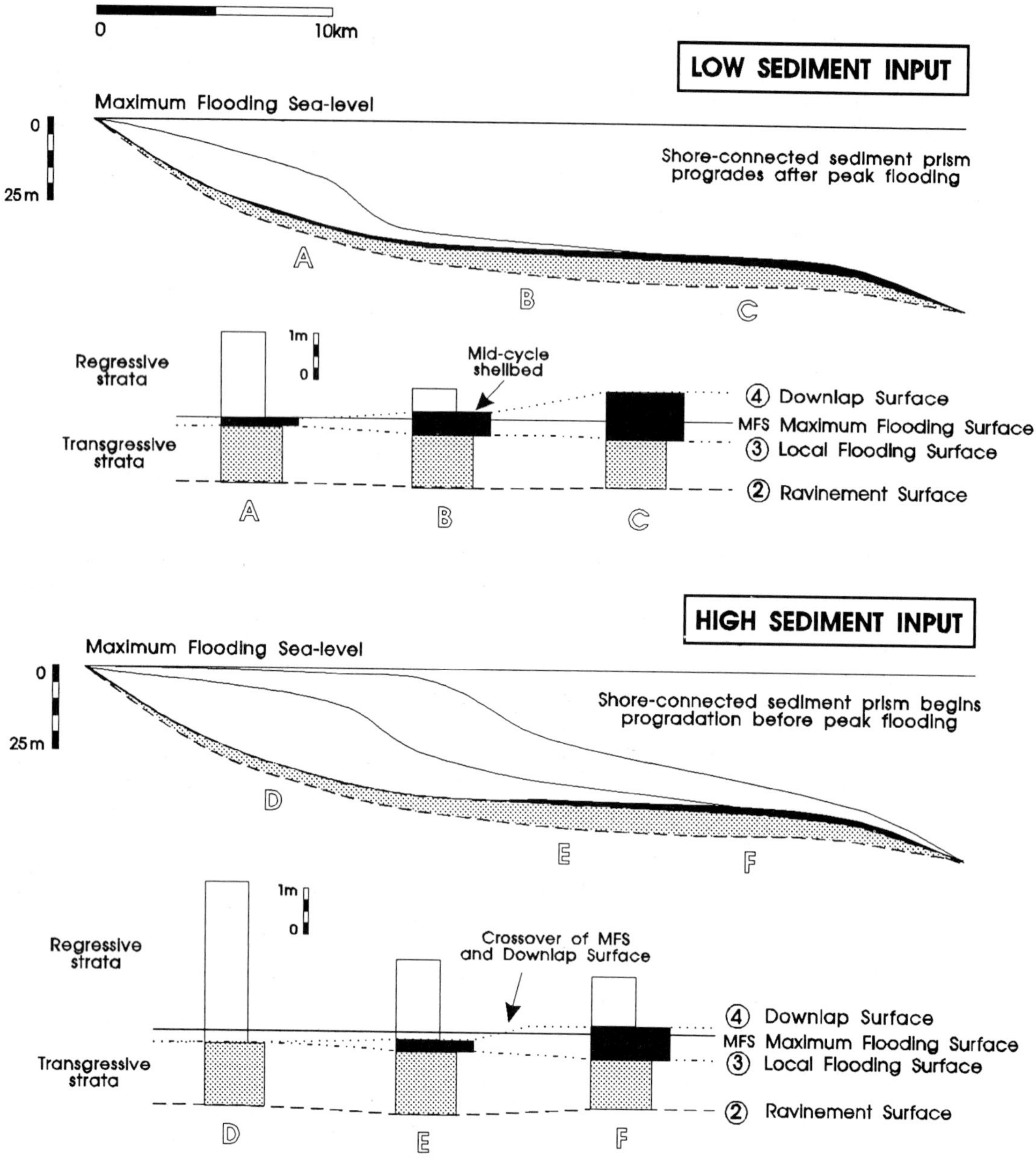

Fig. 11. Cartoons summarizing the relationship between the theoretical maximum flooding surface, the mid-cycle condensed shellbed (black) and identifiable disconformities in outcrop which equate with the ravinement (2), local flooding (3) and downlap (4) surfaces respectively (see also Fig. 12). Upper: columns A, B and C as developed at successively more offshore positions under a regime of low sediment input. Lower: columns D, E and F developed at similar shelf stations to A, B and C, but under conditions of high sediment input. Note that in this case the maximum flooding surface may come to be located within the lower parts of the regressive section. For full explanation, see text.

parautochthonous of Kidwell, 1986) of shallow infaunal and epifaunal invertebrates, dominated by Mollusca, and in facies Zm-3 including clumps of oysters and bryozoans. The biofacies and thicknesses present in particular condensed shellbeds is a reflection of their initial depth of drowning, and of the length of time that the sea floor lay exposed before the arrival of the blanketing sediment of the shelf siltstone facies association (Fig. 11). In the ideal case, and for a given cycle length, the closer to shore that a condensed shellbed develops:

1 the thinner the shellbed, because of the relatively shorter period of time for which the sea floor lies uncovered by prograding, shore-derived sediment; and

2 the less diverse the fauna, and the more shallow-water, inner shelf elements it contains, because benthic faunal diversity increases towards the outer edge of the shelf (e.g. Probert *et al.*, 1979; Probert & Wilson, 1984).

A particularly important implication of the pattern displayed by the Wanganui cyclothems is that a mid-cycle condensed shellbed of facies Sz-3 or Sz-4 up to 30 cm thick can form in only a few thousand years. This inference can be made because (i) cyclothems 1–6 were probably each deposited during a 40-ka obliquity cycle, associated with a sea-level fluctuation of about 100 m (e.g. Ruddiman *et al.*, 1989); (ii) the water depth at the maximum flooding surface for any cyclothem is estimated to have been about 20–40 m (Carter *et al.*, 1991); and (iii) therefore each cyclothem as represented in outcrop corresponds to at most 30 ka of sedimentation, which period has to encompass deposition of the highstand and low-stand systems tracts as well as the development of the condensed sequence shellbed. Similar reasoning indicates that the thicker condensed shellbeds of cyclothems 8–10 (100-ka cycles) were probably deposited in a few tens of thousands of years, which longer timespan may be a prerequisite for the development of facies Zm-3 shellbeds.

Given these interpretations, it is entirely possible that other repetitive mid-cycle shellbeds in the geological column also represent a duration of a only few thousand years, for example the cyclic shellbeds of the Belemnite Marls of the Dorset Lias (Weedon, 1985).

Relationship between transgressive, downlap and maximum flooding surfaces

The low resolution of the seismic records from which the concept of sequence stratigraphy arose has led to ambiguities in use and definition of the terms transgressive surface, maximum flooding surface and downlap surface. Though the original concept (Vail *et al.*, 1977, p. 78, fig. 13a) of a single, mid-cycle unconformity surface, extending from basin margin to centre, has now been refined, misinterpretations remain. For instance, it is not always true that 'the top of the transgressive systems tract is the downlap surface' which 'marks the change from a retrogradational to an aggradational parasequence set and is the surface of maximum flooding' (van Wagoner *et al.*, 1988, p. 44; shown diagrammatically by Loutit *et al.*, 1988, fig. 9). Relationships observed on seismic sections may have made it useful to include the condensed shellbed in the top of the transgressive systems tract, but we prefer in this paper to treat the mid-cycle shellbed as a discrete unit located at the junction between the transgressive and highstand systems tracts.

The stratigraphy of the Wanganui cyclothems is known at a much higher level of resolution than seismic studies, and shows the consistent presence of at least four surfaces of stratal discontinuity within a typical sequence (see Embry, 1990). These surfaces (Figs 9, 11 and 12) are shown below.

Unconformity 4	submarine, starved	downlap surface
Unconformity 3	submarine, transgressed	mid-cycle local flooding
Unconformity 2	shoreface	ravinement
Unconformity 1	subaerial	sequence boundary

Towards basin margins, the sequence boundary (unconformity 1) and the mid-cycle flooding surface (unconformity 3) are separated by a thin sequence of transgressive sediment, within which may occur one or more ravinement or diastem surfaces, depending upon the palaeogeography of the particular locality (see Nummedal & Swift, 1987, pp. 246–8). In the offshore basin, however, unconformities 1, 2 and 3 may all be coincident, or nearly so (Fig. 12), particularly on a rapidly transgressed part of the shelf or in areas of low sediment supply (see Loutit *et al.*, 1988, fig. 30). Unconformity 4 corresponds to the top of the mid-cycle condensed shellbed which usually accumulates on all parts of the transgressed shelf lying outside the contemporary shore-connected clastic wedge. The condensed section below unconformity 4 may therefore thicken, and certainly represents more elapsed time, in a sea-

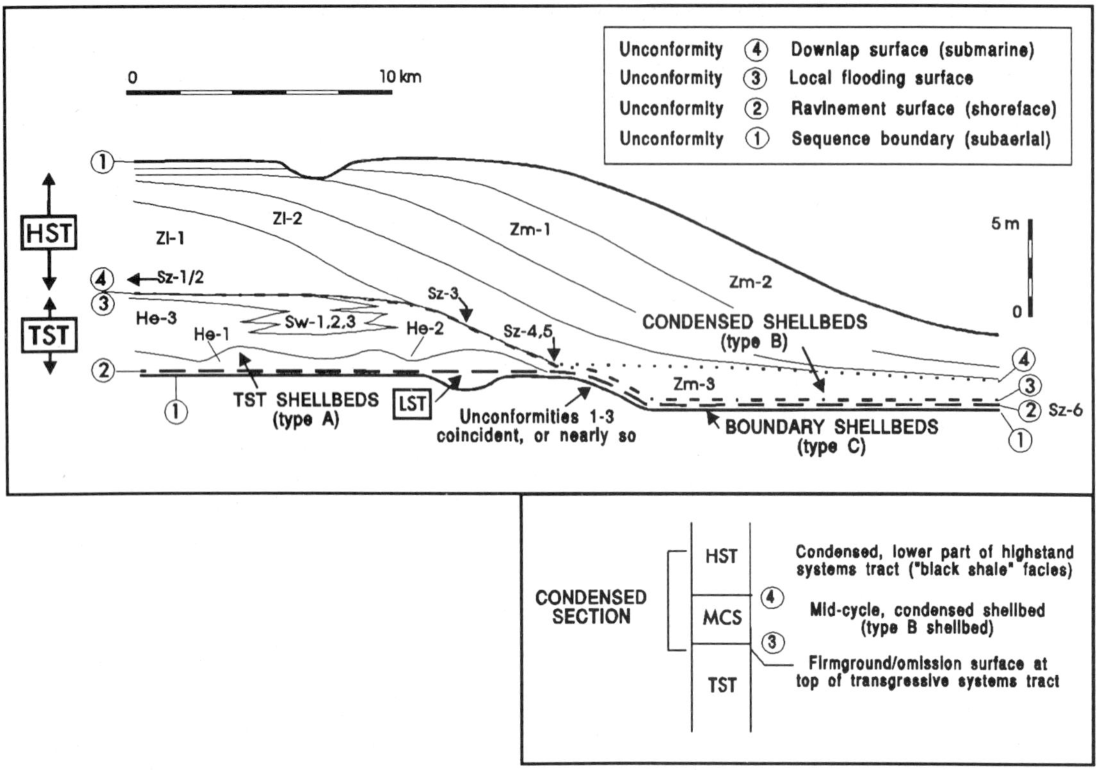

Fig. 12. Summary diagram of the facies, unconformities and systems tracts present within the Wanganui cyclothems. Refer also to Table 1 and Figs 3 and 11.

wards direction. The smothering of a condensed section by terrigenous material marks the arrival of shore-derived clastic sediment, and unconformity 4 accordingly corresponds to the downlap surface. The age of the downlap surface (as marked by the age of the sediment immediately above it) is not isochronous, but becomes younger seawards.

The four surfaces discussed above represent physical features which can be directly identified in outcrop, and which generally correspond also to unconformities. In contrast, the maximum flooding surface is a conceptual surface, corresponding to the time of maximum shoreline transgression within a cycle. Accordingly, the maximum flooding surface is in general not identifiable in outcrop (see Haywick, 1990) and may lie anywhere between the mid-cycle flooding surface (= unconformity 3) and a position low in the high-stand systems tract (see Fig. 11, and further discussion below).

Maximum flooding surfaces

In the Exxon sequence stratigraphic model, the maximum flooding surface represents the shelf-wide bedding surface which corresponds to the time of maximum shoreline transgression during a rising cycle of sea level. The maximum flooding surface therefore coincides with 'the change from a retrogradational to an aggradational parasequence set' (van Wagoner *et al.*, 1988). Depending upon the relative balance between rate of sea-level change, rate of sediment supply and palaeogeographic location, there are three limiting cases for the position of the maximum flooding surface (Fig. 11).

Maximum flooding surface at base of condensed shellbed

In the case of a nearshore location immediately outside the shore-connected sediment wedge, the transgressive surface and the maximum flooding surface may coincide, i.e. there is insufficient time between local transgression and peak flooding to allow any thickness of shellbed to develop (e.g. Fig. 11, positions landward of point A). In such cases, the condensed shellbed will generally be thin, or even represented by a single layer, shell pavement on the top of the underlying transgressive systems tract.

Maximum flooding surface within condensed shellbed

For locations a little further seaward on the inner shelf, and for relatively low terrigenous sediment input, there will be time for a condensed shellbed to develop after drowning and before the shoreline reaches its maximum position. The shellbed will then continue to accumulate during the ensuing sea-level pause or fall, and during the initial progradation of shore-derived silts of the high-stand systems tract. In this scenario, the maximum flooding surface will be located somewhere within the condensed sequence shellbed (e.g. Fig. 11, positions B, C, F). Note that only at particular (and unusual) locations where a moderate sediment input is exactly counterbalanced with the turnaround of the sea-level cycle will the maximum flooding surface lie at the top of the condensed shellbed (e.g. Fig. 11, point A).

Maximum flooding surface above the condensed shellbed

In circumstances of high sediment supply, progradation of the 'high-stand' systems tract may commence before the point of maximum sea-level rise, particularly where the last parts of the sea-level rise are markedly slowed. In such cases, the prograding silts may start to blanket an offshore condensed shellbed before the time of peak sea level, and the maximum flooding surface will be located within the basal part of the 'high-stand' systems tract (e.g. Fig. 11, points D, E). Haywick (1990) has described such a situation within Plio-Pleistocene cyclothems on the eastern side of North Island, New Zealand.

High-stand systems tracts

High-stand systems tract sediments form during the high-stand and early fall of the ideal sea-level cycle (Vail, 1987). The high-stand systems tract of each Wanganui sequence consists of shelf siltstone facies up to 28.2 m thick. Though individual mudstones show evidence for a shallowing-upward character (e.g. cycles 2, 4–6), the Wanganui cyclothems were apparently located too far seawards of the contemporary high-stand shoreline for shoreface or intertidal facies to be represented in the uppermost high-stand systems tract, with the possible exception of the ?distal shoreface Omapu Shellbed in cyclothem 4. However, regressive shoreline deposits within the high-stand systems tract are known from correlative mid-Pleistocene cyclothems in the Rangitikei Valley (Abbott, in prep.), and in older Plio-Pleistocene cyclothems in Hawkes Bay (Haywick, 1990; Haywick *et al.*, 1991). (It is, of course, possible that a small thickness of comparable facies did develop within some of the Wanganui cyclothems, but was later removed by erosion at the superjacent sequence boundary/ravinement surface.)

The high-stand systems tract siltstones in the Wanganui section are therefore interpreted as the seaward parts of regressive, mud-dominated, shore-connected sediment wedges analogous to those found within coastal bights on the modern New Zealand shelf (e.g. Carter *et al.*, 1985), or on the mud-dominated shelf off northern California (Leithold, 1989).

IMPORTANCE OF SHELLBED TYPE FOR CHARACTERIZING SEQUENCES

The shellbed types preserved within the Wanganui coastal section are a powerful indicator of sequence architecture, particularly when considered in association with their enclosing sedimentary facies. Following Carter *et al.* (1991), we recognize two main types of shellbed, termed types A and B, and associated respectively with transgressive systems tracts and mid-cycle flooding surfaces. We also introduce the term type C shellbed for infaunal communities preserved at sequence/ravinement boundaries (Table 3).

Type A shellbeds

Type A shellbeds are deposited in association with an abundant supply of terrigenous sediment (Table 3), i.e. within the body of either a transgressive or a high-stand systems tract. In Pleistocene strata, type A shellbeds can be subdivided according to the known environment of deposition of their constituent fauna (Carter *et al.*, 1991, table 4). However, we discuss here only the two major shellbed varieties which are potentially recognizable also in much older rocks.

Type A1 shellbeds are always associated with abundant pebbly or coarse sandy sediment, and comprise mostly transported and worn shells (sedimentological type of shellbed of Kidwell, 1986). Intertidal and shallow-marine invertebrate species predominate, with densely packed shellbed fabrics. Sedimentary structures are abundant, particularly cross-bedding, and bivalve shells are single-valved and concentrated parallel to the bedding (e.g. Beckvar & Kidwell, 1988, fig. 6; Fig. 5 this paper).

Type A2 shellbeds generally occur as shell-lags and small channel fills within well-sorted fine sand. Molluscan species are preserved *in situ* or near *situm*, and sometimes include double-valved bivalves. They represent the preservation of estuarine (*Austrovenus stutchburyi*, *Paphies australe*), beach (*Dosinia* spp., *Paphies subtriangulata*, *Umbonium* or *Mactra* spp.) and shallow shoreface (*Divaricella*, *Scalpomactra*, *Serratina* or *Zethalia*) environments.

Type A1 and A2 shellbeds are associated with transgressive systems tracts at Wanganui.

Type B shellbeds

Shellbeds associated with conditions of relative sediment starvation that form near the maximum flooding surface of each sequence correspond to type B mid-cycle condensed shellbeds. Typically, therefore, type B shellbeds accumulate on the transgressed shelf, seawards of any contemporaneous shore-connected terrigenous sediment prism. Varieties of type B shellbed that are preserved in the silty sandstone association (e.g. facies Sz-4, Sz-5) sometimes show parallel orientation of bivalve shells, or crude bedding, and therefore probably accumulated on the inner shelf at the feather-edge of any shore-connected sediment body and at depths within the reach of wave or tidal currents (Fig. 12) (see Wilson, 1986). However, the shell-rich silty sandstones (Sz-4, Sz-5) and siltstones (facies Zm-3) are entirely unsorted and accumulated at depths below normal wave-base (Fig. 12).

Type C shellbeds

We introduce here the term 'type C shellbed' for infaunal shellbeds associated with sequence boundaries and/or ravinement surfaces (Table 3). Type C beds occur as single *in situ* layers of sediment-burrowing or rock-boring species, and mark the movement of a zone of shoreline erosion across a shallow, wave-planed cohesive substrate or soft rock platform. At Wanganui, type C shellbeds of the pholad *Barnea similis* mark the basal sequence/ravinement boundaries of cyclothems 1–4 and 6–8.

RELATIONSHIP BETWEEN SEQUENCE THICKNESS AND CYCLE DURATION

Cyclothems 1–6 at Wanganui were deposited during periods of sea-level fluctuation controlled primarily by a 40-ka (tilt) forcing mechanism, whereas cyclothems 7–10 were deposited under a 100-ka (eccentricity) regime (see Ruddiman *et al.*, 1989). In favourable ocean-basin settings, where rates of subsidence and pelagic sediment accumulation are approximately constant, it is possible to distinguish the signature of Milankovitch cycles using power spectrum analysis of the oxygen isotope record for example.

For continental margin locations such as Wanganui, however, sequence and systems tracts thicknesses are the result of complex interaction between rates of tectonic subsidence (or uplift) and rates of sediment supply, as well as sea-level fluctuation (see Galloway, 1989). Moreover, whilst the climatic/sea-level cycles may be viewed as global, and are now well determined for the Pleistocene, subsidence and sediment supply vary from place to place and are not independent variables. It is therefore, in general, not possible to identify the exact part played by each of these three variables in controlling the thickness of continental margin cyclothems (see Kendall & Lerche, 1988).

Acknowledging this problem, the systems tracts within the Wanganui section nonetheless display a number of stratigraphic or sedimentological patterns which can be related plausibly to one or more of the three main parameters which together controlled sediment deposition.

Sediment supply and rate of sea-level rise

Within the 10 transgressive systems tracts at Wanganui, the Kaimatira Formation (in cyclothem 4, 11.1 m thick) and Seafield Sand (in cyclothem 8, 20.5 m thick) stand out as atypically thick (Fig. 3).

Regionally, the Kaimatira Formation (=Potaka Pumice, Seward, 1976) has a pumiceous provenance, corresponding to a pulse of volcanism in the hinterland. Though only the top metre of the formation is conspicuously pumiceous at the coast, it is possible that the abnormal thickness of the Kaimatira Formation compared with the other coastal cyclothems is partly the result of an increased sediment supply to the shoreline during cyclothem 4.

The Seafield Sand is associated with the most varied, and thickest, progradational paracycle within a transgressive systems tract at Wanganui (Fig. 3). The anomalous thickness again suggests an enhanced sediment supply, but would also be consistent with a significant slowing or pause in the rate of sea-level rise during the stage 16 to 15 transgression or the early part of the stage 15 interglacial. This pause in sea-level rise occurred at the time when the transgressing shoreline happened to be in the vicinity of the present Wanganui coast. The Seafield Sand, and perhaps also the similar but thinner Shakespeare Cliff Sand, therefore might be interpreted as mid-Pleistocene examples of the 'drowned' shelf shorelines that are known to characterize the last postglacial transgression (stage 2 to 1; e.g. Sanders & Kumar, 1975; Carter *et al.*, 1985).

Length of climatic cycle

Transgressive systems tract

Leaving aside the Kaimatira and Seafield Formations as atypical, and the Butler's Formation because of its incompleteness, the thickness relations of the transgressive systems tracts at Wanganui are as follows:

Cycles 1–3, 5	Range 1.1–2.75 m	Mean 1.96 m
Cycles 6–7, 9–10	Range 2.8–8.0 m	Mean 4.88 m

The ratio between the length of the 40-ka and 100-ka climatic cycles (0.4) is closely similar to the ratio between the average transgressive sediment thicknesses for cyclothems 1–6 and 7–10 (0.4). This relationship suggests that the Wanganui section received a relatively constant supply of sediment during the transgressive part of successive mid-Pleistocene interglacials (apart from cycles 4 and 8). Such a constant supply might suggest that the sediment in successive cycles was derived largely or entirely from coastal erosion, which is itself controlled by the rate of sea-level rise (G. Postma, personal communication). Be that as it may, the thickness of each transgressive systems tract apparently bears a direct relationship to the length of its equivalent climatic forcing cycle.

Mid-cycle condensed shellbeds

Similar relationships exist between the thicknesses of the mid-cycle condensed shellbeds, namely:

Cycles 2–5	Range 10–30 cm	Mean 16 cm
Cycles 6–10	Range 0.25–4.5 m	Mean 1.65 m

These figures demonstrate that the condensed shellbeds associated with the 100-ka cycle are about an order of magnitude thicker than those associated with the 40-ka cycle. This increased thickness is probably partly a function of the length of the forcing climatic cycle, and partly a function of the magnitude of the associated sea-level oscillation (see below).

Total cyclothem thickness

Despite the relatively thick transgressive systems tract in cyclothems 4 and 8, the thickness of the whole cyclothem is in each case not inconsistent with its timespan. In fact, for all cyclothems except those of cycles 2 and 7, a consistent relationship exists between total cyclothem thickness and duration as judged by correlation to the isotope stage time scale (Fig. 2). On this criterion, however, cyclothem 7 is too thin for its duration, and cyclothem 2 is too thick (Fig. 2).

Magnitude of sea-level oscillations

The change in frequency from 40-ka to 100-ka climatic cycles which occurred at *c*.0.93 Ma (Fig. 2) was accompanied by an increased severity of climatic fluctuation and therefore probably by an increased magnitude of glacial/interglacial sea-level change. Prior to stage 25, paired isotope stages characteristically display 40-ka long fluctuations equivalent to 100 m or less sea-level change, whereas for stages 25–22 (as a group), and later couplets, the changing isotope ratios are consistent with 100-ka-long and *c*.120–140-m-magnitude cycles.

Such a difference in magnitude of the sea-level cycle can be predicted to have three main stratigraphic effects, all of which are observed at Wanganui.

1 Other factors remaining equal, the mid-cycle condensed shellbeds in 100-ka cyclothems will be deposited in water a few tens of metres deeper than the shellbeds in 40-ka cyclothems. This is borne out at Wanganui by the restriction of deeper water shellbed facies Sz-4, and particularly Zm-3, to cyclothems 7–10 (Fig. 3).

2 Regressive systems tracts deposited during a 100-ka cycle will be prograding from a more distant shoreline than their 40-ka cycle equivalents. High-stand facies directly overlying the mid-cycle shellbed should therefore represent a more slowly deposited and offshore shelf environment in a 100-ka cyclothem, and a faster accumulating and nearer shore environment in a 40-ka cyclothem. At Wanganui this may be manifested in the offshore facies Zm-2 which characterizes the Shakespeare Cliff Siltstone (cycle 9, 100 ka), and the shore-proximal facies Zl-1 and Zl-2 which characterize the Lower Okehu Siltstone (cycle 2, 40 ka years).

3 Given that the maximum-flooding shoreline was more distant for 100-ka cycles, their mid-cycle condensed shellbeds should not only be of deeper water aspect (1 above), but also relatively thicker, than their 40-ka counterparts, because they will have lain exposed on the sea floor for longer before being covered by the advancing clinoforms of the high-stand systems tract (see Fig. 11). Such an increased thickness is seen in the mid-cycle shellbeds of cyclothems 8–10 at Wanganui, and perhaps especially in the 4.5-m-thick Tainui Shellbed (cyclothem 9).

Rates of tectonic deformation

The sequence boundary at the base of the Butler's Formation (cyclothem 1) corresponds to a regional unconformity related to increase in rhyolitic volcanic activity in the central North Island (Fleming, 1953, p. 175). Since the sediments of the underlying Maxwell Group contain lignites of probable coastal plain origin, significant subsidence of the Wanganui coastal region must have commenced immediately prior to the sea-level high-stand represented by cyclothem 1. The absence of any identifiable high-stand systems tract sediment in cyclothem 1 suggests that by then there had been insufficient subsidence to create significant accommodation, and that the Butler's Formation accumulated immediately adjacent to the shoreline.

However, by the time of cyclothem 2, about 80 ka later, continuing subsidence had created enough accommodation for the deposition of almost 20 m of high-stand, regressive sediment (=Lower Okehu Siltstone). This indicates a minimum subsidence rate of 0.4 m/ka, which is not significantly different from the rate of tectonic uplift at the Wanganui coast today (e.g. Pillans, 1986). Between cyclothems 2 and 8, high-stand systems tracts follow a thinning trend, apart from the anomalously thick Upper Westmere Siltstone of cyclothem 6 (Fig. 3). This trend could be interpreted as marking regular and continual subsidence between *c*.1.1 and *c*.0.60 Ma, resulting in a more offshore (thinner) transect through the high-stand systems tract of each successive cyclothem.

The high-stand systems tract thinning-trend in the coast section reverses at about cyclothem 8, after which high-stand deposits become thicker in successive cyclothems (Fig. 3). This regressive trend can be interpreted as marking steady tectonic uplift between about 0.60–0.35 Ma, and culminates in the presence of shoreface sediments within the sediments of isotope stage 9 exposed at Landguard Bluff, south of the main Wanganui section (Fleming, 1953; Beu & Edwards, 1984). The regressive trend is also supported by the fact that, in contrast to cyclothems 1–7, cyclothems 8–10 all contain well-sorted, ocean beach or shoreface facies associations in their transgressive systems tract, consistent with an abundant sand delivery from the hinterland.

Although not uniquely constrained, the available evidence therefore suggests that the main Wanganui coast section was deposited during a tectonic cycle which commenced subsidence at *c*.1.1 Ma and reversed about 0.60 Ma, leading to uplift between *c*.0.60(–0.35)–0 Ma. For sustained rates of subsidence of 0.25 or 0.5 m/ka, and assuming (i) no great movement in the position of the axis of tilting, and (ii) a total compaction of about 10%, the sea-floor depth at the maximum flooding surface of cyclothem 7 would have been 27 m or 55 m deep respectively. The shallower end of this range of figures agrees well with independent depth estimates based on sedimentological and faunal evidence.

CONCLUSIONS

Our analysis of 10 mid-Pleistocene cyclothems from Wanganui suggests a number of conclusions relevant to sequence stratigraphic analysis generally.

1 Cyclic sea-level changes such as those of the mid-Pleistocene are accompanied by the deposition of stratigraphic sequences closely similar to those predicted by the Exxon sequence stratigraphic model. This is so even though the 40 (sixth-order) or 100-ka (fifth-order) span of Pleistocene sequences is up to two orders of magnitude less than the 3–5-Ma span of the third-order sequences that comprise the basis for the Exxon sea-level curve.

2 Wanganui fifth- and sixth-order sequences display a highly characteristic internal facies architecture. The transgressive systems tract is marked by nearshore shelf and shoreface sediments of markedly heterolithic nature. Accompanying type A shellbeds are cross-bedded coquina lenses which include *remanié* forms, and faunal assemblages characterized by intertidal or shoreface molluscan species. The local flooding surface lies at the base of a condensed mid-cycle type B shellbed, rich in *in situ* molluscs of offshore shelf habitat. Each sequence terminates with sparsely fossiliferous, massive, bioturbated siltstones of the high-stand systems tract which are separated from the subjacent mid-cycle shellbed across the downlap surface.

3 Sequence boundaries coincide with sharp ravinement surfaces, often penetrated by marine pholad borings (type C shellbed), and sometimes overlain by a transgressive lag gravel of rolled, blackened shells and intrabasinal pebbles (basal type A shellbed). Soils and other non-marine facies do not occur in the coast section, presumably having been eroded by marine planation, but are known at sequence boundaries further inland on the palaeocoastal plain. Two important conclusions, therefore, are: (i) that soils developed during glacial low-stand emergence are not good sequence boundary markers over the large areas of shelf subjected to transgressive marine ravinement; and (ii) that cycle-bounding unconformities often encompass two distinct surfaces, namely the sequence boundary and the ravinement surface, and may in some cases incorporate the local flooding surface (=submarine unconformity), which marks the base of the mid-cycle condensed shellbed, as well.

4 Where both are present as separate stratigraphic features, the distinction between a sequence boundary and a mid-cycle flooding surface is most easily made on the basis of the type of shellbed associated with each. The sequence boundary is marked by an *in situ* type C pholad shellbed, and by overlying transgressive systems tract shellbeds; these type A shellbeds are characterized by moderately sorted sandy-gravelly sediment containing mostly reworked shells, and commonly include species of estuarine or intertidal habitat. In contrast, mid-cycle condensed shellbeds contain an *in situ* or near *situm* fauna of well-preserved offshore molluscs set in a poorly sorted sandy–muddy matrix (type B shellbed). Since these shellbed types are partly characterized by the habitat of their faunas, they will be less easily recognized in strata older than Cenozoic.

5 Typical type B condensed shellbeds are associated with local flooding surfaces within sixth-order sequences, i.e. within eustatic cycles as short as 40 ka, which indicates that a type B shellbed can form in only a few thousand years. Thus the absence of one or more faunal zones is not a necessary condition for the recognition of a type B shellbed in the ancient record.

6 A number of significant differences exist between the sytems tracts of cyclothems deposited during 100-ka high-amplitude and 40-ka moderate-amplitude climatic cycles. Notably, 100-ka cyclothems contain: (i) thicker transgressive systems tracts; (ii) thicker mid-cycle condensed shellbeds; (iii) deeper water (more offshore) faunas in their mid-cycle condensed shellbeds; and (iv) deeper water (more offshore) sedimentary facies in their high-stand systems tracts.

ACKNOWLEDGEMENTS

We acknowledge gratefully the help and hospitality of Bill and Audrey Cox and Glenys Wilson of Mowhanau Beach during fieldwork, and the permission of A.R. Edwards (Stratigraphic Solutions) to cite unpublished nannofossil results from the Lower Castlecliff Siltstone. This research has been financially supported by Special Research Grants from James Cook University, and by Australian Research Council grant #88/2245. We thank also Drs Ed Clifton, Brad Pillans and George Postma for their constructive critical readings of the draft manuscript.

REFERENCES

ABBOTT, S.T. (1992) Relocation of the mid-Pleistocene Waiomio Shellbed (Putikian, *c*.500 ka), and definition of the Waiomio Formation, Rangitikei Valley, Wanganui Basin. *Alcheringa* **16**, 171–180.

ABBOTT, S.T. *Sedimentary analysis of mid-Pleistocene cyclothems from Wanganui Basin, New Zealand*. PhD dissertation, James Cook University of North Queensland, Townsville (in prep.).

ANDERTON, P.W. (1981) Structure and evolution of the South Wanganui Basin, New Zealand. *N. Z. J. Geol. Geophys*. **24**, 39–63.

BAUM, G.R. & VAIL, P.R. (1988) Sequence stratigraphic concepts applied to Paleogene outcrops, Gulf and Atlantic Basins. In: *Sea-level Changes: an Integrated Approach* (Eds Wilgus, C.K., Hastings, B.S., Posamentier, H., van Wagoner, J., Ross, C.A. & Kendall, C.G. St C.). Soc. Econ. Paleont. Mineral. Spec. Publ. 42, pp. 39–45.

BECKVAR, N. & KIDWELL, S.M. (1988) Hiatal shell concentrations, sequence analysis, and sealevel history of a Pleistocene coastal alluvial fan, Punta Chueca, Sonora. *Lethaia* **21**, 257–270.

Beggs, J.M. (1990) Seismic stratigraphy of the Plio-Pleistocene giant foresets, Western Platform, Taranaki Basin. N. Z. In: *Oil Explor. Conf. (1989, Queenstown) Proc.* Vol. 1, pp. 201–207. Ministry of Commerce, Wellington.

Berger, A. (1989) The spectral characteristics of Pre-Quaternary climate records, an example of the relationship between astronomical theory and geosciencies. In: *Climate and Geosciences* (Eds Berger, A., Schneider, S. & Duplessy, J.C.). Reidel Publ. Co., Dordrecht, pp. 47–76.

Berger, A., Loutre, M.F. & Dehant, V. (1989) Pre-Quaternary Milankovitch frequencies. *Nature* **342**, 133.

Beu, A.G. & Edwards, A.R. (1984) New Zealand Pleistocene and Late Pliocene glacio-eustatic cycles. *Palaeogeogr. Palaeoclimatol. Palaeoecol.* **46**, 119–142.

Boardman, D.R. & Heckel, P.H. (1989) Glacial–eustatic sea-level curve for early Late Pennsylvanian sequence in north-central Texas and biostratigraphic correlation with curve for midcontinent North America. *Geology* **17**, 802–805.

Boyd, R., Suter, J. & Penland, S. (1989) Relation of sequence stratigraphy to modern sedimentary environments. *Geology* **17**, 926–929.

Carter, R.M. (1985) The mid-Oligocene Marshall Paraconformity, New Zealand: coincidence with global eustatic sea-level fall or rise? *J. Geol.* **93**, 359–371.

Carter, R.M., Carter L. Williams, J. & Landis, C.A. (1985) *Modern and relict sedimentation on the Otago continental shelf.* N. Z. Oceanog. Inst. Mem. 93.

Carter, R.M., Abbott, S.T., Fulthorpe, C.S., Haywick, D.W. & Henderson, R.A. (1991) Application of global sea-level and sequence stratigraphic models in southern hemisphere Neogene strata from New Zealand. In: *Sedimentation, Tectonics and Eustasy: Sea-level Changes at Active Margins* (Ed. MacDonald, D.I.M.). Spec. Publ. Int. Ass. Sedim. 12, pp. 41–65.

Christie-Blick, N., Grotzinger, J.P. & von der Borch, C.C. (1988) Sequence stratigraphy in Proterozoic successions. *Geology* **16**, 100–104.

Clifton, H.E., Hunter, R.E. & Gardner, J.V. (1988) Analysis of eustatic, tectonic and sedimentologic influences on transgressive and regressive cycles in the Upper Cenozoic Merced Formation, San Francisco, California. In: *New Perspectives in Basin Analysis* (Eds Kleinspehn, K.L. & Paola, C.), pp. 109–128.

Collier, R.E.L.L. (1990) Eustatic and tectonic controls upon Quaternary coastal sedimentation in the Corinth Basin, Greece. *J. Geol. Soc. Lond.* **147**, 301–314.

Cotter, E. (1988) Hierarchy of sea-level cycles in the medial Silurian siliciclastic succession of Pennsylvania. *Geology* **16**, 242–245.

Curray, J.R. (1964) Transgressions and regressions. In: *Papers in Marine Geology, Shepard Commemorative Volume* (Ed. Miller, R.C.). Macmillan, New York, pp. 175–203.

Demarest, J.M. & Kraft, J.C. (1987) Stratigraphic record of Quaternary sea levels: implications for more ancient strata. In: *Sea-level Fluctuation and Coastal Evolution* (Eds Nummedal, D., Pilkey, O.H. & Howard, J.D.). Soc. Econ. Paleont. Mineral. Spec. Publ. 41, pp. 241–259.

DeMenocal, P.B., Ruddiman, W.F. & Kent, D.V. (1990) Depth of post-depositional remanence acquisition in deep-sea sediments: a case study of the Brunhes–Matuyama reversal and oxygen isotopic Stage 19. *Earth Planet. Sci. Lett.* **99**, 1–13.

Dudley, W.C. & Nelson, C.S. 1989 Quaternary surface-water stable isotope signal from calcareous nannofossils at DSDP Site 593, southern Tasman Sea. *Mar. Micropaleontol.* **13**, 353–373.

Embry, A.F. (1990) A tectonic origin for third-order depositional sequences in extensional basins – implications for basin modeling. In: *Quantitative Dynamic Stratigraphy* (Ed. Cross, T.A.). Prentice Hall, New Jersey, pp. 491–501.

Eriksson, K.A. & Simpson, E.L. (1990) Recognition of high-frequency sea-level fluctuations in Proterozoic siliciclastic tidal deposits, Mount Isa, Australia. *Geology* **18**, 474–477.

Feeley, M.H., Moore, T.C., Loutit, T.S. & Bryant, W.R. (1990) Sequence stratigraphy of Mississippi Fan related to oxygen isotope sea level index. *Bull. Am. Assoc. Petrol. Geol.* **74**, 407–424.

Fleming, C.A. (1953) The geology of the Wanganui Subdivision, Waverley and Wanganui sheet districts (N137 & N138). *N. Z. Geol. Surv. Bull.* **52**, 1–362.

Fulthorpe, C.S. (1991) Geological controls on seismic sequence resolution. *Geology* **19**, 61–65.

Fulthorpe, C.S. & Carter, R.M. (1989) Test of seismic sequence methodology on a southern hemisphere passive margin: the Canterbury Basin, New Zealand. *Mar. Petrol. Geol.* **6**, 348–359.

Galloway, W.E. (1989) Genetic stratigraphic sequences in basin analysis I: architecture and genesis of flooding-surface bounded depositional units. *Bull. Am. Assoc. Petrol. Geol.* **73**, 125–142.

Hallam, A. (1969) A pyritised limestone hardground in the Lower Jurassic of Dorset (England). *Sedimentology* **12**, 231–240.

Hallam, A. (1988) A reevaluation of Jurassic eustasy in the light of new data and the revised Exxon curve. In: *Sea-level Changes: an Integrated Approach* (Eds Wilgus, C.K., Hastings, B.S., Posamentier, H., van Wagoner, J., Ross, C.A. & Kendall, C.G. St C.). Soc. Econ. Paleont. Mineral. Spec. Publ. 42, pp. 261–273.

Haq, B.U., Hardenbol, J. & Vail, P.R. (1988) Mesozoic and Cenozoic chronostratigraphy and cycles of sea-level change. In: *Sea-level Changes: an Integrated Approach* (Eds Wilgus, C.K., Hastings, B.S., Posamentier, H., van Wagoner, J., Ross, C.A. & Kendall, C.G. St C.). Soc. Econ. Paleont. Mineral. Spec. Publ. 42, pp. 71–108.

Hays, J.D., Imbrie, J. & Shackleton, N.J. (1976) Variations in the earth's orbit: pacemaker of the ice ages. *Science* **194**, 1121–1132.

Haywick, D.W. (1990) *Stratigraphy, sedimentology, palaeocology and diagenesis of the Petane Group (Plio-Pleistocene) in Central Hawke's Bay, New Zealand.* Unpublished PhD dissertation, James Cook University of North Queensland, Townsville.

Haywick, D.W., Lowe, D.A., Beu, A.G., Henderson, R.A. & Carter, R.M. (1991) Plio-Pleistocene (Nukumaruan) lithostratigraphy of the Tangoio block, and origin of sedimentary cyclicity, central Hawke's Bay,

New Zealand. *N. Z. J. Geol. Geophys.* **34**, 213–225.

HESSELBO, S.P., COE, A.L. & JENKYNS, H.C. (1990a) *Sequence stratigraphy of the Jurassic of Dorset: a field guide*. Int. Assoc. Sed., Nottingham Meeting.

HESSELBO, S.P., COE, A.E. & JENKYNS, H.C. (1990b) Recognition and documentation of depositional sequences from outcrop: an example from the Aptian and Albian on the eastern margin of the Wessex Basin. *J. Geol. Soc. Lond.* **147**, 549–559.

HOUSE, M.R. (1985) A new approach to an absolute timescale from measurements of orbital cycles and sedimentary microrhythms. *Nature* **316**, 721–725.

House, M.R. (1989) *Geology of the Dorset Coast*. The Geologists Association, London.

HOWARD, J.D. & FREY, R.W. (1973) Characteristic physical and biogenic sedimentary structures in Georgia estuaries. *Bull. Am. Assoc. Petrol. Geol.* **57**, 1169–1184.

HUBBARD, R.J. (1988) Age and significance of sequence boundaries on Jurassic and Early Cretaceous rifted continental margins. *Bull. Am. Assoc. Petrol. Geol.* **72**, 49–72.

IMBRIE, J., HAYS, J., MARTINSON, D., MCINTYRE, A., MIX, A., MORLEY, J. *et al.* (1984) The orbital theory of Pleistocene climate: support from a revised chronology of the marine $\delta^{18}O$ record. In: *Milankovitch and Climate*, Part 1 (Eds Berger, A., Imbrie, J., Hays, J., Kukla, G. & Saltzman, B.). Reidel Publ. Co., Dordrecht, pp. 269–305.

JOYCE, J.E., TJALSMA, L.R.C. & PRUTZMAN, J.M. (1990) High-resolution stable isotope record and spectral analysis for the last 5.35 M.Y.: Ocean Drilling Program site 625 northeast Gulf of Mexico. *Paleoceanography* **5**, 507–529.

KAMP, P.J.J. (1978) *Stratigraphy and sedimentology of conglomerates in the Pleistocene Kidnappers Group, Hawke's Bay*. Unpublished MSc thesis, Waikato University, Hamilton.

KENDALL, C.G. St C. & LERCHE, I. (1988) The rise and fall of eustasy. In: *Sea-level Changes: an Integrated Approach* (Eds Wilgus, C.K., Hastings, B.S., Posamentier, H., van Wagoner, J., Ross, C.A & Kendall, C.G. St C.). Soc. Econ. Paleont. Mineral. Spec. Publ. 42, pp. 3–17.

KIDWELL, S.M. (1984) Outcrop features and origin of basin margin unconformities in the lower Chesapeake Group (Miocene), Atlantic coastal plain. In: *Interregional Unconformities and Hydrocarbon Accumulation* (Ed. Schlee, J.S.). Am. Assoc. Petrol. Geol. Mem. 36, pp. 37–58.

KIDWELL, S.M. (1986) Models for fossil concentrations: paleobiologic implications. *Paleobiology* **12**, 6–24.

KOMINZ, M.A. & BOND, G.C. (1990) A new method of testing periodicity in cyclic sediments: application to the Newark Supergroup. *Earth Planet. Sci. Lett.* **98**, 233–244.

LANG, W.D., SPATH, L.F., COX, L.R. & MUIR-WOOD, H.M. (1928) The Belemnite Marls of Charmouth, a series in the Lias of the Dorset coast. *Q. J. Geol. Soc. Lond.* **84**, 179–257.

LEITHOLD, E.L. (1989) Depositional processes on an ancient and modern muddy shelf, northern California. *Sedimentology* **36**, 179–202.

LOUTIT, T.S., HARDENBOL, J. & VAIL, P.R. (1988) Condensed sections: the key to age determination and correlation of continental margin sequences. In: *Sea-level Changes: an Integrated Approach* (Eds Wilgus, C.K., Hastings, B.S., Posamentier, H., van Wagoner, J., Ross, C.A. & Kendall, C.G. St C.). Soc. Econ. Paleont. Mineral. Spec. Publ. 42, pp. 183–213.

LOWRIE, A. & MCDANIEL-LOWRIE, M.L. (1985) Application of Pleistocene climate models to Gulf Coast stratigraphy. *Gulf Coast Assoc. Geol. Soc., Trans.* **35**, 201–208.

MCKNIGHT, D.G. (1969) Infaunal benthic communities of the New Zealand continental shelf. *N. Z. J. Mar. Freshw, Res.* **3**, 409–444.

NELSON, C.S., HENDY, C.H., JARRETT, G.R. & CUTHBERTSON, A.M. (1985) Near-synchroneity of New Zealand alpine glaciations and Northern Hemisphere continental glaciations during the past 750 kyr. *Nature* **318**, 361–363.

NUMMEDAL, D. & SWIFT, D.J.P. (1987) Transgressive stratigraphy at sequence-bounding unconformities: some principles derived from Holocene and Cretaceous examples. In: *Sea-level Fluctuation and Coastal Evolution* (Eds Nummedal, D., Pilkey, O.H. & Howard, J.D.). Soc. Econ. Paleont. Mineral. Spec. Publ. 41, pp. 241–259.

PEMBERTON, S.G. & FREY, R.W. (1985) The *Glossifungites* ichnofacies: modern examples from the Georgia coast, USA. In: *Biogenic Structures: their Use in Interpreting Depositional Environments* (Ed. Curran, H.A.). Soc. Econ. Paleont. Mineral. Spec. Publ. 35, pp. 237–259.

PILAAR, W.F.H. & WAKEFIELD, L.L. (1978) Structural and stratigraphic evolution of the Taranaki Basin, offshore North Island, New Zealand. *J. Assoc. Petrol. Explor. Aust.* **18**, 93–101.

PILLANS, B. (1986) A late Quaternary uplift map for North Island, New Zealand. *Roy. Soc. N. Z., Bull.* **24**, 409–417.

PILLANS, B., ABBOTT, S.T., WILSON, G. & ROBERTS, A. Magnetostratigraphic studies of lower-mid Pleistocene sediments in the Turakina, Whangaehu and Rangitikei valleys, Wanganui Basin, New Zealand (in prep.).

PLINT, A.G. (1988) Global eustasy and the Eocene sequence in the Hampshire Basin, England. *Basin Res.* **1**, 11–22.

PROBERT, P.K. & WILSON, J.B. (1984) Infaunal communities on the Otago shelf. *Estuar. Coastal Shelf Sci.* **19**, 373–391.

PROBERT, P.K., BATHAM, E.J. & WILSON, J.B. (1979) Epibenthic macrofauna off southeastern New Zealand and mid-shelf bryozoan dominance. *N. Z. J. Mar. Freshw. Res.* **13**, 379–392.

RAYMO, M.E., RUDDIMAN, W.F., BACKMAN, J., CLEMENT, B.M. & MARTINSON, D.G. (1989) Late Pliocene variation in Northern Hemisphere ice sheets and North Atlantic deep water circulation. *Paleoceanography* **4**, 413–446.

REINECK, H.E. & SINGH, I.B. (1986) *Depositional Sedimentary Environments*, 2nd edn. Springer-Verlag, Berlin.

RUDDIMAN, W.E., RAYMO, M.E., MARTINSON, D.G., CLEMENT, B.M. & BACKMAN, J. (1989) Pleistocene evolution: Northern Hemisphere ice sheets and North Atlantic Ocean. *Paleoceanography* **4**, 353–412.

SANDERS, K.E. & KUMAR, N. (1975) Evidence of shoreface retreat and in-place 'drowning' during Holocene submergence of barriers, shelf off Fire Island, New York. *Geol. Soc. Am. Bull.* **86**, 65–76.

SEWARD, D. (1976) Tephrostratigaphy of the marine sediments in the Wanganui Basin, New Zealand. *N. Z. J. Geol. Geophys.* **19**, 9–20.

SHACKLETON, N.J. & OPDYKE, N.D. (1973) Oxygen isotope and palaeo-magnetic stratigraphy of equatorial Pacific core V28-238: oxygen isotope temperatures and ice volumes in a 10^5 and 10^6 year scale. *Quat. Res.* **3**, 39–55.

SHACKLETON, N.J., BERGER, A. & PELTIER, W.R. (1990) An alternative astronomical calibration of the lower Pleistocene timescale based on ODP Site 677. *Trans. R. Soc. Edinburgh* **81**, 251–261.

STERN, T.A. & DAVEY, F.J. (1989) Crustal structure and origin of basins formed behind the Hikurangi subduction zone, New Zealand. In: *Origin and Evolution of Sedimentary Basins and their Energy and Mineral Resources* (Ed. Price, R.A.). Int. Union Geodesy Geophys. and Am. Geophys. Union, Geophys. Monogr. 48, pp. 73–85.

STERN, T.A. & DAVEY, F.J. (1990) Deep seismic expression of a foreland basin: Taranaki basin, New Zealand. *Geology* **18**, 979–982.

STEVENS, K.F. & VELLA, P. (1981) Palaeoclimatic interpretation of stable isotope ratios in molluscan fossils from middle Pleistocene marine strata, Wanganui, New Zealand. *Palaeogeogr. Palaeoclimatol. Palaeoecol.* **34**, 257–265.

SWIFT, D.J.P. (1968) Coastal erosion and transgressive stratigraphy. *J. Geol.* **76**, 444–456.

SWIFT, D.J.P. & RICE, D.R. (1984) Sand bodies on muddy shelves: a model for sedimentation in the western interior Cretaceous seaway, North America. In: *Siliciclastic Sediments* (Eds Tillman, R.W. & Siemers, C.T.). Soc. Econ. Paleont. Mineral. Spec. Publ. 34, pp. 43–62.

TURNER, G.M. & KAMP, P.J.J. (1990) Paleomagnetic location of the Jaramillo Subchron and Brunhes–Matuyama transition in the Castlecliffian Stratotype section, Wanganui Basin, New Zealand, *Earth Planet. Sci. Lett.* **100**, 42–50.

VAIL, P. (1987) Seismic stratigraphy interpretation using sequence stratigraphy, part 1: seismic stratigraphy interpretation procedure. In: *Atlas of Seismic Stratigraphy* (Ed. Bally, A.W.). Am. Assoc. Petrol. Geol. Stud. Geol. 27, pp. 1–10.

VAIL, P.R., MITCHUM, R.M. & THOMPSON, S. (1977) Seismic stratigraphy and global changes of sea level, part 4: global cycles of relative changes of sea level. In: *Seismic Stratigraphy, II: an Integrated Approach to Hydrocarbon Exploration* (Eds Berg, R.O. & Woolverton, O.G.) *Am. Assoc. Petrol. Geol. Mem.* **26**, 83–97.

VAIL, P.R., HARDENBOL, J. & TODD, R.G. (1984) Jurassic unconformities, chronostratigraphy, and sea-level changes from seismic stratigraphy and biostratigraphy. In: *Interregional Unconformities and Hydrocarbon Accumulation* (Ed. Schlee, J.S.). Am. Assoc. Petrol. Geol. Mem. 36, pp. 129–144.

VAN WAGONER, J.C., POSAMENTIER, H.W., MITCHUM, R.M., VAIL, P.R., SARG, J.F., LOUTIT, T.S. & HARDENBOL, J. (1988) An overview of the fundamentals of sequence stratigraphy and key definitions. In: *Sea-level Changes: an Integrated Approach* (Eds Wilgus, C.K., Hastings, B.S., Posamentier, H., van Wagoner, J., Ross, C.A. & Kendall, C.G. St C.). Soc. Econ. Paleont. Mineral. Spec. Publ. 42, pp. 39–45.

VERNEKAR, A.D. (1968) Long-period global variations of incoming solar radiation. In: *Research on the Theory of Climate*, Vol. 2. Rept. Travelers Research Center, Hartford, Connecticut.

WEEDON, G.P. (1985) Hemipelagic shelf sedimentation and climatic cycles: the basal Jurassic (Blue Lias) of South Britain. *Earth Planet. Sci. Lett.* **76**, 321–335.

WEEDON, G.P. & JENKYNS, H.C. (1990) Regular and irregular climatic cycles and the Belemnite Marls (Pliensbachian, Lower Jurassic, Wessex Basin). *J. Geol. Soc. Lond.* **147**, 915–918.

WILLIAMS, D.F., THUNELL, R.C., TAPPA, E., DOMENICO, R. & RAFFI, I. (1988) Chronology of the Pleistocene oxygen isotope record: 0–1.88 m.y. BP. *Palaeogeogr. Palaeoclimatol. Palaeoecol.* **64**, 221–240.

WILSON, J.B. (1986) Faunas of tidal current and wave-dominated continental shelves and their use in the recognition of storm deposits. In: *Shelf Sands and Sandstones* (Eds Knight, R.J. & McLean, J.R.). Can. Soc. Petrol. Geol. Mem. 11, pp. 293–301.

Spec. Publs Int. Ass. Sediment. (1994) **19**, 395–411

Cyclic deposition of the Devonian Catskill Delta of the Appalachians, USA

J. VAN TASSELL

Eastern Oregon State College, La Grande, OR 97850-2899, USA

ABSTRACT

The Devonian Catskill Delta complex contains a complex hierarchy of cycles with periods ranging from 20 ka to 3.5 Ma. The origin of these cycles is controversial. Fluvial cycles, shelf storm beds and turbidites were deposited at intervals of 10^2 to 10^3 years along the Catskill Delta margin. Fluvial, shoreline, shelf, slope and basin deposits include cycles with periodicities in the order of 100 ka (sixth-order) and 400 ka (fifth-order) periodicities; 20-ka and 33-ka cycles may be superimposed on the 100-ka cycles, while 0.8–1.5 Ma (fourth-order) cycles and 2 Ma (third-order) cyclic variations in lithology and faunal extinctions may also be present in the Catskill Delta sequence. Some of these depositional cycles have been correlated throughout the Catskill Delta complex and related to global eustatic sea-level fluctuations. Further work is needed to confirm these correlations. Lateral facies shifting, periodic storm activity, tectonism, sea-level fluctuations and climatic change and related variations in sediment input interacted to produce the cycles present in the Catskill Delta complex. The nested hierarchy of cycles present suggests the influence of climate, sea level and sediment input fluctuations triggered by Milankovitch orbital variations ranging from 20 ka to 2.0 Ma.

INTRODUCTION: THE CATSKILL DELTA COMPLEX

The origins of the cycles in the Devonian Catskill Delta sequence of the Appalachians are controversial. Cycles of coarse and fine units in the non-marine portion of the delta were recognized early (Barrell, 1913; Chadwick, 1936) and later interpreted as the result of stream channels shifting across flood plains (Allen & Friend, 1968; Walker & Harms, 1971). Other workers (Goodwin & Anderson, 1985; Brett & Baird, 1986) have focused on allogenic influences on Catskill Delta sediment deposition. Van Tassell (1987) suggested that the deposition of the Catskill Delta complex may have been strongly influenced by 20-ka precession cycles, 40-ka obliquity cycles, and 100-ka, 400-ka and possibly 2.0-Ma orbital eccentricity variations. Episodes of mid-plate thermal uplift and submarine volcanism (Johnson *et al.*, 1985), and episodes of tectonic uplift in the Acadian mountains alternating with tectonically quiescent periods with accompanying variations in rainfall, runoff and clastic sedimentation (Ettensohn, 1985a) have also been called upon to explain the longer period cycles in the Devonian of Euramerica and in the Catskill Delta basinal sequences, respectively. This paper seeks to test further the hypothesis of orbital control on this complexly hierarchical succession.

The Catskill 'Delta' (Barrell, 1913) is actually a series of clastic wedges totalling thousands of metres in thickness. These sediments were eroded from the rising mountains formed during the Acadian orogeny and carried by river systems to the Appalachian foreland basin (Fig. 1). This sedimentary complex was deposited during the Middle and Late Devonian at palaeolatitudes between 15° and 20° S in a seasonally wet and dry (monsoonal) climate (Heckel & Witzke, 1979; Kent & Opdyke, 1985; Woodrow, 1985; Miller & Kent, 1986a, b).

Tectonic activity has played a major role in the deposition of the Catskill Delta sequence. The major disconformity-bounded sequences in the Catskill Delta complex were produced by successive phases of Acadian deformation (Fig. 12). Three and possibly four such tectophases have been noted (Ettensohn, 1985b). Superimposed on these long-term variations are cycles marked at their beginnings by black shales. These key beds include the Marcellus, Geneseo-

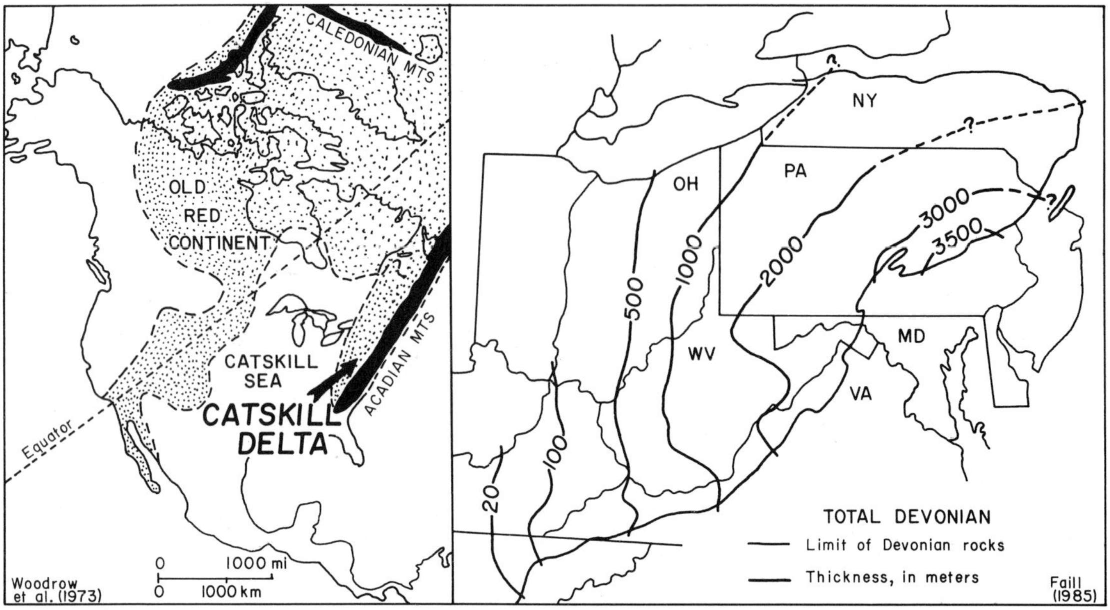

Fig. 1. Middle to Late Devonian palaeogeography, based on Woodrow *et al.* (1973), and isopach map of the Devonian strata in the Appalachian foreland basin, modified from Faill (1985).

Burket, Middlesex, Rhinestreet and Dunkirk black shales. These shales have been traced from New York and Ohio to Tennessee mainly on the basis of their expression in geological well-logs. In New York, the black shales and their easterly dark-grey shale equivalents have been traced into and through highly fossiliferous basin margin units based both on geophysical data and on exposure data (Woodrow *et al.*, 1988). The first cycle, represented by the Hamilton Group, begins with the Marcellus Shale. The Hamilton actually consists of at least four partial cycles in which basinal environments were completely or locally developed. Four Late Devonian cycles, each lasting approximately 2–3 Ma, were completely developed, beginning with the Geneseo-Burket, Middlesex, Rhinestreet and Dunkirk black shales (Ettensohn, 1985a). The Pipe Creek Shale may mark another cycle.

The bases of the Middlesex, Rhinestreet and Pipe Creek shales have been correlated with major breaks in other successions throughout Euramerica on the basis of conodont biostratigraphy (Johnson *et al.*, 1985), and the major sedimentary rhythms and Frasnian extinction events in the New York sequence have been correlated with marked environmental change and discontinuous sedimentation patterns in Belgium, France and Canada (Fig. 13) plus a rather similar pattern and an abrupt termination by drowning of the stromatoporoid reef complexes in western Australia. The synchrony of these changes remains untested, however, since conodont and ammonoid dating are both incomplete (House, 1985).

An aerial view of the Catskill Delta complex (Fig. 2) may have revealed a highland area with limited plant cover on the mountain slopes, making them vulnerable to rapid erosion. Alluvial fans abutted the highlands, and braided streams and sinuous channels separated by broad interfluves flowed across the adjacent alluvial plain. Terrestrial plants and animals flourished, at least locally, near streams and strand. The sinuous channels migrated laterally by erosion and avulsion, but most interfluves existed for as long as 20–30 ka, sufficient time to permit the development of carbonate soils. Seasonal rainfall meant that there were wide fluctuations of the water table and major variations of the flow in the streams during the year (Woodrow, 1985; Woodrow *et al.*, 1988).

The Catskill Delta margin included a wide range of shifting coastal environments, including tidal flats and channels, beaches, distributary mouth bars, interdistributary areas, delta platforms, delta front and prodelta environments, and shelf deposits (Sevon, 1985). River flooding and storm wave action played an important role in the deposition of nearshore sand

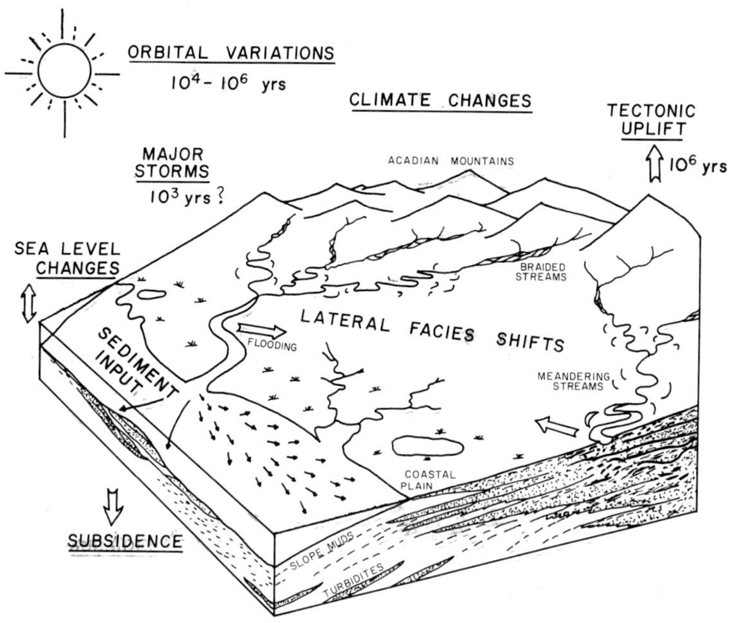

Fig. 2. Upper Devonian sedimentation model for south-central Pennsylvania, modified from Walker & Harms (1971), and factors producing depositional cycles in the Catskill Delta complex.

layers (van Tassell, 1987). Wave heights during these storms may have reached 4–5 m, in contrast to calmer weather waves with heights of 2 m or less produced by the southeast tradewinds (Craft & Bridge, 1987). Turbidity currents may have originated more or less directly from rivers during their flood stages (Walker, 1971). Some flows could also have been triggered by internal waves along the pycnocline in the density-stratified Catskill Sea (Woodrow, 1985). The Catskill turbidity currents deposited silts and very fine sands with an average bed thickness of 10 cm (Kuenen, 1956; Sutton, 1959; Colton, 1967; Frakes, 1967; Walker, 1967; McIver, 1970). Shifting delta distributaries built and abandoned turbidite lobes, producing an apron of silty and sandy sediments along the slope and base of the delta complex (McIver, 1970; Lundegard *et al.*, 1985). Bioturbated hemi-pelagic mudstone also accumulated on the delta slope lateral to areas of active turbidite deposition. Hemi-pelagic claystone and shale with turbidite silt laminae were deposited at the base of the slope (Glaeser, 1979; Woodrow & Isley, 1983). Black shales accumulated on the anaerobic bottom of the adjacent Catskill Sea basin (McIver, 1970; Ettensohn & Barron, 1981; Lundegard *et al.*, 1985).

The Milankovitch hypothesis of orbital control on climatic change and hence on sedimentation has been quantitatively demonstrated to hold true for several non-glacial periods of Earth's history (e.g. Triassic, Olsen, 1986; mid-Cretaceous, de Boer & Wonders, 1984). Thus, there can be little doubt that such variations, in climate at least, have occurred throughout geological history. I therefore start with the assumption that orbitally forced climatic variation did indeed occur in Devonian time. The question addressed in this paper is: Can we or can we not rule out the hypothesis of direct influence of cyclical climatic variations on the Catskill's multi-order cyclicity? I begin by describing the cycles apparent in the wide range of palaeoenvironments present in the Catskill, I examine the crucial question of their long-range correlatability, I look at the more controversial evidence for longer-period (>1-Ma cyclicity), and conclude with a discussion of the possible range of mechanisms that might be invoked.

NON-MARINE CYCLES

The non-marine facies in the Catskill alluvial plain complex vary from braided stream facies to delta-plain–meandering-stream facies and finally to marine–non-marine transition environments (Sevon, 1985). Early studies of fluvial fining-upward cycles (Barrell, 1913, 1914; Chadwick, 1944; Fletcher, 1963, 1967; Allen & Friend, 1968) have been succeeded by detailed quantitative reconstructions

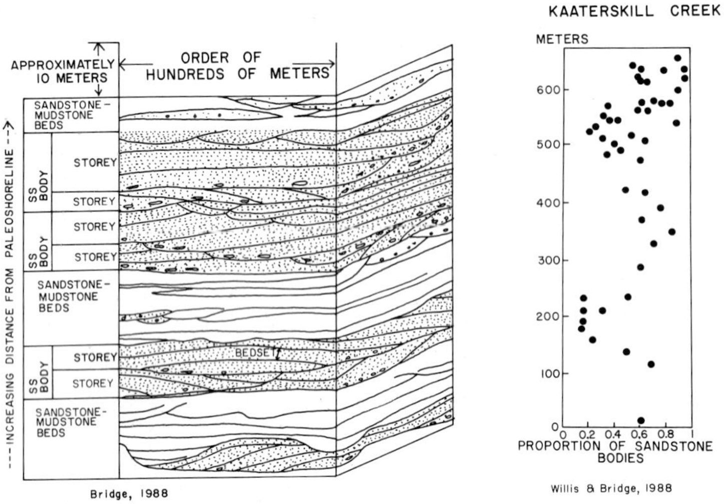

Fig. 3. Schematic bedding geometry of sandstone bodies (stippled) and sandstone–mudstone beds in Devonian fluvial deposits of New York State (Bridge, 1988), and apparent large-scale cyclic variations in the proportion of sandstone bodies in the Givetian Plattekill and Manorkill Formations in the Kaaterskill Creek section, New York (Willis & Bridge, 1988).

of Catskill river systems. Bridge & Gordon (1985), Gordon & Bridge (1987) and Willis & Bridge (1988) have shown that Catskill river deposits include grey to reddish-grey sandstone bodies that are interpreted as the results of aggrading single-channel rivers that migrated laterally within well-defined channel belts (Fig. 3). Each sandstone body consists of one or more storeys defined by a 'major' basal erosion surface and lateral-accretion surfaces that extend from the base to near the top of the storey. The river deposits also include interbedded sandstone, siltstone, mudstone and shale organized into metre-scale fining-upward bedsets. Each bedset is interpreted as the deposit of a single overbank flood on a floodplain. The spatial distribution of sandstone bodies (channel-belt deposits) relative to sandstone–mudstone deposits (overbank deposits) can be explained by periodic channel-belt avulsion during net floodplain deposition (Gordon & Bridge, 1987).

Lithofacies variations in the order of hundreds of metres thick have also been recognized in Catskill alluvial plain rocks (Chadwick, 1944; Fletcher, 1967) but have proven difficult to correlate because of complex lateral lithofacies changes, structural deformation, ambiguous and imprecise definition of formations in their type sections, and lack of biostratigraphically useful fossils (Willis & Bridge, 1988). Evidence of cyclicity is suggested by vertical variations in the proportion of sandstone bodies in Givetian fluvial deposits at Kaaterskill Creek, New York (Willis & Bridge, 1988; Fig. 3). McCave (1969) proposed that two zones with a high proportion of sandstone–mudstone beds at the Kaaterskill Creek section were related to marine transgressions and could be correlated with marine limestones exposed to the northwest. More recent work challenges the association of marine limestones with transgressions (Brett & Baird, 1985). The cycles evident in the proportion of sandstone bodies in the Kaaterskill Creek section do not appear to be present in the Plattekill Creek section 10 km away. Both sections show similar cyclic shifts in palaeocurrent direction, but the data are not sufficient to prove that the Kaaterskill Creek cycles can be traced to the Plattekill Creek section (Willis & Bridge, 1988).

MARINE–NON-MARINE TRANSITION CYCLES

The marine–non-marine transition cycles of the Catskill Delta complex are well preserved in the Irish Valley Member at the base of the Devonian Catskill Formation in the Susquehanna Valley of central Pennsylvania. The 600-m-thick Irish Valley Member was interpreted as a prograding muddy shoreline by Walker (1971) and Walker & Harms (1971). It rests on turbidite and slope deposits and is overlain by a sequence of dominantly non-marine alluvial sediments. The Irish Valley Member contains about 25 repeated facies sequences ('Irish Valley motifs') which vary in thickness from about 4 to 45 m. The motif is interpreted to represent (from the base) marine transgression, non-deposition, winnowing and bioturbation; a marine shoaling phase which is represented dominantly by shale; and,

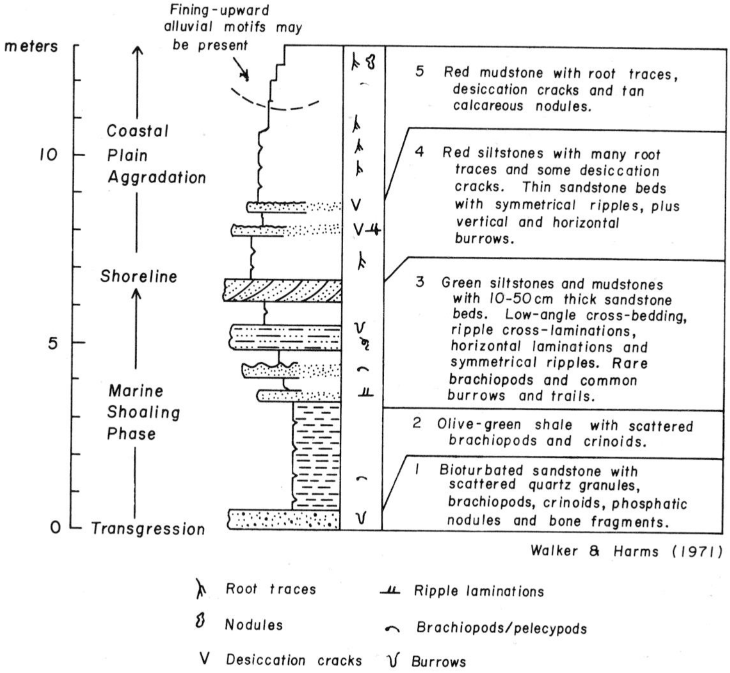

Fig. 4. Idealized 'Irish Valley motif', interpreted to represent one of the oscillations of a prograding muddy shoreline in the central Pennsylvania area (Walker & Harms, 1971).

finally, progradation of a quiet muddy shoreline and the development of an alluvial plain (Figs 4 and 5).

The Irish Valley Member was deposited over an interval of approximately 2.0–2.5 Ma. This suggests that the average depositional period of each of the Irish Valley cycles was approximately 100 ka (van Tassell, 1987).

SHELF AND SLOPE CYCLES

Walker & Harms' (1971) studies of the Irish Valley Member preceded the identification of hummocky cross-stratification in shelf deposits (Hamblin & Walker, 1979, and many others). This discovery paved the way for detailed studies of Catskill Delta shelf sequences such as the shallow-marine shelf strata of the Upper Devonian West Falls Group of south-central New York by Craft & Bridge (1987). The shelf strata of the West Falls Group include centimetre- to decimetre-thick beds of sandstone and siltstone interbedded with mudstone. These beds are arranged into larger-scale coarsening-upward sequences which are typically 15–30 m thick (Fig. 6). The lower part of a sequence is dominated by

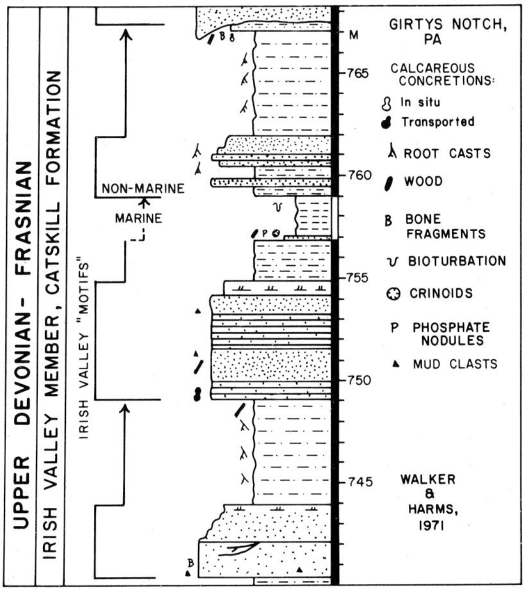

Fig. 5. Interbedding of alluvial and marine facies in the Irish Valley Member of the Catskill Formation at Girtys Notch, Pennsylvania. Modified from Walker & Harms (1971).

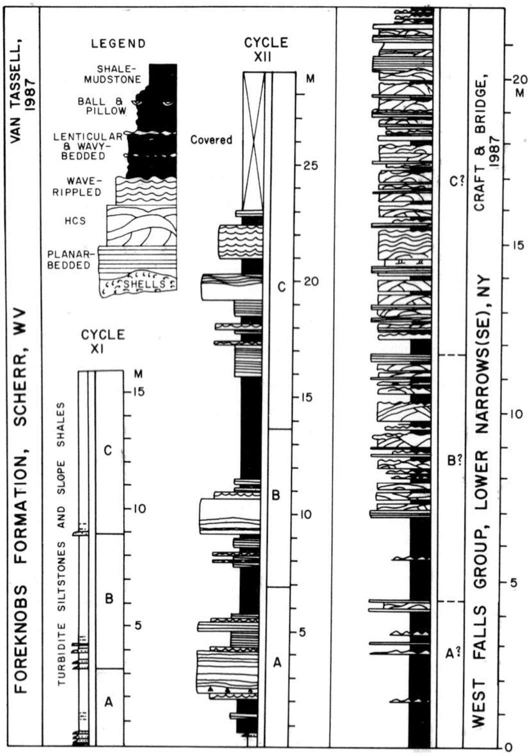

Fig. 6. Representative depositional cycles from the West Falls Group (Craft & Bridge, 1987) of New York and the Foreknobs Formation of West Virginia (van Tassell, 1987). Each cycle represents approximately 100 ka of deposition. Note the distinct fining- or coarsening-upward subcycles in the Foreknobs Formation units.

mudstone and relatively thin beds of bioturbated sandstone or siltstone. These are overlain by progressively thicker and coarser-grained sandstones and thinner, more silty mudstones. The abundance of hummocky cross-stratification and amalgamation of sandstone beds increases upward, along with the size and abundance of plant debris, coquinites, claystone interclasts, and certain bivalves. These sequences are overlain somewhat abruptly by thin bioturbated sandstone and thickening mudstones, making these sedimentary cycles markedly asymmetrical (see Woodrow & Isley, 1983). Each of these sequences records a gradual upward decrease in wave-current strength. These cycles recurred over periods of 10^4–10^5 years, based on the average thickness of the cycles, deposition rates estimated from the 800-m thickness, and the approximately 2–3-Ma depositional period of the West Falls Group (Craft & Bridge, 1987).

The Upper Devonian (Frasnian–Famennian) Brallier, Scherr and Foreknobs Formations of Virginia and West Virginia show a spectrum of cycle patterns ranging from coarsening-upward to fining-upward. One of these cycles, the Back Creek Siltstone Member of the Brallier Formation, consists of three 5–15-m-thick fining-upward packets of olive grey turbidite siltstones and shales which can be traced in a downslope direction down a depositional tongue extending a palinspastic distance of approximately 60 km westward from the Augusta lobe of the Catskill Delta complex. The estimated deposition time of this cycle was approximately 100 ka (van Tassell, 1987).

Turbidite cycles similar to those in the Brallier Formation occur in the overlying Scherr Formation and grade laterally into more proximal portions with a higher percentage of sandstones. Van Tassell (1987, 1988) has shown that the Minnehaha Springs Member at the base of the Scherr Formation is a fining-upward cycle that consists of three fining-upward packets including turbidite siltstones; reddish 'slope' shales probably deposited offshore during major river flooding; lower shoreface, parallel-laminated siltstones; and nearshore (upper shoreface) ripple-bedded and planar-bedded sandstones with flute casts, mudchips and plant fragments (Fig. 7). The estimated deposition time of this cycle, which has been correlated several hundred kilometres across several lobes of the Catskill Delta (Lyke, 1986) and more than 100 km across the outcrop belt into the Appalachian basin (Barrell & Dennison, 1986), is approximately 100 ka (van Tassell, 1987).

The cycle patterns present in the Brallier and Scherr Formations are even more varied. Fining-upward and coarsening-upward cycles both occur, as well as cycles transitional between these two (Fig. 7). In general, cycles consisting mainly of deeper water facies tend to fine upward and are thinner than cycles with shallower water facies, which coarsen upward and are less distinct due to thinning of the shale layers which define the packets recognizable in the deeper water cycles. Cycles which record the transition from nearshore to beach to alluvial environments are also present. Overall, the approximately 20 cycles recognized in the type section of the Foreknobs Formation each had a duration of approximately 100 ka (van Tassell, 1987).

It has proven difficult to establish long-distance correlations of the cycles in the Catskill Delta margin. Prior to the surge of interest in facies vari-

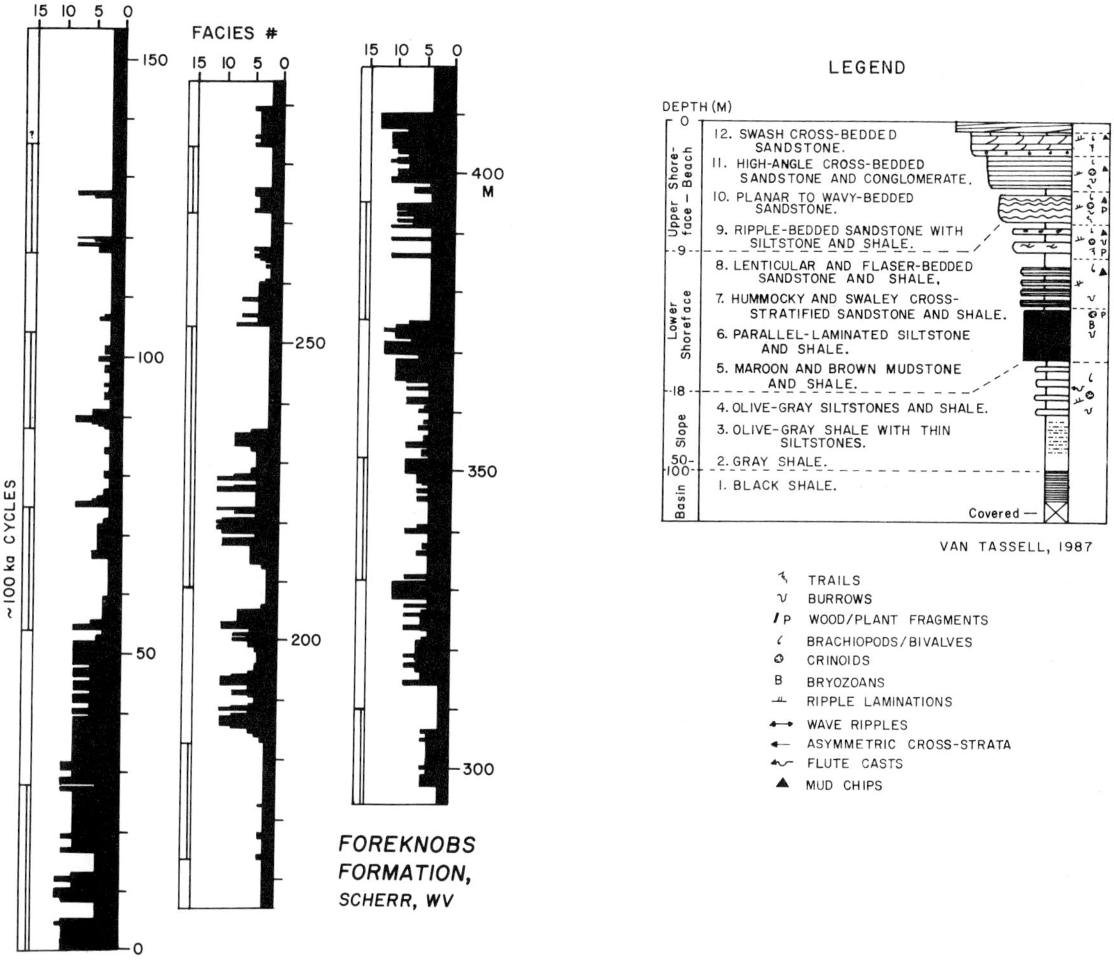

Fig. 7. Facies and 0.1-Ma cycles (1, 2, 3 . . .) in the Foreknobs Formation of West Virginia. From Van Tassell (1987).

ations and facies models in the late 1960s and early 1970s, it was often assumed that units in the Late Devonian Catskill Delta sequence correlated across long distances (e.g. Dyson, 1963; Frakes, 1967). Since the 1970s, however, sedimentologists (e.g. Craft & Bridge, 1987) have criticized the long-distance correlations of Catskill Delta stratigraphers (e.g. Rickard, 1974) because biostratigraphic correlation is very crude and lithostratigraphic correlation is difficult due to the lack of distinctive beds and the overall structural deformation of the region. This change in attitude has roots in Walker & Sutton's (1967) study of the turbidites in the Rock Stream Member of the Sonyea Group in the Finger Lakes area of New York. Before commencing work, Walker & Sutton assumed a simple model of basin filling in which individual beds were more or less continuous and the 10–40 siltstone beds marking the base of the member were correlative. Their careful measurements of sedimentary structures in the turbidites showed that the ABC index values ($A + \frac{1}{2}B$, where A is the percentage of beds in a given interval beginning with Bouma A sequences and B is the percentage of beds beginning with Bouma B units) did not show the expected gradual decrease in successive outcrops in a down-current direction. This led them to propose an alternative model of basin filling in which turbidites were deposited in a series of imbricately-stacked wedges.

Four years later, building on these observations

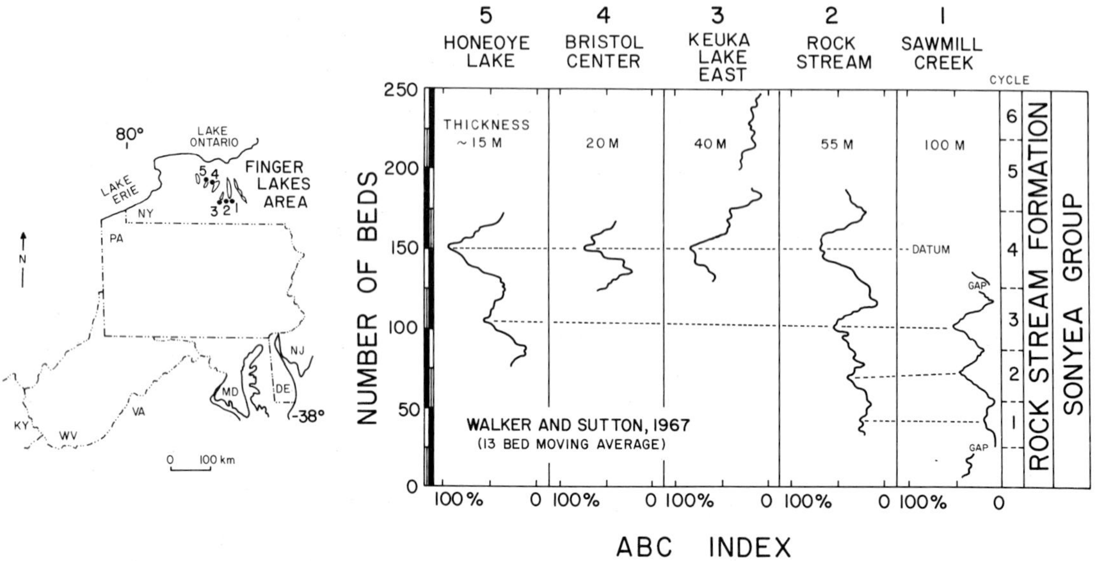

Fig. 8. One possible correlation of turbidite beds in the Upper Devonian Rock Stream Formation of the Sonyea Group in the Finger Lakes area of New York , based on the data of Walker & Sutton (1967).

and stimulated by similar observations by Enos (1969), Walker (1971) pointed out that it was not possible to substantiate the 10 turbidite siltstone and 10 shale members recognized by Frakes (1967) in the marine slope sequences below the non-marine Catskill Formation in the central Pennsylvania area. Walker (1971) concluded that, in places, Frakes' distinction between siltstone and shale appeared arbitrary and correlation of the units for 250 km proved difficult because the units were not distinctive in the field.

Although it is not possible to correlate individual beds, careful study of Walker & Sutton's (1967) data and Walker's (1971) measurements of turbidites at Shamokin Dam, Pennsylvania and Woodmont, Maryland, suggests that there may be rhythmic patterns of deposition in these sequences (Figs 8 and 9). The number of beds per cycle ranges from 50 to 200, with an overall average of approximately 100 beds per cycle. Preliminary studies suggest that these rhythmic patterns can be pieced together in the scattered outcrops in the Finger Lakes area and matched with similar patterns in outcrops of the Upper Devonian (Frasnian) Brallier Formation sequences at Shamokin Dam, Pennsylvania, and Woodmont, Maryland. This suggests that the overall cyclic deposition in all three areas may have had a similar pattern but does not prove that individual cycles in each area correlate with specific cycles in other areas. Additional study is needed to confirm or reject this possibility.

Filer (1991) has recently correlated distal coarse-grained tongues of sediment (including parts of the Foreknobs Formation) produced by widely separated depositional systems along the Catskill Delta Margin by tracing their signature on geophysical logs from wells penetrating the Frasnian Rhinestreet black shales and the overlying grey shales of the Java Formation and Angola Shale, and also the Famennian Huron (Dunkirk) black shales. This study suggested the possible synchrony of small-scale progradational events throughout the entire Appalachian basin.

A CROSS-SECTION ACROSS THE CATSKILL BASIN

Studies of the Upper Eifelian to Middle Givetian Hamilton Group (Brett & Baird, 1985, 1986) provide a glimpse of the lateral variation of facies within cycles and the overall hierarchy of cycles evident in a cross-section across the Catskill basin. The Hamilton Group is transitional between the platform carbonates of the early Middle Devonian and the progradational deposits of the Late Devonian Catskill

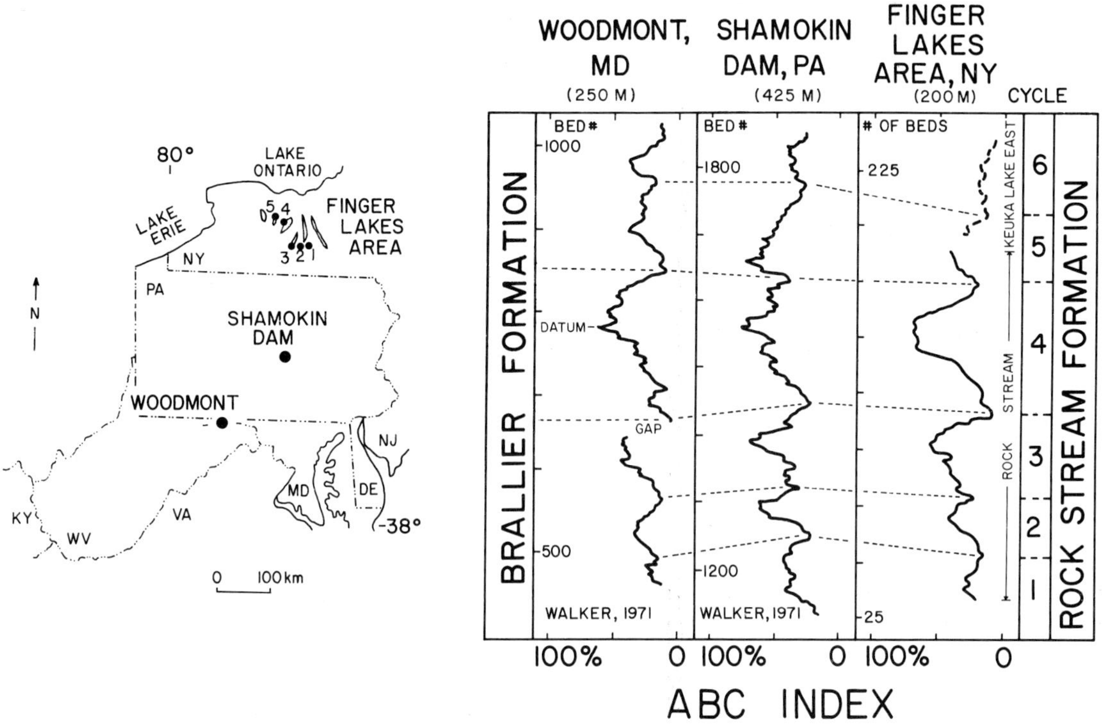

Fig. 9. Comparison of the long-term cyclic depositional patterns of the Rock Stream Formation turbidites with the depositional patterns evident in Walker's (1971) measurements of turbidites in the Upper Devonian Brallier Formation of Maryland and Pennsylvania. Dashed lines show the boundaries of the depositional cycles with similar patterns; correlation of individual cycles with specific cycles in the other regions has not been established. Each cycle represents approximately 100 ka of deposition.

Delta. The detrital sediments in this 90–200-m eastward-thickening wedge of shale, siltstone, sandstone and thin carbonates record the first evidence of the advance of the Catskill Delta complex. The sequence grades westward from alluvial and coastal deposits into marine siltstones and sandstones and then into basinal medium to dark grey mudrocks and thin concretionary limestones. These thin 0.5–2-m-thick carbonate layers were interpreted as isochronous, nearly basin-wide layers formed during coastal entrapment of terrigenous sediment following episodes of rapid eustatic sea-level rise (Cooper, 1957; Johnson & Friedman, 1969; McCave, 1969, 1973). More recent work (Brett & Baird, 1985) has demonstrated that these limestone members contain evidence for deposition in turbulent shallow-shelf settings. The presence of graded coquinites, planar to hummocky cross-stratification, and large overturned coral heads in the sequences supports a storm (tempestite) interpretation. The major, richly fossiliferous carbonates thus appear to represent deposits formed during periods of shoaling rather than deepening.

Brett & Baird (1986) recognized small-scale (0.5–3 m) upward-coarsening cycles on the east side of their study area, closer to the delta margin (Fig. 10). In more basinward regions, the minor upward-coarsening cycles appear to be recorded in the form of thin, 0.5–1-m-scale alternations of fossiliferous, concretionary carbonate beds, each about 20–30 cm thick, alternating with non-calcareous mudstone. Tentative correlation of some of the minor upward-shallowing cycles in New York State has been established, although it has not been possible to trace many of these minor cycles through the intervening fine-grained sediments of the basin which separated the two areas. Based on subdivision of larger packages, these small-scale cycles are estimated to represent 100–125-ka fluctuations (sixth-order cycles of Busch & Rollins, 1984).

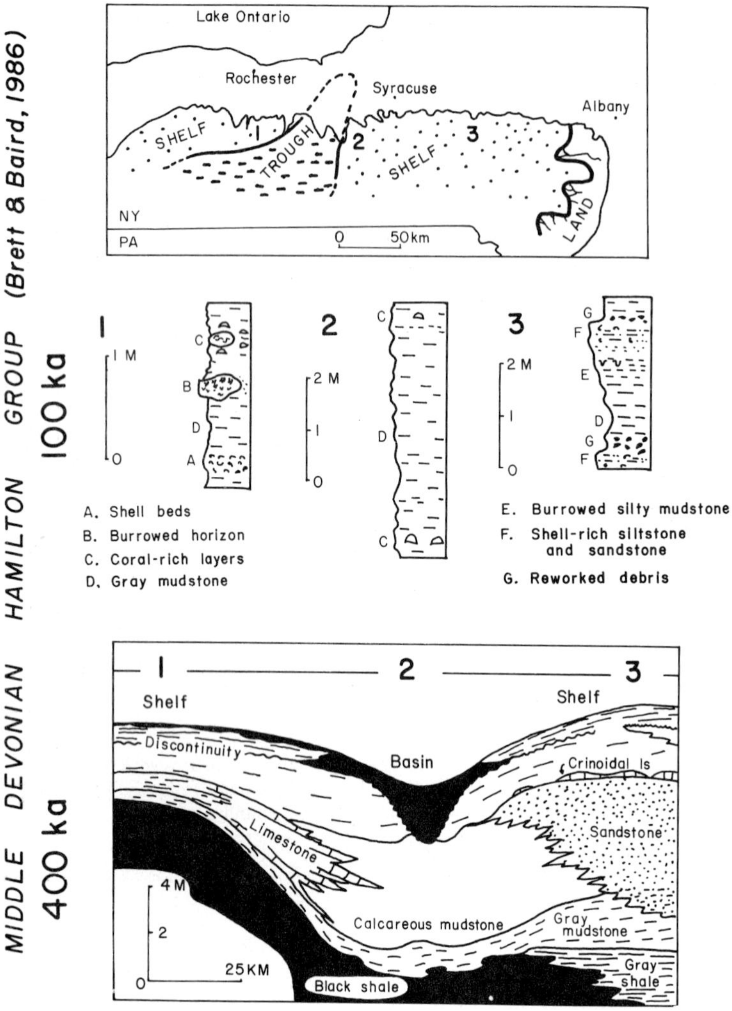

Fig. 10. Regional stratigraphy of the Upper Hamilton Group (Brett & Baird, 1986). Note the lateral variation of the 100-ka cycles, including cycles with numerous thin limestone beds in the Genessee Valley area of New York (1); cycles dominated by mudstone in the Finger Lakes (Cayuga Valley) area of west-central New York (2); and the numerous upward regressive hemicycles dominated by siltstones in the central New York (Tully Valley) area (3). A representative Hamilton Group 400-ka cycle, the Centerfield Limestone and equivalent Chenango Sandstone, in western, west-central, and central New York clearly shows the lateral variation of facies within cycles (Brett & Baird, 1985).

Larger, distinctly cyclic packages up to tens of metres thick (Fig. 10) also form a dominant motif in the Hamilton Group. These major cycles are similar in nature to the smaller cycles but are distinctly thicker, record a much broader range of facies from base to top, and are more laterally extensive. The cycles grade eastward from black clay–shale/mudstone–packstone and grainstone cycles in western New York through transitional calcareous silty mudstones into distinct upward-coarsening shale–sandstone cycles in areas closest to the delta margin. All of the major shale–limestone cycles in western New York are directly correlative with upward-coarsening sequences in the east. In all, about 14–15 subequally spaced, major transgressive–regressive cycles are now recognized over the 6–7-Ma span of the Hamilton Group, suggesting that these cycles record oscillations on the order of 400–500-ka duration (Brett & Baird, 1986). They are thus very similar to fifth-order allocycles such as the mid-continent cyclothems described by Heckel (1977, 1986).

The Hamilton Group has also been previously subdivided into three fourth-order (0.8–1.5 Ma) cycles (Johnson *et al.*, 1985) and itself constitutes a part of a third-order (8–10 Ma) cycle (terminology of Vail *et al.*, 1977). These observations hint at the longer-term depositional rhythms present in the Catskill Delta succession.

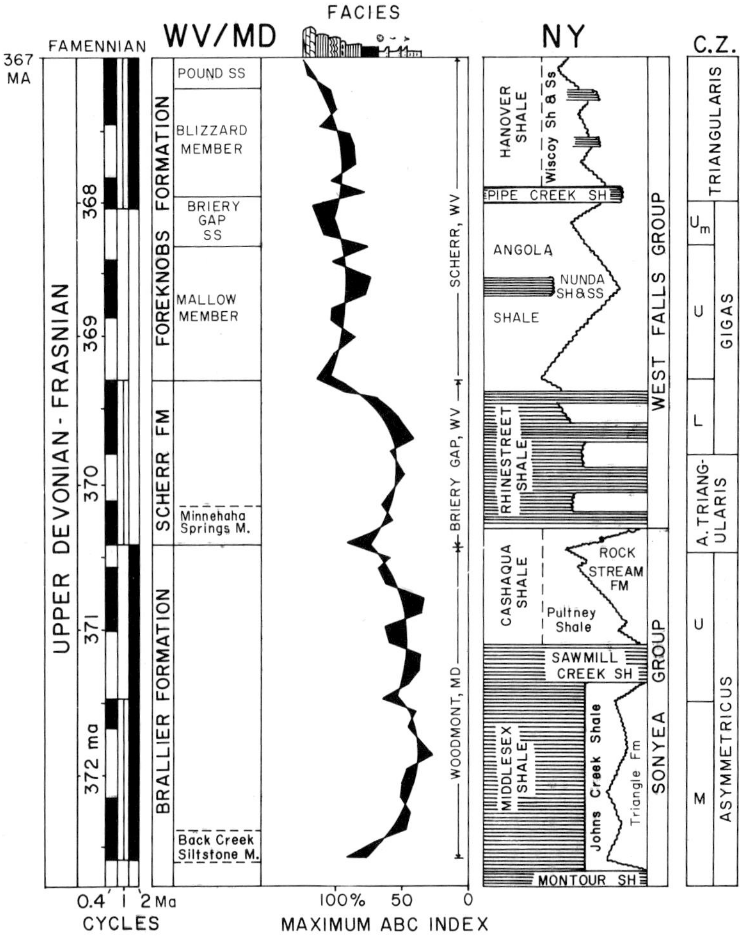

Fig. 11. Long-term depositional cycles in the Frasnian Catskill Delta complex based on the hypothesis that 0.1-Ma variations are a persistent feature of the depositional sequence. Variations on either side of the approximately 1-Ma depositional cycles in the west Virginia–Maryland area are shaded to emphasize the shorter-period 400-ka cycles. Facies symbols are shown in Fig. 7. The stratigraphy of the New York succession is modified from Rickard (1975). The age of the Frasnian–Famennian boundary is from Harland *et al.* (1982). C.Z., conodont zone.

LONGER-PERIOD CYCLES

Van Tassell (1987) pointed out that the recognition of 100-ka cycles in the Catskill Delta succession leads to a provisional way to subdivide the larger radiometrically dated intervals of the complex into smaller subdivisions until better age-dating becomes available. This involves: (i) using the cycles to estimate geological time in approximately 0.1-Ma increments below the Frasnian–Famennian boundary; (ii) plotting the facies present at the point of maximum regression or the maximum ABC index values in each of the cycles in the sequence; and (iii) arranging the sections in stratigraphic order. These plots filter out the effects of the approximately 0.1-Ma and shorter cycles, allowing longer-period rhythms in the succession to be observed more readily.

The resulting curves (Fig. 11), based on ABC index fluctuations at Woodmont, Maryland, measured by Walker (1971) and cycles identified by van Tassell (1987, 1988) at Briery Gap Run and Scherr, West Virginia, indicate a strong rhythm with an average period of approximately 1.2 Ma in the sequence. Deviations on either side of this 1.2-Ma cycle highlight the presence of shorter-period rhythms with recurrence intervals of approximately 0.4 Ma in the succession. In addition, a less obvious cycle with a period of approximately 2 Ma is evident if the asymmetry of individual 1.2-Ma cycles is carefully examined. Also evident is an overall long-term upward-shoaling trend which van Tassell (1987) speculated might be related to sediment input triggered by the third tectophase of the Acadian orogeny. These trends are paralleled by similar depositional variations in the Late Devonian stratigraphy of New York.

ORIGIN OF THE CYCLES

There is clearly a hierarchy of cycles present in the Catskill Delta complex. Delta distributary shifting, turbidite deposition (Lundegard *et al.*, 1985; Van Tassell, 1987), and formation of shelf storm beds (Duke, 1985; Craft & Bridge, 1987) occurred at intervals of 10^2–10^3 years. Studies of fluvial deposits (Schumm, 1968; Woodrow *et al.*, 1973; Willis & Bridge, 1988) indicate the presence of cycles with similar periodicities. Lateral shifting of river channels, delta lobes, and other depositional features helped to produce these cycles. This facies shifting may have been related to fluctuations in storm activity representing climatic fluctuations with a recurrence interval of 1–3 ka. Cycles with periods between 1 and 3 ka have been recognized in a variety of terrestrial and marine environments. This evidence includes oxygen isotope variations in Quaternary ice cores from Antarctica and in deep-sea cores (Dansgaard *et al.*, 1984, Pestiaux *et al.*, 1987), the advance and retreat patterns of North American and European glaciers (Denton & Karlen, 1973), varved slope sediments produced by wind-controlled upwelling off the northern California coast (Anderson *et al.*, 1989), Permian deep-water evaporitic varves in the Delaware Basin interpreted as the result of periodic freshening of the evaporite basin (Anderson, 1982), and storm cycles in the Mississippian ramp-slope deposits of southwest Montana (Elrick *et al.*, 1991). The one factor that could be common to changes of similar periodicity in such diverse environments is climatic change. The cause of these short-term palaeoclimatic fluctuations is not well understood, though one suggestion is that they reflect harmonics or combination tones of the 19–34-ka (precession) and 41-ka (obliquity) orbital-forcing periods or periodic changes in solar activity (Elrick *et al.*, 1991).

Cycles with 100-ka (sixth-order) and 400-ka (fifth-order) periodicities are also present in the shoreline, shelf and basinal Catskill deposits (Brett & Baird, 1986; van Tassell, 1987). It is possible that 100-ka cycles are present in Catskill fluvial deposits, but the evidence is inconclusive. Given that a number of factors influence the proportions of overbank deposits in Catskill fluvial sequences and the observation that there is no reason to presume that these controlling factors would act the same way at all locations on a broad alluvial plain (Gordon & Bridge, 1987), it is not surprising that it is difficult to correlate cyclic trends in Catskill fluvial sequences.

The origin of these longer cycles is controversial. Facies migration has been commonly cited as one possible mechanism. Craft & Bridge (1987) called

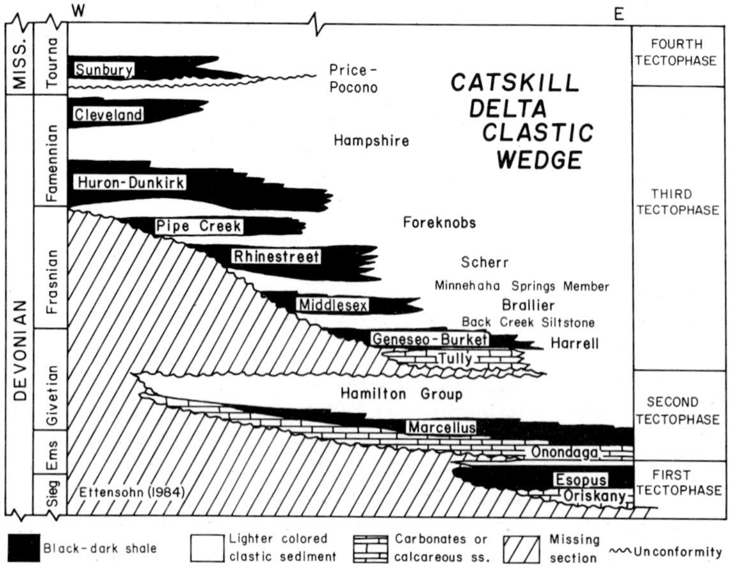

Fig. 12. Large-scale stratigraphic cycles in the Devonian Catskill Delta succession, modified from Ettensohn (1984). Note the cyclic repetition of black shales and the four tectonic episodes (tectophases) indicated by prominent unconformities in the sequence.

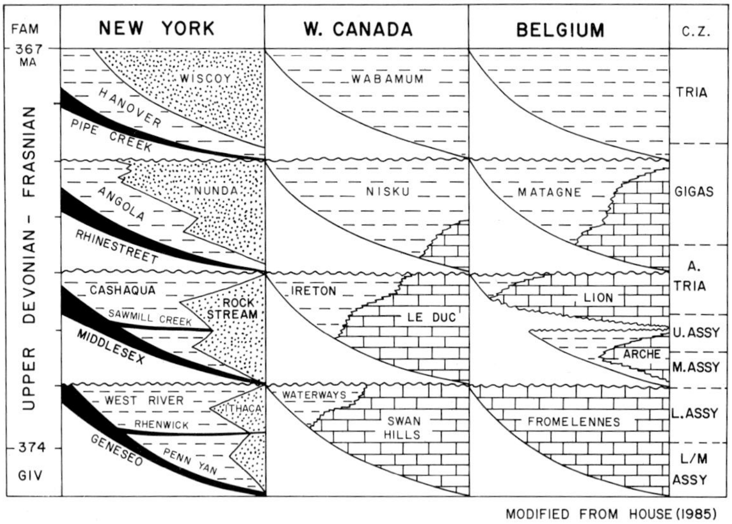

Fig. 13. Possible global correlations of New York Catskill Delta major transgressive–regressive cycles with sequences in Belgium and western Canada, modified from House (1985). Note the strong 1- and 2-Ma cycles and the overall progradation of clastic sediments in the New York area and decline of carbonate sedimentation in the western Canada and Belgium areas.

upon delta lobe progradation and/or the migration of shelf sand ridges associated with sand progradation from river mouths to explain coarsening-upward sequences in Catskill Delta shelf deposits. Walker & Harms (1971) speculated that the repetition of Irish Valley motifs might be the result of the diversion of coastal rivers to a distant part of the coastal plain, combined with regional subsidence and compaction, followed by regression and the formation of a new motif when longshore drift once again provided sediment at a rate exceeding subsidence. As an alternative, abrupt transgressions might be caused by intermittent episodes of tectonic subsidence, perhaps related to tectonic movements in the nearby newly uplifted mountains, but Walker & Harms (1971) pointed out that known rates of tectonic movement are slow compared with the incidence of river deposits and Gordon & Bridge (1987) noted that depositional and tectonic events need to be dated more precisely than at present to test any theory of the relative timing and rate of deposition and subsidence at a point (cf. Beaumont, 1981; Quinlan & Beaumont, 1984; Blair & Bilodeau, 1988; Heller *et al.*, 1988). Willis & Bridge (1988) pointed out, however, that tectonic variations appear to be the most important factor controlling upsection increases in the proportion, thickness and mean grain size of single storey fluvial sandstone bodies. This suggests that local 'cyclic' changes in the proportions of sandstone bodies may reflect either slight changes in the ratio of avulsion rate to deposition, or changes in the width of the effective floodplain relative to channel-belt widths, or local tilting of the floodplain. This has not been proven, however.

A third possible mechanism is eustatic changes in sea level, but Walker & Harms (1971) found no convincing evidence for eustatic movements as a controlling mechanism and both Gordon & Bridge (1987) and Willis & Bridge (1988) suggest that the lack of clear evidence of worldwide sea-level variations and the apparent increase in the magnitude of lithofacies variations associated with the transgression–regression cycles in exposures further away from the shoreline makes it unwise at present to propose eustatic sea-level variations as the major cause of cyclicity during the Givetian and Frasnian (cf. Dennison & Head, 1975; Vail *et al.*, 1977; Johnson *et al.*, 1985).

A fourth possible hypothesis relates variations in the rates of sediment accumulation on the Catskill Delta to fluctuations in climate and rainfall in the source area and the accompanying fluctuations in sediment yield. Such control of pulses of coarse-grained sediment supply has been demonstrated for Pleistocene alluvial fan settings by Frostick & Read (1989) and Nemec & Postma (1993), and for the Eocene South Pyrenean Tremp-Graus foreland basin by de Boer *et al.* (1991). Gordon & Bridge (1987) and Willis & Bridge (1988) argued that there is little clear evidence for climatic change as an explanation for the large-scale variations in Catskill fluvial deposition and that the stratigraphic variations in calcrete development and channel size documented can easily be explained without recourse to climatic change. The possibility of periodic climate change should not be ruled out, however, until detailed studies of the palaeosols are completed. Indeed, given that the palaeogeographic position of the Catskill Delta was consistent with a monsoonal type climate, I prefer to take the view that orbitally forced climatic change is a hypothesis worthy of further testing in the Catskill.

The possibility that pulses of coarse-grained sedimentation were produced by climatic variations related to orbital forcing has been suggested for the Pleistocene by Fischer *et al.* (1990), for Eocene fan delta systems by de Boer *et al.* (1991), and for Devonian fluvial facies in East Greenland by Olsen (1990, pp. 429–438, this volume). Several authors (see Kutzbach & Otto-Bliesner, 1982) have shown that precipitation/evaporation rates may vary strongly in response to forcing by the precession cycle and examples of precession-driven alternations of dry and wet periods have been cited (van Houten, 1964; Anderson, 1984; Olsen, 1986).

The 100-ka (sixth-order) cycles with possible superimposed 20-ka and 33-ka cycles (van Tassell, 1987); 400-ka (fifth-order), 1.2-Ma (fourth-order); and weak 2.0-Ma (third-order) cycles in the Catskill Delta succession correspond well with the periodicities and relative amplitudes of the orbital eccentricity cycles calculated by Berger (1976, 1978; Table 1). This suggests that sediment input variations related to climate variations triggered by orbital changes may have had a major influence on the deposition of the Catskill Delta sequence. This hypothesis explains the observation that the major sedimentary rhythms of the Devonian are composed of microrhythms often sharing microfacies elements of the larger rhythms (House, 1985; Brett & Baird, 1986). It also explains the repetition of cycles throughout the majority of the Givetian and Frasnian Catskill Delta succession and the occurrence of cycles in a wide range of terrestrial and marine depositional environments. Proof of Milankovitch-type forcing will depend on confirming the correlation and synchrony of small-scale progradational events throughout the entire Catskill Delta complex and on a global scale. Greatly improved dating of the cycles and time-series analysis will be necessary to do this and to sort out the effects of lateral facies shifting, tectonism, eustatic sea-level variations, and climate changes, all working simultaneously to produce the complex sedimentary patterns present in the Catskill Delta complex (see Klein & Willard, 1989).

Table 1. Selected main terms in the series expansion of orbital eccentricity (Berger, 1978)

Approximate time	Period (yr)	Amplitude
100 ka	94 945	−0.0087
	99 590	0.0067
	102 384	−0.0016
	102 535	−0.0025
	107 807	−0.0013
	123 297	−0.0074
	131 248	0.0058
	136 618	−0.0016
500 ka	412 885	0.0110
	603 630	0.0020
1.5 Ma	1 306 618	0.0023
2 Ma	2 035 441	−0.0047

CONCLUSIONS

The Catskill Delta complex includes a hierarchy of cycles with periodicities ranging from 1 ka to 20, 33, 100 and 400 ka, to 1.2 and 2 Ma, and longer cycles. The origins of these complex cycles remain to be proven. The shortest period cycles may be related to climatic variations triggered by harmonics or combination tones of the precession and obliquity cycles, or by periodic changes in solar activity. The origins of the cycles with periods from 20 ka to 2 Ma have been controversial. Facies shifting associated with delta progradation and regional tectonism, subsidence and compaction; tectonic movements; eustatic sea-level changes; and climatic variations have all been cited as possible mechanisms. Tentative suggestions of regional correlation of individual cycles, the correspondence of the periodicities of the cycles in the Catskill Delta sequence with known orbital variations, the overall pattern of microrhythms often sharing microfacies elements of the larger rhythms on which they are superimposed, and the repetition of cycles throughout the greater part of the Catskill Delta succession and in a wide range of depositional environments is evidence that sediment input variations related to climatic changes (perhaps shifting of monsoonal belts) triggered by orbital forcing may have played a significant role in the deposition of the sequence in addition to lateral facies shifting, eustatic sea-level changes and tectonism. The influence of major tectonic episodes is clearly recorded in the major unconformity-bounded sequences in the Catskill Delta complex.

Greatly improved dating of the cycles, improved

stratigraphic correlation, and time-series analysis will be necessary to sort out the effects of lateral facies shifting, climatic changes, eustatic sea-level variations, and tectonism, all working simultaneously to produce the complex sedimentary patterns present in the Catskill Delta complex. After a century of research, our study of the Catskill Delta sequence has only just begun.

ACKNOWLEDGEMENTS

The author has benefited greatly from the criticism and suggestions for improvement made over the past decade by Poppe de Boer, David Smith, John Bridge, John Dennison, Carlton Brett, Don Woodrow, Roger Walker, Ed Cotter, Ed Clifton, Frank Ettensohn, Robert Suchecki, Charles Harris, Jim Smith and George McGhee, Jr.

REFERENCES

Allen, J.R.L. & Friend, P.F. (1968) Deposition of the Catskill facies, Appalachian regions, with notes on some other Old Red Sandstone basins. In: *Late Paleozoic and Mesozoic Continental Sedimentation, Northeastern North America* (Ed. Klein, G. DeV.). Geol. Soc. Am. Spec. Pap. 160, pp. 21–24.

Anderson, R.Y. (1982) A long geoclimate record from the Permian. *J. Geophys. Res.* **87**, 7285–7294.

Anderson, R.Y. (1984) Orbital forcing of evaporite sedimentation. In: *Milankovitch and Climate*, Part 1 (Eds Berger, A., Imbrie, J., Hays, J., Kukla, G. & Saltzman, B.). Reidel Publ. Co., Dordrecht, pp. 147–162.

Anderson, R.Y., Gardner, J.V. & Hemphill-Haley, E. (1989) Variability of the late Pleistocene–early Holocene oxygen-minimum zone off northern California. In: *Aspects of Climate Variability in the Pacific and Western Americas* (Ed. Petersen, D.H.). Am. Geophys. Union Monogr. 55, pp. 75–84.

Barrell, J. (1913) The Upper Devonian delta of the Appalachian geosyncline: Part 1. The delta and its relations to the interior sea. *Am. J. Sci.* **36**, 429–472.

Barrell, J. (1914) The Upper Devonian delta of the Appalachian geosyncline, Part 2. *Am. J. Sci.* **37**, 86–109.

Barrell, S.M. & Dennison, J.M. (1986) Northwest–southeast stratigraphic cross section of Devonian Catskill Delta in East-central West Virginia and adjacent Virginia. In: *Appalachian Basin Industrial Associates Fall Meeting Program 11*, pp. 7–32.

Beaumont, C. (1981) Foreland basins. *Geophys. J. R. Astron. Soc.* **65**, 291–329.

Berger, A.L. (1976) Obliquity and precession for the last 5 000 000 years. *Astron. Astrophys.* **51**, 127–135.

Berger, A. (1978) Long-term variations of daily insolation and Quaternary climatic changes. *J. Atmos. Sci.* **35**, 2362–2367.

Blair, T.C. & Bilodeau, W.L. (1988) Development of tectonic cyclothems in rift, pull-apart, and foreland basins: Sedimentary response to episodic tectonism. *Geology* **16**, 517–520.

Brett, C.E. & Baird, G.C. (1985) Carbonate–shale cycles in the Middle Devonian of New York: an evaluation of models for the origin of limestones in terrigenous shelf sequences. *Geology* **13**, 324–327.

Brett, C.E. & Baird, G.C. (1986) Symmetrical and upward shallowing cycles in the Middle Devonian of New York State and their implications for the Punctuated Aggradational Cycle hypothesis. *Paleoceanography* **1**, 431–445.

Bridge, J.S. (1988) *Devonian fluvial deposits of the western Catskill region of New York State*. Soc. Econ. Paleont. Mineral., Eastern Section, 1988 Annual Field Trip Guidebook.

Bridge, J.S. & Gordon, E.A. (1985) Quantitative interpretation of ancient river systems in the Oneonta Formation, Catskill magnafacies. In: *The Catskill Delta* (Eds Woodrow, D.L. & Sevon, W.D.). Geol. Soc. Am. Spec. Pap. 201, pp. 163–182.

Busch, R.M. & Rollins, H.B. (1984) Correlation of Carboniferous strata using a hierarchy of transgressive–regressive units. *Geology* **12**, 471–474.

Chadwick, G.H. (1936) *History and value of the name 'Catskill' in geology*. New York State Museum Bulletin 307.

Chadwick, G.H. (1944) *Geology of the Catskill and Kaaterskill quadrangles, part 2*. New York State Museum Bulletin 336.

Colton, G.W. (1967) Late Devonian current directions in western New York with special reference to *Fucoides graphica*. *J. Geol.* **75**, 11–22.

Cooper, G.A. (1957) Paleoecology of the Middle Devonian of eastern and central United States. In: *Treatise on Marine Ecology and Paleoecology, Part 2, Paleoecology* (Ed. Ladd, H.S.). Geol. Soc. Am. Mem. 67, pp. 249–278.

Craft, J.H. & Bridge, J.S. (1987) Shallow-marine sedimentary processes in the Late Devonian Catskill Sea, New York State. *Geol. Soc. Am. Bull.* **98**, 338–355.

Dansgaard, W., Johnsen, S.J., Clausen, H.B., Dahl-Jensen, D., Gundestrup, N., Hammer, C.H. & Oescheger, H. (1984) North Atlantic oscillations revealed by deep Greenland ice cores. In: *Climate Processes and Climate Sensitivity* (Eds Hanse, J.E. & Takahashi, T.). Am. Geophys. Union Monogr. 29, pp. 288–298.

De Boer, P.L. & Wonders, A.A.H. (1984) Astronomically induced bedding in Cretaceous pelagic sediments near Moria (Italy). In: *Milankovitch and Climate*, Part 1 (Eds Berger, A., Imbrie, J., Hays, J., Kukla, G. & Saltzman, B.). Reidel Publ. Co., Dordrecht, pp. 177–190.

De Boer, P.L., Pragt, J.S.J. & Oost, A.P. (1991) Vertically persistent sedimentary facies boundaries along growth anticlines in the thrust-sheet-top South Pyrenean Tremp-Graus foreland basin. *Basin Res.* **3**, 63–78.

Dennison, J.M. (1985) Catskill Delta shallow marine strata. In: *The Catskill Delta* (Eds Woodrow, D.L. & Sevon, W.D.). Geol. Soc. Am. Spec. Pap. 201, pp. 91–106.

Dennison, J.M. & Head, J.W. (1975) Sea level variations

interpreted from the Appalachian Basin Silurian and Devonian. *Am. J. Sci.* **275**, 1089–1120.
DENTON, G.H. & KARLEN, W. (1973) Holocene climatic variations – their pattern and possible cause. *Quat. Res.* **3**, 155–205.
DUKE, W.L. (1985) Hummocky cross-stratification, tropical hurricanes, and intense winter storms. *Sedimentology* **32**, 167–194.
DYSON, J.L. (1963) *Geology and mineral resources of the New Bloomfield quadrangle*. Pennsylvania Geological Survey Atlas 137A.
ELRICK, M., READ, J.F. & CORUH, C. (1991) Short-term paleoclimatic fluctuations expressed in lower Mississippian ramp-slope deposits, southwestern Montana. *Geology* **19**, 799–802.
ENOS, P. (1969) Anatomy of a flysch. *J. Sedim. Petrol.* **39**, 680–723.
ETTENSOHN, F.R. (1984) The nature and effects of water depth during deposition of the Devonian–Mississippian black shale. In: *Proceedings, 1984 Eastern Oil Shale Symposium*. Kentucky Department of Energy and University of Kentucky Institute for Mining and Minerals Research, Lexington, pp. 333–346.
ETTENSOHN, F.R. (1985a) Controls on development of Catskill Delta complex basin-facies. In: *The Catskill Delta* (Eds Woodrow, D.L. & Sevon, W.D.). Geol. Soc. Am. Spec. Pap. 201, pp. 65–77.
ETTENSOHN, F.R. (1985b) The Catskill Delta complex and the Acadian orogeny: a model. In: *The Catskill Delta* (Eds Woodrow, D.L. & Sevon, W.D.). Geol. Soc. Am. Spec. Pap. 201, pp. 39–49.
ETTENSOHN, F.R. & BARRON, L.S. (1981) Depositional model for the Devonian–Mississippian black shales of North America: a paleoclimatic–paleogeographic approach. In: *Geol. Soc. Am. Cincinnati 1981 Field Trip Guidebook, vol. II: Economic Geology, Structure* (Ed. Roberts, T.G.). Am. Geol. Inst., Falls Church, pp. 344–361.
FAILL, R.T. (1985) The Acadian orogeny and the Catskill Delta. In: *The Catskill Delta* (Eds Woodrow, D.L. & Sevon, W.D.). Geol. Soc. Am. Spec. Pap. 201, pp. 15–37.
FILER, J.K. (1991) Small scale sequences within a portion of the Upper Devonian clastic wedge, Appalachian Basin – origins and implications. *Geol. Soc. Am. Abstracts with Programs* **23**, 28.
FISCHER, A.G., DE BOER, P.L. & PREMOLI SILVA, I. (1990) Cyclostratigraphy. In: *Cretaceous Resources, Events and Rhythms* (Eds Ginsburg, R.N. & Beaudoin, B.). Kluwer Academic Publishers, Dordrecht, pp. 139–172.
FLETCHER, F.W. (1963) Regional stratigraphy of Middle and Upper Devonian nonmarine rocks in southeastern New York. *Pennsylvanian Geol. Surv. Bull.* **G-39**, 25–41.
FLETCHER, F.W. (1967) *Middle and Upper Devonian clastics of the Catskill Front, New York*. New York Geological Association Guidebook 39, pp. C1–C23.
FRAKES, L.A. (1967) Stratigraphy of the Devonian Trimmers Rock in eastern Pennsylvania. *Pennsylvania Geol. Surv. Bull.* **G51**.
FROSTICK, L.E. & REID, I. (1989) Climatic versus tectonic controls of fan sequences: lessons from the Dead Sea, Israel. *J. Geol. Soc. Lond.* **146**, 527–538.
GLAESER, J.D. (1979) Catskill delta slope sediments in the central Appalachian basin: source deposits and reservoir deposits. In: *Geology of Continental Slopes* (Eds Doyle, L.J. & Pilkey, O.H.). Soc. Econ. Paleont. Mineral. Spec. Publ. 27, pp. 343–357.
GOODWIN, P.W. & ANDERSON, E.J. (1985) Punctuated aggradational cycles: General hypothesis of episodic stratigraphic accumulation. *J. Geol.* **93**, 515–533.
GORDON, E.A. & BRIDGE, J.S. (1987) Evolution of Catskill (Upper Devonian) river systems. *J. Sedim. Petrol.* **57**, 234–249.
HAMBLIN, D.P. & WALKER, R.G. (1979) Storm-dominated shallow-marine deposits: The Fernie-Kootenay (Jurassic) transition, southern Rocky Mountains. *Can. J. Earth Sci.* **26**, 1673–1689.
HARLAND, W.B., COX, A.V., LLEWELLYN, R.G., PICTON, C.A.G., SMITH, A.G. & WALTERS, R. (1982) *A Geologic Time Scale*. Cambridge University Press, Cambridge.
HECKEL, P.H. (1977) Origin of phosphatic black shale facies in Pennsylvanian cyclothems in midcontinent North America. *Bull. Am. Assoc. Petrol. Geol.* **61**, 1045–1068.
HECKEL, P.H. (1986) Sea-level curve for Pennsylvanian eustatic marine transgressive–regressive cycles along midcontinent outcrop belt, North America. *Geology* **14**, 330–334.
HECKEL, P.H. & WITZKE, B.J. (1979) Devonian world paleogeography determined from the distribution of carbonates and related lithic paleoclimatic indicators. In: *The Devonian System* (Eds House, M.R., Scrutton, C.T. & Bassett, M.G.). Special Papers in Paleontology 23, pp. 99–123.
HELLER, P.L., ANGEVINE, C.L., WINSLOW, N.S. & PAOLA, C. (1988) Two-phase stratigraphic model of foreland-basin sequences. *Geology* **16**, 501–504.
HOUSE, M.R. (1985) Correlation of mid-Paleozoic ammonoid evolutionary events with global sedimentary perturbations. *Nature* **313**, 17–22.
JOHNSON, J.G., KLAPPER, G. & SANDBERG, C.A. (1985) Devonian eustatic fluctuations in Euramerica. *Geol. Soc. Am. Bull.* **96**, 567–587.
JOHNSON, K.G. & FRIEDMAN, G.M. (1969) The Tully clastic correlatives (Upper Devonian) of New York State: a model for recognition of alluvial, dune(?), tidal, nearshore (bar and lagoon) and offshore sedimentary environments in a tectonic delta complex. *J. Sedim. Petrol.* **39**, 451–485.
KENT, D.V. & OPDYKE, N.D. (1985) Multicomponent magnetisations from the Mississippian Mauch Chunk Formation of the Central Appalachians and their tectonic implications. *J. Geophys. Res.* **90**, 5371–5383.
KLEIN, G. DE V. & WILLARD, D.A. (1989) Origin of the Pennsylvania coal-bearing cyclothems of North America. *Geology* **17**, 152–155.
KUENEN, PH.H. (1956) Problematic origin of the Naples rocks around Ithaca, New York. *Geol. Mijnbouw* **18**, 277–283.
KUTZBACH, J.E. & OTTO-BLIESNER, B.L. (1982) The sensitivity of the African–Asian monsoonal climate to orbital parameter changes for 9000 years BP in a low-resolution general circulation model. *J. Atmos. Sci.* **39**, 1177–1188.
LUNDEGARD, P.D., SAMUELS, N.D. & PRYOR, W.A. (1985) Upper Devonian turbidite sequence, central and southern Appalachian basin: contrasts with submarine fan deposits. Geol. Soc. Am. Spec. Pap. 201, pp. 107–121.

LYKE, W.L. (1986) The stratigraphy, paleogeography, depositional environment, faunal communities, and general petrology of the Minnehaha Member of the Scherr Formation, an upper Devonian turbidite sequence, central Appalachians. *Southeastern Geol.* **26**, 173–192.

MCCAVE, I.N. (1969) Correlation of marine and nonmarine strata with example from Devonian of New York. *Bull. Am. Assoc. Petrol. Geol.* **53**, 155–162.

MCCAVE, I.N. (1973) The sedimentology of a transgression: Portland Point and Catskill Members (Middle Devonian), New York State. *J. Sedim. Petrol.* **43**, 484–504.

MCIVER, N.L. (1970) Appalachian turbidites. In: *Studies in Appalachian Geology, Central and Southern* (Eds Fisher, G.W. & Pettijohn, F.J.). Interscience, New York, pp. 69–81.

MILLER, J.D. & KENT, D.V. (1986a) Paleomagnetism of the Upper Devonian Catskill Formation from the southern limb of the Pennsylvania salient: possible evidence of oroclinal rotation. *Geophys. Res. Lett.* **13**, 1173–1176.

MILLER, J.D. & KENT, D.V. (1986b) Synfolding and prefolding magnetizations in the Upper Devonian Catskill Formation of Eastern Pennsylvania: implications for the tectonic history of Acadia. *J. Geophys. Res.* **91**, 12791–12803.

NEMEC, W. & POSTMA, G. (1993) Quaternary alluvial fans in southwestern Crete: sedimentation processes and geomorphic evolution. In: *Alluvial Sedimentation* (Eds Marzo, M. & Puigdefábregas, C.). Spec. Publ. Int. Assoc. Sedimentol. **17**, 235–276.

OLSEN, H. (1990) Astronomical forcing of meandering river behavior: Milankovitch cycles in Devonian of East Greenland. *Palaeogeogr. Palaeoclimatol. Palaeoecol.* **79**, 99–115.

OLSEN, P.E. (1986) A 40-million year lake record of Early Mesozoic orbital climate forcing. *Science* **234**, 842–848.

PESTIAUX, P., DUPLESSY, J.C. & BERGER, A. (1987) Paleoclimatic variability at frequencies ranging from 1 cycle/10 000 years to 1 cycle/1000 years: evidence for nonlinear behaviour of the climate system. In: *Climate – History, Periodicity and Predictability* (Eds Rampino, M.R. *et al.*). Van Nostrand Reinhold, New York, pp. 285–299.

QUINLAN, G.M. & BEAUMONT, C. (1984) Appalachian thrusting, lithospheric flexure, and the Paleozoic stratigraphy of the Eastern Interior of North America. *Can. J. Earth Sci.* **21**, 973–996.

RICKARD, L.V. (1975) *Correlation of the Silurian and Devonian rocks in New York State.* New York State Museum and Science Service, Geological Survey Map and Chart Series 4.

SCHUMM, S.A. (1968) Speculations concerning paleohydrologic controls of terrestrial sedimentation. *Geol. Soc. Am. Bull.* **79**, 1573–1578.

SEVON, W.D. (1985) Nonmarine facies of the Middle and Late Devonian Catskill coastal alluvial plain. In: *The Catskill Delta* (Eds Woodrow, D.L. & Sevon, W.D.). Geol. Soc. Am. Spec. Pap. 201, pp. 79–90.

SUTTON, R.G. (1959) Use of flute casts in stratigraphic correlation. *Bull. Am. Assoc. Petrol. Geol.* **43**, 230–237.

VAIL, P.R., MITCHUM, R.M., Jr, & THOMPSON, S., III (1977) Seismic stratigraphy and global changes of sea level, 4. Global cycles of relative changes of sea level. In: *Seismic Stratigraphy – Applications to Hydrocarbon Exploration* (Ed. Payton, C.E.). Am. Assoc. Petrol. Geol. Mem. 26, pp. 83–97.

VAN HOUTEN, F.B. (1964) Cyclic lacustrine sedimentation, Upper Triassic Lockatong Formation, central New Jersey and adjacent Pennsylvania. *Kansas Geol. Surv. Bull.* **169**, 497–531.

VAN TASSELL, J. (1987) Upper Devonian Catskill Delta margin cyclic sedimentation: Brallier, Scherr, and Foreknobs Formations of Virginia and West Virginia. *Geol. Soc. Am. Bull.* **99**, 414–426.

VAN TASSELL, J. (1988) Upper Devonian Catskill Delta Milankovitch cycles. In: *Geologic Field Guide, Devonian Delta, East-Central West Virginia and adjacent Virginia* (Ed. Dennison, J.M.). Appalachian Geological Society, pp. 77–84.

WALKER, R.G. (1967) Turbidite sedimentary structures and their relationship to proximal and distal depositional environments. *J. Sedim. Petrol.* **37**, 25–43.

WALKER, R.G. (1971) Non-deltaic depositional environments in the Catskill clastic wedge (Upper Devonian) of central Pennsylvania. *Geol. Soc. Am. Bull.* **82**, 1305–1326.

WALKER, R.G. & HARMS, J.C. (1971) The 'Catskill Delta', a prograding muddy shoreline in central Pennsylvania. *J. Geol.* **79**, 281–299.

WALKER, R.G. & SUTTON, R.G. (1967) Quantitative analysis of turbidites in the Upper Devonian Sonyea Group, New York. *J. Sedim. Petrol.* **37**, 1012–1022.

WILLIS, B.J. & BRIDGE, J.S. (1988) Evolution of Catskill river systems, New York State. In: *Devonian of the World* (Eds McMillan, N.J., Embry, A.F., & Glass, D.J.). Can. Soc. Petrol. Geol. Mem. 14 (1), pp. 85–106.

WOODROW, D.L. (1985) Paleogeography, paleoclimate, and sedimentary processes of the Late Devonian Catskill Delta. In: *The Catskill Delta* (Eds Woodrow, D.L. & Sevon, W.D.). Geol. Soc. Am. Spec. Pap. 201, pp. 51–64.

WOODROW, D.L. & ISLEY, A.M. (1983) Facies, topography, and sedimentary processes in the Catskill Sea (Devonian), New York and Pennsylvania. *Geol. Soc. Am. Bull.* **94**, 459–470.

WOODROW, D.L., FLETCHER, F.W. & AHRNSBRAK, W.F. (1973) Paleogeography and paleoclimate at the deposition sites of the Devonian Catskill and Old Red facies. *Geol. Soc. Am. Bull.* **84**, 3051–3063.

WOODROW, D.L., DENNISON, J.M., ETTENSOHN, F.R., SEVON, W.T. & KIRCHGASSER, W.T. (1988) Middle and Upper Devonian stratigraphy and paleogeography of the central and southern Appalachians and eastern midcontinent, USA. In: *Devonian of the World* (Eds McMillan, N.J., Embry, A.F., & Glass, D.J.). Can. Soc. Petrol. Geol. Mem. 14 (1), pp. 277–301.

Spec. Publs Int. Ass. Sediment. (1994) **19**, 413–428

High-frequency, glacial–eustatic sequences in early Namurian coal-bearing fluviodeltaic deposits, central Scotland

W.A. READ

Department of Geology, University of Keele, Keele, Staffordshire ST5 5BG, UK

ABSTRACT

Sedimentation in the upper part of the Pendleian Limestone Coal Group within the Kilsyth Trough and Kincardine Basin, situated in the northern part of the Midland Valley of Scotland, was influenced by a variety of superimposed controls. These included tectonic uplift of the Highland source area and local autocyclic sedimentary processes, as well as widespread, probably eustatic, sea-level oscillations with two or more periodicities. 'Long' allocycles produced major marine transgressions, at intervals of probably somewhat less than 1 Ma. Between major transgressions, fluviodeltaic sedimentation was generally controlled by high-frequency, probably glacial–eustatic, 'short' allocycles, with a periodicity within the lower range of Milankovitch orbital parameters. These produced a remarkably uniform, 'layer-cake' succession in distal environments relatively remote from strong fluvial influences.

In the 'short' allocycles the rate of eustatic fall outpaced tectonic subsidence so that, in terms of the Exxon Production Research Company's concepts of sequence stratigraphy, they represent high-frequency type 1 sequences, which can be divided into high-stand, low-stand and transgressive systems tracts. Glacial low-stands are usually represented only by incised fluvial channels and the whole succession is high-stand dominated.

INTRODUCTION

Recent research into sedimentary cyclicity has been somewhat hindered by a lack of communication between researchers in the overlapping fields of traditional Carboniferous cyclical sedimentation, sequence stratigraphy and orbitally forced cyclicity. The present paper attempts to integrate concepts in the first two fields and touches on the third, using as an example an early Namurian fluviodeltaic succession composed of particularly well-developed, high-frequency allocycles. This paper is also intended to serve as a precursor to quantitative research, which is currently being undertaken by G.P. Weedon in collaboration with the author.

Carboniferous cyclical sedimentation

With regard to widely effective allocyclic controls, debate still continues on the relative importance of widespread tectonic movements, as first proposed by Phillips (1836) and followed by Weller (1930), and eustatic, particularly glacial–eustatic, sea-level oscillations, as first proposed by Wanless & Shephard (1936). However, both schools of thought now admit the influence of local, autocyclic sedimentary controls such as delta switching and channel avulsion (Moore, 1958; Coleman & Gagliano, 1964; Ferm, 1970). Eustatic mechanisms have recently received renewed support (Heckel, 1986; Leeder, 1988; Read & Forsyth, 1989). Ramsbottom (1977a,b) grouped British Namurian cycles into irregular, unconformity-bounded bundles termed mesothems, but this concept has found little support from subsequent workers (Holdsworth & Collinson, 1988; Read, 1991; Martinsen, 1993).

Sequence stratigraphy

The Exxon Production Research (EPR) Company's concept of sequence stratigraphy postulates that most Phanerozoic coastal sedimentation on passive margins was eustatically controlled. EPR sequence stratigraphy concentrates on the *rates* of eustatic rise and fall relative to the rate of tectonic subsidence and on the accommodation space available for sedimentation. Tectonic subsidence is assumed to increase uniformly basinwards from a hinge line. Other assumptions generally include an open-coast setting with a wide continental shelf composed of thick, unlithified sediments and, tacitly, a 'greenhouse' global climate without continental ice caps.

Symmetrical sinusoidal oscillations in sea level are thought to produce sequences, which are themselves composed of subordinate, progradational parasequences. Systems tracts, with different stacking patterns of parasequences, represent different stages in the eustatic curve (Posamentier *et al.*, 1988; Posamentier & Vail, 1988; Van Wagoner *et al.*, 1988).

Computer simulation studies, which incorporated the above assumptions (Jervey, 1988) have suggested that maximum flooding occurs not at the absolute eustatic high, but shortly after the *R* inflexion point, which marks the maximum rate of sea-level rise on the eustatic curve. Fluviodeltaic aggradation followed by progradation can start after the period of maximum flooding and progradation can continue until the *F* inflexion point, which marks the maximum rate of eustatic fall (Posamentier *et al.*, 1988; Posamentier & Vail, 1988). During a single eustatically induced cycle, fluviodeltaic systems can prograde basinwards during both the high-stand and the low-stand (Van Wagoner *et al.*, 1988).

Eustatic oscillatory cycles are grouped into a hierarchy of different frequencies which are superimposed one upon the other. Third-order sequences, produced by third-order cycles which have frequencies of between 0.5 and 5 Ma (generally between 1 and 2 Ma), are considered to be the basic units of EPR sequence stratigraphy. Fourth- and fifth-order cycles, with frequencies within the range of Milankovitch orbital parameters, are superimposed upon third-order cycles, but their effects (in the form of high-frequency sequences) are detectable only in areas of rapid sedimentation (Mitchum & Van Wagoner, 1991).

For detailed definitions and descriptions of EPR sequence stratigraphy the reader is referred to the papers listed above, together with others in the Wilgus *et al.* (1988) volume, plus the subsequent papers by Van Wagoner *et al.* (1990) and Haq (1991).

Orbitally forced cyclicity

Spectral analysis (see Fischer *et al.*, 1990; de Boer, 1991a) has demonstrated that sedimentation may be affected by climatic changes which are themselves linked to orbital parameters. Thus orbitally forced cyclicity, if linked to an adequate framework of radiometric and magnetostratigraphic dates, has the potential to produce a framework of high-resolution dating and, ultimately, even a global cyclostratigraphy.

To date, this approach has generally been applied to basinal areas of quiet water sedimentation. Coastal and fluviodeltaic siliciclastic environments have generally been dismissed as unsuitable because of 'noise' effects such as local variations in sediment supply and tectonic subsidence which could induce autocyclicity (Algeo & Wilkinson, 1988; de Boer *et al.*, 1991). However, recent research suggests that orbitally forced eustatic cycles really can be detected in such environments. For example, Mitchum & Van Wagoner (1991) have identified eustatic cycles, which they consider to be orbitally forced, with frequencies within the lower range of Milankovitch orbital parameters, in rapidly deposited coastal and fluviodeltaic successions in the Eocene of the Gulf of Mexico.

Pleistocene glacial–eustatic cycles have certainly been identified in ice-free, middle and low latitude, paralic and shelf environments such as the Gulf of Mexico (Morton & Price, 1987; Suter *et al.*, 1987). Carter *et al.* (1991) have managed to identify Plio-Pleistocene glacial–eustatic cycles in shelf deposits, even in tectonically disturbed areas close to an active plate margin in the North Island of New Zealand.

Radiometric dates and cycle frequencies

The framework of radiometric dates for the Namurian and its constituent stages is, as yet, imprecise and unreliable. For example, Hess & Lippolt (1986) have calculated the durations of the Namurian and Namurian A to be 11 and 7 Ma, respectively, whereas Menning (1989) quotes durations of 13 and 7 Ma and Harland *et al.* (1990) durations of 14.6 and 10.1 Ma. Further errors arise through attempting to link stage boundaries, via biozones, to radiometrically dated samples with uncertain biostratigraphic horizons (Miall, 1991).

Considerable errors are thus inevitable if any attempt is made to calculate cycle frequencies from

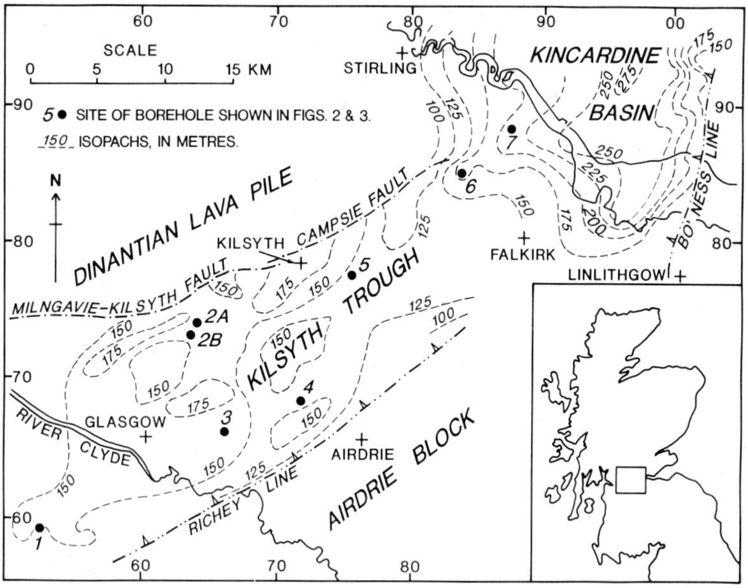

Fig. 1. Sketch map of the area studied showing: (i) major structural elements (see Read, 1988); (ii) isopachs in metres of the upper part of the Limestone Coal Group (between the top of the Black Metals and the base of the Index Limestone); and (iii) sites of sections illustrated in Figs 2 and 3, etc. Numbers along margins refer to British National Grid 100-km squares NS and NT.

such an imprecise dating framework and yet more errors will arise if not all of the cycles have been preserved in the sedimentary record. These problems have recently been discussed with respect to US Pennsylvanian cyclothems by Klein (1990, 1991), de Boer (1991b), Heckel (1991) and Langenheim (1991). Algeo & Wilkinson (1988) have also argued that, because of the restricted range of rates of tectonic subsidence, calculations of the frequencies of all cycles 1–20 m thick, regardless of their origin, may tend to fall within the range of Milankovitch periodicities.

In the following account, Harland *et al.*'s (1990, fig. 1) estimates of the duration of the Pendleian and Arnsbergian stages of the Namurian have provisionally been accepted, albeit only as first approximations. Although they are known to be subject to rather wide error margins, these estimates nevertheless help to quantify such factors as overall rates of tectonic subsidence.

AREA AND SUCCESSION STUDIED

An increasing number of workers (e.g. Klein & Willard, 1989; Martinsen, 1993) have become aware that different mechanisms have controlled cyclicity in specific areas at specific times and that more than one mechanism may have been operating simultaneously. Such a situation almost certainly existed during Namurian A in the Midland Valley of Scotland (Read & Forsyth, 1989), in the north of which the study area (Fig. 1) is located. During the Namurian, tectonic movements took place not only in the Highland source area, but also sporadically within some basinal areas within the Midland Valley (Read, 1988, 1989a,b). In addition, there is strong evidence of local sedimentary autocyclicity in parts of the succession that were fluvially influenced (Read & Dean, 1976, 1982). Nevertheless, during the Pendleian the dominant control within the study area was allocyclic (Read & Forsyth, 1989, 1991).

Using the analogy of communication theory, the Scottish Namurian A succession may be compared with a series of superimposed 'signals' plus a generous amount of random 'noise'. Thus, in order to study the eustatic 'signal', it is necessary to consider the superimposed 'signals', such as contemporaneous tectonic activity, so that these may be filtered out.

Read & Forsyth (1989, 1991) have recently completed a detailed investigation of the study interval so, in order to avoid unnecessary repetition of previously published material, the reader is referred to these papers for supporting evidence and additional information.

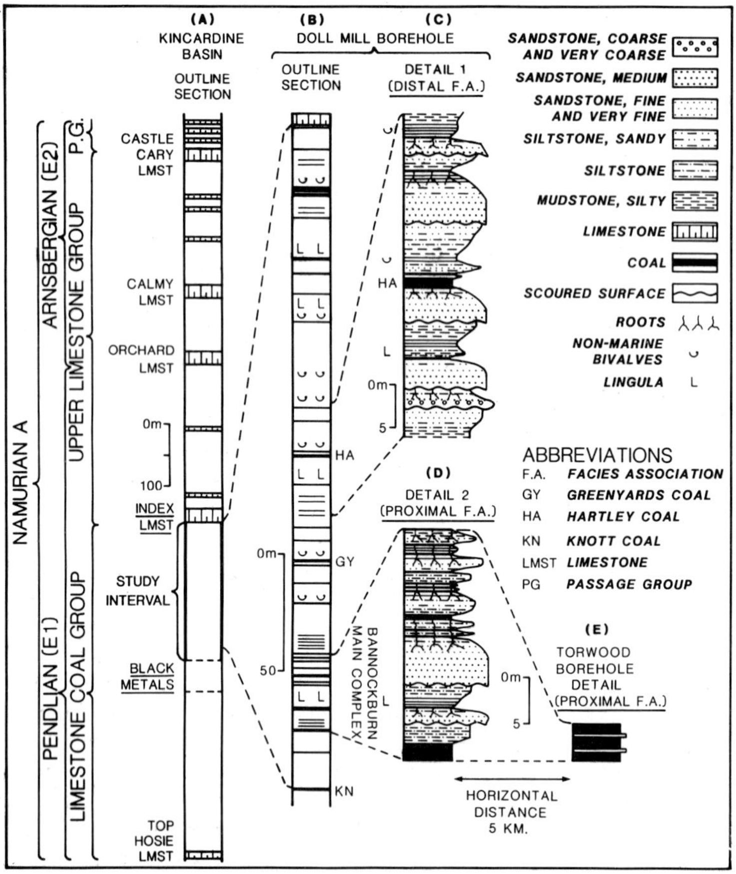

Fig. 2. (A) Outline vertical section of the Namurian A succession in the Kincardine Basin, showing chronostratigraphical and lithostratigraphical divisions and named horizons of the major marine transgressions which mark 'long' allocycles. (B) Skeletal section of the upper part of the Limestone Coal Group in Doll Mill Borehole, 1955 (marked 7 on Fig. 1). (C) Detail of a part of the succession shown in (B) in which the distal facies association, composed of 'short' allocycles, is dominant. (D) Detail of another part of the section shown in (B) in which the largely fluvial, autocyclic, proximal facies association is dominant. (E) Same stratigraphical interval as (D) in Torwood Borehole, 1960 (marked 6 on Fig. 1). Note abrupt lateral variation within the proximal facies association over only about 5 km.

Sources of information

Because of a thick cover of Quaternary deposits, most of the basic information on Upper Carboniferous strata in the Midland Valley comes from continuously cored, mineral-exploration boreholes. The present study concentrates on a part of the succession and an area where this borehole information is both abundant and reliable, namely the Glasgow–Stirling area (Fig. 1) and the upper part of the Limestone Coal Group, between the major marine transgressions marked by the thick 'Black Metals' marine mudstones (Read, 1965) and the Index Limestone (Fig. 2).

The interval studied is early Namurian (Pendleian, E_1a) in age and is composed of fluviodeltaic deposits, including thick coals which have been extensively mined. These deposits mark one of the earlier examples of the 'coal-measures' facies that became so widespread in the Westphalian of northwest Europe, but were more strongly marine influenced than most of these Westphalian deposits.

Correlation

Detailed correlation has been facilitated by the network of closely spaced, geologist-examined, cored boreholes (supplemented by the examination of underground coal workings) and by the remarkable lateral continuity of many of the lithological members, including Lingula bands and coals. Full details have been provided by Forsyth & Read (1962).

Palaeogeographical and structural setting

During the Namurian, Scotland lay on the southern margin of the 'Laurussian' continent of Rowley *et al.* (1985), in near-equatorial latitudes, with open sea both to the east and the west. The prevailing climate was either monsoonal or 'ever-wet' equatorial (Read, 1988).

Thermal subsidence had largely succeeded earlier Dinantian rifting along ENE–WSW lines, which had been accompanied by the effusion of alkali basalts. Some component of right-lateral strike-slip, which was concentrated along a few lineaments, including the Campsie Fault (Read, 1988, 1989b), was superimposed on this subsidence. Such an overall transtensional stress system probably accounted for the subsidence of major north–south downfolds, including the Kincardine Basin, and ENE–WSW half-grabens and grabens, including the Kilsyth Trough (Fig. 1). This basin and trough constitute the major structural units within the study area. Both were partly enclosed by slowly subsiding 'highs', including contemporaneous and earlier volcanic piles, which constrained most siliciclastic sediments to enter the basin from the northwest and the northeast and subsequently to be transported WSW down the trough (Read, 1988, fig. 16.8). Isopachs of the study interval (Fig. 1) clearly define both basin and trough. The main source of the siliciclastic sediments probably lay only a few tens of kilometres to the north, in the Proterozoic metamorphic rocks of the southern Scottish Highlands.

Alkali-basaltic activity continued into the early Namurian along a roughly north–south monoclinal flexure known as the Bo'ness Line (Fig. 1; Read, 1988). Subaerial lavas south of Linlithgow passed northwards along this line into subaqueous lavas and then into a zone of explosive vents.

Major marine incursions

During Namurian A, fluviodeltaic sedimentation in the Midland Valley was regularly interrupted by widespread major marine incursions, each of which has been interpreted by Read & Forsyth (1989) as the maximum flooding episode of a 'long' allocycle. These incursions include the Black Metals and the Index Limestone that bound the top and bottom of the study interval (Fig. 2). In the Upper Limestone Group, which overlies the Limestone Coal Group and which was subject to stronger marine influences, the maximum flooding period of each of the major incursions was generally represented by a relatively thick and laterally persistent marine limestone containing a varied shelly fauna.

Fluviodeltaic facies associations

Within the study area, the upper part of the Limestone Coal Group is composed of two contrasting facies associations – distal and proximal – with different facies geometries which reflect different underlying controls. The volumetrically dominant distal association (Fig. 2C), with which this paper is largely concerned, is characterized by a remarkably regular, laterally persistent 'layer-cake' succession. This reflects the underlying control of high-frequency 'short' allocycles (Read & Forsyth, 1989). Most of these allocycles can be traced through the greater part of the Kilsyth Trough and into the western flank of the Kincardine Basin (Fig. 3; Forsyth & Read, 1962, plate III). Some coals and paleosols and the mudstones that overlie them may be traced over areas of more than 850 km^2.

Coals are laterally persistent, but seldom more than 1 m thick. Within the distal facies association, coals generally contain few clastic partings or splits. Some seams are sharply overlain by a few centimetres of sandy or silty sediment containing broken fragments of the quasi-marine brachiopod *Lingula*. However, most coals are sharply succeeded by dark-grey mudstones, a high proportion of which contain either *Lingula* or brackish-water 'non-marine' bivalves. This abrupt break between the coal and the overlying mudstones indicates rapid flooding and reflects the basic asymmetry of the 'short' allocycles (Read & Forsyth, 1989). The mudstones coarsen upwards, through laminated and commonly bioturbated siltstones and sandy siltstones (interpreted as delta-front and bayfill deposits), into laterally persistent sheet sandstones, which are generally less than 5 m thick.

Many of these sheet sandstones, particularly in sections near to the somewhat constricted eastern end of the Kilsyth Trough and the adjacent southwest corner of the Kincardine Basin, have a characteristic two-storey profile. In such sandstones, a finer-grained, upward-coarsening, locally bioturbated lower portion is truncated erosively by a coarser-grained, upward-fining, cross-stratified, upper portion (Fig. 2C). These two portions have been interpreted as deltaic and fluvial, repectively, by Read & Forsyth (1989). The upper portions grade up into paleosols. Local minor, channel-like, erosive scours less than 5 m deep at the bases of some of the sheet sandstones probably mark distributary channels.

Other, rarer, more deeply eroded scours mark the bases of sporadic, upward-fining channel sandstones. These channels, which are commonly more than 10 m deep and have fairly straight courses (see Read, 1959, fig. 6), cut right through the sheet

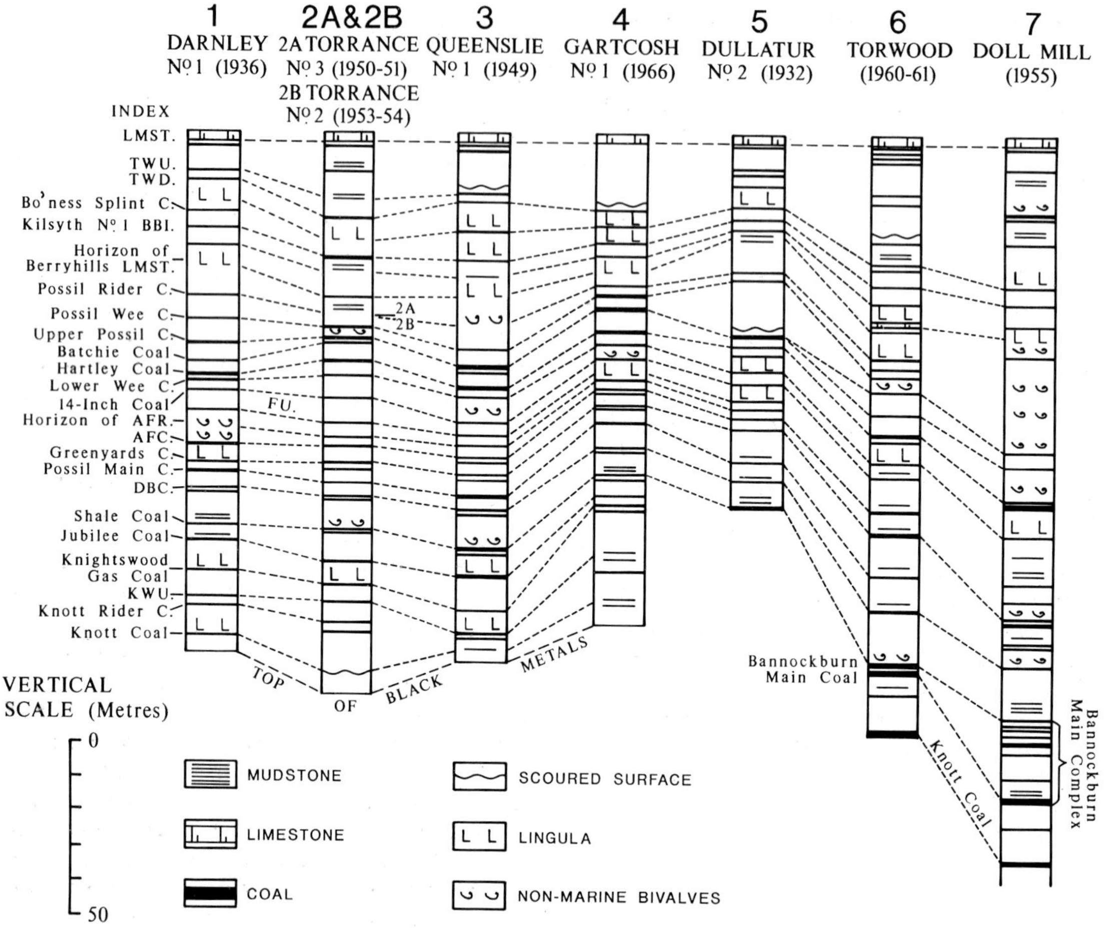

Fig. 3. Comparative skeletal vertical sections of the upper part of the Limestone Coal Group in the Kilsyth Trough and western flank of the Kincardine Basin, showing named coals and other horizons marking 'short' allocycles. For sites of boreholes, see Fig. 1. AFC, Ashfield Coking Coal; AFR, Ashfield Rider Coal; BBI, Blackband Ironstone; C, Coal; DBC, Dumbreck Cloven Coal; FU, 14-inch Ironstone; KWU, Knightswood Under Coal; LMST, Limestone; TWD, Twechar Dirty Coal; TWU, Twechar Under Coal.

sandstones and the underlying mudstones and coals. They have been interpreted by Read & Forsyth (1991) as the fills of fluvial channels incised during periods of lowered base level.

The paleosols (seatearths) which cap the sheet sandstones are generally thin, immature and gley-like. However, a few thicker and more mature paleosols show evidence of leaching and partial oxidation, suggesting periods of subaerial exposure.

In sharp contrast to the distal association, the proximal facies association, which is best developed in 'upstream' areas, particularly in the Kincardine Basin, is highly variable over short distances (Fig. 2D,E; Read & Forsyth, 1989, fig. 4). Coals may be locally almost 5 m thick, but they are laterally impersistent and are commonly split by abruptly thickening and coarsening siliciclastic wedges, thought to be crevasse splays associated with contemporaneous thick, erosive-based, multi-storey, fluvial-channel sandstones. These sandstones may be more than 40 m thick. They generally have very coarse-grained or pebbly bases and they seem to have coexisted with coal-forming peats (Read & Forsyth, 1989, fig. 4). Other thinner, finer-grained, upward-coarsening sandstones probably represent minor lacustrine deltas. Mudstones are seldom laterally extensive and only rarely contain *Lingula* or 'non-marine' bivalves. A detailed description of the abrupt local variations in facies within a wedge of typical proximal facies association strata has been given by Read (1961).

The distal association has been interpreted by Read & Forsyth (1989, 1991) as being marine influenced and dominantly deltaic, although the upper storeys of sheet sandstones and the more deeply cut channel-fill sandstones are thought to be fluvial. By contrast, the depositional environments of the proximal association are thought to be dominantly freshwater and fluvial. The distal association is strongly dominant in the west of the Kilsyth Trough but, as the study interval is traced northeastwards and upstream towards the main sources of siliciclastic input, wedges of the proximal association appear. These wedges thicken and become more numerous eastwards so that they constitute much of the study interval on the eastern flank of the Kincardine Basin and hence disrupt the 'layer-cake' lithostratigraphy (Read & Forsyth, 1989, fig. 2). In some stratigraphical intervals it is possible to demonstrate that deposits of the distal facies association pass upstream into much more variable deposits belonging to the proximal facies association (Read & Forsyth, 1989, fig. 4).

ANALYSIS OF CONTROLS

Tectonic controls

Tectonism in source areas

Uplift and erosion of the closely adjacent Highland source area most probably caused the influx of siliciclastic sediment into the central and eastern parts of the Midland Valley, which initiated the widespread fluviodeltaic environments of the Limestone Coal Group at the beginning of the Namurian (Read, 1988). This movement seems to have been part of a widespread regional uplift of the whole Highland Massif (Dewey, 1982; Watson, 1985). More local episodes of uplift and erosion within parts of the Highlands, which may have been associated with isostatic adjustments, seem to be the most probable explanation for the intermittent, irregularly spaced, local influxes of the fluvially dominated, proximal facies association (Read & Forsyth, 1989).

Basin subsidence

Considerable tectonic subsidence must have taken place during the early Pendleian in both the Kincardine Basin and the Kilsyth Trough, as demonstrated by the isopachs in Fig. 1. The Kincardine Basin was the most rapidly subsiding Namurian structure in Scotland and contains up to 1200 m of Namurian A sediments (Fig. 2A). If Harland *et al.*'s (1990) estimate of 4.6 Ma for the combined duration of the Pendleian plus the Arnsbergian is taken to be correct, and if it is further assumed that the sediments preserved in the Kincardine Basin represent much the greater part of this time interval, this would imply an overall rate of basin subsidence and sediment accumulation equivalent to about 0.3 m/ka of consolidated sediment. If allowance is made for compaction (see Leeder, 1988), this figure must approach the unusually rapid rates of subsidence and sedimentation attained during the Eocene and Plio-Pleistocene in the Gulf of Mexico Basin (Mitchum & Van Wagoner, 1991, table 1).

Intermittent basin subsidence was formerly postulated as the underlying cause of cyclicity in the Limestone Coal Group (e.g. Richey, 1937), but few modern workers still support this theory. Even if basin subsidence had been intermittent, and we have no evidence for this, movement would probably have taken place at irregular time intervals and shifting localities within the Kilsyth Trough and Kincardine Basin. Such irregular tectonic subsidence is unlikely to have produced the remarkably regular, laterally persistent succession seen in the distal facies association in the Kilsyth Trough.

The rapidly subsiding Kincardine Basin is known to have functioned as an effective sediment trap during the deposition of the Black Metals (Read, 1965). This function continued into the study interval, because most of the wedges of proximal facies were 'trapped' within the Kincardine Basin and only about three or four penetrated, at irregular intervals, downstream into the Kilsyth Trough.

Intra-plate stresses

Cloetingh (1988, fig. 2) has suggested that relative changes in sea level of about 10 m, taking place over a time scale of about 1 Ma or even less, could result from widespread tectonic movements linked to variations in intra-plate stresses of the order of a few hundred bars. During the Namurian, stress changes of this magnitude could have been associated with the Variscan orogenic belt which lay a few hundreds of kilometres south of Scotland. However G. de V. Klein (personal communication, 1991) has suggested that Cloetingh's approach may yield displacement values which are too high, because his method is biased towards younger crust and fails to account adequately for interface effects between the asthenosphere and the lithosphere. In addition, there is no reason to suppose that these widespread tectonic movements would necessarily have occurred at such regular time intervals as either the 'long' or the 'short' allocycles.

Volcanic controls

No traces of any volcano-tectonic cycles associated with the filling and emptying of volcanic magma chambers, like those reported by Sloan & Williams (1991), have been detected in the Scottish Namurian succession. The chief control exerted by both the contemporaneous, and the earlier, pre-Namurian, volcanic piles was to form 'highs' which constrained the paths by which siliciclastic sediments could enter and leave the Kincardine Basin and the Kilsyth Trough (see Read, 1988).

Autocyclic, sedimentary controls

The wedges of laterally variable, dominantly fluvial deposits of the proximal facies association constitute the principal source of 'interference' that confuses the 'signal' provided by the 'short' allocycles of the distal facies association. This is largely because facies geometry within the proximal association was dominantly controlled by purely local, largely fluvial, sedimentary processes such as channel avulsion, meander migration and the infilling of relatively small floodbasin lakes by local freshwater deltas. Such local processes affected areas that were relatively small in proportion to the Kincardine Basin and the Kilsyth Trough. As mentioned earlier, these intermittent fluvial influxes were most probably caused by local uplift and erosion within parts of the Highland source area, followed by the regrading of river profiles (Schumm, 1977).

Read & Dean (1967, 1976, 1982) demonstrated that such fluvially influenced intervals in the Scottish Namurian succession tend to be autocyclic. They also discovered that the number of local fluvial autocycles, as delimited by horizons of vegetation colonization, tends to be linearly correlated with the total thickness of the interval and hence with net local subsidence. Accordingly, Read & Forsyth (1989, fig. 5) have used linear regression lines that show this relationship to demonstrate that autocyclicity, associated with influxes of the dominantly fluvial proximal association, is much more important in the Kincardine Basin than in the Kilsyth Trough.

Larger scale autocyclic processes such as the switching of major delta lobes (see Coleman & Gagliano, 1964) were unimportant, because both the Kincardine Basin and the Kilsyth Trough were closely confined by an adjacent low-subsidence area (Figs 1 and 4a; Read, 1988, figs 16.7 and 16.8). Thus there were no neighbouring rapidly subsiding areas into which the deltas could switch.

Namurian sea-level oscillations and orbital forcing

In the preceding account an attempt has been made to identify the possible effects of various tectonic, volcanic and local autocyclic sedimentary processes, so that these effects may, hopefully, be filtered out from the 'signals' of the allocyclic processes that dominantly controlled sedimentation in the study interval, particularly in the Kilsyth Trough.

Three orders of allocycles – 'long', 'intermediate' and 'short' – can be recognized within the Scottish Namurian A succession as a whole and Read (1988) has argued that all three were eustatically controlled. Of these, the 'intermediate' order of allocycles, which controlled the recurrence of the thinner limestones in the Upper Limestone Group (the unnamed limestones in Fig. 2A), cannot easily be recognized within the study interval, although they may possibly be represented by the more persistent *Lingula* bands described by Forsyth (1979). Thus the 'intermediate' allocycles are not considered further in this account but will be described in a separate paper.

'Long' allocycles

The six major transgressions named in Fig. 2A, which are known to have invaded the Midland Valley during the Pendleian and Arnsbergian, mark the maximum flooding periods of six 'long' allocycles. Only five of these allocycles are complete, because the sixth is truncated by the very widespread Mississippian–Pennsylvanian hiatus (Read, 1989a). These major Namurian marine transgressions can be traced into the southern Pennines, where they are represented by persistent 'black shales' containing distinctive thick-shelled ammonoids (Ramsbottom, 1977a, b) and at least two of them seem to be present in the Donetz Basin (Aisenverg *et al.*, 1979). Thus they mark episodes of very widespread flooding, comparable to those represented by the most persistent marine bands in the Westphalian, which have long been regarded as eustatically controlled (see Leeder, 1988).

If a rough estimate of about 4 Ma is accepted for the time interval represented by the five complete 'long' allocycles, simple division would suggest a recurrence interval of about 800 ka. However, as Klein (1991) and de Boer (1991b) have pointed out, this takes no account of possible hiatuses, so the true interval may have been somewhat shorter. Nevertheless, despite these possible errors, the frequency range of the 'long' allocycles almost certainly falls within that of the third-order eustatic cycles (0.5–5 Ma) of EPR sequence stratigraphy (Mitchum & Van Wagoner, 1991) and thus outside that of the Milankovitch orbital parameters which are known to be most effective at climatic forcing (Berger, 1988).

EPR third-order cycles cannot be glacial–eustatic as they can occur during periods of global 'greenhouse' climates (Ziegler *et al.*, 1987) when major ice caps are unlikely to be present (see Haq *et al.*, 1988). Carter *et al.* (1991, table 2) have recently considered the origin of fundamental cycles of sea-level fluctuation and concluded that those within the frequency range of 0.3–5 Ma include two subgroups, both of which represent eustasy of an unknown type.

'Short' allocycles

The 'short' allocycles within the study area are remarkable for their uniformity, regularity and lateral persistence over distances of more than 50 km (Fig. 3; Forsyth & Read, 1962, plate III). These features have been interpreted by Read & Forsyth (1989, 1991) as favouring a eustatic, rather than a tectonic, controlling mechanism. Estimating their frequency raises even greater difficulties than estimating that of the 'long' allocycles, which themselves may not necessarily have occurred at precisely regular intervals. In view of the inadequate framework of radiometric dates for the Pendleian (Harland *et al.*, 1990), all that can be done at present is to make an imprecise estimate by assuming that the stratigraphic interval between the Black Metals and the Index Limestone represents a timespan of approximately 800 ka. As about 25 'short' allocycles can be identified within this interval over a wide area, simple division would give a recurrence period of about 30 ka. Because unknown hiatuses may be present (see de Boer, 1991b), e.g. by 'short' allocycles having been 'drowned out' by the major marine transgressions below and above the study interval, the true frequency was probably somewhat shorter.

The inaccuracy of this estimate makes it impossible to determine whether the frequency of the 'short' allocycles falls within the restricted ranges of either the precession or the obliquity parameters. Berger (1989) has calculated average values of 17.4 and 20.7 ka for precession cycles and 34.3 and 42.9 ka for obliquity cycles during the Late Carboniferous, so that the 'short' allocycles may have been controlled by either of these two orbital parameters. (It seems much less likely that they were controlled by the 100 ka eccentricity parameter.) Nevertheless the frequency of the 'short' allocycles falls within the Late Carboniferous range of the Milankovitch orbital parameters which are known to have strongly influenced Pleistocene climate and glaciation (Berger, 1988, 1989).

The frequency of the 'short' allocycles also falls within the same range as Mitchum & Van Wagoner's (1991) high-frequency (fourth and fifth order) eustatic cycles. Sequences and parasequences resulting from such cycles are preserved only in areas of rapid sedimentation where the sedimentary record is sensitive to relatively minor eustatic fluctuations. Significantly, early Namurian sedimentation was particularly rapid in the Kincardine Basin and the Kilsyth Trough.

De Boer (1991a) has shown that the grouping of cycles into bundles of four or five may indicate orbital forcing, with the modulation of precession cycles by eccentricity, and has applied this approach (de Boer, 1991b) to the US mid-continent Pennsylvanian cycles described by Heckel (1986, 1991). Some faint traces of bundles of three to five cycles may possibly be present in Sections 6 and 7 in Fig. 3 (see also Forsyth & Read, 1962, plate III), but both sections, particularly the latter, have been partly disrupted by wedges of the proximal facies association.

There is also strong circumstantial evidence favouring orbital forcing of global climates as the underlying control for the 'short' allocycles. Veevers & Powell (1987) have identified contemporaneous, continental-scale, Carboniferous ice caps in Gondwanaland. This major, prolonged, glacial episode, which started to affect Australia in the Namurian, coincided approximately with the widespread appearance in Euramerica of allocycles, which Veevers & Powell (1987) considered to be linked to glacial–eustatic sea-level fluctuations. Growth and decay of the Namurian Gondwana ice cap are likely to have been forced by the same family of orbital parameters as affected the Pleistocene ice caps and thus to have resulted in similar glacial–eustatic falls and rises in sea level. However, with different configurations of oceans and continents, the reinforcement mechanisms (Berger, 1988) may have differed. These mechanisms may also have varied within the Carboniferous itself.

Accordingly, the best readily available analogues for Late Carboniferous allocyclic successions in paralic, shelf and shelf-edge environments should be sought in Quaternary successions in similar environments beyond the immediate sedimentational and isostatic effects of glaciation. Read & Forsyth (1989, 1991) have already pointed out the basic underlying similarities between the 'short' allocycles and the glacial–eustatic control of paralic and shallow marine Quaternary cycles in areas like the Gulf of Mexico (see Morton & Price, 1987; Suter *et al.*, 1987).

Although Milankovitch orbital cycles are not themselves asymmetrical, ice sheets tend to decay more rapidly than they grow (Berger, 1988), thus giving rise to markedly asymmetrical glacial–eustatic cycles of the type illustrated by Williams (1988, figs 5 and 6). For late Pleistocene glacial–eustatic cycles in the Gulf of Mexico, Williams (1988) has quoted rapid rates of sea-level fall of about 1.5 m/ka and very rapid rates of sea-level rise of as much as 15 m/ka.

Significantly, the 'short' allocycles are, like the glacial–eustatic Pleistocene sedimentary cycles, markedly asymmetrical, further suggesting that they too may have been controlled by glacial–eustatic cycles. Thus early Namurian rates of fall and rise in sea level may have been comparable to those found in the Pleistocene. Such rates of fall would easily have outpaced the relatively rapid rate of tectonic subsidence (roughly equivalent to 0.3 m/ka of consolidated sediment) in the early Namurian Kincardine Basin, even if a generous allowance is made for compaction.

A glacial–eustatic model would conveniently explain why the more deeply eroded (10 m+) channels in the distal facies association cut right through the sheet sandstones, because if these deeper channels were incised during glacial low-stands, the rate of eustatic fall is likely to have outpaced that of tectonic subsidence. It would also explain the periods of subaerial exposure and partial oxidation, which are indicated by the occasionally preserved remnants of mature paleosols that cap some sheet sandstones. Very rapid rates of postglacial sea-level rise would similarly explain how coal-forming Namurian domed peats could have been rapidly overwhelmed by brackish water, even though the equatorial ombotrophic raised mires (rain-fed domed peats) which are their closest modern equivalents (Moore, 1987) may have an initial rate of accumulation of as high as 4.8 m/ka (Anderson, 1983).

In summary, although it cannot be proved that the frequency of the 'short' allocycles falls within the restricted frequency bands of either precession or obliquity (Algeo & Wilkinson, 1988; Klein, 1990, 1991, de Boer, 1991b) this frequency certainly seems to lie within the lower range of Milankovitch orbital parameters. Namurian continental glaciation is likely to have been accompanied by glacial–eustatic oscillations in sea level (Veevers & Powell, 1987), which, like their Pleistocene counterparts, would have had frequencies within that range (Williams, 1988). This, coupled with the asymmetry, regularity and lateral persistence of the 'short' allocycles, plus their basic similarity to Quaternary glacial–eustatic cycles deposited in similar paralic environments, provides strong circumstantial evidence for glacial–eustatic control. Such a control would obviously have been most effective in areas like the western Kilsyth Trough that lay furthest away from the immediate sources of siliciclastic input and closest to marine influences. Significantly, it is here that the 'short' allocycles of the distal facies association are best developed.

Climatic controls

One scale of orbitally induced climatic changes is obviously likely to have been linked with glacial–interglacial oscillations in major continental ice caps and hence with glacial–eustatic sea-level cycles. If, as Ziegler *et al.* (1987) have suggested, rainfall increased in equatorial latitudes during glacial periods, the increased runoff would have reinforced the effects of increased overall stream gradients caused by glacial low-stands during the deposition of the 'short' allocycles (Read & Forsyth, 1989). Thus orbitally forced climatic changes comparable to those described by de Boer *et al.* (1991) may well have reinforced the effects of glacial–eustatic sea-level oscillations. Cecil (1990) has postulated Carboniferous climatic changes on a variety of time scales in the eastern USA, which affected vegetation cover, runoff and erosion. Long-term climatic changes causing increased runoff may have been partly responsible for the increased siliciclastic input which marks the beginning of the Limestone Coal Group.

NAMURIAN SEQUENCE STRATIGRAPHY

A significant recent development of the EPR concept is the abandonment of any implicit time ranges for sequences and parasequences. Thus it is now realized that a sequence in one area may be the time equivalent of a parasequence in another (Van Wagoner *et al.*, 1990). Also it is now known that in areas of rapid sedimentation, high-frequency cycles can produce type 1 sequences, complete with erosive bases and fluvial channels incised into subaerially exposed parts of the shelf (Haq, 1991; Mitchum & Van Wagoner, 1991). Furthermore, progradational, aggradational and retrogradational patterns of *sequences* are now known to be grouped into sequence sets (analogous to systems tracts) and composite sequences (sequences of sequences). These concepts help to explain how Read & Forsyth (1991) and Martinsen (1993) have used the term 'sequence' to apply to allocycles with markedly different frequencies.

Although the rate of tectonic subsidence was rapid, especially in the Kincardine Basin, it was nevertheless outpaced by the rate of eustatic fall, so that fluvial channels could be incised during low-stands. In terms of EPR sequence stratigraphy, the 'short' allocycles are thus equivalent to high-frequency, type 1 sequences. However, they probably reflect asymmetrical, glacial–eustatic oscillations, rather than symmetrical, sinusoidal oscillations, in sea level. The whole study interval, between the Black Metals and the Index Limestone, can be regarded as a composite sequence.

Figure 4b–f attempts to illustrate the depositional phases of a typical distal-association 'short' allocycle, similar to those in Fig. 2C, in terms of EPR high-stand, low-stand and transgressive systems tracts (HSTs, LSTs and TSTs). Figure 5 shows where these phases and systems tracts are thought to have fallen on an asymmetrical, glacial–eustatic sea-level curve. Maximum flooding probably occurred just after the *R* inflexion point and shortly after this, early in the HST, delta lobes, having filled the Kincardine Basin, started to prograde westwards down the Kilsyth Trough (Fig. 4b). Later in the HST, rivers replaced deltas, particularly after eustatic sea level had started to fall. These rivers would have scoured down into the earlier sandy deltaic deposits, thus creating the upper, fluviatile, storey of the two-storey sheet sandstones (Fig. 4c). Fluvial deposition probably ceased abruptly after the *F* inflexion point and during the following low-stand fluvial channels were incised into, and through, the sheet-like, fluviodeltaic deposits of the preceding HST (Fig. 4d). These 10 m+ deep incised channels would be the sole representatives of the LST.

After the absolute eustatic low had passed, the sea started to rise again, flooding the subaerially eroded surface of the preceding HST and initiating the TST. The sandstone fills of the incised channels would belong to the LST, or to the earliest portions of the succeeding TST. Extensive coal-forming peats would correspond to a rather later stage in the TST, when gradients were reduced, rivers were ponded back and the water table started to rise on the remnant surface of the underlying HST (Fig. 4e). Initially peat growth was probably rapid but, as rain-fed, domed peats (ombotrophic raised mires) formed and the available plant nutrients decreased, vegetation growth and peat accumulation slowed. The dominantly Lycopsid flora was finally killed off by the accelerating rise in sea level, reinforced by continued tectonic subsidence, later in the TST (Fig. 4f). The nearest equivalent to a ravinement or transgressive surface of marine erosion is the sharp break, immediately above the coal and below the thin, occasionally preserved, sandy bed containing broken *Lingula* fragments, which reflected the reworking of earlier sandy fluviodeltaic deposits. The mudstone containing *Lingula* above the coal would then represent the condensed section formed during maximum flooding, shortly after the *R* inflexion point.

If the above synthesis is correct, we would expect a hiatus just below each coal, because the peats presumably formed on the eroded and subaerially weathered surface of the preceding HST. Such hiatuses are probably reflected by the rarely preserved remnants of leached, relatively mature paleosols.

The bulk of the sediments in the study interval belong to HSTs, whereas TSTs and, even more, LSTs are poorly represented. This dominance of HSTs probably reflects the laterally confined

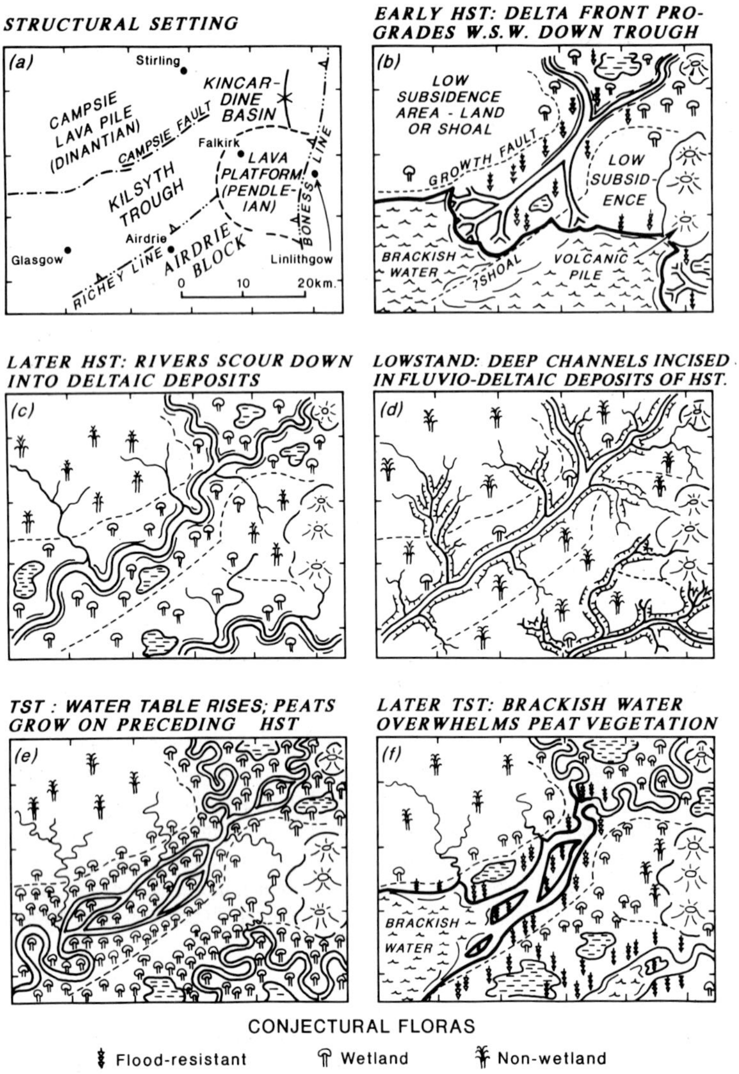

Fig. 4. (a) Structural setting of the study area. (b–f) Tentative reconstruction of successive stages (with their appropriate systems tracts) during the deposition of a single 'short', glacial–eustatic allocycle (high-frequency, type 1 sequence) in the distal facies association. See Fig. 5 for where these stages are thought to have lain on a glacial–eustatic sea-level curve. Details of the vegetation types, particularly those on the low-subsidence areas, are largely speculative. Local sand input from the sides of the Kilsyth Trough has been omitted in order to simplify the diagram. HST, high-stand systems tract; TST, transgressive systems tract.

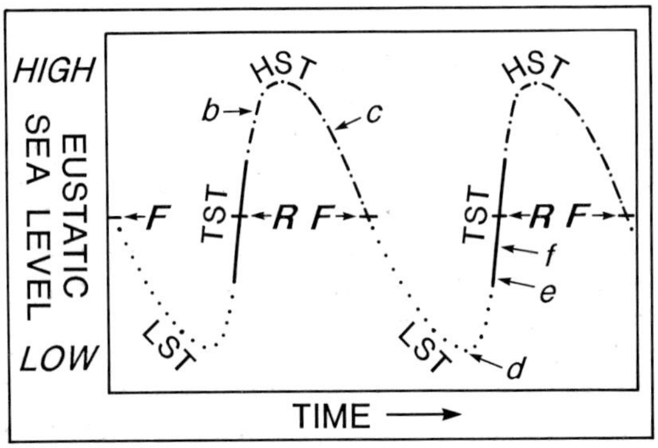

Fig. 5. Suggested relationships between stages in an asymmetrical, glacial–eustatic curve of sea-level change and systems tracts in a high-frequency sequence, equivalent to a 'short' allocycle. Letters b–f refer to the stages illustrated in Fig. 4b–f. *F* and *R* refer to the inflexion points that mark the maximum rate of eustatic fall and rise, respectively. HST, high-stand systems tract; LST, low-stand systems tract; TST, transgressive systems tract.

geographical setting and the close proximity of the Highland source area. In the slightly younger Namurian (Millstone Grit) succession further south in the northern English Pennines, which was related to a separate drainage system, both LSTs and tectonic influences are more obvious (Martinsen, 1993). Still further south in the southern Pennines, most of the Millstone Grit succession is low-stand dominated, with well-developed basin-floor fans, slope fans and prograding low-stand fluviodeltaic wedges (Read, 1991).

Following the recognition of high-frequency sequences, the allocyclic distal facies association sediments in the study interval can be fitted fairly easily into the revised framework of systems tracts, sequences and composite sequences envisaged by Haq (1991) and Mitchum & Van Wagoner (1991), despite a more laterally confined geographical setting and a much more complex pattern of tectonic subsidence than those set out in the simplified models of Jervey (1988), Posamentier *et al.* (1988) and Posamentier & Vail (1988). However, the overlying Upper Limestone Group, which will be the subject of a future paper, fits much less easily. Furthermore, much of the tectonically and fluvially influenced Passage Group of the Midland Valley (Read, 1989a) cannot readily be fitted into the EPR framework. Similarly, Martinsen (1993) has found that parts of the Millstone Grit succession in the northern Pennines do not fit into this framework.

Thus it would be premature at the present time to attempt to upgrade the concepts of EPR sequence stratigraphy into a rigid set of rules and to expect all Carboniferous cyclical successions to conform to such rules. One particularly controversial point concerns the most convenient horizon at which to draw the sequence boundary. Traditionally, the base of each 'short' Limestone Coal Group allocycle has been drawn immediately above the coal, thus utilizing a geographically continuous horizon which is easily recognized in natural sections, underground sections and boreholes (Read & Dean, 1967). This horizon also corresponds closely with the surface of maximum flooding. The latter has been chosen by Galloway (1989) as the base of his genetic stratigraphic sequence, which he has proposed as a substitute for the EPR sequence. However, as Haq (1991) has pointed out, the erosive, unconformable base of an EPR type 1 sequence is more easily recognized on seismic reflection profiles and is therefore generally a more convenient horizon for hydrocarbon exploration.

CONCLUSIONS

Deposition in the upper part of the fluviodeltaic Limestone Coal Group, between the Black Metals and the Index Limestone, was influenced by superimposed eustatic, tectonic and autocyclic sedimentary controls. Nevertheless, uniform, regular, laterally persistent 'short' allocycles make up the bulk of the succession, especially within the Kilsyth Trough. These allocycles are strongly asymmetrical and they most probably reflect glacial–eustatic oscillations in sea level with a frequency within the lower range of Milankovitch orbital parameters. In terms of EPR sequence stratigraphy, they can best be regarded as high-frequency, type 1 sequences, which may be divided into systems tracts. Similarly, the whole interval between the Black Metals and Index Limestone can be regarded as a composite sequence. This interval is high-stand dominated and low-stands are represented only by incised fluvial channels.

'Long' allocycles, probably with a recurrence interval of rather less than 1 Ma, gave rise to a series of major marine incursions which interrupted fluviodeltaic sedimentation. Their cause is as yet unknown.

Regional tectonic uplift and erosion of the Highland Massif at the beginning of the Namurian probably gave rise to the major influx of siliciclastic sediment which initiated the fluviodeltaic depositional environments of the Limestone Coal Group. Phases of more local uplift and erosion, followed by the regrading of stream profiles, were probably responsible for wedges of highly variable, strongly autocyclic, dominantly fluvial strata. These wedges appeared at irregular intervals and interrupted the regular 'layer-cake' lithostratigraphy produced by the succession of 'short' allocycles. However, most of the wedges were trapped in the rapidly subsiding Kincardine Basin.

ACKNOWLEDGEMENTS

Most of the basic information used in this study comes from non-confidential borehole records held in the Edinburgh office of the British Geological Survey, whose assistance is gratefully acknowledged. The author also thanks Ian Forsyth, Gilbert Kelling and Michael Whateley for reading an earlier draft of this paper and making constructive suggestions for its improvement. Ole Martinsen, as referee, and Poppe de Boer, as editor, are similarly thanked for their suggestions, as well as for supplying the author with advance copies of their forthcoming papers.

REFERENCES

AISENVERG, D.E., BRAZHNIKOVA, N.E., VASSILYUK, N.P., REITLINGER, E.A., FOMINA, E.V. & EINOR, O.L. (1979) The Serpukhovian Stage of the Lower Carboniferous of the USSR. In: *The Carboniferous of the USSR* (Ed. Wagner, R.H.). Occas. Publs Yorks. Geol. Soc. 4, pp. 109–124.

ALGEO, T.J. & WILKINSON, B.R. (1988) Periodicity of mesoscale Phanerozoic sedimentary cycles and the role of Milankovitch orbital modulation. *J. Geol.* **96**, 313–322.

ANDERSON, J.A.E. (1983) The structure and development of tropical peat swamps in western Malesia. In: *Ecosystems of the World, 4B, Mires: Swamp, Bog, Fen and Moor* (Ed. Gore, A.J.P.). Elsevier, Amsterdam, pp. 181–199.

BERGER, A. (1988) Milankovitch theory and climate. *Rev. Geophys.* **26**, 624–657.

BERGER, A. (1989) The spectral characteristics of pre-Quaternary climatic records, an example of the relationship between the astronomical theory and geo-sciences. In: *Climate and the Geo-Sciences, a Challenge for Sciences and Society in the 21st Century* (Eds Berger, A., Schneider, S. & Duplessy, J.Cl.). Kluwer, Dordrecht, pp. 47–76.

CARTER, R.M., ABBOTT, S.T., FULTHORPE, C.S., HAYWICK, D.W. & HENDERSON, R.A. (1991) Application of global sea-level and sequence-stratigraphic models in Southern Hemisphere Neogene strata from New Zealand. In: *Sedimentation, Tectonics and Eustasy. Sea-level Changes at Active Margins* (Ed. Macdonald, D.I.M.). Spec. Publs Int. Ass. Sediment. 12, pp. 41–65.

CECIL, C.B. (1990) Paleoclimate controls on stratigraphic repetition of chemical and siliciclastic rocks. *Geology* **18**, 533–536.

CLOETINGH, S. (1988) Intraplate stresses: a tectonic cause for third-order cycles in apparent sea level? In: *Sea-level Changes: an Integrated Approach* (Eds Wilgus, C.K., Hastings, B.S., Posamentier, H., Van Wagoner, J., Ross, C.A. & Kendall, C.G. St C.). Soc. Econ. Paleont. Mineral. Spec. Publs 42, pp. 19–29.

COLEMAN, J.M. & GAGLIANO, S.M. (1964) Cyclic sedimentation in the Mississippi River deltaic plain. *Trans. Gulf Coast Ass. Geol. Socs* **14**, 67–80.

DE BOER, P.L. (1991a) Astronomical cycles reflected in sediments. *Zbl. Geol. Paläont.* Teil I, H8, 911–930.

DE BOER, P.L. (1991b) Comment – Pennsylvanian time scales and cycle periods. *Geology* **19**, 408–409.

DE BOER, P.L., PRAGT, J.S.J. & OOST, A.P. (1991) Vertically persistent sedimentary facies boundaries along growth anticlines in the thrust-sheet-top South Pyrenean Tremp-Graus foreland basin. *Basin Res.* **3**, 63–78.

DEWEY, J.F. (1982) Plate tectonics and the evolution of the British Isles. *J. Geol. Soc. Lond.* **139**, 371–412.

FERM, J.C. (1970) Allegheny deltaic deposits. In: *Deltaic Sedimentation – Ancient and Modern* (Ed. Morgan, J.P.). Soc. Econ. Paleont. Mineral. Spec. Publ. 15, pp. 246–255.

FISCHER, A.G., DE BOER, P.L. & PREMOLI SILVA, I. (1990) Cyclostratigraphy. In: *Cretaceous Resources, Events and Rhythms* (Eds Ginsburg, R.N. & Beaudoin, B.). Kluwer Academic Publishers, Dordrecht, pp. 139–172.

FORSYTH, I.H. (1979) *The Lingula bands in the upper part of the Limestone Coal Group (E_1 Stage of the Namurian) in the Glasgow district.* Rep. Inst. Geol. Sci. 78/9.

FORSYTH, I.H. & READ, W.A. (1962) The correlation of the Limestone Coal Group above the Kilsyth Coking Coal in the Glasgow–Stirling region. *Bull. Geol. Surv. Great Britain* **19**, 29–52.

GALLOWAY, W.E. (1989) Genetic stratigraphic sequences in basin analysis I: architecture and genesis of flooding-surface bounded depositional units. *Bull. Am. Assoc. Petrol. Geol.* **73**, 125–142.

HAQ, B.U. (1991) Sequence stratigraphy, sea-level change, and significance for the deep sea. In: *Sedimentation, Tectonics and Eustasy. Sea-level Changes at Active Margins* (Ed. Macdonald, D.I.M.). Spec. Publs Int. Ass. Sediment. 12, pp. 3–39.

HAQ, B.U., HARDENBOL, J. & VAIL, P.R. (1988) Mesozoic and Cenozoic chronostratigraphy and eustatic cycles. In: *Sea-level Changes: an Integrated Approach* (Eds Wilgus, C.K., Hastings, B.S., Posamentier, H., Van Wagoner, J., Ross, C.A. & Kendall, C.G. St C.). Soc. Econ. Paleont. Mineral. Spec. Publ. 42, pp. 71–108.

HARLAND, W.B., ARMSTRONG, R.L., COX, A.V., CRAIG, L.E., SMITH, A.G. & SMITH, D.G. (1990) *A Geologic Time Scale 1989.* Cambridge University Press, Cambridge.

HECKEL, P.H. (1986) Sea-level curve for Pennsylvanian eustatic marine transgressive–regressive depositional cycles along the mid-continent outcrop belt. *Geology* **14**, 330–334.

HECKEL, P.H. (1991) Comment – Pennsylvanian time scales and cycle periods. *Geology* **19**, 406–407.

HESS, J.C. & LIPPOLT, H.J. (1986) $^{40}Ar/^{39}Ar$ ages of tonstein and tuff sanidines: new calibration points for the improvement of the Upper Carboniferous time scale. *Isotope Geoscience* **59**, 143–154.

HOLDSWORTH, B.K. & COLLINSON, J.D. (1988) Millstone Grit cyclicity revisited. In: *Sedimentation in a Synorogenic Basin Complex. The Upper Carboniferous of Northwest Europe* (Eds Besly, B.M. & Kelling, G.). Blackie, Glasgow, pp. 132–152.

JERVEY, M.T. (1988) Quantitative geological modelling of siliciclastic rock sequences and their seismic expression. In: *Sea-level Changes: an Integrated Approach* (Eds Wilgus, C.K., Hastings, B.S., Posamentier, H., Van Wagoner, J., Ross, C.A. & Kendall, C.G. St C.). Soc. Econ. Paleont. Mineral. Spec. Publ. 42, pp. 47–69.

KLEIN, G. DE V. (1990) Pennsylvanian timescales and cycle periods. *Geology* **18**, 455–457.

KLEIN, G. DE V. (1991). Replies to comments – Pennsylvanian time scales and cycle periods. *Geology* **19**, 405–410.

KLEIN, G. DE V. & WILLARD, D.A. (1989) Origin of the Pennsylvanian coal-bearing cyclothems of North America. *Geology* **17**, 152–155.

LANGENHEIM, R.L. (1991) Comment – Pennsylvanian time scales and cycle periods. *Geology* **19**, 405.

LEEDER, M.R. (1988) Recent developments in Carboni-

ferous geology: a critical review with implications for the British Isles and N.W. Europe. *Proc. Geol. Assoc.* **99**, 73–100.
MARTINSEN, O.J. (1993) Namurian (late Carboniferous) depositional systems of the Craven–Askrigg area, northern England: implications for sequence stratigraphic models. In: *Sequence Stratigraphy and Facies Associations* (Eds Posamentier, H.W., Summerhayes, C.P., Haq, B.U. & Allen, G.P.). Spec. Publs Int. Ass. Sediment. 18, pp. 247–282.
MENNING, M. (1989), A synopsis of numerical time scales, 1917–1986. *Episodes* **12**, 3–5.
MIALL, A.D. (1991) Stratigraphic sequences and their chronostratigraphic correlation. *J. Sedim. Petrol.* **61**, 497–505.
MITCHUM, R.M. & VAN WAGONER, J.C. (1991) High-frequency sequences and their stacking patterns: sequence-stratigraphic evidence of high-frequency eustatic cycles. *Sedim. Geol.* **70**, 131–160.
MOORE, D. (1958) The Yoredale Series of Upper Wensleydale and adjacent parts of northwest Yorkshire. *Proc. Yorks. Geol. Soc.* **31**, 91–148.
MOORE, P.D. (1987) Ecological and hydrological aspects of peat formation. In: *Coal and Coal-bearing Strata: Recent Advances* (Ed. Scott, A.C.). Geol. Soc. Lond. Spec. Publs 32, pp. 7–15.
MORTON, R.A. & PRICE, W.A. (1987) Late Quaternary sea-level fluctuations and sedimentary phases of the Texas coastal plain and shelf. In: *Sea-level Fluctuations and Coastal Evolution* (Eds Nummedal, D., Pilkey, O.H. & Howard, J.D.). Soc. Econ. Paleont. Mineral. Spec. Publs 41, pp. 181–198.
PHILLIPS, J. (1836) *Illustrations of the Geology of Yorkshire. Part II. The Mountain Limestone District.* John Murray, London.
POSAMENTIER, H.W. & VAIL, P.R. (1988) Eustatic controls on clastic deposition II – sequence and tract models. In: *Sea-level Changes: an Integrated Approach* (Eds Wilgus, C.K., Hastings, B.S., Posamentier, H., Van Wagoner, J., Ross, C.A. & Kendall, C.G. St C.). Soc. Econ. Paleont. Mineral. Spec. Publs 42, pp. 125–154.
POSAMENTIER, H.W., JERVEY, M.T. & VAIL, P.R. (1988) Eustatic controls on clastic deposition I – conceptual framework. In: *Sea-level Changes: an Integrated Approach* (Eds Wilgus, C.K., Hastings, B.S., Posamentier, H., Van Wagoner, J., Ross, C.A. & Kendall, C.G. St C.). Soc. Econ. Paleont. Mineral. Spec. Publs 42, pp. 109–124.
RAMSBOTTOM, W.H.C. (1977a) Major cycles of transgression and regression (mesothems) in the Namurian. *Proc. Yorks. Geol. Soc.* **41**, 261–291.
RAMSBOTTOM, W.H.C. (1977b) Correlation of the Scottish Upper Limestone Group (Namurian) with that of the North of England. *Scott. J. Geol.* **13**, 327–330.
READ, W.A. (1959) *The economic geology of the Stirling and Clackmannan Coalfield, Scotland. Area south of the River Forth.* Coalfield Pap. Geol. Surv. 2, HMSO, London.
READ, W.A. (1961) Aberrant cyclic sedimentation in the Limestone Coal Group of the Stirling Coalfield. *Trans. Edinburgh Geol. Soc.* **18**, 271–292.
READ, W.A. (1965) Shoreward facies changes and their relation to cyclical sedimentation in part of the Namurian east of Stirling, Scotland. *Scott. J. Geol.* **1**, 69–96.
READ, W.A. (1988) Controls on Silesian sedimentation in the Midland Valley of Scotland. In: *Sedimentation in a Synorogenic Basin Complex. The Upper Carboniferous of Northwest Europe* (Eds Besly, B.M. & Kelling, G.). Blackie, Glasgow, pp. 222–241.
READ, W.A. (1989a) The interplay of sedimentation, volcanicity and tectonics in the Passage Group (Arnsbergian, E_2 to Westphalian A) in the Midland Valley of Scotland. In: *The Role of Tectonics in Devonian and Carboniferous Sedimentation in the British Isles* (Eds Arthurton, R.S., Gutteridge, P. & Nolan, S.C.). Occas. Publs Yorks. Geol. Soc. 6, pp. 143–152.
READ, W.A. (1989b) Sedimentary evidence for a major subsurface fracture system linking the eastern Campsie and the eastern Ochil faults. *Scott. J. Geol.* **25**, 187–200.
READ, W.A. (1991) The Millstone Grit of the southern Pennines viewed in the light of eustatically controlled sequence stratigraphy. *Geol. J.* **26**, 157–165.
READ, W.A. & DEAN, J.M. (1967) A quantitative study of a sequence of coal-bearing cycles in the Namurian of Central Scotland, 1. *Sedimentology* **9**, 137–156.
READ, W.A. & DEAN, J.M. (1976) Cycles and subsidence: their relationship in different sedimentary and tectonic environments in the Scottish Carboniferous. *Sedimentology* **23**, 107–120.
READ, W.A. & DEAN, J.M. (1982) Quantitative relationships between numbers of fluvial cycles, bulk lithological composition and net subsidence in a Scottish Namurian basin. *Sedimentology* **29**, 181–200.
READ, W.A. & FORSYTH, I.H. (1989) Allocycles and autocycles in the upper part of the Limestone Coal Group (Pendleian E_1) in the Glasgow–Stirling region of the Midland Valley of Scotland. *Geol. J.* **24**, 121–137.
READ, W.A. & FORSYTH, I.H. (1991) Allocycles in the Upper part of the Limestone Coal Group (Pendleian, E_1) of the Glasgow–Stirling Region viewed in the light of sequence stratigraphy. *Geol. J.* **26**, 85–89.
RICHEY, J.E. (1937). Areas of sedimentation of Lower Carboniferous age in the Midland Valley of Scotland. In: *Mem. Geol. Surv. Summ. Prog. for 1935*, pp. 93–110.
ROWLEY, D.B., RAYMOND, A., PARRISH, J.T., LOTTES, A.L., SCOTESE, C.R. & ZIEGLER, A.M. (1985). Carboniferous paleogeographic, phytogenic and paleoclimatic reconstructions. In: *Paleoclimate Controls on Coal Resources in the Pennsylvanian System of North America* (Eds Phillips, T.L. & Cecil, C.B.). *Int. J. Coal. Geol.* **5**, 7–42.
SCHUMM, S.A. (1977) *The Fluvial System.* Wiley, New York.
SLOAN, R.J. & WILLIAMS, B.J.P. (1991) Volcano-tectonic control of offshore to tidal-flat regressive cycles from the Dunquin Group (Silurian) of southwest Ireland. In: *Sedimentation, Tectonics and Eustasy. Sea-level Changes at Active Margins* (Ed. Macdonald, D.I.M.). Spec. Publs Int. Ass. Sediment. 12, pp. 105–119.
SUTER, J.R., BERRYHILL, H.R. & PENLAND, S. (1987) Late Quaternary sea-level fluctuations and depositional sequences, southwest Louisiana continental shelf. In: *Sea-level Fluctuations and Coastal Evolution* (Eds Nummedal, D., Pilkey, O.H. & Howard, J.D.). Soc. Econ. Paleont. Mineral. Spec. Publs 41, pp. 199–219.
VAN WAGONER, J.C., POSAMENTIER, H.W., MITCHUM,

R.M., Vail, P.R., Sarg, J.F., Loutit, T.S. & Hardenbol, J. (1988) An overview of the fundamentals of sequence stratigraphy and key definitions. In: *Sea-level Changes: an Integrated Approach* (Eds Wilgus, C.K., Hastings, B.S., Posamentier, H., Van Wagoner, J., Ross, C.A. & Kendall, C.G. St C.). Soc. Econ. Paleont. Mineral. Spec. Publs 42, pp. 39–45.

Van Wagoner, J.C., Mitchum, R., Campion, K.M. & Rahmanian, V.D. (1990) *Siliciclastic Sequence Stratigraphy in Well Logs, Cores and Outcrops*. Am. Assoc. Petrol. Geol. Methods in Exploration Series 7.

Veevers, J.J. & Powell, C. McA (1987) Late Paleozoic glacial episodes in Gondwanaland reflected in transgressive–regressive depositional sequences in Euramerica. *Geol. Soc. Am. Bull.* **98**, 475–487.

Wanless, H.R. & Shepard, F.P. (1936) Sea-level and climatic changes related to late Paleozoic cycles. *Geol. Soc. Am. Bull.* **47**, 1177–1206.

Watson, J. (1985) Northern Scotland as an Atlantic–North Sea divide. *J. Geol. Soc. Lond.* **142**, 221–243.

Weller, J.M. (1930) Cyclical sedimentation of the Pennsylvanian period and its significance. *J. Geol.* **38**, 97–135.

Wilgus, C.K., Hastings, B.S., Ross, C.A., Posamentier, H., Van Wagoner, J. & Kendall, C.G. St C. (Eds) (1988) *Sea-level Changes: an Integrated Approach*. Soc. Econ. Paleont. Mineral. Spec. Publs 42.

Williams, D.F. (1988) Evidence for and against sea-level changes from the stable isotopic record of the Cenozoic. In: *Sea-level Changes: an Integrated Approach* (Eds Wilgus, C.K., Hastings, B.S., Ross, C.A., Posamentier, H., Van Wagoner, J., Ross, C.A. & Kendall, C.G. St C.). Soc. Econ. Paleont. Mineral. Spec. Publs 42, pp. 31–36.

Ziegler, A.M., Raymond, A.L., Gierlowski, T.C., Horrel, M.A., Rowley, D.B. & Lottes, A.L. (1987) Coal, climate and terrestrial productivity: the present and early Cretaceous compared. In: *Coal and Coal-bearing Strata: Recent Advances* (Ed. Scott, A.C.). Geol. Soc. Lond. Spec. Publs 32, pp. 25–49.

Spec. Publs Int. Ass. Sediment. (1994) **19**, 429–438

Orbital forcing on continental depositional systems — lacustrine and fluvial cyclicity in the Devonian of East Greenland

H. OLSEN

Department of Geology and Geotechnical Engineering, Danmarks Tekniske Højskole, build. 204, DK-2800 Lyngby, Denmark

ABSTRACT

Remarkable cyclicity in a lacustrine succession, the Wimans Bjerg Formation in the Devonian basin in East Greenland, has previously been pointed out by Nicholson & Friend. The present study of the Wimans Bjerg Formation has been carried out to outline the processes leading to the cyclicity.

The Wimans Bjerg Formation is composed of two facies, laminated siltstones and brecciated to massive mudstones, and exhibits a pronounced cyclicity. Individual cycles average 3.6 m in thickness and consist of a unit dominated by laminated siltstones and a unit exclusively composed of brecciated/massive mudstones. A higher order cyclicity is also observed in the variation in thickness of the laminated siltstone units. These mega-cycles are 18.7 m thick on average.

The Wimans Bjerg Formation is interpreted in terms of a playa environment. The brecciated/massive mudstones indicate periods of exposed playa mudflat conditions. The laminated siltstone beds reflect the existence of ephemeral lakes on the playa mudflat.

The cyclicity indicates regular variations in environmental conditions between periods with frequent ephemeral lake expansions and contractions and periods exclusively characterized by playa mudflats. The higher order of cyclicity indicates long-term variations in the periods of punctuated ephemeral lake conditions.

In this paper a comparison is made between the playa lake cyclicity of the Wimans Bjerg Formation, and the cyclicity of the slightly older fluvial Andersson Land Formation in the East Greenland basin. Both units indicate a ratio of mega-cycle thickness to cycle thickness in the order of 5.1–5.2. The cyclic variations may in both cases be explained by cyclic variations in precipitation and could reflect orbital forcing of climate, i.e. astronomical periods of 17.0–20.1 ka and 94.9 ka, controlled by precession and eccentricity cycles.

INTRODUCTION

The Milankovitch theory of climatic change and control on depositional systems is well established (e.g. this volume). However, the literature seems to be biased by marine, mainly pelagic, examples and convincing continental examples are rare (e.g. Olsen, 1986; Astin, 1990; Olsen, 1990).

In this paper I will outline the cyclic development of two continental depositional systems and relate the cyclicity to orbital forcing. In this way I wish to stress that Milankovitch cyclicity is worth looking for in continental settings and that such cyclicity may be a powerful tool in estimation of sedimentation rate and the establishment of a chronostratigraphy in a palaeontologically and radiometrically poorly dated succession.

SETTING

The Devonian basin in East Greenland (Fig. 1) developed as a major intramontane depositional site following the Caledonian orogeny and accumulated more than 8 km of continental sediments during the late Middle to Late Devonian (Givetian to Famennian) (Olsen & Larsen, 1993a). The basin is considered to have been initiated in response to

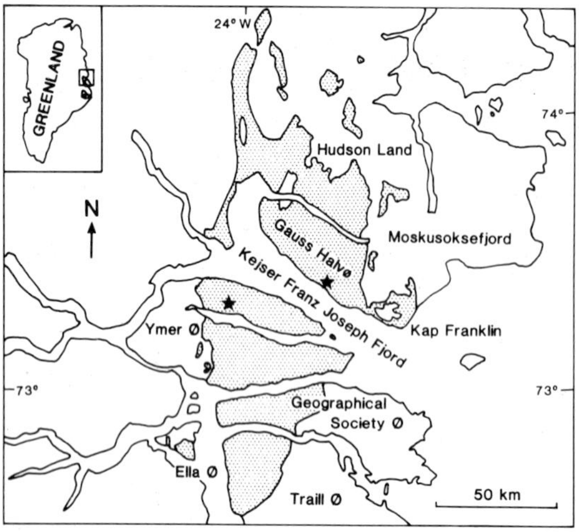

Fig. 1. Geographical position and outcrops of Devonian sediments in northeast Greenland. The star in Ymer Ø indicates the studied section near the Zoologdalen in the Andersson Land Formation. The star in Gauss Halvø indicates Stensiö Bjerg, where the Wimans Bjerg Formation was studied in detail.

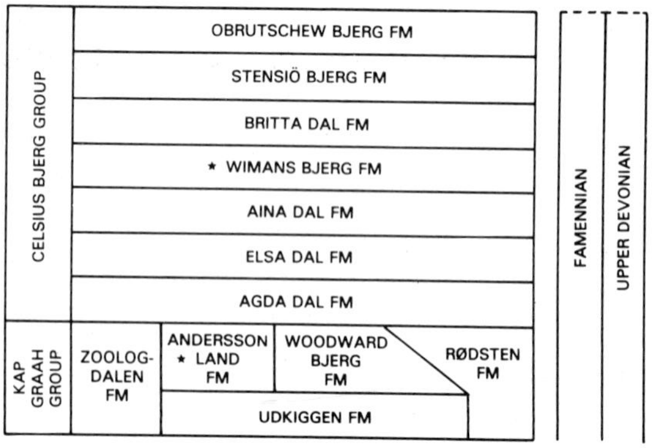

Fig. 2. Lithostratigraphy after Olsen & Larsen (1993a). The Andersson Land Formation and Wimans Bjerg Formation are the subjects of this paper.

extensional collapse of an overthickened Caledonian welt (McClay *et al.*, 1986; Norton *et al.*, 1987; Larsen & Bengaard, 1991). Later phases in basin evolution were characterized by complex deformations, related to a transition into a strike-slip basin (Friend *et al.*, 1983; Larsen & Olsen, 1991). The tectonic regimes exerted major controls on the depositional systems but climatic changes also played an important role (Olsen, 1990, 1993; Olsen & Larsen, 1993b).

INVESTIGATIONS

The present study includes two Upper Devonian (Famennian) sedimentary units, the Andersson Land and Wimans Bjerg Formations (Fig. 2) defined by Olsen & Larsen (1993a). The purpose of the investigation is to outline the cyclic build-up of the deposits and to analyse whether the cyclicities are related to orbital control of the depositional systems. The sediments of the Andersson Land Formation have recently been interpreted as exhibiting Milankovitch-type cyclicity (Olsen, 1990). The formation is fluvial and probably the first example of orbital forcing on a fluvial depositional system. To support this interpretation a comparative study of the lacustrine Wimans Bjerg Formation has been carried out. A succession measured in great detail by Nicholson & Friend (1976) was revisited in order to gain an understanding of the apparent cyclicity discovered by the previous investigators.

In the following the deposits of the two formations are described briefly and the cyclicities observed in the successions are compared.

ANDERSSON LAND FORMATION

The Andersson Land Formation was studied west of Zoologdalen in Ymer Ø (Fig. 1). Here it is *c.*1000 m thick and composed of metre-scale-thick tabular sandstone bodies alternating with siltstone-dominated units (Fig. 3A, B). The two facies associations are equally important. The sandstone bodies are 1.5–15 m thick and composed of fine to medium sand. Trough cross-bedding dominates but parallel lamination and cross-lamination form important constituents.

The sediment bodies are single-storey (most common) or multi-storey. Individual storeys are commonly fining upwards and composed of decimetre-scale fining-upward bedsets, unidirectionally inclined at *c.*7° (Fig. 3A). Palaeocurrents are perpendicular to the inclined (lateral accretion) bedding.

The siltstones are massive, horizontally laminated and brecciated. They are closely associated with sheets and lenses of very fine to fine sandstones less than 1 m thick (Fig. 3B). Desiccation cracks and rootlet horizons are common. The sandstones are mainly composed of intricately interbedded festoon

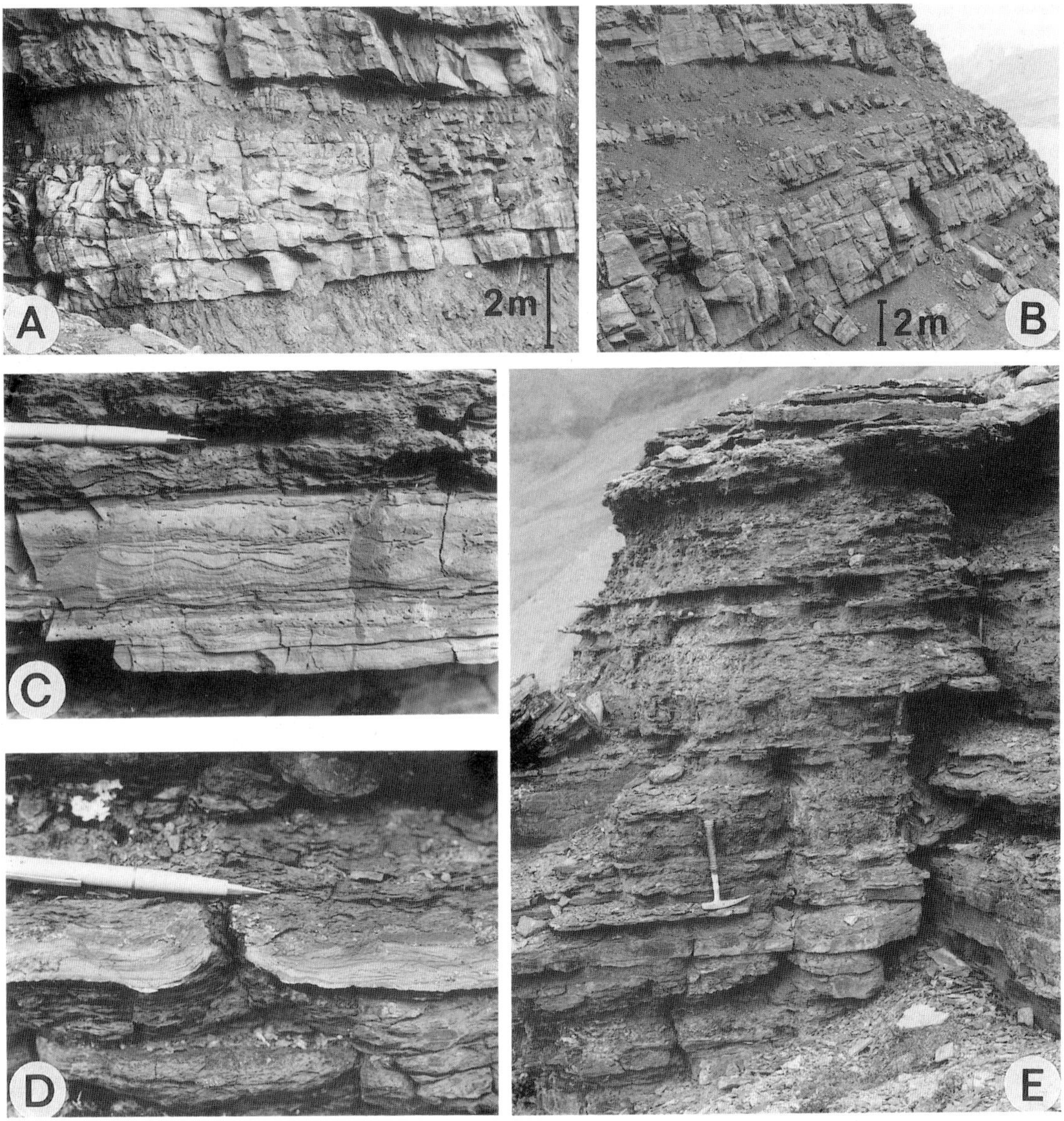

Fig. 3. Field appearance of the Andersson Land Formation (A and B) and the Wimans Bjerg Formation (C, D and E). (A) Multi-storey sandstone body (three stories) deposited as point bars in a meandering river. Notice lateral accretion bedding in left side of uppermost storey grading into a channel plug in the right side. (B) A thick tabular sandstone body (single storey) of point bar origin is embedded in siltstones deposited in flood basins. Notice the thin crevasse splay sandstone in the upper part. (C) Ephemeral lake deposit from the Wimans Bjerg Formation composed of a bedset of six superimposed thin beds, reflecting rapid (?seasonal) expansion and contraction of lakes. Notice the wave ripple lamination. The bedset is overlain by brecciated to massive mudstone of playa mudflat origin (above pencil). (D) A single laminated siltstone bed exhibiting fining upwards and a prominent desiccation crack. This ephemeral lake bed is embedded in brecciated playa mudflat mudstones. (E) Ephemeral lake complex with bedsets and single beds of ephemeral lake deposits and intercalated playa mudflat beds.

cross-bedding, parallel lamination, cross-lamination and scour-and-fills.

The formation is interpreted as a meander belt succession (Olsen, 1990, 1993). The characteristics of the sandstone bodies resemble other inferred ancient point bar deposits (e.g. Leeder, 1973; Puigdefábregas, 1973; Puigdefábregas & van Vliet, 1978; Bridge & Diemer, 1983), and they are accordingly interpreted as having been laid down as point bars of a meandering river. The multi-storey nature of some sandstone bodies reflects the downstream migration of meanders in association with net aggradation. The development of bedsets indicates repeated flooding episodes during which three-dimensional dunes dominated the bar surface.

The siltstones with associated thin sandstone beds are typical flood basin, crevasse splay and crevasse channel deposits (e.g. Farrell, 1987).

WIMANS BJERG FORMATION

The Wimans Bjerg Formation was studied at Stensiö Bjerg in Gauss Halvø (Fig. 1). The formation is composed of two almost equally important facies, laminated siltstones and brecciated to massive mudstones (Figs 3C–E and 4). Locally (forming less than 2% of the total) channel-shaped siltstone and very fine sandstone units occur in the succession.

The laminated siltstones are composed of centimetre-scale-thick beds of siltstone or silty very fine sandstone with sharp and commonly erosive bases. The beds occur singly (Fig. 3D) or stacked in bedsets of two to ten beds (Fig. 3C). Individual beds are dominated by wave ripple lamination and commonly exhibit a more or less complete sequence of parallel lamination, cross-lamination and parallel draped lamination, associated with a fining-upward lithology. Desiccation cracks commonly commence from the top of beds or occur within the draped lamination.

The second important facies is composed of brecciated and apparently massive silty mudstones (Fig. 3C, D). The brecciated mudstones are characterized by a patchy fabric riddled with superimposed desiccation cracks. In less densely brecciated beds individual cracks can be recognized as desiccation cracks and on rare bedding planes polygonal crack patterns can be observed. The majority of beds seem quite massive when looked at with the naked eye. However, thin-section examination always indicates the presence of submillimetre-scale cracks.

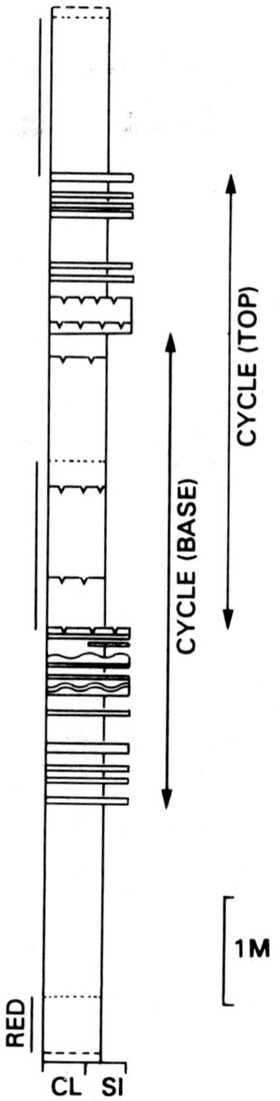

Fig. 4. Characteristic succession from the section studied in the Wimans Bjerg Formation (modified from Nicholson & Friend, 1976). The sediments are composed of ephemeral lake complexes of interbedded laminated siltstones and brecciated to massive mudstones, alternating with playa mudflat units entirely composed of brecciated to massive mudstones. The two types of cycles used for the statistics in Fig. 6 are defined.

The cracks are filled with clay and diagenetic dolomite and quartz. Dolomite crystal rhombs and nodules of dolomite and quartz abound in the massive and brecciated mudstones. Occasionally enterolithic folded dolomitic layers occur.

The Wimans Bjerg Formation is interpreted in terms of a playa environment (Nicholson & Friend,

1976; Olsen, 1993). The laminated siltstones and sandstones resemble type L3 of Smoot (1991) and are likewise interpreted to indicate deposition in shallow lakes experiencing rapid lake-level fluctuations. Emergence of the bed surface occurred frequently as indicated by the desiccation cracks. The fining-upward sequences resemble small-scale sequences described by de Raaf *et al.* (1977) from a shallow marine environment and they are likewise interpreted as having formed during falling energy conditions. The sequences were probably formed in response to heavy (?seasonal) rainstorms and rapid expansions of ephemeral lakes. The subsequent quiescence resulted in deposition from suspension (draped lamination). During dry periods following the rainstorms the lakes contracted, sometimes in a fluctuating manner, and desiccation cracks were formed. These small-scale sequences, accordingly, reflect climatic variations probably on a seasonal scale.

The brecciated mudstones closely resemble modern playa mudflat deposits (Hardie *et al.*, 1978; Demicco & Kordesh, 1986; Smoot, 1991). The brecciated fabric of modern playa muds is due to repeated wetting and drying of clays (Smoot, 1981). Crystals, nodules and layers of gypsum may grow by displacement of the host sediment, or crystals and other evaporite minerals may fill or partly fill desiccation cracks (Hardie *et al.*, 1978). Ancient examples of similar deposits have likewise been interpreted in terms of playa mudflat deposition (disrupted mudstones of Demicco & Kordesh, 1986; crump mudstone fabric of Smoot & Olsen, 1988; type L4 of Smoot, 1991). Accordingly, the brecciated and apparently massive mudstones are interpreted as playa mudflat deposits, probably representing longer timespans per thickness unit than the laminated siltstones. Nodules, crystals and enterolithic folded layers of dolomite and quartz are interpreted as gypsum/anhydrite precipitates, now replaced by dolomite or quartz.

The rare channel-shaped siltstones and sandstones exhibit internal desiccation cracks and they are interpreted as ephemeral stream deposits (Olsen, 1993).

CYCLICITY

Andersson Land Formation

The sandstone bodies (point bars) in the Andersson Land Formation exhibit a remarkable cyclicity (Olsen, 1990). Plots of the thickness of sandstone bodies are shown in Fig. 5a for a 280-m-thick well-exposed ravine section. The position of the sandstone top is used because this reflects the stratigraphic level of the meander belt surface at a given time. The thickness variation occurs in two scales: small (*c.*20 m thick) cycles of decreasing to increasing thickness values (or increasing to decreasing) and a *c.*100 m mega-cycle of decreasing to increasing values of the largest sandstone body in each cycle (Fig. 5a). Even though some error is involved because some of the sandstone bodies are multi-storey, there exists a close relationship between sandstone body thickness and computed river discharge in this section (Olsen, 1990). Accordingly, the cyclic development of sandstone body thickness reflects cyclic changes in the meandering river discharge.

Above the well-exposed 280-m-thick succession from Fig. 5a, a *c.*600-m-thick succession of similar but less well exposed deposits occurs in a mountain ridge exposure. Prominent cuestas occur at regular intervals, either as isolated sandstone bodies or as bundles of two to four major sandstone bodies (Fig. 5b). These sandstone bodies correspond to the major sandstone bodies forming the mega-cyclicity in the well-exposed succession.

The stratigraphic distances between the largest sandstone bodies in successive cycles, *cycle (max)*, and between the smallest sandstone bodies in successive cycles, *cycle (min)*, were measured in the ravine section. The cycle thickness clusters around a mean of 20.0 m (Fig. 6a). The stratigraphic distances between successive bundles (midpoint of bundle or top of isolated sandstone body) were measured along the mountain ridge by aid of a computer assisted stereoscopic study of aerial photographs. These measurements cluster around a mean of 101 m (Fig. 6b).

Wimans Bjerg Formation

The playa deposits at Stensiö Bjerg also exhibit a cyclicity as pointed out by Nicholson & Friend (1976). The cyclicity is defined by two components (Fig. 4): playa mudflat units exclusively (or nearly so) composed of brecciated and massive mudstones (commonly exhibiting an upward transition from red to grey), and ephemeral lake complexes composed of single beds and bedsets of laminated siltstones with varying amounts of interbedded brecciated/massive mudstones (Fig. 3E). The cycle thickness may then be defined as the stratigraphic distance

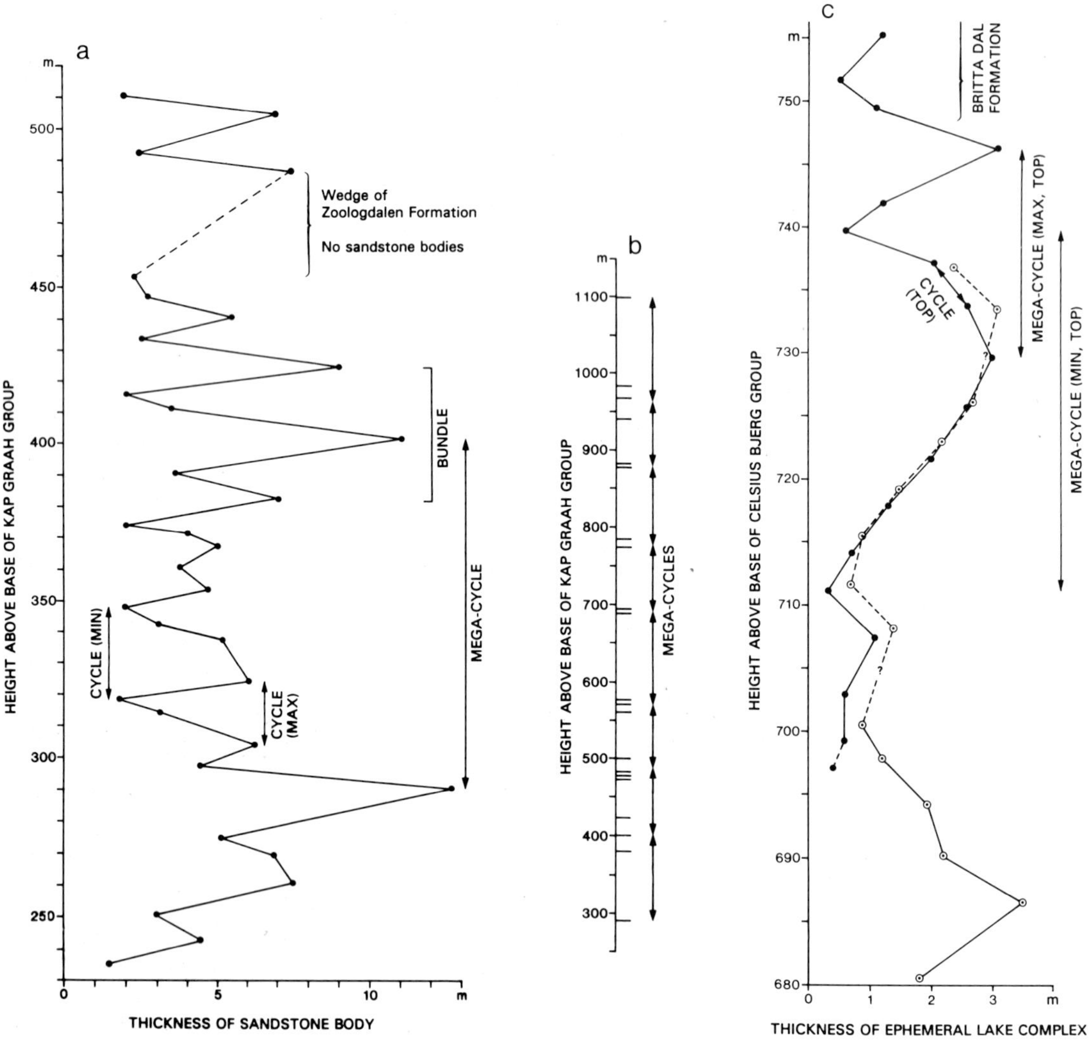

Fig. 5. (a) Plots of sandstone body thickness (plotted at the top of bodies) from the ravine section in the Andersson Land Formation. Definitions of cycle (min), cycle (max), mega-cycle and bundle are provided. (b) Plots of position of the top of major sandstone bodies exposed as prominent sandstone cuestas along a mountain ridge. The mega-cycles defined on this basis are shown. The lowermost mega-cycle is the one shown in (a). (c) Plots of thickness variations of the ephemeral lake complexes in two overlapping sections in the Wimans Bjerg Formation (and lowermost Britta Dal Formation). Notice the good agreement between the two sections. To avoid bias in the statistics due to the overlap, only the data represented by full lines are used. The plots are based on data from Nicholson & Friend (1976, fig. 55). Here the tops of ephemeral lake complexes are indicated. Cycle (top), mega-cycle (min, top) and mega-cycle (max, top) are defined. Although the maximum around 708 m is weakly developed it is present in both sections and regarded as significant.

between the bases of successive complexes. This is called a *cycle*(*base*) in the figures. Alternatively the cycle thickness may be defined as the stratigraphic distance between the tops of successive complexes (=*cycle*(*top*)). The cycles record prolonged periods of exposed playa mudflat conditions (thick brecciated/massive mudstones), alternating with periods in which playa mudflat conditions were punctuated by occasional and repeated expansions/contractions of ephemeral lakes (complexes of brecciated/massive mudstones plus single beds and bedsets of laminated siltstones). The thickness of

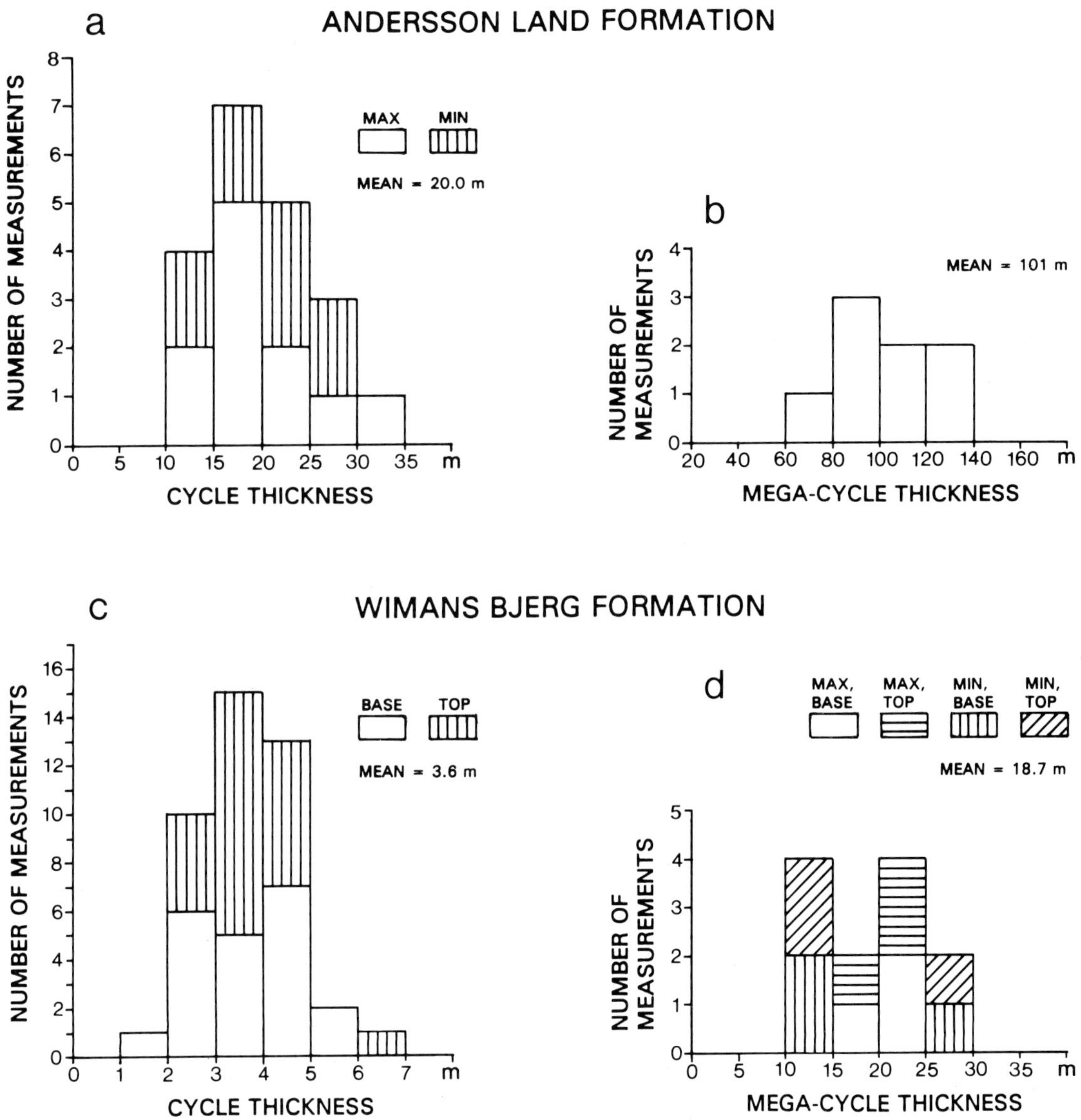

Fig. 6. Distribution of thickness variation in cycles and mega-cycles from the Andersson Land Formation (a and b) and Wimans Bjerg Formation (c and d). See text and Figs 4 and 5 for explanation of cycle types.

cycles exhibits a mean of 3.6 m (Fig. 6c). When plotting the thickness of the ephemeral lake complexes in each cycle an apparent mega-cyclicity is also present (Fig. 5c). Four types of mega-cycles are used for the statistics in Fig. 6d, depending on whether the position of the top or base of ephemeral lake complexes has been used and whether the position of maximum or minimum values has been used. This is done to increase the statistical validity of the relatively sparse data set. The mega-cycle thickness varies systematically with a distance between successive maxima or minima of 18.7 m (mean) (see Fig. 6d).

Causes of cyclicity

The changing size of the meandering river, reflected in the Andersson Land Formation, was shown to be governed by systematic variations in precipitation (Olsen, 1990). Likewise, the alternation between periods of permanent playa mudflat conditions and periods punctuated by ephemeral lake expansions/contractions could be explained very well by a systematic pattern in precipitation variation.

Taking the thickness of ephemeral lake complexes as a record of the duration of the period in which ephemeral lakes regularly existed, the systematic variation (mega-cyclicity) in the thickness of complexes may indicate longer term climatic changes.

The East Greenland basin lay at *c.*10° N latitude during the Famennian (Late Devonian), forming part of the equatorial Laurasia continent, or Old Red continent (e.g. van der Voo, 1988). East Greenland should, accordingly, be viewed in the context of a monsoonal or monsoon-like climate (Olsen, 1990, 1993). Therefore, the cyclicity may reflect cyclic controls on the monsoonal climate.

Although the statistical background for the mega-cyclicity is weak, the comparison of the two formations shows that it may be significant. The ratio of mega-cycle thickness to cycle thickness in the Andersson Land Formation (using mean values) is 5.1. The ratio in the Wimans Bjerg Formation is 5.2, i.e. very close to the Andersson Land Formation. Pure coincidence seems unlikely to explain this similarity.

These ratios are comparable to the ratios of the eccentricity period of 94.9 ka to the two main precession periods, 17.0 and 20.1 ka (interpolation of the palaeoperiods calculated by Berger *et al.*, 1989, to the Famennian at 365 Ma BP). These ratios are 4.7 and 5.6, respectively (arithmetic mean = 5.2). Accordingly, the cyclicity at two orders of magnitude in both formations may reflect climatic variations governed by precession and eccentricity cycles. As seen below, using these orbital periods the computed net sedimentation rates for the two formations correspond quite well with the estimate of average net sedimentation rate in the Famennian (Olsen, 1990). The data from East Greenland accordingly support the prediction of Rossignol-Strick (1983), for example, that the precession and eccentricity cycles exercise the most important orbital forcing on low-latitude (monsoonal) climates. Among other ratios of orbital periods the one closest to 5.1–5.2 is the 412.1-ka eccentricity period to the 94.9-ka eccentricity period (=4.3). This ratio, accordingly, does not fit the present data, and an influence by the 412.1-ka eccentricity cycle is excluded as an explanation for the observed cyclicity.

Applications of the cyclicity

The average net sedimentation rate in the central part of the basin in the Famennian has been estimated at around 48 cm/ka (Olsen, 1990). The inferred orbital or Milankovitch-type cyclicity may be used for comparison of net sedimentation rate and probable subsidence rate stratigraphically and laterally in the basin. The cycle thicknesses in the Andersson Land Formation and inferred precession and eccentricity cycle periods indicate a net sedimentation rate of *c.*106 cm/ka at that locality. The cycle thicknesses and associated cycle periods in the Wimans Bjerg Formation indicate a net sedimentation rate in the order of 20 cm/ka.

Within the Wimans Bjerg Formation lateral variations also occur. At Obrutschew Bjerg in eastern Gauss Halvø the cycles are on a 1-m scale compared to the 3.6 m at Stensiö Bjerg. This implies a sedimentation rate of 5 cm/ka, and reflects the much lower sedimentation rate near the basin margin.

Perspectives

The stratigraphy in the Devonian basin in East Greenland is, in common with many other continental basins, hampered by the scarcity of palaeontological and palynological material for biostratigraphic control. However, with the aid of Milankovitch cyclicity observed in the sedimentary pile a good approach to a chronostratigraphy may be achieved.

If Milankovitch cyclicity is developed in several levels and laterally widespread within these levels, sedimentation rate curves may be established. Thereby, stratigraphic extrapolations above or below well-dated horizons, either of palaeontological/palynological or radiometric nature, may be performed with some accuracy.

In the East Greenland basin the Ankerbjergselv Formation is palaeontologically dated as Givetian (Olsen & Larsen, 1993a). In the north-central part of the basin this formation exhibits a cyclicity on a 20-m scale, defined by the alternations of green and red units associated with facies changes between perennial and ephemeral type depositional systems (Olsen, 1993). If these cycles are supposed to be similar to the precession-controlled cycles in the two Famennian formations, a chronostratigraphic subdivision of the intervening, poorly dated deposits may be attempted. A detailed study of the cyclic development of the Ankerbjergselv Formation is, however, needed prior to this.

CONCLUSIONS

Two formations in the Devonian basin in East Greenland both exhibit cyclicity at two orders of magnitude.

The ratio of mega-cycle thickness to cycle thickness is 5.1 and 5.2 in the two formations respectively, corresponding to variations in precipitation governed by precession period(s) of 17.0 and/or 20.1 ka, and modulated by an eccentricity period of 94.9 ka.

The Milankovitch cyclicity may be used as a tool for subsidence rate calculation and chronostratigraphic subdivision of the continental deposits which are in other ways poorly dated.

ACKNOWLEDGEMENTS

The study was financially supported by grants from the Carlsberg Foundation and carried out mainly at the Geological Survey of Greenland. Vibeke Knudsen and Jette Halskov drafted the figures and Anne-Lene Felland typed the manuscript. The paper is published with the approval of the Geological Survey of Greenland.

REFERENCES

Astin, T.R. (1990) The Devonian lacustrine sediments of Orkney, Scotland: implications for climate cyclicity, basin structure and maturation history. *J. Geol. Soc. Lond.* **147**, 141–151.

Berger, A., Loutre, M.F. & Dehant, V. (1989) Astronomical frequencies for pre-Quaternary palaeoclimate studies. *Terra Nova* **1**, 474–479.

Bridge, J.S. & Diemer, J.A. (1983) Quantitative interpretation of an evolving ancient river system. *Sedimentology* **30**, 599–623.

Demicco, R.V. & Kordesh, E.G. (1986) Facies sequences of a semi-arid closed basin: the Lower Jurassic East Berlin Formation of the Hartford Basin, New England, USA. *Sedimentology* **33**, 107–188.

Farrel, K.M. (1987) Sedimentology and facies architecture of overbank deposits of the Mississippi River, False River region, Louisiana. In: *Recent Developments in Fluvial Sedimentology* (Eds Ethridge, F.G., Flores, R.M. & Harvey, M.D.). Soc. Econ. Paleont. Mineral. Spec. Publ. 39, pp. 111–120.

Friend, P.F., Alexander-Marrack, P.D., Allen, K.C., Nicholson, J. & Yeats, A.K. (1983) Devonian sediments of East Greenland. Review of results. *Meddr. Grønland* **206**, (VI).

Hardie, L.A., Smoot, J.P. & Eugster, H.P. (1978) Saline lakes and their deposits, a sedimentological approach. In: *Modern and Ancient Lake Sediments* (Eds Matter, A. & Tucker, M.E.). Spec. Publs Int. Ass. Sediment. 2, pp. 7–41.

Larsen, P.-H. & Bengaard, H.-J. (1991) Devonian basin initiation in East Geenland: a result of sinistral wrench faulting and Caledonian extensional collapse. *J. Geol. Soc. Lond.* **148**, 355–368.

Larsen, P.-H. & Olsen, H. (1991) The Devonian basin project, North-East Greenland – a summary. *Rapp. Grønlands Geol. Unders.* **145**, 108–113.

Leeder, M.R. (1973) Sedimentology and palaeogeography of the Upper Old Red Sandstone in the Scottish Border Basin. *Scott. J. Geol.* **9**, 117–144.

McClay, K.T., Norton, G.M., Coney, P. & Davis, G.H. (1986) Collapse of the Caledonian orogen and the Old Red Sandstone. *Nature* **323**, 147–149.

Nicholson, J. & Friend, P.F. (1976) Devonian sediments of East Greenland. The central sequence, Kap Graah Group and Mount Celsius Supergroup. *Meddr. Grønland* **206** (V), 117 pp.

Norton, G.M., McClay, K.R. & Way, N.A. (1987) Tectonic evolution of Devonian basins in northern Scotland and southern Norway. *Norsk Geol. Tidsskr.* **67**, 323–338.

Olsen, H. (1990) Astronomical forcing of meandering river behaviour: Milankovitch cycles in Devonian of East Greenland. *Palaeogeogr. Palaeoclimatol. Palaeoecol.* **79**, 99–115.

Olsen, H. (1993) Sedimentary basin analysis of the continental Devonian basin in North-East Greenland. *Bull. Grønl. Geol. Unders.* (in press).

Olsen, H. & Larsen, P.-H. (1993a) Lithostratigraphy of the continental Devonian sediments in North-East Greenland. *Bull. Grønl. Geol. Unders.* **165**, 111 pp.

Olsen, H. & Larsen, P.-H. (1993b) Structural and climatic controls on fluvial depositional systems – Devonian, East Greenland. In: *Alluvial Sedimentation* (Eds Marzo, M. & Puigdefábregas, C.). Spec. Publs Int. Ass. Sediment. 17, pp. 401–424.

Olsen, P.E. (1986) A 40-million-year lake record of early Mesozoic orbital climatic forcing. *Science* **234**, 842–848.

Puigdefábregas, C. (1973) Miocene point-bar deposits in the Ebro Basin, Northern Spain. *Sedimentology* **20**, 133–144.

Puigdefábregas, C. & Van Vliet, A. (1978) Meandering stream deposits from the Tertiary of the Southern Pyrenees. In: *Fluvial Sedimentology* (Ed. Miall, A.D.). Mem. Can. Soc. Petrol. Geol. 5, pp. 469–486.

Raaf, J.F.M. de, Boersma, J.R. & Gelder, A. van (1977) Wave-generated structures and sequences from a shallow marine succession, Lower Carboniferous, County Cork, Ireland. *Sedimentology* **24**, 451–483.

Rossignol-Strick, M. (1983) African monsoons, an immediate climatic response to orbital insolation. *Nature* **304**, 46–49.

SMOOT, J.P. (1981) Subaerial exposure criteria in modern playa and mud cracks (abstract). *Bull. Am. Assoc. Petrol. Geol.* **65**, 994.

SMOOT, J.P. (1991) Sedimentary facies and depositional envirionments of early Mesozoic Newark Supergroup basins, eastern North America. *Palaeogeogr. Palaeoclimatol. Palaeoecol.* **84**, 369–423.

SMOOT, J.P. & OLSEN, P.E. (1988) Massive mudstones in basin analysis and palaeoclimatic interpretation of the Newark Supergroup. In: *Triassic–Jurassic Rifting: North America and North Africa* (Ed. Manspeizer, W.). Elsevier, Amsterdam, pp. 249–274.

VAN DER VOO, R. (1988) Paleozoic paleogeography of North America, Gondwana, and intervening displaced terranes: Comparisons of paleomagnetism with paleoclimatology and biogeographical patterns. *Geol. Soc. Am. Bull.* **100**, 311–324.

Spec. Publs Int. Ass. Sediment. (1994) **19**, 439–457

Climatic controls on ancient desert sedimentation: some late Palaeozoic and Mesozoic examples from NW Europe and the Western Interior of the USA

L.B. CLEMMENSEN*, I.E.I. ØXNEVAD†§ *and* P.L. DE BOER‡

* *Geologisk Institut, Københavns Universitet, DK-1350 København K, Denmark;*
† *Geologisk Institutt, Avd. A, Universitetet i Bergen, 5001 Bergen, Norway; and* ‡ *Comparative Sedimentology Division, Institute of Earth Sciences, P.O. Box 80.021, NL-3508 TA Utrecht, The Netherlands*

ABSTRACT

This study of ancient desert deposits tests the possibility that cyclicity in aeolian and associated deposits, formed at low palaeolatitudes on the megacontinent of Pangaea during the Early Permian to Early Jurassic, was orbitally controlled.

The deposits investigated are characterized by vertically stacked sedimentary cycles. These cycles are simple, or more commonly composite, and range in thickness from about 2 m to more than 400 m. Their composition indicates repeated climatic fluctuations between arid and humid conditions, and the thickness and thickness distribution of these cycles in combination with their palaeolatitudinal position fit a model in which Milankovitch-type climatic fluctuations would have produced them. Some of the formations studied seem to be composed of stacked precession cycles (20 ka), which form part of thicker eccentricity cycles (100 and 400 ka). Other formations display only one order of cyclicity which complicates correlation with specific orbital cycles. If orbital forcing is accepted for these successions, ancient accumulation rates can be calculated; the resulting figures compare very well with values from modern basins. The lack of good time control in most of the basins studied makes it impossible to prove definitely the origin of the cycles. Milankovitch theory, however, is at least a promising candidate for explanation of the cyclic nature of ancient desert deposits.

INTRODUCTION

Orbital variations affecting the intensity and distribution of insolation have been proposed by several authors as the main cause of Quaternary arid–humid fluctuations and related sedimentary cycles in subtropical desert belts (e.g. Talbot, 1985; Perlmutter & Matthews, 1989; Miller *et al.*, 1991; Spaulding, 1991).

It has been speculated that similar climatic cycles are typical of subtropical desert basins of all ages, in non-glacial periods also, when the amplifying effect of polar ice caps was absent, and variations in monsoonal intensity, precipitation–evaporation ratios, etc., were forced directly by orbital parameters (Talbot, 1985; Lancaster, 1990).

In this paper we investigate a number of ancient subtropical desert deposits formed on the megacontinent Pangaea, and test the possibility that these deposits contain sedimentary cycles of climatic origin in response to orbital forcing.

§ Present address: Rogaland Research, P.O. Box 2503 Ullandhaug, N-4004 Stavanger, Norway.

THEORY

Climate is becoming increasingly accepted as an important feature of cyclic stratigraphic sequences in lacustrine and alluvial settings (e.g. Olsen, 1986; Olsen, 1990). So far, however, few papers have discussed the origin of ancient aeolian or partly aeolian deposits in the light of climatic change (e.g. Clemmensen *et al.*, 1989; Clemmensen & Tirsgaard, 1990; Clemmensen & Hegner, 1991; Yang & Baumfalk, pp. 47–61, this volume).

Table 1. Climatic characteristics of modern cyclostratigraphic belts. After Perlmutter & Matthews (1989)

Cyclostratigraphic belt	Climatic maximum	Climatic minimum
8	Polar, humid–subhumid–dry	Polar, dry–arid
7B	Temperate, humid	Polar, dry
7A	Temperate, subhumid	Polar, subhumid
6	Temperate, dry	Polar, humid
5	Temperate, arid	Temperate, humid
4B	Tropical, dry	Temperate, subhumid
4A	Tropical, subhumid	Temperate, dry
3	Tropical, humid	Temperate, arid
2B	Tropical, humid	Tropical, dry
2A	Tropical, very humid	Tropical, subhumid
1	Tropical, very humid	Tropical, very humid

In a recent paper, Perlmutter & Matthews (1989) established a model for cyclostratigraphic analyses of ancient clastic sediments with emphasis on continental depositional systems. This cyclostratigraphic model integrates Milankovitch-induced, short-term climatic change (20–400 ka) with the long-term tectonic evolution (millions of years) of basins in order to interpret and predict the stratigraphic record.

In their model, principal controls on short-term terrestrial sedimentation in a particular basin are: (i) the palaeolatitude of the basin which determines in which cyclostratigraphic belt it lies; and (ii) the latitudinal shifts of climatic zones with time, caused by Milankovitch oscillations. From the palaeolatitude of a given basin, in combination with the land–sea distribution and topography of the Earth's surface, the cyclostratigraphic belt and expected climatic change during a Milankovitch cycle can be estimated.

Ten modern latitudinal cyclostratigraphic belts are defined by Perlmutter & Matthews (1989; p. 459, this volume), each belt exhibiting well-defined climatic and environmental end-members during a complete Milankovitch cycle (Table 1). Cyclostratigraphic belt 3 (15–25° N), for example, experiences a climatic change from tropical humid to subtropical (or temperate) arid and back during a Milankovitch cycle. This is a pronounced climatic change that influences weathering, erosion, sediment transport and the character of the depositional environment. A number of idealized sedimentary cycles for continental basins in cyclostratigraphic belt 3 can accordingly be defined, their exact nature depending on tectonic and palaeogeographic conditions. These cycles would probably be composed of alternating aeolian and fluvial, or aeolian and lacustrine facies or of stacked aeolian units separated by well-developed deflation or stabilization surfaces.

The distribution of cyclostratigraphic belts in the geological past when continent distribution, sea levels and orography were all different is known only in limited detail. Kutzbach & Gallimore (1989) attempted to reconstruct Late Permian and Triassic climates on the Pangaean megacontinent. Based on low-resolution atmospheric circulation models, their results indicate that the equatorial zones and the continental interior of the megacontinent had strong monsoonal circulation. In spite of this, the interior of the continent (up to *c.*40°) was characterized by year-round or seasonal aridity, and light rainfall occurred only near 10° N in connection with the summer monsoons. Kutzbach & Gallimore (1989) state that it is likely that orbital variations modulated the Pangaean monsoons. It is therefore likely that in some periods monsoonal rainfall was more intense and widespread, and that climatic contrasts were better developed on the megacontinent. If this is correct Pangaea should be divisible into a number of cyclostratigraphic belts.

The formations studied all occur in basins situated at low palaeolatitudes. Judging from palaeolatitude alone it is inferred that these basins were located in cyclostratigraphic belts 2 or 3 (Table 2). Direct comparison with modern cyclostratigraphic belts 2 and 3 would therefore indicate that all basins went through repeated arid–humid (belt 3), dry–humid (belt 2B), or subhumid–very humid (belt 2A) climatic fluctuations (Table 2). The concept of global cyclostratigraphy therefore provides a technique for predicting the climate of a specific basin. These predictions can then be compared to the climatic conditions deduced from the stratigraphy and facies

Table 2. Late Palaeozoic and Mesozoic formations studied, and expected climatic characteristics

Formation	Age	Palaeolatitude	Assumed cyclostratigraphic belt*	Predicted climatic change	Sedimentary cycles	Deduced climatic change
Corrie Sandstone	Early Permian	*c*.13° N†	2B	Dry–humid	Ae–(F)	Arid–humid
Dawlish Sands	Late Permian	*c*.8° N†	2A	Subhumid–very humid	Ae–F	Arid–humid
Exmouth Sandstone and Mudstone Fm	Late Permian	*c*.11° N‡	2B	Dry–humid	Ae–F + L	Arid–humid
Helsby Sandstone Fm	Early Triassic (Scythian)	*c*.20° N‡	3	Arid–humid	Ae–F	Arid–humid
Middle Buntsandstein	Early Triassic (Scythian)	*c*.20° N*	3	Arid–humid	Ae–L	Arid–humid
Wingate Sandstone	Early Jurassic	*c*.17° N§	3	Arid–humid	Ae–(F + L)	Arid–humid

* After Perlmutter & Matthews (1989).
† From Glennie (1984).
‡ From Smith *et al.* (1981).
§ From Parrish & Peterson (1988).
Ae, aeolian facies; F, fluvial facies; L, lacustrine facies.

characteristics of the basin deposits (see Matthews & Perlmutter, pp. 459–481, this volume).

The expected thickness of the sedimentary cycles, which consist of sediments and hiatuses formed during alternating more arid and more humid climate conditions, depends on the duration of the Milankovitch cycle and the accumulation rate in the basin. Aeolian *accumulation rate* reflects the net deposition through time (thousands of years to millions of years), and is a function of (i) sand availability and wind regime (controlling the sand budget in the basin), and (ii) basin subsidence and groundwater level (controlling the preservation potential of the aeolian accumulation) (see Kocurek & Havholm, 1993). Aeolian *sedimentation rate* reflects the deposition of sand on dunes or on sand sheets in shorter periods. It is a function of the sand-saturation level and transport capacity of the airflow. Accumulation rate is used here as the basic parameter; sedimentation rate is less useful because it is strongly dependent on the time scale in question (hours, days, years, etc.).

Assuming an average accumulation rate of 1 m/ka, a composite pattern of sedimentary cycles with periods between a few metres and a few hundred metres in thickness would originate. Thus, in a subtropical basin having an average accumulation rate of 1 m/ka, the present-day orbital constants would ideally produce a composite pattern of sedimentary cycles, reflecting periodic changes in aridity/humidity, with frequencies of, on average, 19, 23, 95, 125 and/or 413 m, reflecting the main astronomical periods associated with precession and eccentricity (see Berger, 1988). The obliquity cycle (41, 54 ka) is largely of influence at high latitudes, while at low latitudes its influence might be felt only during periods of ice caps through the coupling of ice volume, global sea level and low-latitude climates. Of course the different orbital cycles interfere, and, in the hypothetical case of a net accumulation rate of 1 m/ka, we may see precession cycles, averaging about 20 m in thickness, grouped into larger cycles of about 100 and 400 m. Only in the case of long and detailed data sets such as are available for example from Pleistocene deep-marine deposits, may statistical techniques allow the recognition of the main 19- and 23-ka precession cycles and of the 95- and 123-ka eccentricity cycles. The shortening of the Earth–Moon distance back in time induces a shortening of the fundamental astronomical periods for the obliquity and the precession (Berger *et al.*, 1989). Thus, in the Early Permian, the predominant precession cycles had periodicities of 17.6 and 21 ka, the obliquity cycles had periodicities of 35.1 and 44.3 ka, while the lengths of the eccentricity cycles were largely unchanged (Berger *et al.*, 1989). As a result, in a Permian low-latitude basin affected by orbitally induced changes of climate, and with again a hypothetical average accumulation rate of 1 m/ka, one would expect a sequence of sedimentary cycles consisting of aeolian or mixed aeolian and aqueous deposits and possibly hiatuses, formed during cyclic arid–humid climate fluctuations, with thicknesses of, on average, 19.5, 40, 100 and/or 413 m in the ideal case. Time control will very rarely be sufficient to detect these frequency changes easily, but because the eccentricity frequencies are constant through time and the precession frequencies increase back in time, it may well be possible to recognize changing interference patterns (bundling) between the precession and eccentricity cycles.

EXAMPLES

The examples discussed (Fig. 1, Table 2) are the Lower Permian Corrie Sandstone, UK (Clemmensen & Abrahamsen, 1983; Clemmensen & Hegner, 1991), the Lower Permian Dawlish Sands, UK (Laming, 1966; Øxnevad, 1991), the Upper Permian Exmouth Sandstone and Mudstone Formation, UK (Mader & Laming, 1985; Øxnevad, 1991), the Lower Triassic Helsby Sandstone Formation, UK (Thompson, 1970; Øxnevad, 1991), the Lower Triassic Middle Buntsandstein, Germany (Clemmensen, 1991), and the Lower Jurassic Wingate Sandstone, USA (Clemmensen & Blakey, 1989). These aeolian or mixed aeolian and aqueous formations with intercalated hiatuses apparently formed during repeated arid–humid climate fluctuations. Here we intend to discuss further the climatic aspects of these ancient desert deposits, and in particular to test the possibility that the dynamics of these deposits were controlled by astronomically induced climate fluctuations.

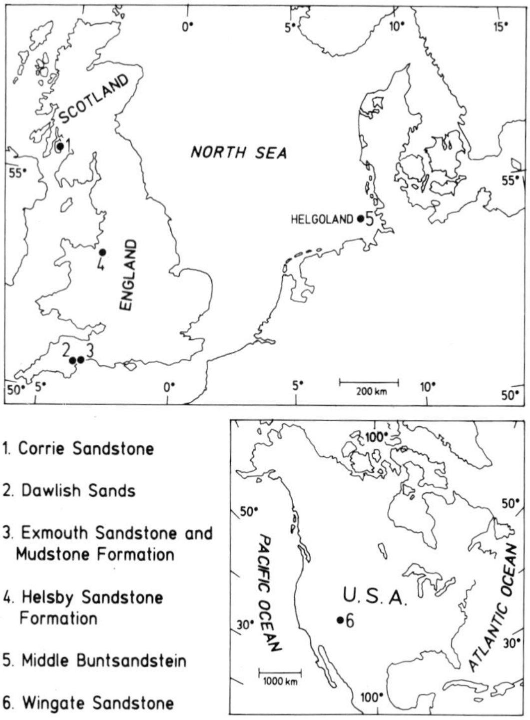

Fig. 1. Location of the late Palaeozoic and Mesozoic formations studied.

Corrie Sandstone (Permian, Scotland)

The Corrie Sandstone (aeolian) and the closely associated Brodick Breccias (fluvial and alluvial fan) form the basal part of the New Red Sandstone succession on the island of Arran in southwestern Scotland. The age of the formation is supposedly Early Permian ('Rotliegend'), but no age-diagnostic fossils have been found in the sandstones. The Corrie Sandstone crops out in the present foreshore in two areas on northeastern Arran (Clemmensen & Abrahamsen, 1983). The thickness of the formation is at least 700 m (Fig. 2). The aeolian and associated alluvial fan deposits most likely formed in a pull-apart basin having a high subsidence rate. The Corrie Sandstone seems to represent a number of relatively small crescentic draas or crescentic dunes. These bedforms were transverse to the resultant sand drift from the NE, but there is evidence to suggest seasonal wind shifts from E to NE, and SW. The aeolian bedforms apparently formed part of a relatively small erg that was flanked at the northwestern side by well-developed alluvial fans. The intermontane basin was situated at *c.*13° N palaeolatitude (Glennie, 1984; Clemmensen & Hegner, 1991; Table 2).

Considering the palaeolatitude only, the basin was probably situated in an Early Permian cyclostratigraphic belt 2B (Tables 1 and 2), characterized by seasonal climate changes, as well as Milankovitch-type climate oscillations between tropical dry and tropical humid. Due to the configuration of Pangaea, the overall climate of the Permian belt 2 may, however, have been significantly more arid than the modern cyclostratigraphic belt 2 (see Kutzbach & Gallimore, 1989).

Description of the sedimentary succession

The Corrie Sandstone is composed of 34 basic aeolian units. Each unit is bounded below and above by deflation surfaces. The basal parts of the aeolian units are characterized by flat-bedded aeolian strata and the upper parts consist of cross-bedded aeolian deposits (Figs 2 and 3). The basic units (termed a, b, c, d, etc. in Fig. 2) have an average thickness of

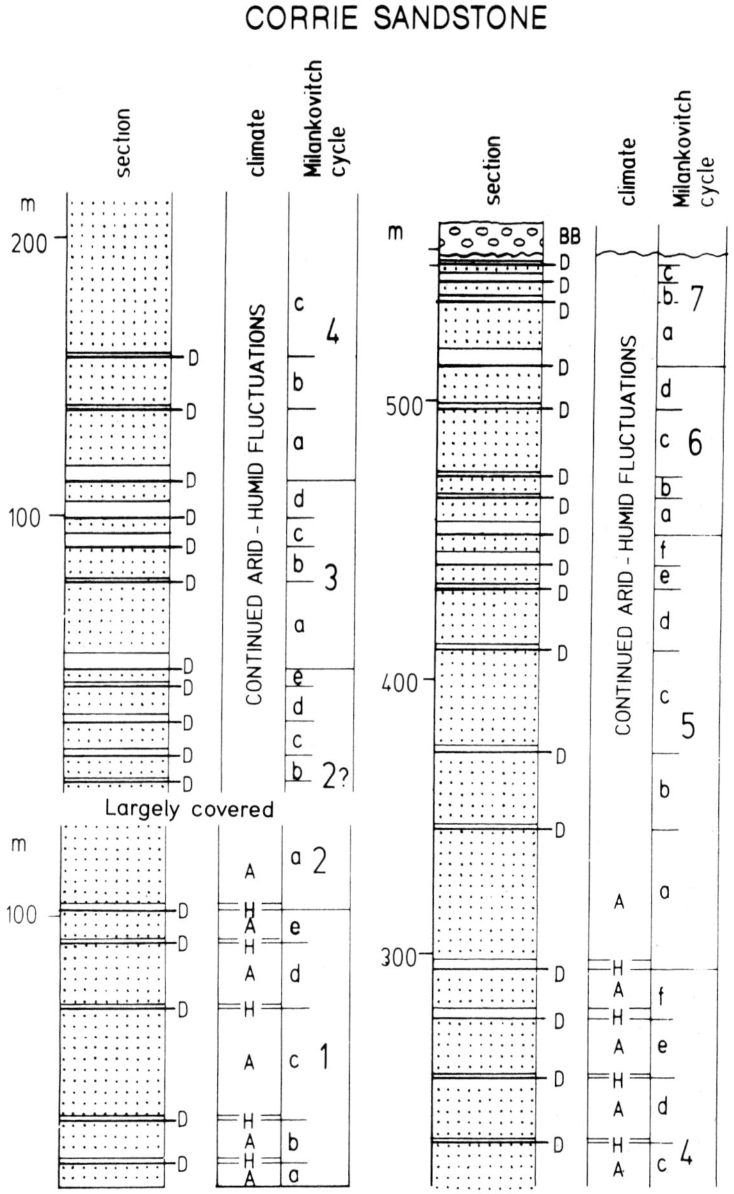

Fig. 2. Simplified sedimentary section of the Corrie Sandstone (Permian), Arran, Scotland. For explanation of the symbols see Fig. 3. The stratigraphic thickness of the covered part of the section is difficult to estimate, due to faulting, but may represent little additional stratigraphic thickness. Section based on Clemmensen & Hegner (1991).

22 m, but individual units vary in thickness from 3 to 70 m. The composition of these basic units thus varies considerably. Thin units (less than 10 m thick) are composed of basal flat-bedded aeolian strata and overlying aeolian strata of nearly equal thickness. Thick units (more than 30 m) are dominated by cross-bedded aeolian strata, and contain thin flat-bedded aeolian strata only at the base. Based on vertical variations in the composition and thickness of the basic aeolian units it seems possible to define one or perhaps two orders of composite (mega) units. The thin composite units (termed 1, 2, 3, etc. in Fig. 2) have an average thickness of 110 m, and unit thickness varies between 62 and 184 m. It should be noted, however, that the thickness of composite unit 2 is uncertain. Finally the succession may contain two, partly preserved, thick composite units up to 440 m thick (see also Clemmensen & Hegner, 1991).

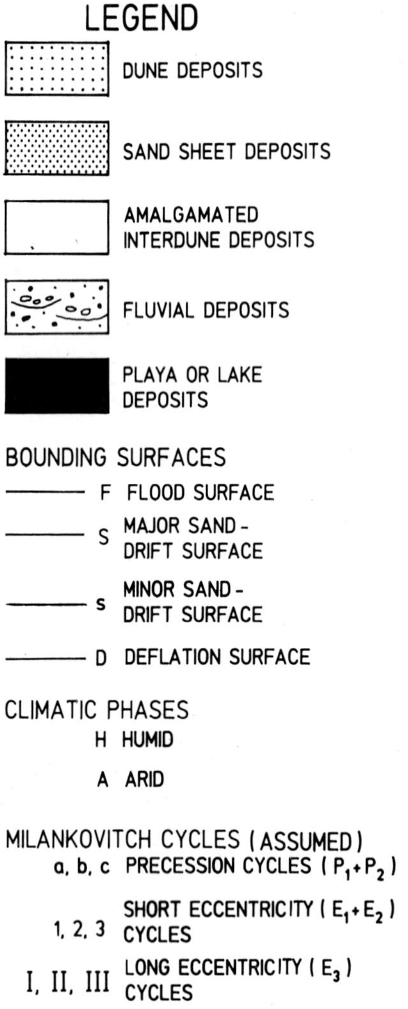

Fig. 3. Legend for Fig. 2 and Figs 4–8.

The ratio between the average thickness of the units is 1:5:(20).

Interpretation of sedimentary succession

The stacking of aeolian units with consistent internal composition suggests a cyclic development of the Corrie Sandstone. The basic aeolian cycles and the associated deflation surfaces reflect alternating periods of erg construction and erg destruction (Fig. 2). Because of the inland position of the basin, erg dynamics were most likely related to climatic variation controlling precipitation, sediment yield to the basin and the nature of aeolian processes in the erg. Erg construction, reflecting a positive sand budget, most probably occurred during arid climatic intervals characterized by a large amount of dry sand available for dune building. Erg destruction, reflecting a negative sand budget and the formation of a deflation surface, probably took place during continued arid or arid–humid transitional climates characterized by a limited amount of dry sand being available for dune building. Humid and transitional climatic humid–arid intervals characterized by intermediate sand supply are probably represented by amalgamated interdune deposits. In the erg margin the humid intervals are represented by alluvial fan sandstones and conglomerates and the arid intervals by aeolian deposits (see Clemmensen & Abrahamsen, 1983). The two orders of composite units suggest that climatic fluctuations also took place on a longer time scale.

The hindcast of climate based on the principles of global cyclostratigraphy indicates alternating periods of arid and humid climate linked to orbital variations (Table 2). The cyclic nature of the Corrie Sandstone and the character of the units are consistent with a system of astronomically induced climatic control on sedimentation (see Clemmensen & Hegner, 1991). Thus the basic sedimentary cycles would represent 20-ka precession (P1 + P2) cycles, the thin composite cycles 100-ka eccentricity (E1 + E2) cycles and the highest order cycles 400-ka eccentricity cycles (E3) (Fig. 2). The basic climatic fluctuations between arid and humid probably record astronomical cycles causing variations in the intensity of monsoonal circulation (Kutzbach & Otto-Bliesner, 1982) and the position of the caloric equator (see de Boer & Smith, pp. 15–24, this volume). These basic climatic fluctuations apparently were modulated by longer term climatic fluctuations that may have been amplified by the waxing and waning of Gondwana ice caps (see Crowley, 1983; Glennie, 1983, 1984).

Implications

The hypothesis of orbital control on the stratigraphy of the Corrie Sandstone allows an estimate of ancient aeolian accumulation rate. The deposition of the Corrie Sandstone thus would have been relatively rapid, and would have covered seven 100-ka eccentricity cycles, i.e. 700 ka. Of course, this interpretation assumes that no major hiatus occurs within the Corrie Sandstone. The related aeolian accumulation

Table 3. Sedimentary cycles, inferred Milankovitch cycle and deduced aeolian accumulation rate

Formation	Cycle thickness (m)	Inferred Milankovitch cycle	Deduced average aeolian accumulation rate* (m/ka)
Corrie Sandstone	22 110 (440)	P1 + P2 E1 + E2 (E3)	2.2
Dawlish Sands	10	P1 + P2	Max. 0.7
Exmouth Sandstone and Mudstone	1.8 9	P1 + P2 E1 + E2	Max. 0.2
Helsby Sandstone	13	P1 + P2	Max. 0.9
Middle Buntsandstein	2.0 11.5 (35.5)	P1 + P2 E1 + E2 (E3)	Max. 0.1
Wingate Sandstone	20	E1 + E2 or E3	0.4 or 0.1

* Aeolian accumulation is considered to take place only during the arid half of each Milankovitch cycle.
P1 + P2, precession cycles; E1 + E2, short eccentricity cycles; E3, long eccentricity cycle.

Table 4. Modern sedimentation rates

Jafurah sand sea, Saudi Arabia	Alexandria coastal dune field, South Africa	Coastal dune field, Vejers, Denmark
0.2 m/ka*	1.5 m/ka†	Min. 5.5 m/ka‡

* Fryberger *et al.* (1984).
† Illenberger & Rust (1988).
‡ Clemmensen, unpublished data.

rate must have been relatively high. Assuming that aeolian sedimentation and accumulation occurred (intermittently) only during the dry half of a Milankovitch cycle (*c.*10 ka in the case of the precession cycle), we arrive at an average accumulation rate of 2.2 m/ka (Table 3). Most probably the cross-bedded aeolian sand (dune or draa deposits), and the flat-bedded aeolian sand (amalgamated interdune or interdraa deposits) record different accumulation rates. In this way the thick basic cycles dominated by cross-bedded sand, and the thin basic cycles with a larger proportion of flat-bedded aeolian sand, may actually represent equal amounts of depositional time. The average accumulation rate of the cycles in the Corrie Sandstone is high and comparable to the highest modern aeolian accumulation rates recorded from high-energy coastal dune systems (Table 4). Nevertheless, it should be stressed that aeolian sedimentation (and accumulation) was not continuous, but was punctuated by periods of non-deposition or erosion. Thus the deduced ancient accumulation rates fall within a range that is very reasonable when compared to modern values. Obviously this is a necessary, but not in itself sufficient requirement for acceptance of orbital forcing.

Preservation of the aeolian cycles in the Corrie Sandstone is thought primarily to record a high and steady subsidence rate, combined with intermittent rises of the groundwater level during the humid climatic intervals.

Dawlish Sands

The Dawlish Sands form the upper part of the Lower Permian New Red Sandstone section in South Devon, southwest England. The age of the formation is early Late Permian. This age is based on the stratigraphic subdivision of the British New Red Sandstone sequence, which is primarily based on lithology, due to the low content of age-diagnostic fossils.

The Dawlish Sands have a maximum thickness of 360 m. The formation is composed of alternating red sandstones and breccias and is interpreted as being of mixed aeolian and fluvial origin. These sediments accumulated in the Dorset Basin, a relatively small intermontane basin within the Variscan mountain belt, formed through crustal extension and possibly additional lithospheric thinning (Chadwick, 1986). It was actively subsiding throughout the Permian with maximum rates of subsidence during the Early Permian and declining during the Late Permian.

The Dawlish Sands accumulated basinward of a gradually retreating alluvial fan complex situated along the western margin of the Dorset Basin (Laming, 1966). During periods of high sediment transfer through the alluvial fans towards the basin, the Dawlish Formation fluvial sands and breccias were deposited. The intervening aeolian sandstones were deposited in low-relief sand sheets and as barchanoid dunes, the latter migrating towards the northwest in response to southeasterly winds. A major source for the aeolian sands was probably local alluvium.

The palaeolatitudinal position of the Dorset Basin was approximately 8° N, i.e. in a Permian cyclostratigraphic belt 2A (Tables 1 and 2). In this latitudinal belt, Milankovitch-type climatic oscillations between tropical humid and tropical subhumid (Tables 1 and 2) are expected, controlling the sediment input, distribution and deposition on time scales corresponding to the length of the orbital cycles. However, given a generally dry Pangaean interior, combined with the fact that the Dorset Basin was located in the rain shadow of the Variscan mountains with respect to southeasterly winds, the climatic conditions may have been considerable drier than in the modern belt 2A.

Description of sedimentary succession

In coastal cliffs northeast of Dawlish on the coast of South Devon, the uppermost part of the Dawlish Formation is exposed (Fig. 4). The exposure reveals a 36-m-thick section. It shows a pattern of pale red aeolian sandstones alternating with dark red fluvial sandstones and breccias. The aeolian and the fluvial deposits are separated by surfaces that were identified as sand-drift and flood-type surfaces (see Clemmensen & Tirsgaard, 1990). In the section studied three or four basic fluvial–aeolian units are recognized. The units range in thickness from 7 to 11 m and they are separated by flood surfaces (Fig. 4).

Interpretation of sedimentary succession

The pattern of alternating fluvial and aeolian deposits suggests a cyclic development of the Dawlish Sands in this area. The fluvial–aeolian cycles are suggested to represent climatic oscillations with pronounced variations in humidity through each cycle (Fig. 4). During humid periods, sediments were transferred through the alluvial fans by a combination of sheet flows and stream flows towards the central part of the Dorset Basin. During more arid

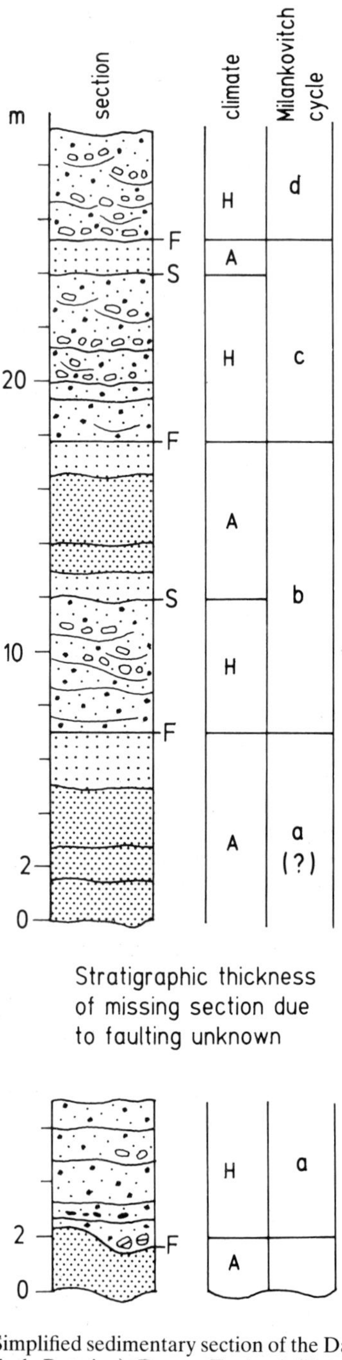

Fig. 4. Simplified sedimentary section of the Dawlish Sands (Early Permian), Devon, England. The aeolian deposits are 1.4 to ≥7 m thick. For legend see Fig. 3.

and windy periods, the local alluvium was winnowed and redeposited in sand sheets and dunes, whereas very little sediment was supplied from the basin margins. This estimate of climatic conditions is consistent with the above hindcast of climate based on the principles of global cyclostratigraphy.

The simple character of the Dawlish Sands depositional cycles prevents a straightforward correlation with specific orbital cycles. However, given that average sediment accumulation rates within the range of 0.1–1.0 m/ka are applicable to the Dorset Basin, we suggest that the climatic oscillations were controlled by the 20-ka precession (P1 + P2) cycle. Cycle thicknesses of 7–11 m thus imply sediment accumulation rates ranging from 0.35 to 0.55 m/ka. Alternatively, the Dawlish Sands depositional cycles may represent eccentricity cycles. This would give sediment accumulation rates of 0.07–0.11 m/ka. Although these numbers may be compatible with the basin subsidence rate proposed by Chadwick (1986), they are low for a tectonically active basin, and we suggest that subsidence rates may have been higher, at least during deposition of the Dawlish Sands. Moreover, the presence of only eccentricity-induced cycles would not logically fit to climatic changes in low latitudes, but rather to high-latitude (ice-cap) effects (see Oerlemans, 1980). In low latitudes, the effect of the cycle of precession is indeed of major influence and is superimposed upon the (100- and 400-ka) eccentricity cycle. In the exceptional case that the effect of the precessional cycle would surpass, by repetition, a certain threshold only during eccentricity maxima every 100 or 400 ka, precessional effects might be obscured in low-latitude climate systems, thus leaving the eccentricity effect as the only one to be recognized. It is unlikely, however, that equilibria allowing such a suppression of the precessional signal would continue long enough to allow the recording of several successive eccentricity maxima on their own.

Implications

Following the above interpretation, the uppermost part of the Dawlish Sands would have been deposited during three or four orbital cycles. The cycles lasted 20 ka (precession, P1 + P2) or (less likely) 100 ka (eccentricity, E1 + E2), implying that the uppermost part of the Dawlish Formation accumulated within less than 80 ka (or 400 ka). This may beg the question of total depositional time for the Dawlish Formation. However, the uppermost part of the Dawlish Sands exposed near Dawlish is not necessarily representative for the entire formation in terms of mode of deposition. Therefore these estimates of depositional time may not be extrapolated beyond the limits of the section studied.

Aeolian accumulation rates can be deduced on the basis of the type of orbital cycles recorded and the preserved thickness of the aeolian deposits. If the cycles indeed cover 20 ka, this implies aeolian accumulation rates of 0.14 to $\geqslant$0.7 m/ka. This is compatible with values recorded in modern aeolian environments (Table 4). If the cycles were 100 ka each, this would give aeolian accumulation rates of 0.03 to $\geqslant$0.14 m/ka. This number is close to values recorded in modern sand sheets in central Sahara (Øxnevad, 1991). However, the Saharan sand sheets differ from those of the Dawlish Formation in that source material is scarce. Moreover, they rest upon stable cratonic basement, and practically no accommodation space is created by subsidence. The higher values for aeolian accumulation are reasonable in a tectonically active setting, such as the Dorset Basin. Therefore, we suggest that the Dawlish Sands record deposition during three or four precession cycles.

Exmouth Sandstone and Mudstone Formation

The Exmouth Formation represents the lower Upper Permian of the New Red Sandstone in the Dorset Basin (South Devon). The formation is approximately 255 m thick and consists of red soft sandstones and interbedded dark red mudstones. These are largely of fluvial origin (Mader & Laming, 1985), but they also include some aeolian deposits.

The Exmouth Formation sediments accumulated in the intermontane Dorset Basin. During the early Late Permian fault activity in the Dorset Basin was low, and basin subsidence was largely controlled by thermal relaxation (Chadwick, 1986). Alluvial fans, which were active along the basin margins during the early Permian, had retreated and sandy braidplains, muddy floodplains and/or playa lakes were established practically throughout the basin.

During the final stages of Exmouth Formation sedimentation, aeolian deposition replaced the deposition of fluvial and playa sediments (Øxnevad, 1991). The aeolian sands accumulated in barchanoid dunes and low-relief sand sheets, forming three distinct depositional units that are separated by deflation surfaces. Each surface marks the termination of an aeolian accumulative event, or more specifically the transition from deposition to deflation. Deposition of the Exmouth Formation aeolian strata was finally terminated by a transition to the accumulation of mud in a vast playa setting (see Holm *et al.*, 1986; Øxnevad, 1991).

In the Late Permian, the Dorset Basin was located at 11°N (Table 1). This palaeolatitudinal position corresponds to a location within the Permian cyclostratigraphic belt 2B (see Perlmutter & Matthews, 1989). Dry–humid orbitally forced climatic oscillations (Tables 1 and 2) are predicted in this climatic belt (see Perlmutter & Matthews, 1989), affecting sediment input and distribution within sedimentary basins.

Description of sedimentary succession

The uppermost 9 m of the Exmouth Sandstone Formation is exposed in coastal cliffs along the beach near Exmouth in South Devon (Fig. 5). The cliff section displays a vertical section upwards from dark red mudstones and white sandstones, through red silty sandstones, to red, fine- to medium-grained sandstones. This reflects a transition from muddy playas and sandy ephemeral streams through mixed aeolian–aqueous sandflat to aeolian barchanoid dunes and sand sheets. At the top of the cliffs, dark red mudstones with interbedded white sandstones mark the termination of the Exmouth Formation and a transition to the overlying formation.

Several depositional units have been identified, based on major bounding surfaces and vertical changes in lithology. In the lower part of the section, which is composed of waterlain mudstone and sandstone and minor aeolian strata, the exact temporal relationship between muddy playa, sandy ephemeral stream and wind-ripple deposits is not known. This is primarily due to a lack of lateral control of facies development. However, it is possible that the observed changes between the playa, ephemeral stream and aeolian deposits record true changes in style of deposition through time. They would thus form two depositional units: a lower unit composed of muddy playa to sandy ephemeral-stream deposits and a second unit consisting of muddy playa to wind-ripple sand-sheet deposits.

The remaining part of the section contains three genetically distinct aeolian deposits: a lower barchanoid dune deposit and two succeeding sand-sheet deposits. The dune deposit is composed of tabular and wedge-shaped cross-strata sets. These directly overlie the underlying sandflat deposits, separated by a sand-drift surface (see Clemmensen & Tirsgaard,

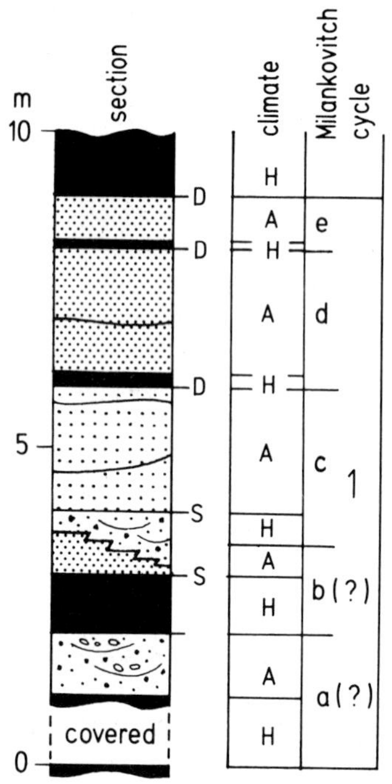

Fig. 5. Simplified sedimentary section of the Exmouth Sandstone Formation (Late Permian), Devon, England. For legend see Fig. 3.

1990). The sand-sheet deposits are composed primarily of low-angle wind-ripple strata. Each aeolian deposit is bounded above by a surface that is laterally extensive, non-climbing and broadly undulating with a maximum relief of 1.5 m. The surface relief is partly filled with dark red mudstones with thinly interbedded aeolian sands, reflecting deposition of mud in ephemeral lakes, muddy playas with temporary incursions of wind-blown sand. These lakes/playas were established in shallow surface depressions, and were fed by ephemeral streams, probably in combination with groundwater. The surfaces themselves appear to have been formed by aeolian deflation and were thereafter stabilized, prior to the establishment of lakes/playas and surface drainage systems. The aeolian deposits and the intervening deflation surfaces define three depositional units, with each unit showing a transition from aqueous to aeolian sedimentation.

Thus, in total, the uppermost part of the Exmouth Formation seems to contain five sedimentary units. The units show temporal variations in style of deposition within an overall trend towards increasing aeolian sediment accumulation. The units vary in thickness laterally, and may eventually pinch out completely. This is largely due to the gentle undulations along the top of the aeolian units. The average unit thickness is 1.9 m, with a maximum of 3.2 m.

Interpretation of sedimentary succession

The pattern of alternating aeolian and aqueous deposition suggests a cyclic nature for the uppermost part of the Exmouth Formation (Fig. 5). The sedimentary development recorded is suggested to represent a long-term climatic trend from humid to dry conditions, on which are superimposed shorter term climatic oscillations. This trend was finally terminated by a return to more permanently wet conditions, as is indicated by the playa mudstones of the succeeding formation (Øxnevad, 1991). We tentatively suggest that the shifts between muddy and sandy aqueous sedimentation reflect changing climatic conditions, which controlled the type of weathering processes taking place in the source area. Humid climatic conditions would favour chemical weathering and thus production of muddy sediments which would be transferred basinward and eventually accumulate in basinal lows. In contrast, a dry climate would favour physical weathering and thus the production of sands that were transported to the basin by ephemeral streams. A similar climatic signal is thought to be recorded by the aqueous–aeolian depositional cycles. During humid climatic phases, chemical weathering produced muddy sediments that were supplied to basinal areas. Here they accumulated in playas/ephemeral lakes in basinal lows that were fed by a combination of surface flow and subsurface groundwater. Apparently, the water table fluctuated in time, causing episodic desiccation of the muddy lake bed/playa surface upon which thin layers of wind-blown sand were deposited. During arid and windy phases, aeolian sands were deposited, forming barchanoid dunes and low-relief sand sheets. The aeolian sands were probably sourced from local sand-rich alluvium deposited during preceding humid phases. The actual onset of climatic change from dry to humid conditions is taken at the deflation surfaces. They represent the termination of the dry-period aeolian accumulation and thus the boundary between the individual humid–arid cycles. The transition from accumulation to deflation probably results from depletion of source material, possibly combined with changes in wind regime at the transition from dry to humid climate.

The inferred climatic fluctuations deduced from the nature of the sedimentary cycles correspond well to arid–humid climatic changes, which are predicted to have taken place during a Milankovitch cycle in a sedimentary basin located within a cyclostratigraphic belt 2B (Tables 1 and 2). Given that the uppermost deposits of the Exmouth Formation reflect five humid–dry climatic cycles, superimposed on a longer-term humid to dry trend, we suggest that the section represents vertically stacked 20-ka precession (P1 + P2) cycles deposited within one 100-ka eccentricity (E1 + E2) cycle.

Implications

Based on the above interpretations the average sediment accumulation rate for the uppermost Exmouth Formation would have been 0.09 m/ka. If this value is used as an estimator for the duration of the entire Exmouth Sandstone Formation (thickness about 250 m), the formation would have accumulated within 2.5–3.0 Ma, which is not unreasonable.

The preserved thickness of the aeolian sandstone units averages 1.8 m, with 3.2 m as the maximum value. Based on the duration of a precession cycle (20 ka), and again assuming deposition during only half of the precession cycle, this implies an average aeolian accumulation rate of 0.18 m/ka and a maximum value of 0.32 m/ka. These numbers are reasonable if compared to values recorded in modern aeolian environments (Table 4).

Helsby Sandstone Formation

The Helsby Sandstone Formation forms part of the Lower Triassic New Red Sandstone in the Cheshire Basin of the English Midlands. The formation is 20–250 m thick, and comprises a group of interbedded red, soft sandstones and pebbly sandstones of mixed fluvial–aeolian origin.

The Cheshire Basin is a relatively small fault bounded structure, located approximately 20° N of the Triassic equator in the northern foreland of the Variscan mountain belt. Basin development was initiated in the Early Permian by lithospheric extension, and subsidence continued throughout the Permian and Triassic (Holloway, 1985; Chadwick, 1986). In the Early Triassic, the basin became connected with the New Red Sandstone basins of southern England, because of increased subsidence in basin margin areas. This resulted in significant fluvial input into the basin from the south. Deposition during the early Triassic was characterized by mixed fluvial and aeolian sedimentation. This continued until the Mid/Late Triassic when marine waters spread over the basin, marking the termination of continental New Red Sandstone deposition in the area (Øxnevad, 1991).

The sandstones of the Helsby Formation are divided into three main members: a lower member composed of aeolian and interbedded fluvial deposits, a middle member largely consisting of fluvial strata, and an upper member predominantly comprising aeolian deposits (see Thompson, 1970). The lateral and vertical distribution of these sediments within the Cheshire Basin indicates that fluvial input took place through three main river systems, draining into the basin from the south, whereas aeolian processes dominated in the adjacent areas.

The Cheshire Basin was located within cyclostratigraphic belt 3 during the Early Triassic, implying climatic fluctuations between temperate arid and tropical humid (Tables 1 and 2).

Description of sedimentary succession

The lower and middle parts of the Helsby Formation are exposed in red sandstone cliffs in the central part of the Cheshire Basin. The composite vertical section measures approximately 80 m (Fig. 6). It is composed of red soft sandstones and pebbly sandstones, forming alternating aeolian and fluvial units. These are bounded by flat, planar sand-drift surfaces and erosional flood surfaces (see Clemmensen & Tirsgaard, 1990).

The fluvial sandstone units range in thickness from 4 to 10 m, with an average of 7.5 m (Fig. 6). The sandstones are dark red or brownish in colour, pebbly and medium-grained to very coarse-grained. They are massive or trough cross-bedded, with cross-strata dip directions indicating fluvial transport towards the north. The aeolian units are 4–9 m thick, with an average thickness of 6.5 m. The aeolian sandstones are red, soft and fine- to medium-grained. Cross-strata sets are composed of sandflow cross-stratification and wind-ripple laminations, and suggest deposition in westward-migrating, transverse, sinuous-crested dunes. Bounding surfaces and associated horizontal deposits separating the dune cross-sets indicate wet or damp interdune areas, although dry deflationary interdunes probably also occurred. Relatively wet conditions were probably maintained by a shallow groundwater table and/or intermittent surface water flow.

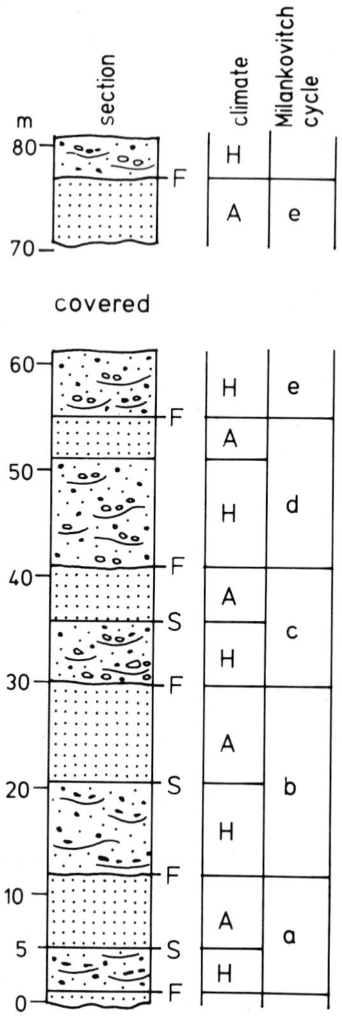

Fig. 6. Simplified sedimentary section of the Helsby Sandstone Formation (Early Triassic), Cheshire, England.

Interpretation of the sedimentary succession

Based on the vertical organization of the fluvial and aeolian sandstone units, we suggest that they represent five vertically stacked depositional cycles. Each cycle comprises a fluvial and an overlying aeolian unit, and flood surfaces mark the cycle boundaries. Cycle thicknesses range from 11 to 22 m, with an average of 13 m. The cycles represent temporal variations in depositional conditions in the central part of the Cheshire Basin near one of the main points of fluvial access to the basin. Similar fluvial–aeolian depositional cycles have not been observed in the areas beyond the main river courses. In these areas, equivalent parts of the Helsby Formation consist predominantly of aeolian sandstones. Whether depositional cycles corresponding to those in the central part of the basin are recorded in these aeolian strata is not known, because of the limited extent of the outcrops.

The Helsby Formation fluvial–aeolian depositional cycles recorded in the central part of the Cheshire Basin are here suggested to represent arid–humid climatic oscillations (Fig. 6). Such oscillations would affect fluvial flow and sediment supply into the basin, as well as depositional processes within the basin. The fluvial pebbly sandstones are most probably related to periods of significant fluvial flow within the main river systems, being controlled by rainfall in southern source areas and by local precipitation. Termination of fluvial activity is marked by the sand-drift surfaces, suggesting the onset of more arid conditions (see Clemmensen & Tirsgaard, 1990). These surfaces suggest exposure of the local alluvium to aeolian processes, implying winnowing and redeposition by easterly winds into low-relief sand sheets and westward-migrating transverse dunes.

Thus, the inferred Helsby Formation arid–humid climatic fluctuations are consistent with the predicted pattern of climatic changes taking place throughout a Milankovitch cycle within cyclostratigraphic belt 3.

Implications

Given an average cycle thickness of 13 m, we tentatively suggest a record of vertically stacked precession cycles (P1 + P2). An average thickness of 6.5 m for aeolian units implies average aeolian accumulation rates of 0.65 m/ka (maximum 0.9 m/ka). These values are reasonable for a tectonically active basin. The implication of this interpretation is that the section of the Helsby Formation studied was deposited during five precession cycles, i.e. within a period of 100 ka.

The above aeolian accumulation rates are reasonable if compared to values derived from modern aeolian environments (Table 4).

Middle Buntsandstein

Lower Triassic strata of the Middle Buntsandstein (Bunter Sandstone Formation) are exposed on the small island of Helgoland in the German North Sea. The exposed section corresponds to the uppermost Volpriehausen, Detfurth, Hardegsen and Solling Folgen (Binot & Röhling, 1988). The Triassic section on Helgoland is correlated, on the basis of sparse vertebrate fossils and gamma-ray measurements, with the Middle Bunter of the northwest German Basin (Binot & Röhling, 1988).

Helgoland is situated in the geographic centre of the northwest German Basin. Maximum subsidence occurred in grabens north of Helgoland (Binot & Röhling, 1988).

The Middle Buntsandstein is thought to represent a large desert-lake system (Clemmensen, 1979, 1991). The lake dried up periodically and the exposed lake bottom was covered by aeolian sand-sheet deposits.

The desert basin was situated at *c.*20° N, in Early Triassic cyclostratigraphic belt 3 (Tables 1 and 2), thus implying alternating temperate arid and tropical humid climatic fluctuations.

Description of sedimentary succession

The most conspicuous sedimentary units in the uppermost Volpriehausen–Folgen and lowermost Detfurth–Folgen are composed of well-developed sand-drift surfaces, lowermost aeolian sand-sheet deposits (up to 1 m thick) and uppermost lacustrine red beds (Fig. 7). Cycle thickness is *c.*11.5 m. Commonly these composite units can be divided into five to six basic sedimentary units, each about 2 m thick and initiated by a minor sand-drift surface and thin or poorly preserved aeolian deposits, and overlain by lacustrine deposits (Fig. 7). This subdivision is also detectable on gamma-ray logs (see Binot & Röhling, 1988). In parts of the Middle Buntsandstein thick composite units have been recognized. In the Detfurth–Folgen two such sedimentary units *c.*35 m thick are identified (Binot & Röhling, 1988). The thickness ratio between the various cycles is 1:5.8:17.8 (Clemmensen, 1991).

Interpretation of sedimentary succession

The alternating pattern of aeolian and subaqueous deposition suggests a cyclic development of the Volpriehausen and Detfurth – Folgen.

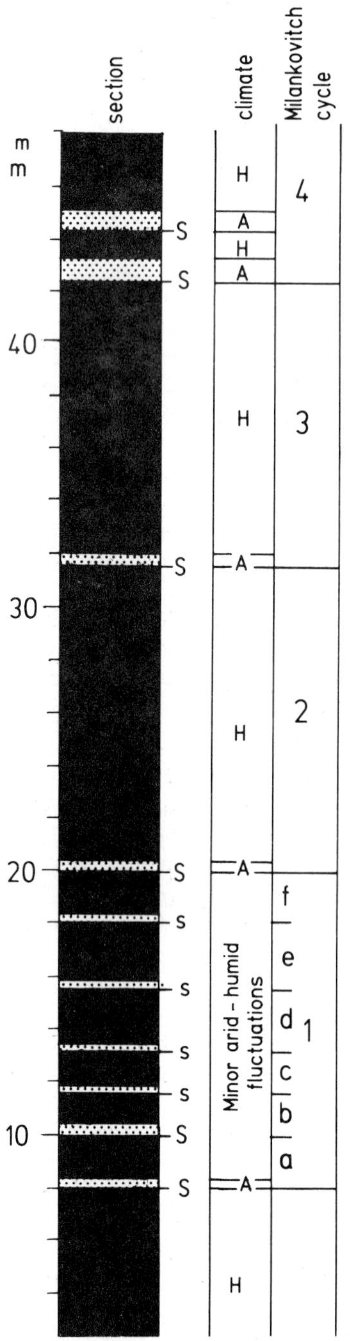

Fig. 7. Simplified sedimentary section of the Middle Buntsandstein (Early Triassic), Helgoland, Germany. Basic cycles are shown only in composite cycle 1. Based on Clemmensen & Tirsgaard (1990).

Each basic sedimentary cycle records a phase of high lake-stand and lacustrine sedimentation (humid phase), following a phase of lake exposure, aeolian sand-drift and sand-sheet formation (arid phase). The composite nature of the cyclicity indicates that climatic fluctuations between arid and humid also took place on a longer time scale. This is consistent with the predictions of climate based on global cyclostratigraphy (Tables 1 and 2). We therefore hypothesize that the basic cycles represent 20 ka precession cycles (P1 + P2), the thin composite cycles are 100-ka eccentricity cycles (E1 + E2) (Fig. 7), and the thick composite cycles are 400-ka eccentricity cycles (E3) (see Binot & Röhling, 1988). We propose that the climatic fluctuations are related to the intensity of monsoonal circulation and hence to the amount of precipitation in the basin (see Kutzbach & Otto-Bliesner, 1982; Olsen, 1986; Spaulding, 1991).

Implications

The interpretation of the above sedimentary cycles as orbitally forced suggests that the *c*.975-m-thick Buntsandstein (Scythian) section in the Helgoland area corresponds to 9.75 Ma. This figure fits very well with the latest dating of the Scythian to 10 Ma by Cowie & Bassett (1989), but not quite so well with the 4 Ma of Harland *et al.* (1990). The value of 4 Ma of the latter authors, however, 'is based on chron interpolation because of a lack of significant age data in the relative stratigraphic interval'.

Assuming for the sake of simplicity that aeolian activity on Helgoland occurred in the arid half of each precession cycle (10 ka), and that the basal sand-drift surface covers only very little time, an average aeolian accumulation rate of at most 0.1 m/ka is deduced (Table 3). This figure is relatively low and certainly lower than the aeolian accumulation rate deduced from the Lower Triassic Helsby Sandstone Formation (Table 3). A comparison with modern aeolian depositional systems shows that sand sheets are less stable terrains, and occur in zones of deflation and transport (Fryberger *et al.*, 1984).

A climatic origin for the sedimentary cycles in the Middle Buntsandstein predicts that the sedimentary cycles should be of large geographical extent, and Binot & Röhling (1988) have indeed been able to trace the composite sedimentary cycles across large parts of the basin.

Wingate Sandstone

The Wingate Sandstone (Lukachukai Member) of the western interior of the USA is composed of large-scale cross-bedded erg deposits (Blakey *et al.*, 1988; Clemmensen & Blakey, 1989). The erg deposits of the Wingate Sandstone intertongue with erg-margin deposits in the Dinosaur Canyon Member of the Moenave Formation towards the southwest (Clemmensen *et al.*, 1989).

The age of the Wingate Sandstone (from scarce palaeontological data and regional stratigraphy) is probably Early Jurassic (Sinemurian to Pliensbachian) (Blakey *et al.*, 1988).

The Wingate Sandstone crops out in southern and eastern Utah and in adjacent areas of Arizona and Colorado. The information presented below is based mainly on studies in northeastern Arizona.

The Wingate Sandstone formed in a large cratonic basin. The formation represents several closely related erg systems having a maximum extent of 110 000 km^2 or more. The ergs formed in an inland desert basin, and there is no evidence of marine deposits in the erg margins.

The desert basin was situated *c.*17° N, in Early Jurassic cyclostratigraphic belt 3, predicting climatic fluctuations between temperate arid and tropical humid (Tables 1 and 2). The dunes in the ergs apparently formed in a directionally varying wind flow with prevailing winds from the NW and periodically strong winds from the SW and possibly also from the NE. Resultant sand transport was towards the SE and the dunes seem to have been of oblique nature.

Description of sedimentary succession

The Wingate Sandstone in northeastern Arizona is composed of five superimposed erg units (Fig. 8) (Clemmensen & Blakey, 1989). The cycles have an average thickness of 23 m and vary in thickness between 15 and 35 m. They have basal erg-order deflation surfaces, a lower, relatively thin, sand sheet or amalgamated interdune deposit, and an upper cross-bedded dune deposit. Recent studies of Loope & Simpson (1992) indicate that amalgamated interdune deposits are more common and more thickly developed in eastern Utah. The uppermost erg unit in the Wingate Sandstone has been truncated erosively and is covered by fluvial deposits of the Kayenta Formation.

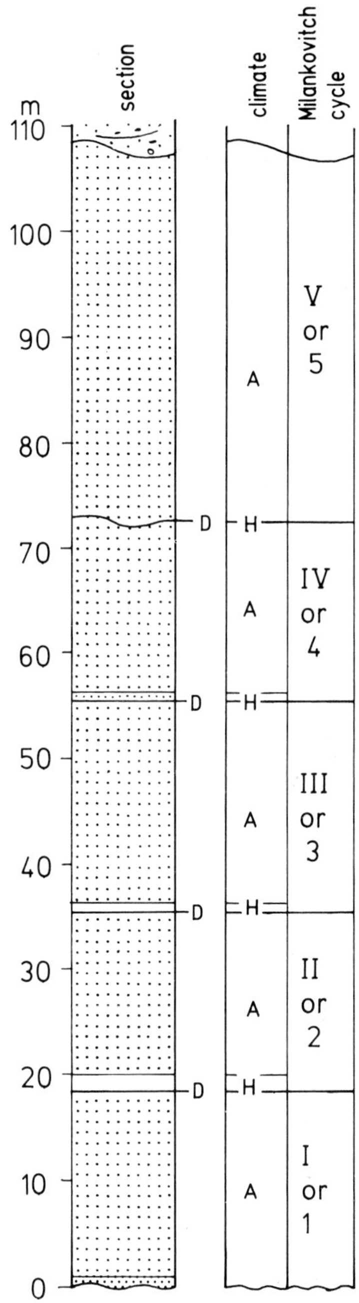

Fig. 8. Simplified sedimentary section of the Wingate Sandstone (Early Jurassic), Arizona, USA. Based on Clemmensen & Blakey (1989).

Interpretation of the sedimentary succession

The stacking of aeolian units with a similar composition and thickness suggests a cyclic nature for the deposition of the Wingate Sandstone.

The sedimentary cycles are thought to be of climatic origin, and are interpreted as recording alternating periods of erg construction and erg destruction or erg stabilization (Fig. 8). Erg construction (positive sand budget; deposition of thick dune sets) probably took place during arid climatic intervals when a good supply of sand was deflated from unconsolidated erg-margin fluvial deposits. Erg destruction (negative sand budget; formation of deflation surface) took place during continued arid conditions, or during climatic changes from arid to humid. Erg initiation (near-neutral sand budget; formation of sand sheets or amalgamated interdune deposits) most likely records the onset of a humid climate. Thus, the climatic conditions deduced are consistent with climatic conditions predicted from the concept of global cyclostratigraphy (Tables 1 and 2). The climatic fluctuations are thus thought to be controlled by astronomically induced variations in the intensity of monsoonal circulation, and related amounts of precipitation in the basin and source areas. Because of the simple nature of the cycles it is difficult to identify them precisely with the predicted Milankovitch periods. The size of the erg (long reconstitution time) might indicate that the sedimentary cycles record 100-ka eccentricity (E1 + E2) or 400-ka eccentricity (E3) cycles rather than precession (P1 + P2) cycles (Fig. 8), but this is speculative.

Implications

The interpretation of the sedimentary cycles as eccentricity cycles implies that the average aeolian accumulation rate in the erg was either 0.1 m/ka (400-ka cycles) or 0.4 m/ka (100-ka cycles) (Table 3). An interpretation of the aeolian cycles as precession cycles would yield an average accumulation rate of 2.3 m/ka. All these values are very reasonable when compared to figures on modern aeolian accumulation rates (Table 4).

Following the arguments of Clemmensen & Blakey (1989), deposition of one sedimentary cycle, 20 m thick and with an upwind extension of 375 km, would require a period of *c.*270 ka. Therefore, we prefer an interpretation of the series in terms of eccentricity cycles. This would lead to a timespan for the entire Wingate Sandstone of either *c.*500 ka (100-ka cycles) or *c.*2 Ma (400-ka cycles). In the case of the availability of local sources, however, the vast amount of sand contained in each erg unit of the series could have been deposited in a shorter timespan thus reducing the required time interval calculated by Clemmensen & Blakey (1989), but even then it is difficult to envisage that units were deposited in about 20 ka. Therefore, on the basis of sand budget considerations, the interpretation of the sedimentary cycles in terms of eccentricity cycles appears sound. This interpretation seems to be in conflict with the observation that at low latitudes direct precession-related climatic effects are expected to be dominant (see Berger, 1978). A recording of only the eccentricity signal in low latitudes may occur when the effect of the precessional cycle is pushed over a certain threshold only during periods with extreme eccentricity. A possibility would be that sea-level fluctuations generated at high latitudes, in response to the eccentricity cycle (polar sea-ice), exerted influences at low latitudes (e.g. precipitation, groundwater level).

With the interpretation of the erg units as a response to eccentricity cycles, the shorter-term precession fluctuations would most likely be contained in the erg-order deflation surfaces, or in the amalgamated interdune deposits at the base of the aeolian cycles.

DISCUSSION AND CONCLUSIONS

The Early Permian to Early Jurassic formations discussed above represent a variety of ancient desert environments which formed on the Pangaea megacontinent at low palaeolatitudes in the Northern Hemisphere (8–20° N) corresponding to the modern cyclostratigraphic belts 2 and 3 of Perlmutter & Matthews (1989).

The climates of this ancient megacontinent have been treated in several papers (e.g. Robinson, 1973; Parrish & Peterson, 1988; Kutzbach & Gallimore, 1989). In the most recent reconstruction, Kutzbach & Gallimore (1989) suggest that large parts of northern Pangaea experienced large-scale summer monsoon circulation during the Late Permian and Triassic. Year-round dry or seasonal aridity zones were simulated in their model for all but the eastern coastal regions, the tropical western coast, and the regions poleward of 40° latitude. The above Late

Permian, Triassic and Early Jurassic formations all occur in the year-round dry or seasonally arid zone of Kutzbach & Gallimore (1989). Variations in the nature of these ancient desert deposits (e.g. desert-lake system or erg-system) seem to reflect variations in palaeotopography or palaeogeography rather than variations in absolute humidity across the continent.

Climatically formed sedimentary cycles, simple or composite, are shown to occur in the successions studied. Individual cycles consist of superimposed aeolian and fluvial, or aeolian and lacustrine facies, or of aeolian sediments separated by deflation surfaces. The thickness of the cycles varies from *c*.2 m to more than 400 m. Climatic estimates based on the composition of the sedimentary cycles are generally consistent with estimates obtained using the cyclostratigraphic model of Perlmutter & Matthews (1989).

Most of the ancient desert deposits are inferred to consist of stacked precession cycles (*c*.20 ka). In some formations precession cycles apparently form part of thicker eccentricity cycles (100 and 400 ka). One formation seems to be primarily composed of eccentricity cycles (100 or 400 ka); precession cycles may here be concealed in erg-order deflation surfaces or in amalgamated interdune deposits.

The cyclic nature of the deposits studied is consistent with repeated fluctuations in climate between relatively arid and relatively humid. Because of a general lack of chronostratigraphic control, the climatic patterns thus deduced cannot be assigned definitely to orbital forcing in most examples. Only in the case of the Middle Buntsandstein of Helgoland is there good time control supporting the orbital forcing interpretation. In the remaining examples absolute dating is poor. It thus could be argued that the deduced climatic fluctuations were of a random nature (see Algeo & Wilkinson, 1988). However, the character of the cycles found in the field match with what would be expected on the basis of the present knowledge of orbital forcing and desert climate sedimentation. A definite proof of the Milankovitch theory in these examples has to await, for example, magnetostratigraphic investigations.

Alternative explanations of the cyclic nature of the examples discussed could involve tectonics or eustasy. Tectonics normally control the long-term stratigraphic evolution of a basin, but are not likely to have a pulsating nature that would produce long series of tectonic events each affecting the dynamics of the depositional systems in exactly the same way. Because of the inland nature of the examples, eustasy can be excluded as a direct control. Indirectly, however, changes in global sea level may have controlled the climate in the desert basins.

The preservation of the precession cycles records a relatively high and steady subsidence rate in the relevant basins, sufficient sediment supply, and depositional systems with a relatively short response time. The preservation of only the eccentricity cycles probably records a depositional system, such as an extensive erg, with a relatively long response time in which only extreme precession influences during eccentricity maxima left clear signals in the sediment. Thus, studies of vertical sequences in ancient desert deposits may eventually lead to the definition of cyclostratigraphic units (see Yang & Baumfalk, p. 47, this volume).

The main conclusion of this paper is that no features of the formations examined are in conflict with the idea of orbital control. This implies that, although the interior of the Pangaean continent was on the average hot and arid (see Kutzbach & Gallimore, 1989), the larger part (at least the interior of the northern part) would have experienced climatic fluctuations between relatively more humid and relatively more arid due to cyclic orbital influences. Arid intervals may be compared to the climatic minimum phases and humid intervals to the climatic maximum phases of Perlmutter & Matthews (1989).

In periods with polar ice caps (i.e. during the early Permian) climatic contrasts on Pangaea were probably amplified. This climatic situation was not considered by Kutzbach & Gallimore (1989), but it is to be expected that the climates in low latitudes were significantly influenced by the waxing and waning of polar ice caps (Glennie, 1984). During climatic minimum phases, growing ice caps pushed the Hadley cells equatorwards and they were compressed. During climatic maximum phases, ice caps decayed and the Hadley cells expanded. Sedimentary basins lying in areas corresponding to the modern cyclostratigraphic belt 2 or 3 would accordingly experience alternating arid (climatic minimum phase) and humid (climatic maximum phase) periods.

The identification of Milankovitch cycles in aeolian or mixed aeolian–aqueous desert basins is important, and it opens up a method of dating such deposits, which commonly occur in basins with poor biostratigraphic and chronostratigraphic control.

ACKNOWLEDGEMENTS

The authors thank K.W. Glennie, D.G. Smith, M.R. Talbot and an anonymous reviewer for constructive comments on the manuscript. D. Pugliese typed the manuscript and R. Madsen drafted the figures.

REFERENCES

Algeo, T.J. & Wilkinson, B.H. (1988) Periodicity of mesoscale Phanerozoic sedimentary cycles and the role of Milankovitch orbital modulation. *J. Geol.* **96**, 313–322.

Berger, A. (1978) Long term variations of daily insolation and Quaternary climatic changes. *J. Atmos. Sci.* **35**, 2362–2367.

Berger, A. (1988) Milankovitch theory and climate. *Rev. Geophys.* **26**, 624–657.

Berger, A., Loutre, M.F. & Dehant, V. (1989) Astronomical frequencies for pre-Quaternary paleoclimate studies. *Terra Nova* **1**, 474–479.

Binot, F. & Röhling, H.-G. (1988) Lithostratigraphie und natürliche Gammastrahlung des Mittleren Buntsandsteins von Helgoland – Ein Vergleich mit der Nordseebohrung J/18-1. *Z. Dt. Geol. Ges.* **139**, 33–49.

Blakey, R.C., Peterson, F. & Kocurek, G. (1988) Synthesis of late Paleozoic and Mesozoic eolian deposits of the Western Interior of the United States. *Sedim. Geol.* **56**, 3–125.

Chadwick, R.A. (1986) Extension tectonics in the Wessex Basin, southern England. *J. Geol. Soc. Lond.* **143**, 465–488.

Clemmensen, L.B. (1979) Triassic lacustrine red-beds and palaeoclimate: the 'Buntsandstein' of Helgoland and the Malmros Klint Member of East Greenland. *Geol. Rundsch.* **68**, 748–774.

Clemmensen, L.B. (1991) Controls on aeolian sand-sheet formation exemplified by the Lower Triassic of Helgoland. In: *Aeolian Grain Transport II: The Erosional Environment* (Eds Barndorf-Nielsen, O.E. & Willets, B.B.). *Acta Mechanica*, Suppl. 2, pp. 161–170.

Clemmensen, L.B. & Abrahamsen, K. (1983) Aeolian stratification and facies association in desert sediments, Arran Basin (Permian), Scotland. *Sedimentology* **30**, 311–339.

Clemmensen, L.B. & Blakey, R.C. (1989) Erg deposits in the Lower Jurassic Wingate Sandstone, northeastern Arizona: oblique dune sedimentation. *Sedimentology* **36**, 449–470.

Clemmensen, L.B. & Hegner, J. (1991) Eolian sequence and erg dynamics, the Permian Corrie Sandstone, Scotland. *J. Sedim. Petrol.* **61**, 768–774.

Clemmensen, L.B. & Tirsgaard, H. (1990) Sand-drift surfaces: A neglected bounding surface. *Geology* **18**, 1142–1145.

Clemmensen, L.B., Olsen, H. & Blakey, R.C. (1989) Erg-margin deposits in the Lower Jurassic Moenave Formation and Wingate Sandstone, southern Utah. *Geol. Soc. Am. Bull.* **101**, 759–773.

Cowie, J.W. & Bassett, M.G. (1989) Global stratigraphic chart. *Episodes*, **12**(2), Suppl.

Crowley, T.J. (1983) The geological record of climatic change. *Rev. Geophys. Space Phys.* **21**, 828–877.

Fryberger, S.G., al-Sari, A.M., Clisham, T.J., Rizvi, S.A.R. & al-Hinai, K.G. (1984) Wind sedimentation in the Jafurah sand sea, Saudi Arabia. *Sedimentology* **31**, 413–431.

Glennie, K.W. (1983) Lower Permian Rotliegend desert sedimentation in the North Sea. In: *Eolian Sediments and Processes* (Eds Brookfield, M.E. & Ahlbrandt, T.S.). Elsevier, Amsterdam, pp. 521–541.

Glennie, K.W. (1984) Early Permian Rotliegend. In: *Introduction to the Petroleum Geology of the North Sea* (Ed. Glennie, K.W.). Blackwell Scientific Publications, Oxford, pp. 41–60.

Harland, W.B., Armstrong, R.L., Cox, A.V., Craig, L.E., Smith, A.G. & Smith, D.G. (1990) *A Geologic Time Scale 1989*. Cambridge University Press, Cambridge.

Holloway, T. (1985) Triassic: Sherwood Sandstone Group (excluding the Kinnerton Sandstone Formation). In: *Atlas of Onshore Sedimentary Basins in England and Wales: Post-Carboniferous Tectonics and Stratigraphy* (Ed. Whittaker, A.). Blackie and Son Ltd, Glasgow, pp. 31–38.

Holm, K., Talbot, M.R., Williams, M.A.J. & Adamson, D.A. (1986) Sedimentation in low relief desert margin systems: the Triassic of south-west England and the Quaternary west-central New South Wales, Australia. In: *Desert Sediments – Ancient and Modern.* Geol. Soc. Lond., Special Scientific Meeting (Abstract volume).

Illenberger, U.K. & Rust, I.G. (1988) A sand budget for the Alexandria coastal dune field, South Africa. *Sedimentology* **35**, 513–521.

Kocurek, G. & Havholm, K. (1993) Eolian event stratigraphy – a conceptual framework. In: *Recent Advances in and Applications of Siliclastic Sequence Stratigraphy* (Eds Weimer, P. & Posamentier, H.). Am. Assoc. Petrol. Geol. Mem. (in press).

Kutzbach, J.E. & Gallimore, R.G. (1989) Pangaean climates: megamonsoons of the megacontinent. *J. Geophys. Res.* **94**, 3341–3357.

Kutzbach, J.E. & Otto-Bliesner, B.L. (1982) The sensitivity of the African–Asian monsoonal climate to orbital parameter changes from 9000 years BP in a low-resolution general climatic circulation model. *J. Atmos. Sci.* **39**, 1177–1188.

Laming, D.J.C. (1966) Imbrication, paleocurrents and other sedimentary features in the Lower New Red Sandstone, Devonshire, England. *J. Sedim. Petrol.* **31**, 128–132.

Lancaster, N. (1990) Paleoclimatic evidence from sand seas. *Palaeogeogr. Palaeoclimatol. Palaeoecol.* **76**, 279–290.

Loope, D.B. & Simpson, E.L. (1992) Significance of thin sets of cross-strata. *J. Sedim. Petrol.* **62**, 849–859.

Mader, D. & Laming, D.J.C. (1985) Braidplain and alluvial-fan environmental history and climatological evolution controlling origin and destruction of eolian dune fields and governing overprinting of sand seas and river plains by calcrete pedogenesis in the Permian and Triassic of South Devon (England). In: *Aspects of Fluvial*

Sedimentation in the Lower Triassic Buntsandstein of Europe (Ed. Mader, D.). Lecture Notes in Earth Sciences, 4. Springer Verlag, Berlin, pp. 519–528.

Miller, G.H., Wendorf, F., Ernst, R., Schild, R., Close, A.E., Freidman, I. & Schwarcz, H.P. (1991) Dating lacustrine episodes in eastern Sahara by the epimerization of isoleucine in ostrich eggshells. *Palaeogeogr. Palaeoclimatol. Palaeoecol.* **84**, 175–189.

Oerlemans, J. (1980) Model experiments on the 100 000-yr glacial cycle. *Nature* **287**, 430–432.

Olsen, P.E. (1986) A 40 million year lake record of early Mesozoic orbital climatic forcing. *Science* **234**, 842–848.

Olsen, H. (1990) Astronomical forcing of meandering river behaviour: Milankovitch cycles in Devonian of East Greenland. *Palaeogeogr. Palaeoclimatol. Palaeoecol.* **79**, 99–115.

Øxnevad, I.E.I. (1991) *Eolian and mixed eolian–aqueous sedimentation in modern and ancient sub-tropical desert basins. Examples from the Sahara and the Permo-Triassic of NW Europe*. Dr Scient. thesis (unpublished), Bergens University.

Parrish, J.T. & Peterson, F. (1988) Wind directions predicted from global circulation models and wind directions determined from eolian sandstones of the western United States – a comparison. In: *Late Paleozoic and Mesozoic Eolian Deposits of the Western Interior of the United States* (Ed. Kocurek, G.). *Sedim. Geol.* **56**, 261–282.

Perlmutter, M.A. & Matthews, M.D. (1989) Global cyclostratigraphy – A model. In: *Quantitative Dynamic Stratigraphy* (Ed. Cross, T.A.). Prentice Hall, Englewood Cliffs, New Jersey, pp. 233–260.

Robinson, P.L. (1973) Paleoclimatology and continental drift. In: *Implications of Continental Drift to Earth Sciences* (Eds Tarling, D.H. & Runcorn, S.K.). Academic Press, London, pp. 451–476.

Smith, A.G., Hurley, A.M. & Briden, J.C. (1981) *Phanerozoic Paleocontinental World Maps*. Cambridge University Press, Cambridge.

Spaulding, W.G. (1991) Pluvial climatic episodes in North America and North Africa: types and correlation with global climate. *Palaeogeogr. Palaeoclimatol. Palaeoecol.* **84**, 217–227.

Talbot, M.R. (1985) Major bounding surfaces in eolian sandstones – a climatic model. *Sedimentology* **32**, 257–265.

Thompson, D.B. (1970) The stratigraphy of the so-called Keuper Sandstone Formation (Scythian – ? Anisian) in the Permo-Triassic Cheshire Basin. *Q. J. Geol. Soc. Lond.* **126**, 151–181.

Spec. Publs Int. Ass. Sediment. (1994) **19**, 459–481

Global cyclostratigraphy: an application to the Eocene Green River Basin

M.D. MATTHEWS *and* M.A. PERLMUTTER

Texaco Exploration and Producing Technology Division, 3901 Briarpark, Houston, TX 77042, USA

ABSTRACT

Global cyclostratigraphy is a process-response model that hindcasts a basin's stratigraphic character by integrating climate and tectonics into a globally synchronous framework. The evolution of depositional environments throughout a basin's history is recreated by superimposing short- to long-term, orbitally driven, climatically controlled stratigraphic variations onto the very long-term, tectonically controlled evolution of basin morphology.

The global nature of this model predicts an interesting difference in the timing of sediment delivery to lacustrine and marine systems in mid-latitudes. In lacustrine systems, runoff and sediment flux are directly linked to lake level. However, in marine systems the phase relationship of runoff and sediment flux to sea level is latitudinally dependent because changes in the sum of worldwide runoff (especially ice storage/melting) affect sea level while changes in local runoff affect sedimentation.

The application of global cyclostratigraphic concepts to the Eocene Green River Basin predicts the general stratigraphic succession and characteristics of the basin. The general evolution of the stratigraphy from fluvial (Wasatch Formation) to lacustrine (Green River Formation) to fluvial (Bridger Formation), and the higher frequency variations within these formations (deep lake–shallow lake, oil shale–trona/marl, braided–meandering rivers) confirm the model's power to infer and provide a context for otherwise conflicting geological observations.

INTRODUCTION

Global cyclostratigraphy is a conceptual/semi-quantitative model that assesses the stratigraphic architecture of a basin. Stratigraphic development is influenced by the interaction of forces that act on many different time scales. To aid in understanding these interactions we have used the following classification:

1 <20 ka: seasonal and very short term;
2 20–50 ka: short term (orbital);
3 50–500 ka: intermediate term (orbital);
4 500–2500 ka: long term (orbital interactions);
5 >1000 ka: very long term (tectonic).

The stratigraphy of a basin is similar to a tapestry that records the history of processes influencing sediment delivery to the basin and sediment distribution within the basin (Fig. 1). The very long-term evolution of basin stratigraphy (fluvial or lacustrine/marine) is determined by the dynamic balance between the ability of the provenance area to supply sediment (production rate and transport capacity) and the ability of the basin to accommodate sediment (subsidence). Thus tectonic conditions control the shape (warp) of the stratigraphic tapestry. The pattern (woof) filling the stratigraphic tapestry is a result of variations in sediment supply, transport and depositional base level. These factors appear to be driven by variations in heat received from the Sun due to orbital variations (Milankovitch, 1941). Stratigraphic variation due to seasonal and very short-term causes, such as sudden tectonic events (fault motion) and autocyclic phenomena (avulsion), are not included in the model because their causes and effects are local.

The core of global cyclostratigraphy is understanding the stratigraphic consequences of the global climate pattern in individual basins. This modelling technique permits stratigraphy to be hindcast for regions where there are little or no direct data. Hindcasting requires an understanding of the spatial variation of the short- to long-term global climate

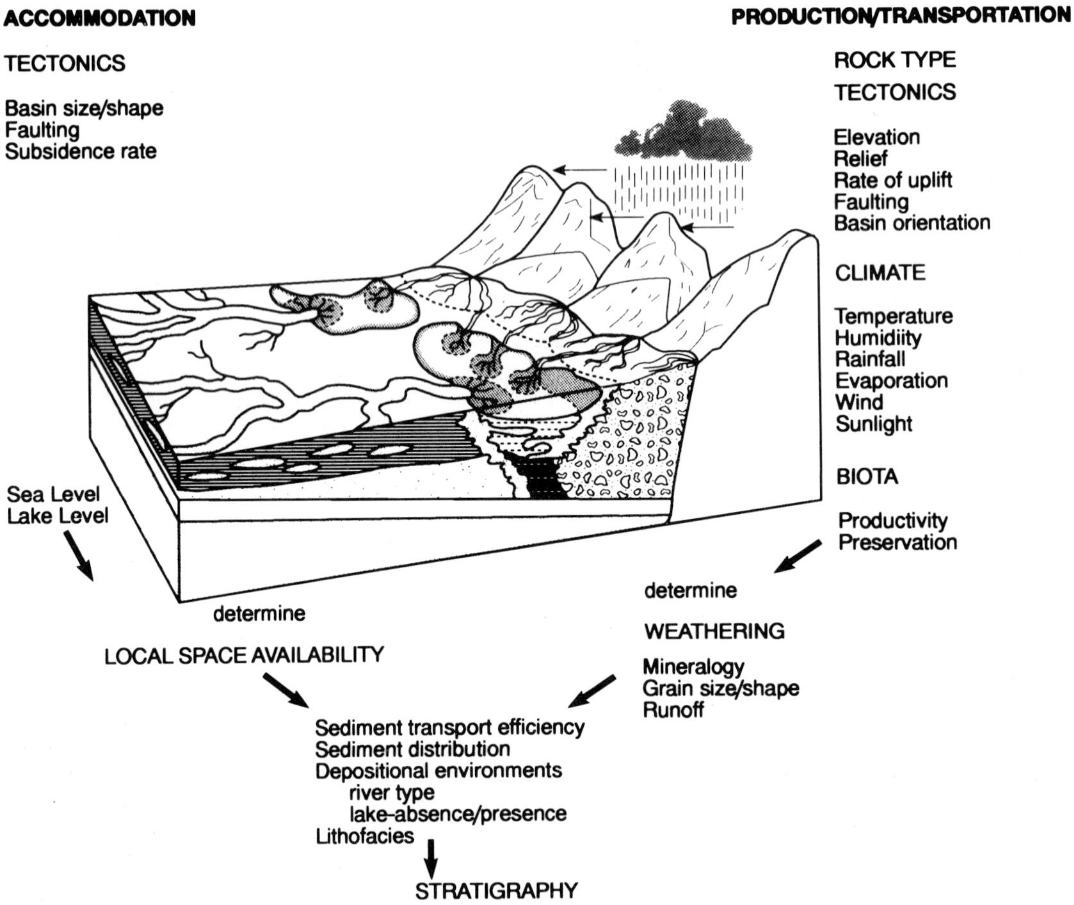

Fig. 1. Factors affecting the production, transport, accommodation and deposition of sediment in a basin.

pattern for a particular geological epoch or age. Climate patterns are derived using a combination of dynamic and conceptual climate models, constrained by the distribution of palaeoclimate indicators for that time period. Sediment yield as a function of climate is modelled by evaluating the effects of provenance, temperature, humidity, runoff and soil binding on sediment production and transport (Perlmutter & Matthews, 1990).

The purpose of this paper is to demonstrate the use of global cyclostratigraphic principles both in hindcasting basin stratigraphy in an unexplored basin and in better understanding the evolution of sequences within a well-studied basin. The first part of the text reviews the principles of global cyclostratigraphy, followed by a brief summary of the tectonic evolution of a foreland basin and its expected impact on the very long-term stratigraphic pattern of the basin. The next section summarizes the global cyclostratigraphy hindcast of Green River Basin stratigraphy, followed by a discussion of previous work on the stratigraphy of the Green River Formation with an emphasis on estimations of climatic conditions within and adjacent to the basin. The accuracy and limitation of the global cyclostratigraphic hindcast stratigraphy and how this analysis technique affects the interpretation of the known stratigraphy are then discussed.

PRESENT GLOBAL CLIMATIC PATTERNS

The factors determining the climatic conditions within a basin are separated into first-order effects

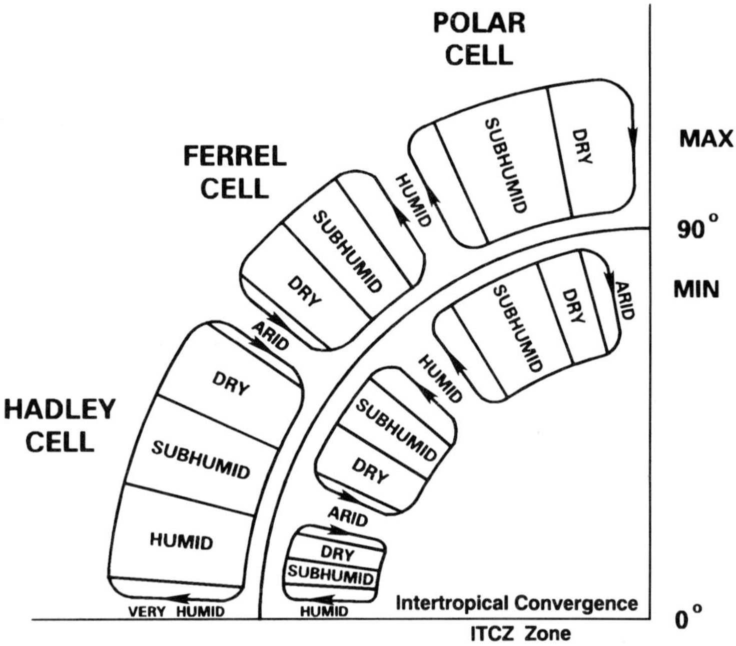

Fig. 2. Hadley circulation at climatic maximum and minimum showing the mean associated humidity effects and the expected changes in relative cell size due to changes in the amount and distribution of heat received from the Sun.

(latitudinal, driven by the Hadley circulation of the atmosphere) and second-order regional/local effects (azonal modifications).

Hadley circulation (zonal)

Global climate generally exhibits a latitudinally zoned pattern (Miller, 1966) arising from the thermal gradient between the equator and the poles and the variation of precipitation and evaporation (humidity/runoff) related to the Hadley circulation (Fig. 2). Global temperature changes from one season to the next affect the relative sizes of the Hadley circulation cells (Polar, Ferrel and Hadley), causing them to shift latitudinal position. This shift results in seasonal changes in humidity and runoff (Lockwood, 1979).

Modification of zonal climate patterns (azonal)

The previous discussion assumes that climatic patterns are zonal and can be defined exclusively by latitude. Although true in many cases, there are significant exceptions (Fig. 3). For example, the seasonal shift of the thermal equator interacts with large landmasses in the mid-latitudes to intensify summer heating and winter cooling of the atmosphere, relative to the ocean. This seasonal differential heating results in a significant poleward shift of the junction of the Northern and Southern hemispheres' Hadley cells (intertropical convergence zone or ITCZ), creating monsoonal climates near the equator, shifting and compressing climates in adjacent poleward regions (Fig. 2).

Additional azonal modifications of climate occur near the margins of land and sea, and farther inland, as a result of the following.

1 Circulation around ocean-centred, mid-latitude, high-pressure cells. The magnitude and nature of this effect is dependent on wind direction, coast line orientation and oceanic currents. For simplicity the examples below assume that a high-pressure cell is centred off an extensive, essentially straight coastline.

(a) East sides of continents: winds are onshore in lower mid-latitudes and offshore in higher mid-latitudes.

(b) West sides of continents: winds are onshore in higher mid-latitudes and offshore in lower mid-latitudes.

(c) North sides of continents: winds are onshore on eastern portions of continents in the Northern Hemisphere and on the western portions in the Southern Hemisphere. Winds are offshore in

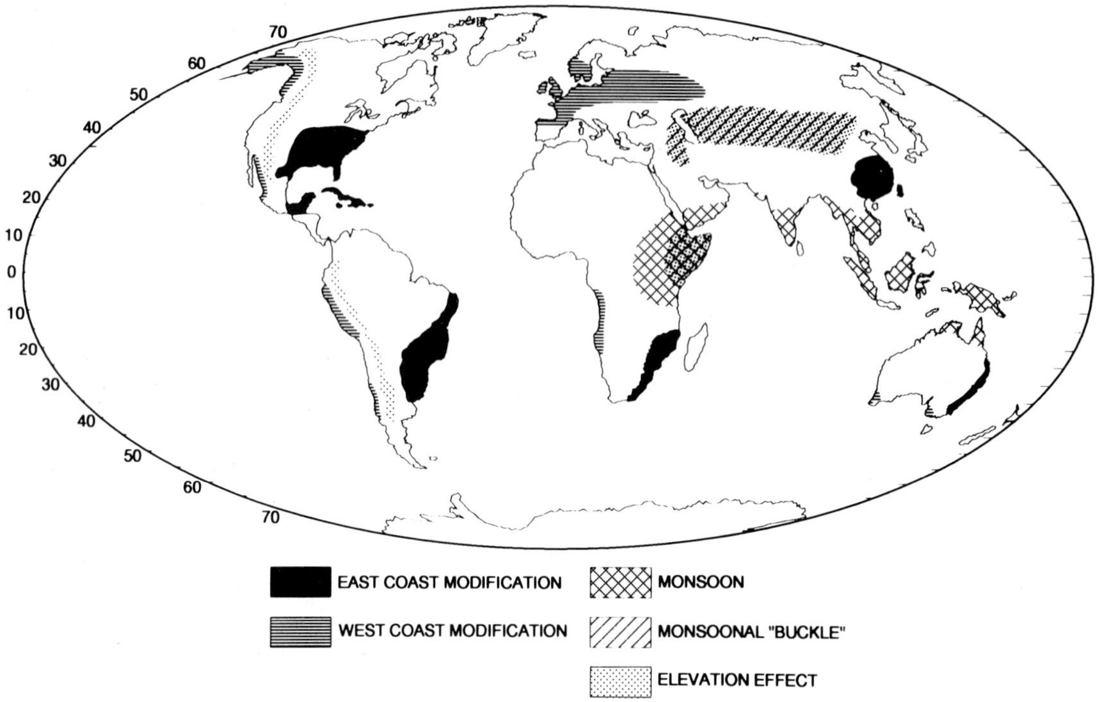

Fig. 3. Present geographical distribution of major deviations from a simple zonal climate.

western portions of continents in the Northern Hemisphere and in eastern portions in the Southern Hemisphere.

(d) South sides of continents: winds are onshore in western areas of the Northern Hemisphere and eastern areas in the Southern Hemisphere. Winds are offshore in eastern areas of the Northern Hemisphere and western areas in the Southern Hemisphere.

2 Ocean currents. The magnitude of the effect of currents is a function of water temperature, the distribution of land and sea, wind direction and sea level.

(a) Warm (surface) boundary currents: presently provide heat and moisture onshore to lower mid-latitude eastern coasts and upper mid-latitude western coasts.

(b) Cool (surface) boundary currents: cool and provide additional precipitation to mid-latitude western coasts by decreasing relative humidity.

(c) Upwelling of cold, deep currents: cool and provide additional precipitation to western coasts, depending on the wind pattern.

3 Elevation. The magnitude of the effect of the vertical motion of air as it crosses mountain ranges is dependent on the initial air temperature, humidity, relief and lapse rate (decrease in temperature with elevation).

(a) Windward sides: the increase in altitude causes air to cool and increase in relative humidity, producing wetter conditions.

(b) Leeward sides: the decrease in altitude causes air to warm and decrease in relative humidity, producing drier conditions.

PALAEOGLOBAL CLIMATIC PATTERNS

In the geological past, climatic change must have arisen due to variations in the amount and distribution of heat received from the Sun, much as the Earth now responds to the seasonal changes discussed above. This section reviews the causes of the changes in insolation and discusses the complexity of the non-linear climatic response of the Earth.

Milankovitch-induced climate change

Changes in the amount/distribution of incoming solar radiation are primarily due to the interaction of three quasi-periodic elements of the Earth's motion about the Sun (Milankovitch, 1941; Berger, 1984). Although the current values of the orbital factors are 23 and 19 ka for precession, 41 ka for obliquity, and 413, 123 and 95 ka for eccentricity, the values are not constant but have changed their period and phase relationships over geological time (Berger & Loutre, pp. 15–24, this volume).

Constructive and destructive interference between the orbital parameters modulates climate and its distribution over the globe. The interaction of the primary (short- to intermediate-term) Milankovitch factors has been calculated to produce long-term periodicity in January solar input at 70° S latitude and 30° N latitude (Matthews & Frohlich, 1991). These long periodicities were used as input by Matthews & Frohlich (1991) to a non-linear ice response model producing periodicities of 1600–2400 ka. The two extreme climates that occur within this time interval we call the climatic maximum end-member and climatic minimum end-member. Interpretation of published and proprietary palaeoclimatic indicators suggests that zonally constrained climatic variations occur back through the Late Proterozoic.

Nested within this long-term envelope are higher frequency climatic reversals directly linked to the fundamental Milankovitch elements. These reversals are responsible for the most noticeable stratigraphic changes. The magnitude of the climatic change associated with these reversals is variable and depends on the non-linear response of the climate system to the smaller changes in solar heating. Thus there are 'lesser' climatic maxima and minima that occur within the time frame of tens to hundreds of thousands of years. For example, if the long-term climate is tropical/humid at the climatic maximum end-member and temperate/arid at the climatic minimum end-member (1600–2400 ka time frame), the short- and intermediate-term climatic cycles may be bounded between more moderate extremes, such as tropical/humid to tropical/dry, tropical/dry to temperate/arid or even the high end of tropical/subhumid to the low end of tropical/subhumid. An example of this is the 'little ice age' that began about 2000 BC (Matthes, 1939) and ended around AD 1800 to 1940 (Lamb, 1961). The 'little ice age' does not represent the coldest glacial period or the warmest interglacial but rather a minor Milankovitch-driven fluctuation where the climate reversed from growing cooler to growing warmer on the back of a larger amplitude, longer term trend.

Continental glaciers cause a significant drop in polar temperature (Budyko, 1969). The increased temperature difference between the poles and the equator would be expected to significantly affect the circulation of heat in the atmosphere and the oceans, selectively deepening climatic minima associated with continental glaciers. From our examination of the data, either the added effect of continental glaciation on global climate is lost within the noise of Milankovitch-associated changes, or glaciation was the rule in the geological past rather than the exception (contrary to the majority of current opinion). We favour glaciers being present at all, or almost all, times when there is a land mass at the pole, with sea ice occurring when a land mass was not at a pole, based on five lines of evidence.

1 Comparisons of the latitudinal pattern of the range of palaeoclimatic indicators from 'greenhouse' conditions (e.g. Cretaceous and Eocene) with those of 'icehouse' conditions (e.g. Pleistocene and Permian) suggest that the latitudinal pattern of end-member climatic shifts is approximately equal over periods of millions of years, regardless of geological time period.

2 Rapid shifts in the coastal onlap curve (Haq *et al.*, 1988) throughout the geological record are most easily explained by changes in glacial volume (Pitman & Golovchenko, 1983).

3 The restricted temperature range for the existence of life suggests that ice may have acted as a thermal buffer to sea-water temperature (King, 1961).

4 Climate modelling suggests that mean annual temperatures below freezing would be expected within polar continental interiors during the Cretaceous (Barron *et al.*, 1981).

5 Glacio-marine drift and Glendonites occur in Cretaceous polar sediments (Kemper, 1987).

Development of global climate patterns over the course of a Milankovitch cycle (zonal)

The sizes and latitudinal positions of the three Hadley circulation cells shifts as Milankovitch oscillations cause the absolute and seasonal distribution of insolation to vary (Frakes, 1979; Lockwood, 1980; Glennie, 1984; Perlmutter & Matthews, 1990). Changes in Hadley circulation force climates to migrate over relatively large distances across the face of the Earth (up to 30° latitude) in geologically

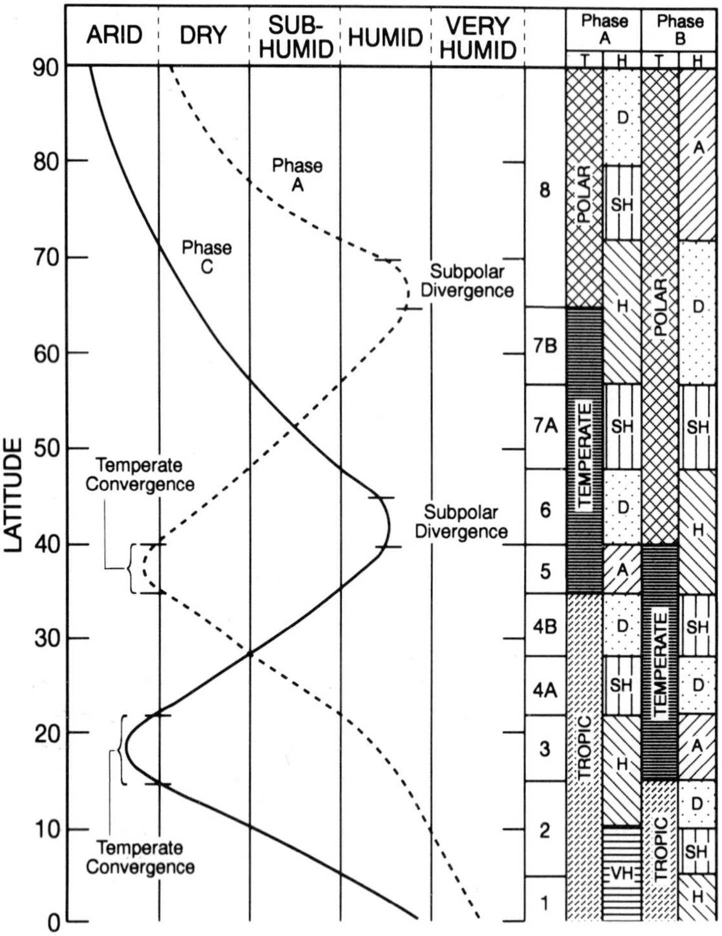

Fig. 4. Climatic change and cyclostratigraphic belts. Vertical scale is latitude (north or south). Horizontal scale is relative humidity. The dashed line represents conditions during the climatic maximum end-member (phase A) and the solid line represents conditions during the climatic minimum end-member (phase C). Latitudinal ranges for maximum and minimum conditions of temperature and humidity are shown to the right. These climatic extremes are summarized into 10 global cyclostratigraphic belts. For example, the extreme climates of belt 3 vary from tropical/humid during the climatic maximum end-member, through tropical/subhumid and tropical/dry during global cooling, to temperate/arid during the climatic minimum end-member, and back through tropical/dry and tropical/subhumid during the warming back to the climatic maximum end-member when the climate is once again tropical/humid. It should be remembered that these climates need to be corrected for non-zonal effects and represent the greatest climatic shifts expected. In many cases the interaction of the Milankovitch parameters will be such that the climatic shifts will be less by varying amounts. Note that for simplicity the diagrams shown ignore any phase relationships between the Northern and Southern hemispheres that may arise due to the precession of the equinoxes.

short periods of time (half a precession cycle, less than 10 ka). The magnitude of this change is well documented for the Pleistocene (Penck, 1914; Sarnthein & Diester-Haass, 1977; Street & Grove, 1979). Ten cyclostratigraphic belts (Fig. 4) are identified from the latitudinal distribution of climate end-member conditions for the last glacial/interglacial and a compilation of climatic indicators from other time periods.

Estimating the climatic range also permits analysis of the sequence of climates in any particular location. For convenience, a complete climate cycle is considered as a cosine function and divided into six phases: phase A, the climatic maximum; phases B1 and B2, the cooling transitions; phase C, the climatic minimum; and phases D1 and D2, the warming transitions (Fig. 5).

The sequence of climates over a cycle of any duration is different in each belt. Although the overall climate of the Earth is commonly expected to be cooler and drier during a climatic minimum, and warmer and wetter during a climatic maximum, the general pattern of humidity (wetness) does not hold for the mid-latitudes because the Ferrel cell circulation is opposite to that of the Hadley and Polar cells (Fig. 2). Ferrel cell circulation moves upwelled air equatorward rather than poleward, causing the wettest regions in the Ferrel cell to be poleward rather than equatorward. Consequently, an equatorward shift (change from global maximum

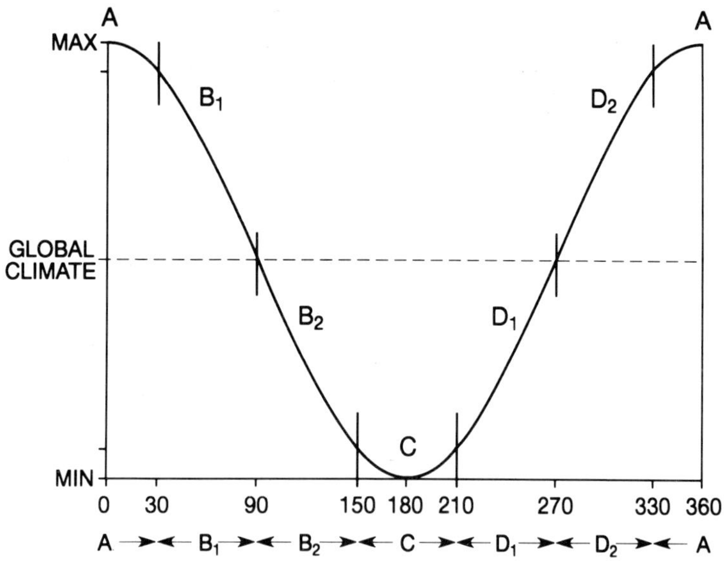

Fig. 5. Idealized Milankovitch cycle showing the associated climatic phases during an astronomical cycle.

to global minimum) in the position of the Ferrel cell causes some mid-latitude regions to become more humid. In belt 3 (15–20° latitude, Fig. 4) the climate is tropical/humid during the climatic maximum end-member (phase A) and temperate/arid during the climatic minimum end-member (phase C). However, in mid-latitude belt 5 (35–40° latitude) the humidity conditions are reversed, temperate/arid during phase A and temperate/humid during phase C. In belt 6 (40–48° latitude) the most humid conditions occur during climatic transition phases B2 and D1, while in belt 7a (48–57° latitude) the most humid conditions occur during the transitional phases B1 and D2.

The direct linkage of climatic change to Milankovitch-driven variation in insolation should result in a precession-driven 180° phase shift between the climate of the Southern and Northern Hemispheres. It appears, however, that the linkage may not be direct. The Pleistocene has sufficient time resolution to estimate independently the phase relationships of the glaciers in both hemispheres and they appear to have behaved synchronously (Broecker & Denton, 1989). This synchronous pattern may be linked to a non-linear behaviour of the global climate system related to heat transfer in the ocean system and continental configuration (Broecker & Denton, 1989). Whether the climate systems are always in or out of phase in the two hemispheres has not been resolved with sufficient accuracy for other time periods because an independent age dating technique of sufficient resolution is not available.

The focus of global cyclostratigraphy, however, is on predicting the statistical range of stratigraphic patterns in individual basins and understanding the differences and similarities of stratigraphic patterns in geographically separate basins of the same age. The model makes no attempt to resolve specific stratigraphic events, and in particular it is not concerned with stratigraphic resolution of the phase differences in the two hemispheres.

Modification of zonal climate patterns over the course of a Milankovitch cycle (azonal)

Changes in insolation during a Milankovitch cycle cause the effects of the azonal climatic modifications to intensify or to diminish. The monsoonal pattern associated with the shift of the ITCZ is greatest during the climatic maximum because of enhanced heating in mid-latitudes, and is least during the climatic minimum because of reduced heating at mid-latitudes.

Overall, onshore winds in mid-latitude areas are expected to be strongest during the climatic maximum due to enhanced continental low pressure, whereas offshore winds should be weakest. Onshore winds in mid-latitudes are expected to be weakest during the climatic minimum as a result of enhanced continental high pressure, and offshore winds should be strongest. Ocean boundary currents will be

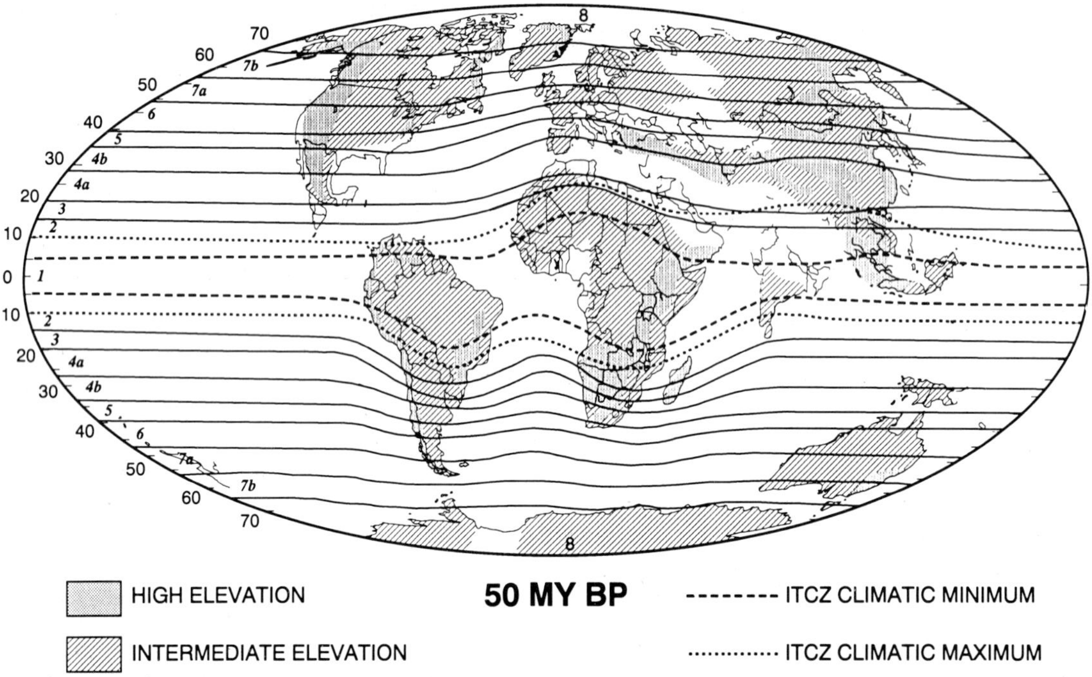

Fig. 6. Global cyclostratigraphic belts for the Eocene. Note that for simplicity the diagrams shown ignore any phase relationships between the Northern and Southern hemispheres that may arise due to the precession of the equinoxes.

warmer during the climatic maximum and gyres will shift poleward, causing upwelling zones to shift poleward. The opposite occurs during the climatic minimum, with cooler currents and upwelling zones shifting equatorward. The effects of elevation are greatest during the climatic minimum because lapse rate is at a maximum, compressing and lowering the vertical sequence of climates. Lapse rate is lowest during the climatic maximum, causing the vertical sequence of climates to expand and shift toward higher altitudes.

Global cyclostratigraphic belt maps (Fig. 6) display the seasonal extremes of the ITCZ at both the climatic maximum end-member and the minimum end-member. Compression and shifting of climates are accounted for by a poleward shift of the belt positions in regions affected by large continents. Once a belt map is constructed for a particular geological time, the belt climate (sequence of climates a particular region or basin will encounter during a Milankovitch cycle) can be assigned. Belt climates reproduce the conditions in centres of continents and oceans without need of modifications. However, in continental regions with significant azonal effects, belt climates must be modified by increasing or decreasing temperature and/or humidity (Table 1).

MILANKOVITCH CLIMATIC PATTERNS AND SEDIMENT FLUX

In each belt, the progression of different climate patterns over a Milankovitch cycle has definite, observable effects on the timing of runoff and sediment input with respect to short-term changes in lake and sea level. In lacustrine systems, a change in runoff affects lake level and sediment yield in the same manner; both are highest during the wettest phases. However, for marine systems it is more complex. Runoff, sediment yield and sea level are not always in phase.

Lacustrine basins

In lacustrine basins, changes in runoff directly affect lake level. Sediment transport is highest when lake (base) level is rising or at high-stand because of high runoff and sediment flushing (the mobilization of sediment stored in drainage areas during previous phases of low runoff, Perlmutter & Matthews, 1990). However, conditions of increasing and maximum runoff occur at different phases of the Milankovitch cycle in different belts because of the reverse humidity gradient associated with the Ferrel cell (Fig. 2). As a consequence, highest lake level and maximum sediment transport also occur at different phases of a Milankovitch cycle, depending only on the sequence of climates from maximum to minimum Milankovitch conditions. To a first approximation (Table 1), high runoff and lake levels are expected during or near: (i) the climatic maximum (phase A) in belts 1–4a, and 7b–8; (ii) the climatic minimum (phase C) in belts 4b and 5; and (iii) the transition climates in belts 6 (phases B2 and D1) and 7a (phases B1 and D2). In addition, belts 4a (phase C) and 4b (phase A) experience secondary peaks in sediment yield and lake level.

When sedimentation patterns, such as stratal geometry, are used to correlate events in continental basins that reside in different cyclostratigraphic belts, it is important to recognize that there may be a phase shift between them, on a Milankovitch time scale. This affects the interpretation of fourth and higher order seismic-sequence stratigraphic boundaries (Vail *et al.*, 1977), and possibly the third-order sequences as well.

Marine basins

In marine basins, the timing of climatically induced changes in continental runoff (described above) is not always in phase with changes in glacio-eustasy. Thus, the relationship between the timing of sediment yield to continental margins and onlap/offlap patterns will not be globally uniform. The common assumption is that the increase in accommodation space on the shelf and in alluvial valleys during rising sea level and high-stand results in a starved condition in adjacent deep basins and development of a condensed section. Similarly, falling sea level and low-stand conditions are assumed to result in stream entrenchment accompanied by higher sedimentation rates in the deep basins (Posamentier *et al.*, 1988; Posamentier & Vail, 1988). We suggest that this condition is most appropriate for belts 4b to 6 (mid-latitudes) where climates become moister, and sediment transport capability increases during falling sea level and low-stand (phases B and C) with dryer climates during rising sea level and high-stand (phases D and A), as shown in Fig. 4 and Table 1. The other belts, in polar and equatorial regions, have a 180° phase shift in the relationship of sediment transport capacity and sea level. Accordingly, the sedimentary record of an area where fluvial sediment input becomes minimal as sea level falls (arid conditions) and increases significantly as sea level rises (more humid conditions) will be quite different from the record where fluvial sediment input becomes high as sea level falls (humid conditions) and decreases significantly as sea level rises (arid conditions) (Perlmutter & Matthews, 1989; Perlmutter *et al.*, 1991). The geographical pattern of climatic change has significant potential to influence the interpretation of third-, fourth- and fifth-order seismic-sequence stratigraphic boundaries.

TECTONICS AND ACCOMMODATION SPACE IN A FORELAND BASIN

The general characteristics of a stratigraphic package are determined by the combination of accommodation space and sediment supply. The spatial variation of the rate of change of subsidence places an upper limit on the amount of sediment (maximum fill potential) that can accumulate at a given location with a basin during a time of interest. The rate of change of sediment delivery to a basin and its distribution within the basin control the realization of this potential.

The integration of these factors can be understood by considering the evolution of a foreland basin. Subsidence is asymmetric, with the deepest point of the basin near the thrust mass (Price, 1973; Beaumont, 1981). Overall, basin subsidence begins slowly and increases as the load of the thrust mass increases, eventually reaching a maximum and gradually decreasing as thrusting ceases and erosion removes the excess load (Beaumont, 1981; Jordan, 1981).

Sediment supply may exceed accommodation during the initial phase as the thrust mass is building to maximum thickness. If discharge is sufficient, river systems bypass excess sediment out of the basin, selectively removing fine-grained material. As the basin matures and loading and thrusting reach a maximum, the rate of formation of accommodation space may exceed the rate of sediment input. This condition causes interior drainage, trapping all transported water and sediment within the basin and forming a lake if runoff is sufficient. Lakes commonly occupy the most rapidly subsiding portion

Table 1. End-member climatic sequences for each global cyclostratigraphic belt showing east and west coast modifications. East coast effects carry moisture onto the continents. West coast effects carry moisture onto the continents and moderate polar temperatures. The extent of inland penetration of these effects depends on the strength of the winds and the location of mountain barriers. The difference in the effects as a function of phase is due to the change in oceanic circulation pattern

Coastal effect	Belt 1 0–5°	Belt 2 5–15°	Belt 3 15–22°	Belt 4a 22–28°	Belt 4b 28–35°	Belt 5 35–40°	Belt 6 40–48°	Belt 7a 48–57°	Belt 7b 57–65°	Belt 8 65–90°
Climatic maximum phase A										
East				Tropical Humid	Tropical Humid	Temperate Humid	Temperate Humid	Temperate Humid		
Base climate	Tropical Very humid	Tropical Very humid to humid	Tropical Humid	Tropical Subhumid	Tropical Dry	Temperate Arid	Temperate Dry	Temperate Subhumid	Temperate Humid	Polar Humid to dry
West						Temperate Subhumid	Temperate Subhumid	Temperate Humid		
Cooling transition phase B1										
East			Tropical Humid	Tropical Humid	Temperate Humid	Temperate Humid	Temperate Humid			
Base climate	Tropical Very humid	Tropical Humid	Tropical Subhumid	Tropical Dry	Temperate Arid	Temperate Dry	Temperate Subhumid	Temperate Humid	Polar Humid	Polar Subhumid to arid
West					Temperate Dry	Temperate Subhumid	Temperate Humid		Temperate Humid	
Cooling transition phase B2										
East		Tropical Humid	Tropical Humid	Temperate Humid	Temperate Humid	Temperate Humid				
Base climate	Tropical Humid	Tropical Subhumid	Tropical Dry	Temperate Arid	Temperate Dry	Temperate Subhumid	Temperate Humid	Polar Humid	Polar Subhumid	Polar Dry to arid
West								Temperate Humid		

Climatic minimum phase C										
East		Tropical Humid Subhumid	Temperate Subhumid	Temperate Subhumid						
Base climate	Tropical Humid	Tropical Subhumid to dry	Temperate Arid	Temperate Dry	Temperate Subhumid	Temperate Humid	Polar Humid	Polar Subhumid	Polar Dry	Polar Arid
West							Temperate Humid			
Warming transition phase D1										
East		Tropical Humid	Tropical Humid	Temperate Humid	Temperate Humid	Temperate Humid				
Base climate	Tropical Humid	Tropical Subhumid	Tropical Dry	Temperate Arid	Temperate Dry	Temperate Subhumid	Temperate Humid	Polar Humid	Polar Subhumid	Polar Dry to arid
West									Temperate Humid	
Warming transition phase D2										
East			Tropical Humid	Tropical Humid	Temperate Humid	Temperate Humid	Temperate Humid			
Base climate	Tropical Very humid	Tropical Humid	Tropical Subhumid	Tropical Dry	Temperate Arid	Temperate Dry	Temperate Subhumid	Temperate Humid	Polar Humid	Polar Subhumid to arid
West					Temperate Dry	Temperate Subhumid	Temperate Humid		Temperate Humid	

of the basin between the thrust and the peripheral upwarp. However, the position of the shoreline and maximum depth of the lake can be shifted locally in regions or during intervals of high sediment supply. After thrusting is complete, subsidence is controlled by sediment loading and the basin gradually returns to a condition where sediment supply exceeds accommodation, forcing lacustrine regimes to evolve into fluvial systems.

Thus, on a very long-term time scale the general stratigraphic architecture of a foreland basin progresses from (i) external drainage to (ii) internal drainage and back to (iii) external drainage. If sufficient runoff is available, a continental basin centre will evolve from a fluvial system to a lacustrine system back to fluvial system. If, however, runoff is low, the sediment influx will also be low and likely much less than the available accommodation space. This condition can result in drainage that remains internal until a change occurs in either the tectonic regime or long-term climate. Under these circumstances, stratigraphy will be produced by mass wasting, ephemeral streams, evaporation and aeolian transport.

GREEN RIVER BASIN

The Green River Basin is located near the junction of Utah, Colorado and Wyoming in the USA. It is a foreland basin that developed in the early Tertiary in response to crustal loading associated with the formation of the Uinta Mountains. Basin subsidence continued intermittently throughout the Eocene. The basin was surrounded by uplifted Precambrian, Palaeozoic and Mesozoic rocks that rose about 2000 m above sea level. Volcanic activity began early and increased in frequency and intensity until the end of the Eocene when basin development ceased and regional erosion began to dissect the basin (Bradley, 1964).

Global cyclostratigraphic analysis

Global cyclostratigraphy provides a technique to evaluate the depositional responses and likely stratigraphic patterns that are expected to occur in different parts of a basin. Model predictions are iteratively refined by comparison to observed data and tested away from points of control. To illustrate the procedure, climatic maximum and minimum end-members and transitions of the Green River Basin are hindcast, using global cyclostratigraphic concepts, and expected stratigraphic responses to these climatic shifts are summarized. These predictions are then compared to the climatic conditions deduced from the basin sedimentary succession and the stratigraphic character of the Green River Formation. Global cyclostratigraphic analysis is based on the dynamic response of sedimentological processes to climatic change. In this respect it considers the sedimentological system to be out of equilibrium with climate, but moving in the direction of equilibrium. The exception to this dynamic condition occurs within phases A and C. These phases represent the time at which the climatic change slows down and reverses direction. At, or just after, the reversal the sedimentological condition must also reverse. Between these two reversals climate and the processes of sedimentation must be at equilibrium. Thus phases A and C are considered to be approximately equilibrium conditions and phases B(1 and 2) and D(1 and 2) dynamically strive towards equilibrium.

Average climate: hindcast

The Green River Basin (Fig. 7) was filled with sediment during the Eocene. It was located at about 45° N, placing it in cyclostratigraphic belt 6 (Fig. 6). This mid-latitude position with the attendant pattern of Ferrel cell circulation causes high humidity/runoff to occur with low temperature, 180° out of phase with respect to the usual occurrence of high humidity with high temperature (Fig. 4). The belt climate for the Basin (phase A, temperate/dry; phase B1, temperate/subhumid; phase B2, temperate/humid; phase C, polar/humid; phase D1, temperate/humid; and phase D2, temperate/subhumid) was further corrected for (i) prevailing wind directions (wet off the Pacific, dry off the continent); (ii) orographic effects associated with the bordering highlands (precipitating moisture with elevation increase); and (iii) proximity to a large body of water (moderating temperature and acting as a moisture source when the wind was off the water). These azonal modifications produced climatic conditions within the Green River Basin which are estimated to have varied from warm temperate/arid during the climatic maximum end-member (phase A), with dry summer winds from the west-southwest and dry winter winds from the north-northwest (Fig. 8a), to temperate/subhumid during the climatic minimum end-member (phase C), with dry summer winds out of the north-northwest and wet winter winds out of the west (Fig. 8b). The climatic transitions (phases B and D) were

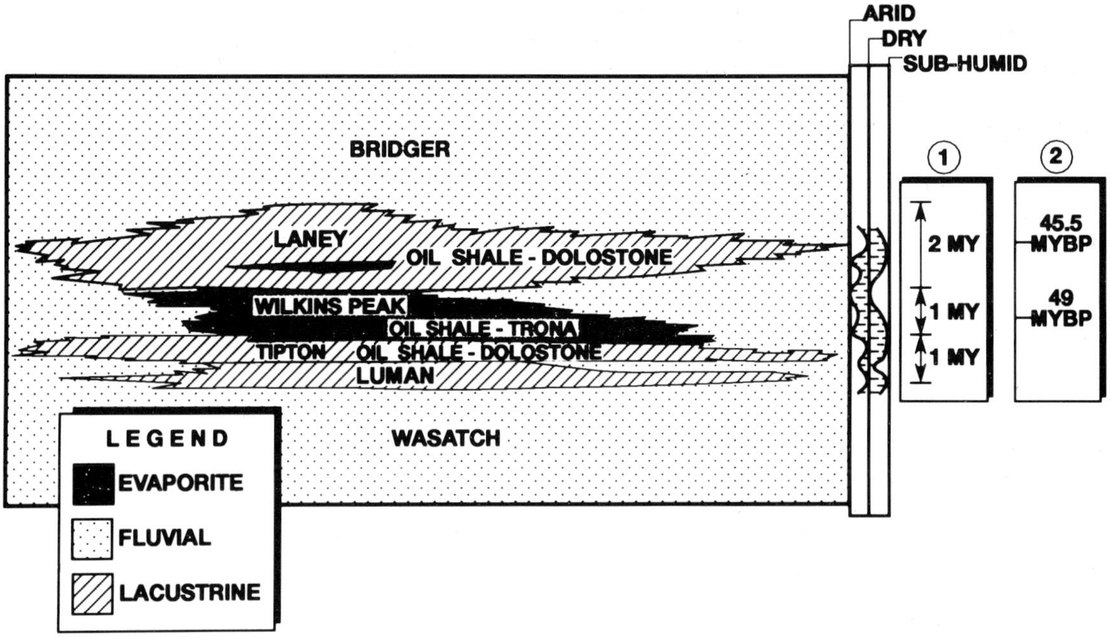

Fig. 7. Schematic section of the Green River Formation and its relationship to the Bridger and Wasatch formations. Facies distributions within the Green River Formation show their relative size and extent throughout the basin (after Bradley & Eugster, 1969). Estimated climatic maxima and minima are indicated, based on extent and salinity of the lake. Note that the climates are interpreted to be constrained by the climatic minimum and maximum end-member climates and that the envelope of short intermediate-term climates is not always symmetrical. Estimations of stratigraphic duration by Bradley (1963) and potassium–argon age dates from Surdam & Stanley (1980).

temperate/dry. Short- and long-term climates are therefore constrained within these bounds but may not always reach these bounds.

Analogies to today's climatic zonations conservatively suggest that mean precipitation varied from less than 20 cm/yr at the climatic maximum end-member to 80–120 cm/yr at the climatic minimum end-member. Mean evaporation is expected to have exceeded precipitation during the climatic maximum end-member and to have been 60–100 cm/yr during the climatic minimum end-member. Mean temperatures are estimated to rarely have reached freezing during the climatic maximum end-member, averaging 17–21°C annually, although winter temperatures below zero were likely during the climatic minimum end-member, with annual averages estimated at 9–14°C.

Depositional systems and stratigraphy: hindcasts

For the purpose of this example, the major provenance areas shedding sediment to the basin are assumed to (i) contain similar parent rock, (ii) be at about the same elevation, and (iii) have analogous climatic patterns.

Phase A. At the climatic maximum end-member, conditions are hindcast as temperate/arid (Fig. 9a). Physical processes governed weathering (Perlmutter & Matthews, 1992), producing predominantly coarse sediments (sand-size or greater). However, the lack of water in the system limited the rate of chemical breakdown of unstable minerals, sediment production, and transport from provenance areas. Consequently, sands were arkosic or contained abundant rock fragments, the volume of clay was relatively small, and sediment was stored in provenance areas and on basin margins. Infrequent precipitation and the abundance of coarse-grained sediments resulted in ephemeral braided rivers dominating the landscape and relatively steep alluvial fans near mountain fronts. Maximum rates of evaporation, coupled with minimum rates of runoff, limited the areal extent of the lake and minimized transport of sedi-

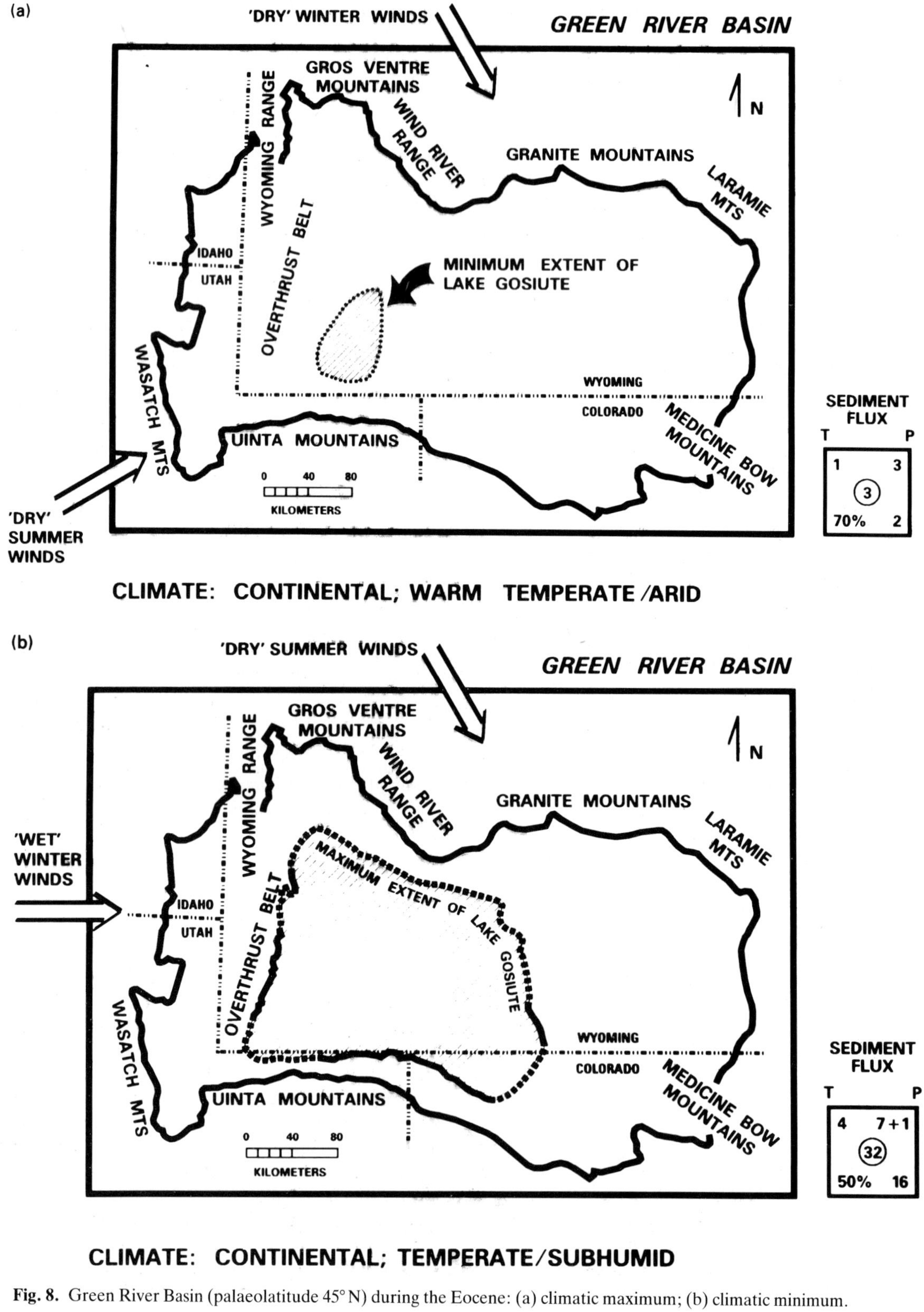

Fig. 8. Green River Basin (palaeolatitude 45° N) during the Eocene: (a) climatic maximum; (b) climatic minimum.

ment to the basin centre, enhancing the potential for chemically deposited carbonates and evaporites.

Phases B1 and B2. The climate during the transition from maximum to minimum is hindcast as temperate/dry (Fig. 9b). Early in the transition in phase B1, physical weathering dominated. As the cycle progressed to phase B2, physical weathering gradually decreased as chemical weathering increased. This change in the mechanism of weathering gradually decreased the proportion of feldspars and rock fragments in the sand and increased the volume of clay. Runoff increased from phase A through phases B1 and B2, flushing into the basin the sediment previously stored in provenance areas and basin margins during phase A. Alluvial fans expanded in volume and fan surfaces decreased in steepness. The early stages of the flush phase may have been accompanied by active downcutting and channel scour in marginal portions of the basin. The abundant coarse material available during the flush caused rivers to braid. Increasing runoff permitted the lake to expand. Locally, the lake margin shifted away from the basin margin as a result of the growth of alluvial fans. Increased sediment transport, production of clay-sized material, and freshening of the lake gradually shifted deposition in the lake from chemical to clastic.

Phase C. At the climatic minimum end-member, conditions are hindcast as temperate/subhumid (Fig. 9c). Chemical weathering slightly exceeded physical weathering, producing substantially more clays and increasing the weathering of feldspars and rock fragments compared to the previous phases. Alluvial fans were at their smallest volume due to conversion of coarse feldspars to fine clay, and at their lowest slope due to increased runoff. Steady runoff and increased suspended load of the rivers caused perennial meandering rivers to evolve. Although the total dissolved load of rivers was high because of increased chemical weathering, concentration was low due to higher runoff. Lakes achieved their maximum extent, becoming completely fresh. Preservation potential of organic material was relatively high due to high rates of production and transport of clay minerals. Warm summer temperatures resulted in a high rate of organic productivity. Winter overturning of the water column resulted in oxidation of the organics at the sediment–water interface. Together these two processes resulted in alternating organic rich and lean varved sediments.

Phases D1 and D2. The climate during the transition from climatic maximum to minimum is hindcast as temperate/dry (Fig. 9d). Early in the transition in phase D1, chemical weathering dominated. As the cycle progressed to phase D2, physical weathering gradually increased as chemical weathering decreased, gradually increasing the proportion of feldspars and rock fragments in the sands and decreasing the volume of shale. As runoff decreased, the rivers became underfit and braided as transport efficiency and production of fine material decreased and the percentage of coarse sediments increased. Alluvial fans expanded and surfaces became steeper as transport efficiency was reduced. The lake contracted as runoff decreased and evaporation potential increased. The drop in base level resulted in local fluvial downcutting near the lake margin. As the lake evaporated, it became saline and organic production decreased. Eventually salts were precipitated in the lake and organic preservation ceased as phase A was reached.

Previous work

The Wasatch, Green River and Bridger Formations (Fig. 7), deposited during the Eocene, reflect the dominant depositional style at different stages of the very long-term tectonic evolution of the basin (Bradley, 1964). Each of these formations displays depositional cycles at shorter frequencies, characteristic of Milankovitch-driven climate changes (Bradley, 1929, 1963; Deardoff & Mannion, 1971; Eugster & Hardie, 1975; Mauger, 1977; O'Neill *et al.*, 1981; Roehler, 1982; Bennett, 1989).

Average climate

The climate within the basin during the deposition of the Green River Formation has been characterized as varying from dry–arid/subtropical in the 'dryer' times to humid/subtropical in the 'wetter times' (Surdam & Stanley, 1979), with the dominant climate being similar to that of the modern day southeastern area of the USA (Bradley, 1929). Knowlton (1923) and Brown (1929) identified two populations of flora, one representing a warm, moist climate and the second a cooler and perhaps somewhat drier climate. These disparate flora were assumed to coexist in time and were respectively interpreted as arising from a mixture of warm wet lowland and cool dry upland terrains. Evidence for a

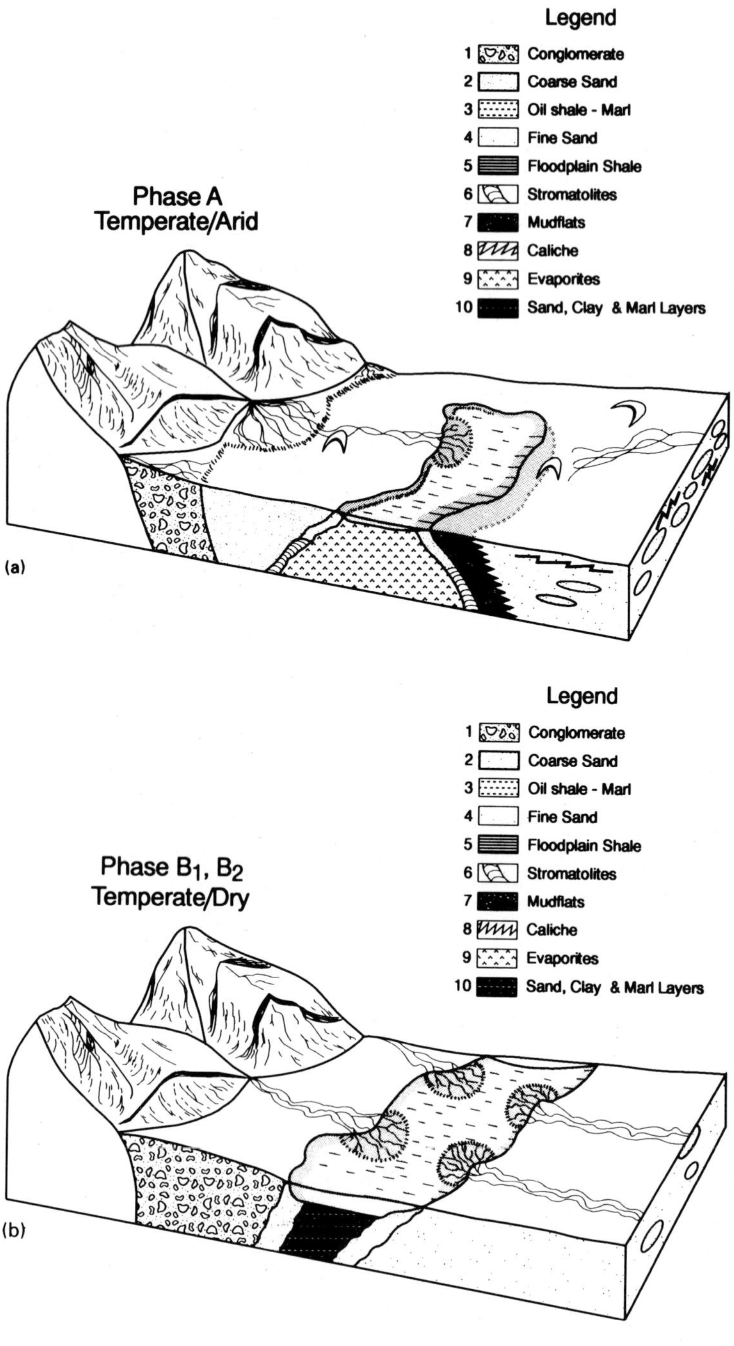

Fig. 9. Diagrammatic representation of the depositional systems of the Green River Basin during (a) phase A; (b) phase B1 and B2; (c) phase C; (d) phase D1 and D2.

drier lowland climate comes from a florule indicative of a subhumid savannah (Bradley & Eugster, 1969) and the presence of *Ephedra* pollen (Bradley, 1963) indicative of a hot dry climate similar to that of Southern California. The presence of a dry–arid climate is also indicated by the occurrence of extensive evaporites.

MacGintie (1969) estimated a mean annual temperature of 15–20°C, with no seasonal frost. Estimates of precipitation range from 60–75 cm/yr

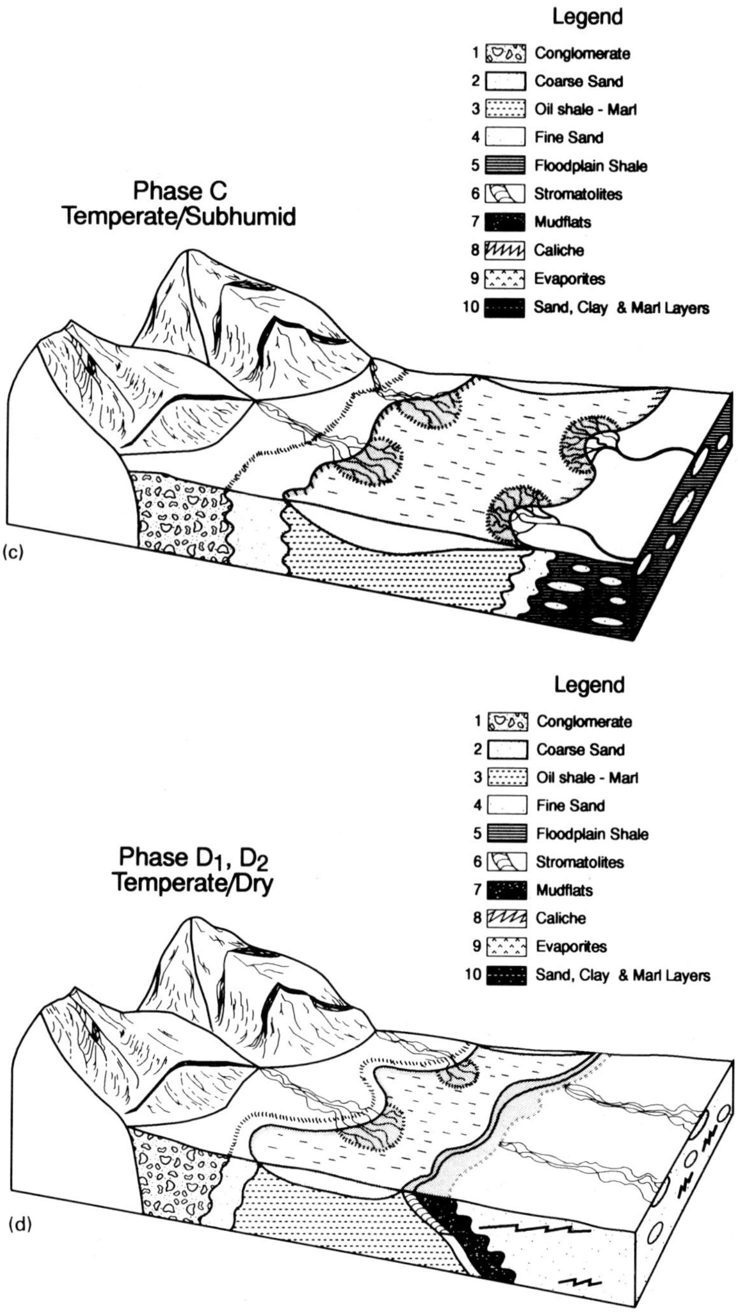

Fig. 9. *Continued.*

(MacGintie, 1969) to 85–95 cm/yr (Bradley, 1963) within the basin. Higher rates, 115 cm/yr, have been estimated for the watershed (MacGintie, 1969). Evaporation rates of 75–130 cm/yr have been proposed for the arid times (Bradley, 1963).

Depositional systems, stratigraphy and sedimentary cycles

The very long-term stratigraphic evolution of the Green River Basin followed the overall evolution of a foreland basin previously outlined and is diagrammatically represented in Fig. 7. Initially, sediment supply to the basin exceeded subsidence, resulting in sediment bypass and deposition of the fluvial Wasatch Formation. As subsidence increased, internal drainage developed, depositing the lacustrine Green River Formation in the deepest portion of the basin. Finally as subsidence decreased, conditions of sediment bypass deposited the fluvial Bridger Formation.

The Green River Formation is subdivided into four members (Fig. 7). The Luman Tongue, upper Tipton Member, as well as the lower and upper Laney Member were deposited during maximum lake extent ('freshwater' facies). The lower Tipton Member, Wilkins Peak Member, and the middle Laney Member were deposited during minimum lake extent ('evaporite' facies). All members of the Green River Formation contain well-known cyclic depositional sequences recorded in the lacustrine portions of the basin. Early in the twentieth century, Bradley (1929) recognized oil shale/marlstone couplets about 3 m thick, representing a duration of about 21 ka. He interpreted oil shales as indicative of a climate with a short, hot summer and a long, cool winter and the marls as being associated with a long, warm summer with a short, mild winter. Deardoff & Mannion (1971) described cyclic sequences in the Wilkins Peak Member: oil shale (wet climate); overlain by trona (arid climate); followed by fluvial sand–marlstone (increased transportation energy); and back to the oil shale. These units range from a few centimetres to about 15 m.

Despite thorough research, the exact periodicity of these cycles is still uncertain because of the uncertain duration of the stratigraphic units (Fig. 7). For example, Bradley (1963) estimated that the Wilkins Peak Member spanned approximately 1 Ma, based on accumulation rates and observed thickness, and rounding off to the nearest 0.5 Ma. Using this duration and Eugster & Hardie's (1975) estimate of at least 50 wet/dry oscillations during the deposition of the Wilkins Peak Member, the periodicity is calculated to be about 20 ka, close to the frequency of the precession cycle. However, the uncertainty implies that it could be as little as 10 ka or as much as 30 ka. In contrast to this Roehler (1982) estimated 1.5 Ma for the longest cycle and 100 ka for the shortest cycle, based on 70–75 cycles of oil shale and evaporite beds in the Wilkins Peak Member and an estimated time interval for these cycles was roughly based on potassium–argon dates on biotites in tuffs and numbers of varves. Thus reoccurrence time estimates range from very short term to long term. The primary cause of this uncertainty is believed to be a combination of insufficient time resolution in age dating techniques, non-constant depositional rates and missing section.

The precession cycle has also been recognized as the dominant periodicity recorded in varved lake sediments of the Green River Formation (Bradley, 1929). A power spectral analysis of variations in organic content estimated from sonic logs of this section shows power at the Milankovitch frequencies of roughly 20, 40 and 100 ka (Bennett, 1989). The age control on Bennett's technique is based on the extraction of non-uniform depositional rates, which are statistically estimated and related to the overall depositional rate. Other estimates of the dominant Milankovitch frequency include the 40-ka obliquity cycle based on a radiometric age date of the Laney Member (Mauger, 1977; O'Neill *et al.*, 1981).

Discussion

Average climate

Published interpretations of climatic conditions during deposition of the Green River Formation are generally consistent with estimates obtained using global cyclostratigraphic principles. Estimates of temperature ranges for the basin are consistent with the values we suggest for the climatic maximum end-member, and estimates of precipitation ranges more closely approximate values we suggest for the climatic minimum end-member.

Given the assumption of a relatively constant climate, floral evidence of warm climates is commonly assumed to represent the basin floor, and cooler climates are generally attributed to higher elevations in the drainage basin. Global cyclostratigraphic concepts suggest that the mixture of warm and cool flora may also arise due to climatic change, with the disparate floral evidence representing distinct climates that occurred sequentially rather than simultaneously.

Long-term changes in precipitation/evaporation within the Eocene Green River Basin are primarily estimated by changes in lake extent and changes in salinity from fresh to evaporitic, including the deposition of trona (Bradley & Eugster, 1969; Surdam & Stanley, 1979; and others). Based on the presence of *Ephedra* pollen (hot, dry climate) in both the traditional evaporative facies, such as the Wilkins Peak Member, and the traditional freshwater facies, such as the Luman Tongue Member (Bradley, 1963), we inferred that dry climates existed within the basin at both relative high-stand and low-stand conditions, supporting the idea of a more rapid climatic signal than the one associated with the progression from wet Luman to dry Wilkins Peak and wet Laney.

Depositional systems and stratigraphy

The very long-term evolution of the Green River Basin sedimentation (fluvial, lacustrine, fluvial) is in good agreement with the tectonic evolutionary component of the global cyclostratigraphic model. Within the Green River Formation we believe there is a complex interaction of the intermediate- and long-term Milankovitch parameters that results in the alternation of freshwater-dominated facies and evaporite-dominated facies on a time scale of 400 ka and longer, based on lake extent variation (Fig. 7). The climatic shifts responsible for these facies changes are bounded by the global cyclostratigraphic maximum and minimum end-member climates. Climatic cycling due to the shorter term Milankovitch parameters is constrained within the drift of these longer term climatic shifts, resulting in the deposition of thin drier facies within the predominantly freshwater sections and fresher water facies within the predominantly dry facies.

An example of the short-term, climatically driven, facies variation within the longer term freshwater-dominated facies is provided by the oil shale/marlstone couplets of the Luman Tongue described by Bradley (1929) as being related to the precession cycle. These facies variations record climatic minimum and maximum variations contained within the envelope of the longer term climatic cycles. During deposition of the fresher water oil shales, the subhumid global cyclostratigraphic climatic minimum was reached, providing an abundance of fresh water to the lake, encouraging organic production and transporting clay minerals to the basin centre. The marlstones represent deposition during maximum climatic conditions. However, the global climatic maximum end-member conditions (arid) are not believed to have been achieved due to the complex interaction of the long-term Milankovitch parameters, limiting the maximum climate during deposition of the Luman Tongue to temperate/dry (Fig. 7). These evaporative conditions resulted in decreasing chemical weathering and clastic transport, deposition of carbonate, and decreasing organic content in the lake. The restricted climatic range, subhumid to dry, did not permit the lake to become hypersaline because the drier periods were of insufficient duration and aridity to significantly evaporate the water provided during the subhumid phase. If subsidence is sufficient, a deep lake may exist under these conditions during the wetter phases.

Short-term cyclic deposition within the long-term evaporative facies is exemplified by the oilshale/trona/sandstone triplets of the Wilkins Peak Member described by Deardoff & Mannion (1971). Oil shale records the minimum climate (wetter), trona beds record the maximum climate (dryer), and the sandstone overlying the trona records the increased transport efficiency of the flush phase during the transition from maximum to minimum. The long-term Milankovitch interactions are inferred to result in a progressive decrease with time of the climatic maximum from the global climatic maximum end-member climate of temperate/subhumid to temperate/dry (Fig. 7). The climatic minimum is believed to have remained near the global climatic minimum end-member climate of temperate/arid. In addition to providing the conditions to deposit oil shale, the subhumid and dry climates provided sufficient mass transport of dissolved solids to permit thick evaporites to form during the arid climatic maximum. The increased proportion of 'dry' time, as well as the greater aridity, caused the lake to become supersaturated, precipitating trona. During these times the water balance of the lake is believed to have been such that a persistent deep lake could not form.

Although cyclicity of facies is well documented in the central portions of ancient lakes, the stratigraphic representation of the subhumid to arid/dry transition climate has not been specifically identified. Our modelling suggests that in the lake facies, the stratigraphic response to a wet-to-arid climatic transition is gradational and thin, the result of limited clastic transport to the basin centre caused by severely reduced runoff and the decreasing production of fine-grained sediment.

Corresponding changes in sedimentation in areas marginal to lakes have also gone unrecognized. Patterns in these areas are more difficult to recognize because: (i) these areas represent more diverse depositional conditions; (ii) they contain a higher diversity of lithologies; and (iii) they have a high proportion of missing section caused by local erosion and reworking.

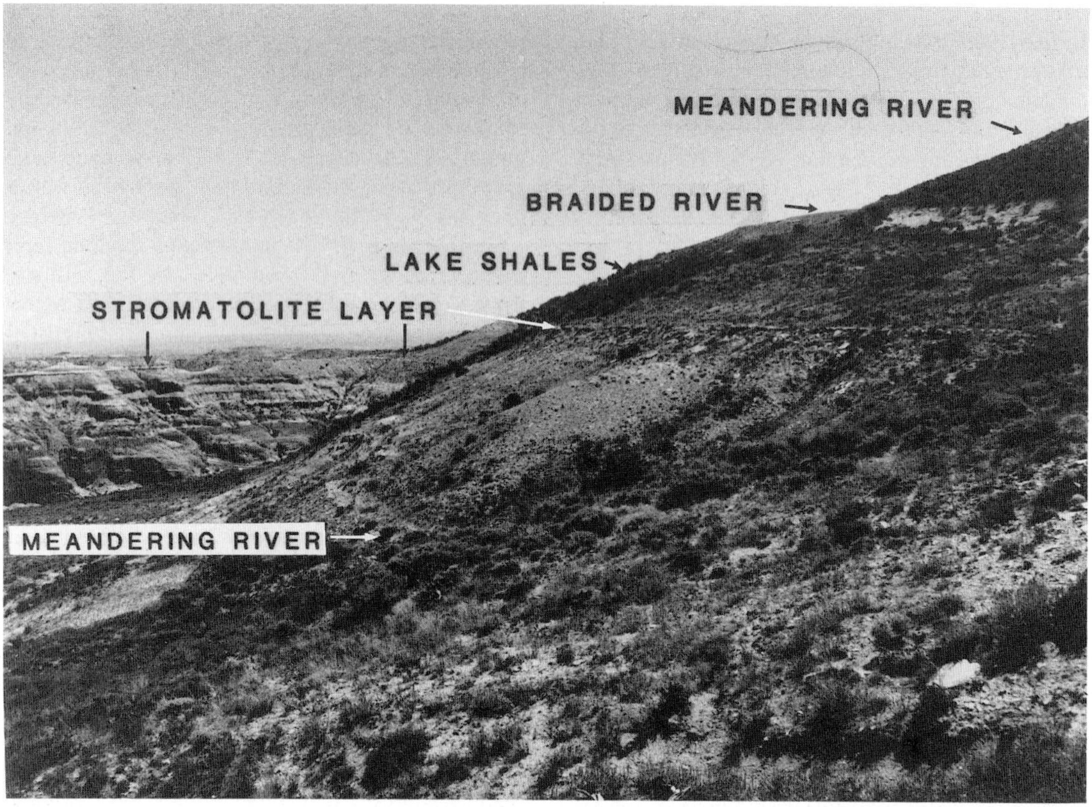

(a)

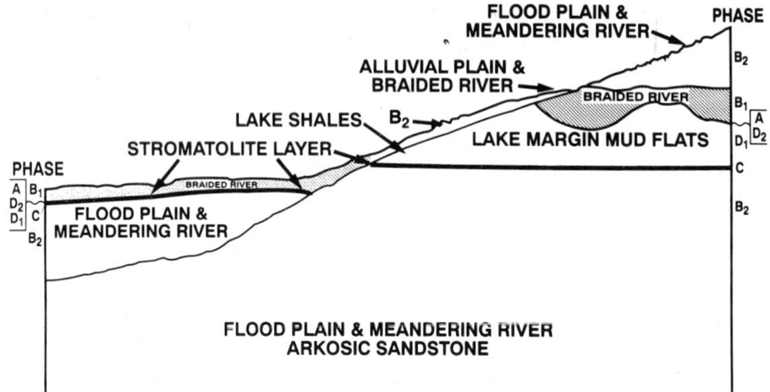

(b)

Fig. 10. (a) View of the Cathedral Bluff Tongue of the Wasatch Formation intercalated with the Laney Member of the Green River Formation showing cyclicity near the lake margin. The section is about 40 m. A complete cycle is estimated to be about 55 m high. (b) Diagrammatic representation of (a) showing the sections attributed to the appropriate phases. Note that erosion removes material deposited during phases A, D2 and parts of D1. Note also that the portion of the section attributed to phase B2 constitutes the majority of the section.

A typical sedimentological succession at a position slightly lakeward of the high-stand Green River Formation shoreline, looking toward the basin margin, is shown in Fig. 10. At the climatic minimum (phase C), deposition at this location is characterized by the oil shale facies. Montmorillonitic clays in the oil shales (Surdam & Stanley, 1980) support a wet weathering environment consistent with a high-stand lake. As the climate cycled back toward the maximum, the lake level fell (early D1) and brought lake-margin stromatolites, linked to falling lake level (Surdam & Wray, 1976), to this location. Green, lake-margin, mudflats replaced the stromatolites as lake level continued to fall (phase D1). Initially, basin margin sediments were reworked as base level dropped. However, as runoff waned, sediment production and transportation rates decreased, and weathered sediments became coarser. Lake-margin sediments developed mud cracks, Magadi-type cherts and caliche crusts, as the environment became arid (Surdam & Stanley, 1979). Reworking by ephemeral streams resulted in the formation of rip-up clasts (phase D2 and A). Locally the stratigraphic response to dropping base level was also regressive, downstepping deltas.

As the climate cycled from the climatic maximum (low-stand, phase A) towards the climatic minimum (high-stand, phase C), runoff increased (phase B1). Extensive local erosion occurred during the early part of phase B1, removing much of phases D2 and A as well as portions of the mudflat deposits formed in phase D1. Rivers eroded down to the stromatolite layers, but only rarely eroded through them. Fluvial 'flush' deposits sit on top of early erosional events. These deposits are sourced from the mobilization of both alluvial fan material on basin margins and sediment stored in drainage basins during phase A. A dry climate for phase B1 (flush) sediments is supported by the tendency of sands associated with rising lake level to be arkosic (Surdam & Stanley, 1980). Red floodplain deposits (associated with braided rivers) became more prevalent towards the end of phase B1. As the climate became wetter and cooler, stored sediment was exhausted and weathering increasingly altered feldspars to clay, causing the braided rivers of phase B1 to evolve into the meandering rivers of phase B2. As the lake continued to expand, the alluvial plain associated with the meandering rivers gave way to carbonate flats and then to mudflats adjacent to the lake margin. As the climatic minimum returned, the lake expanded once more, and oil shales were again deposited at this location.

SUMMARY

The global cyclostratigraphic model integrates the effects of climate and tectonics to estimate and understand the stratigraphic record. Tectonic controls on very long-term basin evolution are considered to affect potential sediment supply to a basin and the ability of the basin to accept it. Tectonic evolution controls the general stratigraphic configuration and style (lacustrine vs. fluvial). Short-, intermediate- and long-term climate affects the type of sediment produced, sediment transport potential, and sediment sorting and distribution within the basin. Climatic cycling governs the short-term stratigraphic facies relationships and changes, as well as the alternation of the dominantly freshwater facies and evaporite-dominated facies through the long-term interaction of the Milankovitch parameters. The particular strength of global cyclostratigraphy is the use of a globally synchronous, relational framework to characterize the dynamic range of stratigraphic relationships within a basin, permitting the range and style of stratigraphic variation to be hindcast.

Geological estimates of the general climatic setting of the Green River Basin (temperature, precipitation and evaporation) confirm the climate reconstructions of the model. More importantly, the model explains the climatic variability observed within the basin and links this variation to the stratigraphic patterns of the basin. The model predicts that the lake volume and configuration varied considerably and periodically at a variety of time scales, suggesting that interpretations of both a deep lake and a shallow lake model may be correct, depending on the time interval of study. The dynamic nature of the model also enables the worker to place a variety of disparate observations into a time sequence that forms a coherent pattern, permitting the incorporation of seemingly contradictory climatic and depositional evidence and points out the potential usefulness of increased stratigraphic resolution.

REFERENCES

Barron, E.J., Thompson, S.L. & Schneider, S.H. (1981) An ice-free Cretaceous? Results from climate model simulations. *Science* **212**, 501–508.

Beaumont, C. (1981) Foreland basins. *Geophys. J. R. Astron. Soc.* **65**, 291–329.

Bennett, M.W. (1989) *Milankovitch cyclicity in the Paleocene–Eocene Green River Formation, Utah and Colorado*. M.S. thesis, University of Utah.

Berger, A. (1984) Accuracy and frequency stability of the Earth's orbital elements during the Quaternary. In: *Milankovitch and Climate* (Eds Berger, A., Imbrie, J., Hays, J., Kukla, G. & Saltzman, B.). Reidel Publ. Co., Dordrecht, pp. 3–41.

Bradley, W.H. (1929) The varves and climate of the Green River Epoch. *US Geol. Surv. Prof. Pap.* **154**, 86–110.

Bradley, W.H. (1963) Paleolimnology. In: *Limnology in North America* (Ed. Frey, D.G.). University of Wisconsin Press, pp. 621–652.

Bradley, W.H. (1964) Geology of Green River Formation and associated Eocene Rocks in Southwestern Wyoming and adjacent parts of Colorado and Utah. *US Geol. Surv. Prof. Pap.* **496-A**.

Bradley, W.H. & Eugster, H.P. (1969) Geology and paleolimnology of the trona deposits and associated authigenic minerals of the Green River Formation of Wyoming. *US Geol. Surv. Prof. Pap.* **496-B**.

Broecker, W.S. & Denton, G.H. (1989) The role of ocean–atmosphere reorganizations in glacial cycles. *Geochim. Cosmochim. Acta* **53**, 2465–2501.

Brown, R.W. (1929) Additions to the flora of the Green River Formation. *US Geol. Surv. Prof. Pap.* **154**, 279–293.

Budyko, M.I. (1969) The effect of solar radiation variations on the climate of the Earth. *Tellus* **21**, 611–619.

Deardorff, D.L. & Mannion, L.E. (1971) Wyoming trona deposits. *Contrib. Geol.* **10**, 25–38.

Eugster, H.P. & Hardie, L.A. (1975) Sedimentation in an ancient playa-lake complex: The Wilkins Peak Member of the Green River Formation of Wyoming. *Geol. Soc. Am. Bull.* **86**, 319–334.

Frakes, L.A. (1979) *Climates Through Geologic Time*. Elsevier, New York.

Glennie, K.W. (1984) Early Permian–Rotliegend. In: *Introduction to the Petroleum Geology of the North Sea* (Ed. Glennie, K.W.). Blackwell Scientific Publications, Oxford, pp. 41–60.

Haq, B.U., Hardenbol, J. & Vail, P.R. (1988) Mesozoic and Cenozoic chronostratigraphy and cycles of sea level changes. In: *Sea-level Changes – An Integrated Approach* (Eds Wilgus, C.K., Hastings, B.S., Posamentier, H., Van Wagoner, J., Ross, C.A. & Kendall, C.G. St C.). Soc. Econ. Paleont. Mineral. Spec. Publs 42, pp. 71–108.

Jordan, T.E. (1981) Thrust loads and foreland basin development, Cretaceous, Western United States. *Am. Assoc. Petrol. Geol.* **65**, 2506–2520.

Kemper, E. (1987) Das Klima der Kreide-Zeit. *Geol. Jb. A* **96**, 5–185.

King, L.C. (1961) The paleoclimatology of Gondwanaland during the Paleozoic and Mesozoic eras. In: *Descriptive Paleoclimatology* (Ed. Nairn, A.E.M.). Interscience Publishers, London, pp. 307–331.

Knowlton, F.H. (1923) Revision of the flora of the Green River Formation. *US Geol. Surv. Prof. Pap.* **154**, 133–197.

Lamb, H.H. (1961) Fundamentals of climate. In: *Descriptive Paleoclimatology* (Ed. Nairn, A.E.M.). Interscience Publishers, London, pp. 8–44.

Lockwood, J.G. (1979) *Causes of Climate*. John Wiley & Sons, New York.

Lockwood, J.G. (1980) Milankovitch theory and ice ages. *Prog. Phys. Geogr.* **4**, 79–87.

MacGintie, H.D. (1969) *Eocene Green River Flora of Northwestern Colorado and Northwestern Utah*. California University Publications, Geological Sciences, 83.

Matthes, F.E. (1939) Report of Committee on glaciers April 1939. *Am. Geophys. Union Trans.* part 4, 518–523.

Matthews, R.K. & Frohlich, C. (1991) Orbital forcing of low-frequency glacioeustasy. *J. Geophys. Res.* **96**(B4), 6797–6803.

Mauger, R.L. (1977) K–Ar ages of biotites from tuffs in Eocene rocks of the Green River, Washakie and Uinta basins, Utah, Wyoming and Colorado. *Univ. Wyoming Contrib. Geol.* **15**, 17–42.

Milankovitch, M.M. (1941) *Canon of Insolation and the Ice Age Problem*. Transactions of Royal Serbian Academy, Beograd (former Yugoslavia). (English translation by Israel program for Scientific Translation and published for the US Department of Commerce and the National Science Foundation.)

Miller, A. (1966) *Meteorology*. Merrill Publ. Co., Columbus.

O'Neill, W.A., Sutter, J.F. & Stanley, K.O. (1981) Sedimentation of Rich Oil Shale Sequences in Eocene Lakes Gosiute and Uinta, Wyoming and Utah. In: *Abstracts Rocky Mountain Section G.S.A. April 16–17*, Vol. 13(4), p. 222.

Penck, A. (1914) The shifting of climatic belts. *Scott. Geogr. Mag.* **30**, 281–293.

Perlmutter, M.A. & Matthews, M.D. (1989) Global cyclostratigraphy: Effects on the timing of sediment delivery to continental margins relative to sea level. In: *Proceedings of the AGU Chapman Conference on Long Term Sea Level Changes, Snow Bird, Utah, April 17–20*.

Perlmutter, M.A. & Matthews, M.D. (1990) Global cyclostratigraphy – a model. In: *Quantitative Dynamic Stratigraphy* (Ed. Cross, T.A.). Prentice Hall, New Jersey, pp. 233–260.

Perlmutter, M.A. & Matthews, M.D. (1992) Global cyclostratigraphy. In: *Encyclopedia of Earth Systems Science* (Ed. Nierenberg, W.A.). Academic Press, California, pp. 379–393.

Perlmutter, M.A., Liro, L.M. & Matthews, M.D. (1991) A global cyclostratigraphic evaluation of sediment flux to continental margins. In: Soc. Econ. Paleont. Mineral. First Annual Theme Meeting, Continental Margins, pp. 26–27.

Pitman, W.C. & Golovchenko, X. (1983) The effect of sealevel changes on the shelf edge and slope of passive margin. In: *The Shelfbreak: Critical Interface on Continental Margins* (Eds Stanley, D.J. & Moore, G.T.). Soc. Econ. Paleont. Mineral. Spec. Publs 33, pp. 41–58.

Posamentier, H.W. & Vail, P.R. (1988) Eustatic controls on clastic deposition II – sequence and system tract models. In: *Sea Level Changes – An Integrated Approach* (Eds Wilgus, C.K., Hastings, B.S., Posamentier, H.W., van Wagoner, J., Ross, C.A. & Kendall, C.G. St C.). Soc. Econ. Paleont. Mineral. Spec. Publs 42, pp. 125–154.

Posamentier, H.W., Jervey, M.T. & Vail, P.R. (1988) Eustatic controls on clastic deposition I – conceptual framework. In: *Sea Level Changes – An Integrated Approach* (Eds Wilgus, C.K., Hastings, B.S., Posamentier, H.W., Ross, C. & Kendall, C.G. St C.). Soc. Econ. Paleont. Mineral. Spec. Publs 42, pp. 109–124.

Price, R.A. (1973) Large-scale gravitational flow of supracrustal rocks, Southern Canadian Rockies. In: *Gravity and Tectonics* (Eds DeJong, K.A. & Scholten, R.). Interscience, John Wiley & Sons, New York, pp. 491–502.

Roehler, H.W. (1982) Physical evidence for saline cycles

of deposition in Eocene Lake Gosiute in Southwest Wyoming. *Bull. Am. Assoc. Petrol. Geol.* **66**, 1698–1699.

SARNTHEIN, S. & DIESTER-HAASS, L. (1977) Eolian-sand turbidities. *J. Sedim. Petrol.* **47**, 868–890.

STREET, F.A. & GROVE, A.T. (1979) Global maps of lake level fluctuations since 30 000 yr BP. *Quat. Res.* **12**, 83–118.

SURDAM, R.C. & STANLEY, K.O. (1979) Lacustrine sedimentation during the culminating phase of Eocene Lake Gosiute, Wyoming (Green River Formation). *Geol. Soc. Am. Bull.* **90**, 93–110.

SURDAM, R.C. & STANLEY, K.O. (1980) The stratigraphy and sedimentologic framework of the Green River Formation, Wyoming. In: *Wyoming Geol. Assoc. Guidebook, Stratigraphy of Wyoming*, pp. 205–221.

SURDAM, R.C. & WRAY, J.L. (1976) Lacustrine stromatolites, Eocene Green River Formation, Wyoming. In: *Stromatolites* (Ed. Walter, M.R.). Elsevier, Amsterdam, pp. 535–541.

VAIL, P.R., MITCHUM, R.M. Jr & THOMPSON, S. III (1977) Seismic stratigraphy and global changes in sea level, part 3, relative changes in sea level from coastal onlap. In: *Seismic Stratigraphy – Applications to Hydrocarbon Exploration* (Ed. Payton, C.E.). Am. Assoc. Petrol. Geol. Mem. 26, pp. 63–81.

Spec. Publs Int. Ass. Sediment. (1994) **19**, 483–507

Reading orbital signals distorted by sedimentation: models and examples

T.D. HERBERT

Geological Research Division, A-015, Scripps Institution of Oceanography, La Jolla, CA 92093, USA

ABSTRACT

A stratigraphic column never gives a precisely linear representation of fluxes of sediment components over time. Weight percent measures generally bear a non-linear relationship to accumulation rates, and variations in accumulation rates also distort the depth axis. Orbital climatic signals in sediments provide a good template for measuring such patterns of distortion. A simple two-component model, in which both components may be time-variant, is developed to study stratigraphic patterns of cyclic sedimentation. Four types of harmonic distortions are recognized: *self modulation*, *cyclical modulation*, *linear modulation* and *step-function modulation*. By properly sampling cyclic waveforms, it may be possible to solve for the dynamics of sedimentation at very high resolution.

The highest frequency distortions are generated by changes in sediment accumulation caused by the climatic effects of a particular orbital cycle. In the depth domain, these generate skewed distributions of the measured variables, and in the frequency domain they generate harmonics of the fundamental frequency. Modelling shows that it is possible to determine which component was more time-variant over the course of a climatic cycle by looking at the convexity/concavity of the waveform. The concept of 'accumulation' versus 'dilution' cycles in carbonate sediments is rigorously defined by critical ratios of model parameters.

Longer term changes in sedimentation rate modulate the stratigraphic frequency of orbital signals. A bewildering splitting of spectral peaks may occur even in the absence of 'noise'. Models show that the highest frequency sedimentary cycles are preferentially distorted, in accord with geological observations. Distortions require that the usual definition of the Nyquist frequency, the minimum sampling rate, be greatly modified. One approach to measuring changes in sedimentation rates is moving-window spectral analysis, whereby only subsets of the time series are considered for each spectral estimation; another is through bedding thickness analysis.

Effects similar to those simulated are observed in real sedimentary records. Late Pleistocene carbonate cycles from the equatorial Pacific seem to result from out-of-phase variations in carbonate and non-carbonate accumulation. Cyclical trends in sedimentation rate may be caused by the climatic effects of low-frequency orbital cycles. Carbonate bedding cycles in the Albian of Italy show probable eccentricity modulation at periods of about 100 and 400 ka, and 100-ka carbonate cycles show modulation at a period of about 400 ka. Increases in sedimentation rate detected by moving-window spectral analysis correspond to increases in mean carbonate content, suggesting a simple relationship between the two variables. Nearly linear frequency modulations caused by changes in sedimentation rate can be seen in moving-window spectral analysis of physical property data of equatorial Pacific and Atlantic cores over the past 5 Ma. Step-function changes in sedimentation rate may be rare, but occur at the Cretaceous/Tertiary (K/T) boundary, where precessional carbonate cycles record an abrupt twofold decrease in sedimentation rate in the earliest Tertiary.

INTRODUCTION

Climatic fluctuations often leave an imprint in the deep-sea record by changing the relative proportions of the major sediment components. Thus ever since the classic work of Arrhenius (1952) in the central equatorial Pacific, studies have shown a link between fluctuations in carbonate, opal and detrital fractions of pelagic sediments and the glacial–interglacial cycles of the Pliocene and Pleistocene (Hays *et al.*,

1969; Dunn & Moore, 1981; Ruddiman *et al.*, 1986; deMenocal *et al.*, 1991). Spectral analyses of climate records over the past few millions of years typically show features matching quasi-periodicities in the Earth's orbital geometry (Imbrie *et al.*, 1984). If the accumulation of major components of the sediment are varying over time, as they must be to produce the observed records, sedimentary climatic series will have inherent distortions in the translation from time to stratigraphic depth. The usual assumption of a linear relation of depth to time will only be true in the most fortuitous case, for example if a decrease in carbonate accumulation is exactly counterbalanced by an increase in opal or detrital flux. Unresolved time–depth distortions may badly complicate frequency analyses by smearing and splitting spectral peaks (Schiffelbein & Dorman, 1986; Weedon, 1989). Furthermore, the pace of accumulation rates should give useful palaeoclimatic information if it can be quantified.

Significant changes in deep-sea sedimentation rates on the orbital (10^4–10^5 years) time scale can be documented in the relatively well dated late Pleistocene. By tuning core time scales so that maxima and minima of oxygen isotopic variations align with inferred orbital forcing, Martinson *et al.* (1987) derived a sedimentation rate curve for the stacked benthic isotope record of Pisias *et al.* (1984) that varies by a factor of four (their fig. 19). Both higher frequency (*c.*20–40 ka) and lower frequency trends are visible, although the dominant fluctuations in sedimentation rate are in-phase with the 100-ka glacial cycles of the late Pleistocene. Hovan *et al.* (1989) used orbital tuning of an oxygen isotope record to estimate large glacial–interglacial variations in dust blown eastward from the Chinese loess deposits to the deep Pacific. Thorium-230 age profiling is an alternative approach to estimating carbonate and non-carbonate fluxes in the ocean over the last glacial cycle (Bacon, 1984; Francois *et al.*, 1990; Yang *et al.*, 1990). Atlantic carbonate cycles above the lysocline seem to be caused by twofold changes in carbonate production acting in opposition to two- to threefold variations in detrital input (Bacon, 1984; Francois *et al.*, 1990). Pacific carbonate changes over the past 18 ka may result from two- to threefold increases in dissolution rate from glacial to interglacial periods, superimposed on nearly constant non-carbonate accumulation (Yang *et al.*, 1990).

The problem of accurately reconstructing the sedimentary response to orbital forcing is still far from solved, however. The approach taken in this paper is one particularly appropriate to the analysis of cyclic sedimentation in sediments older than 3 Ma, where the accuracy of celestial mechanical calculations is no longer good enough to provide the exact match between absolute age and sedimentary cycles that is possible in the late Pliocene and Pleistocene (see Imbrie *et al.*, 1984; Martinson *et al.*, 1987; Shackleton *et al.*, 1990), and time control through bio/magnetostratigraphy also diminishes in precision. I shall argue that several regular patterns of distorted cyclical waveforms may be observed in deep-sea sediments, each giving valuable information on sedimentary dynamics at different time scales.

The model that follows has been developed to describe cyclic variations in *intensive* variables: measures such as calcium carbonate percentage, or percentage of a given fossil species, that inherently depend on the amount of the sediment constituent or fossil. Variations in intensive measures nearly always require changes in sedimentation rate. While variations in *extensive* measures such as carbon and oxygen isotopic ratios imply changes in the environment that might modulate sediment accumulation, the connection is less direct. Variables such as carbonate content are also easily measured, and importantly, persist into older portions of the geological record (for a summary, see Fischer, 1986) where, because of the small primary signal in the absence of ice volume changes, or because of diagenetic alteration, isotopic measures may not be sensitive recorders of climatic variability.

Any model simplifies reality. However, a modelling approach can add perspective to a subject as complex as climatic frequency analysis of sediments, and guide sampling strategies. The model presented is an improvement on the usual assumption that one sedimentary component is invariant (e.g. Dean *et al.*, 1981; Ricken, 1991). Each discussion of modes of depth and amplitude distortions is accompanied by examples from the geological record. I am glad to acknowledge the insights of a number of other studies which have considered the problem of recovering time–depth relations in sedimentary sequences (de Boer, 1982; House, 1985; Schiffelbein & Dorman, 1986; Martinson *et al.*, 1987; Park & Herbert, 1987; Weedon, 1989; Kominz & Bond, 1990; Ricken, 1991; Schwarzacher, 1991).

AMPLITUDE AND FREQUENCY MODULATIONS DURING SEDIMENTATION

One never directly measures the fluxes (units of $g/cm^2/ka$) of components against time in a sedimentary section. Rather, the data consist of the relative abundance of a component, or the ratios of components, against depth. Fluxes can be estimated only on time scales equal to or longer than the spacing of reliable chronostratigraphic tie points, typically every 10^5 to 10^6 years. The model calculations below show that observable features in well-sampled cyclic series, such as the difference between maxima and minima (weight percent amplitude) of sedimentary cycles, the waveform of the cycles, and spectral features revealed in frequency analyses, may be useful in defining the dynamics of the sedimentary system more precisely, and on shorter time scales, than is possible by conventional stratigraphic methods alone.

Any non-linearity in the recording of palaeoclimatic variations in sediments results in frequency distortion of the initial signal. Systematic time–depth distortions of a periodic signal can be modelled as frequency modulation, while the transform between flux and weight percent variations can be idealized as amplitude modulation (Schwarzacher, 1991). Frequency modulation occurs when changing sedimentation rate modifies the stratigraphic representation of a periodic signal. In this case, the accumulation rate of the sediment column has the general form

$$ACC(x) = M + S\sin[(1 + V(t))\omega t],$$

where $ACC(x)$, the instantaneous flux of component A at level x, varies around a mean value M with amplitude S. Sedimentation rate oscillates according to the basic frequency ω and an additional sedimentation rate function $V(t)$. An amplitude modulated signal has the form

$$A(t) = M + [S + V(t)]\sin(\omega t)$$

The amplitude S around the mean of the signal $A(t)$ is modulated by the function $V(t)$. One can distinguish at least four simple distortions of both frequency and amplitude in the sedimentary recording of palaeoclimatic oscillations: *self modulation*, *cyclic modulation* by frequencies lower than ω, *linear modulation* and *step-function modulation*.

We will consider the simplest, but surprisingly common, case of sedimentary variability: a two-component system. The two mathematical components may each be composed of several covarying minerals or fossil groups. For example, in Mid-Cretaceous cyclic sediments of Italy, biogenic silica and carbonate covary in a nearly linear fashion (Herbert *et al.*, 1986). In equatorial Pacific sediments of Pleistocene age, most variance can be described as a mixing between a pure carbonate and an opal-plus-detrital end-member.

Averaged over some time interval, each component will have a mean flux, designated M_A and M_B respectively, and a sensitivity to orbital forcing, S_A and S_B (the sensitivity terms could of course be split into coefficients for precessional, obliquity and eccentricity forcing). Considering first the simplest case of cyclic sedimentation, the flux of component A ($g/cm^2/ka$) is:

$$F_A(t) = M_A + S_A\sin(\omega t) \quad (1)$$

and that of component B is

$$F_B(t) = M_B + S_B\sin(\omega t + \psi_B) \quad (2)$$

If the variable being measured is intensive, such as weight percent of a mineral, then

$$\%A(t) = \frac{F_A(t)}{F_A(t) + F_B(t)} \times 100 \quad (3)$$

For simplicity, the mean flux of component A, M_A is fixed at 1 arbitrary unit; the sensitivities of the two components, S_A and S_B, can range from 0 to 1 unit. Fluxes are prevented from becoming negative by resetting negative values to zero. The ψ_B term contains the phasing of the flux variations of B relative to A, which for simplicity we will assume to be either in-phase ($\psi_B = 0°$), or exactly out-of-phase ($\psi_B = 180°$). Component A will be equated with carbonate and component B with the non-carbonate fraction, although any variable that might be related to changes in accumulation rate, such as oxygen or carbon isotopic values or faunal or floral ratios, could be represented in a similar manner. Extensive palaeoclimatic variables, such as isotopic ratios, will not, however, exhibit the amplitude modulated effects discussed below.

By allowing the accumulation of the two components to build a sediment column, we can better understand the types of time–depth distortion that naturally arise as a result of climatic variation. The sedimentation rate (cm/ka) is the sum of the two fluxes modified by their grain densities (ρ_i) and characteristic porosities (ϕ_i):

$$Sr(t) = \sum \frac{F_i}{P_i(1 - \phi_i)} \quad (4)$$

In non-siliceous settings, the grain densities of carbonate and non-carbonate fractions are nearly equal (2.6–2.7 g/cm^3), but very opal-rich sediments may have grain densities of 2.3–2.5 g/cm^3 (Wilkens & Langseth, 1983). The porosity of soft sediment shows an inverse relationship to carbonate content in most pelagic settings (Mayer, 1991; Herbert & Mayer, 1991, and references therein). In non-compacted sediments, porosity variations are large enough to cause considerable time–depth distortion, even under conditions of constant total mass flux: 1 g of pure carbonate material (ϕ = 55%) will produce a 0.9-cm layer, while 1 g of clay-rich sediment (ϕ = 87%) will produce a layer 2.85 cm thick. Relative porosity differences decrease during compaction, so that sediment thickness is nearly proportional to mass in older, more deeply buried sediments.

Orbitally driven sedimentary series are likely to exhibit multiple waveform distortions over the range of orbital periods of 21–400 ka. Long-term drifts in fluxes of components due to tectonic or evolutionary developments may cause non-periodic modulations. The following sections divide the distortions into different classes, and discuss the implications of each mode for the interpretation of frequency spectra.

Within-waveform distortion

We first consider the simple case outlined in equations (1)–(4), which describe how a sinusoidal climatic variation may be translated into a stratigraphic signal. Both amplitude and frequency modulation

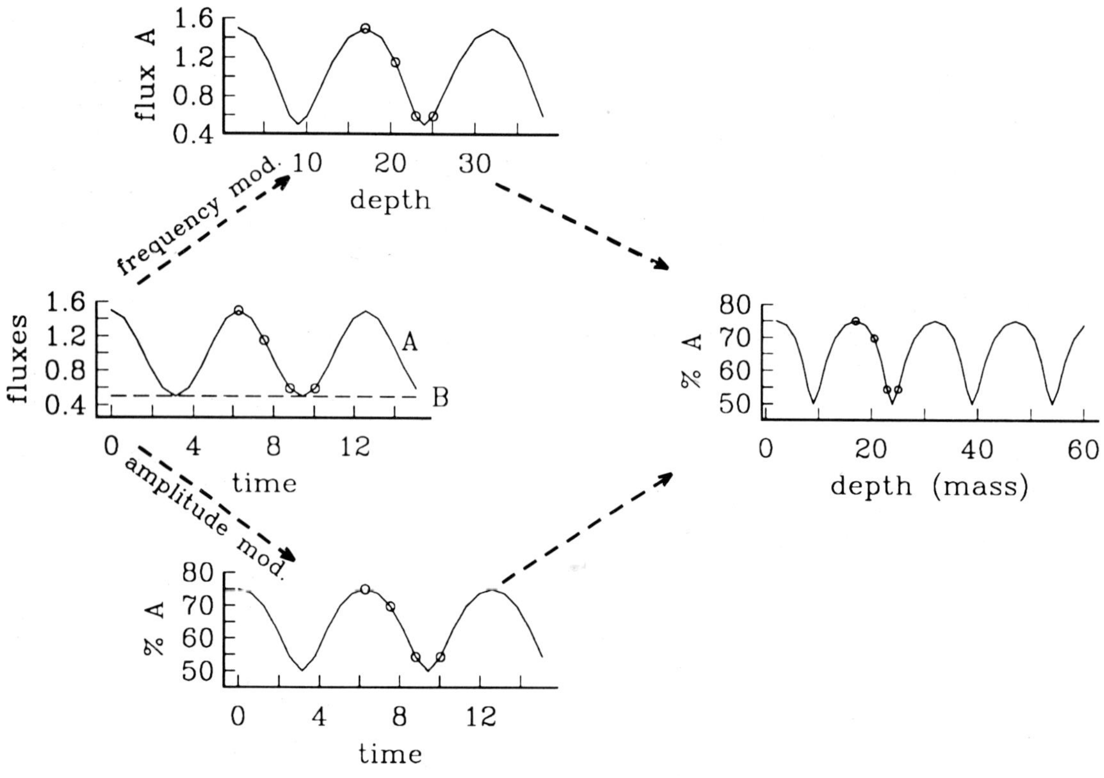

Fig. 1. A climatically driven sinusoidal variation in flux of component *A* (leftmost curve) will be distorted in the sediment column in amplitude by the weight percent measure (lower diagonal) and in spacing by changing sedimentation rate (upper diagonal) to give a sedimentary representation against depth, the combination of the two translations. The flux of component *B* has been held constant in this case; the mean component *A* percentage is 66.7. Component *A* will be equated with carbonate for remaining figures, component *B* with insoluble residue. The amplitudes and waveforms of sedimentary cycles will depend on the sensitivities and phase relations of the fluxes of carbonate and non-carbonate material, but will nearly always lead to distortions of the primary climatic frequencies.

caused by the cyclic oscillation itself may lead to distortions in the stratigraphic column. Amplitude distortion is likely to result whenever closed sum data are collected. For example, carbonate weight percent typically bears a non-linear relationship to variations in carbonate flux (Dean *et al.*, 1981). The degree of non-linearity of the stratigraphic series will depend on the range of carbonate content encountered. Normalizing carbonate content to a time-invariant component restores linearity, but requires correct identification of the appropriate time-invariant component.

Within-waveform amplitude and frequency distortion in sedimentary series create artefacts that must be considered in order to interpret spectra correctly. A simple example illustrates how sedimentary non-linearities tend to take amplitude away from the forcing frequencies. The climatically varying signal is shown in Fig. 1 as a function of time. The flux of component *B* is held constant for simplicity, the mean fraction of component *A* is 66.7% and the relative variation of *A* about its mean is 50%. All spectra shown will be linear amplitude Fourier spectra, and are calculated with the Thomson (1982) multiple taper technique. In addition to its other virtues, the *F*-test ratio of signal to noise estimated by the method is useful in measuring statistical degradation of primary frequencies by distortions. Amplitude distortion due to the non-linearity of the weight percent measure is shown *as a function of time* (Fig. 1) so that we may ignore the effects of changes in sedimentation rate. The Fourier representation of the amplitude distorted waveform contains the fundamental frequency plus integer multiples of the fundamental termed harmonics (Fig. 2A). *F*-test values of the fundamental are not degraded by this type of distortion. As distortion increases (increasing non-linearity), the number of harmonic terms grows, as does their amplitude relative to the fundamental (Schiffelbein & Dorman, 1986).

Instantaneous sedimentation rate in the example above varies by a factor of two, distorting the time–depth relationship in a predictable way (Fig. 1). In this example, the dry bulk density and porosity of both components are taken to be equal for simplicity. The bulk accumulation rate follows the flux of component *A*. The minima of the weight percent curve are therefore condensed, while the peaks are stretched along the depth axis. The spectral effect (Fig. 2B) is again to add distortion in the form of harmonics (Schiffelbein & Dorman, 1986; Weedon, 1989) the largest of which in this case is about 16% of the fundamental. Spectral analysis of the weight percent series, sampled as a function of depth, shows that the combined amplitude and depth distortions are nearly additive: the relative peak height of the first harmonic now rises to 30% of the fundamental peak (Fig. 2C). The calculated height of the fundamental is reduced by 9% relative to its true amplitude. Sedimentary distortions therefore not only lower the amplitude of fundamental climatic frequencies, but also introduce spurious high-frequency peaks. This effect will be particularly important when multiple periodicities are present in the data, as we shall see later.

More interesting and realistic synthetic series can be constructed when both sedimentary components are allowed to vary over time. We will consider cases where the accumulation of both components oscillates in-phase or exactly out-of-phase. Out-of-phase relationships imply opposing responses to the orbital forcing, not different response times of the

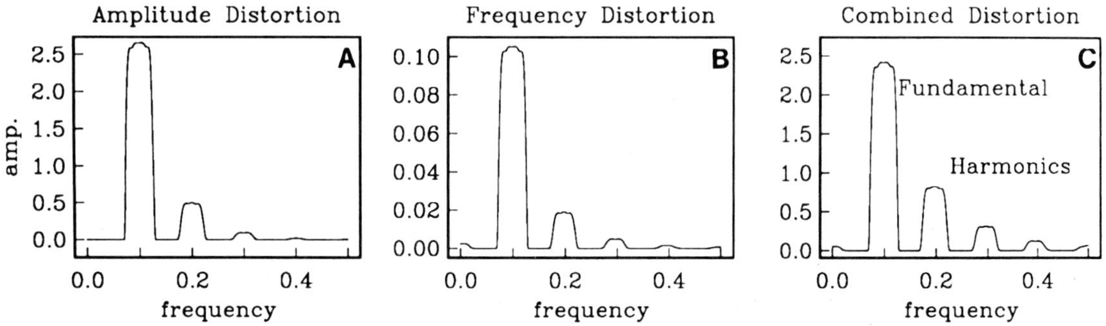

Fig. 2. Amplitude spectra calculated from the time and depth series of Fig. 1. Sedimentary non-linearities generate harmonic distortion of the fundamental palaeoclimatic frequency (note that distortion is additive).

two components; the more complicated question of time lags in a sedimentary system is deferred.

Covariance of the two sedimentary components tends to mask substantial changes in their accumulation when one relies on a weight percent measure. The critical control of both the amplitude (difference between weight percent maxima and minima) of the time series, and the shape of cyclic waveforms, is the ratio of the two sensitivities (S_A/S_B) relative to ratio of the mean fluxes of the two components (M_A/M_B). As noted previously by House (1985) and Ricken (1991), the sensitivity of weight percent carbonate as a palaeoclimatic proxy will vary depending on the mean carbonate content of the section. The amplitudes of carbonate cycles under conditions of a mean ratio of carbonate to non-carbonate of 2:1 with components A and B fluctuating in-phase is summarized in Fig. 3, as functions of relative sensitivities of the two components at fixed values of the carbonate sensitivity (S_A/M_A). Generally, the amplitude of the carbonate weight percent cycle increases with increasing carbonate sensitivity relative to its mean flux. However, the amplitude of the cycle also depends on the relative sensitivities of the two components, and decreases to zero when $S_A/S_B = M_A/M_B$. There is little increase in amplitude sensitivity at high values of S_A/S_B. To the left of the node of zero weight percent amplitude curves the sensitivity of the percent measure may show an abrupt change in slope due to truncation of the weight percent curve at 100% component A.

A critical value of the relative climatic sensitivities of the two components separates two types of weight percent waveforms. This value coincides with a node of exactly zero weight percent sensitivity to changes in the fluxes of the two components, no matter how large they may be. When $S_A/S_B > M_A/M_B$, the troughs of the weight percent curve will have a convex-up appearance (Fig. 4). Most of the waveform distortion is caused by changes in the flux of variable A, and weight percent A varies positively, but non-linearly, with the flux of A. I define this waveform as an 'accumulation' pattern for variable A. When $S_A/S_B < M_A/M_B$, defined here as a 'dilution' regime, the weight percent series of component A will have broad concave-up troughs and narrow peaks (Fig. 4). The major cause of the waveform distortion is the changing flux of the 'dilutant', component B. Now, the weight percent of component A is *inversely* related to its flux. The ratio S_A/S_B to M_A/M_B is therefore an important diagnostic of different sedimentary regimes.

If the accumulation of two components oscillates in opposition, the weight percent measure becomes

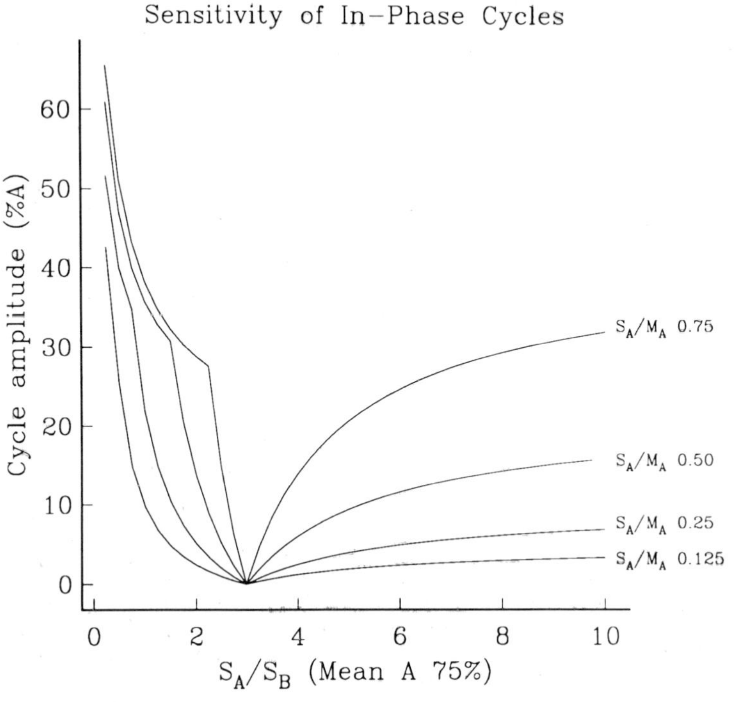

Fig. 3. Difference between maxima and minima of a sedimentary cycle (amplitude in percent component A), as a function of the relative sensitivity of the two components to cyclical forcing (S_A/S_B), for isopleths of constant S_A/M_A, the climatic sensitivity of component A. Both components vary in-phase. Average composition of sediment is 75% component A. Note that there is a node of zero sensitivity of the amplitude measure, despite the fact that the flux of A is varying continuously.

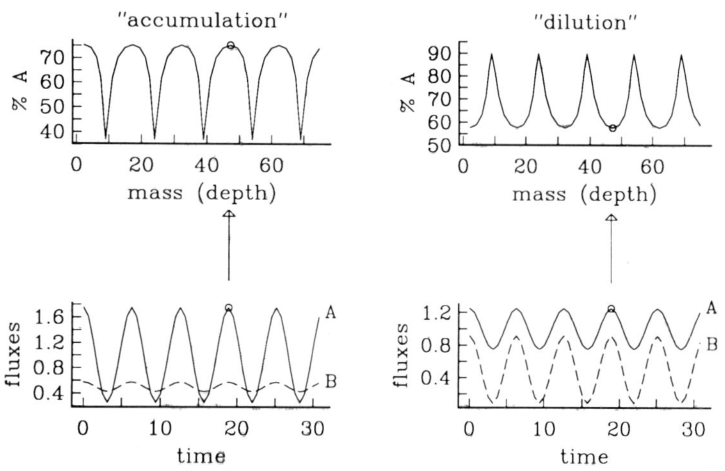

Fig. 4. Changes in the $\%A$ waveform occur across the critical ratio $S_A/S_B = M_A/M_B$ when both components vary in phase. Accumulation cycles ($S_A/S_B > M_A/M_B$) have a convex-upward waveform, with the flux of A (solid curve) varying positively with $\%A$. The flux of component B is shown as a dashed line. Convex-upward waveforms of $\%A$ also occur when component B is invariant. In dilution cycles ($S_A/S_B < M_A/M_B$) the flux of A is anticorrelated to $\%A$, and the $\%A$ (carbonate content) waveform is concave upward. Circles show how a representative point of sediment history is translated to stratigraphic record. Similar changes in waveform were simulated by Decker (1991).

highly sensitive to climatic cycles. Much less time–depth distortion is involved, however, because the decrease in stratigraphic build-up caused by the lesser flux of component A during one phase of the climatic cycle is at least partly compensated by the increased flux of component B. Percentage curves will be more nearly linearly related to fluxes; in fact at the condition $S_A/S_B = 1.0$ the percentage measure varies linearly with flux (Fig. 5) and the depth scale moves evenly with time.

In contrast to the in-phase cycling of components, a weight percent change in component A in out-of-phase cycles is always positively correlated to the flux of component A. When $S_A/S_B < 1.0$ the oscil-

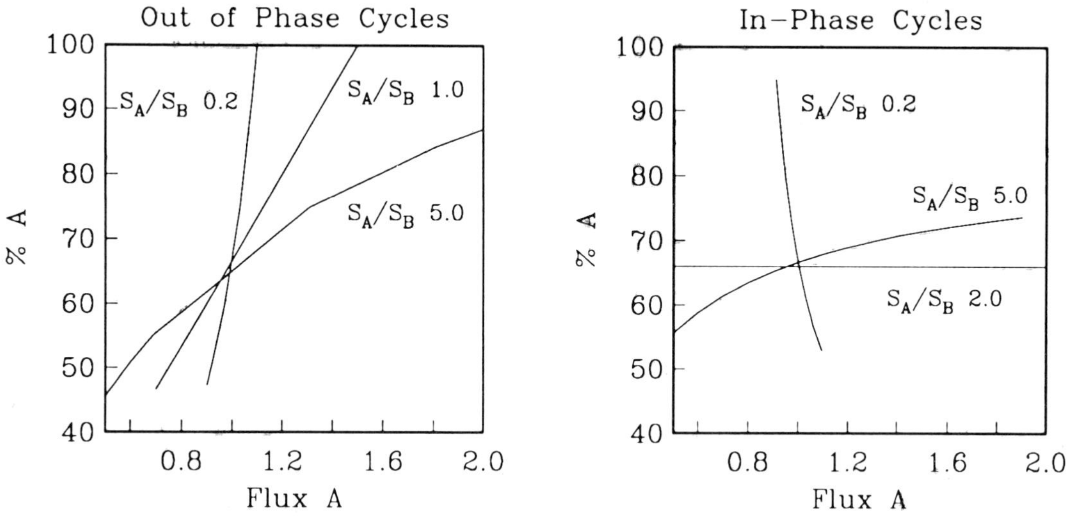

Fig. 5. When the fluxes of the two components are out of phase, the relationship between the percent measure and flux is always positive. Shown here are calculations for a 2:1 mean ratio of components $A:B$; different average compositions would yield weight percent–flux relationships for a particular value S_A/S_B. The case $S_A = S_B$ divides out-of-phase dilution cycles ($S_A < S_B$) from accumulation cycles ($S_A > S_B$). When both phases covary, the condition $S_A/S_B < M_A/M_B$ corresponds to dilution regime for variable A, and $S_A/S_B > M_A/M_B$ corresponds to an accumulation regime for variable A. Note that in the last case, the flux of A is anticorrelated to its weight percent.

lations may again be termed dilution cycles. In such a regime, the weight percent measure is highly sensitive to small changes in the flux of A, but time–depth distortions are driven by the accumulation of component B. When $S_A/S_B > 1.0$ the cycles become driven by the accumulation of component A. In comparison to in-phase variance of sedimentary components, the signature of out-of-phase cycles will therefore be large peak-to-trough differences in weight percent measures, with reduced harmonic distortion of periodicities. At $S_A/S_B < 1.0$, curves of percentage A will be concave up; the waveform reverses at $S_A/S_B > 1.0$.

Multiple modulations

Many palaeoclimatic time series may have several scales of cyclicity that influence sedimentation, resulting in a multiplicity of amplitude and, more importantly, time–depth distortions. For example, the major 100-ka cycle in carbonate content of the equatorial Pacific (Chuey *et al.*, 1987; Farrell & Prell, 1989) almost certainly distorts the representation of the shorter obliquity and precessional cycles in the sediment column. Multiple modulations, in contrast to the previous examples, can actually obscure primary orbital periodicities in the high-frequency end of the spectrum.

The presence of multiple frequencies can be modelled by:

$$A(t) = M_A + \Sigma S_{Ai} \sin(\omega_i t + \psi_i)$$

and

$$B(t) = M_B + \Sigma S_{Bi} \sin(\omega_i t + \psi_i)$$

where each component has a climatic sensitivity associated with each cyclical forcing function ω_i.

Amplitude modulation of the weight percent measure of higher frequency cycles occurs when its relative sensitivity S_A/M_A varies along the stratigraphic axis. Consider, for example, a sediment with a mean carbonate content of 70% containing a precessionally driven carbonate cycle intermingled with variations caused by the eccentricity cycle. If the low frequency diminishes the background carbonate flux to 50 units (carbonate 62.5%), a 5-unit change in carbonate accumulation due to precession may lead to a weight percent change of 5% carbonate. In contrast, the same high-frequency change in carbonate will make only a 2% difference to the weight percent when the low frequency boosts the background carbonate accumulation to 90 units. More generally, the sensitivity of weight percent A at a high frequency H varies cyclically if there is a lower frequency L according to the ratio:

$$\frac{S_{AH}}{M_A + S_{AL}\,(\sin \omega_L t)}$$

This means that the amplitude of the high-frequency weight percent waveform will be modulated along depth, as will its shape (i.e. harmonic structure).

Periodic oscillations in sedimentation rate will also stretch and shrink the local distance between maxima and minima of shorter cycles. In the case of purely intra-waveform distortion, the depth representation of the sedimentary cycle is at least phase-locked to the forcing function in time, i.e. maxima and minima line up, with a constant scale factor, with the input. In spectral terms, the fundamental frequency comes through clearly despite self modulation, and the F-test measure of signal to noise shows little degradation from the linear flux function. However, when longer term cycles in sedimentation modulate the local sedimentation rate, these simple relations no longer hold for shorter periodicities.

Multiply modulated time series will therefore exhibit complex spectral characteristics in the high-frequency end of the spectrum. As before, spectral distortions caused by multiple amplitude and frequency modulation of high-frequency cycles are illustrated separately using a simple example of a sediment with a mean content of 75% carbonate, a constant influx of non-carbonate material and a carbonate climatic sensitivity of 22.5% relative to the mean at each climatic period (Fig. 6). Longer period modulations are chosen at 1.9, 3.2, 5.1 and 19 times the period of the short cycle, in order to simulate the potential effect of obliquity and the 109-ka and 410-ka components of eccentricity on the detection of a precessional cycle. Figure 6 illustrates the flux–amplitude and time–depth distortions suffered by a 21-ka precessional cycle from 41-ka and 100-ka cycles in accumulation. The longer period oscillations modulate local sedimentation rate by a factor of two. The time–depth distortions are clearly visible as condensed cycles at low values of the flux of A, while the amplitude distortion manifests itself by changes of a factor of two in the difference between maxima and minima of the 21-ka weight percent cycle.

In the frequency domain (Fig. 7) the time–depth distortions act to produce artefacts at combination tones ($f_1 \pm nf_2$), where f_1 is the shorter period, f_2 is the modulating period, and n is an integer

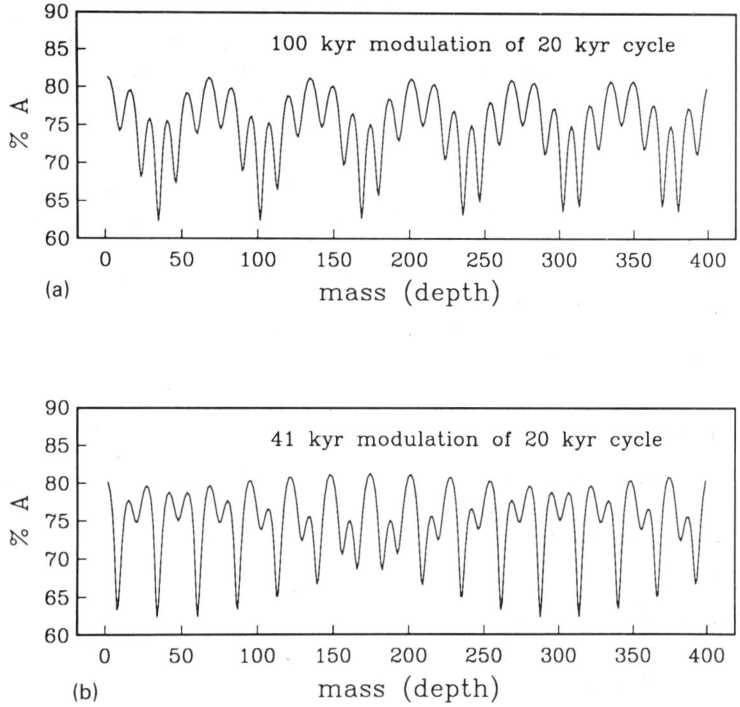

Fig. 6. A shorter cycle (modelled precession) is distorted during sedimentation both in weight percent amplitude (difference between maxima and minima) and in frequency (variable spatial period) when lower frequency climatic cycles influence accumulation (example a: eccentricity; example b: tilt). The flux of component *B* has been held constant; *A* varies by 22.5% around the mean at both precession and the lower frequency. Note that the lower frequency changes in flux modulate both the amplitude and spacing of the shorter cycle. Mass/depth scale is arbitrary.

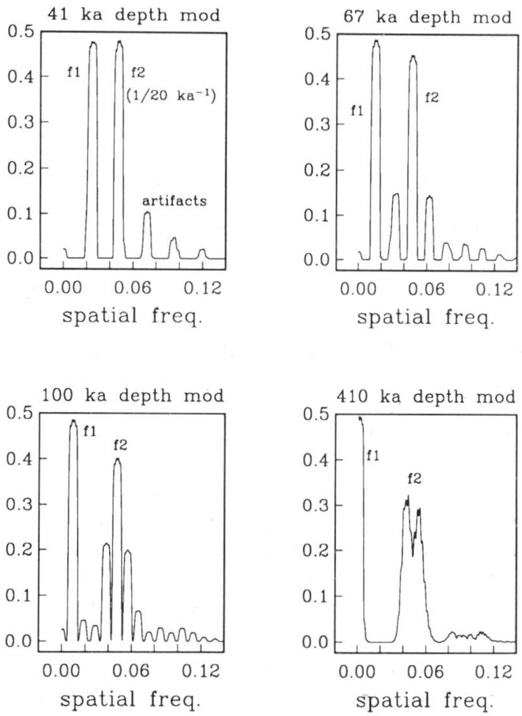

Fig. 7. Spectral artefacts introduced by multiple time–depth (frequency) distortions in sedimentation. A 20-ka signal (f_2) is distorted by a periodic variation in sedimentation rate at a lower frequency (f_1), as in Fig. 6. The modulating frequency f_1 is allowed to vary through the Milankovitch band, while f_2 stays fixed at the precessional period. Both cycles produce the same variance in flux and accumulation rate, and should have the same spectral amplitude. The sedimentary representation of the high-frequency climatic signal becomes progressively more distorted as the modulating period lengthens, as seen by the growth of spectral artefacts. Note the defocusing of the f_2 ($1/20\,ka^{-1}$) frequency, and the lowering of its peak height.

(Weedon, 1989). An intermediate modulating frequency (67 ka) has been added to show the transition of spectral behaviour from shorter to longer modulating periods. Note that the effect of modulation at two to three times the short period has little effect on its detection. As the modulating period grows longer, the amplitude of the high-frequency side lobes grows, until at the *c*.400-ka frequency modulation, they obscure the primary precessional periodicity. The high-frequency peaks now take on the ragged look of a continuous, quasi-periodic spectrum. Further simulations show that the degree of peak splitting depends both on the period of the longer cycle relative to the carrier period, and on the degree of distortion, i.e. on the long-period change in sedimentation rate. Analysis of the signal-to-noise ratio of the model precession cycle, as measured by the *F*-test value at its mean frequency, shows degradation by factors of 5 to 1000, reflecting the fact that the distorted short cycle is no longer recognized as a phase-coherent sinusoid.

In contrast to the time–depth distortion, amplitude modulation affects spectra less severely, a result to be expected from theory (Schwarzacher, 1991). The spectrum suffers ($f_1 \pm nf_2$) distortion, but the peak height of f_1 is unaffected. As the modulating period grows, the ($f_1 \pm f_2$) term becomes closer and closer to f_1, and the distortions fold into the primary frequency. However, multiple frequency modulation simulations with other values of key parameters show significant shifts in the high-frequency end of the spectrum due to amplitude modulation of the weight percent signal. The net effect of distortions is always to split high-frequency peaks and to generate side lobes that may be difficult to distinguish from the 'background' spectrum.

When additional longer modulating frequencies in sedimentation are added, the distortions of a high-frequency term are multiplicative. For example, local sedimentation rates on the precessional time scale would vary by a factor of two if both obliquity and eccentricity sensitivities of 12.5% modulate sedimentation rate under a mean regime of 75% carbonate. One simulation (Fig. 8A) shows the disastrous time–depth distortion of multiple periodicities, here chosen at 5.1 and 19 times the short period, with the relative sensitivity set to 33% for each sinusoid. The spectrum (Fig. 8B) shows that the energy associated with the high-frequency peak spreads out over approximately four times its undistorted width. Again, it is the detection of the highest frequency sinusoid that suffers the most from sedimentation rate distortions.

Long-term, non-periodic modulation

Real sedimentary systems do not simply oscillate at Milankovitch periodicities. Rather, inputs of components to the sedimentary column evolve on time scales of 10^5 years and longer due to variations in tectonic and geochemical cycles. The effects of long-term internal variations of the system might be direct, for example a change in nutrient flux to the oceans could produce a greater flux of biogenic hard parts to the pelagic abyss, or they might be indirect, such as long-term modifications of ocean circulation and inter-basinal changes in carbonate saturation. Long-term variations in sediment flux will produce changes in both the spatial frequency and weight percent amplitude of sedimentary cycles. For simplicity, I assume that the pattern is linear, but the

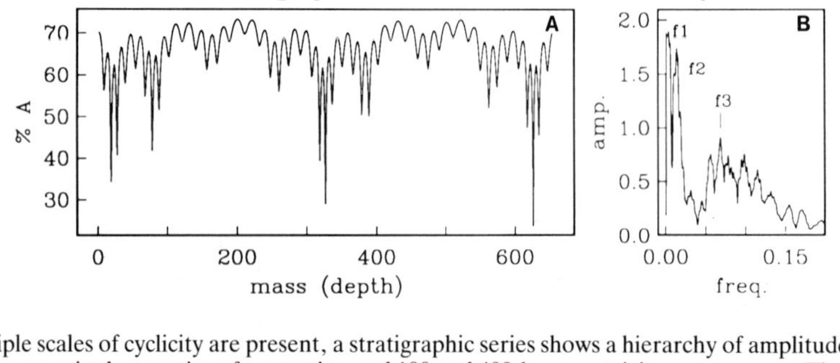

Fig. 8. When multiple scales of cyclicity are present, a stratigraphic series shows a hierarchy of amplitude and spatial distortions. Input frequencies have ratios of precession and 100 and 400-ka eccentricity components. The relative variance at each frequency is identical, and the input Fourier spectrum consists of three spectral lines. The highest frequency cycle (f_3) undergoes the most intense distortion and peak broadening. Note that even in a purely deterministic sedimentary series, it may be hard to use conventional spectral analysis techniques to recognize high-frequency periodicities.

theory is easily extended to a low-order polynomial or other smooth representation of a trend.

Continuing with the two-component model, we now represent the accumulation of both components as having a sinusoidal and a linearly varying component.

$$A(t) = M_A(t) + S_A\sin(\omega t)$$

$$B(t) = M_B(t) + S_B\sin(\omega t + \psi_B)$$

In each case, the trend in accumulation will be of the form

$$M_i(t) = M_i(t_0) + k_i t$$

where k_i is a coefficient that may be either negative or positive, depending on whether accumulation rates of the component increase or decrease with time from the initial point t_0. Both the spatial period and the concentration amplitude of the system will evolve slowly if the absolute value of $k\Delta t/M_i(t_0)$ is not too large over the time interval Δt.

Two concepts need to be emphasized to recognize periodicity in such a system: the local stratigraphic period, and the local amplitude of the percent measure. If sedimentation rate evolves gradually, cycle spacing will be relatively constant over some interval of time longer than one cycle period. However, over greater intervals, cycle spacing will drift toward shorter repeat lengths (higher spatial frequency) in more condensed portions of the record, and greater repeat lengths in regions of increased sedimentation rate. Such changes are illustrated in Fig. 9, which shows a gradual decrease in cycle spacing down-core.

Amplitude distortion may also occur if there are long-term trends in the accumulation of one or more sedimentary components. Such distortions could arise if there are changes, presumably unrelated to orbital forcing, in the factors controlling the accumulation of sediment, e.g. changes in deep ocean circulation, in geochemical cycles, or in the erosion rates of detrital source areas. Amplitude modulation will probably also arise in long time series from real changes in the sensitivity of the Earth's climatic components to orbital forcing. If the changes in M_A and M_B are large enough, then the ratio of S_A/M_A, the relative sensitivity of the concentration of A to cyclic forcing, will change along the stratigraphic dimension as well, causing the difference between maxima and minima of a cycle to evolve.

Such drifting of cycle spacing and amplitude can severely impair statistical recognition of sedimentary cycles when the frequency content of the entire series is analysed. Simulations using the F-test measure show that drift modulations are far more destructive to statistical recognition of cyclicity than are cyclic and within-waveform modulations. One manifestation of the distortion is a leakage of spectral energy away from the central peak heights (Weedon, 1989). Maximum spectral energy will be at the mean stratigraphic frequencies of the sedi-

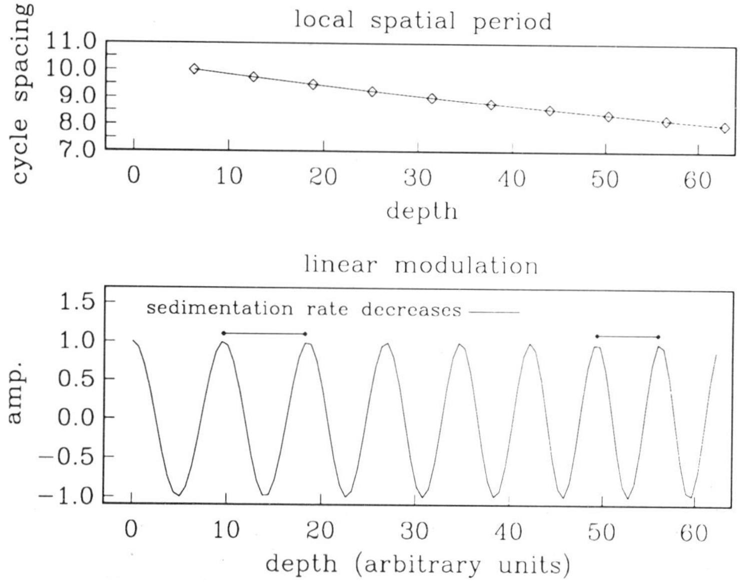

Fig. 9. With progressive condensation, a sedimentary cycle of fixed duration shifts to shorter stratigraphic spacing. The condensation can be measured by continuously monitoring the distance between successive maxima (the local spatial period).

mentary oscillations. Non-stationarity in amplitude can be represented in Fourier terms as a damped, or growing, sine wave. The effect is again to spread energy around the fundamental frequencies (see a text on Fourier representation of signals, e.g. McGillem & Cooper, 1984, Ch. 3).

Step-function changes in sedimentation rate

Abrupt changes in mean accumulation rate are probably rare in pelagic sedimentation, but are conceivable in at least two circumstances. The first is a true catastrophic change in sedimentation, such as caused by a rapid reorganization of the climatic factors controlling sediment accumulation in a particular location. An apparent step-function change in sedimentation might also result from an erosional hiatus, which could remove evidence of a transitional sedimentary regime. Figure 10 illustrates the resulting pattern in cyclic sedimentation.

HOW DO WE RECOGNIZE DETERMINISTIC DISTORTIONS?

The distortion of periodic signals by sedimentation carries implications both for sampling strategies and for methods of spectral analysis. Without due consideration of sampling problems, it will not be possible to recover the true dynamics of sediment accumulation on short time scales. Uncritical spectral analysis of long sedimentary series could confuse as much as it enlightens.

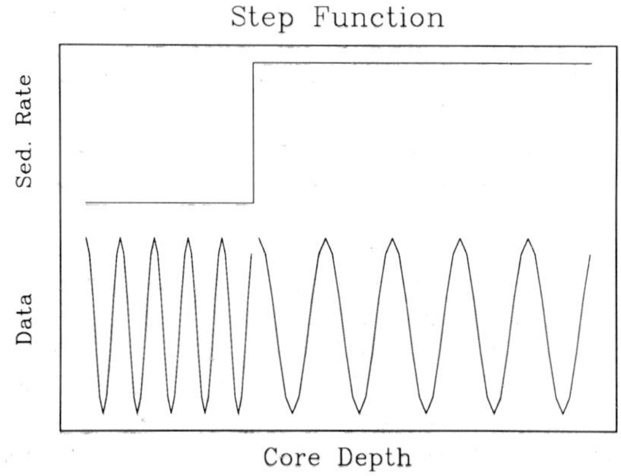

Fig. 10. An abrupt discontinuity in cycle spacing along a stratigraphic sequence may indicate a jump to higher sedimentation rates. The jump could be due either to a rapid palaeoclimatic change, or to the failure to record a time of transition (hiatus).

Spectral analysis of waveform distortion suggests a redefinition of the minimum sampling density needed to recover palaeoclimatic variations accurately in a stratigraphic series. The minimum sample interval, termed the Nyquist frequency (see, for example, McGillem & Cooper, 1984), would normally be considered to be one-half the shortest significant orbital period in a 'Milankovitch'-dominated series. Typically, one would calculate the average sedimentation rate, and sample so as to take an average two or more samples per precessional (*c.*20 ka) cycle. Such sampling will be just adequate to define periodicities in a uniformly accumulating sediment. Ignoring the Nyquist requirement may result in 'aliasing' – the spurious transference of high-frequency energy into lower frequency regions of the spectrum.

However, many sedimentary settings will be characterized by significant variations in sedimentation rate over short distances. The minimum sampling interval therefore needs to be redefined as the *local spatial Nyquist*, the maximal distance between successive samples that will not degrade the accuracy of the resulting spectrum. The local Nyquist interval will vary along section according to sedimentation rate in a manner never exactly known.

Because we have seen that distortions in waveform carry meaningful information on changing fluxes, and because these distortions are represented in Fourier terms by the generation of harmonics, sampling strategies must be improved. The minimum sampling rate required to capture the first harmonic term of a distorted signal is one-quarter the fundamental period (5 ka for precession). Much of the shape information describing a waveform may lie in higher harmonics, requiring greater sampling intensity. Concerns for preventing aliasing and for recovering waveform distortion therefore dictate that time series should be sampled at a nominal spacing of eight times or better the shortest period expected (presumably precession). Additional modulations of the spatial Nyquist by low frequency (cyclic or drifts type) in sedimentation rate must also be considered.

The approach taken by Schiffelbein & Dorman (1986) uses within-waveform distortions to look for systematic changes in sedimentation rate. These authors modelled deterministic time–depth stretching with a distortion parameter K ($K = 1$ gives a 10% distortion, $K = 5$ gives a 50% distortion, etc.). K can be measured in a well-sampled waveform 'by dividing the length of a trough by the length of a peak, measured at zero crossings'. If one assumes

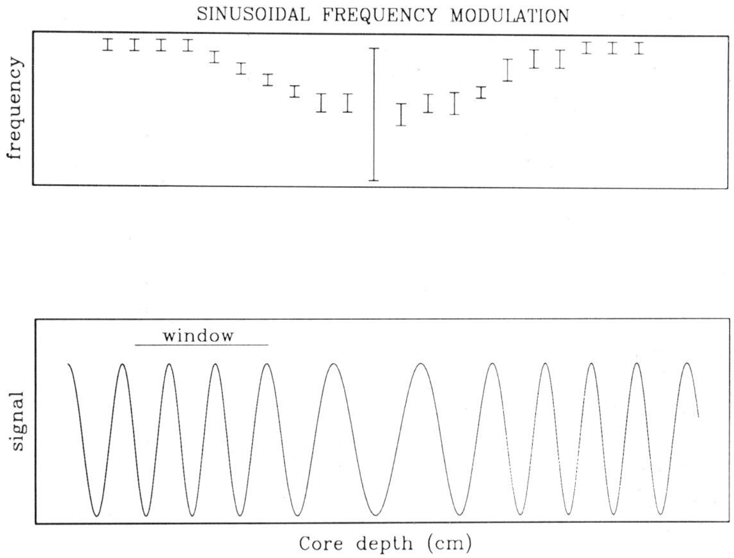

Fig. 11. Cyclical modulation of sedimentation rate stretches and condenses a sedimentary cycle along depth. This might occur to high-frequency (precession or obliquity) cycles when lower frequency orbital cycles are present. The distortions can be recovered if smaller units of data are analysed sequentially for frequency content down-section (moving-window analysis; window length indicated by horizontal bar). Each analysis yields an estimate of the local spatial frequency of the distorted cycle, displayed in the upper panel (the large error bar in the centre is due to inadequate frequency resolution where the cycle is most strongly stretched). Spatial frequency of the sinusoid is highest where sedimentation rate is lowest, and lowest where accumulation rate increases.

that K is a linear function of a climatic variable, it is possible to determine the optimum K by a criterion of spectral simplicity: the spectrum with the fewest, best resolved peaks is superior to a highly complex one (Schiffelbein & Dorman, 1986). K can then be interpreted in terms of the two-component model. However, caution must be applied in using this approach with intensive variables such as carbonate content, which so often bear a non-linear relation to sedimentation rate. A less elegant but more robust method is to forward-model carbonate cycles until the requisite amplitude and depth features are simulated. The shape of the waveform, together with the amplitude, should be diagnostic of both relative variance and phase relations of a two-component system.

Moving-window spectral analysis may be an effective technique for removing time–depth distortions on a scale of several cycles or longer (Park and Herbert, 1987). If sedimentation rate evolves in some systematic way over the course of a record, shorter intervals of the data have less variance in accumulation rate than does the entire series. Moving-window analysis dissects a longer record into smaller pieces, each analysed for frequency content. An overlap of 25–50% between succeeding windows assures that real signals will be continuously traceable along depth. The local spatial period of an orbital cycle, divided by its absolute period, gives the local sedimentation rate. An example of moving-window analysis of a cyclically modulated signal is shown in Fig. 11. The choice of an appropriate window length will be an empirical process, and should balance the need for resolution (shorter windows) with the need to build up a reliable signal. Window lengths corresponding to 200–600 ka work best in my experience.

Ideally, the spatial window length should be varied along the stratigraphic column to adjust to variable sedimentation rate. This would prevent the occasional broad error bars on frequency determinations seen in Fig. 11 that occur when the fixed window is too short to resolve the highest sedimentation rate cycles of the record. Frequency detection would be improved by iteratively modifying the window lengths according to the frequency analysis or to independent estimates of sedimentation rate – shortening the spatial window where the section is more condensed, and lengthening it where the series is more expanded.

Criteria exist to prevent overly enthusiastic application of this method to data of poor quality. First, one should have some independent time estimates in the section, so that there is little chance of 'locking onto' the wrong orbital frequency, and so that moving-window age–depth models can be compared at tie points for consistency with tested time scales. Second, the assumption that peaks can be correlated from spectrum to spectrum needs to be demonstrated. A particularly stringent prediction of the frequency modulation model exists if multiple orbital peaks are present in a stratigraphic series.

According to the frequency modulation model, the different spectral peaks associated with Milankovitch cycles will show coherent spatial frequency shifts as sedimentation rate changes. Thus, a decrease in sedimentation rate by 25% should find spatial periodicities associated with eccentricity, obliquity and precession, condensed by the same factor.

The distortions introduced by sedimentary non-linearities may occasionally be so great that moving-window analysis will not resolve them. It should be particularly difficult to detect precessional carbonate cycles through spectral analysis, because their spacing and amplitude will be modulated by every longer period change in sedimentation. A remedy for this situation has already been applied by outcrop geologists – bedding thickness time series (Schwarzacher, 1954; de Boer, 1982; Schwarzacher & Fischer, 1982). In this case, one assumes that a sinusoidal signal is present, and measures the sequential variations in the thickness of beds, which can be interpreted as a running measure of local sedimentation rate. The time series is then tested for modulation structure, which may be either cyclical or trend-like. Random modulation of bedding will give a noisy spectrum, and the hypothesis of high-frequency cyclicity cannot be confirmed. However, highly structured spectra of bedding thickness are unlikely to be accidental, and can be taken as evidence for a distorted, but recoverable, high-frequency signal.

Two approaches to the spectral analysis of sequential spacing variations may be termed *stratigraphic* and *metronomic*. The stratigraphic method records sequential thickness variations as a function of depth; in other words, it maintains the constant time–depth assumption. Figure 12A shows the application of this method to spacing of the shortest cycles shown in Fig. 8. Clearly, if there are changes in sedimentation rate, these will be incorporated as distortions into the time model for spectral analysis. The frequencies f_1 and f_2 that modulate the spacing of the short cycle are revealed, but in a smeared

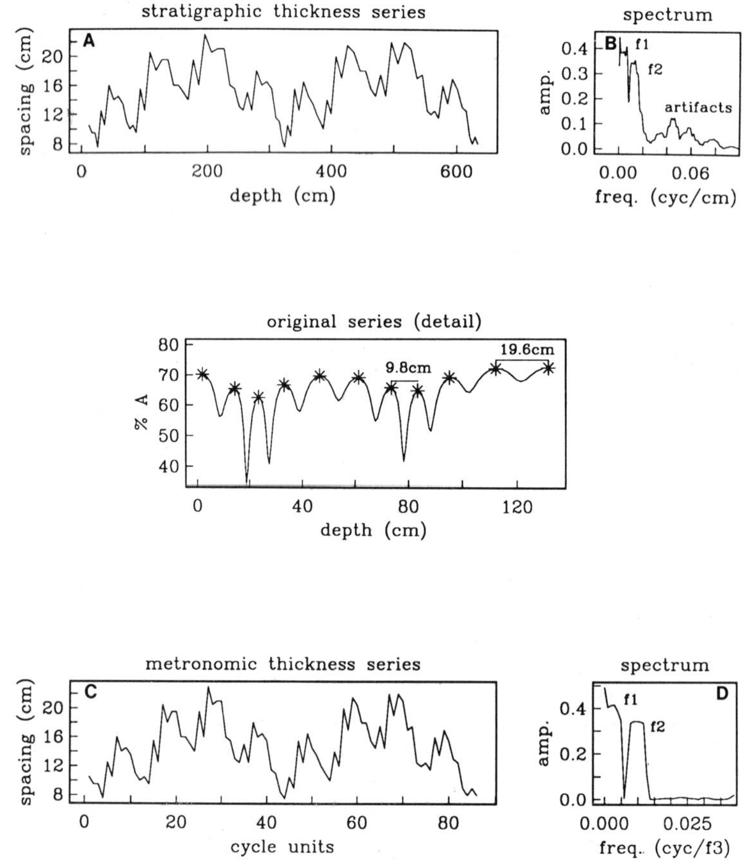

Fig. 12. Cyclic modulations of highly distorted high-frequency cycles may be demonstrated by showing patterns in sequential spacing of maxima and minima of a lithological curve ('original series', a detail of Fig. 8). Recording sequential spacing variations as a function of stratigraphic position (stratigraphic thickness series) assumes that there are no changes in sedimentation rate along the depth axis. Points are then interpolated at constant stratigraphic spacing for frequency analysis. The metronomic method assumes that high-frequency cycles (couplets) mark off equal time, and that thickness measurements are therefore already equally spaced at $1/f_3$. A frequency modulation spectrum of the bedding series with concentrations of energy at discrete frequencies is an indirect proof of the existence of a high-frequency cycle (recall the inability of conventional spectral analysis to resolve f_3 in an ordinary amplitude spectrum, Fig. 8). Periodicity at frequency f_3 ($1/20\ \text{ka}^{-1}$) can be inferred from periodic modulations of its spacing (f_1 = 400-ka cycle; f_2 = 100-ka cycle). In sequences with good signal-to-noise ratio, the metronomic assumption yields cleaner, more accurate spectra than does the stratigraphic assumption.

form (Fig. 12B). A more correct, but risky, approach, is to assume that the signal being measured is a true chronometer, and therefore that each peak represents a constant time increment (Fig. 12C). The time axis becomes then simply the number of short cycle units. Because orbital cycles are not perfect metronomes (see Herbert & D'Hondt, 1990), some error is inherent in this assumption. In addition, mistakes in picking cycle units will propagate time errors throughout the series.

In sedimentary series with a good signal, the metronomic assumption will recover bedding modulating frequencies with far better precision. The statistical superiority of the metronomic approach is demonstrated in Fig. 12D, where both modulating frequencies are well resolved, and have far higher *F*-test values than do their counterparts in the stratigraphic time series.

GEOLOGICAL EXAMPLES

Waveform analysis of equatorial Pacific carbonate cycles

The classic Pleistocene carbonate cycles of the equatorial Pacific (Arrhenius, 1952; Berger, 1973; Farrell & Prell, 1989) have generally been interpreted as a turning on and off of carbonate accumulation, either by varying the productivity of calcareous organisms, or by their preservation on the sea floor. Following either interpretation, one would expect carbonate content to be the major control on bulk accumulation rate. Figure 13 displays two examples of Brunhes chron carbonate records from moderately deep-water depth. Core ERDC 130P is a low-resolution record from the Ontong-Java Plateau (0°02′ N, 161°55′ E, 4112 m water depth) sampled by Schiffelbein (1984). Sample resolution is sufficient only to recognize the dominant 100-ka cycle in carbonate, which appears to have a convex waveform. Core RC11-210 from the central equatorial Pacific (1°49′ N, 140°03′ W, 4420 m depth) has been studied in more detail for its record of isotopic and mineralogical changes (Chuey *et al.*, 1987; Rea *et al.*, 1991). The striking similarity in waveform of the two carbonate records argues for a climatic control on sedimentation over a broad region (Hays *et al.*, 1969).

Inspection of the RC11-210 carbonate curve shows a complex interaction of cyclicities. Higher frequency, generally low amplitude, short cycles ride on the dominant late Pleistocene 100-ka oscillation. The amplitude of the 100-ka cycle at this location is about 10 weight percent carbonate. The short cycles appear to match the precessional periodicity at

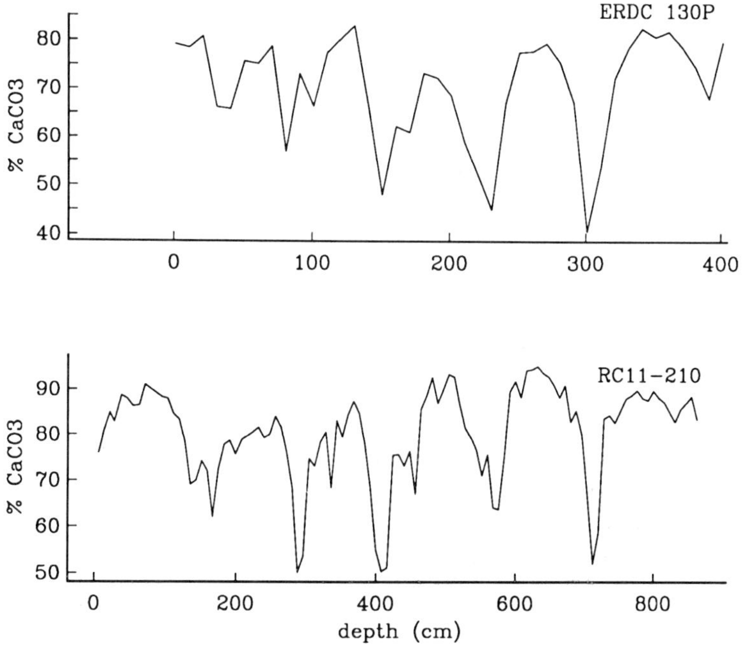

Fig. 13. Similarities in carbonate waveform from sites at equivalent water depths in the Pacific over the past 600 ka. Core ERDC 130P is from the Ontong-Java Plateau (Schiffelbein, 1984), core RC11-210 is from the central equatorial Pacific (Chuey *et al.*, 1987). Sampling density is about 10 cm (12 ka on average) for ERDC 130P, and 8 cm (5 ka on average for RC11-210). The upper 50 ka of sedimentation appear missing from ERDC 130P. Note the sharp spikes of low carbonate content during interglacial stages.

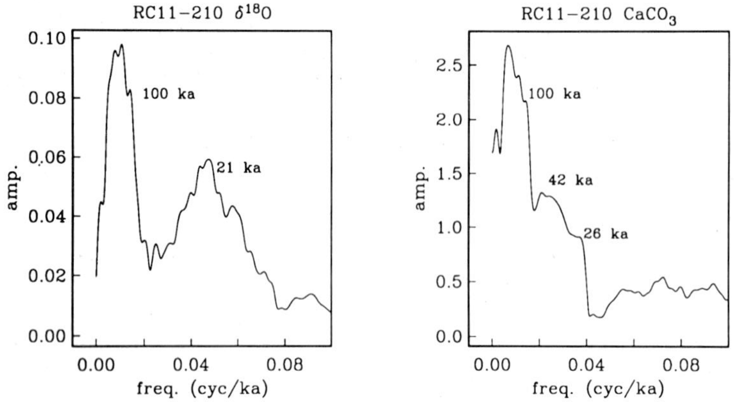

Fig. 14. Comparison of spectra of $\delta^{18}O$ and carbonate records over the last 425 ka from core RC11-210. Apparent spectral energy at 26–27 ka in carbonate spectrum probably is distorted representation of precessional period. Similar spectral artefacts can be generated by large amplitude modulations of a model 20-ka cycle by 100-ka carbonate accumulation rate cycles. Such distortions do not occur when extensive (concentration independent) signals such as $\delta^{18}O$ are used.

about 20 ka. Large excursions occur in the troughs of the 100-ka cycle, corresponding in time to about one-half of a precessional cycle. If the carbonate lows can be interpreted as dissolution maxima (Berger, 1973), these deviations give the appearance of pulses of intense corrosion of carbonate.

Spectra of carbonate content and planktonic $\delta^{18}O$ over the past 435 ka, converted to a *mass* scale from dry bulk-density data of Chuey *et al.* (1987), show the important differences between spectral analysis of intensive and extensive palaeoclimatic variables (Fig. 14). The match of spectral peaks to numerical periods assumes constant mass accumulation rate between the core top and a $\delta^{18}O$ maximum at 435 ka. Peaks at 100 and 21 ka dominate the $\delta^{18}O$ spectrum. No significant 41-ka energy is present. The greater peak width of the *c*.21-ka energy is attributable to greater time–depth distortion suffered by the shorter cycle. Most of the isotopic variance clearly occurs at orbital periodicities. Interpretation of the carbonate spectrum is much more complex beyond the dominant 100-ka peak. Significant peaks occur at about 40, 26.5 and 13 ka. The first appears to be associated with obliquity, and the latter two to be connected to precession in distorted form.

Much of the distortion of the precessional cycle in the carbonate record of RC11-210 may be due to aliasing of precessional harmonic terms into lower frequencies by the 8-cm sampling interval. Comparison of two records of another central equatorial Pacific core, DSDP Site 572 bolsters this hypothesis. Figure 15 shows discretely sampled carbonate data, spaced every 10 cm (Farrell & Prell, 1991) compared with gamma-ray attenuation porosity evaluator (GRAPE) wet bulk-density determinations taken at 1-cm spacing. The similarity of the curves reflects

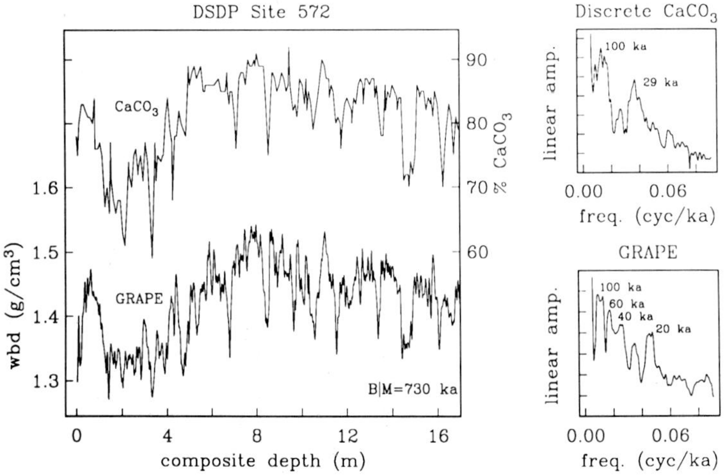

Fig. 15. Comparison of effects of 10-cm (discrete carbonate) and 1-cm (GRAPE wet bulk-density) sample interval on detection of higher Milankovitch frequencies. Although the 10-cm sample interval is nominally twice the Nyquist rate, it smears obliquity and precessional energies into a spurious 29-ka peak. The GRAPE spectrum resolves both cycles. Despite the apparent similarities of the two curves as a function of depth, spectral analysis indicates that sampling at 10 cm is inadequate to capture both amplitude and frequency distortions of the shorter cycles.

the dependency of wet bulk-density on carbonate content in uncompacted sediments (Herbert & Mayer, 1991; Mayer, 1991). Spectral analyses of the two records show the surprising improvement brought about by more intensive sampling. The Fourier spectrum of the more coarsely sampled carbonate curve suggests that outside of the major 100-ka peak, the only major periodicity occurs at about 29 ka, a poor match to any important orbital period. In contrast, the GRAPE curve produces spectra with clearly resolved peaks near the frequencies of obliquity and precession. Despite the fact that the carbonate sampling interval is on average 5 ka, it may fall below the Nyquist frequency for resolving the precessional waveform if sedimentation rate varies by a factor of three to four along core depth. Furthermore, the harmonic terms describing the 20-ka weight percent waveform are certainly undersampled. The result may be to fold spurious energy back into lower frequencies, giving the 29-ka spectral peak.

Can the carbonate oscillations be described satisfactorily with a simple linear model? Constraints for core RC11-210 come from the weighted mean carbonate content of 80%, the 100-ka cycle amplitude of about 10% around the mean, and the large excursion of a single 20-ka cycle during interglacial periods. Simulations with constant non-carbonate flux will not plausibly satisfy the data. For example, the carbonate sensitivity relative to its mean (S_{A100}/M_A) would have to be equal to 1.0 to produce the 100-ka carbonate maxima of 90%. This would also require the minima to reach 0%, a simulation clearly lacking in realism. Furthermore, too much condensation appears in the troughs of the 100-ka cycle for this to be true.

More satisfactory results can be achieved if equatorial Pacific Pleistocene carbonate and non-carbonate fluxes can be anticorrelated on both 100-ka and 20-ka wavelengths. Anti-correlation allows larger weight percent sensitivity to smaller relative variations in carbonate flux. Limits on the model parameters come from matching the amplitude variations in percent carbonate to a reasonable degree, and satisfying the result from orbital tuning (Chuey *et al.*, 1987) that sedimentation rate must be lower by a factor of two during glacial maxima. A satisfactory simulation was achieved with two sinusoids at a frequency ratio of 5:1, with carbonate and non-carbonate accumulation out of phase at both periodicities, $S_{A100} = 0.5$, $S_{A100}/S_{B100} = 10.0$, $S_{A20} = 0.15$ and $S_{A20}/S_{B20} = 10.0$ (Fig. 16). A more realistic simulation would include the amplitude modulation of the precessional cycle, which is smaller during glacial maxima and larger during interglacials.

The crucial sensitivity of the sedimentary system

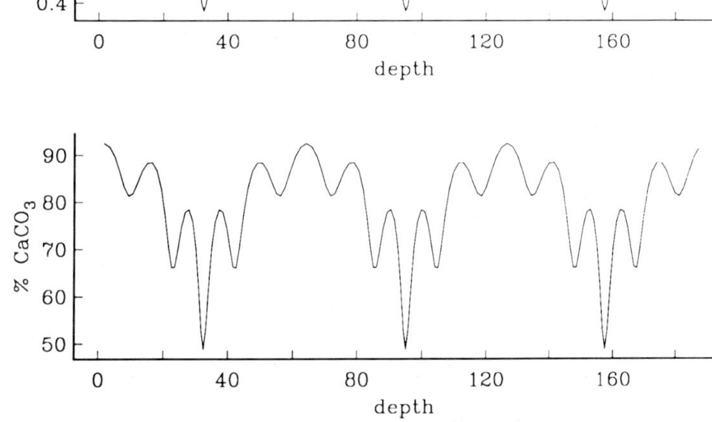

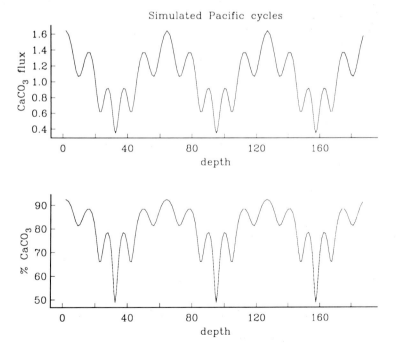

Fig. 16. Lower panel: simulation of equatorial Pacific carbonate cycles, following parameters discussed in text (depth is arbitrary). The essential features are the large amplitude changes of the 21-ka cycle from trough to peak of the 100-ka cycle, the relatively symmetrical variation of percent carbonate around a mean content of 80% on the 100-ka time scale, and the relatively small changes along core in bedding thickness. These features argue for out-of-phase variance of carbonate and non-carbonate fluxes, with carbonate variance 5–10 times greater than the insoluble component. The frequency spectrum of the weight percent curve (not shown) shows that the modelled precessional energy shifts to lower frequencies ($1/26–1/27\ ka^{-1}$) due to amplitude distortion.

gained by allowing the flux of the non-carbonate fraction to vary, albeit subtly, is one lesson of the simulation. Perhaps equally important is the fact that one abrupt carbonate minimum per 100-ka cycle can arise out of a purely linear system. This is a consequence of the non-linearity of the weight percent measure to equal changes in carbonate accumulation on the precessional time scale: the ratio

$$\frac{S_{A20}}{M_A + S_{A100}\sin(100t + \psi)}$$

governs the local amplitude of the 20-ka carbonate cycle. When the 100-ka cycle causes carbonate flux to rise substantially (positive values of $S_{A100}\sin(100t + \psi)$), high-frequency changes in carbonate flux produce little weight percent response. The converse applies during periods of low carbonate flux during interglacial stages. Carbonate minima therefore may not necessarily record anomalously intense carbonate dissolution 'events'.

Cyclical frequency modulations

Beautiful examples of cyclically modulated changes in sedimentation rate and carbonate waveform come from 'bundled' sequences of the Early and Mid-Cretaceous of Italy and France (de Boer, 1982; Schwarzacher and Fischer, 1982; Ricken, 1986; Herbert *et al.*, 1987). Bedding couplets, defined by changes in carbonate content and redox state, appear to be grouped into units of four to six by a repetitive thickening and thinning pattern termed 'bundling'. Fourier analysis of sequential variations in bedding thickness of the couplets shows a modulation frequency equivalent to the bundle cycle (de Boer, 1982; Schwarzacher and Fischer, 1982). These workers proposed that bedding and bundling cycles reflected precessional and eccentricity climate variations, respectively.

The drilling of a core near the town of Piobbico in the Marche region (Tornaghi *et al.*, 1989) has permitted detailed geochemical analysis of Aptian to Albian cyclic sedimentation (Herbert *et al.*, 1986). Percent calcium carbonate and aluminium time series, sampled at an average resolution of 1.5 cm (3 ka at mean sedimentation rates), show high-frequency variations that weather into 10–12-cm bedding couplets in outcrop, and longer period alternations at about 55-cm and 2.5-m spacing (Fig. 17). The 55-cm oscillation corresponds to the bundle rhythm described by de Boer, Schwarzacher and Fischer in strata just above the core section. Bio-

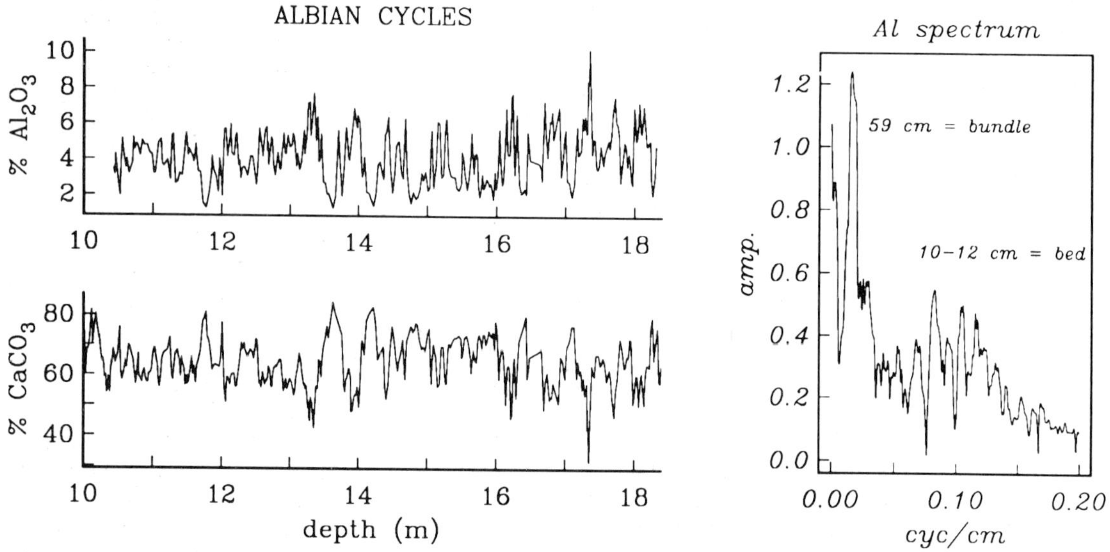

Fig. 17. Carbonate and aluminium determinations taken at 1.5-cm intervals in upper Albian (*T. breggiensis* foraminiferal zone) sediments of Piobbico core. Biostratigraphically estimated sedimentation rate is 0.4–0.6 cm/ka. The spectrum of aluminium content shows a marked periodicity at 59 cm (98–150 ka), corresponding to the bundling modulation of carbonate couplets observed in the field (de Boer, 1982; Schwarzacher & Fischer, 1982). Although high-frequency sedimentary variations consistently fall into a 5 : 1 ratio to the 59-cm cycle, as would precessional cycles embedded in an eccentricity matrix, their spectral representation is badly smeared.

stratigraphically estimated sedimentation rates (4–5 m/Ma) suggest a match of the bedding cycle to the precessional rhythm, and the 55-cm and 2.5-m cycles to the short and long cycles of eccentricity. Spectral analysis shows that a cycle with a mean spacing of 59 cm is by far the most clearly defined statistically (Fig. 17). Note that the ragged appearance of the high-frequency peaks in comparison to the 55–60-cm cycle is reminiscent of spectral behaviour in simulations of mixed periodicities (see Fig. 8).

Modulations in the spacing of bedding and bundling of Albian carbonate variations appear to reflect stretching and shrinking by the other sinusoidal terms. Smooth variations of the local period of the *c*.55-cm carbonate cycle were detected by a 1.25-m moving-window analysis of Park & Herbert (1987). Rather than looking for a maximum in the amplitude spectrum, an *F*-test measure of signal-to-noise ratio was used to identify highly significant spectral lines. Spatial frequency shifts of the bundle carbonate cycle follow a *c*.2.5-m oscillation, with an approximately twofold maximal difference in sedimentation rate implied. Shorter period cycles (obliquity, 20-cm period; precession, 11-cm period) did not produce consistently high *F*-tests along-core, and were not tracked by this analysis.

Following the discovery that aluminium data may be more regular, I have repeated the moving-window analysis with the latter variable and looked to the amplitude spectrum rather than to the *F*-test values to define the bedding (presumed precessional) cycle. A higher frequency peak corresponding to the bedding cycle can indeed be traced down-core (Fig. 18). Still higher frequency variations in sedimentation rate can be resolved by measuring the distance between maxima and minima of the geochemical time series, as shown for a model signal in Fig. 10. As previously discussed, this representation assumes that the bedding is a high-frequency signal that is periodically modulated. Sequential variations in maxima and minima were calculated from the first derivative of a smooth spline function fit to the aluminium data. The resulting thickness series bears a strong resemblance to the carbonate curve, as would be the case if carbonate content is the major control on sediment accumulation rate (Fig. 19). Beds are stretched by a factor of two to four from low to high carbonate portions of the 55–60-cm bundle rhythm.

The strong modulation periodicity of the Albian carbonate bedding pattern, whether calculated as a function of depth (stratigraphic assumption), or as an assumed function of time (metronomic assumption), argues for the existence of a genuine high-frequency (*c*.20 ka) cycle, whose amplitude and spacing are modulated by slower cycles of sedimentation. In accord with the deterministic distortion model, the metronomic assumption produces a cleaner spectrum with higher *F*-test values at spectral maxima than does the stratigraphic assumption (not shown). The bundling peak of the metronomic bedding spectrum is composed of two well-defined lines at precisely the ratio of twin components detected previously after tuning the depth scale of the carbonate data to emphasize the bundling cycle (Park & Herbert, 1987). A low-frequency stretching term at 1/23 the bedding frequency matches the *c*.2.5-m geochemical oscillation already observed. The twin bundling frequency lines match the ratio of

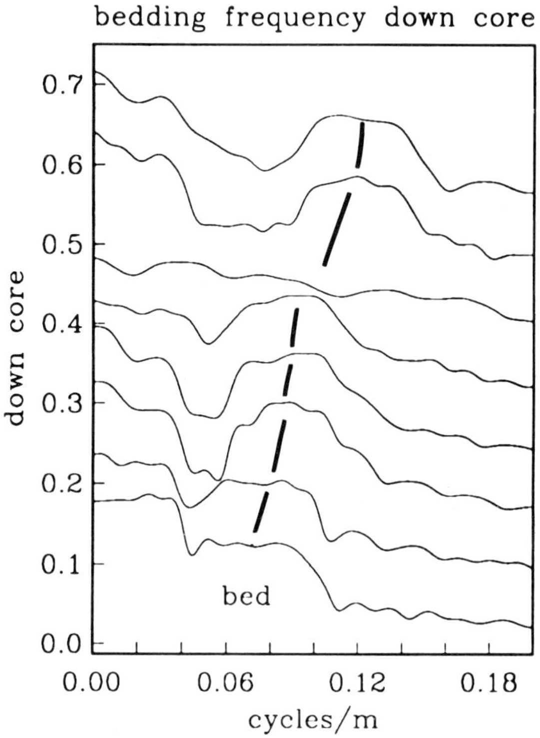

Fig. 18. Example of tracking frequency shifts of an Albian bedding cycle down-core by moving-window analysis. Each trace is a spectral estimate of 1 m of the aluminium time series. A mean frequency for the bedding cycle of 0.085–0.09 cycles/cm is suggested by eyeing the data series. The bedding rhythm shifts over this 2.5-m interval to higher spatial frequency, interpreted as slowing of sedimentation rate. The low-frequency peak, poorly resolved by the short window, is the *c*.60-cm, 100-ka peak.

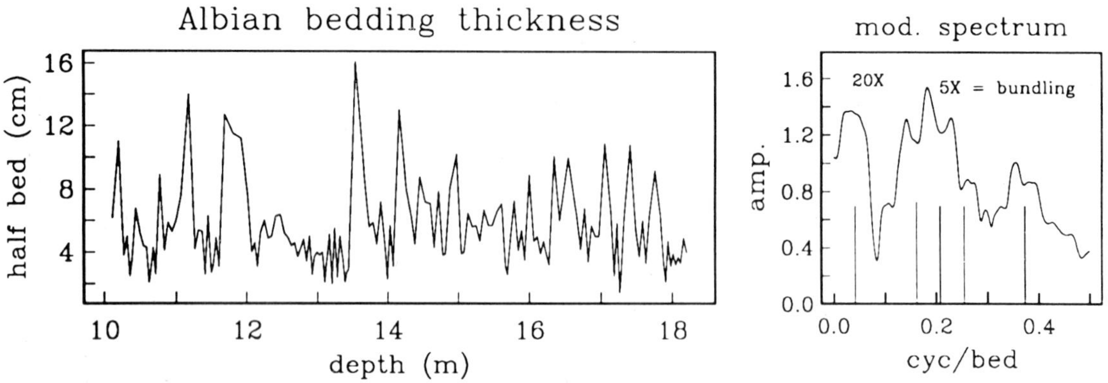

Fig. 19. Sequential variations in spacing between maxima and minima (half bedding cycle) of aluminium data from the Albian of the Piobbico core. Thick beds correspond to high carbonate portions of the bundle (100 ka) cycle. Thin beds coincide with black shale intervals. The spectrum of the thickness series (metronomic assumption shown) shows strong thickness modulation at periods approximately 20 and 5 times the bedding cycle, or at about 400 and 100 ka. Vertical bars positioned at frequencies passing 95% confidence test for line behaviour correspond closely to components of eccentricity. Eccentricity-driven variations in sedimentation rate therefore distort the accumulation of precessional cycles in a regular way.

the 97-ka and 127-ka components of eccentricity to within 2%; a lower frequency is a near match to the 413-ka-long cycle of eccentricity. Scaled to the mean bundling frequency, taken to be the 109-ka average of the short cycle of eccentricity, the bedding cycle has a numerical period of 19.8 ka. The expected value for the Albian precessional period is 20.2 ka (extrapolated from Berger *et al.*, 1989).

Long-term modulations

Nearly continuous gamma-ray attenuation porosity evaluator (GRAPE) measurements taken by the Deep Sea Drilling Project and Ocean Drilling Programs provide climatic time series from Pleistocene to Miocene age sampled at a typical resolution of 1 cm or better (Herbert & Mayer, 1991; Mayer, 1991). Pacific data sets from Legs 85 and 110 show high-frequency oscillation superimposed on the long-term dissolution cycles described by Hays *et al.* (1969). Equatorial Atlantic GRAPE data from Leg 108 record high-frequency variations riding on a long-term decrease of carbonate content from the present into the late Miocene.

Both legs were plagued by problems in magnetostratigraphic dating of pre-Pleistocene intervals, presumably from diagenetic loss of magnetic minerals (Weinreich, 1985; Tauxe *et al.*, 1989). Biostratigraphic data were therefore used to define sediment accumulation rates in many cases. Moving-window analysis (window length 400–500 ka) of representative Pacific and Atlantic GRAPE time series indicates that the frequency modulation model is an alternative method capable of giving very precise sedimentation rate estimates (Figs 20 and 21).

Well-calibrated Brunhes and Matuyama chron orbital peaks can be traced continuously down-core. Their spacing is observed to shift smoothly, as would be expected from gradual changes in sedimentation rate. Sequential spectra show the coherent spatial frequency modulations of orbital peaks predicted from smoothly varying sedimentation rate, most strikingly in the data from Site 573. The ratios of palaeoclimatic frequencies stay locked to orbital values, while their spatial values adjust to changes in sedimentation rate. Local sedimentation rate can be scaled to the orbital model to a precision of a few percent (Figs 20 and 21) by maximizing the match of spatial periodicities to eccentricity, obliquity and precessional numerical periods. The trend at Site 573 is consistent with increasing sedimentation rate down-core (stretching of spatial periods), while Site 661 sediments show frequency modulations indicative of a long-term condensation from the Pleistocene to late Miocene.

There is a sound basis for preferring the sedimentation estimates given by the frequency modulation model over biostratigraphy on time scales shorter than 1 Ma. Comparable time resolution from biostratigraphy requires an aggregate uncertainty of a few tens of thousands of years on datum events used to define local sedimentation rates. Such pre-

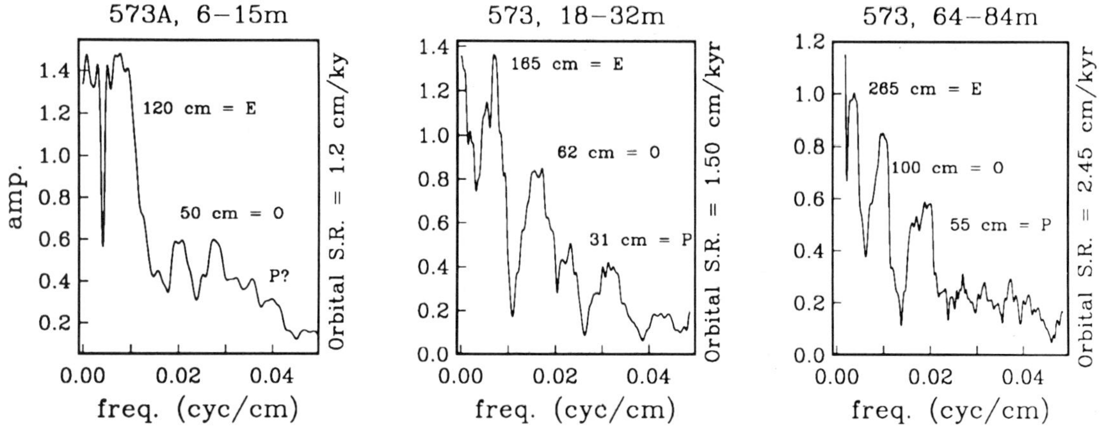

Fig. 20. Example of moving-window detection of an increase in sedimentation rate down-core (from Site 573, equatorial Pacific). Orbital frequencies are continuously traceable down-core, but show gradual changes in spatial frequency. Note that the coherent shifts of the three orbital peaks are consistent with an increase in sedimentation rate of a factor of two with depth. Sedimentation rate that gives optimal match to orbital periods indicated for each window.

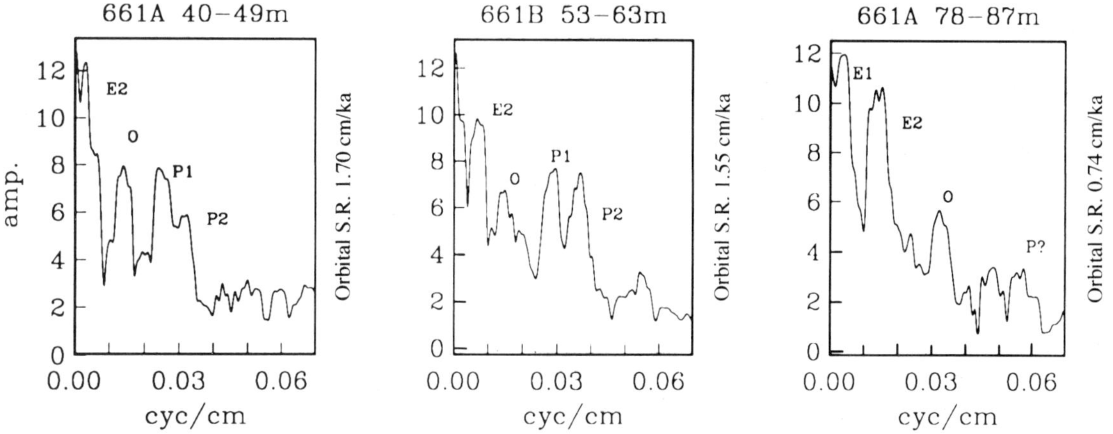

Fig. 21. Sequential shifts in spatial periods of orbital cycles down-core at ODP Site 661 (equatorial Atlantic) are consistent with slowing of sedimentation from upper Pliocene to early Miocene. E2, 400-ka component of eccentricity; E1, 100-ka component of eccentricity; O, obliquity; P1, 23-ka component of precession; P2, 19-ka component of precession.

cision seems over optimistic for several reasons. First, the ages of biostratigraphic events in pre-Pleistocene sediments are derived from linear interpolation within polarity intervals of a few reference sections (e.g. Berggren *et al.*, 1985). Unresolved shifts in sedimentation rate within the polarity zone at the reference section(s), undetected hiatuses or reworking and lack of regional or global synchroneity will bias the age assigned to an event. Second, the resolution of even highly reliable biostratigraphic 'events' may require more intense sampling over a transitional period than is usual (e.g. Thierstein *et al.*, 1977). The orbital model may be vulnerable in different ways: it may not recognize a hiatus if there is no shift in sedimentation rate across the missing interval. It seems appropriate, then, to recognize the limits of biostratigraphic and orbital dating, and to use both in a complementary fashion.

Step functions

A step-function change in the spacing of precessionally driven carbonate cycles has been detected coincident with the Cretaceous/Tertiary (K/T) boundary

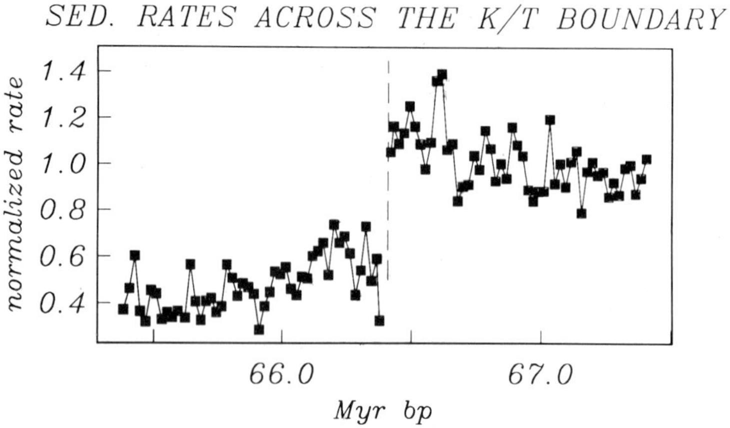

Fig. 22. Normalized variations in sedimentation rate across the K/T boundary in South Atlantic DSDP sites (Herbert and D'Hondt, 1990). The curve compiles sequential changes in the spacing of 20-ka carbonate cycles at five drill sites, normalized at each site to their Maastrichtian mean before averaging between locations.

in the South Atlantic (Herbert & D'Hondt, 1990). This study began as an effort to calibrate carbonate cycles with the palaeomagnetic reversal series of the Late Cretaceous and early Tertiary. A sequential plot of the local spacing of carbonate cycles at a number of DSDP drilling locations shows oscillations coherent with palaeomagnetically deduced changes in accumulation rate. The mean period of the variations is 23.5 ± 3.5 ka according to the Berggren *et al.* (1985) time scale, in good agreement with the expected precessional period of 21 ka. An abrupt drop in cycle spacing at the time of the K/T transition is apparent in a low-resolution compilation of the data (Herbert & D'Hondt, 1990).

A detailed composite time series of sedimentation rate was obtained by measuring the sequential variations in thickness of the cycles across the boundary (Fig. 22). The variations should reflect changes in sedimentation rate measured with a 21-ka yardstick. Magnetic reversal boundaries and the K/T boundary clay were used to correlate between holes. Cycle spacing was normalized at each site to its Maastrichtian mean prior to averaging the data between sites (for example, the spacing of individual carbonate cycles at Site 528 was normalized to the mean Maastrichtian period of 33 cm). Measured spacing of the carbonate rhythms up to the boundary clay shows no sign of a decrease in sedimentation rate prior to the boundary. A very rapid drop in linear sedimentation rate occurs over a timespan that could range from less than 1 to up to 60 000 years. The uncertainty in timing refers to possible errors in the Berggren *et al.* (1985) estimate of the duration of the C29R magnetochrom; the shut-down in sedimentation probably occurred over less than one precessional cycle. Of additional interest are the apparent continuing decline of sedimentation rate into the early Palaeocene, and the very long recovery time of the South Atlantic sedimentary system following the K/T crisis.

CONCLUSIONS

The model of cyclic sedimentation given above has necessarily been simple. Effects on signal preservation such as bioturbational mixing have been ignored, although they would not be difficult to simulate (e.g. Ripepe & Fischer, 1991). Stochastic processes such as slumping and turbidite deposition have been ignored, as have diagenetic modifications of sedimentary cycles during burial. Perhaps more importantly, real changes in the sensitivity of a particular sedimentary variable to orbital forcing over time have not been considered. Such changes, for example the growth in 100-ka response of carbonate and $\delta^{18}O$ across the Matuyama/Brunhes chron transition of the Pleistocene (Ruddiman *et al.*, 1986), have undoubtedly occurred many times in climatic history.

Nevertheless, the simple model presented helps to sharpen the analysis of sedimentary cycles and to suggest improvements in sampling methods. Sedimentary waveforms in a two-component system may go through transitions in behaviour that depend on the ratios of the relative sensitivities of the two components, and their phase relationships. This helps to explain in a rigorous manner why some facies show cyclicity better than others (see also House, 1985; Ricken, 1991). The sensitivity of weight percent measures when components vary in-phase may be low to non-existent; a considerable climatic sensitivity to orbital forcing may thus be masked in many pelagic settings. In contrast, out-of-phase responses of components produce large cyclical changes in sediment composition. It appears likely that more carbonate cycles result from opposite responses of non-carbonate and carbonate accumulation to orbital pacing than has been previously recognized.

We have also seen that the waveform of sedimentary cycles may help us to deduce the time variability of a two-component system. Close attention to the sedimentary waveform, both in amplitude and depth aspects, should make it possible to relax the unrealistic assumption that one component (usually taken to be clay flux) was time-invariant. Diagnostic waveforms can only be seen if the sediment series is sampled adequately, however. Comparison of 1-cm versus 10-cm sampling intervals on the late Pleistocene carbonate record of DSDP Site 572 shows that a sampling rate on average better than the conventional Nyquist requirement may still yield inaccurate frequency spectra. Much denser sampling may be necessary to account adequately for both frequency and amplitude distortions. Collecting a minimum of eight samples per high-frequency cycle will not only result in more reliable spectra of 'untuned' records, but also enable the deconvolution process either as outlined in preceding sections or by the linear algebra method of Kominz & Bond (1990).

Simulations of time–depth distortion also demonstrate just how easily high-frequency climatic signals may be distorted in geological records. In fact, it is commonly difficult to resolve precessional cycles by spectral analysis of 'untuned' data, whether the data set comes from $\delta^{18}O$ variations in the Pleistocene, or from carbonate couplets of Cretaceous age.

It is important to distinguish 'cyclostratigraphy' from circular reasoning. Large variations in intensive sediment variables require time–depth and amplitude distortions in all but the most fortuitous cases. The standard application of Fourier analysis to sequential variations in the sedimentary column assumes a perfect time–depth match. This is a model itself, and a geologically unlikely one. Carrying out spectral analyses does not imply that palaeoceanographers should abandon their geological expertise – it is not more 'objective' to pretend that we are engineers with no knowledge of the sedimentary system. Goodness of fit to data, and consistency of predictions with observable features of the sediment such as fossil preservation, species composition and mineralogy should provide rigorous tests for the usefulness of the orbital model of sedimentation in each geological instance.

ACKNOWLEDGEMENTS

I thank A.G. Fischer, J. Park, W. Ricken and M. Ripepe for helpful comments on an earlier version of this paper. Research was funded by the Petroleum Research Fund of the American Chemical Society and NSF award OCE90-12818.

REFERENCES

Arrhenius, G. (1952) Sediment cores from the East Pacific. *Rep. Swed. Deep Sea Exped.* **5**, 1–227.

Bacon, M.P. (1984) Glacial to interglacial changes in carbonate and clay sedimentation in the Atlantic estimated from thorium-230 measurements. *Isot. Geosci.* **2**, 97–111.

Berger, A.L. (1978) *A simple algorithm to compute long-term variations in daily or monthly insolation*. Inst. Astron. Geophys. G. Lemaitre, Contr. 18.

Berger, A.L., Loutre, M.F. & Dehant, V. (1989) Influence of the changing lunar orbit on the astronomical frequencies of pre-Quaternary insolation patterns. *Paleoceanography* **4**, 555–564.

Berger, W.H. (1973) Deep-sea carbonates: Pleistocene dissolution cycles. *J. Foram. Res.* **3**, 187–195.

Berggren, W.A., Kent, D.V., Flynn, J.J. & Van Couvering, J.A. (1985) Cenozoic geochronology. *Geol. Soc. Am. Bull.* **96**, 1407–1418.

Chuey, J.M., Rea, D.K. & Pisias, N.G. (1987) Late Pleistocene paleoclimatology of the central equatorial Pacific: a quantitative record of eolian and carbonate deposition. *Quat. Res.* **28**, 323–339.

Dean, W.E., Gardner, J.V. & Cepek, P. (1981) Tertiary carbonate-dissolution cycles on the Sierra Leone Rise, eastern equatorial Atlantic Ocean. *Mar. Geol.* **39**, 81–101.

De Boer, P.L. (1982) Cyclicity and storage of organic matter in Middle Cretaceous pelagic sediments. In: *Cyclic and Event Stratification* (Eds Einsele, G. & Seilacher, A.). Springer-Verlag, Berlin, pp. 456–475.

Decker, K. (1991) Rhythmic bedding in siliceous sediments – an overview. In: *Cycles and Events in Stratigraphy* (Eds Einsele, G., Ricken, W. & Seilacher, A.). Springer-Verlag, Berlin, pp. 167–187.

DeMenocal, P., Bloemendal, J. & King, J. (1991) A rock-magnetic record of monsoonal dust deposition to the Arabian Sea: Evidence for a shift in the mode of deposition at 2.4 Ma. In: *Sci. Res. Ocean Drill. Prog., 117* (Eds Prell, W.L. & Niitsuma, N.). College Station, TX (Ocean Drilling Program), pp. 389–407.

Dunn, D.A. & Moore, T.C. (1981) Late Miocene–Pliocene (magnetic Epoch 9–Gilbert magnetic epoch) calcium carbonate stratigraphy of the equatorial Pacific Ocean. *Geol. Soc. Am. Bull.* **92**, 408–452.

FARRELL, J.W. & PRELL, W.L. (1989) Climatic change and $CaCO_3$ preservation: an 800 000 year bathymetric reconstruction from the central equatorial Pacific Ocean. *Paleoceanography* **4**, 447–466.

FARRELL, J.W. & PRELL, W.L. (1991) Pacific $CaCO_3$ preservation and $\delta^{18}O$ since 4 Ma: Paleoceanic and paleoclimatic implications. *Paleoceanography* **6**, 485–498.

FISCHER, A.G. (1986) Climatic rhythms recorded in strata. *Annu. Rev. Earth Planet. Sci.* **14**, 351–376.

FRANCOIS, R., BACON, M.P. & SUMAN, D.O. (1990) Thorium 230 profiling in deep-sea sediments: high resolution records of flux and dissolution of carbonate in the Equatorial Atlantic during the last 24 000 years. *Paleoceanography* **5**, 761–787.

HAYS, J.D., SAITO, T., OPDYKE, N.D. & BURCKLE, L.H. (1969) Pliocene–Pleistocene sediments of the equatorial Pacific: their paleomagnetic, biostratigraphic, and climatic record. *Geol. Soc. Am. Bull.* **80**, 1481–1514.

HERBERT, T.D. & D'HONDT, S.L. (1990) Precessional climate cyclicity in late Cretaceous–early Tertiary marine sediments: a high resolution chronometer of Cretaceous–Tertiary boundary events. *Earth Planet. Sci. Lett.* **99**, 263–275.

HERBERT, T.D. & MAYER, L.A. (1991) Long climatic time series from sediment physical property measurements. *J. Sedim. Petrol.* **61**, 1089–1108.

HERBERT, T.D., STALLARD, R.F. & FISCHER, A.G. (1986) Anoxic events, productivity rhythms, and the orbital signature in a mid-Cretaceous pelagic core. *Paleoceanography* **1**, 495–506.

HOUSE, M.R. (1985) A new approach to an absolute timescale from measurements of orbital cycles and sedimentary microrhythms. *Nature* **315**, 721–725.

HOVAN, S.A., REA, D.K., PISIAS, N.G. & SHACKLETON, N.J. (1989) Direct link between China loess and marine $\delta^{18}O$ records: eolian flux to the north Pacific. *Nature* **340**, 294–296.

IMBRIE, J., HAYS, J.D., MARTINSON, D.G., MCINTYRE, A., MIX, A.C., MORLEY, J.J., PISIAS, N.G. & SHACKLETON, N.J. (1984) The orbital theory of Pleistocene climate: support from a revised chronology of the marine ^{18}O record. In: *Milankovitch and Climate*, Part 1 (Eds Berger. A., Imbrie, J., Hays, J., Kukla, G. & Saltzman, B.). Reidel Publ. Co., Dordrecht, pp. 269–305.

KOMINZ, M.A. & BOND, G.C. (1990) A new method for testing periodicity in cyclic sediments: application to the Newark Supergroup. *Earth Planet. Sci. Lett.* **98**, 233–244.

MARTINSON, D.G., PISIAS, N.G., HAYS, J.D., IMBRIE, J., MOORE, T.C. Jr & SHACKLETON, N.J. (1987) Age dating and the orbital theory of the Ice Ages: Development of a high-resolution 0 to 300 000 year chronostratigraphy. *Quat. Res.* **27**, 1–29.

MAYER, L.A. (1991) Extraction of high-resolution carbonate data for paleoclimate reconstruction. *Nature* **352**, 148–151.

MCGILLEM, C.D. & COOPER, G.R. (1984) *Continuous and Discrete Signal and System Analysis*. Holt, Rinehart and Winston, New York.

PARK, J. & HERBERT, T.D. (1987) Hunting for paleoclimatic periodicities in a sedimentary series with uncertain time scale. *J. Geophys. Res.* **92**, 14027–14040.

PISIAS, N.G., MARTINSON, D.G., MOORE, T.C. Jr, SHACKLETON, N.J., PRELL, W., HAYS, J.D. & BODEN, G. (1984) High resolution stratigraphic correlation of benthic oxygen isotopic records spanning the last 300 000 years. *Mar. Geol.* **46**, 217–233.

REA, D.K., PISIAS, N.G. & NEWBERRY, T. (1991) Late Pleistocene paleoclimatology of the central equatorial Pacific: flux patterns of biogenic sediments. *Paleoceanography* **16**, 227–244.

RICKEN, W. (1986) *Diagenetic Bedding*. Lecture Notes in Earth Sciences. Springer-Verlag, Berlin.

RICKEN, W. (1991) Variation of sedimentation rates in rhythmically bedded sediments: Distinction between depositional types. In: *Cycles and Events in Stratigraphy* (Eds Einsele, G., Ricken, W. & Seilacher, A.). Springer-Verlag, Berlin, pp. 167–187.

RIPEPE, M. & FISCHER, A.G. (1991) Stratigraphic rhythms synthesized from orbital variations. In: *Sedimentary Modeling: Computer Simulations and Methods for Improved Parameter Definition*. Kansas Geol. Survey **233**, 335–344.

RUDDIMAN, W.F., RAYMO, M. & MCINTYRE A. (1986) Matuyama 41 000 year cycles: North Atlantic Ocean and northern hemisphere ice sheets. *Earth Planet. Sci. Lett.* **80**, 117–129.

SCHIFFELBEIN, P. (1984) *Stable isotope sytematics in Pleistocene deep-sea sediment records*. PhD dissertation, University of California San Diego, La Jolla.

SCHIFFELBEIN, P. & DORMAN, L. (1986) Spectral effects of time–depth non-linearities in deep sea sediment records: a demodulation technique for realigning time and depth scales. *J. Geophys. Res.* **91**, 3821–3835.

SCHWARZACHER, W. (1954) Die Grosrhythmik des Dachsteinkalkes von Lofer. *Tscher. Mineral. Petrogr. Mitt.* ser. 3, v. 4, 44–54.

SCHWARZACHER, W. (1991) Milankovitch cycles and the measurement of time. In: *Cycles and Events in Stratigraphy* (Eds Einsele, G., Ricken, W. & Seilacher, A.). Springer-Verlag, Berlin, pp. 855–863.

SCHWARZACHER, W. & FISCHER, A.G. (1982) Limestone–shale bedding and perturbations in the earth's orbit. In: *Cyclic and Event Stratification* (Eds Einsele, G. & Seilacher, A.). Springer-Verlag, Berlin, pp. 72–95.

SHACKLETON, N.J., BERGER, A. & PELTIER, W.R. (1990) An alternative astronomical calibration of the lower Pleistocene timescale based on ODP Site 677. *Trans. R. Soc. Edinburgh* **81**, 251–261.

TAUXE, L., VALET, J.-P. & BLOEMENDAL, J. (1989) Magnetostratigraphy of Leg 108 advanced hydraulic piston cores. In: *Proc. Ocean Drilling Program, Scientific Results, 108* (Eds Ruddiman, W., Sarnthein, M. *et al.*). Ocean Drilling Program, College Station, TX, pp. 429–440.

THIERSTEIN, H.R., GEITZNAUER, K.R., MOLFINO, B. & SHACKLETON, N.J. (1977) Global synchroneity of late Quaternary coccolith datum levels: validation by oxygen isotopes. *Geology* **5**, 400–404.

THOMSON, D.J. (1982) Spectrum estimation and harmonic analysis. *Proc. IEEE* **70**, 1055–1096.

TORNAGHI, M.E., PREMOLI SILVA, I. & RIPEPE, M. (1989) Lithostratigraphy and planktonic foraminiferal biostratigraphy of the Aptian–Albian 'Scisti a Fucoidi' in the Piobbico core, Marche, Italy: background for cyclostratigraphy. *Riv. It. Paleont. Strat.* **95**, 223–264.

WEEDON, G.P. (1989) The detection and illustration of regular sedimentary cycles using Walsh power spectra and filtering, with examples from the Lias of Switzerland. *J. Geol. Soc. Lond.* **146**, 133–144.

WEINREICH, N. (1985) Paleomagnetism of Deep Sea Drilling Project Leg 85 sediments: Neogene magnetostratigraphy and tectonic history of the central Pacific. *Init. Rep. DSDP* **85**, 849–901.

WILKENS, R.H. & LANGSETH, M.G. (1983) Shipboard physical properties measurements. *Init. Rep. DSDP* **70**, 659–674.

YANG, Y.-L., ELDERFIELD, H. & IVANOVICH, M. (1990) Glacial to Holocene changes in carbonate and clay sedimentation in the Equatorial Pacific Ocean estimated from Thorium 230 profiles. *Paleoceanography* **5**, 789–809.

Spec. Publs Int. Ass. Sediment. (1994) **19**, 509–529

The effect of orbital cycles on Late and Middle Cretaceous climate: a comparative general circulation model study

J. PARK* *and* R.J. OGLESBY†‡

* *Department of Geology and Geophysics, Yale University, P.O. Box 6666, New Haven, CT 06511, USA; and*

† *Department of Geological Sciences, Brown University, P.O. Box 1846, Providence, RI 02912-1846, USA*

ABSTRACT

We use sets of climate simulations made with the NCAR (National Center for Atmospheric Research) atmospheric general circulation model (GCM) CCM1 to perform a comparative sensitivity study using two different reconstructions of Cretaceous orography and sea-surface temperatures (SSTs). One set of boundary conditions is appropriate for 100 Ma (Mid-Cretaceous) and has SSTs determined largely by the local surface energy balance as computed in a previous GCM simulation of the Mid-Cretaceous climate. The other set of boundary conditions is appropriate for 70 Ma (Late Cretaceous) and has SSTs determined largely by palaeotemperature data drawn from that time period, which suggest a weak meridional (north–south) SST gradient. No land or sea ice is allowed in either set of climate simulations. We estimated the linear sensitivity of key model variables to precession and obliquity as functions of latitude and longitude. The mean climates for the two reconstructions differ greatly. The hydrological cycle in the 100-Ma simulations shares two critical features with that of the present-day Earth: (i) a symmetric pattern of zonally averaged net evaporation minus precipitation ($E - P$) about the equator, indicative of a low-latitude Hadley-cell circulation and the intertropical convergence zone (ITCZ); and (ii) a west–east/arid–wet asymmetry induced by the summer monsoon over large low-latitude land masses. Both of these features are absent or greatly modified in the 70-Ma simulations, and we infer that they are inhibited by the weak meridional SST gradient from the equator to roughly 40° latitude in the summer hemisphere. The atmospheric circulation in the 70-Ma reconstruction is very sluggish relative to the 100-Ma circulation. Orbital insolation variations in the 100-Ma simulations amplify or diminish monsoonal pressure anomalies and the mean wind flow around them. Associated with these are large changes in the atmospheric hydrological cycle. Large shifts in the evaporation–precipitation balance ($E - P$) over the newly rifted equatorial South Atlantic and East Tethys can be related speculatively to large-scale changes in ocean circulation and the development of deep-water anoxia. Insolation-induced changes in the monsoonal pressure anomalies in the 70-Ma simulations affect the hydrological cycle less strongly at low latitudes. We infer that the weak meridional SST gradient in the low-latitude summer hemisphere inhibits a large response by the hydrological cycle in this region. However, warmer mid-latitude SSTs in the 70-Ma simulations raise the specific humidity of mid-latitude atmospheric flows, thereby increasing the potential for a large hydrological response to precession and obliquity over North America and proto-Europe, where many examples of cyclic limestone sequences can be found. The presence of epicontinental seas, however, inhibits the formation of a summertime monsoon in North America for our 70-Ma simulations, resulting in a weak response in this region to Milankovitch insolation fluctuations. Our conclusion that the SST profile controls the model behaviour is supported by zonal averages of the model's response to orbital insolation fluctuations, which are small relative to zonal averages of the model's 'mean-state' Cretaceous climate.

‡ Present address: Department of Earth and Atmospheric Sciences, Purdue University, West Lafayette, IN 47907, USA.

INTRODUCTION

Alternating lithologies in shelf and pelagic sediments comprise the principal record of Milankovitch rhythms in the Cretaceous. The causative factors behind these sedimentary rhythms are still a matter of debate, but most scenarios depend critically on fluctuations in the global hydrological cycle. An atmospheric general circulation model (GCM) can be used to investigate changes in the atmospheric hydrological cycle that influence regional and global changes in oceanic salinity, circulation and anoxic versus aerated conditions. The geological reconstruction of past climates proceeds from deducing the factors that govern the deposition of individual sedimentary sequences to forming hypotheses about how contemporaneous depositional environments are connected in a global framework. Much of the geological evidence from a given time period can be fit into more than one global climate scenario. Numerical climate models can be used to examine whether hypothesized global climates are consistent with our current understanding of atmospheric and oceanic dynamics. Some of the most comprehensive results can be obtained from atmospheric GCMs, which can simulate three-dimensional wind patterns, temperatures, evaporation and precipitation as equilibrated to a particular set of boundary conditions and forcing. A detailed comparison of model results with the geological record for the Pleistocene is possible with a wide range of proxy evidence, from which past climate during the ice ages can be inferred (e.g. Prell & Kutzbach, 1987; Oglesby *et al.*, 1989). In earlier periods of Earth history, however, constraints on boundary conditions such as sea surface temperature (SST), ice cover and albedo are more limited, and only broad features of the global climate can be investigated. Studies of this type may use idealized boundary conditions that broadly represent an epoch in Earth history (Kutzbach & Gallimore, 1989; Kutzbach *et al.*, 1990), or may vary some boundary conditions systematically (continent placement, ice cover, etc.) to assess the relative influences on climate (Barron & Washington, 1984; Oglesby, 1989).

Climate cycles associated with the Milankovitch orbital insolation variations have been inferred to govern bedded sedimentary sequences from periods ranging from the Precambrian to the Recent (Hays *et al.*, 1976; Imbrie *et al.*, 1984; Grotzinger, 1986; Herbert & Fischer, 1986; Mead *et al.*, 1986; Olsen, 1986; ROCC, 1986; Park & Herbert, 1987; and many others). The nature of the response to orbital insolation cycles can vary greatly, depending on the mean global climate, and the specific sedimentary conditions. In the Pleistocene, orbital insolation variations are thought to influence the growth and decay of continental ice sheets. The albedo feedback associated with large ice sheets is thought to amplify greatly the Earth's response to orbital cycles, which generate large but seasonally opposing variations in solar energy input (5–10% about the mean). In the Cretaceous, orbital cycles appear to influence the timing of short-term ocean anoxic events, as well as cyclic fluctuations in terrigenous input to shelf environments. A major challenge to the climate modeller is to account for such profound changes in the Earth system without the aid of significant ice-albedo feedback. In one sense, however, the presumed absence of widespread continental ice is a partial blessing, because different GCMs do not agree in their predictions of ice sheet development (e.g. Rind *et al.*, 1989; Oglesby, 1990). Moreover, the dynamics of ice sheet growth and decay in the late Pleistocene, which exhibits a 100-ka quasi-period phase-locked to orbital eccentricity, is better modelled by non-linear dynamical models (e.g. Maasch & Saltzman, 1990). Non-linear dynamics may also govern Cretaceous depositional systems, perhaps causing the preponderance of 100-ka cyclicity in some limestone sequences (e.g. Park & Herbert, 1987). Our study cannot model such dynamics, but can suggest how orbital insolation cycles associated with precession and obliquity may have affected the Cretaceous hydrological cycle and ocean circulation.

An explanation of the mean Cretaceous climate is also an unsolved problem. Mean global temperature is thought to have been ≥6.5°C greater in the Mid-Cretaceous (100 Ma) than at present (Barron, 1983), and the palaeofloral record suggests year-round frost-free conditions at many high-latitude locations. There is no evidence for widespread continental glaciation, though evidence of any kind is currently scarce for the interiors of Antarctica and Siberia. Barron & Washington (1984) argued, using the National Center for Atmospheric Research (NCAR) model CCM0, that a combination of continental rearrangement and the prescription that the Earth have no sea ice were sufficient to explain much but not all of the difference in global mean temperatures. Some authors (e.g. Barron & Washington, 1985; Berner, 1990) have suggested that the additional required warming is due to enhanced greenhouse effects caused by elevated levels of atmospheric carbon dioxide (CO_2). However, even with enhanced greenhouse warming, atmospheric GCMs predict below-freezing mean mid-winter temperatures in many continental interiors (Barron & Washington, 1985; Oglesby & Saltzman, 1992). This result appears to contradict palaeofloral constraints in some regions, e.g. the US Western Interior, but is consistent with evidence for (at least) transient snow and ice (Frakes & Francis, 1988; Rich *et al.*, 1988; Gregory *et al.*, 1989; Sloan & Barron, 1990) in the southern high latitudes.

In previous work (Oglesby & Park, 1989; Park & Oglesby, 1990, 1991) we examined the response of the Mid-Cretaceous (100 Ma) climate to orbital insolation changes using the NCAR Community Climate Model CCM1 with prescribed boundary conditions close to those of the Barron & Washington (1984) study. A large response to insolation changes was noted in the evaporation minus precipitation ($E - P$) balance over many regions, suggesting that the model's response to orbital insolation cycles may be sufficient to explain many of the cyclical beddings observed in the sedimentary record. In this paper we contrast our earlier work with a similar set of climate sensitivity experiments that included a 70-Ma continental reconstruction compiled by the University of Chicago Paleoatlas Project, and made available to us by M. Horrell and F. Ziegler. This Late Cretaceous reconstruction has boundary conditions that differ greatly from those of the 100-Ma reconstruction, which was based on Barron *et al.* (1981) and Tarling (1978). We cite four possible comparisons between the reconstructions (Fig. 1).

1 The Late Cretaceous reconstruction includes substantial areas of epicontinental sea within North America. The 100-Ma reconstruction has no epicontinental sea. The presence of an interior seaway inhibits the formation of a mid-continental monsoonal pressure anomaly.

2 Our previous work suggests that wide swings in the $E - P$ balance occurred over the newly rifted South Atlantic Ocean, creating conditions favourable for cyclic anoxia in a silled basin in a manner long hypothesized by sedimentologists (e.g. de Boer & Wonders, 1984), and quantified in the model of Sarmiento *et al.* (1988a). The particular placement of a narrow marine basin at low latitudes between two continents appears to be responsible for the fluctuating $E - P$ balance. Cyclic bedding is observed in deep ocean sedimentary cores from this region both in the Mid-Cretaceous (e.g. Dean *et al.*, 1984) and across the Cretaceous–Tertiary boundary (Herbert & D'Hondt, 1990). The latter sequence was deposited 35-Ma later than our 100-Ma reconstruction, at a time when the South Atlantic was considerably wider. How do the model results depend on the specific width of the South Atlantic?

3 Barron & Peterson (1990) used an oceanic GCM, with the surface winds and $E - P$ balance specified from a previous atmospheric GCM run, to investigate ocean circulation at 100 Ma. They identified eastern Tethys as a possible source of warm, saline deep water to the Mid-Cretaceous ocean. Our 100-Ma sensitivity study identified a considerable $E - P$ response over easternmost Tethys to orbital insolation changes. Taken together, these results are compatible with the cyclic generation of warm, saline, oxygen-depleted deep water at low latitudes, leading to cyclic anoxia in Tethys and perhaps the wider ocean. Warm saline deep water has been proposed as a possible contributor to global warmth through the Cretaceous into the Eocene (e.g. Brass *et al.*, 1982). Does the response of the GCM in the Tethys region to orbital forcing remain fairly constant, or are there significant differences between the Mid-Cretaceous and the Late Cretaceous?

4 The relative warmth of the Cretaceous occurs principally at high latitudes, suggesting an enhanced equator-to-pole transport of heat. As discussed by Barron (1987), warmth at high latitudes can be maintained by atmospheric transport of latent heat in the form of water vapour from low to high latitudes, or by the poleward migration of warm deep ocean waters formed by enhanced evaporation at low latitudes. Oglesby & Saltzman (1990) examined the latter scenario with a simplified zonally symmetric climate model, and found that a prescription that deep ocean water be everywhere ≥15°C forced high-latitude SST to be ≥10°C with a relatively weak equator-to-pole SST gradient. Deep ocean water masses of this warmth are consistent with Late Cretaceous oxygen isotope palaeotemperatures from *Inoceramus* fossils (Saltzman & Barron, 1982). Late Cretaceous deep-water palaeotemperatures determined from benthic foraminifera are somewhat cooler, in the range 4–9°C, though still greater than present-day values (Barrera *et al.*, 1987; Barrera & Huber, 1990). The equator-to-pole SST gradient in the Late Cretaceous reconstruction is derived from various geological palaeotemperature indicators, and is weak in the summer hemisphere of the model, consistent with the poleward migration of warm deep ocean waters. The SSTs computed by Barron & Washington (1984) and used for our 100-Ma study have a relatively large meridional (north–south) temperature gradient, with fairly cool high-latitude temperatures. What impact could a change in the meridional structure of SSTs have on the mean Cretaceous climate and hence on the response to orbital insolation changes?

We describe the modelling procedure in the next section. In the third section we compare the 100-Ma and 70-Ma climate states and their responses to orbital insolation changes, while in the final section we discuss the implications of this comparison for Cretaceous sedimentary cycles. Data constraints on detailed Cretaceous boundary conditions are weak and the underlying causes of the epoch's unusual warmth are yet debated. Therefore, we cannot claim that our comparative sensitivity study models the long-term evolution of the Cretaceous climate system with better than conjectural accuracy. Rather, by isolating those factors in the mean climate that are conducive to the formation of cyclic sedimentation, we hope to offer evidence in favour of particular Cretaceous climate scenarios. Our experiments suggest that a weak meridional SST gradient affects the modelled atmospheric hydrological cycle more profoundly than does the change in orography from the Mid- to the Late Cretaceous. We also find that large variations in the mean climate state strongly affect the model's response to orbital insolation fluctuations.

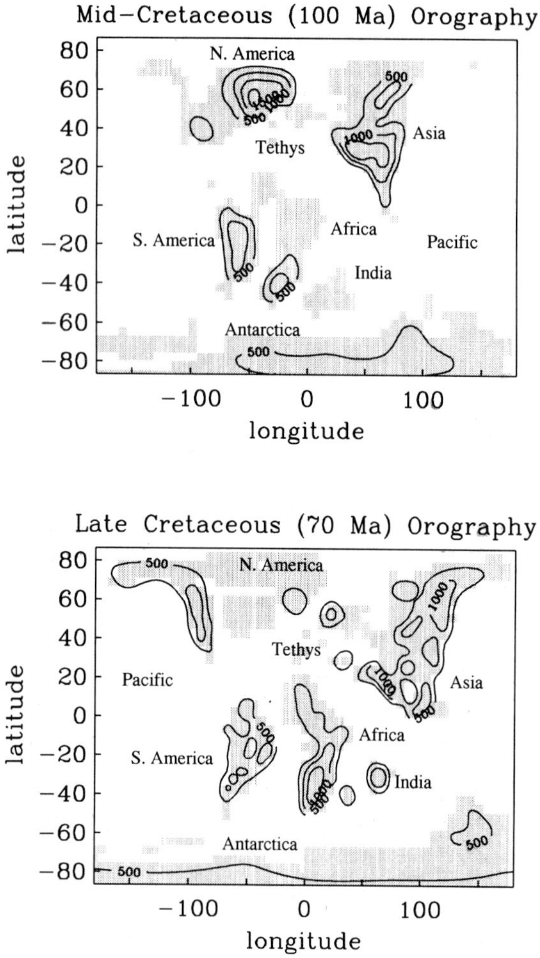

Fig. 1. Continental reconstruction for (a) the Mid-Cretaceous (100 Ma) and (b) the Late Cretaceous (70 Ma), both projected onto the CCM grid. The resolution is 4.5° latitude by 7.5° longitude. Land areas are shaded. Palaeotopography, expressed by the circulation model as a spherical harmonic expansion, is contoured in metres.

EXPERIMENT DESCRIPTION

Surface boundary conditions and modelling assumptions

The model employed for our study is the NCAR CCM, version CCM1. Blackmon (1986) gives an account of the model development and a discussion of its capabilities. Oglesby & Park (1989) and Park & Oglesby (1991) give a thorough description of how the model was employed in the 100-Ma Cretaceous simulations. The model can be used to compute, for a specified set of boundary conditions and external forcing, the atmospheric winds, pressures, temperatures and humidities as prognostic quantities, and, in doing so, computes many other quantities diagnostically, e.g. surface temperatures, precipitation, clouds, radiative fluxes, etc. In our version of CCM1, surface temperatures are calculated over land (and sea ice if any is prescribed) via a surface energy budget, but are held fixed (i.e. prescribed) over ocean due to the large heat capacity of water. All land surfaces in our simulations were defined as 'grassland', which has albedo and surface wetness values that are between those of deserts and forest and are similar to those for 'bare soil' used in earlier Cretaceous GCM studies (e.g. Barron & Washington, 1984; Glancy *et al.*, 1986). Evaporation is calculated from the local surface heat balance using surface moisture values to determine the flux of latent heat from the surface, thereby providing the atmosphere with water vapour. Precipitation is computed as the moisture condensed due to the mutual adjustment (convective or stable as appropriate) of the temperature and moisture fields. A smoother time-average of the evaporation minus precipitation ($E - P$) balance can be made from the divergence of the moisture transport $\boldsymbol{VQ}$. We integrate $\overline{\nabla \cdot \boldsymbol{VQ}}$ vertically from the surface to the 500-mb pressure level (the overbar denotes time-average), above which the atmosphere contains very little moisture. The divergence of moisture transport is equal to the rate at which water exits a 'gridpoint' on the Earth's surface, roughly a rectangle 4.5° in latitude and 7.5° in longitude. A gridpoint is a 'source' of water for the atmosphere if $\overline{\nabla \cdot \boldsymbol{VQ}} > 0$ (i.e. evaporation exceeds precipitation), and is a 'sink' if $\overline{\nabla \cdot \boldsymbol{VQ}} < 0$. In a truly equilibrium model, $E - P$ and $\overline{\nabla \cdot \boldsymbol{VQ}}$ are equivalent at each gridpoint – differences arise from different methods of computation. Both wind velocity $\boldsymbol{V}$ and specific humidity Q are computed spectrally, so that large-scale features in $\overline{\nabla \cdot \boldsymbol{VQ}}$ are relatively easy to identify as compared to the gridpoint-by-gridpoint subtraction of the precipitation field from the evaporation field, which is more subject to stochastic fluctuations.

We chose to run CCM1 in perpetual-season mode to perform the large number of simulations desired for our sensitivity experiment. In seasonal-cycle mode, the solar declination angle in CCM1 varies according to the calendar day, and the model requires a long integration, typically five model years or longer, to reach quasi-equilibrium. In perpetual-season mode, the solar declination angle is held fixed at a single value appropriate for a chosen day and the model typically reaches quasi-equilibrium within 60 model days. Since orbital insolation variations are primarily a summer/winter effect, two days were chosen for our study: January 15 (perpetual January) yielding mid-winter/summer climate statistics for the Northern/Southern Hemisphere, and July 16 (perpetual July) yielding mid-summer/winter climate statistics for the Northern/Southern Hemisphere. Since the model runs are 150 days or longer, we fix surface moisture values over land to avoid runaway summer temperature increases in the continental interiors. (Once a gridpoint becomes parched, all insolation energy that reaches the surface goes into heat.) Though the Cretaceous is considered more 'equable' than the present, some seasonal swing in the SST is likely. In the 100-Ma simulations, we imposed an arbitrary but plausible seasonal SST variation equal to approximately half of what occurs at present. We used separate winter and summer reconstructions for the 70-Ma simulations. We neglect SST perturbations associated with orbital variations. Annual-average precession anomalies are everywhere zero, making this a reasonable assumption for precession (Kutzbach & Gallimore, 1988). Annual-average obliquity anomalies vanish in the global average, but increasing δ shifts insolation from low to high latitudes. Seasonal-cycle simulations may be needed to assess fully any latitudinal shift in SST, or similar climatic changes, caused by obliquity variations.

SST in the 100-Ma simulations is a zonal average of SST (i.e. averaged over longitude) derived by Barron & Washington (1984) using a 'swamp ocean'. A 'swamp ocean' is an infinite moisture source without any heat capacity, and can only be used with annually-averaged or equinoctial insolation. Ocean heat transport was not represented, and SST at a given model gridpoint is determined by the local energy balance. Therefore, the variation of SST with latitude for the 100-Ma experiment resembles the variation of insolation with latitude, with a pronounced equatorial peak (Fig. 2). Some poleward heat transport by the ocean is implicit in the prescription of no sea ice at high latitudes, though this thermal energy has no explicit source in the Barron & Washington (1984) simulation. The Late Cretaceous 70-Ma SST in the University of Chicago Paleoatlas reconstruction is compiled from near-coastal marine and terrestrial flora. The latitudinal variation of these SSTs is consistent with a greatly enhanced oceanic poleward transport of heat relative to the 100-Ma reconstruction: equatorial SSTs are depressed while mid- and high-latitude SSTs are greatly elevated, especially in summer. No explicit physical basis for this SST distribution is given, and it may be possible only on an Earth with enhanced greenhouse warming. However, greenhouse warming, by itself, will tend to warm low-latitude regions as well as high-latitude regions, though probably not as much (e.g. Crowley, 1990; Mitchell, 1990).

Although there is isotopic evidence that equatorial SST declined from the Middle to the Late Cretaceous (Douglas & Savin, 1975), we do not interpret the differences in SST shown in Fig. 2 as representing accurately the evolution of Cretaceous conditions. Neither the currently available isotopic palaeotemperatures nor the 'swamp ocean' calculations are sufficiently reliable. Rather, we interpret the differing SST profiles as representing plausible, but distinctly different, mean states of the global

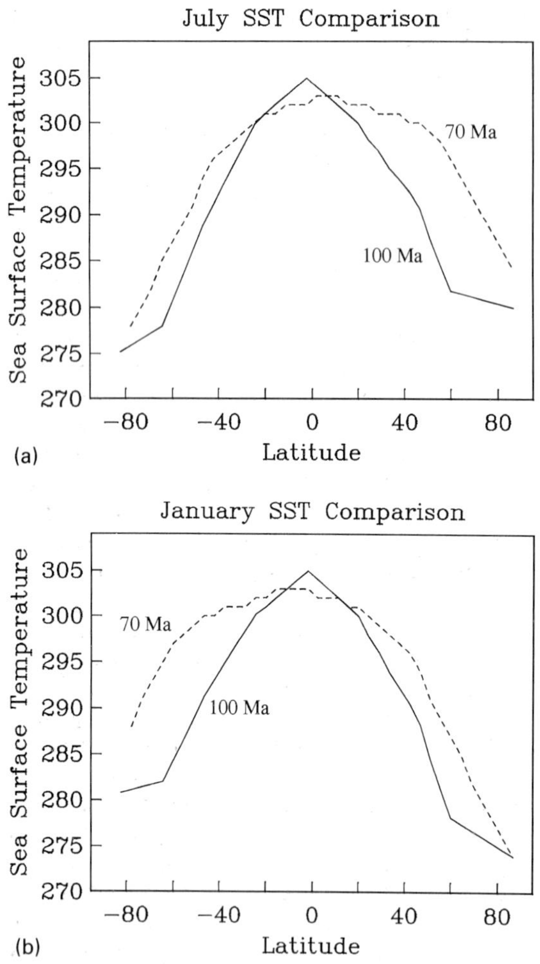

Fig. 2. Zonally averaged SST, in K, for the 100-Ma (solid line) and 70-Ma (dashed line) simulations for (a) July and (b) January.

ocean in the Cretaceous. These different ocean 'states', as seen below, lead to markedly different atmospheric circulations, and therefore influence strongly the model response to orbital cycles. Although most authors believe that Cretaceous CO_2 levels were elevated, we used the present-day atmospheric concentrations in our simulations. Our motivation was twofold. First, the SSTs that we obtained from the Barron & Washington (1984) 'swamp ocean' simulations were calculated with present-day CO_2 levels – using these SSTs in climate simulations with elevated CO_2 would be inconsistent. Second, the critical monsoonal dynamics that respond to orbital insolation changes appear to be governed mostly by temperature gradients between land and ocean regions, rather than by the average global temperature. To first order, changes in CO_2 cause spatially homogeneous changes in temperature, with little difference between land and ocean regions (Oglesby & Saltzman, 1992). A greenhouse-warmed atmosphere could hold and transport greater quantities of water vapour, thereby amplifying the hydrological response in our simulations. However, as seen below in our comparison of the 70-Ma and 100-Ma climate reconstructions, temperature gradients appear to control the *pattern* of water vapour transport.

Any interpretation of our simulations must carry the proviso that the NCAR CCM has been 'tuned' to replicate present-day atmospheric conditions, and there is no guarantee that it replicates Cretaceous atmospheric conditions equally well. Moreover, several recent comparisons of present-day CCM1 simulations with global climate data sets draw attention to shortcomings in the CCM1 simulation of the present-day climate. Comparisons with Earth Radiation Budget Experiment (ERBE) observations suggest that the CCM1 misrepresents the radiative balance over some tropical regions (Kiehl & Ramanathan, 1990). When compared with climate analyses from the European Centre for Medium Range Weather Forecasts (ECMWF), both CCM0 and CCM1 exhibit a negative temperature bias of 3–4 K in the low troposphere at low latitudes, and underestimate the vigour of the time-averaged meridional Hadley cell by about a factor of two (Randel & Williamson, 1990). Comparisons of this kind (termed 'model validation experiments') aid the development of more accurate GCMs, but also suggest that some conclusions about Cretaceous climate response from a CCM1 study may not carry over to improved versions of the CCM. We can hope that at least broad features in our simulations will be reliable indicators of climate response for an Earth with given boundary conditions, with small-scale features and specific details in the simulations remaining highly uncertain. In addition, variability exists in the model results, as only a quasi-equilibrium is obtained. This variability, which may be due to high-frequency (daily or shorter) or low-frequency (decadal scale) quasi-oscillations in the model equations, can be treated as 'noise'. To be considered significant, any model response or 'signal' must exceed this intrinsic 'noise' variability.

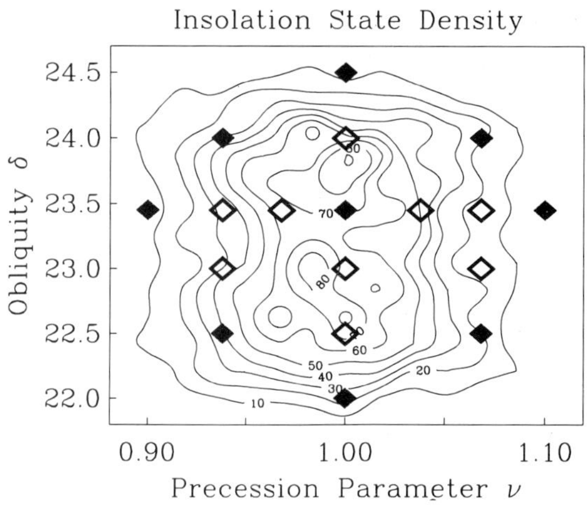

Fig. 3. Range of precession and obliquity values in our simulations. Precession modulates the Earth–Sun distance, which modulates the effective solar constant by a factor ν, shown in the figure. The symbols indicate the 18 insolation states (ν,δ) used in our simulations. All symbols indicate states run with 100-Ma boundary conditions, while only states at the filled symbols were run with 70-Ma boundary conditions. The contours trace a two-dimensional histogram of the insolation states predicted for the past 10 Ma by the astronomical formulas of Berger (1978).

Orbital parameters and model sensitivity coefficients

In CCM1, the obliquity angle δ can be varied by altering a data statement within the computer program. The present-day obliquity is $\delta_o = 23.45°$, with a predicted range of (roughly) 22.0–24.5°. The precessional index p is defined by $p = e \sin \omega(t)$, where e is the eccentricity of the Earth's solar orbit and $\omega(t)$ is the celestial longitude of perihelion, measured counter-clockwise from the vernal equinox. The solar insolation received by the Earth, for a given value of the precessional index, depends on the instantaneous Earth–Sun distance, which varies continuously for an elliptical orbit. The insolation received, relative to that of a perfectly circular orbit, can be parameterized by an 'eccentricity factor' ν that multiplies directly the mean solar insolation (prescribed as 1370 W/m^2 for CCM1). The maximum deviation of ν from unity is roughly 10%, or approximately twice the perturbation to e. Maximum seasonal insolation perturbations occur when the mid-winter and mid-summer equinoxes coincide with perihelion and aphelion of the Earth's orbit. Therefore, although orbital eccentricity governs the maximum possible insolation perturbation, the principal cycles in the mid-summer/mid-winter insolation perturbations are those associated with the precession of the equinoxes.

We define an 'insolation state' (ν, δ) by a choice of eccentricity factor ν and obliquity δ. The case $(\nu_o, \delta_o) = (1.0, 23.45°)$ is used as a mean state for the experiments. We ran 900-model-day control experiments with the 100-Ma reconstruction at this state for January and for July, and model simulation experiments at 17 other insolation states for 150 days (Fig. 3). The first 60 days of each 150-day simulation were allowed for the model to equilibrate with the specified forcing, yielding one 90-day time-averaged sample to compare with the control. These time averages represent 'climatic means' for important variables such as surface temperature and precipitation, and are used for all subsequent analysis. (What we actually do is process 'daily weather' patterns calculated by the model much as present-day daily weather observations are processed to obtain climatic statistics.) The 900-day controls yielded six 90-day samples, each separated by 60 days, providing for an estimate of intrinsic model variability. Statistical comparisons of the 17 'perturbed' states with the control simulation (Park & Oglesby, 1991) indicated that, aside from the intrinsic model variability, the response of the model climate to variations in ν and δ is dominantly linear. Although the internal dynamics represented by the GCM are highly non-linear, time averages of model variables, such as precipitation, did not exhibit chaotic or runaway feedback behaviour. We attribute the linearity of the model response in part to our prescription that our simulations be ice-free. (Ice sheets in GCM simulations tend to grow or wither with time, depending on the particular computer code.) The absence of a significant non-linear response convinced us that fewer insolation states were necessary to determine model sensitivity and intrinsic variability. We therefore used only nine insolation states, as indicated in Fig. 3, to determine model response using the 70-Ma reconstruction.

We estimated the linear sensitivity of time-averaged model variables to orbital forcing by fitting a plane

$$f(\nu, \delta) = m_0 + m_\nu \frac{(\nu - 1)}{0.2} + m_\delta \frac{(\delta - \delta_o)}{2.5} \quad (1)$$

to values obtained from the model runs in each season. Note that this equation corrects a typographical error in Park & Oglesby (1991). If the model response to ν and δ is linear, the regression constant m_0 is equal to the mean-state value of the model variable, obviating the need for a long control simulation. The numerical factors in equation (1) normalize the sensitivity coefficients m_ν and m_δ so that they represent the change in the time-averaged model variable over the entire range of ν (0.9–1.1) and δ (22.0–24.5°), respectively. We fit the plane $f(\nu, \delta)$ to the simulation results at every gridpoint by least-squares, performing a crude local average with adjacent gridpoints to suppress small-scale variations.

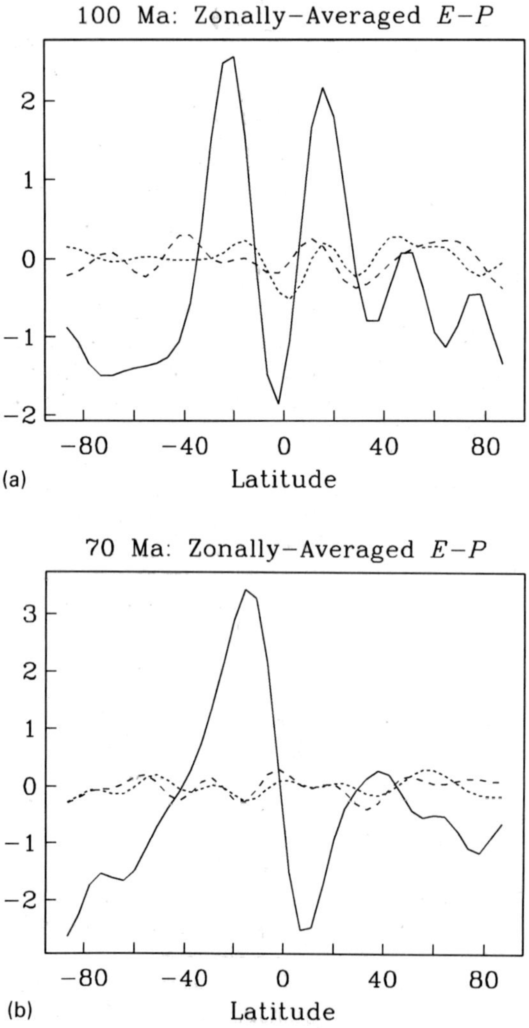

Fig. 4. Zonally averaged $E - P$ for mean state m_0 (solid lines), obliquity sensitivity m_δ (dashed lines) and precession sensitivity m_ν (dotted lines) for July for (a) 100 Ma and (b) 70 Ma. Units are in 10^{-8} m/s, or about 2.5 cm/month.

We estimated the uncertainty in the linear sensitivity coefficients m_ν and m_δ in three ways: (i) with a jackknife over insolation states (Efron, 1982; Park & Oglesby, 1991); (ii) using the least-squares residuals, or misfits, from the linear regression computation; and (iii) through the control standard deviations (for the 100-Ma case only). Though independent of each other, these three measures of uncertainty agree well, reflecting overall the intrinsic model variability, which can be highly non-uniform over the Earth's surface. If model sensitivity to ν and δ is linear, the regression constant m_0 can be taken as the mean-state value of the model variable. Statistical tests reported by Park & Oglesby (1991) on the 100-Ma simulations suggest that, with our prescribed boundary conditions, the non-linear model response to the orbital parameters is negligible.

The geographical dependence of the uncertainty of the linear sensitivity coefficients m_ν and m_δ varies greatly among model variables. We refer the interested reader to Park & Oglesby (1991), which presents contour plots of the jackknife uncertainty for selected model variables. The greatest uncertainty in surface pressure occurs in the wintertime mid-latitudes, where migrating winter storms introduce a large stochastic component to the time-averaged atmospheric pressure. Owing to larger uncertainties, a robust response by the surface pressure to orbital variations is difficult to identify in the wintertime mid-latitudes. Uncertainties in the orbital response of $E - P$, in contrast, are largest in the equatorial intertropical convergerce zone (ITCZ). The linear sensitivity coefficient of $E - P$ to orbital cycles exceeds its uncertainty by a significant amount at several locations in the ITCZ, particularly over the proto-South Atlantic between Africa and South America, where we propose a causal connection with cyclic limestone sequences observed in DSDP cores retrieved from Cretaceous deposits.

COMPARISON OF MODEL RESPONSE FOR 100-Ma AND 70-Ma SIMULATIONS

Most scenarios for the formation of Cretaceous sedimentary bedding cycles involve the atmospheric hydrological cycle and hence the large-scale transport of moisture through the atmosphere. Climatically, this transport of moisture expresses itself through 'sources' and 'sinks', that is, the net evaporation minus precipitation balance. If a region (e.g. model gridpoint) experiences net evaporation it is a source of atmospheric water vapour, while if it experiences net precipitation it is a sink. Figure 4 shows zonally averaged m_0, m_v (precession) and m_δ (obliquity) for $E - P$ for July for the 100-Ma and 70-Ma simulations, which represent the zonal-average mean-state hydrological balance along with the zonal-average sensitivity of the model to precession and obliquity changes. The zonal averages of m_v and m_δ for $E - P$ are small in both sets of simulations, when compared to the mean-state zonal averages m_0. This is largely a consequence of our not allowing SST to vary with either precession or obliquity. For precession, this approximation is probably very good, as insolation variations at any point on the Earth's surface are close to zero over an annual cycle and the relatively large thermal inertia of the ocean mixed-layer will prevent much change in SST (Kutzbach & Gallimore, 1988). For obliquity, the approximation is less certain. Though globally averaged insolation anomalies vanish over an annual cycle, a redistribution of insolation with latitude does occur as a result of the obliquity cycle. Unlike precession, obliquity also affects the timing and effective length of the seasons. For these reasons, models with interactively computed SST will ultimately be required to assess fully the effects of obliquity-induced insolation changes. Given that $m_\delta < m_v$ for most model variables almost everywhere on the Earth's surface, we expect that our fixed-SST assumption is a workable first-order approximation.

The 100-Ma mean state (m_0) shows an $E - P$ pattern qualitatively similar to the present day, with a precipitation maximum near the equator, reflecting the ITCZ , flanked by regions of strong net evaporation, reflecting the semi-permanent subtropical high-pressure zones (many of the world's deserts occur at these latitudes). These features can be thought of as components of a largely convection-driven Hadley cell, with ascending motion and precipitation in the fairly narrow band of the ITCZ and descent of warm dry air in the subtropics. At higher latitudes, the winter (Southern) Hemisphere shows net precipitation everywhere, indicative of low insolation (and hence reduced evaporation) as well as precipitation from cyclonic storm activity. The Northern (summer) Hemisphere also generally has net precipitation, with maxima at 35° N and 60° N and minima at 45° N and 70° N, most probably reflecting specific details of continental placement and monsoonal flows to the south and cyclonic activity to the north.

These features are highly modified or absent in the 70-Ma July simulations. A net precipitation maximum is still found near the equator (though broadened and shifted 10° to the north), indicative of an ITCZ that provides more overall rainfall than at 100 Ma. However, the summertime subtropical evaporation peak is almost completely absent; its only expression is a thin band near 40° N where zonally averaged evaporation and precipitation are closely balanced. The Southern (winter) Hemisphere has a large, broad evaporative peak that stretches from the equator to 40° S. These differences are almost certainly related to differences in SST, as zonal averages for the January simulations (not shown) are nearly a mirror image of those for July (shown in Fig. 4). The 100-Ma simulations have a fairly strong meridional (north–south) SST gradient in both winter and summer, while the 70-Ma simulations have a fairly strong meridional SST gradient in winter but a weak meridional SST gradient in summer. The $E - P$ balances in the winter hemisphere for the two reconstructions are more like each other, reflecting the more-similar SST gradients.

These zonally averaged results imply that the mean states of the Cretaceous climate, as determined by the particular choice of boundary conditions, are very different. Other aspects of model dynamics are affected by the lack of a summertime SST gradient in the 70-Ma reconstruction. The weak meridional SST gradient appears to reduce the intensity of the overall atmospheric circulation. The easterly low-latitude trade winds are very weak in the summer hemisphere. The model wind field at the 500-mb pressure level is a good proxy for the strength of the mean atmospheric circulation. Figure 5 graphs this quantity for the July mean state for both 100-Ma and 70-Ma reconstructions. The mean atmospheric flow is westerly, for the most part, with the most significant variations in the summer hemisphere. In the 100-Ma mean state, the summertime atmospheric flow is weak only over portions of North America and east-central Asia. In the 70-Ma mean state, however, the atmospheric flow is weak over much of the zonal band between the equator and 40° N, as well as the northeastern portions of the northern continents. A similar pattern prevails in the Southern Hemisphere in the January mean state, suggesting that the weak SST gradient

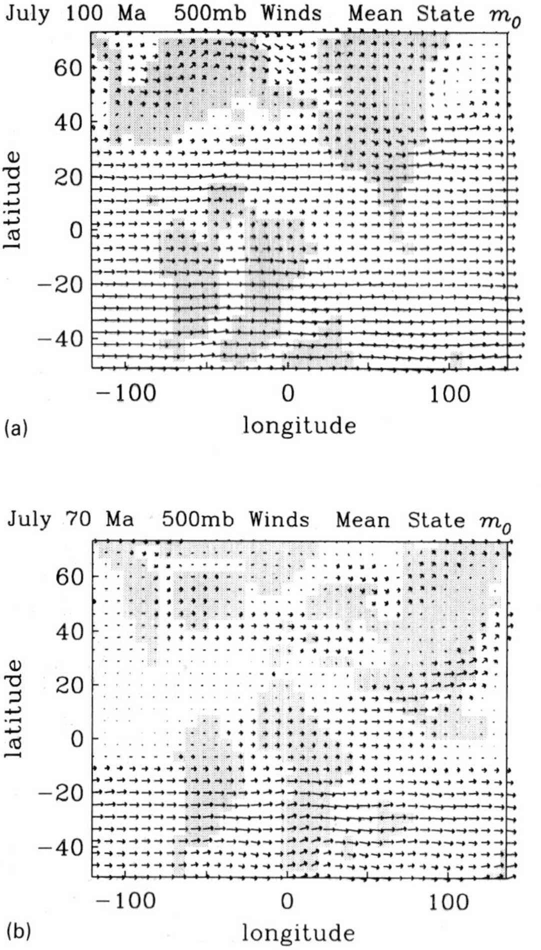

Fig. 5. Mean-state upper-level winds (at the 500-mb pressure level), expressed as vector arrows, for July for (a) 100 Ma and (b) 70 Ma. Length of the arrow denotes the relative magnitude of the wind.

has generated a 'sluggish' mean atmospheric circulation.

The presence of land masses disrupts the zonal symmetry of the Earth's surface properties, and hence the largely zonally symmetric circulation of a landless planet. Continental areas have little effective heat capacity relative to the oceanic mixed layer, so that land surface temperature responds more rapidly than SST to insolation fluctuations. Circulation features associated with this land–ocean heating asymmetry are termed 'monsoonal' effects. In summer, elevated temperature over continental interiors leads to rising air and low-pressure anomalies. Moist air is drawn inland from surrounding ocean surfaces, rises and sheds its water over the continental regions, leading to a preponderance of precipitation over evaporation. In winter, continental interiors are relatively cold, leading to high-pressure anomalies and descending air over the continents. Dry air from the continental interior is transported from land to sea regions. Two other important modifications occur. The first is due to the Coriolis force associated with the Earth's rotation. Coriolis force prevents air from flowing directly from regions of high pressure to regions of low pressure (i.e. from land to ocean or vice-versa), but instead deflects the airflow into the familiar cyclonic or anticyclonic flow around centres of low and high pressure, respectively. Second, the presence of elevated topography tends to amplify monsoonal dynamics. In summertime both the relative surface heating and the uplift of moist air can be enhanced over high-altitude regions, leading to a stronger monsoonal circulation. In winter, cold-air damming by the topographic heights, along with enhanced snowcover, can lead to larger continental high-pressure anomalies and an increased land–ocean pressure gradient. It is these monsoonal circulations that are most likely to respond strongly to orbital insolation variations.

Park & Oglesby (1991) presented results of the 100-Ma sensitivity study largely with contour maps of the mean state and the linear sensitivity coefficients m_ν and m_δ. In this paper, we compare the results from the 100-Ma study with those from the 70-Ma sensitivity study using insolation states with widely differing orbital parameters. We do not show results of individual 150-day model runs, but rather show contour plots of model parameters as predicted from the linear relation summarized in equation (1). In this manner we suppress the intrinsic model variability in order to highlight the underlying linear response of the model to insolation variations. Figure 6 shows July surface pressure values for the 70-Ma and 100-Ma reconstructions, as predicted by equation (1), for extremal insolation states (ν, δ) = (1.075, 22.0°) and (0.925, 24.5°). Several important differences are evident. Both the 70-Ma and 100-Ma simulations show a strong low-pressure region over Asia, induced by summertime heating. The low-pressure cell over Asia is considerably deeper in the 70-Ma simulation than in the 100-Ma simulation. However, in both cases a peak response of about 4 mb occurs as precession and obliquity are varied. The presence of the North American Interior Sea-

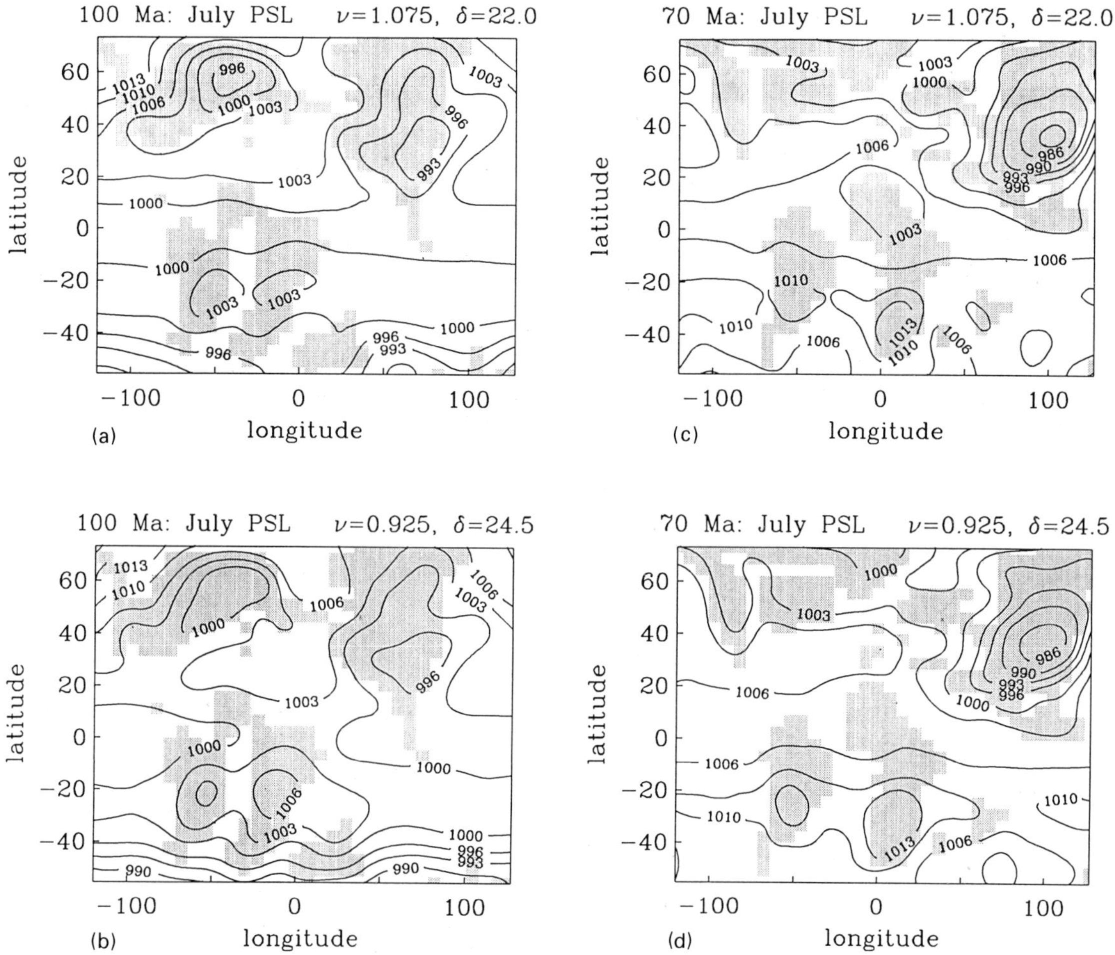

Fig. 6. Sea-level pressure in millibars (mb) for July for (a) 100 Ma with $\nu = 1.075$ and $\delta = 22.0$; (b) 100 Ma with $\nu = 0.925$ and $\delta = 24.5$; (c) 70 Ma with $\nu = 1.075$ and $\delta = 22.0$; (d) 70 Ma with $\nu = 0.925$ and $\delta = 24.5$; Plots are constructed using the mean state m_0 and linear sensitivity coefficients m_δ and m_ν using formula (1), rather than individual model runs, in order to suppress stochastic variation.

way in the 70-Ma reconstruction apparently inhibits the development of a fixed low-pressure system over summertime North America, by diminishing the overall extent of continental heating. With the absence of the seaway in the 100-Ma reconstruction, a strong low-pressure cell is seen over North America.

The intensity of pressure anomalies influences the strength of mean winds and moisture transport, while the global pattern of pressure anomalies influences the direction of these flows. In July simulations with the 100-Ma reconstruction, clear differences in the pattern of continental pressure anomalies occur as the insolation state varies. Summertime (July) low-pressure anomalies over North America and Asia weaken considerably at (0.925, 24.5°), where insolation, and therefore continental heating, is relatively low. In winter (July) Southern Hemisphere, (ν, δ) = (1.075, 22.0°) corresponds to near-maximal insolation. The wintertime monsoonal pressure high over Africa and South America, induced by continental cooling, is nearly absent compared to the high-pressure anomaly at (0.925, 24.5°), which corresponds to near-minimal wintertime insolation in the Southern Hemisphere. In the 70-Ma Northern Hemisphere summer, little absolute pressure variation with insolation state is

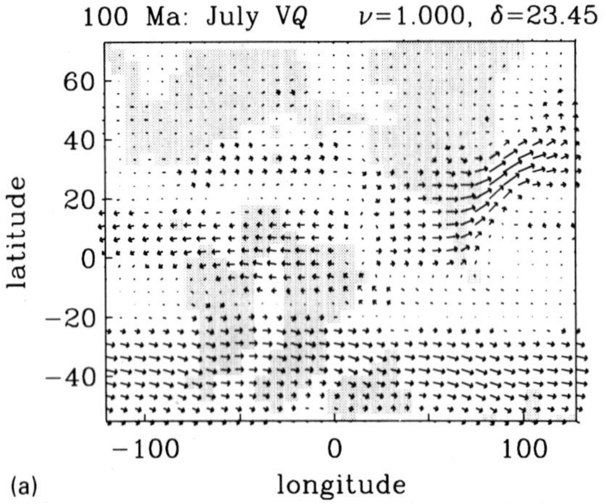

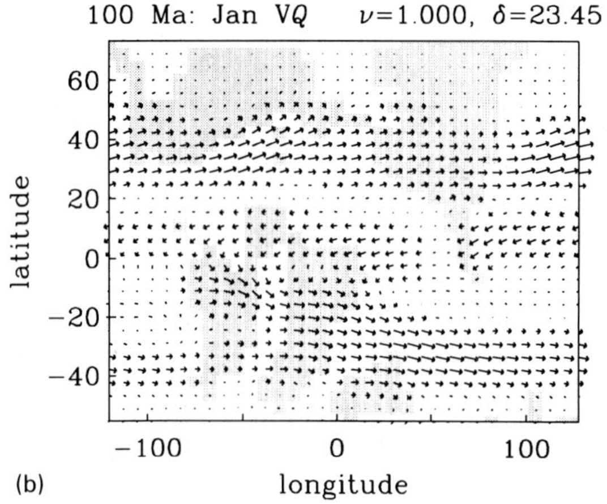

Fig. 7. Mean-state moisture transport $\overline{VQ}$ vertically averaged from the surface to 500 mb and plotted as vector arrows for (a) 100 Ma July and (b) 100 Ma January. Length of the arrow denotes relative magnitude of the moisture transport.

seen over North America, which is expected given its minimal monsoonal low in the mean state. Asia experiences fluctuations in pressure that are similar to those seen in the 100-Ma simulations in an absolute sense, but that are smaller relative to the mean-state pressure anomaly. A key difference is found in the Southern Hemisphere. At 70 Ma, the monsoonal high-pressure anomalies over Africa and South America vary in intensity with the insolation state, but remain strong features of the general circulation. Little variation occurs in the direction of winds and moisture transport as the insolation state varies. The largest change in the pattern of the 70-Ma pressure anomalies occurs at large ν, where a low-pressure anomaly develops over West Africa, which lies partly north of the equator. At 100 Ma, the changes in pressure patterns are more marked, suggesting larger variations in the direction of winds and moisture transport as the insolation state varies.

The moisture transport field VQ is derived from the horizontal component of the wind V times the specific humidity Q, measured as grams of water per kilogram of air. Figure 7 shows the time-averaged mean-state moisture transport field $\overline{VQ}$, integrated vertically from the surface to 500-mb height, for perpetual January and July for the mean state m_0 for the 100-Ma reconstruction. Figure 8 shows the mean-state $\overline{VQ}$ for the 70-Ma reconstruction. Mid-latitude westerly flow is clearly evident in the 100-Ma January mean state, as is a low-latitude easterly flow of moisture (due to the 'trade winds'). In the other graphs in Figs 7 and 8, the continents perturb greatly this largely zonally symmetric pattern of $\overline{VQ}$. The cyclonic flow over summertime proto-Asia causes the mean moisture flow over East Tethys in July to shift from easterly to westerly in both 70-Ma and 100-Ma mean states. In the 70-Ma January mean state, the high-pressure anticyclonic flow over Asia deflects moisture away from the continental interior, leading to a large region with near-zero precipitation, while a large westerly transport of moisture occurs in the 100-Ma January mean state. Smaller such deflections are evident in the July 70-Ma mean state over the southern continents at 20° S.

Changes in the pattern of the July pressure field in response to insolation changes in the 100-Ma simulations lead to large-scale redistributions of atmospheric moisture. Figure 9 shows contour plots of $E - P$ for the 100-Ma July simulations, estimated from the time-averaged divergence of moisture transport $\overline{\nabla \cdot VQ}$ and equation (1), at the same insolation states shown above for surface pressure. At (ν, δ) = (1.075, 22.0°), the enhanced Asian monsoon pulls moisture out of equatorial East Tethys, tending to weaken the ITCZ precipitation maximum in this region. The mean northerly airflow west of the Asian pressure minimum parches central Asia, erasing a region of negative $E - P$ (net precipitation) in the proto-Himalayas in the mean-state simulation. The Pacific coast of Asia is wet, completing the correspondence with the west–east/arid–wet asymmetry of the summer monsoon over present-day Asia. Significant changes in $E - P$ over North America are restricted to its southwest quadrant, where a rough balance between evaporation and

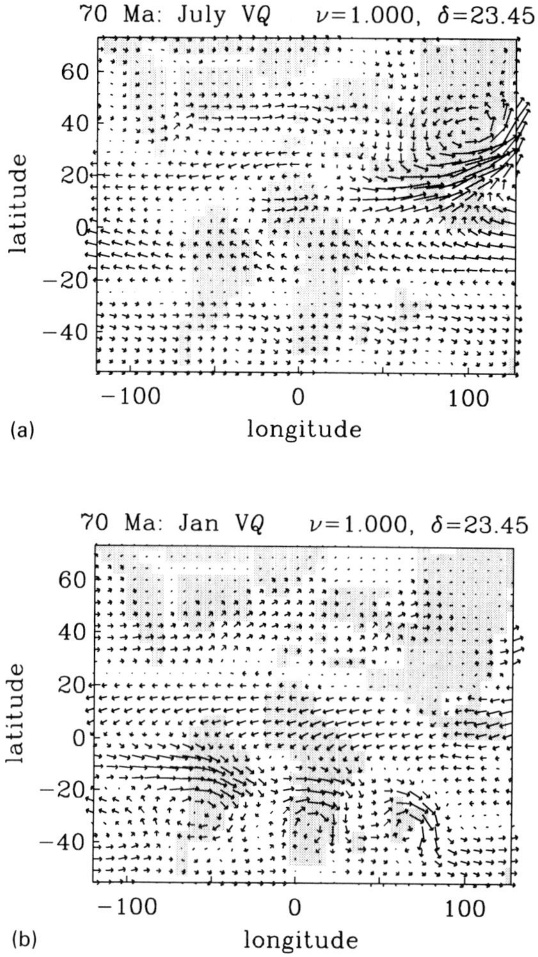

Fig. 8. Mean-state moisture transport $\overline{VQ}$ vertically averaged from the surface to 500 mb and plotted as vector arrows for (a) 70 Ma July and (b) 70 Ma January. Length of the arrow denotes relative magnitude of the moisture transport.

precipitation at (ν, δ) = (0.925, 24.5°) changes to a statistically significant, though local, net precipitation maximum. A large response in the 100-Ma simulations is seen over the southern continents. At (ν, δ) = (1.075, 22.0°), the wintertime monsoonal high is too weak to disrupt greatly the zonal ITCZ rainfall pattern, leaving northern Africa wet. At near-minimal winter insolation, on the other hand, the counterclockwise windflow around the strengthened southern high-pressure anomalies sweeps precipitation away from northern Africa, and intensifies greatly the evaporation over the equatorial South Atlantic.

Such large-scale changes in $E - P$ are much less evident in the 70-Ma sensitivity study (e.g. Fig. 10). Surface area with $E - P < 0$ is largely restricted to continental gridpoints in the climate simulations. The response to ν and δ is likewise dominated by continental locations, and acts principally to modulate the size of the $E - P$ minima over the

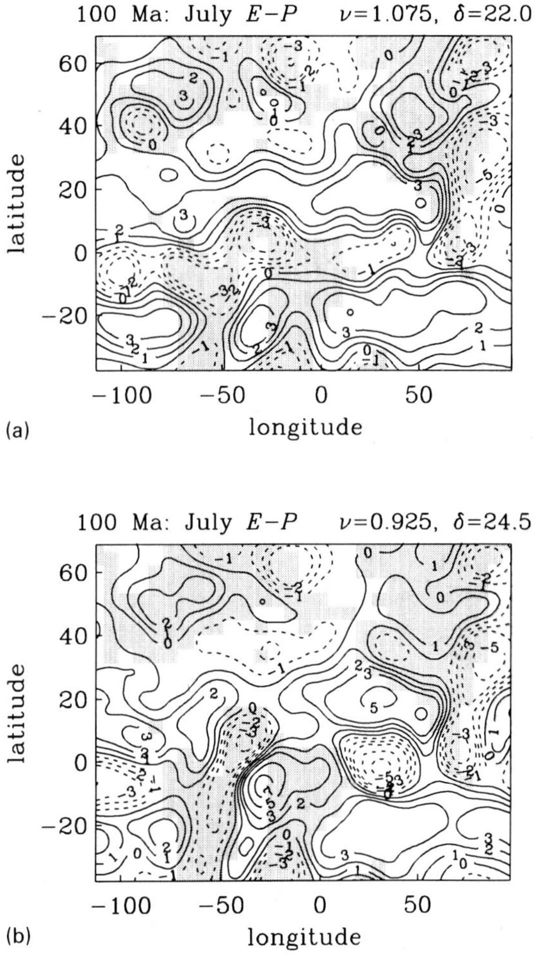

Fig. 9. Evaporation minus precipitation $\overline{(E - P)}$, estimated from vertically integrated $\overline{\nabla \cdot VQ}$, for the 100-Ma July simulations, for (a) the insolation state with ν = 1.075 and δ = 22.0 and (b) the insolation state with ν = 0.925 and δ = 24.5. The insolation states correspond to maximum and minimum winter insolation in the Southern Hemisphere. Plots are constructed using the mean state m_0 and linear sensitivity coefficients m_δ and m_ν using formula (1), rather than individual model runs, in order to suppress stochastic variation. Note that the Tethys region is emphasized. Contours are in 10^{-8} m/s or about 2.5 cm/month.

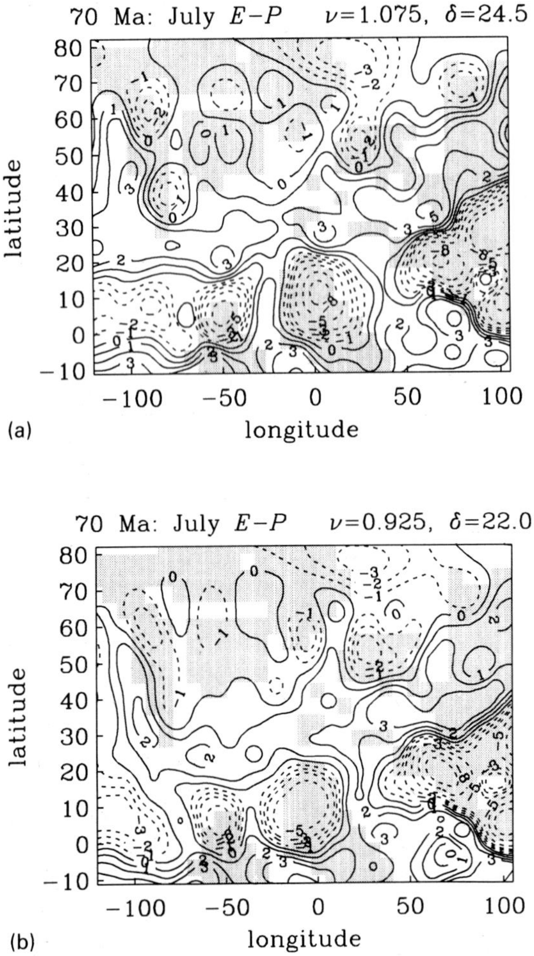

Fig. 10. $E - P$ for the 70-Ma July simulations, estimated from vertically integrated $\overline{\nabla \cdot VQ}$, for (a) the insolation state with $\nu = 1.075$ and $\delta = 24.5$ and (b) the insolation state with $\nu = 0.925$ and $\delta = 22.0$. The insolation states correspond to maximum and minimum summer insolation in the Northern Hemisphere. Contours are in 10^{-8} m/s or about 2.5 cm/month.

continental interiors. These changes are fairly small in a relative sense, and do not greatly change the spatial pattern of the hydrologic balance. The summertime monsoon does not exhibit the west–east/arid–wet asymmetry over continents as seen in the 100-Ma simulations. The cause of this behaviour is likely to be the weak gradient in SST prescribed by the University of Chicago 70-Ma reconstruction. The cyclonic mean wind flow around the Asian low-pressure anomaly transports moisture from East Tethys northward along the Pacific coast of Asia. This flow has warmer, moister air from the equatorial region moving poleward and to the east. If the SST gradient is weak, the net loss of moisture by the eastern arm of the monsoon decreases. The western arm of the summertime monsoon transports cooler, drier air southward, where it warms and evaporates water from the Earth's surface. If the SST gradient is weak, this northerly airflow experiences less heating and consequently generates less net evaporation from the surface. The contrast between strong and

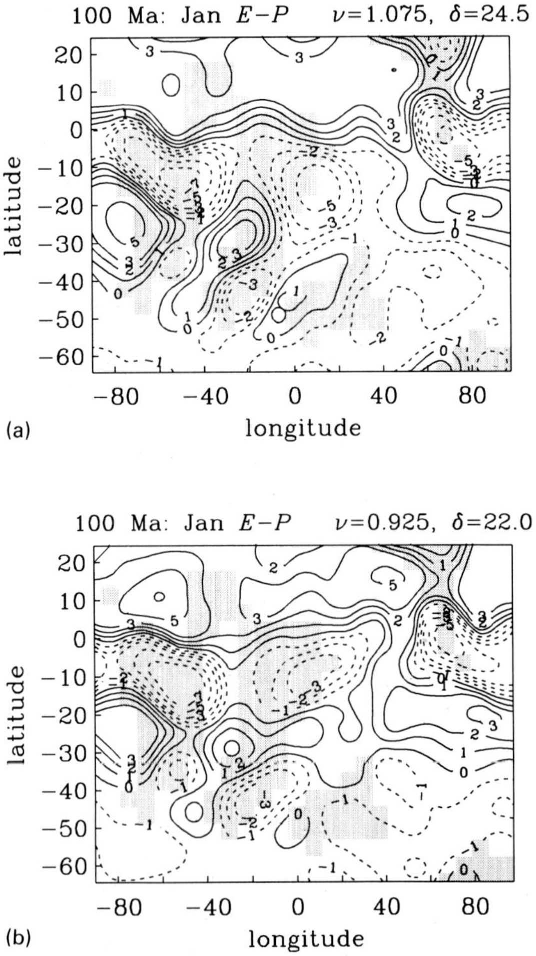

Fig. 11. $E - P$ for the 100-Ma January simulations, estimated from vertically integrated $\overline{\nabla \cdot VQ}$, for (a) the insolation state with $\nu = 1.075$ and $\delta = 24.5$ and (b) the insolation state with $\nu = 0.925$ and $\delta = 22.0$. The insolation states correspond to maximum and minimum summer insolation in the Southern Hemisphere. Contours are in 10^{-8} m/s or about 2.5 cm/month.

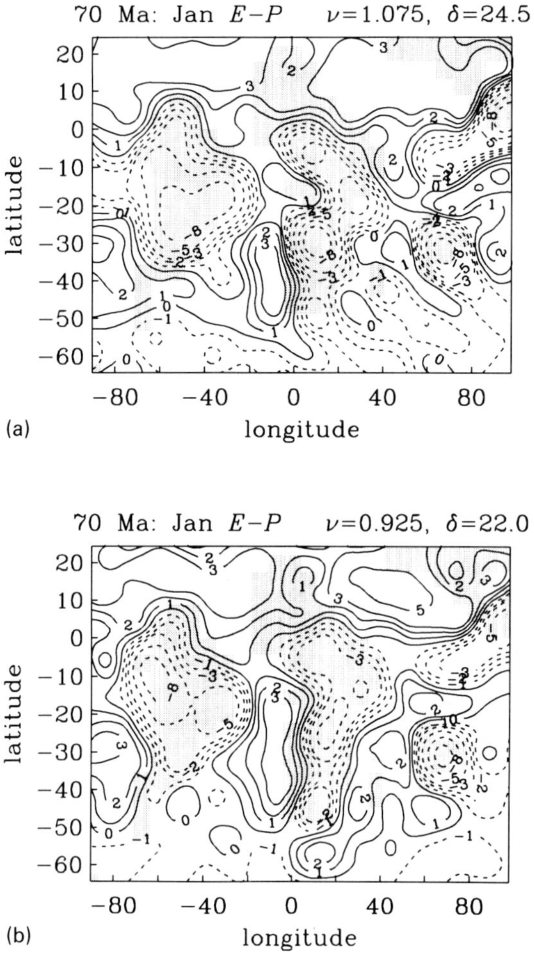

Fig. 12. $E - P$ for the 70-Ma January simulations, estimated from vertically integrated $\overline{\nabla \cdot VQ}$, for (a) the insolation state with $\nu = 1.075$ and $\delta = 24.5$ and (b) the insolation state with $\nu = 0.925$ and $\delta = 22.0$. The insolation states correspond to maximum and minimum summer insolation in the Southern Hemisphere. Contours are in 10^{-8} m/s or about 2.5 cm/month.

weak SST gradients can be seen for the Asian monsoon in Figs 7 and 8. In the 100-Ma simulations (strong SST gradient), $\overline{VQ}$ is large only in the southern half of the monsoonal flow. In the 70-Ma simulations (weak SST gradient), $\overline{VQ}$ is significant in the northern half of the monsoon as well. The principal net moisture transport in the 70-Ma simulations is the inland flow generated by ascending air in the continental interior, leading to a fairly symmetric $E - P$ minimum over low-latitude continental regions.

These contrasting behaviours are also evident in the January simulations over the southern continents (Figs 11 and 12). Choosing insolation states corresponding to maximum and minimum insolation for southern summer, we observe a pronounced west–east asymmetry over mid-latitude South America, Africa and India in the 100-Ma simulations that is magnified by changing the orbital parameters in order to increase Southern Hemisphere insolation. In the 70-Ma simulations, the west–east asymmetry is weak, restricted principally to the west coasts of Africa and proto-Chile, and is not magnified with increasing insolation.

The above emphasis on the meridional SST gradient suggests that large cyclic changes in $E - P$ in the 70-Ma reconstruction should occur in the 40–50° latitude band in summertime, where the SST gradient becomes substantial. Figure 10 shows $E - P$ for July over North America and proto-Europe for the 70-Ma reconstruction. North America is the site of many examples of Cretaceous sedimentary cycles, and thus is a region of particular interest. The interior of North America shifts from weakly wet to a mixture of wet and arid regions as summertime insolation increases from $(\nu, \delta) = (0.925, 22.0°)$ to $(1.075, 24.5°)$, but the absence of a monsoonal low-pressure anomaly weakens the dynamical response to the orbital parameters. By contrast, the intensification of the Asian monsoonal low leads to an increased moisture 'source' over model gridpoints corresponding to European Russia in the 30–50° N latitude range, many of which are parameterized as shallow seas in the University of Chicago reconstruction. The increased net evaporation in this region of the model contrasts with an absence of monsoon-induced aridity on the low-latitude west coast of Asia. Contour plots of m_ν (not shown) reveal that low-latitude west Asia becomes less moist with increasing ν, but the evaporation–precipitation balance $E - P$ remains negative for the full range of the orbital parameters.

DISCUSSION

A comparison of our 70-Ma and 100-Ma climate sensitivity experiments reveals differences that can be related to differences in the prescribed boundary conditions. The largest differences appear to arise from conditions that are currently poorly constrained by data, such as the meridional gradient of SST, rather than by those that are better constrained, such as the widening of the South Atlantic. Our experiments cannot therefore be interpreted as a guide to how the response of the Cretaceous climate to Milankovitch rhythms changed from the Albian to the Maastrichtian, but rather as a guide to how two different hypotheses for mean Cretaceous climate conditions would influence the orbital response, and thereby the deposition of cyclic limestone/marl and limestone/black-shale sequences. It is arguable that continental drift, as expressed in the continent–ocean pattern, is only one of many factors governing the evolution of global climate. With equivalent land-albedo and surface-wetness values and with land and sea ice proscribed in both sets of simulations, the most important difference in our experiment is the meridional profile of SST, and, implicitly, deep-ocean circulation. SSTs for the 100-Ma Barron & Washington (1984) reconstruction are consistent with a world ocean in which SST is largely determined locally. The warm mid-latitude SST and shallow meridional SST gradient prescribed by the 70-Ma University of Chicago reconstruction require maintenance by warm, presumably saline, water at depth, similar to conditions modelled by Oglesby & Saltzman (1990). In addition to this global boundary condition, the 70-Ma reconstruction prescribes a North American Seaway explicitly, and omits a 1000-m elevated plateau on eastern North America that is posited in the 100-Ma reconstruction.

As discussed previously (Oglesby & Park, 1989; Park & Oglesby, 1990, 1991), insolation fluctuations associated with precession and obliquity in the 100-Ma reconstruction are associated with large changes in the atmospheric monsoonal circulation. These changes effect large shifts in the evaporation minus precipitation balance $E - P$ over many regions in the model where a causal connection with the deposition of cyclic carbonates is plausible. A shift from strongly arid to strongly moist conditions is evident over the presumed drainage basin of the proto-South Atlantic, where Cretaceous cyclic limestone/marl and black-shale sequences have been drilled by the Deep-Sea Drilling Project (Melguen, 1979; Borella, 1984; Dean *et al.*, 1984; Stow & Dean, 1984). Oglesby & Park (1989) suggested that these sequences of the South Atlantic were generated in a manner similar to the formation of Eastern Mediterranean sapropels, in which shifts between lagoonal and estuarine circulation in restricted oceanic basins cause shifts in oxic–anoxic conditions (e.g. de Boer & Wonders, 1984; ROCC, 1986; Rossignol-Strick, 1987; Sarmiento *et al.*, 1988a). Cyclic anoxia in the global ocean could be generated by a large model response over regions of deep-water formation (e.g. Sarmiento *et al.*, 1988b). Barron & Peterson (1990) ran an ocean circulation model with the Barron & Washington (1984) mid-Cretaceous boundary conditions and noted that, although the largest source of deep water in their simulation lay in the high-latitude North Pacific, a significant amount of deep water derives from an area of strongly positive $E - P$ in East Tethys off the coast of Asia. Using similar boundary conditions, orbital cycles generate large fluctuations in net evaporation over East Tethys in both seasons, as monsoonal pressure anomalies over Asia and Africa amplify or diminish, with resultant effects on moisture transport. This behaviour is consistent with the cyclic generation of warm, salty deep ocean water at low latitudes, which is a proposed mechanism for episodic global ocean anoxia (e.g. ROCC, 1986). A weak response of $E - P$ to obliquity (though statistically significant in the analysis of our model experiments) is observed in July over the Western Interior of North America. In this region in January we also partially confirm the observations of Glancy *et al.* (1986) in an earlier GCM experiment: the Tethys coast of North America experiences greater precipitation at larger values of the obliquity angle δ, due probably to increased mid-latitude storminess.

It is more difficult to identify possible links between sedimentary cycles and the GCM sensitivity experiment using the 70-Ma reconstruction. The model response to precession and obliquity at low latitudes appears mainly to modulate intense precipitation over the summertime continental interiors, whose western halves lack the present-day monsoonal tendency towards aridity. We infer that the weak SST gradient, which leads to less net poleward atmospheric transport of moisture from low latitudes, is the main factor behind this behaviour. At summertime mid-latitudes, where the SST gradient becomes substantial, a greater response by the hydrological cycle to insolation changes is evident. Model gridpoints corresponding, for the most part, to present-day European Russia experience an increase in net evaporation in summer with increasing insolation. As much of this area is shallow sea in the University of Chicago reconstruction, a cyclic shift from lagoonal to estuarine conditions may be consistent with our simulations, similar to the behaviour noted in the South Atlantic in the 100-Ma simulations. There are major Cretaceous bedded limestone sequences near the southern Russian city of Kislovodsk (A.L. Lerner-Lam, personal communication), so an orbital response in this region is not unexpected. The presence of epicontinental seas in the 70-Ma reconstruction of North America inhibits the formation of a summertime monsoon over the land mass, and so weakens its response to orbital insolation changes. However, there is a weak shift from moist to arid conditions over much of the continent with increasing insolation, coupled with a weak intensification of net-precipitation anomalies on the western shore of the seaway. By itself, this change is much less significant than are other responses by the model; however, it is of interest due to the common occurrence of Cretaceous cyclical sediment sequences in North America.

The climate of the Cretaceous has been considered to be more 'equable' than at present (e.g. Axelrod, 1984), with weaker seasonality and widespread frost-free winters at many high-latitude locations. Weak seasonality and warm high-latitude SST are often taken as indicators of a sluggish global atmospheric circulation. Figure 5 demonstrates how the circulation intensity in our simulations weakens as the meridional temperature gradient weakens, leaving us with the familiar paradox of how to transport sufficient heat poleward with a 'sluggish' circulation (e.g. Barron, 1987). Abundant palaeofloral evidence points toward warm high-latitude SST as an important factor moderating seasonality in coastal regions (e.g. Parrish & Spicer, 1988), and poleward heat transport by the ocean is implied by warmer high-latitude SST. However, 'equability' in the continental interiors seems to be precluded by atmospheric modelling studies (Schneider *et al.*, 1985; Sloan & Barron, 1990).

Because the response to precession and obliquity insolation changes by the Earth's climate depends largely on seasonal effects, the extensive record of limestone/marl sequences would seem to argue against an 'equable' or 'sluggish' transport of water vapour by the atmosphere. The robust and changeable low-latitude dynamics in the 100-Ma simulations are inconsistent with a sluggish overall transport of either air or water vapour. Water-vapour transport in the 70-Ma simulations is also large, as their weaker summertime atmospheric circulations are largely compensated by an increase in specific humidity Q over the warmer mid-latitude sea-surface. In fact, the mean-state moisture transport $\overline{VQ}$ for the 70-Ma simulations is larger than $\overline{VQ}$ for the 100-Ma simulations (Figs 7 and 8). Therefore a 'sluggish' atmospheric circulation does not necessarily imply a 'sluggish' hydrological circulation.

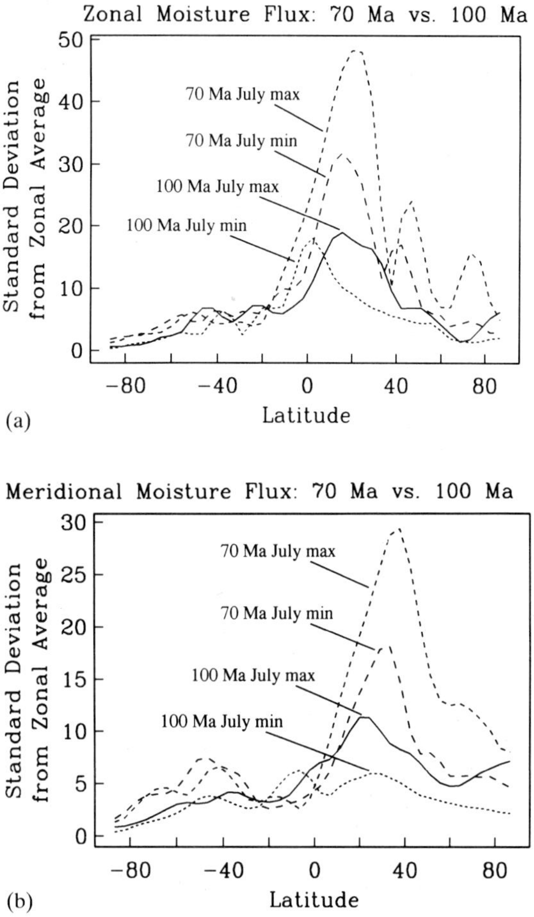

Fig. 13. Standard deviation from a zonal average of the moisture flux $\overline{VQ}$ vertically averaged from the surface to 500 mb for 100 Ma and 70 Ma for (a) the meridional (north–south) moisture flux (vQ) and (b) the zonal (east–west) moisture flux (uQ). Units are in (g/kg)m/s.

One measure of 'vigour' in the hydrological circulation is the standard deviation of time-averaged model hydrological variables from their zonal averages. Figure 13 compares values of this deviation for the time-averaged moisture flux $\overline{VQ}$ for the July 70-Ma and 100-Ma reconstructions and insolation states $(\nu, \delta) = (1.075, 24.5°)$ and $(0.925, 22.0°)$, corresponding to near-maximal and near-minimal insolation in the summer (Northern) Hemisphere. The climate states are constructed, as in earlier figures, from the linear sensitivity relation (1), from which we compute zonal (east–west) averages and the standard deviations about these averages. Overall there is a 60–80% increase in the variability of $\overline{VQ}$ with longitude as summertime insolation goes from minimal to maximal values, in marked contrast to the weak sensitivity of the zonal mean to insolation perturbations. The zone of maximal variability in the zonal moisture flow shifts from the equator to 20° N latitude for the 100-Ma reconstruction, reflecting the shift in relative strength between the high-African and low-Asian pressure anomalies. There is a large variability in $\overline{VQ}$ within zonal bands in the 70-Ma climate states. Warmer mid-latitude SST in the 70-Ma reconstruction and, hence, warmer air over ocean gridpoints in the model allow the atmosphere to hold and transport more moisture. However, the zonal averages of $\overline{VQ}$ for the two sets of simulations have roughly comparable magnitude, so the variability of the 70-Ma moisture transport relative to its mean value is larger overall than that of the 100-Ma simulations. Note the large peak at 40° N latitude in the variability of the meridional moisture transport for the 70-Ma simulations. A similar peak at 35° S occurs in an identical calculation performed for the January simulations – this suggests a causal link with the latitude at which the summertime meridional SST gradient becomes substantial, rather than with model orography, e.g. the position of North America.

It may be possible to use the above contrasts to investigate different hypotheses for the mean Cretaceous climate. For instance, the prevalence of large shifts in the E − P balance over ocean regions in the 100-Ma reconstruction suggests that the boundary conditions of this 'world' are more consistent with the production of warm saline deep ocean water at low latitudes and the development of cyclic global anoxia. However, in seasonal-cycle simulations using CCM1 (Oglesby & Saltzman, 1992) the net evaporation over the present-day Arabian Sea is greater than in the 100-Ma East Tethys. The Arabian Sea is not a major source of deep water to the present-day world ocean, primarily because the volume of warm, saline high density water produced is relatively small compared to the volume of cold, dense deep water from multiple high-latitude sources. In the ocean modelling study of Barron & Peterson (1990), frost-free polar SST, in effect, weakens high-latitude sources of deep water, so that East Tethys makes a significant contribution. A 'world' with a weaker SST gradient might find it yet more difficult to form deep water at high latitudes, lowering the threshold for generating deep water at low latitudes. Therefore, we cannot conclude that the smaller net evaporation overålow-latitude seas (e.g. Tethys) in the 70-Ma 'world' precludes the production of low-latitude deep water, or that the smaller orbitally induced $E - P$ fluctuations would not generate cycles in its production. A more extensive modelling study, involving ocean circulation models, is required to resolve this question more fully.

The mean climate at 70 Ma has been reconstructed by the University of Chicago Paleoatlas project from palaeofloral and shallow-marine data. Low-latitude Asia in this reconstruction is characterized by desert flora, suggesting arid conditions. In our January simulations for the 70-Ma orography and SST, a strong wintertime high-pressure anomaly over Asia leaves this region with only minimal rainfall, but in July low-latitude Asia is very wet from the Tethys to the Pacific. This seems to contradict the palaeofloral data. The west–east/arid–wet monsoon pattern in the 100-Ma simulations would be more consistent with aridity in much of low-latitude Asia. This appears to be evidence against a weak meridional SST gradient in the waters around Asia, at least on the Pacific side where warm mid-latitude SST would enhance the evaporation and transport of moisture in the north and west arms of the cyclonic monsoonal flow. Further simulations are needed, however, to examine the effect of desert albedo and surface wetness over Asia on the orbital response. A key negative feature of our 100-Ma sensitivity study is the weak response to v and δ on the northern shore of Tethys, the location of many examples of bedded sediments thought to be orbitally controlled. Although the response of our particular 70-Ma simulations was also not strong, Fig. 13 suggests that the northward shift of the meridional SST gradient increases the potential for large, orbitally modulated variations in the mid-latitude hydrological cycle. It may be necessary to posit large variations in SST with longitude to accommodate both aridity in low-latitude Asia and orbital cycles along the north shore of Tethys. For example, warm mid-latitude SST may have existed only in the Tethys region and the South Atlantic, where tectonic barriers could have impeded deep-water flow, while the Pacific had an SST profile that more resembles the locally determined profile of Barron & Washington (1984). A thorough investigation of such a hypothesis will probably challenge the next generation of coupled ocean–atmosphere GCMs.

Although progress in the study of Cretaceous climate dynamics can probably be made by testing simple hypotheses about its surface boundary conditions, the best approach is to collect and incorporate constraints from the geological record. The palaeofloral reconstruction for the 70-Ma orography was not used in this comparative study, and an assessment of its effects is the subject of ongoing study (Horrell & Oglesby, 1990). The coverage of the current $\delta^{18}O$ data set for ocean palaeotemperatures is sparse, and contains unknown biases from salinity, diagenesis and vital effects. Individual fossil palaeotemperatures from the Mesozoic can be highly uncertain, and the global picture afforded by the sparse coverage of the world ocean is more so, especially for times before the Maastrichtian. Careful inspection of individual foraminiferal tests with electron microscopy has been used to screen out cases of obvious carbonate diagenesis (e.g. Barrera & Huber, 1990), which should improve the accuracy of estimates. Since atmospheric dynamics are driven largely by temperature gradients, estimates of relative variations in palaeotemperature across ocean basins are at least as helpful as absolute palaeotemperatures. Such estimates may emerge more confidently from a more densely sampled data set.

ACKNOWLEDGEMENTS

We thank Mark Horrell and Fred Ziegler for providing the 70-Ma orography compiled by the University of Chicago Paleoatlas Project, and Mark for helpful comments on its data constraints. This work was supported by NCAR grant SCD-36211005, and NSF grants EAR-8657206 (J.P.) and ATM-8802630 (R.O.). R.O. also received support from the Doherty Oceanography Endowment at Brown University.

REFERENCES

Axelrod, D.I. (1984) An interpretation of Cretaceous and Tertiary biota in polar regions. *Palaeogeogr. Palaeoclimatol. Palaeoecol.* **45**, 105–147.

Barrera, E. & Huber, B.T. (1990) Evolution of Antarctic waters during the Maastrichtian: Foraminifer oxygen and carbon isotope ratios, Leg 113. In: *Proc. Ocean Drilling Program, Scient. Results*, **113**, Ocean Drilling Program, College Station, TX, pp. 813–827.

Barrera, E., Huber, B.T., Savin, S.M. & Webb, P.-N. (1987) Antarctic marine temperatures: late Campanian through early Paleocene. *Paleoceanography* **2**, 21–47.

Barron, E.J. (1983) A warm, equable Cretaceous: The nature of the problem. *Earth Sci. Rev.* **19**, 305–338.

Barron, E.J. (1987) Eocene equator-to-pole surface temperatures: A significant climate problem? *Paleoceanography* **2**, 729–739.

Barron, E.J. & Peterson, W.H. (1990) Mid-Cretaceous ocean circulation: Results from model sensitivity studies. *Paleoceanography* **5**, 319–337.

Barron, E.J. & Washington, W.M. (1984) The role of geographic variables in explaining paleoclimates: Results from Cretaceous climate model simulations. *J. Geophys. Res.* **89**, 1267–1279.

Barron, E.J. & Washington, W.M. (1985) Warm Cretaceous climates: High atmospheric CO_2 as a plausible mechanism. In: *The Carbon Cycle and Atmospheric CO_2: Natural Variations Archean to Present* (Eds Sundquist, E.T. & Broecker, W.S.). Am. Geophys. Union, Washington DC, pp. 546–553.

Barron, E.J., Harrison, C.G.A., Sloan, J.L. & Hay, W.W. (1981) Paleogeography, 180 million years ago to the present. *Ecologae Geol. Helv.* **74**, 443–470.

Berger, A. (1978) *A simple algorithm to compute long-term variations of daily or monthly insolation*. Inst. Astron. Geophys. G. Lemaitre Contrib. 18.

Berner, R.A. (1990) Atmospheric carbon dioxide levels over Phanerozoic time. *Science* **249**, 1382–1386.

Blackmon, M.L. (1986) Building, testing and using a general circulation model. In: *Large-Scale Transport Processes in Oceans and Atmosphere* (Eds Willebrand, J. & Anderson, D.L.T.). Reidel Publ. Co., Dordrecht, pp. 1–70.

Borella, P.E. (1984) Sedimentology, petrology, and cyclic sedimentation patterns, Walvis Ridge transect, Leg 74, Deep Sea Drilling Project. *Init. Rep. DSDP* **74**, 645–662.

Brass, G.W., Southam, J.R. & Peterson, W.H. (1982) Warm saline bottom water in the ancient ocean. *Nature* **296**, 620–623.

Crowley, T.J. (1990) Are there any satisfactory geologic analogues for a future greenhouse warming? *J. Climate* 3, 1282–1292.

Dean, W.E., Arthur, M.A. & Stow, D.A.V. (1984) Origin and geochemistry of Cretaceous deep-sea black shales and multicolored claystones, with emphasis on Deep Sea Drilling Project Site 530, southern Angola Basin. *Init. Rep. DSDP* **75**, 819–844.

De Boer, P.L. & Wonders, A.A.H. (1984) Orbitally induced rhythmic bedding in Cretaceous pelagic sediments near Moria (Italy). In: *Milankovitch and Climate*, Part 1 (Eds Berger, A., Imbrie, J., Hays, J., Kukla, G. & Saltzman, B.). Reidel Publ. Co., Dordrecht, pp. 177–190.

Douglas, R. & Savin, S. (1975) Oxygen and carbon isotope analysis of Tertiary and Cretaceous microfossils from Shatsky Rise and other sites in the North Pacific Ocean. *Init. Rep. DSDP* **32**, 509–520.

Efron, B. (1982) *The Jackknife, the Bootstrap, and Other Resampling Plans*. Society for Industrial and Applied Mathematics, Philadelphia.

FRAKES, L.A. & FRANCIS, J.E. (1988) A guide to Phanerozoic cold polar climates from high-latitude ice-rafting in the Cretaceous. *Nature* **333**, 547–549.

GLANCY, T.J., BARRON, E.J. & ARTHUR, M.A. (1986) An initial study of the sensitivity of modeled Cretaceous climate to cyclical insolation forcing. *Paleoceanography* **1**, 523–537.

GREGORY, R.T., DOUTHITT, C.B., DUDDY, I.R., RICH, P.V. & RICH, T.H. (1989) Oxygen isotopic composition of carbonate concretions from the lower Cretaceous of Victoria, Australia: Implications for the evolution of meteoric waters on Australian continent in a paleopolar environment. *Earth Planet. Sci. Lett.* **92**, 27–42.

GROTZINGER, J.P. (1986) Upward shallowing platform cycles: A response to 2.2 billion years of low-amplitude, high-frequency (Milankovitch band) sea level oscillations. *Paleoceanography* **1**, 403–416.

HAYS, J.D., IMBRIE, J.I. & SHACKLETON, N.J. (1976) Variations in the earth's orbit: pacemaker of the ice ages. *Science* **194**, 1121–1132.

HERBERT, T.D. & D'HONDT, S.L. (1990) Precessional climate cyclicity in Late Cretaceous–Early Tertiary marine sediments: a high resolution chronometer of Cretaceous–Tertiary boundary events. *Earth Planet. Sci. Lett.* **99**, 263–275.

HERBERT, T.D. & FISCHER, A.G. (1986) Milankovitch climatic origin of mid-Cretaceous black shale rhythms in central Italy. *Nature* **321**, 739–743.

HORRELL, M.A. & OGLESBY, R.J. (1990) Could Cretaceous high-latitude forests have contributed to polar warmth? Implications from a general circulation model study (abstract). *Eos Trans. AGU* **71**, 473.

IMBRIE, J., HAYS, J.D., MARTINSON, D.G., MCINTYRE, A., MIX, A.C., MORLEY, J.J. *et al.* (1984) The orbital theory of Pleistocene climate: support from a revised chronology of the marine $\delta(^{18}O)$ record. In: *Milankovitch and Climate*, Part 1 (Eds Berger, A., Imbrie, J., Hays, J., Kukla, G. & Saltzman, B.). Reidel Publ. Co., Dordrecht, pp. 269–306.

KIEHL, J.T. & RAMANATHAN, V. (1990) Comparison of cloud forcing derived from the Earth Radiation Budget Experiment with that simulated by the NCAR Community Climate Model. *J. Geophys. Res.* **95**, 11679–11698.

KUTZBACH, J.E. & GALLIMORE, R.G. (1988) Sensitivity of a coupled atmosphere mixed-layer ocean model to changes in orbital forcing at 9000 yr BP. *J. Geophys. Res.* **93**, 803–821.

KUTZBACH, J.E. & GALLIMORE, R.G. (1989) Pangean climates: Megamonsoons of the megacontinent. *J. Geophys. Res.* **94**, 3341–3357.

KUTZBACH, J.E., GUETTER, P.J. & WASHINGTON, W.M. (1990) Simulated circulation of an idealized ocean for Pangean time. *Paleoceanography* **5**, 299–317.

MAASCH, K.A. & SALTZMAN, B. (1990) A low-order dynamical model of global climate variability over the full Pleistocene. *J. Geophys. Res.* **95**, 1955–1963.

MEAD, G.A., TAUXE, L. & LABRECQUE, J.L. (1986) Oligocene paleoceanography of the South Atlantic: Paleoclimatic implications of sediment accumulation rates and magnetic susceptibility measurements. *Paleoceanography* **1**, 273–284.

MELGUEN, M. (1979) Facies evolution, carbonate dissolution cycles in sediments from the eastern South Atlantic (DSDP Leg 40) since the early Cretaceous. *Init. Rep. DSDP* **40**, 981–1024.

MITCHELL, J.F.B. (1990) Greenhouse warming: Is the mid-Holocene a good analogue? *J. Climate* **3**, 1177–1192.

OGLESBY, R.J. (1989) A GCM study of Antarctic glaciation. *Climate Dynamics* **3**, 135–156.

OGLESBY, R.J. (1990) Sensitivity of glaciation to initial snow cover, CO_2, snow albedo, and oceanic roughness in the NCAR CCM. *Climate Dynamics* **4**, 219–235.

OGLESBY, R.J. & PARK, J. (1989) The effect of precessional insolation changes on Cretaceous climate and cyclic sedimentation. *J. Geophys. Res.* **94**, 14793–14816.

OGLESBY, R.J. & SALTZMAN, B. (1990) Extending the EBM: the effect of deep ocean temperature on climate with applications to the Cretaceous. *Palaeogeogr. Palaeoclimatol. Palaeoecol.* **82**, 237–259.

OGLESBY, R.J. & SALTZMAN, B. (1992) Equilibrium climate statistics of a GCM as a function of atmospheric CO_2: I. Geographic distributions of primary variables. *J. Climate* **5**, 66–92.

OGLESBY, R.J., MAASCH, K.A. & SALTZMAN, B. (1989) Glacial meltwater cooling of the Gulf of Mexico: GCM implications for Holocene and present-day climates. *Climate Dynamics* **3**, 115–133.

OLSEN, P.E. (1986) A 40-million-year lake record of early Mesozoic orbital climate forcing. *Science* **234**, 842–844.

PARK, J. & HERBERT, T.D. (1987) Hunting for paleoclimatic periodicities in a geologic time series with an uncertain time scale. *J. Geophys. Res.* **92**, 14027–14040.

PARK, J. & OGLESBY, R.J. (1990) A comparison of precession and obliquity effects in a Cretaceous paleoclimate simulation. *Geophys. Res. Lett* **17**, 1929–1932.

PARK, J. & OGLESBY, R.J. (1991) Milankovitch rhythms in the Cretaceous: A GCM modelling study. *Palaeogeogr. Palaeoclimatol. Palaeoecol.* **90**, 329–355.

PARRISH, J.T. & SPICER, R.A. (1988) Late Cretaceous terrestrial vegetation: A near-polar temperature curve. *Geology* **16**, 22–25.

PRELL, W.L. & KUTZBACH, J.E. (1987) Variability of the monsoon over the past 150 000 years: Comparison of observed and simulated paleoclimatic time series. *J. Geophys. Res.* **92**, 8411–8425.

RANDEL, W.J. & WILLIAMSON, D.L. (1990) A comparison of the climate simulated by the NCAR Community Climate Model (CCM1:R15) with ECMWF analyses. *J. Climate* **3**, 608–633.

RICH, P.V., RICH, T.H., WAGSTAFF, B.E., MCEWEN-MASON, J., DOUTHITT, C.B., GREGORY, R.T. & FELTON, E.A. (1988) Evidence for low temperatures and biologic diversity in Cretaceous high latitudes of Australia. *Science* **242**, 1403–1406.

RIND, D., POTEET, D. & KUKLA, G. (1989) Can Milankovitch orbital variations initiate the growth of ice sheets in a general circulation model? *J. Geophys. Res.* **94**, 12851–12871.

ROCC (RESEARCH ON CRETACEOUS CYCLES GROUP) (1986) Rhythmic bedding in Upper Cretaceous pelagic carbonate sequences: Varying sedimentary response to climate forcing. *Geology* **14**, 153–156.

ROSSIGNOL-STRICK, M. (1987) Rainy periods and bottom water stagnation initiating brine accumulation and metal concentrations: 1. the late Quaternary. *Paleoceano-*

graphy **2**, 333–360.

SALTZMAN, E. & BARRON, E.J. (1982) Deep circulation in the late Cretaceous: Oxygen isotope paleotemperatures from *Inoceramus* remains in DSDP cores. *Palaeogeogr. Palaeoclimatol. Palaeoecol.* **40**, 167–182.

SARMIENTO, J.L., HERBERT, T.D. & TOGGWEILER, J.R. (1988a) Mediterranean nutrient balance and episodes of anoxia. *Global Biogeochem. Cycles* **2**, 427–444.

SARMIENTO, J.L., HERBERT, T.D. & TOGGWEILER, J.R. (1988b) Causes of anoxia in the world ocean. *Global Biogeochem. Cycles* **2**, 115–128.

SCHNEIDER, S.H., THOMPSON, S.L. & BARRON, E.J. (1985) Mid-Cretaceous continental surface temperatures: Are high CO_2 concentrations needed to simulate above freezing winter conditions? In: *The Carbon Cycle and Atmospheric CO_2 Natural Variations Archean to Present* (Eds Sundquist, E.T. & Broecker, W.S.). Am. Geophys. Union, Washington DC, pp. 554–559.

SLOAN, L.C. & BARRON, E.J. (1990) 'Equable' climates during Earth history? *Geology* **18**, 489–492.

STOW, D.A.V & DEAN, W.E. (1984) Middle Cretaceous black shales at site 530 in the southeastern Angola Basin. *Init. Rep. DSDP* **75**, 809–817.

TARLING, D.H. (1978) The geological–geophysical framework of ice ages. In: *Climate Change* (Ed. Gribben, J.). Cambridge University Press, New York, pp. 3–24.

Spec. Publs Int. Ass. Sediment. (1994) **19**, 531–544

Cyclicity or chaos? Orbital forcing versus non-linear dynamics

D.G. SMITH

Petroconsultants (UK) Ltd, Europa House, 266 Upper Richmond Road, Putney, London SW15 6TQ, UK

ABSTRACT

Chaos theory (or non-linear dynamics) predicts complex yet non-random output from natural systems in which feedback mechanisms are important. It therefore competes with the Milankovitch hypothesis as an explanation for sedimentary cyclicity. In this paper, the stratigraphic record is presented as the product of a highly complex system of interacting influences, which can be expected to show some of the characteristics of dynamical systems. These include recurrence, fractal-like distribution of stratal properties, low-dimensionality, approximate coincidence in time of diverse phenomena, and problems of distinguishing cause and effect.

Analytical methods to distinguish system-generated information from noise in observational time-series data are at an early stage of development. Graphical methods, however, lend some support to the hypothesis that the 'Stratigraphy Machine' has all the characteristics of a non-linear dynamical system. Such systems are usefully conceptualized as comprising coupled oscillators, like a forced pendulum. As such, they are capable of damping out an input forcing signal such as the predicted orbitally mediated changes in solar insolation. Given suitable tuned strength of coupling, they are equally capable of resonating to the same input frequencies. A dynamical systems view of the stratigraphic record, in which the Milankovitch signal may at different times be (i) unambiguously encoded, (ii) cryptic but suspected, or (iii) absent, is therefore qualitatively consistent with the available evidence. Non-linear dynamics could prove to be the reason for the preservation of the orbital signal (and of many other phenomena) in the stratigraphic record. Quantitative demonstration can in future be approached through modelling and simulation.

INTRODUCTION

Many, perhaps most, major scientific advances are made through interdisciplinary cross-fertilization. Milankovitch was not an earth scientist, but the papers in this volume testify to the depth of his influence on our science. This paper describes a deliberate attempt to expose the discipline of stratigraphy to the newly developing multidisciplinary concepts of non-linear dynamics, or chaos theory. The aim is to explore the possibility that 'cyclicity' at all scales in the stratigraphic record might be a product of the internal workings of what I shall call the 'Stratigraphy Machine', and not simply a direct response to a simple input 'forcing' function.

Sequence stratigraphy has of course brought about great changes in stratigraphic thinking, in particular a greater awareness of the role that models necessarily play in stratigraphic interpretation. The problem with sequence stratigraphy is that the theory has been developed almost entirely from the data, which makes it difficult to test the theory from the data – hence the difficult question of sea level versus tectonics as the driving force. Does non-linear dynamics have the potential to provide a rationale for the way that stratigraphy works at a more fundamental level? Could it provide us with a new approach to questions of cause and effect in stratigraphy, at all scales?

This paper will review some of the criticisms that have been levelled at the Milankovitch hypothesis, that is, the hypothesis that orbitally controlled variations in incoming solar energy have a measurable impact on the stratigraphic record, especially the pre-Quaternary record. I next review the essential links in the Milankovitch chain, from variable insolation at the top of the atmosphere, to its influence on the entire hydrosphere, to the recording of the resulting changes in the stratigraphic record.

I then introduce my concept of the Stratigraphy Machine, the complex system of interacting processes whose output is the stratigraphic record. Emphasis is placed on the interdependence and feedback relationships among the variables, rather than on the individual variables themselves. Chaos theory will then be introduced, albeit at a superficial level, but sufficient to outline its key predictions should it indeed apply to the stratigraphic record. I then look at the extent to which stratigraphic phenomena meet those predictions, and finally return to the question of cyclicity in the Milankovitch waveband and its relationship to astronomical forcing.

Others have explored the value of the chaos paradigm in other areas of the earth sciences (see especially the recent collection of papers edited by Middleton, 1991), and both Tetzlaff (1990) and Slingerland (1990) have commented on its implications for predictability in quantitative dynamic stratigraphy. There is inevitably some overlap with the subject matter of this paper, which is nevertheless included in this volume because of the particular relevance to the quasi-periodic nature of orbital forcing.

PROBLEMS WITH ORBITAL FORCING

Problems with cyclicity

In the context of the subject matter of this volume, geologists are largely polarized into two camps: those for whom orbital forcing of stratal cyclicity is almost an article of faith, and those for whom the problems of proving it beyond all reasonable doubt are just too severe. The problem of precise age calibration of sedimentary cycles at Milankovitch frequencies forces reliance on cycle thickness as a proxy for time. Periodicities are measured in centimetres rather than years, and 'period' ratios are actually thickness ratios. The case for periodicity in time rests on Sander's Rule, that 'equally spaced beds represent equal time intervals and are therefore time-periodic, but beds which are not of equal thickness do not permit the exclusion of rhythmic events as their cause' (see Schwarzacher, 1985, p. 382). Algeo & Wilkinson (1988) provided an excellent summary of the arguments against orbital forcing, supported by a substantial data set of cycles of a wide range of dimensions and ages, largely from shallow-marine environments. Their key lines of attack can be summarized as follows.

1 Milankovitch-band periodicity need not equate to orbital forcing; alternative mechanisms such as autocyclicity are available.

2 Regularity of cycle thickness is a function of constraints on rates of subsidence.

3 Calculated periods of 'mesoscale' sedimentary cycles almost inevitably fall within the Milankovitch frequency band whatever their cause.

4 The extreme difficulty of relating rock thickness to time is a major obstacle to demonstrating rhythmicity in time.

5 Estimates of cycle periods are generally too imprecise to demonstrate correspondence to the predicted orbital periods with any confidence.

6 Cycle periods estimated from their own data set fail to cluster around the predicted periodicities.

7 Ratios between periodicities (whether in terms of thickness or time) are of no value in demonstrating orbital control, as too many ratios are available between pairs of periodicities predicted from the astronomical theory.

It will be seen below that an even greater problem with the Milankovitch hypothesis is that of passing the incoming orbital signal through the web of intercoupled mechanisms that constitute the strata-forming process, and into some encoded form in the strata themselves. Non-linear dynamics may provide the only reasonable explanation of how, under certain conditions at least, this is possible at all.

Alternative causes of cyclicity

In addition to Algeo & Wilkinson's objections to the interpretation of cyclic phenomena in the stratigraphic record, there are multiple mechanisms available that could individually cause regular repetitions of strata; their collective effects are considered below, in a later section of this paper. Four principal mechanisms are:

1 tectonics, in the form of episodic subsidence;

2 sediment supply, which might fluctuate because of episodic uplift of the source area or because of changing erosion rates, e.g. through climatic change;

3 sediment type (lithofacies), which might vary due to changes in water depth and/or distance offshore, or oceanographic variables such as temperature and circulation;

4 accommodation, the space available beneath

some kind of base level, which can vary because of either subsidence or sea-level changes.

The arguments presented below in favour of a non-linear dynamical model for the accumulation of strata lend weight to the view that it is unreasonable to look for a single cause for cyclicity, because of the impossibility of separating the effects of the different mechanisms that are available. Furthermore, it is likely that the interactions between the various mechanisms are at least as important as each individual mechanism. This is a viewpoint that is absent from much of the recent literature on tectonics versus sea level as the primary control on sequence stratigraphy.

THE MILANKOVITCH MACHINE

We can think of the Milankovitch theory of orbitally forced climatic change in terms of three links of a chain. First, the theory in the strict sense predicts (i) changes in the amount of solar energy (insolation) received at the top of the Earth's atmosphere, and (ii) the variation in insolation with latitude. The second link in the chain is the prediction that changes in climate are linked in some way to changes in insolation. Insolation is the major energy input to the atmosphere/ocean system, whose motions, modified by the Earth's daily rotation, are manifested as climate. While the changes in insolation can be predicted quantitatively and with considerable confidence, prediction of the resulting climatic changes is quite a different matter. It is possible to model the processes in the atmosphere in a variety of different ways, but, as followers of weather forecasts are all too well aware, confidence in the result is another matter. Two particular sources of uncertainty are relevant to this paper. The first is the likely importance of threshold effects in climatic change; this is the source of much of the present concern about anthropogenic enhancement of the greenhouse effect, for instance. There is abundant and increasing evidence for the suddenness of climatic change, at various time scales, in the historical and geological past (Crowley & North, 1988). The second source of uncertainty brings us to the central theme of deterministic chaos, and that is sensitive dependence on initial conditions. The same model may yield very different results from initial conditions that differ from one another by apparently insignificantly small amounts. Thus, (i) limitations on weather forecasting will always be imposed by the impossibility of specifying the initial conditions to infinite precision, and (ii) even very simple deterministic models can be made to yield surprisingly unpredictable results, as illustrated below.

The third essential link in the Milankovitch chain is the recording of climatic change in the sedimentary record. There are indeed sedimentary rhythms that record the predicted periodicities remarkably faithfully. However, the mechanism by which the Milankovitch signal is encoded in sediment generally remains obscure. Rocks do not record sunshine, so there is no possibility of a direct link between changes in insolation and changes in some lithological property. It is more likely that changes in insolation lead to changes in one or more aspects of the *processes* that result in stratal accumulation; the orbital signal goes into a black box, out of which is extruded the stratigraphic record. The black box is what I refer to as the Stratigraphy Machine, and the workings of its contents remain largely mysterious.

THE STRATIGRAPHY MACHINE

Equations of stratigraphy?

It is well known that the long-term addition of sediment to the stratigraphic record depends on certain apparently simple factors: we need a supply of sediment, and somewhere to put it:

Stratal accumulation =
f(Sediment supply, Accommodation)

Accommodation space can be thought of in terms of a base level, a surface below which sediment will accumulate, and above which it will be transported somewhere else. If such a basin is not to fill up too quickly, we need a mechanism for subsidence, which we can provide through a combination of tectonic and thermal processes in the lithosphere. We can provide some additional accommodation space by altering the base level, by raising sea level in the case of marine environments, or through tectonothermal uplift and subsidence in the non-marine realm. Cant (1989) has explored the implications of the following relationship, expressed as the controls on changes in water depth:

(Subsidence increment)
+ (Change in eustatic sea level)
− (Thickness of sediment deposited)
= (Change in water depth)

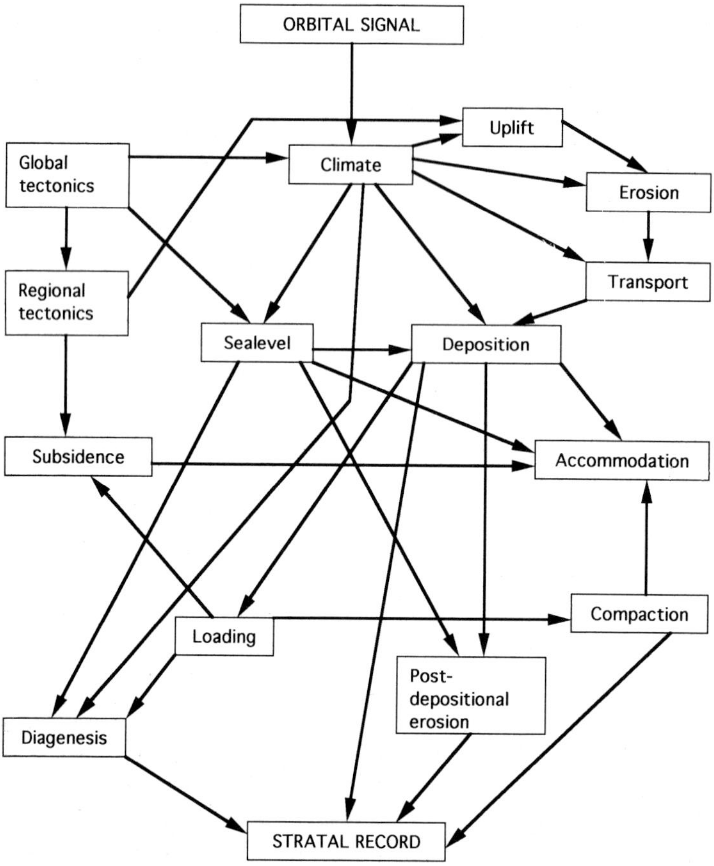

Fig. 1. The Stratigraphy Machine, showing the complex system of interactions and feedbacks through which the orbital signal (top) must pass in order to stand a chance of being encoded in strata (bottom). Deposition requires a source of sediment and somewhere to put it. The source is an area of uplift, which is primarily a function of regional tectonics, though climate may also have an influence. Erosion, to generate sediment, is strongly dependent on climate, which also influences the transport of sediment. Climate is also an influence on depositional environment, along with sea level, itself partly a function of climate, mainly through glaciation, which is linked to uplift and to global tectonics. Sea level is also controlled by global tectonics. Deposition is further influenced by accommodation (the space available beneath some kind of base level, which might be sea level itself, or storm wave-base). There is a negative feedback loop between accommodation and deposition, but a positive one involving sediment compaction as a result of loading; loading also acts to increase subsidence and hence accommodation. Other post-depositional processes include erosion (redistribution of sediment), partly influenced by changes in sea level/base level, and diagenesis, influenced by factors including sea level, compaction and climate.

Such a model is useful, not least because the relationships as expressed are linear, and substitution of known or assumed values might appear to yield a unique answer. However, these processes are not independent of one another, or of additional variables such as climate and topography (Fig. 1). For instance, sediment accumulation itself causes subsidence; erosion requires uplift but also brings about further uplift through isostatic adjustment; uplift of the source and subsidence of the basin are coupled through mass balance considerations; the rate of erosion is controlled by climate as well as by topography, but topography in turn is a major influence on climate. Sea-level changes affect subsidence through isostasy, and may be brought about by global tectonics or by climatic change. Global tectonic cycles probably influence long-term climatic change as well as long-term sea-level change through the greenhouse–icehouse cycle (Fischer, 1984; Worsley *et al.*, 1984) and are probably linked, through mantle processes, to episodes of volcanic as well as tectonic activity.

It is clearly unrealistic to attempt to decouple any of these processes from any of the others, yet the majority of models of stratal accumulation are forced to do just that because of our present lack of understanding (i) of the relative strengths of the couplings and interactions and (ii) of the characteristic time scales of many of the processes.

Some important characteristics of strata

Although we know too little as yet to propose a fully coupled model of the processes that together create the stratigraphic record, we can at least consider the conceptual implications of the Stratigraphy Machine. The following characteristics should be borne in mind when, later in this paper, we look at the stratigraphic record in the light of the predictions of non-linear dynamics for just such a feedback-dominated system. All of the following pose questions for the processes of stratal accumulation, and thus are precisely those that tend to be avoided in the classical approach to stratigraphy, in which understanding of process is expected to be a passive end-product of meticulous description.

Order combined with unpredictability

Sedimentary strata are highly structured. We have well-ordered distributions of grain shapes, compositions and sizes at the small scale; bedding, sorting and sedimentary structures at intermediate scales; and mappable formations or seismic sequences grouped into basins at larger scales. On the other hand, strata are not so highly ordered as to be predictable beyond the immediately accessible; prediction from the known to the unknown is the major preoccupation of exploration geology and the failure rate is high.

Fractal properties

The stratigraphic record is remarkable for its ability to record processes and events over a huge range of both length and time scales, from the existence of a coccolith (length $c.10^{-5}$ m, lifespan $c.10^{0}$ a) or the impact of a raindrop ($c.10^{-2}$ m, $c.10^{-8}$ a), to the long-term advance and retreat of the sea across the North American continent ($c.10^{7}$ m, $c.10^{8}$ a). At least some important processes are rather independent of scale, such as deposition of sediment to fill a standing body of water, which can take place in a puddle, a lake, or the Gulf of Mexico. As a result, the stratigraphic record is not only 'more gap than record' (Ager, 1981) but is more gap the closer you look at it, like the Cantor Set of fractal mathematics (Plotnick, 1986). Stratigraphic completeness is thus a function of the scale of observation (Sadler, 1981; Sadler & Strauss, 1990). The fractal model leads to another useful result, through the fact that fractal functions are not differentiable; the concept of slope at a point has no meaning, because reduction of the interval over which the slope is calculated reveals more detail, not less. The lesson for stratigraphy is that 'accumulation rates' calculated by dividing net thickness by estimated time duration have to be interpreted with great care.

Problems of cause and effect

The empirical nature of classical stratigraphic correlation is due to the general difficulty of distinguishing cause and effect in stratigraphy. Tops of biozones frequently coincide with lithological changes: did the implied change in environment cause the organisms to move somewhere else, or was their evolution causally linked to the same changes? Can sea level really be singled out as an essentially independent cause of depositional packaging? Is it right to look for a single cause of the Cretaceous–Tertiary (K–T) boundary phenomena (extinctions, volcanism, impact debris), or is it more likely that they are linked to each other in some more dynamic way (Shaw, 1987a)?

Reduction of variables

The number of variables conceptually available as controls on stratigraphic processes is almost infinite. Each 'variable' in Fig. 1 subsumes many others: how many variables are needed to specify climate, for instance? Yet our perception in any given real geological situation is that a very small number of variables need be considered. The essence of the arguments over eustatic versus alternative controls on depositional sequences is that it is quite reasonable to argue that a single variable, eustatic sea level, has a large measure of control over a wide range of depositional settings. The opposing camp of course are equally certain that tectonic effects are dominant. The point here is not which side is correct, but that the stratigraphic record behaves as though only a very few of the available variables are in overall control.

Milankovitch cyclicity is an example of this effect. In many settings, the Milankovitch climatic signal that we know to exist is masked by the effects of other variables. In other settings it apparently comes through clearly, with all of its relative periodicities intact, and the other variables are reduced to a background role with only minor influence on the structure of the resulting strata.

Recurrence

Given the rather limited number of commonly occurring lithologies (shale, sandstone, limestone and a few odds and ends), recurrence is an inevitable characteristic of the stratigraphic record (Zeller, 1964). There is, however, an array of recurrence intervals and of degrees of regularity that is less easily predicted from classical models. At the scale of field and laboratory observation, there are spatial recurrences that allow us to classify stratigraphic phenomena in terms of grain size, lithological composition, maturity, bedding relationships such as coarsening-up cycles, and so on.

Temporal recurrence is nearer to the subject matter of this volume. It too covers a wide range of scales, from tidal cycles and annual varves at short time scales and a high degree of periodicity, through the mesoscale cycles of the Milankovitch waveband, to cycles of continental assembly and breakup at much longer time scales and less regular periodicity. At the bottom end of the periodicity scale come the unique events such as those characterized by Ager (1981) under the Principle of the Persistence of Facies; the Rhaetic Bone Beds, basal Cambrian

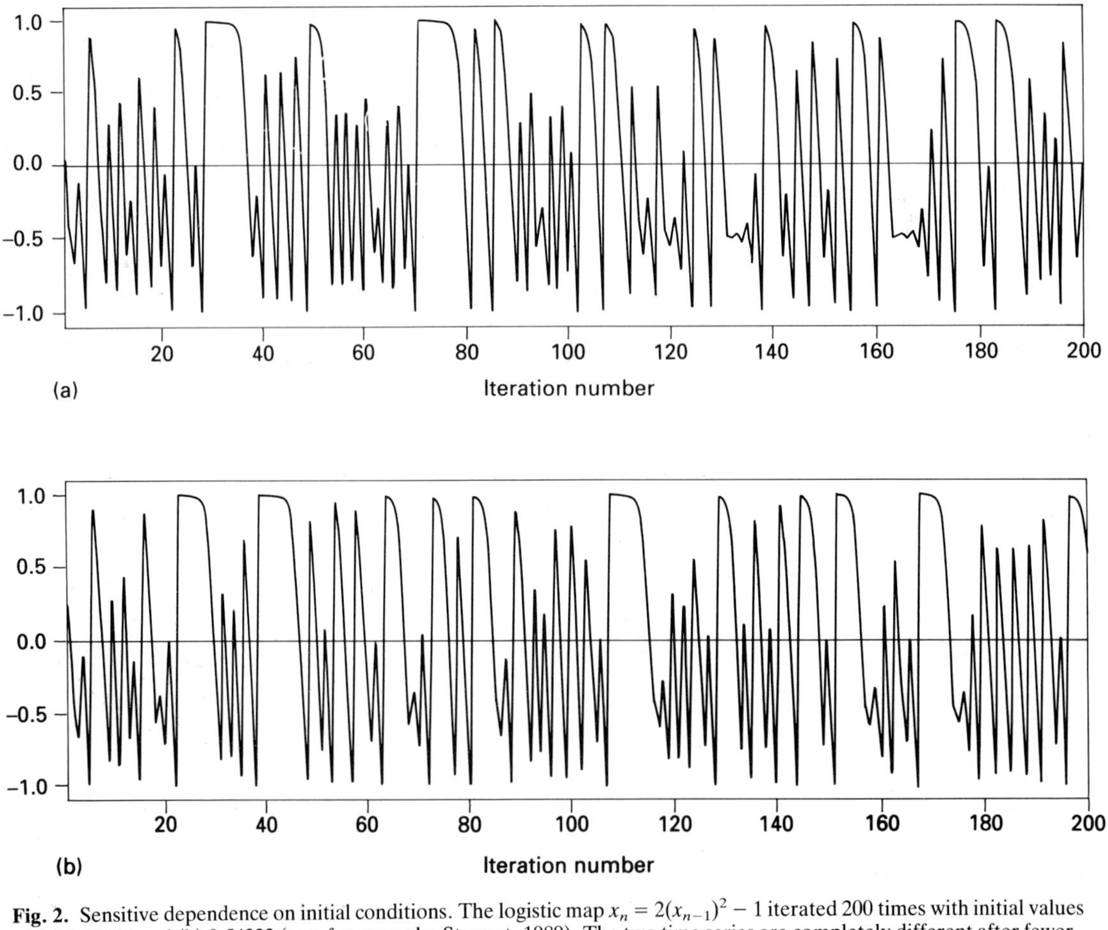

Fig. 2. Sensitive dependence on initial conditions. The logistic map $x_n = 2(x_{n-1})^2 - 1$ iterated 200 times with initial values of (a) 0.54321 and (b) 0.54322 (see, for example, Stewart, 1989). The two time series are completely different after fewer than 20 iterations.

quartzites, Urgonian Limestones, Cretaceous Chalks and so on, that seem to represent a concerted swing of a number of environmental variables into an unusual if not unique field of stability.

Summary

Thus, the stratigraphic record is simultaneously characterized by order and randomness, uniqueness and recurrence, periodicity and aperiodicity, all of these features applying at all of a very wide range of length and time scales. It is an unfortunate fact that classical geological training, reinforced by the dictates of stratigraphic codes, conspires to conceal these most interesting and relevant features of the stratigraphic record. We now turn to the topical subject of non-linear dynamics, for insights into the above characteristics of the Stratigraphy Machine and its output.

NON-LINEAR DYNAMICS

The topic of non-linear dynamics and chaos has recently been introduced in a geological context by Middleton (1991). General accounts of chaos theory can be found in more popular works such as Gleick (1987) and Stewart (1989). Two key aspects of

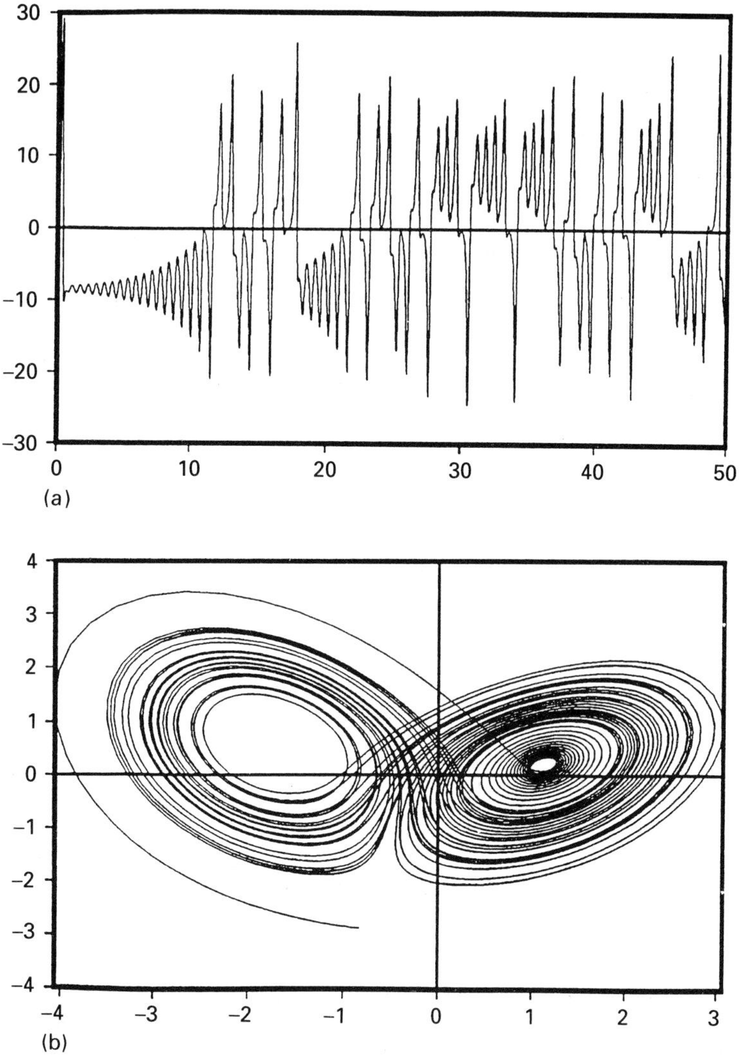

Fig. 3. (a) Time series of the variable y, and (b) two-dimensional projection of successive points in x, y, z-space, from iterative solution of Lorenz's equations (Lorenz, 1963; Goodings, 1991a) describing convective flow. The equations are $dx/dt = -\sigma(x - y)$, $dy/dt = rx - y - xz$, $dz/dt = xy - bz$. The values of the three constants used are $\sigma = 10$, $b = 8/3$, $r = 28$. Initial values of the variables are $x = 0$, $y = 1$, $z = 0$. The trajectory in (b) is not strictly a strange attractor, as it is constructed from a succession of discrete points, but it is a fair representation of one.

'chaos' are illustrated here: Fig. 2 illustrates sensitive dependence on initial conditions, using a simple difference equation; Fig. 3 introduces the state-space plot by comparing the elegant trajectory described by Lorenz's equations (Lorenz, 1963) with the chaotic nature of the corresponding time series.

The strange attractor approximated by iterating Lorenz's equations (Fig. 3b) shows the evolution of that particular system through time, but without using time as one of the graph's axes – time can be thought of as running along the trajectory. A very remarkable theorem, due to F. Takens and described by Stewart (1989) and Goodings (1991b), proves that an approximation to a system's attractor can be constructed by an apparently completely different route. All that is needed is the time series of a single variable (i.e. successive values of a single variable measured at equal time intervals), such as Lorenz's y-variable plotted in Fig. 3a. For a two-dimensional trajectory, successive values in the time series are simply plotted against one another, $y(t)$ against $y(t-1)$. For a higher (n) dimensional plot, the variables would be $y(t)$, $y(t-1)$, $y(t-2)$, . . . , $y(t-n+1)$. The resulting trajectory clearly has the property of representing successive time steps in the system's evolution.

Such a direct approach to a system's attractor invites experimentation with geological data. Three examples are given in Figs 4–6, using data at successively longer time scales (and with successively

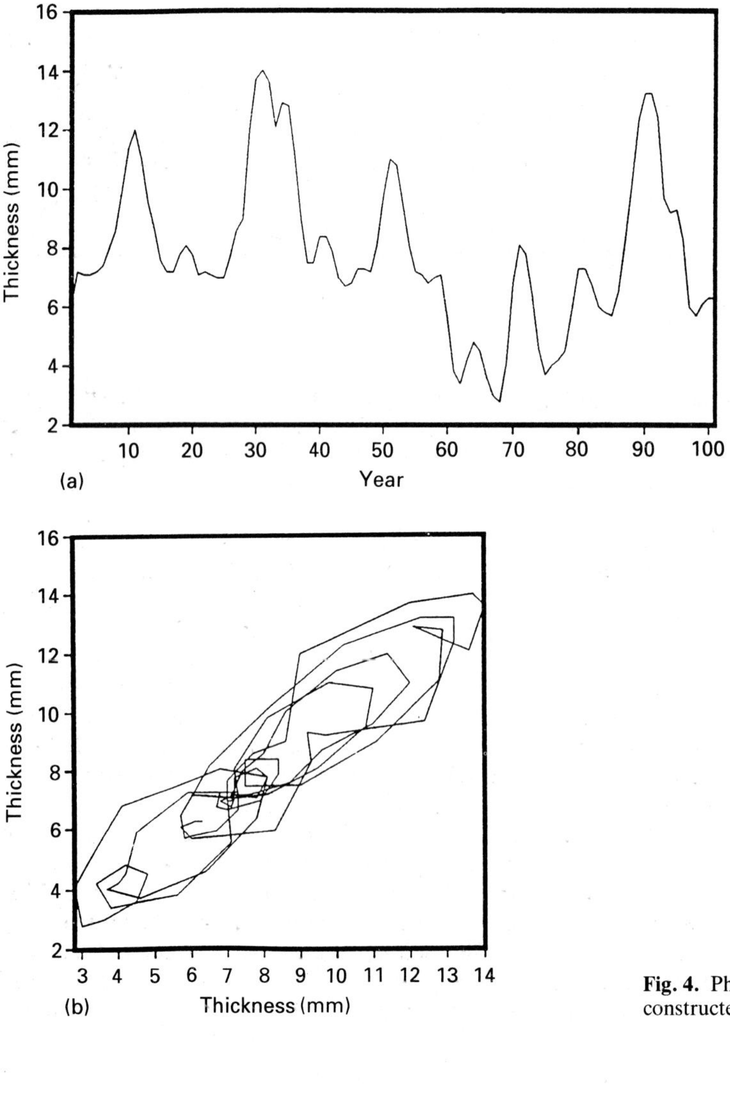

Fig. 4. Phase portrait (b) of annual varve data (a), constructed as described in the text.

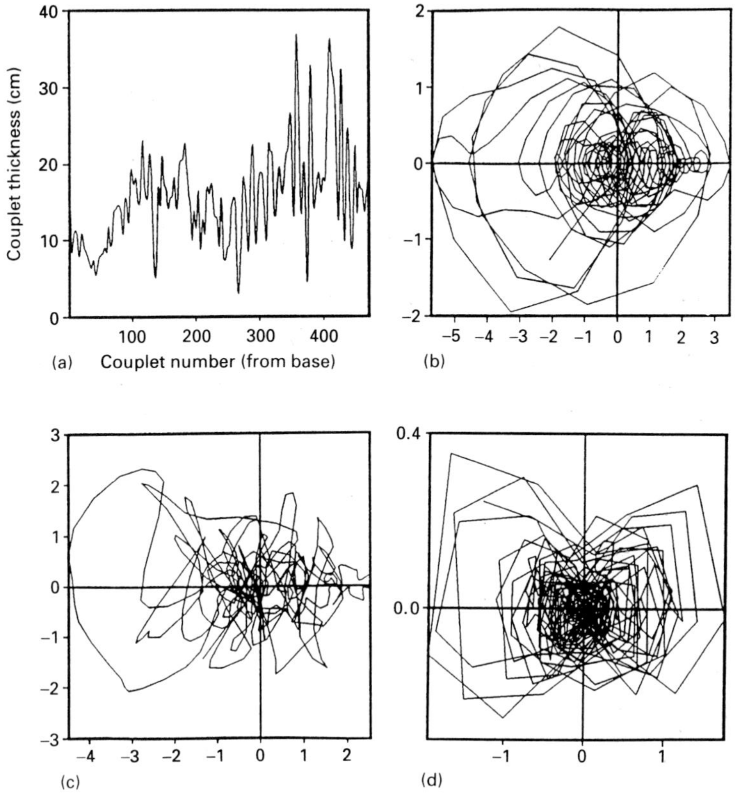

Fig. 5. (a) Cenomanian couplet thicknesses, from de Boer (1983). (b)–(d) Three orthogonal views of a three-dimensional phase portrait of the same data, constructed as described in the text.

greater relaxation of the strict requirement for equal spacing in time of the data points). The first example is shown in Fig. 4. The data set comprises thickness measurements (in millimetres) of a set of varves (annual increments of lake-floor sediment typical of a periglacial environment; data from Davis, 1973). The result is a graph of each year's increment plotted against that of the previous year.

There is considerable order in the data, whose trajectory visits a rather restricted and well-organized part of the available 'state-space'. There is thus considerable correlation between successive annual sediment increments (random, uncorrelated data would simply fill the available state-space). Secondly, there is a central cluster of 'preferred' values about which the trajectory revolves. Third, there are two larger circuits, one revolving around smaller thickness values and the other around larger values. As the raw data are quite simple and have a recurrent pattern that appears to include the *c*.22-year sunspot cycle, it is easy enough to see the relationship between oscillations in the original data and the two parts of the figure-of-eight of the state-space trajectory. As the data set is very small, it is of course very likely that it does not explore all the parts of state-space that a larger data set would reach.

A second example is shown in Fig. 5. Here, the time series illustrated (Fig. 5a) comprises the thickness (in centimetres) of about 460 successive couplets of pelagic marl and limestone of Cenomanian age in central Italy (data from de Boer, 1983 and personal communication). It is likely that the couplets were controlled by the precessional cycles of the Earth's orbit and are thus of equal duration (about 21 ka), but this is not critical to the method. Because the resulting figure is evidently more complex than Fig. 4b, I have attempted to illustrate its three-dimensional nature. I first smoothed the data with a three-point moving average to allow for imprecision of measurement (while being aware that smoothing will introduce a degree of correlation between successive points). I then took the smoothed data and two successive lags of itself, and plotted these as the

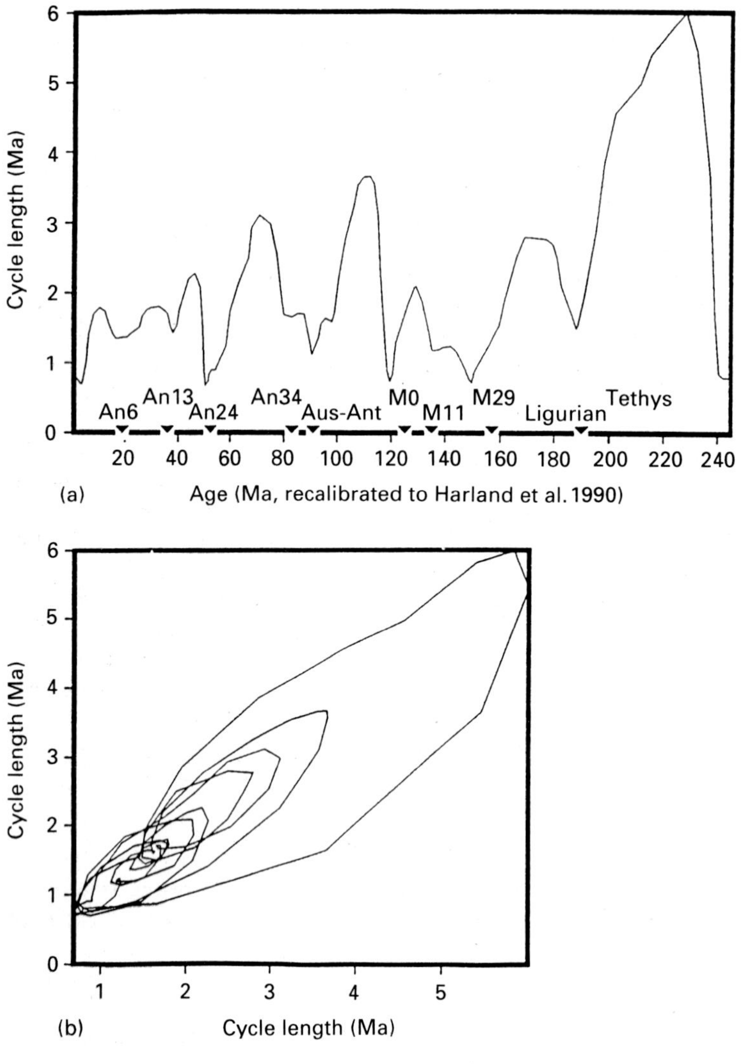

Fig. 6. (a) Haq *et al.* (1988) 'sequence' duration data, with ages recalibrated to the Harland *et al.* (1990) time scale. Data have been smoothed with a three-point moving average to allow for the rounding of sequence boundary ages, and to emphasize the longer term (*c.*10 Ma) variations in sequence length, also showing ages of magnetic chrons at which major reorganization of tectonic plate motions took place. (b) Phase portrait constructed by plotting duration of sequence $n + 1$ against duration of sequence n.

three 'variables'. They are plotted in pairs in Fig. 5b–d, to give three orthogonal views of the resulting three-dimensional state portrait. Although the figure described is complex and not particularly pretty, it can at least be seen that, like Fig. 4b, it explores only part of the available state-space, it has some definite shape to it, and moreover its shape in the three views is different.

A third example (Fig. 6) plots the length of successive 'global cycles' from the chart of Haq *et al.* (1988). I have again smoothed the data a little, to allow for the fact that many of the cycle boundary ages are rounded to the nearest 0.2, 0.5 or 1.0 Ma. The trajectory again implies a possibly surprising degree of order in the data (not all of which can be ascribed to the smoothing). Referring to the time-series plot of the cycle-length data, the tendency of the trajectory to follow one of a rather small number of loops is an expression of the pseudoperiodicity in the data itself. One is drawn to the rather close coincidence of the major troughs in the time series with periods of major plate tectonic reorganization around times of significant continental rifting events (which are indicated, in terms of magnetic anomaly numbers, along the bottom of the figure). The example demonstrates the kind of coupling that can appear between superficially unrelated data sets. This is in keeping with the properties of dynamical systems; it might indeed suggest that the dynamics of the sequence-generating system are rejuvenated into higher frequency activity at irregular intervals by whatever combination of events are involved in major events of continental rifting.

The most important fact that emerges from these three graphical experiments is that the trajectory in each case visits only a restricted part of the available state-space. This is suggestive (although not proof) of control by a non-linear dynamical system that is governed by a chaotic attractor. The relative simplicity of the resulting figure in each case suggests that the underlying attractor can be represented in only two or three dimensions. This result is comparable to that obtained for the dripping tap experiment, which was devised to test the same graphical method of reconstructing an attractor from time-series data (Crutchfield *et al.*, 1986). Phase portraits of this kind may well be useful for exploring geological time series that are insufficiently periodic to yield useful information through spectral methods.

IS THE STRATIGRAPHIC RECORD CHAOTIC?

Dynamical systems share a number of common characteristics. Many of these are shared by non-chaotic natural systems, so we cannot diagnose chaos by demonstrating their presence. It is nevertheless instructive to review the properties of the stratigraphic record, and what we know of its generative system, in the light of the properties predicted by non-linear dynamics.

The state of a dynamical system at any time t is governed by its state at some previous time $t - \Delta t$, and is not predetermined by its state at some initial time t_0. Contrast this with the trajectory of a bullet in classical mechanics, which is completely determined by its initial velocity and direction. This 'recursiveness' is modelled by the use of iteration in the many mapping functions (e.g. the logistic map; Fig. 2) that have been used to explore the properties of chaos. It results in the sensitive dependence on initial conditions that is so characteristic of chaos. In turn, this implies lack of predictability over anything but the shortest of time scales. Lack of predictability is certainly a characteristic of the stratigraphic record; prediction ahead of the drill is the major preoccupation of exploration geology, and it continues to be conducted with the poorly quantifiable levels of confidence of long-range weather forecasting. Basin-modelling as presently practised in quantitative dynamic stratigraphy tends to be essentially deterministic – a given set of input parameters will always yield the same stratigraphy. This is intuitively unrealistic: who would care to predict the detailed stratigraphy that will eventually result from the complete filling of a presently active sedimentary basin?

Dynamical systems are distinct from those of classical mechanics in requiring a throughput of energy; they are dissipative, not frictionless. The Stratigraphy Machine is typical of natural systems in requiring such an external source of energy to drive it. Energy is available from three sources: (i) the kinetic energy of crustal movements (ultimately powered by heat-flow and mantle convection) which create the topographic gradients between sediment source and basin; (ii) the solar power required to drive the weathering–transport–deposition cycle; and (iii) gravitational energy supplied by the dynamics of the solar system.

Self-similarity, the similarity of appearance of a function or phenomenon at many different scales, is characteristic of the output of non-linear systems. Of turbulence, Lewis Fry Richardson (a pioneer investigator of fractal phenomena) said:

> *Big whorls have little whorls*
> *Which feed on their velocity*
> *And little whorls have lesser whorls*
> *And so on to viscosity.*

Not only is turbulence itself relevant to sedimentary processes, but many of the other classical examples of self-similarity are also drawn from the earth sciences. The Gutenberg–Richter law of the inverse relationship between earthquake magnitude and frequency defines a $1/f$, or red-noise, power spectrum (Mandelbrot & Wallis, 1969; Bak & Chen, 1991; Weedon, 1993). The coastline of Great Britain, which gets longer with every decrease in the scale used to measure it, is readily perceived as self-similar. In stratigraphy, many (though by no means all) phenomena are rather independent of scale, and this is reflected in the appropriateness of the hierarchical approach to their classification.

Avalanches at all scales from piles of sand at their critical resting angle are the model for the newly described phenomenon of self-organized criticality, a state towards which systems that are not in equilibrium may evolve (Bak *et al.*, 1988; Bak & Chen, 1991). Such systems generate fractally distributed output, possibly implying that self-organized criticality is the normal state for many natural systems.

Relevance to systems of sedimentation is intuitively obvious and suggests considerable scope for modelling.

Recurrence, or pseudoperiodic repetition, is characteristic of chaotic systems; the examples illustrated in Figs 2 and 3 show the characteristic tendency towards the repetition of patterns, yet without true periodicity. The output is not chaotic in the sense of lacking structure, only in the sense of lacking predictability. We know the kinds of patterns that recur in our weather, for instance, but cannot predict precisely when the next front will arrive. Upward fining or coarsening cycles are among many common signatures in the stratigraphic record, but we cannot specify precisely how they will stack up, or how they will group themselves into higher order 'cycles'.

Low-dimensionality, or reduction of many input variables to a rather small number of apparently controlling parameters, is a convenient attribute of natural systems. It allows description of such systems in terms of a much smaller number of variables than might otherwise be the case. One does not need measurements of large numbers of variables at closely spaced stations all over the Earth in order to make reasonable descriptions, and even short-term predictions, of the weather. A few measurements of temperature and pressure, together with some knowledge of past behaviour, may be enough. The reason is the degree of coupling, or feedback, between variables, that ensures that there will be much parallelism of behaviour among them. Such coupling is central to non-linear dynamical systems (witness the dependence of each variable on both of the others in Lorenz's equations in the caption to Fig. 3). An important implication of this is the loss of distinction between cause and effect.

Low-dimensionality is manifested in stratigraphy by the fact that quite reasonable modelling can be achieved with very low numbers of variables, whereas the total number of 'available' variables is conceptually very large. Conversely, the danger lies in assuming that modelling can proceed by decoupling variables and placing them on either side of an equation. Thus, sediment supply, thermally-driven subsidence and eustatic sea level are not mutually independent (Fig. 1). Sediment supply is related to uplift in the source area, which is related to basin subsidence through both thermal and mass-balance considerations. Subsidence is related through isostasy to sediment input as well as to some underlying thermal mechanism. Eustasy is related to some combination of global-scale tectonic processes and to climatic change. Climate further influences sediment supply through both erosion and transport. None of these relationships is simple, and many of them involve significant (but poorly understood) time lags. None of the variables can be decoupled and used as the primary input to a modelling equation.

At a strictly qualitative level therefore, the processes that allow strata to accumulate, and the form of the strata themselves, are fully consistent with the properties of a dynamical system. The more fundamental question of what is driving it all is beyond the scope of this paper, but the Gaia concept of Lovelock (1979) and the referenced papers of H.R. Shaw provide interesting further reading on this matter.

DISCUSSION

Non-linear dynamics thus has considerable relevance for stratigraphy, including the question of cyclicity and its possible relationship to the Milankovitch hypothesis. The Stratigraphy Machine, the dynamic system of coupled variables whose output is the stratigraphic record, clearly has all the characteristics of non-linear dynamics, from consideration of both its workings and its output. An immediate consequence of accepting the stratigraphic record as the product of such a system is that we are required to rethink some of our most treasured prejudices, particularly concerning causality. Yes, sea level is implicated in driving depositional sequences, but it is probably more truthful to regard both sea level and the stratigraphic record as non-independent outcomes of the complexities illustrated (and oversimplified) in Fig. 1. Yes, the orbital forcing function is an influence on global climate throughout geological time, but getting it into the stratigraphic record is another matter. Indeed, even the orbital frequencies themselves are not completely external to the Earth, because the planetary orbits are responses to resonances that relate to the original segregation of matter in the solar system among and within the individual planets and their satellites (Shaw, 1989).

There is also, I believe, a message here regarding that most empirical of sciences, stratigraphic correlation. Non-linear dynamics suggests to me that correlation is less a matter of coincidence than of coupling, with its implication of common origin in the internal dynamics of the system. Correlations that 'work', like stratigraphic boundary phenomena that seem to go together, have good reason to do so, without necessarily implying precise synchrony.

However, we are still only at the stage of qualitative concepts. Given the difficulty of finding analytical methods of deriving meaningful parameters from the noisy data of stratigraphy, the next step in making use of non-linear dynamics will be through numerical and graphical modelling. Very simple models can be run to generate data whose statistical properties can then be compared with real data. An example is Huang & Turcotte's (1990) model of the earthquake swarms generated by coupled fault systems. Another profitable approach is that of Shaw & Chouet (1988) and Shaw (1988), who adapted two very simple chaos-generating equations (the logistic and sine-circle maps) for modelling the behaviour of magma chambers; these mathematically simple functions are capable of generating considerable richness of chaotic output. Shaw's (1987b) investigation of the history and prediction of Hawaiian volcanism also illustrates the value of graphical experiments on simple attractors. The major barrier to similar modelling of stratigraphic systems is the lack of hard information on the relative rates of the critical processes involved. Gaffin & Maasch (1991) have taken a major step in the right direction; their non-linear model of stratal accumulation generates shifts in coastal onlap with no change in sea level. Critical to the non-linearity of their model is the incorporation of delays, specifically in the isostatic response to loading; many linear models assume instantaneous response. Critically, though, Gaffin and Maasch lacked the hard data on process rates that would have allowed a fully quantitative model.

Finally, does the non-linear dynamic approach answer the question posed by the title of this paper? It does not do so in any definite sense, but I believe that it at least points the way forward. It suggests, firstly, that we should not necessarily look for a single explanation of why and when the Milankovitch signal becomes such an overriding influence on the sedimentary record that the forcing frequencies can be extracted from the record by classical time-series methods. (I assume that readers of this volume will be in no doubt that there is now a growing number of such cases.) Non-linear systems can resonate to an incoming signal, or damp it out completely; the stronger the coupling, the greater the range of frequencies over which the system will lock on to the incoming signal and amplify it (Bak, 1986). Numerical investigations by Shaw (1987a, 1989) have suggested phase-locking over a wide range of time scales for a range of phenomena from extraterrestrial impacts to biological evolution. In the case of the pelagic environment, repository of so many faithful records of the orbital frequencies, we can imagine that the accommodation oscillator is switched off (accommodation space being virtually infinite), thus reducing this part of the Stratigraphy Machine to even greater simplicity, and increasing the chances of phase-locking with the incoming orbital signal. We can probably learn much from this and other systems in which the Milankovitch signal appears to increase and decrease in intensity, such as the Lias of southern Britain (Weedon & Jenkyns, 1990).

In conclusion, chaos need not displace Milankovitch from his seat on the bandwagon, but rather it provides us with a quite new and potentially revolutionary paradigm with which to challenge many of stratigraphy's concepts and practices.

ACKNOWLEDGEMENTS

Ilfryn Price suggested this line of research and fired my enthusiasm for it. Poppe de Boer invited me to present a lecture on the subject at a symposium in Utrecht, at which Graham Weedon urged me to submit a written version. I thank all of them for their support, and Gerard Middleton, Robin Whatley and Poppe de Boer for suggesting significant improvements to the text.

REFERENCES

AGER, D.V. (1981) *The Nature of the Stratigraphic Record*, 2nd edn. Macmillan, London.

ALGEO, T.J. & WILKINSON, B.H. (1988) Periodicity of mesoscale Phanerozoic sedimentary cycles and the role of Milankovitch orbital modulation. *J. Geol.* **96**, 313–322.

BAK, P. (1986) The Devil's Staircase. *Physics Today*, December, 38–45.

BAK, P. & CHEN, K. (1991) Self-organised criticality. *Sci. Am.* **Jan**, 26–33.

BAK, P., TANG, C. & WIESENFELD, K. (1988) Self-organised criticality. *Phys. Rev. A.* **38**, 364–374.

CANT, D.J. (1989) Simple equations of sedimentation: applications to sequence stratigraphy. *Basin Res.* **2**, 73–81.

CROWLEY, T.J. & NORTH, G.R. (1988) Abrupt climate change and extinction events in Earth history. *Science* **240**, 996–1002.

Crutchfield, J.P., Farmer, J.D., Packard, N.H. & Shaw, R.S. (1986) Chaos. *Sci. Am.* **Dec**, 38–49.

Davis, J.C. (1973) *Statistics and Data Analysis in Geology*. John Wiley & Sons, Chichester.

De Boer, P.L. (1983) Aspects of Middle Cretaceous pelagic sedimentation in southern Europe. *Geol. Ultraiectina* **31**.

Fischer, A.G. (1984) The two Phanerozoic supercycles. In: *Catastrophes in Earth History* (Eds Berggren, W.A. & Van Couvering, J.A.). Princeton University Press, pp. 129–150.

Gaffin, S.R. & Maasch, K.A. (1991) Anomalous cyclicity in climate and stratigraphy and modelling nonlinear oscillations. *J. Geophys. Res.* **96**, 6701–6711.

Gleick, J. (1987) *Chaos: Making a New Science*. Viking, New York.

Goodings, D. (1991a) Nonlinear differential equations and attractors. In: *Nonlinear Dynamics. Chaos and Fractals, with Applications to Geological Systems* (Ed. Middleton, G.V.). Geol. Assoc. Can., Short Course Notes 9, pp. 23–33.

Goodings, D. (1991b) Chaos in a time series. In: *Nonlinear Dynamics, Chaos and Fractals, with Applications to Geological Systems*. (Ed. Middleton, G.V.). Geol. Assoc. Can., Short Course Notes 9, pp. 35–46.

Haq, B.U., Hardenbol, J. & Vail, P.R. (1988) Mesozoic and Cenozoic chronostratigraphy and cycles of sealevel change. In: *Sea-Level Changes: an Integrated Approach* (Eds Wilgus, C.K., Hastings, B.S., Posamentier, H., Van Wagoner, J., Ross, C.A. & Kendall, C.G. St C.). Soc. Econ. Paleont. Mineral. Spec. Publs. 42, pp. 71–108.

Harland, W.B., Armstrong, R.L., Cox, A.V., Craig, L.E., Smith, A.G. & Smith, D.G. (1990) *A Geologic Time Scale 1989*. Cambridge University Press, Cambridge.

Huang, J. & Turcotte, D.L. (1990) Evidence for chaotic fault interactions in the seismicity of the San Andreas fault and Nankai trough. *Nature* **348**, 234–236 (and commentary by C. Scholz, pp. 197–198).

Lorenz, E.N. (1963) Deterministic non-periodic flows. *J. Atmos. Sci.* **20**, 130–141.

Lovelock, J.E. (1979) *Gaia. A New Look at Life on Earth*. Oxford University Press, Oxford.

Mandelbrot, B.B. & Wallis, J.R. (1969) Some long-run properties of geophysical records. *Water Resources Res.* **5**, 321–340.

Middleton, G.V. (Ed.) (1991) *Nonlinear Dynamics, Chaos and Fractals, with Applications to Geological Systems*. Geol. Assoc. Can., Short Course Notes 9.

Plotnick, R.E. (1986) A fractal model for the distribution of stratigraphic hiatuses. *J. Geol.* **94**, 885–890.

Sadler, P.M. (1981) Sediment accumulation rates and the completeness of stratigraphic sections. *J. Geol.* **89**, 569–584.

Sadler, P.M. & Strauss, D.J. (1990) Estimation of completeness of stratigraphical sections using empirical data theoretical models. *J. Geol. Soc. Lond.* **147**, 471–485.

Schwarzacher, W. (1985) Principles of quantitative lithostratigraphy – the treatment of single sections. In: *Quantitative Stratigraphy* (Eds Gradstein, F.M., Agterberg, F.P., Brower, J.C. & Schwarzacher, W.S.). UNESCO, Paris and Reidel Publ. Co., Dordrecht, pp. 361–386.

Shaw, H.R. (1987a) The periodic structure of the natural record, and nonlinear dynamics. *EOS* **68**, 1651–1662.

Shaw, H.R. (1987b) Uniqueness of volcanic systems. *US Geol. Surv. Prof. Pap.* **1350**, 1357–1394.

Shaw, H.R. (1988) Mathematical attractor theory and plutonic–volcanic episodicity. In: *Modelling of Volcanic Processes* (Eds King, C.-Y. & Scarpa, R.). Friedr. Vieweg, Braunschwieg/Wiesbaden, pp. 162–206.

Shaw, H.R. (1989) *Terrestrial–cosmological correlations in evolutionary processes*. USGS Open File Report 88-43.

Shaw, H.R. & Chouet, B. (1988) *Application of nonlinear dynamics to the history of seismic tremor at Kilauea volcano, Hawaii*. USGS Open File Report 88-539.

Slingerland, R. (1990) Predictability and chaos in quantitative dynamic stratigraphy. In: *Quantitative Dynamic Stratigraphy* (Ed. Cross, T.A.). Prentice Hall, New Jersey, pp. 45–53.

Stewart, I. (1989) *Does God Play Dice? The Mathematics of Chaos*. Penguin, Middlesex.

Tetzlaff, D.M. (1990) Limits to the predictive ability of dynamic models that simulate clastic sedimentation. In: *Quantitative Dynamic Stratigraphy* (Ed. Cross, T.A.). Prentice Hall, New Jersey, pp. 55–65.

Weedon, G.P. (1993) The recognition and stratigraphic implications of orbital forcing of climate and sedimentary cycles. *Sediment. Rev.* **1**, 31–50.

Weedon, G.P. & Jenkins, H.C. (1990) Regular and irregular climatic cycles and the Belemnite Marls (Pliensbachian, Lower Jurassic, Wessex Basin). *J. Geol. Soc. Lond.* **147**, 915–918.

Worsley, T.R., Nance, D. & Moody. J.B. (1984) Global tectonics and eustasy for the past 2 billion years. *Mar. Geol.* **58**, 373–400.

Zeller, E.J. (1964) Cycles and psychology. *Kansas Geol. Surv. Bull.* **169**, 631–636.

Index

References to figures appear in *italic type*.
References to tables appear in **bold type**.